Table of Contents

UNIT PRICES

GENERAL REQUIREMENTS

SITE CONSTRUCTION

CONCRETE

METALS

WOOD & PLASTICS

THERMAL & MOISTURE PROTECTION

FINISHES

SPECIALTIES

EQUIPMENT

SPECIAL CONSTRUCTION

CONVEYING SYSTEMS

MECHANICAL

ELECTRICAL

SQUARE FOOT

ASSEMBLIES

SERVICES

BUILDING SITEWORK

REFERENCE INFORMATION

REFERENCE NUMBERS

CREWS

COST INDEXES

INDEX

iii

Related Means Products and Services

Melville J. Mossman, P.E., Senior Editor of this cost data book, suggests the following Means products and services as **companion** information resources to *Plumbing Cost Data*:

Construction Cost Data Books

Mechanical Cost Data 2005
Electrical Cost Data 2005

Reference Books

Green Building: Project Planning & Cost Estimating
Plumbing Estimating, 2nd Edition
Mechanical Estimating, 3rd Edition
Electrical Estimating, 3rd Edition
Cyberplaces 2nd Ed.: The Internet Guide for Architects, Engineers, Contractors & Facility Managers
HVAC: Design Criteria, Options, Selection, 2nd Edition
HVAC Systems Evaluation
Understanding Building Automation Systems

Seminars and In-House Training

Mechanical & Electrical Estimating

Means on the Internet

Visit Means at **http://www.rsmeans.com.** The site contains useful *interactive* cost and reference material. Request or download *FREE* estimating software demos. Visit our bookstore for convenient ordering, plus learn more about new publications and companion products.

Means Data on CD-ROM

Get the information found in Means traditional cost books on Means new, easy to use CD-ROM, Means *CostWorks 2005*. Means *CostWorks* users can now enhance their Costlist with the Means *CostWorks Estimator*. For more information see the special full color brochure inserted in this book.

Means Consulting Services

Means Consulting Group provides a number of construction cost-related services including: Cost Estimates, Customized Databases, Benchmark Studies, Estimating Audits, Feasibility Studies, Litigation Support, Staff Assessments and Customized Training.

Means Data for Computer Estimating

Construction costs for software applications.

Over 25 unit price and assemblies cost databases are available through a number of leading estimating and facilities management software providers (listed below). For more information see the "yellow" pages at the back of this publication.

- 3D International
- 4Clicks-Solutions, LLC
- Aepco, Inc.
- Applied Flow Technology
- ArenaSoft Estimating
- ARES Corporation
- Barchan Associates
- BSD – Building Systems Design, Inc.
- CMS – Computerized Micro Solutions
- Corecon Technologies, Inc.
- CorVet Systems
- Discover Software
- Estimating Systems, Inc.
- Entech Engineering
- FAMIS Software, Inc.
- ISES Corporation
- Magellan K-12
- Maximus Asset Solutions
- MC^2 – Management Computer Controls
- Neddam Software Technologies
- Prime Time
- Quest Solutions, Inc.
- RIB Software (Americas), Inc.
- Shaw Beneco Enterprises, Inc.
- Schwab Technology Solutions, Inc.
- Timberline Software Corp.
- TMA Systems, LLC
- US Cost, Inc.
- Vanderweil Facility Advisors
- Vertigraph, Inc.

How the Book Is Built: An Overview

A Powerful Construction Tool

You have in your hands one of the most powerful construction tools available today. A successful project is built on the foundation of an accurate and dependable estimate. This book will enable you to construct just such an estimate.

For the casual user the book is designed to be:

- quickly and easily understood so you can get right to your estimate
- filled with valuable information so you can understand the necessary factors that go into the cost estimate

For the regular user, the book is designed to be:

- a handy desk reference that can be quickly referred to for key costs
- a comprehensive, fully reliable source of current construction costs and productivity rates, so you'll be prepared to estimate any project
- a source book for preliminary project cost, product selections, and alternate materials and methods

To meet all of these requirements we have organized the book into the following clearly defined sections.

How To Use the Book: The Details

This section contains an in-depth explanation of how the book is arranged . . . and how you can use it to determine a reliable construction cost estimate. It includes information about how we develop our cost figures and how to completely prepare your estimate.

Unit Price Section

All cost data has been divided into the 16 divisions according to the MasterFormat system of classification and numbering as developed by the Construction Specifications Institute (CSI) and Construction Specifications Canada (CSC). For a listing of these divisions and an outline of their subdivisions, see the Unit Price Section Table of Contents.

Estimating tips are included at the beginning of each division.

Division 17: Quick Project Estimates:

In addition to the 16 Unit Price Divisions there is a S.F. (Square Foot) and C.F. (Cubic Foot) Cost Division, Division 17. It contains costs for 59 different building types that allow you to quickly make a rough estimate for the overall cost of a project or its major components.

Assemblies Section

The cost data in this section has been organized in an "Assemblies" format. These assemblies are the functional elements of a building and are arranged according to the 7 divisions of the UNIFORMAT II classification system. For a complete explanation of a typical "Assemblies" page, see "How To Use the Assemblies Cost Tables."

Reference Section

This section includes information on Reference Numbers, Change Orders, Crew Listings, Historical Cost Indexes, City Cost Indexes, Location Factors and a listing of Abbreviations. It is visually identified by a vertical gray bar on the edge of pages.

Reference Numbers: Following selected items and major classifications throughout the book are "reference numbers" shown in bold squares. These numbers refer you to related information in other sections.

In these other sections, you'll find related tables, explanations, and estimating information. Also included are alternate pricing methods, and technical data, along with information on design and economy in construction. You'll also find helpful tips on estimating and construction.

Change Orders: This section includes information on the factors that influence the pricing of change orders.

Crew Listings: This section lists all the crews referenced. For the purposes of this book, a crew is composed of more than one trade classification and/or the addition of equipment to any trade classification. Power equipment is included in the cost of the crew. Costs are shown both with bare labor rates and with the installing contractor's overhead and profit added. For each, the total crew cost per eight-hour day and the composite cost per labor-hour are listed.

Historical Cost Indexes: These indexes provide you with data to adjust construction costs over time. If you know costs for a past project, you can use these indexes to estimate the cost to construct the same project today.

City Cost Indexes: Obviously, costs vary depending on the regional economy. You can adjust the "national average" costs in this book to 316 major cities throughout the U.S. and Canada by using the data in this section. How to use information is included.

Location Factors, to quickly adjust the data to over 930 zip code areas, are included.

Abbreviations: A listing of the abbreviations and the terms they represent, is included.

Index

A comprehensive listing of all terms and subjects in this book to help you find what you need quickly.

The Scope of This Book

This book is designed to be comprehensive and easy to use. To that end we have made certain assumptions:

1. We have established material prices based on a "national average."
2. We have computed labor costs based on a 30-city "national average" of union wage rates.
3. We have targeted the data for projects of a certain size range.

For a more detailed explanation of how the cost data is developed, see "How To Use the Book: The Details."

Project Size

This book is aimed primarily at commercial and industrial projects costing $1,000,000 and up, or large multi-family housing projects. Costs are primarily for new construction or major renovation of buildings rather than repairs or minor alterations.

With reasonable exercise of judgment the figures can be used for any building work. *However, for civil engineering structures such as bridges, dams, highways, or the like, please refer to **Means Heavy Construction Cost Data.***

How to Use the Book: The Details

What's Behind the Numbers? The Development of Cost Data

The staff at RSMeans continuously monitors developments in the construction industry in order to ensure reliable, thorough and up-to-date cost information.

While *overall* construction costs may vary relative to general economic conditions, price fluctuations within the industry are dependent upon many factors. Individual price variations may, in fact, be opposite to overall economic trends. Therefore, costs are continually monitored and complete updates are published yearly. Also, new items are frequently added in response to changes in materials and methods.

Costs—$ (U.S.)

All costs represent U.S. national averages and are given in U.S. dollars. The Means City Cost Indexes can be used to adjust costs to a particular location. The City Cost Indexes for Canada can be used to adjust U.S. national averages to local costs in Canadian dollars.

Material Costs

The RSMeans staff contacts manufacturers, dealers, distributors, and contractors all across the U.S. and Canada to determine national average material costs. If you have access to current material costs for your specific location, you may wish to make adjustments to reflect differences from the national average. Included within material costs are fasteners for a normal installation. RSMeans engineers use manufacturers' recommendations, written specifications and/or standard construction practice for size and spacing of fasteners. Adjustments to material costs may be required for your specific application or location. Material costs do not include sales tax.

Labor Costs

Labor costs are based on the average of wage rates from 30 major U.S. cities. Rates are determined from labor union agreements or prevailing wages for construction trades for the current year. Rates along with overhead and profit markups are listed on the inside back cover of this book.

- If wage rates in your area vary from those used in this book, or if rate increases are expected within a given year, labor costs should be adjusted accordingly.

Labor costs reflect productivity based on actual working conditions. These figures include time spent during a normal workday on tasks other than actual installation, such as material receiving and handling, mobilization at site, site movement, breaks, and cleanup.

Productivity data is developed over an extended period so as not to be influenced by abnormal variations and reflects a typical average.

Equipment Costs

Equipment costs include not only rental, but also operating costs for equipment under normal use. The operating costs include parts and labor for routine servicing such as repair and replacement of pumps, filters and worn lines. Normal operating expendables such as fuel, lubricants, tires and electricity (where applicable) are also included. Extraordinary operating expendables with highly variable wear patterns such as diamond bits and blades are excluded. These costs are included under materials. Equipment rental rates are obtained from industry sources throughout North America—contractors, suppliers, dealers, manufacturers, and distributors.

Crew Equipment Cost/Day—The power equipment required for each crew is included in the crew cost. The daily cost for crew equipment is based on dividing the weekly bare rental rate by 5 (number of working days per week), and then adding the hourly operating cost times 8 (hours per day). This "Crew Equipment Cost/Day" is listed in Subdivision 01590.

Mobilization/Demobolization—The cost to move construction equipment from an equipment yard or rental company to the job site and back again is not included in equipment costs. Mobilization (to the site) and demolization (from the site) costs can be found in Section 02305-250. If a piece of equipment is already at the job site, it is not appropriate to utlize mob/demob costs again in an estimate.

General Conditions

Cost data in this book is presented in two ways: Bare Costs and Total Cost including O&P (Overhead and Profit). General Conditions, when applicable, should also be added to the Total Cost including O&P. The costs for General Conditions are listed in Division 1 of the Unit Price Section and the Reference Section of this book. General Conditions for the *Installing Contractor* may range from 0% to 10% of the Total Cost including O&P. For the *General* or *Prime Contractor*, costs for General Conditions may range from 5% to 15% of the Total Cost including O&P, with a figure of 10% as the most typical allowance.

Overhead and Profit

Total Cost including O&P for the *Installing Contractor* is shown in the last column on both the Unit Price and the Assemblies pages of this book. This figure is the sum of the bare material cost plus 10% for profit, the base labor cost plus total overhead and profit, and the bare equipment cost plus 10% for profit. Details for the calculation of Overhead and Profit on labor are shown on the inside back cover and in the Reference Section of this book. (See the "How to Use the Unit Price Pages" for an example of this calculation).

Factors Affecting Costs

Costs can vary depending upon a number of variables. Here's how we have handled the main factors affecting costs.

Quality—The prices for materials and the workmanship upon which productivity is based represent sound construction work. They are also in line with U.S. government specifications.

Overtime—We have made no allowance for overtime. If you anticipate premium time or work beyond normal working hours, be sure to make an appropriate adjustment to your labor costs.

Productivity—The productivity, daily output, and labor-hour figures for each line item are based on working an eight-hour day in daylight hours in moderate temperatures. For work that extends beyond normal work hours or is performed under adverse conditions, productivity may decrease. (See the section in "How To Use the Unit Price Pages" for more on productivity.)

Size of Project—The size, scope of work, and type of construction project will have a significant impact on cost. Economies of scale can reduce costs for large projects. Unit costs can often run higher for small projects. Costs in this book are intended for the size and type of project as previously described in "How the Book Is Built: An Overview." Costs for projects of a significantly different size or type should be adjusted accordingly.

Location—Material prices in this book are for metropolitan areas. However, in dense urban areas, traffic and site storage limitations may increase costs. Beyond a 20-mile radius of large cities, extra trucking or transportation charges may also increase the material costs slightly. On the other hand, lower wage rates may be in effect. Be sure to consider both these factors when preparing an estimate, particularly if the job site is located in a central city or remote rural location.

In addition, highly specialized subcontract items may require travel and per diem expenses for mechanics.

Other factors—
- season of year
- contractor management
- weather conditions
- local union restrictions
- building code requirements
- availability of:
  - adequate energy
  - skilled labor
  - building materials
- owner's special requirements/ restrictions
- safety requirements
- environmental considerations

Unpredictable Factors—General business conditions influence "in-place" costs of all items. Substitute materials and construction methods may have to be employed. These may affect the installed cost and/or life cycle costs. Such factors may be difficult to evaluate and cannot necessarily be predicted on the basis of the job's location in a particular section of the country. Thus, where these factors apply, you may find significant, but unavoidable cost variations for which you will have to apply a measure of judgment to your estimate.

Rounding of Costs

In general, all unit prices in excess of $5.00 have been rounded to make them easier to use and still maintain adequate precision of the results. The rounding rules we have chosen are in the following table.

Prices from ...	Rounded to the nearest ...
$.01 to $5.00	$.01
$5.01 to $20.00	$.05
$20.01 to $100.00	$.50
$100.01 to $300.00	$1.00
$300.01 to $1,000.00	$5.00
$1,000.01 to $10,000.00	$25.00
$10,000.01 to $50,000.00	$100.00
$50,000.01 and above	$500.00

Contingencies

Contingencies: The allowance for contingencies generally is to provide for unforeseen construction difficulties. On alterations or repair jobs 20% is none too much. If drawings are final and only field contingencies are being considered, 2% or 3% is probably sufficient and often nothing need be added. As far as the contract is concerned, future changes in plans will be covered by extras. The contractor should consider inflationary price trends and possible material shortages during the course of the job. If drawings are not complete or approved or a budget cost wanted, it is wise to add 5% to 10%. Contingencies, then, are a matter of judgment. Additional allowances are shown in Division 01250-200 for contingencies and job

conditions and Division 01250-400 for factors to convert prices for repair and remodeling jobs.

Important Estimating Considerations

One reason for listing a job size "minimum" is to ensure that the construction craftsmen are productive for 8 hours a day on a continuous basis. Otherwise, tasks that only require 6 to 7 hours could be billed as an 8-hour day, thus providing an erroneous productivity estimate. The "productivity," or daily output of each craftsman, includes mobilization and cleanup time, break time, plan layout time, as well as an allowance to carry stock from the storage trailer or location on the job site up to 200' into the building and on the first or second floor. If material has to be transported over greater distances or to higher floors, an additional allowance should be considered by the estimator. An allowance has also been included in the piping and fittings installation time for a leak check and minor tightening.

Equipment installation time includes all applicable items following: positioning, leveling and securing the unit in place, connecting all associated piping, ducts, vents, etc., which shall have been estimated separately, connecting to an adjacent power source, filling/bleeding, startup, adjusting the controls up and down to assure proper response, setting the integral controls/valves/ regulators/thermostats for proper operation (does not include external building type control systems, DDC systems, etc.), explaining/training owner's operator, and warranty. A reasonable breakdown of the labor costs would be as follows:

1. Movement into building, installation/ setting of equipment — 35%
2. Connection to piping/duct/power, etc. — 25%
3. Filling/flushing/cleaning/touchup, etc. — 15%
4. Startup/running adjustments — 5%
5. Training owner's rep. — 5%
6. Warranty/call back/service — 15%

Note: For an explanation of "Selective Demolition" versus "Removal for Replacement" see reference number R15050-720.

Note that cranes or other lifting equipment are not included on any lines in Division 15. For example, if a crane is required to lift a heavy piece of pipe into place high above a gym floor, or to put a rooftop unit on the roof of a four-story building, etc., it must be added. Due to the potential for extreme variation—from nothing additional required, to a major crane or helicopter—we feel that including a nominal amount for "lifting contingency" would be useless and detract from the accuracy of the estimate. When using equipment rental from Means do not forget to include the cost of the operator(s).

Estimating Labor-Hours

The labor-hours expressed in this publication are based on Average Installation time, using an efficiency level of approximately 60-65% (see item 7) which has been found reasonable and acceptable by many contractors.

The book uses this national efficiency average to establish a consistent benchmark.

For bid situations, adjustments to this efficiency level should be the responsibility of the contractor bidding the project.

The unit labor-hour is divided in the following manner. A typical day for a journeyman might be:

1.	Study Plans	3%	14.4 min.
2.	Material Procurement	3%	14.4 min.
3.	Receiving and Storing	3%	14.4 min
4.	Mobilization	5%	24.0 min.
5.	Site Movement	5%	24.0 min.
6.	Layout and Marking	8%	38.4 min.
7.	Actual Installation	64%	307.2 min.
8.	Cleanup	3%	14.4 min.
9.	Breaks, Non-Productive	6%	28.8 min.
		100%	480.0 min.

If any of the percentages expressed in this breakdown do not apply to the particular work or project situation, then that percentage or a portion of it may be deducted from or added to labor hours.

Final Checklist

Estimating can be a straightforward process provided you remember the basics. Here's a checklist of some of the items you should remember to do before completing your estimate.

Did you remember to . . .

- factor in the City Cost Index for your locale
- take into consideration which items have been marked up and by how much
- mark up the entire estimate sufficiently for your purposes
- read the background information on techniques and technical matters that could impact your project time span and cost
- include all components of your project in the final estimate
- double check your figures to be sure of your accuracy
- for more information, please see "Tips for Accurate Estimating," R01100-005 in the Reference Section
- call RSMeans if you have any questions about your estimate or the data you've found in our publications

Remember, RSMeans stands behind its publications. If you have any questions about your estimate . . . about the costs you've used from our books . . . or even about the technical aspects of the job that may affect your estimate, feel free to call the RSMeans editors at 1-800-334-3509.

Unit Price Section

Table of Contents

How to Use the Unit Price Pages

The following is a detailed explanation of a sample entry in the Unit Price Section. Next to each bold number below is the item being described with appropriate component of the sample entry following in parenthesis. Some prices are listed as bare costs, others as costs that include overhead and profit of the installing contractor. In most cases, if the work is to be subcontracted, the general contractor will need to add an additional markup (RSMeans suggests using 10%) to the figures in the column "Total Incl. O&P."

Division Number/Title (03300/Cast-In-Place Concrete)

Use the Unit Price Section Table of Contents to locate specific items. The sections are classified according to the CSI MasterFormat (1995 Edition).

Line Numbers (03310 240 3900)

Each unit price line item has been assigned a unique 12-digit code based on the CSI MasterFormat classification.

- Level One - CSI-MasterFormat Division
- Level Two - CSI

03300
03310-240-3900

- Means 12-digit Line Number
- Level Four - Means
- Level Three - CSI

Description (Concrete-In-Place, etc.)

Each line item is described in detail. Sub-items and additional sizes are indented beneath the appropriate line items. The first line or two after the main item (in boldface) may contain descriptive information that pertains to all line items beneath this boldface listing.

Reference Number Information

R03310 -010 You'll see reference numbers shown in bold rectangles at the beginning of some sections. These refer to related items in the Reference Section, visually identified by a vertical gray bar on the edge of pages.

The relation may be: (1) an estimating procedure that should be read before estimating, (2) an alternate pricing method, or (3) technical information.

The "R" designates the Reference Section. The numbers refer to the MasterFormat classification system.

It is strongly recommended that you review all reference numbers that appear within the section in which you are working.

Note: Not all reference numbers appear in all Means publications.

03300 | Cast-In-Place Concrete

03310 | Structural Concrete

			CREW	DAILY OUTPUT	LABOR-HOURS	UNIT	MAT.	LABOR	EQUIP.	TOTAL	TOTAL INCL O&P	
240	0010	**CONCRETE IN PLACE** Including forms (4 uses) reinforcing										240
	0050	steel and finishing unless otherwise indicated	R03310 -010									
	0300	Beams, 5 kip per L.F., 10' span	R03310 -100	C-14A	15	.04	C.Y.	284		46.50	770.50	1,050
	0350	25' span	"	18.55	10.782		214	370	39.50	623.50	860
	3850	Over 5 C.Y.		81.04	1.382		226	45	.29	271.29	320
	3900	Footings, strip, 18" x 9", unreinforced		40	2.800		101	91.50	.59	193.09	255
	3920	18" x 9", reinforced		35	3.200		120	105	.67	225.67	297
	3925	20" x 10", unreinforced		45	2.489		98	81.50	.52	180.02	236
	3930	20" x 10", reinforced		40	2.800		114	91.50	.59	206.09	269
	3935	24" x 12", unreinforced		55	2.036		97	66.50	.43	163.93	
	3940	24" x 12", reinforced		48	2.333		113	76	.49	189.49	2
	3945	36" x 12", unreinforced		70	1.600		93	52.50	.34	145.84	184
	3950	36" x 12", reinforced		60	1.867		108	61	.39	169.39	214
	4000	Foundation mat, under 10 C.Y.		38.67	2.896		164	94.50	.61	259.11	330

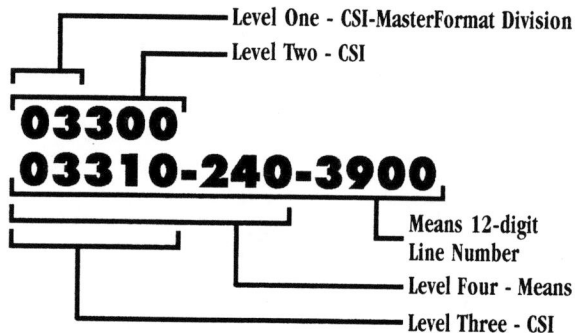

2

5 Crew (C-14C)

The "Crew" column designates the typical trade or crew used to install the item. If an installation can be accomplished by one trade and requires no power equipment, that trade and the number of workers are listed (for example, "2 Carpenters"). If an installation requires a composite crew, a crew code designation is listed (for example, "C-14C"). You'll find full details on all composite crews in the Crew Listings.

- For a complete list of all trades utilized in this book and their abbreviations, see the inside back cover.

Crew No.	Bare Costs		Incl. Subs O & P		Cost Per Labor-Hour	
Crew C-14C	Hr.	Daily	Hr.	Daily	Bare Costs	Incl. O&P
1 Carpenter Foreman (out)	$36.25	$290.00	$56.45	$451.60	$32.66	$51.14
6 Carpenters	34.25	1644.00	53.35	2560.80		
2 Rodmen (reinf.)	37.95	607.20	62.60	1001.60		
4 Laborers	26.70	854.40	41.55	1329.60		
1 Cement Finisher	32.85	262.80	48.35	386.80		
1 Gas Engine Vibrator		24.00		26.40	.21	.24
112 L.H., Daily Totals		$3682.40		$5756.80	$32.87	$51.38

6 Productivity: Daily Output (40.0)/Labor-Hours (2.80)

The "Daily Output" represents the typical number of units the designated crew will install in a normal 8-hour day. To find out the number of days the given crew would require to complete the installation, divide your quantity by the daily output. For example:

Quantity	÷	Daily Output	=	Duration
100 C.Y.	÷	40.0/ Crew Day	=	2.50 Crew Days

The "Labor-Hours" figure represents the number of labor-hours required to install one unit of work. To find out the number of labor-hours required for your particular task, multiply the quantity of the item times the number of labor-hours shown. For example:

Quantity	x	Productivity Rate	=	Duration
100 C.Y.	x	2.80 Labor-Hours/ C.Y.	=	280 Labor-Hours

7 Unit (C.Y.)

The abbreviated designation indicates the unit of measure upon which the price, production, and crew are based (C.Y. = Cubic Yard). For a complete listing of abbreviations refer to the Abbreviations Listing in the Reference Section of this book.

8 Bare Costs:

Mat. (Bare Material Cost) (101)

The unit material cost is the "bare" material cost with no overhead and profit included. *Costs shown reflect national average material prices for January of the current year and include delivery to the job site. No sales taxes are included.*

Labor (91.50)

The unit labor cost is derived by multiplying bare labor-hour costs for Crew C-14C by labor-hour units. The bare labor-hour cost is found in the Crew Section under C-14C. (If a trade is listed, the hourly labor cost—the wage rate—is found on the inside back cover.)

Labor-Hour Cost Crew C-14C	x	Labor-Hour Units	=	Labor
$32.66	x	2.80	=	$91.00

Equip. (Equipment) (.59)

Equipment costs for each crew are listed in the description of each crew. Tools or equipment whose value justifies purchase or ownership by a contractor are considered overhead as shown on the inside back cover. The unit equipment cost is derived by multiplying the bare equipment hourly cost by the labor-hour units.

Equipment Cost Crew C-14C	x	Labor-Hour Units	=	Equip.
.21	x	2.80	=	$.59

Total (193.09)

The total of the bare costs is the arithmetic total of the three previous columns: mat., labor, and equip.

Material	+	Labor	+	Equip.	=	Total
$101	+	$91.50	+	$.59	=	$193.09

9 Total Costs Including O&P

This figure is the sum of the bare material cost plus 10% for profit; the bare labor cost plus total overhead and profit (per the inside back cover or, if a crew is listed, from the crew listings); and the bare equipment cost plus 10% for profit.

Material is Bare Material Cost + 10% = 101 + 10.10	=	$111.10
Labor for Crew C-14C = Labor-Hour Cost (51.14) x Labor-Hour Units (2.80)	=	$143.19
Equip. is Bare Equip. Cost + 10% = .59 + .06	=	$ .65
Total (Rounded)	=	$255

Division 1
General Requirements

Estimating Tips

The General Requirements of any contract are very important to both the bidder and the owner. These lay the ground rules under which the contract will be executed and have a significant influence on the cost of operations. Therefore, it is extremely important to thoroughly read and understand the General Requirements both before preparing an estimate and when the estimate is complete, to ascertain that nothing in the contract is overlooked. Caution should be exercised when applying items listed in Division 1 to an estimate. Many of the items are included in the unit prices listed in the other divisions such as mark-ups on labor and company overhead.

01200 Price & Payment Procedures

- When estimating historic preservation projects (depending on the condition of the existing structure and the owner's requirements), a 15-20% contingency or allowance is recommended, regardless of the stage of the drawings.

01300 Administrative Requirements

- Before determining a final cost estimate, it is a good practice to review all the items listed in subdivision 01300 to make final adjustments for items that may need customizing to specific job conditions.
- Historic preservation projects may require specialty labor and methods, as well as extra time to protect existing materials that must be preserved and/or restored. Some additional expenses may be incurred in architectural fees for facility surveys and other special inspections and analyses.
- Requirements for initial and periodic submittals can represent a significant cost to the General Requirements of a job. Thoroughly check the submittal specifications when estimating a project to determine any costs that should be included.

01400 Quality Requirements

- All projects will require some degree of Quality Control. This cost is not included in the unit cost of construction listed in each division. Depending upon the terms of the contract, the various costs of inspection and testing can be the responsibility of either the owner or the contractor. Be sure to include the required costs in your estimate.

01500 Temporary Facilities & Controls

- Barricades, access roads, safety nets, scaffolding, security and many more requirements for the execution of a safe project are elements of direct cost. These costs can easily be overlooked when preparing an estimate. When looking through the major classifications of this subdivision, determine which items apply to each division in your estimate.

01590 Equipment Rental

- This subdivision contains transportation, handling, storage, protection and product options and substitutions. Listed in this cost manual are average equipment rental rates for all types of equipment. This is useful information when estimating the time and materials requirement of any particular operation in order to establish a unit or total cost.
- A good rule of thumb is that weekly rental is 3 times daily rental and that monthly rental is 3 times weekly rental.
- The figures in the column for Crew Equipment Cost represent the rental rate used in determining the daily cost of equipment in a crew. It is calculated by dividing the weekly rate by 5 days and adding the hourly operating cost times 8 hours.

01740 Execution Requirements

- When preparing an estimate, read the specifications to determine the requirements for Contract Closeout thoroughly. Final cleaning, record documentation, operation and maintenance data, warranties and bonds, and spare parts and maintenance materials can all be elements of cost for the completion of a contract. Do not overlook these in your estimate.

01830 Operations & Maintenance

- If maintenance and repair are included in your contract, they require special attention. To estimate the cost to remove and replace any unit usually requires a site visit to determine the accessibility and the specific difficulty at that location. Obstructions, dust control, safety, and often overtime hours must be considered when preparing your estimate.

Reference Numbers

Reference numbers are shown in bold squares at the beginning of some major classifications. These numbers refer to related items in the Reference Section. The reference information may be an estimating procedure, an alternate pricing method or technical information.

Note: Not all subdivisions listed here necessarily appear in this publication.

01107	Professional Consultant	CREW	DAILY OUTPUT	LABOR-HOURS	UNIT	MAT.	LABOR	EQUIP.	TOTAL	TOTAL INCL O&P	
100	**0011**	**ARCHITECTURAL FEES**										**100**
	0020	For new construction										
	0060	Minimum				Project					4.90%	
	0090	Maximum									16%	
	0100	For alteration work, to $500,000, add to fee									50%	
	0150	Over $500,000, add to fee				↓					25%	
200	**0010**	**CONSTRUCTION MANAGEMENT FEES** $1,000,000 job, minimum				Project					4.50%	**200**
	0050	Maximum									7.50%	
	0300	$5,000,000 job, minimum									2.50%	
	0350	Maximum				↓					4%	
300	**0010**	**ENGINEERING FEES**										**300**
	0020	Educational planning consultant, minimum	R01107 -030			Project					.50%	
	0100	Maximum				"					2.50%	
	0200	Electrical, minimum				Contrct					4.10%	
	0300	Maximum									10.10%	
	0400	Elevator & conveying systems, minimum									2.50%	
	0500	Maximum									5%	
	0600	Food service & kitchen equipment, minimum									8%	
	0700	Maximum									12%	
	0800	Landscaping & site development, minimum									2.50%	
	0900	Maximum									6%	
	1000	Mechanical (plumbing & HVAC), minimum									4.10%	
	1100	Maximum				↓					10.10%	
	1200	Structural, minimum				Project					1%	
	1300	Maximum				"					2.50%	
700	**0010**	**SURVEYING** Conventional, topographical, minimum	A-7	3.30	7.273	Acre	16.30	269	18.40	303.70	455	**700**
	0100	Maximum	A-8	.60	53.333		49	1,925	101	2,075	3,125	
	0300	Lot location and lines, minimum, for large quantities	A-7	2	12		25.50	445	30.50	501	745	
	0320	Average	"	1.25	19.200		46	710	48.50	804.50	1,200	
	0400	Maximum, for small quantities	A-8	1	32	↓	73.50	1,150	61	1,284.50	1,925	
	0600	Monuments, 3' long	A-7	10	2.400	Ea.	20	88.50	6.05	114.55	166	
	0800	Property lines, perimeter, cleared land	"	1,000	.024	L.F.	.03	.89	.06	.98	1.47	
	0900	Wooded land	A-8	875	.037	"	.05	1.31	.07	1.43	2.16	
	1500	Aerial surveying, including ground control, minimum fee, 10 acres				Total					5,700	
	1510	100 acres									9,500	
	1550	From existing photography, deduct				↓					1,370	
	1600	2' contours, 10 acres				Acre					460	
	1650	20 acres									315	
	1800	50 acres									95	
	1850	100 acres									85	
	2000	1000 acres									17.85	
	2050	10,000 acres				↓					11.50	
	2150	For 1' contours and										
	2160	dense urban areas, add to above				Acre					40%	

Important: See the Reference Section for critical supporting data - Reference Nos., Crews, & City Cost Indexes

		01250	Contract Modification Procedures	CREW	DAILY OUTPUT	LABOR-HOURS	UNIT	2005 BARE COSTS				TOTAL INCL O&P	
								MAT.	LABOR	EQUIP.	TOTAL		
200	0010		**CONTINGENCIES** for estimate at conceptual stage				Project					20%	200
	0050		Schematic stage									15%	
	0100		Preliminary working drawing stage (Design Dev.)									10%	
	0150		Final working drawing stage				↓					3%	
300	0010		**CREWS** For building construction, see How To Use This Book										300
400	0010		**FACTORS** Cost adjustments R01250 -010										400
	0100		Add to construction costs for particular job requirements										
	0500		Cut & patch to match existing construction, add, minimum				Costs	2%	3%				
	0550		Maximum					5%	9%				
	0800		Dust protection, add, minimum					1%	2%				
	0850		Maximum					4%	11%				
	1100		Equipment usage curtailment, add, minimum					1%	1%				
	1150		Maximum					3%	10%				
	1400		Material handling & storage limitation, add, minimum					1%	1%				
	1450		Maximum					6%	7%				
	1700		Protection of existing work, add, minimum					2%	2%				
	1750		Maximum					5%	7%				
	2000		Shift work requirements, add, minimum						5%				
	2050		Maximum						30%				
	2300		Temporary shoring and bracing, add, minimum					2%	5%				
	2350		Maximum					5%	12%				
	2400		Work inside prisons and high security areas, add, minimum						30%				
	2450		Maximum				↓		50%				
500	0010		**JOB CONDITIONS** Modifications to total										500
	0020		project cost summaries										
	0100		Economic conditions, favorable, deduct				Project					2%	
	0200		Unfavorable, add									5%	
	0300		Hoisting conditions, favorable, deduct									2%	
	0400		Unfavorable, add									5%	
	0700		Labor availability, surplus, deduct									1%	
	0800		Shortage, add									10%	
	0900		Material storage area, available, deduct									1%	
	1000		Not available, add									2%	
	1100		Subcontractor availability, surplus, deduct									5%	
	1200		Shortage, add									12%	
	1300		Work space, available, deduct									2%	
	1400		Not available, add				↓					5%	
600	0010		**OVERTIME** For early completion of projects or where R01100 -110										600
	0020		labor shortages exist, add to usual labor, up to				Costs		100%				

		01255	Cost Indexes										
200	0010		**CONSTRUCTION COST INDEX** (Reference) over 930 zip code locations in										200
	0020		The U.S. and Canada, total bldg cost, min. (Clarksdale, MS)				%					66.80%	
	0050		Average									100%	
	0100		Maximum (New York, NY)				↓					132.40%	
400	0010		**HISTORICAL COST INDEXES** (Reference) Back to 1955										400
500	0010		**LABOR INDEX** (Reference) For over 930 zip code locations in										500
	0020		the U.S. and Canada, minimum (Clarksdale, MS)				%		31.70%				
	0050		Average						100%				
	0100		Maximum (New York, NY)				↓		162.30%				
600	0011		**MATERIAL INDEX** For over 930 zip code locations in										600
	0020		the U.S. and Canada, minimum (Elizabethtown, KY)				%	91.10%					
	0040		Average					100%					
	0060		Maximum (Ketchikan, AK)				↓	143.10%					

01200 | Price & Payment Procedures

01290	Payment Procedures		CREW	DAILY OUTPUT	LABOR-HOURS	UNIT	2005 BARE COSTS				TOTAL INCL O&P	
							MAT.	LABOR	EQUIP.	TOTAL		
800	0010	**TAXES** Sales tax, State, average	R01100 -090				%	4.85%					800
	0050	Maximum						7.25%					
	0200	Social Security, on first $87,900 of wages	R01100 -100						7.65%				
	0300	Unemployment, MA, combined Federal and State, minimum							1.92%				
	0350	Average							6.20%				
	0400	Maximum							11.76%				

01300 | Administrative Requirements

01310	Project Management/Coordination		CREW	DAILY OUTPUT	LABOR-HOURS	UNIT	2005 BARE COSTS				TOTAL INCL O&P	
							MAT.	LABOR	EQUIP.	TOTAL		
150	0010	**PERMITS** Rule of thumb, most cities, minimum					Job					.50%	150
	0100	Maximum					"					2%	
200	0010	**PERFORMANCE BOND** For buildings, minimum	R01100 -080				Job					.60%	200
	0100	Maximum					"					2.50%	
350	0010	**INSURANCE** Builders risk, standard, minimum	R01100 -040				Job					.24%	350
	0050	Maximum										.64%	
	0200	All-risk type, minimum	R01100 -060									.25%	
	0250	Maximum										.62%	
	0400	Contractor's equipment floater, minimum					Value					.50%	
	0450	Maximum					"					1.50%	
	0600	Public liability, average					Job					2.02%	
	0800	Workers' compensation & employer's liability, average											
	0850	by trade, carpentry, general					Payroll		18.42%				
	1000	Electrical							6.46%				
	1150	Insulation							15.53%				
	1450	Plumbing							7.96%				
	1550	Sheet metal work (HVAC)							11.46%				
400	0010	**MAIN OFFICE EXPENSE** Average for General Contractors											400
	0020	As a percentage of their annual volume											
	0125	Annual volume under 1 million dollars					% Vol.				13.60%		
	0145	Up to 2.5 million dollars									8%		
	0150	Up to 4.0 million dollars									6.80%		
	0200	Up to 7.0 million dollars									5.60%		
	0250	Up to 10 million dollars									5.10%		
	0300	Over 10 million dollars									3.90%		
500	0010	**MARK-UP** For General Contractors for change	R01100 -070										500
	0100	of scope of job as bid											
	0200	Extra work, by subcontractors, add					%					10%	
	0250	By General Contractor, add										15%	
	0400	Omitted work, by subcontractors, deduct all but										5%	
	0450	By General Contractor, deduct all but										7.50%	
	0600	Overtime work, by subcontractors, add										15%	
	0650	By General Contractor, add										10%	
	1000	Installing contractors, on his own labor, minimum							46.90%				
	1100	Maximum							87.30%				
600	0010	**OVERHEAD** As percent of direct costs, minimum					%				5%		600
	0050	Average									13%		
	0100	Maximum									30%		

Important: See the Reference Section for critical supporting data - Reference Nos., Crews, & City Cost Indexes

01300 | Administrative Requirements

620

01310	Project Management/Coordination	CREW	DAILY OUTPUT	LABOR-HOURS	UNIT	2005 BARE COSTS				TOTAL INCL O&P
						MAT.	LABOR	EQUIP.	TOTAL	
0010	**OVERHEAD & PROFIT** Allowance to add to items in this	R01100 -070								
0020	book that do not include Subs O&P, average				%				25%	
0100	Allowance to add to items in this book that									
0110	do include Subs O&P, minimum				%					5%
0150	Average									10%
0200	Maximum									15%
0300	Typical, by size of project, under $100,000									30%
0350	$500,000 project									25%
0400	$2,000,000 project									20%
0450	Over $10,000,000 project									15%
0010	**FIELD PERSONNEL** Clerk average				Week		320		320	500
0100	Field engineer, minimum						765		765	1,200
0120	Average						995		995	1,550
0140	Maximum						1,150		1,150	1,800
0160	General purpose laborer, average						1,050		1,050	1,625
0180	Project manager, minimum						1,425		1,425	2,225
0200	Average						1,625		1,625	2,525
0220	Maximum						1,850		1,850	2,875
0240	Superintendent, minimum						1,375		1,375	2,150
0260	Average						1,500		1,500	2,325
0280	Maximum						1,725		1,725	2,675
0290	Timekeeper, average						890		890	1,375

700

01321 | Construction Photos

		CREW	DAILY OUTPUT	LABOR-HOURS	UNIT	MAT.	LABOR	EQUIP.	TOTAL	TOTAL INCL O&P
0010	**PHOTOGRAPHS** 8" x 10", 4 shots, 2 prints ea., std. mounting				Set	292			292	320
0100	Hinged linen mounts					315			315	350
0200	8" x 10", 4 shots, 2 prints each, in color					340			340	370
0300	For I.D. slugs, add to all above					4.09			4.09	4.50
1500	Time lapse equipment, camera and projector, buy					3,675			3,675	4,050
1550	Rent per month					545			545	600
1700	Cameraman and film, including processing, B.&W.				Day	1,350			1,350	1,475
1720	Color				"	1,350			1,350	1,475

500

01500 | Temporary Facilities & Controls

01510	Temporary Utilities	CREW	DAILY OUTPUT	LABOR-HOURS	UNIT	2005 BARE COSTS				TOTAL INCL O&P
						MAT.	LABOR	EQUIP.	TOTAL	
0010	**TEMPORARY UTILITIES**									
0100	Heat, incl. fuel and operation, per week, 12 hrs. per day	1 Skwk	100	.080	CSF Flr	6.25	2.79		9.04	11.25
0200	24 hrs. per day	"	60	.133		9.40	4.65		14.05	17.60
0350	Lighting, incl. service lamps, wiring & outlets, minimum	1 Elec	34	.235		2.32	9.60		11.92	16.80
0360	Maximum	"	17	.471		5.05	19.20		24.25	34
0400	Power for temp lighting only, per month, min/month 6.6 KWH								.75	1.18
0450	Maximum/month 23.6 KWH								2.85	3.14
0600	Power for job duration incl. elevator, etc., minimum								47	51.70
0650	Maximum								110	121
1000	Toilet, portable, see division 01590-400									

800

01520 | Construction Facilities

		CREW	DAILY OUTPUT	LABOR-HOURS	UNIT	MAT.	LABOR	EQUIP.	TOTAL	TOTAL INCL O&P
0010	**OFFICE** Trailer, furnished, no hookups, 20' x 8', buy	2 Skwk	1	16	Ea.	6,200	560		6,760	7,700
0250	Rent per month					154			154	169

500

01520 | Construction Facilities

		CREW	DAILY OUTPUT	LABOR-HOURS	UNIT	MAT.	LABOR	EQUIP.	TOTAL	TOTAL INCL O&P	
500	0300	32' x 8', buy	2 Skwk	.70	22.857	Ea.	9,250	795		10,045	11,500	**500**
	0350	Rent per month					163			163	180	
	0400	50' x 10', buy	2 Skwk	.60	26.667		17,000	930		17,930	20,200	
	0450	Rent per month					254			254	280	
	0500	50' x 12', buy	2 Skwk	.50	32		20,300	1,125		21,425	24,000	
	0550	Rent per month					286			286	315	
	0700	For air conditioning, rent per month, add				↓	39.50			39.50	43.50	
	0800	For delivery, add per mile				Mile	1.53			1.53	1.68	
	1000	Portable buildings, prefab, on skids, economy, 8' x 8'	2 Carp	265	.060	S.F.	81.50	2.07		83.57	93	
	1100	Deluxe, 8' x 12'	"	150	.107	"	88.50	3.65		92.15	103	
	1200	Storage boxes, 20' x 8', buy	2 Skwk	1.80	8.889	Ea.	3,175	310		3,485	3,975	
	1250	Rent per month					74.50			74.50	82	
	1300	40' x 8', buy	2 Skwk	1.40	11.429		4,250	400		4,650	5,300	
	1350	Rent per month				↓	105			105	116	

01530 | Temporary Construction

		CREW	DAILY OUTPUT	LABOR-HOURS	UNIT	MAT.	LABOR	EQUIP.	TOTAL	TOTAL INCL O&P	
700	0010	**PROTECTION** Stair tread, 2" x 12" planks, 1 use	1 Carp	75	.107	Tread	4.80	3.65		8.45	11	**700**
	0100	Exterior plywood, 1/2" thick, 1 use		65	.123		1.86	4.22		6.08	8.60	
	0200	3/4" thick, 1 use	↓	60	.133	↓	2.82	4.57		7.39	10.20	
900	0010	**WINTER PROTECTION** Reinforced plastic on wood										**900**
	0100	framing to close openings	2 Clab	750	.021	S.F.	.38	.57		.95	1.31	
	0200	Tarpaulins hung over scaffolding, 8 uses, not incl. scaffolding		1,500	.011		.18	.28		.46	.64	
	0250	Tarpaulin polyester reinf. w/ integral fastening system 11 mils thick		1,600	.010		.78	.27		1.05	1.28	
	0300	Prefab fiberglass panels, steel frame, 8 uses	↓	1,200	.013	↓	.74	.36		1.10	1.36	

01540 | Construction Aids

		CREW	DAILY OUTPUT	LABOR-HOURS	UNIT	MAT.	LABOR	EQUIP.	TOTAL	TOTAL INCL O&P	
750	0010	**SCAFFOLDING**										**750**
	0015	Steel tubular, reg, rent/mo, no plank, incl erect or dismantle	R01540 -100									
	0090	Building exterior, wall face, 1 to 5 stories, 6'-4" x 5' frames	3 Carp	24	1	C.S.F.	24.50	34.50		59	80	
	0200	6 to 12 stories	4 Carp	21.20	1.509		24.50	51.50		76	107	
	0310	13 to 20 stories	5 Carp	20	2		24.50	68.50		93	134	
	0460	Building interior, wall face area, up to 16' high	3 Carp	25	.960		24.50	33		57.50	77.50	
	0560	16' to 40' high		23	1.043	↓	24.50	35.50		60	82	
	0800	Building interior floor area, up to 30' high	↓	312	.077	C.C.F.	2.57	2.63		5.20	6.95	
	0900	Over 30' high	4 Carp	275	.116	"	2.57	3.99		6.56	9.05	
	0910	Steel tubular, heavy duty shoring, buy										
	0920	Frames 5' high 2' wide				Ea.	82.50			82.50	91	
	0925	5' high 4' wide					93.50			93.50	103	
	0930	6' high 2' wide					94.50			94.50	104	
	0935	6' high 4' wide				↓	111			111	122	
	0940	Accessories										
	0945	Cross braces				Ea.	17.60			17.60	19.35	
	0950	U-head, 8" x 8"					19.25			19.25	21	
	0955	J-head, 4" x 8"					14.10			14.10	15.50	
	0960	Base plate, 8" x 8"					15.60			15.60	17.20	
	0965	Leveling jack				↓	33.50			33.50	37	
	1000	Steel tubular, regular, buy										
	1100	Frames 3' high 5' wide				Ea.	64			64	70	
	1150	5' high 5' wide					73.50			73.50	81	
	1200	6'-4" high 5' wide					92.50			92.50	102	
	1350	7'-6" high 6' wide					160			160	175	
	1500	Accessories cross braces					16.50			16.50	18.15	
	1550	Guardrail post					16.50			16.50	18.15	
	1600	Guardrail 7' section				↓	8			8	8.80	

Important: See the Reference Section for critical supporting data - Reference Nos., Crews, & City Cost Indexes

01540 | Construction Aids

		CREW	DAILY OUTPUT	LABOR HOURS	UNIT	2005 BARE COSTS MAT.	LABOR	EQUIP.	TOTAL	TOTAL INCL O&P	
750	1650	Screw jacks & plates				Ea.	26.50			26.50	29	**750**
	1700	Sidearm brackets	R01540 -100				31			31	34	
	1750	8″ casters					36.50			36.50	40	
	1800	Plank 2″ x 10″ x 16′-0″					47			47	51.50	
	1900	Stairway section					270			270	296	
	1910	Stairway starter bar					32			32	35	
	1920	Stairway inside handrail					58.50			58.50	64	
	1930	Stairway outside handrail					80.50			80.50	88.50	
	1940	Walk-thru frame guardrail				▼	40.50			40.50	44.50	
	2000	Steel tubular, regular, rent/mo.										
	2100	Frames 3′ high 5′ wide				Ea.	3.75			3.75	4.13	
	2150	5′ high 5′ wide					3.75			3.75	4.13	
	2200	6′-4″ high 5′ wide					3.75			3.75	4.13	
	2250	7′-6″ high 6′ wide					7			7	7.70	
	2500	Accessories, cross braces					.60			.60	.66	
	2550	Guardrail post					1			1	1.10	
	2600	Guardrail 7′ section					.75			.75	.83	
	2650	Screw jacks & plates					1.50			1.50	1.65	
	2700	Sidearm brackets					1.50			1.50	1.65	
	2750	8″ casters					6			6	6.60	
	2800	Outrigger for rolling tower					3			3	3.30	
	2850	Plank 2″ x 10″ x 16′-0″					5			5	5.50	
	2900	Stairway section					10			10	11	
	2910	Stairway starter bar					.10			.10	.11	
	2920	Stairway inside handrail					5			5	5.50	
	2930	Stairway outside handrail					5			5	5.50	
	2940	Walk-thru frame guardrail				▼	2			2	2.20	
	3000	Steel tubular, heavy duty shoring, rent/mo.										
	3250	5′ high 2′ & 4′ wide				Ea.	5			5	5.50	
	3300	6′ high 2′ & 4′ wide					5			5	5.50	
	3500	Accessories, cross braces					1			1	1.10	
	3600	U - head, 8″ x 8″					1			1	1.10	
	3650	J - head, 4″ x 8″					1			1	1.10	
	3700	Base plate, 8″ x 8″					1			1	1.10	
	3750	Leveling jack					2			2	2.20	
	5700	Planks, 2x10x16′-0″, labor only, erect or remove to 50′ H	3 Carp	144	.167			5.70		5.70	8.90	
	5800	Over 50′ high	4 Carp	160	.200	▼		6.85		6.85	10.65	
	6000	Heavy duty shoring for elevated slab forms to 8′-2″ high, floor area										
	6010	incl. erection or dismantle labor										
	6100	1 use/month	4 Carp	36	.889	C.S.F.	29.50	30.50		60	80	
	6150	2 uses/month	″	36	.889	″	14.80	30.50		45.30	64	
	6500	To 14′-8″ high										
	6600	1 use/month	4 Carp	18	1.778	C.S.F.	43	61		104	143	
	6650	2 uses/month	″	18	1.778	″	21.50	61		82.50	119	
755	0010	**SCAFFOLDING SPECIALTIES**										**75**
	1200	Sidewalk bridge, heavy duty steel posts & beams, including										
	1210	parapet protection & waterproofing										
	1220	8′ to 10′ wide, 2 posts	3 Carp	15	1.600	L.F.	34.50	55		89.50	124	
	1230	3 posts	″	10	2.400	″	53.50	82		135.50	187	
	1500	Sidewalk bridge using tubular steel										
	1510	scaffold frames, including planking	3 Carp	45	.533	L.F.	4.72	18.25		22.97	33.50	
	1600	For 2 uses per month, deduct from all above					50%					
	1700	For 1 use every 2 months, add to all above					100%					
	1900	Catwalks, 20″ wide, no guardrails, 7′ span, buy				Ea.	138			138	151	
	2000	10′ span, buy				″	176			176	194	
	2800	Hand winch-operated masons scaffolding, no plank										

01540 | Construction Aids

		CREW	DAILY OUTPUT	LABOR-HOURS	UNIT	2005 BARE COSTS MAT.	LABOR	EQUIP.	TOTAL	TOTAL INCL O&P	
755	2810	plank moving not required										**755**
	2900	98' long, 10'-6" high, buy				Ea.	28,900			28,900	31,700	
	3000	Rent per month					1,150			1,150	1,275	
	3100	28'-6" high, buy					35,400			35,400	38,900	
	3200	Rent per month					1,425			1,425	1,550	
	3400	196' long, 28'-6" high, buy					68,500			68,500	75,000	
	3500	Rent per month					2,725			2,725	3,000	
	3600	64'-6" high, buy					93,500			93,500	103,000	
	3700	Rent per month					3,750			3,750	4,125	
	3720	Putlog, standard, 8' span, with hangers, buy					67			67	74	
	3730	Rent per month					10			10	11	
	3750	12' span, buy					101			101	111	
	3755	Rent per month					15			15	16.50	
	3760	Trussed type, 16' span, buy					231			231	254	
	3770	Rent per month					20			20	22	
	3790	22' span, buy					277			277	305	
	3795	Rent per month					30			30	33	
	3800	Rolling ladders with handrails, 30" wide, buy, 2 step					173			173	191	
	4000	7 step					585			585	645	
	4050	10 step					765			765	845	
	4100	Rolling towers, buy, 5' wide, 7' long, 10' high					1,250			1,250	1,375	
	4200	For 5' high added sections, to buy, add					209			209	230	
	4300	Complete incl. wheels, railings, outriggers,										
	4350	21' high, to buy				Ea.	2,125			2,125	2,325	
	4400	Rent/month				"	159			159	175	
800	0010	**TARPAULINS** Cotton duck, 10 oz. to 13.13 oz. per S.Y., minimum				S.F.	.49			.49	.54	**800**
	0050	Maximum					.58			.58	.64	
	0100	Polyvinyl coated nylon, 14 oz. to 18 oz., minimum					.48			.48	.53	
	0150	Maximum					.68			.68	.75	
	0200	Reinforced polyethylene 3 mils thick, white					.11			.11	.12	
	0300	4 mils thick, white, clear or black					.14			.14	.15	
	0400	5.5 mils thick, clear					.19			.19	.21	
	0500	White, fire retardant					.18			.18	.20	
	0600	7.5 mils, oil resistant, fire retardant					.19			.19	.21	
	0700	8.5 mils, black					.24			.24	.26	
	0710	Woven polyethylene, 6 mils thick					.48			.48	.53	
	0730	Polyester reinforced w/ integral fastening system 11 mils thick					1.07			1.07	1.18	
	0740	Mylar polyester, non-reinforced, 7 mils thick					1.17			1.17	1.29	
820	0010	**SMALL TOOLS** As % of contractor's work, minimum				Total					.50%	**820**
	0100	Maximum				"					2%	

01550 | Vehicular Access & Parking

		CREW	DAILY OUTPUT	LABOR-HOURS	UNIT	2005 BARE COSTS MAT.	LABOR	EQUIP.	TOTAL	TOTAL INCL O&P	
700	0010	**ROADS AND SIDEWALKS** Temporary										**700**
	0050	Roads, gravel fill, no surfacing, 4" gravel depth	B-14	715	.067	S.Y.	3.23	1.89	.30	5.42	6.80	
	0100	8" gravel depth	"	615	.078	"	6.45	2.19	.35	8.99	10.90	
	1000	Ramp, 3/4" plywood on 2" x 6" joists, 16" O.C.	2 Carp	300	.053	S.F.	1.48	1.83		3.31	4.47	
	1100	On 2" x 10" joists, 16" O.C.	"	275	.058		2.11	1.99		4.10	5.40	
	2200	Sidewalks, 2" x 12" planks, 2 uses	1 Carp	350	.023		.80	.78		1.58	2.10	
	2300	Exterior plywood, 2 uses, 1/2" thick		750	.011		.31	.37		.68	.91	
	2400	5/8" thick		650	.012		.39	.42		.81	1.09	
	2500	3/4" thick		600	.013		.47	.46		.93	1.23	

Important: See the Reference Section for critical supporting data - Reference Nos., Crews, & City Cost Indexes

01560	Barriers & Enclosures	CREW	DAILY OUTPUT	LABOR-HOURS	UNIT	2005 BARE COSTS				TOTAL INCL O&P	
						MAT.	LABOR	EQUIP.	TOTAL		
100 0010	**BARRICADES** 5' high, 3 rail @ 2" x 8", fixed	2 Carp	30	.533	L.F.	11	18.25		29.25	40.50	**10**
0150	Movable		20	.800		11	27.50		38.50	54.50	
1000	Guardrail, wooden, 3' high, 1" x 6", on 2" x 4" posts		200	.080		1.02	2.74		3.76	5.40	
1100	2" x 6", on 4" x 4" posts	↓	165	.097		1.85	3.32		5.17	7.20	
1200	Portable metal with base pads, buy					15.50			15.50	17.05	
1250	Typical installation, assume 10 reuses	2 Carp	600	.027	↓	1.60	.91		2.51	3.18	
1300	Barricade tape, polyethelyne, 7 mil, 3" wide x 500' long roll				Ea.	25			25	27.50	
250 0010	**TEMPORARY FENCING** Chain link, 11 ga, 5' high	2 Clab	400	.040	L.F.	4.65	1.07		5.72	6.75	**25**
0100	6' high		300	.053		4.48	1.42		5.90	7.15	
0200	Rented chain link, 6' high, to 1000' (up to 12 mo.)		400	.040		2.94	1.07		4.01	4.89	
0250	Over 1000' (up to 12 mo.)	↓	300	.053		2.13	1.42		3.55	4.56	
0350	Plywood, painted, 2" x 4" frame, 4' high	A-4	135	.178		5.30	5.85		11.15	14.85	
0400	4" x 4" frame, 8' high	"	110	.218		9.90	7.20		17.10	22	
0500	Wire mesh on 4" x 4" posts, 4' high	2 Carp	100	.160		7.90	5.50		13.40	17.25	
0550	8' high	"	80	.200	↓	12.10	6.85		18.95	24	
400 0010	**TEMPORARY CONSTRUCTION** See also division 01530										**40**
800 0010	**WATCHMAN** Service, monthly basis, uniformed person, minimum				Hr.					9	**80**
0100	Maximum									16	
0200	Person and command dog, minimum									12	
0300	Maximum				↓					17.70	
0500	Sentry dog, leased, with job patrol (yard dog), 1 dog				Week					230	
0600	2 dogs				"					300	
0800	Purchase, trained sentry dog, minimum				Ea.					1,000	
0900	Maximum				"					2,500	

01580	Project Signs										
700 0010	**SIGNS** High Intensity reflectorized, no posts, buy				S.F.	16.40			16.40	18	**70**

01590 | Equipment Rental

		UNIT	HOURLY OPER. COST	RENT PER DAY	RENT PER WEEK	RENT PER MONTH	CREW EQUIPMENT COST/DAY	
100	0010	**CONCRETE EQUIPMENT RENTAL**							**100**
	0100	without operators							
	0150	For batch plant, see div. 01590-500							
	0200	Bucket, concrete lightweight, 1/2 C.Y.	Ea.	.55	15.35	46	138	13.60	
	0300	1 C.Y.		.60	18.65	56	168	16	
	0400	1-1/2 C.Y.		.75	25.50	76	228	21.20	
	0500	2 C.Y.		.80	30.50	91	273	24.60	
	0580	8 C.Y.		4.40	200	600	1,800	155.20	
	0600	Cart, concrete, self propelled, operator walking, 10 C.F.		2.10	55	165	495	49.80	
	0700	Operator riding, 18 C.F.		3.35	83.50	250	750	76.80	
	0800	Conveyer for concrete, portable, gas, 16" wide, 26' long		7.35	117	350	1,050	128.80	
	0900	46' long		7.70	142	425	1,275	146.60	
	1000	56' long		7.85	150	450	1,350	152.80	
	1100	Core drill, electric, 2-1/2 H.P., 1" to 8" bit diameter		1.61	62.50	187	560	50.30	
	1150	11 HP, 8" to 18" cores		6.61	83.50	250.80	750	103.05	
	1200	Finisher, concrete floor, gas, riding trowel, 48" diameter		4.95	86.50	260	780	91.60	
	1300	Gas, manual, 3 blade, 36" trowel		1.05	16.65	50	150	18.40	
	1400	4 blade, 48" trowel		1.50	21	63	189	24.60	
	1500	Float, hand-operated (Bull float) 48" wide		.08	13.35	40	120	8.65	
	1570	Curb builder, 14 H.P., gas, single screw		9.50	195	585	1,750	193	
	1590	Double screw		9.90	228	685	2,050	216.20	
	1600	Grinder, concrete and terrazzo, electric, floor		2.10	88.50	266	800	70	
	1700	Wall grinder		1.05	44.50	133	400	35	
	1800	Mixer, powered, mortar and concrete, gas, 6 C.F., 18 H.P.		5.25	102	305	915	103	
	1900	10 C.F., 25 H.P.		6.40	120	360	1,075	123.20	
	2000	16 C.F.		6.70	143	430	1,300	139.60	
	2100	Concrete, stationary, tilt drum, 2 C.Y.		5.25	200	600	1,800	162	
	2120	Pump, concrete, truck mounted 4" line 80' boom		21.65	885	2,655	7,975	704.20	
	2140	5" line, 110' boom		28.05	1,175	3,510	10,500	926.40	
	2160	Mud jack, 50 C.F. per hr.		4.82	115	346	1,050	107.75	
	2180	225 C.F. per hr.		7.56	154	461.60	1,375	152.80	
	2190	Shotcrete pump rig, 12 CY/hr		11.10	225	675	2,025	223.80	
	2600	Saw, concrete, manual, gas, 18 H.P.		3.60	35	105	315	49.80	
	2650	Self-propelled, gas, 30 H.P.		7.05	96.50	290	870	114.40	
	2700	Vibrators, concrete, electric, 60 cycle, 2 H.P.		.37	10.65	32	96	9.35	
	2800	3 H.P.		.56	15.65	47	141	13.90	
	2900	Gas engine, 5 H.P.		.95	18.35	55	165	18.60	
	3000	8 H.P.		1.35	22	66	198	24	
	3050	Vibrating screed, gas engine, 8 H.P.		1.68	49.50	148	445	43.05	
	3100	Concrete transit mixer, hydraulic drive							
	3120	6 x 4, 250 H.P., 8 C.Y., rear discharge		34.95	550	1,650	4,950	609.60	
	3200	Front discharge		40.65	670	2,015	6,050	728.20	
	3300	6 x 6, 285 H.P., 12 C.Y., rear discharge		40.60	670	2,010	6,025	726.80	
	3400	Front discharge		41.55	680	2,040	6,125	740.40	
200	0010	**EARTHWORK EQUIPMENT RENTAL** Without operators							**200**
	0040	Aggregate spreader, push type 8' to 12' wide	Ea.	1.80	34.50	103	310	35	
	0045	Tailgate type, 8' wide	"	1.75	31.50	95	285	33	
	0050	Augers for vertical drilling							
	0055	Earth auger, truck-mounted, for fence & sign posts	Ea.	8.30	485	1,455	4,375	357.40	
	0060	For borings and monitoring wells		29.20	620	1,855	5,575	604.60	
	0070	Earth auger, portable, trailer mounted		1.55	22.50	67	201	25.80	
	0075	Earth auger, truck-mounted, for caissons, water wells, utility poles		153.10	3,250	9,770	29,300	3,179	
	0080	Auger, horizontal boring machine, 12" to 36" diameter, 45 H.P.		15.85	182	545	1,625	235.80	
	0090	12" to 48" diameter, 65 H.P.		22.30	325	970	2,900	372.40	
	0095	Auger, for fence posts, gas engine, hand held		.30	4.33	13	39	5	
	0100	Excavator, diesel hydraulic, crawler mounted, 1/2 C.Y. cap.		15.80	345	1,030	3,100	332.40	
	0120	5/8 C.Y. capacity		19.30	465	1,390	4,175	432.40	
	0140	3/4 C.Y. capacity		22.50	490	1,470	4,400	474	

Important: See the Reference Section for critical supporting data - Reference Nos., Crews, & City Cost Indexes

01590 | Equipment Rental

		UNIT	HOURLY OPER. COST	RENT PER DAY	RENT PER WEEK	RENT PER MONTH	CREW EQUIPMENT COST/DAY
200	0150	1 C.Y. capacity	Ea.	27.10	570	1,705	5,125	557.80	200
	0200	1-1/2 C.Y. capacity		33.55	755	2,260	6,775	720.40
	0300	2 C.Y. capacity		42.35	950	2,855	8,575	909.80
	0320	2-1/2 C.Y. capacity		55.75	1,275	3,845	11,500	1,215
	0340	3-1/2 C.Y. capacity		94.45	2,100	6,320	19,000	2,020
	0341	Attachments						
	0342	Bucket thumbs		2.50	208	625	1,875	145
	0345	Grapples		1	200	600.40	1,800	128.10
	0350	Gradall type, truck mounted, 3 ton @ 15' radius, 5/8 C.Y.		38.35	885	2,655	7,975	837.80
	0370	1 C.Y. capacity		44.25	1,025	3,085	9,250	971
	0400	Backhoe-loader, 40 to 45 H.P., 5/8 C.Y. capacity		8.40	177	530	1,600	173.20
	0450	45 H.P. to 60 H.P., 3/4 C.Y. capacity		10.60	218	655	1,975	215.80
	0460	80 H.P., 1-1/4 C.Y. capacity		13.30	250	750	2,250	256.40
	0470	112 H.P., 1-1/2 C.Y. capacity		18.10	390	1,165	3,500	377.80
	0480	Attachments						
	0482	Compactor, 20,000 lb		4.25	117	350	1,050	104
	0485	Hydraulic hammer, 750 ft-lbs		1.95	68.50	205	615	56.60
	0486	Hydraulic hammer, 1200 ft-lbs		4	133	400	1,200	112
	0500	Brush chipper, gas engine, 6" cutter head, 35 H.P.		5.90	93.50	280	840	103.20
	0550	12" cutter head, 130 H.P.		9.35	150	450	1,350	164.80
	0600	15" cutter head, 165 H.P.		13.40	158	475	1,425	202.20
	0750	Bucket, clamshell, general purpose, 3/8 C.Y.		1	35	105	315	29
	0800	1/2 C.Y.		1.10	41.50	125	375	33.80
	0850	3/4 C.Y.		1.25	51.50	155	465	41
	0900	1 C.Y.		1.30	55	165	495	43.40
	0950	1-1/2 C.Y.		2.05	75	225	675	61.40
	1000	2 C.Y.		2.15	85	255	765	68.20
	1010	Bucket, dragline, medium duty, 1/2 C.Y.		.60	22.50	67	201	18.20
	1020	3/4 C.Y.		.60	23.50	70	210	18.80
	1030	1 C.Y.		.65	25.50	76	228	20.40
	1040	1-1/2 C.Y.		.95	38.50	115	345	30.60
	1050	2 C.Y.		1.05	43.50	130	390	34.40
	1070	3 C.Y.		1.55	58.50	175	525	47.40
	1200	Compactor, manually guided 2-drum vibratory smooth roller, 7.5 H.P.		4.60	143	430	1,300	122.80
	1250	Rammer compactor, gas, 1000 lb. blow		1.60	36.50	110	330	34.80
	1300	Vibratory plate, gas, 18" plate, 3000 lb. blow		1.50	22	66	198	25.20
	1350	21" plate, 5000 lb. blow		1.80	37.50	112	335	36.80
	1370	Curb builder/extruder, 14 H.P., gas, single screw		9.50	197	590	1,775	194
	1390	Double screw		9.90	228	685	2,050	216.20
	1500	Disc harrow attachment, for tractor		.36	60.50	181	545	39.10
	1750	Extractor, piling, see lines 2500 to 2750						
	1810	Feller buncher, shearing & accumulating trees, 100 H.P.	Ea.	19.80	455	1,370	4,100	432.40
	1860	Grader, self-propelled, 25,000 lb.		18.20	400	1,205	3,625	386.60
	1910	30,000 lb.		20.70	485	1,455	4,375	456.60
	1920	40,000 lb.		30.50	730	2,190	6,575	682
	1930	55,000 lb.		40.25	1,025	3,080	9,250	938
	1950	Hammer, pavement demo., hyd., gas, self-prop., 1000 to 1250 lb.		20.15	390	1,170	3,500	395.20
	2000	Diesel 1300 to 1500 lb.		27.80	590	1,765	5,300	575.40
	2050	Pile driving hammer, steam or air, 4150 ft.-lb. @ 225 BPM		6.35	275	825	2,475	215.80
	2100	8750 ft.-lb. @ 145 BPM		8.25	450	1,350	4,050	336
	2150	15,000 ft.-lb. @ 60 BPM		8.60	485	1,455	4,375	359.80
	2200	24,450 ft.-lb. @ 111 BPM		11.35	535	1,605	4,825	411.80
	2250	Leads, 15,000 ft.-lb. hammers	L.F.	.03	1.69	5.06	15.20	1.25
	2300	24,450 ft.-lb. hammers and heavier	"	.05	2.67	8	24	2
	2350	Diesel type hammer, 22,400 ft.-lb.	Ea.	24	620	1,860	5,575	564
	2400	41,300 ft.-lb.		32.20	675	2,025	6,075	662.60
	2450	141,000 ft.-lb.		60.90	1,475	4,430	13,300	1,373
	2500	Vib. elec. hammer/extractor, 200 KW diesel generator, 34 H.P.		26.25	655	1,965	5,900	603

01590 | Equipment Rental

		UNIT	HOURLY OPER. COST	RENT PER DAY	RENT PER WEEK	RENT PER MONTH	CREW EQUIPMENT COST/DAY	
200	**2550**	80 H.P.	Ea.	44.80	960	2,885	8,650	935.40	**200**
	2600	150 H.P.		64	1,475	4,455	13,400	1,403	
	2700	Extractor, steam or air, 700 ft.-lb.		14.50	410	1,235	3,700	363	
	2750	1000 ft.-lb.		16.60	515	1,545	4,625	441.80	
	2800	Log chipper, up to 22" diam, 600 H.P.		38.53	1,250	3,770	11,300	1,062	
	2850	Logger, for skidding & stacking logs, 150 H.P.		33.80	790	2,375	7,125	745.40	
	2900	Rake, spring tooth, with tractor		7.99	213	638	1,925	191.50	
	3000	Roller, vibratory, tandem, smooth drum, 20 H.P.		5.05	112	335	1,000	107.40	
	3050	35 H.P.		7.15	203	610	1,825	179.20	
	3100	Towed type vibratory compactor, smooth drum, 50 H.P.		31.95	560	1,680	5,050	591.60	
	3150	Sheepsfoot, 50 H.P.		32.65	580	1,745	5,225	610.20	
	3170	Landfill compactor, 220 HP		46.75	1,150	3,470	10,400	1,068	
	3200	Pneumatic tire roller, 80 H.P.		8.35	292	875	2,625	241.80	
	3250	120 H.P.		13.60	505	1,515	4,550	411.80	
	3300	Sheepsfoot vibratory roller, 200 H.P.		35.35	855	2,560	7,675	794.80	
	3320	340 H.P.		48.90	1,225	3,670	11,000	1,125	
	3350	Smooth drum vibratory roller, 75 H.P.		13.60	395	1,180	3,550	344.80	
	3400	125 H.P.		17.15	495	1,480	4,450	433.20	
	3410	Rotary mower, brush, 60", with tractor		11.35	230	690	2,075	228.80	
	3450	Scrapers, towed type, 9 to 12 C.Y. capacity		4.12	196	589	1,775	150.75	
	3500	12 to 17 C.Y. capacity		1.51	262	785.60	2,350	169.20	
	3550	Scrapers, self-propelled, 4 x 4 drive, 2 engine, 14 C.Y. capacity		82.40	1,500	4,465	13,400	1,552	
	3600	2 engine, 24 C.Y. capacity		119.45	2,275	6,805	20,400	2,317	
	3640	32 - 44 C.Y. capacity		142.65	2,650	7,935	23,800	2,728	
	3650	Self-loading, 11 C.Y. capacity		40.55	825	2,475	7,425	819.40	
	3700	22 C.Y. capacity		76.40	1,725	5,185	15,600	1,648	
	3710	Screening plant 110 H.P. w/ 5' x 10' screen		21.75	370	1,105	3,325	395	
	3720	5' x 16' screen		23.80	465	1,390	4,175	468.40	
	3850	Shovels, see Cranes division 01590-600							
	3860	Shovel/backhoe bucket, 1/2 C.Y.	Ea.	1.80	55	165	495	47.40	
	3870	3/4 C.Y.		1.85	61.50	185	555	51.80	
	3880	1 C.Y.		1.95	71.50	215	645	58.60	
	3890	1-1/2 C.Y.		2.10	85	255	765	67.80	
	3910	3 C.Y.		2.40	118	355	1,075	90.20	
	3950	Stump chipper, 18" deep, 30 H.P.		4.56	57	171	515	70.70	
	4110	Tractor, crawler, with bulldozer, torque converter, diesel 80 H.P.		16.10	310	930	2,800	314.80	
	4150	105 H.P.		22.10	460	1,385	4,150	453.80	
	4200	140 H.P.		27.25	590	1,770	5,300	572	
	4260	200 H.P.		40.95	985	2,960	8,875	919.60	
	4310	300 H.P.		53.10	1,275	3,850	11,600	1,195	
	4360	410 H.P.		71.95	1,575	4,740	14,200	1,524	
	4370	500 H.P.		95.25	2,125	6,365	19,100	2,035	
	4380	700 H.P.		142.20	3,325	10,005	30,000	3,139	
	4400	Loader, crawler, torque conv., diesel, 1-1/2 C.Y., 80 H.P.		14.95	320	965	2,900	312.60	
	4450	1-1/2 to 1-3/4 C.Y., 95 H.P.		17.35	385	1,150	3,450	368.80	
	4510	1-3/4 to 2-1/4 C.Y., 130 H.P.		23.60	605	1,815	5,450	551.80	
	4530	2-1/2 to 3-1/4 C.Y., 190 H.P.		35.50	840	2,520	7,550	788	
	4560	3-1/2 to 5 C.Y., 275 H.P.		46.95	1,200	3,590	10,800	1,094	
	4610	Tractor loader, wheel, torque conv., 4 x 4, 1 to 1-1/4 C.Y., 65 H.P.		9.85	192	575	1,725	193.80	
	4620	1-1/2 to 1-3/4 C.Y., 80 H.P.		12.50	235	705	2,125	241	
	4650	1-3/4 to 2 C.Y., 100 H.P.		14.20	278	835	2,500	280.60	
	4710	2-1/2 to 3-1/2 C.Y., 130 H.P.		15.25	305	915	2,750	305	
	4730	3 to 4-1/2 C.Y., 170 H.P.		20.95	465	1,390	4,175	445.60	
	4760	5-1/4 to 5-3/4 C.Y., 270 H.P.		34.55	705	2,120	6,350	700.40	
	4810	7 to 8 C.Y., 375 H.P.		58.95	1,250	3,755	11,300	1,223	
	4870	12-1/2 C.Y., 690 H.P.		82.30	1,950	5,860	17,600	1,830	
	4880	Wheeled, skid steer, 10 C.F., 30 H.P. gas		8.80	138	415	1,250	153.40	
	4890	1 C.Y., 78 H.P., diesel		10.35	192	575	1,725	197.80	

Important: See the Reference Section for critical supporting data - Reference Nos., Crews, & City Cost Indexes

01590 | Equipment Rental

			UNIT	HOURLY OPER. COST	RENT PER DAY	RENT PER WEEK	RENT PER MONTH	CREW EQUIPMENT COST/DAY	
200	4891	Attachments for all skid steer loaders							**200**
	4892	Auger	Ea.	.40	66.50	200	600	43.20	
	4893	Backhoe		.65	109	326	980	70.40	
	4894	Broom		.63	105	314	940	67.85	
	4895	Forks		.22	37.50	112	335	24.15	
	4896	Grapple		.50	83	249	745	53.80	
	4897	Concrete hammer		.98	163	489	1,475	105.65	
	4898	Tree spade		1	167	500	1,500	108	
	4899	Trencher		.68	114	342	1,025	73.85	
	4900	Trencher, chain, boom type, gas, operator walking, 12 H.P.		2.75	43.50	130	390	48	
	4910	Operator riding, 40 H.P.		8.55	242	725	2,175	213.40	
	5000	Wheel type, diesel, 4' deep, 12" wide		46.90	735	2,205	6,625	816.20	
	5100	Diesel, 6' deep, 20" wide		63.75	1,575	4,750	14,300	1,460	
	5150	Ladder type, diesel, 5' deep, 8" wide		23.10	705	2,115	6,350	607.80	
	5200	Diesel, 8' deep, 16" wide		56.80	1,600	4,830	14,500	1,420	
	5210	Tree spade, self-propelled		9.60	267	800	2,400	236.80	
	5250	Truck, dump, tandem, 12 ton payload		20.45	272	815	2,450	326.60	
	5300	Three axle dump, 16 ton payload		27.95	420	1,265	3,800	476.60	
	5350	Dump trailer only, rear dump, 16-1/2 C.Y.		4.25	115	345	1,025	103	
	5400	20 C.Y.		4.60	132	395	1,175	115.80	
	5450	Flatbed, single axle, 1-1/2 ton rating		11.85	56.50	170	510	128.80	
	5500	3 ton rating		15.15	91.50	275	825	176.20	
	5550	Off highway rear dump, 25 ton capacity		40.40	980	2,945	8,825	912.20	
	5600	35 ton capacity		41.30	1,000	3,015	9,050	933.40	
	5610	50 ton capacity		53.15	1,250	3,780	11,300	1,181	
	5620	65 ton capacity		56.85	1,375	4,105	12,300	1,276	
	5630	100 ton capacity		73	1,775	5,325	16,000	1,649	
	6000	Vibratory plow, 25 H.P., walking		3.90	58.50	175	525	66.20	
400	0010	**GENERAL EQUIPMENT RENTAL** Without operators							**400**
	0150	Aerial lift, scissor type, to 15' high, 1000 lb. cap., electric	Ea.	2.30	41.50	125	375	43.40	
	0160	To 25' high, 2000 lb. capacity		2.70	60	180	540	57.60	
	0170	Telescoping boom to 40' high, 500 lb. capacity, gas		11.45	268	805	2,425	252.60	
	0180	To 45' high, 500 lb. capacity		12.30	310	925	2,775	283.40	
	0190	To 60' high, 600 lb. capacity		14.25	410	1,230	3,700	360	
	0195	Air compressor, portable, 6.5 CFM, electric		.42	14.65	44	132	12.15	
	0196	Gasoline		.51	22	66	198	17.30	
	0200	Air compressor, portable, gas engine, 60 C.F.M.		4.50	32.50	97	291	55.40	
	0300	160 C.F.M.		7.80	41.50	125	375	87.40	
	0400	Diesel engine, rotary screw, 250 C.F.M.		7.80	83.50	250	750	112.40	
	0500	365 C.F.M.		10.35	107	320	960	146.80	
	0550	450 C.F.M.		12.55	127	380	1,150	176.40	
	0600	600 C.F.M.		21.35	212	635	1,900	297.80	
	0700	750 C.F.M.		23.05	222	665	2,000	317.40	
	0800	For silenced models, small sizes, add		3%	5%	5%	5%		
	0900	Large sizes, add		5%	7%	7%	7%		
	0920	Air tools and accessories							
	0930	Breaker, pavement, 60 lb.	Ea.	.40	11	33	99	9.80	
	0940	80 lb.		.40	12.65	38	114	10.80	
	0950	Drills, hand (jackhammer) 65 lb.		.45	15	45	135	12.60	
	0960	Track or wagon, swing boom, 4" drifter		35.05	640	1,920	5,750	664.40	
	0970	5" drifter		47.70	735	2,205	6,625	822.60	
	0975	Track mounted quarry drill, 6" diameter drill		50.45	795	2,380	7,150	879.60	
	0980	Dust control per drill		.78	12	36	108	13.45	
	0990	Hammer, chipping, 12 lb.		.40	21.50	64	192	16	
	1000	Hose, air with couplings, 50' long, 3/4" diameter		.03	5.65	17	51	3.65	
	1100	1" diameter		.03	5.65	17	51	3.65	
	1200	1-1/2" diameter		.04	7.35	22	66	4.70	
	1300	2" diameter		.10	17.35	52	156	11.20	

01590 | Equipment Rental

		UNIT	HOURLY OPER. COST	RENT PER DAY	RENT PER WEEK	RENT PER MONTH	CREW EQUIPMENT COST/DAY	
400	1400	2-1/2" diameter	Ea.	.12	19.65	59	177	12.75	400
	1410	3" diameter		.16	27.50	82	246	17.70	
	1450	Drill, steel, 7/8" x 2'		.05	8	24	72	5.20	
	1460	7/8" x 6'		.06	9.35	28	84	6.10	
	1520	Moil points		.03	4.33	13	39	2.85	
	1525	Pneumatic nailer w/accessories		.41	27	81	243	19.50	
	1530	Sheeting driver for 60 lb. breaker		.10	6.95	20.80	62.50	4.95	
	1540	For 90 lb. breaker		.15	10	30	90	7.20	
	1550	Spade, 25 lb.		.35	6.35	19	57	6.60	
	1560	Tamper, single, 35 lb.		.58	38.50	115	345	27.65	
	1570	Triple, 140 lb.		.87	57.50	173	520	41.55	
	1580	Wrenches, impact, air powered, up to 3/4" bolt		.25	7.35	22	66	6.40	
	1590	Up to 1-1/4" bolt		.35	15.35	46	138	12	
	1600	Barricades, barrels, reflectorized, 1 to 50 barrels		.02	2.60	7.80	23.50	1.70	
	1610	100 to 200 barrels		.01	1.93	5.80	17.40	1.25	
	1620	Barrels with flashers, 1 to 50 barrels		.02	3.27	9.80	29.50	2.10	
	1630	100 to 200 barrels		.02	2.60	7.80	23.50	1.70	
	1640	Barrels with steady burn type C lights		.03	4.33	13	39	2.85	
	1650	Illuminated board, trailer mounted, with generator		.65	117	350	1,050	75.20	
	1670	Portable barricade, stock, with flashers, 1 to 6 units		.02	3.27	9.80	29.50	2.10	
	1680	25 to 50 units		.02	3.03	9.10	27.50	2	
	1690	Butt fusion machine, electric		22.35	425	1,270	3,800	432.80	
	1695	Electro fusion machine		8.85	168	505	1,525	171.80	
	1700	Carts, brick, hand powered, 1000 lb. capacity		.30	49.50	148	445	32	
	1800	Gas engine, 1500 lb., 7-1/2' lift		2.77	94	282	845	78.55	
	1822	Dehumidifier, medium, 6 lb/hr, 150 CFM		.68	41.50	124	370	30.25	
	1824	Large, 18 lb/hr, 600 CFM		1.36	82.50	248	745	60.50	
	1830	Distributor, asphalt, trailer mtd, 2000 gal., 38 H.P. diesel		7.35	243	730	2,200	204.80	
	1840	3000 gal., 38 H.P. diesel		8.35	278	835	2,500	233.80	
	1850	Drill, rotary hammer, electric, 1-1/2" diameter		.40	24.50	73	219	17.80	
	1860	Carbide bit for above		.03	5.35	16	48	3.45	
	1865	Rotary, crawler, 250 H.P.		87.35	1,600	4,835	14,500	1,666	
	1870	Emulsion sprayer, 65 gal., 5 H.P. gas engine		1.80	72	216	650	57.60	
	1880	200 gal., 5 H.P. engine		4.85	120	360	1,075	110.80	
	1920	Floodlight, mercury vapor, or quartz, on tripod							
	1930	1000 watt	Ea.	.30	12	36	108	9.60	
	1940	2000 watt		.52	22	66	198	17.35	
	1950	Floodlights, trailer mounted with generator, 1 - 300 watt light		2.45	65	195	585	58.60	
	1960	2 - 1000 watt lights		3.30	108	325	975	91.40	
	2000	4 - 300 watt lights		2.75	76.50	230	690	68	
	2020	Forklift, wheeled, for brick, 18', 3000 lb., 2 wheel drive, gas		13.70	177	530	1,600	215.60	
	2040	28', 4000 lb., 4 wheel drive, diesel		11.75	232	695	2,075	233	
	2050	For rough terrain, 8000 lb., 16' lift, 68 H.P.		15.70	360	1,075	3,225	340.60	
	2060	For plant, 4 T. capacity, 80 H.P., 2 wheel drive, gas		7.80	83.50	250	750	112.40	
	2080	10 T. capacity, 120 H.P., 2 wheel drive, diesel		11.95	157	470	1,400	189.60	
	2100	Generator, electric, gas engine, 1.5 KW to 3 KW		1.75	15	45	135	23	
	2200	5 KW		2.40	21.50	65	195	32.20	
	2300	10 KW		3.95	46.50	140	420	59.60	
	2400	25 KW		7.65	71.50	215	645	104.20	
	2500	Diesel engine, 20 KW		5.95	61.50	185	555	84.60	
	2600	50 KW		11.05	68.50	205	615	129.40	
	2700	100 KW		16.20	78.50	235	705	176.60	
	2800	250 KW		45.65	142	425	1,275	450.20	
	2850	Hammer, hydraulic, for mounting on boom, to 500 ft.-lb.		1.75	63.50	190	570	52	
	2860	1000 ft.-lb.		3.10	102	305	915	85.80	
	2900	Heaters, space, oil or electric, 50 MBH		.92	9.65	29	87	13.15	
	3000	100 MBH		1.66	13.65	41	123	21.50	
	3100	300 MBH		5.33	35	105	315	63.65	

Important: See the Reference Section for critical supporting data - Reference Nos., Crews, & City Cost Indexes

01590 | Equipment Rental

			UNIT	HOURLY OPER. COST	RENT PER DAY	RENT PER WEEK	RENT PER MONTH	CREW EQUIPMENT COST/DAY	
400	3150	500 MBH	Ea.	10.70	50	150	450	115.60	400
	3200	Hose, water, suction with coupling, 20' long, 2" diameter		.02	4	12	36	2.55	
	3210	3" diameter		.03	6.65	20	60	4.25	
	3220	4" diameter		.03	8	24	72	5.05	
	3230	6" diameter		.10	20	60	180	12.80	
	3240	8" diameter		.25	41.50	125	375	27	
	3250	Discharge hose with coupling, 50' long, 2" diameter		.01	3	9	27	1.90	
	3260	3" diameter		.01	4	12	36	2.50	
	3270	4" diameter		.02	5.65	17	51	3.55	
	3280	6" diameter		.06	14	42	126	8.90	
	3290	8" diameter		.31	52	156	470	33.70	
	3295	Insulation blower		.11	7	21	63	5.10	
	3300	Ladders, extension type, 16' to 36' long		.17	28.50	86	258	18.55	
	3400	40' to 60' long		.26	36.50	110	330	24.10	
	3405	Lance for cutting concrete		2.78	105	315	945	85.25	
	3407	Lawn mower, rotary, 22", 5HP		.97	27	81	243	23.95	
	3408	48" self propelled		3.06	100	300	900	84.50	
	3410	Level, laser type, for pipe and sewer leveling		1.27	84.50	253	760	60.75	
	3430	Electronic		.76	50.50	151	455	36.30	
	3440	Laser type, rotating beam for grade control		1.04	69	207	620	49.70	
	3460	Builders level with tripod and rod		.07	12.35	37	111	7.95	
	3500	Light towers, towable, with diesel generator, 2000 watt		2.75	76.50	230	690	68	
	3600	4000 watt		3.30	108	325	975	91.40	
	3700	Mixer, powered, plaster and mortar, 6 C.F., 7 H.P.		1.30	26.50	80	240	26.40	
	3800	10 C.F., 9 H.P.		1.55	41.50	125	375	37.40	
	3850	Nailer, pneumatic		.41	27	81	243	19.50	
	3900	Paint sprayers complete, 8 CFM		.69	45.50	137	410	32.90	
	4000	17 CFM		1.10	73.50	220	660	52.80	
	4020	Pavers, bituminous, rubber tires, 8' wide, 50 H.P., diesel		26.80	775	2,320	6,950	678.40	
	4030	10' wide, 150 H.P.		59.45	1,300	3,895	11,700	1,255	
	4050	Crawler, 8' wide, 100 H.P., diesel		55.60	1,450	4,335	13,000	1,312	
	4060	10' wide, 150 H.P.		66.95	1,875	5,590	16,800	1,654	
	4070	Concrete paver, 12' to 24' wide, 250 H.P.		57.90	1,325	3,995	12,000	1,262	
	4080	Placer-spreader-trimmer, 24' wide, 300 H.P.		87.60	2,025	6,045	18,100	1,910	
	4100	Pump, centrifugal gas pump, 1-1/2", 4 MGPH		2.60	36.50	110	330	42.80	
	4200	2", 8 MGPH		3.25	41.50	125	375	51	
	4300	3", 15 MGPH		3.45	43.50	130	390	53.60	
	4400	6", 90 MGPH		15.75	152	455	1,375	217	
	4500	Submersible electric pump, 1-1/4", 55 GPM		.35	17	51	153	13	
	4600	1-1/2", 83 GPM		.41	19.65	59	177	15.10	
	4700	2", 120 GPM		.43	24.50	73	219	18.05	
	4800	3", 300 GPM		.90	33.50	100	300	27.20	
	4900	4", 560 GPM		6.16	135	405	1,225	130.30	
	5000	6", 1590 GPM		9.27	200	600	1,800	194.15	
	5100	Diaphragm pump, gas, single, 1-1/2" diameter		.78	38.50	116	350	29.45	
	5200	2" diameter		2.70	48.50	145	435	50.60	
	5300	3" diameter		2.75	48.50	145	435	51	
	5400	Double, 4" diameter		3.55	76.50	230	690	74.40	
	5500	Trash pump, self-priming, gas, 2" diameter		2.55	26.50	80	240	36.40	
	5600	Diesel, 4" diameter		6.05	53.50	160	480	80.40	
	5650	Diesel, 6" diameter		18.35	112	335	1,000	213.80	
	5655	Grout Pump		8.30	58.50	175	525	101.40	
	5660	Rollers, see division 01590-200							
	5700	Salamanders, L.P. gas fired, 100,000 BTU	Ea.	1.66	10.65	32	96	19.70	
	5705	50,000 BTU		1.25	7.65	23	69	14.60	
	5720	Sandblaster, portable, open top, 3 C.F. capacity		.40	20	60	180	15.20	
	5730	6 C.F. capacity		.65	29	87	261	22.60	
	5740	Accessories for above		.11	18.35	55	165	11.90	

01590 | Equipment Rental

		UNIT	HOURLY OPER. COST	RENT PER DAY	RENT PER WEEK	RENT PER MONTH	CREW EQUIPMENT COST/DAY	
400								400
5750	Sander, floor	Ea.	.72	17	51	153	15.95	
5760	Edger		.71	25	75	225	20.70	
5800	Saw, chain, gas engine, 18" long		1.20	16	48	144	19.20	
5900	36" long		.55	48.50	145	435	33.40	
5950	60" long		.55	50	150	450	34.40	
6000	Masonry, table mounted, 14" diameter, 5 H.P.		1.26	54.50	163	490	42.70	
6050	Portable cut-off, 8 H.P.		1.30	28.50	85	255	27.40	
6100	Circular, hand held, electric, 7-1/4" diameter		.20	7.35	22	66	6	
6200	12" diameter		.27	11	33	99	8.75	
6250	Wall saw, w/hydraulic power, 10 H.P		2.04	95	284.80	855	73.30	
6275	Shot blaster, walk behind, 20" wide		1.42	545	1,635	4,900	338.35	
6300	Steam cleaner, 100 gallons per hour		2.25	61.50	185	555	55	
6310	200 gallons per hour		3	75	225	675	69	
6340	Tar Kettle/Pot, 400 gallon		2.69	51.50	155	465	52.50	
6350	Torch, cutting, acetylene-oxygen, 150' hose		1.50	10	30	90	18	
6360	Hourly operating cost includes tips and gas		8.10				64.80	
6410	Toilet, portable chemical		.11	17.65	53	159	11.50	
6420	Recycle flush type		.13	21.50	65	195	14.05	
6430	Toilet, fresh water flush, garden hose,		.15	24.50	73	219	15.80	
6440	Hoisted, non-flush, for high rise		.13	21.50	64	192	13.85	
6450	Toilet, trailers, minimum		.22	36.50	110	330	23.75	
6460	Maximum		.66	110	330	990	71.30	
6465	Tractor, farm with attachment	▼	10.25	222	665	2,000	215	
6470	Trailer, office, see division 01520-500							
6500	Trailers, platform, flush deck, 2 axle, 25 ton capacity	Ea.	4.25	90	270	810	88	
6600	40 ton capacity		5.55	125	375	1,125	119.40	
6700	3 axle, 50 ton capacity		6	138	415	1,250	131	
6800	75 ton capacity		7.45	182	545	1,625	168.60	
6810	Trailer mounted cable reel for H.V. line work		4.47	213	639	1,925	163.55	
6820	Trailer mounted cable tensioning rig		8.75	415	1,250	3,750	320	
6830	Cable pulling rig	▼	55.44	2,375	7,120	21,400	1,868	
6850	Trailer, storage, see division 01520-500							
6900	Water tank, engine driven discharge, 5000 gallons	Ea.	5.60	120	360	1,075	116.80	
6925	10,000 gallons		7.70	168	505	1,525	162.60	
6950	Water truck, off highway, 6000 gallons		49.25	700	2,095	6,275	813	
7010	Tram car for H.V. line work, powered, 2 conductor		5.73	116	347	1,050	115.25	
7020	Transit (builder's level) with tripod		.07	12.35	37	111	7.95	
7030	Trench box, 3000 lbs. 6'x8'		.67	90.50	272	815	59.75	
7040	7200 lbs. 6'x20'		.82	136	409	1,225	88.35	
7050	8000 lbs., 8' x 16'		.88	147	442	1,325	95.45	
7060	9500 lbs., 8'x20'		1.19	199	597	1,800	128.90	
7065	11,000 lbs., 8'x24'		1.34	224	671	2,025	144.90	
7070	12,000 lbs., 10' x 20'		1.77	294	883	2,650	190.75	
7100	Truck, pickup, 3/4 ton, 2 wheel drive		5.75	53.50	160	480	78	
7200	4 wheel drive		5.90	63.50	190	570	85.20	
7250	Crew carrier, 9 passenger		5.44	92	276.40	830	98.80	
7290	Tool van, 24,000 G.V.W.		8.40	91.50	274.80	825	122.15	
7300	Tractor, 4 x 2, 30 ton capacity, 195 H.P.		13.65	165	495	1,475	208.20	
7410	250 H.P.		18.20	243	730	2,200	291.60	
7500	6 x 2, 40 ton capacity, 240 H.P.		17.45	270	810	2,425	301.60	
7600	6 x 4, 45 ton capacity, 240 H.P.		22.20	293	880	2,650	353.60	
7620	Vacuum truck, hazardous material, 2500 gallon		6.51	300	902	2,700	232.50	
7625	5,000 gallon		11.37	400	1,202.80	3,600	331.50	
7640	Tractor, with A frame, boom and winch, 225 H.P.		15.75	223	670	2,000	260	
7650	Vacuum, H.E.P.A., 16 gal., wet/dry		.27	24	72	216	16.55	
7655	55 gal, wet/dry		.60	36	108	325	26.40	
7660	Water tank, portable	▼	1	9.35	28	84	13.60	
7690	Large production vacuum loader, 3150 CFM		15	600	1,800	5,400	480	

Important: See the Reference Section for critical supporting data - Reference Nos., Crews, & City Cost Indexes

01590 | Equipment Rental

			UNIT	HOURLY OPER. COST	RENT PER DAY	RENT PER WEEK	RENT PER MONTH	CREW EQUIPMENT COST/DAY	
400	7700	Welder, electric, 200 amp	Ea.	3.79	60.50	181	545	66.50	400
	7800	300 amp		5.28	64	192	575	80.65	
	7900	Gas engine, 200 amp		5.70	28	84	252	62.40	
	8000	300 amp		7.70	32.50	98	294	81.20	
	8100	Wheelbarrow, any size		.06	10.65	32	96	6.90	
	8200	Wrecking ball, 4000 lb.		1.85	68.50	205	615	55.80	
500	0010	**HIGHWAY EQUIPMENT RENTAL**							500
	0050	Asphalt batch plant, portable drum mixer, 100 ton/hr.	Ea.	55.75	1,275	3,850	11,600	1,216	
	0060	200 ton/hr.		61.85	1,350	4,045	12,100	1,304	
	0070	300 ton/hr.		72.05	1,600	4,785	14,400	1,533	
	0100	Backhoe attachment, long stick, up to 185 HP, 10.5' long		.30	20	60	180	14.40	
	0140	Up to 250 HP, 12' long		.33	21.50	65	195	15.65	
	0180	Over 250 HP, 15' long		.43	28.50	85	255	20.45	
	0200	Special dipper arm, up to 100 HP, 32' long		.88	58.50	175	525	42.05	
	0240	Over 100 HP, 33' long		1.10	73.50	220	660	52.80	
	0300	Concrete batch plant, portable, electric, 200 CY/Hr		10.40	570	1,715	5,150	426.20	
	0500	Grader attachment, ripper/scarifier, rear mounted							
	0520	Up to 135 HP	Ea.	2.75	58.50	175	525	57	
	0540	Up to 180 HP		3.30	75	225	675	71.40	
	0580	Up to 250 HP		3.65	85	255	765	80.20	
	0700	Pvmt. removal bucket, for hyd. excavator, up to 90 HP		1.40	43.50	130	390	37.20	
	0740	Up to 200 HP		1.60	65	195	585	51.80	
	0780	Over 200 HP		1.70	76.50	230	690	59.60	
	0900	Aggregate spreader, self-propelled, 187 HP		37.15	790	2,370	7,100	771.20	
	1000	Chemical spreader, 3 C.Y.		2.20	61.50	185	555	54.60	
	1900	Hammermill, traveling, 250 HP		39.98	1,700	5,120	15,400	1,344	
	2000	Horizontal borer, 3" diam, 13 HP gas driven		3.95	51.50	155	465	62.60	
	2200	Hydromulchers, gas power, 3000 gal., for truck mounting		10.50	187	560	1,675	196	
	2400	Joint & crack cleaner, walk behind, 25 HP		2.05	46.50	140	420	44.40	
	2500	Filler, trailer mounted, 400 gal., 20 HP		6	182	545	1,625	157	
	3000	Paint striper, self propelled, double line, 30 HP		5.15	155	465	1,400	134.20	
	3200	Post drivers, 6" I-Beam frame, for truck mounting		8.15	415	1,250	3,750	315.20	
	3400	Road sweeper, self propelled, 8' wide, 90 HP		23.50	430	1,285	3,850	445	
	4000	Road mixer, self-propelled, 130 HP		29.05	570	1,715	5,150	575.40	
	4100	310 HP		54	1,850	5,570	16,700	1,546	
	4200	Cold mix paver, incl pug mill and bitumen tank,							
	4220	165 HP	Ea.	66.95	1,875	5,600	16,800	1,656	
	4250	Paver, asphalt, wheel or crawler, 130 H.P., diesel		65.65	1,775	5,325	16,000	1,590	
	4300	Paver, road widener, gas 1' to 6', 67 HP		29.65	630	1,890	5,675	615.20	
	4400	Diesel, 2' to 14', 88 HP		39.95	955	2,870	8,600	893.60	
	4600	Slipform pavers, curb and gutter, 2 track, 75 HP		24.85	640	1,920	5,750	582.80	
	4700	4 track, 165 HP		34.75	745	2,235	6,700	725	
	4800	Median barrier, 215 HP		35.35	775	2,325	6,975	747.80	
	4901	Trailer, low bed, 75 ton capacity		7.90	178	535	1,600	170.20	
	5000	Road planer, walk behind, 10" cutting width, 10 HP		2	26	78	234	31.60	
	5100	Self propelled, 12" cutting width, 64 HP		5.30	195	585	1,750	159.40	
	5200	Pavement profiler, 4' to 6' wide, 450 HP		146.05	2,675	8,040	24,100	2,776	
	5300	8' to 10' wide, 750 HP		233.70	4,000	11,980	35,900	4,266	
	5400	Roadway plate, steel, 1"x8'x20'		.06	10	30	90	6.50	
	5600	Stabilizer, self-propelled, 150 HP		27.40	540	1,625	4,875	544.20	
	5700	310 HP		44.55	1,150	3,445	10,300	1,045	
	5800	Striper, thermal, truck mounted 120 gal. paint, 150 H.P.		32.40	485	1,450	4,350	549.20	
	6000	Tar kettle, 330 gal., trailer mounted		2.37	36.50	110	330	40.95	
	7000	Tunnel locomotive, diesel, 8 to 12 ton		20.95	540	1,615	4,850	490.60	
	7005	Electric, 10 ton		20.65	610	1,830	5,500	531.20	
	7010	Muck cars, 1/2 C.Y. capacity		1.50	20	60	180	24	
	7020	1 C.Y. capacity		1.70	28	84	252	30.40	
	7030	2 C.Y. capacity		1.85	32.50	98	294	34.40	

01590 | Equipment Rental

		UNIT	HOURLY OPER. COST	RENT PER DAY	RENT PER WEEK	RENT PER MONTH	CREW EQUIPMENT COST/DAY	
500	7040	Side dump, 2 C.Y. capacity	Ea.	2	40	120	360	40	**500**
	7050	3 C.Y. capacity		2.70	45	135	405	48.60	
	7060	5 C.Y. capacity		3.85	58.50	175	525	65.80	
	7100	Ventilating blower for tunnel, 7-1/2 H.P.		1.22	44	132	395	36.15	
	7110	10 H.P.		1.41	45.50	137	410	38.70	
	7120	20 H.P.		2.30	56.50	170	510	52.40	
	7140	40 H.P.		4.02	83.50	250	750	82.15	
	7160	60 H.P.		6.09	127	380	1,150	124.70	
	7175	75 H.P.		7.79	168	505	1,525	163.30	
	7180	200 H.P.		17.45	245	735	2,200	286.60	
	7800	Windrow loader, elevating		33.40	875	2,620	7,850	791.20	
600	0010	**LIFTING AND HOISTING EQUIPMENT RENTAL**							**600**
	0100	without operators							
	0120	Aerial lift truck, 2 person, to 80'	Ea.	18.60	580	1,740	5,225	496.80	
	0140	Boom work platform, 40' snorkel		9.80	203	610	1,825	200.40	
	0150	Crane, flatbed mntd, 3 ton cap.		12.45	183	550	1,650	209.60	
	0200	Crane, climbing, 106' jib, 6000 lb. capacity, 410 FPM		43.35	1,350	4,070	12,200	1,161	
	0300	101' jib, 10,250 lb. capacity, 270 FPM		48.75	1,725	5,150	15,500	1,420	
	0400	Tower, static, 130' high, 106' jib,							
	0500	6200 lb. capacity at 400 FPM	Ea.	46.50	1,575	4,700	14,100	1,312	
	0600	Crawler mounted, lattice boom, 1/2 C.Y., 15 tons at 12' radius		20.76	570	1,710	5,125	508.10	
	0700	3/4 C.Y., 20 tons at 12' radius		27.68	680	2,045	6,125	630.45	
	0800	1 C.Y., 25 tons at 12' radius		36.90	895	2,690	8,075	833.20	
	0900	1-1/2 C.Y., 40 tons at 12' radius		41.15	930	2,795	8,375	888.20	
	1000	2 C.Y., 50 tons at 12' radius		51.35	1,325	3,975	11,900	1,206	
	1100	3 C.Y., 75 tons at 12' radius		49.10	1,300	3,885	11,700	1,170	
	1200	100 ton capacity, 60' boom		62.80	1,700	5,085	15,300	1,519	
	1300	165 ton capacity, 60' boom		92.20	2,175	6,560	19,700	2,050	
	1400	200 ton capacity, 70' boom		98.10	2,350	7,015	21,000	2,188	
	1500	350 ton capacity, 80' boom		141.90	3,325	9,945	29,800	3,124	
	1600	Truck mounted, lattice boom, 6 x 4, 20 tons at 10' radius		27.76	960	2,875	8,625	797.10	
	1700	25 tons at 10' radius		29.66	1,025	3,070	9,200	851.30	
	1800	8 x 4, 30 tons at 10' radius		36.66	1,075	3,260	9,775	945.30	
	1900	40 tons at 12' radius		32.74	1,150	3,450	10,400	951.90	
	2000	8 x 4, 60 tons at 15' radius		39.81	1,250	3,770	11,300	1,072	
	2050	82 tons at 15' radius		46.50	1,500	4,470	13,400	1,266	
	2100	90 tons at 15' radius		44.08	1,625	4,860	14,600	1,325	
	2200	115 tons at 15' radius		50.55	1,800	5,400	16,200	1,484	
	2300	150 tons at 18' radius		43.10	1,825	5,500	16,500	1,445	
	2350	165 tons at 18' radius		77.25	2,150	6,470	19,400	1,912	
	2400	Truck mounted, hydraulic, 12 ton capacity		32.45	570	1,715	5,150	602.60	
	2500	25 ton capacity		32.60	595	1,780	5,350	616.80	
	2550	33 ton capacity		33.45	625	1,870	5,600	641.60	
	2560	40 ton capacity		31.85	595	1,790	5,375	612.80	
	2600	55 ton capacity		47.05	865	2,595	7,775	895.40	
	2700	80 ton capacity		56	910	2,735	8,200	995	
	2720	100 ton capacity		80.75	2,325	6,945	20,800	2,035	
	2740	120 ton capacity		85.55	2,525	7,550	22,700	2,194	
	2760	150 ton capacity		104.65	3,175	9,535	28,600	2,744	
	2800	Self-propelled, 4 x 4, with telescoping boom, 5 ton		15.50	320	965	2,900	317	
	2900	12-1/2 ton capacity		25.10	510	1,525	4,575	505.80	
	3000	15 ton capacity		25.70	585	1,755	5,275	556.60	
	3050	20 ton capacity		26.80	610	1,835	5,500	581.40	
	3100	25 ton capacity		27.25	595	1,785	5,350	575	
	3150	40 ton capacity		45.20	845	2,540	7,625	869.60	
	3200	Derricks, guy, 20 ton capacity, 60' boom, 75' mast		12.72	330	996	3,000	300.95	
	3300	100' boom, 115' mast		20.60	570	1,710	5,125	506.80	

R01590 -100

Important: See the Reference Section for critical supporting data - Reference Nos., Crews, & City Cost Indexes

01590 | Equipment Rental

			UNIT	HOURLY OPER. COST	RENT PER DAY	RENT PER WEEK	RENT PER MONTH	CREW EQUIPMENT COST/DAY	
600	3400	Stiffleg, 20 ton capacity, 70' boom, 37' mast	Ea.	14.71	425	1,280	3,850	373.70	600
	3500	100' boom, 47' mast		23.19	695	2,080	6,250	601.50	
	3550	Helicopter, small, lift to 1250 lbs. maximum, w/pilot		67.85	2,675	8,050	24,200	2,153	
	3600	Hoists, chain type, overhead, manual, 3/4 ton		.10	2	6	18	2	
	3900	10 ton		.60	13	39	117	12.60	
	4000	Hoist and tower, 5000 lb. cap., portable electric, 40' high		4.19	191	574	1,725	148.30	
	4100	For each added 10' section, add		.09	15	45	135	9.70	
	4200	Hoist and single tubular tower, 5000 lb. electric, 100' high		5.66	267	801	2,400	205.50	
	4300	For each added 6'-6" section, add		.15	25	75	225	16.20	
	4400	Hoist and double tubular tower, 5000 lb., 100' high		6.06	294	882	2,650	224.90	
	4500	For each added 6'-6" section, add		.17	28.50	85	255	18.35	
	4550	Hoist and tower, mast type, 6000 lb., 100' high		6.56	305	915	2,750	235.50	
	4570	For each added 10' section, add		.11	18.35	55	165	11.90	
	4600	Hoist and tower, personnel, electric, 2000 lb., 100' @ 125 FPM		13.52	815	2,440	7,325	596.15	
	4700	3000 lb., 100' @ 200 FPM		15.40	915	2,750	8,250	673.20	
	4800	3000 lb., 150' @ 300 FPM		17.10	1,025	3,090	9,275	754.80	
	4900	4000 lb., 100' @ 300 FPM		17.73	1,050	3,150	9,450	771.85	
	5000	6000 lb., 100' @ 275 FPM		19.14	1,100	3,300	9,900	813.10	
	5100	For added heights up to 500', add	L.F.	.01	1.67	5	15	1.10	
	5200	Jacks, hydraulic, 20 ton	Ea.	.05	6.35	19	57	4.20	
	5500	100 ton	"	.30	18.65	56	168	13.60	
	6000	Jacks, hydraulic, climbing with 50' jackrods							
	6010	and control consoles, minimum 3 mo. rental							
	6100	30 ton capacity	Ea.	1.65	110	329	985	79	
	6150	For each added 10' jackrod section, add		.05	3.33	10	30	2.40	
	6300	50 ton capacity		2.65	177	530	1,600	127.20	
	6350	For each added 10' jackrod section, add		.06	4	12	36	2.90	
	6500	125 ton capacity		6.95	465	1,390	4,175	333.60	
	6550	For each added 10' jackrod section, add		.48	31.50	95	285	22.85	
	6600	Cable jack, 10 ton capacity with 200' cable		1.38	91.50	275	825	66.05	
	6650	For each added 50' of cable, add		.15	10	30	90	7.20	
700	0010	**WELLPOINT EQUIPMENT RENTAL** See also division 02240							70
	0020	Based on 2 months rental							
	0100	Combination jetting & wellpoint pump, 60 H.P. diesel	Ea.	9.17	272	817	2,450	236.75	
	0200	High pressure gas jet pump, 200 H.P., 300 psi	"	16.39	233	698	2,100	270.70	
	0300	Discharge pipe, 8" diameter	L.F.	.01	.44	1.32	3.96	.35	
	0350	12" diameter		.01	.65	1.96	5.90	.45	
	0400	Header pipe, flows up to 150 G.P.M., 4" diameter		.01	.40	1.20	3.60	.30	
	0500	400 G.P.M., 6" diameter		.01	.47	1.42	4.26	.35	
	0600	800 G.P.M., 8" diameter		.01	.65	1.96	5.90	.45	
	0700	1500 G.P.M., 10" diameter		.01	.69	2.06	6.20	.50	
	0800	2500 G.P.M., 12" diameter		.02	1.30	3.89	11.65	.95	
	0900	4500 G.P.M., 16" diameter		.02	1.66	4.98	14.95	1.15	
	0950	For quick coupling aluminum and plastic pipe, add		.03	1.72	5.15	15.45	1.25	
	1100	Wellpoint, 25' long, with fittings & riser pipe, 1-1/2" or 2" diameter	Ea.	.05	3.43	10.28	31	2.45	
	1200	Wellpoint pump, diesel powered, 4" diameter, 20 H.P.		4.45	157	471	1,425	129.80	
	1300	6" diameter, 30 H.P.		5.81	195	584	1,750	163.30	
	1400	8" suction, 40 H.P.		7.91	267	801	2,400	223.50	
	1500	10" suction, 75 H.P.		10.86	310	936	2,800	274.10	
	1600	12" suction, 100 H.P.		16.25	500	1,500	4,500	430	
	1700	12" suction, 175 H.P.		21.61	550	1,650	4,950	502.90	
800	0010	**MARINE EQUIPMENT RENTAL**							80
	0200	Barge, 400 Ton, 30' wide x 90' long	Ea.	16.15	240	720	2,150	273.20	
	0240	800 Ton, 45' wide x 90' long		26.65	345	1,030	3,100	419.20	
	2000	Tugboat, diesel, 100 HP		16.95	168	505	1,525	236.60	
	2040	250 HP		31.45	315	945	2,825	440.60	
	2080	380 HP		73.10	925	2,780	8,350	1,141	

R01590 -100

R02240 -900

01740 | Cleaning

			DAILY OUTPUT	LABOR-HOURS	UNIT	2005 BARE COSTS				TOTAL INCL O&P	
						MAT.	LABOR	EQUIP.	TOTAL		
500	0010	**CLEANING UP** After job completion, allow, minimum			Job					.30%	500
	0040	Maximum			"					1%	
	0050	Cleanup of floor area, continuous, per day, during const.	A-5	24	.750	M.S.F.	1.70	20	1.34	23.04	34.50
	0100	Final by GC at end of job	"	11.50	1.565	"	2.65	42	2.80	47.45	71
	1000	Mechanical demolition, see division 02220									

01800 | Facility Operation

01810 | Commissioning

			CREW	DAILY OUTPUT	LABOR-HOURS	UNIT	2005 BARE COSTS				TOTAL INCL O&P	
							MAT.	LABOR	EQUIP.	TOTAL		
100	0010	**COMMISSIONING** Including documentation of design intent										100
	0100	performance verification, O&M, training, min				Project					.50%	
	0150	Maximum				"					.75%	

For information about Means Estimating Seminars, see yellow pages 12 and 13 in back of book

Division 2
Site Construction

Estimating Tips

02200 Site Preparation
- If possible visit the site and take an inventory of the type, quantity and size of the trees. Certain trees may have a landscape resale value or firewood value. Stump disposal can be very expensive, particularly if they cannot be buried at the site. Consider using a bulldozer in lieu of hand cutting trees.
- Estimators should visit the site to determine the need for haul road, access, storage of materials, and security considerations. When estimating for access roads on unstable soil, consider using a geotextile stabilization fabric. It can greatly reduce the quantity of crushed stone or gravel. Sites of limited size and access can cause cost overruns due to lost productivity. Theft and damage is another consideration if the location is isolated. A temporary fence or security guards may be required. Investigate the site thoroughly.

02210 Subsurface Investigation
In preparing estimates on structures involving earthwork or foundations, all information concerning soil characteristics should be obtained. Look particularly for hazardous waste, evidence of prior dumping of debris, and previous stream beds.

02220 Selective Demolition
The costs shown for selective demolition do not include rubbish handling or disposal. These items should be estimated separately using Means data or other sources.
- Historic preservation often requires that the contractor remove materials from the existing structure, rehab them and replace them. The estimator must be aware of any related measures and precautions that must be taken when doing selective demolition, and cutting and patching. Requirements may include special handling and storage, as well as security.
- In addition to Section 02220, you can find selective demolition items in each division. Example: Roofing demolition is in division 7.

02300 Earthwork
- Estimating the actual cost of performing earthwork requires careful consideration of the variables involved. This includes items such as type of soil, whether or not water will be encountered, dewatering, whether or not banks need bracing, disposal of excavated earth, length of haul to fill or spoil sites, etc. If the project has large quantities of cut or fill, consider raising or lowering the site to reduce costs while paying close attention to the effect on site drainage and utilities if doing this.
- If the project has large quantities of fill, creating a borrow pit on the site can significantly lower the costs.
- It is very important to consider what time of year the project is scheduled for completion. Bad weather can create large cost overruns from dewatering, site repair and lost productivity from cold weather.

02500 Utility Services
02600 Drainage & Containment
- Never assume that the water, sewer and drainage lines will go in at the early stages of the project. Consider the site access needs before dividing the site in half with open trenches, loose pipe, and machinery obstructions. Always inspect the site to establish that the site drawings are complete. Check off all existing utilities on your drawing as you locate them. If you find any discrepancies, mark up the site plan for further research. Differing site conditions can be very costly if discovered later in the project.
- See also Section 02955 for restoration of pipe where removal/replacement may be undesirable. Use of new types of piping materials can reduce the overall project cost. Owners/design engineers should consider the installing construction as a valuable source of current information on piping products that could lead to significant utility cost savings.

02700 Bases, Ballasts, Pavements/Appurtenances
- When estimating paving, keep in mind the project schedule. If an asphaltic paving project is in a colder climate and runs through to the spring, consider placing the base course in the autumn, then topping it in the spring just prior to completion. This could save considerable costs in spring repair. Keep in mind that prices for asphalt and concrete are generally higher in the cold seasons.
- See also Sections 02960/02965.

02900 Planting
- The timing of planting and guarantee specifications often dictate the costs for establishing tree and shrub growth and a stand of grass or ground cover. Establish the work performance schedule to coincide with the local planting season. Maintenance and growth guarantees can add from 20% to 100% to the total landscaping cost. The cost to replace trees and shrubs can be as high as 5% of the total cost depending on the planting zone, soil conditions and time of year.

02960 & 02965 Flexible Pavement Surfacing Recovery
- Recycling of asphalt pavement is becoming very popular and is an alternative to removal and replacement of asphalt pavement. It can be a good value engineering proposal if removed pavement can be recycled either at the site or another site that is reasonably close to the project site.

Reference Numbers
Reference numbers are shown in bold squares at the beginning of some major classifications. These numbers refer to related items in the Reference Section. The reference information may be an estimating procedure, an alternate pricing method or technical information.

Note: Not all subdivisions listed here necessarily appear in this publication.

02080	Utility Materials	CREW	DAILY OUTPUT	LABOR-HOURS	UNIT	2005 BARE COSTS				TOTAL INCL O&P
						MAT.	LABOR	EQUIP.	TOTAL	
100	**0010 FIRE HYDRANTS**	G3010 -400								**100**
	0020 Mechanical joints unless otherwise noted									
	1000 Fire hydrants, two way; excavation and backfill not incl.									
	1100 4-1/2" valve size, depth 2'-0"	B-21	10	2.800	Ea.	1,100	86.50	15.85	1,202.35	1,350
	1200 4'-6"		9	3.111		1,250	96	17.60	1,363.60	1,525
	1260 6'-0"		7	4		1,200	124	22.50	1,346.50	1,550
	1340 8'-0"		6	4.667		1,375	144	26.50	1,545.50	1,750
	1420 10'-0"		5	5.600		1,450	173	31.50	1,654.50	1,900
	2000 5-1/4" valve size, depth 2'-0"		10	2.800		865	86.50	15.85	967.35	1,100
	2080 4'-0"		9	3.111		975	96	17.60	1,088.60	1,250
	2160 6'-0"		7	4		1,100	124	22.50	1,246.50	1,425
	2240 8'-0"		6	4.667		1,200	144	26.50	1,370.50	1,575
	2320 10'-0"		5	5.600		1,325	173	31.50	1,529.50	1,750
	2350 For threeway valves, add					7%				
	2400 Lower barrel extensions with stems, 1'-0"	B-20	14	1.714		345	51.50		396.50	460
	2440 2'-0"		13	1.846		420	55.50		475.50	545
	2480 3'-0"		12	2		495	60		555	640
	2520 4'-0"		10	2.400		570	72		642	735
	5000 Indicator post									
	5020 Adjustable, valve size 4" to 14", 4' bury	B-21	10	2.800	Ea.	380	86.50	15.85	482.35	565
	5060 8' bury		7	4		485	124	22.50	631.50	745
	5080 10' bury		6	4.667		495	144	26.50	665.50	795
	5100 12' bury		5	5.600		745	173	31.50	949.50	1,125
	5120 14' bury		4	7		815	216	39.50	1,070.50	1,275
	5500 Non-adjustable, valve size 4" to 14", 3' bury		10	2.800		380	86.50	15.85	482.35	565
	5520 3'-6" bury		10	2.800		380	86.50	15.85	482.35	565
	5540 4' bury		9	3.111		380	96	17.60	493.60	585
400	**0010 UTILITY BOXES** Precast concrete, 6" thick	F1030 -400								**400**
	0050 5' x 10' x 6' high, I.D.	B-13	2	28	Ea.	1,525	805	310	2,640	3,275
	0100 6' x 10' x 6' high, I.D.		2	28		1,575	805	310	2,690	3,325
	0150 5' x 12' x 6' high, I.D.		2	28		1,675	805	310	2,790	3,450
	0200 6' x 12' x 6' high, I.D.		1.80	31.111		1,875	895	345	3,115	3,825
	0250 6' x 13' x 6' high, I.D.		1.50	37.333		2,475	1,075	410	3,960	4,825
	0300 8' x 14' x 7' high, I.D.		1	56		2,675	1,600	615	4,890	6,075
	0350 Hand hole, precast concrete, 1-1/2" thick									
	0400 1'-0" x 2'-0" x 1'-9", I.D., light duty	B-1	4	6	Ea.	260	164		424	540
	0450 4'-6" x 3'-2" x 2'-0", O.D., heavy duty	B-6	3	8	"	800	231	72	1,103	1,325
500	**0010 VALVES** Water distribution, see also div. 15110									**500**
	3000 Butterfly valves with boxes, cast iron, mech. jt.									
	3100 4" diameter	B-20	6	4	Ea.	214	120		334	420
	3140 6" diameter	"	5	4.800		246	144		390	495
	3180 8" diameter	B-21	4	7		298	216	39.50	553.50	705
	3300 10" diameter		3.50	8		440	247	45.50	732.50	915
	3340 12" diameter		3	9.333		570	288	53	911	1,125
	3400 14" diameter		2	14		980	435	79	1,494	1,825
	3440 16" diameter		2	14		1,325	435	79	1,839	2,200
	3460 18" diameter		1.50	18.667		1,650	575	106	2,331	2,800
	3480 20" diameter		1	28		2,075	865	158	3,098	3,800
	3500 24" diameter		.50	56		2,875	1,725	315	4,915	6,200
	3600 With lever operator									
	3610 4" diameter	B-20	6	4	Ea.	116	120		236	315
	3614 6" diameter	"	5	4.800		162	144		306	405
	3616 8" diameter	B-21	4	7		235	216	39.50	490.50	640
	3618 10" diameter		3.50	8		375	247	45.50	667.50	850
	3620 12" diameter		3	9.333		700	288	53	1,041	1,275

Important: See the Reference Section for critical supporting data - Reference Nos., Crews, & City Cost Indexes

02080	Utility Materials	CREW	DAILY OUTPUT	LABOR-HOURS	UNIT	MAT.	LABOR	EQUIP.	TOTAL	TOTAL INCL O&P
						2005 BARE COSTS				
500 3622	14" diameter	B-21	2	14	Ea.	1,125	435	79	1,639	1,975
3624	16" diameter		2	14		1,525	435	79	2,039	2,425
3626	18" diameter		1.50	18.667		1,750	575	106	2,431	2,925
3628	20" diameter		1	28		2,325	865	158	3,348	4,075
3630	24" diameter	↓	.50	56	↓	2,500	1,725	315	4,540	5,775
3700	Check valves, flanged									
3710	4" diameter	B-20	6	4	Ea.	425	120		545	650
3714	6" diameter	"	5	4.800		720	144		864	1,025
3716	8" diameter	B-21	4	7		1,475	216	39.50	1,730.50	1,975
3718	10" diameter		3.50	8		2,325	247	45.50	2,617.50	3,000
3720	12" diameter		3	9.333		3,575	288	53	3,916	4,450
3722	14" diameter		2	14		5,725	435	79	6,239	7,050
3724	16" diameter		2	14		7,900	435	79	8,414	9,425
3726	18" diameter		1.50	18.667		11,700	575	106	12,381	13,800
3728	20" diameter		1	28		13,600	865	158	14,623	16,400
3730	24" diameter	↓	.50	56	↓	19,500	1,725	315	21,540	24,500
3800	Gate valves, C.I., 250 PSI, mechanical joint, w/boxes									
3810	4" diameter	B-21A	8	5	Ea.	166	165	63.50	394.50	505
3814	6" diameter		6.80	5.882		214	194	74.50	482.50	615
3816	8" diameter		5.60	7.143		340	236	90.50	666.50	830
3818	10" diameter		4.80	8.333		530	275	105	910	1,125
3820	12" diameter	↓	4.08	9.804		730	325	124	1,179	1,425
3822	14" diameter	B-21	2	14		760	435	79	1,274	1,600
3824	16" diameter		1	28		900	865	158	1,923	2,500
3826	18" diameter		.80	35		1,100	1,075	198	2,373	3,125
3828	20" diameter		.80	35		1,400	1,075	198	2,673	3,450
3830	24" diameter		.50	56		1,725	1,725	315	3,765	4,925
3831	30" diameter		.35	80		2,075	2,475	455	5,005	6,600
3832	36" diameter	↓	.30	93.333		2,475	2,875	530	5,880	7,775
3880	Sleeve, for tapping mains, 8" x 4", add					505			505	555
3884	10" x 6", add					1,025			1,025	1,125
3888	12" x 6", add					1,025			1,025	1,125
3892	12" x 8", add				↓	1,100			1,100	1,200
3900	Globe valves, flanged, iron body, class 125									
3910	4" diameter	B-20	10	2.400	Ea.	1,125	72		1,197	1,350
3914	6" diameter	"	9	2.667		2,175	80		2,255	2,525
3916	8" diameter	B-21	6	4.667		4,050	144	26.50	4,220.50	4,700
3918	10" diameter		5	5.600		6,375	173	31.50	6,579.50	7,300
3920	12" diameter		4	7		8,900	216	39.50	9,155.50	10,200
3922	14" diameter		3	9.333		9,550	288	53	9,891	11,000
3924	16" diameter		2	14		8,025	435	79	8,539	9,575
3926	18" diameter		.80	35		8,825	1,075	198	10,098	11,600
3928	20" diameter		.60	46.667		8,700	1,450	264	10,414	12,100
3930	24" diameter	↓	.50	56	↓	10,300	1,725	315	12,340	14,400
600 0010	**UTILITY ACCESSORIES**									
0400	Underground tape, detectable, reinforced, alum. foil core, 2"	1 Clab	150	.053	C.L.F.	1.43	1.42		2.85	3.79
0500	6"	"	140	.057	"	3.58	1.53		5.11	6.30

02110 | Hazmat Removal & Handling

		CREW	DAILY OUTPUT	LABOR-HOURS	UNIT	2005 BARE COSTS				TOTAL INCL O&P
						MAT.	LABOR	EQUIP.	TOTAL	
300	**0010**	**HAZARDOUS WASTE CLEANUP/PICKUP/DISPOSAL**									**300**
	0100	For contractor equipment, i.e. dozer,									
	0110	Front end loader, dump truck, etc., see div. 01590-200									
	1000	Solid pickup									
	1100	55 gal. drums				Ea.					220
	1120	Bulk material, minimum				Ton					165
	1130	Maximum				"					550
	1200	Transportation to disposal site									
	1220	Truckload = 80 drums or 25 C.Y. or 18 tons									
	1260	Minimum				Mile					2.50
	1270	Maximum				"					4.40
	3000	Liquid pickup, vacuum truck, stainless steel tank									
	3100	Minimum charge, 4 hours									
	3110	1 compartment, 2200 gallon				Hr.					110
	3120	2 compartment, 5000 gallon				"					110
	3400	Transportation in 6900 gallon bulk truck				Mile					4.75
	3410	In teflon lined truck				"					5.50
	5000	Heavy sludge or dry vacuumable material				Hr.					110
	6000	Dumpsite disposal charge, minimum				Ton					110
	6020	Maximum				"					440

02115 | Underground Storage Tank Removal

		CREW	DAILY OUTPUT	LABOR-HOURS	UNIT	MAT.	LABOR	EQUIP.	TOTAL	TOTAL INCL O&P
200	**0010**	**REMOVAL OF UNDERGROUND STORAGE TANKS**			R02115 -200						**200**
	0011	Petroleum storage tanks, non-leaking									
	0100	Excavate & load onto trailer									
	0110	3000 gal. to 5000 gal. tank	B-14	4	12	Ea.	335	54		389	580
	0120	6000 gal to 8000 gal tank	B-3A	3	13.333		375	240		615	845
	0130	9000 gal to 12000 gal tank	"	2	20		565	360		925	1,275
	0190	Known leaking tank add				%				100%	100%
	0200	Remove sludge, water and remaining product from tank bottom									
	0201	of tank with vacuum truck									
	0300	3000 gal to 5000 gal tank	A-13	5	1.600	Ea.		53	96	149	186
	0310	6000 gal to 8000 gal tank		4	2			66	120	186	232
	0320	9000 gal to 12000 gal tank		3	2.667			88	160	248	310
	0390	Dispose of sludge off-site, average				Gal.					4.40
	0400	Insert inert solid CO2 "dry ice" into tank									
	0401	For cleaning/transporting tanks (1.5 lbs./100 gal. cap)	1 Clab	500	.016	Lb.	1.34	.43		1.77	2.13
	1020	Haul tank to certified salvage dump, 100 miles round trip									
	1023	3000 gal. to 5000 gal. tank				Ea.				550	690
	1026	6000 gal. to 8000 gal. tank								650	825
	1029	9,000 gal. to 12,000 gal. tank								875	1,100
	1100	Disposal of contaminated soil to landfill									
	1110	Minimum				C.Y.					110
	1111	Maximum				"					330
	1120	Disposal of contaminated soil to									
	1121	bituminous concrete batch plant									
	1130	Minimum				C.Y.					55
	1131	Maximum				"					110
	2010	Decontamination of soil on site incl poly tarp on top/bottom									
	2011	Soil containment berm, and chemical treatment									
	2020	Minimum	B-11C	100	.160	C.Y.	5.60	4.91	2.16	12.67	16
	2021	Maximum	"	100	.160		7.25	4.91	2.16	14.32	17.85
	2050	Disposal of decontaminated soil, minimum									66
	2055	Maximum									135

Important: See the Reference Section for critical supporting data - Reference Nos., Crews, & City Cost Indexes

02210 | Subsurface Investigation

		CREW	DAILY OUTPUT	LABOR-HOURS	UNIT	2005 BARE COSTS				TOTAL INCL O&P
						MAT.	LABOR	EQUIP.	TOTAL	
200	0010	**CORE DRILLING** Reinforced concrete slab, up to 6" thick slab									200
	0020	Including bit, layout and set up									
	0100	1" diameter core	B-89A	28	.571	Ea.	2.40	17.60	3.68	23.68	34
	0150	Each added inch thick, add		300	.053		.43	1.64	.34	2.41	3.40
	0300	3" diameter core		23	.696		5.35	21.50	4.48	31.33	44.50
	0350	Each added inch thick, add		186	.086		.96	2.65	.55	4.16	5.80
	0500	4" diameter core		19	.842		5.35	26	5.40	36.75	52.50
	0550	Each added inch thick, add		170	.094		1.21	2.90	.61	4.72	6.50
	0700	6" diameter core		14	1.143		8.80	35	7.35	51.15	72.50
	0750	Each added inch thick, add		140	.114		1.49	3.52	.74	5.75	7.90
	0900	8" diameter core		11	1.455		12	45	9.35	66.35	93
	0950	Each added inch thick, add		95	.168		2.02	5.20	1.08	8.30	11.45
	1100	10" diameter core		10	1.600		16	49.50	10.30	75.80	105
	1150	Each added inch thick, add		80	.200		2.65	6.15	1.29	10.09	13.95
	1300	12" diameter core		9	1.778		19.20	54.50	11.45	85.15	119
	1350	Each added inch thick, add		68	.235		3.18	7.25	1.52	11.95	16.40
	1500	14" diameter core		7	2.286		23.50	70.50	14.70	108.70	151
	1550	Each added inch thick, add		55	.291		4.04	8.95	1.87	14.86	20.50
	1700	18" diameter core		4	4		30	123	26	179	254
	1750	Each added inch thick, add	↓	28	.571		5.30	17.60	3.68	26.58	37.50
	1760	For horizontal holes, add to above				↓				30%	30%
	1770	Prestressed hollow core plank, 6" thick									
	1780	1" diameter core	B-89A	52	.308	Ea.	1.59	9.45	1.98	13.02	18.70
	1790	Each added inch thick, add		350	.046		.28	1.41	.29	1.98	2.82
	1800	3" diameter core		50	.320		3.51	9.85	2.06	15.42	21.50
	1810	Each added inch thick, add		240	.067		.58	2.05	.43	3.06	4.30
	1820	4" diameter core		48	.333		4.67	10.25	2.15	17.07	23.50
	1830	Each added inch thick, add		216	.074		.81	2.28	.48	3.57	4.96
	1840	6" diameter core		44	.364		5.80	11.20	2.34	19.34	26.50
	1850	Each added inch thick, add		175	.091		.96	2.81	.59	4.36	6.10
	1860	8" diameter core		32	.500		7.75	15.40	3.22	26.37	36
	1870	Each added inch thick, add		118	.136		1.34	4.17	.87	6.38	8.95
	1880	10" diameter core		28	.571		10.45	17.60	3.68	31.73	43
	1890	Each added inch thick, add		99	.162		1.44	4.97	1.04	7.45	10.45
	1900	12" diameter core		22	.727		12.75	22.50	4.68	39.93	54
	1910	Each added inch thick, add	↓	85	.188	↓	2.12	5.80	1.21	9.13	12.65
	1950	Minimum charge for above, 3" diameter core		7	2.286	Total		70.50	14.70	85.20	125
	2000	4" diameter core		6.80	2.353			72.50	15.15	87.65	130
	2050	6" diameter core		6	2.667			82	17.15	99.15	147
	2100	8" diameter core		5.50	2.909			89.50	18.75	108.25	160
	2150	10" diameter core		4.75	3.368			104	21.50	125.50	185
	2200	12" diameter core		3.90	4.103			126	26.50	152.50	225
	2250	14" diameter core		3.38	4.734			146	30.50	176.50	261
	2300	18" diameter core	↓	3.15	5.079	↓		156	32.50	188.50	279

02220 | Site Demolition

			CREW	DAILY OUTPUT	LABOR-HOURS	UNIT	MAT.	LABOR	EQUIP.	TOTAL	TOTAL INCL O&P
240	0010	**MINOR SITE DEMOLITION** R02220 -510										24
	0015	No hauling, abandon catch basin or manhole	B-6	7	3.429	Ea.		99	31	130	186
	0020	Remove existing catch basin or manhole, masonry		4	6			173	54	227	325
	0030	Catch basin or manhole frames and covers, stored		13	1.846			53	16.60	69.60	100
	0040	Remove and reset	↓	7	3.429			99	31	130	186
	0900	Hydrants, fire, remove only	B-21A	5	8			264	101	365	510
	0950	Remove and reset	"	2	20	↓		660	253	913	1,275
	2900	Pipe removal, sewer/water, no excavation, 12" diameter	B-6	175	.137	L.F.		3.95	1.23	5.18	7.45
	2930	15"-18" diameter		150	.160			4.61	1.44	6.05	8.70
	2960	21"-24" diameter	↓	120	.200	↓		5.75	1.80	7.55	10.85

02220 | Site Demolition

		CREW	DAILY OUTPUT	LABOR-HOURS	UNIT	2005 BARE COSTS				TOTAL INCL O&P	
						MAT.	LABOR	EQUIP.	TOTAL		
240	3000	27"-36" diameter *R02220-510*	B-6	90	.267	L.F.		7.70	2.40	10.10	14.45	**240**
	3200	Steel, welded connections, 4" diameter		160	.150			4.32	1.35	5.67	8.15	
	3300	10" diameter		80	.300	↓		8.65	2.70	11.35	16.25	
	4000	Sidewalk removal, bituminous, 2-1/2" thick		325	.074	S.Y.		2.13	.66	2.79	4	
	4100	Concrete, plain, 4"	↓	160	.150	"		4.32	1.35	5.67	8.15	
250	0010	**DEMOLISH, REMOVE PAVEMENT AND CURB** *R02220-510*										**250**
	5010	Pavement removal, bituminous roads, 3" thick	B-38	690	.058	S.Y.		1.74	1.20	2.94	3.98	
	5050	4" to 6" thick	"	420	.095	"		2.85	1.96	4.81	6.55	
330	0010	**SELECTIVE DEMOLITION, DUMP CHARGES** *R02220-510*										**330**
	0100	Building construction materials				Ton					70	
	0300	Rubbish only									60	
	0500	Reclamation station, usual charge	↓			↓					85	
350	0010	**SELECTIVE DEMOLITION, RUBBISH HANDLING** *R02220-510*										**350**
	0020	The following are to be added to the demolition prices										
	0400	Chute, circular, prefabricated steel, 18" diameter	B-1	40	.600	L.F.	28	16.40		44.40	56.50	
	0440	30" diameter	"	30	.800	"	37.50	22		59.50	75.50	
	0725	Dumpster, weekly rental, 1 dump/week, 20 C.Y. capacity (8 Tons)				Week					440	
	0800	30 C.Y. capacity (10 Tons)									665	
	0840	40 C.Y. capacity (13 Tons)				↓					805	
	1000	Dust partition, 6 mil polyethylene, 1" x 3" frame	2 Carp	2,000	.008	S.F.	.17	.27		.44	.62	
	1080	2" x 4" frame	"	2,000	.008	"	.28	.27		.55	.74	
	2000	Load, haul, and dump, 50' haul	2 Clab	24	.667	C.Y.		17.80		17.80	27.50	
	2040	100' haul		16.50	.970			26		26	40.50	
	2080	Over 100' haul, add per 100 L.F.		35.50	.451			12.05		12.05	18.75	
	2120	In elevators, per 10 floors, add	↓	140	.114			3.05		3.05	4.75	
	3000	Loading & trucking, including 2 mile haul, chute loaded	B-16	45	.711			19.50	10.60	30.10	41.50	
	3040	Hand loading truck, 50' haul	"	48	.667			18.25	9.95	28.20	39.50	
	3080	Machine loading truck	B-17	120	.267			7.60	4.52	12.12	16.60	
	5000	Haul, per mile, up to 8 C.Y. truck	B-34B	1,165	.007			.19	.41	.60	.74	
	5100	Over 8 C.Y. truck	"	1,550	.005	↓		.14	.31	.45	.56	
370	0010	**SELECTIVE DEMOLITION, TORCH CUTTING** *R02220-510*										**370**
	0020	Steel, 1" thick plate	1 Clab	360	.022	L.F.	.18	.59		.77	1.12	
	0040	1" diameter bar	"	210	.038	Ea.		1.02		1.02	1.58	
	1000	Oxygen lance cutting, reinforced concrete walls										
	1040	12" to 16" thick walls	1 Clab	10	.800	L.F.		21.50		21.50	33	
	1080	24" thick walls	"	6	1.333	"		35.50		35.50	55.50	

02240 | Dewatering

		CREW	DAILY OUTPUT	LABOR-HOURS	UNIT	MAT.	LABOR	EQUIP.	TOTAL	TOTAL INCL O&P	
500	0010	**DEWATERING**										**500**
	0020	Excavate drainage trench, 2' wide, 2' deep	B-11C	90	.178	C.Y.		5.45	2.40	7.85	11	
	0100	2' wide, 3' deep, with backhoe loader	"	135	.119			3.64	1.60	5.24	7.30	
	0200	Excavate sump pits by hand, light soil	1 Clab	7.10	1.127			30		30	47	
	0300	Heavy soil	"	3.50	2.286	↓		61		61	95	
	0500	Pumping 8 hr., attended 2 hrs. per day, including 20 L.F.										
	0550	of suction hose & 100 L.F. discharge hose										
	0600	2" diaphragm pump used for 8 hours	B-10H	4	3	Day		96	14.25	110.25	162	
	0650	4" diaphragm pump used for 8 hours	B-10I	4	3			96	21.50	117.50	170	
	0800	8 hrs. attended, 2" diaphragm pump	B-10H	1	12			385	57	442	650	
	0900	3" centrifugal pump	B-10J	1	12			385	63	448	655	
	1000	4" diaphragm pump	B-10I	1	12			385	86.50	471.50	680	
	1100	6" centrifugal pump	B-10K	1	12	↓		385	248	633	855	
	1300	CMP, incl. excavation 3' deep, 12" diameter	B-6	115	.209	L.F.	9.65	6	1.88	17.53	22	
	1400	18" diameter		100	.240	"	12	6.90	2.16	21.06	26	
	1600	Sump hole construction, incl. excavation and gravel, pit	↓	1,250	.019	C.F.	.70	.55	.17	1.42	1.81	

Important: See the Reference Section for critical supporting data - Reference Nos., Crews, & City Cost Indexes

02240 | Dewatering

			DAILY OUTPUT	LABOR-HOURS	UNIT	MAT.	LABOR	EQUIP.	TOTAL	TOTAL INCL O&P	
							2005 BARE COSTS				
			CREW								
500	1700	With 12" gravel collar, 12" pipe, corrugated, 16 ga.	B-6	70	.343	L.F.	15.10	9.90	3.08	28.08	35	50
	1800	15" pipe, corrugated, 16 ga.		55	.436		19.30	12.60	3.92	35.82	44.50	
	1900	18" pipe, corrugated, 16 ga.		50	.480		22.50	13.85	4.32	40.67	51.50	
	2000	24" pipe, corrugated, 14 ga.		40	.600		27	17.30	5.40	49.70	62.50	
	2200	Wood lining, up to 4' x 4', add		300	.080	SFCA	12.80	2.31	.72	15.83	18.40	
	9950	See div. 02240-900 for wellpoints										
	9960	See div. 02240-700 for deep well systems										
700	0010	**WELLS** For dewatering 10' to 20' deep, 2' diameter										700
	0020	with steel casing, minimum	B-6	165	.145	V.L.F.	2.35	4.19	1.31	7.85	10.50	
	0050	Average		98	.245		4.68	7.05	2.20	13.93	18.40	
	0100	Maximum		49	.490		12.50	14.10	4.40	31	40	
	0300	For pumps for dewatering, see division 01590-400										
	0500	For domestic water wells, see division 02520-510										
900	0010	**WELLPOINTS** For wellpoint equipment rental, see div. 01590-700										900
	0100	Installation and removal of single stage system										
	0110	Labor only, .75 labor-hours per L.F., minimum	1 Clab	10.70	.748	LF Hdr		19.95		19.95	31	
	0200	2.0 labor-hours per L.F., maximum	"	4	2	"		53.50		53.50	83	
	0400	Pump operation, 4 @ 6 hr. shifts										
	0410	Per 24 hour day	4 Eqlt	1.27	25.197	Day		835		835	1,250	
	0500	Per 168 hour week, 160 hr. straight, 8 hr. double time		.18	177	Week		5,875		5,875	8,850	
	0550	Per 4.3 week month		.04	800	Month		26,400		26,400	39,800	
	0600	Complete installation, operation, equipment rental, fuel &										
	0610	removal of system with 2" wellpoints 5' O.C.										
	0700	100' long header, 6" diameter, first month	4 Eqlt	3.23	9.907	LF Hdr	125	325		450	630	
	0800	Thereafter, per month		4.13	7.748		99.50	256		355.50	495	
	1000	200' long header, 8" diameter, first month		6	5.333		110	176		286	385	
	1100	Thereafter, per month		8.39	3.814		56	126		182	252	
	1300	500' long header, 8" diameter, first month		10.63	3.010		43.50	99.50		143	198	
	1400	Thereafter, per month		20.91	1.530		31	50.50		81.50	111	
	1600	1,000' long header, 10" diameter, first month		11.62	2.754		37.50	91		128.50	178	
	1700	Thereafter, per month		41.81	.765		18.70	25.50		44.20	58.50	
	1900	Note: above figures include pumping 168 hrs. per week										
	1910	and include the pump operator and one stand-by pump.										

R02240 -900

02305 | Equipment

			DAILY OUTPUT	LABOR-HOURS	UNIT	MAT.	LABOR	EQUIP.	TOTAL	TOTAL INCL O&P	
							2005 BARE COSTS				
			CREW								
250	0010	**MOBILIZATION OR DEMOB.** (One or the other, unless noted)										250
	0015	Up to 25 mi haul dist (50 mi round trip for mob/demob crew)										
	0020	Dozer, loader, backhoe, excav., grader, paver, roller, 70 to 150 H.P.	B-34N	4	2	Ea.		55	112	167	207	
	0100	Above 150 HP	B-34K	3	2.667			73.50	175	248.50	305	
	0900	Shovel or dragline, 3/4 C.Y.	"	3.60	2.222			61	146	207	254	
	1100	Small equipment, placed in rear of, or towed by pickup truck	A-3A	8	1			27	10.65	37.65	52.50	
	1150	Equip up to 70 HP, on flatbed trailer behind pickup truck	A-3D	4	2			53.50	43.50	97	129	
	2000	Mob & demob truck-mounted crane up to 75 ton, driver only	1 EQHV	3.60	2.222			80		80	120	
	2100	Crane, truck-mounted, over 75 ton	A-3E	2.50	6.400			203	34	237	350	
	2200	Crawler-mounted, up to 75 ton	A-3F	2	8			254	278	532	690	
	2300	Over 75 ton	A-3G	1.50	10.667			340	405	745	960	
	2500	For each additional 5 miles haul distance, add						10%	10%			

02305 | Equipment

		CREW	DAILY OUTPUT	LABOR-HOURS	UNIT	2005 BARE COSTS MAT.	LABOR	EQUIP.	TOTAL	TOTAL INCL O&P	
250	3000	For large pieces of equipment, allow for assembly/knockdown										**250**
	3100	For mob/demob of micro-tunneling equip, see section 02441-400										

02315 | Excavation and Fill

		CREW	DAILY OUTPUT	LABOR-HOURS	UNIT	2005 BARE COSTS MAT.	LABOR	EQUIP.	TOTAL	TOTAL INCL O&P	
210	0010	**BORROW, LOADING AND/OR SPREADING**										**210**
	0020	Material only, bank run gravel				Ton	12.70			12.70	14	
	0500	Haul 2 mi. spread, 200HP dozer, bank run gravel	B-15	1,100	.025			.75	1.70	2.45	3.01	
	1000	Hand spread, bank run gravel	A-5	33	.545			14.55	.98	15.53	23.50	
	1800	Delivery charge, minimum 20 tons, 1 hr round trip, add	B-34B	130	.062			1.70	3.67	5.37	6.60	
	1820	1-1/2 hr. round trip, add		93	.086			2.37	5.15	7.52	9.25	
	1840	2 hr. round trip, add		65	.123			3.39	7.35	10.74	13.20	
490	0010	**HAULING**, excavated or borrow, loose cubic yards										**490**
	0012	no loading included, highway haulers										
	0020	6 C.Y. dump truck, 1/4 mile round trip, 5.0 loads/hr.	B-34A	195	.041	L.C.Y.		1.13	1.68	2.81	3.56	
	0030	1/2 mile round trip, 4.1 loads/hr.		160	.050			1.38	2.04	3.42	4.35	
	0040	1 mile round trip, 3.3 loads/hr. R01590 -100		130	.062			1.70	2.51	4.21	5.35	
	0100	2 mile round trip, 2.6 loads/hr.		100	.080			2.20	3.27	5.47	6.95	
	0150	3 mile round trip, 2.1 loads/hr.		80	.100			2.76	4.08	6.84	8.70	
	0200	4 mile round trip, 1.8 loads/hr.		70	.114			3.15	4.67	7.82	9.95	
	0310	12 C.Y. dump truck, 1/4 mile round trip 3.7 loads/hr.	B-34B	288	.028			.77	1.66	2.43	2.99	
	0400	2 mile round trip, 2.2 loads/hr.		180	.044			1.22	2.65	3.87	4.78	
	0450	3 mile round trip, 1.9 loads/hr.		170	.047			1.30	2.80	4.10	5.05	
	1300	Hauling in medium traffic, add								20%	20%	
	1400	Heavy traffic, add								30%	30%	
	1600	Grading at dump, or embankment if required, by dozer	B-10B	1,000	.012			.38	.92	1.30	1.59	
	1800	Spotter at fill or cut, if required	1 Clab	8	1	Hr.		26.50		26.50	41.50	
520	0010	**FILL**, spread dumped material, no compaction										**520**
	0020	By dozer, no compaction	B-10B	1,000	.012	L.C.Y.		.38	.92	1.30	1.59	
	0100	By hand	1 Clab	12	.667	"		17.80		17.80	27.50	
	0500	Gravel fill, compacted, under floor slabs, 4" deep	B-37	10,000	.005	S.F.	.16	.13	.01	.30	.39	
	0600	6" deep		8,600	.006		.24	.16	.01	.41	.51	
	0700	9" deep		7,200	.007		.39	.19	.01	.59	.74	
	0800	12" deep		6,000	.008		.55	.22	.02	.79	.98	
	1000	Alternate pricing method, 4" deep		120	.400	E.C.Y.	11.80	11.25	.90	23.95	31.50	
	1100	6" deep		160	.300		11.80	8.45	.67	20.92	27	
	1200	9" deep		200	.240		11.80	6.75	.54	19.09	24	
	1300	12" deep		220	.218		11.80	6.15	.49	18.44	23	
	1400	Granular fill				L.C.Y.	7.90			7.90	8.65	
610	0010	**EXCAVATING, TRENCH** or continuous footing, common earth G1030 805										**610**
	0020	No sheeting or dewatering included										
	0050	1' to 4' deep, 3/8 C.Y. tractor loader/backhoe	B-11C	150	.107	B.C.Y.		3.27	1.44	4.71	6.60	
	0060	1/2 C.Y. tractor loader/backhoe	B-11M	200	.080			2.45	1.28	3.73	5.15	
	0090	4' to 6' deep, 1/2 C.Y. tractor loader/backhoe	"	200	.080			2.45	1.28	3.73	5.15	
	0100	5/8 C.Y. hydraulic backhoe	B-12Q	250	.064			2	1.73	3.73	4.96	
	0300	1/2 C.Y. hydraulic excavator, truck mounted	B-12J	200	.080			2.50	4.19	6.69	8.45	
	0500	6' to 10' deep, 3/4 C.Y. hydraulic backhoe, 6' to 10' deep	B-12F	225	.071			2.23	2.11	4.34	5.70	
	0600	1 C.Y. hydraulic excavator, truck mounted	B-12K	400	.040			1.25	2.43	3.68	4.58	
	0900	10' to 14' deep, 3/4 C.Y. hydraulic backhoe	B-12F	200	.080			2.50	2.37	4.87	6.45	
	1000	1-1/2 C.Y. hydraulic backhoe	B-12B	540	.030			.93	1.33	2.26	2.89	
	1300	14' to 20' deep, 1 C.Y. hydraulic backhoe	B-12A	320	.050			1.57	1.74	3.31	4.31	
	1400	By hand with pick and shovel 2' to 6' deep, light soil	1 Clab	8	1			26.50		26.50	41.50	
	1500	Heavy soil	"	4	2			53.50		53.50	83	
	1700	For tamping backfilled trenches, air tamp, add	A-1G	100	.080	E.C.Y.		2.14	.35	2.49	3.70	
	1900	Vibrating plate, add	B-18	180	.133	"		3.65	.20	3.85	5.95	

Important: See the Reference Section for critical supporting data - Reference Nos., Crews, & City Cost Indexes

02315 | Excavation and Fill

			CREW	DAILY OUTPUT	LABOR-HOURS	UNIT	MAT.	LABOR	EQUIP.	TOTAL	TOTAL INCL O&P	
610	2100	Trim sides and bottom for concrete pours, common earth	B-18	1,500	.016	S.F.		.44	.02	.46	.71	61
	2300	Hardpan	↓	600	.040	"		1.09	.06	1.15	1.77	
620	0010	**EXCAVATING, UTILITY TRENCH** Common earth [G1030 805]										62
	0050	Trenching with chain trencher, 12 H.P., operator walking										
	0100	4" wide trench, 12" deep	B-53	800	.010	L.F.		.33	.06	.39	.57	
	0150	18" deep		750	.011			.35	.06	.41	.60	
	0200	24" deep		700	.011			.38	.07	.45	.65	
	0300	6" wide trench, 12" deep		650	.012			.41	.07	.48	.69	
	0350	18" deep		600	.013			.44	.08	.52	.75	
	0400	24" deep		550	.015			.48	.09	.57	.82	
	0450	36" deep		450	.018			.59	.11	.70	1.01	
	0600	8" wide trench, 12" deep		475	.017			.56	.10	.66	.95	
	0650	18" deep		400	.020			.66	.12	.78	1.13	
	0700	24" deep		350	.023			.76	.14	.90	1.29	
	0750	36" deep	↓	300	.027	↓		.88	.16	1.04	1.51	
	1000	Backfill by hand including compaction, add										
	1050	4" wide trench, 12" deep	A-1G	800	.010	L.F.		.27	.04	.31	.47	
	1100	18" deep		530	.015			.40	.07	.47	.70	
	1150	24" deep		400	.020			.53	.09	.62	.93	
	1300	6" wide trench, 12" deep		540	.015			.40	.06	.46	.69	
	1350	18" deep		405	.020			.53	.09	.62	.91	
	1400	24" deep		270	.030			.79	.13	.92	1.37	
	1450	36" deep		180	.044			1.19	.19	1.38	2.06	
	1600	8" wide trench, 12" deep		400	.020			.53	.09	.62	.93	
	1650	18" deep		265	.030			.81	.13	.94	1.39	
	1700	24" deep		200	.040			1.07	.17	1.24	1.85	
	1750	36" deep	↓	135	.059	↓		1.58	.26	1.84	2.74	
	2000	Chain trencher, 40 H.P. operator riding										
	2050	6" wide trench and backfill, 12" deep	B-54	1,200	.007	L.F.		.22	.18	.40	.53	
	2100	18" deep		1,000	.008			.26	.21	.47	.63	
	2150	24" deep		975	.008			.27	.22	.49	.65	
	2200	36" deep		900	.009			.29	.24	.53	.70	
	2250	48" deep		750	.011			.35	.28	.63	.84	
	2300	60" deep		650	.012			.41	.33	.74	.97	
	2400	8" wide trench and backfill, 12" deep		1,000	.008			.26	.21	.47	.63	
	2450	18" deep		950	.008			.28	.22	.50	.67	
	2500	24" deep		900	.009			.29	.24	.53	.70	
	2550	36" deep		800	.010			.33	.27	.60	.79	
	2600	48" deep		650	.012			.41	.33	.74	.97	
	2700	12" wide trench and backfill, 12" deep		975	.008			.27	.22	.49	.65	
	2750	18" deep		860	.009			.31	.25	.56	.73	
	2800	24" deep		800	.010			.33	.27	.60	.79	
	2850	36" deep		725	.011			.36	.29	.65	.87	
	3000	16" wide trench and backfill, 12" deep		835	.010			.32	.26	.58	.76	
	3050	18" deep		750	.011			.35	.28	.63	.84	
	3100	24" deep	↓	700	.011	↓		.38	.31	.69	.91	
	3200	Compaction with vibratory plate, add								50%	50%	
	5100	Hand excavate and trim for pipe bells after trench excavation										
	5200	8" pipe	1 Clab	155	.052	L.F.		1.38		1.38	2.14	
	5300	18" pipe	"	130	.062	"		1.64		1.64	2.56	
640	0010	**UTILITY BEDDING** For pipe and conduit, not incl. compaction [G1030 805]										6
	0050	Crushed or screened bank run gravel	B-6	150	.160	L.C.Y.	20	4.61	1.44	26.05	30.50	
	0100	Crushed stone 3/4" to 1/2"		150	.160		27.50	4.61	1.44	33.55	38.50	
	0200	Sand, dead or bank	↓	150	.160	↓	4.13	4.61	1.44	10.18	13.20	
	0500	Compacting bedding in trench	A-1D	90	.089	E.C.Y.		2.37	.28	2.65	4	
	0600	If material source exceeds 2 miles, add for extra mileage.				L.C.Y.						

For expanded coverage of these items see Means Heavy Construction Cost Data 2005

		02315	Excavation and Fill	CREW	DAILY OUTPUT	LABOR-HOURS	UNIT	2005 BARE COSTS MAT.	LABOR	EQUIP.	TOTAL	TOTAL INCL O&P	
640	0610		See 02315-490-0010 for hauling mileage add.				L.C.Y.						640

		02441	Microtunneling	CREW	DAILY OUTPUT	LABOR-HOURS	UNIT	2005 BARE COSTS MAT.	LABOR	EQUIP.	TOTAL	TOTAL INCL O&P	
400	0010		**MICROTUNNELING** Not including excavation, backfill, shoring,										400
	0020		or dewatering, average 50'/day, slurry method										
	0100		24" to 48" outside diameter, minimum				L.F.					640	
	0110		Adverse conditions, add				%					50%	
	1000		Rent microtunneling machine, average monthly lease				Month					85,500	
	1010		Operating technician				Day					640	
	1100		Mobilization and demobilization, minimum				Job					42,800	
	1110		Maximum				"					430,000	

		02445	Boring or Jacking Conduits	CREW	DAILY OUTPUT	LABOR-HOURS	UNIT	2005 BARE COSTS MAT.	LABOR	EQUIP.	TOTAL	TOTAL INCL O&P	
300	0010		**HORIZONTAL BORING** Casing only, 100' minimum,										300
	0020		not incl. jacking pits or dewatering										
	0100		Roadwork, 1/2" thick wall, 24" diameter casing	B-42	20	3.200	L.F.	67	96	53.50	216.50	284	
	0200		36" diameter		16	4		107	120	67	294	380	
	0300		48" diameter		15	4.267		157	128	71.50	356.50	455	
	0500		Railroad work, 24" diameter		15	4.267		67	128	71.50	266.50	355	
	0600		36" diameter		14	4.571		107	137	76.50	320.50	415	
	0700		48" diameter		12	5.333		157	160	89	406	520	
	0900		For ledge, add								155	190	
	1000		Small diameter boring, 3", sandy soil	B-82	900	.018		18.45	.53	.07	19.05	21.50	
	1040		Rocky soil	"	500	.032		18.45	.96	.13	19.54	22	
	1100		Prepare jacking pits, incl. mobilization & demobilization, minimum				Ea.				2,850	3,375	
	1101		Maximum				"				16,100	19,300	

		02510	Water Distribution	CREW	DAILY OUTPUT	LABOR-HOURS	UNIT	2005 BARE COSTS MAT.	LABOR	EQUIP.	TOTAL	TOTAL INCL O&P	
600	0010		**VALVES**										600
	9000		Valves, gate valve, N.R.S. post type, 4" diameter	B-21	32	.875	Ea.	365	27	4.95	396.95	450	
	9020		6" diameter		20	1.400		460	43.50	7.90	511.40	585	
	9040		8" diameter		16	1.750		705	54	9.90	768.90	870	
	9060		10" diameter		16	1.750		1,150	54	9.90	1,213.90	1,350	
	9080		12" diameter		13	2.154		1,425	66.50	12.20	1,503.70	1,675	
	9100		14" diameter		11	2.545		1,650	78.50	14.40	1,742.90	1,975	
	9120		O.S.&Y., 4" diameter		32	.875		291	27	4.95	322.95	365	
	9140		6" diameter		20	1.400		425	43.50	7.90	476.40	540	
	9160		8" diameter		16	1.750		695	54	9.90	758.90	860	

Important: See the Reference Section for critical supporting data - Reference Nos., Crews, & City Cost Indexes

02510 | Water Distribution

			CREW	DAILY OUTPUT	LABOR-HOURS	UNIT	2005 BARE COSTS				TOTAL INCL O&P
							MAT.	LABOR	EQUIP.	TOTAL	
600	9180	10" diameter	B-21	16	1.750	Ea.	1,125	54	9.90	1,188.90	1,350
	9200	12" diameter		13	2.154		1,475	66.50	12.20	1,553.70	1,750
	9220	14" diameter	▼	11	2.545		4,150	78.50	14.40	4,242.90	4,725
	9400	Check valves, rubber disc, 2-1/2" diameter	B-20	44	.545		350	16.40		366.40	410
	9420	3" diameter	"	38	.632		350	19		369	415
	9440	4" diameter	B-21	32	.875		430	27	4.95	461.95	520
	9480	6" diameter		20	1.400		650	43.50	7.90	701.40	790
	9500	8" diameter		16	1.750		1,050	54	9.90	1,113.90	1,250
	9520	10" diameter		16	1.750		1,700	54	9.90	1,763.90	1,975
	9540	12" diameter		13	2.154		2,675	66.50	12.20	2,753.70	3,075
	9542	14" diameter		11	2.545		3,775	78.50	14.40	3,867.90	4,300
	9700	Detector check valves, reducing, 4" diameter		32	.875		705	27	4.95	736.95	820
	9720	6" diameter		20	1.400		1,075	43.50	7.90	1,126.40	1,275
	9740	8" diameter		16	1.750		1,775	54	9.90	1,838.90	2,050
	9760	10" diameter		16	1.750		3,500	54	9.90	3,563.90	3,950
	9800	Galvanized, 4" diameter		32	.875		920	27	4.95	951.95	1,075
	9820	6" diameter		20	1.400		1,350	43.50	7.90	1,401.40	1,550
	9840	8" diameter		16	1.750		2,150	54	9.90	2,213.90	2,450
	9860	10" diameter	▼	16	1.750	▼	4,250	54	9.90	4,313.90	4,775
710	0010	**TAPPING, CROSSES AND SLEEVES**									
	4000	Drill and tap pressurized main (labor only)									
	4100	6" main, 1" to 2" service	Q-1	3	5.333	Ea.		196		196	295
	4150	8" main, 1" to 2" service	"	2.75	5.818	"		214		214	320
	4500	Tap and insert gate valve									
	4600	8" main, 4" branch	B-21	3.20	8.750	Ea.		270	49.50	319.50	475
	4650	6" branch		2.70	10.370			320	58.50	378.50	560
	4651	Piping, drill, tap & insert gate valve, 8" main, 6" branch		2.70	10.370			320	58.50	378.50	560
	4700	10" Main, 4" branch		2.70	10.370			320	58.50	378.50	560
	4750	6" branch		2.35	11.915			370	67.50	437.50	645
	4800	12" main, 6" branch		2.35	11.915			370	67.50	437.50	645
	4850	8" branch		2.35	11.915			370	67.50	437.50	645
	7020	Crosses, 4" x 4"		37	.757		455	23.50	4.28	482.78	540
	7030	6" x 4"		25	1.120		540	34.50	6.35	580.85	650
	7040	6" x 6"		25	1.120		540	34.50	6.35	580.85	650
	7060	8" x 6"		21	1.333		660	41	7.55	708.55	795
	7080	8" x 8"		21	1.333		715	41	7.55	763.55	855
	7100	10" x 6"		21	1.333		1,300	41	7.55	1,348.55	1,525
	7120	10" x 10"		21	1.333		1,425	41	7.55	1,473.55	1,625
	7140	12" x 6"		18	1.556		1,300	48	8.80	1,356.80	1,525
	7160	12" x 12"		18	1.556		1,650	48	8.80	1,706.80	1,900
	7180	14" x 6"		16	1.750		3,275	54	9.90	3,338.90	3,700
	7200	14" x 14"		16	1.750		3,450	54	9.90	3,513.90	3,900
	7220	16" x 6"		14	2		3,500	62	11.30	3,573.30	3,950
	7240	16" x 10"		14	2		3,550	62	11.30	3,623.30	4,000
	7260	16" x 16"		14	2		3,725	62	11.30	3,798.30	4,200
	7280	18" x 6"		10	2.800		5,225	86.50	15.85	5,327.35	5,900
	7300	18" x 12"		10	2.800		5,275	86.50	15.85	5,377.35	5,950
	7320	18" x 18"		10	2.800		4,825	86.50	15.85	4,927.35	5,450
	7340	20" x 6"		8	3.500		4,225	108	19.80	4,352.80	4,850
	7360	20" x 12"		8	3.500		4,525	108	19.80	4,652.80	5,200
	7380	20" x 20"		8	3.500		6,625	108	19.80	6,752.80	7,500
	7400	24" x 6"		6	4.667		5,475	144	26.50	5,645.50	6,275
	7420	24" x 12"		6	4.667		5,550	144	26.50	5,720.50	6,350
	7440	24" x 18"		6	4.667		8,275	144	26.50	8,445.50	9,350
	7460	24" x 24"	▼	6	4.667	▼	8,625	144	26.50	8,795.50	9,725

For expanded coverage of these items see *Means Heavy Construction Cost Data 2005*

02510	Water Distribution	CREW	DAILY OUTPUT	LABOR-HOURS	UNIT	MAT.	LABOR	EQUIP.	TOTAL	TOTAL INCL O&P	
710							2005 BARE COSTS				**710**
7600	Cut-in sleeves with rubber gaskets, 4"	B-21	18	1.556	Ea.	150	48	8.80	206.80	249	
7620	6"		12	2.333		191	72	13.20	276.20	335	
7640	8"		10	2.800		260	86.50	15.85	362.35	435	
7660	10"		10	2.800		360	86.50	15.85	462.35	545	
7680	12"		9	3.111		430	96	17.60	543.60	640	
7800	Cut-in valves with rubber gaskets, 4"		18	1.556		380	48	8.80	436.80	505	
7820	6"		12	2.333		515	72	13.20	600.20	690	
7840	8"		10	2.800		795	86.50	15.85	897.35	1,025	
7860	10"		10	2.800		995	86.50	15.85	1,097.35	1,250	
7880	12"		9	3.111		1,200	96	17.60	1,313.60	1,500	
7900	Tapping Valve 4 inch, MJ, ductile iron		18	1.556		380	48	8.80	436.80	505	
7920	Tapping Valve 6 inch, MJ, ductile iron		12	2.333		455	72	13.20	540.20	625	
8000	Sleeves with rubber gaskets, 4" x 4"		37	.757		775	23.50	4.28	802.78	895	
8010	6" x 4"		25	1.120		435	34.50	6.35	475.85	535	
8020	6" x 6"		25	1.120		470	34.50	6.35	510.85	575	
8030	8" x 4"		21	1.333		470	41	7.55	518.55	590	
8040	8" x 6"		21	1.333		520	41	7.55	568.55	645	
8060	8" x 8"		21	1.333		610	41	7.55	658.55	740	
8070	10" x 4"		21	1.333		475	41	7.55	523.55	595	
8080	10" x 6"		21	1.333		530	41	7.55	578.55	655	
8090	10" x 8"		21	1.333		890	41	7.55	938.55	1,050	
8100	10" x 10"		21	1.333		1,000	41	7.55	1,048.55	1,175	
8110	12" x 4"		18	1.556		515	48	8.80	571.80	650	
8120	12" x 6"		18	1.556		805	48	8.80	861.80	970	
8130	12" x 8"		18	1.556		715	48	8.80	771.80	870	
8135	12" x 10"		18	1.556		985	48	8.80	1,041.80	1,150	
8140	12" x 12"		18	1.556		1,200	48	8.80	1,256.80	1,375	
8160	14" x 6"		16	1.750		2,600	54	9.90	2,663.90	2,975	
8180	14" x 14"		16	1.750		2,775	54	9.90	2,838.90	3,150	
8200	16" x 6"		14	2		2,800	62	11.30	2,873.30	3,175	
8220	16" x 10"		14	2		2,850	62	11.30	2,923.30	3,225	
8240	16" x 16"		14	2		2,975	62	11.30	3,048.30	3,375	
8260	18" x 6"		10	2.800		4,175	86.50	15.85	4,277.35	4,750	
8280	18" x 12"		10	2.800		4,225	86.50	15.85	4,327.35	4,775	
8300	18" x 18"		10	2.800		4,325	86.50	15.85	4,427.35	4,925	
8320	20" x 6"		8	3.500		3,375	108	19.80	3,502.80	3,925	
8340	20" x 12"		8	3.500		3,625	108	19.80	3,752.80	4,200	
8360	20" x 20"		8	3.500		5,300	108	19.80	5,427.80	6,025	
8380	24" x 6"		6	4.667		4,375	144	26.50	4,545.50	5,050	
8400	24" x 12"		6	4.667		4,450	144	26.50	4,620.50	5,125	
8420	24" x 18"		6	4.667		6,625	144	26.50	6,795.50	7,525	
8440	24" x 24"		6	4.667		6,900	144	26.50	7,070.50	7,825	
8800	Curb box, 6' long	B-20	20	1.200		69.50	36		105.50	133	
8820	8' long	"	18	1.333		81	40		121	152	
730	**WATER SUPPLY, DUCTILE IRON PIPE** cement lined										**730**
0010											
0020	Not including excavation or backfill										
2000	Pipe, class 50 water piping, 18' lengths										
2020	Mechanical joint, 4" diameter	B-21A	200	.200	L.F.	12.50	6.60	2.53	21.63	26.50	
2040	6" diameter		160	.250		14.50	8.25	3.16	25.91	32	
2060	8" diameter		133.33	.300		16.05	9.90	3.80	29.75	37	
2080	10" diameter		114.29	.350		21.50	11.55	4.43	37.48	46.50	
2100	12" diameter		105.26	.380		26.50	12.55	4.81	43.86	54	
2120	14" diameter		100	.400		34	13.20	5.05	52.25	63	
2140	16" diameter		72.73	.550		37	18.15	6.95	62.10	76	
2160	18" diameter		68.97	.580		46.50	19.10	7.35	72.95	88.50	
2170	20" diameter		57.14	.700		54.50	23	8.85	86.35	105	

Important: See the Reference Section for critical supporting data - Reference Nos., Crews, & City Cost Indexes

02510 | Water Distribution

		CREW	DAILY OUTPUT	LABOR-HOURS	UNIT	2005 BARE COSTS MAT.	LABOR	EQUIP.	TOTAL	TOTAL INCL O&P	
730	2180	24" diameter	B-21A	47.06	.850	L.F.	70	28	10.75	108.75	131	730
	3000	Tyton, Push-on joint, 4" diameter		400	.100		7.35	3.30	1.27	11.92	14.50	
	3020	6" diameter		333.33	.120		8.40	3.96	1.52	13.88	16.85	
	3040	8" diameter		200	.200		11.50	6.60	2.53	20.63	25.50	
	3060	10" diameter		181.82	.220		18.15	7.25	2.78	28.18	34	
	3080	12" diameter		160	.250		19.15	8.25	3.16	30.56	37	
	3100	14" diameter		133.33	.300		21	9.90	3.80	34.70	42	
	3120	16" diameter		114.29	.350		29.50	11.55	4.43	45.48	55	
	3140	18" diameter		100	.400		32.50	13.20	5.05	50.75	61.50	
	3160	20" diameter		88.89	.450		36	14.85	5.70	56.55	68.50	
	3180	24" diameter		76.92	.520		47	17.15	6.60	70.75	85	
	8000	Fittings, mechanical joint										
	8006	90° bend, 4" diameter	B-20A	16	2	Ea.	126	64.50		190.50	237	
	8020	6" diameter		12.80	2.500		169	80.50		249.50	310	
	8040	8" diameter		10.67	2.999		245	96.50		341.50	420	
	8060	10" diameter	B-21A	11.43	3.500		480	115	44.50	639.50	755	
	8080	12" diameter		10.53	3.799		495	125	48	668	790	
	8100	14" diameter		10	4		685	132	50.50	867.50	1,000	
	8120	16" diameter		7.27	5.502		735	181	69.50	985.50	1,175	
	8140	18" diameter		6.90	5.797		1,000	191	73.50	1,264.50	1,475	
	8160	20" diameter		5.71	7.005		2,550	231	88.50	2,869.50	3,250	
	8180	24" diameter		4.70	8.511		3,150	281	108	3,539	4,000	
	8200	Wye or tee, 4" diameter	B-20A	10.67	2.999		153	96.50		249.50	315	
	8220	6" diameter		8.53	3.751		192	121		313	395	
	8240	8" diameter		7.11	4.501		284	145		429	535	
	8260	10" diameter	B-21A	7.62	5.249		490	173	66.50	729.50	870	
	8280	12" diameter		7.02	5.698		610	188	72	870	1,050	
	8300	14" diameter		6.67	5.997		1,250	198	76	1,524	1,750	
	8320	16" diameter		4.85	8.247		1,300	272	104	1,676	1,950	
	8340	18" diameter		4.60	8.696		2,650	287	110	3,047	3,450	
	8360	20" diameter		3.81	10.499		3,375	345	133	3,853	4,375	
	8380	24" diameter		3.14	12.739		3,850	420	161	4,431	5,075	
	8398	45° bends, 4" diameter	B-20A	16	2		133	64.50		197.50	245	
	8400	6" diameter	"	12.80	2.500		146	80.50		226.50	284	
	8410	12" diameter	B-21A	10.53	3.799		355	125	48	528	635	
	8420	16" diameter		7.27	5.502		460	181	69.50	710.50	860	
	8430	20" diameter		5.71	7.005		2,125	231	88.50	2,444.50	2,800	
	8440	24" diameter		4.70	8.511		2,575	281	108	2,964	3,375	
	8450	Decreaser, 6" x 4" diameter	B-20A	14.22	2.250		104	72.50		176.50	226	
	8460	8" x 6" diameter	"	11.64	2.749		213	88.50		301.50	370	
	8470	10" x 6 " diameter	B-21A	13.33	3.001		217	99	38	354	430	
	8480	12" x 6" diameter		12.70	3.150		245	104	40	389	470	
	8490	16" x 6" diameter		10	4		885	132	50.50	1,067.50	1,225	
	8500	20" x 6" diameter		8.42	4.751		1,400	157	60	1,617	1,850	
	8550	Butterfly valves with boxes, cast iron										
	8560	4" diameter	B-20	6	4	Ea.	760	120		880	1,025	
	8580	8" diameter	B-21	4	7		760	216	39.50	1,015.50	1,225	
	8590	10" diameter		3.50	8		1,050	247	45.50	1,342.50	1,575	
	8600	12" diameter		3	9.333		1,350	288	53	1,691	1,975	
	8610	14" diameter		2	14		1,875	435	79	2,389	2,800	
	8620	16" diameter		2	14		2,400	435	79	2,914	3,400	
	9600	Steel sleeve and tap, 4" diameter	B-20	3	8		430	241		671	845	
	9620	6" diameter		2	12		505	360		865	1,125	
	9630	8" diameter		2	12		680	360		1,040	1,300	
740	0010	**WATER SUPPLY, POLYETHYLENE PIPE, C901**										74
	0020	Not including excavation or backfill										

For expanded coverage of these items see Means Heavy Construction Cost Data 2005

02510 | Water Distribution

		CREW	DAILY OUTPUT	LABOR-HOURS	UNIT	2005 BARE COSTS				TOTAL INCL O&P	
						MAT.	LABOR	EQUIP.	TOTAL		
740	1000	Piping, 160 P.S.I., 3/4" diameter	B-20	525	.046	L.F.	.31	1.38		1.69	2.48	**740**
	1120	1" diameter		485	.049		.39	1.49		1.88	2.75
	1140	1-1/2" diameter		450	.053		.71	1.60		2.31	3.28
	1160	2" diameter		365	.066		1.09	1.98		3.07	4.28
	2000	Fittings, insert type, nylon, 160 & 250 psi, cold water									
	2220	Clamp ring, stainless steel, 3/4" diameter	B-20	345	.070	Ea.	.99	2.09		3.08	4.35
	2240	1" diameter		321	.075		1.04	2.25		3.29	4.64
	2260	1-1/2" diameter		285	.084		1.29	2.53		3.82	5.35
	2280	2" diameter		255	.094		1.45	2.83		4.28	6
	2300	Coupling, 3/4" diameter		66	.364		.72	10.95		11.67	17.85
	2320	1" diameter		57	.421		.96	12.65		13.61	21
	2340	1-1/2" diameter		51	.471		2.24	14.15		16.39	24.50
	2360	2" diameter		48	.500		2.69	15.05		17.74	26.50
	2400	Elbow, 90°, 3/4" diameter		66	.364		1.28	10.95		12.23	18.45
	2420	1" diameter		57	.421		1.37	12.65		14.02	21
	2440	1-1/2" diameter		51	.471		3.41	14.15		17.56	26
	2460	2" diameter		48	.500		5.70	15.05		20.75	30
750	0010	**WATER SUPPLY, POLYVINYL CHLORIDE PIPE**										**750**
	0020	Not including excavation or backfill, unless specified									
	2100	AWWA Class 160, S.D.R. 26, 1-1/2" diameter	B-20	750	.032	L.F.	.45	.96		1.41	2
	2120	2" diameter		686	.035		1.09	1.05		2.14	2.84
	2140	2-1/2" diameter		500	.048		1.62	1.44		3.06	4.03
	2160	3" diameter		430	.056		2.32	1.68		4	5.15
	2180	4" diameter		375	.064		3.78	1.93		5.71	7.15
	2200	6" diameter		316	.076		8.15	2.28		10.43	12.50
	2210	8" diameter		260	.092		13.80	2.78		16.58	19.50
	3010	AWWA C905, PR 100, DR 41									
	3030	14" diameter	B-20A	213	.150	L.F.	12.60	4.84		17.44	21.50
	3040	16" diameter		200	.160		16.70	5.15		21.85	26
	3050	18" diameter		160	.200		21	6.45		27.45	33
	3060	20" diameter		133	.241		26	7.75		33.75	41
	3070	24" diameter		107	.299		36.50	9.65		46.15	55
	3080	30" diameter		80	.400		56.50	12.90		69.40	82
	3090	36" diameter		80	.400		80	12.90		92.90	108
	3100	42" diameter		60	.533		109	17.20		126.20	146
	3200	48" diameter		60	.533		143	17.20		160.20	183
	3960	Pressure pipe, class 200, SDR 21, 3/4"		1,340	.024		.20	.77		.97	1.39
	3980	1"		1,100	.029		.41	.94		1.35	1.88
	4000	1-1/2"		920	.035		.86	1.12		1.98	2.66
	4010	2"		840	.038		1.32	1.23		2.55	3.32
	4020	2-1/2"		600	.053		1.96	1.72		3.68	4.78
	4030	3"		520	.062		2.80	1.98		4.78	6.10
	4040	4"		460	.070		4.61	2.24		6.85	8.45
	4050	6"		380	.084		10	2.71		12.71	15.15
	4060	8"		320	.100		16.85	3.22		20.07	23.50
	4090	Including trenching to 3' deep, 3/4"	Q-1C	300	.080		.20	2.89	4.73	7.82	9.75
	4100	1"		280	.086		.41	3.09	5.05	8.55	10.70
	4110	1-1/2"		260	.092		.86	3.33	5.45	9.64	11.95
	4120	2"		220	.109		1.32	3.93	6.45	11.70	14.45
	4130	2-1/2"		200	.120		1.96	4.33	7.10	13.39	16.45
	4140	3"		175	.137		2.80	4.95	8.10	15.85	19.50
	4150	4"		150	.160		4.61	5.75	9.45	19.81	24
	4160	6"		125	.192		10	6.95	11.35	28.30	34
	4165	Fittings									
	4170	Elbow, 90°, 3/4"	B-20A	114	.281	Ea.	.58	9.05		9.63	14.45

Important: See the Reference Section for critical supporting data - Reference Nos., Crews, & City Cost Indexes

02510 | Water Distribution

		CREW	DAILY OUTPUT	LABOR-HOURS	UNIT	2005 BARE COSTS				TOTAL INCL O&P
						MAT.	LABOR	EQUIP.	TOTAL	
4180	1"	B-20A	100	.320	Ea.	1.05	10.30		11.35	16.90
4190	1-1/2"		80	.400		1.84	12.90		14.74	21.50
4200	2"		72	.444		3.30	14.35		17.65	25.50
4210	3"		46	.696		7.20	22.50		29.70	42
4220	4"		36	.889		21.50	28.50		50	67.50
4230	6"		24	1.333		49.50	43		92.50	120
4240	8"		14	2.286		92	73.50		165.50	213
4250	Elbow, 45°, 3/4"		114	.281		.71	9.05		9.76	14.60
4260	1"		100	.320		.87	10.30		11.17	16.70
4270	1-1/2"		80	.400		15	12.90		27.90	36
4280	2"		72	.444		10.75	14.35		25.10	34
4290	2-1/2"		54	.593		11.50	19.10		30.60	41.50
4300	3"		46	.696		14.55	22.50		37.05	50
4310	4"		36	.889		26.50	28.50		55	73
4320	6"		24	1.333		49	43		92	119
4330	8"		14	2.286		95	73.50		168.50	217
4340	Tee, 3/4"		76	.421		.38	13.55		13.93	21
4350	1"		66	.485		.63	15.65		16.28	24.50
4360	1-1/2"		54	.593		18.85	19.10		37.95	49.50
4370	2"		48	.667		22.50	21.50		44	57.50
4380	2-1/2"		36	.889		25	28.50		53.50	71
4390	3"		30	1.067		26	34.50		60.50	81
4400	4"		24	1.333		40.50	43		83.50	110
4410	6"		14.80	2.162		90	69.50		159.50	205
4420	8"		9	3.556		159	115		274	350
4430	Coupling, 3/4"		114	.281		.22	9.05		9.27	14.05
4440	1"		100	.320		.35	10.30		10.65	16.15
4450	1-1/2"		80	.400		7.50	12.90		20.40	28
4460	2"		72	.444		7.65	14.35		22	30.50
4470	2-1/2"		54	.593		8.30	19.10		27.40	38
4480	3"		46	.696		10.45	22.50		32.95	45.50
4490	4"		36	.889		19	28.50		47.50	64.50
4500	6"		24	1.333		29	43		72	97.50
4510	8"		14	2.286		60	73.50		133.50	178
4520	Class 150, SDR 18, AWWA C900, 4"		380	.084	L.F.	2.38	2.71		5.09	6.75
4530	6"		316	.101		4.70	3.26		7.96	10.15
4540	8"		264	.121		8.15	3.91		12.06	14.90
4550	10"		220	.145		12.30	4.69		16.99	20.50
4560	12"		186	.172		17.40	5.55		22.95	27.50
8000	Fittings with rubber gasket									
8003	Class 150, D.R. 18									
8006	90° Bend , 4" diameter	B-20	100	.240	Ea.	39.50	7.20		46.70	55
8020	6" diameter		90	.267		70	8		78	89.50
8040	8" diameter		80	.300		136	9		145	163
8060	10" diameter		50	.480		235	14.45		249.45	282
8080	12" diameter		30	.800		300	24		324	370
8100	Tee, 4" diameter		90	.267		54.50	8		62.50	72
8120	6" diameter		80	.300		121	9		130	147
8140	8" diameter		70	.343		173	10.30		183.30	206
8160	10" diameter		40	.600		196	18.05		214.05	243
8180	12" diameter		20	1.200		420	36		456	515
8200	45° Bend, 4" diameter		100	.240		38.50	7.20		45.70	54
8220	6" diameter		90	.267		68.50	8		76.50	87.50
8240	8" diameter		50	.480		130	14.45		144.45	166
8260	10" diameter		50	.480		199	14.45		213.45	241
8280	12" diameter		30	.800		258	24		282	320

For expanded coverage of these items see Means Heavy Construction Cost Data 2005

02510 | Water Distribution

		CREW	DAILY OUTPUT	LABOR-HOURS	UNIT	MAT.	LABOR	EQUIP.	TOTAL	TOTAL INCL O&P	
750							2005 BARE COSTS					750
8300	Reducing tee 6"x4"	B-20	100	.240	Ea.	118	7.20		125.20	141	
8320	8" x 6"		90	.267		210	8		218	244	
8330	10" x 6"		90	.267		270	8		278	310	
8340	10"x8"		90	.267		281	8		289	325	
8350	12" x 6"		90	.267		320	8		328	365	
8360	12" x 8"		90	.267		335	8		343	385	
8400	Tapped service tee (threaded type) 6" x 6" x 3/4		100	.240		42	7.20		49.20	57.50	
8420	6" x 6" x 3/4"		90	.267		42	8		50	58.50	
8430	6" x 6" x 1"		90	.267		42	8		50	58.50	
8440	6" x 6" x 1 1/2"		90	.267		42	8		50	58.50	
8450	6" x 6" x 2"		90	.267		42	8		50	58.50	
8460	8" x 8" x 3/4"		90	.267		177	8		185	208	
8470	8" x 8" x 1"		90	.267		177	8		185	208	
8480	8" x 8" x 1 1/2"		90	.267		177	8		185	208	
8490	8" x 8" x 2"		90	.267		177	8		185	208	
8500	Repair coupling 4"		100	.240		25	7.20		32.20	39	
8520	6" diameter		90	.267		38.50	8		46.50	54.50	
8540	8" diameter		50	.480		84.50	14.45		98.95	116	
8560	10" diameter		50	.480		143	14.45		157.45	180	
8580	12" diameter		50	.480		209	14.45		223.45	253	
8600	Plug end 4"		100	.240		21	7.20		28.20	35	
8620	6" diameter		90	.267		38	8		46	54.50	
8640	8" diameter		50	.480		65	14.45		79.45	94	
8660	10" diameter		50	.480		68.50	14.45		82.95	97.50	
8680	12" diameter	▼	50	.480	▼	84.50	14.45		98.95	116	
760												760
0010	**WATER SUPPLY, HDPE**, butt fusion joints, SDR 21, 40' lengths										
0100	4" diameter	B-22A	400	.095	L.F.	1.84	2.88	1.76	6.48	8.40	
0200	6" diameter		380	.100		2.65	3.03	1.85	7.53	9.65	
0300	8" diameter		320	.119		4.48	3.60	2.20	10.28	12.90	
0400	10" diameter		300	.127		6.95	3.84	2.34	13.13	16.20	
0500	12" diameter		260	.146		9.80	4.43	2.70	16.93	20.50	
0600	14" diameter		220	.173		11.80	5.25	3.19	20.24	24.50	
0700	16" diameter		180	.211		15.45	6.40	3.90	25.75	31	
0800	18" diameter		140	.271		19.50	8.20	5	32.70	39.50	
0900	24" diameter	▼	100	.380	▼	35	11.50	7.05	53.55	63.50	
1000	Fittings										
1100	Elbows, 90 degrees										
1200	4" diameter	B-22B	32	.500	Ea.	25.50	15.40	5.35	46.25	58	
1300	6" diameter		28	.571		67	17.60	6.15	90.75	108	
1400	8" diameter		24	.667		168	20.50	7.15	195.65	225	
1500	10" diameter		18	.889		290	27.50	9.55	327.05	375	
1600	12" diameter		12	1.333		490	41	14.30	545.30	620	
1700	14" diameter		9	1.778		490	54.50	19.10	563.60	640	
1800	16" diameter		6	2.667		590	82	28.50	700.50	805	
1900	18" diameter		4	4		795	123	43	961	1,125	
2000	24" diameter	▼	3	5.333	▼	1,625	164	57.50	1,846.50	2,100	
2100	Tees										
2200	4" diameter	B-22B	30	.533	Ea.	35.50	16.40	5.75	57.65	71	
2300	6" diameter		26	.615		93	18.95	6.60	118.55	140	
2400	8" diameter		22	.727		236	22.50	7.80	266.30	305	
2500	10" diameter		15	1.067		310	33	11.45	354.45	410	
2600	12" diameter		10	1.600		425	49.50	17.20	491.70	565	
2700	14" diameter		8	2		505	61.50	21.50	588	675	
2800	16" diameter		6	2.667		590	82	28.50	700.50	810	
2900	18" diameter	▼	4	4		800	123	43	966	1,125	

Important: See the Reference Section for critical supporting data - Reference Nos., Crews, & City Cost Indexes

			DAILY	LABOR-		2005 BARE COSTS				TOTAL	
02510	**Water Distribution**	CREW	OUTPUT	HOURS	UNIT	MAT.	LABOR	EQUIP.	TOTAL	INCL O&P	
760	3000	24" diameter	B-22B	2	8	Ea.	1,650	246	86	1,982	2,300	760
780	0010	**WATER SUPPLY, COPPER PIPE**										780
	0020	Not including excavation or backfill										
	2000	Tubing, type K, 20' joints, 3/4" diameter	Q-1	400	.040	L.F.	2.70	1.47		4.17	5.20	
	2200	1" diameter		320	.050		3.57	1.84		5.41	6.70	
	3000	1-1/2" diameter		265	.060		5.85	2.22		8.07	9.80	
	3020	2" diameter		230	.070		9.10	2.56		11.66	13.85	
	3040	2-1/2" diameter		146	.110		13.25	4.03		17.28	20.50	
	3060	3" diameter		134	.119		18.50	4.39		22.89	27	
	4012	4" diameter	▼	95	.168		30.50	6.20		36.70	43.50	
	4016	6" diameter	Q-2	95	.253	▼	93	9.65		102.65	117	
	5000	Tubing, type L										
	5108	2" diameter	Q-1	105	.152	L.F.	7.55	5.60		13.15	16.70	
	6010	3" diameter		140	.114		15	4.20		19.20	23	
	6012	4" diameter	▼	95	.168		25	6.20		31.20	37	
	6016	6" diameter	Q-2	100	.240	▼	86	9.15		95.15	108	
	7020	Fittings, brass, corporation stops, 3/4" diameter	1 Plum	19	.421	Ea.	16.65	17.20		33.85	44.50	
	7040	1" diameter		16	.500		25.50	20.50		46	58.50	
	7060	1-1/2" diameter		13	.615		76	25		101	122	
	7080	2" diameter		11	.727		126	29.50		155.50	184	
	7100	Curb stops, 3/4" diameter		19	.421		36	17.20		53.20	65.50	
	7120	1" diameter		16	.500		49.50	20.50		70	85	
	7140	1-1/2" diameter		13	.615		110	25		135	159	
	7160	2" diameter	▼	11	.727	▼	180	29.50		209.50	243	

			DAILY	LABOR-		2005 BARE COSTS				TOTAL	
02520	**Wells**	CREW	OUTPUT	HOURS	UNIT	MAT.	LABOR	EQUIP.	TOTAL	INCL O&P	
510	0010	**WELLS & ACCESSORIES**, domestic										51
	0100	Drilled, 4" to 6" diameter	B-23	120	.333	L.F.		9.05	28	37.05	45	
	0200	8" diameter	"	95.20	.420	"		11.40	35	46.40	56.50	
	0400	Gravel pack well, 40' deep, incl. gravel & casing, complete										
	0500	24" diameter casing x 18" diameter screen	B-23	.13	307	Total	22,900	8,350	25,800	57,050	66,500	
	0600	36" diameter casing x 18" diameter screen		.12	333	"	24,700	9,025	28,000	61,725	72,000	
	0800	Observation wells, 1-1/4" riser pipe	▼	163	.245	V.L.F.	12.60	6.65	20.50	39.75	47	
	0900	For flush Buffalo roadway box, add	1 Skwk	16.60	.482	Ea.	34.50	16.80		51.30	64	
	1200	Test well, 2-1/2" diameter, up to 50' deep (15 to 50 GPM)	B-23	1.51	26.490	"	515	720	2,225	3,460	4,150	
	1300	Over 50' deep, add	"	121.80	.328	L.F.	13.75	8.90	27.50	50.15	59.50	
	1500	Pumps, installed in wells to 100' deep, 4" submersible										
	1510	1/2 H.P.	Q-1	3.22	4.969	Ea.	330	183		513	640	
	1520	3/4 H.P.		2.66	6.015		390	221		611	765	
	1600	1 H.P.	▼	2.29	6.987		415	257		672	840	
	1700	1-1/2 H.P.	Q-22	1.60	10		1,075	370	375	1,820	2,150	
	1800	2 H.P.		1.33	12.030		1,250	440	455	2,145	2,550	
	1900	3 H.P.		1.14	14.035		1,425	515	530	2,470	2,925	
	2000	5 H.P.		1.14	14.035		1,950	515	530	2,995	3,500	
	3000	Pump, 6" submersible, 25' to 150' deep, 25 H.P., 249 to 297 GPM		.89	17.978		4,250	660	675	5,585	6,425	
	3100	25' to 500' deep, 30 H.P., 100 to 300 GPM	▼	.73	21.918	▼	4,850	805	825	6,480	7,425	
	8000	Steel well casing	B-23A	3,020	.008	Lb.	.59	.24	1.08	1.91	2.21	
	8110	Well screen assembly, stainless steel, 2" diameter		273	.088	L.F.	47.50	2.64	11.95	62.09	69	
	8120	3" diameter		253	.095		65.50	2.85	12.85	81.20	90.50	
	8130	4" diameter		200	.120		75	3.60	16.30	94.90	106	
	8140	5" diameter		168	.143		87.50	4.29	19.40	111.19	125	
	8150	6" diameter		126	.190		105	5.70	26	136.70	152	
	8160	8" diameter		98.50	.244		137	7.30	33	177.30	198	
	8170	10" diameter	▼	73	.329	▼	172	9.85	44.50	226.35	253	

02520 | Wells

		CREW	DAILY OUTPUT	LABOR-HOURS	UNIT	2005 BARE COSTS				TOTAL INCL O&P	
						MAT.	LABOR	EQUIP.	TOTAL		
510	8180	12" diameter	B-23A	62.50	.384	L.F.	202	11.55	52	265.55	297	510
	8190	14" diameter		54.30	.442		229	13.25	60	302.25	340	
	8200	16" diameter		48.30	.497		253	14.90	67.50	335.40	375	
	8210	18" diameter		39.20	.612		315	18.40	83	416.40	465	
	8220	20" diameter		31.20	.769		360	23	104	487	545	
	8230	24" diameter		23.80	1.008		445	30.50	137	612.50	690	
	8240	26" diameter		21	1.143		500	34.50	155	689.50	775	
	8300	Slotted PVC, 1-1/4" diameter		521	.046		1.69	1.38	6.25	9.32	10.90	
	8310	1-1/2" diameter		488	.049		2.46	1.48	6.65	10.59	12.35	
	8320	2" diameter		273	.088		3.41	2.64	11.95	18	21	
	8330	3" diameter		253	.095		3.62	2.85	12.85	19.32	22.50	
	8340	4" diameter		200	.120		4.17	3.60	16.30	24.07	28	
	8350	5" diameter		168	.143		4.38	4.29	19.40	28.07	33	
	8360	6" diameter		126	.190		6.10	5.70	26	37.80	44	
	8370	8" diameter	▼	98.50	.244		9.20	7.30	33	49.50	58	
	8400	Artificial gravel pack, 2" screen, 6" casing	B-23B	174	.138		2.77	4.14	19.95	26.86	31.50	
	8405	8" casing		111	.216		3.78	6.50	31.50	41.78	48.50	
	8410	10" casing		74.50	.322		4.75	9.65	46.50	60.90	71.50	
	8415	12" casing		60	.400		6.40	12	58	76.40	89	
	8420	14" casing		50.20	.478		8.30	14.35	69	91.65	107	
	8425	16" casing		40.70	.590		11.45	17.70	85.50	114.65	134	
	8430	18" casing		36	.667		13.30	20	96.50	129.80	152	
	8435	20" casing		29.50	.814		15.30	24.50	118	157.80	184	
	8440	24" casing		25.70	.934		17	28	135	180	211	
	8445	26" casing		24.60	.976		19	29.50	141	189.50	221	
	8450	30" casing		20	1.200		22	36	174	232	271	
	8455	36" casing		16.40	1.463	▼	23.50	44	212	279.50	325	
	8500	Develop well		8	3	Hr.	215	90	435	740	855	
	8550	Pump test well		8	3		57	90	435	582	680	
	8560	Standby well	B-23A	8	3		55.50	90	405	550.50	650	
	8570	Standby, drill rig		8	3	▼		90	405	495	590	
	8580	Surface seal well, concrete filled	▼	1	24	Ea.	540	720	3,250	4,510	5,275	
	8590	Well test pump, install & remove	B-23	1	40			1,075	3,350	4,425	5,375	
	8600	Well sterilization, chlorine	2 Clab	1	16	▼	400	425		825	1,100	
	9950	See div. 02240-900 for wellpoints										
	9960	See div. 02240-700 for drainage wells										
520	0010	**WATER SUPPLY WELLS, PUMPS** with pressure control										**520**
	1000	Deep well, jet, 42 gal. galvanized tank										
	1040	3/4 HP	1 Plum	.80	10	Ea.	585	410		995	1,250	
	3000	Shallow well, jet, 30 gal. galvanized tank										
	3040	1/2 HP	1 Plum	2	4	Ea.	370	163		533	655	

02530 | Sanitary Sewerage

		CREW	DAILY OUTPUT	LABOR-HOURS	UNIT	2005 BARE COSTS				TOTAL INCL O&P	
705	0010	**SEWAGE COLLECTION, VALVES**										**705**
	1000	Backwater sewer line valve										
	1010	Offset type, bronze swing check assy and bronze cover										
	1020	B & S connections										
	1040	2" size	2 Skwk	14	1.143	Ea.	282	40		322	370	
	1050	3" size		12.50	1.280		415	44.50		459.50	525	
	1060	4" size	▼	11	1.455		525	50.50		575.50	660	
	1070	6" size	3 Skwk	15	1.600		755	56		811	915	
	1080	8" size	"	7.50	3.200	▼	1,075	112		1,187	1,350	
	1110	Backwater drainage control										
	1120	Offset type, w/bronze manually operated shear gate										
	1130	B & S connections										

Important: See the Reference Section for critical supporting data - Reference Nos., Crews, & City Cost Indexes

02530 | Sanitary Sewerage

			CREW	DAILY OUTPUT	LABOR-HOURS	UNIT	2005 BARE COSTS				TOTAL INCL O&P	
							MAT.	LABOR	EQUIP.	TOTAL		
705	1160	4" size	2 Skwk	11	1.455	Ea.	815	50.50		865.50	975	705
	1170	6" size	3 Skwk	15	1.600	"	1,300	56		1,356	1,525	
730	0010	**SEWAGE COLLECTION, CONCRETE PIPE**										730
	0020	See 02630-530 for sewage/drainage collection, concrete pipe										
780	0010	**SEWAGE COLLECTION, POLYVINYL CHLORIDE PIPE** G3020 110										780
	0020	Not including excavation or backfill										
	2000	10' lengths, S.D.R. 35, B&S, 4" diameter	B-20	375	.064	L.F.	1.92	1.93		3.85	5.10	
	2040	6" diameter	"	350	.069		3.44	2.06		5.50	7	
	2120	10" diameter	B-21	330	.085		8.75	2.62	.48	11.85	14.25	
	2160	12" diameter		320	.087		9.75	2.70	.50	12.95	15.50	
	2200	15" diameter	↓	190	.147		14.75	4.56	.83	20.14	24	
	4000	Piping, DWV PVC,no exc/bkfill, 10' L, Sch 40, 4" dia	B-20	375	.064		1.88	1.93		3.81	5.05	
	4010	6" dia		350	.069		4.06	2.06		6.12	7.70	
	4020	8" dia	↓	335	.072	↓	12.50	2.15		14.65	17.10	

02540 | Septic Tank Systems

			CREW	DAILY OUTPUT	LABOR-HOURS	UNIT	MAT.	LABOR	EQUIP.	TOTAL	TOTAL INCL O&P	
400	0010	**SEPTIC TANKS**										400
	0015	Septic tanks, not incl exc or piping, precast, 1,000 gal G3020 300	B-21	8	3.500	Ea.	560	108	19.80	687.80	805	
	0100	2,000 gallon	"	5	5.600		1,100	173	31.50	1,304.50	1,525	
	0200	5,000 gallon	B-13	3.50	16		5,625	460	176	6,261	7,100	
	0300	15,000 gallon, 4 piece	B-13B	1.70	32.941		15,600	950	525	17,075	19,100	
	0400	25,000 gallon, 4 piece		1.10	50.909		33,000	1,475	815	35,290	39,400	
	0500	40,000 gallon, 4 piece	↓	.80	70		38,800	2,025	1,125	41,950	47,000	
	0520	50,000 gallon, 5 piece	B-13C	.60	93.333		44,600	2,675	2,525	49,800	56,000	
	0540	75,000 gallon, cast in place	C-14C	.25	448		54,500	14,600	94	69,194	83,000	
	0560	100,000 gallon	"	.15	746		67,000	24,400	157	91,557	112,500	
	0600	High density polyethylene, 1,000 gallon	B-21	6	4.667		865	144	26.50	1,035.50	1,200	
	0700	1,500 gallon	"	4	7		1,125	216	39.50	1,380.50	1,600	
	1000	Distribution boxes, concrete, 7 outlets	2 Clab	16	1		104	26.50		130.50	156	
	1100	9 outlets	"	8	2		281	53.50		334.50	395	
	1150	Leaching field chambers, 13' x 3'-7" x 1'-4", standard	B-13	16	3.500		565	101	38.50	704.50	820	
	1200	Heavy duty, 8' x 4' x 1'-6"		14	4		390	115	44	549	655	
	1300	13' x 3'-9" x 1'-6"		12	4.667		1,375	134	51.50	1,560.50	1,775	
	1350	20' x 4' x 1'-6"	↓	5	11.200		910	320	123	1,353	1,625	
	1400	Leaching pit, precast concrete, 3' diameter, 3' deep	B-21	8	3.500		370	108	19.80	497.80	595	
	1500	6' diameter, 3' section		4.70	5.957		645	184	33.50	862.50	1,025	
	2000	Velocity reducing pit, precast conc., 6' diameter, 3' deep	↓	4.70	5.957	↓	310	184	33.50	527.50	665	
	2200	Excavation for septic tank, 3/4 C.Y. backhoe	B-12F	145	.110	C.Y.		3.45	3.27	6.72	8.90	
	2400	4' trench for disposal field, 3/4 C.Y. backhoe	"	335	.048	L.F.		1.49	1.42	2.91	3.84	
	2600	Gravel fill, run of bank	B-6	150	.160	C.Y.	16.15	4.61	1.44	22.20	26.50	
	2800	Crushed stone, 3/4"	"	150	.160	"	24	4.61	1.44	30.05	34.50	

02550 | Piped Energy Distribution

			CREW	DAILY OUTPUT	LABOR-HOURS	UNIT	MAT.	LABOR	EQUIP.	TOTAL	TOTAL INCL O&P	
200	0010	**PIPE CONDUIT, PREFABRICATED / PREINSULATED** R15100 -070										20
	0020	Does not include trenching, fittings or crane.										
	0300	For cathodic protection, add 12 to 14%										
	0310	of total built-up price (casing plus service pipe)										
	0580	Polyurethane insulated system, 250°F. max. temp.										
	0620	Black steel service pipe, standard wt., 1/2" insulation										
	0660	3/4" diam. pipe size	Q-17	54	.296	L.F.	24.50	10.90	1.49	36.89	45	
	0670	1" diam. pipe size		50	.320		27	11.80	1.61	40.41	49	
	0680	1-1/4" diam. pipe size		47	.340		30	12.55	1.72	44.27	53.50	
	0690	1-1/2" diam. pipe size		45	.356		32.50	13.10	1.79	47.39	57.50	
	0700	2" diam. pipe size		42	.381		34	14.05	1.92	49.97	60	
	0710	2-1/2" diam. pipe size	↓	34	.471	↓	34.50	17.35	2.37	54.22	66	

02550 | Piped Energy Distribution

			CREW	DAILY OUTPUT	LABOR-HOURS	UNIT	2005 BARE COSTS MAT.	LABOR	EQUIP.	TOTAL	TOTAL INCL O&P	
200	0720	3" diam. pipe size	Q-17	28	.571	L.F.	39.50	21	2.88	63.38	78	**200**
	0730	4" diam. pipe size		22	.727		50	27	3.67	80.67	99.50	
	0740	5" diam. pipe size		18	.889		64	33	4.48	101.48	124	
	0750	6" diam. pipe size	Q-18	23	1.043		74	40	3.51	117.51	145	
	0760	8" diam. pipe size		19	1.263		109	48.50	4.24	161.74	197	
	0770	10" diam. pipe size		16	1.500		143	57.50	5.05	205.55	250	
	0780	12" diam. pipe size		13	1.846		178	70.50	6.20	254.70	310	
	0790	14" diam. pipe size		11	2.182		199	83.50	7.35	289.85	350	
	0800	16" diam. pipe size		10	2.400		228	91.50	8.05	327.55	400	
	0810	18" diam. pipe size		8	3		262	115	10.10	387.10	470	
	0820	20" diam. pipe size		7	3.429		292	131	11.50	434.50	530	
	0830	24" diam. pipe size		6	4		355	153	13.45	521.45	640	
	0900	For 1" thick insulation, add					10%					
	0940	For 1-1/2" thick insulation, add					13%					
	0980	For 2" thick insulation, add					20%					
	1500	Gland seal for system, 3/4" diam. pipe size	Q-17	32	.500	Ea.	395	18.45	2.52	415.97	465	
	1510	1" diam. pipe size		32	.500		395	18.45	2.52	415.97	465	
	1540	1-1/4" diam. pipe size		30	.533		425	19.65	2.69	447.34	495	
	1550	1-1/2" diam. pipe size		30	.533		425	19.65	2.69	447.34	495	
	1560	2" diam. pipe size		28	.571		505	21	2.88	528.88	590	
	1570	2-1/2" diam. pipe size		26	.615		545	22.50	3.10	570.60	635	
	1580	3" diam. pipe size		24	.667		585	24.50	3.36	612.86	680	
	1590	4" diam. pipe size		22	.727		690	27	3.67	720.67	805	
	1600	5" diam. pipe size		19	.842		850	31	4.24	885.24	985	
	1610	6" diam. pipe size	Q-18	26	.923		905	35.50	3.10	943.60	1,050	
	1620	8" diam. pipe size		25	.960		1,050	36.50	3.23	1,089.73	1,200	
	1630	10" diam. pipe size		23	1.043		1,275	40	3.51	1,318.51	1,475	
	1640	12" diam. pipe size		21	1.143		1,400	43.50	3.84	1,447.34	1,625	
	1650	14" diam. pipe size		19	1.263		1,575	48.50	4.24	1,627.74	1,825	
	1660	16" diam. pipe size		18	1.333		1,850	51	4.48	1,905.48	2,100	
	1670	18" diam. pipe size		16	1.500		1,975	57.50	5.05	2,037.55	2,275	
	1680	20" diam. pipe size		14	1.714		2,250	65.50	5.75	2,321.25	2,575	
	1690	24" diam. pipe size		12	2		2,500	76.50	6.70	2,583.20	2,875	
	2000	Elbow, 45° for system										
	2020	3/4" diam. pipe size	Q-17	14	1.143	Ea.	250	42	5.75	297.75	345	
	2040	1" diam. pipe size		13	1.231		257	45.50	6.20	308.70	360	
	2050	1-1/4" diam. pipe size		11	1.455		292	53.50	7.35	352.85	410	
	2060	1-1/2" diam. pipe size		9	1.778		305	65.50	8.95	379.45	445	
	2070	2" diam. pipe size		6	2.667		320	98.50	13.45	431.95	515	
	2080	2-1/2" diam. pipe size		4	4		345	147	20	512	625	
	2090	3" diam. pipe size		3.50	4.571		400	168	23	591	720	
	2100	4" diam. pipe size		3	5.333		465	197	27	689	835	
	2110	5" diam. pipe size		2.80	5.714		595	211	29	835	1,000	
	2120	6" diam. pipe size	Q-18	4	6		675	229	20	924	1,100	
	2130	8" diam. pipe size		3	8		980	305	27	1,312	1,575	
	2140	10" diam. pipe size		2.40	10		1,250	380	33.50	1,663.50	1,975	
	2150	12" diam. pipe size		2	12		1,650	460	40.50	2,150.50	2,550	
	2160	14" diam. pipe size		1.80	13.333		2,050	510	45	2,605	3,075	
	2170	16" diam. pipe size		1.60	15		2,425	575	50.50	3,050.50	3,600	
	2180	18" diam. pipe size		1.30	18.462		3,050	705	62	3,817	4,475	
	2190	20" diam. pipe size		1	24		3,850	915	80.50	4,845.50	5,700	
	2200	24" diam. pipe size		.70	34.286		4,825	1,300	115	6,240	7,425	
	2260	For elbow, 90°, add					25%					
	2300	For tee, straight, add					85%	30%				
	2340	For tee, reducing, add					170%	30%				
	2380	For weldolet, straight, add					50%					

R15100-070

Important: See the Reference Section for critical supporting data - Reference Nos., Crews, & City Cost Indexes

02550 | Piped Energy Distribution

		CREW	DAILY OUTPUT	LABOR-HOURS	UNIT	2005 BARE COSTS				TOTAL INCL O&P
						MAT.	LABOR	EQUIP.	TOTAL	
200	2400	Polyurethane insulation, 1" R15100 -070									**200**
	2410	FRP carrier and casing								
	2420	4"	Q-5	18	.889	L.F.	37.50	33		70.50	90.50
	2422	6"	Q-6	23	1.043		65	40		105	132
	2424	8"		19	1.263		106	48.50		154.50	190
	2426	10"		16	1.500		143	57.50		200.50	244
	2428	12"		13	1.846		186	70.50		256.50	310
	2430	FRP carrier and PVC casing								
	2440	4"	Q-5	18	.889	L.F.	14.80	33		47.80	65.50
	2444	8"	Q-6	19	1.263		31	48.50		79.50	107
	2446	10"		16	1.500		44.50	57.50		102	135
	2448	12"		13	1.846		57	70.50		127.50	169
	2450	PVC carrier and casing								
	2460	4"	Q-1	36	.444	L.F.	7.30	16.35		23.65	32.50
	2462	6"	"	29	.552		10.25	20.50		30.75	42
	2464	8"	Q-2	36	.667		14.25	25.50		39.75	53.50
	2466	10"		32	.750		19.50	28.50		48	64.50
	2468	12"		31	.774		22.50	29.50		52	69.50
	2800	Calcium silicate insulated system, high temp. (1200°F)								
	2840	Steel casing with protective exterior coating								
	2850	6-5/8" diameter	Q-18	52	.462	L.F.	45.50	17.65	1.55	64.70	78.50
	2860	8-5/8" diameter		50	.480		50	18.35	1.61	69.96	84.50
	2870	10-3/4" diameter		47	.511		58.50	19.50	1.72	79.72	96
	2880	12-3/4" diameter		44	.545		63.50	21	1.83	86.33	104
	2890	14" diameter		41	.585		71.50	22.50	1.97	95.97	115
	2900	16" diameter		39	.615		77	23.50	2.07	102.57	122
	2910	18" diameter		36	.667		85	25.50	2.24	112.74	134
	2920	20" diameter		34	.706		95.50	27	2.37	124.87	148
	2930	22" diameter		32	.750		133	28.50	2.52	164.02	192
	2940	24" diameter		29	.828		151	31.50	2.78	185.28	217
	2950	26" diameter		26	.923		173	35.50	3.10	211.60	246
	2960	28" diameter		23	1.043		215	40	3.51	258.51	300
	2970	30" diameter		21	1.143		228	43.50	3.84	275.34	320
	2980	32" diameter		19	1.263		257	48.50	4.24	309.74	360
	2990	34" diameter		18	1.333		260	51	4.48	315.48	365
	3000	36" diameter		16	1.500		279	57.50	5.05	341.55	395
	3040	For multi-pipe casings, add					10%				
	3060	For oversize casings, add					2%				
	3400	Steel casing gland seal, single pipe								
	3420	6-5/8" diameter	Q-18	25	.960	Ea.	635	36.50	3.23	674.73	760
	3440	8-5/8" diameter		23	1.043		745	40	3.51	788.51	885
	3450	10-3/4" diameter		21	1.143		835	43.50	3.84	882.34	990
	3460	12-3/4" diameter		19	1.263		995	48.50	4.24	1,047.74	1,175
	3470	14" diameter		17	1.412		1,100	54	4.74	1,158.74	1,275
	3480	16" diameter		16	1.500		1,275	57.50	5.05	1,337.55	1,500
	3490	18" diameter		15	1.600		1,425	61	5.40	1,491.40	1,650
	3500	20" diameter		13	1.846		1,575	70.50	6.20	1,651.70	1,850
	3510	22" diameter		12	2		1,775	76.50	6.70	1,858.20	2,075
	3520	24" diameter		11	2.182		1,975	83.50	7.35	2,065.85	2,300
	3530	26" diameter		10	2.400		2,250	91.50	8.05	2,349.55	2,625
	3540	28" diameter		9.50	2.526		2,550	96.50	8.50	2,655	2,950
	3550	30" diameter		9	2.667		2,600	102	8.95	2,710.95	3,025
	3560	32" diameter		8.50	2.824		2,925	108	9.50	3,042.50	3,375
	3570	34" diameter		8	3		3,175	115	10.10	3,300.10	3,675
	3580	36" diameter		7	3.429		3,375	131	11.50	3,517.50	3,925
	3620	For multi-pipe casings, add					5%				

02550	Piped Energy Distribution	CREW	DAILY OUTPUT	LABOR-HOURS	UNIT	2005 BARE COSTS				TOTAL INCL O&P
						MAT.	LABOR	EQUIP.	TOTAL	
4000	Steel casing anchors, single pipe R15100-070	Q-18	8	3	Ea.	570	115	10.10	695.10	815
4020	6-5/8" diameter									
4040	8-5/8" diameter		7.50	3.200		595	122	10.75	727.75	850
4050	10-3/4" diameter		7	3.429		785	131	11.50	927.50	1,075
4060	12-3/4" diameter		6.50	3.692		835	141	12.40	988.40	1,150
4070	14" diameter		6	4		980	153	13.45	1,146.45	1,325
4080	16" diameter		5.50	4.364		1,150	167	14.65	1,331.65	1,525
4090	18" diameter		5	4.800		1,275	183	16.15	1,474.15	1,725
4100	20" diameter		4.50	5.333		1,425	204	17.90	1,646.90	1,875
4110	22" diameter		4	6		1,575	229	20	1,824	2,100
4120	24" diameter		3.50	6.857		1,725	262	23	2,010	2,325
4130	26" diameter		3	8		1,950	305	27	2,282	2,650
4140	28" diameter		2.50	9.600		2,125	365	32.50	2,522.50	2,925
4150	30" diameter		2	12		2,275	460	40.50	2,775.50	3,225
4160	32" diameter		1.50	16		2,725	610	54	3,389	3,975
4170	34" diameter		1	24		3,050	915	80.50	4,045.50	4,825
4180	36" diameter	▼	1	24		3,325	915	80.50	4,320.50	5,125
4220	For multi-pipe, add				▼	5%	20%			
4800	Steel casing elbow									
4820	6-5/8" diameter	Q-18	15	1.600	Ea.	785	61	5.40	851.40	960
4830	8-5/8" diameter		15	1.600		835	61	5.40	901.40	1,025
4850	10-3/4" diameter		14	1.714		1,000	65.50	5.75	1,071.25	1,200
4860	12-3/4" diameter		13	1.846		1,200	70.50	6.20	1,276.70	1,450
4870	14" diameter		12	2		1,275	76.50	6.70	1,358.20	1,525
4880	16" diameter		11	2.182		1,400	83.50	7.35	1,490.85	1,650
4890	18" diameter		10	2.400		1,600	91.50	8.05	1,699.55	1,900
4900	20" diameter		9	2.667		1,700	102	8.95	1,810.95	2,050
4910	22" diameter		8	3		1,825	115	10.10	1,950.10	2,200
4920	24" diameter		7	3.429		2,025	131	11.50	2,167.50	2,425
4930	26" diameter		6	4		2,200	153	13.45	2,366.45	2,675
4940	28" diameter		5	4.800		2,375	183	16.15	2,574.15	2,925
4950	30" diameter		4	6		2,400	229	20	2,649	3,000
4960	32" diameter		3	8		2,725	305	27	3,057	3,500
4970	34" diameter		2	12		2,975	460	40.50	3,475.50	4,000
4980	36" diameter	▼	2	12	▼	3,175	460	40.50	3,675.50	4,225
5500	Black steel service pipe, std. wt., 1" thick insulation									
5510	3/4" diameter pipe size	Q-17	54	.296	L.F.	19.90	10.90	1.49	32.29	40
5540	1" diameter pipe size		50	.320		20.50	11.80	1.61	33.91	42.50
5550	1-1/4" diameter pipe size		47	.340		23	12.55	1.72	37.27	46
5560	1-1/2" diameter pipe size		45	.356		25.50	13.10	1.79	40.39	49.50
5570	2" diameter pipe size		42	.381		28	14.05	1.92	43.97	54
5580	2-1/2" diameter pipe size		34	.471		29.50	17.35	2.37	49.22	61
5590	3" diameter pipe size		28	.571		33.50	21	2.88	57.38	71.50
5600	4" diameter pipe size		22	.727		43	27	3.67	73.67	92
5610	5" diameter pipe size	▼	18	.889		58.50	33	4.48	95.98	118
5620	6" diameter pipe size	Q-18	23	1.043	▼	63.50	40	3.51	107.01	134
6000	Black steel service pipe, std. wt., 1-1/2" thick insul.									
6010	3/4" diameter pipe size	Q-17	54	.296	L.F.	20	10.90	1.49	32.39	40
6040	1" diameter pipe size		50	.320		22.50	11.80	1.61	35.91	44.50
6050	1-1/4" diameter pipe size		47	.340		25.50	12.55	1.72	39.77	48.50
6060	1-1/2" diameter pipe size		45	.356		27.50	13.10	1.79	42.39	52
6070	2" diameter pipe size		42	.381		30	14.05	1.92	45.97	56
6080	2-1/2" diameter pipe size		34	.471		32	17.35	2.37	51.72	63.50
6090	3" diameter pipe size		28	.571		36	21	2.88	59.88	74
6100	4" diameter pipe size		22	.727		46	27	3.67	76.67	95
6110	5" diameter pipe size	▼	18	.889	▼	58.50	33	4.48	95.98	118

Important: See the Reference Section for critical supporting data - Reference Nos., Crews, & City Cost Indexes

02550 | Piped Energy Distribution

			CREW	DAILY OUTPUT	LABOR-HOURS	UNIT	MAT.	LABOR	EQUIP.	TOTAL	TOTAL INCL O&P
								2005 BARE COSTS			
200	6120	6" diameter pipe size	Q-18	23	1.043	L.F.	66.50	40	3.51	110.01	137
	6130	8" diameter pipe size		19	1.263		95.50	48.50	4.24	148.24	182
	6140	10" diameter pipe size		16	1.500		125	57.50	5.05	187.55	229
	6150	12" diameter pipe size		13	1.846		149	70.50	6.20	225.70	277
	6190	For 2" thick insulation, add					15%				
	6220	For 2-1/2" thick insulation, add					25%				
	6260	For 3" thick insulation, add					30%				
	6800	Black steel service pipe, ex. hvy. wt., 1" thick insul.									
	6820	3/4" diameter pipe size	Q-17	50	.320	L.F.	21	11.80	1.61	34.41	42.50
	6840	1" diameter pipe size		47	.340		22.50	12.55	1.72	36.77	45
	6850	1-1/4" diameter pipe size		44	.364		25.50	13.40	1.83	40.73	50.50
	6860	1-1/2" diameter pipe size		42	.381		27.50	14.05	1.92	43.47	53
	6870	2" diameter pipe size		40	.400		29	14.75	2.02	45.77	56
	6880	2-1/2" diameter pipe size		31	.516		36	19	2.60	57.60	71
	6890	3" diameter pipe size		27	.593		40.50	22	2.99	65.49	81
	6900	4" diameter pipe size		21	.762		53	28	3.84	84.84	105
	6910	5" diameter pipe size		17	.941		74.50	34.50	4.74	113.74	139
	6920	6" diameter pipe size	Q-18	22	1.091		82.50	41.50	3.67	127.67	157
	7400	Black steel service pipe, ex. hvy. wt., 1-1/2" thick insul.									
	7420	3/4" diameter pipe size	Q-17	50	.320	L.F.	21	11.80	1.61	34.41	42.50
	7440	1" diameter pipe size		47	.340		24	12.55	1.72	38.27	47
	7450	1-1/4" diameter pipe size		44	.364		27.50	13.40	1.83	42.73	52.50
	7460	1-1/2" diameter pipe size		42	.381		30.50	14.05	1.92	46.47	56.50
	7470	2" diameter pipe size		40	.400		30.50	14.75	2.02	47.27	57.50
	7480	2-1/2" diameter pipe size		31	.516		36.50	19	2.60	58.10	71.50
	7490	3" diameter pipe size		27	.593		43	22	2.99	67.99	83.50
	7500	4" diameter pipe size		21	.762		56	28	3.84	87.84	108
	7510	5" diameter pipe size		17	.941		77	34.50	4.74	116.24	142
	7520	6" diameter pipe size	Q-18	22	1.091		85	41.50	3.67	130.17	160
	7530	8" diameter pipe size		18	1.333		127	51	4.48	182.48	221
	7540	10" diameter pipe size		15	1.600		151	61	5.40	217.40	264
	7550	12" diameter pipe size		13	1.846		186	70.50	6.20	262.70	315
	7590	For 2" thick insulation, add					13%				
	7640	For 2-1/2" thick insulation, add					18%				
	7680	For 3" thick insulation, add					24%				
	9000	Combined steam pipe with condensate return									
	9010	8" and 4" in 24" case	Q-18	100	.240	L.F.	268	9.15	.81	277.96	310
	9020	6" and 3" in 20" case		104	.231		183	8.80	.78	192.58	216
	9030	3" and 1-1/2" in 16" case		110	.218		137	8.35	.73	146.08	164
	9040	2" and 1-1/4" in 12-3/4" case		114	.211		119	8.05	.71	127.76	144
	9050	1-1/2" and 1-1/4" in 10-3/4" case		116	.207		104	7.90	.70	112.60	128
	9100	Steam pipe only (no return)									
	9110	6" in 18" case	Q-18	108	.222	L.F.	151	8.50	.75	160.25	180
	9120	4" in 14" case		112	.214		119	8.20	.72	127.92	144
	9130	3" in 14" case		114	.211		110	8.05	.71	118.76	134
	9140	2-1/2" in 12-3/4" case		118	.203		103	7.75	.68	111.43	125
	9150	2" in 12-3/4" case		122	.197		98.50	7.50	.66	106.66	120
464	0010	**PIPING, GAS SERVICE & DISTRIBUTION, POLYETHYLENE**									
	0020	not including excavation or backfill									
	1000	60 psi coils, comp cplg @ 100', 1/2" diameter, SDR 9.3	B-20A	608	.053	L.F.	.47	1.70		2.17	3.11
	1040	1-1/4" diameter, SDR 11		544	.059		.87	1.90		2.77	3.85
	1100	2" diameter, SDR 11		488	.066		1.08	2.11		3.19	4.42
	1160	3" diameter, SDR 11		408	.078		2.25	2.53		4.78	6.35
	1500	60 PSI 40' joints with coupling, 3" diameter, SDR 11	B-21A	408	.098		2.25	3.23	1.24	6.72	8.75
	1540	4" diameter, SDR 11		352	.114		5.15	3.75	1.44	10.34	13

(Row 6120 notes: R15100 -070)

			DAILY	LABOR-		2005 BARE COSTS				TOTAL	
	02550 | **Piped Energy Distribution**	CREW	OUTPUT	HOURS	UNIT	MAT.	LABOR	EQUIP.	TOTAL	INCL O&P	
464	1600	6" diameter, SDR 11	B-21A	328	.122	L.F.	16.10	4.02	1.54	21.66	25.50	**464**
	1640	8" diameter, SDR 11	↓	272	.147	↓	22	4.85	1.86	28.71	33.50	
550	0010	**GAS STATION PRODUCT LINE**										**550**
	0020	Primary containment pipe, fiberglass-reinforced										
	0030	Plastic pipe 15' & 30' lengths										
	0040	2" diameter	Q-6	425	.056	L.F.	3.23	2.16		5.39	6.80	
	0050	3" diameter		400	.060		4.23	2.29		6.52	8.10	
	0060	4" diameter	↓	375	.064	↓	5.45	2.45		7.90	9.70	
	0100	Fittings										
	0110	Elbows, 90° & 45°, bell-ends, 2"	Q-6	24	1	Ea.	32.50	38		70.50	93.50	
	0120	3" diameter		22	1.091		34.50	41.50		76	100	
	0130	4" diameter		20	1.200		45.50	46		91.50	119	
	0200	Tees, bell ends, 2"		21	1.143		39.50	43.50		83	109	
	0210	3" diameter		18	1.333		40	51		91	121	
	0230	Flanges bell ends, 2"		24	1		13.10	38		51.10	72	
	0240	3" diameter		22	1.091		16.50	41.50		58	80.50	
	0250	4" diameter		20	1.200		22.50	46		68.50	93.50	
	0260	Sleeve couplings, 2"		21	1.143		8.35	43.50		51.85	74.50	
	0270	3" diameter		18	1.333		11.85	51		62.85	89.50	
	0280	4" diameter		15	1.600		16.35	61		77.35	110	
	0290	Threaded adapters 2"		21	1.143		10.95	43.50		54.45	77.50	
	0300	3" diameter		18	1.333		19.25	51		70.25	97.50	
	0310	4" diameter		15	1.600		26	61		87	121	
	0320	Reducers, 2"		27	.889		14.60	34		48.60	67	
	0330	3" diameter		22	1.091		16.90	41.50		58.40	81	
	0340	4" diameter	↓	20	1.200	↓	22	46		68	93	
	1010	Gas station product line for secondary containment (double wall)										
	1100	Fiberglass reinforced plastic pipe 25' lengths										
	1120	Pipe, plain end, 3"	Q-6	375	.064	L.F.	5.55	2.45		8	9.80	
	1130	4" diameter		350	.069		9.10	2.62		11.72	13.95	
	1140	5" diameter		325	.074		11.85	2.82		14.67	17.25	
	1150	6" diameter	↓	300	.080	↓	12.20	3.06		15.26	18	
	1200	Fittings										
	1230	Elbows, 90° & 45°, 3"	Q-6	18	1.333	Ea.	40	51		91	121	
	1240	4" diameter		16	1.500		69.50	57.50		127	163	
	1250	5" diameter		14	1.714		162	65.50		227.50	277	
	1260	6" diameter		12	2		164	76.50		240.50	295	
	1270	Tees, 3"		15	1.600		59	61		120	157	
	1280	4" diameter		12	2		87	76.50		163.50	211	
	1290	5" diameter		9	2.667		176	102		278	345	
	1300	6" diameter		6	4		184	153		337	430	
	1310	Couplings, 3"		18	1.333		28	51		79	108	
	1320	4" diameter		16	1.500		72.50	57.50		130	166	
	1330	5" diameter		14	1.714		150	65.50		215.50	265	
	1340	6" diameter		12	2		156	76.50		232.50	287	
	1350	Cross-over nipples, 3"		18	1.333		6.40	51		57.40	83.50	
	1360	4" diameter		16	1.500		7.50	57.50		65	94.50	
	1370	5" diameter		14	1.714		11.15	65.50		76.65	111	
	1380	6" diameter		12	2		11.70	76.50		88.20	128	
	1400	Telescoping, reducers, concentric 4" x 3"		18	1.333		21.50	51		72.50	100	
	1410	5" x 4"		17	1.412		55.50	54		109.50	143	
	1420	6" x 5"	↓	16	1.500	↓	134	57.50		191.50	233	

Important: See the Reference Section for critical supporting data - Reference Nos., Crews, & City Cost Indexes

02620 | Subdrainage

		CREW	DAILY OUTPUT	LABOR-HOURS	UNIT	2005 BARE COSTS				TOTAL INCL O&P
						MAT.	LABOR	EQUIP.	TOTAL	
610	**0010 PIPING, SUBDRAINAGE, CONCRETE** G1030 805									**61**
	0021 Not including excavation and backfill									
	3000 Porous wall concrete underdrain, std. strength, 4" diameter	B-20	335	.072	L.F.	1.95	2.15		4.10	5.50
	3020 6" diameter	"	315	.076		2.54	2.29		4.83	6.35
	3040 8" diameter	B-21	310	.090		3.13	2.79	.51	6.43	8.30
	3100 18" diameter	"	165	.170		10.05	5.25	.96	16.26	20
	4000 Extra strength, 6" diameter	B-20	315	.076		2.57	2.29		4.86	6.40
	4020 8" diameter	B-21	310	.090		3.86	2.79	.51	7.16	9.10
	4040 10" diameter		285	.098		7.70	3.04	.56	11.30	13.80
	4060 12" diameter		230	.122		8.35	3.76	.69	12.80	15.80
	4080 15" diameter		200	.140		9.25	4.33	.79	14.37	17.75
	4100 18" diameter		165	.170		13.50	5.25	.96	19.71	24
620	**0010 PIPING, SUBDRAINAGE, CORRUGATED METAL** G1030 805									**62**
	0021 Not including excavation and backfill									
	2010 Aluminum, perforated									
	2020 6" diameter, 18 ga.	B-14	380	.126	L.F.	2.76	3.55	.57	6.88	9.15
	2200 8" diameter, 16 ga.		370	.130		3.94	3.64	.58	8.16	10.60
	2220 10" diameter, 16 ga.		360	.133		4.93	3.75	.60	9.28	11.85
	2240 12" diameter, 16 ga.		285	.168		5.50	4.73	.76	10.99	14.20
	2260 18" diameter, 16 ga.		205	.234		8.25	6.60	1.05	15.90	20.50
	3000 Uncoated galvanized, perforated									
	3020 6" diameter, 18 ga.	B-20	380	.063	L.F.	4.37	1.90		6.27	7.75
	3200 8" diameter, 16 ga.	"	370	.065		6	1.95		7.95	9.65
	3220 10" diameter, 16 ga.	B-21	360	.078		9	2.40	.44	11.84	14.10
	3240 12" diameter, 16 ga.		285	.098		9.45	3.04	.56	13.05	15.70
	3260 18" diameter, 16 ga.		205	.137		14.40	4.22	.77	19.39	23.50
	4000 Steel, perforated, asphalt coated									
	4020 6" diameter 18 ga.	B-20	380	.063	L.F.	3.49	1.90		5.39	6.80
	4030 8" diameter 18 ga	"	370	.065		5.45	1.95		7.40	9.05
	4040 10" diameter 16 ga	B-21	360	.078		6.30	2.40	.44	9.14	11.10
	4050 12" diameter 16 ga		285	.098		7.20	3.04	.56	10.80	13.25
	4060 18" diameter 16 ga		205	.137		9.85	4.22	.77	14.84	18.20
630	**0010 PIPING, SUBDRAINAGE, PLASTIC**									**63**
	0020 Not including excavation and backfill									
	1110 10"				Ea.	3.01			3.01	3.31
	2100 Perforated PVC, 4" diameter	B-14	314	.153	L.F.	.71	4.29	.69	5.69	8.20
	2110 6" diameter		300	.160		1.31	4.49	.72	6.52	9.20
	2120 8" diameter		290	.166		1.49	4.65	.74	6.88	9.65
	2130 10" diameter		280	.171		2.31	4.82	.77	7.90	10.85
	2140 12" diameter		270	.178		3.20	4.99	.80	8.99	12.10

02630 | Storm Drainage

		CREW	DAILY OUTPUT	LABOR-HOURS	UNIT	MAT.	LABOR	EQUIP.	TOTAL	TOTAL INCL O&P
110	**0010 CATCH BASIN GRATES AND FRAMES** not including footing, excavation									**11**
	1600 Frames & covers, C.I., 24" square, 500 lb.	B-6	7.80	3.077	Ea.	239	88.50	27.50	355	430
	1700 26" D shape, 600 lb.		7	3.429		380	99	31	510	605
	1800 Light traffic, 18" diameter, 100 lb.		10	2.400		128	69	21.50	218.50	271
	1900 24" diameter, 300 lb.		8.70	2.759		196	79.50	25	300.50	365
	2000 36" diameter, 900 lb.		5.80	4.138		425	119	37	581	690
	2100 Heavy traffic, 24" diameter, 400 lb.		7.80	3.077		193	88.50	27.50	309	380
	2200 36" diameter, 1150 lb.		3	8		610	231	72	913	1,100
	2300 Mass. State standard, 26" diameter, 475 lb.		7	3.429		465	99	31	595	695
	2400 30" diameter, 620 lb.		7	3.429		325	99	31	455	540
	2500 Watertight, 24" diameter, 350 lb.		7.80	3.077		325	88.50	27.50	441	520
	2600 26" diameter, 500 lb.		7	3.429		310	99	31	440	525
	2700 32" diameter, 575 lb.		6	4		665	115	36	816	945
	2800 3 piece cover & frame, 10" deep,									

For expanded coverage of these items see *Means Heavy Construction Cost Data 2005*

02630 | Storm Drainage

		CREW	DAILY OUTPUT	LABOR-HOURS	UNIT	2005 BARE COSTS				TOTAL INCL O&P	
						MAT.	LABOR	EQUIP.	TOTAL		
110	2900	1200 lbs., for heavy equipment	B-6	3	8	Ea.	990	231	72	1,293	1,525	**110**
	3000	Raised for paving 1-1/4" to 2" high,										
	3100	4 piece expansion ring										
	3200	20" to 26" diameter	1 Clab	3	2.667	Ea.	114	71		185	237	
	3300	30" to 36" diameter	"	3	2.667	"	159	71		230	286	
	3320	Frames and covers, existing, raised for paving, 2", including										
	3340	row of brick, concrete collar, up to 12" wide frame	B-6	18	1.333	Ea.	35.50	38.50	12	86	112	
	3360	20" to 26" wide frame		11	2.182		56.50	63	19.60	139.10	180	
	3380	30" to 36" wide frame	↓	9	2.667		70	77	24	171	222	
	3400	Inverts, single channel brick	D-1	3	5.333		78.50	166		244.50	340	
	3500	Concrete		5	3.200		61	99.50		160.50	220	
	3600	Triple channel, brick		2	8		119	249		368	510	
	3700	Concrete	↓	3	5.333	↓	105	166		271	370	
400	0010	**STORM DRAINAGE MANHOLES, FRAMES & COVERS** not including										**400**
	0020	footing, excavation, backfill (See line items for frame & cover)										
	0050	Brick, 4' inside diameter, 4' deep	D-1	1	16	Ea.	320	495		815	1,100	
	0100	6' deep		.70	22.857		445	710		1,155	1,575	
	0150	8' deep	↓	.50	32	↓	565	995		1,560	2,150	
	0200	For depths over 8', add		4	4	V.L.F.	174	124		298	380	
	0400	Concrete blocks (radial), 4' I.D., 4' deep		1.50	10.667	Ea.	255	330		585	785	
	0500	6' deep		1	16		335	495		830	1,125	
	0600	8' deep	↓	.70	22.857	↓	415	710		1,125	1,525	
	0700	For depths over 8', add		5.50	2.909	V.L.F.	41.50	90.50		132	184	
	0800	Concrete, cast in place, 4' x 4', 8" thick, 4' deep	C-14H	2	24	Ea.	445	810	12	1,267	1,750	
	0900	6' deep		1.50	32		650	1,075	16	1,741	2,400	
	1000	8' deep	↓	1	48	↓	920	1,625	24	2,569	3,575	
	1100	For depths over 8', add		8	6	V.L.F.	107	202	3	312	435	
	1110	Precast, 4' I.D., 4' deep	B-22	4.10	7.317	Ea.	675	229	58	962	1,175	
	1120	6' deep		3	10		870	315	79.50	1,264.50	1,525	
	1130	8' deep	↓	2	15	↓	1,025	470	119	1,614	1,975	
	1140	For depths over 8', add		16	1.875	V.L.F.	143	58.50	14.85	216.35	264	
	1150	5' I.D., 4' deep	B-6	3	8	Ea.	720	231	72	1,023	1,225	
	1160	6' deep		2	12		970	345	108	1,423	1,725	
	1170	8' deep	↓	1.50	16	↓	1,225	460	144	1,829	2,225	
	1180	For depths over 8', add		12	2	V.L.F.	159	57.50	18	234.50	283	
	1190	6' I.D., 4' deep		2	12	Ea.	1,175	345	108	1,628	1,950	
	1200	6' deep		1.50	16		1,525	460	144	2,129	2,550	
	1210	8' deep	↓	1	24	↓	1,875	690	216	2,781	3,375	
	1220	For depths over 8', add		8	3	V.L.F.	246	86.50	27	359.50	435	
	1250	Slab tops, precast, 8" thick										
	1300	4' diameter manhole	B-6	8	3	Ea.	165	86.50	27	278.50	345	
	1400	5' diameter manhole		7.50	3.200		325	92	29	446	535	
	1500	6' diameter manhole	↓	7	3.429		405	99	31	535	630	
	3800	Steps, heavyweight cast iron, 7" x 9"	1 Bric	40	.200		12.20	7.05		19.25	24	
	3900	8" x 9"		40	.200		18.35	7.05		25.40	31	
	3928	12" x 10-1/2"		40	.200		16.40	7.05		23.45	29	
	4000	Standard sizes, galvanized steel		40	.200		15.30	7.05		22.35	27.50	
	4100	Aluminum	↓	40	.200	↓	18.90	7.05		25.95	32	
510	0010	**PIPING, STORM DRAINAGE, CORRUGATED METAL** G3020 110										**510**
	0020	Not including excavation or backfill										
	2000	Corrugated metal pipe, galvanized and coated										
	2020	Bituminous coated with paved invert, 20' lengths										
	2040	8" diameter, 16 ga.	B-14	330	.145	L.F.	8.95	4.09	.65	13.69	16.85	
	2060	10" diameter, 16 ga.	↓	260	.185	↓	10.75	5.20	.83	16.78	20.50	

Important: See the Reference Section for critical supporting data - Reference Nos., Crews, & City Cost Indexes

	02630	Storm Drainage	CREW	DAILY OUTPUT	LABOR-HOURS	UNIT	2005 BARE COSTS				TOTAL INCL O&P	
							MAT.	LABOR	EQUIP.	TOTAL		
510	2080	12″ diameter, 16 ga.	B-14	210	.229	L.F.	12.85	6.40	1.03	20.28	25	510
	2100	15″ diameter, 16 ga.		200	.240		15.65	6.75	1.08	23.48	29	
	2120	18″ diameter, 16 ga.		190	.253		20	7.10	1.14	28.24	34.50	
	2140	24″ diameter, 14 ga.		160	.300		24.50	8.45	1.35	34.30	41.50	
	2160	30″ diameter, 14 ga.	B-13	120	.467		32.50	13.45	5.15	51.10	61.50	
	2180	36″ diameter, 12 ga.		120	.467		47.50	13.45	5.15	66.10	78.50	
	2200	48″ diameter, 12 ga.		100	.560		72.50	16.10	6.15	94.75	111	
	2220	60″ diameter, 10 ga.	B-13B	75	.747		93.50	21.50	11.95	126.95	149	
	2240	72″ diameter, 8 ga.	″	45	1.244		140	36	19.90	195.90	231	
	2250	End sections, 8″ diameter, 16 ga.	B-14	20	2.400	Ea.	52.50	67.50	10.80	130.80	173	
	2255	10″ diameter, 16 ga.		20	2.400		59.50	67.50	10.80	137.80	181	
	2260	12″ diameter, 16 ga.		18	2.667		73.50	75	12	160.50	210	
	2265	15″ diameter, 16 ga.		18	2.667		91	75	12	178	229	
	2270	18″ diameter, 16 ga.		16	3		115	84.50	13.50	213	271	
	2275	24″ diameter, 16 ga.	B-13	16	3.500		164	101	38.50	303.50	380	
	2280	30″ diameter, 16 ga.		14	4		310	115	44	469	570	
	2285	36″ diameter, 14 ga.		14	4		410	115	44	569	675	
	2290	48″ diameter, 14 ga.		10	5.600		850	161	61.50	1,072.50	1,250	
	2292	60″ diameter, 14 ga.		6	9.333		1,175	269	103	1,547	1,800	
	2294	72″ diameter, 14 ga.	B-13B	5	11.200		2,025	320	179	2,524	2,925	
	2342	18″ diameter, 16 ga.	B-14	20	2.400		174	67.50	10.80	252.30	310	
	2344	24″ diameter, 14 ga.		16	3		243	84.50	13.50	341	410	
	2346	30″ diameter, 14 ga.		15	3.200		300	90	14.40	404.40	485	
	2348	36″ diameter, 14 ga.	B-13	15	3.733		405	107	41	553	655	
	2350	48″ diameter, 12 ga.	″	12	4.667		580	134	51.50	765.50	905	
	2352	60″ diameter, 10 ga.	B-13B	10	5.600		780	161	89.50	1,030.50	1,200	
	2354	72″ diameter, 10 ga.	″	6	9.333		965	269	149	1,383	1,650	
	2410	18″ diameter, 16 ga.	B-14	16	3		232	84.50	13.50	330	400	
	2412	24″ diameter, 14 ga.	″	16	3		350	84.50	13.50	448	530	
	2414	30″ diameter, 14 ga.	B-13	12	4.667		455	134	51.50	640.50	765	
	2416	36″ diameter, 14 ga.		11	5.091		560	147	56	763	905	
	2418	48″ diameter, 12 ga.		10	5.600		845	161	61.50	1,067.50	1,250	
	2420	60″ diameter, 10 ga.	B-13B	8	7		1,225	202	112	1,539	1,775	
	2422	72″ diameter, 10 ga.	″	5	11.200		1,475	320	179	1,974	2,325	
	2500	Galvanized, uncoated, 20′ lengths										
	2520	8″ diameter, 16 ga.	B-14	355	.135	L.F.	6.90	3.80	.61	11.31	14.10	
	2540	10″ diameter, 16 ga.		280	.171		7.60	4.82	.77	13.19	16.65	
	2560	12″ diameter, 16 ga.		220	.218		8.65	6.15	.98	15.78	20	
	2580	15″ diameter, 16 ga.		220	.218		11	6.15	.98	18.13	22.50	
	2600	18″ diameter, 16 ga.		205	.234		14.55	6.60	1.05	22.20	27.50	
	2620	24″ diameter, 14 ga.		175	.274		21	7.70	1.23	29.93	36.50	
	2640	30″ diameter, 14 ga.	B-13	130	.431		27	12.40	4.74	44.14	54	
	2660	36″ diameter, 12 ga.		130	.431		44	12.40	4.74	61.14	73	
	2680	48″ diameter, 12 ga.		110	.509		59	14.65	5.60	79.25	93.50	
	2690	60″ diameter, 10 ga.	B-13B	78	.718		93	20.50	11.50	125	147	
	2695	72″ diameter, 10 ga.	″	60	.933		117	27	14.90	158.90	187	
	2711	Bends or elbows, 12″ diameter, 16 ga.	B-14	30	1.600	Ea.	104	45	7.20	156.20	192	
	2712	15″ diameter, 16 ga.		25.04	1.917		128	54	8.65	190.65	234	
	2714	18″ diameter, 16 ga.		20	2.400		144	67.50	10.80	222.30	275	
	2716	24″ diameter, 14 ga.		16	3		210	84.50	13.50	308	375	
	2718	30″ diameter, 14 ga.		15	3.200		263	90	14.40	367.40	445	
	2720	36″ diameter, 14 ga.	B-13	15	3.733		360	107	41	508	605	
	2722	48″ diameter, 12 ga.		12	4.667		495	134	51.50	680.50	810	
	2724	60″ diameter, 10 ga.		10	5.600		760	161	61.50	982.50	1,150	
	2726	72″ diameter, 10 ga.		6	9.333		940	269	103	1,312	1,550	
	2728	Wyes or tees, 12″ diameter, 16 ga.	B-14	22.48	2.135		138	60	9.60	207.60	256	

02630	Storm Drainage	CREW	DAILY OUTPUT	LABOR-HOURS	UNIT	MAT.	2005 BARE COSTS LABOR	EQUIP.	TOTAL	TOTAL INCL O&P	
510	2730	18″ diameter, 16 ga.	B-14	15	3.200	Ea.	201	90	14.40	305.40	375	510
	2732	24″ diameter, 14 ga.		15	3.200		310	90	14.40	414.40	500	
	2734	30″ diameter, 14 ga.	↓	14	3.429		410	96.50	15.45	521.95	615	
	2736	36″ diameter, 14 ga.	B-13	14	4		530	115	44	689	810	
	2738	48″ diameter, 12 ga.		12	4.667		780	134	51.50	965.50	1,125	
	2740	60″ diameter, 10 ga.		10	5.600		1,150	161	61.50	1,372.50	1,575	
	2742	72″ diameter, 10 ga.	↓	6	9.333		1,350	269	103	1,722	2,000	
	2780	End sections, 8″ diameter	B-14	24	2		71.50	56	9	136.50	175	
	2785	10″ diameter		22	2.182		73	61.50	9.80	144.30	186	
	2790	12″ diameter		35	1.371		82.50	38.50	6.15	127.15	157	
	2800	18″ diameter	↓	30	1.600		98	45	7.20	150.20	185	
	2810	24″ diameter	B-13	25	2.240		143	64.50	24.50	232	284	
	2820	30″ diameter		25	2.240		272	64.50	24.50	361	425	
	2825	36″ diameter		20	2.800		360	80.50	31	471.50	555	
	2830	48″ diameter	↓	10	5.600		745	161	61.50	967.50	1,125	
	2835	60″ diameter	B-13B	5	11.200		970	320	179	1,469	1,775	
	2840	72″ diameter	″	4	14		1,750	405	224	2,379	2,800	
	2850	Couplings, 12″ diameter					16			16	17.60	
	2855	18″ diameter					23.50			23.50	25.50	
	2860	24″ diameter					31.50			31.50	34.50	
	2865	30″ diameter					38.50			38.50	42.50	
	2870	36″ diameter					54			54	59.50	
	2875	48″ diameter					73.50			73.50	81	
	2880	60″ diameter					116			116	127	
	2885	72″ diameter	↓				160			160	176	
530	0010	**SEWAGE/DRAINAGE COLLECTION, CONCRETE PIPE** G3020 110										530
	0020	Not including excavation or backfill										
	1000	Non-reinforced pipe, extra strength, B&S or T&G joints										
	1010	6″ diameter	B-14	265.04	.181	L.F.	4.13	5.10	.82	10.05	13.30	
	1020	8″ diameter		224	.214		4.54	6	.96	11.50	15.35	
	1030	10″ diameter		216	.222		5.05	6.25	1	12.30	16.30	
	1040	12″ diameter		200	.240		6.20	6.75	1.08	14.03	18.45	
	1050	15″ diameter		180	.267		7.25	7.50	1.20	15.95	21	
	1060	18″ diameter		144	.333		8.90	9.35	1.50	19.75	26	
	1070	21″ diameter		112	.429		10.95	12.05	1.93	24.93	33	
	1080	24″ diameter	↓	100	.480	↓	13.40	13.50	2.16	29.06	38	
	2000	Reinforced culvert, class 3, no gaskets										
	2010	12″ diameter	B-14	150	.320	L.F.	11.20	9	1.44	21.64	28	
	2020	15″ diameter		150	.320		14.35	9	1.44	24.79	31.50	
	2030	18″ diameter		132	.364		15.10	10.20	1.64	26.94	34	
	2035	21″ diameter		120	.400		19.60	11.25	1.80	32.65	41	
	2040	24″ diameter	↓	100	.480		22.50	13.50	2.16	38.16	48	
	2045	27″ diameter	B-13	92	.609		28	17.50	6.70	52.20	65.50	
	2050	30″ diameter		88	.636		30.50	18.30	7	55.80	69	
	2060	36″ diameter	↓	72	.778		43.50	22.50	8.55	74.55	92	
	2070	42″ diameter	B-13B	72	.778		59.50	22.50	12.45	94.45	114	
	2080	48″ diameter		64	.875		73.50	25	14	112.50	135	
	2090	60″ diameter		48	1.167		118	33.50	18.65	170.15	201	
	2100	72″ diameter		40	1.400		166	40.50	22.50	229	270	
	2120	84″ diameter		32	1.750		281	50.50	28	359.50	420	
	2140	96″ diameter	↓	24	2.333		335	67	37.50	439.50	515	
	2200	With gaskets, class 3, 12″ diameter	B-21	168	.167		12.25	5.15	.94	18.34	22.50	
	2220	15″ diameter		160	.175		14.70	5.40	.99	21.09	25.50	
	2230	18″ diameter		152	.184		18.40	5.70	1.04	25.14	30.50	
	2240	24″ diameter	↓	136	.206	↓	27.50	6.35	1.17	35.02	41.50	

Important: See the Reference Section for critical supporting data - Reference Nos., Crews, & City Cost Indexes

		02630	Storm Drainage	CREW	DAILY OUTPUT	LABOR-HOURS	UNIT	MAT.	2005 BARE COSTS LABOR	EQUIP.	TOTAL	TOTAL INCL O&P	
530	2260		30" diameter	B-13	88	.636	L.F.	37	18.30	7	62.30	76	53
	2270		36" diameter	"	72	.778		55	22.50	8.55	86.05	104	
	2290		48" diameter	B-13B	64	.875		90	25	14	129	153	
	2310		72" diameter	"	40	1.400		233	40.50	22.50	296	345	
	2330		Flared ends, 6'-1" long, 12" diameter	B-21	190	.147		34	4.56	.83	39.39	45	
	2340		15" diameter		155	.181		38.50	5.60	1.02	45.12	52.50	
	2400		6'-2" long, 18" diameter		122	.230		40	7.10	1.30	48.40	56.50	
	2420		24" diameter	↓	88	.318		46.50	9.85	1.80	58.15	68	
	2440		36" diameter	B-13	60	.933		84.50	27	10.30	121.80	145	
	3080		Radius pipe, add to pipe prices, 12" to 60" diameter					50%					
	3090		Over 60" diameter, add				↓	20%					
	3500		Reinforced elliptical, 8' lengths, C507 class 3										
	3520		14" x 23" inside, round equivalent 18" diameter	B-21	82	.341	L.F.	26.50	10.55	1.93	38.98	47.50	
	3530		24" x 38" inside, round equivalent 30" diameter	B-13	58	.966		47	28	10.65	85.65	106	
	3540		29" x 45" inside, round equivalent 36" diameter		52	1.077		60.50	31	11.85	103.35	128	
	3550		38" x 60" inside, round equivalent 48" diameter		38	1.474		79	42.50	16.25	137.75	170	
	3560		48" x 76" inside, round equivalent 60" diameter		26	2.154		120	62	23.50	205.50	254	
	3570		58" x 91" inside, round equivalent 72" diameter	↓	22	2.545	↓	171	73.50	28	272.50	335	
	3780		Concrete slotted pipe, class 4 mortar joint										
	3800		12" diameter	B-21	168	.167	L.F.	13.15	5.15	.94	19.24	23.50	
	3840		18" diameter	"	152	.184	"	20.50	5.70	1.04	27.24	32.50	
	3900		Class 4 O-ring										
	3940		12" diameter	B-21	168	.167	L.F.	13.75	5.15	.94	19.84	24	
	3960		18" diameter	"	152	.184	"	18.45	5.70	1.04	25.19	30.50	
810	0010	**OVAL ARCH CULVERTS**											81
	3000		Corrugated galvanized or aluminum, coated & paved										
	3020		17" x 13", 16 ga., 15" equivalent	B-14	200	.240	L.F.	26.50	6.75	1.08	34.33	40.50	
	3040		21" x 15", 16 ga., 18" equivalent		150	.320		34	9	1.44	44.44	53	
	3060		28" x 20", 14 ga., 24" equivalent		125	.384		49	10.80	1.73	61.53	72	
	3080		35" x 24", 14 ga., 30" equivalent	↓	100	.480		59.50	13.50	2.16	75.16	89	
	3100		42" x 29", 12 ga., 36" equivalent	B-13	100	.560		88.50	16.10	6.15	110.75	129	
	3120		49" x 33", 12 ga., 42" equivalent		90	.622		107	17.90	6.85	131.75	152	
	3140		57" x 38", 12 ga., 48" equivalent	↓	75	.747	↓	119	21.50	8.20	148.70	173	
	3160		Steel, plain oval arch culverts, plain										
	3180		17" x 13", 16 ga., 15" equivalent	B-14	225	.213	L.F.	14.30	6	.96	21.26	26	
	3200		21" x 15", 16 ga., 18" equivalent		175	.274		16.95	7.70	1.23	25.88	32	
	3220		28" x 20", 14 ga., 24" equivalent	↓	150	.320		27	9	1.44	37.44	45.50	
	3240		35" x 24", 14 ga., 30" equivalent	B-13	108	.519		34	14.95	5.70	54.65	67	
	3260		42" x 29", 12 ga., 36" equivalent		108	.519		56.50	14.95	5.70	77.15	91.50	
	3280		49" x 33", 12 ga., 42" equivalent		92	.609		66.50	17.50	6.70	90.70	108	
	3300		57" x 38", 12 ga., 48" equivalent	↓	75	.747	↓	58.50	21.50	8.20	88.20	107	
	3320		End sections, 17" x 13"		22	2.545	Ea.	78.50	73.50	28	180	230	
	3340		42" x 29"	↓	17	3.294	"	320	95	36.50	451.50	540	
	3360		Multi-plate arch, steel	B-20	1,690	.014	Lb.	.90	.43		1.33	1.65	

		02740	Flexible Pavement	CREW	DAILY OUTPUT	LABOR-HOURS	UNIT	MAT.	2005 BARE COSTS LABOR	EQUIP.	TOTAL	TOTAL INCL O&P	
310	0010	**ASPHALTIC CONCRETE PAVEMENT, HIGHWAYS**											31
	0020		and large paved areas										

For expanded coverage of these items see *Means Heavy Construction Cost Data 2005*

02740 | Flexible Pavement

		CREW	DAILY OUTPUT	LABOR-HOURS	UNIT	MAT.	LABOR	EQUIP.	TOTAL	TOTAL INCL O&P	
310							2005 BARE COSTS					**310**
0080	Binder course, 1-1/2" thick	B-25	7,725	.011	S.Y.	2.63	.33	.26	3.22	3.69	
0120	2" thick		6,345	.014		3.50	.40	.32	4.22	4.82	
0160	3" thick		4,905	.018		5.20	.52	.41	6.13	6.95	
0200	4" thick	↓	4,140	.021		6.95	.62	.49	8.06	9.15	
0300	Wearing course, 1" thick	B-25B	10,575	.009		1.86	.27	.21	2.34	2.68	
0340	1-1/2" thick		7,725	.012		2.83	.37	.28	3.48	3.98	
0380	2" thick		6,345	.015		3.80	.45	.35	4.60	5.25	
0420	2-1/2" thick		5,480	.018		4.69	.52	.40	5.61	6.40	
0460	3" thick	↓	4,900	.020		5.60	.58	.45	6.63	7.55	
0500	Open graded friction course	B-25C	5,000	.010	↓	1.66	.28	.35	2.29	2.66	
0800	Alternate method of figuring paving costs										
0810	Binder course, 1-1/2" thick	B-25	630	.140	Ton	34.50	4.06	3.19	41.75	48	
0811	2" thick		690	.128		34.50	3.71	2.91	41.12	47	
0812	3" thick		800	.110		34.50	3.20	2.51	40.21	45.50	
0813	4" thick	↓	850	.104		34.50	3.01	2.37	39.88	45	
0850	Wearing course, 1" thick	B-25B	575	.167		34.50	4.93	3.81	43.24	49.50	
0851	1-1/2" thick		630	.152		34.50	4.50	3.48	42.48	48.50	
0852	2" thick		690	.139		34.50	4.11	3.17	41.78	48	
0853	2-1/2" thick		745	.129		34.50	3.80	2.94	41.24	47	
0854	3" thick	↓	800	.120		34.50	3.54	2.74	40.78	46.50	
1000	Pavement replacement over trench, 2" thick	B-37	90	.533	S.Y.	3.87	15	1.19	20.06	28.50	
1050	4" thick		70	.686		7.65	19.25	1.54	28.44	40	
1080	6" thick	↓	55	.873	↓	12.20	24.50	1.95	38.65	53.50	
1200	For paving projects 300 tons or less add for trucking										
1300	See 02315-490-0010 for hauling costs										
315	0011	**ASPHALTIC CONCRETE PAVEMENT, LOTS & DRIVEWAYS**										**315**
0020	6" stone base, 2" binder course, 1" topping	B-25C	9,000	.005	S.F.	1.38	.16	.20	1.74	1.98	
0300	Binder course, 1-1/2" thick		35,000	.001		.31	.04	.05	.40	.46	
0400	2" thick		25,000	.002		.41	.06	.07	.54	.62	
0500	3" thick		15,000	.003		.63	.10	.12	.85	.97	
0600	4" thick		10,800	.004		.82	.13	.16	1.11	1.28	
0800	Sand finish course, 3/4" thick		41,000	.001		.19	.03	.04	.26	.31	
0900	1" thick	↓	34,000	.001		.23	.04	.05	.32	.38	
1000	Fill pot holes, hot mix, 2" thick	B-16	4,200	.008		.42	.21	.11	.74	.90	
1100	4" thick		3,500	.009		.61	.25	.14	1	1.21	
1120	6" thick	↓	3,100	.010		.82	.28	.15	1.25	1.51	
1140	Cold patch, 2" thick	B-51	3,000	.016		.50	.43	.04	.97	1.27	
1160	4" thick		2,700	.018		.95	.48	.05	1.48	1.85	
1180	6" thick	↓	1,900	.025	↓	1.48	.68	.07	2.23	2.76	

02810 | Irrigation System

		CREW	DAILY OUTPUT	LABOR-HOURS	UNIT	MAT.	LABOR	EQUIP.	TOTAL	TOTAL INCL O&P	
							2005 BARE COSTS					
200	0010	**SUBSURFACE DRIP IRRIGATION**, looped grid, pressure compensating										**200**
0100	Preinserted emitter, line, hand bury, irregular area, small	3 Skwk	1,200	.020	L.F.	.34	.70		1.04	1.45	
0150	Medium		1,800	.013		.34	.46		.80	1.09	
0200	Large		2,520	.010		.34	.33		.67	.89	
0250	Rectangular area, small		2,040	.012		.34	.41		.75	1.01	
0300	Medium	↓	2,640	.009	↓	.34	.32		.66	.86	

Important: See the Reference Section for critical supporting data - Reference Nos., Crews, & City Cost Indexes

02810	Irrigation System	CREW	DAILY OUTPUT	LABOR-HOURS	UNIT	2005 BARE COSTS MAT.	LABOR	EQUIP.	TOTAL	TOTAL INCL O&P	
200											**200**
0350	Large	3 Skwk	3,600	.007	L.F.	.34	.23		.57	.73	
0400	Install in trench, irregular area, small		4,050	.006		.34	.21		.55	.69	
0450	Medium		7,488	.003		.34	.11		.45	.54	
0500	Large		16,560	.001		.34	.05		.39	.45	
0550	Rectangular area, small		8,100	.003		.34	.10		.44	.53	
0600	Medium		21,960	.001		.34	.04		.38	.43	
0650	Large		33,264	.001		.34	.03		.37	.41	
0700	Trenching and backfill	B-53	500	.016			.53	.10	.63	.91	
0750	For vinyl tubing, 1/4", add to above						10%				
0800	Vinyl tubing, 1/4", material only					.06			.06	.07	
0850	Supply tubing, 1/2", material only, 100' coil					.12			.12	.13	
0900	500' coil					.11			.11	.12	
0950	Compression fittings	1 Skwk	90	.089	Ea.	.71	3.10		3.81	5.60	
1000	Barbed fittings, 1/4"		360	.022		.17	.77		.94	1.39	
1100	Flush risers		60	.133		.78	4.65		5.43	8.10	
1150	Flush ends, figure eight		180	.044		.14	1.55		1.69	2.56	
1200	Ball valve, 4-1/2"		20	.400		6.05	13.95		20	28	
1250	4-3/4"		20	.400		7.15	13.95		21.10	29.50	
1300	Auto flush, spring loaded		90	.089		1.93	3.10		5.03	6.95	
1350	Volumetric		90	.089		6.05	3.10		9.15	11.50	
1400	Air relief valve, inline with compensation tee, 1/2"		45	.178		6.50	6.20		12.70	16.80	
1450	1"		30	.267		18.10	9.30		27.40	34.50	
1500	Round box for flush ends, 6"		30	.267		3.58	9.30		12.88	18.40	
1550	Fertilizer injector, non-proportional		4	2		3.48	69.50		72.98	112	
1600	Screen filter, 3/4" screen		12	.667		23	23		46	61.50	
1650	1" disk		8	1		60.50	35		95.50	121	
1700	1-1/2" disk		4	2		144	69.50		213.50	267	
1750	2" disk		3	2.667		289	93		382	465	
1800	Typical installation 18" O.C., small, minimum				S.F.				.75	.86	
1850	Maximum								1.39	1.60	
1900	Large, minimum								.55	.64	
2000	Maximum								.92	1.07	
2100	For non-pressure compensating systems, deduct								10%	10%	
300	**0010** **SPRINKLER IRRIGATION SYSTEM** For lawns [G2050 710]										**30**
0100	Golf course with fully automatic system	C-17	.05	1,600	9 holes	82,500	56,500		139,000	178,500	
0200	24' diam. head at 15' O.C incl. piping, auto oper., minimum	B-20	70	.343	Head	18.20	10.30		28.50	36	
0300	Maximum		40	.600		42	18.05		60.05	74	
0500	60' diam. head at 40' O.C. incl. piping, auto oper., minimum		28	.857		55	26		81	101	
0600	Maximum		23	1.043		154	31.50		185.50	219	
0800	Residential system, custom, 1" supply		2,000	.012	S.F.	.27	.36		.63	.86	
0900	1-1/2" supply		1,800	.013	"	.32	.40		.72	.97	
0990	For renovation work, add to above						50%				
1020	Pop up spray head w/risers, hi-pop, full circle pattern, 4"	2 Skwk	76	.211	Ea.	3.36	7.35		10.71	15.10	
1030	1/2 circle pattern		76	.211		3.43	7.35		10.78	15.15	
1040	6", full circle pattern		76	.211		8.60	7.35		15.95	21	
1050	1/2 circle pattern		76	.211		8.60	7.35		15.95	21	
1060	12", full circle pattern		76	.211		12.25	7.35		19.60	25	
1070	1/2 circle pattern		76	.211		12.30	7.35		19.65	25	
1080	Pop up bubbler head w/risers, hi-pop bubbler head, 4"		76	.211		3.50	7.35		10.85	15.25	
1090	6"		76	.211		8.60	7.35		15.95	21	
1100	12"		76	.211		12.25	7.35		19.60	25	
1110	Impact full/part circle sprinklers, 28'-54' 25-60 PSI		37	.432		11.45	15.05		26.50	36	
1120	Spaced 37'-49' @ 25-50 PSI		37	.432		25	15.05		40.05	51	
1130	Spaced 43'-61' @ 30-60 PSI		37	.432		45.50	15.05		60.55	73.50	
1140	Spaced 54'-78' @ 40-80 PSI		37	.432		105	15.05		120.05	139	

For expanded coverage of these items see Means Site Work and Landscape Cost Data 2005

02810	Irrigation System	CREW	DAILY OUTPUT	LABOR-HOURS	UNIT	2005 BARE COSTS				TOTAL INCL O&P
						MAT.	LABOR	EQUIP.	TOTAL	
1145	Impact rotor pop-up full/part commercial circle sprinklers									
1150	Spaced 42'-65' 35-80 PSI	2 Skwk	25	.640	Ea.	23	22.50		45.50	60
1160	Spaced 48'-76' 45-85 PSI	"	25	.640	"	8.85	22.50		31.35	44
1165	Impact rotor pop-up part. circle comm., 53'-75', 55-100 PSI, w/ acc									
1170	Plastic case, metal cover	2 Skwk	25	.640	Ea.	148	22.50		170.50	198
1180	Rubber cover		25	.640		112	22.50		134.50	158
1190	Iron case, metal cover		22	.727		136	25.50		161.50	189
1200	Rubber cover		22	.727		147	25.50		172.50	201
1250	Plastic case, 2 nozzle, metal cover		25	.640		108	22.50		130.50	154
1260	Rubber cover		25	.640		115	22.50		137.50	161
1270	Iron case, 2 nozzle, metal cover		22	.727		157	25.50		182.50	213
1280	Rubber cover		22	.727		164	25.50		189.50	220
1282	Impact rotor pop-up full circle comm., 39'-99', 30-100 PSI									
1284	Plastic case, metal cover	2 Skwk	25	.640	Ea.	113	22.50		135.50	159
1286	Rubber cover		25	.640		125	22.50		147.50	173
1288	Iron case, metal cover		22	.727		162	25.50		187.50	219
1290	Rubber cover		22	.727		171	25.50		196.50	228
1292	Plastic case, 2 nozzle, metal cover		22	.727		122	25.50		147.50	174
1294	Rubber cover		22	.727		126	25.50		151.50	179
1296	Iron case, 2 nozzle, metal cover		20	.800		157	28		185	216
1298	Rubber cover		20	.800		167	28		195	228
1305	Electric remote control valve, plastic, 3/4"		18	.889		17.05	31		48.05	67
1310	1"		18	.889		30	31		61	81
1320	1-1/2"		18	.889		58	31		89	112
1330	2"		18	.889		85.50	31		116.50	142
1335	Quick coupling valves, brass, locking cover									
1340	Inlet coupling valve, 3/4"	2 Skwk	18.75	.853	Ea.	42	29.50		71.50	92.50
1350	1"		18.75	.853		51	29.50		80.50	103
1360	Controller valve boxes, 6" round boxes		18.75	.853		3.73	29.50		33.23	50.50
1370	10" round boxes		14.25	1.123		8.15	39		47.15	70
1380	12" square box		9.75	1.641		16.40	57		73.40	107
1388	Electromech. control, 14 day 3-60 min, auto start to 23/day									
1390	4 station	2 Skwk	1.04	15.385	Ea.	177	535		712	1,025
1400	7 station		.64	25		229	870		1,099	1,600
1410	12 station		.40	40		284	1,400		1,684	2,475
1420	Dual programs, 18 station		.24	66.667		1,400	2,325		3,725	5,175
1430	23 station		.16	100		1,825	3,475		5,300	7,450
1435	Backflow preventer, bronze, 0-175 PSI, w/valves, test cocks									
1440	3/4"	2 Skwk	2	8	Ea.	115	279		394	560
1450	1"		2	8		120	279		399	565
1460	1-1/2"		2	8		239	279		518	695
1470	2"		2	8		249	279		528	710
1475	Pressure vacuum breaker, brass, 15-150 PSI									
1480	3/4"	2 Skwk	2	8	Ea.	69.50	279		348.50	510
1490	1"		2	8		91	279		370	535
1500	1-1/2"		2	8		180	279		459	635
1510	2"		2	8		167	279		446	620
2000	Pop-up spray, head & nozzle, low/medium volume, plastic	1 Skwk	30	.267		3.40	9.30		12.70	18.20
2200	Brass, economy		30	.267		7.45	9.30		16.75	22.50
2400	Heavy duty		30	.267		12.50	9.30		21.80	28
3000	Riser mounted spray, head, low/medium volume, plastic		30	.267		2.27	9.30		11.57	16.95
3200	Brass		30	.267		3.11	9.30		12.41	17.85
4000	Pop-up impact sprinkler, body & case, plastic low/medium volume		30	.267		17.60	9.30		26.90	34
4200	Brass, high/medium volume		30	.267		32.50	9.30		41.80	50
4400	High volume		30	.267		92	9.30		101.30	115
5000	Quick coupling valve and key, brass, thread type		25	.320		40.50	11.15		51.65	62

Important: See the Reference Section for critical supporting data - Reference Nos., Crews, & City Cost Indexes

02810	Irrigation System	CREW	DAILY OUTPUT	LABOR-HOURS	UNIT	2005 BARE COSTS MAT.	LABOR	EQUIP.	TOTAL	TOTAL INCL O&P	
300	5200	Lug type	1 Skwk	25	.320	Ea.	43.50	11.15		54.65	65	**30**
	5400	Pop-up gear drive sprinkler, nozzle & case plastic, medium volume	↓	30	.267	↓	20.50	9.30		29.80	37	
	6000	Riser mounted gear drive sprinkler, nozzle and case										
	6200	Plastic, medium volume	1 Skwk	30	.267	Ea.	11.95	9.30		21.25	27.50	
	7000	Riser mounted impact sprinkler, body, part or										
	7200	Full circle plastic, low/medium volume	1 Skwk	25	.320	Ea.	11.95	11.15		23.10	30.50	
	7400	Brass, low/medium volume	↓	25	.320	↓	27	11.15		38.15	47	
	7600	Medium volume		25	.320		32.50	11.15		43.65	53	
	7800	Female thread	↓	25	.320	↓	65	11.15		76.15	89	
	8000	Riser mounted impact sprinkler, body, full circle only										
	8200	Plastic, low/medium volume	1 Skwk	25	.320	Ea.	7.25	11.15		18.40	25.50	
	8400	Brass, low/medium volume		25	.320		15.55	11.15		26.70	34.50	
	8600	High volume		25	.320		57	11.15		68.15	80	
	8800	Very high volume	↓	25	.320	↓	101	11.15		112.15	129	

02955	Restoration of Underground Piping	CREW	DAILY OUTPUT	LABOR-HOURS	UNIT	2005 BARE COSTS MAT.	LABOR	EQUIP.	TOTAL	TOTAL INCL O&P	
210	0010	**PIPE INTERNAL CLEANING & INSPECTION**										**21**
	0100	Cleaning, pressure pipe systems										
	0120	Pig method, lengths 1000' to 10,000'										
	0140	4" diameter thru 24" diameter, minimum				L.F.				2.14	2.46	
	0160	Maximum				"				5.35	6.40	
	6000	Sewage/sanitary systems										
	6100	Power rodder with header & cutters										
	6110	Mobilization charge, minimum				Total				320	375	
	6120	Mobilization charge, maximum				"				750	850	
	6140	4" diameter				L.F.				1.07	1.23	
	6150	6" diameter								1.34	1.55	
	6160	8" diameter								1.60	1.87	
	6170	10" diameter								1.87	2.14	
	6180	12" diameter								2.03	2.35	
	6190	14" diameter								2.14	2.46	
	6200	16" diameter								2.41	2.78	
	6210	18" diameter								2.68	3.05	
	6220	20" diameter								2.78	3.21	
	6230	24" diameter								3.10	3.58	
	6240	30" diameter								3.75	4.28	
	6250	36" diameter								4.28	4.92	
	6260	48" diameter								4.60	5.30	
	6270	60" diameter								6	6.90	
	6280	72" diameter				↓				6.65	7.65	
	9000	Inspection, television camera with film										
	9060	500 linear feet				Total				1,075	1,230	
220	0010	**PIPEBURSTING**, 300' runs, replace with HDPE pipe										**2**
	0020	Not including excavation, backfill, shoring, or dewatering										
	0100	6" to 15" diameter, minimum				L.F.					80	
	0200	Maximum									160	
	0300	18" to 36" diameter, minimum									187	
	0400	Maximum				↓					320	

For expanded coverage of these items see *Means Site Work and Landscape Cost Data 2005*

02955	Restoration of Underground Piping	CREW	DAILY OUTPUT	LABOR-HOURS	UNIT	2005 BARE COSTS				TOTAL INCL O&P	
						MAT.	LABOR	EQUIP.	TOTAL		
220	0500	Mobilize and demobilize, minimum				Job					2,675	**220**
	0600	Maximum				"					26,800	
230	0010	**LINING PIPE** with cement, incl. bypass and cleaning										**230**
	0020	Less than 10,000 L.F., urban, 6" to 10"	C-17E	130	.615	L.F.	6.90	21.50	.61	29.01	41.50	
	0200	24" to 36"		90	.889		10.95	31.50	.88	43.33	61.50	
	0300	48" to 72"		80	1		17.45	35.50	.99	53.94	75.50	
240	0010	**HDPE PIPE LINING,** excludes cleaning and video inspection										**240**
	0020	Pipe relined with one pipe size smaller than original (4" for 6")										
	0100	6" diameter, original size	B-6B	600	.080	L.F.	1.84	2.19	1.46	5.49	7.05	
	0150	8" diameter, original size		600	.080		2.65	2.19	1.46	6.30	7.95	
	0200	10" diameter, original size		600	.080		4.48	2.19	1.46	8.13	9.95	
	0250	12" diameter, original size		400	.120		6.95	3.28	2.20	12.43	15.15	
	0300	14" diameter, original size		400	.120		9.80	3.28	2.20	15.28	18.30	
310	0010	**CORROSION RESISTANCE** Wrap & coat, add to pipe, 4" dia.				L.F.	1.46			1.46	1.61	**310**
	0020	5" diameter					1.53			1.53	1.68	
	0040	6" diameter					1.53			1.53	1.68	
	0060	8" diameter					2.37			2.37	2.61	
	0080	10" diameter					3.54			3.54	3.89	
	0100	12" diameter					3.54			3.54	3.89	
	0120	14" diameter					5.20			5.20	5.70	
	0140	16" diameter					5.20			5.20	5.70	
	0160	18" diameter					5.25			5.25	5.75	
	0180	20" diameter					5.50			5.50	6.05	
	0200	24" diameter					6.70			6.70	7.35	
	0220	Small diameter pipe, 1" diameter, add					1.10			1.10	1.21	
	0240	2" diameter					1.21			1.21	1.33	
	0260	2-1/2" diameter					1.33			1.33	1.46	
	0280	3" diameter					1.33			1.33	1.46	
	0300	Fittings, field covered, add				S.F.	7			7	7.70	
	0500	Coating, bituminous, per diameter inch, 1 coat, add				L.F.	.27			.27	.30	
	0540	3 coat					.42			.42	.46	
	0560	Coal tar epoxy, per diameter inch, 1 coat, add					.16			.16	.18	
	0600	3 coat					.34			.34	.37	
	1000	Polyethylene H.D. extruded, .025" thk., 1/2" diameter add					.24			.24	.26	
	1020	3/4" diameter					.25			.25	.28	
	1040	1" diameter					.28			.28	.31	
	1060	1-1/4" diameter					.34			.34	.37	
	1080	1-1/2" diameter					.38			.38	.42	
	1100	.030" thk., 2" diameter					.41			.41	.45	
	1120	2-1/2" diameter					.48			.48	.53	
	1140	.035" thk., 3" diameter					.60			.60	.66	
	1160	3-1/2" diameter					.68			.68	.75	
	1180	4" diameter					.76			.76	.84	
	1200	.040" thk, 5" diameter					1.05			1.05	1.16	
	1220	6" diameter					1.09			1.09	1.20	
	1240	8" diameter					1.43			1.43	1.57	
	1260	10" diameter					1.74			1.74	1.91	
	1280	12" diameter					2.08			2.08	2.29	
	1300	.060" thk., 14" diameter					2.87			2.87	3.16	
	1320	16" diameter					3.36			3.36	3.70	
	1340	18" diameter					3.97			3.97	4.37	
	1360	20" diameter					4.46			4.46	4.91	
	1380	Fittings, field wrapped, add				S.F.	4.61			4.61	5.05	
400	0010	**PIPE REPAIR**										**400**
	0020	Not including excavation or backfill										

Important: See the Reference Section for critical supporting data - Reference Nos., Crews, & City Cost Indexes

02955	Restoration of Underground Piping	CREW	DAILY OUTPUT	LABOR-HOURS	UNIT	2005 BARE COSTS				TOTAL INCL O&P
						MAT.	LABOR	EQUIP.	TOTAL	
400 0100	Clamp, stainless steel, lightweight, for steel pipe									400
0110	3" long, 1/2" diameter pipe	1 Plum	34	.235	Ea.	10.45	9.60		20.05	26
0120	3/4" diameter pipe		32	.250		10.80	10.20		21	27
0130	1" diameter pipe		30	.267		11.60	10.90		22.50	29
0140	1-1/4" diameter pipe		28	.286		12.30	11.65		23.95	31
0150	1-1/2" diameter pipe		26	.308		12.70	12.55		25.25	33
0160	2" diameter pipe		24	.333		13.85	13.60		27.45	36
0170	2-1/2" diameter pipe		23	.348		15.10	14.20		29.30	38
0180	3" diameter pipe		22	.364		16.65	14.85		31.50	41
0190	3-1/2" diameter pipe	↓	21	.381		17.85	15.55		33.40	43
0200	4" diameter pipe	B-20	56	.429		18.65	12.90		31.55	40.50
0210	5" diameter pipe		53	.453		21.50	13.60		35.10	44.50
0220	6" diameter pipe		48	.500		24	15.05		39.05	49.50
0230	8" diameter pipe		30	.800		28	24		52	68
0240	10" diameter pipe		28	.857		71.50	26		97.50	119
0250	12" diameter pipe		24	1		76	30		106	131
0260	14" diameter pipe		22	1.091		79.50	33		112.50	138
0270	16" diameter pipe		20	1.200		85.50	36		121.50	150
0280	18" diameter pipe		18	1.333		90.50	40		130.50	162
0290	20" diameter pipe		16	1.500		100	45		145	180
0300	24" diameter pipe	↓	14	1.714		111	51.50		162.50	203
0360	For 6" long, add					100%	40%			
0370	For 9" long, add					200%	100%			
0380	For 12" long, add					300%	150%			
0390	For 18" long, add				↓	500%	200%			
0400	Pipe Freezing for live repairs of systems 3/8 inch to 6 inch									
0410	Note: Pipe Freezing can also used to install a valve into a live system									
0420	Pipe Freezing each side 3/8 inch	2 Skwk	8	2	Ea.	500	69.50		569.50	660
0425	Pipe Freezing each side 3/8 inch , second location same kit		8	2		19	69.50		88.50	129
0430	Pipe Freezing each side 3/4 inch		8	2		430	69.50		499.50	585
0435	Pipe Freezing each side 3/4 inch , second location same kit		8	2		19	69.50		88.50	129
0440	Pipe Freezing each side 1 1/2 inch		6	2.667		430	93		523	620
0445	Pipe Freezing each side 1 1/2 inch , second location same kit		6	2.667		19	93		112	166
0450	Pipe Freezing each side 2 inch		6	2.667		635	93		728	845
0455	Pipe Freezing each side 2 inch , second location same kit		6	2.667		19	93		112	166
0460	Pipe Freezing each side 2 1/2 inch-3 inch		6	2.667		890	93		983	1,125
0465	Pipe Freezing each side 2 1/2-3 inch , second loc. same kit		6	2.667		38	93		131	187
0470	Pipe Freezing each side 4 inch		4	4		1,450	139		1,589	1,800
0475	Pipe Freezing each side 4 inch , second location same kit		4	4		60	139		199	283
0480	Pipe Freezing each side 5-6 inch		4	4		3,475	139		3,614	4,050
0485	Pipe Freezing each side 5-6 inch , second location same kit	↓	4	4		180	139		319	415
0490	Pipe Freezing extra 20 lb CO2 cyl. (3/8 to 2 " - 1 ea, 3 " -2 ea)					189			189	208
0500	Pipe Freezing extra 50 lb CO2 cylinders (4 " - 2 ea, 5-6 " -6 ea)				↓	425			425	465
1000	Clamp, stainless steel, with threaded service tap									
1040	Full seal for iron, steel, PVC pipe									
1100	6" long, 2" diameter pipe	1 Plum	17	.471	Ea.	135	19.20		154.20	177
1110	2-1/2" diameter pipe		16	.500		143	20.50		163.50	188
1120	3" diameter pipe		15.60	.513		152	21		173	199
1130	3-1/2" diameter pipe	↓	15	.533		159	22		181	207
1140	4" diameter pipe	B-20	40	.600		174	18.05		192.05	220
1150	6" diameter pipe		34	.706		201	21		222	255
1160	8" diameter pipe		21	1.143		251	34.50		285.50	330
1170	10" diameter pipe		20	1.200		278	36		314	360
1180	12" diameter pipe	↓	17	1.412	↓	315	42.50		357.50	415

For expanded coverage of these items see *Means Site Work and Landscape Cost Data 2005*

02955	Restoration of Underground Piping	CREW	DAILY OUTPUT	LABOR-HOURS	UNIT	2005 BARE COSTS				TOTAL INCL O&P
						MAT.	LABOR	EQUIP.	TOTAL	
1205	For 9" long, add				Ea.	20%	45%			
1210	For 12" long, add					40%	80%			
1220	For 18" long, add				↓	70%	110%			
1600	Clamp, stainless steel, single section									
1640	Full seal for iron, steel, PVC pipe									
1700	6" long, 2" diameter pipe	1 Plum	17	.471	Ea.	70	19.20		89.20	106
1710	2-1/2" diameter pipe		16	.500		76	20.50		96.50	115
1720	3" diameter pipe		15.60	.513		79.50	21		100.50	119
1730	3-1/2" diameter pipe	↓	15	.533		87	22		109	129
1740	4" diameter pipe	B-20	40	.600		95	18.05		113.05	133
1750	6" diameter pipe		34	.706		119	21		140	164
1760	8" diameter pipe		21	1.143		159	34.50		193.50	228
1770	10" diameter pipe		20	1.200		182	36		218	257
1780	12" diameter pipe	↓	17	1.412		198	42.50		240.50	284
1800	For 9" long, add					40%	45%			
1810	For 12" long, add					60%	80%			
1820	For 18" long, add				↓	120%	110%			
2000	Clamp, stainless steel, two section									
2040	Full seal, for iron, steel, PVC pipe									
2100	6" long, 4" diameter pipe	B-20	24	1	Ea.	135	30		165	195
2110	6" diameter pipe		20	1.200		159	36		195	230
2120	8" diameter pipe		13	1.846		174	55.50		229.50	279
2130	10" diameter pipe		12	2		254	60		314	375
2140	12" diameter pipe		10	2.400		278	72		350	415
2200	9" long, 4" diameter pipe		16	1.500		174	45		219	262
2210	6" diameter pipe		13	1.846		198	55.50		253.50	305
2220	8" diameter pipe		9	2.667		206	80		286	350
2230	10" diameter pipe		8	3		285	90		375	455
2240	12" diameter pipe		7	3.429		315	103		418	510
2250	14" diameter pipe		6.40	3.750		475	113		588	700
2260	16" diameter pipe		6	4		500	120		620	735
2270	18" diameter pipe		5	4.800		515	144		659	790
2280	20" diameter pipe		4.60	5.217		555	157		712	855
2290	24" diameter pipe	↓	4	6		950	180		1,130	1,325
2320	For 12" long, add to 9"					15%	25%			
2330	For 18" long, add to 9"				↓	70%	55%			
8000	For internal cleaning and inspection, see Div. 02955-210									
8100	For pipe testing, see Div. 15955-700									

For information about Means Estimating Seminars, see yellow pages 12 and 13 in back of book

Division 3
Concrete

Estimating Tips

General

- Carefully check all the plans and specifications. Concrete often appears on drawings other than structural drawings, including mechanical and electrical drawings for equipment pads. The cost of cutting and patching is often difficult to estimate. See Subdivisions 02220 and 03055 for demolition costs.
- Always obtain concrete prices from suppliers near the job site. A volume discount can often be negotiated depending upon competition in the area. Remember to add for waste, particularly for slabs and footings on grade.

03100 Concrete Forms & Accessories

- A primary cost for concrete construction is forming. Most jobs today are constructed with prefabricated forms. The selection of the forms best suited for the job and the total square feet of forms required for efficient concrete forming and placing are key elements in estimating concrete construction. Enough forms must be available for erection to make efficient use of the concrete placing equipment and crew.
- Concrete accessories for forming and placing depend upon the systems used. Study the plans and specifications to assure that all special accessory requirements have been included in the cost estimate such as anchor bolts, inserts and hangers.

03200 Concrete Reinforcement

- Ascertain that the reinforcing steel supplier has included all accessories, cutting, bending and an allowance for lapping, splicing and waste. A good rule of thumb is 10% for lapping, splicing and waste. Also, 10% waste should be allowed for welded wire fabric.

03300 Cast-in-Place Concrete

- When estimating structural concrete, pay particular attention to requirements for concrete additives, curing methods and surface treatments. Special consideration for climate, hot or cold, must be included in your estimate. Be sure to include requirements for concrete placing equipment and concrete finishing.

03400 Precast Concrete
03500 Cementitious Decks & Toppings

- The cost of hauling precast concrete structural members is often an important factor. For this reason, it is important to get a quote from the nearest supplier. It may become economically feasible to set up precasting beds on the site if the hauling costs are prohibitive.

Reference Numbers

Reference numbers are shown in bold squares at the beginning of some major classifications. These numbers refer to related items in the Reference Section. The reference information may be an estimating procedure, an alternate pricing method or technical information.

Note: Not all subdivisions listed here necessarily appear in this publication.

03110 | Structural C.I.P. Forms

			CREW	DAILY OUTPUT	LABOR-HOURS	UNIT	2005 BARE COSTS				TOTAL INCL O&P	
							MAT.	LABOR	EQUIP.	TOTAL		
425	0010	FORMS IN PLACE, EQUIPMENT FOUNDATIONS job built										425
	0020	1 use	C-2	160	.300	SFCA	2.67	10		12.67	18.50	
	0050	2 use		190	.253		1.47	8.40		9.87	14.70	
	0100	3 use		200	.240		1.07	8		9.07	13.60	
	0150	4 use	▼	205	.234	▼	.87	7.80		8.67	13.10	
430	0010	FORMS IN PLACE, FOOTINGS Continuous wall, plywood, 1 use	C-1	375	.085	SFCA	2.31	2.76		5.07	6.85	430
	0050	2 use		440	.073		1.27	2.35		3.62	5.05	
	0100	3 use		470	.068		.92	2.20		3.12	4.45	
	0150	4 use		485	.066		.75	2.14		2.89	4.16	
	5000	Spread footings, job-built lumber, 1 use		305	.105		1.66	3.40		5.06	7.15	
	5050	2 use		371	.086		.92	2.79		3.71	5.35	
	5100	3 use		401	.080		.67	2.58		3.25	4.75	
	5150	4 use	▼	414	.077	▼	.54	2.50		3.04	4.49	
445	0010	FORMS IN PLACE, SLAB ON GRADE										445
	3000	Edge forms, wood, 4 use, on grade, to 6" high	C-1	600	.053	L.F.	.27	1.73		2	2.98	
	6000	Trench forms in floor, wood, 1 use		160	.200	SFCA	2.20	6.45		8.65	12.50	
	6050	2 use		175	.183		1.22	5.90		7.12	10.55	
	6100	3 use		180	.178		.89	5.75		6.64	9.95	
	6150	4 use	▼	185	.173	▼	.72	5.60		6.32	9.50	
	8760	Void form, corrugated fiberboard, 6" x 12", 10' long	▼	240	.133	S.F.	9.20	4.31		13.51	16.80	

03150 | Concrete Accessories

			CREW	DAILY OUTPUT	LABOR-HOURS	UNIT	MAT.	LABOR	EQUIP.	TOTAL	TOTAL INCL O&P	
620	0010	ACCESSORIES, SLEEVES AND CHASES										620
	0100	Plastic, 1 use, 9" long, 2" diameter	1 Carp	100	.080	Ea.	.53	2.74		3.27	4.85	
	0150	4" diameter		90	.089		1.56	3.04		4.60	6.45	
	0200	6" diameter		75	.107		2.75	3.65		6.40	8.75	
	0250	12" diameter		60	.133		18.05	4.57		22.62	27	
	5000	Sheet metal, 2" diameter		100	.080		.68	2.74		3.42	5	
	5100	4" diameter		90	.089		.85	3.04		3.89	5.70	
	5150	6" diameter		75	.107		1.23	3.65		4.88	7.05	
	5200	12" diameter		60	.133		2.46	4.57		7.03	9.80	
	6000	Steel pipe, 2" diameter		100	.080		3.49	2.74		6.23	8.10	
	6100	4" diameter		90	.089		11.90	3.04		14.94	17.80	
	6150	6" diameter		75	.107		25.50	3.65		29.15	34	
	6200	12" diameter	▼	60	.133	▼	64	4.57		68.57	77.50	

03200 | Concrete Reinforcement

03210 | Reinforcing Steel

			CREW	DAILY OUTPUT	LABOR-HOURS	UNIT	2005 BARE COSTS				TOTAL INCL O&P	
							MAT.	LABOR	EQUIP.	TOTAL		
600	0010	REINFORCING IN PLACE A615 Grade 60, incl. access. labor										600
	0502	Footings, #4 to #7	4 Rodm	4,200	.008	Lb.	.42	.29		.71	.94	
	0552	#8 to #18		7,200	.004		.42	.17		.59	.74	
	0602	Slab on grade, #3 to #7	▼	4,200	.008	▼	.40	.29		.69	.92	
	3000	For epoxy dowel anchoring, see Div 05090-300										

Important: See the Reference Section for critical supporting data - Reference Nos., Crews, & City Cost Indexes

03220 | Welded Wire Fabric

		CREW	DAILY OUTPUT	LABOR-HOURS	UNIT	2005 BARE COSTS				TOTAL INCL O&P
						MAT.	LABOR	EQUIP.	TOTAL	
200	0010	**WELDED WIRE FABRIC** ASTM A185										200
	0040	Reinforcing sheets, 6x6-W1.4xW1.4				S.F.	.44			.44	.48

03300 | Cast-In-Place Concrete

03310 | Structural Concrete

			CREW	DAILY OUTPUT	LABOR-HOURS	UNIT	2005 BARE COSTS				TOTAL INCL O&P
							MAT.	LABOR	EQUIP.	TOTAL	
220	0010	**CONCRETE, READY MIX** Normal weight										220
	0020	2000 psi				C.Y.	77.50			77.50	85.50
	0100	2500 psi					79.50			79.50	87.50
	0150	3000 psi					81			81	89
	0200	3500 psi					82			82	90
	0300	4000 psi					84			84	92.50
	1000	For high early strength cement, add					10%				
	2000	For all lightweight aggregate, add					45%				
240	0010	**CONCRETE IN PLACE** Including forms (4 uses), reinforcing										240
	0050	steel and finishing unless otherwise indicated									
	3800	Footings, spread under 1 C.Y.	C-14C	38.07	2.942	C.Y.	162	96	.62	258.62	330
	3850	Over 5 C.Y.		81.04	1.382		226	45	.29	271.29	320
	3900	Footings, strip, 18" x 9", unreinforced		40	2.800		101	91.50	.59	193.09	255
	3920	18" x 9", reinforced		35	3.200		120	105	.67	225.67	297
	3925	20" x 10", unreinforced		45	2.489		98	81.50	.52	180.02	236
	3930	20" x 10", reinforced		40	2.800		114	91.50	.59	206.09	269
	3935	24" x 12", unreinforced		55	2.036		97	66.50	.43	163.93	211
	3940	24" x 12", reinforced		48	2.333		113	76	.49	189.49	244
	3945	36" x 12", unreinforced		70	1.600		93	52.50	.34	145.84	184
	3950	36" x 12", reinforced		60	1.867		108	61	.39	169.39	214
	4000	Foundation mat, under 10 C.Y.		38.67	2.896		164	94.50	.61	259.11	330
	4050	Over 20 C.Y.		56.40	1.986		144	65	.42	209.42	260
	4650	Slab on grade, not including finish, 4" thick	C-14E	60.75	1.449		104	48.50	.39	152.89	192
	4700	6" thick	"	92	.957		98	32	.26	130.26	159
700	0010	**PLACING CONCRETE** and vibrating, including labor & equipment										700
	1900	Footings, continuous, shallow, direct chute	C-6	120	.400	C.Y.		11.20	.40	11.60	17.70
	1950	Pumped	C-20	150	.427			12.25	5	17.25	24.50
	2000	With crane and bucket	C-7	90	.800			23	10.65	33.65	47.50
	2100	Footings, continuous, deep, direct chute	C-6	140	.343			9.60	.34	9.94	15.20
	2150	Pumped	C-20	160	.400			11.50	4.70	16.20	23
	2200	With crane and bucket	C-7	110	.655			18.90	8.75	27.65	38.50
	2400	Footings, spread, under 1 C.Y., direct chute	C-6	55	.873			24.50	.87	25.37	38.50
	2450	Pumped	C-20	65	.985			28.50	11.55	40.05	56.50
	2500	With crane and bucket	C-7	45	1.600			46	21.50	67.50	94.50
	2600	Over 5 C.Y., direct chute	C-6	120	.400			11.20	.40	11.60	17.70
	2650	Pumped	C-20	150	.427			12.25	5	17.25	24.50
	2700	With crane and bucket	C-7	100	.720			21	9.60	30.60	42.50
	2900	Foundation mats, over 20 C.Y., direct chute	C-6	350	.137			3.85	.14	3.99	6.05
	2950	Pumped	C-20	400	.160			4.59	1.88	6.47	9.10
	3000	With crane and bucket	C-7	300	.240			6.95	3.20	10.15	14.10

03350 | Concrete Finishing

			CREW	DAILY OUTPUT	LABOR-HOURS	UNIT	2005 BARE COSTS				TOTAL INCL O&P
							MAT.	LABOR	EQUIP.	TOTAL	
300	0010	**FINISHING FLOORS** Monolithic, screed finish	1 Cefi	900	.009	S.F.		.29		.29	.43	300
	0100	Screed and bull float (darby) finish		725	.011			.36		.36	.53

For expanded coverage of these items see Means Concrete & Masonry Cost Data 2005

			DAILY	LABOR-		\multicolumn{4}{c}{2005 BARE COSTS}	TOTAL	
03350		**Concrete Finishing**	CREW	OUTPUT	HOURS	UNIT	MAT.	LABOR	EQUIP.	TOTAL	INCL O&P	
300	0150	Screed, float, and broom finish	1 Cefi	630	.013	S.F.		.42		.42	.61	**300**
	0200	Screed, float, and hand trowel		600	.013			.44		.44	.64	
	0250	Machine trowel	↓	550	.015	↓		.48		.48	.70	
325	0010	**CONTROL JOINT**, concrete floor slab										**325**
	0100	Sawcut in green concrete										
	0120	1" depth	C-27	2,000	.008	L.F.		.26	.06	.32	.44	
	0140	1-1/2" depth		1,800	.009			.29	.06	.35	.50	
	0160	2" depth	↓	1,600	.010			.33	.07	.40	.56	
	0200	Clean out control joint of debris	C-28	6,000	.001	↓		.04		.04	.06	
	0300	Joint sealant										
	0320	Backer rod, polyethylene, 1/4" diameter	1 Cefi	460	.017	L.F.	.02	.57		.59	.86	
	0340	Sealant, polyurethane										
	0360	1/4" x 1/4" (308 LF/Gal)	1 Cefi	270	.030	L.F.	.15	.97		1.12	1.60	
	0380	1/4" x 1/2" (154 LF/Gal)	"	255	.031	"	.31	1.03		1.34	1.86	

For information about Means Estimating Seminars, see yellow pages 12 and 13 in back of book

Division 5
Metals

Estimating Tips

05050 Basic Metal Materials & Methods

• Nuts, bolts, washers, connection angles and plates can add a significant amount to both the tonnage of a structural steel job as well as the estimated cost. As a rule of thumb add 10% to the total weight to account for these accessories.

• Type 2 steel construction, commonly referred to as "simple construction," consists generally of field bolted connections with lateral bracing supplied by other elements of the building, such as masonry walls or x-bracing. The estimator should be aware, however, that shop connections may be accomplished by welding or bolting. The method may be particular to the fabrication shop and may have an impact on the estimated cost.

05200 Metal Joists

• In any given project the total weight of open web steel joists is determined by the loads to be supported and the design. However, economies can be realized in minimizing the amount of labor used to place the joists. This is done by maximizing the joist spacing and therefore minimizing the number of joists required to be installed on the job. Certain spacings and locations may be required by the design, but in other cases maximizing the spacing and keeping it as uniform as possible will keep the costs down.

05300 Metal Deck

• The takeoff and estimating of metal deck involves more than simply the area of the floor or roof and the type of deck specified or shown on the drawings. Many different sizes and types of openings may exist. Small openings for individual pipes or conduits may be drilled after the floor/roof is installed, but larger openings may require special deck lengths as well as reinforcing or structural support. The estimator should determine who will be supplying this reinforcing. Additionally, some deck terminations are part of the deck package, such as screed angles and pour stops, and others will be part of the steel contract, such as angles attached to structural members and cast-in-place angles and plates. The estimator must ensure that all pieces are accounted for in the complete estimate.

05500 Metal Fabrications

• The most economical steel stairs are those that use common materials, standard details and most importantly, a uniform and relatively simple method of field assembly. Commonly available A36 channels and plates are very good choices for the main stringers of the stairs, as are angles and tees for the carrier members. Risers and treads are usually made by specialty shops, and it is most economical to use a typical detail in as many places as possible. The stairs should be pre-assembled and shipped directly to the site. The field connections should be simple and straightforward to be accomplished efficiently and with a minimum of equipment and labor.

Reference Numbers

Reference numbers are shown in bold squares at the beginning of some major classifications. These numbers refer to related items in the Reference Section. The reference information may be an estimating procedure, an alternate pricing method or technical information.

Note: Not all subdivisions listed here necessarily appear in this publication.

05090	Metal Fastenings	CREW	DAILY OUTPUT	LABOR-HOURS	UNIT	2005 BARE COSTS				TOTAL INCL O&P
						MAT.	LABOR	EQUIP.	TOTAL	
080	0010	**ANCHOR BOLTS**									080
	0100	J-type, incl. hex nut & washer, 1/2" diameter x 6" long	2 Carp	70	.229	Ea.	.94	7.85		8.79	13.25
	0110	12" long		65	.246		1.17	8.45		9.62	14.45
	0120	18" long		60	.267		1.52	9.15		10.67	15.90
	0130	3/4" diameter x 8" long		50	.320		1.39	10.95		12.34	18.60
	0140	12" long		45	.356		1.74	12.20		13.94	21
	0150	18" long		40	.400		2.26	13.70		15.96	24
	0160	1" diameter x 12" long		35	.457		3.32	15.65		18.97	28
	0170	18" long		30	.533		4	18.25		22.25	33
	0180	24" long		25	.640		4.88	22		26.88	39.50
	0190	36" long		20	.800		6.70	27.50		34.20	50
	0200	1-1/2" diameter x 18" long		22	.727		11.85	25		36.85	52
	0210	24" long		16	1		14.10	34.50		48.60	69
	0300	L-type, incl. hex nut & washer, 3/4" diameter x 12" long		45	.356		1.31	12.20		13.51	20.50
	0310	18" long		40	.400		1.67	13.70		15.37	23.50
	0320	24" long		35	.457		2.04	15.65		17.69	26.50
	0330	30" long		30	.533		2.59	18.25		20.84	31.50
	0340	36" long		25	.640		2.96	22		24.96	37.50
	0350	1" diameter x 12" long		35	.457		2.28	15.65		17.93	27
	0360	18" long		30	.533		2.84	18.25		21.09	31.50
	0370	24" long		25	.640		3.50	22		25.50	38
	0380	30" long		23	.696		4.13	24		28.13	41.50
	0390	36" long		20	.800		4.73	27.50		32.23	47.50
	0400	42" long		18	.889		5.75	30.50		36.25	54
	0410	48" long		15	1.067		6.45	36.50		42.95	64
	0420	1-1/4" diameter x 18" long		25	.640		5.10	22		27.10	39.50
	0430	24" long		22	.727		6.05	25		31.05	45.50
	0440	30" long		20	.800		7	27.50		34.50	50
	0450	36" long		18	.889		8	30.50		38.50	56.50
	0460	42" long		16	1		9	34.50		43.50	63.50
	0470	48" long		14	1.143		10.30	39		49.30	72.50
	0480	54" long		12	1.333		12.15	45.50		57.65	84.50
	0490	60" long		10	1.600		13.35	55		68.35	100
	0500	1-1/2" diameter x 18" long		22	.727		7.85	25		32.85	47.50
	0510	24" long		19	.842		9.15	29		38.15	55
	0520	30" long		17	.941		10.35	32		42.35	61.50
	0530	36" long		16	1		11.90	34.50		46.40	66.50
	0540	42" long		15	1.067		13.55	36.50		50.05	72
	0550	48" long		13	1.231		15.25	42		57.25	82.50
	0560	54" long		11	1.455		18.55	50		68.55	98
	0570	60" long		9	1.778		20.50	61		81.50	118
	0580	1-3/4" diameter x 18" long		20	.800		11.85	27.50		39.35	55.50
	0590	24" long		18	.889		13.90	30.50		44.40	63
	0600	30" long		17	.941		16.20	32		48.20	68
	0610	36" long		16	1		18.45	34.50		52.95	74
	0620	42" long		14	1.143		20.50	39		59.50	84
	0630	48" long		12	1.333		23	45.50		68.50	96
	0640	54" long		10	1.600		28.50	55		83.50	117
	0650	60" long		8	2		30.50	68.50		99	141
	0660	2" diameter x 24" long		17	.941		17.70	32		49.70	69.50
	0670	30" long		15	1.067		19.95	36.50		56.45	79
	0680	36" long		13	1.231		22	42		64	89.50
	0690	42" long		11	1.455		24.50	50		74.50	105
	0700	48" long		10	1.600		28	55		83	117
	0710	54" long		9	1.778		33.50	61		94.50	132
	0720	60" long		8	2		36	68.50		104.50	147

05090	Metal Fastenings	CREW	DAILY OUTPUT	LABOR-HOURS	UNIT	2005 BARE COSTS				TOTAL INCL O&P
						MAT.	LABOR	EQUIP.	TOTAL	
080 0730	66" long	2 Carp	7	2.286	Ea.	38.50	78.50		117	165	**080**
0740	72" long	↓	6	2.667		42	91.50		133.50	189
0990	For galvanized, add				↓	75%				
300 0010	**CHEMICAL ANCHORS**, Includes layout & drilling										**300**
1430	Chemical anchor, w/rod & epoxy cartridge, 3/4" diam. x 9-1/2" long	B-89A	27	.593	Ea.	11.65	18.25	3.82	33.72	45.50
1435	1" diameter x 11-3/4" long		24	.667		22.50	20.50	4.29	47.29	61.50
1440	1-1/4" diameter x 14" long		21	.762		43	23.50	4.91	71.41	89
1445	1-3/4" diameter x 15" long		20	.800		81	24.50	5.15	110.65	134
1450	18" long		17	.941		97.50	29	6.05	132.55	159
1455	2" diameter x 18" long		16	1		124	31	6.45	161.45	192
1460	24" long	↓	15	1.067	↓	162	33	6.85	201.85	238
1500	Chemical anchoring, epoxy cartridge, excludes layout, drilling, fastener									
1530	For fastener 3/4" dia x 6" embedment	B-89A	27	.593	Ea.	4.58	18.25	3.82	26.65	38
1535	1" dia x 8" embedment		24	.667		6.85	20.50	4.29	31.64	44.50
1540	1-1/4" dia x 10" embedment		21	.762		13.75	23.50	4.91	42.16	57
1545	1-3/4" dia x 12" embedment		20	.800		23	24.50	5.15	52.65	69
1550	14" embedment		17	.941		27.50	29	6.05	62.55	81.50
1555	2" dia x 12" embedment		16	1		36.50	31	6.45	73.95	95.50
1560	18" embedment	↓	15	1.067	↓	46	33	6.85	85.85	109
340 0010	**DRILLING** For anchors, up to 4" deep, incl. bit and layout										**340**
0050	in concrete or brick walls and floors, no anchor									
0100	Holes, 1/4" diameter	1 Carp	75	.107	Ea.	.08	3.65		3.73	5.80
0150	For each additional inch of depth, add		430	.019		.02	.64		.66	1.01
0200	3/8" diameter		63	.127		.08	4.35		4.43	6.85
0250	For each additional inch of depth, add		340	.024		.02	.81		.83	1.28
0300	1/2" diameter		50	.160		.08	5.50		5.58	8.65
0350	For each additional inch of depth, add		250	.032		.02	1.10		1.12	1.73
0400	5/8" diameter		48	.167		.15	5.70		5.85	9.05
0450	For each additional inch of depth, add		240	.033		.04	1.14		1.18	1.82
0500	3/4" diameter		45	.178		.18	6.10		6.28	9.70
0550	For each additional inch of depth, add		220	.036		.04	1.25		1.29	1.99
0600	7/8" diameter		43	.186		.21	6.35		6.56	10.20
0650	For each additional inch of depth, add		210	.038		.05	1.30		1.35	2.09
0700	1" diameter		40	.200		.24	6.85		7.09	10.90
0750	For each additional inch of depth, add		190	.042		.06	1.44		1.50	2.32
0800	1-1/4" diameter		38	.211		.35	7.20		7.55	11.65
0850	For each additional inch of depth, add		180	.044		.09	1.52		1.61	2.47
0900	1-1/2" diameter		35	.229		.53	7.85		8.38	12.80
0950	For each additional inch of depth, add	↓	165	.048	↓	.13	1.66		1.79	2.73
1000	For ceiling installations, add						40%			
1100	Drilling & layout for drywall/plaster walls, up to 1" deep, no anchor									
1200	Holes, 1/4" diameter	1 Carp	150	.053	Ea.	.01	1.83		1.84	2.86
1300	3/8" diameter		140	.057		.01	1.96		1.97	3.06
1400	1/2" diameter		130	.062		.01	2.11		2.12	3.29
1500	3/4" diameter		120	.067		.02	2.28		2.30	3.58
1600	1" diameter		110	.073		.03	2.49		2.52	3.91
1700	1-1/4" diameter		100	.080		.04	2.74		2.78	4.32
1800	1-1/2" diameter	↓	90	.089	↓	.07	3.04		3.11	4.81
1900	For ceiling installations, add						40%			
1910	Drilling & layout for steel, up to 1/4" deep, no anchor									
1920	Holes, 1/4" diameter	1 Sswk	112	.071	Ea.	.11	2.73		2.84	5
1925	For each additional 1/4" depth, add		336	.024		.11	.91		1.02	1.75
1930	3/8" diameter		104	.077		.13	2.93		3.06	5.40
1935	For each additional 1/4" depth, add		312	.026		.13	.98		1.11	1.89
1940	1/2" diameter	↓	96	.083	↓	.14	3.18		3.32	5.85

05090 | Metal Fastenings

		CREW	DAILY OUTPUT	LABOR-HOURS	UNIT	MAT.	2005 BARE COSTS LABOR	EQUIP.	TOTAL	TOTAL INCL O&P	
340											**340**
1945	For each additional 1/4" depth, add	1 Sswk	288	.028	Ea.	.14	1.06		1.20	2.06	
1950	5/8" diameter		88	.091		.24	3.47		3.71	6.45	
1955	For each additional 1/4" depth, add		264	.030		.24	1.16		1.40	2.33	
1960	3/4" diameter		80	.100		.26	3.82		4.08	7.15	
1965	For each additional 1/4" depth, add		240	.033		.26	1.27		1.53	2.56	
1970	7/8" diameter		72	.111		.31	4.24		4.55	7.95	
1975	For each additional 1/4" depth, add		216	.037		.31	1.41		1.72	2.87	
1980	1" diameter		64	.125		.35	4.77		5.12	8.95	
1985	For each additional 1/4" depth, add	▼	192	.042		.35	1.59		1.94	3.23	
1990	For drilling up, add				▼		40%				
380	0010 **EXPANSION ANCHORS** & shields										**380**
0100	Bolt anchors for concrete, brick or stone, no layout and drilling										
0200	Expansion shields, zinc, 1/4" diameter, 1-5/16" long, single	1 Carp	90	.089	Ea.	1.01	3.04		4.05	5.85	
0300	1-3/8" long, double		85	.094		1.11	3.22		4.33	6.20	
0400	3/8" diameter, 1-1/2" long, single		85	.094		1.66	3.22		4.88	6.85	
0500	2" long, double		80	.100		2.05	3.43		5.48	7.60	
0600	1/2" diameter, 2-1/16" long, single		80	.100		2.75	3.43		6.18	8.40	
0700	2-1/2" long, double		75	.107		2.65	3.65		6.30	8.60	
0800	5/8" diameter, 2-5/8" long, single		75	.107		3.93	3.65		7.58	10	
0900	2-3/4" long, double		70	.114		3.93	3.91		7.84	10.40	
1000	3/4" diameter, 2-3/4" long, single		70	.114		5.85	3.91		9.76	12.50	
1100	3-15/16" long, double		65	.123		7.80	4.22		12.02	15.10	
1500	Self drilling anchor, snap-off, for 1/4" diameter bolt		26	.308		.83	10.55		11.38	17.30	
1600	3/8" diameter bolt		23	.348		1.20	11.90		13.10	19.85	
1700	1/2" diameter bolt		20	.400		1.84	13.70		15.54	23.50	
1800	5/8" diameter bolt		18	.444		3.08	15.20		18.28	27	
1900	3/4" diameter bolt	▼	16	.500	▼	5.20	17.15		22.35	32	
2100	Hollow wall anchors for gypsum wall board, plaster or tile										
2300	1/8" diameter, short	1 Carp	160	.050	Ea.	.26	1.71		1.97	2.96	
2400	Long		150	.053		.31	1.83		2.14	3.19	
2500	3/16" diameter, short		150	.053		.55	1.83		2.38	3.46	
2600	Long		140	.057		.59	1.96		2.55	3.70	
2700	1/4" diameter, short		140	.057		.68	1.96		2.64	3.80	
2800	Long		130	.062		.76	2.11		2.87	4.12	
3000	Toggle bolts, bright steel, 1/8" diameter, 2" long		85	.094		.25	3.22		3.47	5.30	
3100	4" long		80	.100		.37	3.43		3.80	5.75	
3200	3/16" diameter, 3" long		80	.100		.42	3.43		3.85	5.80	
3300	6" long		75	.107		.58	3.65		4.23	6.35	
3400	1/4" diameter, 3" long		75	.107		.47	3.65		4.12	6.20	
3500	6" long		70	.114		.66	3.91		4.57	6.85	
3600	3/8" diameter, 3" long		70	.114		.90	3.91		4.81	7.10	
3700	6" long		60	.133		1.58	4.57		6.15	8.85	
3800	1/2" diameter, 4" long		60	.133		2.33	4.57		6.90	9.65	
3900	6" long	▼	50	.160	▼	3.85	5.50		9.35	12.80	
4000	Nailing anchors										
4100	Nylon nailing anchor, 1/4" diameter, 1" long	1 Carp	3.20	2.500	C	18	85.50		103.50	153	
4200	1-1/2" long		2.80	2.857		23	98		121	178	
4300	2" long		2.40	3.333		38.50	114		152.50	221	
4400	Metal nailing anchor, 1/4" diameter, 1" long		3.20	2.500		27.50	85.50		113	164	
4500	1-1/2" long		2.80	2.857		37.50	98		135.50	194	
4600	2" long	▼	2.40	3.333	▼	48	114		162	231	
5000	Screw anchors for concrete, masonry,										
5100	stone & tile, no layout or drilling included										
5200	Jute fiber, #6, #8, & #10, 1" long	1 Carp	240	.033	Ea.	.24	1.14		1.38	2.04	
5300	#12, 1-1/2" long		200	.040		.35	1.37		1.72	2.52	
5400	#14, 2" long	▼	160	.050	▼	.55	1.71		2.26	3.28	

Important: See the Reference Section for critical supporting data - Reference Nos., Crews, & City Cost Indexes

05090 | Metal Fastenings

		CREW	DAILY OUTPUT	LABOR-HOURS	UNIT	2005 BARE COSTS				TOTAL INCL O&P
						MAT.	LABOR	EQUIP.	TOTAL	
380	5500	#16, 2" long	1 Carp	150	.053	Ea.	.57	1.83		2.40	3.48
	5600	#20, 2" long		140	.057		.91	1.96		2.87	4.05
	5700	Lag screw shields, 1/4" diameter, short		90	.089		.44	3.04		3.48	5.20
	5800	Long		85	.094		.51	3.22		3.73	5.55
	5900	3/8" diameter, short		85	.094		.80	3.22		4.02	5.90
	6000	Long		80	.100		.94	3.43		4.37	6.40
	6100	1/2" diameter, short		80	.100		1.11	3.43		4.54	6.55
	6200	Long		75	.107		1.39	3.65		5.04	7.25
	6300	3/4" diameter, short		70	.114		3.12	3.91		7.03	9.55
	6400	Long		65	.123		3.78	4.22		8	10.70
	6600	Lead, #6 & #8, 3/4" long		260	.031		.16	1.05		1.21	1.82
	6700	#10 - #14, 1-1/2" long		200	.040		.23	1.37		1.60	2.38
	6800	#16 & #18, 1-1/2" long		160	.050		.31	1.71		2.02	3.01
	6900	Plastic, #6 & #8, 3/4" long		260	.031		.10	1.05		1.15	1.75
	7000	#8 & #10, 7/8" long		240	.033		.04	1.14		1.18	1.82
	7100	#10 & #12, 1" long		220	.036		.13	1.25		1.38	2.08
	7200	#14 & #16, 1-1/2" long	▼	160	.050	▼	.07	1.71		1.78	2.75
	8000	Wedge anchors, not including layout or drilling									
	8050	Carbon steel, 1/4" diameter, 1-3/4" long	1 Carp	150	.053	Ea.	.42	1.83		2.25	3.31
	8100	3 1/4" long		140	.057		.55	1.96		2.51	3.66
	8150	3/8" diameter, 2-1/4" long		145	.055		.63	1.89		2.52	3.63
	8200	5" long		140	.057		1.10	1.96		3.06	4.26
	8250	1/2" diameter, 2-3/4" long		140	.057		.96	1.96		2.92	4.11
	8300	7" long		125	.064		1.65	2.19		3.84	5.20
	8350	5/8" diameter, 3-1/2" long		130	.062		1.90	2.11		4.01	5.35
	8400	8-1/2" long		115	.070		4.05	2.38		6.43	8.15
	8450	3/4" diameter, 4-1/4" long		115	.070		2.29	2.38		4.67	6.25
	8500	10" long		95	.084		5.20	2.88		8.08	10.25
	8550	1" diameter, 6" long		100	.080		7.70	2.74		10.44	12.70
	8575	9" long		85	.094		10	3.22		13.22	16
	8600	12" long		75	.107		10.80	3.65		14.45	17.55
	8650	1-1/4" diameter, 9" long		70	.114		14	3.91		17.91	21.50
	8700	12" long	▼	60	.133	▼	17.90	4.57		22.47	27
	8750	For type 303 stainless steel, add					350%				
	8800	For type 316 stainless steel, add					450%				
	8950	Self-drilling concrete screw, hex washer head, 3/16" dia x 1-3/4" long	1 Carp	300	.027	Ea.	.32	.91		1.23	1.77
	8960	2-1/4" long		250	.032		.49	1.10		1.59	2.25
	8970	Phillips flat head, 3/16" dia x 1-3/4" long		300	.027		.33	.91		1.24	1.78
	8980	2-1/4" long	▼	250	.032	▼	.48	1.10		1.58	2.24
460	0010	**LAG SCREWS**									
	0020	Steel, 1/4" diameter, 2" long	1 Carp	200	.040	Ea.	.08	1.37		1.45	2.22
	0100	3/8" diameter, 3" long		150	.053		.23	1.83		2.06	3.10
	0200	1/2" diameter, 3" long		130	.062		.38	2.11		2.49	3.70
	0300	5/8" diameter, 3" long	▼	120	.067	▼	.75	2.28		3.03	4.39
500	0010	**MACHINE SCREWS**									
	0020	Steel, round head, #8 x 1" long				C	2.28			2.28	2.51
	0110	#8 x 2" long					4.96			4.96	5.45
	0200	#10 x 1" long					3.25			3.25	3.58
	0300	#10 x 2" long				▼	6.05			6.05	6.70
540	0010	**MACHINERY ANCHORS**, heavy duty, incl. sleeve, floating base nut,									
	0020	lower stud & coupling nut, fiber plug, connecting stud, washer & nut.									
	0030	For flush mounted embedment in poured concrete heavy equip. pads.									
	0200	Material only, 1/2" diameter stud & bolt				Ea.	52.50			52.50	58

05090	Metal Fastenings	CREW	DAILY OUTPUT	LABOR-HOURS	UNIT	2005 BARE COSTS				TOTAL INCL O&P	
						MAT.	LABOR	EQUIP.	TOTAL		
540											**540**
0300	5/8" diameter				Ea.	58.50			58.50	64	
0500	3/4" diameter					67.50			67.50	74	
0600	7/8" diameter					73.50			73.50	81	
0800	1" diameter					77.50			77.50	85	
0900	1-1/4" diameter				▼	103			103	113	
580	0010	**POWDER ACTUATED** Tools & fasteners									**580**
0020	Stud driver, .22 caliber, buy, minimum				Ea.	310			310	340	
0100	Maximum				"	495			495	545	
0300	Powder charges for above, low velocity				C	16.60			16.60	18.25	
0400	Standard velocity					23.50			23.50	26	
0600	Drive pins & studs, 1/4" & 3/8" diam., to 3" long, minimum					11.85			11.85	13.05	
0700	Maximum				▼	46.50			46.50	51	
0800	Pneumatic stud driver for 1/8" diameter studs				Ea.	2,150			2,150	2,350	
0900	Drive pins for above, 1/2" to 3/4" long				M	490			490	540	
600	0010	**RIVETS**									**600**
0100	Aluminum rivet & mandrel, 1/2" grip length x 1/8" diameter				C	5			5	5.50	
0200	3/16" diameter					7.70			7.70	8.45	
0300	Aluminum rivet, steel mandrel, 1/8" diameter					7.30			7.30	8.05	
0400	3/16" diameter					6.70			6.70	7.35	
0500	Copper rivet, steel mandrel, 1/8" diameter					6.65			6.65	7.30	
0600	Monel rivet, steel mandrel, 1/8" diameter					23.50			23.50	26	
0700	3/16" diameter					68			68	75	
0800	Stainless rivet & mandrel, 1/8" diameter					11.90			11.90	13.05	
0900	3/16" diameter					22			22	24.50	
1000	Stainless rivet, steel mandrel, 1/8" diameter					9.25			9.25	10.20	
1100	3/16" diameter					16.75			16.75	18.45	
1200	Steel rivet and mandrel, 1/8" diameter					5.75			5.75	6.30	
1300	3/16" diameter				▼	8.60			8.60	9.50	
1400	Hand riveting tool, minimum				Ea.	111			111	122	
1500	Maximum					212			212	233	
1600	Power riveting tool, minimum					800			800	880	
1700	Maximum				▼	2,025			2,025	2,225	
880	0010	**WELD ROD**									**880**
0020	Steel, type 6011, 1/8" dia, less than 500#				Lb.	1.72			1.72	1.89	
0100	500# to 2,000#					1.55			1.55	1.71	
0200	2,000# to 5,000#					1.46			1.46	1.60	
0300	5/32" diameter, less than 500#					1.65			1.65	1.82	
0310	500# to 2,000#					1.49			1.49	1.64	
0320	2,000# to 5,000#					1.40			1.40	1.54	
0400	3/16" dia, less than 500#					1.68			1.68	1.84	
0500	500# to 2,000#					1.51			1.51	1.66	
0600	2,000# to 5,000#					1.42			1.42	1.56	
0620	Steel, type 6010, 1/8" dia, less than 500#					1.90			1.90	2.09	
0630	500# to 2,000#					1.71			1.71	1.88	
0640	2,000# to 5,000#					1.61			1.61	1.77	
0650	Steel, type 7018 Low Hydrogen, 1/8" dia, less than 500#					1.62			1.62	1.78	
0660	500# to 2,000#					1.46			1.46	1.61	
0670	2,000# to 5,000#					1.37			1.37	1.51	
0700	Steel, type 7024 Jet Weld, 1/8" dia, less than 500#					1.72			1.72	1.89	
0710	500# to 2,000#					1.55			1.55	1.71	
0720	2,000# to 5,000#					1.46			1.46	1.60	
1550	Aluminum, type 4043 TIG, 1/8" dia, less than 10#					5.95			5.95	6.55	
1560	10# to 60#					5.35			5.35	5.90	
1570	Over 60#				▼	5.05			5.05	5.55	

05090 | Metal Fastenings

		CREW	DAILY OUTPUT	LABOR-HOURS	UNIT	MAT.	LABOR	EQUIP.	TOTAL	TOTAL INCL O&P	
880	1600	Aluminum, type 5356 TIG, 1/8" dia, less than 10#				Lb.	6.60			6.60	7.25	880
	1610	10# to 60#					5.95			5.95	6.55	
	1620	Over 60#					5.60			5.60	6.15	
	1900	Cast iron, type 8 Nickel, 1/8" dia, less than 500#					20.50			20.50	22.50	
	1910	500# to 1,000#					18.35			18.35	20	
	1920	Over 1,000#					17.25			17.25	18.95	
	2000	Stainless steel, type 316/316L, 1/8" dia, less than 500#					9.40			9.40	10.30	
	2100	500# to 1000#					8.45			8.45	9.30	
	2220	Over 1000#				Lb.	7.95			7.95	8.75	
900	0010	**WELDING STRUCTURAL** R05090-520										900
	0020	Field welding, 1/8" E6011, cost per welder, no oper. engr	E-14	8	1	Hr.	3.44	40	10.15	53.59	87	
	0200	With 1/2 operating engineer	E-13	8	1.500		3.44	56.50	10.15	70.09	112	
	0300	With 1 operating engineer	E-12	8	2		3.44	73	10.15	86.59	137	
	0500	With no operating engineer, 2# weld rod per ton	E-14	8	1	Ton	3.44	40	10.15	53.59	87	
	0600	8# E6011 per ton	"	2	4		13.75	161	40.50	215.25	345	
	0800	With one operating engineer per welder, 2# E6011 per ton	E-12	8	2		3.44	73	10.15	86.59	137	
	0900	8# E6011 per ton	"	2	8		13.75	293	40.50	347.25	545	
	1200	Continuous fillet, stick welding, incl. equipment										
	1300	Single pass, 1/8" thick, 0.1#/L.F.	E-14	150	.053	L.F.	.17	2.14	.54	2.85	4.62	
	1400	3/16" thick, 0.2#/L.F.		75	.107		.34	4.28	1.08	5.70	9.20	
	1500	1/4" thick, 0.3#/L.F.		50	.160		.52	6.40	1.62	8.54	13.85	
	1610	5/16" thick, 0.4#/L.F.		38	.211		.69	8.45	2.14	11.28	18.25	
	1800	3 passes, 3/8" thick, 0.5#/L.F.		30	.267		.86	10.70	2.71	14.27	23	
	2010	4 passes, 1/2" thick, 0.7#/L.F.		22	.364		1.20	14.60	3.69	19.49	31.50	
	2200	5 to 6 passes, 3/4" thick, 1.3#/L.F.		12	.667		2.24	27	6.75	35.99	58	
	2400	8 to 11 passes, 1" thick, 2.4#/L.F.		6	1.333		4.13	53.50	13.55	71.18	115	
	2600	For all position welding, add, minimum						20%				
	2700	Maximum						300%				
	2900	For semi-automatic welding, deduct, minimum						5%				
	3000	Maximum						15%				
	4000	Cleaning and welding plates, bars, or rods										
	4010	to existing beams, columns, or trusses	E-14	12	.667	L.F.	.86	27	6.75	34.61	56.50	

05100 | Structural Metal Framing

05120 | Structural Steel

		CREW	DAILY OUTPUT	LABOR-HOURS	UNIT	MAT.	LABOR	EQUIP.	TOTAL	TOTAL INCL O&P	
440	0010	**LIGHTWEIGHT FRAMING**										440
	0400	Angle framing, field fabricated, 4" and larger	E-3	440	.055	Lb.	.51	2.12	.18	2.81	4.55	
	0450	Less than 4" angles		265	.091		.53	3.52	.31	4.36	7.20	
	0600	Channel framing, field fabricated, 8" and larger		500	.048		.53	1.86	.16	2.55	4.09	
	0650	Less than 8" channels		335	.072		.53	2.78	.24	3.55	5.85	
	1000	Continuous slotted channel framing system, shop fab, min	2 Sswk	2,400	.007		2.71	.25		2.96	3.44	
	1200	Maximum	"	1,600	.010		3.06	.38		3.44	4.05	
	1250	Plate & bar stock for reinforcing beams and trusses					.96			.96	1.06	
	1300	Cross bracing, rods, shop fabricated, 3/4" diameter	E-3	700	.034		1.05	1.33	.12	2.50	3.67	
	1310	7/8" diameter		850	.028		1.05	1.10	.10	2.25	3.23	
	1320	1" diameter		1,000	.024		1.05	.93	.08	2.06	2.92	
	1330	Angle, 5" x 5" x 3/8"		2,800	.009		1.05	.33	.03	1.41	1.79	

05100 | Structural Metal Framing

05120 | Structural Steel

			CREW	DAILY OUTPUT	LABOR-HOURS	UNIT	2005 BARE COSTS				TOTAL INCL O&P	
							MAT.	LABOR	EQUIP.	TOTAL		
440	1350	Hanging lintels, shop fabricated, average	E-3	850	.028	Lb.	1.05	1.10	.10	2.25	3.23	440
	1380	Roof frames, shop fabricated, 3'-0" square, 5' span	E-2	4,200	.013		1.05	.49	.32	1.86	2.36	
	1400	Tie rod, not upset, 1-1/2" to 4" diameter, with turnbuckle	2 Sswk	800	.020		1.14	.76		1.90	2.62	
	1420	No turnbuckle	↓	700	.023		1.09	.87		1.96	2.76	
	1500	Upset, 1-3/4" to 4" diameter, with turnbuckle		800	.020		1.14	.76		1.90	2.62	
	1520	No turnbuckle	↓	700	.023	↓	1.09	.87		1.96	2.76	
520	0010	**PIPE SUPPORT FRAMING**										520
	0020	Under 10#/L.F.	E-4	3,900	.008	Lb.	1.17	.32	.02	1.51	1.88	
	0200	10.1 to 15#/L.F.		4,300	.007		1.16	.29	.02	1.47	1.80	
	0400	15.1 to 20#/L.F.		4,800	.007		1.14	.26	.02	1.42	1.73	
	0600	Over 20#/L.F.	↓	5,400	.006	↓	1.12	.23	.02	1.37	1.66	
640	0010	**STRUCTURAL STEEL MEMBERS**										640
	0020	Shop fab'd for 100-ton, 1-2 story project, bolted conn's.										
	0100	W 6 x 9	E-2	600	.093	L.F.	8.65	3.45	2.21	14.31	17.90	
	0300	W 8 x 10	"	600	.093		9.65	3.45	2.21	15.31	18.95	
	8490	For projects 75 to 99 tons, add					10%					
	8492	50 to 74 tons, add					20%					
	8494	25 to 49 tons, add					30%	10%				
	8496	10 to 24 tons, add					50%	25%				
	8498	2 to 9 tons, add					75%	50%				
	8499	Less than 2 tons, add				↓	100%	100%				

05500 | Metal Fabrications

05540 | Floor Plates

			CREW	DAILY OUTPUT	LABOR-HOURS	UNIT	2005 BARE COSTS				TOTAL INCL O&P	
							MAT.	LABOR	EQUIP.	TOTAL		
200	0010	**CHECKERED PLATE**, field fabricated										200
	0020	1/4" & 3/8", 2000 to 5000 S.F., bolted	E-4	2,900	.011	Lb.	.76	.43	.03	1.22	1.63	
	0100	Welded	↓	4,400	.007	"	.71	.28	.02	1.01	1.30	
	0300	Pit or trench cover and frame, 1/4" plate, 2' to 3' wide	↓	100	.320	S.F.	20	12.35	.81	33.16	45	
	0400	For galvanizing, add				Lb.	.41			.41	.45	
	0500	Platforms, 1/4" plate, no handrails included, rectangular	E-4	4,200	.008		1.21	.29	.02	1.52	1.88	
	0600	Circular	"	2,500	.013	↓	1.69	.49	.03	2.21	2.79	

For information about Means Estimating Seminars, see yellow pages 12 and 13 in back of book

Important: See the Reference Section for critical supporting data - Reference Nos., Crews, & City Cost Indexes

Division 6
Wood & Plastics

Estimating Tips

06050 Basic Wood & Plastic Materials & Methods

- Common to any wood framed structure are the accessory connector items such as screws, nails, adhesives, hangers, connector plates, straps, angles and holdowns. For typical wood framed buildings, such as residential projects, the aggregate total for these items can be significant, especially in areas where seismic loading is a concern. For floor and wall framing, the material cost is based on 10 to 25 lbs. per MBF. Holdowns, hangers and other connectors should be taken off by the piece.

06100 Rough Carpentry

- Lumber is a traded commodity and therefore sensitive to supply and demand in the marketplace. Even in "budgetary" estimating of wood framed projects, it is advisable to call local suppliers for the latest market pricing.
- Common quantity units for wood framed projects are "thousand board feet" (MBF). A board foot is a volume of wood, $1'' \times 1' \times 1'$, or 144 cubic inches. Board foot quantities are generally calculated using nominal material dimensions—dressed sizes are ignored. Board foot per lineal foot of any stick of lumber can be calculated by dividing the nominal cross sectional area by 12. As an example, 2,000 lineal feet of 2 x 12 equates to 4 MBF by dividing the nominal area, 2 x 12, by 12, which equals 2, and multiplying by 2,000 to give 4,000 board feet. This simple rule applies to all nominal dimensioned lumber.
- Waste is an issue of concern at the quantity takeoff for any area of construction. Framing lumber is sold in even foot lengths, i.e., 10', 12', 14', 16', and depending on spans, wall heights and the grade of lumber, waste is inevitable. A rule of thumb for lumber waste is 5% to 10% depending on material quality and the complexity of the framing.
- Wood in various forms and shapes is used in many projects, even where the main structural framing is steel, concrete or masonry. Plywood as a back-up partition material and 2x boards used as blocking and cant strips around roof edges are two common examples. The estimator should ensure that the costs of all wood materials are included in the final estimate.

06200 Finish Carpentry

- It is necessary to consider the grade of workmanship when estimating labor costs for erecting millwork and interior finish. In practice, there are three grades: premium, custom and economy. The Means daily output for base and case moldings is in the range of 200 to 250 L.F. per carpenter per day. This is appropriate for most average custom grade projects. For premium projects an adjustment to productivity of 25% to 50% should be made depending on the complexity of the job.

Reference Numbers

Reference numbers are shown in bold squares at the beginning of some major classifications. These numbers refer to related items in the Reference Section. The reference information may be an estimating procedure, an alternate pricing method or technical information.

Note: Not all subdivisions listed here necessarily appear in this publication.

06110 | Wood Framing

		CREW	DAILY OUTPUT	LABOR-HOURS	UNIT	2005 BARE COSTS				TOTAL INCL O&P
						MAT.	LABOR	EQUIP.	TOTAL	
545	**0010 FRAMING, MISCELLANEOUS**									**545**
8500	Firestops, 2" x 4"	2 Carp	.51	31.373	M.B.F.	510	1,075		1,585	2,225
8505	Pneumatic nailed		.62	25.806		510	885		1,395	1,925
8520	2" x 6"		.60	26.667		535	915		1,450	2,025
8525	Pneumatic nailed		.73	21.858		535	750		1,285	1,775
8540	2" x 8"		.60	26.667		620	915		1,535	2,100
8560	2" x 12"		.70	22.857		800	785		1,585	2,100
8600	Nailers, treated, wood construction, 2" x 4"		.53	30.189		730	1,025		1,755	2,400
8605	Pneumatic nailed		.64	25.157		730	860		1,590	2,150
8620	2" x 6"		.75	21.333		770	730		1,500	2,000
8625	Pneumatic nailed	▼	.90	17.778	▼	770	610		1,380	1,800

06160 | Sheathing

		CREW	DAILY OUTPUT	LABOR-HOURS	UNIT	2005 BARE COSTS				TOTAL INCL O&P
						MAT.	LABOR	EQUIP.	TOTAL	
800	**0010 SHEATHING** Plywood on roof, CDX									**800**
0050	3/8" thick	2 Carp	1,525	.010	S.F.	.65	.36		1.01	1.28
0055	Pneumatic nailed		1,860	.009		.65	.29		.94	1.18
0200	5/8" thick		1,300	.012		.78	.42		1.20	1.52
0205	Pneumatic nailed		1,586	.010		.78	.35		1.13	1.40
0300	3/4" thick		1,200	.013		.94	.46		1.40	1.74
0305	Pneumatic nailed		1,464	.011		.94	.37		1.31	1.61
0500	Plywood on walls with exterior CDX, 3/8" thick		1,200	.013		.65	.46		1.11	1.43
0505	Pneumatic nailed		1,488	.011		.65	.37		1.02	1.29
0700	5/8" thick		1,050	.015		.78	.52		1.30	1.67
0705	Pneumatic nailed		1,302	.012		.78	.42		1.20	1.52
0800	3/4" thick		975	.016		.94	.56		1.50	1.91
0805	Pneumatic nailed	▼	1,209	.013	▼	.94	.45		1.39	1.74

06400 | Architectural Woodwork

06410 | Custom Cabinets

		CREW	DAILY OUTPUT	LABOR-HOURS	UNIT	2005 BARE COSTS				TOTAL INCL O&P
						MAT.	LABOR	EQUIP.	TOTAL	
100	**0010 CABINETS** Corner china cabinets, stock pine,									**100**
0020	80" high, unfinished, minimum	2 Carp	6.60	2.424	Ea.	455	83		538	630
0100	Maximum	"	4.40	3.636	"	1,000	125		1,125	1,300
0700	Kitchen base cabinets, hardwood, not incl. counter tops,									
0710	24" deep, 35" high, prefinished									
0800	One top drawer, one door below, 12" wide	2 Carp	24.80	.645	Ea.	132	22		154	181
0840	18" wide		23.30	.687		197	23.50		220.50	253
0880	24" wide		22.30	.717		236	24.50		260.50	298
1000	Four drawers, 12" wide		24.80	.645		315	22		337	380
1040	18" wide		23.30	.687		269	23.50		292.50	335
1060	24" wide		22.30	.717		292	24.50		316.50	360
1200	Two top drawers, two doors below, 27" wide		22	.727		261	25		286	325
1260	36" wide		20.30	.788		300	27		327	370
1300	48" wide		18.90	.847		345	29		374	425
1500	Range or sink base, two doors below, 30" wide		21.40	.748		231	25.50		256.50	294
1540	36" wide		20.30	.788		259	27		286	325
1580	48" wide	▼	18.90	.847		291	29		320	365
1800	For sink front units, deduct				▼	50			50	55

Important: See the Reference Section for critical supporting data - Reference Nos., Crews, & City Cost Indexes

06410		Custom Cabinets	CREW	DAILY OUTPUT	LABOR-HOURS	UNIT	2005 BARE COSTS				TOTAL INCL O&P	
							MAT.	LABOR	EQUIP.	TOTAL		
100	9000	For deluxe models of all cabinets, add					40%					10
	9500	For custom built in place, add					25%	10%				
	9550	Rule of thumb, kitchen cabinets not including										
	9560	appliances & counter top, minimum	2 Carp	30	.533	L.F.	92	18.25		110.25	130	
	9600	Maximum	"	25	.640	"	240	22		262	298	
400	0010	**VANITIES**										40
	8000	Vanity bases, 2 doors, 30" high, 21" deep, 24" wide	2 Carp	20	.800	Ea.	177	27.50		204.50	237	
	8050	30" wide		16	1		203	34.50		237.50	277	
	8100	36" wide		13.33	1.200		271	41		312	360	
	8150	48" wide	▼	11.43	1.400		325	48		373	430	
	9000	For deluxe models of all vanities, add to above					40%					
	9500	For custom built in place, add to above				▼	25%	10%				

For information about Means Estimating Seminars, see yellow pages 12 and 13 in back of book

Division Notes

	CREW	DAILY OUTPUT	LABOR-HOURS	UNIT	2005 BARE COSTS				TOTAL INCL O&P
					MAT.	LABOR	EQUIP.	TOTAL	

Division 7
Thermal & Moisture Protection

Estimating Tips

07100 Dampproofing & Waterproofing

- Be sure of the job specifications before pricing this subdivision. The difference in cost between waterproofing and dampproofing can be great. Waterproofing will hold back standing water. Dampproofing prevents the transmission of water vapor. Also included in this section are vapor retarding membranes.

07200 Thermal Protection

- Insulation and fireproofing products are measured by area, thickness, volume or R value. Specifications may only give what the specific R value should be in a certain situation. The estimator may need to choose the type of insulation to meet that R value.

07300 Shingles, Roof Tiles & Roof Coverings
07400 Roofing & Siding Panels

- Many roofing and siding products are bought and sold by the square. One square is equal to an area that measures 100 square feet.

This simple change in unit of measure could create a large error if the estimator is not observant. Accessories necessary for a complete installation must be figured into any calculations for both material and labor.

07500 Membrane Roofing
07600 Flashing & Sheet Metal
07700 Roof Specialties & Accessories

- The items in these subdivisions compose a roofing system. No one component completes the installation and all must be estimated. Built-up or single ply membrane roofing systems are made up of many products and installation trades. Wood blocking at roof perimeters or penetrations, parapet coverings, reglets, roof drains, gutters, downspouts, sheet metal flashing, skylights, smoke vents or roof hatches all need to be considered along with the roofing material. Several different installation trades will need to work together on the roofing system. Inherent difficulties in the scheduling and coordination of various trades must be accounted for when estimating labor costs.

07900 Joint Sealers

- To complete the weather-tight shell the sealants and caulkings must be estimated. Where different materials meet—at expansion joints, at flashing penetrations, and at hundreds of other locations throughout a construction project—they provide another line of defense against water penetration. Often, an entire system is based on the proper location and placement of caulking or sealants. The detail drawings that are included as part of a set of architectural plans, show typical locations for these materials. When caulking or sealants are shown at typical locations, this means the estimator must include them for all the locations where this detail is applicable. Be careful to keep different types of sealants separate, and remember to consider backer rods and primers if necessary.

Reference Numbers

Reference numbers are shown in bold squares at the beginning of some major classifications. These numbers refer to related items in the Reference Section. The reference information may be an estimating procedure, an alternate pricing method or technical information.

Note: Not all subdivisions listed here necessarily appear in this publication.

	07650	**Flexible Flashing**	CREW	DAILY OUTPUT	LABOR-HOURS	UNIT	2005 BARE COSTS				TOTAL INCL O&P	
							MAT.	LABOR	EQUIP.	TOTAL		
600	0010	**FLASHING** Aluminum, mill finish, .013" thick	1 Rofc	145	.055	S.F.	.35	1.62		1.97	3.16	**600**
	0030	.016" thick		145	.055		.51	1.62		2.13	3.33	
	0060	.019" thick		145	.055		.65	1.62		2.27	3.49	
	0100	.032" thick		145	.055		1.06	1.62		2.68	3.94	
	0200	.040" thick		145	.055		1.45	1.62		3.07	4.37	
	0300	.050" thick		145	.055		1.84	1.62		3.46	4.79	
	0325	Mill finish 5" x 7" step flashing, .016" thick		1,920	.004	Ea.	.10	.12		.22	.32	
	0350	Mill finish 12" x 12" step flashing, .016" thick		1,600	.005	"	.40	.15		.55	.69	
	0400	Painted finish, add				S.F.	.24			.24	.26	
	0500	Fabric-backed 2 sides, .004" thick	1 Rofc	330	.024		.93	.71		1.64	2.24	
	0700	.005" thick		330	.024		1.09	.71		1.80	2.42	
	0750	Mastic-backed, self adhesive		460	.017		2.26	.51		2.77	3.36	
	0800	Mastic-coated 2 sides, .004" thick		330	.024		.93	.71		1.64	2.24	
	1000	.005" thick		330	.024		1.09	.71		1.80	2.42	
	1100	.016" thick		330	.024		1.20	.71		1.91	2.54	
	1300	Asphalt flashing cement, 5 gallon				Gal.	3.98			3.98	4.38	
	1600	Copper, 16 oz, sheets, under 1000 lbs.	1 Rofc	115	.070	S.F.	2.62	2.05		4.67	6.35	
	1700	Over 4000 lbs.		155	.052		2.91	1.52		4.43	5.80	
	1900	20 oz sheets, under 1000 lbs.		110	.073		3.90	2.14		6.04	7.95	
	2000	Over 4000 lbs.		145	.055		3.62	1.62		5.24	6.75	
	2200	24 oz sheets, under 1000 lbs.		105	.076		4.69	2.24		6.93	8.95	
	2300	Over 4000 lbs.		135	.059		4.34	1.75		6.09	7.75	
	2500	32 oz sheets, under 1000 lbs.		100	.080		6.20	2.36		8.56	10.85	
	2600	Over 4000 lbs.		130	.062		5.80	1.81		7.61	9.50	
	2800	Copper, paperbacked 1 side, 2 oz		330	.024		.86	.71		1.57	2.17	
	2900	3 oz		330	.024		1.12	.71		1.83	2.45	
	3100	Paperbacked 2 sides, 2 oz		330	.024		.86	.71		1.57	2.17	
	3150	3 oz		330	.024		1.11	.71		1.82	2.44	
	3200	5 oz		330	.024		1.66	.71		2.37	3.05	
	3250	7 oz		330	.024		2.72	.71		3.43	4.21	
	3400	Mastic-backed 2 sides, copper, 2 oz		330	.024		1.02	.71		1.73	2.34	
	3500	3 oz		330	.024		1.26	.71		1.97	2.61	
	3700	5 oz		330	.024		1.85	.71		2.56	3.26	
	3800	Fabric-backed 2 sides, copper, 2 oz		330	.024		1.09	.71		1.80	2.42	
	4000	3 oz		330	.024		1.40	.71		2.11	2.76	
	4100	5 oz		330	.024		1.90	.71		2.61	3.31	
	4300	Copper-clad stainless steel, .015" thick, under 500 lbs.		115	.070		3.16	2.05		5.21	6.95	
	4400	Over 2000 lbs.		155	.052		3.05	1.52		4.57	5.95	
	4600	.018" thick, under 500 lbs.		100	.080		4.18	2.36		6.54	8.60	
	4700	Over 2000 lbs.		145	.055		3.06	1.62		4.68	6.15	
	5800	Lead, 2.5 lb. per SF, up to 12" wide		135	.059		2.94	1.75		4.69	6.20	
	5900	Over 12" wide		135	.059		3.25	1.75		5	6.55	
	8500	Shower pan, bituminous membrane, 7 oz		155	.052		1.08	1.52		2.60	3.78	
	8550	3 ply copper and fabric, 3 oz		155	.052		1.63	1.52		3.15	4.38	
	8600	7 oz		155	.052		3.37	1.52		4.89	6.30	
	8650	Copper, 16 oz		100	.080		3.11	2.36		5.47	7.45	
	8700	Lead on copper and fabric, 5 oz		155	.052		1.62	1.52		3.14	4.37	
	8800	7 oz		155	.052		2.93	1.52		4.45	5.80	
	8850	Polyvinyl chloride, .030" thick		160	.050		.30	1.47		1.77	2.84	
	8900	Stainless steel sheets, 32 ga, .010" thick		155	.052		2.16	1.52		3.68	4.97	
	9000	28 ga, .015" thick		155	.052		2.68	1.52		4.20	5.55	
	9100	26 ga, .018" thick		155	.052		3.25	1.52		4.77	6.15	
	9200	24 ga, .025" thick		155	.052		4.22	1.52		5.74	7.25	
	9400	Terne coated stainless steel, .015" thick, 28 ga		155	.052		4.05	1.52		5.57	7.05	
	9500	.018" thick, 26 ga		155	.052		4.57	1.52		6.09	7.65	
	9600	Zinc and copper alloy (brass), .020" thick		155	.052		3.26	1.52		4.78	6.20	

Important: See the Reference Section for critical supporting data - Reference Nos., Crews, & City Cost Indexes

07600 | Flashing and Sheet Metal

07650	Flexible Flashing	CREW	DAILY OUTPUT	LABOR-HOURS	UNIT	2005 BARE COSTS MAT.	LABOR	EQUIP.	TOTAL	TOTAL INCL O&P	
600	9700	.027" thick	1 Rofc	155	.052	S.F.	4.37	1.52		5.89	7.40	600
	9800	.032" thick		155	.052		5.10	1.52		6.62	8.20	
	9900	.040" thick	↓	155	.052	↓	6.20	1.52		7.72	9.45	

07720	Roof Accessories										
560	0010	**SNOW GUARDS**										560
	0100	Slate & asphalt shingle roofs	1 Rofc	160	.050	Ea.	8	1.47		9.47	11.30	
	0200	Standing seam metal roofs		48	.167		12.25	4.91		17.16	22	
	0300	Surface mount for metal roofs	↓	48	.167	↓	6.75	4.91		11.66	15.80	

07800 | Fire and Smoke Protection

07840	Firestopping	CREW	DAILY OUTPUT	LABOR-HOURS	UNIT	2005 BARE COSTS MAT.	LABOR	EQUIP.	TOTAL	TOTAL INCL O&P	
100	0010	**FIRESTOPPING** R07800 -030										100
	0100	Metallic piping, non insulated										
	0110	Through walls, 2" diameter	1 Carp	16	.500	Ea.	9.95	17.15		27.10	37.50	
	0120	4" diameter		14	.571		15.20	19.55		34.75	47.50	
	0130	6" diameter		12	.667		20.50	23		43.50	58	
	0140	12" diameter		10	.800		36.50	27.50		64	82.50	
	0150	Through floors, 2" diameter		32	.250		6.05	8.55		14.60	20	
	0160	4" diameter		28	.286		8.70	9.80		18.50	25	
	0170	6" diameter		24	.333		11.40	11.40		22.80	30.50	
	0180	12" diameter	↓	20	.400	↓	19.25	13.70		32.95	42.50	
	0190	Metallic piping, insulated										
	0200	Through walls, 2" diameter	1 Carp	16	.500	Ea.	14.15	17.15		31.30	42	
	0210	4" diameter		14	.571		19.40	19.55		38.95	52	
	0220	6" diameter		12	.667		24.50	23		47.50	62.50	
	0230	12" diameter		10	.800		40.50	27.50		68	87	
	0240	Through floors, 2" diameter		32	.250		10.20	8.55		18.75	24.50	
	0250	4" diameter		28	.286		12.85	9.80		22.65	29.50	
	0260	6" diameter		24	.333		15.60	11.40		27	35	
	0270	12" diameter	↓	20	.400	↓	19.25	13.70		32.95	42.50	
	0280	Non metallic piping, non insulated										
	0290	Through walls, 2" diameter	1 Carp	12	.667	Ea.	41	23		64	81	
	0300	4" diameter		10	.800		51.50	27.50		79	99.50	
	0310	6" diameter		8	1		72	34.50		106.50	133	
	0330	Through floors, 2" diameter		16	.500		32	17.15		49.15	62	
	0340	4" diameter		6	1.333		40	45.50		85.50	115	
	0350	6" diameter	↓	6	1.333	↓	48	45.50		93.50	124	
	0370	Ductwork, insulated & non insulated, round										
	0380	Through walls, 6" diameter	1 Carp	12	.667	Ea.	21	23		44	58.50	
	0390	12" diameter		10	.800		41.50	27.50		69	88.50	
	0400	18" diameter		8	1		67.50	34.50		102	128	
	0410	Through floors, 6" diameter		16	.500		11.45	17.15		28.60	39	
	0420	12" diameter		14	.571		21	19.55		40.55	53.50	
	0430	18" diameter	↓	12	.667	↓	36.50	23		59.50	75.50	
	0440	Ductwork, insulated & non insulated, rectangular										
	0450	With stiffener/closure angle, through walls, 6" x 12"	1 Carp	8	1	Ea.	17.35	34.50		51.85	72.50	
	0460	12" x 24"	↓	6	1.333	↓	23	45.50		68.50	96.50	

7

07840 | Firestopping

		CREW	DAILY OUTPUT	LABOR-HOURS	UNIT	2005 BARE COSTS				TOTAL INCL O&P
						MAT.	LABOR	EQUIP.	TOTAL	
100 0470	24" x 48"	1 Carp	4	2	Ea.	65.50	68.50		134	179 **100**
0480	With stiffener/closure angle, through floors, 6" x 12" R07800-030		10	.800		9.35	27.50		36.85	53
0490	12" x 24"		8	1		16.85	34.50		51.35	72
0500	24" x 48"		6	1.333		33	45.50		78.50	108
0510	Multi trade openings									
0520	Through walls, 6" x 12"	1 Carp	2	4	Ea.	36.50	137		173.50	253
0530	12" x 24"	"	1	8		147	274		421	585
0540	24" x 48"	2 Carp	1	16		585	550		1,135	1,500
0550	48" x 96"	"	.75	21.333		2,350	730		3,080	3,750
0560	Through floors, 6" x 12"	1 Carp	2	4		36.50	137		173.50	253
0570	12" x 24"	"	1	8		147	274		421	585
0580	24" x 48"	2 Carp	.75	21.333		585	730		1,315	1,800
0590	48" x 96"	"	.50	32		2,350	1,100		3,450	4,300
0600	Structural penetrations, through walls									
0610	Steel beams, W8 x 10	1 Carp	8	1	Ea.	23	34.50		57.50	78.50
0620	W12 x 14		6	1.333		36.50	45.50		82	111
0630	W21 x 44		5	1.600		73	55		128	166
0640	W36 x 135		3	2.667		177	91.50		268.50	335
0650	Bar joists, 18" deep		6	1.333		33.50	45.50		79	108
0660	24" deep		6	1.333		41.50	45.50		87	117
0670	36" deep		5	1.600		62.50	55		117.50	154
0680	48" deep		4	2		73	68.50		141.50	187
0690	Construction joints, floor slab at exterior wall									
0700	Precast, brick, block or drywall exterior									
0710	2" wide joint	1 Carp	125	.064	L.F.	5.20	2.19		7.39	9.10
0720	4" wide joint	"	75	.107	"	10.40	3.65		14.05	17.15
0730	Metal panel, glass or curtain wall exterior									
0740	2" wide joint	1 Carp	40	.200	L.F.	12.30	6.85		19.15	24
0750	4" wide joint	"	25	.320	"	16.80	10.95		27.75	35.50
0760	Floor slab to drywall partition									
0770	Flat joint	1 Carp	100	.080	L.F.	5.10	2.74		7.84	9.85
0780	Fluted joint		50	.160		10.40	5.50		15.90	20
0790	Etched fluted joint		75	.107		6.75	3.65		10.40	13.15
0800	Floor slab to concrete/masonry partition									
0810	Flat joint	1 Carp	75	.107	L.F.	11.45	3.65		15.10	18.30
0820	Fluted joint	"	50	.160	"	13.55	5.50		19.05	23.50
0830	Concrete/CMU wall joints									
0840	1" wide	1 Carp	100	.080	L.F.	6.25	2.74		8.99	11.10
0850	2" wide		75	.107		11.45	3.65		15.10	18.30
0860	4" wide		50	.160		22	5.50		27.50	32.50
0870	Concrete/CMU floor joints									
0880	1" wide	1 Carp	200	.040	L.F.	3.12	1.37		4.49	5.55
0890	2" wide		150	.053		5.70	1.83		7.53	9.15
0900	4" wide		100	.080		10.90	2.74		13.64	16.25

07900 | Joint Sealers

07920 | Joint Sealants

		CREW	DAILY OUTPUT	LABOR-HOURS	UNIT	2005 BARE COSTS				TOTAL INCL O&P
						MAT.	LABOR	EQUIP.	TOTAL	
800 0010	**CAULKING AND SEALANTS**									**800**
0020	Acoustical sealant, elastomeric, cartridges				Ea.	2.21			2.21	2.43

			CREW	DAILY OUTPUT	LABOR-HOURS	UNIT	2005 BARE COSTS				TOTAL INCL O&P	
	07920	**Joint Sealants**					MAT.	LABOR	EQUIP.	TOTAL		
800	0030	Backer rod, polyethylene, 1/4" diameter	1 Bric	4.60	1.739	C.L.F.	2.12	61.50		63.62	96	**800**
	0050	1/2" diameter		4.60	1.739		3.39	61.50		64.89	97	
	0070	3/4" diameter		4.60	1.739		5.80	61.50		67.30	100	
	0090	1" diameter	↓	4.60	1.739	↓	9.35	61.50		70.85	104	
	0100	Acrylic latex caulk, white										
	0200	11 fl. oz cartridge				Ea.	1.88			1.88	2.07	
	0500	1/4" x 1/2"	1 Bric	248	.032	L.F.	.15	1.14		1.29	1.90	
	0600	1/2" x 1/2"		250	.032		.31	1.13		1.44	2.06	
	0800	3/4" x 3/4"		230	.035		.69	1.23		1.92	2.63	
	0900	3/4" x 1"		200	.040		.92	1.41		2.33	3.16	
	1000	1" x 1"	↓	180	.044	↓	1.15	1.57		2.72	3.66	
	1400	Butyl based, bulk				Gal.	22			22	24.50	
	1500	Cartridges				"	27			27	29.50	
	1700	Bulk, in place 1/4" x 1/2", 154 L.F./gal.	1 Bric	230	.035	L.F.	.14	1.23		1.37	2.03	
	1800	1/2" x 1/2", 77 L.F./gal.	"	180	.044	"	.29	1.57		1.86	2.71	
	2000	Latex acrylic based, bulk				Gal.	23			23	25.50	
	2100	Cartridges					26			26	28.50	
	2300	Polysulfide compounds, 1 component, bulk				↓	44			44	48	
	2600	1 or 2 component, in place, 1/4" x 1/4", 308 L.F./gal.	1 Bric	145	.055	L.F.	.14	1.94		2.08	3.12	
	2700	1/2" x 1/4", 154 L.F./gal.		135	.059		.28	2.09		2.37	3.49	
	2900	3/4" x 3/8", 68 L.F./gal.		130	.062		.64	2.17		2.81	4.01	
	3000	1" x 1/2", 38 L.F./gal.	↓	130	.062	↓	1.15	2.17		3.32	4.57	
	3200	Polyurethane, 1 or 2 component				Gal.	47.50			47.50	52.50	
	3300	Cartridges				"	46.50			46.50	51.50	
	3500	Bulk, in place, 1/4" x 1/4"	1 Bric	150	.053	L.F.	.15	1.88		2.03	3.03	
	3600	1/2" x 1/4"		145	.055		.31	1.94		2.25	3.30	
	3800	3/4" x 3/8", 68 L.F./gal.		130	.062		.70	2.17		2.87	4.07	
	3900	1" x 1/2"	↓	110	.073	↓	1.24	2.56		3.80	5.25	
	4100	Silicone rubber, bulk				Gal.	34.50			34.50	38	
	4200	Cartridges				"	40.50			40.50	44.50	
	4400	Neoprene gaskets, closed cell, adhesive, 1/8" x 3/8"	1 Bric	240	.033	L.F.	.20	1.17		1.37	2.01	
	4500	1/4" x 3/4"		215	.037		.48	1.31		1.79	2.53	
	4700	1/2" x 1"		200	.040		1.40	1.41		2.81	3.69	
	4800	3/4" x 1-1/2"	↓	165	.048	↓	2.91	1.71		4.62	5.80	
	5500	Resin epoxy coating, 2 component, heavy duty				Gal.	26			26	28.50	
	5800	Tapes, sealant, P.V.C. foam adhesive, 1/16" x 1/4"				C.L.F.	4.60			4.60	5.05	
	5900	1/16" x 1/2"					6.80			6.80	7.50	
	5950	1/16" x 1"					11.30			11.30	12.45	
	6000	1/8" x 1/2"				↓	7.65			7.65	8.40	
	6200	Urethane foam, 2 component, handy pack, 1 C.F.				Ea.	27.50			27.50	30.50	
	6300	50.0 C.F. pack				C.F.	14.05			14.05	15.45	

For information about Means Estimating Seminars, see yellow pages 12 and 13 in back of book

		CREW	DAILY OUTPUT	LABOR-HOURS	UNIT	2005 BARE COSTS				TOTAL INCL O&P
						MAT.	LABOR	EQUIP.	TOTAL	

Division 9
Finishes

Estimating Tips

General

- Room Finish Schedule: A complete set of plans should contain a room finish schedule. If one is not available, it would be well worth the time and effort to put one together. A room finish schedule should contain the room number, room name (for clarity), floor materials, base materials, wainscot materials, wainscot height, wall materials (for each wall), ceiling materials, ceiling height and special instructions.
- Surplus Finishes: Review the specifications to determine if there is any requirement to provide certain amounts of extra materials for the owner's maintenance department. In some cases the owner may require a substantial amount of materials, especially when it is a special order item or long lead time item.

09200 Plaster & Gypsum Board

- Lath is estimated by the square yard for both gypsum and metal lath, plus usually 5% allowance for waste. Furring, channels and accessories are measured by the linear foot. An extra foot should be allowed for each accessory miter or stop.
- Plaster is also estimated by the square yard. Deductions for openings vary by preference, from zero deduction to 50% of all openings over 2 feet in width. Some estimators deduct a percentage of the total yardage for openings. The estimator should allow one extra square foot for each linear foot of horizontal interior or exterior angle located below the ceiling level. Also, double the areas of small radius work.
- Each room should be measured, perimeter times maximum wall height. Floors and ceiling areas are equal to length times width.
- Drywall accessories, studs, track, and acoustical caulking are all measured by the linear foot. Drywall taping is figured by the square foot. Gypsum wallboard is estimated by the square foot. No material deductions should be made for door or window openings under 32 S.F. Coreboard can be obtained in a 1" thickness for solid wall and shaft work. Additions should be made to price out the inside or outside corners.
- Different types of partition construction should be listed separately on the quantity sheets. There may be walls with studs of various widths, double studded, and similar or dissimilar surface materials. Shaft work is usually different construction from surrounding partitions requiring separate quantities and pricing of the work.

09300 Tile
09400 Terrazzo

- Tile and terrazzo areas are taken off on a square foot basis. Trim and base materials are measured by the linear foot. Accent tiles are listed per each. Two basic methods of installation are used. Mud set is approximately 30% more expensive than the thin set. In terrazzo work, be sure to include the linear footage of embedded decorative strips, grounds, machine rubbing and power cleanup.

09600 Flooring

- Wood flooring is available in strip, parquet, or block configuration. The latter two types are set in adhesives with quantities estimated by the square foot. The laying pattern will influence labor costs and material waste. In addition to the material and labor for laying wood floors, the estimator must make allowances for sanding and finishing these areas unless the flooring is prefinished.
- Most of the various types of flooring are all measured on a square foot basis. Base is measured by the linear foot. If adhesive materials are to be quantified, they are estimated at a specified coverage rate by the gallon depending upon the specified type and the manufacturer's recommendations.
- Sheet flooring is measured by the square yard. Roll widths vary, so consideration should be given to use the most economical width, as waste must be figured into the total quantity. Consider also the installation methods available, direct glue down or stretched.

09700 Wall Finishes

- Wall coverings are estimated by the square foot. The area to be covered is measured, length by height of wall above baseboards, to calculate the square footage of each wall. This figure is divided by the number of square feet in the single roll which is being used. Deduct, in full, the areas of openings such as doors and windows. Where a pattern match is required allow 25%-30% waste. One gallon of paste should be sufficient to hang 12 single rolls of light to medium weight paper.

09800 Acoustical Treatment

- Acoustical systems fall into several categories. The takeoff of these materials should be by the square foot of area with a 5% allowance for waste. Do not forget about scaffolding, if applicable, when estimating these systems.

09900 Paints & Coatings

- A major portion of the work in painting involves surface preparation. Be sure to include cleaning, sanding, filling and masking costs in the estimate.
- Painting is one area where bids vary to a greater extent than almost any other section of a project. This arises from the many methods of measuring surfaces to be painted. The estimator should check the plans and specifications carefully to be sure of the required number of coats.
- Protection of adjacent surfaces is not included in painting costs. When considering the method of paint application, an important factor is the amount of protection and masking required. These must be estimated separately and may be the determining factor in choosing the method of application.

Reference Numbers

Reference numbers are shown in bold squares at the beginning of some major classifications. These numbers refer to related items in the Reference Section. The reference information may be an estimating procedure, an alternate pricing method or technical information.

Note: Not all subdivisions listed here necessarily appear in this publication.

09910	Paints	CREW	DAILY OUTPUT	LABOR-HOURS	UNIT	2005 BARE COSTS				TOTAL INCL O&P
						MAT.	LABOR	EQUIP.	TOTAL	
630 0010	**MISCELLANEOUS, INTERIOR**									**630**
3800	Grilles, per side, oil base, primer coat, brushwork	1 Pord	520	.015	S.F.	.10	.47		.57	.82
3850	Spray		1,140	.007		.10	.21		.31	.43
3880	Paint 1 coat, brushwork		520	.015		.11	.47		.58	.83
3900	Spray		1,140	.007		.12	.21		.33	.45
3920	Paint 2 coats, brushwork		325	.025		.21	.75		.96	1.36
3940	Spray		650	.012		.24	.38		.62	.83
3950	Prime & paint 1 coat		325	.025		.20	.75		.95	1.35
3960	Prime & paint 2 coats		270	.030		.20	.91		1.11	1.58
4250	Paint 1 coat, brushwork	2 Pord	1,300	.012	L.F.	.11	.38		.49	.69
4500	Louvers, one side, primer, brushwork	1 Pord	524	.015	S.F.	.06	.47		.53	.76
4520	Paint one coat, brushwork		520	.015		.06	.47		.53	.77
4530	Spray		1,140	.007		.06	.21		.27	.39
4540	Paint two coats, brushwork		325	.025		.11	.75		.86	1.25
4550	Spray		650	.012		.12	.38		.50	.70
4560	Paint three coats, brushwork		270	.030		.16	.91		1.07	1.54
4570	Spray		500	.016		.18	.49		.67	.94
5000	Pipe, to 4" diameter, primer or sealer coat, oil base, brushwork	2 Pord	1,250	.013	L.F.	.06	.39		.45	.66
5100	Spray		2,165	.007		.06	.23		.29	.40
5200	Paint 1 coat, brushwork		1,250	.013		.06	.39		.45	.66
5300	Spray		2,165	.007		.06	.23		.29	.40
5350	Paint 2 coats, brushwork		775	.021		.11	.63		.74	1.07
5400	Spray		1,240	.013		.13	.39		.52	.73
5420	Paint 3 coats, brushwork		775	.021		.17	.63		.80	1.14
5450	To 8" diameter, primer or sealer coat, brushwork		620	.026		.12	.79		.91	1.32
5500	Spray		1,085	.015		.20	.45		.65	.90
5550	Paint 1 coat, brushwork		620	.026		.17	.79		.96	1.38
5600	Spray		1,085	.015		.19	.45		.64	.89
5650	Paint 2 coats, brushwork		385	.042		.23	1.27		1.50	2.16
5700	Spray		620	.026		.25	.79		1.04	1.47
5720	Paint 3 coats, brushwork		385	.042		.34	1.27		1.61	2.28
5750	To 12" diameter, primer or sealer coat, brushwork		415	.039		.18	1.18		1.36	1.97
5800	Spray		725	.022		.21	.68		.89	1.25
5850	Paint 1 coat, brushwork		415	.039		.17	1.18		1.35	1.96
6000	Spray		725	.022		.19	.68		.87	1.23
6200	Paint 2 coats, brushwork		260	.062		.34	1.88		2.22	3.20
6250	Spray		415	.039		.38	1.18		1.56	2.18
6270	Paint 3 coats, brushwork		260	.062		.50	1.88		2.38	3.38
6300	To 16" diameter, primer or sealer coat, brushwork		310	.052		.24	1.58		1.82	2.64
6350	Spray		540	.030		.27	.91		1.18	1.66
6400	Paint 1 coat, brushwork		310	.052		.23	1.58		1.81	2.63
6450	Spray		540	.030		.26	.91		1.17	1.65
6500	Paint 2 coats, brushwork		195	.082		.45	2.51		2.96	4.27
6550	Spray		310	.052		.50	1.58		2.08	2.92
6600	Radiators, per side, primer, brushwork	1 Pord	520	.015	S.F.	.06	.47		.53	.77
6620	Paint one coat, brushwork		520	.015		.05	.47		.52	.77
6640	Paint two coats, brushwork		340	.024		.11	.72		.83	1.20
6660	Paint three coats, brushwork		283	.028		.16	.87		1.03	1.48

For information about Means Estimating Seminars, see yellow pages 12 and 13 in back of book

Division 10 Specialties

Estimating Tips

General
- The items in this division are usually priced per square foot or each.
- Many items in Division 10 require some type of support system or special anchors that are not usually furnished with the item. The required anchors must be added to the estimate in the appropriate division.
- Some items in Division 10 such as lockers, may require assembly before installation. Verify the amount of assembly required. Assembly can often exceed installation time.

10150 Compartments & Cubicles
- Support angles and blocking are not included in the installation of toilet compartments, shower/dressing compartments or cubicles. Appropriate line items from Divisions 5 or 6 may need to be added to support the installations.
- Toilet partitions are priced by the stall. A stall consists of a side wall, pilaster and door with hardware. Toilet tissue holders and grab bars are extra.

10600 Partitions
- The required acoustical rating of a folding partition can have a significant impact on costs. Verify the sound transmission coefficient rating of the panel priced to the specification requirements.

10800 Toilet/Bath/Laundry Accessories
- Grab bar installation does not include supplemental blocking or backing to support the required load. When grab bars are installed at an existing facility provisions must be made to attach the grab bars to solid structure.

Reference Numbers
Reference numbers are shown in bold squares at the beginning of some major classifications. These numbers refer to related items in the Reference Section. The reference information may be an estimating procedure, an alternate pricing method or technical information.

Note: Not all subdivisions listed here necessarily appear in this publication.

10160 | Metal Toilet Compartments

		CREW	DAILY OUTPUT	LABOR-HOURS	UNIT	MAT.	LABOR	EQUIP.	TOTAL	TOTAL INCL O&P
100								2005 BARE COSTS		**100**
0010	**TOILET PARTITIONS, METAL**									
0110	Cubicles, ceiling hung									
0200	Painted metal	2 Carp	4	4	Ea.	455	137		592	715
0500	Stainless steel	"	4	4		1,175	137		1,312	1,525
0600	For handicap units, incl. 52" grab bars, add				▼	320			320	355
0900	Floor and ceiling anchored									
1000	Painted metal	2 Carp	5	3.200	Ea.	415	110		525	630
1300	Stainless steel	"	5	3.200		1,350	110		1,460	1,675
1400	For handicap units, incl. 52" grab bars, add				▼	287			287	315
1610	Floor mounted									
1700	Painted metal	2 Carp	7	2.286	Ea.	465	78.50		543.50	635
2000	Stainless steel	"	7	2.286		1,450	78.50		1,528.50	1,700
2100	For handicap units, incl. 52" grab bars, add					287			287	315
2200	For juvenile units, deduct				▼	39.50			39.50	43.50
2450	Floor mounted, headrail braced									
2500	Painted metal	2 Carp	6	2.667	Ea.	415	91.50		506.50	600
2800	Stainless steel	"	6	2.667		1,350	91.50		1,441.50	1,625
2900	For handicap units, incl. 52" grab bars, add					287			287	315
3000	Wall hung partitions, painted metal	2 Carp	7	2.286		525	78.50		603.50	700
3300	Stainless steel	"	7	2.286		1,325	78.50		1,403.50	1,575
3400	For handicap units, incl. 52" grab bars, add				▼	287			287	315
4000	Screens, entrance, floor mounted, 58" high, 48" wide									
4200	Painted metal	2 Carp	15	1.067	Ea.	188	36.50		224.50	264
4500	Stainless steel	"	15	1.067	"	715	36.50		751.50	845
4650	Urinal screen, 18" wide									
4700	Painted metal	2 Carp	8	2	Ea.	156	68.50		224.50	279
5000	Stainless steel	"	8	2	"	515	68.50		583.50	670
5100	Floor mounted, head rail braced									
5300	Painted metal	2 Carp	8	2	Ea.	172	68.50		240.50	296
5600	Stainless steel	"	8	2	"	600	68.50		668.50	765
5750	Pilaster, flush									
5800	Painted metal	2 Carp	10	1.600	Ea.	252	55		307	365
6100	Stainless steel		10	1.600		450	55		505	580
6300	Urinal screen, post braced, painted metal		10	1.600		260	55		315	370
6600	Stainless steel	▼	10	1.600	▼	450	55		505	580
6700	Wall hung, bracket supported									
6800	Painted metal	2 Carp	10	1.600	Ea.	262	55		317	375
7100	Stainless steel		10	1.600		410	55		465	535
7400	Flange supported, painted metal		10	1.600		193	55		248	298
7700	Stainless steel		10	1.600		475	55		530	610
7800	Wedge type, painted metal		10	1.600		228	55		283	335
8100	Stainless steel	▼	10	1.600	▼	490	55		545	625

10165 | Plastic Laminate Toilet Compartment

		CREW	DAILY OUTPUT	LABOR-HOURS	UNIT	MAT.	LABOR	EQUIP.	TOTAL	TOTAL INCL O&P
100	0010 **TOILET PARTITIONS, PLASTIC LAMINATE**									**100**
0110	Cubicles, ceiling hung									
0300	Plastic laminate on particle board	2 Carp	4	4	Ea.	590	137		727	865
0600	For handicap units, incl. 52" grab bars, add				"	320			320	355
0900	Floor and ceiling anchored									
1100	Plastic laminate on particle board	2 Carp	5	3.200	Ea.	665	110		775	900
1400	For handicap units, incl. 52" grab bars, add				"	287			287	315
1610	Floor mounted									
1800	Plastic laminate on particle board	2 Carp	7	2.286	Ea.	585	78.50		663.50	760
2450	Floor mounted, headrail braced									
2600	Plastic laminate on particle board	2 Carp	6	2.667	Ea.	770	91.50		861.50	990
3400	For handicap units, incl. 52" grab bars, add				▼	287			287	315

Important: See the Reference Section for critical supporting data - Reference Nos., Crews, & City Cost Indexes

10165 | Plastic Laminate Toilet Compartment

		CREW	DAILY OUTPUT	LABOR-HOURS	UNIT	MAT.	LABOR	EQUIP.	TOTAL	TOTAL INCL O&P
100 4300	Entrance screen, floor mtd., plas. lam., 58″ high, 48″ wide	2 Carp	15	1.067	Ea.	435	36.50		471.50	530
4800	Urinal screen, 18″ wide, ceiling braced, plastic laminate		8	2		293	68.50		361.50	425
5400	Floor mounted, headrail braced		8	2		355	68.50		423.50	495
5900	Pilaster, flush, plastic laminate		10	1.600		360	55		415	480
6400	Post braced, plastic laminate		10	1.600		360	55		415	480
6700	Wall hung, bracket supported									
6900	Plastic laminate on particle board	2 Carp	10	1.600	Ea.	167	55		222	269
7450	Flange supported									
7500	Plastic laminate on particle board	2 Carp	10	1.600	Ea.	395	55		450	520

10170 | Plastic Toilet Compartments

		CREW	DAILY OUTPUT	LABOR-HOURS	UNIT	MAT.	LABOR	EQUIP.	TOTAL	TOTAL INCL O&P
100 0010	**TOILET PARTITIONS, PLASTIC**									
0110	Cubicles, ceiling hung									
0250	Phenolic	2 Carp	4	4	Ea.	875	137		1,012	1,175
0600	For handicap units, incl. 52″ grab bars, add				″	320			320	355
0900	Floor and ceiling anchored									
1050	Phenolic	2 Carp	5	3.200	Ea.	920	110		1,030	1,200
1400	For handicap units, incl. 52″ grab bars, add				″	287			287	315
1610	Floor mounted									
1750	Phenolic	2 Carp	7	2.286	Ea.	990	78.50		1,068.50	1,225
2100	For handicap units, incl. 52″ grab bars, add					287			287	315
2200	For juvenile units, deduct					39.50			39.50	43.50
2450	Floor mounted, headrail braced									
2550	Phenolic	2 Carp	6	2.667	Ea.	890	91.50		981.50	1,125

10180 | Stone Toilet Compartments

		CREW	DAILY OUTPUT	LABOR-HOURS	UNIT	MAT.	LABOR	EQUIP.	TOTAL	TOTAL INCL O&P
100 0010	**TOILET PARTITIONS, STONE**									
0100	Cubicles, ceiling hung, marble	2 Marb	2	8	Ea.	1,475	266		1,741	2,025
0600	For handicap units, incl. 52″ grab bars, add					320			320	355
0800	Floor & ceiling anchored, marble	2 Marb	2.50	6.400		1,600	212		1,812	2,100
1400	For handicap units, incl. 52″ grab bars, add					287			287	315
1600	Floor mounted, marble	2 Marb	3	5.333		945	177		1,122	1,325
2400	Floor mounted, headrail braced, marble	″	3	5.333		905	177		1,082	1,275
2900	For handicap units, incl. 52″ grab bars, add					287			287	315
4100	Entrance screen, floor mounted marble, 58″ high, 48″ wide	2 Marb	9	1.778		595	59		654	745
4600	Urinal screen, 18″ wide, ceiling braced, marble	D-1	6	2.667		595	83		678	780
5100	Floor mounted, head rail braced									
5200	Marble	D-1	6	2.667	Ea.	520	83		603	700
5700	Pilaster, flush, marble		9	1.778		660	55.50		715.50	815
6200	Post braced, marble		9	1.778		655	55.50		710.50	805

10185 | Shower/Dressing Compartments

		CREW	DAILY OUTPUT	LABOR-HOURS	UNIT	MAT.	LABOR	EQUIP.	TOTAL	TOTAL INCL O&P
100 0010	**PARTITIONS, SHOWER** Floor mounted, no plumbing									
0100	Cabinet, incl. base, no door, painted steel, 1″ thick walls	2 Shee	5	3.200	Ea.	730	129		859	1,000
0300	With door, fiberglass		4.50	3.556		590	143		733	870
0600	Galvanized and painted steel, 1″ thick walls		5	3.200		770	129		899	1,050
0800	Stall, 1″ thick wall, no base, enameled steel		5	3.200		830	129		959	1,125
1200	Stainless steel		5	3.200		1,325	129		1,454	1,675
1400	For double entry type, no doors, deduct					10%				
1500	Circular fiberglass, cabinet 36″ diameter,	2 Shee	4	4		605	161		766	910
1700	One piece, 36″ diameter, less door		4	4		510	161		671	805
1800	With door		3.50	4.571		840	184		1,024	1,200
2000	Curved shell shower, no door needed		3	5.333		730	214		944	1,125
2300	For fiberglass seat, add to both above					104			104	114

10185 | Shower/Dressing Compartments

		CREW	DAILY OUTPUT	LABOR-HOURS	UNIT	MAT.	LABOR	EQUIP.	TOTAL	TOTAL INCL O&P	
100	2400	Glass stalls, with doors, no receptors, chrome on brass	2 Shee	3	5.333	Ea.	1,200	214		1,414	1,650	**100**
	2700	Anodized aluminum	"	4	4		830	161		991	1,150	
	2900	Marble shower stall, stock design, with shower door	2 Marb	1.20	13.333		1,850	445		2,295	2,700	
	3000	With curtain		1.30	12.308		1,625	410		2,035	2,425	
	3200	Receptors, precast terrazzo, 32" x 32"		14	1.143		225	38		263	305	
	3300	48" x 34"		9.50	1.684		385	56		441	510	
	3500	Plastic, simulated terrazzo receptor, 32" x 32"		14	1.143		95	38		133	163	
	3600	32" x 48"		12	1.333		140	44.50		184.50	222	
	3800	Precast concrete, colors, 32" x 32"		14	1.143		188	38		226	265	
	3900	48" x 48"	▼	8	2		200	66.50		266.50	320	
	4100	Shower doors, economy plastic, 24" wide	1 Shee	9	.889		104	35.50		139.50	169	
	4200	Tempered glass door, economy		8	1		182	40		222	262	
	4400	Folding, tempered glass, aluminum frame		6	1.333		315	53.50		368.50	430	
	4500	Sliding, tempered glass, 48" opening		6	1.333		207	53.50		260.50	310	
	4700	Deluxe, tempered glass, chrome on brass frame, minimum		8	1		257	40		297	345	
	4800	Maximum		1	8		700	320		1,020	1,275	
	4850	On anodized aluminum frame, minimum		2	4		122	161		283	380	
	4900	Maximum	▼	1	8	▼	410	320		730	950	
	5100	Shower enclosure, tempered glass, anodized alum. frame										
	5120	2 panel & door, corner unit, 32" x 32"	1 Shee	2	4	Ea.	410	161		571	700	
	5140	Neo-angle corner unit, 16" x 24" x 16"	"	2	4		755	161		916	1,075	
	5200	Shower surround, 3 wall, polypropylene, 32" x 32"	1 Carp	4	2		276	68.50		344.50	410	
	5220	PVC, 32" x 32"		4	2		259	68.50		327.50	390	
	5240	Fiberglass		4	2		300	68.50		368.50	435	
	5250	2 wall, polypropylene, 32" x 32"		4	2		215	68.50		283.50	345	
	5270	PVC		4	2		267	68.50		335.50	400	
	5290	Fiberglass	▼	4	2		297	68.50		365.50	430	
	5300	Tub doors, tempered glass & frame, minimum	1 Shee	8	1		175	40		215	255	
	5400	Maximum		6	1.333		405	53.50		458.50	530	
	5600	Chrome plated, brass frame, minimum		8	1		229	40		269	315	
	5700	Maximum		6	1.333		455	53.50		508.50	585	
	5900	Tub/shower enclosure, temp. glass, alum. frame, minimum		2	4		310	161		471	585	
	6200	Maximum		1.50	5.333		605	214		819	995	
	6500	On chrome-plated brass frame, minimum		2	4		430	161		591	720	
	6600	Maximum	▼	1.50	5.333		885	214		1,099	1,300	
	6800	Tub surround, 3 wall, polypropylene	1 Carp	4	2		174	68.50		242.50	298	
	6900	PVC		4	2		264	68.50		332.50	395	
	7000	Fiberglass, minimum		4	2		294	68.50		362.50	430	
	7100	Maximum	▼	3	2.667	▼	505	91.50		596.50	695	

10520 | Fire Protection Specialties

10525 | Fire Prot. Specialties

		CREW	DAILY OUTPUT	LABOR-HOURS	UNIT	MAT.	LABOR	EQUIP.	TOTAL	TOTAL INCL O&P	
200	0010	**FIRE EQUIPMENT CABINETS** Not equipped, 20 ga. steel box,										**200**
	0040	recessed, D.S. glass in door, box size given [D4020 310]										
	1000	Portable extinguisher, single, 8" x 12" x 27", alum. door & frame	Q-12	8	2	Ea.	84.50	73.50		158	204	
	1100	Steel door and frame		8	2		62.50	73.50		136	180	
	1200	Stainless steel door and frame		8	2		129	73.50		202.50	253	
	2000	Portable extinguisher, large, 8" x 12" x 36", alum. door & frame	▼	8	2		144	73.50		217.50	269	

Important: See the Reference Section for critical supporting data - Reference Nos., Crews, & City Cost Indexes

10525 | Fire Prot. Specialties

			CREW	DAILY OUTPUT	LABOR-HOURS	UNIT	2005 BARE COSTS MAT.	LABOR	EQUIP.	TOTAL	TOTAL INCL O&P
200	2100	Steel door and frame	Q-12	8	2	Ea.	93.50	73.50		167	214
	2200	Stainless steel door and frame		8	2		148	73.50		221.50	274
	2500	8" x 16" x 38", aluminum door & frame		8	2		162	73.50		235.50	289
	2600	Steel door and frame		8	2		145	73.50		218.50	270
	3000	Hose rack assy., 1-1/2" valve & 100' hose, 24" x 40" x 5-1/2"									
	3100	Aluminum door and frame	Q-12	6	2.667	Ea.	206	98		304	375
	3200	Steel door and frame		6	2.667		134	98		232	296
	3300	Stainless steel door and frame		6	2.667		268	98		366	445
	4000	Hose rack assy., 2-1/2" x 1-1/2" valve, 100' hose, 24" x 40" x 8"									
	4100	Aluminum door and frame	Q-12	6	2.667	Ea.	208	98		306	375
	4200	Steel door and frame		6	2.667		141	98		239	305
	4300	Stainless steel door and frame		6	2.667		272	98		370	445
	5000	Hose rack assy., 2-1/2" x 1-1/2" valve, 100' hose									
	5010	and extinguisher, 30" x 40" x 8"									
	5100	Aluminum door and frame	Q-12	5	3.200	Ea.	265	117		382	470
	5200	Steel door and frame		5	3.200		154	117		271	345
	5300	Stainless steel door and frame		5	3.200		291	117		408	495
	6000	Hose rack assy., 1-1/2" valve, 100' hose									
	6010	and 2-1/2" FD valve, 24" x 44" x 8"									
	6100	Aluminum door and frame	Q-12	5	3.200	Ea.	236	117		353	435
	6200	Steel door and frame		5	3.200		151	117		268	345
	6300	Stainless steel door and frame		5	3.200		305	117		422	515
	7000	Hose rack assy., 1-1/2" valve & 100' hose, 2-1/2" FD valve									
	7010	and extinguisher, 30" x 44" x 8"									
	7100	Aluminum door and frame	Q-12	5	3.200	Ea.	279	117		396	480
	7200	Steel door and frame		5	3.200		283	117		400	485
	7300	Stainless steel door and frame		5	3.200		370	117		487	585
	8000	Valve cabinet for 2-1/2" FD angle valve, 18" x 18" x 8"									
	8100	Aluminum door and frame	Q-12	12	1.333	Ea.	92	49		141	175
	8200	Steel door and frame		12	1.333		69.50	49		118.50	151
	8300	Stainless steel door and frame		12	1.333		121	49		170	207
300	0010	**FIRE EXTINGUISHERS**									
	0120	CO2, portable with swivel horn, 5 lb.				Ea.	130			130	143
	0140	With hose and "H" horn, 10 lb.					160			160	176
	0160	15 lb.					190			190	209
	0180	20 lb.					220			220	242
	0360	Wheeled type, cart mounted, 50 lb.					1,200			1,200	1,325
	0400	100 lb.					1,525			1,525	1,675
	1000	Dry chemical, pressurized									
	1040	Standard type, portable, painted, 2-1/2 lb.				Ea.	27.50			27.50	30.50
	1060	5 lb.					60			60	66
	1080	10 lb.					80			80	88
	1100	20 lb.					110			110	121
	1120	30 lb.					157			157	173
	1300	Standard type, wheeled, 150 lb.					1,600			1,600	1,750
	2000	ABC all purpose type, portable, 2-1/2 lb.					30			30	33
	2060	5 lb.					45			45	49.50
	2080	9-1/2 lb.					68			68	75
	2100	20 lb.					98			98	108
	2300	Wheeled, 45 lb.					650			650	715
	2360	150 lb.					1,600			1,600	1,750
	3000	Dry chemical, outside cartridge to -65°F, painted, 9 lb.					200			200	220
	3060	26 lb.					250			250	275
	5000	Pressurized water, 2-1/2 gallon, stainless steel					85			85	93.50
	5060	With anti-freeze					135			135	149

Reference notes in table: D4020 330, D4020 410, R10520 -310, R10520 -320

		10525	Fire Prot. Specialties	CREW	DAILY OUTPUT	LABOR-HOURS	UNIT	2005 BARE COSTS				TOTAL INCL O&P	
								MAT.	LABOR	EQUIP.	TOTAL		
300	9400		Installation of extinguishers, 12 or more, on wood	1 Carp	30	.267	Ea.		9.15		9.15	14.25	300
	9420		On masonry or concrete	"	15	.533	↓		18.25		18.25	28.50	

		10810	Toilet Accessories	CREW	DAILY OUTPUT	LABOR-HOURS	UNIT	2005 BARE COSTS				TOTAL INCL O&P	
								MAT.	LABOR	EQUIP.	TOTAL		
100	0010		COMMERCIAL TOILET ACCESSORIES										100
	0200		Curtain rod, stainless steel, 5' long, 1" diameter	1 Carp	13	.615	Ea.	30.50	21		51.50	66.50	
	0300		1-1/4" diameter		13	.615		29	21		50	64.50	
	0400		Diaper changing station, horizontal, wall mounted, plastic	↓	10	.800	↓	174	27.50		201.50	235	
	0500		Dispenser units, combined soap & towel dispensers,										
	0510		mirror and shelf, flush mounted	1 Carp	10	.800	Ea.	330	27.50		357.50	410	
	0600		Towel dispenser and waste receptacle,										
	0610		18 gallon capacity	1 Carp	10	.800	Ea.	256	27.50		283.50	325	
	0800		Grab bar, straight, 1-1/4" diameter, stainless steel, 18" long		24	.333		18.75	11.40		30.15	38.50	
	0900		24" long		23	.348		21	11.90		32.90	41.50	
	1000		30" long		22	.364		21	12.45		33.45	43	
	1100		36" long		20	.400		19.75	13.70		33.45	43	
	1105		42" long		20	.400		20.50	13.70		34.20	44	
	1200		1-1/2" diameter, 24" long		23	.348		50.50	11.90		62.40	74	
	1300		36" long		20	.400		57	13.70		70.70	84	
	1310		42" long		18	.444		48	15.20		63.20	76.50	
	1500		Tub bar, 1-1/4" diameter, 24" x 36"		14	.571		87.50	19.55		107.05	127	
	1600		Plus vertical arm		12	.667		86.50	23		109.50	131	
	1900		End tub bar, 1" diameter, 90° angle, 16" x 32"		12	.667		118	23		141	166	
	2300		Hand dryer, surface mounted, electric, 115 volt, 20 amp		4	2		485	68.50		553.50	635	
	2400		230 volt, 10 amp		4	2		485	68.50		553.50	635	
	2600		Hat and coat strip, stainless steel, 4 hook, 36" long		24	.333		48	11.40		59.40	71	
	2700		6 hook, 60" long		20	.400		84	13.70		97.70	114	
	3000		Mirror, with stainless steel 3/4" square frame, 18" x 24"		20	.400		65.50	13.70		79.20	93.50	
	3100		36" x 24"		15	.533		104	18.25		122.25	143	
	3200		48" x 24"		10	.800		159	27.50		186.50	217	
	3300		72" x 24"		6	1.333		196	45.50		241.50	287	
	3500		With 5" stainless steel shelf, 18" x 24"		20	.400		114	13.70		127.70	147	
	3600		36" x 24"		15	.533		161	18.25		179.25	206	
	3700		48" x 24"		10	.800		171	27.50		198.50	231	
	3800		72" x 24"		6	1.333		286	45.50		331.50	385	
	4100		Mop holder strip, stainless steel, 5 holders, 48" long		20	.400		80.50	13.70		94.20	110	
	4200		Napkin/tampon dispenser, recessed		15	.533		310	18.25		328.25	375	
	4300		Robe hook, single, regular		36	.222		4.95	7.60		12.55	17.30	
	4400		Heavy duty, concealed mounting		36	.222		11.10	7.60		18.70	24	
	4600		Soap dispenser, chrome, surface mounted, liquid		20	.400		44	13.70		57.70	69.50	
	4700		Powder		20	.400		42	13.70		55.70	67.50	
	5000		Recessed stainless steel, liquid		10	.800		143	27.50		170.50	200	
	5100		Powder		10	.800		174	27.50		201.50	234	
	5300		Soap tank, stainless steel, 1 gallon		10	.800		159	27.50		186.50	218	
	5400		5 gallon		5	1.600		219	55		274	325	
	5600		Shelf, stainless steel, 5" wide, 18 ga., 24" long		24	.333		41.50	11.40		52.90	64	
	5700		48" long		16	.500		76.50	17.15		93.65	111	
	5800		8" wide shelf, 18 ga., 24" long	↓	22	.364	↓	49	12.45		61.45	73.50	

Important: See the Reference Section for critical supporting data - Reference Nos., Crews, & City Cost Indexes

			DAILY	LABOR-		2005 BARE COSTS				TOTAL	
10810	**Toilet Accessories**	CREW	OUTPUT	HOURS	UNIT	MAT.	LABOR	EQUIP.	TOTAL	INCL O&P	
100	5900	48" long	1 Carp	14	.571	Ea.	108	19.55		127.55	150	100
	6000	Toilet seat cover dispenser, stainless steel, recessed		20	.400		102	13.70		115.70	134	
	6050	Surface mounted		15	.533		26	18.25		44.25	57	
	6100	Toilet tissue dispenser, surface mounted, SS, single roll		30	.267		10.80	9.15		19.95	26	
	6200	Double roll		24	.333		15.25	11.40		26.65	34.50	
	6400	Towel bar, stainless steel, 18" long		23	.348		35	11.90		46.90	57	
	6500	30" long		21	.381		56.50	13.05		69.55	83	
	6700	Towel dispenser, stainless steel, surface mounted		16	.500		41	17.15		58.15	72	
	6800	Flush mounted, recessed		10	.800		172	27.50		199.50	232	
	7000	Towel holder, hotel type, 2 guest size		20	.400		15.40	13.70		29.10	38.50	
	7200	Towel shelf, stainless steel, 24" long, 8" wide		20	.400		48.50	13.70		62.20	75	
	7400	Tumbler holder, tumbler only		30	.267		25	9.15		34.15	42	
	7500	Soap, tumbler & toothbrush		30	.267		22.50	9.15		31.65	39.50	
	7700	Wall urn ash receiver, surface mount, 11" long		12	.667		113	23		136	160	
	7800	7-1/2", long		18	.444		83	15.20		98.20	115	
	8000	Waste receptacles, stainless steel, with top, 13 gallon		10	.800		187	27.50		214.50	248	
	8100	36 gallon	▼	8	1	▼	345	34.50		379.50	435	

10820	**Bath Accessories**										
400	0010	**MEDICINE CABINETS** With mirror, st. st. frame, 16" x 22", unlighted	1 Carp	14	.571	Ea.	69.50	19.55		89.05	107	400
	0100	Wood frame		14	.571		96.50	19.55		116.05	137	
	0300	Sliding mirror doors, 20" x 16" x 4-3/4", unlighted		7	1.143		86.50	39		125.50	156	
	0400	24" x 19" x 8-1/2", lighted		5	1.600		136	55		191	236	
	0600	Triple door, 30" x 32", unlighted, plywood body		7	1.143		214	39		253	297	
	0700	Steel body		7	1.143		282	39		321	370	
	0900	Oak door, wood body, beveled mirror, single door		7	1.143		125	39		164	199	
	1000	Double door		6	1.333		320	45.50		365.50	420	
	1200	Hotel cabinets, stainless, with lower shelf, unlighted		10	.800		173	27.50		200.50	233	
	1300	Lighted	▼	5	1.600	▼	256	55		311	370	

For information about Means Estimating Seminars, see yellow pages 12 and 13 in back of book

Division Notes

	CREW	DAILY OUTPUT	LABOR-HOURS	UNIT	2005 BARE COSTS				TOTAL INCL O&P
					MAT.	LABOR	EQUIP.	TOTAL	

Division 11
Equipment

Estimating Tips

General

- The items in this division are usually priced per square foot or each. Many of these items are purchased by the owner for installation by the contractor. Check the specifications for responsibilities, and include time for receiving, storage, installation and mechanical and electrical hook-ups in the appropriate divisions.

- Many items in Division 11 require some type of support system that is not usually furnished with the item. Examples of these systems include blocking for the attachment of casework and support angles for ceiling hung projection screens. The required blocking or supports must be added to the estimate in the appropriate division.

- Some items in Division 11 may require assembly or electrical hook-ups. Verify the amount of assembly required or the need for a hard electrical connection and add the appropriate costs.

Reference Numbers

Reference numbers are shown in bold squares at the beginning of some major classifications. These numbers refer to related items in the Reference Section. The reference information may be an estimating procedure, an alternate pricing method or technical information.

Note: Not all subdivisions listed here necessarily appear in this publication.

11010 | Maintenance Equipment

11013	Floor/Wall Cleaning Equipment	CREW	DAILY OUTPUT	LABOR-HOURS	UNIT	2005 BARE COSTS				TOTAL INCL O&P	
						MAT.	LABOR	EQUIP.	TOTAL		
800	0010 **VACUUM CLEANING**										800
	0020 Central, 3 inlet, residential	1 Skwk	.90	8.889	Total	610	310		920	1,150	
	0200 Commercial		.70	11.429		1,150	400		1,550	1,900	
	0400 5 inlet system, residential		.50	16		925	560		1,485	1,900	
	0600 7 inlet system, commercial		.40	20		1,025	695		1,720	2,225	
	0800 9 inlet system, residential		.30	26.667		1,300	930		2,230	2,875	
	4010 Rule of thumb: First 1200 S.F., installed									1,125	
	4020 For each additional S.F., add				S.F.					.18	

11040 | Ecclesiastical Equipment

11041	Ecclesiastical Equipment	CREW	DAILY OUTPUT	LABOR-HOURS	UNIT	2005 BARE COSTS				TOTAL INCL O&P	
						MAT.	LABOR	EQUIP.	TOTAL		
250	0010 **CHURCH EQUIPMENT**										250
	0020 Altar, wood, custom design, plain	1 Carp	1.40	5.714	Ea.	1,825	196		2,021	2,300	
	0050 Deluxe	"	.20	40	"	8,800	1,375		10,175	11,800	
	0150 Baptistry, fiberglass, 3'-6" deep, x 13'-7" long,										
	0160 steps at both ends, incl. plumbing, minimum	L-8	1	20	Ea.	2,775	710		3,485	4,150	
	0200 Maximum	"	.70	28.571		5,400	1,025		6,425	7,525	
	0250 Add for filter, heater and lights					1,175			1,175	1,300	

11100 | Mercantile Equipment

11102	Barber Shop Equipment	CREW	DAILY OUTPUT	LABOR-HOURS	UNIT	2005 BARE COSTS				TOTAL INCL O&P	
						MAT.	LABOR	EQUIP.	TOTAL		
150	0010 **BARBER EQUIPMENT**										150
	0020 Chair, hydraulic, movable, minimum	1 Carp	24	.333	Ea.	420	11.40		431.40	480	
	0050 Maximum	"	16	.500		2,750	17.15		2,767.15	3,050	
	0500 Sink, hair washing basin, rough plumbing not incl.	1 Plum	8	1		279	41		320	365	
	1000 Sterilizer, liquid solution for tools					128			128	140	

11110 | Commercial Laundry & Dry Cleaning Equipment

11119	Laundry Cleaning	CREW	DAILY OUTPUT	LABOR-HOURS	UNIT	2005 BARE COSTS				TOTAL INCL O&P	
						MAT.	LABOR	EQUIP.	TOTAL		
450	0010 **LAUNDRY EQUIPMENT** Not incl. rough-in										450
	0500 Dryers, gas fired residential, 16 lb. capacity, average	1 Plum	3	2.667	Ea.	565	109		674	785	
	1000 Commercial, 30 lb. capacity, coin operated, single		3	2.667		2,575	109		2,684	3,000	
	1100 Double stacked		2	4		5,525	163		5,688	6,325	

Important: See the Reference Section for critical supporting data - Reference Nos., Crews, & City Cost Indexes

11110 | Commercial Laundry & Dry Cleaning Equipment

11119	Laundry Cleaning	CREW	DAILY OUTPUT	LABOR-HOURS	UNIT	2005 BARE COSTS				TOTAL INCL O&P
						MAT.	LABOR	EQUIP.	TOTAL	
450	1500 Industrial, 30 lb. capacity	1 Plum	2	4	Ea.	2,200	163		2,363	2,650
	1600 50 lb. capacity		1.70	4.706		2,875	192		3,067	3,450
	5000 Washers, residential, 4 cycle, average		3	2.667		645	109		754	875
	5300 Commercial, coin operated, average	↓	3	2.667		1,050	109		1,159	1,325
	6000 Combination washer/extractor, 20 lb. capacity	L-6	1.50	8		3,725	325		4,050	4,600
	6100 30 lb. capacity		.80	15		7,200	610		7,810	8,850
	6200 50 lb. capacity		.68	17.647		8,950	720		9,670	10,900
	6300 75 lb. capacity		.30	40		18,500	1,625		20,125	22,800
	6350 125 lb. capacity	↓	.16	75	↓	22,400	3,050		25,450	29,300

11140 | Vehicle Service Equipment

11141	Service Station Equipment	CREW	DAILY OUTPUT	LABOR-HOURS	UNIT	2005 BARE COSTS				TOTAL INCL O&P
						MAT.	LABOR	EQUIP.	TOTAL	
100	0010 **COMPRESSED AIR EQUIPMENT**									
	0030 Compressors, electric, 1-1/2 H.P., standard controls	L-4	1.50	16	Ea.	296	510		806	1,125
	0550 Dual controls		1.50	16		475	510		985	1,325
	0600 5 H.P., 115/230 volt, standard controls		1	24		1,700	760		2,460	3,050
	0650 Dual controls	↓	1	24	↓	1,800	760		2,560	3,175
200	0010 **FUEL DISPENSING EQUIPMENT**									
	1100 Product dispenser with vapor recovery for 6 nozzles, installed, not									
	1110 including piping to storage tanks				Ea.	18,000			18,000	19,800
300	0010 **LUBRICATION EQUIPMENT**									
	3000 Lube equipment, 3 reel type, with pumps, not including piping	L-4	.50	48	Set	6,825	1,525		8,350	9,875
	3700 Pump lubrication, pneumatic, not incl. air compressor									
	3710 Oil/gear lube	Q-1	9.60	1.667	Ea.	615	61.50		676.50	770
	3720 Grease	"	9.60	1.667	"	640	61.50		701.50	795
400	0010 **SPRAY PAINTING EQUIPMENT**									
	4000 Spray painting booth, 26' long, complete	L-4	.40	60	Ea.	12,700	1,900		14,600	16,900

11190 | Detention Equipment

11191	Detention Equipment	CREW	DAILY OUTPUT	LABOR-HOURS	UNIT	2005 BARE COSTS				TOTAL INCL O&P
						MAT.	LABOR	EQUIP.	TOTAL	
150	0010 **DETENTION EQUIPMENT**									
	0500 Bar front, rolling, 7/8" bars, 4" O.C., 7' high, 5' wide, with hardware	E-4	2	16	Ea.	4,600	620	40.50	5,260.50	6,225
	2000 Cells, prefab., 5' to 6' wide, 7' to 8' high, 7' to 8' deep,									
	2010 bar front, cot, not incl. plumbing	E-4	1.50	21.333	Ea.	7,600	825	54	8,479	9,900
	3000 Toilet apparatus including wash basin, average	L-8	1.50	13.333	"	2,850	475		3,325	3,850

11310 | Sewage & Sludge Pumps

			CREW	DAILY OUTPUT	LABOR-HOURS	UNIT	2005 BARE COSTS				TOTAL INCL O&P
							MAT.	LABOR	EQUIP.	TOTAL	
350	0010	**PUMPS, PNEUMATIC EJECTOR**										**350**
	0020	With steel receiver, level controls, inlet/outlet gate/check valves									
	0030	Cross connect. not incl. compressor, fittings or piping									
	0040	Duplex									
	0050	30 GPM	Q-2	1.70	14.118	Ea.	14,400	540		14,940	16,600
	0060	50 GPM		1.56	15.385		20,300	585		20,885	23,200
	0070	100 GPM		1.30	18.462		24,500	705		25,205	28,100
	0080	150 GPM		1.10	21.818		47,300	830		48,130	53,500
	0090	200 GPM		.85	28.235		60,000	1,075		61,075	67,500
	0100	250 GPM	Q-3	.91	35.165		77,500	1,375		78,875	87,000
	0110	300 GPM	"	.57	56.140		94,500	2,175		96,675	107,500
700	0010	**SEWAGE PUMPING STATIONS** Prefabricated steel, concrete										**700**
	0020	or fiberglass, 200 GPM	C-17D	.17	494	Total	34,100	17,400	2,925	54,425	68,000
	0200	1,000 GPM		.07	1,200		48,100	42,300	7,100	97,500	126,500
	0500	Add for generator unit, 200 GPM, steel		.34	247		24,600	8,725	1,475	34,800	42,200
	0600	Concrete		.51	164		15,600	5,800	975	22,375	27,300
	1000	Add for generator unit, 1,000 GPM, steel		.30	280		26,200	9,875	1,650	37,725	45,900
	1200	Concrete		.38	221		22,100	7,800	1,300	31,200	37,900
	1500	For drilled water well, if required, add	B-23	.50	80		6,375	2,175	6,700	15,250	17,800

11390 | Pkg Sewage Treat Plants

			CREW	DAILY OUTPUT	LABOR-HOURS	UNIT	2005 BARE COSTS				TOTAL INCL O&P
							MAT.	LABOR	EQUIP.	TOTAL	
200	0010	**SEWAGE TREATMENT PLANTS,** not incl fencing or external piping										**200**
	0020	Steel packaged, blown air aeration plants									
	0100	1,000 GPD				Gal.				16.85	19.40
	0200	5,000 GPD								11.25	12.90
	0300	15,000 GPD								6.20	7.10
	0400	30,000 GPD								5.85	6.70
	0500	50,000 GPD								4.50	5.05
	0600	100,000 GPD								3.95	4.50
	0700	200,000 GPD								2.81	3.23
	0800	500,000 GPD								2.75	3.15
	1000	Concrete, extended aeration, primary and secondary treatment									
	1010	10,000 GPD				Gal.				12.35	14.65
	1100	30,000 GPD								6.20	7.15
	1200	50,000 GPD								5.05	5.85
	1400	100,000 GPD								3.94	4.55
	1500	500,000 GPD								2,081	3.26
	1700	Municipal wastewater treatment facility									
	1720	1.0 MGD				Gal.				4.83	5.55
	1740	1.5 MGD								4.78	5.50
	1760	2.0 MGD								4.10	4.73
	1780	3.0 MGD								3.20	3.71
	1800	5.0 MGD								2.92	3.37
	2000	Holding tank system, not incl. excavation or backfill									
	2010	Recirculating chemical water closet	2 Plum	4	4	Ea.	750	163		913	1,075
	2100	For voltage converter, add	"	16	1		199	41		240	281
	2200	For high level alarm, add	1 Plum	7.80	1.026		114	42		156	188
900	0010	**WASTEWATER TREATMENT SYSTEM** Fiberglass, 1,000 gallon	B-21	1.29	21.705	Ea.	2,925	670	123	3,718	4,400	**900**
	0100	1,500 gallon	"	1.03	27.184	"	6,400	840	154	7,394	8,525

Important: See the Reference Section for critical supporting data - Reference Nos., Crews, & City Cost Indexes

		11405	Food Storage Equipment	CREW	DAILY OUTPUT	LABOR-HOURS	UNIT	2005 BARE COSTS MAT.	LABOR	EQUIP.	TOTAL	TOTAL INCL O&P	
110	0010		**FOOD STORAGE EQUIPMENT**										11
	2350		Cooler, reach-in, beverage, 6' long	Q-1	6	2.667	Ea.	3,550	98		3,648	4,075	
	4300		Freezers, reach-in, 44 C.F.		4	4		8,125	147		8,272	9,175	
	4500		68 C.F.		3	5.333		9,150	196		9,346	10,400	
	8300		Refrigerators, reach-in type, 44 C.F.		5	3.200		4,950	118		5,068	5,625	
	8310		With glass doors, 68 C.F.	↓	4	4	↓	7,150	147		7,297	8,075	

		11420	Food Cooking Equipment										
110	0010		**COOKING EQUIPMENT**										11
	0020		Bake oven, gas, one section	Q-1	8	2	Ea.	4,225	73.50		4,298.50	4,750	
	0300		Two sections		7	2.286		8,625	84		8,709	9,625	
	0600		Three sections	↓	6	2.667		12,200	98		12,298	13,600	
	0900		Electric convection, single deck	L-7	4	7		4,525	231		4,756	5,325	
	6350		Kettle, w/steam jacket, tilting, w/positive lock, SS, 20 gallons		7	4		5,350	132		5,482	6,100	
	6600		60 gallons	↓	6	4.667	↓	6,825	154		6,979	7,775	

		11425	Hood and Ventilation Equipment										
110	0010		**HOOD & VENTILATION EQUIPMENT**										11
	7950		Hood fire protection system, minimum	Q-1	3	5.333	Ea.	3,225	196		3,421	3,850	
	8050		Maximum	"	1	16	"	24,100	590		24,690	27,500	

		11435	Ice Machines										
110	0010		**ICE MACHINES**										11
	5800		Ice cube maker, 50 pounds per day	Q-1	6	2.667	Ea.	1,325	98		1,423	1,600	
	6050		500 pounds per day	"	4	4	"	2,875	147		3,022	3,400	

		11440	Cleaning and Disposal Equipment										
110	0010		**CLEANING & DISPOSAL EQUIPMENT**										11
	2700		Dishwasher, commercial, rack type										
	2720		10 to 12 racks per hour	Q-1	3.20	5	Ea.	3,650	184		3,834	4,300	
	2750		Semi-automatic 38 to 50 racks per hour	"	1.30	12.308		6,550	455		7,005	7,875	
	2800		Automatic, 190 to 230 racks per hour	L-6	.35	34.286		8,200	1,400		9,600	11,100	
	2820		235 to 275 racks per hour		.25	48		19,000	1,950		20,950	23,800	
	2840		8,750 to 12,500 dishes per hour	↓	.10	120	↓	41,200	4,900		46,100	52,500	

11450 | Residential Equipment

		11454	Residential Appliances	CREW	DAILY OUTPUT	LABOR-HOURS	UNIT	2005 BARE COSTS MAT.	LABOR	EQUIP.	TOTAL	TOTAL INCL O&P	
500	0010		**RESIDENTIAL APPLIANCES**										50
	0020		Cooking range, 30" free standing, 1 oven, minimum	2 Clab	10	1.600	Ea.	241	42.50		283.50	330	
	0050		Maximum		4	4		1,475	107		1,582	1,800	
	0150		2 oven, minimum		10	1.600		1,475	42.50		1,517.50	1,700	
	0200		Maximum	↓	10	1.600		1,550	42.50		1,592.50	1,775	
	2450		Dehumidifier, portable, automatic, 15 pint					149			149	164	
	2550		40 pint					166			166	183	
	2750		Dishwasher, built-in, 2 cycles, minimum	L-1	4	4	↓	248	163		411	515	

11454	Residential Appliances	CREW	DAILY OUTPUT	LABOR-HOURS	UNIT	2005 BARE COSTS				TOTAL INCL O&P	
						MAT.	LABOR	EQUIP.	TOTAL		
500 2800	Maximum	L-1	2	8	Ea.	289	325		614	810	**500**
2950	4 or more cycles, minimum		4	4		262	163		425	530	
2960	Average		4	4		350	163		513	630	
3000	Maximum	↓	2	8		540	325		865	1,075	
3200	Dryer, automatic, minimum	L-2	3	5.333		282	159		441	560	
3250	Maximum	"	2	8		790	239		1,029	1,250	
3300	Garbage disposal, sink type, minimum	L-1	10	1.600		42.50	65.50		108	144	
3350	Maximum	"	10	1.600		145	65.50		210.50	258	
3550	Heater, electric, built-in, 1250 watt, ceiling type, minimum	1 Elec	4	2		72	81.50		153.50	201	
3600	Maximum		3	2.667		118	109		227	292	
3700	Wall type, minimum		4	2		102	81.50		183.50	234	
3750	Maximum		3	2.667		136	109		245	310	
3900	1500 watt wall type, with blower		4	2		127	81.50		208.50	260	
3950	3000 watt	↓	3	2.667		258	109		367	445	
4150	Hood for range, 2 speed, vented, 30" wide, minimum	L-3	5	3.200		37.50	120		157.50	224	
4200	Maximum		3	5.333		595	199		794	960	
4300	42" wide, minimum		5	3.200		225	120		345	430	
4330	Custom		5	3.200		625	120		745	870	
4350	Maximum	↓	3	5.333		760	199		959	1,150	
4500	For ventless hood, 2 speed, add					15			15	16.50	
4650	For vented 1 speed, deduct from maximum					39			39	43	
4850	Humidifier, portable, 8 gallons per day					149			149	164	
5000	15 gallons per day					179			179	197	
5200	Icemaker, automatic, 20 lb. per day	1 Plum	7	1.143		360	46.50		406.50	465	
5350	51 lb. per day		2	4		1,000	163		1,163	1,350	
6400	Sump pump cellar drainer, pedestal, 1/3 H.P., molded PVC base		3	2.667		87	109		196	260	
6450	Solid brass	↓	2	4	↓	179	163		342	445	
6460	Sump pump, see also division 15440-940										
6650	Washing machine, automatic, minimum	1 Plum	3	2.667	Ea.	279	109		388	470	
6700	Maximum	"	1	8		1,025	325		1,350	1,625	
6900	Water heater, electric, glass lined, 30 gallon, minimum	L-1	5	3.200		252	131		383	470	
6950	Maximum		3	5.333		350	218		568	710	
7100	80 gallon, minimum		2	8		485	325		810	1,025	
7150	Maximum	↓	1	16		670	655		1,325	1,725	
7180	Water heater, gas, glass lined, 30 gallon, minimum	2 Plum	5	3.200		345	131		476	575	
7220	Maximum		3	5.333		480	218		698	855	
7260	50 gallon, minimum		2.50	6.400		480	261		741	925	
7300	Maximum		1.50	10.667	↓	665	435		1,100	1,400	
7310	Water heater, see also division 15480-200										
7350	Water softener, automatic, to 30 grains per gallon	2 Plum	5	3.200	Ea.	430	131		561	670	
7400	To 100 grains per gallon	"	4	4		600	163		763	905	
7450	Vent kits for dryers	1 Carp	10	.800	↓	12.65	27.50		40.15	56.50	

11460	Unit Kitchens										
100 0010	**UNIT KITCHENS**										**100**
1500	Combination range, refrigerator and sink, 30" wide, minimum	L-1	2	8	Ea.	730	325		1,055	1,300	
1550	Maximum		1	16		1,450	655		2,105	2,575	
1570	60" wide, average		1.40	11.429		2,400	465		2,865	3,350	
1590	72" wide, average		1.20	13.333		2,725	545		3,270	3,825	
1600	Office model, 48" wide		2	8		2,050	325		2,375	2,775	
1620	Refrigerator and sink only	↓	2.40	6.667	↓	2,050	272		2,322	2,675	
1640	Combination range, refrigerator, sink, microwave										
1660	oven and ice maker	L-1	.80	20	Ea.	4,025	815		4,840	5,650	

11471 | Darkroom Processing

		CREW	DAILY OUTPUT	LABOR-HOURS	UNIT	2005 BARE COSTS				TOTAL INCL O&P
						MAT.	LABOR	EQUIP.	TOTAL	
700	0010	**DARKROOM EQUIPMENT**									700
	0020	Developing sink, 5" deep, 24" x 48"	Q-1	2	8	Ea.	4,225	294		4,519	5,100
	0050	48" x 52"		1.70	9.412		4,300	345		4,645	5,250
	0200	10" deep, 24" x 48"		1.70	9.412		5,275	345		5,620	6,325
	0250	24" x 108"		1.50	10.667		1,900	390		2,290	2,700
	3500	Washers, round, minimum sheet 11" x 14"		2	8		2,600	294		2,894	3,325
	3550	Maximum sheet 20" x 24"		1	16		3,050	590		3,640	4,250
	3800	Square, minimum sheet 20" x 24"		1	16		2,750	590		3,340	3,900
	3900	Maximum sheet 50" x 56"	▼	.80	20	▼	4,275	735		5,010	5,800
	4500	Combination tank sink, tray sink, washers, with									
	4510	dry side tables, average	Q-1	.45	35.556	Ea.	8,100	1,300		9,400	10,900

11520 | Industrial Equipment

		CREW	DAILY OUTPUT	LABOR-HOURS	UNIT	2005 BARE COSTS				TOTAL INCL O&P
						MAT.	LABOR	EQUIP.	TOTAL	
250	0010	**DUST COLLECTION SYSTEMS** Commercial / industrial									250
	0120	Central vacuum units									
	0130	Includes stand, filters and motorized shaker									
	0200	500 CFM, 10" inlet, 2 HP	Q-20	2.40	8.333	Ea.	3,475	310		3,785	4,300
	0220	1000 CFM, 10" inlet, 3 HP		2.20	9.091		3,625	335		3,960	4,525
	0240	1500 CFM, 10" inlet, 5 HP		2	10		3,700	370		4,070	4,650
	0260	3000 CFM, 13" inlet, 10 HP		1.50	13.333		9,700	495		10,195	11,500
	0280	5000 CFM, 16" inlet, 2 @ 10 HP	▼	1	20	▼	10,200	740		10,940	12,400
	1000	Vacuum tubing, galvanized									
	1100	2-1/8" OD, 16 ga.	Q-9	440	.036	L.F.	1.84	1.32		3.16	4.04
	1110	2-1/2" OD, 16 ga.		420	.038		2.13	1.38		3.51	4.46
	1120	3" OD, 16 ga.		400	.040		2.71	1.45		4.16	5.20
	1130	3-1/2" OD, 16 ga.		380	.042		3.84	1.52		5.36	6.55
	1140	4" OD, 16 ga.		360	.044		3.93	1.61		5.54	6.80
	1150	5" OD, 14 ga.		320	.050		7.25	1.81		9.06	10.80
	1160	6" OD, 14 ga.		280	.057		8.15	2.07		10.22	12.15
	1170	8" OD, 14 ga.		200	.080		12.35	2.89		15.24	18
	1180	10" OD, 12 ga.		160	.100		12.35	3.62		15.97	19.10
	1190	12" OD, 12 ga.		120	.133		35.50	4.82		40.32	46.50
	1200	14" OD, 12 ga.	▼	80	.200	▼	39	7.25		46.25	54
	1940	Hose, flexible wire reinforced rubber									
	1956	3" dia.	Q-9	400	.040	L.F.	5.85	1.45		7.30	8.70
	1960	4" dia.		360	.044		7.90	1.61		9.51	11.15
	1970	5" dia.		320	.050		9.10	1.81		10.91	12.80
	1980	6" dia.	▼	280	.057	▼	10.50	2.07		12.57	14.75
	2000	90° Elbow, slip fit									
	2110	2-1/8" dia.	Q-9	70	.229	Ea.	7.70	8.25		15.95	21
	2120	2-1/2" dia.		65	.246		10.75	8.90		19.65	25.50
	2130	3" dia.		60	.267		14.55	9.65		24.20	31
	2140	3-1/2" dia.		55	.291		18	10.55		28.55	36
	2150	4" dia.		50	.320		22.50	11.60		34.10	42.50
	2160	5" dia.		45	.356		42.50	12.85		55.35	66.50
	2170	6" dia.		40	.400		58.50	14.45		72.95	87
	2180	8" dia.	▼	30	.533	▼	111	19.30		130.30	152

		CREW	DAILY OUTPUT	LABOR-HOURS	UNIT	2005 BARE COSTS				TOTAL INCL O&P
	11520 | **Industrial Equipment**					MAT.	LABOR	EQUIP.	TOTAL	
250	2400	45° Elbow, slip fit									**250**
	2410	2-1/8" dia.	Q-9	70	.229	Ea.	6.75	8.25		15	20
	2420	2-1/2" dia.		65	.246		9.90	8.90		18.80	24.50
	2430	3" dia.		60	.267		11.95	9.65		21.60	28
	2440	3-1/2" dia.		55	.291		15.05	10.55		25.60	33
	2450	4" dia.		50	.320		19.50	11.60		31.10	39.50
	2460	5" dia.		45	.356		33	12.85		45.85	56
	2470	6" dia.		40	.400		44.50	14.45		58.95	71.50
	2480	8" dia.	▼	35	.457	▼	87.50	16.55		104.05	122
	2800	90° TY, slip fit thru 6" dia									
	2810	2-1/8" dia.	Q-9	42	.381	Ea.	15.05	13.80		28.85	37.50
	2820	2-1/2" dia.		39	.410		19.55	14.85		34.40	44.50
	2830	3" dia.		36	.444		26	16.10		42.10	53
	2840	3-1/2" dia.		33	.485		33	17.55		50.55	63.50
	2850	4" dia.		30	.533		47.50	19.30		66.80	81.50
	2860	5" dia.		27	.593		90	21.50		111.50	132
	2870	6" dia.	▼	24	.667		123	24		147	172
	2880	8" dia., butt end					260			260	286
	2890	10" dia., butt end					470			470	515
	2900	12" dia., butt end					470			470	515
	2910	14" dia., butt end					855			855	940
	2920	6" x 4" dia., butt end					74.50			74.50	82
	2930	8" x 4" dia., butt end					141			141	156
	2940	10" x 4" dia., butt end					185			185	204
	2950	12" x 4" dia., butt end				▼	213			213	234
	3100	90° Elbow, butt end segmented									
	3110	8" dia., butt end, segmented				Ea.	168			168	185
	3120	10" dia., butt end, segmented					276			276	305
	3130	12" dia., butt end, segmented					350			350	385
	3140	14" dia., butt end, segmented				▼	440			440	480
	3200	45° Elbow, butt end segmented									
	3210	8" dia., butt end, segmented				Ea.	143			143	157
	3220	10" dia., butt end, segmented					199			199	219
	3230	12" dia., butt end, segmented					204			204	225
	3240	14" dia., butt end, segmented				▼	251			251	276
	3400	All butt end fittings require one coupling per joint.									
	3410	Labor for fitting included with couplings.									
	3460	Compression coupling, galvanized, neoprene gasket									
	3470	2-1/8" dia.	Q-9	44	.364	Ea.	11.70	13.15		24.85	33
	3480	2-1/2" dia.		44	.364		11.70	13.15		24.85	33
	3490	3" dia.		38	.421		15.35	15.25		30.60	40.50
	3500	3-1/2" dia.		35	.457		17.20	16.55		33.75	44.50
	3510	4" dia.		33	.485		18.70	17.55		36.25	47.50
	3520	5" dia.		29	.552		21.50	19.95		41.45	54
	3530	6" dia.		26	.615		25	22.50		47.50	62
	3540	8" dia.		22	.727		45.50	26.50		72	90.50
	3550	10" dia.		20	.800		66.50	29		95.50	118
	3560	12" dia.		18	.889		82.50	32		114.50	141
	3570	14" dia.	▼	16	1	▼	117	36		153	185
	3800	Air gate valves, galvanized									
	3810	2-1/8" dia.	Q-9	30	.533	Ea.	89	19.30		108.30	128
	3820	2-1/2" dia.		28	.571		93.50	20.50		114	135
	3830	3" dia.		26	.615		109	22.50		131.50	154
	3840	4" dia.		23	.696		142	25		167	195
	3850	6" dia.	▼	18	.889	▼	196	32		228	266

Important: See the Reference Section for critical supporting data - Reference Nos., Crews, & City Cost Indexes

11600 | Laboratory Equipment

11620 | Laboratory Equipment

		CREW	DAILY OUTPUT	LABOR-HOURS	UNIT	2005 BARE COSTS MAT.	LABOR	EQUIP.	TOTAL	TOTAL INCL O&P
110	0010	**LABORATORY EQUIPMENT**									110
	0670	Service fixtures, average				Ea.	44			44	48.50
	0680	For sink assembly with hot and cold water, add	1 Plum	1.40	5.714		590	233		823	1,000
	0700	Glassware washer, undercounter, minimum	L-1	1.80	8.889		6,200	365		6,565	7,375
	0710	Maximum	"	1	16		8,175	655		8,830	9,975
	1200	Refrigerator, blood bank, 28.6 C.F. emergency signal					6,750			6,750	7,400
	1210	Reach-in, 16.9 C.F.					6,900			6,900	7,575
	1400	Safety equipment, eye wash, hand held					294			294	325
	1450	Deluge shower					565			565	625
	1600	Sink, one piece plastic, flask wash, hose, free standing	1 Plum	1.60	5		1,500	204		1,704	1,950
	1610	Epoxy resin sink, 25" x 16" x 10"		2	4		166	163		329	430
	1850	Utensil washer-sanitizer	↓	2	4	↓	8,975	163		9,138	10,100
	8000	Alternate pricing method: as percent of lab furniture									
	8050	Installation, not incl. plumbing & duct work				% Furn.					22%
	8100	Plumbing, final connections, simple system									10%
	8110	Moderately complex system									15%
	8120	Complex system									20%
	8150	Electrical, simple system									10%
	8160	Moderately complex system									20%
	8170	Complex system				↓					35%

11700 | Medical Equipment

11710 | Medical Sterilizing Equipment

		CREW	DAILY OUTPUT	LABOR-HOURS	UNIT	2005 BARE COSTS MAT.	LABOR	EQUIP.	TOTAL	TOTAL INCL O&P
110	0010	**MEDICAL STERILIZING EQUIPMENT**									110
	0700	Distiller, water, steam heated, 50 gal. capacity	1 Plum	1.40	5.714	Ea.	15,200	233		15,433	17,100
	5600	Sterilizers, floor loading, 26" x 62" x 42", single door, steam					130,500			130,500	143,500
	5650	Double door, steam					167,500			167,500	184,500
	5800	General purpose, 20" x 20" x 38", single door					12,600			12,600	13,800
	6000	Portable, counter top, steam, minimum					3,125			3,125	3,450
	6020	Maximum					4,900			4,900	5,375
	6050	Portable, counter top, gas, 17"x15"x32-1/2"					32,300			32,300	35,500
	6150	Manual washer/sterilizer, 16"x16"x26"	1 Plum	2	4	↓	44,300	163		44,463	48,900
	6200	Steam generators, electric 10 KW to 180 KW, freestanding									
	6250	Minimum	1 Elec	3	2.667	Ea.	6,575	109		6,684	7,400
	6300	Maximum	"	.70	11.429		23,700	465		24,165	26,800
	8200	Bed pan washer-sanitizer	1 Plum	2	4	↓	5,875	163		6,038	6,700

11730 | Patient Care Equipment

		CREW	DAILY OUTPUT	LABOR-HOURS	UNIT	2005 BARE COSTS MAT.	LABOR	EQUIP.	TOTAL	TOTAL INCL O&P
110	0010	**PATIENT CARE EQUIPMENT**									11
	1800	Heat therapy unit, humidified, 26" x 78" x 28"				Ea.	2,800			2,800	3,075
	2100	Hubbard tank with accessories, stainless steel,									
	2110	125 GPM at 45 psi water pressure				Ea.	21,600			21,600	23,800
	2150	For electric overhead hoist, add					2,350			2,350	2,600
	3600	Paraffin bath, 126°F, auto controlled					1,250			1,250	1,375
	4600	Station, dietary, medium, with ice					13,200			13,200	14,500
	8400	Whirlpool bath, mobile, sst, 18" x 24" x 60"					3,775			3,775	4,150
	8450	Fixed, incl. mixing valves	1 Plum	2	4	↓	7,575	163		7,738	8,575

11740 | Dental Equipment

			CREW	DAILY OUTPUT	LABOR-HOURS	UNIT	2005 BARE COSTS				TOTAL INCL O&P	
							MAT.	LABOR	EQUIP.	TOTAL		
200	0010	**DENTAL EQUIPMENT**										200
	0020	Central suction system, minimum	1 Plum	1.20	6.667	Ea.	2,900	272		3,172	3,575	
	0100	Maximum	"	.90	8.889		22,300	365		22,665	25,000	
	0300	Air compressor, minimum	1 Skwk	.80	10		3,175	350		3,525	4,025	
	0400	Maximum		.50	16		22,000	560		22,560	25,100	
	0600	Chair, electric or hydraulic, minimum		.50	16		3,625	560		4,185	4,850	
	0700	Maximum		.25	32		7,675	1,125		8,800	10,200	
	2000	Light, ceiling mounted, minimum		8	1		1,875	35		1,910	2,125	
	2100	Maximum	↓	8	1		3,800	35		3,835	4,225	
	2200	Unit light, minimum	2 Skwk	5.33	3.002		1,400	105		1,505	1,725	
	2210	Maximum		5.33	3.002		4,600	105		4,705	5,250	
	2220	Track light, minimum		3.20	5		2,500	174		2,674	3,025	
	2230	Maximum	↓	3.20	5		5,475	174		5,649	6,300	
	2300	Sterilizers, steam portable, minimum					3,175			3,175	3,500	
	2350	Maximum					11,000			11,000	12,100	
	2600	Steam, institutional					13,700			13,700	15,000	
	2650	Dry heat, electric, portable, 3 trays				↓	1,075			1,075	1,175	

11760 | Operating Room Equipment

			CREW	DAILY OUTPUT	LABOR-HOURS	UNIT	MAT.	LABOR	EQUIP.	TOTAL	TOTAL INCL O&P	
110	0010	**OPERATING ROOM EQUIPMENT**										110
	5000	Scrub, surgical, stainless steel, single station, minimum	1 Plum	3	2.667	Ea.	2,125	109		2,234	2,525	
	5100	Maximum				"	7,075			7,075	7,775	

11781 | Mortuary Equipment

			CREW	DAILY OUTPUT	LABOR-HOURS	UNIT	MAT.	LABOR	EQUIP.	TOTAL	TOTAL INCL O&P	
110	0010	**MORTUARY EQUIPMENT**										110
	0015	Autopsy table, standard	1 Plum	1	8	Ea.	7,275	325		7,600	8,525	
	0020	Deluxe	"	.60	13.333	"	13,400	545		13,945	15,600	

For information about Means Estimating Seminars, see yellow pages 12 and 13 in back of book

Important: See the Reference Section for critical supporting data - Reference Nos., Crews, & City Cost Indexes

Division 13
Special Construction

Estimating Tips

General

- The items and systems in this division are usually estimated, purchased, supplied and installed as a unit by one or more subcontractors. The estimator must ensure that all parties are operating from the same set of specifications and assumptions and that all necessary items are estimated and will be provided. Many times the complex items and systems are covered but the more common ones such as excavation or a crane are overlooked for the very reason that everyone assumes nobody could miss them. The estimator should be the central focus and be able to ensure that all systems are complete.

- Another area where problems can develop in this division is at the interface between systems. The estimator must ensure, for instance, that anchor bolts, nuts and washers are estimated and included for the air-supported structures and pre-engineered buildings to be bolted to their foundations.

Utility supply is a common area where essential items or pieces of equipment can be missed or overlooked due to the fact that each subcontractor may feel it is the others' responsibility. The estimator should also be aware of certain items which may be supplied as part of a package but installed by others, and ensure that the installing contractor's estimate includes the cost of installation. Conversely, the estimator must also ensure that items are not costed by two different subcontractors, resulting in an inflated overall estimate.

13120 Pre-Engineered Structures

- The foundations and floor slab, as well as rough mechanical and electrical, should be estimated, as this work is required for the assembly and erection of the structure. Generally, as noted in the book, the pre-engineered building comes as a shell and additional features, such as windows and doors, must be included by the estimator. Here again, the estimator must have a clear understanding of the scope of each portion of the work and all the necessary interfaces.

13200 Storage Tanks

- The prices in this subdivision for above and below ground storage tanks do not include foundations or hold-down slabs. The estimator should refer to Divisions 2 and 3 for foundation system pricing. In addition to the foundations, required tank accessories such as tank gauges, leak detection devices, and additional manholes and piping must be added to the tank prices.

Reference Numbers

Reference numbers are shown in bold squares at the beginning of some major classifications. These numbers refer to related items in the Reference Section. The reference information may be an estimating procedure, an alternate pricing method or technical information.

Note: Not all subdivisions listed here necessarily appear in this publication.

13035 | Special Purpose Rooms

			CREW	DAILY OUTPUT	LABOR-HOURS	UNIT	2005 BARE COSTS				TOTAL INCL O&P	
							MAT.	LABOR	EQUIP.	TOTAL		
800	0010	SAUNA Prefabricated, incl. heater & controls, 7' high, 6' x 4', C/C	L-7	2.20	12.727	Ea.	3,525	420		3,945	4,525	800
	0050	6' x 4', C/P		2	14		3,275	460		3,735	4,325	
	0400	6' x 5', C/C		2	14		3,950	460		4,410	5,075	
	0450	6' x 5', C/P		2	14		3,700	460		4,160	4,800	
	0600	6' x 6', C/C		1.80	15.556		4,200	515		4,715	5,425	
	0650	6' x 6', C/P		1.80	15.556		3,925	515		4,440	5,125	
	0800	6' x 9', C/C		1.60	17.500		5,275	580		5,855	6,700	
	0850	6' x 9', C/P		1.60	17.500		5,000	580		5,580	6,400	
	1000	8' x 12', C/C		1.10	25.455		8,175	840		9,015	10,300	
	1050	8' x 12', C/P		1.10	25.455		7,500	840		8,340	9,550	
	1200	8' x 8', C/C		1.40	20		6,225	660		6,885	7,875	
	1250	8' x 8', C/P		1.40	20		5,850	660		6,510	7,475	
	1400	8' x 10', C/C		1.20	23.333		6,925	770		7,695	8,800	
	1450	8' x 10', C/P		1.20	23.333		6,425	770		7,195	8,275	
	1600	10' x 12', C/C		1	28		8,650	925		9,575	11,000	
	1650	10' x 12', C/P	↓	1	28		7,825	925		8,750	10,100	
	2500	Heaters only (incl. above), wall mounted, to 200 C.F.					460			460	505	
	2750	To 300 C.F.					555			555	615	
	3000	Floor standing, to 720 C.F., 10,000 watts, w/controls	1 Elec	3	2.667		1,450	109		1,559	1,725	
	3250	To 1,000 C.F., 16,000 watts	"	3	2.667	↓	1,500	109		1,609	1,800	
940	0010	STEAM BATH Heater, timer & head, single, to 140 C.F.	1 Plum	1.20	6.667	Ea.	975	272		1,247	1,475	940
	0500	To 300 C.F.		1.10	7.273		1,100	297		1,397	1,650	
	1000	Commercial size, with blow-down assembly, to 800 C.F.		.90	8.889		3,800	365		4,165	4,725	
	1500	To 2500 C.F.	↓	.80	10		5,975	410		6,385	7,200	
	2000	Multiple, motels, apts., 2 baths, w/ blow-down assm., 500 C.F.	Q-1	1.30	12.308		3,750	455		4,205	4,800	
	2500	4 baths	"	.70	22.857		4,075	840		4,915	5,775	
	2700	Conversion unit for residential tub, including door				↓	3,075			3,075	3,375	

13128 | Pre-Engineered Structures

			CREW	DAILY OUTPUT	LABOR-HOURS	UNIT	2005 BARE COSTS				TOTAL INCL O&P	
							MAT.	LABOR	EQUIP.	TOTAL		
200	0010	COMFORT STATIONS Prefab., stock, w/doors, windows & fixt.										200
	0100	Not incl. interior finish or electrical										
	0300	Mobile, on steel frame, minimum				S.F.	36.50			36.50	40	
	0350	Maximum					57.50			57.50	63.50	
	0400	Permanent, including concrete slab, minimum	B-12J	50	.320		142	10	16.75	168.75	190	
	0500	Maximum	"	43	.372	↓	206	11.65	19.50	237.15	266	
	0600	Alternate pricing method, mobile, minimum				Fixture	1,775			1,775	1,950	
	0650	Maximum					2,650			2,650	2,925	
	0700	Permanent, minimum	B-12J	.70	22.857		10,200	715	1,200	12,115	13,600	
	0750	Maximum	"	.50	32	↓	17,000	1,000	1,675	19,675	22,100	

13151	Swimming Pools	CREW	DAILY OUTPUT	LABOR-HOURS	UNIT	2005 BARE COSTS				TOTAL INCL O&P
						MAT.	LABOR	EQUIP.	TOTAL	
200 0010	**SWIMMING POOLS** Residential in-ground, vinyl lined, concrete sides									**200**
0020	Sides including equipment, sand bottom	B-52	300	.187	SF Surf	10.60	5.85	1.31	17.76	22
0100	Metal or polystyrene sides	B-14	410	.117	↓	8.85	3.29	.53	12.67	15.45
0200	Add for vermiculite bottom				↓	.68			.68	.75
0500	Gunite bottom and sides, white plaster finish									
0600	12' x 30' pool	B-52	145	.386	SF Surf	17.55	12.10	2.72	32.37	41
0720	16' x 32' pool	↓	155	.361	↓	15.85	11.35	2.54	29.74	38
0750	20' x 40' pool	↓	250	.224	↓	14.15	7.05	1.58	22.78	28
0810	Concrete bottom and sides, tile finish									
0820	12' x 30' pool	B-52	80	.700	SF Surf	17.75	22	4.93	44.68	59
0830	16' x 32' pool	↓	95	.589		14.65	18.50	4.15	37.30	49
0840	20' x 40' pool	↓	130	.431	↓	11.65	13.50	3.03	28.18	37
1100	Motel, gunite with plaster finish, incl. medium									
1150	capacity filtration & chlorination	B-52	115	.487	SF Surf	21.50	15.30	3.43	40.23	51.50
1200	Municipal, gunite with plaster finish, incl. high									
1250	capacity filtration & chlorination	B-52	100	.560	SF Surf	28	17.60	3.94	49.54	62.50
1350	Add for formed gutters				L.F.	48.50			48.50	53.50
1360	Add for stainless steel gutters				"	144			144	158
1700	Filtration and deck equipment only, as % of total				Total				20%	20%
1800	Deck equipment, rule of thumb, 20' x 40' pool				SF Pool					1.30
1900	5000 S.F. pool				"					1.90
3000	Painting pools, preparation + 3 coats, 20' x 40' pool, epoxy	2 Pord	.33	48.485	Total	605	1,475		2,080	2,900
3100	Rubber base paint, 18 gallons	"	.33	48.485	↓	460	1,475		1,935	2,725
3500	42' x 82' pool, 75 gallons, epoxy paint	3 Pord	.14	171		2,550	5,250		7,800	10,700
3600	Rubber base paint	"	.14	171	↓	1,975	5,250		7,225	10,100
700 0010	**SWIMMING POOL EQUIPMENT** Diving stand, stainless steel, 3 meter	2 Carp	.40	40	Ea.	4,925	1,375		6,300	7,550 **700**
0300	1 meter	"	2.70	5.926	"	3,650	203		3,853	4,325
0900	Filter system, sand or diatomite type, incl. pump, 6,000 gal./hr.	2 Plum	1.80	8.889	Total	1,125	365		1,490	1,775
1020	Add for chlorination system, 800 S.F. pool	↓	3	5.333	Ea.	203	218		421	550
1040	5,000 S.F. pool	↓	3	5.333	"	1,150	218		1,368	1,600
1100	Gutter system, stainless steel, with grating, stock,									
1110	contains supply and drainage system	E-1	20	1.200	L.F.	170	44.50	4.06	218.56	267
1120	Integral gutter and 5' high wall system, stainless steel	"	10	2.400	"	254	89	8.10	351.10	440
2100	Lights, underwater, 12 volt, with transformer, 300 watt	1 Elec	.40	20	Ea.	145	815		960	1,375
2200	110 volt, 500 watt, standard		.40	20		130	815		945	1,375
2400	Low water cutoff type	↓	.40	20	↓	135	815		950	1,375
2800	Heaters, see division 15510-880									

13201	Storage Tanks	CREW	DAILY OUTPUT	LABOR-HOURS	UNIT	2005 BARE COSTS				TOTAL INCL O&P
						MAT.	LABOR	EQUIP.	TOTAL	
300 0010	**GROUND TANKS** Not incl. pipe or pumps, prestress conc., 250,000 gal.				Ea.					284,200 **30**
3000	Steel, storage, above ground, including cradles, coating,									
3020	fittings, not including fdn, pumps or piping									
3040	Single wall, interior, 275 gallon	Q-5	5	3.200	Ea.	305	118		423	515
3060	550 gallon	"	2.70	5.926		1,400	218		1,618	1,850
3080	1,000 gallon	Q-7	5	6.400		2,200	250		2,450	2,800
3100	1,500 gallon		4.75	6.737		3,325	263		3,588	4,050
3120	2,000 gallon	↓	4.60	6.957	↓	3,800	272		4,072	4,600

13201	Storage Tanks	CREW	DAILY OUTPUT	LABOR-HOURS	UNIT	2005 BARE COSTS				TOTAL INCL O&P	
						MAT.	LABOR	EQUIP.	TOTAL		
300 3320	Double wall, 500 gallon capacity	Q-5	2.40	6.667	Ea.	2,425	246		2,671	3,050	**300**
3330	2000 gallon capacity	Q-7	4.15	7.711		5,575	300		5,875	6,575	
3340	4000 gallon capacity		3.60	8.889		9,925	345		10,270	11,400	
3350	6000 gallon capacity		2.40	13.333		11,700	520		12,220	13,700	
3360	8000 gallon capacity		2	16		15,000	625		15,625	17,400	
3370	10000 gallon capacity		1.80	17.778		16,600	695		17,295	19,300	
3380	15000 gallon capacity		1.50	21.333		25,200	835		26,035	29,000	
3390	20000 gallon capacity		1.30	24.615		28,600	960		29,560	33,000	
3400	25000 gallon capacity		1.15	27.826		34,800	1,075		35,875	39,900	
3410	30000 gallon capacity	↓	1	32	↓	38,200	1,250		39,450	43,900	
800 0010	**UNDERGROUND STORAGE TANKS**										**800**
0210	Fiberglass, underground, single wall, U.L. listed, not including										
0220	manway or hold-down strap										
0225	550 gallon capacity	Q-5	2.67	5.993	Ea.	1,500	221		1,721	1,975	
0230	1,000 gallon capacity	"	2.46	6.504		1,875	240		2,115	2,425	
0240	2,000 gallon capacity	Q-7	4.57	7.002		2,525	273		2,798	3,175	
0250	4,000 gallon capacity		3.55	9.014		3,650	350		4,000	4,550	
0260	6,000 gallon capacity		2.67	11.985		4,075	470		4,545	5,175	
0270	8,000 gallon capacity		2.29	13.974		4,950	545		5,495	6,275	
0280	10,000 gallon capacity		2	16		5,550	625		6,175	7,050	
0282	12,000 gallon capacity		1.88	17.021		7,150	665		7,815	8,875	
0284	15,000 gallon capacity		1.68	19.048		8,075	745		8,820	10,000	
0290	20,000 gallon capacity		1.45	22.069		11,200	860		12,060	13,600	
0300	25,000 gallon capacity		1.28	25		17,500	975		18,475	20,800	
0320	30,000 gallon capacity		1.14	28.070		21,900	1,100		23,000	25,800	
0340	40,000 gallon capacity		.89	35.955		30,800	1,400		32,200	35,900	
0360	48,000 gallon capacity	↓	.81	39.506		50,500	1,550		52,050	58,000	
0500	For manway, fittings and hold-downs, add					20%	15%				
0600	For manways, add					850			850	935	
1000	For helical heating coil, add	Q-5	2.50	6.400	↓	2,225	236		2,461	2,800	
1020	Fiberglass, underground, double wall, U.L. listed										
1030	includes manways, not incl. hold-down straps										
1040	600 gallon capacity	Q-5	2.42	6.612	Ea.	3,325	244		3,569	4,025	
1050	1,000 gallon capacity	"	2.25	7.111		4,650	262		4,912	5,525	
1060	2,500 gallon capacity	Q-7	4.16	7.692		6,800	300		7,100	7,950	
1070	3,000 gallon capacity		3.90	8.205		7,325	320		7,645	8,550	
1080	4,000 gallon capacity		3.64	8.791		8,700	345		9,045	10,100	
1090	6,000 gallon capacity		2.42	13.223		9,725	515		10,240	11,500	
1100	8,000 gallon capacity		2.08	15.385		10,800	600		11,400	12,800	
1110	10,000 gallon capacity		1.82	17.582		12,000	685		12,685	14,200	
1120	12,000 gallon capacity		1.70	18.824		14,800	735		15,535	17,400	
1122	15,000 gallon capacity		1.52	21.053		20,500	820		21,320	23,700	
1124	20,000 gallon capacity		1.33	24.060		24,700	940		25,640	28,600	
1126	25,000 gallon capacity		1.16	27.586		32,500	1,075		33,575	37,400	
1128	30,000 gallon capacity	↓	1.03	31.068	↓	39,000	1,225		40,225	44,600	
1140	For hold-down straps, add					2%	10%				
1150	For hold-downs 500-4000 gal, add	Q-7	16	2	Set	237	78		315	375	
1160	For hold-downs 5000-15000 gal, add		8	4		475	156		631	755	
1170	For hold-downs 20,000 gal, add		5.33	6.004		710	234		944	1,125	
1180	For hold-downs 25,000 gal, add		4	8		945	310		1,255	1,525	
1190	For hold-downs 30,000 gal, add	↓	2.60	12.308	↓	1,425	480		1,905	2,275	
2210	Fiberglass, underground, single wall, U.L. listed, including										
2220	hold-down straps, no manways										
2225	550 gallon capacity	Q-5	2	8	Ea.	1,725	295		2,020	2,350	
2230	1,000 gallon capacity	"	1.88	8.511		2,125	315		2,440	2,800	
2240	2,000 gallon capacity	Q-7	3.55	9.014	↓	2,750	350		3,100	3,550	

Important: See the Reference Section for critical supporting data - Reference Nos., Crews, & City Cost Indexes

	13201	Storage Tanks	CREW	DAILY OUTPUT	LABOR-HOURS	UNIT	2005 BARE COSTS				TOTAL INCL O&P	
							MAT.	LABOR	EQUIP.	TOTAL		
800	2250	4,000 gallon capacity	Q-7	2.90	11.034	Ea.	3,900	430		4,330	4,925	800
	2260	6,000 gallon capacity		2	16		4,550	625		5,175	5,950	
	2270	8,000 gallon capacity		1.78	17.978		5,425	700		6,125	7,025	
	2280	10,000 gallon capacity		1.60	20		6,025	780		6,805	7,800	
	2282	12,000 gallon capacity		1.52	21.053		7,625	820		8,445	9,625	
	2284	15,000 gallon capacity		1.39	23.022		8,550	900		9,450	10,800	
	2290	20,000 gallon capacity		1.14	28.070		11,900	1,100		13,000	14,800	
	2300	25,000 gallon capacity		.96	33.333		18,500	1,300		19,800	22,300	
	2320	30,000 gallon capacity	▼	.80	40	▼	23,400	1,550		24,950	28,100	
	3020	Fiberglass, underground, double wall, U.L. listed										
	3030	includes manways and hold-down straps										
	3040	600 gallon capacity	Q-5	1.86	8.602	Ea.	3,550	315		3,865	4,400	
	3050	1,000 gallon capacity	"	1.70	9.412		4,875	345		5,220	5,900	
	3060	2,500 gallon capacity	Q-7	3.29	9.726		7,050	380		7,430	8,325	
	3070	3,000 gallon capacity		3.13	10.224		7,575	400		7,975	8,925	
	3080	4,000 gallon capacity		2.93	10.921		8,950	425		9,375	10,500	
	3090	6,000 gallon capacity		1.86	17.204		10,200	670		10,870	12,200	
	3100	8,000 gallon capacity		1.65	19.394		11,300	755		12,055	13,500	
	3110	10,000 gallon capacity		1.48	21.622		12,500	845		13,345	15,100	
	3120	12,000 gallon capacity		1.40	22.857		15,300	890		16,190	18,200	
	3122	15,000 gallon capacity		1.28	25		20,900	975		21,875	24,500	
	3124	20,000 gallon capacity		1.06	30.189		25,400	1,175		26,575	29,800	
	3126	25,000 gallon capacity		.90	35.556		33,500	1,400		34,900	38,900	
	3128	30,000 gallon capacity	▼	.74	43.243	▼	40,400	1,700		42,100	46,900	
	5000	Steel underground, sti-P3, set in place, not incl. hold-down bars.										
	5500	Excavation, pad, pumps and piping not included										
	5510	Single wall, 500 gallon capacity, 7 gauge shell	Q-5	2.70	5.926	Ea.	1,000	218		1,218	1,425	
	5520	1,000 gallon capacity, 7 gauge shell	"	2.50	6.400		1,650	236		1,886	2,150	
	5530	2,000 gallon capacity, 1/4" thick shell	Q-7	4.60	6.957		3,025	272		3,297	3,725	
	5535	2,500 gallon capacity, 7 gauge shell	Q-5	3	5.333		3,675	197		3,872	4,350	
	5540	5,000 gallon capacity, 1/4" thick shell	Q-7	3.20	10		6,850	390		7,240	8,100	
	5560	10,000 gallon capacity, 1/4" thick shell		2	16		11,100	625		11,725	13,200	
	5580	15,000 gallon capacity, 5/16" thick shell		1.70	18.824		13,200	735		13,935	15,700	
	5600	20,000 gallon capacity, 5/16" thick shell		1.50	21.333		17,800	835		18,635	20,800	
	5610	25,000 gallon capacity, 3/8" thick shell		1.30	24.615		21,500	960		22,460	25,200	
	5620	30,000 gallon capacity, 3/8" thick shell		1.10	29.091		28,600	1,125		29,725	33,200	
	5630	40,000 gallon capacity, 3/8" thick shell		.90	35.556		37,400	1,400		38,800	43,200	
	5640	50,000 gallon capacity, 3/8" thick shell	▼	.80	40	▼	46,800	1,550		48,350	54,000	
	6200	Steel, underground, 360°, double wall, U.L. listed,										
	6210	with sti-P3 corrosion protection,										
	6220	(dielectric coating, cathodic protection, electrical										
	6230	isolation) 30 year warranty,										
	6240	not incl. manholes or hold-downs.										
	6250	500 gallon capacity	Q-5	2.40	6.667	Ea.	3,025	246		3,271	3,700	
	6260	1,000 gallon capactiy	"	2.25	7.111		4,400	262		4,662	5,250	
	6270	2,000 gallon capacity	Q-7	4.16	7.692		5,800	300		6,100	6,825	
	6280	3,000 gallon capacity		3.90	8.205		6,400	320		6,720	7,525	
	6290	4,000 gallon capacity		3.64	8.791		8,150	345		8,495	9,500	
	6300	5,000 gallon capacity		2.91	10.997		10,700	430		11,130	12,300	
	6310	6,000 gallon capacity		2.42	13.223		12,700	515		13,215	14,700	
	6320	8,000 gallon capacity		2.08	15.385		13,200	600		13,800	15,400	
	6330	10,000 gallon capacity		1.82	17.582		14,900	685		15,585	17,400	
	6340	12,000 gallon capacity		1.70	18.824		18,000	735		18,735	20,800	
	6350	15,000 gallon capacity		1.33	24.060		19,900	940		20,840	23,200	
	6360	20,000 gallon capacity		1.33	24.060		23,900	940		24,840	27,700	
	6370	25,000 gallon capacity	▼	1.16	27.586	▼	40,100	1,075		41,175	45,700	

13201	Storage Tanks	CREW	DAILY OUTPUT	LABOR-HOURS	UNIT	MAT.	LABOR	EQUIP.	TOTAL	TOTAL INCL O&P	
						2005 BARE COSTS					
800											800
6380	30,000 gallon capacity	Q-7	1.03	31.068	Ea.	48,300	1,225		49,525	55,000	
6390	40,000 gallon capacity		.80	40		62,000	1,550		63,550	70,500	
6395	50,000 gallon capacity		.73	43.836	↓	79,500	1,700		81,200	90,000	
6400	For hold-downs 500-2000 gal, add		16	2	Set	370	78		448	525	
6410	For hold-downs 3000-6000 gal, add		12	2.667		460	104		564	660	
6420	For hold-downs 8000-12,000 gal, add		11	2.909		1,175	114		1,289	1,475	
6430	For hold-downs 15,000 gal, add		9	3.556		1,475	139		1,614	1,825	
6440	For hold-downs 20,000 gal, add		8	4		1,650	156		1,806	2,025	
6450	For hold-downs 20,000 gal plus, add	↓	6	5.333	↓	1,650	208		1,858	2,125	
6500	For manways, add				Ea.	640			640	705	
6600	In place with hold-downs										
6652	550 gallon capacity	Q-5	1.84	8.696	Ea.	3,400	320		3,720	4,200	

13281	Hazardous Material Remediation	CREW	DAILY OUTPUT	LABOR-HOURS	UNIT	MAT.	LABOR	EQUIP.	TOTAL	TOTAL INCL O&P	
						2005 BARE COSTS					
120	0010	**BULK ASBESTOS REMOVAL**										120
0020	Includes disposable tools and 2 suits and 1 respirator filter/day/worker										
0200	Boiler insulation	A-9	480	.133	S.F.	.44	4.95		5.39	8.30	
0210	With metal lath add				%				50%		
0300	Boiler breeching or flue insulation	A-9	520	.123	S.F.	.35	4.57		4.92	7.60	
0310	For active boiler, add				%				100%		
0400	Duct or AHU insulation	A-10B	440	.073	S.F.	.21	2.71		2.92	4.50	
0500	Duct vibration isolation joints, up to 24 Sq. In. duct	A-9	56	1.143	Ea.	3.24	42.50		45.74	70.50	
0520	25 Sq. In. to 48 Sq. In. duct		48	1.333		3.78	49.50		53.28	82	
0530	49 Sq. In. to 76 Sq. In. duct		40	1.600	↓	4.54	59.50		64.04	99	
0600	Pipe insulation, air cell type, up to 4" diameter pipe		900	.071	L.F.	.20	2.64		2.84	4.39	
0610	4" to 8" diameter pipe		800	.080		.23	2.97		3.20	4.94	
0620	10" to 12" diameter pipe		700	.091		.26	3.40		3.66	5.65	
0630	14" to 16" diameter pipe		550	.116	↓	.33	4.32		4.65	7.15	
0650	Over 16" diameter pipe		650	.098	S.F.	.28	3.66		3.94	6.05	
0700	With glove bag up to 3" diameter pipe		300	.213	L.F.	3.15	7.95		11.10	15.95	
1000	Pipe fitting insulation up to 4" diameter pipe		320	.200	Ea.	.57	7.45		8.02	12.35	
1100	6" to 8" diameter pipe		304	.211		.60	7.80		8.40	13	
1110	10" to 12" diameter pipe		192	.333		.95	12.40		13.35	20.50	
1120	14" to 16" diameter pipe		128	.500	↓	1.42	18.60		20.02	31	
1130	Over 16" diameter pipe		176	.364	S.F.	1.03	13.50		14.53	22.50	
1200	With glove bag, up to 8" diameter pipe		100	.640	L.F.	6.55	24		30.55	44.50	
2000	Scrape foam fireproofing from flat surface		2,400	.027	S.F.	.08	.99		1.07	1.64	
2100	Irregular surfaces		1,200	.053		.15	1.98		2.13	3.30	
3000	Remove cementitious material from flat surface		1,800	.036		.10	1.32		1.42	2.20	
3100	Irregular surface		1,400	.046	↓	.13	1.70		1.83	2.82	
6000	Remove contaminated soil from crawl space by hand	↓	400	.160	C.F.	.45	5.95		6.40	9.90	
6100	With large production vacuum loader	A-12	700	.091	"	.26	3.40	.69	4.35	6.40	
7000	Radiator backing, not including radiator removal	A-9	1,200	.053	S.F.	.15	1.98		2.13	3.30	
9000	For type C (supplied air) respirator equipment, add				%					10%	
125	0010	**ASBESTOS ABATEMENT WORK AREA** Containment and preparation.										125
0100	Pre-cleaning, HEPA vacuum and wet wipe, flat surfaces	A-10	12,000	.005	S.F.	.02	.20		.22	.33	
0200	Protect carpeted area, 2 layers 6 mil poly on 3/4" plywood	"	1,000	.064		1.50	2.38		3.88	5.40	
0300	Separation barrier, 2" x 4" @ 16", 1/2" plywood ea. side, 8' high	2 Carp	400	.040	↓	1.25	1.37		2.62	3.51	

Important: See the Reference Section for critical supporting data - Reference Nos., Crews, & City Cost Indexes

		13281	Hazardous Material Remediation	CREW	DAILY OUTPUT	LABOR-HOURS	UNIT	2005 BARE COSTS				TOTAL INCL O&P	
								MAT.	LABOR	EQUIP.	TOTAL		
125	0310		12' high	2 Carp	320	.050	S.F.	1.40	1.71		3.11	4.21	125
	0320		16' high		200	.080		1.50	2.74		4.24	5.90	
	0400		Personnel decontam. chamber, 2" x 4" @ 16", 3/4" ply ea. side		280	.057		2.50	1.96		4.46	5.80	
	0450		Waste decontam. chamber, 2" x 4" studs @ 16", 3/4" ply ea. side		360	.044		3	1.52		4.52	5.65	
	0500		Cover surfaces with polyethelene sheeting										
	0501		Including glue and tape										
	0550		Floors, each layer, 6 mil	A-10	8,000	.008	S.F.	2.85	.30		3.15	3.61	
	0551		4 mil		9,000	.007		1.94	.26		2.20	2.55	
	0560		Walls, each layer, 6 mil		6,000	.011		2.84	.40		3.24	3.75	
	0561		4 mil		7,000	.009		1.95	.34		2.29	2.69	
	0570		For heights above 12', add						20%				
	0575		For heights above 20', add						30%				
	0580		For fire retardant poly, add					100%					
	0590		For large open areas, deduct					10%	20%				
	0600		Seal floor penetrations with foam firestop to 36 Sq. In.	2 Carp	200	.080	Ea.	6.25	2.74		8.99	11.15	
	0610		36 Sq. In. to 72 Sq. In.		125	.128		12.50	4.38		16.88	20.50	
	0615		72 Sq. In. to 144 Sq. In.		80	.200		25	6.85		31.85	38	
	0620		Wall penetrations, to 36 square inches		180	.089		6.25	3.04		9.29	11.65	
	0630		36 Sq. In. to 72 Sq. In.		100	.160		12.50	5.50		18	22.50	
	0640		72 Sq. In. to 144 Sq. In.		60	.267		25	9.15		34.15	42	
	0800		Caulk seams with latex	1 Carp	230	.035	L.F.	.15	1.19		1.34	2.03	
	0900		Set up neg. air machine, 1-2k C.F.M. /25 M.C.F. volume	1 Asbe	4.30	1.860	Ea.		69		69	109	
130	0010		**DEMOLITION IN ASBESTOS CONTAMINATED AREA**										130
	0200		Ceiling, including suspension system, plaster and lath	A-9	2,100	.030	S.F.	.09	1.13		1.22	1.89	
	0210		Finished plaster, leaving wire lath		585	.109		.31	4.07		4.38	6.75	
	0220		Suspended acoustical tile		3,500	.018		.05	.68		.73	1.13	
	0230		Concealed tile grid system		3,000	.021		.06	.79		.85	1.32	
	0240		Metal pan grid system		1,500	.043		.12	1.59		1.71	2.63	
	0250		Gypsum board		2,500	.026		.07	.95		1.02	1.58	
	0260		Lighting fixtures up to 2' x 4'		72	.889	Ea.	2.52	33		35.52	55	
	0400		Partitions, non load bearing										
	0410		Plaster, lath, and studs	A-9	690	.093	S.F.	.81	3.45		4.26	6.35	
	0450		Gypsum board and studs	"	1,390	.046	"	.13	1.71		1.84	2.84	
	9000		For type C (supplied air) respirator equipment, add				%					10%	
135	0010		**ASBESTOS ABATEMENT EQUIPMENT** and supplies, buy R13281-120										13
	0200		Air filtration device, 2000 C.F.M.				Ea.	2,500			2,500	2,750	
	0250		Large volume air sampling pump, minimum					365			365	400	
	0260		Maximum					680			680	745	
	0300		Airless sprayer unit, 2 gun					2,000			2,000	2,200	
	0350		Light stand, 500 watt					250			250	275	
	0400		Personal respirators										
	0410		Negative pressure, 1/2 face, dual operation, min.				Ea.	22.50			22.50	25	
	0420		Maximum					24			24	26.50	
	0450		P.A.P.R., full face, minimum					400			400	440	
	0460		Maximum					700			700	770	
	0470		Supplied air, full face, incl. air line, minimum					450			450	495	
	0480		Maximum					600			600	660	
	0500		Personnel sampling pump, minimum					450			450	495	
	0510		Maximum					750			750	825	
	1500		Power panel, 20 unit, incl. G.F.I.					1,800			1,800	1,975	
	1600		Shower unit, including pump and filters					1,125			1,125	1,250	
	1700		Supplied air system (type C)					10,000			10,000	11,000	
	1750		Vacuum cleaner, HEPA, 16 gal., stainless steel, wet/dry					1,000			1,000	1,100	
	1760		55 gallon					2,200			2,200	2,425	
	1800		Vacuum loader, 9-18 ton/hr					90,000			90,000	99,000	
	1900		Water atomizer unit, including 55 gal. drum					230			230	253	

	13281	Hazardous Material Remediation	CREW	DAILY OUTPUT	LABOR-HOURS	UNIT	2005 BARE COSTS MAT.	LABOR	EQUIP.	TOTAL	TOTAL INCL O&P	
135	2000	Worker prot., whole body, foot, head cover & gloves, plastic R13281-120				Ea.	5.50			5.50	6.05	135
	2500	Respirator, single use					10.50			10.50	11.55	
	2550	Cartridge for respirator					11.70			11.70	12.85	
	2570	Glove bag, 7 mil, 50" x 64"					8.50			8.50	9.35	
	2580	10 mil, 44" x 60"					8.40			8.40	9.25	
	3000	HEPA vacuum for work area, minimun					1,050			1,050	1,150	
	6000	Disposable polyethelene bags, 6 mil, 3 C.F.					1.15			1.15	1.27	
	6300	Disposable fiber drums, 3 C.F.					6.50			6.50	7.15	
	6400	Pressure sensitive caution lables, 3" x 5"					.88			.88	.97	
	6450	11" x 17"					6.50			6.50	7.15	
	6500	Negative air machine, 1800 C.F.M.					775			775	855	
140	0010	**DECONTAMINATION CONTAINMENT AREA DEMOLITION** and clean-up										140
	0100	Spray exposed substrate with surfactant (bridging)										
	0200	Flat surfaces	A-9	6,000	.011	S.F.	.36	.40		.76	1.03	
	0250	Irregular surfaces		4,000	.016	"	.31	.59		.90	1.28	
	0300	Pipes, beams, and columns		2,000	.032	L.F.	.56	1.19		1.75	2.50	
	1000	Spray encapsulate polyethelene sheeting		8,000	.008	S.F.	.31	.30		.61	.81	
	1100	Roll down polyethelene sheeting		8,000	.008	"		.30		.30	.47	
	1500	Bag polyethelene sheeting		400	.160	Ea.	.77	5.95		6.72	10.25	
	2000	Fine clean exposed substrate, with nylon brush		2,400	.027	S.F.		.99		.99	1.56	
	2500	Wet wipe substrate		4,800	.013			.50		.50	.78	
	2600	Vacuum surfaces, fine brush		6,400	.010			.37		.37	.59	
	3000	Structural demolition										
	3100	Wood stud walls	A-9	2,800	.023	S.F.		.85		.85	1.34	
	3500	Window manifolds, not incl. window replacement		4,200	.015			.57		.57	.89	
	3600	Plywood carpet protection		2,000	.032			1.19		1.19	1.88	
	4000	Remove custom decontamination facility	A-10A	8	3	Ea.	15	112		127	193	
	4100	Remove portable decontamination facility	3 Asbe	12	2	"	12.50	74		86.50	131	
	5000	HEPA vacuum, shampoo carpeting	A-9	4,800	.013	S.F.	.05	.50		.55	.84	
	9000	Final cleaning of protected surfaces	A-10A	8,000	.003	"		.11		.11	.18	
145	0010	**OSHA TESTING**										145
	0100	Certified technician, minimum				Day					300	
	0110	Maximum				"					500	
	0200	Personal sampling, PCM analysis, NIOSH 7400, minimum	1 Asbe	8	1	Ea.	2.75	37		39.75	61.50	
	0210	Maximum	"	4	2	"	3	74		77	120	
	0300	Industrial hygenist, minimum				Day					400	
	0310	Maximum				"					550	
	1000	Cleaned area samples	1 Asbe	8	1	Ea.	2.63	37		39.63	61.50	
	1100	PCM air sample analysis, NIOSH 7400, minimum		8	1		30	37		67	91.50	
	1110	Maximum		4	2		3.09	74		77.09	120	
	1200	TEM air sample analysis, NIOSH 7402, minimum								100	125	
	1210	Maximum								400	500	
150	0010	**ENCAPSULATION WITH SEALANTS**										150
	0100	Ceilings and walls, minimum	A-9	21,000	.003	S.F.	.27	.11		.38	.48	
	0110	Maximum		10,600	.006	"	.42	.22		.64	.81	
	0300	Pipes to 12" diameter including minor repairs, minimum		800	.080	L.F.	.37	2.97		3.34	5.10	
	0310	Maximum		400	.160	"	1.04	5.95		6.99	10.55	
155	0010	**WASTE PACKAGING, HANDLING, & DISPOSAL**										155
	0100	Collect and bag bulk material, 3 C.F. bags, by hand	A-9	400	.160	Ea.	1.15	5.95		7.10	10.65	
	0200	Large production vacuum loader	A-12	880	.073		.80	2.70	.55	4.05	5.75	
	1000	Double bag and decontaminate	A-9	960	.067		2.30	2.48		4.78	6.45	
	2000	Containerize bagged material in drums, per 3 C.F. drum	"	800	.080		6.50	2.97		9.47	11.85	
	3000	Cart bags 50' to dumpster	2 Asbe	400	.040			1.48		1.48	2.34	

Important: See the Reference Section for critical supporting data - Reference Nos., Crews, & City Cost Indexes

		13281 | **Hazardous Material Remediation**	CREW	DAILY OUTPUT	LABOR-HOURS	UNIT	2005 BARE COSTS				TOTAL INCL O&P	
							MAT.	LABOR	EQUIP.	TOTAL		
155	5000	Disposal charges, not including haul, minimum				C.Y.					50	**155**
	5020	Maximum				"					175	
	5100	Remove refrigerant from system	1 Plum	40	.200	Lb.		8.15		8.15	12.30	
	9000	For type C (supplied air) respirator equipment, add				%					10%	
600	0010	**MOLD ABATEMENT WORK AREA** Containment and preparation										**600**
	0100	Pre-cleaning, HEPA vacuum and wet wipe, flat surfaces	A-10	12,000	.005	S.F.	.02	.20		.22	.33	
	0300	Separation barrier, 2" x 4" @ 16", 1/2" plywood ea. side, 8' high	2 Carp	400	.040		1.25	1.37		2.62	3.51	
	0310	12' high		320	.050		1.40	1.71		3.11	4.21	
	0320	16' high		200	.080		1.50	2.74		4.24	5.90	
	0400	Personnel decontam. chamber, 2" x 4" @ 16", 3/4" ply ea. side		280	.057		2.50	1.96		4.46	5.80	
	0450	Waste decontam. chamber, 2" x 4" studs @ 16", 3/4" ply each side	▼	360	.044	▼	3	1.52		4.52	5.65	
	0500	Cover surfaces with polyethelene sheeting										
	0501	Including glue and tape										
	0550	Floors, each layer, 6 mil	A-10	8,000	.008	S.F.	2.85	.30		3.15	3.61	
	0551	4 mil		9,000	.007		1.94	.26		2.20	2.55	
	0560	Walls, each layer, 6 mil		6,000	.011		2.84	.40		3.24	3.75	
	0561	4 mil	▼	7,000	.009	▼	1.95	.34		2.29	2.69	
	0570	For heights above 12', add						20%				
	0575	For heights above 20', add						30%				
	0580	For fire retardant poly, add					100%					
	0590	For large open areas, deduct					10%	20%				
	0600	Seal floor penetrations with foam firestop to 36 sq in	2 Carp	200	.080	Ea.	6.25	2.74		8.99	11.15	
	0610	36 sq in to 72 sq in		125	.128		12.50	4.38		16.88	20.50	
	0615	72 sq in to 144 sq in		80	.200		25	6.85		31.85	38	
	0620	Wall penetrations, to 36 square inches		180	.089		6.25	3.04		9.29	11.65	
	0630	36 Sq. in. to 72 sq. in.		100	.160		12.50	5.50		18	22.50	
	0640	72 Sq. in. to 144 sq. in.	▼	60	.267	▼	25	9.15		34.15	42	
	0800	Caulk seams with latex	1 Carp	230	.035	L.F.	.15	1.19		1.34	2.03	
	0900	Set up neg. air machine, 1-2k C.F.M. /25 M.C.F. volume	1 Asbe	4.30	1.860	Ea.		69		69	109	
610	0010	**DEMOLITION IN MOLD CONTAMINATED AREA**										**61**
	0200	Ceiling, including suspension system, plaster and lath	A-9	2,100	.030	S.F.	.09	1.13		1.22	1.89	
	0210	Finished plaster, leaving wire lath		585	.109		.31	4.07		4.38	6.75	
	0220	Suspended acoustical tile		3,500	.018		.05	.68		.73	1.13	
	0230	Concealed tile grid system		3,000	.021		.06	.79		.85	1.32	
	0240	Metal pan grid system		1,500	.043		.12	1.59		1.71	2.63	
	0250	Gypsum board		2,500	.026		.07	.95		1.02	1.58	
	0255	Plywood		2,500	.026	▼	.07	.95		1.02	1.58	
	0260	Lighting fixtures up to 2' x 4'	▼	72	.889	Ea.	2.52	33		35.52	55	
	0400	Partitions, non load bearing										
	0410	Plaster, lath, and studs	A-9	690	.093	S.F.	.81	3.45		4.26	6.35	
	0450	Gypsum board and studs		1,390	.046		.13	1.71		1.84	2.84	
	0465	Carpet & pad		1,390	.046	▼	.13	1.71		1.84	2.84	
	0600	Pipe insulation, air cell type, up to 4" diameter pipe		900	.071	L.F.	.20	2.64		2.84	4.39	
	0610	4" to 8" diameter pipe		800	.080		.23	2.97		3.20	4.94	
	0620	10" to 12" diameter pipe		700	.091		.26	3.40		3.66	5.65	
	0630	14" to 16" diameter pipe		550	.116	▼	.33	4.32		4.65	7.15	
	0650	Over 16" diameter pipe	▼	650	.098	S.F.	.28	3.66		3.94	6.05	
	9000	For type C (supplied air) respirator equipment, add				%					10%	

		13630	Solar Collector Components		CREW	DAILY OUTPUT	LABOR-HOURS	UNIT	2005 BARE COSTS				TOTAL INCL O&P	
									MAT.	LABOR	EQUIP.	TOTAL		
200	0010	**SOLAR ENERGY**		D2020 265										200
	0020	System/Package prices, not including connecting												
	0030	pipe, insulation, or special heating/plumbing fixtures		D2020 270										
	0150	For solar ultraviolet pipe insulation see Div. 15086												
	0500	Hot water, standard package, low temperature		D2020 275										
	0540	1 collector, circulator, fittings, 65 gal. tank			Q-1	.50	32	Ea.	1,450	1,175		2,625	3,350	
	0580	2 collectors, circulator, fittings, 120 gal. tank		D2020 280		.40	40		2,050	1,475		3,525	4,450	
	0620	3 collectors, circulator, fittings, 120 gal. tank				.34	47.059		2,475	1,725		4,200	5,325	
	0700	Medium temperature package		D2020 285										
	0720	1 collector, circulator, fittings, 80 gal. tank			Q-1	.50	32	Ea.	1,550	1,175		2,725	3,475	
	0740	2 collectors, circulator, fittings, 120 gal. tank		D2020 290		.40	40		2,200	1,475		3,675	4,625	
	0780	3 collectors, circulator, fittings, 120 gal. tank				.30	53.333		4,000	1,950		5,950	7,350	
	0980	For each additional 120 gal. tank, add		D2020 295					700			700	770	
	2250	Controller, liquid temperature			1 Plum	5	1.600		87.50	65.50		153	194	
	2300	Circulators, air		D3010 650										
	2310	Blowers												
	2330	100-300 S.F. system, 1/10 HP		D3010 660	Q-9	16	1	Ea.	136	36		172	206	
	2340	300-500 S.F. system, 1/5 HP				15	1.067		188	38.50		226.50	266	
	2350	Two speed, 100-300 S.F., 1/10 HP		D 3010 675		14	1.143		136	41.50		177.50	214	
	2400	Reversible fan, 20" diameter, 2 speed				18	.889		108	32		140	169	
	2520	Space & DHW system, less duct work		R13600 -610		.50	32		1,475	1,150		2,625	3,400	
	2550	Booster fan 6" diameter, 120 CFM				16	1		35	36		71	94	
	2570	6" diameter, 225 CFM				16	1		43	36		79	103	
	2580	8" diameter, 150 CFM				16	1		39.50	36		75.50	99	
	2590	8" diameter, 310 CFM				14	1.143		60.50	41.50		102	130	
	2600	8" diameter, 425 CFM				14	1.143		68	41.50		109.50	139	
	2650	Rheostat				32	.500		14.80	18.10		32.90	44.50	
	2660	Shutter/damper				12	1.333		39	48		87	117	
	2670	Shutter motor				16	1		92.50	36		128.50	158	
	2800	Circulators, liquid, 1/25 HP, 5.3 GPM			Q-1	14	1.143		147	42		189	225	
	2820	1/20 HP, 17 GPM				12	1.333		154	49		203	244	
	2850	1/20 HP, 17 GPM, stainless steel				12	1.333		243	49		292	340	
	2870	1/12 HP, 30 GPM				10	1.600		330	59		389	455	
	3000	Collector panels, air with aluminum absorber plate												
	3010	Wall or roof mount												
	3040	Flat black, plastic glazing												
	3080	4' x 8'			Q-9	6	2.667	Ea.	635	96.50		731.50	845	
	3100	4' x 10'				5	3.200	"	780	116		896	1,050	
	3200	Flush roof mount, 10' to 16' x 22" wide				96	.167	L.F.	52	6.05		58.05	66.50	
	3210	Manifold, by L.F. width of collectors				160	.100	"	27	3.62		30.62	35	
	3300	Collector panels, liquid with copper absorber plate												
	3320	Black chrome, tempered glass glazing												
	3330	Alum. frame, 4' x 8', 5/32" single glazing			Q-1	9.50	1.684	Ea.	545	62		607	695	
	3390	Alum. frame, 4' x 10', 5/32" single glazing				6	2.667		650	98		748	855	
	3450	Flat black, alum. frame, 3' x 8'				9	1.778		430	65.50		495.50	575	
	3500	4' x 8.5'				5.50	2.909		490	107		597	700	
	3520	4' x 10.5'				10	1.600		595	59		654	745	
	3540	4' x 12.5'				5	3.200		735	118		853	985	
	3550	Liquid with fin tube absorber plate												
	3560	Alum. frame 4' x 8' tempered glass			Q-1	10	1.600	Ea.	495	59		554	635	
	3580	Liquid with vacuum tubes, 4' x 6'-10"				9	1.778		880	65.50		945.50	1,075	
	3600	Liquid, full wetted, plastic, alum. frame, 3' x 10'				5	3.200		195	118		313	390	
	3650	Collector panel mounting, flat roof or ground rack				7	2.286		45	84		129	176	
	3670	Roof clamps				70	.229	Set	1.61	8.40		10.01	14.40	

Important: See the Reference Section for critical supporting data - Reference Nos., Crews, & City Cost Indexes

13630 | Solar Collector Components

		CREW	DAILY OUTPUT	LABOR-HOURS	UNIT	MAT.	LABOR	EQUIP.	TOTAL	TOTAL INCL O&P	
200	3700	Roof strap, teflon	1 Plum	205	.039	L.F.	9.50	1.59		11.09	12.85	200
	3900	Differential controller with two sensors *(R13600-610)*										
	3930	Thermostat, hard wired	1 Plum	8	1	Ea.	87.50	41		128.50	158	
	3950	Line cord and receptacle		12	.667		83.50	27		110.50	133	
	4050	Pool valve system		2.50	3.200		215	131		346	430	
	4070	With 12 VAC actuator		2	4		276	163		439	550	
	4080	Pool pump system, 2" pipe size		6	1.333		172	54.50		226.50	272	
	4100	Five station with digital read-out		3	2.667		218	109		327	405	
	4150	Sensors										
	4200	Brass plug, 1/2" MPT	1 Plum	32	.250	Ea.	17.95	10.20		28.15	35	
	4210	Brass plug, reversed		32	.250		24.50	10.20		34.70	42.50	
	4220	Freeze prevention		32	.250		22	10.20		32.20	39.50	
	4240	Screw attached		32	.250		10.95	10.20		21.15	27.50	
	4250	Brass, immersion		32	.250		26.50	10.20		36.70	44.50	
	4300	Heat exchanger										
	4330	Fluid to air coil, up flow, 45 MBH	Q-1	4	4	Ea.	295	147		442	545	
	4380	70 MBH		3.50	4.571		335	168		503	620	
	4400	80 MBH		3	5.333		445	196		641	785	
	4580	Fluid to fluid package includes two circulating pumps										
	4590	expansion tank, check valve, relief valve										
	4600	controller, high temperature cutoff and sensors	Q-1	2.50	6.400	Ea.	695	235		930	1,125	
	4650	Heat transfer fluid										
	4700	Propylene glycol, inhibited anti-freeze	1 Plum	28	.286	Gal.	8.80	11.65		20.45	27.50	
	4800	Solar storage tanks, knocked down										
	4810	Air, galvanized steel clad, double wall, 4" fiberglass										
	4820	insulation, 20 Mil PVC lining										
	4870	4' high, 4' x 4', = 64 C.F./450 gallons	Q-9	2	8	Ea.	2,000	289		2,289	2,675	
	4880	4' x 8' = 128 C.F./900 gallons		1.50	10.667		2,600	385		2,985	3,450	
	4890	4' x 12' = 190 C.F./1300 gallons		1.30	12.308		3,475	445		3,920	4,500	
	4900	8' x 8' = 250 C.F./1700 gallons		1	16		3,875	580		4,455	5,175	
	5010	6'-3" high, 7' x 7' = 306 C.F./2000 gallons	Q-10	1.20	20		9,450	750		10,200	11,600	
	5020	7' x 10'-6" = 459 C.F./3000 gallons		.80	30		11,900	1,125		13,025	14,800	
	5030	7' x 14' = 613 C.F./4000 gallons		.60	40		13,900	1,500		15,400	17,600	
	5040	10'-6" x 10'-6" = 689 C.F./4500 gallons		.50	48		14,400	1,800		16,200	18,600	
	5050	10'-6" x 14' = 919 C.F./6000 gallons		.40	60		16,900	2,250		19,150	22,100	
	5060	14' x 14' = 1225 C.F./8000 gallons	Q-11	.40	80		19,800	3,075		22,875	26,500	
	5070	14' x 17'-6" = 1531 C.F./10,000 gallons		.30	106		22,300	4,075		26,375	30,800	
	5080	17'-6" x 17'-6" = 1914 C.F./12,500 gallons		.25	128		26,200	4,900		31,100	36,500	
	5090	17'-6" x 21' = 2297 C.F./15,000 gallons		.20	160		29,300	6,125		35,425	41,600	
	5100	21' x 21' = 2756 C.F./18,000 gallons		.18	177		33,500	6,800		40,300	47,400	
	5120	30 Mil reinforced Chemflex lining,										
	5140	4' high, 4' x 4' = 64 C.F./450 gallons	Q-9	2	8	Ea.	2,050	289		2,339	2,700	
	5150	4' x 8' = 128 C.F./900 gallons		1.50	10.667		3,075	385		3,460	4,000	
	5160	4' x 12' = 190 C.F./1300 gallons		1.30	12.308		4,075	445		4,520	5,175	
	5170	8' x 8' = 250 C.F./1700 gallons		1	16		4,550	580		5,130	5,925	
	5190	6'-3" high, 7' x 7' = 306 C.F./2000 gallons	Q-10	1.20	20		10,100	750		10,850	12,300	
	5200	7' x 10'-6" = 459 C.F./3000 gallons		.80	30		12,300	1,125		13,425	15,200	
	5210	7' x 14' = 613 C.F./4000 gallons		.60	40		14,300	1,500		15,800	18,000	
	5220	10'-6" x 10'-6" = 689 C.F./4500 gallons		.50	48		14,800	1,800		16,600	19,000	
	5230	10'-6" x 14' = 919 C.F./6000 gallons		.40	60		17,300	2,250		19,550	22,600	
	5240	14' x 14' = 1225 C.F./8000 gallons	Q-11	.40	80		20,300	3,075		23,375	27,000	
	5250	14' x 17'-6" = 1531 C.F./10,000 gallons		.30	106		22,700	4,075		26,775	31,300	
	5260	17'-6" x 17'-6" = 1914 C.F./12,500 gallons		.25	128		26,800	4,900		31,700	37,000	
	5270	17'-6" x 21' = 2297 C.F./15,000 gallons		.20	160		30,100	6,125		36,225	42,500	
	5280	21' x 21' = 2756 C.F./18,000 gallons		.18	177		34,500	6,800		41,300	48,500	
	5290	30 Mil reinforced Hypalon lining, add					.02%					

	13630	Solar Collector Components	CREW	DAILY OUTPUT	LABOR-HOURS	UNIT	2005 BARE COSTS MAT.	LABOR	EQUIP.	TOTAL	TOTAL INCL O&P	
200	7000	Solar control valves and vents										200
	7050	Air purger, 1" pipe size [R13600-610]	1 Plum	12	.667	Ea.	43.50	27		70.50	88.50	
	7070	Air eliminator, automatic 3/4" size		32	.250		28.50	10.20		38.70	47	
	7090	Air vent, automatic, 1/8" fitting		32	.250		10.20	10.20		20.40	26.50	
	7100	Manual, 1/8" NPT		32	.250		2.24	10.20		12.44	17.80	
	7120	Backflow preventer, 1/2" pipe size		16	.500		58.50	20.50		79	94.50	
	7130	3/4" pipe size		16	.500		62.50	20.50		83	99	
	7150	Balancing valve, 3/4" pipe size		20	.400		26	16.35		42.35	53	
	7180	Draindown valve, 1/2" copper tube		9	.889		202	36.50		238.50	277	
	7200	Flow control valve, 1/2" pipe size		22	.364		58.50	14.85		73.35	87	
	7220	Expansion tank, up to 5 gal.		32	.250		61.50	10.20		71.70	83	
	7250	Hydronic controller (aquastat)		8	1		43.50	41		84.50	110	
	7400	Pressure gauge, 2" dial		32	.250		22	10.20		32.20	40	
	7450	Relief valve, temp. and pressure 3/4" pipe size		30	.267		9.10	10.90		20	26.50	
	7500	Solenoid valve, normally closed										
	7520	Brass, 3/4" NPT, 24V	1 Plum	9	.889	Ea.	212	36.50		248.50	288	
	7530	1" NPT, 24V		9	.889		212	36.50		248.50	288	
	7750	Vacuum relief valve, 3/4" pipe size		32	.250		26.50	10.20		36.70	44.50	
	7800	Thermometers										
	7820	Digital temperature monitoring, 4 locations	1 Plum	2.50	3.200	Ea.	128	131		259	335	
	7900	Upright, 1/2" NPT		8	1		24	41		65	88	
	7970	Remote probe, 2" dial		8	1		30.50	41		71.50	95	
	7990	Stem, 2" dial, 9" stem		16	.500		20	20.50		40.50	52.50	
	8250	Water storage tank with heat exchanger and electric element										
	8270	66 gal. with 2" x 2 lb. density insulation	1 Plum	1.60	5	Ea.	730	204		934	1,100	
	8300	80 gal. with 2" x 2 lb. density insulation		1.60	5		835	204		1,039	1,225	
	8380	120 gal. with 2" x 2 lb. density insulation		1.40	5.714		940	233		1,173	1,375	
	8400	120 gal. with 2" x 2 lb. density insul., 40 S.F. heat coil		1.40	5.714		1,075	233		1,308	1,550	
	8500	Water storage module, plastic										
	8600	Tubular, 12" diameter, 4' high	1 Carp	48	.167	Ea.	80	5.70		85.70	97.50	
	8610	12" diameter, 8' high		40	.200		122	6.85		128.85	146	
	8620	18" diameter, 5' high		38	.211		134	7.20		141.20	159	
	8630	18" diameter, 10' high		32	.250		176	8.55		184.55	207	
	8640	58" diameter, 5' high	2 Carp	32	.500		490	17.15		507.15	560	
	8650	Cap, 12" diameter					12.65			12.65	13.95	
	8660	18" diameter					15.85			15.85	17.40	

	13720	Detection & Alarm										
065	0010	**DETECTION SYSTEMS**, not including wires & conduits										065
	0100	Burglar alarm, battery operated, mechanical trigger	1 Elec	4	2	Ea.	254	81.50		335.50	400	
	0200	Electrical trigger		4	2		305	81.50		386.50	455	
	0400	For outside key control, add		8	1		72	41		113	140	
	0600	For remote signaling circuitry, add		8	1		114	41		155	187	
	0800	Card reader, flush type, standard		2.70	2.963		850	121		971	1,125	
	1000	Multi-code		2.70	2.963		1,100	121		1,221	1,375	
	3594	Fire, alarm control panel										
	3600	4 zone	2 Elec	2	8	Ea.	940	325		1,265	1,500	
	3800	8 zone		1	16		1,400	650		2,050	2,525	
	4000	12 zone		.67	23.988		1,825	980		2,805	3,450	
	4020	Alarm device	1 Elec	8	1		122	41		163	195	
	4050	Actuating device		8	1		292	41		333	380	
	4200	Battery and rack		4	2		690	81.50		771.50	880	
	4400	Automatic charger		8	1		445	41		486	550	
	4600	Signal bell		8	1		49.50	41		90.50	115	
	4800	Trouble buzzer or manual station		8	1		37	41		78	101	
	5000	Detector, rate of rise		8	1		33.50	41		74.50	97.50	

FIRE

Important: See the Reference Section for critical supporting data - Reference Nos., Crews, & City Cost Indexes

13720	Detection & Alarm	CREW	DAILY OUTPUT	LABOR-HOURS	UNIT	2005 BARE COSTS				TOTAL INCL O&P	
						MAT.	LABOR	EQUIP.	TOTAL		
065	5100	Fixed temperature	1 Elec	8	1	Ea.	28	41		69	91.50	065
5200	Smoke detector, ceiling type		6.20	1.290		75	52.50		127.50	161
5400	Duct type		3.20	2.500		250	102		352	425
5600	Strobe and horn		5.30	1.509		95	61.50		156.50	197
5800	Fire alarm horn		6.70	1.194		36.50	48.50		85	113
6000	Door holder, electro-magnetic		4	2		77.50	81.50		159	207
6200	Combination holder and closer		3.20	2.500		430	102		532	625
6400	Code transmitter		4	2		690	81.50		771.50	880
6600	Drill switch		8	1		86.50	41		127.50	156
6800	Master box		2.70	2.963		3,100	121		3,221	3,575
7000	Break glass station		8	1		50	41		91	116
7800	Remote annunciator, 8 zone lamp		1.80	4.444		175	181		356	465
8000	12 zone lamp	2 Elec	2.60	6.154		300	251		551	705
8200	16 zone lamp	"	2.20	7.273		300	296		596	770

13838	Pneumatic/Electric Controls	CREW	DAILY OUTPUT	LABOR-HOURS	UNIT	2005 BARE COSTS				TOTAL INCL O&P	
						MAT.	LABOR	EQUIP.	TOTAL		
200	0010	**CONTROL COMPONENTS**										200
2000	Gauges, pressure or vacuum									
2100	2″ diameter dial	1 Stpi	32	.250	Ea.	8.35	10.25		18.60	24.50
2200	2-1/2″ diameter dial		32	.250		9.10	10.25		19.35	25.50
2300	3-1/2″ diameter dial		32	.250		20	10.25		30.25	37.50
2400	4-1/2″ diameter dial		32	.250		30	10.25		40.25	48.50
2700	Flanged iron case, black ring									
2800	3-1/2″ diameter dial	1 Stpi	32	.250	Ea.	63.50	10.25		73.75	85
2900	4-1/2″ diameter dial		32	.250		86.50	10.25		96.75	110
3000	6″ diameter dial		32	.250		68.50	10.25		78.75	91
3010	Steel case, 0 - 300 psi									
3012	2″ diam. dial	1 Stpi	16	.500	Ea.	18.15	20.50		38.65	51
3014	4″ diam. dial	"	16	.500	"	42.50	20.50		63	77.50
3020	Aluminum case, 0 - 300 psi									
3022	3-1/2″ diam. dial	1 Stpi	16	.500	Ea.	56	20.50		76.50	92.50
3024	4-1/2″ diam. dial		16	.500		71	20.50		91.50	110
3026	6″ diam. dial		16	.500		118	20.50		138.50	161
3028	8-1/2″ diam. dial		16	.500		281	20.50		301.50	340
3030	Brass case, 0 - 300 psi									
3032	2″ diam. dial	1 Stpi	16	.500	Ea.	23.50	20.50		44	57
3034	4-1/2″ diam. dial	"	16	.500	"	53.50	20.50		74	90
3040	Steel case, high pressure, 0 -10,000 psi									
3042	4-1/2″ diam. dial	1 Stpi	16	.500	Ea.	134	20.50		154.50	178
3044	6-1/2″ diam. dial		16	.500		215	20.50		235.50	267
3046	8-1/2″ diam. dial		16	.500		390	20.50		410.50	460
3080	Pressure gauge, differential, magnehelic									
3084	0 - 2″ W.C., with air filter kit	1 Stpi	6	1.333	Ea.	79.50	54.50		134	169
3300	For compound pressure-vacuum, add					18%				
4000	Thermometers									
4100	Dial type, 3-1/2″ diameter, vapor type, union connection	1 Stpi	32	.250	Ea.	96.50	10.25		106.75	121
4120	Liquid type, union connection		32	.250		135	10.25		145.25	163
4500	Stem type, 6-1/2″ case, 2″ stem, 1/2″ NPT		32	.250		55	10.25		65.25	76

		13838	Pneumatic/Electric Controls	CREW	DAILY OUTPUT	LABOR-HOURS	UNIT	2005 BARE COSTS				TOTAL INCL O&P	
								MAT.	LABOR	EQUIP.	TOTAL		
200	4520		4" stem, 1/2" NPT	1 Stpi	32	.250	Ea.	75.50	10.25		85.75	98.50	200
	4600		9" case, 3-1/2" stem, 3/4" NPT		28	.286		90	11.70		101.70	117	
	4620		6" stem, 3/4" NPT		28	.286		96	11.70		107.70	123	
	4640		8" stem, 3/4" NPT		28	.286		118	11.70		129.70	148	
	4660		12" stem, 1" NPT	↓	26	.308	↓	130	12.60		142.60	162	
	4670		Bi-metal, dial type, steel case brass stem										
	4672		2" dial, 4" - 9" stem	1 Stpi	16	.500	Ea.	29.50	20.50		50	63.50	
	4673		2-1/2" dial, 4" - 9" stem		16	.500		34	20.50		54.50	68	
	4674		3-1/2" dial, 4" - 9" stem	↓	16	.500	↓	33	20.50		53.50	67	
	4680		Mercury filled, industrial, union connection type										
	4682		Angle stem, 7" scale	1 Stpi	16	.500	Ea.	74.50	20.50		95	113	
	4683		9" scale		16	.500		109	20.50		129.50	151	
	4684		12" scale		16	.500		121	20.50		141.50	164	
	4686		Straight stem, 7" scale		16	.500		96	20.50		116.50	136	
	4687		9" scale		16	.500		90	20.50		110.50	130	
	4688		12" scale	↓	16	.500	↓	110	20.50		130.50	152	
	4690		Mercury filled, industrial, separable socket type										
	4692		Angle stem, with socket, 7" scale	1 Stpi	16	.500	Ea.	64.50	20.50		85	102	
	4693		9" scale		16	.500		77	20.50		97.50	116	
	4694		12" scale		16	.500		156	20.50		176.50	202	
	4696		Straight stem, with socket, 7" scale		16	.500		64.50	20.50		85	102	
	4697		9" scale		16	.500		77	20.50		97.50	116	
	4698		12" scale	↓	16	.500	↓	90.50	20.50		111	131	
	6000		Valves, motorized zone										
	6100		Sweat connections, 1/2" C x C	1 Stpi	20	.400	Ea.	67.50	16.40		83.90	99	
	6110		3/4" C x C		20	.400		72.50	16.40		88.90	104	
	6120		1" C x C		19	.421		84.50	17.25		101.75	119	
	6140		1/2" C x C, with end switch, 2 wire		20	.400		71.50	16.40		87.90	103	
	6150		3/4" C x C, with end switch, 2 wire		20	.400		76	16.40		92.40	108	
	6160		1" C x C, with end switch, 2 wire		19	.421		86	17.25		103.25	121	
	6180		1-1/4" C x C, w/end switch, 2 wire	↓	15	.533	↓	103	22		125	147	

		13851	Detection & Alarm	CREW	DAILY OUTPUT	LABOR-HOURS	UNIT	2005 BARE COSTS				TOTAL INCL O&P	
								MAT.	LABOR	EQUIP.	TOTAL		
350	0010		**TANK LEAK DETECTION SYSTEMS** Liquid and vapor										350
	0100		For hydrocarbons and hazardous liquids/vapors										
	0120		Controller, data acquisition, incl. printer, modem, RS232 port										
	0140		24 channel, for use with all probes				Ea.	2,100			2,100	2,325	
	0160		9 channel, for external monitoring				"	1,625			1,625	1,800	
	0200		Probes										
	0210		Well monitoring										
	0220		Liquid phase detection				Ea.	735			735	805	
	0230		Hydrocarbon vapor, fixed position					675			675	745	
	0240		Hydrocarbon vapor, float mounted					600			600	660	
	0250		Both liquid and vapor hydrocarbon				↓	910			910	1,000	
	0300		Secondary containment, liquid phase										
	0310		Pipe trench/manway sump				Ea.	325			325	360	
	0320		Double wall pipe and manual sump				↓	320			320	350	

Important: See the Reference Section for critical supporting data - Reference Nos., Crews, & City Cost Indexes

13851		Detection & Alarm	CREW	DAILY OUTPUT	LABOR-HOURS	UNIT	2005 BARE COSTS				TOTAL INCL O&P	
							MAT.	LABOR	EQUIP.	TOTAL		
350	0330	Double wall fiberglass annular space				Ea.	295			295	325	35
	0340	Double wall steel tank annular space				↓	295			295	325	
	0500	Accessories										
	0510	Modem, non-dedicated phone line				Ea.	264			264	290	
	0600	Monitoring, internal										
	0610	Automatic tank gauge, incl. overfill				Ea.	1,050			1,050	1,150	
	0620	Product line				"	1,125			1,125	1,250	
	0700	Monitoring, special										
	0710	Cathodic protection				Ea.	630			630	690	
	0720	Annular space chemical monitor				"	855			855	940	

13910		Basic Fire Protection Matl/Methd	CREW	DAILY OUTPUT	LABOR-HOURS	UNIT	2005 BARE COSTS				TOTAL INCL O&P	
							MAT.	LABOR	EQUIP.	TOTAL		
400	0010	**FIRE HOSE AND EQUIPMENT** D4020 310										40
	0200	Adapters, rough brass, straight hose threads										
	0220	One piece, female to male, rocker lugs D4020 330										
	0240	1" x 1"				Ea.	26.50			26.50	29	
	0260	1-1/2" x 1" D4020 410					26			26	28.50	
	0280	1-1/2" x 1-1/2"					9.20			9.20	10.10	
	0300	2" x 1-1/2" R10520 -310					47			47	51.50	
	0320	2" x 2"					47			47	51.50	
	0340	2-1/2" x 1-1/2" R10520 -320					31			31	34	
	0360	2-1/2" x 2"					21			21	23	
	0380	2-1/2" x 2-1/2"					14.60			14.60	16.10	
	0400	3" x 2-1/2"					39			39	43	
	0420	3" x 3"					104			104	114	
	0500	For polished brass, add					50%					
	0520	For polished chrome, add				↓	75%					
	0700	One piece, female to male, hexagon										
	0740	1-1/2" x 3/4"				Ea.	25			25	27.50	
	0760	2" x 1-1/2"					46			46	50.50	
	0780	2-1/2" x 1"					88			88	97	
	0800	2-1/2" x 1-1/2"					31.50			31.50	34.50	
	0820	2-1/2" x 2"					20.50			20.50	22.50	
	0840	3" x 2-1/2"					38			38	42	
	0900	For polished chrome, add				↓	75%					
	1100	Swivel, female to female, pin lugs										
	1120	1-1/2" x 1-1/2"				Ea.	34			34	37	
	1200	2-1/2" x 2-1/2"					66.50			66.50	73.50	
	1260	For polished brass, add					50%					
	1280	For polished chrome, add				↓	75%					
	1400	Couplings, sngl & dbl jacket, pin lug or rocker lug, cast brass										
	1410	1-1/2"				Ea.	29.50			29.50	32	
	1420	2-1/2"				"	38.50			38.50	42.50	
	1500	For polished brass, add					20%					
	1520	For polished chrome, add					40%					
	1580	Reducing, F x M, interior installation, cast brass										
	1590	2" x 1-1/2"				Ea.	37			37	40.50	
	1600	2-1/2" x 1-1/2"				↓	8.25			8.25	9.10	

13910	Basic Fire Protection Matl/Methd	CREW	DAILY OUTPUT	LABOR-HOURS	UNIT	2005 BARE COSTS				TOTAL INCL O&P
						MAT.	LABOR	EQUIP.	TOTAL	
1680	For polished brass, add	R10520-310			Ea.	50%				
1720	For polished chrome, add					75%				
1900	Escutcheon plate, for angle valves, polished brass, 1-1/2"	R10520-320				10.90			10.90	12
1920	2-1/2"					26			26	29
1940	3"					24			24	26.50
1980	For polished chrome, add					15%				
2200	Hose, less couplings									
2260	Synthetic jacket, lined, 300 lb. test, 1-1/2" diameter	Q-12	2,600	.006	L.F.	1.62	.23		1.85	2.12
2280	2-1/2" diameter		2,200	.007		2.70	.27		2.97	3.37
2360	High strength, 500 lb. test, 1-1/2" diameter		2,600	.006		1.68	.23		1.91	2.19
2380	2-1/2" diameter		2,200	.007		2.91	.27		3.18	3.60
2600	Hose rack, swinging, for 1-1/2" diameter hose,									
2620	Enameled steel, 50' & 75' lengths of hose	Q-12	20	.800	Ea.	33	29.50		62.50	81
2640	100' and 125' lengths of hose		20	.800		33	29.50		62.50	81
2680	Chrome plated, 50' and 75' lengths of hose		20	.800		51	29.50		80.50	101
2700	100' and 125' lengths of hose		20	.800		51	29.50		80.50	101
2780	For hose rack nipple, 1-1/2" polished brass, add					15.90			15.90	17.50
2820	2-1/2" polished brass, add					26.50			26.50	29
2840	1-1/2" polished chrome, add					16.50			16.50	18.15
2860	2-1/2" polished chrome, add					29			29	31.50
2990	Hose reel, swinging, for 1-1/2" polyester neoprene lined hose									
3000	50' long	Q-12	14	1.143	Ea.	75.50	42		117.50	147
3020	100' long		14	1.143		102	42		144	177
3060	For 2-1/2" cotton rubber hose, 75' long		14	1.143		118	42		160	194
3100	150' long		14	1.143		138	42		180	216
3750	Hydrants, wall, w/caps, single, flush, polished brass									
3800	2-1/2" x 2-1/2"	Q-12	5	3.200	Ea.	120	117		237	310
3840	2-1/2" x 3"		5	3.200		161	117		278	355
3860	3" x 3"		4.80	3.333		196	122		318	400
3900	For polished chrome, add					20%				
3950	Double, flush, polished brass									
4000	2-1/2" x 2-1/2" x 4"	Q-12	5	3.200	Ea.	320	117		437	530
4040	2-1/2" x 2-1/2" x 6"		4.60	3.478		465	128		593	705
4080	3" x 3" x 4"		4.90	3.265		680	120		800	930
4120	3" x 3" x 6"		4.50	3.556		705	130		835	970
4200	For polished chrome, add					10%				
4350	Double, projecting, polished brass									
4400	2-1/2" x 2-1/2" x 4"	Q-12	5	3.200	Ea.	148	117		265	340
4450	2-1/2" x 2-1/2" x 6"	"	4.60	3.478	"	286	128		414	510
4460	Valve control, dbl. flush/projecting hydrant, cap &									
4470	chain, ext. rod & cplg., escutcheon, polished brass	Q-12	8	2	Ea.	174	73.50		247.50	300
4480	Four-way square, flush, polished brass									
4540	2-1/2"(4) x 6"	Q-12	3.60	4.444	Ea.	1,725	163		1,888	2,150
5000	Nipples, straight hose to tapered iron pipe, brass									
5060	Female to female, 1-1/2" x 1-1/2"				Ea.	13.45			13.45	14.80
5080	2" x 2"					15.50			15.50	17.05
5100	2-1/2" x 2-1/2"					18.75			18.75	20.50
5200	Double male or male to female, 1" x 1"					23.50			23.50	25.50
5220	1-1/2" x 1"					34			34	37.50
5230	1-1/2" x 1-1/2"					7.50			7.50	8.25
5260	2" x 1-1/2"					37.50			37.50	41.50
5270	2" x 2"					47			47	51.50
5280	2-1/2" x 1-1/2"					31			31	34
5300	2-1/2" x 2"					25			25	27.50
5310	2-1/2" x 2-1/2"					14.60			14.60	16.10
5340	For polished chrome, add					75%				

Important: See the Reference Section for critical supporting data - Reference Nos., Crews, & City Cost Indexes

13910 | Basic Fire Protection Matl/Methd

		CREW	DAILY OUTPUT	LABOR-HOURS	UNIT	2005 BARE COSTS MAT.	LABOR	EQUIP.	TOTAL	TOTAL INCL O&P
400 5600	Nozzles, brass R10520-310									**400**
5620	Adjustable fog, 3/4" booster line				Ea.	70.50			70.50	77.50
5630	1" booster line R10520-320					87			87	95.50
5640	1-1/2" leader line					91			91	100
5660	2-1/2" direct connection					178			178	196
5680	2-1/2" playpipe nozzle					147			147	162
5780	For chrome plated, add					8%				
5850	Electrical fire, adjustable fog, no shock									
5900	1-1/2"				Ea.	268			268	295
5920	2-1/2"					365			365	400
5980	For polished chrome, add					6%				
6200	Heavy duty, comb. adj. fog and str. stream, with handle									
6210	1" booster line				Ea.	256			256	282
6240	1-1/2"					253			253	279
6260	2-1/2", for playpipe					310			310	340
6280	2-1/2" direct connection					385			385	420
6300	2-1/2" playpipe combination					520			520	570
6480	For polished chrome, add					7%				
6500	Plain fog, polished brass, 1-1/2"					48			48	53
6540	Chrome plated, 1-1/2"					48.50			48.50	53.50
6700	Plain stream, polished brass, 1-1/2" x 10"					26.50			26.50	29.50
6760	2-1/2" x 15" x 7/8" or 1-1/2"					48.50			48.50	53
6860	For polished chrome, add					20%				
7000	Underwriters playpipe, 2-1/2" x 30" with 1-1/8" tip				Ea.	182			182	201
7040	Less tip				"	144			144	159
7140	Standpipe connections, wall, w/plugs & chains									
7160	Single, flush, brass, 2-1/2" x 2-1/2"	Q-12	5	3.200	Ea.	88.50	117		205.50	275
7180	2-1/2" x 3"	"	5	3.200	"	92	117		209	278
7240	For polished chrome, add					15%				
7280	Double, flush, polished brass									
7300	2-1/2" x 2-1/2" x 4"	Q-12	5	3.200	Ea.	297	117		414	500
7330	2-1/2" x 2-1/2" x 6"		4.60	3.478		405	128		533	640
7340	3" x 3" x 4"		4.90	3.265		530	120		650	760
7370	3" x 3" x 6"		4.50	3.556		630	130		760	890
7400	For polished chrome, add					15%				
7440	For sill cock combination, add				Ea.	50			50	55.50
7580	Double projecting, polished brass									
7600	2-1/2" x 2-1/2" x 4"	Q-12	5	3.200	Ea.	273	117		390	475
7630	2-1/2" x 2-1/2" x 6"	"	4.60	3.478	"	450	128		578	690
7680	For polished chrome, add					15%				
7900	Three way, flush, polished brass									
7920	2-1/2" (3) x 4"	Q-12	4.80	3.333	Ea.	930	122		1,052	1,200
7930	2-1/2" (3) x 6"	"	4.80	3.333		940	122		1,062	1,200
8000	For polished chrome, add					9%				
8020	Three way, projecting, polished brass									
8040	2-1/2"(3) x 4"	Q-12	4.80	3.333	Ea.	620	122		742	865
8070	2-1/2" (3) x 6"	"	4.60	3.478		620	128		748	875
8100	For polished chrome, add					12%				
8200	Four way, square, flush, polished brass,									
8240	2-1/2"(4) x 6"	Q-12	3.60	4.444	Ea.	1,625	163		1,788	2,050
8300	For polished chrome, add				"	10%				
8550	Wall, vertical, flush, cast brass									
8600	Two way, 2-1/2" x 2-1/2" x 4"	Q-12	5	3.200	Ea.	665	117		782	905
8660	Four way, 2-1/2"(4) x 6"		3.80	4.211		1,475	154		1,629	1,850
8680	Six way, 2-1/2"(6) x 6"		3.40	4.706		1,975	173		2,148	2,425
8700	For polished chrome, add					10%				

			R10520	CREW	DAILY OUTPUT	LABOR-HOURS	UNIT	2005 BARE COSTS				TOTAL INCL O&P
	13910	**Basic Fire Protection Matl/Methd**						MAT.	LABOR	EQUIP.	TOTAL	
400	8800	Sidewalk siamese unit, polished brass, two way	R10520 -310	Q-12	2.50	6.400	Ea.	315	235		550	700
	8820	2-1/2" x 2-1/2" x 4"										
	8850	2-1/2" x 2-1/2" x 6"	R10520 -320		2	8		525	293		818	1,025
	8860	3" x 3" x 4"			2.50	6.400		510	235		745	915
	8890	3" x 3" x 6"			2	8		690	293		983	1,200
	8940	For polished chrome, add						12%				
	9100	Sidewalk siamese unit, polished brass, three way										
	9120	2-1/2" x 2-1/2" x 2-1/2" x 6"		Q-12	2	8	Ea.	755	293		1,048	1,275
	9160	For polished chrome, add						15%				
	9200	Storage house, hose only, primed steel						475			475	525
	9220	Aluminum						1,050			1,050	1,150
	9280	Hose and hydrant house, primed steel						645			645	705
	9300	Aluminum						935			935	1,025
	9340	Tools, crowbar and brackets		1 Carp	12	.667		66.50	23		89.50	109
	9360	Combination hydrant wrench and spanner						17.80			17.80	19.60
	9380	Fire axe and brackets										
	9400	6 lb.		1 Carp	12	.667	Ea.	68.50	23		91.50	111
800	0010	**FIRE VALVES**	D4020 310									
	0020	Angle, combination pressure adjust/restricting, rough brass										
	0030	1-1/2"	D4020 330	1 Spri	12	.667	Ea.	47	27		74	93
	0040	2-1/2"		"	7	1.143	"	99.50	46.50		146	181
	0042	Nonpressure adjustable/restricting, rough brass										
	0044	1-1/2"		1 Spri	12	.667	Ea.	30	27		57	74
	0046	2-1/2"	D4020 410	"	7	1.143	"	51.50	46.50		98	127
	0050	For polished brass, add						30%				
	0060	For polished chrome, add						40%				
	0080	Wheel handle, 300 lb., 1-1/2"	R10520 -310	1 Spri	12	.667	Ea.	30	27		57	74
	0090	2-1/2"	R10520 -320	"	7	1.143	"	51.50	46.50		98	127
	0100	For polished brass, add						35%				
	0110	For polished chrome, add						50%				
	1000	Ball drip, automatic, rough brass, 1/2"		1 Spri	20	.400	Ea.	8.60	16.30		24.90	34
	1010	3/4"		"	20	.400	"	9.75	16.30		26.05	35.50
	1100	Ball, 175 lb., sprinkler system, FM/UL, threaded, bronze										
	1120	Slow close										
	1150	1" size		1 Spri	19	.421	Ea.	84	17.15		101.15	119
	1160	1-1/4" size			15	.533		91	21.50		112.50	133
	1170	1-1/2" size			13	.615		115	25		140	165
	1180	2" size			11	.727		145	29.50		174.50	205
	1190	2-1/2" size		Q-12	15	1.067		196	39		235	275
	1230	For supervisory switch kit, all sizes										
	1240	One circuit, add		1 Spri	48	.167	Ea.	63.50	6.80		70.30	80.50
	1280	Quarter turn for trim										
	1300	1/2" size		1 Spri	22	.364	Ea.	11.15	14.80		25.95	35
	1310	3/4" size			20	.400		11.95	16.30		28.25	37.50
	1320	1" size			19	.421		13.30	17.15		30.45	40.50
	1330	1-1/4" size			15	.533		22	21.50		43.50	57
	1340	1-1/2" size			13	.615		27.50	25		52.50	68
	1350	2" size			11	.727		32.50	29.50		62	81
	1400	Caps, polished brass with chain, 3/4"						23.50			23.50	25.50
	1420	1"						29.50			29.50	32.50
	1440	1-1/2"						7.85			7.85	8.65
	1460	2-1/2"						12.35			12.35	13.60
	1480	3"						18.35			18.35	20
	2000	Foam, control valve, 3"		1 Spri	6	1.333		1,175	54.50		1,229.50	1,375
	2020	Supply valve, 2-1/2"			7	1.143		44	46.50		90.50	119

120

13910 | Basic Fire Protection Matl/Methd

			CREW	DAILY OUTPUT	LABOR-HOURS	UNIT	MAT.	LABOR	EQUIP.	TOTAL	TOTAL INCL O&P	
800	2040	Proportioner, 8"	R10520 -310	1 Spri	2	4	Ea.	1,375	163		1,538	1,750	800
	2060	Oscillating foam monitor		Q-12	5.33	3.002		6,700	110		6,810	7,525	
	3000	Gate, hose, wheel handle, N.R.S., rough brass, 1-1/2"	R10520 -320	1 Spri	12	.667		75	27		102	124	
	3040	2-1/2", 300 lb.		"	7	1.143		102	46.50		148.50	183	
	3080	For polished brass, add						40%					
	3090	For polished chrome, add						50%					
	3800	Hydrant, screw type, crank handle, brass											
	3840	2-1/2" size		Q-12	11	1.455	Ea.	199	53.50		252.50	300	
	3880	For chrome, same price											
	4200	Hydrolator, vent and draining, rough brass, 1-1/2"		1 Spri	12	.667	Ea.	41.50	27		68.50	87	
	4280	For polished brass, add						50%					
	4290	For polished chrome, add						90%					
	5000	Pressure restricting, adjustable rough brass, 1-1/2"		1 Spri	12	.667		70.50	27		97.50	119	
	5020	2-1/2"		"	7	1.143		99	46.50		145.50	180	
	5080	For polished brass, add						30%					
	5090	For polished chrome, add						45%					
	6000	Roof manifold, horiz., brass, without valves & caps											
	6040	2-1/2" x 2-1/2" x 4"		Q-12	4.80	3.333	Ea.	103	122		225	298	
	6060	2-1/2" x 2-1/2" x 6"			4.60	3.478		122	128		250	325	
	6080	2-1/2" x 2-1/2" x 2-1/2" x 4"			4.60	3.478		166	128		294	375	
	6090	2-1/2" x 2-1/2" x 2-1/2" x 6"			4.60	3.478		171	128		299	380	
	7000	Sprinkler line tester, cast brass						17.75			17.75	19.55	
	8000	Wye, leader line, ball type, swivel female x male x male											
	8040	2-1/2" x 1-1/2" x 1-1/2" polished brass					Ea.	252			252	277	
	8060	2-1/2" x 1-1/2" x 1-1/2" polished chrome					"	259			259	285	

13920 | Fire Pumps

			CREW	DAILY OUTPUT	LABOR-HOURS	UNIT	MAT.	LABOR	EQUIP.	TOTAL	TOTAL INCL O&P	
400	0010	**FIRE PUMPS** Including controller, fittings and relief valve									400
	0030	Diesel									
	0050	500 GPM, 50 psi, 27 HP, 4" pump	Q-13	.64	50	Ea.	48,700	1,950		50,650	56,500	
	0100	500 GPM, 100 psi, 62 HP, 4" pump		.60	53.333		51,500	2,075		53,575	59,500	
	0150	500 GPM, 125 psi, 78 HP, 4" pump		.56	57.143		52,000	2,225		54,225	61,000	
	0200	750 GPM, 50 psi, 44 HP, 5" pump		.60	53.333		52,000	2,075		54,075	60,000	
	0250	750 GPM, 100 psi, 80 HP, 4" pump		.56	57.143		54,500	2,225		56,725	63,000	
	0300	750 GPM, 165 psi, 203 HP, 5" pump		.52	61.538		56,000	2,400		58,400	65,500	
	0350	1000 GPM, 50 psi, 48 HP, 5" pump		.58	55.172		52,500	2,150		54,650	61,000	
	0400	1000 GPM, 100 psi, 89 HP, 4" pump		.56	57.143		55,500	2,225		57,725	65,000	
	0450	1000 GPM, 150 psi, 148 HP, 4" pump		.48	66.667		55,000	2,600		57,600	64,500	
	0470	1000 GPM, 200 psi, 280 HP, 5" pump		.40	80		69,000	3,100		72,100	80,500	
	0480	1250 GPM, 75 psi, 75 HP, 5" pump		.54	59.259		53,000	2,300		55,300	61,500	
	0500	1500 GPM, 50 psi, 66 HP, 6" pump		.50	64		56,500	2,475		58,975	66,000	
	0550	1500 GPM, 100 psi, 140 HP, 6" pump		.46	69.565		57,500	2,700		60,200	67,500	
	0600	1500 GPM, 150 psi, 228 HP, 6" pump		.42	76.190		65,000	2,950		67,950	76,000	
	0650	1500 GPM, 200 psi, 279 HP, 6" pump		.38	84.211		102,000	3,275		105,275	117,000	
	0700	2,000 GPM, 100 psi, 167 HP, 6" pump		.34	94.118		59,000	3,650		62,650	70,500	
	0750	2000 GPM, 150 psi, 284 HP, 6"pump		.30	106		80,000	4,150		84,150	94,500	
	0800	2500 GPM, 100 psi, 213 HP, 8" pump		.32	100		75,500	3,875		79,375	89,000	
	0820	2500 GPM, 150 psi, 365 HP, 8" pump		.26	123		88,500	4,775		93,275	104,000	
	0850	3000 GPM, 100 psi, 250 HP, 8" pump		.28	114		87,000	4,450		91,450	102,000	
	0900	3000 GPM, 150 psi, 384 HP, 10" pump		.20	160		101,000	6,225		107,225	121,000	
	0950	3500 GPM, 100 psi, 300 HP, 10" pump		.24	133		104,000	5,175		109,175	122,500	
	1000	3500 GPM., 150 psi, 518 HP, 10" pump		.20	160		147,000	6,225		153,225	171,000	
	3000	Electric									
	3100	250 GPM, 55 psi, 15 HP, 3,550 RPM, 2" pump	Q-13	.70	45.714	Ea.	11,400	1,775		13,175	15,300	
	3200	500 GPM, 50 psi, 27 HP, 1770 RPM, 4" pump		.68	47.059		14,200	1,825		16,025	18,400	

13920 | Fire Pumps

		CREW	DAILY OUTPUT	LABOR-HOURS	UNIT	2005 BARE COSTS				TOTAL INCL O&P
						MAT.	LABOR	EQUIP.	TOTAL	
400											400
3250	500 GPM, 100 psi, 47 HP, 3550 RPM, 3" pump	Q-13	.66	48.485	Ea.	17,200	1,875		19,075	21,800
3300	500 GPM, 125 psi, 64 HP, 3550 RPM, 3" pump		.62	51.613		18,800	2,000		20,800	23,700
3350	750 GPM, 50 psi, 44 HP, 1,770 RPM, 5" pump		.64	50		20,700	1,950		22,650	25,700
3400	750 GPM, 100 psi, 66 HP, 3550 RPM, 4" pump		.58	55.172		18,300	2,150		20,450	23,500
3450	750 GPM, 165 psi, 120 HP, 3550 RPM, 4" pump		.56	57.143		23,900	2,225		26,125	29,700
3500	1000 GPM, 50 psi, 48 HP 1770 RPM, 5" pump		.60	53.333		20,700	2,075		22,775	25,900
3550	1000 GPM, 100 psi, 86 HP, 3550 RPM, 5" pump		.54	59.259		23,200	2,300		25,500	29,000
3600	1000 GPM, 150 psi, 142 HP, 3550 RPM, 5" pump		.50	64		26,300	2,475		28,775	32,700
3650	1000 GPM, 200 psi, 245 HP, 1770 RPM, 6" pump		.36	88.889		57,000	3,450		60,450	67,500
3660	1250 GPM, 75 psi, 75 HP, 1770 RPM, 5" pump		.55	58.182		21,300	2,250		23,550	26,800
3700	1500 GPM, 50 psi, 66 HP, 1770 RPM, 6" pump		.50	64		22,300	2,475		24,775	28,300
3750	1500 GPM, 100 psi, 139 HP, 1770 RPM, 6" pump		.46	69.565		25,200	2,700		27,900	31,800
3800	1500 GPM, 150 psi, 200 HP, 1770 RPM, 6" pump		.36	88.889		53,500	3,450		56,950	64,000
3850	1500 GPM, 200 psi, 279 HP, 1770 RPM, 6" pump		.32	100		58,500	3,875		62,375	70,500
3900	2000 GPM, 100 psi, 167 HP, 1770 RPM, 6" pump		.34	94.118		32,200	3,650		35,850	41,000
3950	2000 GPM, 150 psi, 292 HP, 1770 RPM, 6" pump		.28	114		43,200	4,450		47,650	54,000
4000	2500 GPM, 100 psi, 213 HP, 1770 RPM, 8" pump		.30	106		35,300	4,150		39,450	45,100
4040	2500 GPM, 135 psi, 339 HP, 1770 RPM, 8" pump		.26	123		50,500	4,775		55,275	62,500
4100	3000 GPM, 100 psi, 250 HP, 1770 RPM, 8" pump		.28	114		42,300	4,450		46,750	53,000
4150	3000 GPM, 140 psi, 428 HP, 1770 RPM, 10" pump		.24	133		67,000	5,175		72,175	82,000
4200	3500 GPM, 100 psi, 300 HP, 1770 RPM, 10" pump		.26	123		52,500	4,775		57,275	64,500
4250	3500 GPM, 140 psi, 450 HP, 1770 RPM, 10" pump		.24	133		67,000	5,175		72,175	82,000
5000	For jockey pump 1", 3 HP, with control, add	Q-12	2	8		2,550	293		2,843	3,250

13930 | Wet-Pipe Fire Supp. Sprinklers

			CREW	DAILY OUTPUT	LABOR-HOURS	UNIT	MAT.	LABOR	EQUIP.	TOTAL	TOTAL INCL O&P
400												400
0010	**SPRINKLER SYSTEM COMPONENTS**										
0600	Accelerator	D4010 305	1 Spri	8	1	Ea.	380	41		421	480
0800	Air compressor for dry pipe system, automatic, complete	D4010 400									
0820	280 gal. system capacity, 3/4 HP		1 Spri	1.30	6.154	Ea.	720	251		971	1,175
0860	520 gal. system capacity, 1 HP	D4010 415		1.30	6.154		850	251		1,101	1,325
0910	650 gal. system capacity, 1-1/2 HP			1.30	6.154		865	251		1,116	1,325
0920	790 gal. system capacity, 2 HP	D4010 425		1.30	6.154		900	251		1,151	1,375
0960	Air pressure maintenance control			24	.333		157	13.60		170.60	194
1100	Alarm, electric pressure switch (circuit closer)	D4010 435		26	.308		148	12.55		160.55	182
1140	For explosion proof, max 20 PSI, contacts close or open			26	.308		296	12.55		308.55	345
1220	Water motor, complete with gong	R10520 -110		4	2		188	81.50		269.50	330
1400	Deluge system, pressured monitoring panel, 120V			18	.444		885	18.10		903.10	1,000
1600	Dehydrator package, incl. valves and nipples	R10520 -120		12	.667		325	27		352	395
1800	Firecycle system, controls, includes panel,										
1820	batteries, solenoid valves and pressure switches	R10520 -130	Q-13	1	32	Ea.	8,800	1,250		10,050	11,600
1980	Detector		1 Spri	16	.500		325	20.50		345.50	385
2000	Release, emergency, manual, for hydraulic or pneumatic system			12	.667		99	27		126	150
2060	Release, thermostatic, for hydraulic or pneumatic release line			20	.400		300	16.30		316.30	355
2200	Sprinkler cabinets, 6 head capacity			16	.500		35	20.50		55.50	69.50
2260	12 head capacity			16	.500		37	20.50		57.50	72
2340	Sprinkler head escutcheons, standard, brass tone, 1" size			40	.200		3.44	8.15		11.59	16.10
2360	Chrome, 1" size			40	.200		1.79	8.15		9.94	14.25
2400	Recessed type, brass tone			40	.200		4.44	8.15		12.59	17.20
2440	Chrome or white enamel			40	.200		2	8.15		10.15	14.50
2600	Sprinkler heads, not including supply piping										
2640	Dry, pendent, 1/2" orifice, 3/4" or 1" NPT										
2660	1/2" to 6" length		1 Spri	14	.571	Ea.	36.50	23.50		60	75
2670	6-1/4" to 9" length			14	.571		43	23.50		66.50	82.50
2680	9-1/4" to 12" length			14	.571		43	23.50		66.50	82.50
2690	12-1/4" to 15" length			14	.571		50.50	23.50		74	91

Important: See the Reference Section for critical supporting data - Reference Nos., Crews, & City Cost Indexes

13930 | Wet-Pipe Fire Supp. Sprinklers

		DAILY	LABOR-		2005 BARE COSTS				TOTAL
	CREW	OUTPUT	HOURS	UNIT	MAT.	LABOR	EQUIP.	TOTAL	INCL O&P
400 2700	15-1/4" to 18" length R10520-110	1 Spri	14	.571	Ea.	53.50	23.50		77	94 **400**
2710	18-1/4" to 21" length		13	.615		56	25		81	100
2720	21-1/4" to 24" length R10520-120		13	.615		60	25		85	104
2730	24-1/4" to 27" length		13	.615		62	25		87	106
2740	27-1/4" to 30" length R10520-130		13	.615		64.50	25		89.50	109
2750	30-1/4" to 33" length		13	.615		67	25		92	112
2760	33-1/4" to 36" length		13	.615		70	25		95	115
2780	36-1/4" to 39" length		12	.667		73	27		100	121
2790	39-1/4" to 42" length	▼	12	.667		74.50	27		101.50	123
2800	For each inch or fraction, add					.94			.94	1.03
3600	Foam-water, pendent or upright, 1/2" NPT	1 Spri	12	.667	▼	56	27		83	103
3700	Standard spray, pendent or upright, brass, 135° to 286°F									
3720	1/2" NPT, 3/8" orifice	1 Spri	16	.500	Ea.	6.75	20.50		27.25	38.50
3730	1/2" NPT, 7/16" orifice		16	.500		6.75	20.50		27.25	38.50
3740	1/2" NPT, 1/2" orifice		16	.500		5.10	20.50		25.60	36.50
3760	1/2" NPT, 17/32" orifice		16	.500		5.85	20.50		26.35	37.50
3780	3/4" NPT, 17/32" orifice	▼	16	.500	▼	5.55	20.50		26.05	37
3800	For open sprinklers, deduct					15%				
3840	For chrome, add				Ea.	1.28			1.28	1.41
3860	For wax and lead coating, add					5.70			5.70	6.30
3880	For wax coating, add					2.81			2.81	3.09
3900	For lead coating, add					3.38			3.38	3.72
3920	For 360°F, same cost									
3930	For 400°F, add				Ea.	5.70			5.70	6.30
3940	For 500°F, add				"	5.70			5.70	6.30
4200	Sidewall, vertical brass, 135 -286°F									
4240	1/2" NPT, 1/2" orifice	1 Spri	16	.500	Ea.	7.80	20.50		28.30	39.50
4280	3/4" NPT, 17/32" orifice	"	16	.500		8.30	20.50		28.80	40
4360	For satin chrome, add				▼	1.13			1.13	1.24
4400	For 360°F, same cost									
4410	For 400°F, add				Ea.	5.70			5.70	6.30
4420	For 500°F, add				"	7.80			7.80	8.60
4500	Sidewall, horizontal, brass, 135° to 286°F									
4520	1/2" NPT, 1/2" orifice	1 Spri	16	.500	Ea.	6.75	20.50		27.25	38.50
4540	For 360°F, same cost									
4800	Recessed pendent, brass, 135° to 286°F									
4820	1/2" NPT, 3/8" orifice	1 Spri	10	.800	Ea.	16.10	32.50		48.60	67
4830	1/2" NPT, 7/16" orifice		10	.800		11.45	32.50		43.95	61.50
4840	1/2" NPT, 1/2" orifice		10	.800		13.50	32.50		46	64
4860	1/2" NPT, 17/32" orifice	▼	10	.800		10.90	32.50		43.40	61
4900	For satin chrome, add				▼	.84			.84	.92
5000	Recessed-vertical sidewall, brass, 135°-286°F									
5020	1/2" NPT, 3/8" orifice	1 Spri	10	.800	Ea.	12.50	32.50		45	63
5030	1/2" NPT, 7/16" orifice		10	.800		11.95	32.50		44.45	62
5040	1/2" NPT, 1/2" orifice	▼	10	.800	▼	11.45	32.50		43.95	61.50
5100	For bright nickel, same cost									
5600	Concealed, complete with cover plate									
5620	1/2" NPT, 1/2" orifice, 135°F to 212°F	1 Spri	9	.889	Ea.	15.60	36		51.60	71.50
5800	Window, brass, 1/2" NPT, 1/4" orifice		16	.500		15.60	20.50		36.10	48
5810	1/2" NPT, 5/16" orifice		16	.500		15.60	20.50		36.10	48
5820	1/2" NPT, 3/8" orifice		16	.500		15.60	20.50		36.10	48
5830	1/2" NPT, 7/16" orifice		16	.500		15.60	20.50		36.10	48
5840	1/2" NPT, 1/2" orifice	▼	16	.500		16.60	20.50		37.10	49.50
5860	For polished chrome, add					2.10			2.10	2.31
5880	3/4" NPT, 5/8" orifice	1 Spri	16	.500		17.30	20.50		37.80	50
5890	3/4 NPT, 3/4" orifice	"	16	.500	▼	17.30	20.50		37.80	50

13930	Wet-Pipe Fire Supp. Sprinklers		CREW	DAILY OUTPUT	LABOR-HOURS	UNIT	2005 BARE COSTS				TOTAL INCL O&P
							MAT.	LABOR	EQUIP.	TOTAL	
6000	Sprinkler head guards, bright zinc, 1/2" NPT	R10520-110				Ea.	2.80			2.80	3.08
6020	Bright zinc, 3/4" NPT						2.80			2.80	3.08
6025	Residential sprinkler components, (one and two family)	R10520-120									
6026	Water motor alarm, with strainer		1 Spri	4	2	Ea.	189	81.50		270.50	330
6027	Fast response, glass bulb, 135° to 155° f.	R10520-130									
6028	1/2" NPT, pendent, brass		1 Spri	16	.500	Ea.	10.25	20.50		30.75	42.50
6029	1/2" NPT, sidewall, brass			16	.500		10.75	20.50		31.25	43
6030	1/2" NPT, pendent, brass, extended coverage			16	.500		10.75	20.50		31.25	43
6031	1/2" NPT, sidewall, brass, extended coverage			16	.500		10.25	20.50		30.75	42.50
6032	3/4" NPT sidewall, brass, extended coverage			16	.500		10.75	20.50		31.25	43
6033	For chrome, add						15%				
6034	For polyester/teflon coating add						20%				
6100	Sprinkler head wrenches, standard head					Ea.	11.70			11.70	12.85
6120	Recessed head						18.20			18.20	20
6160	Tamper switch, (valve supervisory switch)		1 Spri	16	.500		73.50	20.50		94	112
6200	Valves										
6210	Alarm, includes										
6220	retard chamber, trim, gauges, alarm line strainer										
6260	3" size		Q-12	3	5.333	Ea.	850	196		1,046	1,225
6280	4" size		"	2	8		860	293		1,153	1,400
6300	6" size		Q-13	4	8		955	310		1,265	1,525
6320	8" size		"	3	10.667		1,125	415		1,540	1,875
6500	Check, swing, C.I. body, brass fittings, auto. ball drip										
6520	4" size		Q-12	3	5.333	Ea.	163	196		359	475
6540	6" size		Q-13	4	8		300	310		610	800
6580	8" size		"	3	10.667		570	415		985	1,250
6800	Check, wafer, butterfly type, C.I. body, bronze fittings										
6820	4" size		Q-12	4	4	Ea.	172	147		319	410
6840	6" size		Q-13	5.50	5.818		296	226		522	665
6860	8" size			5	6.400		530	249		779	960
6880	10" size			4.50	7.111		805	276		1,081	1,300
7000	Deluge, assembly, incl. trim, pressure										
7020	operated relief, emergency release, gauges										
7040	2" size		Q-12	2	8	Ea.	1,025	293		1,318	1,575
7060	3" size			1.50	10.667		1,225	390		1,615	1,950
7080	4" size			1	16		1,475	585		2,060	2,500
7100	6" size		Q-13	1.80	17.778		1,875	690		2,565	3,100
7800	Pneumatic actuator, bronze, required on all										
7820	pneumatic release systems, any size deluge		1 Spri	18	.444	Ea.	196	18.10		214.10	244
8000	Dry pipe air check valve, 3" size		Q-12	2	8		1,175	293		1,468	1,750
8200	Dry pipe valve, incl. trim and gauges, 3" size			2	8		1,250	293		1,543	1,825
8220	4" size			1	16		1,300	585		1,885	2,300
8240	6" size		Q-13	2	16		1,525	620		2,145	2,625
8280	For accelerator trim with gauges, add		1 Spri	8	1		425	41		466	525
8400	Firecycle package, includes swing check										
8420	and flow control valves with required trim										
8440	2" size		Q-12	2	8	Ea.	1,325	293		1,618	1,925
8460	3" size			1.50	10.667		2,050	390		2,440	2,850
8480	4" size			1	16		2,300	585		2,885	3,400
8500	6" size		Q-13	1.40	22.857		3,100	890		3,990	4,775
8800	Flow control valve, includes trim and gauges, 2" size		Q-12	2	8		1,175	293		1,468	1,725
8820	3" size		"	1.50	10.667		1,375	390		1,765	2,100
8840	4" size		Q-13	2.80	11.429		1,625	445		2,070	2,475
8860	6" size		"	2	16		1,725	620		2,345	2,850
9200	Pressure operated relief valve, brass body		1 Spri	18	.444		128	18.10		146.10	169
9600	Waterflow indicator, with recycling retard and										

Important: See the Reference Section for critical supporting data - Reference Nos., Crews, & City Cost Indexes

13930	Wet-Pipe Fire Supp. Sprinklers	CREW	DAILY OUTPUT	LABOR-HOURS	UNIT	2005 BARE COSTS				TOTAL INCL O&P
						MAT.	LABOR	EQUIP.	TOTAL	
400	9610	two single pole retard switches, 2″ thru 6″ pipe size R10520 -110	1 Spri	8	1	Ea.	118	41		159	192	40

13960	CO2 Fire Extinguishing									
200	0010	**AUTOMATIC FIRE SUPPRESSION SYSTEMS** D4090 410									20
	0040	For detectors and control stations, see division 13850									
	0100	Control panel, single zone with batteries (2 zones det., 1 suppr.)	1 Elec	1	8	Ea.	1,275	325		1,600	1,875
	0150	Multizone (4) with batteries (8 zones det., 4 suppr.)	″	.50	16		2,450	650		3,100	3,675
	1000	Dispersion nozzle, CO2, 3″ x 5″	1 Plum	18	.444		50	18.15		68.15	82.50
	1100	FM200, 1-1/2″	″	14	.571		50	23.50		73.50	90
	2000	Extinguisher, CO2 system, high pressure, 75 lb. cylinder	Q-1	6	2.667		950	98		1,048	1,200
	2100	100 lb. cylinder	″	5	3.200		1,025	118		1,143	1,300
	2400	FM200 system, filled, with mounting bracket									
	2460	26 lb. container	Q-1	8	2	Ea.	1,800	73.50		1,873.50	2,075
	2480	44 lb. container		7	2.286		2,400	84		2,484	2,775
	2500	63 lb. container		6	2.667		2,800	98		2,898	3,225
	2520	101 lb. container		5	3.200		3,750	118		3,868	4,300
	2540	196 lb. container		4	4		5,800	147		5,947	6,600
	3000	Electro/mechanical release	L-1	4	4		125	163		288	380
	3400	Manual pull station	1 Plum	6	1.333		45	54.50		99.50	132
	4000	Pneumatic damper release	″	8	1		175	41		216	255
	6000	Average FM200 system, minimum				C.F.	1.25			1.25	1.38
	6020	Maximum				″	2.50			2.50	2.75

For information about Means Estimating Seminars, see yellow pages 12 and 13 in back of book

Division Notes

	CREW	DAILY OUTPUT	LABOR-HOURS	UNIT	2005 BARE COSTS				TOTAL INCL O&P
					MAT.	LABOR	EQUIP.	TOTAL	

Division 14
Conveying Systems

Estimating Tips

General

- Many products in Division 14 will require some type of support or blocking for installation not included with the item itself. Examples are supports for conveyors or tube systems, attachment points for lifts, and footings for hoists or cranes. Add these supports in the appropriate division.

14100 Dumbwaiters
14200 Elevators

- Dumbwaiters and elevators are estimated and purchased in a method similar to buying a car. The manufacturer has a base unit with standard features. Added to this base unit price will be whatever options the owner or specifications require. Increased load capacity, additional vertical travel, additional stops, higher speed, and cab finish options are items to be considered. When developing an estimate for dumbwaiters and elevators, remember that some items needed by the installers may have to be included as part of the general contract.

Examples are:
- shaftway
- rail support brackets
- machine room
- electrical supply
- sill angles
- electrical connections
- pits
- roof penthouses
- pit ladders

Check the job specifications and drawings before pricing.

- Installation of elevators and handicapped lifts in historic structures can require significant additional costs. The associated structural requirements may involve cutting into and repairing finishes, mouldings, flooring, etc. The estimator must account for these special conditions.

14300 Escalators & Moving Walks

- Escalators and moving walks are specialty items installed by specialty contractors. There are numerous options associated with these items. For specific options contact a manufacturer or contractor. In a method similar to estimating dumbwaiters and elevators, you should verify the extent of general contract work and add items as necessary.

14400 Lifts
14500 Material Handling
14600 Hoists & Cranes

- Products such as correspondence lifts, conveyors, chutes, pneumatic tube systems, material handling cranes and hoists as well as other items specified in this subdivision may require trained installers. The general contractor might not have any choice as to who will perform the installation or when it will be performed. Long lead times are often required for these products, making early decisions in scheduling necessary.

Reference Numbers

Reference numbers are shown in bold squares at the beginning of some major classifications. These numbers refer to related items in the Reference Section. The reference information may be an estimating procedure, an alternate pricing method or technical information.

Note: Not all subdivisions listed here necessarily appear in this publication.

14450	Vehicle Lifts	CREW	DAILY OUTPUT	LABOR-HOURS	UNIT	2005 BARE COSTS				TOTAL INCL O&P	
						MAT.	LABOR	EQUIP.	TOTAL		
700	0010	**HYDRAULIC LIFT AUTO/TRUCK**									700
	0200	Double post, 9000 lb frame	L-4	1.15	20.870	Ea.	5,150	665		5,815	6,700
	0400	26000 lb frame		.22	109		22,900	3,475		26,375	30,600
	2200	Hoists, single post, 8,000# capacity, swivel arms		.40	60		4,300	1,900		6,200	7,700
	2400	Two posts, adjustable frames, 11,000# capacity		.25	96		5,525	3,050		8,575	10,900
	2500	24,000# capacity		.15	160		7,375	5,075		12,450	16,000
	2700	7,500# capacity, frame supports		.50	48		6,150	1,525		7,675	9,150
	2800	Four post, roll on ramp		.50	48		5,525	1,525		7,050	8,475
	2810	Hydraulic lifts, above ground, 2 post, clear floor, 6000 lb cap		2.67	8.989		5,775	285		6,060	6,800
	2815	9000 lb capacity		2.29	10.480		13,700	335		14,035	15,600
	2820	15,000 lb capacity		2	12		15,500	380		15,880	17,600
	2825	30,000 lb capacity		1.60	15		34,300	475		34,775	38,400
	2830	4 post, ramp style, 25,000 lb capacity		2	12		13,400	380		13,780	15,300
	2835	35,000 lb capacity		1	24		62,500	760		63,260	70,000
	2840	50,000 lb capacity		1	24		70,000	760		70,760	78,000
	2845	75,000 lb capacity		1	24		81,000	760		81,760	90,500
	2850	For drive thru tracks, add, minimum					860			860	945
	2855	Maximum					1,475			1,475	1,625
	2860	Ramp extensions, 3'(set of 2)					705			705	780
	2865	Rolling jack platform					2,450			2,450	2,700
	2870	Elec/hyd jacking beam					6,575			6,575	7,250
	2880	Scissor lift, portable, 6000 lb capacity					6,450			6,450	7,100

14610	Fixed Hoists	CREW	DAILY OUTPUT	LABOR-HOURS	UNIT	2005 BARE COSTS				TOTAL INCL O&P	
						MAT.	LABOR	EQUIP.	TOTAL		
500	0010	**MATERIAL HANDLING**									500
	1500	Cranes, portable hydraulic, floor type, 2,000 lb capacity				Ea.	1,800			1,800	1,975
	1600	4,000 lb capacity					3,100			3,100	3,425
	1800	Movable gantry type, 12' to 15' range, 2,000 lb capacity					2,775			2,775	3,050
	1900	6,000 lb capacity					4,025			4,025	4,425
	2100	Hoists, electric overhead, chain, hook hung, 15' lift, 1 ton cap.					1,825			1,825	2,000
	2200	3 ton capacity					2,275			2,275	2,500
	2500	5 ton capacity					5,000			5,000	5,500
	2600	For hand-pushed trolley, add					15%				
	2700	For geared trolley, add					30%				
	2800	For motor trolley, add					75%				
	3000	For lifts over 15', 1 ton, add				L.F.	17.10			17.10	18.80
	3100	5 ton, add				"	40			40	44
	3300	Lifts, scissor type, portable, electric, 36" high, 2,000 lb				Ea.	3,275			3,275	3,600
	3400	48" high, 4,000 lb				"	3,750			3,750	4,125

For information about Means Estimating Seminars, see yellow pages 12 and 13 in back of book

Division 15
Mechanical

Estimating Tips

15100 Building Services Piping
This subdivision is primarily basic pipe and related materials. The pipe may be used by any of the mechanical disciplines, i.e., plumbing, fire protection, heating, and air conditioning.

- The piping section lists the add to labor for elevated pipe installation. These adds apply to all elevated pipe, fittings, valves, insulation, etc., that are placed above 10' high. CAUTION: the correct percentage may vary for the same pipe. For example, the percentage add for the basic pipe installation should be based on the maximum height that the craftsman must install for that particular section. If the pipe is to be located 14' above the floor but it is suspended on threaded rod from beams, the bottom flange of which is 18' high (4' rods), then the height is actually 18' and the add is 20%. The pipe coverer, however, does not have to go above the 14' and so his add should be 10%.

- Most pipe is priced first as straight pipe with a joint (coupling, weld, etc.) every 10' and a hanger usually every 10'. There are exceptions with hanger spacing such as: for cast iron pipe (5') and plastic pipe (3 per 10'). Following each type of pipe there are several lines listing sizes and the amount to be subtracted to delete couplings and hangers. This is for pipe that is to be buried or supported together on trapeze hangers. The reason that the couplings are deleted is that these runs are usually long and frequently longer lengths of pipe are used. By deleting the couplings the estimator is expected to look up and add back the correct reduced number of couplings.

- When preparing an estimate it may be necessary to approximate the fittings. Fittings usually run between 25% and 50% of the cost of the pipe. The lower percentage is for simpler runs, and the higher number is for complex areas like mechanical rooms.

- For historic restoration projects, the systems must be as invisible as possible, and pathways must be sought for pipes, conduits, and ductwork. While installations in accessible spaces (such as basements and attics) are relatively straightforward to estimate, labor costs may be more difficult to determine when delivery systems must be concealed.

15400 Plumbing Fixtures & Equipment

- Plumbing fixture costs usually require two lines, the fixture itself and its "rough-in, supply and waste".

- In the Assemblies Section (Plumbing D2010) for the desired fixture, the System Components Group in the center of the page shows the fixture itself on the first line while the rest of the list (fittings, pipe, tubing, etc.) will total up to what we refer to in the Unit Price section as "Rough-in, supply, waste and vent". Note that for most fixtures we allow a nominal 5' of tubing to reach from the fixture to a main or riser.

- Remember that gas and oil fired units need venting.

15500 Heat Generation Equipment

- When estimating the cost of an HVAC system, check to see who is responsible for providing and installing the temperature control system. It is possible to overlook controls, assuming that they would be included in the electrical estimate.

- When looking up a boiler be careful on specified capacity. Some manufacturers rate their products on output while others use input.

- Include HVAC insulation for pipe, boiler and duct (wrap and liner).

- Be careful when looking up mechanical items to get the correct pressure rating and connection type (thread, weld, flange).

15700 Heating/Ventilation/ Air Conditioning Equipment

- Combination heating and cooling units are sized by the air conditioning requirements. (See Reference No. R15710-020 for preliminary sizing guide.)

- A ton of air conditioning is nominally 400 CFM.

- Rectangular duct is taken off by the linear foot for each size, but its cost is usually estimated by the pound. Remember that SMACNA standards now base duct on internal pressure.

- Prefabricated duct is estimated and purchased like pipe: straight sections and fittings.

- Note that cranes or other lifting equipment are not included on any lines in Division 15. For example, if a crane is required to lift a heavy piece of pipe into place high above a gym floor, or to put a rooftop unit on the roof of a four-story building, etc., it must be added. Due to the potential for extreme variation—from nothing additional required, to a major crane or helicopter—we feel that including a nominal amount for "lifting contingency" would be useless and detract from the accuracy of the estimate. When using equipment rental from Means do not forget to include the cost of the operator(s).

Reference Numbers
Reference numbers are shown in bold squares at the beginning of some major classifications. These numbers refer to related items in the Reference Section. The reference information may be an estimating procedure, an alternate pricing method or technical information.

Note: Not all subdivisions listed here necessarily appear in this publication.

Note: **i2 Trade Service,** *in part, has been used as a reference source for some of the material prices used in Division 15.*

	15051	Mechanical General	CREW	DAILY OUTPUT	LABOR-HOURS	UNIT	2005 BARE COSTS				TOTAL INCL O&P	
							MAT.	LABOR	EQUIP.	TOTAL		
100	0010	**BOILERS, GENERAL** Prices do not include flue piping, elec. wiring,										100
	0020	gas or oil piping, boiler base, pad, or tankless unless noted										
	0100	Boiler H.P.: 10 KW = 34 lbs/steam/hr = 33,475 BTU/hr.										
	0120											
	0150	To convert SFR to BTU rating: Hot water, 150 x SFR;										
	0160	Forced hot water, 180 x SFR; steam, 240 x SFR										
700	0010	**PIPING** See also divisions 02500 & 02600 for site work R15100 -070										700
	1000	Add to labor for elevated installation										
	1080	10' to 15' high						10%				
	1100	15' to 20' high						20%				
	1120	20' to 25' high						25%				
	1140	25' to 30' high						35%				
	1160	30' to 35' high						40%				
	1180	35' to 40' high						50%				
	1200	Over 40' high						55%				

	15055	Selective Mech Demolition	CREW	DAILY OUTPUT	LABOR-HOURS	UNIT	MAT.	LABOR	EQUIP.	TOTAL	TOTAL INCL O&P	
300	0010	**HVAC DEMOLITION**										300
	0100	Air conditioner, split unit, 3 ton	Q-5	2	8	Ea.		295		295	445	
	0150	Package unit, 3 ton	Q-6	3	8			305		305	460	
	0190	Rooftop, self contained, up to 5 ton	1 Plum	1.20	6.667			272		272	410	
	0250	Air curtain	Q-9	20	.800	L.F.		29		29	44.50	
	0254	Air filters, up thru 16,000 CFM		20	.800	Ea.		29		29	44.50	
	0256	20,000 thru 60,000 CFM		16	1			36		36	55.50	
	0297	Boiler blowdown	Q-5	8	2			73.50		73.50	111	
	0298	Boilers										
	0300	Electric, up thru 148 kW	Q-19	2	12	Ea.		460		460	685	
	0310	150 thru 518 kW	"	1	24			915		915	1,375	
	0320	550 thru 2000 kW	Q-21	.40	80			3,100		3,100	4,650	
	0330	2070 kW and up	"	.30	106			4,150		4,150	6,200	
	0340	Gas and/or oil, up thru 150 MBH	Q-7	2.20	14.545			570		570	855	
	0350	160 thru 2000 MBH		.80	40			1,550		1,550	2,350	
	0360	2100 thru 4500 MBH		.50	64			2,500		2,500	3,750	
	0370	4600 thru 7000 MBH		.30	106			4,175		4,175	6,250	
	0380	7100 thru 12,000 MBH		.16	200			7,800		7,800	11,700	
	0390	12,200 thru 25,000 MBH		.12	266			10,400		10,400	15,600	
	0400	Central station air handler unit, up thru 15 ton	Q-5	1.60	10			370		370	555	
	0410	17.5 thru 30 ton	"	.80	20			735		735	1,100	
	0430	Computer room unit										
	0434	Air cooled split, up thru 10 ton	Q-5	.67	23.881	Ea.		880		880	1,325	
	0436	12 thru 23 ton		.53	30.189			1,100		1,100	1,675	
	0440	Chilled water, up thru 10 ton		1.30	12.308			455		455	680	
	0444	12 thru 23 ton		1	16			590		590	885	
	0450	Glycol system, up thru 10 ton		.53	30.189			1,100		1,100	1,675	
	0454	12 thru 23 ton		.40	40			1,475		1,475	2,225	
	0460	Water cooled, not including condenser, up thru 10 ton		.80	20			735		735	1,100	
	0464	12 thru 23 ton		.60	26.667			985		985	1,475	
	0600	Condenser, up thru 50 ton		1	16			590		590	885	
	0610	51 thru 100 ton	Q-6	.80	30			1,150		1,150	1,725	
	0620	150 thru 1000 ton	"	.16	150			5,725		5,725	8,625	
	0660	Condensing unit, up thru 10 ton	Q-5	1.25	12.800			470		470	710	
	0670	11 thru 50 ton	"	.40	40			1,475		1,475	2,225	
	0680	60 thru 100 ton	Q-6	.30	80			3,050		3,050	4,600	
	0700	Cooling tower, up thru 400 ton		.80	30			1,150		1,150	1,725	
	0710	450 thru 600 ton		.53	45.283			1,725		1,725	2,600	

Important: See the Reference Section for critical supporting data - Reference Nos., Crews, & City Cost Indexes

		CREW	DAILY OUTPUT	LABOR- HOURS	UNIT	2005 BARE COSTS				TOTAL INCL O&P
						MAT.	LABOR	EQUIP.	TOTAL	
0720	700 thru 1300 ton	Q-6	.40	60	Ea.		2,300		2,300	3,450
0780	Dehumidifier, up thru 155 lb./hr.	Q-1	8	2			73.50		73.50	111
0790	240 lb./hr. and up	"	2	8	↓		294		294	440
1560	Ductwork									
1570	Metal, steel, sst, fabricated	Q-9	1,000	.016	Lb.		.58		.58	.89
1580	Aluminum, fabricated		485	.033	"		1.19		1.19	1.84
1590	Spiral, prefabricated		400	.040	L.F.		1.45		1.45	2.23
1600	Fiberglass, prefabricated		400	.040			1.45		1.45	2.23
1610	Flex, prefabricated		500	.032			1.16		1.16	1.78
1620	Glass fiber reinforced plastic, prefabricated	↓	280	.057	↓		2.07		2.07	3.18
1630	Diffusers, registers or grills, up to 20" max dimension	1 Shee	50	.160	Ea.		6.45		6.45	9.90
1640	21 thru 36" max dimension		36	.222			8.95		8.95	13.75
1650	Above 36" max dimension	↓	30	.267			10.70		10.70	16.50
1700	Evaporator, up thru 12,000 BTUH	Q-5	5.30	3.019			111		111	167
1710	12,500 thru 30,000 BTUH	"	2.70	5.926			218		218	330
1720	31,000 BTUH and up	Q-6	1.50	16			610		610	920
1730	Evaporative cooler, up thru 5 H.P.	Q-9	2.70	5.926			214		214	330
1740	10 thru 30 H.P.	"	.67	23.881	↓		865		865	1,325
1750	Exhaust systems									
1760	Exhaust components	1 Shee	8	1	System		40		40	62
1770	Weld fume hoods	"	20	.400	Ea.		16.10		16.10	24.50
2120	Fans, up thru 1 H.P. or 2000 CFM	Q-9	8	2			72.50		72.50	111
2124	2 thru 10 H.P. or 20,000 CFM		5.30	3.019			109		109	168
2128	15 thru 30 H.P. or above 20,000 CFM	↓	4	4			145		145	223
2150	Fan coil air conditioner, chilled water, up thru 7.5 ton	Q-5	14	1.143			42		42	63.50
2154	Direct expansion, up thru 10 ton		8	2			73.50		73.50	111
2158	11 thru 30 ton	↓	2	8			295		295	445
2170	Flue shutter damper	Q-9	8	2			72.50		72.50	111
2200	Furnace, electric	Q-20	2	10			370		370	565
2300	Gas or oil, under 120 MBH	Q-9	4	4			145		145	223
2340	Over 120 MBH	"	3	5.333			193		193	297
2730	Heating and ventilating unit	Q-5	2.70	5.926			218		218	330
2740	Heater, electric, wall, baseboard and quartz	1 Elec	10	.800			32.50		32.50	48.50
2750	Heater, electric, unit, cabinet, fan and convector	"	8	1			41		41	60.50
2760	Heat exchanger, shell and tube type	Q-5	1.60	10			370		370	555
2770	Plate type	Q-6	.60	40	↓		1,525		1,525	2,300
2810	Heat pump									
2820	Air source, split, up thru 10 ton	Q-5	.90	17.778	Ea.		655		655	985
2830	15 thru 25 ton	Q-6	.80	30			1,150		1,150	1,725
2850	Single package, up thru 12 ton	Q-5	1	16			590		590	885
2860	Water source, split, up thru 15 ton	"	.90	17.778			655		655	985
2870	20 thru 50 ton	Q-6	.80	30			1,150		1,150	1,725
2910	Heat recovery package, up thru 20,000 CFM	Q-5	2	8			295		295	445
2920	25,000 CFM and up		1.20	13.333			490		490	740
2930	Heat transfer package, up thru 130 GPM		.80	20			735		735	1,100
2934	255 thru 800 GPM		.42	38.095			1,400		1,400	2,100
2940	Humidifier		10.60	1.509			55.50		55.50	83.50
2961	Hydronic unit heaters, up thru 200 MBH		14	1.143			42		42	63.50
2962	Above 200 MBH		8	2			73.50		73.50	111
2964	Valance units	↓	32	.500	↓		18.45		18.45	27.50
2966	Radiant floor heating									
2967	System valves, controls, manifolds	Q-5	16	1	Ea.		37		37	55.50
2968	Per room distribution		8	2			73.50		73.50	111
2970	Hydronic heating, baseboard radiation		16	1			37		37	55.50
2976	Convectors and free standing radiators	↓	18	.889			33		33	49
2980	Induced draft fan, up thru 1 H.P.	Q-9	4.60	3.478	↓		126		126	194

15055	Selective Mech Demolition	CREW	DAILY OUTPUT	LABOR-HOURS	UNIT	2005 BARE COSTS				TOTAL INCL O&P
						MAT.	LABOR	EQUIP.	TOTAL	
300 2984	1-1/2 H.P. thru 7-1/2 H.P.	Q-9	2.20	7.273	Ea.		263		263	405 **300**
2988	Infra-red unit	Q-5	16	1	↓		37		37	55.50
2992	Louvers	1 Shee	46	.174	S.F.		7		7	10.75
3000	Mechanical equipment, light items. Unit is weight, not cooling.	Q-5	.90	17.778	Ton		655		655	985
3600	Heavy items	"	1.10	14.545	"		535		535	805
3700	Deduct for salvage (when applicable), minimum				Job					55
3710	Maximum				"					420
3720	Make up air unit, up thru 6000 CFM	Q-5	3	5.333	Ea.		197		197	295
3730	8000 thru 30,000 CFM	"	1.60	10			370		370	555
3740	35,000 thru 75,000 CFM	Q-6	1	24			915		915	1,375
3800	Mixing boxes, constant and VAV	Q-9	18	.889			32		32	49.50
4000	Packaged terminal air conditioner, up thru 18,000 BTUH	Q-5	8	2			73.50		73.50	111
4010	24,000 thru 48,000 BTUH	"	2.80	5.714	↓		211		211	315
5000	Refrigerant compressor, reciprocating or scroll									
5010	Up thru 5 ton	1 Stpi	6	1.333	Ea.		54.50		54.50	82
5020	5.08 thru 10 ton	Q-5	6	2.667			98.50		98.50	148
5030	15 thru 50 ton	"	.40	40			1,475		1,475	2,225
5040	60 thru 130 ton	Q-6	.48	50			1,900		1,900	2,875
5100	Roof top air conditioner, up thru 10 ton	Q-5	1.40	11.429			420		420	635
5110	12-1/2 thru 40 ton	Q-6	1	24			915		915	1,375
5120	50 thru 140 ton	↓	.50	48			1,825		1,825	2,750
5130	150 thru 300 ton	↓	.30	80			3,050		3,050	4,600
6000	Self contained single package air conditioner, up thru 10 ton	Q-5	1.60	10			370		370	555
6010	15 thru 60 ton	Q-6	1.20	20			765		765	1,150
6100	Space heaters, up thru 200 MBH	Q-5	10	1.600			59		59	88.50
6110	Over 200 MBH		5	3.200			118		118	177
6200	Split ductless, both sections	↓	8	2			73.50		73.50	111
6300	Steam condensate meter	1 Stpi	11	.727			30		30	45
6600	Thru-the-wall air conditioner	L-2	8	2	↓		60		60	93
7000	Vent chimney, prefabricated, up thru 12" diameter	Q-9	94	.170	V.L.F.		6.15		6.15	9.45
7010	14" thru 36" diameter	↓	40	.400			14.45		14.45	22.50
7020	38" thru 48" diameter	↓	32	.500			18.10		18.10	28
7030	54" thru 60" diameter	Q-10	14	1.714	↓		64.50		64.50	99
7400	Ventilators, up thru 14" neck diameter	Q-9	58	.276	Ea.		10		10	15.35
7410	16" thru 50" neck diameter		40	.400			14.45		14.45	22.50
7450	Relief vent, up thru 24" x 96"		22	.727			26.50		26.50	40.50
7460	48" x 60" thru 96" x 144"	↓	10	1.600			58		58	89
8000	Water chiller up thru 10 ton	Q-5	2.50	6.400			236		236	355
8010	15 thru 100 ton	Q-6	.48	50			1,900		1,900	2,875
8020	110 thru 500 ton	Q-7	.29	110			4,300		4,300	6,475
8030	600 thru 1000 ton		.23	139			5,425		5,425	8,150
8040	1100 ton and up	↓	.20	160			6,250		6,250	9,375
8400	Window air conditioner	1 Carp	16	.500	↓		17.15		17.15	26.50
600 0010	**PLUMBING DEMOLITION**									**600**
0400	Air compressor, up thru 2 H.P.	Q-1	10	1.600	Ea.		59		59	88.50
0410	3 H.P. thru 7-1/2 H.P.		5.60	2.857			105		105	158
0420	10 H.P. thru 15 H.P.	↓	1.40	11.429			420		420	630
0430	20 H.P. thru 30 H.P.	Q-2	1.30	18.462			705		705	1,050
0500	Backflow preventer, up thru 2" diameter	1 Plum	17	.471			19.20		19.20	29
0510	2-1/2" thru 3" diameter	Q-1	10	1.600			59		59	88.50
0520	4" thru 6" diameter	"	5	3.200			118		118	177
0530	8" thru 10" diameter	Q-2	3	8	↓		305		305	460
0700	Carriers and supports									
0710	Fountains, sinks, lavatories and urinals	1 Plum	14	.571	Ea.		23.50		23.50	35
0720	Water closets	"	12	.667	"		27		27	41

Note for row 0400: R15050 -720

Important: See the Reference Section for critical supporting data - Reference Nos., Crews, & City Cost Indexes

15055 | Selective Mech Demolition

		CREW	DAILY OUTPUT	LABOR-HOURS	UNIT	2005 BARE COSTS				TOTAL INCL O&P
						MAT.	LABOR	EQUIP.	TOTAL	
600	0730	Grinder pump or sewage ejector system	R15050 -720								600
	0732	Simplex	Q-1	7	2.286	Ea.		84		84	126
	0734	Duplex	"	2.80	5.714			210		210	315
	0738	Hot water dispenser	1 Plum	36	.222			9.10		9.10	13.65
	0740	Hydrant, wall		26	.308			12.55		12.55	18.90
	0744	Ground		12	.667			27		27	41
	0750	Hot water dispenser		36	.222			9.10		9.10	13.65
	0760	Cleanouts and drains, up thru 4" pipe diameter		10	.800			32.50		32.50	49
	0764	5" thru 8" pipe diameter	Q-1	10	1.600			59		59	88.50
	0780	Industrial safety fixtures	1 Plum	8	1			41		41	61.50
	1020	Fixtures, including 10' piping									
	1100	Bath tubs, cast iron	1 Plum	4	2	Ea.		81.50		81.50	123
	1120	Fiberglass		6	1.333			54.50		54.50	82
	1140	Steel		5	1.600			65.50		65.50	98
	1150	Bidet	Q-1	7	2.286			84		84	126
	1200	Lavatory, wall hung	1 Plum	10	.800			32.50		32.50	49
	1220	Counter top		8	1			41		41	61.50
	1300	Sink, steel or cast iron, single		8	1			41		41	61.50
	1320	Double		7	1.143			46.50		46.50	70
	1340	Shower, stall and receptor	Q-1	6	2.667			98		98	147
	1350	Group	"	7	2.286			84		84	126
	1400	Water closet, floor mounted	1 Plum	8	1			41		41	61.50
	1420	Wall mounted	"	7	1.143			46.50		46.50	70
	1440	Wash fountain, 36" diameter	Q-2	8	3			114		114	172
	1442	54" diameter	"	7	3.429			131		131	197
	1500	Urinal, floor mounted	1 Plum	4	2			81.50		81.50	123
	1520	Wall mounted	"	7	1.143			46.50		46.50	70
	1590	Whirl pool or hot tub	Q-1	2.60	6.154			226		226	340
	1600	Water fountains, free standing	1 Plum	8	1			41		41	61.50
	1620	Recessed		6	1.333			54.50		54.50	82
	1800	Medical gas specialties		8	1			41		41	61.50
	1900	Piping fittings, single connection, up thru 1-1/2" diameter		30	.267			10.90		10.90	16.35
	1910	2" thru 4" diameter		14	.571			23.50		23.50	35
	1980	Pipe hanger/support removal		80	.100			4.09		4.09	6.15
	1990	Glass pipe with fittings, 1" thru 3" diameter		200	.040	L.F.		1.63		1.63	2.46
	1992	4" thru 6" diameter		150	.053			2.18		2.18	3.27
	2000	Piping, metal, up thru 1-1/2" diameter		200	.040			1.63		1.63	2.46
	2050	2" thru 3-1/2" diameter		150	.053			2.18		2.18	3.27
	2100	4" thru 6" diameter	2 Plum	100	.160			6.55		6.55	9.80
	2150	8" thru 14" diameter	"	60	.267			10.90		10.90	16.35
	2153	16" thru 20" diameter	Q-18	70	.343			13.10	1.15	14.25	21
	2155	24" thru 26" diameter		55	.436			16.70	1.47	18.17	26.50
	2156	30" thru 36" diameter		40	.600			23	2.02	25.02	36.50
	2160	Glass pipe with fittings, up thru 1-1/2" diameter	1 Plum	250	.032			1.31		1.31	1.96
	2162	2" thru 3" diameter	"	200	.040			1.63		1.63	2.46
	2164	4" thru 6" diameter	Q-1	200	.080			2.94		2.94	4.42
	2166	8" thru 14" diameter		150	.107			3.92		3.92	5.90
	2168	16" diameter		100	.160			5.90		5.90	8.85
	2170	Prison fixtures, lavatory or sink		18	.889	Ea.		32.50		32.50	49
	2172	Shower		5.60	2.857			105		105	158
	2174	Urinal or water closet		13	1.231			45.50		45.50	68
	2180	Pumps, all fractional horse-power		12	1.333			49		49	73.50
	2184	1 H.P. thru 5 H.P.		6	2.667			98		98	147
	2186	7-1/2 H.P. thru 15 H.P.		2.50	6.400			235		235	355
	2188	20 H.P. thru 25 H.P.	Q-2	4	6			229		229	345
	2190	30 H.P. thru 60 H.P.		.80	30			1,150		1,150	1,725

15055	Selective Mech Demolition	CREW	DAILY OUTPUT	LABOR-HOURS	UNIT	MAT.	LABOR	EQUIP.	TOTAL	TOTAL INCL O&P	
							2005 BARE COSTS				
600	2192	75 H.P. thru 100 H.P. R15050-720	Q-2	.60	40	Ea.		1,525		1,525	2,300	**600**
	2194	150 H.P.	↓	.50	48			1,825		1,825	2,750	
	2198	Pump, sump or submersible	1 Plum	12	.667			27		27	41	
	2200	Receptors and interceptors, up thru 20 GPM	"	8	1			41		41	61.50	
	2204	25 thru 100 GPM	Q-1	6	2.667			98		98	147	
	2208	125 thru 300 GPM	"	2.40	6.667			245		245	370	
	2211	325 thru 500 GPM	Q-2	2.60	9.231	↓		350		350	530	
	2212	Deduct for salvage, aluminum scrap				Ton					525	
	2214	Brass scrap									420	
	2216	Copper scrap									1,025	
	2218	Lead scrap									235	
	2220	Steel scrap				↓					65	
	2230	Temperature maintenance cable	1 Plum	1,200	.007	L.F.		.27		.27	.41	
	2250	Water heater, 40 gal.	"	6	1.333	Ea.		54.50		54.50	82	
	3100	Tanks, water heaters and liquid containers										
	3110	Up thru 40 gallons	Q-1	22	.727	Ea.		27		27	40	
	3120	50 thru 120 gallons		14	1.143			42		42	63	
	3130	130 thru 240 gallons		7.60	2.105			77.50		77.50	116	
	3140	250 thru 500 gallons	↓	5.40	2.963			109		109	164	
	3150	600 thru 1000 gallons	Q-2	1.60	15			570		570	860	
	3160	1100 thru 2000 gallons		.70	34.286			1,300		1,300	1,975	
	3170	2100 thru 4000 gallons	↓	.50	48			1,825		1,825	2,750	
	6000	Remove and reset fixtures, minimum	1 Plum	6	1.333			54.50		54.50	82	
	6100	Maximum	"	4	2			81.50		81.50	123	
	9100	Valve, metal, up thru 1-1/2" diameter	1 Stpi	28	.286			11.70		11.70	17.60	
	9110	2" thru 3" diameter	Q-1	11	1.455			53.50		53.50	80.50	
	9120	4" thru 6" diameter	"	8	2			73.50		73.50	111	
	9130	8" thru 14" diameter	Q-2	8	3			114		114	172	
	9140	16" thru 20" diameter		2	12			460		460	690	
	9150	24" diameter	↓	1.20	20			765		765	1,150	
	9200	Valve, plastic, up thru 1-1/2" diameter	1 Plum	42	.190			7.80		7.80	11.70	
	9210	2" thru 3" diameter		15	.533			22		22	33	
	9220	4" thru 6" diameter		12	.667			27		27	41	
	9300	Vent flashing and caps	↓	55	.145			5.95		5.95	8.95	
	9350	Water filter, commercial, 1" thru 1-1/2"	Q-1	2	8			294		294	440	
	9360	2" thru 2-1/2"	"	1.60	10	↓		370		370	555	
	9400	Water heaters (when tank size given see line 15055 600 3000)										
	9410	Up thru 245 GPH	Q-1	2.40	6.667	Ea.		245		245	370	
	9420	250 thru 756 GPH		1.60	10			370		370	555	
	9430	775 thru 1640 GPH	↓	.80	20			735		735	1,100	
	9440	1650 thru 4000 GPH	Q-2	.50	48			1,825		1,825	2,750	
	9470	Water softener	Q-1	2	8	↓		294		294	440	

15060	Hangers & Supports										
300	0010	**PIPE HANGERS AND SUPPORTS**, TYPE numbers per MSS-SP58										**300**
	0050	Brackets										
	0060	Beam side or wall, malleable iron, TYPE 34										
	0070	3/8" threaded rod size	1 Plum	48	.167	Ea.	3.39	6.80		10.19	14	
	0080	1/2" threaded rod size		48	.167		6.55	6.80		13.35	17.45	
	0090	5/8" threaded rod size		48	.167		8.70	6.80		15.50	19.80	
	0100	3/4" threaded rod size		48	.167		10.45	6.80		17.25	22	
	0110	7/8" threaded rod size	↓	48	.167		12.55	6.80		19.35	24	
	0120	For concrete installation, add				↓		30%				
	0150	Wall, welded steel, medium, TYPE 32										
	0160	0 size, 12" wide, 18" deep	1 Plum	34	.235	Ea.	255	9.60		264.60	294	
	0170	1 size, 18" wide 24" deep	↓	34	.235	↓	305	9.60		314.60	350	

Important: See the Reference Section for critical supporting data - Reference Nos., Crews, & City Cost Indexes

	15060	Hangers & Supports	CREW	DAILY OUTPUT	LABOR-HOURS	UNIT	2005 BARE COSTS				TOTAL INCL O&P	
							MAT.	LABOR	EQUIP.	TOTAL		
300	0180	2 size, 24" wide, 30" deep	1 Plum	34	.235	Ea.	375	9.60		384.60	425	300
	0200	Beam attachment, welded, TYPE 22										
	0202	3/8"	Q-15	80	.200	Ea.	15.35	7.35	1.01	23.71	29	
	0203	1/2"		76	.211		16.35	7.75	1.06	25.16	31	
	0204	5/8"		72	.222		17.35	8.15	1.12	26.62	32.50	
	0205	3/4"		68	.235		23.50	8.65	1.19	33.34	40.50	
	0206	7/8"		64	.250		30	9.20	1.26	40.46	48	
	0207	1"		56	.286		45.50	10.50	1.44	57.44	67.50	
	0300	Clamps										
	0310	C-clamp, for mounting on steel beam flange, w/locknut, TYPE 23										
	0320	3/8" threaded rod size	1 Plum	160	.050	Ea.	1.81	2.04		3.85	5.05	
	0330	1/2" threaded rod size		160	.050		2.26	2.04		4.30	5.55	
	0340	5/8" threaded rod size		160	.050		3.46	2.04		5.50	6.90	
	0350	3/4" threaded rod size		160	.050		4.68	2.04		6.72	8.20	
	0352	7/8" threaded rod size		140	.057		16.40	2.33		18.73	21.50	
	0400	High temperature to 1050°F, alloy steel										
	0410	4" pipe size	Q-1	106	.151	Ea.	27.50	5.55		33.05	38.50	
	0420	6" pipe size		106	.151		46	5.55		51.55	59	
	0430	8" pipe size		97	.165		52	6.05		58.05	66	
	0440	10" pipe size		84	.190		70.50	7		77.50	88	
	0450	12" pipe size		72	.222		81.50	8.15		89.65	102	
	0460	14" pipe size		64	.250		171	9.20		180.20	202	
	0470	16" pipe size		56	.286		195	10.50		205.50	230	
	0480	Beam clamp, flange type, TYPE 25										
	0482	For 3/8" bolt	1 Plum	48	.167	Ea.	4.09	6.80		10.89	14.75	
	0483	For 1/2" bolt		44	.182		9.45	7.45		16.90	21.50	
	0484	For 5/8" bolt		40	.200		18	8.15		26.15	32	
	0485	For 3/4" bolt		36	.222		18.35	9.10		27.45	33.50	
	0486	For 1" bolt		32	.250		88	10.20		98.20	112	
	0500	I-beam, for mounting on bottom flange, strap iron, TYPE 21										
	0510	2" flange size	1 Plum	96	.083	Ea.	4.42	3.40		7.82	9.95	
	0520	3" flange size		95	.084		4.56	3.44		8	10.15	
	0530	4" flange size		93	.086		4.67	3.51		8.18	10.45	
	0540	5" flange size		92	.087		4.84	3.55		8.39	10.65	
	0550	6" flange size		90	.089		4.95	3.63		8.58	10.90	
	0560	7" flange size		88	.091		5.10	3.71		8.81	11.25	
	0570	8" flange size		86	.093		5.30	3.80		9.10	11.50	
	0600	One hole, vertical mounting, malleable iron										
	0610	1/2" pipe size	1 Plum	160	.050	Ea.	.76	2.04		2.80	3.91	
	0620	3/4" pipe size		145	.055		.91	2.25		3.16	4.39	
	0630	1" pipe size		136	.059		1.35	2.40		3.75	5.10	
	0640	1-1/4" pipe size		128	.063		1.51	2.55		4.06	5.50	
	0650	1-1/2" pipe size		120	.067		1.69	2.72		4.41	5.95	
	0660	2" pipe size		112	.071		2.45	2.92		5.37	7.10	
	0670	2-1/2" pipe size		104	.077		4.77	3.14		7.91	9.95	
	0680	3" pipe size		96	.083		6.80	3.40		10.20	12.55	
	0690	3-1/2" pipe size		90	.089		6.80	3.63		10.43	12.90	
	0700	4" pipe size		84	.095		10.45	3.89		14.34	17.30	
	0750	Riser or extension pipe, carbon steel, TYPE 8										
	0756	1/2" pipe size	1 Plum	52	.154	Ea.	3.33	6.30		9.63	13.10	
	0760	3/4" pipe size		48	.167		3.33	6.80		10.13	13.90	
	0770	1" pipe size		47	.170		3.35	6.95		10.30	14.15	
	0780	1-1/4" pipe size		46	.174		4.09	7.10		11.19	15.20	
	0790	1-1/2" pipe size		45	.178		4.30	7.25		11.55	15.65	
	0800	2" pipe size		43	.186		4.52	7.60		12.12	16.35	
	0810	2-1/2" pipe size		41	.195		4.77	7.95		12.72	17.25	

15060	Hangers & Supports	CREW	DAILY OUTPUT	LABOR-HOURS	UNIT	MAT.	LABOR	EQUIP.	TOTAL	TOTAL INCL O&P	
300	0820	3" pipe size	1 Plum	40	.200	Ea.	5.20	8.15		13.35	18.05	300
0830	3-1/2" pipe size		39	.205		6.35	8.40		14.75	19.60
0840	4" pipe size		38	.211		6.55	8.60		15.15	20
0850	5" pipe size		37	.216		9.05	8.85		17.90	23.50
0860	6" pipe size		36	.222		10.90	9.10		20	25.50
0870	8" pipe size		34	.235		17.65	9.60		27.25	34
0880	10" pipe size		32	.250		26.50	10.20		36.70	44.50
0890	12" pipe size		28	.286		32	11.65		43.65	53
0900	For plastic coating 3/4 to 4", add					190%				
0910	For copper plating 3/4 to 4", add					58%				
0950	Two piece, complete, carbon steel, medium weight, TYPE 4									
0960	1/2" pipe size	Q-1	137	.117	Ea.	2.32	4.30		6.62	9
0970	3/4" pipe size		134	.119		2.32	4.39		6.71	9.15
0980	1" pipe size		132	.121		2.45	4.46		6.91	9.40
0990	1-1/4" pipe size		130	.123		3.04	4.53		7.57	10.15
1000	1-1/2" pipe size		126	.127		3.04	4.67		7.71	10.35
1010	2" pipe size		124	.129		3.76	4.75		8.51	11.30
1020	2-1/2" pipe size		120	.133		4.05	4.90		8.95	11.80
1030	3" pipe size		117	.137		4.52	5.05		9.57	12.50
1040	3-1/2" pipe size		114	.140		6.40	5.15		11.55	14.80
1050	4" pipe size		110	.145		6.40	5.35		11.75	15.10
1060	5" pipe size		106	.151		9.25	5.55		14.80	18.50
1070	6" pipe size		104	.154		15.60	5.65		21.25	25.50
1080	8" pipe size		100	.160		22.50	5.90		28.40	33.50
1090	10" pipe size		96	.167		33.50	6.15		39.65	46
1100	12" pipe size		89	.180		39	6.60		45.60	53
1110	14" pipe size		82	.195		68	7.20		75.20	86
1120	16" pipe size		68	.235		73.50	8.65		82.15	94
1130	For galvanized, add					45%				
1150	Insert, concrete									
1160	Wedge type, carbon steel body, malleable iron nut									
1170	1/4" threaded rod size	1 Plum	96	.083	Ea.	1.77	3.40		5.17	7.05
1180	3/8" threaded rod size		96	.083		1.81	3.40		5.21	7.10
1190	1/2" threaded rod size		96	.083		1.91	3.40		5.31	7.20
1200	5/8" threaded rod size		96	.083		1.95	3.40		5.35	7.25
1210	3/4" threaded rod size		96	.083		2.30	3.40		5.70	7.65
1220	7/8" threaded rod size		96	.083		2.89	3.40		6.29	8.30
1230	For galvanized, add					1.99			1.99	2.19
1250	Pipe guide sized for insulation									
1260	No. 1, 1" pipe size, 1" thick insulation	1 Stpi	26	.308	Ea.	114	12.60		126.60	144
1270	No. 2, 1-1/4"-2" pipe size, 1" thick insulation		23	.348		134	14.25		148.25	170
1280	No. 3, 1-1/4"-2" pipe size, 1-1/2" thick insulation		21	.381		134	15.60		149.60	172
1290	No. 4, 2-1/2"-3-1/2" pipe size, 1-1/2" thick insulation		18	.444		134	18.20		152.20	176
1300	No. 5, 4"-5" pipe size, 1-1/2" thick insulation		16	.500		149	20.50		169.50	195
1310	No. 6, 5"-6" pipe size, 2" thick insulation	Q-5	21	.762		166	28		194	224
1320	No. 7, 8" pipe size, 2" thick insulation		16	1		290	37		327	375
1330	No. 8, 10" pipe size, 2" thick insulation		12	1.333		550	49		599	680
1340	No. 9, 12" pipe size, 2" thick insulation	Q-6	17	1.412		550	54		604	685
1350	No. 10, 12"-14" pipe size, 2-1/2" thick insulation		16	1.500		715	57.50		772.50	870
1360	No. 11, 16" pipe size, 2-1/2" thick insulation		10.50	2.286		715	87.50		802.50	915
1370	No. 12, 16"-18" pipe size, 3" thick insulation		9	2.667		755	102		857	985
1380	No. 13, 20" pipe size, 3" thick insulation		7.50	3.200		755	122		877	1,025
1390	No. 14, 24" pipe size, 3" thick insulation		7	3.429		965	131		1,096	1,250
1400	Bands									
1410	Adjustable band, carbon steel, for non-insulated pipe, TYPE 7									
1420	1/2" pipe size	Q-1	142	.113	Ea.	.19	4.14		4.33	6.45

Important: See the Reference Section for critical supporting data - Reference Nos., Crews, & City Cost Indexes

		CREW	DAILY OUTPUT	LABOR-HOURS	UNIT	2005 BARE COSTS				TOTAL INCL O&P
15060	**Hangers & Supports**					MAT.	LABOR	EQUIP.	TOTAL	
300 1430	3/4" pipe size	Q-1	140	.114	Ea.	.85	4.20		5.05	7.25	300
1440	1" pipe size		137	.117		.99	4.30		5.29	7.55
1450	1-1/4" pipe size		134	.119		1.03	4.39		5.42	7.75
1460	1-1/2" pipe size		131	.122		1.14	4.49		5.63	8
1470	2" pipe size		129	.124		1.58	4.56		6.14	8.60
1480	2-1/2" pipe size		125	.128		1.65	4.71		6.36	8.90
1490	3" pipe size		122	.131		1.81	4.82		6.63	9.25
1500	3-1/2" pipe size		119	.134		1.96	4.95		6.91	9.60
1510	4" pipe size		114	.140		2.82	5.15		7.97	10.85
1520	5" pipe size		110	.145		2.98	5.35		8.33	11.35
1530	6" pipe size		108	.148		4.13	5.45		9.58	12.75
1540	8" pipe size	▼	104	.154	▼	6.45	5.65		12.10	15.60
1550	For copper plated, add					50%				
1560	For galvanized, add					30%				
1570	For plastic coating, add					30%				
1600	Adjusting nut malleable iron, steel band, TYPE 9									
1610	1/2" pipe size, galvanized band	Q-1	137	.117	Ea.	2.73	4.30		7.03	9.45
1620	3/4" pipe size, galvanized band		135	.119		2.73	4.36		7.09	9.55
1630	1" pipe size, gavanized band		132	.121		2.73	4.46		7.19	9.70
1640	1-1/4" pipe size, galvanized band		129	.124		3.30	4.56		7.86	10.50
1650	1-1/2" pipe size, galvanized band		126	.127		3.30	4.67		7.97	10.65
1660	2" pipe size, galvanized band		124	.129		3.58	4.75		8.33	11.10
1670	2-1/2" pipe size, galvanized band		120	.133		6	4.90		10.90	13.95
1680	3" pipe size, galvanized band		117	.137		6.55	5.05		11.60	14.75
1690	3-1/2" pipe size, galvanized band		114	.140		8.20	5.15		13.35	16.80
1700	4" pipe size, cadmium plated band	▼	110	.145		10.50	5.35		15.85	19.60
1740	For plastic coated band, add					35%				
1750	For completely copper coated, add				▼	45%				
1800	Clevis, adjustable, carbon steel, for non-insulated pipe, TYPE 1									
1810	1/2" pipe size	Q-1	137	.117	Ea.	.34	4.30		4.64	6.80
1820	3/4" pipe size		135	.119		.34	4.36		4.70	6.90
1830	1" pipe size		132	.121		.36	4.46		4.82	7.10
1840	1-1/4" pipe size		129	.124		.41	4.56		4.97	7.30
1850	1-1/2" pipe size		126	.127		.45	4.67		5.12	7.50
1860	2" pipe size		124	.129		.50	4.75		5.25	7.70
1870	2-1/2" pipe size		120	.133		.82	4.90		5.72	8.25
1880	3" pipe size		117	.137		1.03	5.05		6.08	8.70
1890	3-1/2" pipe size		114	.140		1.08	5.15		6.23	8.95
1900	4" pipe size		110	.145		1.24	5.35		6.59	9.40
1910	5" pipe size		106	.151		1.72	5.55		7.27	10.25
1920	6" pipe size		104	.154		1.98	5.65		7.63	10.70
1930	8" pipe size		100	.160		3.26	5.90		9.16	12.45
1940	10" pipe size		96	.167		27	6.15		33.15	39
1950	12" pipe size		89	.180		34	6.60		40.60	47.50
1960	14" pipe size		82	.195		51	7.20		58.20	67.50
1970	16" pipe size		68	.235		53	8.65		61.65	71.50
1971	18" pipe size		54	.296		60.50	10.90		71.40	83
1972	20" pipe size	▼	38	.421	▼	127	15.50		142.50	163
1980	For galvanized, add					66%				
1990	For copper plated 1/2" to 4", add					77%				
2000	For light weight 1/2" to 4", deduct					13%				
2010	Insulated pipe type, 3/4" to 12" pipe, add					180%				
2020	Insulated pipe type, chrome-moly U-strap, add					530%				
2250	Split ring, malleable iron, for non-insulated pipe, TYPE 11									
2260	1/2" pipe size	Q-1	137	.117	Ea.	3.89	4.30		8.19	10.75
2270	3/4" pipe size	▼	135	.119	▼	3.92	4.36		8.28	10.85

15060 | Hangers & Supports

		CREW	DAILY OUTPUT	LABOR-HOURS	UNIT	2005 BARE COSTS				TOTAL INCL O&P
						MAT.	LABOR	EQUIP.	TOTAL	
2280	1" pipe size	Q-1	132	.121	Ea.	4.14	4.46		8.60	11.25
2290	1-1/4" pipe size		129	.124		5.15	4.56		9.71	12.50
2300	1-1/2" pipe size		126	.127		6.25	4.67		10.92	13.85
2310	2" pipe size		124	.129		7.10	4.75		11.85	14.95
2320	2-1/2" pipe size		120	.133		10.25	4.90		15.15	18.60
2330	3" pipe size		117	.137		12.75	5.05		17.80	21.50
2340	3-1/2" pipe size		114	.140		13.05	5.15		18.20	22
2350	4" pipe size		110	.145		13.30	5.35		18.65	22.50
2360	5" pipe size		106	.151		17.70	5.55		23.25	28
2370	6" pipe size		104	.154		39.50	5.65		45.15	52
2380	8" pipe size	▼	100	.160	▼	70	5.90		75.90	86
2390	For copper plated, add					8%				
2500	Washer, flat steel									
2502	3/8"	1 Plum	240	.033	Ea.	.05	1.36		1.41	2.11
2503	1/2"		220	.036		.07	1.49		1.56	2.31
2504	5/8"		200	.040		.15	1.63		1.78	2.63
2505	3/4"		180	.044		.11	1.82		1.93	2.85
2506	7/8"		160	.050		.16	2.04		2.20	3.25
2507	1"		140	.057		.16	2.33		2.49	3.69
2508	1-1/8"	▼	120	.067	▼	.48	2.72		3.20	4.62
2520	Nut, steel, hex									
2522	3/8"	1 Plum	200	.040	Ea.	.15	1.63		1.78	2.63
2523	1/2"		180	.044		.27	1.82		2.09	3.03
2524	5/8"		160	.050		.56	2.04		2.60	3.69
2525	3/4"		140	.057		.76	2.33		3.09	4.35
2526	7/8"		120	.067		1.27	2.72		3.99	5.50
2527	1"		100	.080		2.83	3.27		6.10	8
2528	1-1/8"	▼	80	.100	▼	5.40	4.09		9.49	12.10
2532	Turnbuckle, TYPE 13									
2534	3/8"	1 Plum	80	.100	Ea.	12.30	4.09		16.39	19.65
2535	1/2"		72	.111		15.25	4.54		19.79	23.50
2536	5/8"		64	.125		15.95	5.10		21.05	25
2538	7/8"		48	.167		29.50	6.80		36.30	43
2539	1"		40	.200		37	8.15		45.15	53
2540	1-1/8"	▼	32	.250	▼	67	10.20		77.20	89
2544	Eye rod, welded									
2546	3/8"	1 Plum	56	.143	Ea.	16.30	5.85		22.15	26.50
2547	1/2"		48	.167		24	6.80		30.80	37
2548	5/8"		40	.200		36.50	8.15		44.65	52.50
2549	3/4"		32	.250		46.50	10.20		56.70	66.50
2550	7/8"		24	.333		57	13.60		70.60	83
2551	1"	▼	16	.500	▼	72	20.50		92.50	110
2556	Eye rod, welded, linked									
2558	3/8"	1 Plum	56	.143	Ea.	49.50	5.85		55.35	63.50
2559	1/2"		48	.167		67.50	6.80		74.30	84.50
2560	5/8"		40	.200		97	8.15		105.15	119
2561	3/4"		32	.250		126	10.20		136.20	153
2562	7/8"		24	.333		153	13.60		166.60	190
2563	1"	▼	16	.500	▼	197	20.50		217.50	247
2650	Rods, carbon steel									
2660	Continuous thread									
2670	1/4" thread size	1 Plum	144	.056	L.F.	.25	2.27		2.52	3.69
2680	3/8" thread size		144	.056		.27	2.27		2.54	3.71
2690	1/2" thread size		144	.056		.33	2.27		2.60	3.77
2700	5/8" thread size		144	.056		.50	2.27		2.77	3.96
2710	3/4" thread size	▼	144	.056	▼	1.03	2.27		3.30	4.54

Important: See the Reference Section for critical supporting data - Reference Nos., Crews, & City Cost Indexes

			DAILY	LABOR-		2005 BARE COSTS				TOTAL
15060	**Hangers & Supports**	CREW	OUTPUT	HOURS	UNIT	MAT.	LABOR	EQUIP.	TOTAL	INCL O&P
2720	7/8" thread size	1 Plum	144	.056	L.F.	1.55	2.27		3.82	5.10
2721	1" thread size	Q-1	160	.100		2.59	3.68		6.27	8.40
2722	1-1/8" thread size	"	120	.133		3.96	4.90		8.86	11.70
2725	1/4" thread size, bright finish	1 Plum	144	.056		.79	2.27		3.06	4.28
2726	1/2" thread size, bright finish	"	144	.056		.59	2.27		2.86	4.06
2730	For galvanized, add					40%				
2750	Both ends machine threaded 18" length									
2760	3/8" thread size	1 Plum	240	.033	Ea.	2.06	1.36		3.42	4.32
2770	1/2" thread size		240	.033		3.30	1.36		4.66	5.70
2780	5/8" thread size		240	.033		4.79	1.36		6.15	7.30
2790	3/4" thread size		240	.033		7.30	1.36		8.66	10.05
2800	7/8" thread size		240	.033		16.35	1.36		17.71	20
2810	1" thread size		240	.033		22.50	1.36		23.86	26.50
2820	Rod couplings									
2821	3/8"	1 Plum	60	.133	Ea.	2.23	5.45		7.68	10.65
2822	1/2"		54	.148		2.59	6.05		8.64	11.95
2823	5/8"		48	.167		2.94	6.80		9.74	13.50
2824	3/4"		44	.182		4.04	7.45		11.49	15.60
2825	7/8"		40	.200		4.52	8.15		12.67	17.25
2826	1"		34	.235		6.10	9.60		15.70	21
2827	1-1/8"		30	.267		17.25	10.90		28.15	35.50
2860	Pipe hanger assy, adj. clevis, saddle, rod, clamp, insul. allowance									
2864	1/2" pipe size	Q-5	35	.457	Ea.	17.05	16.85		33.90	44.50
2866	3/4" pipe size		34.80	.460		17.65	16.95		34.60	45
2868	1" pipe size		34.60	.462		31.50	17.05		48.55	60
2869	1-1/4" pipe size		34.30	.466		31.50	17.20		48.70	60.50
2870	1-1/2" pipe size		33.90	.472		32	17.40		49.40	61.50
2872	2" pipe size		33.30	.480		34	17.70		51.70	63.50
2874	2-1/2" pipe size		32.30	.495		39	18.25		57.25	70.50
2876	3" pipe size		31.20	.513		42	18.90		60.90	74.50
2880	4" pipe size		30.70	.521		44	19.20		63.20	77.50
2884	6" pipe size		29.80	.537		53.50	19.80		73.30	88.50
2888	8" pipe size		28	.571		98.50	21		119.50	140
2892	10" pipe size		25.20	.635		182	23.50		205.50	235
2896	12" pipe size		23.20	.690		224	25.50		249.50	285
2900	Rolls									
2910	Adjustable yoke, carbon steel with CI roll, TYPE 43									
2918	2" pipe size	Q-1	140	.114	Ea.	10.70	4.20		14.90	18.05
2920	2-1/2" pipe size		137	.117		10.70	4.30		15	18.20
2930	3" pipe size		131	.122		11.90	4.49		16.39	19.85
2940	3-1/2" pipe size		124	.129		15.80	4.75		20.55	24.50
2950	4" pipe size		117	.137		16.05	5.05		21.10	25
2960	5" pipe size		110	.145		18.75	5.35		24.10	28.50
2970	6" pipe size		104	.154		23.50	5.65		29.15	34
2980	8" pipe size		96	.167		35	6.15		41.15	47.50
2990	10" pipe size		80	.200		41.50	7.35		48.85	56.50
3010	14" pipe size		56	.286		163	10.50		173.50	196
3020	16" pipe size		48	.333		209	12.25		221.25	248
3050	Chair, carbon steel with CI roll									
3060	2" pipe size	1 Plum	68	.118	Ea.	14.40	4.81		19.21	23
3070	2-1/2" pipe size		65	.123		15.35	5.05		20.40	24.50
3080	3" pipe size		62	.129		16.90	5.25		22.15	26.50
3090	3-1/2" pipe size		60	.133		20.50	5.45		25.95	30.50
3100	4" pipe size		58	.138		23	5.65		28.65	33.50
3110	5" pipe size		56	.143		25	5.85		30.85	36.50
3120	6" pipe size		53	.151		34.50	6.15		40.65	47.50

300

15060 | Hangers & Supports

		CREW	DAILY OUTPUT	LABOR-HOURS	UNIT	2005 BARE COSTS				TOTAL INCL O&P
						MAT.	LABOR	EQUIP.	TOTAL	
3130	8" pipe size	1 Plum	50	.160	Ea.	46.50	6.55		53.05	61.50
3140	10" pipe size		48	.167		58.50	6.80		65.30	75
3150	12" pipe size	↓	46	.174		85	7.10		92.10	105
3170	Single pipe roll, (see line 2650 for rods), TYPE 41, 1" pipe size	Q-1	137	.117		7.75	4.30		12.05	15
3180	1-1/4" pipe size		131	.122		7.90	4.49		12.39	15.45
3190	1-1/2" pipe size		129	.124		8.05	4.56		12.61	15.70
3200	2" pipe size		124	.129		8.20	4.75		12.95	16.15
3210	2-1/2" pipe size		118	.136		9.65	4.99		14.64	18.15
3220	3" pipe size		115	.139		11.30	5.10		16.40	20
3230	3-1/2" pipe size		113	.142		12.75	5.20		17.95	22
3240	4" pipe size		112	.143		14.35	5.25		19.60	23.50
3250	5" pipe size		110	.145		15.90	5.35		21.25	25.50
3260	6" pipe size		101	.158		20.50	5.85		26.35	31.50
3270	8" pipe size		90	.178		32	6.55		38.55	45
3280	10" pipe size		80	.200		38	7.35		45.35	52.50
3290	12" pipe size	↓	68	.235	↓	42	8.65		50.65	59
3300	Saddles (add vertical pipe riser, usually 3" diameter)									
3310	Pipe support, complete, adjust., CI saddle, TYPE 36									
3320	2-1/2" pipe size	1 Plum	96	.083	Ea.	183	3.40		186.40	206
3330	3" pipe size		88	.091		185	3.71		188.71	210
3340	3-1/2" pipe size		79	.101		186	4.14		190.14	211
3350	4" pipe size		68	.118		243	4.81		247.81	274
3360	5" pipe size		64	.125		245	5.10		250.10	277
3370	6" pipe size		59	.136		249	5.55		254.55	282
3380	8" pipe size		53	.151		264	6.15		270.15	299
3390	10" pipe size		50	.160		291	6.55		297.55	330
3400	12" pipe size	↓	48	.167	↓	300	6.80		306.80	345
3450	For standard pipe support, one piece, CI deduct					34%				
3460	For stanchion support, CI with steel yoke, deduct					60%				
3500	Multiple pipe support 2' long strut	1 Plum	30	.267	Ea.	4.46	10.90		15.36	21.50
3510	3' long strut		20	.400		6.70	16.35		23.05	32
3520	4' long strut	↓	17	.471		8.90	19.20		28.10	39
3550	Insulation shield 1" thick, 1/2" pipe size, TYPE 40	1 Asbe	100	.080		4.93	2.97		7.90	10.10
3560	3/4" pipe size		100	.080		6.10	2.97		9.07	11.45
3570	1" pipe size		98	.082		6.10	3.03		9.13	11.55
3580	1-1/4" pipe size		98	.082		6.45	3.03		9.48	11.90
3590	1-1/2" pipe size		96	.083		6.45	3.09		9.54	12
3600	2" pipe size		96	.083		6.75	3.09		9.84	12.35
3610	2-1/2" pipe size		94	.085		7	3.16		10.16	12.70
3620	3" pipe size		94	.085		7.45	3.16		10.61	13.15
3630	2" thick, 3-1/2" pipe size		92	.087		8.25	3.23		11.48	14.20
3640	4" pipe size		92	.087		9.70	3.23		12.93	15.80
3650	5" pipe size		90	.089		14.40	3.30		17.70	21
3660	6" pipe size		90	.089		16.45	3.30		19.75	23.50
3670	8" pipe size		88	.091		24.50	3.37		27.87	32.50
3680	10" pipe size		88	.091		36.50	3.37		39.87	45.50
3690	12" pipe size		86	.093		43	3.45		46.45	53
3700	14" pipe size		86	.093		68.50	3.45		71.95	81
3710	16" pipe size		84	.095		72	3.53		75.53	84.50
3720	18" pipe size		84	.095		83.50	3.53		87.03	97.50
3730	20" pipe size		82	.098		91	3.62		94.62	106
3732	24" pipe size		80	.100		160	3.71		163.71	183
3733	30" pipe size		78	.103		207	3.81		210.81	233
3735	36" pipe size	↓	75	.107	↓	220	3.96		223.96	248
3750	Covering protection saddle, TYPE 39									
3760	1" covering size									

15060	Hangers & Supports	CREW	DAILY OUTPUT	LABOR-HOURS	UNIT	MAT.	2005 BARE COSTS LABOR	EQUIP.	TOTAL	TOTAL INCL O&P
3770	3/4" pipe size	1 Plum	68	.118	Ea.	3.65	4.81		8.46	11.20
3780	1" pipe size		68	.118		3.65	4.81		8.46	11.20
3790	1-1/4" pipe size		68	.118		3.65	4.81		8.46	11.20
3800	1-1/2" pipe size		66	.121		3.96	4.95		8.91	11.80
3810	2" pipe size		66	.121		4.04	4.95		8.99	11.90
3820	2-1/2" pipe size		64	.125		5.70	5.10		10.80	13.95
3830	3" pipe size		64	.125		7.30	5.10		12.40	15.75
3840	3-1/2" pipe size		62	.129		7.30	5.25		12.55	15.95
3850	4" pipe size		62	.129		9.20	5.25		14.45	18
3860	5" pipe size		60	.133		9.20	5.45		14.65	18.30
3870	6" pipe size	▼	60	.133	▼	10.95	5.45		16.40	20
3900	1-1/2" covering size									
3910	3/4" pipe size	1 Plum	68	.118	Ea.	11.05	4.81		15.86	19.35
3920	1" pipe size		68	.118		11.05	4.81		15.86	19.35
3930	1-1/4" pipe size		68	.118		11.05	4.81		15.86	19.35
3940	1-1/2" pipe size		66	.121		11.05	4.95		16	19.60
3950	2" pipe size		66	.121		11.80	4.95		16.75	20.50
3960	2-1/2" pipe size		64	.125		14.15	5.10		19.25	23.50
3970	3" pipe size		64	.125		14.15	5.10		19.25	23.50
3980	3-1/2" pipe size		62	.129		14.35	5.25		19.60	23.50
3990	4" pipe size		62	.129		14.35	5.25		19.60	23.50
4000	5" pipe size		60	.133		14.35	5.45		19.80	24
4010	6" pipe size		60	.133		20.50	5.45		25.95	30.50
4020	8" pipe size		58	.138		27	5.65		32.65	38
4022	10" pipe size		56	.143		23.50	5.85		29.35	35
4024	12" pipe size	▼	54	.148	▼	38	6.05		44.05	50.50
4028	2" covering size									
4029	2-1/2" pipe size	1 Plum	62	.129	Ea.	12	5.25		17.25	21
4032	3" pipe size		60	.133		15.95	5.45		21.40	26
4033	4" pipe size		58	.138		15.95	5.65		21.60	26
4034	6" pipe size		56	.143		23	5.85		28.85	34
4035	8" pipe size		54	.148		27	6.05		33.05	38.50
4080	10" pipe size		58	.138		28.50	5.65		34.15	40
4090	12" pipe size		56	.143		42.50	5.85		48.35	55.50
4100	14" pipe size		56	.143		42.50	5.85		48.35	55.50
4110	16" pipe size		54	.148		59.50	6.05		65.55	74.50
4120	18" pipe size		54	.148		59.50	6.05		65.55	74.50
4130	20" pipe size		52	.154		66	6.30		72.30	82
4150	24" pipe size		50	.160		77	6.55		83.55	95
4160	30" pipe size		48	.167		85	6.80		91.80	104
4180	36" pipe size	▼	45	.178	▼	93.50	7.25		100.75	114
4186	2-1/2" covering size									
4187	3" pipe size	1 Plum	58	.138	Ea.	15.55	5.65		21.20	25.50
4188	4" pipe size		56	.143		16.90	5.85		22.75	27.50
4189	6" pipe size		52	.154		26	6.30		32.30	38
4190	8" pipe size		48	.167		31.50	6.80		38.30	45
4191	10" pipe size		44	.182		34.50	7.45		41.95	49
4192	12" pipe size		40	.200		46	8.15		54.15	63
4193	14" pipe size		36	.222		46	9.10		55.10	64
4194	16" pipe size		32	.250		68	10.20		78.20	90
4195	18" pipe size	▼	28	.286	▼	72.50	11.65		84.15	97.50
4200	Sockets									
4210	Rod end, malleable iron, TYPE 16									
4220	1/4" thread size	1 Plum	240	.033	Ea.	1.22	1.36		2.58	3.39
4230	3/8" thread size		240	.033		1.31	1.36		2.67	3.49
4240	1/2" thread size	▼	230	.035	▼	1.57	1.42		2.99	3.87

300

	15060	**Hangers & Supports**	CREW	DAILY OUTPUT	LABOR-HOURS	UNIT	MAT.	LABOR	EQUIP.	TOTAL	TOTAL INCL O&P	
								2005 BARE COSTS				
300	4250	5/8" thread size	1 Plum	225	.036	Ea.	2.84	1.45		4.29	5.30	**300**
	4260	3/4" thread size		220	.036		7.55	1.49		9.04	10.60	
	4270	7/8" thread size	↓	210	.038		9.90	1.56		11.46	13.25	
	4290	Strap, 1/2" pipe size, TYPE 26	Q-1	142	.113		.47	4.14		4.61	6.75	
	4300	3/4" pipe size		140	.114		.51	4.20		4.71	6.85	
	4310	1" pipe size		137	.117		.53	4.30		4.83	7.05	
	4320	1-1/4" pipe size		134	.119		.60	4.39		4.99	7.25	
	4330	1-1/2" pipe size		131	.122		.67	4.49		5.16	7.50	
	4340	2" pipe size		129	.124		.73	4.56		5.29	7.65	
	4350	2-1/2" pipe size		125	.128		1.14	4.71		5.85	8.35	
	4360	3" pipe size		122	.131		1.31	4.82		6.13	8.70	
	4370	3-1/2" pipe size		119	.134		14.30	4.95		19.25	23	
	4380	4" pipe size	↓	114	.140	↓	16	5.15		21.15	25.50	
	4400	U-bolt, carbon steel										
	4410	Standard, with nuts, TYPE 42										
	4420	1/2" pipe size	1 Plum	160	.050	Ea.	.55	2.04		2.59	3.68	
	4430	3/4" pipe size		158	.051		.55	2.07		2.62	3.72	
	4450	1" pipe size		152	.053		.62	2.15		2.77	3.91	
	4460	1-1/4" pipe size		148	.054		.71	2.21		2.92	4.10	
	4470	1-1/2" pipe size		143	.056		.79	2.29		3.08	4.30	
	4480	2" pipe size		139	.058		.86	2.35		3.21	4.48	
	4490	2-1/2" pipe size		134	.060		1.42	2.44		3.86	5.25	
	4500	3" pipe size		128	.063		1.50	2.55		4.05	5.50	
	4510	3-1/2" pipe size		122	.066		1.53	2.68		4.21	5.70	
	4520	4" pipe size		117	.068		1.60	2.79		4.39	5.95	
	4530	5" pipe size		114	.070		1.70	2.87		4.57	6.20	
	4540	6" pipe size		111	.072		2.99	2.94		5.93	7.70	
	4550	8" pipe size		109	.073		3.42	3		6.42	8.25	
	4560	10" pipe size		107	.075		5.50	3.05		8.55	10.65	
	4570	12" pipe size	↓	104	.077	↓	7.35	3.14		10.49	12.75	
	4580	For plastic coating on 1/2" thru 6" size, add						150%					
	4700	U-hook, carbon steel, requires mounting screws or bolts											
	4710	3/4" thru 2" pipe size											
	4720	6" long	1 Plum	96	.083	Ea.	4.09	3.40		7.49	9.60	
	4730	8" long		96	.083		4.36	3.40		7.76	9.90	
	4740	10" long		96	.083		5.50	3.40		8.90	11.15	
	4750	12" long	↓	96	.083		6.10	3.40		9.50	11.80	
	4760	For copper plated, add						50%					
	8000	Pipe clamp, plastic, 1/2" CTS	1 Plum	80	.100		.20	4.09		4.29	6.35	
	8010	3/4" CTS		73	.110		.21	4.48		4.69	7	
	8020	1" CTS		68	.118		.50	4.81		5.31	7.75	
	8080	Economy clamp, 1/4" CTS		175	.046		.05	1.87		1.92	2.87	
	8090	3/8" CTS		168	.048		.06	1.95		2.01	2.99	
	8100	1/2" CTS		160	.050		.06	2.04		2.10	3.14	
	8110	3/4" CTS		145	.055		.12	2.25		2.37	3.52	
	8200	Half clamp, 1/2" CTS		80	.100		.07	4.09		4.16	6.25	
	8210	3/4" CTS		73	.110		.10	4.48		4.58	6.85	
	8300	Suspension clamp, 1/2" CTS		80	.100		.20	4.09		4.29	6.35	
	8310	3/4" CTS		73	.110		.21	4.48		4.69	7	
	8320	1" CTS		68	.118		.50	4.81		5.31	7.75	
	8400	Insulator, 1/2" CTS		80	.100		.34	4.09		4.43	6.50	
	8410	3/4" CTS		73	.110		.35	4.48		4.83	7.15	
	8420	1" CTS		68	.118		.36	4.81		5.17	7.60	
	8500	J hook clamp with nail 1/2" CTS		240	.033		.11	1.36		1.47	2.17	
	8501	3/4" CTS	↓	240	.033	↓	.12	1.36		1.48	2.18	
	8800	Wire cable support system											

Important: See the Reference Section for critical supporting data - Reference Nos., Crews, & City Cost Indexes

15060 | Hangers & Supports

			DAILY	LABOR-		2005 BARE COSTS				TOTAL
		CREW	OUTPUT	HOURS	UNIT	MAT.	LABOR	EQUIP.	TOTAL	INCL O&P
300	8810	Cable with hook terminal and locking device									300
	8830	2 mm, (.079 ") dia cable, (100 lb. cap.)									
	8840	1 m, (3.3') length, with hook	1 Shee	96	.083	Ea.	3.74	3.35		7.09	9.25
	8850	2 m, (6.6') length, with hook		84	.095		4.32	3.83		8.15	10.65
	8860	3 m, (9.9') length, with hook	↓	72	.111		4.90	4.47		9.37	12.25
	8870	5 m, (16.4') length, with hook	Q-9	60	.267		6.15	9.65		15.80	21.50
	8880	10 m, (32.8') length, with hook	"	30	.533	↓	8.90	19.30		28.20	39.50
	8900	3mm, (.118 ") dia cable, (200 lb. cap.)									
	8910	1 m, (3.3') length, with hook	1 Shee	96	.083	Ea.	4.80	3.35		8.15	10.45
	8920	2 m, (6.6') length, with hook		84	.095		5.40	3.83		9.23	11.85
	8930	3 m, (9.9') length, with hook	↓	72	.111		5.95	4.47		10.42	13.40
	8940	5 m, (16.4') length, with hook	Q-9	60	.267		7.35	9.65		17	23
	8950	10 m, (32.8') length, with hook	"	30	.533	↓	10.60	19.30		29.90	41
	9000	Cable system accessories									
	9010	Anchor bolt, 3/8", with nut	1 Shee	140	.057	Ea.	1.47	2.30		3.77	5.15
	9020	Air duct corner protector		160	.050		.64	2.01		2.65	3.79
	9030	Air duct support attachment	↓	140	.057	↓	.24	2.30		2.54	3.79
	9040	Flange clip, hammer-on style									
	9044	For flange thickness 3/32" - 9/64", 160 lb. cap.	1 Shee	180	.044	Ea.	.29	1.79		2.08	3.07
	9048	For flange thickness 1/8" - 1/4", 200 lb. cap.		160	.050		.33	2.01		2.34	3.45
	9052	For flange thickness 5/16" - 1/2", 200 lb. cap.		150	.053		.36	2.14		2.50	3.70
	9056	For flange thickness 9/16" - 3/4", 200 lb. cap.		140	.057	↓	.53	2.30		2.83	4.11
	9060	Wire insulation protection tube	↓	180	.044	L.F.	.41	1.79		2.20	3.20
	9070	Wire cutter				Ea.	40			40	44

15070 | Mech Sound/Vibration/Seismic Ctrl

			DAILY	LABOR-		2005 BARE COSTS				TOTAL
		CREW	OUTPUT	HOURS	UNIT	MAT.	LABOR	EQUIP.	TOTAL	INCL O&P
800	0010	**VIBRATION ABSORBERS**									800
	0100	Hangers, neoprene flex									
	0200	10 - 120 lb. capacity				Ea.	13.55			13.55	14.90
	0220	75 - 550 lb. capacity					17.65			17.65	19.40
	0240	250 - 1100 lb. capacity					37			37	40.50
	0260	1000 - 4000 lb. capacity					59.50			59.50	65.50
	0500	Spring flex, 60 lb. capacity					19			19	21
	0520	450 lb. capacity					23.50			23.50	26
	0540	900 lb. capacity					31			31	34.50
	0560	1100-1300 lb. capacity				↓	31			31	34.50
	0600	Rubber in shear									
	0610	45 - 340 lb., up to 1/2" rod size	1 Stpi	22	.364	Ea.	12	14.90		26.90	35.50
	0620	130 - 700 lb., up to 3/4" rod size	↓	20	.400		27	16.40		43.40	54
	0630	50 - 1000 lb., up to 3/4" rod size	↓	18	.444		31	18.20		49.20	61.50
	1000	Mounts, neoprene, 45 - 380 lb. capacity					10.25			10.25	11.25
	1020	250 - 1100 lb. capacity					26.50			26.50	29.50
	1040	1000 - 4000 lb. capacity					60			60	66
	1100	Spring flex, 60 lb. capacity					37			37	40.50
	1120	165 lb. capacity					37			37	40.50
	1140	260 lb. capacity					37			37	40.50
	1160	450 lb. capacity					48.50			48.50	53.50
	1180	600 lb. capacity					48.50			48.50	53.50
	1200	750 lb. capacity					48.50			48.50	53.50
	1220	900 lb. capacity					49			49	54
	1240	1100 lb. capacity					50.50			50.50	55.50
	1260	1300 lb. capacity					53.50			53.50	59
	1280	1500 lb. capacity					70.50			70.50	77.50
	1300	1800 lb. capacity					74			74	81.50
	1320	2200 lb. capacity					82.50			82.50	90.50
	1340	2600 lb. capacity				↓	84			84	92.50

15070 | Mech Sound/Vibration/Seismic Ctrl

		CREW	DAILY OUTPUT	LABOR-HOURS	UNIT	2005 BARE COSTS				TOTAL INCL O&P
						MAT.	LABOR	EQUIP.	TOTAL	
800	1399	Spring type									800
	1400	2 Piece								
	1410	50 - 1000 lb.	1 Stpi	12	.667	Ea.	102	27.50		129.50	153
	1420	1100 - 1600 lb.	"	12	.667	"	110	27.50		137.50	162
	1500	Double spring open									
	1510	150 - 450 lb.	1 Stpi	24	.333	Ea.	74	13.65		87.65	102
	1520	500 - 1000 lb.		24	.333		74	13.65		87.65	102
	1530	1100 - 1600 lb.		24	.333		74	13.65		87.65	102
	1540	1700 - 2400 lb.		24	.333		80	13.65		93.65	109
	1550	2500 - 3400 lb.		24	.333		136	13.65		149.65	171
	2000	Pads, cork rib, 18" x 18" x 1", 10-50 psi					100			100	110
	2020	18" x 36" x 1", 10-50 psi					201			201	221
	2100	Shear flexible pads, 18" x 18" x 3/8", 20-70 psi					47.50			47.50	52.50
	2120	18" x 36" x 3/8", 20-70 psi					88.50			88.50	97.50
	2150	Laminated neoprene and cork									
	2160	1" thick	1 Stpi	16	.500	S.F.	47	20.50		67.50	82.50
	3000	Note overlap in capacities due to deflections									

15075 | Mechanical Identification

		CREW	DAILY OUTPUT	LABOR-HOURS	UNIT	MAT.	LABOR	EQUIP.	TOTAL	TOTAL INCL O&P
400	0010	**PIPING SYSTEM IDENTIFICATION LABELS**									400
	0100	Indicate contents and flow direction									
	0106	Pipe markers									
	0110	Plastic snap around									
	0114	1/2" pipe	1 Plum	80	.100	Ea.	3.40	4.09		7.49	9.90
	0116	3/4" pipe		80	.100		3.40	4.09		7.49	9.90
	0118	1" pipe		80	.100		3.40	4.09		7.49	9.90
	0120	2" pipe		75	.107		4.40	4.36		8.76	11.40
	0122	3" pipe		70	.114		8.20	4.67		12.87	16
	0124	4" pipe		60	.133		8.20	5.45		13.65	17.20
	0126	6" pipe		60	.133		9	5.45		14.45	18.10
	0128	8" pipe		56	.143		12.30	5.85		18.15	22.50
	0130	10" pipe		56	.143		12.30	5.85		18.15	22.50
	0200	Over 10" pipe size		50	.160		15	6.55		21.55	26.50
	1110	Self adhesive									
	1114	1" pipe	1 Plum	80	.100	Ea.	2.40	4.09		6.49	8.80
	1116	2" pipe		75	.107		2.40	4.36		6.76	9.20
	1118	3" pipe		70	.114		3.80	4.67		8.47	11.20
	1120	4" pipe		60	.133		3.80	5.45		9.25	12.40
	1122	6" pipe		60	.133		3.80	5.45		9.25	12.40
	1124	8" pipe		56	.143		3.80	5.85		9.65	12.95
	1126	10" pipe		56	.143		5.30	5.85		11.15	14.60
	1200	Over 10" pipe size		50	.160		7.30	6.55		13.85	17.85
	2000	Valve tags									
	2010	Numbered plus identifying legend									
	2100	Brass, 2" diameter	1 Plum	40	.200	Ea.	2.10	8.15		10.25	14.60
	2200	Plastic, 1-1/2" Dia.	"	40	.200	"	1.80	8.15		9.95	14.30

15080 | Mechanical Insulation

		CREW	DAILY OUTPUT	LABOR-HOURS	UNIT	MAT.	LABOR	EQUIP.	TOTAL	TOTAL INCL O&P
200	0010	**DUCT INSULATION**									200
	0100	Rule of thumb, as a percentage of total mechanical costs				Job				10%	
	0110	Insulation req'd is based on the surface size/area to be covered									
	3780	Foam, rubber									
	3782	1" thick	1 Stpi	50	.160	S.F.	2.66	6.55		9.21	12.80
	7878	Contact cement, quart can				Ea.	22			22	24

Important: See the Reference Section for critical supporting data - Reference Nos., Crews, & City Cost Indexes

		15080 Mechanical Insulation	CREW	DAILY OUTPUT	LABOR-HOURS	UNIT	MAT.	LABOR	EQUIP.	TOTAL	TOTAL INCL O&P	
							2005 BARE COSTS					
200	9600	Minimum labor/equipment charge	1 Stpi	4	2	Job		82		82	123	**200**
400	0010	**EQUIPMENT INSULATION**										**400**
	0100	Rule of thumb, as a percentage of total mechanical costs				Job				10%		
	0110	Insulation req'd is based on the surface size/area to be covered										
	1000	Boiler, 1-1/2″ calcium silicate, 1/2″ cement finish	Q-14	50	.320	S.F.	3.30	10.70		14	20.50	
	1020	2″ fiberglass	″	80	.200	″	2.03	6.70		8.73	12.80	
	2000	Breeching, 2″ calcium silicate with 1/2″ cement finish, no lath										
	2020	Rectangular	Q-14	42	.381	S.F.	3.87	12.70		16.57	24.50	
	2040	Round	″	38.70	.413	″	4.33	13.80		18.13	27	
	2300	Calcium silicate block, + 200°F to + 1200°F										
	2310	On irregular surfaces, valves and fittings										
	2340	1″ thick	Q-14	30	.533	S.F.	2.70	17.80		20.50	31	
	2360	1-1/2″ thick	″	25	.640	″	3.15	21.50		24.65	37	
	2410	On plane surfaces										
	2420	1″ thick	Q-14	126	.127	S.F.	2.70	4.24		6.94	9.65	
	2430	1-1/2″ thick	″	120	.133	″	3.15	4.45		7.60	10.50	
	2900	Domestic water heater wrap kit										
	2920	1-1/2″ with vinyl jacket, 20-60 gal.	1 Plum	8	1	Ea.	16.60	41		57.60	80	
600	0010	**PIPING INSULATION**										**600**
	0100	Rule of thumb, as a percentage of total mechanical costs				Job				10%		
	0110	Insulation req'd is based on the surface size/area to be covered										
	2930	Insulated protectors, (ADA)										
	2935	For exposed piping under sinks or lavatories.										
	2940	Vinyl coated foam, velcro tabs										
	2945	P Trap, 1-1/4″ or 1-1/2″	1 Plum	32	.250	Ea.	17.15	10.20		27.35	34	
	2960	Valve and supply cover										
	2965	1/2″, 3/8″, and 7/16″ pipe size	1 Plum	32	.250	Ea.	17.15	10.20		27.35	34	
	2970	Extension drain cover										
	2975	1-1/4″, or 1-1/2″ pipe size	1 Plum	32	.250	Ea.	17.60	10.20		27.80	34.50	
	2980	Tailpiece offset (wheelchair)										
	2985	1-1/4″ pipe size	1 Plum	32	.250	Ea.	19.60	10.20		29.80	37	
	4000	Pipe covering (price copper tube one size less than IPS)										
	4040	Air cell, asbestos free corrugated felt, with cover										
	4100	3 ply, 1/2″ iron pipe size	Q-14	145	.110	L.F.	.75	3.69		4.44	6.65	
	4130	3/4″ iron pipe size		145	.110		.85	3.69		4.54	6.75	
	4140	1″ iron pipe size		145	.110		.90	3.69		4.59	6.80	
	4150	1-1/4″ iron pipe size		140	.114		1	3.82		4.82	7.10	
	4160	1-1/2″ iron pipe size		140	.114		1.10	3.82		4.92	7.20	
	4170	2″ iron pipe size		135	.119		1.20	3.96		5.16	7.55	
	4180	2-1/2″ iron pipe size		135	.119		1.35	3.96		5.31	7.75	
	4190	3″ iron pipe size		130	.123		1.50	4.11		5.61	8.15	
	4200	3-1/2″ iron pipe size		130	.123		1.76	4.11		5.87	8.45	
	4210	4″ iron pipe size		125	.128		1.95	4.28		6.23	8.90	
	4220	5″ iron pipe size		120	.133		2.25	4.45		6.70	9.55	
	4230	6″ iron pipe size		115	.139		2.40	4.65		7.05	10	
	4240	8″ iron pipe size		105	.152		3.30	5.10		8.40	11.70	
	4250	10″ iron pipe size		95	.168		3.90	5.65		9.55	13.20	
	4260	12″ iron pipe size		90	.178		4.50	5.95		10.45	14.30	
	4280	Cellular glass, closed cell foam, all service jacket, sealant,										
	4281	working temp. (-450°F to +900°F), 0 water vapor xmission										
	4284	1″ wall,										
	4286	1/2″ iron pipe size	Q-14	120	.133	L.F.	1.65	4.45		6.10	8.85	
	4300	1-1/2″ wall,										
	4301	1″ iron pipe size	Q-14	105	.152	L.F.	2.65	5.10		7.75	10.95	

15080	Mechanical Insulation	CREW	DAILY OUTPUT	LABOR-HOURS	UNIT	MAT.	LABOR	EQUIP.	TOTAL	TOTAL INCL O&P	
						2005 BARE COSTS					
600											**600**
4304	2-1/2" iron pipe size	Q-14	90	.178	L.F.	4.46	5.95		10.41	14.25	
4306	3" iron pipe size		85	.188		4.51	6.30		10.81	14.85	
4308	4" iron pipe size		70	.229		6	7.65		13.65	18.65	
4310	5" iron pipe size		65	.246		6.90	8.20		15.10	20.50	
4320	2" wall,										
4322	1" iron pipe size	Q-14	100	.160	L.F.	3.94	5.35		9.29	12.80	
4324	2-1/2" iron pipe size		85	.188		5.85	6.30		12.15	16.30	
4326	3" iron pipe size		80	.200		6	6.70		12.70	17.15	
4328	4" iron pipe size		65	.246		6.70	8.20		14.90	20.50	
4330	5" iron pipe size		60	.267		8.05	8.90		16.95	23	
4332	6" iron pipe size		50	.320		9.55	10.70		20.25	27.50	
4336	8" iron pipe size		40	.400		12.30	13.35		25.65	34.50	
4338	10" iron pipe size		35	.457		12.75	15.25		28	38	
4350	2-1/2" wall,										
4360	12" iron pipe size	Q-14	32	.500	L.F.	20	16.70		36.70	48.50	
4362	14" iron pipe size	"	28	.571	"	21.50	19.10		40.60	54	
4370	3" wall,										
4378	6" iron pipe size	Q-14	48	.333	L.F.	12.75	11.15		23.90	31.50	
4380	8" iron pipe size		38	.421		14.85	14.05		28.90	38.50	
4382	10" iron pipe size		33	.485		19.65	16.20		35.85	47	
4384	16" iron pipe size		25	.640		25.50	21.50		47	61.50	
4386	18" iron pipe size		22	.727		28	24.50		52.50	69	
4388	20" iron pipe size		20	.800		32.50	26.50		59	78	
4400	3-1/2" wall,										
4412	12" iron pipe size	Q-14	27	.593	L.F.	28.50	19.80		48.30	62.50	
4414	14" iron pipe size	"	25	.640	"	30.50	21.50		52	67	
4430	4" wall,										
4446	16" iron pipe size	Q-14	22	.727	L.F.	31	24.50		55.50	72.50	
4448	18" iron pipe size		20	.800		40	26.50		66.50	86	
4450	20" iron pipe size		18	.889		41.50	29.50		71	92.50	
4480	Fittings, average with fabric and mastic										
4484	1" wall,										
4486	1/2" iron pipe size	1 Asbe	40	.200	Ea.	2.93	7.40		10.33	14.90	
4500	1-1/2" wall,										
4502	1" iron pipe size	1 Asbe	38	.211	Ea.	5.05	7.80		12.85	17.90	
4504	2-1/2" iron pipe size		32	.250		6.25	9.30		15.55	21.50	
4506	3" iron pipe size		30	.267		6.90	9.90		16.80	23	
4508	4" iron pipe size		28	.286		10.60	10.60		21.20	28.50	
4510	5" iron pipe size		24	.333		11.70	12.35		24.05	32.50	
4520	2" wall,										
4522	1" iron pipe size	1 Asbe	36	.222	Ea.	5.85	8.25		14.10	19.40	
4524	2-1/2" iron pipe size		30	.267		6.90	9.90		16.80	23	
4526	3" iron pipe size		28	.286		8.30	10.60		18.90	26	
4528	4" iron pipe size		24	.333		11.15	12.35		23.50	32	
4530	5" iron pipe size		22	.364		14.05	13.50		27.55	37	
4532	6" iron pipe size		20	.400		16.45	14.85		31.30	41.50	
4536	8" iron pipe size		12	.667		25	24.50		49.50	66.50	
4538	10" iron pipe size		8	1		31	37		68	92.50	
4550	2-1/2" wall,										
4560	12" iron pipe size	1 Asbe	6	1.333	Ea.	53	49.50		102.50	137	
4562	14" iron pipe size	"	4	2	"	61.50	74		135.50	185	
4570	3" wall,										
4578	6" iron pipe size	1 Asbe	16	.500	Ea.	19.10	18.55		37.65	50.50	
4580	8" iron pipe size		10	.800		28	29.50		57.50	78	
4582	10" iron pipe size		6	1.333		40	49.50		89.50	122	
4584	16" iron pipe size		2	4		95.50	148		243.50	340	

Important: See the Reference Section for critical supporting data - Reference Nos., Crews, & City Cost Indexes

	**15080	Mechanical Insulation**	CREW	DAILY OUTPUT	LABOR-HOURS	UNIT	MAT.	LABOR	EQUIP.	TOTAL	TOTAL INCL O&P
600	4586	18" iron pipe size	1 Asbe	2	4	Ea.	123	148		271	370
	4588	20" iron pipe size	↓	2	4	↓	145	148		293	395
	4600	3-1/2" wall,									
	4612	12" iron pipe size	1 Asbe	4	2	Ea.	69	74		143	193
	4614	14" iron pipe size	"	2	4	"	76.50	148		224.50	320
	4630	4" wall,									
	4646	16" iron pipe size	1 Asbe	2	4	Ea.	115	148		263	360
	4648	18" iron pipe size		2	4		140	148		288	390
	4650	20" iron pipe size	↓	2	4	↓	163	148		311	415
	4900	Calcium silicate, with cover									
	5100	1" wall, 1/2" iron pipe size	Q-14	170	.094	L.F.	2.18	3.14		5.32	7.35
	5130	3/4" iron pipe size		170	.094		2.19	3.14		5.33	7.35
	5140	1" iron pipe size		170	.094		2.22	3.14		5.36	7.40
	5150	1-1/4" iron pipe size		165	.097		2.37	3.24		5.61	7.70
	5160	1-1/2" iron pipe size		165	.097		2.49	3.24		5.73	7.85
	5170	2" iron pipe size		160	.100		2.53	3.34		5.87	8.05
	5180	2-1/2" iron pipe size		160	.100		2.69	3.34		6.03	8.20
	5190	3" iron pipe size		150	.107		2.89	3.56		6.45	8.80
	5200	4" iron pipe size		140	.114		3.56	3.82		7.38	9.90
	5210	5" iron pipe size		135	.119		3.79	3.96		7.75	10.40
	5220	6" iron pipe size		130	.123		4.01	4.11		8.12	10.90
	5280	1-1/2" wall, 1/2" iron pipe size		150	.107		2.35	3.56		5.91	8.20
	5310	3/4" iron pipe size		150	.107		2.36	3.56		5.92	8.20
	5320	1" iron pipe size		150	.107		2.59	3.56		6.15	8.45
	5330	1-1/4" iron pipe size		145	.110		2.80	3.69		6.49	8.90
	5340	1-1/2" iron pipe size		145	.110		3	3.69		6.69	9.10
	5350	2" iron pipe size		140	.114		3.31	3.82		7.13	9.65
	5360	2-1/2" iron pipe size		140	.114		3.60	3.82		7.42	9.95
	5370	3" iron pipe size		135	.119		3.79	3.96		7.75	10.40
	5380	4" iron pipe size		125	.128		4.39	4.28		8.67	11.60
	5390	5" iron pipe size		120	.133		4.94	4.45		9.39	12.50
	5400	6" iron pipe size		110	.145		5.10	4.86		9.96	13.25
	5460	2" wall, 1/2" iron pipe size		135	.119		3.63	3.96		7.59	10.25
	5490	3/4" iron pipe size		135	.119		3.82	3.96		7.78	10.45
	5500	1" iron pipe size		135	.119		4.01	3.96		7.97	10.65
	5510	1-1/4" iron pipe size		130	.123		4.31	4.11		8.42	11.25
	5520	1-1/2" iron pipe size		130	.123		4.49	4.11		8.60	11.45
	5530	2" iron pipe size		125	.128		4.73	4.28		9.01	11.95
	5540	2-1/2" iron pipe size		125	.128		5.65	4.28		9.93	12.95
	5550	3" iron pipe size		120	.133		5.70	4.45		10.15	13.30
	5560	4" iron pipe size		115	.139		6.55	4.65		11.20	14.55
	5570	5" iron pipe size		110	.145		7.45	4.86		12.31	15.80
	5580	6" iron pipe size	↓	105	.152	↓	8.10	5.10		13.20	16.95
	5600	Calcium silicate, no cover									
	5720	1" wall, 1/2" iron pipe size	Q-14	180	.089	L.F.	2	2.97		4.97	6.90
	5740	3/4" iron pipe size		180	.089		2	2.97		4.97	6.90
	5750	1" iron pipe size		180	.089		1.92	2.97		4.89	6.80
	5760	1-1/4" iron pipe size		175	.091		1.96	3.05		5.01	7
	5770	1-1/2" iron pipe size		175	.091		1.97	3.05		5.02	7
	5780	2" iron pipe size		170	.094		2.27	3.14		5.41	7.45
	5790	2-1/2" iron pipe size		170	.094		2.40	3.14		5.54	7.60
	5800	3" iron pipe size		160	.100		2.58	3.34		5.92	8.10
	5810	4" iron pipe size		150	.107		3.17	3.56		6.73	9.10
	5820	5" iron pipe size		145	.110		3.36	3.69		7.05	9.50
	5830	6" iron pipe size		140	.114		3.53	3.82		7.35	9.90
	5900	1-1/2" wall, 1/2" iron pipe size	↓	160	.100	↓	2.11	3.34		5.45	7.55

	15080	Mechanical Insulation	CREW	DAILY OUTPUT	LABOR-HOURS	UNIT	2005 BARE COSTS				TOTAL INCL O&P	
							MAT.	LABOR	EQUIP.	TOTAL		
600	5920	3/4" iron pipe size	Q-14	160	.100	L.F.	2.16	3.34		5.50	7.65	**600**
	5930	1" iron pipe size		160	.100		2.35	3.34		5.69	7.85	
	5940	1-1/4" iron pipe size		155	.103		2.53	3.45		5.98	8.25	
	5950	1-1/2" iron pipe size		155	.103		2.71	3.45		6.16	8.45	
	5960	2" iron pipe size		150	.107		3	3.56		6.56	8.90	
	5970	2-1/2" iron pipe size		150	.107		3.27	3.56		6.83	9.20	
	5980	3" iron pipe size		145	.110		3.42	3.69		7.11	9.55	
	5990	4" iron pipe size		135	.119		3.78	3.96		7.74	10.40	
	6000	5" iron pipe size		130	.123		4.46	4.11		8.57	11.40	
	6010	6" iron pipe size		120	.133		4.55	4.45		9	12.05	
	6020	7" iron pipe size		115	.139		5.35	4.65		10	13.20	
	6030	8" iron pipe size		105	.152		6.05	5.10		11.15	14.70	
	6040	9" iron pipe size		100	.160		7.25	5.35		12.60	16.40	
	6050	10" iron pipe size		95	.168		8	5.65		13.65	17.70	
	6060	12" iron pipe size		90	.178		9.50	5.95		15.45	19.80	
	6070	14" iron pipe size		85	.188		10.70	6.30		17	21.50	
	6080	16" iron pipe size		80	.200		12	6.70		18.70	24	
	6090	18" iron pipe size		75	.213		15.25	7.15		22.40	28	
	6120	2" wall, 1/2" iron pipe size		145	.110		3.35	3.69		7.04	9.50	
	6140	3/4" iron pipe size		145	.110		3.53	3.69		7.22	9.70	
	6150	1" iron pipe size		145	.110		3.71	3.69		7.40	9.90	
	6160	1-1/4" iron pipe size		140	.114		3.99	3.82		7.81	10.40	
	6170	1-1/2" iron pipe size		140	.114		4.15	3.82		7.97	10.55	
	6180	2" iron pipe size		135	.119		4.37	3.96		8.33	11.05	
	6190	2-1/2" iron pipe size		135	.119		5.25	3.96		9.21	12.05	
	6200	3" iron pipe size		130	.123		5.25	4.11		9.36	12.30	
	6210	4" iron pipe size		125	.128		6.10	4.28		10.38	13.45	
	6220	5" iron pipe size		120	.133		6.95	4.45		11.40	14.70	
	6230	6" iron pipe size		115	.139		7.60	4.65		12.25	15.70	
	6240	7" iron pipe size		110	.145		8.25	4.86		13.11	16.70	
	6250	8" iron pipe size		105	.152		9.05	5.10		14.15	18	
	6260	9" iron pipe size		100	.160		10.25	5.35		15.60	19.75	
	6270	10" iron pipe size		95	.168		11.35	5.65		17	21.50	
	6280	12" iron pipe size		90	.178		12.65	5.95		18.60	23.50	
	6290	14" iron pipe size		85	.188		13.90	6.30		20.20	25	
	6300	16" iron pipe size		80	.200		15.30	6.70		22	27.50	
	6310	18" iron pipe size		75	.213		17.20	7.15		24.35	30	
	6320	20" iron pipe size		65	.246		21	8.20		29.20	36.50	
	6330	22" iron pipe size		60	.267		23.50	8.90		32.40	40	
	6340	24" iron pipe size		55	.291		24	9.70		33.70	42	
	6360	3" wall, 1/2" iron pipe size		115	.139		6.05	4.65		10.70	14.05	
	6380	3/4" iron pipe size		115	.139		6.10	4.65		10.75	14.05	
	6390	1" iron pipe size		115	.139		6.15	4.65		10.80	14.10	
	6400	1-1/4" iron pipe size		110	.145		6.25	4.86		11.11	14.50	
	6410	1-1/2" iron pipe size		110	.145		6.30	4.86		11.16	14.55	
	6420	2" iron pipe size		105	.152		6.70	5.10		11.80	15.40	
	6430	2-1/2" iron pipe size		105	.152		7.65	5.10		12.75	16.45	
	6440	3" iron pipe size		100	.160		7.70	5.35		13.05	16.90	
	6450	4" iron pipe size		95	.168		9.80	5.65		15.45	19.70	
	6460	5" iron pipe size		90	.178		10.65	5.95		16.60	21	
	6470	6" iron pipe size		90	.178		12	5.95		17.95	22.50	
	6480	7" iron pipe size		85	.188		13.35	6.30		19.65	24.50	
	6490	8" iron pipe size		85	.188		14.25	6.30		20.55	25.50	
	6500	9" iron pipe size		80	.200		16.20	6.70		22.90	28.50	
	6510	10" iron pipe size		75	.213		17.25	7.15		24.40	30	
	6520	12" iron pipe size		70	.229		19.05	7.65		26.70	33	

Important: See the Reference Section for critical supporting data - Reference Nos., Crews, & City Cost Indexes

15080 | Mechanical Insulation

		CREW	DAILY OUTPUT	LABOR-HOURS	UNIT	2005 BARE COSTS				TOTAL INCL O&P
						MAT.	LABOR	EQUIP.	TOTAL	
600	6530	14" iron pipe size	Q-14	65	.246	L.F.	19.75	8.20		27.95	35
	6540	16" iron pipe size		60	.267		20.50	8.90		29.40	36.50
	6550	18" iron pipe size		55	.291		26.50	9.70		36.20	45
	6560	20" iron pipe size		50	.320		32	10.70		42.70	52
	6570	22" iron pipe size		45	.356		34.50	11.90		46.40	57
	6580	24" iron pipe size		40	.400		37	13.35		50.35	62
	6600	Fiberglass, with all service jacket									
	6640	1/2" wall, 1/2" iron pipe size	Q-14	250	.064	L.F.	.60	2.14		2.74	4.03
	6660	3/4" iron pipe size		240	.067		.67	2.23		2.90	4.25
	6670	1" iron pipe size		230	.070		.71	2.32		3.03	4.45
	6680	1-1/4" iron pipe size		220	.073		.79	2.43		3.22	4.70
	6690	1-1/2" iron pipe size		220	.073		.83	2.43		3.26	4.74
	6700	2" iron pipe size		210	.076		.91	2.54		3.45	5
	6710	2-1/2" iron pipe size		200	.080		1.04	2.67		3.71	5.35
	6720	3" iron pipe size		190	.084		1.15	2.81		3.96	5.70
	6730	3-1/2" iron pipe size		180	.089		1.24	2.97		4.21	6.05
	6740	4" iron pipe size		160	.100		1.47	3.34		4.81	6.85
	6750	5" iron pipe size		150	.107		1.70	3.56		5.26	7.45
	6760	6" iron pipe size		120	.133		1.80	4.45		6.25	9.05
	6840	1" wall, 1/2" iron pipe size		240	.067		.71	2.23		2.94	4.29
	6860	3/4" iron pipe size		230	.070		.83	2.32		3.15	4.58
	6870	1" iron pipe size		220	.073		.83	2.43		3.26	4.74
	6880	1-1/4" iron pipe size		210	.076		.96	2.54		3.50	5.10
	6890	1-1/2" iron pipe size		210	.076		1.02	2.54		3.56	5.15
	6900	2" iron pipe size		200	.080		1.12	2.67		3.79	5.45
	6910	2-1/2" iron pipe size		190	.084		1.23	2.81		4.04	5.80
	6920	3" iron pipe size		180	.089		1.39	2.97		4.36	6.20
	6930	3-1/2" iron pipe size		170	.094		1.45	3.14		4.59	6.55
	6940	4" iron pipe size		150	.107		1.80	3.56		5.36	7.60
	6950	5" iron pipe size		140	.114		2.07	3.82		5.89	8.30
	6960	6" iron pipe size		120	.133		2.20	4.45		6.65	9.45
	6970	7" iron pipe size		110	.145		2.71	4.86		7.57	10.65
	6980	8" iron pipe size		100	.160		3.14	5.35		8.49	11.90
	6990	9" iron pipe size		90	.178		3.41	5.95		9.36	13.10
	7000	10" iron pipe size		90	.178		3.67	5.95		9.62	13.40
	7010	12" iron pipe size		80	.200		4.17	6.70		10.87	15.15
	7020	14" iron pipe size		80	.200		4.84	6.70		11.54	15.85
	7030	16" iron pipe size		70	.229		5.75	7.65		13.40	18.35
	7040	18" iron pipe size		70	.229		7	7.65		14.65	19.75
	7050	20" iron pipe size		60	.267		7.80	8.90		16.70	22.50
	7060	24" iron pipe size		60	.267		9.55	8.90		18.45	24.50
	7080	1-1/2" wall, 1/2" iron pipe size		230	.070		1.47	2.32		3.79	5.30
	7100	3/4" iron pipe size		220	.073		1.53	2.43		3.96	5.50
	7110	1" iron pipe size		210	.076		1.61	2.54		4.15	5.80
	7120	1-1/4" iron pipe size		200	.080		1.73	2.67		4.40	6.10
	7130	1-1/2" iron pipe size		200	.080		1.82	2.67		4.49	6.20
	7140	2" iron pipe size		190	.084		1.98	2.81		4.79	6.60
	7150	2-1/2" iron pipe size		180	.089		2.18	2.97		5.15	7.10
	7160	3" iron pipe size		170	.094		2.25	3.14		5.39	7.45
	7170	3-1/2" iron pipe size		160	.100		2.47	3.34		5.81	7.95
	7180	4" iron pipe size		140	.114		2.58	3.82		6.40	8.85
	7190	5" iron pipe size		130	.123		2.88	4.11		6.99	9.65
	7200	6" iron pipe size		110	.145		2.96	4.86		7.82	10.90
	7210	7" iron pipe size		100	.160		3.41	5.35		8.76	12.20
	7220	8" iron pipe size		90	.178		3.70	5.95		9.65	13.40
	7230	9" iron pipe size		85	.188		4.22	6.30		10.52	14.55

15080	Mechanical Insulation	CREW	DAILY OUTPUT	LABOR-HOURS	UNIT	2005 BARE COSTS				TOTAL INCL O&P
						MAT.	LABOR	EQUIP.	TOTAL	
600 7240	10" iron pipe size	Q-14	80	.200	L.F.	4.66	6.70		11.36	15.70	600
7250	12" iron pipe size		75	.213		5.10	7.15		12.25	16.85
7260	14" iron pipe size		70	.229		5.85	7.65		13.50	18.50
7270	16" iron pipe size		65	.246		6.60	8.20		14.80	20
7280	18" iron pipe size		60	.267		8.55	8.90		17.45	23.50
7290	20" iron pipe size		55	.291		9.05	9.70		18.75	25.50
7300	24" iron pipe size		50	.320		10.60	10.70		21.30	28.50
7320	2" wall, 1/2" iron pipe size		220	.073		2.29	2.43		4.72	6.35
7340	3/4" iron pipe size		210	.076		2.37	2.54		4.91	6.65
7350	1" iron pipe size		200	.080		2.54	2.67		5.21	7
7360	1-1/4" iron pipe size		190	.084		2.69	2.81		5.50	7.40
7370	1-1/2" iron pipe size		190	.084		2.84	2.81		5.65	7.55
7380	2" iron pipe size		180	.089		2.95	2.97		5.92	7.95
7390	2-1/2" iron pipe size		170	.094		3.16	3.14		6.30	8.45
7400	3" iron pipe size		160	.100		3.41	3.34		6.75	9
7410	3-1/2" iron pipe size		150	.107		3.68	3.56		7.24	9.65
7420	4" iron pipe size		130	.123		3.91	4.11		8.02	10.80
7430	5" iron pipe size		120	.133		4.42	4.45		8.87	11.90
7440	6" iron pipe size		100	.160		4.54	5.35		9.89	13.45
7450	7" iron pipe size		90	.178		5.20	5.95		11.15	15.05
7460	8" iron pipe size		80	.200		5.60	6.70		12.30	16.70
7470	9" iron pipe size		75	.213		6.20	7.15		13.35	18.05
7480	10" iron pipe size		70	.229		6.70	7.65		14.35	19.45
7490	12" iron pipe size		65	.246		7.40	8.20		15.60	21
7500	14" iron pipe size		60	.267		8.35	8.90		17.25	23.50
7510	16" iron pipe size		55	.291		9.50	9.70		19.20	26
7520	18" iron pipe size		50	.320		11.35	10.70		22.05	29.50
7530	20" iron pipe size		45	.356		12.60	11.90		24.50	32.50
7540	24" iron pipe size	▼	40	.400	▼	14.10	13.35		27.45	36.50
7600	For fiberglass with standard canvas jacket, deduct					5%				
7660	For fittings, add 3 L.F. for each fitting									
7680	plus 4 L.F. for each flange of the fitting									
7700	For equipment or congested areas, add						20%			
7720	Finishes, for .010" aluminum jacket, add	Q-14	200	.080	S.F.	.53	2.67		3.20	4.80
7740	For .016" aluminum jacket, add		200	.080		.68	2.67		3.35	4.97
7760	For .010" stainless steel, add	▼	160	.100	▼	1.99	3.34		5.33	7.45
7780	For single layer of felt, add					10%	10%			
7784	Polyethylene tubing flexible closed cell foam, UV resistant									
7828	Standard temperature (-90° to +212° F)									
7830	3/8" wall, 1/8" iron pipe size	1 Asbe	130	.062	L.F.	.41	2.28		2.69	4.05
7831	1/4" iron pipe size		130	.062		.43	2.28		2.71	4.07
7832	3/8" iron pipe size		130	.062		.48	2.28		2.76	4.13
7833	1/2" iron pipe size		126	.063		.53	2.36		2.89	4.30
7834	3/4" iron pipe size		122	.066		.61	2.43		3.04	4.51
7835	1" iron pipe size		120	.067		.66	2.47		3.13	4.63
7836	1-1/4" iron pipe size		118	.068		.87	2.52		3.39	4.93
7837	1-1/2" iron pipe size		118	.068		1.06	2.52		3.58	5.15
7838	2" iron pipe size		116	.069		1.19	2.56		3.75	5.35
7839	2-1/2" iron pipe size		114	.070		1.72	2.60		4.32	6
7840	3" iron pipe size		112	.071		2.26	2.65		4.91	6.65
7842	1/2" wall, 1/8" iron pipe size		120	.067		.60	2.47		3.07	4.56
7843	1/4" iron pipe size		120	.067		.65	2.47		3.12	4.62
7844	3/8" iron pipe size		120	.067		.70	2.47		3.17	4.67
7845	1/2" iron pipe size		118	.068		.75	2.52		3.27	4.80
7846	3/4" iron pipe size		116	.069		.88	2.56		3.44	5
7847	1" iron pipe size	▼	114	.070	▼	.95	2.60		3.55	5.15

Important: See the Reference Section for critical supporting data - Reference Nos., Crews, & City Cost Indexes

15080	Mechanical Insulation	CREW	DAILY OUTPUT	LABOR-HOURS	UNIT	2005 BARE COSTS				TOTAL INCL O&P
						MAT.	LABOR	EQUIP.	TOTAL	
600 7848	1-1/4″ iron pipe size	1 Asbe	112	.071	L.F.	1.19	2.65		3.84	5.50
7849	1-1/2″ iron pipe size		110	.073		1.43	2.70		4.13	5.85
7850	2″ iron pipe size		108	.074		1.67	2.75		4.42	6.20
7851	2-1/2″ iron pipe size		106	.075		2.26	2.80		5.06	6.90
7852	3″ iron pipe size		104	.077		2.83	2.85		5.68	7.60
7853	3-1/2″ iron pipe size		102	.078		3.16	2.91		6.07	8.05
7854	4″ iron pipe size		100	.080		3.92	2.97		6.89	9
7855	3/4″ wall, 1/8″ iron pipe size		110	.073		.92	2.70		3.62	5.25
7856	1/4″ iron pipe size		110	.073		.97	2.70		3.67	5.35
7857	3/8″ iron pipe size		108	.074		1.11	2.75		3.86	5.55
7858	1/2″ iron pipe size		106	.075		1.24	2.80		4.04	5.80
7859	3/4″ iron pipe size		104	.077		1.61	2.85		4.46	6.25
7860	1″ iron pipe size		102	.078		1.85	2.91		4.76	6.65
7861	1-1/4″ iron pipe size		100	.080		2.51	2.97		5.48	7.45
7862	1-1/2″ iron pipe size		100	.080		2.94	2.97		5.91	7.90
7863	2″ iron pipe size		98	.082		3.44	3.03		6.47	8.55
7864	2-1/2″ iron pipe size		96	.083		4.33	3.09		7.42	9.65
7865	3″ iron pipe size		94	.085		5.20	3.16		8.36	10.75
7866	3-1/2″ iron pipe size		92	.087		5.60	3.23		8.83	11.25
7867	4″ iron pipe size		90	.089		6.85	3.30		10.15	12.75
7868	1″ wall, 1/4″ iron pipe size		100	.080		1.67	2.97		4.64	6.50
7869	3/8″ iron pipe size		98	.082		1.74	3.03		4.77	6.70
7870	1/2″ iron pipe size		96	.083		1.85	3.09		4.94	6.90
7871	3/4″ iron pipe size		94	.085		2.42	3.16		5.58	7.65
7872	1″ iron pipe size		92	.087		2.79	3.23		6.02	8.15
7873	1-1/4″ iron pipe size		90	.089		3.18	3.30		6.48	8.70
7874	1-1/2″ iron pipe size		90	.089		3.63	3.30		6.93	9.20
7875	2″ iron pipe size		88	.091		4.77	3.37		8.14	10.55
7876	2-1/2″ iron pipe size		86	.093		6.20	3.45		9.65	12.25
7877	3″ iron pipe size		84	.095		7.30	3.53		10.83	13.60
7878	Contact cement, quart can				Ea.	22			22	24
7879	Rubber tubing, flexible closed cell foam									
7880	3/8″ wall, 1/4″ iron pipe size	1 Asbe	120	.067	L.F.	.28	2.47		2.75	4.21
7900	3/8″ iron pipe size		120	.067		.30	2.47		2.77	4.23
7910	1/2″ iron pipe size		115	.070		.35	2.58		2.93	4.46
7920	3/4″ iron pipe size		115	.070		.40	2.58		2.98	4.51
7930	1″ iron pipe size		110	.073		.49	2.70		3.19	4.80
7940	1-1/4″ iron pipe size		110	.073		.60	2.70		3.30	4.92
7950	1-1/2″ iron pipe size		110	.073		.72	2.70		3.42	5.05
8100	1/2″ wall, 1/4″ iron pipe size		90	.089		.37	3.30		3.67	5.60
8120	3/8″ iron pipe size		90	.089		.49	3.30		3.79	5.75
8130	1/2″ iron pipe size		89	.090		.57	3.33		3.90	5.90
8140	3/4″ iron pipe size		89	.090		.63	3.33		3.96	5.95
8150	1″ iron pipe size		88	.091		.68	3.37		4.05	6.05
8160	1-1/4″ iron pipe size		87	.092		.86	3.41		4.27	6.35
8170	1-1/2″ iron pipe size		87	.092		1.03	3.41		4.44	6.55
8180	2″ iron pipe size		86	.093		1.20	3.45		4.65	6.75
8190	2-1/2″ iron pipe size		86	.093		1.65	3.45		5.10	7.25
8200	3″ iron pipe size		85	.094		1.72	3.49		5.21	7.40
8210	3-1/2″ iron pipe size		85	.094		2.30	3.49		5.79	8.05
8220	4″ iron pipe size		80	.100		2.73	3.71		6.44	8.85
8230	5″ iron pipe size		80	.100		3.75	3.71		7.46	10
8300	3/4″ wall, 1/4″ iron pipe size		90	.089		.67	3.30		3.97	5.95
8320	3/8″ iron pipe size		90	.089		.74	3.30		4.04	6
8330	1/2″ iron pipe size		89	.090		.81	3.33		4.14	6.15
8340	3/4″ iron pipe size		89	.090		.97	3.33		4.30	6.30

	15080	Mechanical Insulation	CREW	DAILY OUTPUT	LABOR-HOURS	UNIT	2005 BARE COSTS				TOTAL INCL O&P
							MAT.	LABOR	EQUIP.	TOTAL	
600	8350	1" iron pipe size	1 Asbe	88	.091	L.F.	1.18	3.37		4.55	6.60	600
	8360	1-1/4" iron pipe size		87	.092		1.36	3.41		4.77	6.90
	8370	1-1/2" iron pipe size		87	.092		2.07	3.41		5.48	7.70
	8380	2" iron pipe size		86	.093		2.43	3.45		5.88	8.10
	8390	2-1/2" iron pipe size		86	.093		2.85	3.45		6.30	8.60
	8400	3" iron pipe size		85	.094		3.32	3.49		6.81	9.15
	8410	3-1/2" iron pipe size		85	.094		4.06	3.49		7.55	9.95
	8420	4" iron pipe size		80	.100		4.64	3.71		8.35	10.95
	8430	5" iron pipe size		80	.100		5.65	3.71		9.36	12.10
	8440	6" iron pipe size		80	.100		6.85	3.71		10.56	13.40
	8444	1" wall, 1/2" iron pipe size		86	.093		1.53	3.45		4.98	7.15
	8445	3/4" iron pipe size		84	.095		1.90	3.53		5.43	7.70
	8446	1" iron pipe size		84	.095		2.30	3.53		5.83	8.15
	8447	1-1/4" iron pipe size		82	.098		2.68	3.62		6.30	8.65
	8448	1-1/2" iron pipe size		82	.098		3.02	3.62		6.64	9
	8449	2" iron pipe size		80	.100		4.70	3.71		8.41	11
	8450	2-1/2" iron pipe size	▼	80	.100	▼	6.15	3.71		9.86	12.60
	8456	Rubber insulation tape, 1/8" x 2" x 30'				Ea.	10.80			10.80	11.90
	8460	Polyolefin tubing, flexible closed cell foam, UV stabilized, work									
	8462	temp.-165° F to +210° F, 0 water vapor transmission.									
	8464	3/8" wall, 1/8" iron pipe size	1 Asbe	140	.057	L.F.	.30	2.12		2.42	3.68
	8466	1/4" iron pipe size		140	.057		.31	2.12		2.43	3.69
	8468	3/8" iron pipe size		140	.057		.35	2.12		2.47	3.74
	8470	1/2" iron pipe size		136	.059		.38	2.18		2.56	3.86
	8472	3/4" iron pipe size		132	.061		.44	2.25		2.69	4.03
	8474	1" iron pipe size		130	.062		.48	2.28		2.76	4.13
	8476	1-1/4" iron pipe size		128	.063		.63	2.32		2.95	4.35
	8478	1-1/2" iron pipe size		128	.063		.76	2.32		3.08	4.50
	8480	2" iron pipe size		126	.063		.86	2.36		3.22	4.67
	8482	2-1/2" iron pipe size		123	.065		1.25	2.41		3.66	5.20
	8484	3" iron pipe size		121	.066		1.63	2.45		4.08	5.65
	8486	4" iron pipe size		118	.068		2.69	2.52		5.21	6.95
	8500	1/2" wall, 1/8" iron pipe size		130	.062		.43	2.28		2.71	4.07
	8502	1/4" iron pipe size		130	.062		.47	2.28		2.75	4.12
	8504	3/8" iron pipe size		130	.062		.50	2.28		2.78	4.15
	8506	1/2" iron pipe size		128	.063		.54	2.32		2.86	4.25
	8508	3/4" iron pipe size		126	.063		.63	2.36		2.99	4.41
	8510	1" iron pipe size		123	.065		.68	2.41		3.09	4.56
	8512	1-1/4" iron pipe size		121	.066		.86	2.45		3.31	4.82
	8514	1-1/2" iron pipe size		119	.067		1.03	2.49		3.52	5.05
	8516	2" iron pipe size		117	.068		1.20	2.54		3.74	5.30
	8518	2-1/2" iron pipe size		114	.070		1.63	2.60		4.23	5.90
	8520	3" iron pipe size		112	.071		2.04	2.65		4.69	6.40
	8522	4" iron pipe size		110	.073		2.82	2.70		5.52	7.35
	8534	3/4" wall, 1/8" iron pipe size		120	.067		.66	2.47		3.13	4.63
	8536	1/4" iron pipe size		120	.067		.70	2.47		3.17	4.67
	8538	3/8" iron pipe size		117	.068		.82	2.54		3.36	4.90
	8540	1/2" iron pipe size		114	.070		.89	2.60		3.49	5.10
	8542	3/4" iron pipe size		112	.071		1.16	2.65		3.81	5.45
	8544	1" iron pipe size		110	.073		1.33	2.70		4.03	5.70
	8546	1-1/4" iron pipe size		108	.074		1.80	2.75		4.55	6.30
	8548	1-1/2" iron pipe size		108	.074		2.12	2.75		4.87	6.65
	8550	2" iron pipe size		106	.075		2.48	2.80		5.28	7.15
	8552	2-1/2" iron pipe size		104	.077		3.12	2.85		5.97	7.95
	8554	3" iron pipe size		102	.078		3.75	2.91		6.66	8.70
	8556	4" iron pipe size	▼	100	.080	▼	4.92	2.97		7.89	10.10

Important: See the Reference Section for critical supporting data - Reference Nos., Crews, & City Cost Indexes

15080 | Mechanical Insulation

		CREW	DAILY OUTPUT	LABOR-HOURS	UNIT	2005 BARE COSTS				TOTAL INCL O&P
						MAT.	LABOR	EQUIP.	TOTAL	
600	8570	1" wall, 1/8" iron pipe size	1 Asbe	110	.073	L.F.	1.20	2.70		3.90	5.60	600
	8572	1/4" iron pipe size		108	.074		1.23	2.75		3.98	5.70
	8574	3/8" iron pipe size		106	.075		1.25	2.80		4.05	5.80
	8576	1/2" iron pipe size		104	.077		1.33	2.85		4.18	5.95
	8578	3/4" iron pipe size		102	.078		1.74	2.91		4.65	6.50
	8580	1" iron pipe size		100	.080		2.01	2.97		4.98	6.90
	8582	1-1/4" iron pipe size		97	.082		2.29	3.06		5.35	7.35
	8584	1-1/2" iron pipe size		97	.082		2.61	3.06		5.67	7.70
	8586	2" iron pipe size		95	.084		3.44	3.12		6.56	8.70
	8588	2-1/2" iron pipe size		93	.086		4.33	3.19		7.52	9.80
	8590	3" iron pipe size	▼	91	.088	▼	5.25	3.26		8.51	10.90
	8600	Seam Roller				Ea.	7.50			7.50	8.25
	8602	Fuse Seal applicator (Butane)					200			200	220
	8604	Fuse Seal 1/2" sticks				▼	.91			.91	1
	8606	Leaktite adhesive				Qt.	15.45			15.45	16.95
	8608	Leaktite adhesive				Gal.	53.50			53.50	58.50
	8610	NOTE: Preslit/preglued vs unslit, same price									
900	0010	**PIPE INSULATION PROTECTIVE JACKETING**										900
	0100	PVC, white, 48" lengths									
	0120	20 mil thick									
	0140	Size based on OD of insulation									
	0150	1-1/2" ID	Q-14	270	.059	L.F.	1.23	1.98		3.21	4.47
	0152	2" ID		260	.062		1.22	2.06		3.28	4.58
	0154	2-1/2" ID		250	.064		1.28	2.14		3.42	4.78
	0156	3" ID		240	.067		1.32	2.23		3.55	4.96
	0158	3-1/2" ID		230	.070		1.43	2.32		3.75	5.25
	0160	4" ID		220	.073		1.48	2.43		3.91	5.45
	0162	4-1/2" ID		210	.076		1.56	2.54		4.10	5.75
	0164	5" ID		200	.080		1.69	2.67		4.36	6.10
	0166	5-1/2" ID		190	.084		1.77	2.81		4.58	6.40
	0168	6" ID		180	.089		1.90	2.97		4.87	6.75
	0170	6-1/2" ID		175	.091		2	3.05		5.05	7
	0172	7" ID		170	.094		2.05	3.14		5.19	7.20
	0174	7-1/2" ID		164	.098		2.15	3.26		5.41	7.50
	0176	8" ID		161	.099		2.25	3.32		5.57	7.75
	0178	8-1/2" ID		158	.101		2.39	3.38		5.77	8
	0180	9" ID		155	.103		2.49	3.45		5.94	8.20
	0182	9-1/2" ID		152	.105		2.57	3.52		6.09	8.40
	0184	10" ID		149	.107		2.62	3.59		6.21	8.55
	0186	10-1/2" ID		146	.110		2.83	3.66		6.49	8.90
	0188	11" ID		143	.112		2.93	3.74		6.67	9.10
	0190	11-1/2" ID		140	.114		3.06	3.82		6.88	9.35
	0192	12" ID		137	.117		3.03	3.90		6.93	9.50
	0194	12-1/2" ID		134	.119		3.21	3.99		7.20	9.85
	0196	13-1/2" ID		132	.121		3.27	4.05		7.32	10
	0198	14" ID		130	.123		3.71	4.11		7.82	10.60
	0200	15" ID		128	.125		3.84	4.18		8.02	10.80
	0202	16" ID		126	.127		3.97	4.24		8.21	11.05
	0204	17" ID		124	.129		4.23	4.31		8.54	11.45
	0206	18" ID		122	.131		4.43	4.38		8.81	11.75
	0208	19" ID		120	.133		4.64	4.45		9.09	12.15
	0210	20" ID		118	.136		5.05	4.53		9.58	12.70
	0212	21" ID		116	.138		5.75	4.61		10.36	13.60
	0214	22" ID		114	.140		5.95	4.69		10.64	13.95
	0216	23" ID	▼	112	.143	▼	6.30	4.77		11.07	14.50

	15080	**Mechanical Insulation**	CREW	DAILY OUTPUT	LABOR-HOURS	UNIT	2005 BARE COSTS MAT.	LABOR	EQUIP.	TOTAL	TOTAL INCL O&P	
900	0218	24" ID	Q-14	110	.145	L.F.	6.45	4.86		11.31	14.75	**900**
	0220	25" ID		108	.148		6.80	4.95		11.75	15.30	
	0222	26" ID		106	.151		7	5.05		12.05	15.65	
	0224	27" ID		104	.154		7.25	5.15		12.40	16.05	
	0226	28" ID		102	.157		7.55	5.25		12.80	16.55	
	0228	29" ID		100	.160		7.75	5.35		13.10	17	
	0230	30" ID		98	.163		8.05	5.45		13.50	17.50	
	0300	For colors, add				Ea.	10%					
	1000	30 mil thick										
	1010	Size based on OD of insulation										
	1020	2" ID	Q-14	260	.062	L.F.	1.34	2.06		3.40	4.71	
	1022	2-1/2" ID		250	.064		1.57	2.14		3.71	5.10	
	1024	3" ID		240	.067		1.71	2.23		3.94	5.40	
	1026	3-1/2" ID		230	.070		1.91	2.32		4.23	5.75	
	1028	4" ID		220	.073		2	2.43		4.43	6.05	
	1030	4-1/2" ID		210	.076		2.17	2.54		4.71	6.40	
	1032	5" ID		200	.080		2.43	2.67		5.10	6.90	
	1034	5-1/2" ID		190	.084		2.62	2.81		5.43	7.30	
	1036	6" ID		180	.089		2.82	2.97		5.79	7.80	
	1038	6-1/2" ID		175	.091		3.02	3.05		6.07	8.15	
	1040	7" ID		170	.094		3.08	3.14		6.22	8.35	
	1042	7-1/2" ID		164	.098		3.19	3.26		6.45	8.65	
	1044	8" ID		161	.099		3.28	3.32		6.60	8.85	
	1046	8-1/2" ID		158	.101		3.48	3.38		6.86	9.20	
	1048	9" ID		155	.103		3.79	3.45		7.24	9.60	
	1050	9-1/2" ID		152	.105		3.88	3.52		7.40	9.80	
	1052	10" ID		149	.107		4.07	3.59		7.66	10.15	
	1054	10-1/2" ID		146	.110		4.31	3.66		7.97	10.55	
	1056	11" ID		143	.112		4.45	3.74		8.19	10.80	
	1058	11-1/2" ID		140	.114		4.59	3.82		8.41	11.05	
	1060	12" ID		137	.117		4.67	3.90		8.57	11.30	
	1062	12-1/2" ID		134	.119		4.81	3.99		8.80	11.60	
	1064	13-1/2" ID		132	.121		5.20	4.05		9.25	12.15	
	1066	14" ID		130	.123		5.65	4.11		9.76	12.70	
	1068	15" ID		128	.125		5.85	4.18		10.03	13.05	
	1070	16" ID		126	.127		6.20	4.24		10.44	13.50	
	1072	17" ID		124	.129		6.40	4.31		10.71	13.85	
	1074	18" ID		122	.131		6.70	4.38		11.08	14.25	
	1076	19" ID		120	.133		6.95	4.45		11.40	14.70	
	1078	20" ID		118	.136		7.25	4.53		11.78	15.10	
	1080	21" ID		116	.138		7.45	4.61		12.06	15.45	
	1082	22" ID		114	.140		7.75	4.69		12.44	15.95	
	1084	23" ID		112	.143		8.05	4.77		12.82	16.40	
	1086	24" ID		110	.145		8.30	4.86		13.16	16.80	
	1088	25" ID		108	.148		8.60	4.95		13.55	17.25	
	1090	26" ID		106	.151		8.85	5.05		13.90	17.70	
	1092	27" ID		104	.154		8.95	5.15		14.10	17.95	
	1094	28" ID		102	.157		9.35	5.25		14.60	18.55	
	1096	29" ID		100	.160		9.70	5.35		15.05	19.15	
	1098	30" ID		98	.163		9.90	5.45		15.35	19.50	
	1300	For colors, add				Ea.	10%					
	2000	PVC, white, fitting covers										
	2020	Fiberglass insulation inserts included with sizes 1-3/4" thru 9-3/4".										
	2030	Size is based on OD of insulation										
	2040	90° Elbow fitting										
	2060	1-3/4"	Q-14	135	.119	Ea.	.59	3.96		4.55	6.90	

Important: See the Reference Section for critical supporting data - Reference Nos., Crews, & City Cost Indexes

15080 | Mechanical Insulation

		CREW	DAILY OUTPUT	LABOR-HOURS	UNIT	2005 BARE COSTS MAT.	LABOR	EQUIP.	TOTAL	TOTAL INCL O&P	
900											900
2062	2"	Q-14	130	.123	Ea.	.66	4.11		4.77	7.25	
2064	2-1/4"		128	.125		.81	4.18		4.99	7.50	
2068	2-1/2"		126	.127		.81	4.24		5.05	7.60	
2070	2-3/4"		123	.130		1.05	4.34		5.39	8	
2072	3"		120	.133		.93	4.45		5.38	8.05	
2074	3-3/8"		116	.138		1.21	4.61		5.82	8.60	
2076	3-3/4"		113	.142		1.17	4.73		5.90	8.75	
2078	4-1/8"		110	.145		1.47	4.86		6.33	9.25	
2080	4-3/4""		105	.152		1.79	5.10		6.89	10	
2082	5-1/4""		100	.160		2.08	5.35		7.43	10.75	
2084	5-3/4""		95	.168		2.91	5.65		8.56	12.10	
2086	6-1/4""		90	.178		5.95	5.95		11.90	15.90	
2088	6-3/4""		87	.184		5.15	6.15		11.30	15.40	
2090	7-1/4""		85	.188		7.95	6.30		14.25	18.60	
2092	7-3/4""		83	.193		6.75	6.45		13.20	17.55	
2094	8-3/4""		80	.200		8.80	6.70		15.50	20	
2096	9-3/4""		77	.208		12.25	6.95		19.20	24.50	
2098	10-7/8"		74	.216		16.65	7.20		23.85	29.50	
2100	11-7/8"		71	.225		18.95	7.55		26.50	33	
2102	12-7/8"		68	.235		23.50	7.85		31.35	38.50	
2104	14-1/8"		66	.242		28	8.10		36.10	44	
2106	15-1/8"		64	.250		31	8.35		39.35	47	
2108	16-1/8"		63	.254		34	8.50		42.50	51	
2110	17-1/8"		62	.258		38.50	8.60		47.10	55.50	
2112	18-1/8"		61	.262		47.50	8.75		56.25	66	
2114	19-1/8"		60	.267		65.50	8.90		74.40	86	
2116	20-1/8"	▼	59	.271	▼	84.50	9.05		93.55	107	
2200	45° Elbow fitting										
2220	1-3/4" thru 9-3/4" same price as 90° Elbow fitting										
2320	10-7/8"	Q-14	74	.216	Ea.	14.50	7.20		21.70	27.50	
2322	11-7/8"		71	.225		16	7.55		23.55	29.50	
2324	12-7/8"		68	.235		21	7.85		28.85	35.50	
2326	14-1/8"		66	.242		24.50	8.10		32.60	40	
2328	15-1/8"		64	.250		26.50	8.35		34.85	42.50	
2330	16-1/8"		63	.254		29.50	8.50		38	46	
2332	17-1/8"		62	.258		32.50	8.60		41.10	49.50	
2334	18-1/8"		61	.262		40	8.75		48.75	58	
2336	19-1/8"		60	.267		58	8.90		66.90	78	
2338	20-1/8"	▼	59	.271	▼	84.50	9.05		93.55	107	
2400	Tee fitting										
2410	1-3/4"	Q-14	96	.167	Ea.	1.02	5.55		6.57	9.90	
2412	2"		94	.170		1.12	5.70		6.82	10.20	
2414	2-1/4"		91	.176		1.27	5.85		7.12	10.65	
2416	2-1/2"		88	.182		1.38	6.05		7.43	11.10	
2418	2-3/4"		85	.188		1.52	6.30		7.82	11.55	
2420	3"		82	.195		1.63	6.50		8.13	12.10	
2422	3-3/8"		79	.203		1.88	6.75		8.63	12.70	
2424	3-3/4"		76	.211		2.17	7.05		9.22	13.50	
2426	4-1/8"		73	.219		2.54	7.30		9.84	14.35	
2428	4-3/4""		70	.229		3.14	7.65		10.79	15.50	
2430	5-1/4""		67	.239		3.75	8		11.75	16.75	
2432	5-3/4""		63	.254		5	8.50		13.50	18.90	
2434	6-1/4""		60	.267		6.60	8.90		15.50	21.50	
2436	6-3/4""		59	.271		8.15	9.05		17.20	23.50	
2438	7-1/4""		57	.281		13.15	9.40		22.55	29.50	
2440	7-3/4""	▼	54	.296	▼	14.45	9.90		24.35	31.50	

15080	Mechanical Insulation	CREW	DAILY OUTPUT	LABOR-HOURS	UNIT	2005 BARE COSTS				TOTAL INCL O&P
						MAT.	LABOR	EQUIP.	TOTAL	
900 2442	8-3/4""	Q-14	52	.308	Ea.	17.50	10.30		27.80	35.50 **900**
2444	9-3/4""		50	.320		20.50	10.70		31.20	39.50
2446	10-7/8"		48	.333		23.50	11.15		34.65	43
2448	11-7/8"		47	.340		24	11.35		35.35	44
2450	12-7/8"		46	.348		28.50	11.60		40.10	49.50
2452	14-1/8"		45	.356		31	11.90		42.90	53
2454	15-1/8"		44	.364		33.50	12.15		45.65	56
2456	16-1/8"		43	.372		36	12.45		48.45	59.50
2458	17-1/8"		42	.381		38.50	12.70		51.20	62.50
2460	18-1/8"		41	.390		42.50	13.05		55.55	67.50
2462	19-1/8"		40	.400		46.50	13.35		59.85	72
2464	20-1/8"	▼	39	.410	▼	99	13.70		112.70	131
4000	Mechanical grooved fitting cover									
4020	90° Elbow fitting									
4030	3/4" & 1"	Q-14	140	.114	Ea.	4.84	3.82		8.66	11.30
4040	1-1/4" & 1-1/2"		135	.119		6.20	3.96		10.16	13.10
4042	2"		130	.123		9.85	4.11		13.96	17.35
4044	2-1/2"		125	.128		10.50	4.28		14.78	18.30
4046	3"		120	.133		11.30	4.45		15.75	19.45
4048	3-1/2"		115	.139		15.90	4.65		20.55	25
4050	4"		110	.145		15.20	4.86		20.06	24.50
4052	5"		100	.160		19.70	5.35		25.05	30
4054	6"		90	.178		26.50	5.95		32.45	39
4056	8"		80	.200		31.50	6.70		38.20	45.50
4058	10"		75	.213		40.50	7.15		47.65	56
4060	12"		68	.235		58.50	7.85		66.35	76.50
4062	14"		65	.246		81	8.20		89.20	102
4064	16"		63	.254		110	8.50		118.50	134
4066	18"	▼	61	.262	▼	141	8.75		149.75	169
4100	45° Elbow fitting									
4120	3/4" & 1"	Q-14	140	.114	Ea.	4.13	3.82		7.95	10.55
4130	1-1/4" & 1-1/2"		135	.119		5.50	3.96		9.46	12.30
4140	2"		130	.123		9	4.11		13.11	16.40
4142	2-1/2"		125	.128		10.05	4.28		14.33	17.80
4144	3"		120	.133		11	4.45		15.45	19.15
4146	3-1/2"		115	.139		14.40	4.65		19.05	23
4148	4"		110	.145		13.30	4.86		18.16	22.50
4150	5"		100	.160		17.80	5.35		23.15	28
4152	6"		90	.178		24.50	5.95		30.45	36.50
4154	8"		80	.200		28	6.70		34.70	41.50
4156	10"		75	.213		35.50	7.15		42.65	50.50
4158	12"		68	.235		50.50	7.85		58.35	68.50
4160	14"		65	.246		57	8.20		65.20	76
4162	16"		63	.254		93.50	8.50		102	116
4164	18"	▼	61	.262	▼	121	8.75		129.75	147
4200	Tee fitting									
4220	3/4" & 1"	Q-14	93	.172	Ea.	6.60	5.75		12.35	16.30
4230	1-1/4" & 1-1/2"		90	.178		8.50	5.95		14.45	18.70
4240	2"		87	.184		15.55	6.15		21.70	27
4242	2-1/2"		84	.190		15.55	6.35		21.90	27
4244	3"		80	.200		16.60	6.70		23.30	29
4246	3-1/2"		77	.208		21.50	6.95		28.45	35
4248	4"		73	.219		22	7.30		29.30	35.50
4250	5"		67	.239		26.50	8		34.50	42
4252	6"		60	.267		31.50	8.90		40.40	49
4254	8"	▼	54	.296	▼	44	9.90		53.90	64

Important: See the Reference Section for critical supporting data - Reference Nos., Crews, & City Cost Indexes

15080	Mechanical Insulation	CREW	DAILY OUTPUT	LABOR-HOURS	UNIT	2005 BARE COSTS				TOTAL INCL O&P
						MAT.	LABOR	EQUIP.	TOTAL	
900 4256	10"	Q-14	50	.320	Ea.	50.50	10.70		61.20	73
4258	12"		46	.348		67	11.60		78.60	92.50
4260	14"		43	.372		95	12.45		107.45	124
4262	16"		42	.381		126	12.70		138.70	159
4264	18"	▼	41	.390	▼	161	13.05		174.05	198

15100 | Building Services Piping

15106	Glass Pipe & Fittings	CREW	DAILY OUTPUT	LABOR-HOURS	UNIT	2005 BARE COSTS				TOTAL INCL O&P
						MAT.	LABOR	EQUIP.	TOTAL	
120 0010	**PIPE, GLASS** Borosilicate, couplings & hangers 10' O.C. R15100-070									
0020	Drainage									
1100	1-1/2" diameter	Q-1	52	.308	L.F.	9.15	11.30		20.45	27
1120	2" diameter		44	.364		12.20	13.35		25.55	33.50
1140	3" diameter		39	.410		16.30	15.10		31.40	40.50
1160	4" diameter		30	.533		29.50	19.60		49.10	62
1180	6" diameter	▼	26	.615	▼	53.50	22.50		76	93
1870	To delete coupling & hanger, subtract									
1880	1-1/2" diam. to 2" diam.					19%	22%			
1890	3" diam. to 6" diam.					20%	17%			
2000	Process supply (pressure), beaded joints									
2040	1/2" diameter	1 Plum	36	.222	L.F.	3.05	9.10		12.15	17
2060	3/4" diameter		31	.258		3.53	10.55		14.08	19.75
2080	1" diameter	▼	27	.296		12.85	12.10		24.95	32.50
2100	1-1/2" diameter	Q-1	47	.340		15.40	12.50		27.90	35.50
2120	2" diameter		39	.410		20.50	15.10		35.60	45.50
2140	3" diameter		34	.471		30	17.30		47.30	58.50
2160	4" diameter		25	.640		41.50	23.50		65	81.50
2180	6" diameter	▼	21	.762	▼	82.50	28		110.50	133
2860	To delete coupling & hanger, subtract									
2870	1/2" diam. to 1" diam.					25%	33%			
2880	1-1/2" diam. to 3" diam.					22%	21%			
2890	4" diam. to 6" diam.					23%	15%			
3000	Beaded joint, armored, translucent									
3040	1/2" diameter	1 Plum	36	.222	L.F.	13.55	9.10		22.65	28.50
3060	3/4" diameter		31	.258		16.10	10.55		26.65	33.50
3080	1" diameter	▼	27	.296		20.50	12.10		32.60	40.50
3100	1-1/2" diameter	Q-1	47	.340		24	12.50		36.50	45
3120	2" diameter		39	.410		31	15.10		46.10	56.50
3140	3" diameter		34	.471		48.50	17.30		65.80	79.50
3160	4" diameter		25	.640		64.50	23.50		88	107
3180	6" diameter	▼	21	.762	▼	125	28		153	180
3710	To delete coupling & hanger, subtract									
3720	1/2" diam. to 3/4" diam.					30%	27%			
3730	1" diam. to 3" diam.					28%	22%			
3740	4" diam. to 6" diam.					24%	15%			
3800	Conical joint, transparent									
3980	6" diameter	Q-1	21	.762	L.F.	570	28		598	670
4500	To delete couplings & hangers, subtract									
4530	6" diam.	▼				22%	26%			

160		**15106**	**Glass Pipe & Fittings**	CREW	DAILY OUTPUT	LABOR-HOURS	UNIT	2005 BARE COSTS				TOTAL INCL O&P	160
								MAT.	LABOR	EQUIP.	TOTAL		
0010	**PIPE, GLASS, FITTINGS**												
0020		Drainage, beaded ends											
0040		Coupling & labor required at joints not incl. in fitting											
0050		price. Add 1 per joint for installed price											
0070		90° Bend or sweep, 1-1/2"				Ea.	19.70			19.70	21.50	
0090		2"					25			25	27.50	
0100		3"					41			41	45	
0110		4"					65.50			65.50	72	
0120		6" (sweep only)				▼	200			200	219	
0200		45° Bend or sweep same as 90°										
0350		Tee, single sanitary, 1-1/2"				Ea.	32			32	35	
0370		2"					32			32	35	
0380		3"					47.50			47.50	52	
0390		4"					85			85	93.50	
0400		6"					228			228	251	
0410		Tee, straight, 1-1/2"					39.50			39.50	43.50	
0430		2"					39.50			39.50	43.50	
0440		3"					57			57	62.50	
0450		4"					100			100	110	
0460		6"				▼	246			246	270	
0500		Coupling, stainless steel, TFE seal ring										
0520		1-1/2"	Q-1	32	.500	Ea.	13.45	18.40		31.85	42.50	
0530		2"		30	.533		17.20	19.60		36.80	48.50	
0540		3"		25	.640		23	23.50		46.50	61	
0550		4"		23	.696		39.50	25.50		65	82	
0560		6"	▼	20	.800	▼	92	29.50		121.50	145	
0600		Coupling, stainless steel, bead to plain end										
0610		1-1/2"	Q-1	36	.444	Ea.	17.40	16.35		33.75	43.50	
0620		2"		34	.471		21.50	17.30		38.80	50	
0630		3"		29	.552		37	20.50		57.50	71	
0640		4"		27	.593		55	22		77	94	
0650		6"	▼	24	.667	▼	188	24.50		212.50	244	
2000		Process supply (pressure), beaded ends										
2050		90° Sweep elbow, 1/2"				Ea.	19.90			19.90	22	
2070		3/4"					23.50			23.50	25.50	
2080		1"					31			31	34	
2090		1-1/2"					32.50			32.50	35.50	
2100		2"					44			44	48	
2120		3"					69			69	76	
2130		4"					97			97	107	
2140		6"				▼	243			243	267	
2150		45° Sweep elbow, same as 90°										
2250		Tee, straight, 1/2"				Ea.	34			34	37	
2270		3/4"					34			34	37	
2280		1"					37.50			37.50	41.50	
2290		1-1/2"					55.50			55.50	61	
2300		2"					65.50			65.50	72	
2310		3"					80			80	88	
2320		4"					135			135	148	
2330		6"				▼	250			250	275	
2350		Coupling, Viton liner, for temperatures to 400°F										
2370		1/2"	Q-1	40	.400	Ea.	28	14.70		42.70	53	
2380		3/4"		37	.432		32	15.90		47.90	59	
2390		1"		35	.457		38	16.80		54.80	67.50	
2400		1-1/2"		32	.500		42	18.40		60.40	74	
2410		2"	▼	30	.533	▼	51.50	19.60		71.10	86.50	

Important: See the Reference Section for critical supporting data - Reference Nos., Crews, & City Cost Indexes

		15106 | Glass Pipe & Fittings	CREW	DAILY OUTPUT	LABOR-HOURS	UNIT	2005 BARE COSTS				TOTAL INCL O&P	
							MAT.	LABOR	EQUIP.	TOTAL		
160	2420	3"	Q-1	25	.640	Ea.	95	23.50		118.50	140	160
	2430	4"		23	.696		114	25.50		139.50	165	
	2440	6"		20	.800		245	29.50		274.50	315	
	2550	For beaded joint armored fittings, add					200%					
	2600	Conical ends. Flange set, gasket & labor not incl. in fitting										
	2620	price. Add 1 per joint for installed price.										
	2650	90° Sweep elbow, 1"				Ea.	53			53	58.50	
	2670	1-1/2"					77			77	84.50	
	2680	2"					89			89	98	
	2690	3"					151			151	166	
	2700	4"					340			340	375	
	2710	6"					645			645	710	
	2750	Cross (straight), add					55%					
	2850	Tee, add					20%					

		15107 | Metal Pipe & Fittings										
220	0010	**PIPE, BRASS** Plain end,										220
	0900	Field threaded, coupling & clevis hanger 10' O.C.										
	0920	Regular weight										
	0980	1/8" diameter	1 Plum	62	.129	L.F.	2.23	5.25		7.48	10.35	
	1000	1/4" diameter		57	.140		2.86	5.75		8.61	11.75	
	1100	3/8" diameter		52	.154		3.47	6.30		9.77	13.25	
	1120	1/2" diameter		48	.167		4.20	6.80		11	14.85	
	1140	3/4" diameter		46	.174		5.75	7.10		12.85	17	
	1160	1" diameter		43	.186		8.30	7.60		15.90	20.50	
	1180	1-1/4" diameter	Q-1	72	.222		12.55	8.15		20.70	26	
	1200	1-1/2" diameter		65	.246		14.90	9.05		23.95	30	
	1220	2" diameter		53	.302		20.50	11.10		31.60	39	
	1240	2-1/2" diameter		41	.390		31	14.35		45.35	56	
	1260	3" diameter		31	.516		44	19		63	77	
	1280	3-1/2" diameter	Q-2	39	.615		73.50	23.50		97	117	
	1300	4" diameter	"	37	.649		71.50	24.50		96	116	
	1930	To delete coupling & hanger, subtract										
	1940	1/8" diam. to 1/2" diam.					14%	46%				
	1950	3/4" diam. to 1-1/2" diam.					8%	47%				
	1960	2" diam. to 4" diam.					10%	37%				
260	0010	**PIPE, BRASS, FITTINGS** Rough bronze, threaded										260
	1000	Standard wt., 90° Elbow										
	1040	1/8"	1 Plum	13	.615	Ea.	7.80	25		32.80	46.50	
	1060	1/4"		13	.615		7.80	25		32.80	46.50	
	1080	3/8"		13	.615		7.80	25		32.80	46.50	
	1100	1/2"		12	.667		7.80	27		34.80	49.50	
	1120	3/4"		11	.727		10.40	29.50		39.90	56	
	1140	1"		10	.800		16.90	32.50		49.40	67.50	
	1160	1-1/4"	Q-1	17	.941		27.50	34.50		62	82	
	1180	1-1/2"		16	1		34	37		71	92.50	
	1200	2"		14	1.143		54.50	42		96.50	123	
	1220	2-1/2"		11	1.455		132	53.50		185.50	226	
	1240	3"		8	2		202	73.50		275.50	335	
	1260	4"	Q-2	11	2.182		410	83		493	575	
	1280	5"		8	3		1,125	114		1,239	1,425	
	1300	6"		7	3.429		1,675	131		1,806	2,025	
	1500	45° Elbow, 1/8"	1 Plum	13	.615		9.55	25		34.55	48.50	
	1540	1/4"		13	.615		9.55	25		34.55	48.50	
	1560	3/8"		13	.615		9.55	25		34.55	48.50	
	1580	1/2"		12	.667		9.55	27		36.55	51.50	

15107 | Metal Pipe & Fittings

		CREW	DAILY OUTPUT	LABOR-HOURS	UNIT	2005 BARE COSTS				TOTAL INCL O&P	
						MAT.	LABOR	EQUIP.	TOTAL		
260	1600	3/4″	1 Plum	11	.727	Ea.	13.55	29.50		43.05	59.50	260
	1620	1″	↓	10	.800		23	32.50		55.50	74.50	
	1640	1-1/4″	Q-1	17	.941		37	34.50		71.50	92.50	
	1660	1-1/2″		16	1		46.50	37		83.50	107	
	1680	2″		14	1.143		75	42		117	146	
	1700	2-1/2″		11	1.455		143	53.50		196.50	238	
	1720	3″	↓	8	2		220	73.50		293.50	355	
	1740	4″	Q-2	11	2.182		505	83		588	680	
	1760	5″		8	3		930	114		1,044	1,200	
	1780	6″	↓	7	3.429		1,300	131		1,431	1,625	
	2000	Tee, 1/8″	1 Plum	9	.889		9.15	36.50		45.65	64.50	
	2040	1/4″		9	.889		9.15	36.50		45.65	64.50	
	2060	3/8″		9	.889		9.15	36.50		45.65	64.50	
	2080	1/2″		8	1		9.15	41		50.15	71.50	
	2100	3/4″		7	1.143		13	46.50		59.50	84.50	
	2120	1″	↓	6	1.333		23.50	54.50		78	108	
	2140	1-1/4″	Q-1	10	1.600		40.50	59		99.50	133	
	2160	1-1/2″		9	1.778		45.50	65.50		111	149	
	2180	2″		8	2		75.50	73.50		149	194	
	2200	2-1/2″		7	2.286		180	84		264	325	
	2220	3″	↓	5	3.200		275	118		393	475	
	2240	4″	Q-2	7	3.429		680	131		811	945	
	2260	5″		5	4.800		1,425	183		1,608	1,850	
	2280	6″	↓	4	6		2,225	229		2,454	2,800	
	2500	Coupling, 1/8″	1 Plum	26	.308		6.50	12.55		19.05	26	
	2540	1/4″		22	.364		6.50	14.85		21.35	29.50	
	2560	3/8″		18	.444		6.50	18.15		24.65	34.50	
	2580	1/2″		15	.533		6.50	22		28.50	40	
	2600	3/4″		14	.571		9.15	23.50		32.65	45	
	2620	1″		13	.615		15.60	25		40.60	55	
	2640	1-1/4″	Q-1	22	.727		26	27		53	68.50	
	2660	1-1/2″		20	.800		34	29.50		63.50	81	
	2680	2″		18	.889		56	32.50		88.50	111	
	2700	2-1/2″		14	1.143		95.50	42		137.50	168	
	2720	3″	↓	10	1.600		132	59		191	234	
	2740	4″	Q-2	12	2		272	76.50		348.50	415	
	2760	5″		10	2.400		510	91.50		601.50	700	
	2780	6″	↓	9	2.667	↓	725	102		827	950	
	3000	Union, 125 lb										
	3020	1/8″	1 Plum	12	.667	Ea.	22	27		49	65.50	
	3040	1/4″		12	.667		22	27		49	65.50	
	3060	3/8″		12	.667		22	27		49	65.50	
	3080	1/2″		11	.727		22	29.50		51.50	69	
	3100	3/4″		10	.800		30.50	32.50		63	82.50	
	3120	1″	↓	9	.889		40	36.50		76.50	98.50	
	3140	1-1/4″	Q-1	16	1		58	37		95	120	
	3160	1-1/2″		15	1.067		69	39		108	135	
	3180	2″		13	1.231		107	45.50		152.50	185	
	3200	2-1/2″		10	1.600		279	59		338	395	
	3220	3″	↓	7	2.286		435	84		519	605	
	3240	4″	Q-2	10	2.400	↓	1,075	91.50		1,166.50	1,350	
320	0010	**PIPE, CAST IRON** Soil, on hangers 5′ O.C.										320
	0020	Single hub, service wt., lead & oakum joints 10′ O.C.										
	2120	2″ diameter	Q-1	63	.254	L.F.	4.63	9.35		13.98	19.15	
	2140	3″ diameter	↓	60	.267	↓	6.35	9.80		16.15	21.50	

R15100 -070

Important: See the Reference Section for critical supporting data - Reference Nos., Crews, & City Cost Indexes

		15107	**Metal Pipe & Fittings**	CREW	DAILY OUTPUT	LABOR-HOURS	UNIT	2005 BARE COSTS				TOTAL INCL O&P
								MAT.	LABOR	EQUIP.	TOTAL	
320	2160		4" diameter R15100 -070	Q-1	55	.291	L.F.	8.10	10.70		18.80	25
	2180		5" diameter	Q-2	76	.316		8.20	12.05		20.25	27
	2200		6" diameter	"	73	.329		13.75	12.55		26.30	34
	2220		8" diameter	Q-3	59	.542		21.50	21		42.50	55
	2240		10" diameter		54	.593		39	23		62	77.50
	2260		12" diameter		48	.667		55.50	26		81.50	100
	2261		15" diameter		40	.800		81	31		112	136
	2320		For service weight, double hub, add					10%				
	2340		For extra heavy, single hub, add					48%	4%			
	2360		For extra heavy, double hub, add					71%	4%			
	2400		Lead for caulking, (1#/dia in)	Q-1	160	.100	Lb.	.96	3.68		4.64	6.60
	2420		Oakum for caulking, (1/8#/dia in)	"	40	.400	"	2.36	14.70		17.06	24.50
	2960		To delete hangers, subtract									
	2970		2" diam. to 4" diam.					16%	19%			
	2980		5" diam. to 8" diam.					14%	14%			
	2990		10" diam. to 15" diam.					13%	19%			
	3000		Single hub, service wt, push-on gasket joints 10' O.C.									
	3010		2" diameter	Q-1	66	.242	L.F.	4.94	8.90		13.84	18.85
	3020		3" diameter		63	.254		6.75	9.35		16.10	21.50
	3030		4" diameter		57	.281		8.65	10.30		18.95	25
	3040		5" diameter	Q-2	79	.304		9	11.60		20.60	27.50
	3050		6" diameter	"	75	.320		14.60	12.20		26.80	34.50
	3060		8" diameter	Q-3	62	.516		23	20		43	55.50
	3070		10" diameter		56	.571		42	22.50		64.50	79.50
	3080		12" diameter		49	.653		59	25.50		84.50	103
	3082		15" diameter		40	.800		85	31		116	141
	3100		For service weight, double hub, add					65%				
	3110		For extra heavy, single hub, add					48%	4%			
	3120		For extra heavy, double hub, add					29%	4%			
	3130		To delete hangers, subtract									
	3140		2" diam. to 4" diam.					12%	21%			
	3150		5" diam. to 8" diam.					10%	16%			
	3160		10" diam. to 15" diam.					9%	21%			
	4000		No hub, couplings 10' O.C.									
	4100		1-1/2" diameter	Q-1	71	.225	L.F.	5.30	8.30		13.60	18.30
	4120		2" diameter		67	.239		5.45	8.80		14.25	19.20
	4140		3" diameter		64	.250		7.15	9.20		16.35	21.50
	4160		4" diameter		58	.276		9.10	10.15		19.25	25.50
	4180		5" diameter	Q-2	83	.289		13.10	11.05		24.15	31
	4200		6" diameter	"	79	.304		16.30	11.60		27.90	35.50
	4220		8" diameter	Q-3	69	.464		25.50	18.05		43.55	55.50
	4240		10" diameter	"	61	.525		46	20.50		66.50	81
	4280		To delete hangers, subtract									
	4290		1-1/2" diam. to 6" diam.					22%	47%			
	4300		8" diam. to 10" diam.					21%	44%			
360	0010		**PIPE, CAST IRON, FITTINGS** Soil									
	0040		Hub and spigot, service weight, lead & oakum joints									
	0080		1/4 bend, 2"	Q-1	16	1	Ea.	9.75	37		46.75	66.50
	0120		3"		14	1.143		13	42		55	77.50
	0140		4"		13	1.231		20.50	45.50		66	90.50
	0160		5"	Q-2	18	1.333		28.50	51		79.50	108
	0180		6"	"	17	1.412		35.50	54		89.50	120
	0200		8"	Q-3	11	2.909		107	113		220	287
	0220		10"		10	3.200		156	125		281	360
	0224		12"		9	3.556		211	138		349	440

		CREW	DAILY OUTPUT	LABOR-HOURS	UNIT	2005 BARE COSTS				TOTAL INCL O&P
						MAT.	LABOR	EQUIP.	TOTAL	
0226	15"	Q-3	7	4.571	Ea.	675	178		853	1,025
0251	Long sweep elbow									
0252	2"	Q-1	16	1	Ea.	13.65	37		50.65	70.50
0253	3"		14	1.143		19.90	42		61.90	85
0254	4"	▼	13	1.231		31	45.50		76.50	102
0255	6"	Q-2	17	1.412		57.50	54		111.50	145
0256	8"	Q-3	11	2.909		141	113		254	325
0257	10"		10	3.200		225	125		350	435
0258	12"		9	3.556		365	138		503	610
0259	15"	▼	7	4.571		610	178		788	945
0266	Closet bend, 3" diameter with flange 10" x 16"	Q-1	14	1.143		53	42		95	121
0268	16"x16"		12	1.333		73	49		122	154
0270	Closet bend, 4" diameter, 1" x 4" ring, 6" x 16"		13	1.231		59	45.50		104.50	133
0280	8" x 16"		13	1.231		52.50	45.50		98	126
0290	10" x 12"		12	1.333		49.50	49		98.50	128
0300	10" x 18"		11	1.455		75	53.50		128.50	163
0310	12" x 16"		11	1.455		59	53.50		112.50	146
0330	16" x 16"		10	1.600		74.50	59		133.50	171
0340	1/8 bend, 2"		16	1		6.95	37		43.95	63
0350	3"		14	1.143		10.90	42		52.90	75
0360	4"	▼	13	1.231		15.85	45.50		61.35	85.50
0380	5"	Q-2	18	1.333		22	51		73	101
0400	6"	"	17	1.412		27	54		81	111
0420	8"	Q-3	11	2.909		80.50	113		193.50	259
0440	10"		10	3.200		116	125		241	315
0460	12"		9	3.556		219	138		357	450
0461	15"	▼	7	4.571		435	178		613	750
0500	Sanitary tee, 2"	Q-1	10	1.600		12.60	59		71.60	102
0540	3"		9	1.778		22	65.50		87.50	123
0620	4"	▼	8	2		27	73.50		100.50	141
0700	5"	Q-2	12	2		53.50	76.50		130	174
0800	6"	"	11	2.182		61	83		144	192
0880	8"	Q-3	7	4.571		161	178		339	445
0881	10"		7	4.571		295	178		473	595
0882	12"		6	5.333		490	208		698	850
0883	15"	▼	4	8	▼	975	310		1,285	1,550
0900	Sanitary tee, tapped									
0901	2" x 2"	Q-1	9	1.778	Ea.	18.10	65.50		83.60	118
0902	3" x 2"		8	2		20.50	73.50		94	134
0903	4" x 2"	▼	7	2.286	▼	25.50	84		109.50	154
0910	Sanitary cross, tapped [double tapped sanitary tee]									
0911	2" x 2"	Q-1	7	2.286	Ea.	21.50	84		105.50	150
0912	3" x 2"		6	2.667		24.50	98		122.50	174
0913	4" x 2"	▼	5	3.200		33	118		151	213
0940	Sanitary Tee, reducing									
0942	3" x 2"	Q-1	10	1.600	Ea.	18.95	59		77.95	110
0943	4" x 3"		9	1.778		24.50	65.50		90	126
0944	4" x 2"	▼	9	1.778		23	65.50		88.50	124
0945	5" x 3"	Q-2	9	2.667		50.50	102		152.50	209
0946	6" x 4"		12	2		69	76.50		145.50	191
0947	6" x 3"		8	3		47	114		161	224
0948	6" x 2"	▼	12.50	1.920		45.50	73		118.50	160
0949	8" x 6"	Q-3	9	3.556		136	138		274	355
0950	8" x 5"		9	3.556		136	138		274	355
0951	8" x 4"		10	3.200		103	125		228	300
0954	10" x 6"	▼	8	4	▼	233	156		389	490

15107 | Metal Pipe & Fittings

		CREW	DAILY OUTPUT	LABOR-HOURS	UNIT	2005 BARE COSTS				TOTAL INCL O&P
						MAT.	LABOR	EQUIP.	TOTAL	
0958	12" x 8"	Q-3	7	4.571	Ea.	420	178		598	735
0962	15" x 10"	↓	5	6.400		735	249		984	1,175
1000	Tee, 2"	Q-1	10	1.600		17.05	59		76.05	107
1060	3"		9	1.778		27.50	65.50		93	129
1120	4"	↓	8	2		35.50	73.50		109	150
1200	5"	Q-2	12	2		53.50	76.50		130	174
1300	6"	"	11	2.182		61	83		144	192
1380	8"	Q-3	7	4.571	↓	161	178		339	445
1400	Combination Y and 1/8 bend									
1420	2"	Q-1	10	1.600	Ea.	17.05	59		76.05	107
1460	3"		9	1.778		26	65.50		91.50	127
1520	4"	↓	8	2		35.50	73.50		109	151
1540	5"	Q-2	12	2		68	76.50		144.50	190
1560	6"		11	2.182		85.50	83		168.50	220
1580	8"	↓	7	3.429		211	131		342	430
1582	12"	Q-3	6	5.333	↓	480	208		688	835
1584	Combination Y & 1/8 bend, reducing									
1586	3" x 2"	Q-1	10	1.600	Ea.	19.45	59		78.45	110
1587	4" x 2"		9.50	1.684		26.50	62		88.50	122
1588	4" x 3"	↓	9	1.778		31	65.50		96.50	133
1589	6" x 2"	Q-2	12.50	1.920		55.50	73		128.50	171
1590	6" x 3"		12	2		59	76.50		135.50	180
1591	6" x 4"	↓	11	2.182		62	83		145	193
1592	8" x 2"	Q-3	11	2.909		105	113		218	285
1593	8" x 4"		10	3.200		103	125		228	300
1594	8" x 6"	↓	9	3.556		142	138		280	365
1600	Double Y, 2"	Q-1	8	2		20.50	73.50		94	134
1610	3"		7	2.286		35.50	84		119.50	165
1620	4"	↓	6.50	2.462		49.50	90.50		140	190
1630	5"	Q-2	9	2.667		83.50	102		185.50	245
1640	6"	"	8	3		127	114		241	310
1650	8"	Q-3	5.50	5.818		310	227		537	680
1660	10"		5	6.400		470	249		719	895
1670	12"	↓	4.50	7.111	↓	630	277		907	1,100
1676	Combination double Y & 1/8 bend									
1678	2"	Q-1	8	2	Ea.	33	73.50		106.50	147
1680	3"		7	2.286		44.50	84		128.50	175
1682	4"	↓	6.50	2.462		73.50	90.50		164	217
1684	6"	Q-2	8	3	↓	253	114		367	450
1690	Combination double Y & 1/8 bend, CI, reducing									
1692	3" x 2"	Q-1	8	2	Ea.	36	73.50		109.50	151
1694	4" x 2"		7	2.286		45.50	84		129.50	176
1696	4" x 3"	↓	7	2.286		50.50	84		134.50	182
1698	6" x 4"	Q-2	9	2.667	↓	166	102		268	335
1700	Double Y, CI, reducing									
1702	3" x 2"	Q-1	8	2	Ea.	30.50	73.50		104	145
1703	4" x 2"		7	2.286		36.50	84		120.50	166
1704	4" x 3"	↓	7	2.286		41	84		125	171
1706	6" x 3"	Q-2	9.50	2.526		88.50	96.50		185	243
1707	6" x 4"	"	9	2.667		87	102		189	249
1708	8" x 4"	Q-3	8.50	3.765		215	147		362	455
1709	8" x 6"		8	4		221	156		377	475
1711	10" x 6"		8	4		365	156		521	640
1712	10" x 8"		6.50	4.923		410	192		602	740
1713	12" x 6"		7.50	4.267		470	166		636	770
1714	12" x 8"	↓	6	5.333	↓	510	208		718	870

15107	Metal Pipe & Fittings	CREW	DAILY OUTPUT	LABOR-HOURS	UNIT	2005 BARE COSTS				TOTAL INCL O&P
						MAT.	LABOR	EQUIP.	TOTAL	
1740	Reducer, 3" x 2"	Q-1	15	1.067	Ea.	9.50	39		48.50	69.50
1750	4" x 2"		14.50	1.103		10	40.50		50.50	72
1760	4" x 3"		14	1.143		12.40	42		54.40	76.50
1770	5" x 2"		14	1.143		16.40	42		58.40	81
1780	5" x 3"		13.50	1.185		16.40	43.50		59.90	83.50
1790	5" x 4"		13	1.231		16.40	45.50		61.90	86
1800	6" x 2"		13.50	1.185		22	43.50		65.50	90
1810	6" x 3"		13	1.231		24	45.50		69.50	94.50
1830	6" x 4"	Q-2	12.50	1.920		25	73		98	138
1840	6" x 5"	Q-1	11	1.455		27	53.50		80.50	110
1880	8" x 3"	Q-2	13.50	1.778		43	68		111	150
1900	8" x 4"		13	1.846		42	70.50		112.50	152
1920	8" x 5"		12	2		43	76.50		119.50	163
1940	8" x 6"		12	2		43	76.50		119.50	162
1942	10" x 4"		12.50	1.920		63	73		136	179
1943	10" x 6"		12	2		66.50	76.50		143	188
1944	10" x 8"		9.50	2.526		70.50	96.50		167	223
1945	12" x 4"		11.50	2.087		102	79.50		181.50	233
1946	12" x 6"		11	2.182		110	83		193	246
1947	12" x 8"		9	2.667		113	102		215	277
1948	12" x 10"		8.50	2.824		115	108		223	288
1949	15" x 6"	Q-3	12	2.667		214	104		318	390
1950	15" x 8"		10	3.200		233	125		358	445
1951	15" x 10"		9.50	3.368		241	131		372	460
1952	15" x 12"		9	3.556		244	138		382	475
1960	Increaser, 2" x 3"	Q-1	15	1.067		21.50	39		60.50	82.50
1980	2" x 4"		14	1.143		21.50	42		63.50	86.50
2000	2" x 5"		13	1.231		26	45.50		71.50	97
2020	3" x 4"		13	1.231		22.50	45.50		68	93
2040	3" x 5"		13	1.231		26	45.50		71.50	96.50
2060	3" x 6"		12	1.333		32.50	49		81.50	109
2070	4" x 5"		13	1.231		26	45.50		71.50	97
2080	4" x 6"		12	1.333		32.50	49		81.50	109
2090	4" x 8"	Q-2	13	1.846		66	70.50		136.50	179
2100	5" x 6"	Q-1	11	1.455		48	53.50		101.50	134
2110	5" x 8"	Q-2	12	2		77.50	76.50		154	201
2120	6" x 8"		12	2		77.50	76.50		154	201
2130	6" x 10"		8	3		133	114		247	320
2140	8" x 10"		6.50	3.692		133	141		274	360
2150	10" x 12"		5.50	4.364		251	166		417	525
2500	Y, 2"	Q-1	10	1.600		11.85	59		70.85	102
2510	3"		9	1.778		22	65.50		87.50	123
2520	4"		8	2		29.50	73.50		103	144
2530	5"	Q-2	12	2		49.50	76.50		126	170
2540	6"	"	11	2.182		67.50	83		150.50	200
2550	8"	Q-3	7	4.571		165	178		343	450
2560	10"		6	5.333		267	208		475	605
2570	12"		5	6.400		460	249		709	880
2580	15"		4	8		975	310		1,285	1,550
2581	Y, reducing									
2582	3" x 2"	Q-1	10	1.600	Ea.	17.10	59		76.10	107
2584	4" x 2"		9.50	1.684		23	62		85	118
2586	4" x 3"		9	1.778		25	65.50		90.50	126
2588	6" x 2"	Q-2	12.50	1.920		43	73		116	157
2590	6" x 3"		12	2		44	76.50		120.50	163
2592	6" x 4"		12	2		45.50	76.50		122	165

Important: See the Reference Section for critical supporting data - Reference Nos., Crews, & City Cost Indexes

15107 | Metal Pipe & Fittings

		CREW	DAILY OUTPUT	LABOR-HOURS	UNIT	2005 BARE COSTS				TOTAL INCL O&P
						MAT.	LABOR	EQUIP.	TOTAL	
360 2594	8" x 2"	Q-3	11	2.909	Ea.	91	113		204	270
2596	8" x 3"		10.50	3.048		96	119		215	283
2598	8" x 4"		10	3.200		82.50	125		207.50	278
2600	8" x 6"		9	3.556		103	138		241	320
2602	10" x 3"		10	3.200		151	125		276	355
2604	10" x 4"		9.50	3.368		149	131		280	360
2606	10" x 6"		8	4		162	156		318	415
2608	10" x 8"		8	4		223	156		379	480
2610	12" x 4"		9	3.556		252	138		390	485
2612	12" x 6"		8.50	3.765		259	147		406	505
2614	12" x 8"		8	4		310	156		466	575
2616	12" x 10"		7.50	4.267		400	166		566	690
2618	15" x 4"		7	4.571		765	178		943	1,125
2620	15" x 6"		6.50	4.923		800	192		992	1,175
2622	15" x 8"		6.50	4.923		815	192		1,007	1,175
2624	15" x 10"		5	6.400		835	249		1,084	1,300
2626	15" x 12"		4.50	7.111		855	277		1,132	1,350
3000	For extra heavy, add					44%	4%			
3600	Hub and spigot, service weight gasket joint									
3605	Note: gaskets and joint labor have									
3606	been included with all listed fittings.									
3610	1/4 bend, 2"	Q-1	20	.800	Ea.	12.80	29.50		42.30	58
3620	3"		17	.941		17.05	34.50		51.55	71
3630	4"		15	1.067		25.50	39		64.50	87
3640	5"	Q-2	21	1.143		36.50	43.50		80	106
3650	6"	"	19	1.263		43.50	48		91.50	121
3660	8"	Q-3	12	2.667		125	104		229	293
3670	10"		11	2.909		183	113		296	370
3680	12"		10	3.200		246	125		371	460
3690	15"		8	4		720	156		876	1,025
3700	Closet bend, 3" diameter with ring 10" x 16"	Q-1	17	.941		57	34.50		91.50	115
3710	16" x 16"		15	1.067		77	39		116	144
3730	Closet bend, 4" diameter, 1" x 4" ring, 6" x 16"		15	1.067		64	39		103	130
3740	8" x 16"		15	1.067		57.50	39		96.50	123
3750	10" x 12"		14	1.143		54.50	42		96.50	123
3760	10" x 18"		13	1.231		80.50	45.50		126	157
3770	12" x 16"		13	1.231		64.50	45.50		110	139
3780	16" x 16"		12	1.333		79.50	49		128.50	161
3800	1/8 bend, 2"		20	.800		10	29.50		39.50	55
3810	3"		17	.941		14.95	34.50		49.45	68.50
3820	4"		15	1.067		21	39		60	82
3830	5"	Q-2	21	1.143		30	43.50		73.50	98.50
3840	6"	"	19	1.263		35	48		83	111
3850	8"	Q-3	12	2.667		98.50	104		202.50	264
3860	10"		11	2.909		143	113		256	325
3870	12"		10	3.200		254	125		379	465
3880	15"		8	4		480	156		636	760
3900	Sanitary Tee, 2"	Q-1	12	1.333		18.70	49		67.70	94
3910	3"		10	1.600		30	59		89	122
3920	4"		9	1.778		37	65.50		102.50	140
3930	5"	Q-2	13	1.846		69	70.50		139.50	182
3940	6"	"	11	2.182		77	83		160	210
3950	8"	Q-3	8	4		197	156		353	450
3980	Tee, 2"	Q-1	12	1.333		23	49		72	99
3990	3"		10	1.600		35.50	59		94.50	128
4000	4"		9	1.778		45.50	65.50		111	149

		CREW	DAILY OUTPUT	LABOR-HOURS	UNIT	2005 BARE COSTS				TOTAL INCL O&P
15107	**Metal Pipe & Fittings**					MAT.	LABOR	EQUIP.	TOTAL	
4010	5"	Q-2	13	1.846	Ea.	69	70.50		139.50	182
4020	6"	"	11	2.182		77	83		160	210
4030	8"	Q-3	8	4	↓	197	156		353	450
4060	Combination Y and 1/8 bend									
4070	2"	Q-1	12	1.333	Ea.	23	49		72	99
4080	3"		10	1.600		34	59		93	126
4090	4"	↓	9	1.778		46	65.50		111.50	149
4100	5"	Q-2	13	1.846		83.50	70.50		154	198
4110	6"	"	11	2.182		102	83		185	237
4120	8"	Q-3	8	4		247	156		403	505
4160	Double Y, 2"	Q-1	10	1.600		29.50	59		88.50	121
4170	3"		8	2		47.50	73.50		121	164
4180	4"	↓	7	2.286		64.50	84		148.50	197
4190	5"	Q-2	10	2.400		107	91.50		198.50	256
4200	6"	"	9	2.667		151	102		253	320
4210	8"	Q-3	6	5.333		365	208		573	710
4220	10"		5	6.400		555	249		804	985
4230	12"	↓	4.50	7.111		735	277		1,012	1,225
4260	Reducer, 3" x 2"	Q-1	17	.941		16.60	34.50		51.10	70.50
4270	4" x 2"		16.50	.970		18.15	35.50		53.65	73.50
4280	4" x 3"		16	1		21.50	37		58.50	79
4290	5" x 2"		16	1		27.50	37		64.50	85.50
4300	5" x 3"		15.50	1.032		28.50	38		66.50	88
4310	5" x 4"		15	1.067		29.50	39		68.50	91.50
4320	6" x 2"		15.50	1.032		33.50	38		71.50	93.50
4330	6" x 3"		15	1.067		36.50	39		75.50	99
4340	6" x 5"	↓	13	1.231		43	45.50		88.50	115
4360	8" x 3"	Q-2	15	1.600		65	61		126	163
4370	8" x 4"		15	1.600		65	61		126	163
4380	8" x 5"		14	1.714		69	65.50		134.50	175
4390	8" x 6"	↓	14	1.714		69	65.50		134.50	175
4430	Increaser, 2" x 3"	Q-1	17	.941		25.50	34.50		60	80
4440	2" x 4"		16	1		26.50	37		63.50	85
4450	2" x 5"		15	1.067		34	39		73	96.50
4460	3" x 4"		15	1.067		27.50	39		66.50	89.50
4470	3" x 5"		15	1.067		34	39		73	96.50
4480	3" x 6"		14	1.143		40.50	42		82.50	108
4490	4" x 5"		15	1.067		34	39		73	96.50
4500	4" x 6"	↓	14	1.143		40.50	42		82.50	108
4510	4" x 8"	Q-2	15	1.600		84	61		145	184
4520	5" x 6"	Q-1	13	1.231		56	45.50		101.50	130
4530	5" x 8"	Q-2	14	1.714		95.50	65.50		161	204
4540	6" x 8"		14	1.714		95.50	65.50		161	204
4550	6" x 10"		10	2.400		160	91.50		251.50	315
4560	8" x 10"		8.50	2.824		160	108		268	340
4570	10" x 12"	↓	7.50	3.200		286	122		408	500
4600	Y, 2"	Q-1	12	1.333		17.90	49		66.90	93
4610	3"		10	1.600		30	59		89	122
4620	4"	↓	9	1.778		39.50	65.50		105	142
4630	5"	Q-2	13	1.846		65	70.50		135.50	178
4640	6"	"	11	2.182		84	83		167	217
4650	8"	Q-3	8	4		201	156		357	455
4660	10"		7	4.571		320	178		498	625
4670	12"	↓	6	5.333	↓	530	208		738	890
4900	For extra heavy, add					44%	4%			
4940	Gasket and making push-on joint									

15107	Metal Pipe & Fittings	CREW	DAILY OUTPUT	LABOR-HOURS	UNIT	2005 BARE COSTS				TOTAL INCL O&P
						MAT.	LABOR	EQUIP.	TOTAL	
360											36
4950	2"	Q-1	40	.400	Ea.	3.04	14.70		17.74	25.50
4960	3"		35	.457		4.05	16.80		20.85	30
4970	4"	▼	32	.500		5.10	18.40		23.50	33
4980	5"	Q-2	43	.558		7.85	21.50		29.35	40.50
4990	6"	"	40	.600		8.15	23		31.15	43.50
5000	8"	Q-3	32	1		17.95	39		56.95	78.50
5010	10"		29	1.103		27.50	43		70.50	94.50
5020	12"		25	1.280		35	50		85	114
5022	15"	▼	21	1.524	▼	42	59.50		101.50	135
5030	Note: gaskets and joint labor have									
5040	Been included with all listed fittings.									
5990	No hub									
6000	Cplg. & labor required at joints not incl. in fitting									
6010	price. Add 1 coupling per joint for installed price									
6020	1/4 Bend, 1-1/2"				Ea.	5.50			5.50	6.05
6060	2"					5.95			5.95	6.55
6080	3"					8.30			8.30	9.10
6120	4"					11.90			11.90	13.10
6140	5"					26.50			26.50	29.50
6160	6"					30			30	33
6180	8"					83			83	91
6181	10"					130			130	143
6184	1/4 Bend, long sweep, 1-1/2"					12.85			12.85	14.10
6186	2"					12.85			12.85	14.10
6188	3"					15.30			15.30	16.85
6189	4"					24.50			24.50	27
6190	5"					45			45	49.50
6191	6"					25			25	27.50
6192	8"					133			133	147
6193	10"					239			239	263
6200	1/8 Bend, 1-1/2"					4.55			4.55	5
6210	2"					5.10			5.10	5.60
6212	3"					6.85			6.85	7.55
6214	4"					8.70			8.70	9.60
6216	5"					18.60			18.60	20.50
6218	6"					20			20	22
6220	8"				▼	58			58	64
6222	10"					108			108	119
6364	Closet flange									
6366	4"				Ea.	7.95			7.95	8.75
6370	Closet bend, no hub									
6376	4" x 16"				Ea.	41.50			41.50	46
6380	Sanitary Tee, tapped, 1-1/2"					10.05			10.05	11.05
6382	2" x 1-1/2"					9.15			9.15	10.05
6384	2"					10.10			10.10	11.10
6386	3" x 2"					14.15			14.15	15.60
6388	3"					26			26	29
6390	4" x 1-1/2"					12.65			12.65	13.90
6392	4" x 2"					14.15			14.15	15.60
6394	6" x 1-1/2"					29.50			29.50	32.50
6396	6" x 2"					30			30	33
6459	Sanitary Tee, 1-1/2"					7.45			7.45	8.15
6460	2"					8.30			8.30	9.10
6470	3"					10.55			10.55	11.60
6472	4"					15.60			15.60	17.15
6474	5"				▼	44.50			44.50	49

15107	Metal Pipe & Fittings	CREW	DAILY OUTPUT	LABOR-HOURS	UNIT	2005 BARE COSTS				TOTAL INCL O&P
						MAT.	LABOR	EQUIP.	TOTAL	
6476	6"				Ea.	45.50			45.50	50
6478	8"					184			184	202
6480	10"				↓	196			196	216
6724	Sanitary Tee, reducing									
6725	3" x 2"				Ea.	9.15			9.15	10.10
6726	4" x 3"					14.35			14.35	15.80
6727	5" x 4"					34.50			34.50	37.50
6728	6" x 3"					33.50			33.50	36.50
6729	8" x 4"					97			97	107
6730	Y, 1-1/2"					7.75			7.75	8.55
6740	2"					7.75			7.75	8.55
6750	3"					11			11	12.10
6760	4"					17.90			17.90	19.70
6762	5"					42			42	46.50
6764	6"					47.50			47.50	52.50
6768	8"					112			112	124
6791	Y, reducing, 3" x 2"					8.30			8.30	9.10
6792	4" x 2"					11.45			11.45	12.60
6793	5" x 2"					26			26	28.50
6794	6" x 2"					28.50			28.50	31.50
6795	6" x 4"					37			37	41
6796	8" x 4"					64			64	70.50
6797	8" x 6"					79			79	86.50
6798	10" x 6"					177			177	195
6799	10" x 8"					213			213	234
6800	Double Y, 2"					11.90			11.90	13.10
6920	3"					22			22	24
7000	4"					45			45	49.50
7100	6"					79			79	87
7120	8"				↓	226			226	248
7200	Combination Y and 1/8 Bend									
7220	1-1/2"				Ea.	8.20			8.20	9
7260	2"					8.70			8.70	9.60
7320	3"					13.25			13.25	14.60
7400	4"					24.50			24.50	26.50
7480	5"					53.50			53.50	58.50
7500	6"					67.50			67.50	74.50
7520	8"					167			167	184
7800	Reducer, 3" x 2"					4.13			4.13	4.54
7820	4" x 2"					6.45			6.45	7.10
7840	4" x 3"				↓	6.45			6.45	7.10
8000	Coupling, standard (by CISPI Mfrs.)									
8020	1-1/2"	Q-1	48	.333	Ea.	5.25	12.25		17.50	24
8040	2"		44	.364		5.25	13.35		18.60	26
8080	3"		38	.421		6.20	15.50		21.70	30.50
8120	4"	↓	33	.485		7.35	17.85		25.20	35
8160	5"	Q-2	44	.545		17.95	21		38.95	51
8180	6"	"	40	.600		18.70	23		41.70	55
8200	8"	Q-3	33	.970		35.50	38		73.50	96
8220	10"	"	26	1.231	↓	48.50	48		96.50	126
8300	Coupling, cast iron clamp & neoprene gasket (by MG)									
8310	1-1/2"	Q-1	48	.333	Ea.	7.75	12.25		20	27
8320	2"		44	.364		8	13.35		21.35	29
8330	3"		38	.421		8.35	15.50		23.85	32.50
8340	4"	↓	33	.485		9.80	17.85		27.65	38
8350	5"	Q-2	44	.545	↓	18.55	21		39.55	52

Important: See the Reference Section for critical supporting data - Reference Nos., Crews, & City Cost Indexes

15107 | Metal Pipe & Fittings

		CREW	DAILY OUTPUT	LABOR-HOURS	UNIT	2005 BARE COSTS				TOTAL INCL O&P
						MAT.	LABOR	EQUIP.	TOTAL	
360										**360**
8360	6"	Q-2	40	.600	Ea.	25.50	23		48.50	62.50
8380	8"	Q-3	33	.970		44.50	38		82.50	106
8400	10"	"	26	1.231	↓	68	48		116	147
8410	Reducing, no hub									
8416	2" x 1-1/2"	Q-1	44	.364	Ea.	9.10	13.35		22.45	30
8600	Coupling, Stainless steel, heavy duty									
8620	1-1/2"	Q-1	48	.333	Ea.	11.10	12.25		23.35	30.50
8630	2"		44	.364		11.85	13.35		25.20	33
8640	2" x 1-1/2"		44	.364		15.60	13.35		28.95	37
8650	3"		38	.421		13.65	15.50		29.15	38.50
8660	4"		33	.485		15.90	17.85		33.75	44.50
8670	4" x 3"		33	.485		24	17.85		41.85	53
8680	5"	Q-2	44	.545		32.50	21		53.50	67.50
8690	6"	"	40	.600		37	23		60	75.50
8700	8"	Q-3	33	.970		76.50	38		114.50	142
8710	10"	"	26	1.231		99	48		147	181
380	0010	**PIPE, CAST IRON/STEEL, FITTINGS** Special								**380**
	0020	Cast iron, drainage, threaded, black								
	0030	90° Elbow, straight								
	0031	1-1/2" pipe size	Q-1	20	.800	Ea.	14.15	29.50		43.65	59.50
	0032	2" pipe size		18	.889		21.50	32.50		54	72.50
	0033	3" pipe size		10	1.600		80	59		139	177
	0034	4" pipe size	↓	6	2.667	↓	127	98		225	286
	0040	90° Long turn elbow, straight								
	0041	1-1/2" pipe size	Q-1	20	.800	Ea.	19.80	29.50		49.30	66
	0042	2" pipe size		18	.889		29.50	32.50		62	81.50
	0043	3" pipe size		10	1.600		86.50	59		145.50	184
	0044	4" pipe size	↓	6	2.667	↓	179	98		277	345
	0050	90° Street elbow, straight								
	0051	1-1/2" pipe size	Q-1	20	.800	Ea.	20	29.50		49.50	66
	0052	2" pipe size	"	18	.889	"	26.50	32.50		59	78.50
	0060	45° Elbow								
	0061	1-1/2" pipe size	Q-1	20	.800	Ea.	13.85	29.50		43.35	59
	0062	2" pipe size		18	.889		19.95	32.50		52.45	71
	0063	3" pipe size		10	1.600		78	59		137	175
	0064	4" pipe size	↓	6	2.667	↓	122	98		220	282
	0070	45° Street elbow								
	0071	1-1/2" pipe size	Q-1	20	.800	Ea.	18.85	29.50		48.35	64.50
	0072	2" pipe size	"	18	.889	"	30.50	32.50		63	82.50
	0092	Tees, straight								
	0093	1-1/2" pipe size	Q-1	13	1.231	Ea.	23.50	45.50		69	93.50
	0094	2" pipe size	"	11	1.455	"	38.50	53.50		92	123
	0100	TY's, straight								
	0101	1-1/2" pipe size	Q-1	13	1.231	Ea.	23	45.50		68.50	93
	0102	2" pipe size		11	1.455		37.50	53.50		91	122
	0103	3" pipe size		6	2.667		135	98		233	295
	0104	4" pipe size	↓	4	4	↓	205	147		352	445
	0120	45° Y branch, straight								
	0121	1-1/2" pipe size	Q-1	13	1.231	Ea.	28	45.50		73.50	99
	0122	2" pipe size		11	1.455		52	53.50		105.50	138
	0123	3" pipe size		6	2.667		175	98		273	340
	0124	4" pipe size	↓	4	4	↓	268	147		415	515
	0147	Double Y branch, straight								
	0148	1-1/2" pipe size	Q-1	10	1.600	Ea.	68	59		127	163
	0149	2" pipe size	"	7	2.286	"	69	84		153	202
	0160	P trap								

	15107	Metal Pipe & Fittings	CREW	DAILY OUTPUT	LABOR-HOURS	UNIT	2005 BARE COSTS				TOTAL INCL O&P	
							MAT.	LABOR	EQUIP.	TOTAL		
380	0161	1-1/2" pipe size	Q-1	15	1.067	Ea.	41.50	39		80.50	105	**380**
	0162	2" pipe size		13	1.231		66.50	45.50		112	141	
	0163	3" pipe size		7	2.286		209	84		293	355	
	0164	4" pipe size	↓	5	3.200	↓	500	118		618	725	
	0180	Tucker connection										
	0181	1-1/2" pipe size	Q-1	24	.667	Ea.	39	24.50		63.50	80	
	0182	2" pipe size	"	21	.762	"	48	28		76	95	
	0205	Cast iron, drainage, threaded, galvanized										
	0206	90° Elbow, straight										
	0207	1-1/2" pipe size	Q-1	20	.800	Ea.	19.85	29.50		49.35	66	
	0208	2" pipe size		18	.889		28.50	32.50		61	80.50	
	0209	3" pipe size		10	1.600		113	59		172	214	
	0210	4" pipe size	↓	6	2.667	↓	220	98		318	390	
	0216	90° Long turn elbow, straight										
	0217	1-1/2" pipe size	Q-1	20	.800	Ea.	22.50	29.50		52	69	
	0218	2" pipe size		18	.889		38.50	32.50		71	91.50	
	0219	3" pipe size		10	1.600		119	59		178	220	
	0220	4" pipe size	↓	6	2.667	↓	200	98		298	365	
	0226	90° Street elbow, straight										
	0227	1-1/2" pipe size	Q-1	20	.800	Ea.	25	29.50		54.50	71.50	
	0228	2" pipe size	"	18	.889	"	39.50	32.50		72	92.50	
	0236	45° Elbow										
	0237	1-1/2" pipe size	Q-1	20	.800	Ea.	19.60	29.50		49.10	65.50	
	0238	2" pipe size		18	.889		28	32.50		60.50	80	
	0239	3" pipe size		10	1.600		97.50	59		156.50	196	
	0240	4" pipe size	↓	6	2.667	↓	159	98		257	320	
	0246	45° Street elbow										
	0247	1-1/2" pipe size	Q-1	20	.800	Ea.	26	29.50		55.50	72.50	
	0248	2" pipe size	"	18	.889	"	40	32.50		72.50	93	
	0268	Tees, straight										
	0269	1-1/2" pipe size	Q-1	13	1.231	Ea.	30	45.50		75.50	101	
	0270	2" pipe size	"	11	1.455	"	50	53.50		103.50	136	
	0276	TY's, straight										
	0277	1-1/2" pipe size	Q-1	13	1.231	Ea.	30	45.50		75.50	101	
	0278	2" pipe size		11	1.455		49.50	53.50		103	135	
	0279	3" pipe size		6	2.667		191	98		289	355	
	0280	4" pipe size	↓	4	4	↓	238	147		385	485	
	0296	45° Y branch, straight										
	0297	1-1/2" pipe size	Q-1	13	1.231	Ea.	33.50	45.50		79	105	
	0298	2" pipe size		11	1.455		67	53.50		120.50	155	
	0299	3" pipe size		6	2.667		231	98		329	400	
	0300	4" pipe size	↓	4	4	↓	365	147		512	620	
	0323	Double Y branch, straight										
	0324	1-1/2" pipe size	Q-1	10	1.600	Ea.	68	59		127	163	
	0325	2" pipe size	"	7	2.286	"	104	84		188	241	
	0336	P Trap										
	0337	1-1/2" pipe size	Q-1	15	1.067	Ea.	54	39		93	119	
	0338	2" pipe size		13	1.231		94.50	45.50		140	172	
	0339	3" pipe size		7	2.286		291	84		375	445	
	0340	4" pipe size	↓	5	3.200	↓	720	118		838	965	
	0356	Tucker connection										
	0357	1-1/2" pipe size	Q-1	24	.667	Ea.	38.50	24.50		63	79.50	
	0358	2" pipe size	"	21	.762	"	51.50	28		79.50	99	
420	0010	**PIPE, COPPER** Solder joints	R15100 -050									**420**
	0100	Solder										

15107 | Metal Pipe & Fittings

		CREW	DAILY OUTPUT	LABOR-HOURS	UNIT	2005 BARE COSTS				TOTAL INCL O&P	
						MAT.	LABOR	EQUIP.	TOTAL		
420	0120	Solder, lead free, roll R15100 -050				Lb.	9.30			9.30	10.25	**420**
	1000	Type K tubing, couplings & clevis hangers 10' O.C.										
	1100	1/4" diameter R15100 -070	1 Plum	84	.095	L.F.	1.37	3.89		5.26	7.35	
	1120	3/8" diameter		82	.098		1.57	3.99		5.56	7.70	
	1140	1/2" diameter		78	.103		1.79	4.19		5.98	8.25	
	1160	5/8" diameter		77	.104		3.05	4.24		7.29	9.75	
	1180	3/4" diameter		74	.108		3.02	4.42		7.44	9.95	
	1200	1" diameter		66	.121		3.97	4.95		8.92	11.80	
	1220	1-1/4" diameter		56	.143		4.91	5.85		10.76	14.15	
	1240	1-1/2" diameter		50	.160		6.50	6.55		13.05	16.95	
	1260	2" diameter		40	.200		9.95	8.15		18.10	23.50	
	1280	2-1/2" diameter	Q-1	60	.267		14.85	9.80		24.65	31	
	1300	3" diameter		54	.296		20.50	10.90		31.40	39.50	
	1320	3-1/2" diameter		42	.381		28	14		42	52	
	1330	4" diameter		38	.421		35	15.50		50.50	62	
	1340	5" diameter		32	.500		91	18.40		109.40	128	
	1360	6" diameter	Q-2	38	.632		133	24		157	182	
	1380	8" diameter	"	34	.706		234	27		261	299	
	1390	For other than full hard temper, add					13%					
	1440	For silver solder, add						15%				
	1800	For medical clean, (oxygen class), add					12%					
	1950	To delete cplgs. & hngrs., 1/4"-1" pipe, subtract					27%	60%				
	1960	1-1/4"-3" pipe, subtract					14%	52%				
	1970	3-1/2"-5" pipe, subtract					10%	60%				
	1980	6"-8" pipe, subtract					19%	53%				
	2000	Type L tubing, couplings & hangers 10' O.C.										
	2100	1/4" diameter	1 Plum	88	.091	L.F.	.94	3.71		4.65	6.65	
	2120	3/8" diameter		84	.095		1.28	3.89		5.17	7.25	
	2140	1/2" diameter		81	.099		1.50	4.03		5.53	7.70	
	2160	5/8" diameter		79	.101		2.23	4.14		6.37	8.65	
	2180	3/4" diameter		76	.105		2.20	4.30		6.50	8.85	
	2200	1" diameter		68	.118		3.10	4.81		7.91	10.60	
	2220	1-1/4" diameter		58	.138		4.25	5.65		9.90	13.15	
	2240	1-1/2" diameter		52	.154		5.40	6.30		11.70	15.40	
	2260	2" diameter		42	.190		8.40	7.80		16.20	21	
	2280	2-1/2" diameter	Q-1	62	.258		12.75	9.50		22.25	28.50	
	2300	3" diameter		56	.286		17.25	10.50		27.75	35	
	2320	3-1/2" diameter		43	.372		23.50	13.70		37.20	46.50	
	2340	4" diameter		39	.410		29.50	15.10		44.60	55	
	2360	5" diameter		34	.471		82.50	17.30		99.80	117	
	2380	6" diameter	Q-2	40	.600		103	23		126	148	
	2400	8" diameter	"	36	.667		180	25.50		205.50	236	
	2410	For other than full hard temper, add					21%					
	2590	For silver solder, add						15%				
	2900	For medical clean, (oxygen class), add					12%					
	2940	To delete cplgs. & hngrs., 1/4"-1" pipe, subtract					37%	63%				
	2960	1-1/4"-3" pipe, subtract					12%	53%				
	2970	3-1/2"-5" pipe, subtract					12%	63%				
	2980	6"-8" pipe, subtract					24%	55%				
	3000	Type M tubing, couplings & hangers 10' O.C.										
	3100	1/4" diameter	1 Plum	90	.089	L.F.	1.04	3.63		4.67	6.60	
	3120	3/8" diameter		87	.092		1.05	3.76		4.81	6.80	
	3140	1/2" diameter		84	.095		1.18	3.89		5.07	7.15	
	3160	5/8" diameter		81	.099		1.71	4.03		5.74	7.95	
	3180	3/4" diameter		78	.103		1.73	4.19		5.92	8.20	
	3200	1" diameter		70	.114		2.38	4.67		7.05	9.60	

			DAILY	LABOR-			2005 BARE COSTS			TOTAL
15107	**Metal Pipe & Fittings**	CREW	OUTPUT	HOURS	UNIT	MAT.	LABOR	EQUIP.	TOTAL	INCL O&P
420	3220	1-1/4" diameter	R15100-050	1 Plum	60	.133	L.F.	3.48	5.45		8.93	12.05	420

		CREW	DAILY OUTPUT	LABOR-HOURS	UNIT	MAT.	LABOR	EQUIP.	TOTAL	TOTAL INCL O&P
3220	1-1/4" diameter R15100-050	1 Plum	60	.133	L.F.	3.48	5.45		8.93	12.05
3240	1-1/2" diameter		54	.148		4.68	6.05		10.73	14.25
3260	2" diameter R15100-070	↓	44	.182		7.30	7.45		14.75	19.20
3280	2-1/2" diameter	Q-1	64	.250		10.95	9.20		20.15	26
3300	3" diameter		58	.276		14.65	10.15		24.80	31.50
3320	3-1/2" diameter		45	.356		21	13.10		34.10	42.50
3340	4" diameter		40	.400		26.50	14.70		41.20	51
3360	5" diameter	↓	36	.444		79.50	16.35		95.85	112
3370	6" diameter	Q-2	42	.571		103	22		125	146
3380	8" diameter	"	38	.632	↓	170	24		194	222
3440	For silver solder, add						15%			
3960	To delete cplgs. & hngrs., 1/4"-1" pipe, subtract					35%	65%			
3970	1-1/4"-3" pipe, subtract					19%	56%			
3980	3-1/2"-5" pipe, subtract					13%	65%			
3990	6"-8" pipe, subtract					28%	58%			
4000	Type DWV tubing, couplings & hangers 10' O.C.									
4100	1-1/4" diameter	1 Plum	60	.133	L.F.	3.67	5.45		9.12	12.25
4120	1-1/2" diameter		54	.148		4.56	6.05		10.61	14.10
4140	2" diameter	↓	44	.182		6.10	7.45		13.55	17.85
4160	3" diameter	Q-1	58	.276		10.90	10.15		21.05	27.50
4180	4" diameter		40	.400		19.15	14.70		33.85	43
4200	5" diameter	↓	36	.444		57	16.35		73.35	87
4220	6" diameter	Q-2	42	.571		81.50	22		103.50	123
4240	8" diameter	"	38	.632	↓	201	24		225	257
4730	To delete cplgs. & hngrs., 1-1/4"-2" pipe, subtract					16%	53%			
4740	2"-4" pipe, subtract					13%	60%			
4750	5"-8" pipe, subtract					23%	58%			
5200	ACR tubing, type L, hard temper, cleaned and									
5220	capped, no couplings or hangers									
5240	3/8" OD				L.F.	.77			.77	.85
5250	1/2" OD					1.22			1.22	1.34
5260	5/8" OD					1.49			1.49	1.64
5270	3/4" OD					2.10			2.10	2.31
5280	7/8" OD					2.34			2.34	2.57
5290	1-1/8" OD					3.37			3.37	3.71
5300	1-3/8" OD					4.54			4.54	4.99
5310	1-5/8" OD					5.80			5.80	6.40
5320	2-1/8" OD					9.15			9.15	10.05
5330	2-5/8" OD					13.60			13.60	15
5340	3-1/8" OD					18.30			18.30	20
5350	3-5/8" OD					24			24	26.50
5360	4-1/8" OD				↓	30.50			30.50	33.50
5380	ACR tubing, type L, annealed, cleaned and capped									
5381	No couplings or hangers									
5384	3/8"	1 Stpi	160	.050	L.F.	.77	2.05		2.82	3.93
5385	1/2"		160	.050		1.22	2.05		3.27	4.42
5386	5/8"		160	.050		1.49	2.05		3.54	4.72
5387	3/4"		130	.062		2.10	2.52		4.62	6.10
5388	7/8"		130	.062		2.34	2.52		4.86	6.35
5389	1-1/8"		115	.070		3.37	2.85		6.22	8
5390	1-3/8"		100	.080		4.54	3.28		7.82	9.90
5391	1-5/8"		90	.089		5.80	3.64		9.44	11.85
5392	2-1/8"	↓	80	.100		9.15	4.10		13.25	16.20
5393	2-5/8"	Q-5	125	.128		13.60	4.72		18.32	22
5394	3-1/8"		105	.152		18.30	5.60		23.90	28.50
5395	4-1/8"	↓	95	.168	↓	30.50	6.20		36.70	43

Important: See the Reference Section for critical supporting data - Reference Nos., Crews, & City Cost Indexes

15107	Metal Pipe & Fittings		CREW	DAILY OUTPUT	LABOR-HOURS	UNIT	2005 BARE COSTS				TOTAL INCL O&P
							MAT.	LABOR	EQUIP.	TOTAL	
420	5800	Refrigeration tubing, dryseal, 50' coils	R15100 -050				Coil	20			20	22
	5840	1/8" OD										
	5850	3/16" OD	R15100 -070					22			22	24.50
	5860	1/4" OD						25			25	27.50
	5870	5/16" OD						32.50			32.50	35.50
	5880	3/8" OD						36.50			36.50	40
	5890	1/2" OD						46.50			46.50	51.50
	5900	5/8" OD						63			63	69.50
	5910	3/4" OD						73.50			73.50	81
	5920	7/8" OD						109			109	120
	5930	1-1/8" OD						160			160	176
	5940	1-3/8" OD						246			246	271
	5950	1-5/8" OD						315			315	350
460	0010	**PIPE, COPPER, FITTINGS** Wrought unless otherwise noted										
	0040	Solder joints, copper x copper										
	0070	90° elbow, 1/4"		1 Plum	22	.364	Ea.	1.57	14.85		16.42	24
	0090	3/8"			22	.364		1.49	14.85		16.34	24
	0100	1/2"			20	.400		.48	16.35		16.83	25
	0110	5/8"			19	.421		2.99	17.20		20.19	29.50
	0120	3/4"			19	.421		1.06	17.20		18.26	27
	0130	1"			16	.500		2.61	20.50		23.11	33.50
	0140	1-1/4"			15	.533		3.94	22		25.94	37.50
	0150	1-1/2"			13	.615		5.55	25		30.55	44
	0160	2"			11	.727		11.20	29.50		40.70	57
	0170	2-1/2"		Q-1	13	1.231		23.50	45.50		69	94
	0180	3"			11	1.455		31.50	53.50		85	115
	0190	3-1/2"			10	1.600		110	59		169	210
	0200	4"			9	1.778		72.50	65.50		138	179
	0210	5"			6	2.667		405	98		503	590
	0220	6"		Q-2	9	2.667		450	102		552	650
	0230	8"		"	8	3		1,675	114		1,789	2,000
	0250	45° elbow, 1/4"		1 Plum	22	.364		2.65	14.85		17.50	25.50
	0270	3/8"			22	.364		2.16	14.85		17.01	25
	0280	1/2"			20	.400		.86	16.35		17.21	25.50
	0290	5/8"			19	.421		4.56	17.20		21.76	31
	0300	3/4"			19	.421		1.52	17.20		18.72	27.50
	0310	1"			16	.500		3.82	20.50		24.32	34.50
	0320	1-1/4"			15	.533		5.45	22		27.45	39
	0330	1-1/2"			13	.615		6.55	25		31.55	45
	0340	2"			11	.727		10.95	29.50		40.45	56.50
	0350	2-1/2"		Q-1	13	1.231		21	45.50		66.50	91
	0360	3"			13	1.231		34.50	45.50		80	106
	0370	3-1/2"			10	1.600		55	59		114	149
	0380	4"			9	1.778		73.50	65.50		139	180
	0390	5"			6	2.667		285	98		383	460
	0400	6"		Q-2	9	2.667		450	102		552	650
	0410	8"		"	8	3		1,525	114		1,639	1,850
	0450	Tee, 1/4"		1 Plum	14	.571		3.30	23.50		26.80	38.50
	0470	3/8"			14	.571		2.53	23.50		26.03	38
	0480	1/2"			13	.615		.81	25		25.81	39
	0490	5/8"			12	.667		5.50	27		32.50	47
	0500	3/4"			12	.667		1.95	27		28.95	43
	0510	1"			10	.800		6	32.50		38.50	55.50
	0520	1-1/4"			9	.889		8.65	36.50		45.15	64
	0530	1-1/2"			8	1		12	41		53	74.50

460

		CREW	DAILY OUTPUT	LABOR-HOURS	UNIT	MAT.	LABOR	EQUIP.	TOTAL	TOTAL INCL O&P
						2005 BARE COSTS				
0540	2"	1 Plum	7	1.143	Ea.	18.75	46.50		65.25	90.50
0550	2-1/2"	Q-1	8	2		42	73.50		115.50	158
0560	3"		7	2.286		64	84		148	197
0570	3-1/2"		6	2.667		186	98		284	350
0580	4"		5	3.200		155	118		273	350
0590	5"		4	4		510	147		657	780
0600	6"	Q-2	6	4		695	153		848	995
0610	8"	"	5	4.800		2,675	183		2,858	3,225
0612	Tee, reducing on the outlet, 1/4"	1 Plum	15	.533		5.70	22		27.70	39.50
0613	3/8"		15	.533		4.99	22		26.99	38.50
0614	1/2"		14	.571		4.38	23.50		27.88	40
0615	5/8"		13	.615		8.85	25		33.85	47.50
0616	3/4"		12	.667		2.82	27		29.82	44
0617	1"		11	.727		5.75	29.50		35.25	51
0618	1-1/4"		10	.800		9.35	32.50		41.85	59.50
0619	1-1/2"		9	.889		9.95	36.50		46.45	65.50
0620	2"		8	1		16.25	41		57.25	79.50
0621	2-1/2"	Q-1	9	1.778		44	65.50		109.50	147
0622	3"		8	2		54	73.50		127.50	171
0623	4"		6	2.667		110	98		208	268
0624	5"		5	3.200		510	118		628	735
0625	6"	Q-2	7	3.429		695	131		826	960
0626	8"	"	6	4		2,675	153		2,828	3,175
0630	Tee, reducing on the run, 1/4"	1 Plum	15	.533		6.65	22		28.65	40.50
0631	3/8"		15	.533		7.55	22		29.55	41.50
0632	1/2"		14	.571		8.90	23.50		32.40	45
0633	5/8"		13	.615		9.40	25		34.40	48.50
0634	3/4"		12	.667		2.30	27		29.30	43.50
0635	1"		11	.727		7.55	29.50		37.05	53
0636	1-1/4"		10	.800		12	32.50		44.50	62
0637	1-1/2"		9	.889		21	36.50		57.50	78
0638	2"		8	1		27	41		68	91.50
0639	2-1/2"	Q-1	9	1.778		66	65.50		131.50	171
0640	3"		8	2		93.50	73.50		167	214
0641	4"		6	2.667		207	98		305	375
0642	5"		5	3.200		480	118		598	705
0643	6"	Q-2	7	3.429		735	131		866	1,000
0644	8"	"	6	4		2,675	153		2,828	3,175
0650	Coupling, 1/4"	1 Plum	24	.333		.37	13.60		13.97	21
0670	3/8"		24	.333		.48	13.60		14.08	21
0680	1/2"		22	.364		.33	14.85		15.18	23
0690	5/8"		21	.381		1.02	15.55		16.57	24.50
0700	3/4"		21	.381		.71	15.55		16.26	24.50
0710	1"		18	.444		1.44	18.15		19.59	29
0715	1-1/4"		17	.471		2.69	19.20		21.89	32
0716	1-1/2"		15	.533		3.55	22		25.55	37
0718	2"		13	.615		5.95	25		30.95	44.50
0721	2-1/2"	Q-1	15	1.067		12.60	39		51.60	73
0722	3"		13	1.231		18.90	45.50		64.40	89
0724	3-1/2"		8	2		36.50	73.50		110	151
0726	4"		7	2.286		40	84		124	170
0728	5"		6	2.667		98	98		196	255
0731	6"	Q-2	8	3		162	114		276	350
0732	8"	"	7	3.429		525	131		656	770
0741	Coupling, reducing, concentric									
0743	1/2"	1 Plum	23	.348	Ea.	.96	14.20		15.16	22.50

460

		CREW	DAILY OUTPUT	LABOR-HOURS	UNIT	MAT.	LABOR	EQUIP.	TOTAL	TOTAL INCL O&P
15107	**Metal Pipe & Fittings**									
0745	3/4"	1 Plum	21.50	.372	Ea.	1.37	15.20		16.57	24.50
0747	1"		19.50	.410		3.05	16.75		19.80	28.50
0748	1-1/4"		18	.444		3.50	18.15		21.65	31.50
0749	1-1/2"		16	.500		6.15	20.50		26.65	37.50
0751	2"		14	.571		9.60	23.50		33.10	45.50
0752	2-1/2"		13	.615		23	25		48	63.50
0753	3"	Q-1	14	1.143		28.50	42		70.50	94.50
0755	4"	"	8	2		61	73.50		134.50	178
0757	5"	Q-2	7.50	3.200		210	122		332	415
0759	6"		7	3.429		345	131		476	575
0761	8"		6.50	3.692		570	141		711	835
0771	Cap, sweat									
0773	1/2"	1 Plum	40	.200	Ea.	.34	8.15		8.49	12.65
0775	3/4"		38	.211		.63	8.60		9.23	13.65
0777	1"		32	.250		1.44	10.20		11.64	16.95
0778	1-1/4"		29	.276		2.09	11.25		13.34	19.25
0779	1-1/2"		26	.308		3.14	12.55		15.69	22.50
0781	2"		22	.364		5.75	14.85		20.60	29
0791	Flange, sweat									
0793	3"	Q-1	22	.727	Ea.	46.50	27		73.50	91
0795	4"		18	.889		65	32.50		97.50	121
0797	5"		12	1.333		121	49		170	207
0799	6"	Q-2	18	1.333		127	51		178	217
0801	8"	"	16	1.500		525	57		582	660
0850	Unions, 1/4"	1 Plum	21	.381		9.50	15.55		25.05	34
0870	3/8"		21	.381		9.60	15.55		25.15	34
0880	1/2"		19	.421		4.89	17.20		22.09	31.50
0890	5/8"		18	.444		22	18.15		40.15	51.50
0900	3/4"		18	.444		6.40	18.15		24.55	34.50
0910	1"		15	.533		11	22		33	45
0920	1-1/4"		14	.571		19.05	23.50		42.55	56
0930	1-1/2"		12	.667		25	27		52	68.50
0940	2"		10	.800		43	32.50		75.50	96
0950	2-1/2"	Q-1	12	1.333		93.50	49		142.50	177
0960	3"	"	10	1.600		242	59		301	355
0980	Adapter, copper x male IPS, 1/4"	1 Plum	20	.400		5.20	16.35		21.55	30
0990	3/8"		20	.400		2.58	16.35		18.93	27.50
1000	1/2"		18	.444		.99	18.15		19.14	28.50
1010	3/4"		17	.471		1.66	19.20		20.86	31
1020	1"		15	.533		4.30	22		26.30	37.50
1030	1-1/4"		13	.615		6.70	25		31.70	45.50
1040	1-1/2"		12	.667		7.70	27		34.70	49.50
1050	2"		11	.727		13	29.50		42.50	59
1060	2-1/2"	Q-1	10.50	1.524		50	56		106	139
1070	3"		10	1.600		66	59		125	161
1080	3-1/2"		9	1.778		86.50	65.50		152	194
1090	4"		8	2		86.50	73.50		160	206
1200	5"		6	2.667		500	98		598	695
1210	6"	Q-2	8.50	2.824		500	108		608	710
1214	Adapter, copper x female IPS									
1216	1/2"	1 Plum	18	.444	Ea.	1.58	18.15		19.73	29
1218	3/4"		17	.471		2.15	19.20		21.35	31.50
1220	1"		15	.533		5.25	22		27.25	39
1221	1/1/4"		13	.615		7.70	25		32.70	46.50
1222	1-1/2"		12	.667		12	27		39	54
1224	2"		11	.727		16.35	29.50		45.85	62.50

460

		CREW	DAILY OUTPUT	LABOR-HOURS	UNIT	MAT.	LABOR	EQUIP.	TOTAL	TOTAL INCL O&P	
	15107 | Metal Pipe & Fittings						2005 BARE COSTS				
460 1250	Cross, 1/2"	1 Plum	10	.800	Ea.	6.90	32.50		39.40	56.50	460
1260	3/4"		9.50	.842		13.40	34.50		47.90	66.50	
1270	1"		8	1		23	41		64	86.50	
1280	1-1/4"		7.50	1.067		32.50	43.50		76	102	
1290	1-1/2"		6.50	1.231		46.50	50.50		97	127	
1300	2"		5.50	1.455		88	59.50		147.50	187	
1310	2-1/2"	Q-1	6.50	2.462		203	90.50		293.50	360	
1320	3"	"	5.50	2.909		256	107		363	440	
1500	Tee fitting, mechanically formed, (Type 1).										
1520	1/2" run size, 3/8" to 1/2" branch size	1 Plum	80	.100	Ea.		4.09		4.09	6.15	
1530	3/4" run size, 3/8" to 3/4" branch size		60	.133			5.45		5.45	8.20	
1540	1" run size, 3/8" to 1" branch size		54	.148			6.05		6.05	9.10	
1550	1-1/4" run size, 3/8" to 1-1/4" branch size		48	.167			6.80		6.80	10.25	
1560	1-1/2" run size, 3/8" to 1-1/2" branch size		40	.200			8.15		8.15	12.30	
1570	2" run size, 3/8" to 2" branch size		35	.229			9.35		9.35	14.05	
1580	2-1/2" run size, 1/2" to 2" branch size		32	.250			10.20		10.20	15.35	
1590	3" run size, 1" to 2" branch size		26	.308			12.55		12.55	18.90	
1600	4" run size, 1" to 2" branch size		24	.333			13.60		13.60	20.50	
1640	Tee fitting, mechanically formed, (Type 2)										
1650	2-1/2" run size, 2-1/2" branch size	1 Plum	12.50	.640	Ea.		26		26	39.50	
1660	3" run size, 2-1/2" to 3" branch size		12	.667			27		27	41	
1670	3-1/2" run size, 2-1/2" to 3-1/2" branch size		11	.727			29.50		29.50	44.50	
1680	4" run size, 2-1/2" to 4" branch size		10.50	.762			31		31	47	
1698	5" run size, 2" to 4" branch size		9.50	.842			34.50		34.50	51.50	
1700	6" run size, 2" to 4" branch size		8.50	.941			38.50		38.50	58	
1710	8" run size, 2" to 4" branch size		7	1.143			46.50		46.50	70	
1800	ACR fittings, OD size										
1802	Tee, straight										
1808	5/8"	1 Stpi	12	.667	Ea.	.81	27.50		28.31	42	
1810	3/4"		12	.667		5.50	27.50		33	47	
1812	7/8"		10	.800		2.07	33		35.07	51.50	
1813	1"		10	.800		14.90	33		47.90	65.50	
1814	1-1/8"		10	.800		6	33		39	55.50	
1816	1-3/8"		9	.889		8.65	36.50		45.15	64	
1818	1-5/8"		8	1		13.35	41		54.35	76	
1820	2-1/8"		7	1.143		21	47		68	93.50	
1822	2-5/8"	Q-5	8	2		42	73.50		115.50	158	
1824	3-1/8"		7	2.286		64	84		148	198	
1826	4-1/8"		5	3.200		155	118		273	350	
1830	90° elbow										
1836	5/8"	1 Stpi	19	.421	Ea.	1.66	17.25		18.91	28	
1838	3/4"		19	.421		3.03	17.25		20.28	29.50	
1840	7/8"		16	.500		3.01	20.50		23.51	34.50	
1842	1-1/8"		16	.500		4.03	20.50		24.53	35.50	
1844	1-3/8"		15	.533		3.94	22		25.94	37.50	
1846	1-5/8"		13	.615		6.15	25		31.15	45	
1848	2-1/8"		11	.727		11.20	30		41.20	57.50	
1850	2-5/8"	Q-5	13	1.231		23.50	45.50		69	94	
1852	3-1/8"		11	1.455		31.50	53.50		85	115	
1854	4-1/8"		9	1.778		80.50	65.50		146	188	
1860	Coupling										
1866	5/8"	1 Stpi	21	.381	Ea.	.37	15.60		15.97	24	
1868	3/4"		21	.381		1.13	15.60		16.73	24.50	
1870	7/8"		18	.444		.71	18.20		18.91	28.50	
1871	1"		18	.444		2.58	18.20		20.78	30.50	
1872	1-1/8"		18	.444		1.53	18.20		19.73	29	

Important: See the Reference Section for critical supporting data - Reference Nos., Crews, & City Cost Indexes

15107 | Metal Pipe & Fittings

		CREW	DAILY OUTPUT	LABOR-HOURS	UNIT	2005 BARE COSTS MAT.	LABOR	EQUIP.	TOTAL	TOTAL INCL O&P	
460	1874	1-3/8"	1 Stpi	17	.471	Ea.	2.69	19.25		21.94	32	460
	1876	1-5/8"		15	.533		3.55	22		25.55	37	
	1878	2-1/8"	↓	13	.615		5.95	25		30.95	44.50	
	1880	2-5/8"	Q-5	15	1.067		13.35	39.50		52.85	73.50	
	1882	3-1/8"	↓	13	1.231		19.30	45.50		64.80	89.50	
	1884	4-1/8"	↓	7	2.286	↓	44	84		128	176	
	2000	DWW, solder joints, copper x copper										
	2030	90° Elbow, 1-1/4"	1 Plum	13	.615	Ea.	4.82	25		29.82	43.50	
	2050	1-1/2"		12	.667		7.20	27		34.20	49	
	2070	2"	↓	10	.800		9.40	32.50		41.90	59.50	
	2090	3"	Q-1	10	1.600		23.50	59		82.50	114	
	2100	4"	"	9	1.778		100	65.50		165.50	209	
	2150	45° Elbow, 1-1/4"	1 Plum	13	.615		3.99	25		28.99	42.50	
	2170	1-1/2"		12	.667		3.30	27		30.30	44.50	
	2180	2"	↓	10	.800		7.60	32.50		40.10	57.50	
	2190	3"	Q-1	10	1.600		16.10	59		75.10	106	
	2200	4"	"	9	1.778		76.50	65.50		142	183	
	2250	Tee, Sanitary, 1-1/4"	1 Plum	9	.889		8.55	36.50		45.05	64	
	2270	1-1/2"		8	1		10.65	41		51.65	73.50	
	2290	2"	↓	7	1.143		12.45	46.50		58.95	83.50	
	2310	3"	Q-1	7	2.286		50	84		134	181	
	2330	4"	"	6	2.667		127	98		225	287	
	2400	Coupling, 1-1/4"	1 Plum	14	.571		2.03	23.50		25.53	37	
	2420	1-1/2"		13	.615		2.53	25		27.53	41	
	2440	2"	↓	11	.727		3.50	29.50		33	48.50	
	2460	3"	Q-1	11	1.455		6.80	53.50		60.30	88	
	2480	4"	"	10	1.600	↓	21.50	59		80.50	113	
	2600	Traps, see division 15155										
	3500	Compression joint fittings										
	3510	As used for plumbing and oil burner work										
	3520	Fitting price includes nuts and sleeves										
	3540	Sleeve, 1/8"				Ea.	.11			.11	.12	
	3550	3/16"					.11			.11	.12	
	3560	1/4"					.05			.05	.06	
	3570	5/16"					.11			.11	.12	
	3580	3/8"					.23			.23	.25	
	3600	1/2"					.31			.31	.34	
	3620	Nut, 1/8"					.17			.17	.19	
	3630	3/16"					.18			.18	.20	
	3640	1/4"					.17			.17	.19	
	3650	5/16"					.25			.25	.28	
	3660	3/8"					.27			.27	.30	
	3670	1/2"					.48			.48	.53	
	3710	Union, 1/8"	1 Plum	26	.308		1.01	12.55		13.56	20	
	3720	3/16"		24	.333		.97	13.60		14.57	21.50	
	3730	1/4"		24	.333		1.02	13.60		14.62	21.50	
	3740	5/16"		23	.348		1.30	14.20		15.50	23	
	3750	3/8"		22	.364		1.25	14.85		16.10	24	
	3760	1/2"		22	.364		1.91	14.85		16.76	24.50	
	3780	5/8"		21	.381		2.12	15.55		17.67	26	
	3820	Union tee, 1/8"		17	.471		3.65	19.20		22.85	33	
	3830	3/16"		16	.500		2.84	20.50		23.34	33.50	
	3840	1/4"		15	.533		2.38	22		24.38	35.50	
	3850	5/16"		15	.533		4.80	22		26.80	38.50	
	3860	3/8"		15	.533		2.75	22		24.75	36	
	3870	1/2"	↓	15	.533	↓	4.62	22		26.62	38	

15107 | Metal Pipe & Fittings

		CREW	DAILY OUTPUT	LABOR-HOURS	UNIT	2005 BARE COSTS				TOTAL INCL O&P	
						MAT.	LABOR	EQUIP.	TOTAL		
460	3910	Union elbow, 1/4"	1 Plum	24	.333	Ea.	1.88	13.60		15.48	22.50	460
	3920	5/16"		23	.348		3.10	14.20		17.30	25	
	3930	3/8"		22	.364		2.24	14.85		17.09	25	
	3940	1/2"		22	.364		3.65	14.85		18.50	26.50	
	3980	Female connector, 1/8"		26	.308		.83	12.55		13.38	19.80	
	4000	3/16" x 1/8"		24	.333		1.30	13.60		14.90	22	
	4010	1/4" x 1/8"		24	.333		.88	13.60		14.48	21.50	
	4020	1/4"		24	.333		1.11	13.60		14.71	21.50	
	4030	3/8" x 1/4"		22	.364		1.23	14.85		16.08	24	
	4040	1/2" x 3/8"		22	.364		1.96	14.85		16.81	24.50	
	4050	5/8" x 1/2"		21	.381		2.46	15.55		18.01	26	
	4090	Male connector, 1/8"		26	.308		.60	12.55		13.15	19.55	
	4100	3/16" x 1/8"		24	.333		.62	13.60		14.22	21	
	4110	1/4" x 1/8"		24	.333		.64	13.60		14.24	21	
	4120	1/4"		24	.333		.64	13.60		14.24	21	
	4130	5/16" x 1/8"		23	.348		.75	14.20		14.95	22.50	
	4140	5/16" x 1/4"		23	.348		.86	14.20		15.06	22.50	
	4150	3/8" x 1/8"		22	.364		.89	14.85		15.74	23.50	
	4160	3/8" x 1/4"		22	.364		1.03	14.85		15.88	23.50	
	4170	3/8"		22	.364		1.26	14.85		16.11	24	
	4180	3/8" x 1/2"		22	.364		1.41	14.85		16.26	24	
	4190	1/2" x 3/8"		22	.364		1.37	14.85		16.22	24	
	4200	1/2"		22	.364		1.91	14.85		16.76	24.50	
	4210	5/8" x 1/2"		21	.381		1.78	15.55		17.33	25.50	
	4240	Male elbow, 1/8"		26	.308		1.30	12.55		13.85	20.50	
	4250	3/16" x 1/8"		24	.333		1.43	13.60		15.03	22	
	4260	1/4" x 1/8"		24	.333		1.08	13.60		14.68	21.50	
	4270	1/4"		24	.333		1.23	13.60		14.83	22	
	4280	3/8" x 1/4"		22	.364		1.91	14.85		16.76	24.50	
	4290	3/8"		22	.364		2.69	14.85		17.54	25.50	
	4300	1/2" x 1/4"		22	.364		2.60	14.85		17.45	25.50	
	4310	1/2" x 3/8"		22	.364		4.07	14.85		18.92	27	
	4340	Female elbow, 1/8"		26	.308		2.78	12.55		15.33	22	
	4350	1/4" x 1/8"		24	.333		1.58	13.60		15.18	22	
	4360	1/4"		24	.333		2.08	13.60		15.68	23	
	4370	3/8" x 1/4"		22	.364		2.06	14.85		16.91	25	
	4380	1/2" x 3/8"		22	.364		3.50	14.85		18.35	26.50	
	4390	1/2"		22	.364		4.71	14.85		19.56	27.50	
	4420	Male run tee, 1/4" x 1/8"		15	.533		1.94	22		23.94	35	
	4430	5/16" x 1/8"		15	.533		3.80	22		25.80	37	
	4440	3/8" x 1/4"		15	.533		3	22		25	36.50	
	4480	1/4" x 1/8"		15	.533		2.25	22		24.25	35.50	
	4490	1/4"		15	.533		2.42	22		24.42	35.50	
	4500	3/8" x 1/4"		15	.533		2.82	22		24.82	36	
	4510	1/2" x 3/8"		15	.533		4.38	22		26.38	38	
	4520	1/2"	▼	15	.533	▼	7	22		29	40.50	
	4800	Flare joint fittings										
	4810	Refrigeration fittings										
	4820	Flare joint nuts and labor not incl. in price. Add 1 nut per jnt.										
	4830	90° Elbow, 1/4"				Ea.	1.86			1.86	2.05	
	4840	3/8"					2.16			2.16	2.38	
	4850	1/2"					3.13			3.13	3.44	
	4860	5/8"					4.58			4.58	5.05	
	4870	3/4"					6.90			6.90	7.60	
	5030	Tee, 1/4"					1.61			1.61	1.77	
	5040	5/16"	▼				2.28			2.28	2.51	

Important: See the Reference Section for critical supporting data - Reference Nos., Crews, & City Cost Indexes

			DAILY	LABOR-			2005 BARE COSTS				TOTAL
15107	**Metal Pipe & Fittings**	CREW	OUTPUT	HOURS	UNIT	MAT.	LABOR	EQUIP.	TOTAL		INCL O&P
460	5050	3/8"				Ea.	2.60			2.60		2.86	460
	5060	1/2"					3.96			3.96		4.36
	5070	5/8"					6.15			6.15		6.75
	5080	3/4"					8.70			8.70		9.55
	5140	Union, 3/16"					1.86			1.86		2.05
	5150	1/4"					.63			.63		.69
	5160	5/16"					1.18			1.18		1.30
	5170	3/8"					1.26			1.26		1.39
	5180	1/2"					1.32			1.32		1.45
	5190	5/8"					2.15			2.15		2.37
	5200	3/4"					3.21			3.21		3.53
	5260	Long flare nut, 3/16"	1 Stpi	42	.190		.78	7.80		8.58		12.55
	5270	1/4"		41	.195		.86	8		8.86		12.95
	5280	5/16"		40	.200		2.11	8.20		10.31		14.60
	5290	3/8"		39	.205		1.18	8.40		9.58		13.95
	5300	1/2"		38	.211		1.35	8.60		9.95		14.45
	5310	5/8"		37	.216		2.87	8.85		11.72		16.45
	5320	3/4"		34	.235		2.45	9.65		12.10		17.20
	5380	Short flare nut, 3/16"		42	.190		1.15	7.80		8.95		12.95
	5390	1/4"		41	.195		.39	8		8.39		12.45
	5400	5/16"		40	.200		.63	8.20		8.83		13
	5410	3/8"		39	.205		.54	8.40		8.94		13.25
	5420	1/2"		38	.211		.79	8.60		9.39		13.80
	5430	5/8"		36	.222		1.11	9.10		10.21		14.90
	5440	3/4"	▼	34	.235		1.52	9.65		11.17		16.15
	5500	90° Elbow flare by MIPS, 1/4"					1.35			1.35		1.49
	5510	3/8"					1.69			1.69		1.86
	5520	1/2"					2.70			2.70		2.97
	5530	5/8"					3.13			3.13		3.44
	5540	3/4"					5.25			5.25		5.80
	5600	Flare by FIPS, 1/4"					4.05			4.05		4.46
	5610	3/8"					2.87			2.87		3.16
	5620	1/2"					4.31			4.31		4.74
	5670	Flare by sweat, 1/4"					1.35			1.35		1.49
	5680	3/8"					1.52			1.52		1.67
	5690	1/2"					2.70			2.70		2.97
	5700	5/8"					3.04			3.04		3.34
	5760	Tee flare by IPS, 1/4"					2.34			2.34		2.57
	5770	3/8"					3.19			3.19		3.51
	5780	1/2"					4.89			4.89		5.40
	5790	5/8"					6.25			6.25		6.90
	5850	Connector, 1/4"					.80			.80		.88
	5860	3/8"					1			1		1.10
	5870	1/2"					1.27			1.27		1.40
	5880	5/8"					1.40			1.40		1.54
	5890	3/4"					4.64			4.64		5.10
	5950	Seal cap, 1/4"					.30			.30		.33
	5960	3/8"					.59			.59		.65
	5970	1/2"					.68			.68		.75
	5980	5/8"					1.01			1.01		1.11
	5990	3/4"				▼	2.11			2.11		2.32
	6000	Water service fittings										
	6010	Flare joints nut and labor are included in the fitting price.										
	6020	90° Elbow, cxc, 3/8"	1 Plum	19	.421	Ea.	13.70	17.20		30.90		41
	6030	1/2"		18	.444		13.70	18.15		31.85		42.50
	6040	3/4"	▼	16	.500	▼	17.45	20.50		37.95		49.50

179

			DAILY	LABOR-			2005 BARE COSTS				TOTAL
15107	**Metal Pipe & Fittings**	CREW	OUTPUT	HOURS	UNIT	MAT.	LABOR	EQUIP.	TOTAL		INCL O&P
460	6050	1"	1 Plum	15	.533	Ea.	33	22		55	69.50	460
	6060	1-1/4"		12	.667		64	27		91	112
	6070	1-1/2"		12	.667		64	27		91	112
	6080	2"		10	.800		114	32.50		146.50	174
	6090	C x MPT, 3/8"		19	.421		9.85	17.20		27.05	37
	6100	1/2"		18	.444		9.85	18.15		28	38.50
	6110	3/4"		16	.500		11.40	20.50		31.90	43
	6120	1"		15	.533		31	22		53	67
	6130	1-1/4"		13	.615		64	25		89	109
	6140	1-1/2"		12	.667		64	27		91	112
	6150	2"		10	.800		89.50	32.50		122	147
	6160	C x FPT, 3/8"		19	.421		10.60	17.20		27.80	37.50
	6170	1/2"		18	.444		10.60	18.15		28.75	39
	6180	3/4"		16	.500		13.10	20.50		33.60	45
	6190	1"		15	.533		30	22		52	66
	6200	1-1/4"		13	.615		30	25		55	71
	6210	1-1/2"		12	.667		138	27		165	193
	6220	2"		10	.800		211	32.50		243.50	282
	6230	Tee, cxcxc, 3/8"		13	.615		20	25		45	60
	6240	1/2"		12	.667		20	27		47	63
	6250	3/4"		11	.727		23.50	29.50		53	70.50
	6260	1"		10	.800		41.50	32.50		74	94.50
	6270	1-1/4"		8	1		109	41		150	181
	6280	1-1/2"		8	1		130	41		171	205
	6290	2"		7	1.143		233	46.50		279.50	325
	6300	Tee, cxcxFPT, 3/8"		13	.615		12.25	25		37.25	51.50
	6310	1/2"		12	.667		18.15	27		45.15	61
	6320	3/4"		11	.727		25.50	29.50		55	73
	6330	Tube nut, cxnut seat, 3/8"		40	.200		4.05	8.15		12.20	16.75
	6340	1/2"		38	.211		4.05	8.60		12.65	17.40
	6350	3/4"		34	.235		4.05	9.60		13.65	18.90
	6360	1"		32	.250		7.35	10.20		17.55	23.50
	6370	1-1/4"		25	.320		32.50	13.05		45.55	55
	6380	Coupling, cxc, 3/8"		19	.421		12.25	17.20		29.45	39.50
	6390	1/2"		18	.444		12.25	18.15		30.40	41
	6400	3/4"		16	.500		15.50	20.50		36	47.50
	6410	1"		15	.533		28.50	22		50.50	64.50
	6420	1-1/4"		12	.667		46.50	27		73.50	92
	6430	1-1/2"		12	.667		70.50	27		97.50	119
	6440	2"		10	.800		106	32.50		138.50	165
	6450	Adapter, cxFPT, 3/8"		19	.421		9.25	17.20		26.45	36
	6460	1/2"		18	.444		9.25	18.15		27.40	37.50
	6470	3/4"		16	.500		11.20	20.50		31.70	43
	6480	1"		15	.533		24	22		46	59.50
	6490	1-1/4"		13	.615		51	25		76	94
	6500	1-1/2"		12	.667		51	27		78	97
	6510	2"		10	.800		66	32.50		98.50	122
	6520	Adapter, cxMPT, 3/8"		19	.421		8.30	17.20		25.50	35
	6530	1/2"		18	.444		8.30	18.15		26.45	36.50
	6540	3/4"		16	.500		11.45	20.50		31.95	43
	6550	1"		15	.533		20	22		42	55
	6560	1-1/4"		13	.615		50.50	25		75.50	93.50
	6570	1-1/2"		12	.667		50.50	27		77.50	96.50
	6580	2"	▼	10	.800	▼	68.50	32.50		101	124
	6990	Polybutylene/polyethylene pipe, See 15108 for plastic ftng.									
	7000	Insert type Brass/copper, 100 psi @ 180°F, CTS									

Important: See the Reference Section for critical supporting data - Reference Nos., Crews, & City Cost Indexes

15107 | Metal Pipe & Fittings

		CREW	DAILY OUTPUT	LABOR-HOURS	UNIT	2005 BARE COSTS MAT.	LABOR	EQUIP.	TOTAL	TOTAL INCL O&P
460	7010	Adapter MPT 3/8" x 3/8" CTS	1 Plum	29	.276	Ea.	.97	11.25		12.22	18
	7020	1/2" x 1/2"		26	.308		.97	12.55		13.52	19.95
	7030	3/4" x 1/2"		26	.308		1.44	12.55		13.99	20.50
	7040	3/4" x 3/4"		25	.320		1.42	13.05		14.47	21
	7050	Adapter CTS 1/2" x 1/2" sweat		24	.333		.38	13.60		13.98	21
	7060	3/4" x 3/4" sweat		22	.364		.47	14.85		15.32	23
	7070	Coupler center set 3/8" CTS		25	.320		.49	13.05		13.54	20
	7080	1/2" CTS		23	.348		.38	14.20		14.58	22
	7090	3/4" CTS		22	.364		.47	14.85		15.32	23
	7100	Elbow 90°, copper 3/8"		25	.320		1.08	13.05		14.13	21
	7110	1/2" CTS		23	.348		.69	14.20		14.89	22.50
	7120	3/4" CTS		22	.364		.84	14.85		15.69	23.50
	7130	Tee copper 3/8" CTS		17	.471		1.28	19.20		20.48	30.50
	7140	1/2" CTS		15	.533		.88	22		22.88	34
	7150	3/4" CTS		14	.571		1.35	23.50		24.85	36.50
	7160	3/8" x 3/8" x 1/2"		16	.500		.88	20.50		21.38	31.50
	7170	1/2" x 3/8" x 1/2"		15	.533		.88	22		22.88	34
	7180	3/4" x 1/2" x 3/4"		14	.571		1.35	23.50		24.85	36.50
480	0010	**PIPE/TUBE, GROOVED-JOINT FOR COPPER**									
	4000	Fittings: coupling and labor required at joints not incl.									
	4001	in fitting price. add 1 per joint for installed price.									
	4010	Coupling, rigid style									
	4018	2" diameter	1 Plum	50	.160	Ea.	8.10	6.55		14.65	18.70
	4020	2-1/2" diameter	Q-1	80	.200		9.20	7.35		16.55	21
	4022	3" diameter		67	.239		10.30	8.80		19.10	24.50
	4024	4" diameter		50	.320		15.80	11.75		27.55	35
	4026	5" diameter		40	.400		27	14.70		41.70	52
	4028	6" diameter	Q-2	50	.480		36	18.30		54.30	67
	4100	Elbow, 90° or 45°									
	4108	2" diameter	1 Plum	25	.320	Ea.	12.35	13.05		25.40	33.50
	4110	2-1/2" diameter	Q-1	40	.400		13.50	14.70		28.20	37
	4112	3" diameter		33	.485		19.10	17.85		36.95	48
	4114	4" diameter		25	.640		44.50	23.50		68	84.50
	4116	5" diameter		20	.800		127	29.50		156.50	184
	4118	6" diameter	Q-2	25	.960		204	36.50		240.50	279
	4200	Tee									
	4208	2" diameter	1 Plum	17	.471	Ea.	20.50	19.20		39.70	52
	4210	2-1/2" diameter	Q-1	27	.593		22	22		44	57
	4212	3" diameter		22	.727		33	27		60	76.50
	4214	4" diameter		17	.941		73	34.50		107.50	132
	4216	5" diameter		13	1.231		205	45.50		250.50	294
	4218	6" diameter	Q-2	17	1.412		254	54		308	360
	4300	Reducer, concentric									
	4310	3" x 2-1/2" diameter	Q-1	35	.457	Ea.	18.05	16.80		34.85	45.50
	4312	4" x 2-1/2" diameter		32	.500		37.50	18.40		55.90	68.50
	4314	4" x 3" diameter		29	.552		37.50	20.50		58	71.50
	4316	5" x 3" diameter		25	.640		106	23.50		129.50	153
	4318	5" x 4" diameter		22	.727		106	27		133	157
	4320	6" x 3" diameter	Q-2	28	.857		115	32.50		147.50	176
	4322	6" x 4" diameter		26	.923		115	35		150	180
	4324	6" x 5" diameter		24	1		115	38		153	185
	4350	Flange, w/groove gasket									
	4351	ANSI class 125 and 150									
	4355	2" diameter	1 Plum	23	.348	Ea.	65.50	14.20		79.70	94
	4356	2-1/2" diameter	Q-1	37	.432		68.50	15.90		84.40	99.50
	4358	3" diameter		31	.516		71.50	19		90.50	108

15107 | Metal Pipe & Fittings

		CREW	DAILY OUTPUT	LABOR-HOURS	UNIT	2005 BARE COSTS				TOTAL INCL O&P
						MAT.	LABOR	EQUIP.	TOTAL	
480											**480**
4360	4" diameter	Q-1	23	.696	Ea.	78.50	25.50		104	125
4362	5" diameter	↓	19	.842		106	31		137	163
4364	6" diameter	Q-2	23	1.043	↓	115	40		155	187
500	**PIPE, CORROSION RESISTANT** No couplings or hangers R15100-070										**500**
0020	Iron alloy, drain, mechanical joint									
1000	1-1/2" diameter	Q-1	70	.229	L.F.	22.50	8.40		30.90	37.50
1100	2" diameter		66	.242		26	8.90		34.90	42
1120	3" diameter		60	.267		36	9.80		45.80	55
1140	4" diameter	↓	52	.308	↓	46.50	11.30		57.80	68
1980	Iron alloy, drain, B&S joint									
2000	2" diameter	Q-1	54	.296	L.F.	28	10.90		38.90	47.50
2100	3" diameter		52	.308		40.50	11.30		51.80	61.50
2120	4" diameter	↓	48	.333		56	12.25		68.25	80
2140	6" diameter	Q-2	59	.407		93.50	15.50		109	127
2160	8" diameter	"	54	.444	↓	169	16.95		185.95	212
2980	Plastic, epoxy, fiberglass filament wound									
3000	2" diameter	Q-1	62	.258	L.F.	5.20	9.50		14.70	19.95
3100	3" diameter		51	.314		7.10	11.55		18.65	25
3120	4" diameter		45	.356		9.40	13.10		22.50	30
3140	6" diameter	↓	32	.500		14.40	18.40		32.80	43.50
3160	8" diameter	Q-2	38	.632		19.80	24		43.80	58
3180	10" diameter		32	.750		34	28.50		62.50	80.50
3200	12" diameter	↓	28	.857	↓	43	32.50		75.50	96.50
3980	Polyester, fiberglass filament wound									
4000	2" diameter	Q-1	62	.258	L.F.	9.85	9.50		19.35	25
4100	3" diameter		51	.314		13.20	11.55		24.75	32
4120	4" diameter		45	.356		14.90	13.10		28	36
4140	6" diameter	↓	32	.500		23.50	18.40		41.90	53.50
4160	8" diameter	Q-2	38	.632		32	24		56	71
4180	10" diameter		32	.750		41	28.50		69.50	88
4200	12" diameter	↓	28	.857	↓	47	32.50		79.50	101
4980	Polypropylene, acid resistant, fire retardant, schedule 40									
5000	1-1/2" diameter	Q-1	68	.235	L.F.	1.13	8.65		9.78	14.25
5100	2" diameter		62	.258		1.53	9.50		11.03	15.95
5120	3" diameter		51	.314		4.28	11.55		15.83	22
5140	4" diameter		45	.356		4.55	13.10		17.65	24.50
5160	6" diameter	↓	32	.500	↓	7.65	18.40		26.05	36
5980	Proxylene, fire retardant, Schedule 40									
6000	1-1/2" diameter	Q-1	68	.235	L.F.	3.87	8.65		12.52	17.25
6100	2" diameter		62	.258		5.30	9.50		14.80	20
6120	3" diameter		51	.314		9.60	11.55		21.15	28
6140	4" diameter		45	.356		13.65	13.10		26.75	34.50
6160	6" diameter	↓	32	.500		24.50	18.40		42.90	54.50
6820	For Schedule 80, add				↓	35%	2%			
9000	For "external pipe protection" (see div.02955-800)									
560	**PIPE, CORROSION RESISTANT, FITTINGS**										**560**
0030	Iron alloy									
0050	Mechanical joint									
0060	1/4 Bend, 1-1/2"	Q-1	12	1.333	Ea.	28	49		77	105
0080	2"		10	1.600		36.50	59		95.50	129
0090	3"		9	1.778		51	65.50		116.50	155
0100	4"		8	2		70.50	73.50		144	189
0110	1/8 Bend, 1-1/2"		12	1.333		28	49		77	105
0130	2"		10	1.600		36.50	59		95.50	129
0140	3"	↓	9	1.778	↓	51	65.50		116.50	155

15107 | Metal Pipe & Fittings

		CREW	DAILY OUTPUT	LABOR-HOURS	UNIT	2005 BARE COSTS				TOTAL INCL O&P
						MAT.	LABOR	EQUIP.	TOTAL	
560	**0150**	4"	Q-1	8	2	Ea.	70.50	73.50		144	189
	0160	Tee and Y, sanitary, straight									
	0170	1-1/2"	Q-1	8	2	Ea.	37.50	73.50		111	152
	0180	2"		7	2.286		49.50	84		133.50	180
	0190	3"		6	2.667		74	98		172	229
	0200	4"		5	3.200		111	118		229	299
	0360	Coupling, 1-1/2"		14	1.143		26.50	42		68.50	92
	0380	2"		12	1.333		30.50	49		79.50	107
	0390	3"		11	1.455		32.50	53.50		86	116
	0400	4"		10	1.600		36.50	59		95.50	129
	0500	Bell & Spigot									
	0510	1/4 and 1/16 bend, 2"	Q-1	16	1	Ea.	39	37		76	98.50
	0520	3"		14	1.143		58	42		100	127
	0530	4"		13	1.231		75	45.50		120.50	151
	0540	6"	Q-2	17	1.412		269	54		323	375
	0550	8"	"	12	2		925	76.50		1,001.50	1,150
	0620	1/8 bend, 2"	Q-1	16	1		39	37		76	98.50
	0640	3"		14	1.143		58.50	42		100.50	128
	0650	4"		13	1.231		75	45.50		120.50	151
	0660	6"	Q-2	17	1.412		181	54		235	280
	0680	8"	"	12	2		535	76.50		611.50	705
	0700	Tee, sanitary, 2"	Q-1	10	1.600		51	59		110	145
	0710	3"		9	1.778		168	65.50		233.50	284
	0720	4"		8	2		187	73.50		260.50	315
	0730	6"	Q-2	11	2.182		310	83		393	465
	0740	8"	"	8	3		925	114		1,039	1,200
	1800	Y, sanitary, 2"	Q-1	10	1.600		50	59		109	144
	1820	3"		9	1.778		85	65.50		150.50	192
	1830	4"		8	2		139	73.50		212.50	264
	1840	6"	Q-2	11	2.182		226	83		309	375
	1850	8"	"	8	3		1,375	114		1,489	1,675
	3000	Epoxy, filament wound									
	3030	Quick-lock joint									
	3040	90° Elbow, 2"	Q-1	28	.571	Ea.	58.50	21		79.50	96
	3060	3"		16	1		67.50	37		104.50	130
	3070	4"		13	1.231		92	45.50		137.50	169
	3080	6"		8	2		134	73.50		207.50	259
	3090	8"	Q-2	9	2.667		248	102		350	425
	3100	10"		7	3.429		310	131		441	535
	3110	12"		6	4		445	153		598	720
	3120	45° Elbow, 2"	Q-1	28	.571		45.50	21		66.50	81.50
	3130	3"		16	1		64.50	37		101.50	126
	3140	4"		13	1.231		88	45.50		133.50	165
	3150	6"		8	2		134	73.50		207.50	259
	3160	8"	Q-2	9	2.667		248	102		350	425
	3170	10"		7	3.429		310	131		441	535
	3180	12"		6	4		445	153		598	720
	3190	Tee, 2"	Q-1	19	.842		140	31		171	200
	3200	3"		11	1.455		169	53.50		222.50	267
	3210	4"		9	1.778		203	65.50		268.50	320
	3220	6"		5	3.200		310	118		428	515
	3230	8"	Q-2	6	4		400	153		553	670
	3240	10"		5	4.800		520	183		703	845
	3250	12"		4	6		755	229		984	1,175
	4000	Polypropylene, acid resistant									
	4020	Non-pressure, electrofusion joints									

15107 | Metal Pipe & Fittings

		DAILY OUTPUT	LABOR-HOURS	UNIT	2005 BARE COSTS				TOTAL INCL O&P
		CREW			MAT.	LABOR	EQUIP.	TOTAL	
560									**560**
4050	1/4 bend, 1-1/2"	1 Plum	16	.500	Ea.	14	20.50		34.50	46
4060	2"	Q-1	28	.571		18	21		39	51.50
4080	3"		17	.941		29.50	34.50		64	84.50
4090	4"		14	1.143		47	42		89	115
4110	6"		8	2		115	73.50		188.50	238
4150	1/4 Bend, long sweep									
4170	1-1/2"	1 Plum	16	.500	Ea.	14.45	20.50		34.95	46.50
4180	2"	Q-1	28	.571		19.75	21		40.75	53
4200	3"		17	.941		34	34.50		68.50	89
4210	4"		14	1.143		48.50	42		90.50	117
4250	1/8 bend, 1-1/2"	1 Plum	16	.500	Ea.	13.75	20.50		34.25	45.50
4260	2"	Q-1	28	.571		16.75	21		37.75	50
4280	3"		17	.941		30.50	34.50		65	86
4290	4"		14	1.143		35	42		77	102
4310	6"		8	2		96.50	73.50		170	217
4400	Tee, sanitary									
4420	1-1/2"	1 Plum	10	.800	Ea.	17.40	32.50		49.90	68
4430	2"	Q-1	17	.941		21	34.50		55.50	75
4450	3"		11	1.455		41.50	53.50		95	127
4460	4"		9	1.778		62.50	65.50		128	167
4480	6"		5	3.200		106	118		224	293
4490	Tee, sanitary reducing, 2" x 2" x 1-1/2"		17	.941		21	34.50		55.50	75
4491	3" x 3" x 1-1/2"		11	1.455		36	53.50		89.50	121
4492	3" x 3" x 2"		11	1.455		39	53.50		92.50	124
4493	4" x 4" x 2"		10	1.600		56	59		115	150
4494	4" x 4" x 3"		9	1.778		58.50	65.50		124	163
4496	6" x 6" x 4"		5	3.200		106	118		224	293
4500	Tee/wye, long turn									
4520	1-1/2"	1 Plum	10	.800	Ea.	23	32.50		55.50	74.50
4530	2"	Q-1	17	.941		31	34.50		65.50	86
4550	3"		11	1.455		49.50	53.50		103	135
4570	4"		9	1.778		70	65.50		135.50	176
4650	Wye 45°, 1-1/2"	1 Plum	10	.800		17.70	32.50		50.20	68.50
4652	2"	Q-1	17	.941		25	34.50		59.50	80
4653	3"		11	1.455		44.50	53.50		98	129
4654	4"		9	1.778		64.50	65.50		130	170
4656	6"		5	3.200		164	118		282	355
4660	Wye, reducing									
4662	2" x 2" x 1-1/2"	Q-1	17	.941	Ea.	25	34.50		59.50	79.50
4666	3" x 3" x 2"		11	1.455		41.50	53.50		95	126
4668	4" x 4" x 2"		10	1.600		59.50	59		118.50	154
4669	4" x 4" x 3"		9	1.778		62	65.50		127.50	167
4671	6" x 6" x 2"		6	2.667		104	98		202	261
4673	6" x 6" x 3"		5.50	2.909		105	107		212	277
4675	6" x 6" x 4"		5	3.200		106	118		224	294
4678	Combination Y & 1/8 bend									
4681	1-1/2"	1 Plum	10	.800	Ea.	23	32.50		55.50	74.50
4683	2"	Q-1	17	.941		31	34.50		65.50	86
4684	3"		11	1.455		49.50	53.50		103	135
4685	4"		9	1.778		70	65.50		135.50	176
4689	Combination Y & 1/8 bend, reducing									
4692	2" x 2" x 1-1/2"	Q-1	17	.941	Ea.	28	34.50		62.50	82.50
4694	3" x 3" x 1-1/2"		12	1.333		41	49		90	119
4695	3" x 3" x 2"		11	1.455		44.50	53.50		98	130
4697	4" x 4" x 2"		10	1.600		61.50	59		120.50	156
4699	4" x 4" x 3"		9	1.778		64.50	65.50		130	170

Important: See the Reference Section for critical supporting data - Reference Nos., Crews, & City Cost Indexes

15107 | Metal Pipe & Fittings

		CREW	DAILY OUTPUT	LABOR-HOURS	UNIT	2005 BARE COSTS				TOTAL INCL O&P
						MAT.	LABOR	EQUIP.	TOTAL	
560										**560**
4710	Hub adapter									
4712	1-1/2"	1 Plum	16	.500	Ea.	16.65	20.50		37.15	49
4713	2"	Q-1	28	.571		17.75	21		38.75	51
4714	3"		17	.941		22.50	34.50		57	76.50
4715	4"	↓	14	1.143	↓	32.50	42		74.50	99
4719	Mechanical joint adapter									
4721	1-1/2"	1 Plum	16	.500	Ea.	16.70	20.50		37.20	49
4722	2"	Q-1	28	.571		17.60	21		38.60	51
4723	3"		17	.941		21.50	34.50		56	76
4724	4"	↓	14	1.143	↓	32	42		74	98.50
4728	Couplings									
4731	1-1/2"	1 Plum	16	.500	Ea.	10.95	20.50		31.45	42.50
4732	2"	Q-1	28	.571		13.65	21		34.65	46.50
4733	3"		17	.941		17.45	34.50		51.95	71
4734	4"		14	1.143		25	42		67	90.50
4736	6"	↓	8	2	↓	39.50	73.50		113	155
620	0010 **PIPE, STEEL**									**620**
0020	All pipe sizes are to Spec. A-53 unless noted otherwise	R15100-070								
0030	Schedule 10, see 15107-690-0500									
0050	Schedule 40, threaded, with couplings, and clevis type									
0060	hangers sized for covering, 10' O.C.									
0540	Black, 1/4" diameter	1 Plum	66	.121	L.F.	2.09	4.95		7.04	9.75
0550	3/8" diameter		65	.123		2.02	5.05		7.07	9.80
0560	1/2" diameter		63	.127		1.28	5.20		6.48	9.20
0570	3/4" diameter		61	.131		1.54	5.35		6.89	9.75
0580	1" diameter	↓	53	.151		2.16	6.15		8.31	11.65
0590	1-1/4" diameter	Q-1	89	.180		2.75	6.60		9.35	12.95
0600	1-1/2" diameter		80	.200		3.16	7.35		10.51	14.55
0610	2" diameter		64	.250		4.13	9.20		13.33	18.35
0620	2-1/2" diameter		50	.320		7	11.75		18.75	25.50
0630	3" diameter		43	.372		9.05	13.70		22.75	30.50
0640	3-1/2" diameter		40	.400		11.95	14.70		26.65	35
0650	4" diameter		36	.444		14.25	16.35		30.60	40
0660	5" diameter	↓	26	.615		26	22.50		48.50	62.50
0670	6" diameter	Q-2	31	.774		32.50	29.50		62	80.50
0680	8" diameter		27	.889		51	34		85	107
0690	10" diameter		23	1.043		78	40		118	146
0700	12" diameter	↓	18	1.333	↓	86	51		137	171
0809	A-106, gr. A/B, seamless w/cplgs. & hangers									
0811	1/4" diameter	1 Plum	66	.121	L.F.	4.93	4.95		9.88	12.85
0812	3/8" diameter		65	.123		3.66	5.05		8.71	11.55
0813	1/2" diameter		63	.127		2.75	5.20		7.95	10.85
0814	3/4" diameter		61	.131		2.25	5.35		7.60	10.50
0815	1" diameter	↓	53	.151		2.95	6.15		9.10	12.50
0816	1-1/4" diameter	Q-1	89	.180		3.67	6.60		10.27	14
0817	1-1/2" diameter		80	.200		4.88	7.35		12.23	16.40
0819	A-53, 2" diameter		64	.250		5.15	9.20		14.35	19.45
0821	2-1/2" diameter		50	.320		9.95	11.75		21.70	28.50
0822	3" diameter		43	.372		12.05	13.70		25.75	34
0823	4" diameter	↓	36	.444	↓	20	16.35		36.35	46.50
1220	To delete coupling & hanger, subtract									
1230	1/4" diam. to 3/4" diam.					31%	56%			
1240	1" diam. to 1-1/2" diam.					23%	51%			
1250	2" diam. to 4" diam.					23%	41%			
1260	5" diam. to 12" diam.					21%	45%			
1290	Galvanized, 1/4" diameter	↓ 1 Plum	66	.121	L.F.	2.36	4.95		7.31	10.05

15107 | Metal Pipe & Fittings

		CREW	DAILY OUTPUT	LABOR-HOURS	UNIT	2005 BARE COSTS				TOTAL INCL O&P
						MAT.	LABOR	EQUIP.	TOTAL	
620	1300	3/8" diameter	1 Plum	65	.123	L.F.	2.69	5.05		7.74	10.50	620
1310	1/2" diameter		63	.127		1.52	5.20		6.72	9.45
1320	3/4" diameter		61	.131		1.76	5.35		7.11	10
1330	1" diameter		53	.151		2.46	6.15		8.61	11.95
1340	1-1/4" diameter	Q-1	89	.180		3.20	6.60		9.80	13.45
1350	1-1/2" diameter		80	.200		3.71	7.35		11.06	15.15
1360	2" diameter		64	.250		4.85	9.20		14.05	19.15
1370	2-1/2" diameter		50	.320		7.90	11.75		19.65	26.50
1380	3" diameter		43	.372		10.20	13.70		23.90	32
1390	3-1/2" diameter		40	.400		16.50	14.70		31.20	40
1400	4" diameter		36	.444		16.65	16.35		33	43
1410	5" diameter		26	.615		27.50	22.50		50	64.50
1420	6" diameter	Q-2	31	.774		38	29.50		67.50	86
1430	8" diameter		27	.889		57	34		91	114
1440	10" diameter		23	1.043		84	40		124	153
1450	12" diameter		18	1.333		115	51		166	204
1750	To delete coupling & hanger, subtract									
1760	1/4" diam. to 3/4" diam.					31%	56%			
1770	1" diam. to 1-1/2" diam.					23%	51%			
1780	2" diam. to 4" diam.					23%	41%			
1790	5" diam. to 12" diam.					21%	45%			
1900	Pipe nipple std blk 2"lg 1/2"	1 Plum	19	.421	Ea.	.68	17.20		17.88	27
1910	Pipe nipple std blk 2"lg 1" diameter	"	15	.533		1.26	22		23.26	34.50
1920	3"lg 2-1/2"	Q-1	16	1		8.35	37		45.35	64.50
2000	Welded, sch. 40, on yoke & roll hangers, sized for covering,									
2010	10' O.C. (no hangers incl. for 14" diam. and up)									
2040	Black, 1" diameter	Q-15	93	.172	L.F.	2.48	6.35	.87	9.70	13.20
2050	1-1/4" diameter		84	.190		3.21	7	.96	11.17	15.15
2060	1-1/2" diameter		76	.211		2.98	7.75	1.06	11.79	16.10
2070	2" diameter		61	.262		4.07	9.65	1.32	15.04	20.50
2080	2-1/2" diameter		47	.340		5.35	12.50	1.72	19.57	26.50
2090	3" diameter		43	.372		6.45	13.70	1.88	22.03	29.50
2100	3-1/2" diameter		39	.410		10.70	15.10	2.07	27.87	36.50
2110	4" diameter		37	.432		12	15.90	2.18	30.08	39.50
2120	5" diameter		32	.500		21	18.40	2.52	41.92	53.50
2130	6" diameter	Q-16	36	.667		26.50	25.50	2.24	54.24	70
2140	8" diameter		29	.828		39	31.50	2.78	73.28	93.50
2150	10" diameter		24	1		64	38	3.36	105.36	132
2160	12" diameter		19	1.263		83.50	48	4.24	135.74	169
2170	14" diameter		15	1.600		36	61	5.40	102.40	137
2180	16" diameter		13	1.846		41.50	70.50	6.20	118.20	158
2190	18" diameter		11	2.182		43.50	83	7.35	133.85	181
2200	20" diameter		9	2.667		61	102	8.95	171.95	230
2220	24" diameter		8	3		63	114	10.10	187.10	252
2230	26" diameter		7.50	3.200		145	122	10.75	277.75	355
2240	30" diameter		6	4		158	153	13.45	324.45	420
2260	36" diameter		4.50	5.333		181	203	17.90	401.90	525
2560	To delete hanger, subtract									
2570	1" diam. to 1-1/2" diam.					15%	34%			
2580	2" diam. to 3-1/2" diam.					9%	21%			
2590	4" diam. to 12" diam.					5%	12%			
3250	Flanged, 150 lb. weld neck, on yoke & roll hangers									
3260	sized for covering, 10' O.C.									
3290	Black, 1" diameter	Q-15	70	.229	L.F.	5.80	8.40	1.15	15.35	20.50
3300	1-1/4" diameter		64	.250		6.50	9.20	1.26	16.96	22.50
3310	1-1/2" diameter		58	.276		6.30	10.15	1.39	17.84	23.50

R15100 -070

Important: See the Reference Section for critical supporting data - Reference Nos., Crews, & City Cost Indexes

15107 | Metal Pipe & Fittings

		CREW	DAILY OUTPUT	LABOR-HOURS	UNIT	2005 BARE COSTS				TOTAL INCL O&P
						MAT.	LABOR	EQUIP.	TOTAL	
620 3320	2" diameter R15100 -070	Q-15	45	.356	L.F.	7.95	13.10	1.79	22.84	30.50 **62**
3330	2-1/2" diameter		36	.444		9.95	16.35	2.24	28.54	38
3340	3" diameter		32	.500		11.80	18.40	2.52	32.72	43.50
3350	3-1/2" diameter		29	.552		18.50	20.50	2.78	41.78	54
3360	4" diameter		26	.615		19.70	22.50	3.10	45.30	59
3370	5" diameter		21	.762		30.50	28	3.84	62.34	79.50
3380	6" diameter	Q-16	25	.960		38	36.50	3.23	77.73	101
3390	8" diameter		19	1.263		57	48	4.24	109.24	140
3400	10" diameter		16	1.500		86.50	57	5.05	148.55	187
3410	12" diameter		14	1.714		127	65.50	5.75	198.25	244
3470	For 300 lb. flanges, add					63%				
3480	For 600 lb. flanges, add					310%				
3960	To delete flanges & hanger, subtract									
3970	1" diam. to 2" diam.					76%	65%			
3980	2-1/2" diam. to 4" diam.					62%	59%			
3990	5" diam. to 12" diam.					60%	46%			
4750	Schedule 80, threaded, with couplings, and clevis type hangers									
4760	sized for covering, 10' O.C.									
4790	Black, 1/4" diameter	1 Plum	54	.148	L.F.	2.59	6.05		8.64	11.95
4800	3/8" diameter		53	.151		2.38	6.15		8.53	11.85
4810	1/2" diameter		52	.154		2.84	6.30		9.14	12.55
4820	3/4" diameter		50	.160		3.75	6.55		10.30	13.90
4830	1" diameter		45	.178		3.83	7.25		11.08	15.10
4840	1-1/4" diameter	Q-1	75	.213		5.25	7.85		13.10	17.55
4850	1-1/2" diameter		69	.232		6.30	8.55		14.85	19.70
4860	2" diameter		56	.286		7.55	10.50		18.05	24
4870	2-1/2" diameter		44	.364		9.85	13.35		23.20	31
4880	3" diameter		38	.421		12.85	15.50		28.35	37.50
4890	3-1/2" diameter		35	.457		18.45	16.80		35.25	46
4900	4" diameter		32	.500		25	18.40		43.40	55
4910	5" diameter		23	.696		28	25.50		53.50	69.50
4920	6" diameter	Q-2	28	.857		48.50	32.50		81	103
4930	8" diameter		23	1.043		92.50	40		132.50	162
4940	10" diameter		20	1.200		119	46		165	200
4950	12" diameter		15	1.600		160	61		221	268
5430	To delete coupling & hanger, subtract									
5440	1/4" diam. to 1/2" diam.					31%	54%			
5450	3/4" diam. to 1-1/2" diam.					28%	49%			
5460	2" diam. to 4" diam.					21%	40%			
5470	5" diam. to 12" diam.					17%	43%			
5510	Galvanized, 1/4" diameter	1 Plum	54	.148	L.F.	4.81	6.05		10.86	14.40
5520	3/8" diameter		53	.151		5.50	6.15		11.65	15.30
5530	1/2" diameter		52	.154		4.90	6.30		11.20	14.85
5540	3/4" diameter		50	.160		4.95	6.55		11.50	15.25
5550	1" diameter		45	.178		6.85	7.25		14.10	18.45
5560	1-1/4" diameter	Q-1	75	.213		8.20	7.85		16.05	21
5570	1-1/2" diameter		69	.232		12.15	8.55		20.70	26
5580	2" diameter		56	.286		13.35	10.50		23.85	30.50
5590	2-1/2" diameter		44	.364		15.45	13.35		28.80	37
5600	3" diameter		38	.421		19.50	15.50		35	45
5610	3-1/2" diameter		35	.457		21.50	16.80		38.30	49
5620	4" diameter		32	.500		30.50	18.40		48.90	61
5630	5" diameter		23	.696		38	25.50		63.50	80.50
5640	6" diameter	Q-2	28	.857		60	32.50		92.50	115
5650	8" diameter		23	1.043		100	40		140	170
5660	10" diameter		20	1.200		117	46		163	198

15107 | Metal Pipe & Fittings

		CREW	DAILY OUTPUT	LABOR-HOURS	UNIT	2005 BARE COSTS				TOTAL INCL O&P
						MAT.	LABOR	EQUIP.	TOTAL	
5670	12" diameter	Q-2	15	1.600	L.F.	157	61		218	265
5930	To delete coupling & hanger, subtract									
5940	1/4" diam. to 1/2" diam.					31%	54%			
5950	3/4" diam. to 1-1/2" diam.					28%	49%			
5960	2" diam. to 4" diam.					21%	40%			
5970	5" diam. to 12" diam.					17%	43%			
6000	Welded, on yoke & roller hangers									
6010	sized for covering, 10' O.C.									
6040	Black, 1" diameter	Q-15	85	.188	L.F.	4.56	6.90	.95	12.41	16.45
6050	1-1/4" diameter		79	.203		6.15	7.45	1.02	14.62	19.05
6060	1-1/2" diameter		72	.222		7.20	8.15	1.12	16.47	21.50
6070	2" diameter		57	.281		8.15	10.30	1.41	19.86	26
6080	2-1/2" diameter		44	.364		10.30	13.35	1.83	25.48	33.50
6090	3" diameter		40	.400		13.05	14.70	2.02	29.77	38.50
6100	3-1/2" diameter		34	.471		18.55	17.30	2.37	38.22	49
6110	4" diameter		33	.485		24.50	17.85	2.44	44.79	56.50
6120	5" diameter		26	.615		28	22.50	3.10	53.60	68
6130	6" diameter	Q-16	30	.800		46.50	30.50	2.69	79.69	100
6140	8" diameter		25	.960		88.50	36.50	3.23	128.23	156
6150	10" diameter		20	1.200		119	46	4.03	169.03	203
6160	12" diameter		15	1.600		160	61	5.40	226.40	273
6540	To delete hanger, subtract									
6550	1" diam. to 1-1/2" diam.					30%	14%			
6560	2" diam. to 3" diam.					23%	9%			
6570	3-1/2" diam. to 5" diam.					12%	6%			
6580	6" diam. to 12" diam.					10%	4%			
7250	Flanged, 300 lb. weld neck, on yoke & roll hangers									
7260	sized for covering, 10' O.C.									
7290	Black, 1" diameter	Q-15	66	.242	L.F.	8.60	8.90	1.22	18.72	24
7300	1-1/4" diameter		61	.262		10.20	9.65	1.32	21.17	27
7310	1-1/2" diameter		54	.296		11.20	10.90	1.49	23.59	30.50
7320	2" diameter		42	.381		12.70	14	1.92	28.62	37
7330	2-1/2" diameter		33	.485		16.20	17.85	2.44	36.49	47.50
7340	3" diameter		29	.552		20.50	20.50	2.78	43.78	56
7350	3-1/2" diameter		24	.667		30	24.50	3.36	57.86	73.50
7360	4" diameter		23	.696		36.50	25.50	3.51	65.51	82.50
7370	5" diameter		19	.842		42	31	4.24	77.24	97
7380	6" diameter	Q-16	23	1.043		66.50	40	3.51	110.01	137
7390	8" diameter		17	1.412		118	54	4.74	176.74	215
7400	10" diameter		14	1.714		164	65.50	5.75	235.25	286
7410	12" diameter		12	2		223	76.50	6.70	306.20	365
7470	For 600 lb. flanges, add					100%				
7940	To delete flanges & hanger, subtract									
7950	1" diam. to 1-1/2" diam.					75%	66%			
7960	2" diam. to 3" diam.					62%	60%			
7970	3-1/2" diam. to 5" diam.					54%	66%			
7980	6" diam. to 12" diam.					55%	62%			
9000	Threading pipe labor, one end, all schedules through 80									
9010	1/4" through 3/4" pipe size	1 Plum	80	.100	Ea.		4.09		4.09	6.15
9020	1" through 2" pipe size		73	.110			4.48		4.48	6.75
9030	2-1/2" pipe size		53	.151			6.15		6.15	9.25
9040	3" pipe size		50	.160			6.55		6.55	9.80
9050	3-1/2" pipe size	Q-1	89	.180			6.60		6.60	9.95
9060	4" pipe size		73	.219			8.05		8.05	12.10
9070	5" pipe size		53	.302			11.10		11.10	16.70
9080	6" pipe size		46	.348			12.80		12.80	19.25

R15100 -070

620

Important: See the Reference Section for critical supporting data - Reference Nos., Crews, & City Cost Indexes

15107 | Metal Pipe & Fittings

		CREW	DAILY OUTPUT	LABOR-HOURS	UNIT	2005 BARE COSTS				TOTAL INCL O&P	
						MAT.	LABOR	EQUIP.	TOTAL		
620	9090	8" pipe size	Q-1	29	.552	Ea.		20.50		20.50	30.50	**620**
	9100	10" pipe size		21	.762			28		28	42	
	9110	12" pipe size	↓	13	1.231	↓		45.50		45.50	68	
	9200	Welding labor per joint										
	9210	Schedule 40,										
	9230	1/2" pipe size	Q-15	32	.500	Ea.		18.40	2.52	20.92	30.50	
	9240	3/4" pipe size		27	.593			22	2.99	24.99	36.50	
	9250	1" pipe size		23	.696			25.50	3.51	29.01	42.50	
	9260	1-1/4" pipe size		20	.800			29.50	4.03	33.53	48.50	
	9270	1-1/2" pipe size		19	.842			31	4.24	35.24	51	
	9280	2" pipe size		16	1			37	5.05	42.05	61	
	9290	2-1/2" pipe size		13	1.231			45.50	6.20	51.70	75	
	9300	3" pipe size		12	1.333			49	6.70	55.70	81	
	9310	4" pipe size		10	1.600			59	8.05	67.05	97.50	
	9320	5" pipe size		9	1.778			65.50	8.95	74.45	108	
	9330	6" pipe size		8	2			73.50	10.10	83.60	122	
	9340	8" pipe size		5	3.200			118	16.15	134.15	195	
	9350	10" pipe size		4	4			147	20	167	243	
	9360	12" pipe size		3	5.333			196	27	223	325	
	9370	14" pipe size		2.60	6.154			226	31	257	375	
	9380	16" pipe size		2.20	7.273			267	36.50	303.50	440	
	9390	18" pipe size		2	8			294	40.50	334.50	485	
	9400	20" pipe size		1.80	8.889			325	45	370	540	
	9410	22" pipe size		1.70	9.412			345	47.50	392.50	570	
	9420	24" pipe size	↓	1.50	10.667	↓		390	54	444	650	
	9450	Schedule 80,										
	9460	1/2" pipe size	Q-15	27	.593	Ea.		22	2.99	24.99	36.50	
	9470	3/4" pipe size		23	.696			25.50	3.51	29.01	42.50	
	9480	1" pipe size		20	.800			29.50	4.03	33.53	48.50	
	9490	1-1/4" pipe size		19	.842			31	4.24	35.24	51	
	9500	1-1/2" pipe size		18	.889			32.50	4.48	36.98	54	
	9510	2" pipe size		15	1.067			39	5.40	44.40	65	
	9520	2-1/2" pipe size		12	1.333			49	6.70	55.70	81	
	9530	3" pipe size		11	1.455			53.50	7.35	60.85	88.50	
	9540	4" pipe size		8	2			73.50	10.10	83.60	122	
	9550	5" pipe size		6	2.667			98	13.45	111.45	162	
	9560	6" pipe size		5	3.200			118	16.15	134.15	195	
	9570	8" pipe size		4	4			147	20	167	243	
	9580	10" pipe size		3	5.333			196	27	223	325	
	9590	12" pipe size	↓	2	8			294	40.50	334.50	485	
	9600	14" pipe size	Q-16	2.60	9.231			350	31	381	565	
	9610	16" pipe size		2.30	10.435			400	35	435	640	
	9620	18" pipe size		2	12			460	40.50	500.50	735	
	9630	20" pipe size		1.80	13.333			510	45	555	815	
	9640	22" pipe size		1.60	15			570	50.50	620.50	915	
	9650	24" pipe size	↓	1.50	16	↓		610	54	664	975	
640	0010	**PIPE, STEEL, FITTINGS** Threaded										**64**
	0020	Cast Iron										
	0040	Standard weight, black										
	0060	90° Elbow, straight										
	0070	1/4"	1 Plum	16	.500	Ea.	4.01	20.50		24.51	35	
	0080	3/8"		16	.500		5.80	20.50		26.30	37	
	0090	1/2"		15	.533		2.54	22		24.54	36	
	0100	3/4"		14	.571		2.65	23.50		26.15	38	
	0110	1"	↓	13	.615		3.14	25		28.14	41.50	
	0120	1-1/4"	Q-1	22	.727	↓	4.46	27		31.46	45	

R15100-070

15107	Metal Pipe & Fittings	CREW	DAILY OUTPUT	LABOR-HOURS	UNIT	MAT.	LABOR	EQUIP.	TOTAL	TOTAL INCL O&P
						2005 BARE COSTS				
0130	1-1/2"	Q-1	20	.800	Ea.	6.15	29.50		35.65	51
0140	2"		18	.889		9.65	32.50		42.15	59.50
0150	2-1/2"		14	1.143		23	42		65	88.50
0160	3"		10	1.600		38	59		97	130
0170	3-1/2"		8	2		103	73.50		176.50	224
0180	4"		6	2.667		70	98		168	224
0190	5"		5	3.200		149	118		267	340
0200	6"	Q-2	7	3.429		193	131		324	410
0210	8"	"	6	4		475	153		628	750
0250	45° Elbow, straight									
0260	1/4"	1 Plum	16	.500	Ea.	4.81	20.50		25.31	36
0270	3/8"		16	.500		5.15	20.50		25.65	36
0280	1/2"		15	.533		3.89	22		25.89	37.50
0300	3/4"		14	.571		3.92	23.50		27.42	39.50
0320	1"		13	.615		4.59	25		29.59	43
0330	1-1/4"	Q-1	22	.727		6.20	27		33.20	47
0340	1-1/2"		20	.800		10.25	29.50		39.75	55.50
0350	2"		18	.889		11.80	32.50		44.30	62
0360	2-1/2"		14	1.143		30.50	42		72.50	97
0370	3"		10	1.600		48.50	59		107.50	142
0380	3-1/2"		8	2		108	73.50		181.50	230
0400	4"		6	2.667		101	98		199	258
0420	5"		5	3.200		222	118		340	420
0440	6"	Q-2	7	3.429		285	131		416	510
0460	8"	"	6	4		570	153		723	855
0500	Tee, straight									
0510	1/4"	1 Plum	10	.800	Ea.	6.30	32.50		38.80	56
0520	3/8"		10	.800		6.10	32.50		38.60	55.50
0530	1/2"		9	.889		3.97	36.50		40.47	59
0540	3/4"		9	.889		4.63	36.50		41.13	59.50
0550	1"		8	1		4.11	41		45.11	66
0560	1-1/4"	Q-1	14	1.143		7.50	42		49.50	71.50
0570	1-1/2"		13	1.231		9.75	45.50		55.25	79
0580	2"		11	1.455		13.60	53.50		67.10	95.50
0590	2-1/2"		9	1.778		35.50	65.50		101	138
0600	3"		6	2.667		54	98		152	207
0610	3-1/2"		5	3.200		109	118		227	297
0620	4"		4	4		106	147		253	335
0630	5"		3	5.333		281	196		477	605
0640	6"	Q-2	4	6		320	229		549	695
0650	8"	"	3	8		720	305		1,025	1,250
0660	Tee, reducing, run or outlet									
0661	1/2"	1 Plum	9	.889	Ea.	8.25	36.50		44.75	63.50
0662	3/4"		9	.889		7.45	36.50		43.95	62.50
0663	1"		8	1		8.15	41		49.15	70.50
0664	1-1/4"	Q-1	14	1.143		15.70	42		57.70	80.50
0665	1-1/2"		13	1.231		17.45	45.50		62.95	87
0666	2"		11	1.455		25	53.50		78.50	108
0667	2-1/2"		9	1.778		46.50	65.50		112	150
0668	3"		6	2.667		70	98		168	224
0669	3-1/2"		5	3.200		126	118		244	315
0670	4"		4	4		143	147		290	380
0671	5"		3	5.333		249	196		445	570
0672	6"	Q-2	4	6		320	229		549	695
0673	8"	"	3	8		420	305		725	925
0674	Reducer, concentric									

Important: See the Reference Section for critical supporting data - Reference Nos., Crews, & City Cost Indexes

15107 | Metal Pipe & Fittings

		CREW	DAILY OUTPUT	LABOR-HOURS	UNIT	2005 BARE COSTS				TOTAL INCL O&P
						MAT.	LABOR	EQUIP.	TOTAL	
640 0675	3/4"	1 Plum	18	.444	Ea.	7.10	18.15		25.25	35.50 **640**
0676	1"	"	15	.533		4.81	22		26.81	38.50
0677	1-1/4"	Q-1	26	.615		13.95	22.50		36.45	49.50
0678	1-1/2"		24	.667		22	24.50		46.50	61
0679	2"		21	.762		25.50	28		53.50	70.50
0680	2-1/2"		18	.889		36.50	32.50		69	89
0681	3"		14	1.143		55.50	42		97.50	124
0682	3-1/2"		12	1.333		96.50	49		145.50	180
0683	4"		10	1.600		103	59		162	203
0684	5"		6	2.667		143	98		241	305
0685	6"	Q-2	8	3		188	114		302	380
0686	8"	"	7	3.429		820	131		951	1,100
0687	Reducer, eccentric									
0688	3/4"	1 Plum	16	.500	Ea.	14.95	20.50		35.45	47
0689	1"	"	14	.571		15.70	23.50		39.20	52.50
0690	1-1/4"	Q-1	25	.640		26	23.50		49.50	64
0691	1-1/2"		22	.727		34	27		61	77.50
0692	2"		20	.800		48	29.50		77.50	96.50
0693	2-1/2"		16	1		66	37		103	128
0694	3"		12	1.333		104	49		153	188
0695	3-1/2"		10	1.600		144	59		203	248
0696	4"		9	1.778		188	65.50		253.50	305
0697	5"		5	3.200		229	118		347	430
0698	6"	Q-2	8	3		291	114		405	490
0699	8"	"	7	3.429		1,125	131		1,256	1,450
0700	Standard weight, galvanized cast iron									
0720	90° Elbow, straight									
0730	1/4"	1 Plum	16	.500	Ea.	9.35	20.50		29.85	41
0740	3/8"		16	.500		9.35	20.50		29.85	41
0750	1/2"		15	.533		7.55	22		29.55	41.50
0760	3/4"		14	.571		10.50	23.50		34	46.50
0770	1"		13	.615		12.10	25		37.10	51.50
0780	1-1/4"	Q-1	22	.727		18.85	27		45.85	60.50
0790	1-1/2"		20	.800		26	29.50		55.50	72.50
0800	2"		18	.889		38	32.50		70.50	91
0810	2-1/2"		14	1.143		78.50	42		120.50	149
0820	3"		10	1.600		119	59		178	220
0830	3-1/2"		8	2		190	73.50		263.50	320
0840	4"		6	2.667		218	98		316	385
0850	5"		5	3.200		440	118		558	660
0860	6"	Q-2	7	3.429		575	131		706	825
0870	8"	"	6	4		1,400	153		1,553	1,775
0900	45° Elbow, straight									
0910	1/4"	1 Plum	16	.500	Ea.	14.25	20.50		34.75	46
0920	3/8"		16	.500		15.30	20.50		35.80	47.50
0930	1/2"		15	.533		10.65	22		32.65	45
0940	3/4"		14	.571		12.25	23.50		35.75	48.50
0950	1"		13	.615		13.60	25		38.60	53
0960	1-1/4"	Q-1	22	.727		18.30	27		45.30	60
0970	1-1/2"		20	.800		31.50	29.50		61	79
0980	2"		18	.889		44.50	32.50		77	98
0990	2-1/2"		14	1.143		87.50	42		129.50	159
1000	3"		10	1.600		139	59		198	242
1010	3-1/2"		8	2		257	73.50		330.50	395
1020	4"		6	2.667		257	98		355	430
1030	5"		5	3.200		655	118		773	900

15107	Metal Pipe & Fittings	CREW	DAILY OUTPUT	LABOR-HOURS	UNIT	2005 BARE COSTS				TOTAL INCL O&P
						MAT.	LABOR	EQUIP.	TOTAL	
640	1040	6"	Q-2	7	3.429	Ea.	845	131		976	1,125	640
1050		8"	"	6	4	↓	1,675	153		1,828	2,075
1100		Tee, straight									
1110		1/4"	1 Plum	10	.800	Ea.	18.60	32.50		51.10	69.50
1120		3/8"		10	.800		11.95	32.50		44.45	62
1130		1/2"		9	.889		11.75	36.50		48.25	67.50
1140		3/4"		9	.889		15.05	36.50		51.55	71
1150		1"	↓	8	1		16.35	41		57.35	79.50
1160		1-1/4"	Q-1	14	1.143		28.50	42		70.50	94.50
1170		1-1/2"		13	1.231		38	45.50		83.50	110
1180		2"		11	1.455		47	53.50		100.50	133
1190		2-1/2"		9	1.778		102	65.50		167.50	211
1200		3"		6	2.667		263	98		361	435
1210		3-1/2"		5	3.200		270	118		388	475
1220		4"		4	4		305	147		452	555
1230		5"	↓	3	5.333		835	196		1,031	1,200
1240		6"	Q-2	4	6		945	229		1,174	1,400
1250		8"	"	3	8	↓	2,150	305		2,455	2,800
1300		Extra heavy weight, black									
1310		Couplings, steel straight									
1320		1/4"	1 Plum	19	.421	Ea.	1.76	17.20		18.96	28
1330		3/8"		19	.421		1.92	17.20		19.12	28
1340		1/2"		19	.421		2.59	17.20		19.79	29
1350		3/4"		18	.444		2.77	18.15		20.92	30.50
1360		1"	↓	15	.533		3.53	22		25.53	37
1370		1-1/4"	Q-1	26	.615		5.65	22.50		28.15	40.50
1380		1-1/2"		24	.667		5.65	24.50		30.15	43.50
1390		2"		21	.762		8.65	28		36.65	51.50
1400		2-1/2"		18	.889		12.80	32.50		45.30	63
1410		3"		14	1.143		15.20	42		57.20	80
1420		3-1/2"		12	1.333		20.50	49		69.50	96
1430		4"		10	1.600		24	59		83	115
1440		5"	↓	6	2.667		34.50	98		132.50	185
1450		6"	Q-2	8	3		42	114		156	219
1460		8"		7	3.429		76.50	131		207.50	281
1470		10"		6	4		120	153		273	360
1480		12"	↓	4	6	↓	161	229		390	520
1510		90° Elbow, straight									
1520		1/2"	1 Plum	15	.533	Ea.	12.60	22		34.60	47
1530		3/4"		14	.571		13.35	23.50		36.85	49.50
1540		1"	↓	13	.615		16.15	25		41.15	56
1550		1-1/4"	Q-1	22	.727		24	27		51	66.50
1560		1-1/2"		20	.800		30	29.50		59.50	77
1580		2"		18	.889		37	32.50		69.50	89.50
1590		2-1/2"		14	1.143		86	42		128	158
1600		3"		10	1.600		114	59		173	214
1610		4"	↓	6	2.667		238	98		336	410
1620		6"	Q-2	7	3.429	↓	620	131		751	875
1650		45° Elbow, straight									
1660		1/2"	1 Plum	15	.533	Ea.	18	22		40	53
1670		3/4"		14	.571		17.35	23.50		40.85	54
1680		1"		13	.615		21	25		46	61
1690		1-1/4"	Q-1	22	.727		34.50	27		61.50	78
1700		1-1/2"		20	.800		38	29.50		67.50	85.50
1710		2"		18	.889		54	32.50		86.50	108
1720		2-1/2"	↓	14	1.143	↓	93	42		135	165

Important: See the Reference Section for critical supporting data - Reference Nos., Crews, & City Cost Indexes

			DAILY	LABOR-		2005 BARE COSTS				TOTAL
15107	**Metal Pipe & Fittings**		CREW	OUTPUT	HOURS	UNIT	MAT.	LABOR	EQUIP.	TOTAL	INCL 0&P
640 1800	Tee, straight										
1810	1/2"		1 Plum	9	.889	Ea.	19.90	36.50		56.40	76.50
1820	3/4"			9	.889		19.90	36.50		56.40	76.50
1830	1"			8	1		24	41		65	88
1840	1-1/4"		Q-1	14	1.143		36	42		78	103
1850	1-1/2"			13	1.231		46	45.50		91.50	119
1860	2"			11	1.455		57	53.50		110.50	144
1870	2-1/2"			9	1.778		122	65.50		187.50	233
1880	3"			6	2.667		166	98		264	330
1890	4"			4	4		325	147		472	575
1900	6"		Q-2	4	6		715	229		944	1,125
4000	Standard weight, black										
4010	Couplings, steel straight, merchants										
4030	1/4"		1 Plum	19	.421	Ea.	.47	17.20		17.67	26.50
4040	3/8"			19	.421		.57	17.20		17.77	26.50
4050	1/2"			19	.421		.61	17.20		17.81	26.50
4060	3/4"			18	.444		.77	18.15		18.92	28.50
4070	1"			15	.533		1.08	22		23.08	34
4080	1-1/4"		Q-1	26	.615		1.38	22.50		23.88	35.50
4090	1-1/2"			24	.667		1.75	24.50		26.25	39
4100	2"			21	.762		2.50	28		30.50	45
4110	2-1/2"			18	.889		7.10	32.50		39.60	57
4120	3"			14	1.143		10	42		52	74
4130	3-1/2"			12	1.333		17.70	49		66.70	93
4140	4"			10	1.600		17.70	59		76.70	108
4150	5"			6	2.667		32.50	98		130.50	183
4160	6"		Q-2	8	3		39	114		153	215
4166	Plug, 1/4"		1 Plum	38	.211		1.31	8.60		9.91	14.40
4167	3/8"			38	.211		1.31	8.60		9.91	14.40
4168	1/2"			38	.211		1.13	8.60		9.73	14.20
4169	3/4"			32	.250		3.42	10.20		13.62	19.10
4170	1"			30	.267		3.65	10.90		14.55	20.50
4171	1-1/4"		Q-1	52	.308		4.19	11.30		15.49	21.50
4172	1-1/2"			48	.333		5.95	12.25		18.20	25
4173	2"			42	.381		7.65	14		21.65	29.50
4176	2-1/2"			36	.444		11.35	16.35		27.70	37
4200	Standard weight, galvanized										
4210	Couplings, steel straight, merchants										
4230	1/4"		1 Plum	19	.421	Ea.	.55	17.20		17.75	26.50
4240	3/8"			19	.421		.70	17.20		17.90	27
4250	1/2"			19	.421		.74	17.20		17.94	27
4260	3/4"			18	.444		.93	18.15		19.08	28.50
4270	1"			15	.533		1.30	22		23.30	34.50
4280	1-1/4"		Q-1	26	.615		1.66	22.50		24.16	36
4290	1-1/2"			24	.667		2.07	24.50		26.57	39.50
4300	2"			21	.762		3.08	28		31.08	45.50
4310	2-1/2"			18	.889		8.75	32.50		41.25	58.50
4320	3"			14	1.143		11.60	42		53.60	76
4330	3-1/2"			12	1.333		20.50	49		69.50	96
4340	4"			10	1.600		20.50	59		79.50	111
4350	5"			6	2.667		38	98		136	189
4360	6"		Q-2	8	3		47	114		161	224
4370	Plug, galvanized, square head										
4374	1/2"		1 Plum	38	.211	Ea.	2.21	8.60		10.81	15.40
4375	3/4"			32	.250		2.40	10.20		12.60	18
4376	1"			30	.267		3.99	10.90		14.89	20.50

15107 | Metal Pipe & Fittings

		CREW	DAILY OUTPUT	LABOR-HOURS	UNIT	MAT.	LABOR	EQUIP.	TOTAL	TOTAL INCL O&P	
										2005 BARE COSTS	
640 4377	1-1/4"	Q-1	52	.308	Ea.	4.31	11.30		15.61	21.50	**640**
4378	1-1/2"		48	.333		5.80	12.25		18.05	25	
4379	2"		42	.381		7.20	14		21.20	29	
4380	2-1/2"		36	.444		15.30	16.35		31.65	41.50	
4381	3"		28	.571		19.55	21		40.55	53	
4382	4"		20	.800		52	29.50		81.50	102	
4384	6"		8	2		86	73.50		159.50	206	
5000	Malleable iron, 150 lb.										
5020	Black										
5040	90° elbow, straight										
5060	1/4"	1 Plum	16	.500	Ea.	1.99	20.50		22.49	32.50	
5070	3/8"		16	.500		1.99	20.50		22.49	32.50	
5080	1/2"		15	.533		1.38	22		23.38	34.50	
5090	3/4"		14	.571		1.67	23.50		25.17	37	
5100	1"		13	.615		2.90	25		27.90	41	
5110	1-1/4"	Q-1	22	.727		4.77	27		31.77	45.50	
5120	1-1/2"		20	.800		6.25	29.50		35.75	51	
5130	2"		18	.889		10.80	32.50		43.30	61	
5140	2-1/2"		14	1.143		24	42		66	89.50	
5150	3"		10	1.600		35.50	59		94.50	128	
5160	3-1/2"		8	2		94	73.50		167.50	215	
5170	4"		6	2.667		75.50	98		173.50	231	
5180	5"		5	3.200		215	118		333	415	
5190	6"	Q-2	7	3.429		255	131		386	480	
5250	45° elbow, straight										
5270	1/4"	1 Plum	16	.500	Ea.	3	20.50		23.50	34	
5280	3/8"		16	.500		3	20.50		23.50	34	
5290	1/2"		15	.533		2.27	22		24.27	35.50	
5300	3/4"		14	.571		2.81	23.50		26.31	38	
5310	1"		13	.615		3.55	25		28.55	42	
5320	1-1/4"	Q-1	22	.727		6.30	27		33.30	47	
5330	1-1/2"		20	.800		7.75	29.50		37.25	52.50	
5340	2"		18	.889		11.75	32.50		44.25	62	
5350	2-1/2"		14	1.143		34	42		76	101	
5360	3"		10	1.600		44	59		103	137	
5370	3-1/2"		8	2		86.50	73.50		160	206	
5380	4"		6	2.667		86.50	98		184.50	242	
5390	5"		5	3.200		340	118		458	550	
5400	6"	Q-2	7	3.429		340	131		471	570	
5450	Tee, straight										
5470	1/4"	1 Plum	10	.800	Ea.	3.36	32.50		35.86	52.50	
5480	3/8"		10	.800		3.36	32.50		35.86	52.50	
5490	1/2"		9	.889		1.84	36.50		38.34	56.50	
5500	3/4"		9	.889		2.65	36.50		39.15	57.50	
5510	1"		8	1		4.52	41		45.52	66.50	
5520	1-1/4"	Q-1	14	1.143		7.35	42		49.35	71	
5530	1-1/2"		13	1.231		9.15	45.50		54.65	78	
5540	2"		11	1.455		15.55	53.50		69.05	97.50	
5550	2-1/2"		9	1.778		33.50	65.50		99	136	
5560	3"		6	2.667		49.50	98		147.50	202	
5570	3-1/2"		5	3.200		119	118		237	310	
5580	4"		4	4		119	147		266	350	
5590	5"		3	5.333		340	196		536	670	
5600	6"	Q-2	4	6		380	229		609	760	
5601	Tee, reducing, on outlet										
5602	1/2"	1 Plum	9	.889	Ea.	4.03	36.50		40.53	59	

Important: See the Reference Section for critical supporting data - Reference Nos., Crews, & City Cost Indexes

	15107	Metal Pipe & Fittings	CREW	DAILY OUTPUT	LABOR-HOURS	UNIT	2005 BARE COSTS				TOTAL INCL O&P
							MAT.	LABOR	EQUIP.	TOTAL	
	5603	3/4"	1 Plum	9	.889	Ea.	4.09	36.50		40.59	59
	5604	1"	↓	8	1		5.15	41		46.15	67
	5605	1-1/4"	Q-1	14	1.143		8.95	42		50.95	73
	5606	1-1/2"		13	1.231		12.90	45.50		58.40	82
	5607	2"		11	1.455		16.60	53.50		70.10	99
	5608	2-1/2"		9	1.778		43.50	65.50		109	147
	5609	3"		6	2.667		69	98		167	223
	5610	3-1/2"		5	3.200		154	118		272	345
	5611	4"		4	4		154	147		301	390
	5612	5"	↓	3	5.333	↓	415	196		611	750
	5650	Coupling									
	5670	1/4"	1 Plum	19	.421	Ea.	2.46	17.20		19.66	28.50
	5680	3/8"		19	.421		2.46	17.20		19.66	28.50
	5690	1/2"		19	.421		1.90	17.20		19.10	28
	5700	3/4"		18	.444		2.23	18.15		20.38	30
	5710	1"	↓	15	.533		3.34	22		25.34	36.50
	5720	1-1/4"	Q-1	26	.615		4.33	22.50		26.83	39
	5730	1-1/2"		24	.667		5.85	24.50		30.35	43.50
	5740	2"		21	.762		8.65	28		36.65	51.50
	5750	2-1/2"		18	.889		24	32.50		56.50	75.50
	5760	3"		14	1.143		32.50	42		74.50	98.50
	5770	3-1/2"		12	1.333		65	49		114	145
	5780	4"		10	1.600		65	59		124	160
	5790	5"	↓	6	2.667		186	98		284	350
	5800	6"	Q-2	8	3		186	114		300	375
	5840	Reducer, concentric, 1/4"	1 Plum	19	.421		2.61	17.20		19.81	29
	5850	3/8"		19	.421		3.33	17.20		20.53	29.50
	5860	1/2"		19	.421		2.64	17.20		19.84	29
	5870	3/4"		16	.500		3.12	20.50		23.62	34
	5880	1"	↓	15	.533		5.25	22		27.25	39
	5890	1-1/4"	Q-1	26	.615		5.90	22.50		28.40	40.50
	5900	1-1/2"		24	.667		8.45	24.50		32.95	46.50
	5910	2"		21	.762		12.20	28		40.20	55.50
	5911	2-1/2"		18	.889		27	32.50		59.50	79
	5912	3"		14	1.143		33	42		75	99.50
	5913	3-1/2"		12	1.333		77	49		126	159
	5914	4"		10	1.600		77	59		136	174
	5915	5"	↓	6	2.667		145	98		243	305
	5916	6"	Q-2	8	3		186	114		300	375
	5981	Bushing, 1/4"	1 Plum	19	.421		2.52	17.20		19.72	29
	5982	3/8"		19	.421		2.52	17.20		19.72	29
	5983	1/2"		19	.421		2.52	17.20		19.72	29
	5984	3/4"		16	.500		2.48	20.50		22.98	33
	5985	1"	↓	15	.533		3.63	22		25.63	37
	5986	1-1/4"	Q-1	26	.615		4.51	22.50		27.01	39
	5987	1-1/2"		24	.667		5.20	24.50		29.70	43
	5988	2"	↓	21	.762		5.60	28		33.60	48
	5989	Cap, 1/4"	1 Plum	38	.211		1.82	8.60		10.42	14.95
	5991	3/8"		38	.211		1.54	8.60		10.14	14.65
	5992	1/2"		38	.211		1.39	8.60		9.99	14.50
	5993	3/4"		32	.250		1.88	10.20		12.08	17.40
	5994	1"	↓	30	.267		2.28	10.90		13.18	18.85
	5995	1-1/4"	Q-1	52	.308		3.01	11.30		14.31	20.50
	5996	1-1/2"		48	.333		4.14	12.25		16.39	23
	5997	2"	↓	42	.381		6.05	14		20.05	27.50
	6000	For galvanized elbows, tees, and couplings add				↓	20%				

15107	Metal Pipe & Fittings	CREW	DAILY OUTPUT	LABOR-HOURS	UNIT	2005 BARE COSTS				TOTAL INCL O&P
						MAT.	LABOR	EQUIP.	TOTAL	
6058	For galvanized reducers, caps and plugs add					20%				
6100	90° elbow, galvanized, 150 lb., reducing									
6110	3/4" x 1/2"	1 Plum	15.40	.519	Ea.	3.75	21		24.75	36
6112	1" x 3/4"		14	.571		4.81	23.50		28.31	40.50
6114	1" x 1/2"	↓	14.50	.552		5.10	22.50		27.60	39.50
6116	1-1/4" x 1"	Q-1	24.20	.661		8.05	24.50		32.55	45.50
6118	1-1/4" x 3/4"		25.40	.630		9.70	23		32.70	45.50
6120	1-1/4" x 1/2"		26.20	.611		7.60	22.50		30.10	42.50
6122	1-1/2" x 1-1/4"		21.60	.741		12.90	27		39.90	55
6124	1-1/2" x 1"		23.50	.681		12.90	25		37.90	51.50
6126	1-1/2" x 3/4"		24.60	.650		12.90	24		36.90	50
6128	2" x 1-1/2"		20.50	.780		15.05	28.50		43.55	59.50
6130	2" x 1-1/4"		21	.762		17.25	28		45.25	61
6132	2" x 1"		22.80	.702		17.90	26		43.90	58.50
6134	2" x 3/4"		23.90	.669		18.25	24.50		42.75	57
6136	2-1/2" x 2"		12.30	1.301		51	48		99	128
6138	2-1/2" x 1-1/2"		12.50	1.280		51	47		98	127
6140	3" x 2-1/2"		8.60	1.860		90	68.50		158.50	202
6142	3" x 2"		11.80	1.356		80.50	50		130.50	164
6144	4" x 3"	↓	8.20	1.951	↓	204	72		276	330
6160	90° elbow, black, 150 lb., reducing									
6170	1" x 3/4"	1 Plum	14	.571	Ea.	3.47	23.50		26.97	39
6174	1-1/2" x 1"	Q-1	23.50	.681		8.25	25		33.25	46.50
6178	1-1/2" x 3/4"		24.60	.650		9.40	24		33.40	46.50
6182	2" x 1-1/2"		20.50	.780		11.90	28.50		40.40	56
6186	2" x 1"		22.80	.702		13.55	26		39.55	54
6190	2" x 3/4"		23.90	.669		14.40	24.50		38.90	53
6194	2-1/2" x 2"	↓	12.30	1.301	↓	32.50	48		80.50	108
7000	Union, with brass seat									
7010	1/4"	1 Plum	15	.533	Ea.	10.70	22		32.70	45
7020	3/8"		15	.533		7.40	22		29.40	41
7030	1/2"		14	.571		6.70	23.50		30.20	42.50
7040	3/4"		13	.615		7	25		32	45.50
7050	1"	↓	12	.667		10	27		37	52
7060	1-1/4"	Q-1	21	.762		13.75	28		41.75	57
7070	1-1/2"		19	.842		17.90	31		48.90	66
7080	2"		17	.941		21	34.50		55.50	75
7090	2-1/2"		13	1.231		62	45.50		107.50	136
7100	3"	↓	9	1.778	↓	74.50	65.50		140	181
7120	Union, galvanized									
7124	1/2"	1 Plum	14	.571	Ea.	7.50	23.50		31	43.50
7125	3/4"		13	.615		8.80	25		33.80	47.50
7126	1"	↓	12	.667		11.50	27		38.50	53.50
7127	1-1/4"	Q-1	21	.762		16.65	28		44.65	60.50
7128	1-1/2"		19	.842		20	31		51	68.50
7129	2"		17	.941		23	34.50		57.50	77.50
7130	2-1/2"		13	1.231		80.50	45.50		126	157
7131	3"	↓	9	1.778	↓	113	65.50		178.50	223
7500	Malleable iron, 300 lb									
7520	Black									
7540	90° Elbow, straight, 1/4"	1 Plum	16	.500	Ea.	7.65	20.50		28.15	39
7560	3/8"		16	.500		6.80	20.50		27.30	38
7570	1/2"		15	.533		8.80	22		30.80	42.50
7580	3/4"		14	.571		9.95	23.50		33.45	46
7590	1"	↓	13	.615		12.75	25		37.75	52
7600	1-1/4"	Q-1	22	.727	↓	18.60	27		45.60	60.50

640

Important: See the Reference Section for critical supporting data - Reference Nos., Crews, & City Cost Indexes

						2005 BARE COSTS				TOTAL
15107	**Metal Pipe & Fittings**	CREW	DAILY OUTPUT	LABOR-HOURS	UNIT	MAT.	LABOR	EQUIP.	TOTAL	INCL O&P
640 7610	1-1/2"	Q-1	20	.800	Ea.	22	29.50		51.50	68
7620	2"		18	.889		32	32.50		64.50	84
7630	2-1/2"		14	1.143		81.50	42		123.50	153
7640	3"		10	1.600		94	59		153	192
7650	4"	↓	6	2.667		214	98		312	385
7700	45° Elbow, straight, 1/4"	1 Plum	16	.500		11.40	20.50		31.90	43
7720	3/8"		16	.500		11.40	20.50		31.90	43
7730	1/2"		15	.533		12.70	22		34.70	47
7740	3/4"		14	.571		14.10	23.50		37.60	50.50
7750	1"	↓	13	.615		15.60	25		40.60	55
7760	1-1/4"	Q-1	22	.727		25	27		52	67.50
7770	1-1/2"		20	.800		32.50	29.50		62	80
7780	2"		18	.889		49.50	32.50		82	103
7790	2-1/2"		14	1.143		107	42		149	181
7800	3"		10	1.600		111	59		170	211
7810	4"	↓	6	2.667		320	98		418	495
7850	Tee, straight, 1/4"	1 Plum	10	.800		9.65	32.50		42.15	59.50
7870	3/8"		10	.800		10.20	32.50		42.70	60.50
7880	1/2"		9	.889		13	36.50		49.50	69
7890	3/4"		9	.889		14.05	36.50		50.55	70
7900	1"	↓	8	1		16.90	41		57.90	80
7910	1-1/4"	Q-1	14	1.143		24	42		66	89.50
7920	1-1/2"		13	1.231		28	45.50		73.50	99
7930	2"		11	1.455		41.50	53.50		95	127
7940	2-1/2"		9	1.778		98.50	65.50		164	207
7950	3"		6	2.667		144	98		242	305
7960	4"	↓	4	4		400	147		547	660
7980	6"	Q-2	4	6		425	229		654	810
7990	8"	"	3	8		665	305		970	1,200
8050	Couplings, straight, 1/4"	1 Plum	19	.421		6.90	17.20		24.10	33.50
8070	3/8"		19	.421		6.90	17.20		24.10	33.50
8080	1/2"		19	.421		7.70	17.20		24.90	34.50
8090	3/4"		18	.444		8.85	18.15		27	37.50
8100	1"	↓	15	.533		10.15	22		32.15	44
8110	1-1/4"	Q-1	26	.615		12.05	22.50		34.55	47.50
8120	1-1/2"		24	.667		18.05	24.50		42.55	57
8130	2"		21	.762		26	28		54	70.50
8140	2-1/2"		18	.889		46.50	32.50		79	100
8150	3"		14	1.143		67	42		109	137
8160	4"	↓	10	1.600	↓	84	59		143	181
8200	Galvanized									
8220	90° Elbow, straight, 1/4"	1 Plum	16	.500	Ea.	12.95	20.50		33.45	45
8222	3/8"		16	.500		13.20	20.50		33.70	45
8224	1/2"		15	.533		15.75	22		37.75	50.50
8226	3/4"		14	.571		17.65	23.50		41.15	54.50
8228	1"	↓	13	.615		22.50	25		47.50	63
8230	1-1/4"	Q-1	22	.727		35.50	27		62.50	79
8232	1-1/2"		20	.800		38.50	29.50		68	86.50
8234	2"		18	.889		65	32.50		97.50	121
8236	2-1/2"		14	1.143		139	42		181	216
8238	3"		10	1.600		170	59		229	276
8240	4"	↓	6	2.667	↓	495	98		593	690
8280	45° Elbow, straight									
8282	1/2"	1 Plum	15	.533	Ea.	23.50	22		45.50	58.50
8284	3/4"		14	.571		27	23.50		50.50	64.50
8286	1"	↓	13	.615	↓	29.50	25		54.50	70.50

15107 | Metal Pipe & Fittings

		CREW	DAILY OUTPUT	LABOR-HOURS	UNIT	2005 BARE COSTS				TOTAL INCL O&P
						MAT.	LABOR	EQUIP.	TOTAL	
640 8288	1-1/4"	Q-1	22	.727	Ea.	46	27		73	90.50
8290	1-1/2"		20	.800		59.50	29.50		89	109
8292	2"	↓	18	.889		82	32.50		114.50	140
8310	Tee, straight, 1/4"	1 Plum	10	.800		18.45	32.50		50.95	69.50
8312	3/8"		10	.800		18.80	32.50		51.30	69.50
8314	1/2"		9	.889		23.50	36.50		60	80.50
8316	3/4"		9	.889		26	36.50		62.50	83
8318	1"	↓	8	1		32	41		73	96.50
8320	1-1/4"	Q-1	14	1.143		45.50	42		87.50	113
8322	1-1/2"		13	1.231		47.50	45.50		93	120
8324	2"		11	1.455		79.50	53.50		133	168
8326	2-1/2"		9	1.778		233	65.50		298.50	355
8328	3"		6	2.667		272	98		370	445
8330	4"	↓	4	4		680	147		827	965
8380	Couplings, straight, 1/4"	1 Plum	19	.421		6.90	17.20		24.10	33.50
8382	3/8"		19	.421		9.90	17.20		27.10	37
8384	1/2"		19	.421		10.20	17.20		27.40	37
8386	3/4"		18	.444		11.50	18.15		29.65	40
8388	1"	↓	15	.533		15.20	22		37.20	49.50
8390	1-1/4"	Q-1	26	.615		19.80	22.50		42.30	56
8392	1-1/2"		24	.667		31.50	24.50		56	72
8394	2"		21	.762		38.50	28		66.50	84
8396	2-1/2"		18	.889		98	32.50		130.50	157
8398	3"		14	1.143		117	42		159	192
8399	4"	↓	10	1.600	↓	124	59		183	226
8529	Black									
9500	Union with brass seat, 1/4"	1 Plum	15	.533	Ea.	14.25	22		36.25	48.50
9530	3/8"		15	.533		14.45	22		36.45	49
9540	1/2"		14	.571		11.70	23.50		35.20	48
9550	3/4"		13	.615		12.95	25		37.95	52
9560	1"	↓	12	.667		16.90	27		43.90	59.50
9570	1-1/4"	Q-1	21	.762		27.50	28		55.50	72.50
9580	1-1/2"		19	.842		29	31		60	78
9590	2"		17	.941		35.50	34.50		70	91.50
9600	2-1/2"		13	1.231		115	45.50		160.50	195
9610	3"		9	1.778		149	65.50		214.50	263
9620	4"	↓	5	3.200		475	118		593	695
9630	Union, all iron, 1/4"	1 Plum	15	.533		24	22		46	59.50
9650	3/8"		15	.533		24	22		46	59.50
9660	1/2"		14	.571		25.50	23.50		49	63
9670	3/4"		13	.615		27.50	25		52.50	68
9680	1"	↓	12	.667		33.50	27		60.50	78
9690	1-1/4"	Q-1	21	.762		51.50	28		79.50	98.50
9700	1-1/2"		19	.842		63	31		94	116
9710	2"		17	.941		79.50	34.50		114	140
9720	2-1/2"		13	1.231		141	45.50		186.50	223
9730	3"	↓	9	1.778		268	65.50		333.50	395
9750	For galvanized unions, add				↓	15%				
660 0010	**PIPE, STEEL, FITTINGS** Flanged, welded and special type									**660**
0020	flanged joints, C.I., standard weight, black. One gasket & bolt									
0040	set, mat'l only, required at each joint, not included (see line 0620)									
0060	90° Elbow, straight, 1-1/2" pipe size	Q-1	14	1.143	Ea.	220	42		262	305
0080	2" pipe size		13	1.231		134	45.50		179.50	215
0090	2-1/2" pipe size		12	1.333		145	49		194	233
0100	3" pipe size		11	1.455		147	53.50		200.50	243
0110	4" pipe size	↓	8	2		149	73.50		222.50	275

Important: See the Reference Section for critical supporting data - Reference Nos., Crews, & City Cost Indexes

		CREW	DAILY OUTPUT	LABOR-HOURS	UNIT	2005 BARE COSTS				TOTAL INCL O&P	
						MAT.	LABOR	EQUIP.	TOTAL		
660	0120	5" pipe size	Q-1	7	2.286	Ea.	355	84		439	515	660
	0130	6" pipe size	Q-2	9	2.667		234	102		336	410
	0140	8" pipe size		8	3		405	114		519	615
	0150	10" pipe size		7	3.429		885	131		1,016	1,175
	0160	12" pipe size		6	4		1,775	153		1,928	2,175
	0171	90° Elbow, reducing									
	0172	2-1/2" by 2" pipe size	Q-1	12	1.333	Ea.	425	49		474	545
	0173	3" by 2-1/2" pipe size		11	1.455		450	53.50		503.50	570
	0174	4" by 3" pipe size		8	2		330	73.50		403.50	475
	0175	5" by 3" pipe size		7	2.286		685	84		769	880
	0176	6" by 4" pipe size	Q-2	9	2.667		410	102		512	605
	0177	8" by 6" pipe size		8	3		600	114		714	830
	0178	10" by 8" pipe size		7	3.429		1,150	131		1,281	1,450
	0179	12" by 10" pipe size		6	4		2,200	153		2,353	2,625
	0200	45° Elbow, straight, 1-1/2" pipe size	Q-1	14	1.143		265	42		307	355
	0220	2" pipe size		13	1.231		194	45.50		239.50	281
	0230	2-1/2" pipe size		12	1.333		145	49		194	233
	0240	3" pipe size		11	1.455		121	53.50		174.50	214
	0250	4" pipe size		8	2		149	73.50		222.50	275
	0260	5" pipe size		7	2.286		355	84		439	515
	0270	6" pipe size	Q-2	9	2.667		234	102		336	410
	0280	8" pipe size		8	3		405	114		519	615
	0290	10" pipe size		7	3.429		885	131		1,016	1,175
	0300	12" pipe size		6	4		1,775	153		1,928	2,175
	0310	Cross, straight									
	0311	2-1/2" pipe size	Q-1	6	2.667	Ea.	425	98		523	615
	0312	3" pipe size		5	3.200		445	118		563	665
	0313	4" pipe size		4	4		585	147		732	860
	0314	5" pipe size		3	5.333		950	196		1,146	1,350
	0315	6" pipe size	Q-2	5	4.800		1,275	183		1,458	1,675
	0316	8" pipe size		4	6		1,975	229		2,204	2,525
	0317	10" pipe size		3	8		2,325	305		2,630	3,025
	0318	12" pipe size		2	12		3,550	460		4,010	4,600
	0350	Tee, straight, 1-1/2" pipe size	Q-1	10	1.600		252	59		311	365
	0370	2" pipe size		9	1.778		147	65.50		212.50	260
	0380	2-1/2" pipe size		8	2		213	73.50		286.50	345
	0390	3" pipe size		7	2.286		150	84		234	291
	0400	4" pipe size		5	3.200		228	118		346	425
	0410	5" pipe size		4	4		610	147		757	890
	0420	6" pipe size	Q-2	6	4		330	153		483	595
	0430	8" pipe size		5	4.800		565	183		748	895
	0440	10" pipe size		4	6		1,525	229		1,754	2,025
	0450	12" pipe size		3	8		2,375	305		2,680	3,075
	0459	Tee, reducing on outlet									
	0460	2-1/2" by 2" pipe size	Q-1	8	2	Ea.	300	73.50		373.50	440
	0461	3" by 2-1/2" pipe size		7	2.286		350	84		434	505
	0462	4" by 3" pipe size		5	3.200		470	118		588	690
	0463	5" by 4" pipe size		4	4		985	147		1,132	1,300
	0464	6" by 4" pipe size	Q-2	6	4		450	153		603	725
	0465	8" by 6" pipe size		5	4.800		705	183		888	1,050
	0466	10" by 8" pipe size		4	6		1,600	229		1,829	2,125
	0467	12" by 10" pipe size		3	8		2,775	305		3,080	3,525
	0476	Reducer, concentric									
	0477	3" by 2-1/2"	Q-1	12	1.333	Ea.	430	49		479	545
	0478	4" by 3"		9	1.778		300	65.50		365.50	430
	0479	5" by 4"		8	2		465	73.50		538.50	620

		CREW	DAILY OUTPUT	LABOR-HOURS	UNIT	MAT.	LABOR	EQUIP.	TOTAL	TOTAL INCL O&P	
660	0480	6" by 4"	Q-2	10	2.400	Ea.	405	91.50		496.50	585	**660**
	0481	8" by 6"		9	2.667		510	102		612	715	
	0482	10" by 8"		8	3		1,050	114		1,164	1,325	
	0483	12" by 10"	↓	7	3.429	↓	1,800	131		1,931	2,175	
	0492	Reducer, eccentric										
	0493	4" by 3"	Q-1	8	2	Ea.	490	73.50		563.50	650	
	0494	5" by 4"	"	7	2.286		725	84		809	920	
	0495	6" by 4"	Q-2	9	2.667		460	102		562	665	
	0496	8" by 6"		8	3		585	114		699	815	
	0497	10" by 8"		7	3.429		1,500	131		1,631	1,850	
	0498	12" by 10"	↓	6	4		2,025	153		2,178	2,450	
	0500	For galvanized elbows and tees, add					100%					
	0520	For extra heavy weight elbows and tees, add					140%					
	0620	Gasket and bolt set, 150#, 1/2" pipe size	1 Plum	20	.400		2.36	16.35		18.71	27	
	0622	3/4" pipe size		19	.421		2.36	17.20		19.56	28.50	
	0624	1" pipe size		18	.444		2.42	18.15		20.57	30	
	0626	1-1/4" pipe size		17	.471		2.54	19.20		21.74	32	
	0628	1-1/2" pipe size		15	.533		2.71	22		24.71	36	
	0630	2" pipe size		13	.615		4.33	25		29.33	43	
	0640	2-1/2" pipe size		12	.667		4.58	27		31.58	46	
	0650	3" pipe size		11	.727		4.80	29.50		34.30	50	
	0660	3-1/2" pipe size		9	.889		8.70	36.50		45.20	64	
	0670	4" pipe size		8	1		8.70	41		49.70	71	
	0680	5" pipe size		7	1.143		12.90	46.50		59.40	84	
	0690	6" pipe size		6	1.333		13.55	54.50		68.05	97	
	0700	8" pipe size		5	1.600		14.45	65.50		79.95	114	
	0710	10" pipe size		4.50	1.778		30	72.50		102.50	142	
	0720	12" pipe size		4.20	1.905		32	78		110	153	
	0730	14" pipe size		4	2		45.50	81.50		127	173	
	0740	16" pipe size		3	2.667		59.50	109		168.50	230	
	0750	18" pipe size		2.70	2.963		83	121		204	274	
	0760	20" pipe size		2.30	3.478		111	142		253	335	
	0780	24" pipe size		1.90	4.211		149	172		321	425	
	0790	26" pipe size		1.60	5		168	204		372	490	
	0810	30" pipe size		1.40	5.714		208	233		441	580	
	0830	36" pipe size	↓	1.10	7.273	↓	256	297		553	725	
	0850	For 300 lb gasket set, add					40%					
	2000	Flanged unions, 125 lb., black, 1/2" pipe size	1 Plum	17	.471	Ea.	34	19.20		53.20	66.50	
	2040	3/4" pipe size		17	.471		45	19.20		64.20	78.50	
	2050	1" pipe size	↓	16	.500		43.50	20.50		64	78.50	
	2060	1-1/4" pipe size	Q-1	28	.571		51.50	21		72.50	88.50	
	2070	1-1/2" pipe size		27	.593		47.50	22		69.50	85.50	
	2080	2" pipe size		26	.615		56	22.50		78.50	95.50	
	2090	2-1/2" pipe size		24	.667		76	24.50		100.50	121	
	2100	3" pipe size		22	.727		86	27		113	135	
	2110	3-1/2" pipe size		18	.889		146	32.50		178.50	209	
	2120	4" pipe size		16	1		117	37		154	184	
	2130	5" pipe size	↓	14	1.143		272	42		314	360	
	2140	6" pipe size	Q-2	19	1.263		255	48		303	355	
	2150	8" pipe size	"	16	1.500		570	57		627	710	
	2200	For galvanized unions, add				↓	150%					
	2290	Threaded flange										
	2300	Cast iron										
	2310	Black, 125 lb., per flange										
	2320	1" pipe size	1 Plum	27	.296	Ea.	19.50	12.10		31.60	39.50	
	2330	1-1/4" pipe size	Q-1	44	.364	↓	21.50	13.35		34.85	44	

Important: See the Reference Section for critical supporting data - Reference Nos., Crews, & City Cost Indexes

15107 | Metal Pipe & Fittings

		CREW	DAILY OUTPUT	LABOR-HOURS	UNIT	2005 BARE COSTS				TOTAL INCL O&P
						MAT.	LABOR	EQUIP.	TOTAL	
660 2340	1-1/2" pipe size	Q-1	40	.400	Ea.	23.50	14.70		38.20	47.50 **660**
2350	2" pipe size		36	.444		21.50	16.35		37.85	48
2360	2-1/2" pipe size		28	.571		25	21		46	59
2370	3" pipe size		20	.800		32.50	29.50		62	79.50
2380	3-1/2" pipe size		16	1		45.50	37		82.50	106
2390	4" pipe size		12	1.333		44	49		93	122
2400	5" pipe size	↓	10	1.600		62	59		121	157
2410	6" pipe size	Q-2	14	1.714		70	65.50		135.50	176
2420	8" pipe size		12	2		110	76.50		186.50	236
2430	10" pipe size		10	2.400		196	91.50		287.50	355
2440	12" pipe size	↓	8	3	↓	420	114		534	630
2460	For galvanized flanges, add					95%				
2490	Blind flange									
2492	Cast iron									
2494	Black, 125 lb., per flange									
2496	1" pipe size	1 Plum	27	.296	Ea.	31	12.10		43.10	52
2500	1-1/2" pipe size	Q-1	40	.400		35	14.70		49.70	60.50
2502	2" pipe size		36	.444		39.50	16.35		55.85	68
2504	2-1/2" pipe size		28	.571		43.50	21		64.50	79
2506	3" pipe size		20	.800		52	29.50		81.50	101
2508	4" pipe size		12	1.333		66.50	49		115.50	147
2510	5" pipe size	↓	10	1.600		108	59		167	208
2512	6" pipe size	Q-2	14	1.714		117	65.50		182.50	228
2514	8" pipe size		12	2		184	76.50		260.50	320
2516	10" pipe size		10	2.400		273	91.50		364.50	440
2518	12" pipe size	↓	8	3	↓	495	114		609	715
2520	For galvanized flanges, add					80%				
2570	Threaded flange									
2580	Forged steel,									
2590	Black 150 lb., per flange									
2600	1/2" pipe size	1 Plum	30	.267	Ea.	13.35	10.90		24.25	31
2610	3/4" pipe size		28	.286		13.35	11.65		25	32
2620	1" pipe size	↓	27	.296		13.35	12.10		25.45	33
2630	1-1/4" pipe size	Q-1	44	.364		13.35	13.35		26.70	34.50
2640	1-1/2" pipe size		40	.400		13.35	14.70		28.05	36.50
2650	2" pipe size		36	.444		16.25	16.35		32.60	42.50
2660	2-1/2" pipe size		28	.571		20	21		41	53.50
2670	3" pipe size		20	.800		21	29.50		50.50	67
2690	4" pipe size		12	1.333		24.50	49		73.50	101
2700	5" pipe size	↓	10	1.600		36.50	59		95.50	129
2710	6" pipe size	Q-2	14	1.714		41	65.50		106.50	144
2720	8" pipe size		12	2		69.50	76.50		146	192
2730	10" pipe size	↓	10	2.400	↓	126	91.50		217.50	277
2860	Black 300 lb., per flange									
2870	1/2" pipe size	1 Plum	30	.267	Ea.	20	10.90		30.90	38.50
2880	3/4" pipe size		28	.286		20	11.65		31.65	39.50
2890	1" pipe size	↓	27	.296		20	12.10		32.10	40
2900	1-1/4" pipe size	Q-1	44	.364		20	13.35		33.35	42
2910	1-1/2" pipe size		40	.400		20	14.70		34.70	44
2920	2" pipe size		36	.444		27.50	16.35		43.85	54.50
2930	2-1/2" pipe size		28	.571		26	21		47	60.50
2940	3" pipe size		20	.800		29	29.50		58.50	76
2960	4" pipe size	↓	12	1.333		47.50	49		96.50	126
2970	6" pipe size	Q-2	14	1.714	↓	81	65.50		146.50	188
3000	Weld joint, butt, carbon steel, standard weight									
3040	90° elbow, long radius									

15107 | Metal Pipe & Fittings

		CREW	DAILY OUTPUT	LABOR-HOURS	UNIT	2005 BARE COSTS				TOTAL INCL O&P
						MAT.	LABOR	EQUIP.	TOTAL	
3050	1/2" pipe size	Q-15	16	1	Ea.	18.90	37	5.05	60.95	82
3060	3/4" pipe size		16	1		18.90	37	5.05	60.95	82
3070	1" pipe size		16	1		9.60	37	5.05	51.65	71.50
3080	1-1/4" pipe size		14	1.143		9.60	42	5.75	57.35	80
3090	1-1/2" pipe size		13	1.231		9.60	45.50	6.20	61.30	85.50
3100	2" pipe size		10	1.600		10.55	59	8.05	77.60	109
3110	2-1/2" pipe size		8	2		12.80	73.50	10.10	96.40	136
3120	3" pipe size		7	2.286		15.35	84	11.50	110.85	156
3130	4" pipe size		5	3.200		25.50	118	16.15	159.65	223
3136	5" pipe size		4	4		54	147	20	221	300
3140	6" pipe size	Q-16	5	4.800		59	183	16.15	258.15	360
3150	8" pipe size		3.75	6.400		111	244	21.50	376.50	510
3160	10" pipe size		3	8		203	305	27	535	715
3170	12" pipe size		2.50	9.600		300	365	32.50	697.50	915
3180	14" pipe size		2	12		395	460	40.50	895.50	1,175
3190	16" pipe size		1.50	16		545	610	54	1,209	1,575
3191	18" pipe size		1.25	19.200		720	730	64.50	1,514.50	1,950
3192	20" pipe size		1.15	20.870		1,025	795	70	1,890	2,400
3194	24" pipe size		1.02	23.529		1,375	895	79	2,349	2,925
3200	45° Elbow, long									
3210	1/2" pipe size	Q-15	16	1	Ea.	26.50	37	5.05	68.55	90
3220	3/4" pipe size		16	1		26.50	37	5.05	68.55	90
3230	1" pipe size		16	1		9.90	37	5.05	51.95	72
3240	1-1/4" pipe size		14	1.143		9.90	42	5.75	57.65	80.50
3250	1-1/2" pipe size		13	1.231		9.90	45.50	6.20	61.60	85.50
3260	2" pipe size		10	1.600		9.90	59	8.05	76.95	108
3270	2-1/2" pipe size		8	2		11.85	73.50	10.10	95.45	135
3280	3" pipe size		7	2.286		12.80	84	11.50	108.30	153
3290	4" pipe size		5	3.200		22.50	118	16.15	156.65	220
3296	5" pipe size		4	4		34.50	147	20	201.50	281
3300	6" pipe size	Q-16	5	4.800		47.50	183	16.15	246.65	345
3310	8" pipe size		3.75	6.400		79.50	244	21.50	345	475
3320	10" pipe size		3	8		144	305	27	476	650
3330	12" pipe size		2.50	9.600		203	365	32.50	600.50	810
3340	14" pipe size		2	12		273	460	40.50	773.50	1,025
3341	16" pipe size		1.50	16		460	610	54	1,124	1,475
3342	18" pipe size		1.25	19.200		655	730	64.50	1,449.50	1,900
3343	20" pipe size		1.15	20.870		680	795	70	1,545	2,025
3345	24" pipe size		1.05	22.857		1,025	870	77	1,972	2,500
3346	26" pipe size		.85	28.235		1,125	1,075	95	2,295	2,975
3347	30" pipe size		.45	53.333		1,250	2,025	179	3,454	4,625
3349	36" pipe size		.38	63.158		1,375	2,400	212	3,987	5,350
3350	Tee, straight									
3352	For reducing tees and concentrics see starting line 4600									
3360	1/2" pipe size	Q-15	10	1.600	Ea.	48.50	59	8.05	115.55	151
3370	3/4" pipe size		10	1.600		48.50	59	8.05	115.55	151
3380	1" pipe size		10	1.600		26	59	8.05	93.05	126
3390	1-1/4" pipe size		9	1.778		30	65.50	8.95	104.45	141
3400	1-1/2" pipe size		8	2		30	73.50	10.10	113.60	155
3410	2" pipe size		6	2.667		25.50	98	13.45	136.95	190
3420	2-1/2" pipe size		5	3.200		35	118	16.15	169.15	233
3430	3" pipe size		4	4		39	147	20	206	286
3440	4" pipe size		3	5.333		55.50	196	27	278.50	385
3446	5" pipe size		2.50	6.400		91	235	32.50	358.50	490
3450	6" pipe size	Q-16	3	8		94.50	305	27	426.50	595
3460	8" pipe size		2.50	9.600		165	365	32.50	562.50	770

Important: See the Reference Section for critical supporting data - Reference Nos., Crews, & City Cost Indexes

			CREW	DAILY OUTPUT	LABOR-HOURS	UNIT	MAT.	LABOR	EQUIP.	TOTAL	TOTAL INCL O&P	
		15107 | Metal Pipe & Fittings						2005 BARE COSTS				
660	3470	10" pipe size	Q-16	2	12	Ea.	297	460	40.50	797.50	1,050	660
	3480	12" pipe size		1.60	15		415	570	50.50	1,035.50	1,375	
	3481	14" pipe size		1.30	18.462		675	705	62	1,442	1,850	
	3482	16" pipe size		1	24		815	915	80.50	1,810.50	2,350	
	3483	18" pipe size		.80	30		1,300	1,150	101	2,551	3,250	
	3484	20" pipe size		.75	32		2,000	1,225	108	3,333	4,150	
	3486	24" pipe size		.70	34.286		2,575	1,300	115	3,990	4,925	
	3487	26" pipe size		.55	43.636		2,825	1,675	147.	4,647	5,775	
	3488	30" pipe size		.30	80		3,125	3,050	269	6,444	8,300	
	3490	36" pipe size	▼	.20	120		3,425	4,575	405	8,405	11,100	
	3491	Eccentric reducer, 1-1/2" pipe size	Q-15	14	1.143		40.50	42	5.75	88.25	114	
	3492	2" pipe size		11	1.455		57.50	53.50	7.35	118.35	152	
	3493	2-1/2" pipe size		9	1.778		69	65.50	8.95	143.45	184	
	3494	3" pipe size		8	2		78	73.50	10.10	161.60	208	
	3495	4" pipe size	▼	6	2.667		84.50	98	13.45	195.95	255	
	3496	6" pipe size	Q-16	5	4.800		148	183	16.15	347.15	455	
	3497	8" pipe size		4	6		170	229	20	419	555	
	3498	10" pipe size		3	8		179	305	27	511	685	
	3499	12" pipe size	▼	2.50	9.600		256	365	32.50	653.50	865	
	3501	Cap, 1-1/2" pipe size	Q-15	28	.571		11.20	21	2.88	35.08	47	
	3502	2" pipe size		22	.727		10.90	27	3.67	41.57	56	
	3503	2-1/2" pipe size		18	.889		11.50	32.50	4.48	48.48	66.50	
	3504	3" pipe size		16	1		11.50	37	5.05	53.55	73.50	
	3505	4" pipe size	▼	12	1.333		16.65	49	6.70	72.35	99	
	3506	6" pipe size	Q-16	10	2.400		29	91.50	8.05	128.55	178	
	3507	8" pipe size		8	3		43.50	114	10.10	167.60	231	
	3508	10" pipe size		6	4		71.50	153	13.45	237.95	320	
	3509	12" pipe size		5	4.800		108	183	16.15	307.15	410	
	3511	14" pipe size		4	6		144	229	20	393	525	
	3512	16" pipe size		4	6		167	229	20	416	550	
	3513	18" pipe size	▼	3	8	▼	229	305	27	561	740	
	3517	Weld joint, butt, carbon steel, extra strong										
	3519	90° elbow, long										
	3520	1/2" pipe size	Q-15	13	1.231	Ea.	24	45.50	6.20	75.70	101	
	3530	3/4" pipe size		12	1.333		24	49	6.70	79.70	107	
	3540	1" pipe size		11	1.455		21.50	53.50	7.35	82.35	112	
	3550	1-1/4" pipe size		10	1.600		21.50	59	8.05	88.55	121	
	3560	1-1/2" pipe size		9	1.778		21.50	65.50	8.95	95.95	132	
	3570	2" pipe size		8	2		33	73.50	10.10	116.60	159	
	3580	2-1/2" pipe size		7	2.286		33	84	11.50	128.50	175	
	3590	3" pipe size		6	2.667		33	98	13.45	144.45	198	
	3600	4" pipe size		4	4		55	147	20	222	305	
	3606	5" pipe size	▼	3.50	4.571		114	168	23	305	405	
	3610	6" pipe size	Q-16	4.50	5.333		128	203	17.90	348.90	465	
	3620	8" pipe size		3.50	6.857		276	261	23	560	725	
	3630	10" pipe size		2.50	9.600		375	365	32.50	772.50	1,000	
	3640	12" pipe size	▼	2.25	10.667	▼	450	405	36	891	1,150	
	3650	45° Elbow, long										
	3660	1/2" pipe size	Q-15	13	1.231	Ea.	16.95	45.50	6.20	68.65	93.50	
	3670	3/4" pipe size		12	1.333		16.95	49	6.70	72.65	99.50	
	3680	1" pipe size		11	1.455		8.95	53.50	7.35	69.80	98.50	
	3690	1-1/4" pipe size		10	1.600		8.95	59	8.05	76	107	
	3700	1-1/2" pipe size		9	1.778		8.95	65.50	8.95	83.40	118	
	3710	2" pipe size		8	2		8.95	73.50	10.10	92.55	132	
	3720	2-1/2" pipe size		7	2.286		13.10	84	11.50	108.60	153	
	3730	3" pipe size	▼	6	2.667	▼	16.30	98	13.45	127.75	180	

15107	Metal Pipe & Fittings	CREW	DAILY OUTPUT	LABOR-HOURS	UNIT	2005 BARE COSTS				TOTAL INCL O&P
						MAT.	LABOR	EQUIP.	TOTAL	
3740	4" pipe size	Q-15	4	4	Ea.	27.50	147	20	194.50	274
3746	5" pipe size	↓	3.50	4.571		59.50	168	23	250.50	345
3750	6" pipe size	Q-16	4.50	5.333		66	203	17.90	286.90	400
3760	8" pipe size		3.50	6.857		125	261	23	409	560
3770	10" pipe size		2.50	9.600		224	365	32.50	621.50	830
3780	12" pipe size	↓	2.25	10.667	↓	315	405	36	756	995
3800	Tee, straight									
3810	1/2" pipe size	Q-15	9	1.778	Ea.	57.50	65.50	8.95	131.95	172
3820	3/4" pipe size		8.50	1.882		57.50	69	9.50	136	178
3830	1" pipe size		8	2		32	73.50	10.10	115.60	157
3840	1-1/4" pipe size		7	2.286		32	84	11.50	127.50	174
3850	1-1/2" pipe size		6	2.667		36	98	13.45	147.45	201
3860	2" pipe size		5	3.200		32	118	16.15	166.15	230
3870	2-1/2" pipe size		4	4		51	147	20	218	300
3880	3" pipe size		3.50	4.571		49.50	168	23	240.50	335
3890	4" pipe size		2.50	6.400		80	235	32.50	347.50	480
3896	5" pipe size	↓	2.25	7.111		163	262	36	461	615
3900	6" pipe size	Q-16	2.25	10.667		136	405	36	577	800
3910	8" pipe size		2	12		255	460	40.50	755.50	1,025
3920	10" pipe size		1.75	13.714		415	525	46	986	1,300
3930	12" pipe size	↓	1.50	16	↓	600	610	54	1,264	1,625
4190	Weld fittings, reducing, standard weight									
4200	Welding ring w/spacer pins, 2" pipe size				Ea.	1.95			1.95	2.15
4210	2-1/2" pipe size					2.35			2.35	2.59
4220	3" pipe size					2.44			2.44	2.68
4230	4" pipe size					2.54			2.54	2.79
4236	5" pipe size					2.99			2.99	3.29
4240	6" pipe size					2.99			2.99	3.29
4250	8" pipe size					3.53			3.53	3.88
4260	10" pipe size					4.32			4.32	4.75
4270	12" pipe size					4.95			4.95	5.45
4280	14" pipe size					5.75			5.75	6.35
4290	16" pipe size					6.50			6.50	7.15
4300	18" pipe size					7.50			7.50	8.25
4310	20" pipe size					7.95			7.95	8.75
4330	24" pipe size					9.25			9.25	10.20
4340	26" pipe size					12.35			12.35	13.60
4350	30" pipe size					12.35			12.35	13.60
4370	36" pipe size				↓	17.75			17.75	19.55
4600	Tee, reducing on outlet									
4601	2-1/2" x 2" pipe size	Q-15	5	3.200	Ea.	47	118	16.15	181.15	246
4602	3" x 2-1/2" pipe size		4	4		51.50	147	20	218.50	300
4603	3-1/2" x 3" pipe size		3.50	4.571		64.50	168	23	255.50	350
4604	4" x 3" pipe size		3	5.333		105	196	27	328	440
4605	5" x 4" pipe size	↓	2.50	6.400		125	235	32.50	392.50	530
4606	6" x 5" pipe size	Q-16	3	8		340	305	27	672	865
4607	8" x 6" pipe size		2.50	9.600		215	365	32.50	612.50	825
4608	10" x 8" pipe size		2	12		365	460	40.50	865.50	1,125
4609	12" x 10" pipe size		1.60	15		535	570	50.50	1,155.50	1,500
4610	16" x 12" pipe size		1.50	16		955	610	54	1,619	2,025
4611	14" x 12" pipe size	↓	1.52	15.789	↓	730	600	53	1,383	1,775
4618	Reducer, concentric									
4619	2-1/2" by 2" pipe size	Q-15	10	1.600	Ea.	17.60	59	8.05	84.65	117
4620	3" by 2-1/2" pipe size		9	1.778		21	65.50	8.95	95.45	131
4621	3-1/2" by 3" pipe size		8	2		37.50	73.50	10.10	121.10	163
4622	4" by 2-1/2" pipe size	↓	7	2.286	↓	22.50	84	11.50	118	164

15107 | Metal Pipe & Fittings

		CREW	DAILY OUTPUT	LABOR-HOURS	UNIT	2005 BARE COSTS				TOTAL INCL O&P
						MAT.	LABOR	EQUIP.	TOTAL	
660	4623	5" by 3" pipe size	Q-15	7	2.286	Ea.	47.50	84	11.50	143	191
	4624	6" by 4" pipe size	Q-16	6	4		39.50	153	13.45	205.95	287
	4625	8" by 6" pipe size		5	4.800		52	183	16.15	251.15	350
	4626	10" by 8" pipe size		4	6		91	229	20	340	465
	4627	12" by 10" pipe size		3	8		113	305	27	445	615
	4660	Reducer, eccentric									
	4662	3" x 2" pipe size	Q-15	8	2	Ea.	24.50	73.50	10.10	108.10	149
	4664	4" x 3" pipe size		6	2.667		30	98	13.45	141.45	194
	4666	4" x 2" pipe size		6	2.667		37	98	13.45	148.45	202
	4670	6" x 4" pipe size	Q-16	5	4.800		62	183	16.15	261.15	360
	4672	6" x 3" pipe size		5	4.800		97.50	183	16.15	296.65	400
	4676	8" x 6" pipe size		4	6		79	229	20	328	455
	4678	8" x 4" pipe size		4	6		145	229	20	394	525
	4682	10" x 8" pipe size		3	8		93.50	305	27	425.50	595
	4684	10" x 6" pipe size		3	8		165	305	27	497	670
	4688	12" x 10" pipe size		2.50	9.600		147	365	32.50	544.50	745
	4690	12" x 8" pipe size		2.50	9.600		230	365	32.50	627.50	840
	4691	14" x 12" pipe size		2.20	10.909		258	415	36.50	709.50	950
	4693	16" x 14" pipe size		1.80	13.333		350	510	45	905	1,200
	4694	16" x 12" pipe size		2	12		465	460	40.50	965.50	1,250
	4696	18" x 16" pipe size		1.60	15		605	570	50.50	1,225.50	1,575
	5630	T-O-L, 1/4" pipe size, nozzle	Q-15	23	.696		4.20	25.50	3.51	33.21	47
	5631	3/8" pipe size, nozzle		23	.696		4.20	25.50	3.51	33.21	47
	5632	1/2" pipe size, nozzle		22	.727		4.20	27	3.67	34.87	48.50
	5633	3/4" pipe size, nozzle		21	.762		4.93	28	3.84	36.77	51.50
	5634	1" pipe size, nozzle		20	.800		5.65	29.50	4.03	39.18	54.50
	5635	1-1/4" pipe size, nozzle		18	.889		8.20	32.50	4.48	45.18	63
	5636	1-1/2" pipe size, nozzle		16	1		8.75	37	5.05	50.80	70.50
	5637	2" pipe size, nozzle		12	1.333		9.90	49	6.70	65.60	92
	5638	2-1/2" pipe size, nozzle		10	1.600		32	59	8.05	99.05	132
	5639	4" pipe size, nozzle		6	2.667		72	98	13.45	183.45	241
	5640	W-O-L, 1/4" pipe size, nozzle		23	.696		8.30	25.50	3.51	37.31	51.50
	5641	3/8" pipe size, nozzle		23	.696		8.30	25.50	3.51	37.31	51.50
	5642	1/2" pipe size, nozzle		22	.727		8.30	27	3.67	38.97	53
	5643	3/4" pipe size, nozzle		21	.762		8.35	28	3.84	40.19	55.50
	5644	1" pipe size, nozzle		20	.800		8.90	29.50	4.03	42.43	58
	5645	1-1/4" pipe size, nozzle		18	.889		10.95	32.50	4.48	47.93	66
	5646	1-1/2" pipe size, nozzle		16	1		10.95	37	5.05	53	73
	5647	2" pipe size, nozzle		12	1.333		11.10	49	6.70	66.80	93
	5648	2-1/2" pipe size, nozzle		10	1.600		25	59	8.05	92.05	125
	5649	3" pipe size, nozzle		8	2		29	73.50	10.10	112.60	154
	5650	4" pipe size, nozzle		6	2.667		36.50	98	13.45	147.95	202
	5651	5" pipe size, nozzle		5	3.200		85.50	118	16.15	219.65	289
	5652	6" pipe size, nozzle		4	4		103	147	20	270	355
	5653	8" pipe size, nozzle		3	5.333		197	196	27	420	540
	5654	10" pipe size, nozzle		2.60	6.154		350	226	31	607	760
	5655	12" pipe size, nozzle		2.20	7.273		640	267	36.50	943.50	1,150
	5656	14" pipe size, nozzle		2	8		1,175	294	40.50	1,509.50	1,775
	5674	S-O-L, 1/4" pipe size, outlet		23	.696		4.20	25.50	3.51	33.21	47
	5675	3/8" pipe size, outlet		23	.696		4.20	25.50	3.51	33.21	47
	5676	1/2" pipe size, outlet		22	.727		4.20	27	3.67	34.87	48.50
	5677	3/4" pipe size, outlet		21	.762		4.93	28	3.84	36.77	51.50
	5678	1" pipe size, outlet		20	.800		5.65	29.50	4.03	39.18	54.50
	5679	1-1/4" pipe size, outlet		18	.889		8.40	32.50	4.48	45.38	63
	5680	1-1/2" pipe size, outlet		16	1		8.75	37	5.05	50.80	70.50
	5681	2" pipe size, outlet		12	1.333		9.90	49	6.70	65.60	92

15107 | Metal Pipe & Fittings

		CREW	DAILY OUTPUT	LABOR-HOURS	UNIT	2005 BARE COSTS				TOTAL INCL O&P
						MAT.	LABOR	EQUIP.	TOTAL	
660	6000	Weld-on flange, forged steel									660
	6020	Slip-on, 150 lb. flange (welded front and back)									
	6050	1/2" pipe size	Q-15	18	.889	Ea.	10.75	32.50	4.48	47.73	66
	6060	3/4" pipe size		18	.889		10.75	32.50	4.48	47.73	66
	6070	1" pipe size		17	.941		10.75	34.50	4.74	49.99	69
	6080	1-1/4" pipe size		16	1		10.75	37	5.05	52.80	73
	6090	1-1/2" pipe size		15	1.067		10.75	39	5.40	55.15	77
	6100	2" pipe size		12	1.333		10	49	6.70	65.70	92
	6110	2-1/2" pipe size		10	1.600		15.60	59	8.05	82.65	115
	6120	3" pipe size		9	1.778		19.15	65.50	8.95	93.60	129
	6130	3-1/2" pipe size		7	2.286		25.50	84	11.50	121	167
	6140	4" pipe size		6	2.667		24	98	13.45	135.45	188
	6150	5" pipe size		5	3.200		28	118	16.15	162.15	226
	6160	6" pipe size	Q-16	6	4		34	153	13.45	200.45	281
	6170	8" pipe size		5	4.800		58	183	16.15	257.15	355
	6180	10" pipe size		4	6		85.50	229	20	334.50	460
	6190	12" pipe size		3	8		133	305	27	465	635
	6191	14" pipe size		2.50	9.600		180	365	32.50	577.50	785
	6192	16" pipe size		1.80	13.333		271	510	45	826	1,125
	6200	300 lb. flange									
	6210	1/2" pipe size	Q-15	17	.941	Ea.	14.35	34.50	4.74	53.59	73
	6220	3/4" pipe size		17	.941		14.35	34.50	4.74	53.59	73
	6230	1" pipe size		16	1		14.35	37	5.05	56.40	77
	6240	1-1/4" pipe size		13	1.231		14.35	45.50	6.20	66.05	90.50
	6250	1-1/2" pipe size		12	1.333		14.35	49	6.70	70.05	96.50
	6260	2" pipe size		11	1.455		15.45	53.50	7.35	76.30	106
	6270	2-1/2" pipe size		9	1.778		22	65.50	8.95	96.45	132
	6280	3" pipe size		7	2.286		30	84	11.50	125.50	172
	6290	4" pipe size		6	2.667		43.50	98	13.45	154.95	210
	6300	5" pipe size		4	4		56	147	20	223	305
	6310	6" pipe size	Q-16	5	4.800		72	183	16.15	271.15	370
	6320	8" pipe size		4	6		117	229	20	366	495
	6330	10" pipe size		3.40	7.059		165	269	23.50	457.50	615
	6340	12" pipe size		2.80	8.571		204	325	29	558	745
	6400	Welding neck, 150 lb. flange									
	6410	1/2" pipe size	Q-15	40	.400	Ea.	15.40	14.70	2.02	32.12	41
	6420	3/4" pipe size		36	.444		15.40	16.35	2.24	33.99	44
	6430	1" pipe size		32	.500		15.40	18.40	2.52	36.32	47
	6440	1-1/4" pipe size		29	.552		15.40	20.50	2.78	38.68	50.50
	6450	1-1/2" pipe size		26	.615		15.40	22.50	3.10	41	54.50
	6460	2" pipe size		20	.800		17.40	29.50	4.03	50.93	67.50
	6470	2-1/2" pipe size		16	1		19.30	37	5.05	61.35	82.50
	6480	3" pipe size		14	1.143		23	42	5.75	70.75	95
	6500	4" pipe size		10	1.600		35	59	8.05	102.05	136
	6510	5" pipe size		8	2		41	73.50	10.10	124.60	167
	6520	6" pipe size	Q-16	10	2.400		50	91.50	8.05	149.55	202
	6530	8" pipe size		7	3.429		82.50	131	11.50	225	300
	6540	10" pipe size		6	4		97	153	13.45	263.45	350
	6550	12" pipe size		5	4.800		175	183	16.15	374.15	485
	6551	14" pipe size		4.50	5.333		244	203	17.90	464.90	595
	6552	16" pipe size		3	8		289	305	27	621	810
	6553	18" pipe size		2.50	9.600		445	365	32.50	842.50	1,075
	6554	20" pipe size		2.30	10.435		535	400	35	970	1,225
	6556	24" pipe size		2	12		725	460	40.50	1,225.50	1,525
	6557	26" pipe size		1.70	14.118		830	540	47.50	1,417.50	1,775
	6558	30" pipe size		.90	26.667		950	1,025	89.50	2,064.50	2,675

Important: See the Reference Section for critical supporting data - Reference Nos., Crews, & City Cost Indexes

15107 | Metal Pipe & Fittings

		CREW	DAILY OUTPUT	LABOR-HOURS	UNIT	2005 BARE COSTS				TOTAL INCL O&P
						MAT.	LABOR	EQUIP.	TOTAL	
6559	36" pipe size	Q-16	.75	32	Ea.	1,100	1,225	108	2,433	3,150
6560	300 lb. flange									
6570	1/2" pipe size	Q-15	36	.444	Ea.	19	16.35	2.24	37.59	48
6580	3/4" pipe size		34	.471		19	17.30	2.37	38.67	49.50
6590	1" pipe size		30	.533		19	19.60	2.69	41.29	53.50
6600	1-1/4" pipe size		28	.571		19	21	2.88	42.88	55.50
6610	1-1/2" pipe size		24	.667		19	24.50	3.36	46.86	61.50
6620	2" pipe size		18	.889		20.50	32.50	4.48	57.48	76.50
6630	2-1/2" pipe size		14	1.143		27.50	42	5.75	75.25	99.50
6640	3" pipe size		12	1.333		35	49	6.70	90.70	119
6650	4" pipe size		8	2		53.50	73.50	10.10	137.10	181
6660	5" pipe size	↓	7	2.286		65	84	11.50	160.50	210
6670	6" pipe size	Q-16	9	2.667		97.50	102	8.95	208.45	270
6680	8" pipe size		6	4		142	153	13.45	308.45	400
6690	10" pipe size		5	4.800		214	183	16.15	413.15	530
6700	12" pipe size	↓	4	6	↓	298	229	20	547	690
7740	Plain ends for plain end pipe, mechanically coupled									
7750	Cplg & labor required at joints not included add 1 per									
7760	joint for installed price, see line 9180									
7770	Malleable iron, painted, unless noted otherwise									
7800	90° Elbow 1"				Ea.	16.75			16.75	18.45
7810	1-1/2"					32			32	35.50
7820	2"					32			32	35.50
7830	2-1/2"					32			32	35.50
7840	3"					38.50			38.50	42.50
7860	4"					56			56	61.50
7870	5" welded steel					101			101	112
7880	6"					80.50			80.50	88.50
7890	8" welded steel					170			170	187
7900	10" welded steel					214			214	235
7910	12" welded steel					272			272	300
7970	45° Elbow 1"					16.75			16.75	18.45
7980	1-1/2"					32			32	35.50
7990	2"					32			32	35.50
8000	2-1/2"					32			32	35.50
8010	3"					38.50			38.50	42.50
8030	4"					56			56	61.50
8040	5" welded steel					101			101	112
8050	6"					80.50			80.50	88.50
8060	8"					93.50			93.50	103
8070	10" welded steel					131			131	144
8080	12" welded steel					169			169	186
8140	Tee, straight 1"					21			21	23
8150	1-1/2"					43			43	47.50
8160	2"					43			43	47.50
8170	2-1/2"					43			43	47.50
8180	3"					62			62	68.50
8200	4"					86.50			86.50	95
8210	5" welded steel					99			99	109
8220	6"					122			122	134
8230	8" welded steel					169			169	186
8240	10" welded steel					258			258	284
8250	12" welded steel				↓	350			350	385
8340	Segmentally welded steel, painted									
8390	Wye 2"				Ea.	66			66	72.50
8400	2-1/2"				↓	66			66	72.50

		15107	Metal Pipe & Fittings	CREW	DAILY OUTPUT	LABOR-HOURS	UNIT	2005 BARE COSTS MAT.	LABOR	EQUIP.	TOTAL	TOTAL INCL O&P	
660	8410		3"				Ea.	74			74	81.50	**660**
	8430		4"					110			110	121	
	8440		5"					134			134	147	
	8450		6"					184			184	203	
	8460		8"					239			239	263	
	8470		10"					350			350	385	
	8480		12"					545			545	595	
	8540		Wye, lateral 2"					70.50			70.50	77.50	
	8550		2-1/2"					70.50			70.50	77.50	
	8560		3"					79			79	86.50	
	8580		4"					119			119	131	
	8590		5"					155			155	170	
	8600		6"					198			198	218	
	8610		8"					288			288	315	
	8620		10"					405			405	450	
	8630		12"					550			550	605	
	8690		Cross, 2"					75			75	82.50	
	8700		2-1/2"					75			75	82.50	
	8710		3"					89			89	98	
	8730		4"					122			122	134	
	8740		5"					174			174	191	
	8750		6"					229			229	252	
	8760		8"					294			294	325	
	8770		10"					430			430	470	
	8780		12"					620			620	680	
	8800		Tees, reducing 2" x 1"					44.50			44.50	49	
	8810		2" x 1-1/2"					44.50			44.50	49	
	8820		3" x 1"					44.50			44.50	49	
	8830		3" x 1-1/2"					44.50			44.50	49	
	8840		3" x 2"					44.50			44.50	49	
	8850		4" x 1"					65			65	71.50	
	8860		4" x 1-1/2"					65			65	71.50	
	8870		4" x 2"					65			65	71.50	
	8880		4" x 2-1/2"					66.50			66.50	73.50	
	8890		4" x 3"					66.50			66.50	73.50	
	8900		6" x 2"					108			108	119	
	8910		6" x 3"					114			114	125	
	8920		6" x 4"					114			114	125	
	8930		8" x 2"					139			139	152	
	8940		8" x 3"					148			148	163	
	8950		8" x 4"					148			148	163	
	8960		8" x 5"					152			152	167	
	8970		8" x 6"					152			152	167	
	8980		10" x 4"					218			218	240	
	8990		10" x 6"					231			231	254	
	9000		10" x 8"					231			231	254	
	9010		12" x 6"					335			335	370	
	9020		12" x 8"					335			335	370	
	9030		12" x 10"				▼	350			350	385	
	9080		Adapter nipples 3" long										
	9090		1"				Ea.	5.85			5.85	6.40	
	9100		1-1/2"					5.85			5.85	6.40	
	9110		2"					5.85			5.85	6.40	
	9120		2-1/2"					6.80			6.80	7.50	
	9130		3"					8.45			8.45	9.30	
	9140		4"				▼	14			14	15.40	

Important: See the Reference Section for critical supporting data - Reference Nos., Crews, & City Cost Indexes

15107 | Metal Pipe & Fittings

		CREW	DAILY OUTPUT	LABOR-HOURS	UNIT	2005 BARE COSTS MAT.	LABOR	EQUIP.	TOTAL	TOTAL INCL O&P
660	9150	6"				Ea.	38			38	41.50
	9180	Coupling, mechanical, plain end pipe to plain end pipe or fitting									
	9190	1"	Q-1	29	.552	Ea.	21.50	20.50		42	54
	9200	1-1/2"		28	.571		21.50	21		42.50	55
	9210	2"		27	.593		21.50	22		43.50	56.50
	9220	2-1/2"		26	.615		21.50	22.50		44	57.50
	9230	3"		25	.640		32	23.50		55.50	70.50
	9240	3-1/2"		24	.667		37	24.50		61.50	77.50
	9250	4"		22	.727		37	27		64	80.50
	9260	5"	Q-2	28	.857		52.50	32.50		85	107
	9270	6"		24	1		64	38		102	128
	9280	8"		19	1.263		114	48		162	198
	9290	10"		16	1.500		148	57		205	249
	9300	12"		12	2		186	76.50		262.50	320
	9310	Outlets for precut holes through pipe wall									
	9331	Strapless type, with gasket									
	9332	4" to 8" pipe x 1/2"	1 Plum	13	.615	Ea.	21	25		46	61
	9333	4" to 8" pipe x 3/4"		13	.615		22.50	25		47.50	63
	9334	10" pipe and larger x 1/2"		11	.727		21	29.50		50.50	67.50
	9335	10" pipe and larger x 3/4"		11	.727		22.50	29.50		52	69.50
	9341	Thermometer wells with gasket									
	9342	4" to 8" pipe, 6" stem	1 Plum	14	.571	Ea.	32	23.50		55.50	70.50
	9343	8" pipe and larger, 6" stem	"	13	.615	"	32	25		57	73.50
	9400	Mechanical joint ends for plain end pipe									
	9410	Malleable iron, black									
	9420	90° Elbows, 1-1/4"	Q-1	29	.552	Ea.	9.90	20.50		30.40	41.50
	9430	1-1/2"		27	.593		11.30	22		33.30	45.50
	9440	2"		24	.667		12.40	24.50		36.90	50.50
	9490	Tee, reducing outlet									
	9510	1-1/4" x 1/2"	Q-1	18	.889	Ea.	5.90	32.50		38.40	55.50
	9520	1-1/4" x 3/4"		18	.889		5.90	32.50		38.40	55.50
	9530	1-1/4" x 1"		18	.889		5.90	32.50		38.40	55.50
	9540	1-1/2" x 1/2"		17	.941		6.25	34.50		40.75	59
	9550	1-1/2" x 3/4"		17	.941		6.25	34.50		40.75	59
	9560	1-1/2" x 1"		17	.941		6.25	34.50		40.75	59
	9570	2" x 1/2"		15	1.067		7.65	39		46.65	67.50
	9580	2" x 3/4"		15	1.067		7.65	39		46.65	67.50
	9590	2" x 1"		15	1.067		7.65	39		46.65	67.50
	9640	Tee, reducing run and outlet									
	9660	1-1/4" x 1" x 1/2"	Q-1	18	.889	Ea.	6.30	32.50		38.80	56
	9670	1-1/4" x 1" x 3/4"		18	.889		6.30	32.50		38.80	56
	9680	1-1/4" x 1" x 1"		18	.889		6.30	32.50		38.80	56
	9690	1-1/2" x 1-1/4" x 1/2"		17	.941		6.70	34.50		41.20	59.50
	9700	1-1/2" x 1-1/4" x 3/4"		17	.941		6.70	34.50		41.20	59.50
	9710	1-1/2" x 1-1/4" x 1"		17	.941		6.70	34.50		41.20	59.50
	9720	2" x 1-1/2" x 1/2"		15	1.067		8.10	39		47.10	68
	9730	2" x 1-1/2" x 3/4"		15	1.067		8.10	39		47.10	68
	9740	2" x 1-1/2" x 1"		15	1.067		8.10	39		47.10	68
	9790	Tee, outlet									
	9810	3" x 1-1/4"	Q-1	29	.552	Ea.	14.25	20.50		34.75	46
	9820	3" x 1-1/2"		28	.571		14.25	21		35.25	47
	9830	3" x 2"		26	.615		15.90	22.50		38.40	51.50
	9840	4" x 1-1/4"		28	.571		17.70	21		38.70	51
	9850	4" x 1-1/2"		26	.615		18.60	22.50		41.10	54.50
	9860	4" x 2"		24	.667		18.60	24.50		43.10	57.50
	9940	For galvanized fittings for plain end pipe, add					20%				

15107	Metal Pipe & Fittings		CREW	DAILY OUTPUT	LABOR-HOURS	UNIT	2005 BARE COSTS				TOTAL INCL O&P
							MAT.	LABOR	EQUIP.	TOTAL	
690	0010	**PIPE, GROOVED-JOINT STEEL FITTINGS & VALVES**	R15100 -070									690
	0012	Fittings are ductile iron. Steel fittings noted.									
	0020	Pipe includes coupling & clevis type hanger 10' O.C.									
	0500	Schedule 10, black									
	0550	2" diameter	1 Plum	43	.186	L.F.	4.10	7.60		11.70	15.90
	0560	2-1/2" diameter	Q-1	61	.262		5.10	9.65		14.75	20
	0570	3" diameter		55	.291		5.90	10.70		16.60	22.50
	0580	3-1/2" diameter		53	.302		7.30	11.10		18.40	25
	0590	4" diameter		49	.327		7.05	12		19.05	26
	0600	5" diameter		40	.400		13.45	14.70		28.15	37
	0610	6" diameter	Q-2	46	.522		13.75	19.90		33.65	45
	0620	8" diameter	"	41	.585		27	22.50		49.50	63
	0700	To delete couplings & hangers, subtract									
	0710	2" diam. to 5" diam.					25%	20%			
	0720	6" diam. to 8" diam.					27%	15%			
	1000	Schedule 40, black									
	1040	3/4" diameter	1 Plum	71	.113	L.F.	1.98	4.60		6.58	9.10
	1050	1" diameter		63	.127		1.85	5.20		7.05	9.85
	1060	1-1/4" diameter		58	.138		2.37	5.65		8.02	11.05
	1070	1-1/2" diameter		51	.157		2.21	6.40		8.61	12.10
	1080	2" diameter		40	.200		3.30	8.15		11.45	15.95
	1090	2-1/2" diameter	Q-1	57	.281		4.80	10.30		15.10	21
	1100	3" diameter		50	.320		6	11.75		17.75	24.50
	1110	4" diameter		45	.356		10.95	13.10		24.05	31.50
	1120	5" diameter		37	.432		19.95	15.90		35.85	46
	1130	6" diameter	Q-2	42	.571		25.50	22		47.50	61
	1140	8" diameter		37	.649		38	24.50		62.50	79
	1150	10" diameter		31	.774		58.50	29.50		88	109
	1160	12" diameter		27	.889		79.50	34		113.50	139
	1170	14" diameter		20	1.200		91	46		137	169
	1180	16" diameter		17	1.412		99.50	54		153.50	191
	1190	18" diameter		14	1.714		160	65.50		225.50	275
	1200	20" diameter		12	2		190	76.50		266.50	325
	1210	24" diameter		10	2.400		224	91.50		315.50	385
	1740	To delete coupling & hanger, subtract									
	1750	3/4" diam. to 2" diam.					65%	27%			
	1760	2-1/2" diam. to 5" diam.					41%	18%			
	1770	6" diam. to 12" diam.					31%	13%			
	1780	14" diam. to 24" diam.					35%	10%			
	1800	Galvanized									
	1840	3/4" diameter	1 Plum	71	.113	L.F.	2.19	4.60		6.79	9.30
	1850	1" diameter		63	.127		2.13	5.20		7.33	10.15
	1860	1-1/4" diameter		58	.138		2.76	5.65		8.41	11.50
	1870	1-1/2" diameter		51	.157		3.18	6.40		9.58	13.15
	1880	2" diameter		40	.200		3.91	8.15		12.06	16.60
	1890	2-1/2" diameter	Q-1	57	.281		5.75	10.30		16.05	22
	1900	3" diameter		50	.320		7.25	11.75		19	25.50
	1910	4" diameter		45	.356		10	13.10		23.10	30.50
	1920	5" diameter		37	.432		26	15.90		41.90	52.50
	1930	6" diameter	Q-2	42	.571		33.50	22		55.50	70
	1940	8" diameter		37	.649		49.50	24.50		74	91.50
	1950	10" diameter		31	.774		75	29.50		104.50	127
	1960	12" diameter		27	.889		103	34		137	164
	2540	To delete coupling & hanger, subtract									
	2550	3/4" diam. to 2" diam.					36%	27%			
	2560	2-1/2" diam. to 5" diam.					19%	18%			

15107 | Metal Pipe & Fittings

		CREW	DAILY OUTPUT	LABOR-HOURS	UNIT	2005 BARE COSTS				TOTAL INCL O&P
						MAT.	LABOR	EQUIP.	TOTAL	
690	2570	6" diam. to 12" diam.			R15100 -070		14%	13%				690
	2600	Schedule 80, black									
	2610	3/4" diameter	1 Plum	65	.123	L.F.	3.99	5.05		9.04	11.95
	2650	1" diameter		61	.131		3.95	5.35		9.30	12.40
	2660	1-1/4" diameter		55	.145		5.30	5.95		11.25	14.80
	2670	1-1/2" diameter		49	.163		6.45	6.65		13.10	17.05
	2680	2" diameter		38	.211		7.45	8.60		16.05	21
	2690	2-1/2" diameter	Q-1	54	.296		9.50	10.90		20.40	27
	2700	3" diameter		48	.333		12.35	12.25		24.60	32
	2710	4" diameter		44	.364		24	13.35		37.35	46
	2720	5" diameter		35	.457		27	16.80		43.80	55
	2730	6" diameter	Q-2	40	.600		44.50	23		67.50	83.50
	2740	8" diameter		35	.686		86.50	26		112.50	135
	2750	10" diameter		29	.828		113	31.50		144.50	172
	2760	12" diameter		24	1		151	38		189	224
	3240	To delete coupling & hanger, subtract									
	3250	3/4" diam. to 2" diam.					30%	25%			
	3260	2-1/2" diam. to 5" diam.					14%	17%			
	3270	6" diam. to 12" diam.					12%	12%			
	3300	Galvanized									
	3310	3/4" diameter	1 Plum	65	.123	L.F.	5.20	5.05		10.25	13.25
	3350	1" diameter		61	.131		6.95	5.35		12.30	15.70
	3360	1-1/4" diameter		55	.145		8.25	5.95		14.20	18.05
	3370	1-1/2" diameter		46	.174		12.30	7.10		19.40	24.50
	3380	2" diameter		38	.211		13.15	8.60		21.75	27.50
	3390	2-1/2" diameter	Q-1	54	.296		14.95	10.90		25.85	33
	3400	3" diameter		48	.333		18.95	12.25		31.20	39.50
	3410	4" diameter		44	.364		29.50	13.35		42.85	52.50
	3420	5" diameter		35	.457		36.50	16.80		53.30	65.50
	3430	6" diameter	Q-2	40	.600		56	23		79	96
	3440	8" diameter		35	.686		93.50	26		119.50	143
	3450	10" diameter		29	.828		110	31.50		141.50	169
	3460	12" diameter		24	1		147	38		185	220
	3920	To delete coupling & hanger, subtract									
	3930	3/4" diam. to 2" diam.					30%	25%			
	3940	2-1/2" diam. to 5" diam.					15%	17%			
	3950	6" diam. to 12" diam.					11%	12%			
	3990	Fittings: cplg. & labor required at joints not incl. in fitting									
	3994	price. Add 1 per joint for installed price.									
	4000	Elbow, 90° or 45°, painted									
	4030	3/4" diameter	1 Plum	50	.160	Ea.	19.10	6.55		25.65	31
	4040	1" diameter		50	.160		10.15	6.55		16.70	21
	4050	1-1/4" diameter		40	.200		10.15	8.15		18.30	23.50
	4060	1-1/2" diameter		33	.242		10.15	9.90		20.05	26
	4070	2" diameter		25	.320		10.15	13.05		23.20	31
	4080	2-1/2" diameter	Q-1	40	.400		10.15	14.70		24.85	33
	4090	3" diameter		33	.485		18	17.85		35.85	47
	4100	4" diameter		25	.640		19.60	23.50		43.10	57
	4110	5" diameter		20	.800		47	29.50		76.50	96
	4120	6" diameter	Q-2	25	.960		55.50	36.50		92	116
	4130	8" diameter		21	1.143		116	43.50		159.50	194
	4140	10" diameter		18	1.333		211	51		262	310
	4150	12" diameter		15	1.600		340	61		401	460
	4170	14" diameter		12	2		605	76.50		681.50	780
	4180	16" diameter		11	2.182		785	83		868	990
	4190	18" diameter	Q-3	15	2.133		1,000	83		1,083	1,225

15107 | Metal Pipe & Fittings

		CREW	DAILY OUTPUT	LABOR-HOURS	UNIT	2005 BARE COSTS				TOTAL INCL O&P
						MAT.	LABOR	EQUIP.	TOTAL	
4200	20" diameter	Q-3	13	2.462	Ea.	1,325	96		1,421	1,600
4210	24" diameter	↓	11	2.909	↓	1,900	113		2,013	2,275
4250	For galvanized elbows, add				↓	26%				
4690	Tee, painted									
4700	3/4" diameter	1 Plum	38	.211	Ea.	20.50	8.60		29.10	35.50
4740	1" diameter		33	.242		15.60	9.90		25.50	32
4750	1-1/4" diameter		27	.296		15.60	12.10		27.70	35.50
4760	1-1/2" diameter		22	.364		15.60	14.85		30.45	39.50
4770	2" diameter	↓	17	.471		15.60	19.20		34.80	46
4780	2-1/2" diameter	Q-1	27	.593		15.60	22		37.60	50
4790	3" diameter		22	.727		22	27		49	64
4800	4" diameter		17	.941		33	34.50		67.50	88.50
4810	5" diameter	↓	13	1.231		76	45.50		121.50	152
4820	6" diameter	Q-2	17	1.412		90	54		144	180
4830	8" diameter		14	1.714		198	65.50		263.50	315
4840	10" diameter		12	2		410	76.50		486.50	570
4850	12" diameter		10	2.400		575	91.50		666.50	770
4851	14" diameter		9	2.667		610	102		712	825
4852	16" diameter	↓	8	3		685	114		799	925
4853	18" diameter	Q-3	11	2.909		860	113		973	1,125
4854	20" diameter		10	3.200		1,225	125		1,350	1,525
4855	24" diameter	↓	8	4		1,875	156		2,031	2,300
4900	For galvanized tees, add				↓	24%				
4906	Couplings, rigid style, painted									
4908	1" diameter	1 Plum	100	.080	Ea.	7.45	3.27		10.72	13.10
4909	1-1/4" diameter		100	.080		7.45	3.27		10.72	13.10
4910	1-1/2" diameter		67	.119		7.45	4.88		12.33	15.55
4912	2" diameter	↓	50	.160		7.65	6.55		14.20	18.25
4914	2-1/2" diameter	Q-1	80	.200		8.80	7.35		16.15	21
4916	3" diameter		67	.239		10.30	8.80		19.10	24.50
4918	4" diameter		50	.320		14.60	11.75		26.35	34
4920	5" diameter	↓	40	.400		18.90	14.70		33.60	43
4922	6" diameter	Q-2	50	.480		25.50	18.30		43.80	55.50
4924	8" diameter		42	.571		40	22		62	76.50
4926	10" diameter		35	.686		71	26		97	118
4928	12" diameter		32	.750		79.50	28.50		108	131
4930	14" diameter		24	1		104	38		142	172
4931	16" diameter		20	1.200		136	46		182	218
4932	18" diameter		18	1.333		157	51		208	250
4933	20" diameter	↓	16	1.500		179	57		236	283
4934	24" diameter	Q-9	13	1.231	↓	275	44.50		319.50	370
4940	Flexible, standard, painted									
4950	3/4" diameter	1 Plum	100	.080	Ea.	5.40	3.27		8.67	10.80
4960	1" diameter		100	.080		5.40	3.27		8.67	10.80
4970	1-1/4" diameter		80	.100		7.15	4.09		11.24	14
4980	1-1/2" diameter		67	.119		7.80	4.88		12.68	15.95
4990	2" diameter	↓	50	.160		8.25	6.55		14.80	18.90
5000	2-1/2" diameter	Q-1	80	.200		9.85	7.35		17.20	22
5010	3" diameter		67	.239		10.90	8.80		19.70	25
5020	3-1/2" diameter		57	.281		15.85	10.30		26.15	33
5030	4" diameter		50	.320		12.95	11.75		24.70	32
5040	5" diameter	↓	40	.400		24.50	14.70		39.20	48.50
5050	6" diameter	Q-2	50	.480		28.50	18.30		46.80	59
5070	8" diameter		42	.571		46.50	22		68.50	84.50
5090	10" diameter		35	.686		77.50	26		103.50	125
5110	12" diameter	↓	32	.750	↓	88.50	28.50		117	140

Important: See the Reference Section for critical supporting data - Reference Nos., Crews, & City Cost Indexes

15107 | Metal Pipe & Fittings

		CREW	DAILY OUTPUT	LABOR-HOURS	UNIT	2005 BARE COSTS MAT.	LABOR	EQUIP.	TOTAL	TOTAL INCL O&P	
690	5120	14" diameter	Q-2	24	1	Ea.	107	38		145	176	**690**
	5130	16" diameter		20	1.200		141	46		187	224	
	5140	18" diameter		18	1.333		165	51		216	258	
	5150	20" diameter		16	1.500		259	57		316	370	
	5160	24" diameter		13	1.846		276	70.50		346.50	410	
	5176	Lightweight style, painted										
	5178	1-1/2" diameter	1 Plum	67	.119	Ea.	6.80	4.88		11.68	14.85	
	5180	2" diameter	"	50	.160		6.95	6.55		13.50	17.45	
	5182	2-1/2" diameter	Q-1	80	.200		8.10	7.35		15.45	19.95	
	5184	3" diameter		67	.239		9.35	8.80		18.15	23.50	
	5186	3-1/2" diameter		57	.281		13.30	10.30		23.60	30	
	5188	4" diameter		50	.320		13.30	11.75		25.05	32.50	
	5190	5" diameter		40	.400		19.10	14.70		33.80	43	
	5192	6" diameter	Q-2	50	.480		23	18.30		41.30	53	
	5194	8" diameter		42	.571		36	22		58	73	
	5196	10" diameter		35	.686		97	26		123	147	
	5198	12" diameter		32	.750		108	28.50		136.50	162	
	5200	For galvanized couplings, add					33%					
	5220	Tee, reducing, painted										
	5225	2" x 1-1/2" diameter	Q-1	38	.421	Ea.	33.50	15.50		49	60.50	
	5226	2-1/2" x 2" diameter		28	.571		33.50	21		54.50	68.50	
	5227	3" x 2-1/2" diameter		23	.696		29.50	25.50		55	71	
	5228	4" x 3" diameter		18	.889		40	32.50		72.50	93	
	5229	5" x 4" diameter		15	1.067		86	39		125	154	
	5230	6" x 4" diameter	Q-2	18	1.333		95	51		146	181	
	5231	8" x 6" diameter		15	1.600		198	61		259	310	
	5232	10" x 8" diameter		13	1.846		260	70.50		330.50	390	
	5233	12" x 10" diameter		11	2.182		395	83		478	560	
	5234	14" x 12" diameter		10	2.400		445	91.50		536.50	625	
	5235	16" x 12" diameter		9	2.667		550	102		652	760	
	5236	18" x 12" diameter	Q-3	12	2.667		660	104		764	880	
	5237	18" x 16" diameter		11	2.909		830	113		943	1,075	
	5238	20" x 16" diameter		10	3.200		1,075	125		1,200	1,375	
	5239	24" x 20" diameter		9	3.556		1,725	138		1,863	2,075	
	5240	Reducer, concentric, painted										
	5241	2-1/2" x 2" diameter	Q-1	43	.372	Ea.	11.80	13.70		25.50	33.50	
	5242	3" x 2-1/2" diameter		35	.457		14.25	16.80		31.05	41	
	5243	4" x 3" diameter		29	.552		17.20	20.50		37.70	49.50	
	5244	5" x 4" diameter		22	.727		24	27		51	66.50	
	5245	6" x 4" diameter	Q-2	26	.923		28	35		63	83.50	
	5246	8" x 6" diameter		23	1.043		72.50	40		112.50	140	
	5247	10" x 8" diameter		20	1.200		148	46		194	232	
	5248	12" x 10" diameter		16	1.500		266	57		323	380	
	5255	Eccentric, painted										
	5256	2-1/2" x 2" diameter	Q-1	42	.381	Ea.	25	14		39	48.50	
	5257	3" x 2-1/2" diameter		34	.471		28.50	17.30		45.80	57.50	
	5258	4" x 3" diameter		28	.571		35	21		56	70	
	5259	5" x 4" diameter		21	.762		47.50	28		75.50	94.50	
	5260	6" x 4" diameter	Q-2	25	.960		55.50	36.50		92	116	
	5261	8" x 6" diameter		22	1.091		112	41.50		153.50	186	
	5262	10" x 8" diameter		19	1.263		310	48		358	415	
	5263	12" x 10" diameter		15	1.600		425	61		486	555	
	5270	Coupling, reducing, painted										
	5272	2" x 1-1/2" diameter	1 Plum	52	.154	Ea.	11.75	6.30		18.05	22.50	
	5274	2-1/2" x 2" diameter	Q-1	82	.195		15.35	7.20		22.55	27.50	
	5276	3" x 2" diameter		69	.232		17.60	8.55		26.15	32	

R15100 -070

15107 | Metal Pipe & Fittings

		CREW	DAILY OUTPUT	LABOR-HOURS	UNIT	2005 BARE COSTS				TOTAL INCL O&P
						MAT.	LABOR	EQUIP.	TOTAL	
5278	4" x 2" diameter	Q-1	52	.308	Ea.	28	11.30		39.30	48
5280	5" x 4" diameter	↓	42	.381		31.50	14		45.50	56
5282	6" x 4" diameter	Q-2	52	.462		48	17.60		65.60	79
5284	8" x 6" diameter	"	44	.545	↓	72	21		93	111
5290	Outlet coupling, painted									
5294	1-1/2" x 1" pipe size	1 Plum	65	.123	Ea.	15	5.05		20.05	24
5296	2" x 1" pipe size	"	48	.167		15.40	6.80		22.20	27
5298	2-1/2" x 1" pipe size	Q-1	78	.205		24	7.55		31.55	38
5300	2-1/2" x 1" pipe size	1 Plum	70	.114		27	4.67		31.67	37
5302	3" x 1" pipe size	Q-1	65	.246		23	9.05		32.05	39
5304	4" x 3/4" pipe size		48	.333		34	12.25		46.25	56
5306	4" x 1-1/2" pipe size	↓	46	.348		48.50	12.80		61.30	72.50
5308	6" x 1-1/2" pipe size	Q-2	44	.545	↓	68.50	21		89.50	107
5750	Flange, w/groove gasket, black steel									
5752	See 15107-660-0620 for gasket & bolt set									
5760	ANSI class 125 and 150, painted									
5780	2" pipe size	1 Plum	23	.348	Ea.	34	14.20		48.20	59
5790	2-1/2" pipe size	Q-1	37	.432		42.50	15.90		58.40	70.50
5800	3" pipe size		31	.516		45.50	19		64.50	79
5820	4" pipe size		23	.696		61.50	25.50		87	106
5830	5" pipe size	↓	19	.842		71	31		102	125
5840	6" pipe size	Q-2	23	1.043		77.50	40		117.50	146
5850	8" pipe size		17	1.412		87.50	54		141.50	178
5860	10" pipe size		14	1.714		139	65.50		204.50	252
5870	12" pipe size		12	2		182	76.50		258.50	315
5880	14" pipe size		10	2.400		415	91.50		506.50	600
5890	16" pipe size		9	2.667		485	102		587	690
5900	18" pipe size		6	4		595	153		748	885
5910	20" pipe size		5	4.800		720	183		903	1,075
5920	24" pipe size	↓	4.50	5.333	↓	920	203		1,123	1,300
5940	ANSI class 350, painted									
5946	2" pipe size	1 Plum	23	.348	Ea.	42.50	14.20		56.70	68.50
5948	2-1/2" pipe size	Q-1	37	.432		49.50	15.90		65.40	78.50
5950	3" pipe size		31	.516		67.50	19		86.50	103
5952	4" pipe size		23	.696		90.50	25.50		116	138
5954	5" pipe size	↓	19	.842		103	31		134	160
5956	6" pipe size	Q-2	23	1.043		119	40		159	191
5958	8" pipe size		17	1.412		137	54		191	232
5960	10" pipe size	↓	14	1.714		219	65.50		284.50	340
5962	12" pipe size	1 Plum	12	.667		233	27		260	298
6100	Coupling, for PVC plastic pipe									
6110	2" diameter	1 Plum	50	.160	Ea.	9.20	6.55		15.75	19.90
6112	2-1/2" diameter	Q-1	80	.200		12.85	7.35		20.20	25
6114	3" diameter		67	.239		15.60	8.80		24.40	30.50
6116	4" diameter	↓	50	.320		20.50	11.75		32.25	40
6118	6" diameter	Q-2	50	.480		33.50	18.30		51.80	64.50
6120	8" diameter		42	.571		55.50	22		77.50	94
6122	10" diameter		35	.686		91	26		117	140
6124	12" diameter	↓	32	.750	↓	115	28.50		143.50	170
7400	Suction diffuser									
7402	Grooved end inlet x flanged outlet									
7410	3" x 3"	Q-1	50	.320	Ea.	355	11.75		366.75	410
7412	4" x 4"		38	.421		480	15.50		495.50	555
7414	5" x 5"	↓	30	.533		560	19.60		579.60	650
7416	6" x 6"	Q-2	38	.632		710	24		734	815
7418	8" x 8"	↓	27	.889	↓	1,325	34		1,359	1,500

R15100-070

690

Important: See the Reference Section for critical supporting data - Reference Nos., Crews, & City Cost Indexes

			DAILY	LABOR-		2005 BARE COSTS				TOTAL
15107	**Metal Pipe & Fittings**	CREW	OUTPUT	HOURS	UNIT	MAT.	LABOR	EQUIP.	TOTAL	INCL O&P
690	7420	10" x 10"	Q-2	20	1.200	Ea.	1,800	46		1,846	2,050
	7422	12" x 12"		16	1.500		2,950	57		3,007	3,325
	7424	14" x 14"		15	1.600		3,725	61		3,786	4,200
	7426	16" x 14"	↓	14	1.714	↓	3,825	65.50		3,890.50	4,325
	7500	Strainer, tee type, painted									
	7506	2" pipe size	1 Plum	38	.211	Ea.	221	8.60		229.60	256
	7508	2-1/2" pipe size	Q-1	62	.258		232	9.50		241.50	269
	7510	3" pipe size		50	.320		260	11.75		271.75	305
	7512	4" pipe size		38	.421		295	15.50		310.50	350
	7514	5" pipe size	↓	30	.533		425	19.60		444.60	500
	7516	6" pipe size	Q-2	38	.632		465	24		489	545
	7518	8" pipe size		27	.889		710	34		744	830
	7520	10" pipe size		20	1.200		1,050	46		1,096	1,225
	7522	12" pipe size		16	1.500		1,350	57		1,407	1,550
	7524	14" pipe size		15	1.600		4,775	61		4,836	5,375
	7526	16" pipe size	↓	14	1.714	↓	5,950	65.50		6,015.50	6,650
	7570	Expansion joint, max. 3" travel									
	7572	2" diameter	1 Plum	38	.211	Ea.	181	8.60		189.60	212
	7574	3" diameter	Q-1	50	.320		213	11.75		224.75	252
	7576	4" diameter	"	38	.421		266	15.50		281.50	315
	7578	6" diameter	Q-2	38	.632	↓	465	24		489	545
	7800	Ball valve w/handle, carbon steel trim									
	7810	1-1/2" pipe size	1 Plum	50	.160	Ea.	63	6.55		69.55	79
	7812	2" pipe size	"	38	.211		69.50	8.60		78.10	89.50
	7814	2-1/2" pipe size	Q-1	62	.258		152	9.50		161.50	181
	7816	3" pipe size		50	.320		246	11.75		257.75	289
	7818	4" pipe size	↓	38	.421		380	15.50		395.50	445
	7820	6" pipe size	Q-2	30	.800	↓	1,175	30.50		1,205.50	1,350
	7830	With gear operator									
	7834	2-1/2" pipe size	Q-1	62	.258	Ea.	310	9.50		319.50	355
	7836	3" pipe size		50	.320		435	11.75		446.75	500
	7838	4" pipe size	↓	38	.421		555	15.50		570.50	640
	7840	6" pipe size	Q-2	30	.800	↓	1,325	30.50		1,355.50	1,500
	7870	Check valve									
	7874	2-1/2" pipe size	Q-1	62	.258	Ea.	109	9.50		118.50	134
	7876	3" pipe size		50	.320		128	11.75		139.75	159
	7878	4" pipe size		38	.421		136	15.50		151.50	173
	7880	5" pipe size	↓	30	.533		226	19.60		245.60	279
	7882	6" pipe size	Q-2	38	.632		268	24		292	330
	7884	8" pipe size		27	.889		365	34		399	455
	7886	10" pipe size		20	1.200		1,050	46		1,096	1,250
	7888	12" pipe size	↓	16	1.500	↓	1,250	57		1,307	1,450
	7900	Plug valve, balancing, w/lever operator									
	7906	3" pipe size	Q-1	50	.320	Ea.	245	11.75		256.75	288
	7908	4" pipe size	"	38	.421		300	15.50		315.50	355
	7909	6" pipe size	Q-2	30	.800	↓	465	30.50		495.50	555
	7916	With gear operator									
	7920	3" pipe size	Q-1	50	.320	Ea.	445	11.75		456.75	510
	7922	4" pipe size	"	38	.421		460	15.50		475.50	530
	7926	8" pipe size	Q-2	27	.889		840	34		874	975
	7928	10" pipe size		20	1.200		1,300	46		1,346	1,500
	7930	12" pipe size	↓	16	1.500	↓	1,975	57		2,032	2,250
	8000	Butterfly valve, 2 position handle, with standard trim									
	8010	1-1/2" pipe size	1 Plum	50	.160	Ea.	96	6.55		102.55	116
	8020	2" pipe size	"	38	.211		96	8.60		104.60	119
	8030	3" pipe size	Q-1	50	.320	↓	138	11.75		149.75	170

R15100
-070

15107 | Metal Pipe & Fittings

		CREW	DAILY OUTPUT	LABOR-HOURS	UNIT	2005 BARE COSTS				TOTAL INCL O&P
						MAT.	LABOR	EQUIP.	TOTAL	
8050	4" pipe size	Q-1	38	.421	Ea.	171	15.50		186.50	212
8070	6" pipe size	Q-2	38	.632		305	24		329	370
8080	8" pipe size		27	.889		435	34		469	525
8090	10" pipe size		20	1.200		625	46		671	760
8200	With stainless steel trim									
8240	1-1/2" pipe size	1 Plum	50	.160	Ea.	122	6.55		128.55	144
8250	2" pipe size	"	38	.211		122	8.60		130.60	147
8270	3" pipe size	Q-1	50	.320		164	11.75		175.75	198
8280	4" pipe size	"	38	.421		178	15.50		193.50	220
8300	6" pipe size	Q-2	38	.632		330	24		354	400
8310	8" pipe size		27	.889		505	34		539	610
8320	10" pipe size		20	1.200		725	46		771	870
8322	12" pipe size		16	1.500		1,150	57		1,207	1,325
8324	14" pipe size		15	1.600		1,500	61		1,561	1,750
8326	16" pipe size		14	1.714		2,050	65.50		2,115.50	2,375
8328	18" pipe size	Q-3	12	2.667		2,525	104		2,629	2,950
8330	20" pipe size		11	2.909		3,425	113		3,538	3,950
8332	24" pipe size		10	3.200		4,375	125		4,500	4,975
8336	Note: sizes 8" up w/manual gear operator									
9000	Cut one groove, labor									
9010	3/4" pipe size	Q-1	152	.105	Ea.		3.87		3.87	5.80
9020	1" pipe size		140	.114			4.20		4.20	6.30
9030	1-1/4" pipe size		124	.129			4.75		4.75	7.15
9040	1-1/2" pipe size		114	.140			5.15		5.15	7.75
9050	2" pipe size		104	.154			5.65		5.65	8.50
9060	2-1/2" pipe size		96	.167			6.15		6.15	9.20
9070	3" pipe size		88	.182			6.70		6.70	10.05
9080	3-1/2" pipe size		83	.193			7.10		7.10	10.65
9090	4" pipe size		78	.205			7.55		7.55	11.35
9100	5" pipe size		72	.222			8.15		8.15	12.30
9110	6" pipe size		70	.229			8.40		8.40	12.65
9120	8" pipe size		54	.296			10.90		10.90	16.40
9130	10" pipe size		38	.421			15.50		15.50	23.50
9140	12" pipe size		30	.533			19.60		19.60	29.50
9150	14" pipe size		20	.800			29.50		29.50	44
9160	16" pipe size		19	.842			31		31	46.50
9170	18" pipe size		18	.889			32.50		32.50	49
9180	20" pipe size		17	.941			34.50		34.50	52
9190	24" pipe size		15	1.067			39		39	59
9210	Roll one groove									
9220	3/4" pipe size	Q-1	266	.060	Ea.		2.21		2.21	3.33
9230	1" pipe size		228	.070			2.58		2.58	3.88
9240	1-1/4" pipe size		200	.080			2.94		2.94	4.42
9250	1-1/2" pipe size		178	.090			3.31		3.31	4.97
9260	2" pipe size		116	.138			5.05		5.05	7.60
9270	2-1/2" pipe size		110	.145			5.35		5.35	8.05
9280	3" pipe size		100	.160			5.90		5.90	8.85
9290	3-1/2" pipe size		94	.170			6.25		6.25	9.40
9300	4" pipe size		86	.186			6.85		6.85	10.30
9310	5" pipe size		84	.190			7		7	10.55
9320	6" pipe size		80	.200			7.35		7.35	11.05
9330	8" pipe size		66	.242			8.90		8.90	13.40
9340	10" pipe size		58	.276			10.15		10.15	15.25
9350	12" pipe size		46	.348			12.80		12.80	19.25
9360	14" pipe size		30	.533			19.60		19.60	29.50
9370	16" pipe size		28	.571			21		21	31.50

R15100 -070

690

Important: See the Reference Section for critical supporting data - Reference Nos., Crews, & City Cost Indexes

15107 | Metal Pipe & Fittings

		CREW	DAILY OUTPUT	LABOR-HOURS	UNIT	2005 BARE COSTS				TOTAL INCL O&P
						MAT.	LABOR	EQUIP.	TOTAL	
690 9380	18" pipe size R15100-070	Q-1	27	.593	Ea.		22		22	33 **690**
9390	20" pipe size		25	.640			23.50		23.50	35.50
9400	24" pipe size		23	.696			25.50		25.50	38.50
920 0010	**PIPE, STAINLESS STEEL** R15100-070									**920**
0020	Welded, with clevis type hangers 10' O.C.									
0500	Schedule 5, type 304									
0540	1/2" diameter	Q-15	128	.125	L.F.	4.89	4.60	.63	10.12	12.95
0550	3/4" diameter		116	.138		5.60	5.05	.70	11.35	14.50
0560	1" diameter		103	.155		6.90	5.70	.78	13.38	17.05
0570	1-1/4" diameter		93	.172		7.95	6.35	.87	15.17	19.20
0580	1-1/2" diameter		85	.188		9.20	6.90	.95	17.05	21.50
0590	2" diameter		69	.232		11.10	8.55	1.17	20.82	26.50
0600	2-1/2" diameter		53	.302		15.60	11.10	1.52	28.22	35.50
0610	3" diameter		48	.333		18.85	12.25	1.68	32.78	41.50
0620	4" diameter		44	.364		24	13.35	1.83	39.18	48.50
0630	5" diameter		36	.444		49	16.35	2.24	67.59	81
0640	6" diameter	Q-16	42	.571		45	22	1.92	68.92	84.50
0650	8" diameter		34	.706		69.50	27	2.37	98.87	120
0660	10" diameter		26	.923		98.50	35	3.10	136.60	164
0670	12" diameter		21	1.143		130	43.50	3.84	177.34	213
0700	To delete hangers, subtract									
0710	1/2" diam. to 1-1/2" diam.					8%	19%			
0720	2" diam. to 5" diam.					4%	9%			
0730	6" diam. to 12" diam.					3%	4%			
0750	For small quantities, add				L.F.	10%				
1250	Schedule 5, type 316									
1290	1/2" diameter	Q-15	128	.125	L.F.	6.25	4.60	.63	11.48	14.50
1300	3/4" diameter		116	.138		7.30	5.05	.70	13.05	16.40
1310	1" diameter		103	.155		8.35	5.70	.78	14.83	18.65
1320	1-1/4" diameter		93	.172		10.10	6.35	.87	17.32	21.50
1330	1-1/2" diameter		85	.188		11.50	6.90	.95	19.35	24
1340	2" diameter		69	.232		14.05	8.55	1.17	23.77	29.50
1350	2-1/2" diameter		53	.302		20.50	11.10	1.52	33.12	41
1360	3" diameter		48	.333		26	12.25	1.68	39.93	49
1370	4" diameter		44	.364		32	13.35	1.83	47.18	57
1380	5" diameter		36	.444		61	16.35	2.24	79.59	94.50
1390	6" diameter	Q-16	42	.571		59.50	22	1.92	83.42	101
1400	8" diameter		34	.706		90	27	2.37	119.37	142
1410	10" diameter		26	.923		131	35	3.10	169.10	200
1420	12" diameter		21	1.143		168	43.50	3.84	215.34	255
1490	For small quantities, add					10%				
1940	To delete hanger, subtract									
1950	1/2" diam. to 1-1/2" diam.					5%	19%			
1960	2" diam. to 5" diam.					3%	9%			
1970	6" diam. to 12" diam.					2%	4%			
2000	Schedule 10, type 304									
2040	1/4" diameter	Q-15	131	.122	L.F.	3.78	4.49	.62	8.89	11.60
2050	3/8" diameter		128	.125		4.01	4.60	.63	9.24	12
2060	1/2" diameter		125	.128		5	4.71	.65	10.36	13.30
2070	3/4" diameter		113	.142		5.85	5.20	.71	11.76	15.10
2080	1" diameter		100	.160		8.60	5.90	.81	15.31	19.20
2090	1-1/4" diameter		91	.176		10.45	6.45	.89	17.79	22
2100	1-1/2" diameter		83	.193		11.50	7.10	.97	19.57	24.50
2110	2" diameter		67	.239		14.35	8.80	1.20	24.35	30.50
2120	2-1/2" diameter		51	.314		17.85	11.55	1.58	30.98	38.50

21

15107	Metal Pipe & Fittings	CREW	DAILY OUTPUT	LABOR-HOURS	UNIT	2005 BARE COSTS				TOTAL INCL O&P
						MAT.	LABOR	EQUIP.	TOTAL	
920 2130	3" diameter	Q-15	46	.348	L.F.	22	12.80	1.75	36.55	45 **920**
2140	4" diameter		42	.381		29	14	1.92	44.92	54.50
2150	5" diameter		35	.457		39	16.80	2.30	58.10	70.50
2160	6" diameter	Q-16	40	.600		45	23	2.02	70.02	86
2170	8" diameter		33	.727		69.50	27.50	2.44	99.44	121
2180	10" diameter		25	.960		98.50	36.50	3.23	138.23	167
2190	12" diameter		21	1.143		123	43.50	3.84	170.34	206
2250	For small quantities, add					10%				
2650	To delete hanger, subtract									
2660	1/4" diam. to 3/4" diam.					9%	22%			
2670	1" diam. to 2" diam.					4%	15%			
2680	2-1/2" diam. to 5" diam.					3%	8%			
2690	6" diam. to 12" diam.					3%	4%			
2750	Schedule 10, type 316									
2790	1/4" diameter	Q-15	131	.122	L.F.	4.30	4.49	.62	9.41	12.15
2800	3/8" diameter		128	.125		4.83	4.60	.63	10.06	12.90
2810	1/2" diameter		125	.128		6.25	4.71	.65	11.61	14.65
2820	3/4" diameter		113	.142		7.35	5.20	.71	13.26	16.70
2830	1" diameter		100	.160		9.75	5.90	.81	16.46	20.50
2840	1-1/4" diameter		91	.176		13.25	6.45	.89	20.59	25
2850	1-1/2" diameter		83	.193		14.85	7.10	.97	22.92	28
2860	2" diameter		67	.239		18.40	8.80	1.20	28.40	34.50
2870	2-1/2" diameter		51	.314		24	11.55	1.58	37.13	45.50
2880	3" diameter		46	.348		30.50	12.80	1.75	45.05	54.50
2890	4" diameter		42	.381		37	14	1.92	52.92	63.50
2900	5" diameter		35	.457		55	16.80	2.30	74.10	88.50
2910	6" diameter	Q-16	40	.600		59.50	23	2.02	84.52	102
2920	8" diameter		33	.727		98	27.50	2.44	127.94	152
2930	10" diameter		25	.960		131	36.50	3.23	170.73	203
2940	12" diameter		21	1.143		164	43.50	3.84	211.34	250
2990	For small quantities, add					10%				
3430	To delete hanger, subtract									
3440	1/4" diam. to 3/4" diam.					6%	22%			
3450	1" diam. to 2" diam.					3%	15%			
3460	2-1/2" diam. to 5" diam.					2%	8%			
3470	6" diam. to 12" diam.					2%	4%			
3500	Threaded, couplings and hangers 10' O.C.									
3520	Schedule 40, type 304									
3540	1/4" diameter	1 Plum	54	.148	L.F.	5.25	6.05		11.30	14.90
3550	3/8" diameter		53	.151		6.10	6.15		12.25	15.95
3560	1/2" diameter		52	.154		7.20	6.30		13.50	17.35
3570	3/4" diameter		51	.157		9.60	6.40		16	20
3580	1" diameter		45	.178		12.05	7.25		19.30	24
3590	1-1/4" diameter	Q-1	76	.211		15.60	7.75		23.35	29
3600	1-1/2" diameter		69	.232		18.25	8.55		26.80	33
3610	2" diameter		57	.281		24.50	10.30		34.80	42.50
3620	2-1/2" diameter		44	.364		42	13.35		55.35	66
3630	3" diameter		38	.421		55	15.50		70.50	84
3640	4" diameter	Q-2	51	.471		76.50	17.95		94.45	111
3740	For small quantities, add					10%				
4200	To delete couplings & hangers, subtract									
4210	1/4" diam. to 3/4" diam.					15%	56%			
4220	1" diam. to 2" diam.					18%	49%			
4230	2-1/2" diam. to 4" diam.					34%	40%			
4250	Schedule 40, type 316									
4290	1/4" diameter	1 Plum	54	.148	L.F.	6.15	6.05		12.20	15.85

R15100 -070

Important: See the Reference Section for critical supporting data - Reference Nos., Crews, & City Cost Indexes

		CREW	DAILY OUTPUT	LABOR-HOURS	UNIT	2005 BARE COSTS				TOTAL INCL O&P	
15107	**Metal Pipe & Fittings**					MAT.	LABOR	EQUIP.	TOTAL		
920 4300	3/8" diameter R15100 -070	1 Plum	53	.151	L.F.	7.35	6.15		13.50	17.35	**920**
4310	1/2" diameter		52	.154		9.50	6.30		15.80	19.95	
4320	3/4" diameter		51	.157		11.80	6.40		18.20	22.50	
4330	1" diameter		45	.178		15.45	7.25		22.70	28	
4340	1-1/4" diameter	Q-1	76	.211		19.90	7.75		27.65	33.50	
4350	1-1/2" diameter		69	.232		23.50	8.55		32.05	38.50	
4360	2" diameter		57	.281		31.50	10.30		41.80	50	
4370	2-1/2" diameter		44	.364		54	13.35		67.35	79.50	
4380	3" diameter		38	.421		73.50	15.50		89	104	
4390	4" diameter	Q-2	51	.471		100	17.95		117.95	137	
4490	For small quantities, add					10%					
4900	To delete couplings & hangers, subtract										
4910	1/4" diam. to 3/4" diam.					12%	56%				
4920	1" diam. to 2" diam.					14%	49%				
4930	2-1/2" diam. to 4" diam.					27%	40%				
5000	Schedule 80, type 304										
5040	1/4" diameter	1 Plum	53	.151	L.F.	7.05	6.15		13.20	17.05	
5050	3/8" diameter		52	.154		8.10	6.30		14.40	18.35	
5060	1/2" diameter		51	.157		10.15	6.40		16.55	21	
5070	3/4" diameter		48	.167		13.20	6.80		20	25	
5080	1" diameter		43	.186		18.30	7.60		25.90	31.50	
5090	1-1/4" diameter	Q-1	73	.219		27	8.05		35.05	41.50	
5100	1-1/2" diameter		67	.239		31	8.80		39.80	47	
5110	2" diameter		54	.296		39.50	10.90		50.40	60	
5190	For small quantities, add					10%					
5700	To delete couplings & hangers, subtract										
5710	1/4" diam. to 3/4" diam.					10%	53%				
5720	1" diam. to 2" diam.					14%	47%				
5750	Schedule 80, type 316										
5790	1/4" diameter	1 Plum	53	.151	L.F.	8.75	6.15		14.90	18.90	
5800	3/8" diameter		52	.154		10	6.30		16.30	20.50	
5810	1/2" diameter		51	.157		14.45	6.40		20.85	25.50	
5820	3/4" diameter		48	.167		16.55	6.80		23.35	28.50	
5830	1" diameter		43	.186		23	7.60		30.60	37	
5840	1-1/4" diameter	Q-1	73	.219		34.50	8.05		42.55	50	
5850	1-1/2" diameter		67	.239		39.50	8.80		48.30	56.50	
5860	2" diameter		54	.296		50.50	10.90		61.40	72.50	
5950	For small quantities, add					10%					
7000	To delete couplings & hangers, subtract										
7010	1/4" diam. to 3/4" diam.					9%	53%				
7020	1" diam. to 2" diam.					14%	47%				
8000	Weld joints with clevis type hangers 10' O.C.										
8010	Schedule 40, type 304										
8050	1/8" pipe size	Q-15	126	.127	L.F.	4.35	4.67	.64	9.66	12.50	
8060	1/4" pipe size		125	.128		4.46	4.71	.65	9.82	12.70	
8070	3/8" pipe size		122	.131		5.10	4.82	.66	10.58	13.60	
8080	1/2" pipe size		118	.136		6.25	4.99	.68	11.92	15.10	
8090	3/4" pipe size		109	.147		7.90	5.40	.74	14.04	17.60	
8100	1" pipe size		95	.168		10	6.20	.85	17.05	21	
8110	1-1/4" pipe size		86	.186		12.40	6.85	.94	20.19	25	
8120	1-1/2" pipe size		78	.205		14.50	7.55	1.03	23.08	28.50	
8130	2" pipe size		62	.258		18.75	9.50	1.30	29.55	36	
8140	2-1/2" pipe size		49	.327		28.50	12	1.65	42.15	51	
8150	3" pipe size		44	.364		36.50	13.35	1.83	51.68	62	
8160	3-1/2" pipe size		44	.364		43	13.35	1.83	58.18	69	
8170	4" pipe size		39	.410		49	15.10	2.07	66.17	79	

15107 | Metal Pipe & Fittings

		CREW	DAILY OUTPUT	LABOR-HOURS	UNIT	2005 BARE COSTS				TOTAL INCL O&P
						MAT.	LABOR	EQUIP.	TOTAL	
8180	5" pipe size	Q-15	32	.500	L.F.	71.50	18.40	2.52	92.42	109
8190	6" pipe size	Q-16	37	.649		85.50	24.50	2.18	112.18	133
8200	8" pipe size		29	.828		138	31.50	2.78	172.28	203
8210	10" pipe size		24	1		236	38	3.36	277.36	320
8220	12" pipe size		20	1.200		267	46	4.03	317.03	365
8300	Schedule 40, type 316									
8310	1/8" pipe size	Q-15	126	.127	L.F.	4.31	4.67	.64	9.62	12.45
8320	1/4" pipe size		125	.128		5.15	4.71	.65	10.51	13.45
8330	3/8" pipe size		122	.131		6.20	4.82	.66	11.68	14.85
8340	1/2" pipe size		118	.136		8.35	4.99	.68	14.02	17.45
8350	3/4" pipe size		109	.147		9.75	5.40	.74	15.89	19.60
8360	1" pipe size		95	.168		12.95	6.20	.85	20	24.50
8370	1-1/4" pipe size		86	.186		16.05	6.85	.94	23.84	29
8380	1-1/2" pipe size		78	.205		18.80	7.55	1.03	27.38	33
8390	2" pipe size		62	.258		24.50	9.50	1.30	35.30	42.50
8400	2-1/2" pipe size		49	.327		37.50	12	1.65	51.15	61.50
8410	3" pipe size		44	.364		51	13.35	1.83	66.18	78
8420	3-1/2" pipe size		44	.364		59	13.35	1.83	74.18	87
8430	4" pipe size		39	.410		67	15.10	2.07	84.17	99
8440	5" pipe size		32	.500		102	18.40	2.52	122.92	142
8450	6" pipe size	Q-16	37	.649		114	24.50	2.18	140.68	164
8460	8" pipe size		29	.828		213	31.50	2.78	247.28	286
8470	10" pipe size		24	1		287	38	3.36	328.36	375
8480	12" pipe size		20	1.200		390	46	4.03	440.03	505
8500	Schedule 80, type 304									
8510	1/4" pipe size	Q-15	110	.145	L.F.	6.15	5.35	.73	12.23	15.60
8520	3/8" pipe size		109	.147		7	5.40	.74	13.14	16.60
8530	1/2" pipe size		106	.151		8.90	5.55	.76	15.21	18.95
8540	3/4" pipe size		96	.167		11.50	6.15	.84	18.49	23
8550	1" pipe size		87	.184		15.45	6.75	.93	23.13	28
8560	1-1/4" pipe size		81	.198		20	7.25	1	28.25	34
8570	1-1/2" pipe size		74	.216		23	7.95	1.09	32.04	38
8580	2" pipe size		58	.276		28.50	10.15	1.39	40.04	48.50
8590	2-1/2" pipe size		46	.348		59	12.80	1.75	73.55	86
8600	3" pipe size		41	.390		79.50	14.35	1.97	95.82	111
8610	4" pipe size		33	.485		124	17.85	2.44	144.29	167
8630	6" pipe size	Q-16	30	.800		274	30.50	2.69	307.19	350
8640	Schedule 80, type 316									
8650	1/4" pipe size	Q-15	110	.145	L.F.	7.80	5.35	.73	13.88	17.40
8660	3/8" pipe size		109	.147		8.95	5.40	.74	15.09	18.70
8670	1/2" pipe size		106	.151		12.95	5.55	.76	19.26	23.50
8680	3/4" pipe size		96	.167		14.35	6.15	.84	21.34	26
8690	1" pipe size		87	.184		19.30	6.75	.93	26.98	32
8700	1-1/4" pipe size		81	.198		26.50	7.25	1	34.75	41
8710	1-1/2" pipe size		74	.216		30	7.95	1.09	39.04	46
8720	2" pipe size		58	.276		37.50	10.15	1.39	49.04	58
8730	2-1/2" pipe size		46	.348		97.50	12.80	1.75	112.05	128
8740	3" pipe size		41	.390		114	14.35	1.97	130.32	149
8760	4" pipe size		33	.485		158	17.85	2.44	178.29	204
8770	6" pipe size	Q-16	30	.800		425	30.50	2.69	458.19	520
8790	Schedule 160, type 304									
8800	1/2" pipe size	Q-15	96	.167	L.F.	12.70	6.15	.84	19.69	24
8810	3/4" pipe size		88	.182		14.95	6.70	.92	22.57	27.50
8820	1" pipe size		80	.200		25.50	7.35	1.01	33.86	40
8840	1-1/2" pipe size		67	.239		35	8.80	1.20	45	53.50
8850	2" pipe size		53	.302		51	11.10	1.52	63.62	74.50

Reference box: R15100 -070

15107 | Metal Pipe & Fittings

		CREW	DAILY OUTPUT	LABOR-HOURS	UNIT	MAT.	2005 BARE COSTS LABOR	EQUIP.	TOTAL	TOTAL INCL O&P	
920	8870	3" pipe size	Q-15	37	.432	L.F.	83.50	15.90	2.18	101.58	118	920
	8900	Schedule 160, type 316										
	8910	1/2" pipe size	Q-15	96	.167	L.F.	19	6.15	.84	25.99	31	
	8920	3/4" pipe size		88	.182		20	6.70	.92	27.62	33	
	8930	1" pipe size		80	.200		28.50	7.35	1.01	36.86	43.50	
	8940	1-1/4" pipe size		73	.219		41	8.05	1.10	50.15	58.50	
	8950	1-1/2" pipe size		67	.239		47	8.80	1.20	57	66	
	8960	2" pipe size		53	.302		65	11.10	1.52	77.62	90	
	8990	4" pipe size	↓	30	.533	↓	164	19.60	2.69	186.29	213	
	9100	Threading pipe labor, sst, one end, schedules 40 & 80										
	9110	1/4" through 3/4" pipe size	1 Plum	61.50	.130	Ea.		5.30		5.30	8	
	9120	1" through 2" pipe size		55.90	.143			5.85		5.85	8.80	
	9130	2-1/2" pipe size		41.50	.193			7.85		7.85	11.85	
	9140	3" pipe size	↓	38.50	.208			8.50		8.50	12.75	
	9150	3-1/2" pipe size	Q-1	68.40	.234			8.60		8.60	12.95	
	9160	4" pipe size		73	.219			8.05		8.05	12.10	
	9170	5" pipe size		40.70	.393			14.45		14.45	21.50	
	9180	6" pipe size		35.40	.452			16.60		16.60	25	
	9190	8" pipe size		22.30	.717			26.50		26.50	39.50	
	9200	10" pipe size		16.10	.994			36.50		36.50	55	
	9210	12" pipe size	↓	12.30	1.301	↓		48		48	72	
	9250	Welding labor per joint for stainless steel										
	9260	Schedule 5 and 10										
	9270	1/4" pipe size	Q-15	36	.444	Ea.		16.35	2.24	18.59	27	
	9280	3/8" pipe size		35	.457			16.80	2.30	19.10	28	
	9290	1/2" pipe size		35	.457			16.80	2.30	19.10	28	
	9300	3/4" pipe size		28	.571			21	2.88	23.88	34.50	
	9310	1" pipe size		25	.640			23.50	3.23	26.73	39	
	9320	1-1/4" pipe size		22	.727			27	3.67	30.67	44	
	9330	1-1/2" pipe size		21	.762			28	3.84	31.84	46	
	9340	2" pipe size		18	.889			32.50	4.48	36.98	54	
	9350	2-1/2" pipe size		12	1.333			49	6.70	55.70	81	
	9360	3" pipe size		9.73	1.644			60.50	8.30	68.80	100	
	9370	4" pipe size		7.37	2.171			80	10.95	90.95	132	
	9380	5" pipe size		6.15	2.602			95.50	13.10	108.60	158	
	9390	6" pipe size		5.71	2.802			103	14.10	117.10	171	
	9400	8" pipe size		3.69	4.336			159	22	181	264	
	9410	10" pipe size		2.91	5.498			202	27.50	229.50	335	
	9420	12" pipe size	↓	2.31	6.926	↓		255	35	290	425	
	9500	Schedule 40										
	9510	1/4" pipe size	Q-15	28	.571	Ea.		21	2.88	23.88	34.50	
	9520	3/8" pipe size		27	.593			22	2.99	24.99	36.50	
	9530	1/2" pipe size		25.40	.630			23	3.17	26.17	38.50	
	9540	3/4" pipe size		22.22	.720			26.50	3.63	30.13	44	
	9550	1" pipe size		20.25	.790			29	3.98	32.98	48	
	9560	1-1/4" pipe size		18.82	.850			31.50	4.28	35.78	51.50	
	9570	1-1/2" pipe size		17.78	.900			33	4.54	37.54	55	
	9580	2" pipe size		15.09	1.060			39	5.35	44.35	64.50	
	9590	2-1/2" pipe size		7.96	2.010			74	10.15	84.15	122	
	9600	3" pipe size		6.43	2.488			91.50	12.55	104.05	152	
	9610	4" pipe size		4.88	3.279			121	16.50	137.50	199	
	9620	5" pipe size		4.26	3.756			138	18.95	156.95	229	
	9630	6" pipe size		3.77	4.244			156	21.50	177.50	259	
	9640	8" pipe size		2.44	6.557			241	33	274	395	
	9650	10" pipe size		1.92	8.333			305	42	347	505	
	9660	12" pipe size	↓	1.52	10.526	↓		385	53	438	640	

R15100 -070

15107 | Metal Pipe & Fittings

			CREW	DAILY OUTPUT	LABOR-HOURS	UNIT	MAT.	LABOR	EQUIP.	TOTAL	TOTAL INCL O&P	
920	9750	Schedule 80										**920**
	9760	1/4" pipe size R15100-070	Q-15	21.55	.742	Ea.	27.50	3.74		31.24	45	
	9770	3/8" pipe size		20.75	.771		28.50	3.89		32.39	47	
	9780	1/2" pipe size		19.54	.819		30	4.13		34.13	50	
	9790	3/4" pipe size		17.09	.936		34.50	4.72		39.22	57	
	9800	1" pipe size		15.58	1.027		38	5.20		43.20	62.50	
	9810	1-1/4" pipe size		14.48	1.105		40.50	5.55		46.05	67	
	9820	1-1/2" pipe size		13.68	1.170		43	5.90		48.90	71	
	9830	2" pipe size		11.61	1.378		50.50	6.95		57.45	83.50	
	9840	2-1/2" pipe size		6.12	2.614		96	13.20		109.20	160	
	9850	3" pipe size		4.94	3.239		119	16.30		135.30	197	
	9860	4" pipe size		3.75	4.267		157	21.50		178.50	260	
	9870	5" pipe size		3.27	4.893		180	24.50		204.50	297	
	9880	6" pipe size		2.90	5.517		203	28		231	335	
	9890	8" pipe size		1.87	8.556		315	43		358	525	
	9900	10" pipe size		1.48	10.811		400	54.50		454.50	660	
	9910	12" pipe size		1.17	13.675		505	69		574	830	
	9920	Schedule 160, 1/2" pipe size		17	.941		34.50	4.74		39.24	57	
	9930	3/4" pipe size		14.81	1.080		39.50	5.45		44.95	65.50	
	9940	1" pipe size		13.50	1.185		43.50	5.95		49.45	72	
	9950	1-1/4" pipe size		12.55	1.275		47	6.45		53.45	77.50	
	9960	1-1/2" pipe size		11.85	1.350		49.50	6.80		56.30	82	
	9970	2" pipe size		10	1.600		59	8.05		67.05	97.50	
	9980	3" pipe size		4.28	3.738		138	18.85		156.85	228	
	9990	4" pipe size		3.25	4.923		181	25		206	300	
960	0010	**PIPE, STAINLESS STEEL, FITTINGS**										**960**
	0100	Butt weld joint, schedule 5, type 304										
	0120	90° Elbow, long										
	0140	1/2"	Q-15	17.50	.914	Ea.	11.05	33.50	4.61	49.16	67.50	
	0150	3/4"		14	1.143		17.35	42	5.75	65.10	88.50	
	0160	1"		12.50	1.280		12.70	47	6.45	66.15	92	
	0170	1-1/4"		11	1.455		20.50	53.50	7.35	81.35	111	
	0180	1-1/2"		10.50	1.524		15.90	56	7.70	79.60	110	
	0190	2"		9	1.778		21	65.50	8.95	95.45	131	
	0200	2-1/2"		6	2.667		30.50	98	13.45	141.95	195	
	0210	3"		4.86	3.292		38.50	121	16.60	176.10	242	
	0220	3-1/2"		4.27	3.747		203	138	18.90	359.90	450	
	0230	4"		3.69	4.336		71	159	22	252	340	
	0240	5"		3.08	5.195		505	191	26	722	875	
	0250	6"	Q-16	4.29	5.594		203	213	18.80	434.80	565	
	0260	8"		2.76	8.696		345	330	29	704	910	
	0270	10"		2.18	11.009		710	420	37	1,167	1,450	
	0280	12"		1.73	13.873		1,125	530	46.50	1,701.50	2,075	
	0320	For schedule 5, type 316, add					30%					
	0600	45° Elbow, long										
	0620	1/2"	Q-15	17.50	.914	Ea.	15.90	33.50	4.61	54.01	73	
	0630	3/4"		14	1.143		15.90	42	5.75	63.65	87	
	0640	1"		12.50	1.280		17.95	47	6.45	71.40	98	
	0650	1-1/4"		11	1.455		30.50	53.50	7.35	91.35	122	
	0660	1-1/2"		10.50	1.524		16.35	56	7.70	80.05	110	
	0670	2"		9	1.778		18.75	65.50	8.95	93.20	129	
	0680	2-1/2"		6	2.667		31	98	13.45	142.45	196	
	0690	3"		4.86	3.292		32	121	16.60	169.60	235	
	0700	3-1/2"		4.27	3.747		203	138	18.90	359.90	450	
	0710	4"		3.69	4.336		56.50	159	22	237.50	325	

15107 | Metal Pipe & Fittings

		CREW	DAILY OUTPUT	LABOR-HOURS	UNIT	2005 BARE COSTS				TOTAL INCL O&P
						MAT.	LABOR	EQUIP.	TOTAL	
0720	5"	Q-15	3.08	5.195	Ea.	355	191	26	572	705
0730	6"	Q-16	4.29	5.594		122	213	18.80	353.80	475
0740	8"		2.76	8.696		212	330	29	571	765
0750	10"		2.18	11.009		475	420	37	932	1,200
0760	12"		1.73	13.873		760	530	46.50	1,336.50	1,675
0800	For schedule 5, type 316, add					25%				
1100	Tee, straight									
1130	1/2"	Q-15	11.66	1.372	Ea.	44.50	50.50	6.90	101.90	133
1140	3/4"		9.33	1.715		44.50	63	8.65	116.15	154
1150	1"		8.33	1.921		42.50	70.50	9.70	122.70	164
1160	1-1/4"		7.33	2.183		73	80.50	11	164.50	213
1170	1-1/2"		7	2.286		42.50	84	11.50	138	186
1180	2"		6	2.667		44.50	98	13.45	155.95	211
1190	2-1/2"		4	4		127	147	20	294	385
1200	3"		3.24	4.938		83	182	25	290	390
1210	3-1/2"		2.85	5.614		222	206	28.50	456.50	585
1220	4"		2.46	6.504		118	239	33	390	525
1230	5"		2	8		475	294	40.50	809.50	1,000
1240	6"	Q-16	2.85	8.421		355	320	28.50	703.50	905
1250	8"		1.84	13.043		810	495	44	1,349	1,700
1260	10"		1.45	16.552		1,275	630	55.50	1,960.50	2,400
1270	12"		1.15	20.870		1,600	795	70	2,465	3,050
1320	For schedule 5, type 316, add					25%				
2000	Butt weld joint, schedule 10, type 304									
2020	90° elbow, long									
2040	1/2"	Q-15	17	.941	Ea.	10.40	34.50	4.74	49.64	68.50
2050	3/4"		14	1.143		10.40	42	5.75	58.15	81
2060	1"		12.50	1.280		10.40	47	6.45	63.85	89.50
2070	1-1/4"		11	1.455		19.45	53.50	7.35	80.30	110
2080	1-1/2"		10.50	1.524		15.15	56	7.70	78.85	109
2090	2"		9	1.778		20.50	65.50	8.95	94.95	131
2100	2-1/2"		6	2.667		45	98	13.45	156.45	211
2110	3"		4.86	3.292		41.50	121	16.60	179.10	246
2120	3-1/2"		4.27	3.747		176	138	18.90	332.90	420
2130	4"		3.69	4.336		52.50	159	22	233.50	320
2140	5"		3.08	5.195		128	191	26	345	455
2150	6"	Q-16	4.29	5.594		110	213	18.80	341.80	460
2160	8"		2.76	8.696		137	330	29	496	685
2170	10"		2.18	11.009		395	420	37	852	1,100
2180	12"		1.73	13.873		495	530	46.50	1,071.50	1,400
2500	45° elbow, long									
2520	1/2"	Q-15	17.50	.914	Ea.	5.25	33.50	4.61	43.36	61.50
2530	3/4"		14	1.143		5.25	42	5.75	53	75
2540	1"		12.50	1.280		6.75	47	6.45	60.20	85.50
2550	1-1/4"		11	1.455		10.50	53.50	7.35	71.35	100
2560	1-1/2"		10.50	1.524		15.15	56	7.70	78.85	109
2570	2"		9	1.778		17.40	65.50	8.95	91.85	128
2580	2-1/2"		6	2.667		12	98	13.45	123.45	175
2590	3"		4.86	3.292		13.50	121	16.60	151.10	215
2600	3-1/2"		4.27	3.747		67	138	18.90	223.90	300
2610	4"		3.69	4.336		21	159	22	202	287
2620	5"		3.08	5.195		104	191	26	321	430
2630	6"	Q-16	4.29	5.594		46.50	213	18.80	278.30	390
2640	8"		2.76	8.696		76.50	330	29	435.50	615
2650	10"		2.18	11.009		179	420	37	636	870
2660	12"		1.73	13.873		273	530	46.50	849.50	1,150

960

15107 | Metal Pipe & Fittings

		CREW	DAILY OUTPUT	LABOR-HOURS	UNIT	MAT.	LABOR	EQUIP.	TOTAL	TOTAL INCL O&P
960							2005 BARE COSTS			**960**
3000	Tee, straight									
3030	1/2"	Q-15	11.66	1.372	Ea.	13.50	50.50	6.90	70.90	98.50
3040	3/4"		9.33	1.715		13.50	63	8.65	85.15	119
3050	1"		8.33	1.921		36.50	70.50	9.70	116.70	157
3060	1-1/4"		7.33	2.183		21	80.50	11	112.50	156
3070	1-1/2"		7	2.286		40	84	11.50	135.50	183
3080	2"		6	2.667		40.50	98	13.45	151.95	206
3090	2-1/2"		4	4		38.50	147	20	205.50	285
3100	3"		3.24	4.938		27	182	25	234	330
3110	3-1/2"		2.85	5.614		40	206	28.50	274.50	385
3120	4"		2.46	6.504		111	239	33	383	520
3130	5"	▼	2	8		128	294	40.50	462.50	625
3140	6"	Q-16	2.85	8.421		107	320	28.50	455.50	635
3150	8"		1.84	13.043		233	495	44	772	1,050
3151	10"		1.45	16.552		385	630	55.50	1,070.50	1,425
3152	12"	▼	1.15	20.870		535	795	70	1,400	1,850
3154	For schedule 10, type 316, add				▼	25%				
3281	Butt weld joint, schedule 40, type 304									
3284	90° Elbow, long, 1/2"	Q-15	12.70	1.260	Ea.	5.25	46.50	6.35	58.10	82.50
3288	3/4"		11.10	1.441		5.25	53	7.25	65.50	93.50
3289	1"		10.13	1.579		6	58	7.95	71.95	103
3290	1-1/4"		9.40	1.702		8.25	62.50	8.60	79.35	113
3300	1-1/2"		8.89	1.800		6.75	66	9.05	81.80	117
3310	2"		7.55	2.119		9.75	78	10.70	98.45	140
3320	2-1/2"		3.98	4.020		19.50	148	20.50	188	266
3330	3"		3.21	4.984		25	183	25	233	330
3340	3-1/2"		2.83	5.654		104	208	28.50	340.50	460
3350	4"		2.44	6.557		46.50	241	33	320.50	450
3360	5"	▼	2.13	7.512		164	276	38	478	635
3370	6"	Q-16	2.83	8.481		145	325	28.50	498.50	675
3380	8"		1.83	13.115		244	500	44	788	1,075
3390	10"		1.44	16.667		510	635	56	1,201	1,575
3400	12"	▼	1.14	21.053	▼	720	805	70.50	1,595.50	2,075
3410	For schedule 40, type 316, add					25%				
3460	45° Elbow, long, 1/2"	Q-15	12.70	1.260	Ea.	6.75	46.50	6.35	59.60	84
3470	3/4"		11.10	1.441		6.75	53	7.25	67	95
3480	1"		10.13	1.579		6.75	58	7.95	72.70	104
3490	1-1/4"		9.40	1.702		10.50	62.50	8.60	81.60	115
3500	1-1/2"		8.89	1.800		8.25	66	9.05	83.30	119
3510	2"		7.55	2.119		15.75	78	10.70	104.45	146
3520	2-1/2"		3.98	4.020		19.50	148	20.50	188	266
3530	3"		3.21	4.984		18	183	25	226	325
3540	3-1/2"		2.83	5.654		106	208	28.50	342.50	465
3550	4"		2.44	6.557		33	241	33	307	435
3560	5"	▼	2.13	7.512		114	276	38	428	580
3570	6"	Q-16	2.83	8.481		85.50	325	28.50	439	610
3580	8"		1.83	13.115		162	500	44	706	975
3590	10"		1.44	16.667		305	635	56	996	1,350
3600	12"	▼	1.14	21.053	▼	415	805	70.50	1,290.50	1,725
3610	For schedule 40, type 316, add					25%				
3660	Tee, straight 1/2"	Q-15	8.46	1.891	Ea.	14.25	69.50	9.55	93.30	131
3670	3/4"		7.40	2.162		14.25	79.50	10.90	104.65	148
3680	1"		6.74	2.374		14.25	87.50	11.95	113.70	160
3690	1-1/4"		6.27	2.552		24	94	12.85	130.85	182
3700	1-1/2"		5.92	2.703		14.25	99.50	13.60	127.35	180
3710	2"	▼	5.03	3.181	▼	15.75	117	16.05	148.80	211

Important: See the Reference Section for critical supporting data - Reference Nos., Crews, & City Cost Indexes

15107 | Metal Pipe & Fittings

		CREW	DAILY OUTPUT	LABOR-HOURS	UNIT	MAT.	LABOR	EQUIP.	TOTAL	TOTAL INCL O&P	
960							2005 BARE COSTS					960
3720	2-1/2"	Q-15	2.65	6.038	Ea.	40	222	30.50	292.50	410	
3730	3"		2.14	7.477		32.50	275	37.50	345	490	
3740	3-1/2"		1.88	8.511		25.50	315	43	383.50	545	
3750	4"		1.62	9.877		25.50	365	50	440.50	630	
3760	5"		1.42	11.268		49	415	57	521	740	
3770	6"	Q-16	1.88	12.766		170	485	43	698	965	
3780	8"		1.22	19.672		340	750	66	1,156	1,575	
3790	10"		.96	25		715	955	84	1,754	2,300	
3800	12"		.76	31.579		2,525	1,200	106	3,831	4,700	
3810	For schedule 40, type 316, add					25%					
3820	Tee, reducing on outlet, 3/4" x 1/2"	Q-15	7.73	2.070	Ea.	55.50	76	10.45	141.95	187	
3822	1" x 1/2"		7.24	2.210		77	81.50	11.15	169.65	219	
3824	1" x 3/4"		6.96	2.299		64.50	84.50	11.60	160.60	210	
3826	1-1/4" x 1"		6.43	2.488		203	91.50	12.55	307.05	375	
3828	1-1/2" x 1/2"		6.58	2.432		107	89.50	12.25	208.75	265	
3830	1-1/2" x 3/4"		6.35	2.520		107	92.50	12.70	212.20	271	
3832	1-1/2" x 1"		6.18	2.589		77	95	13.05	185.05	242	
3834	2" x 1"		5.50	2.909		75	107	14.65	196.65	260	
3836	2" x 1-1/2"		5.30	3.019		70.50	111	15.20	196.70	262	
3838	2-1/2" x 2"		3.15	5.079		182	187	25.50	394.50	510	
3840	3" x 1-1/2"		2.72	5.882		155	216	29.50	400.50	530	
3842	3" x 2"		2.65	6.038		131	222	30.50	383.50	515	
3844	4" x 2"		2.10	7.619		236	280	38.50	554.50	720	
3846	4" x 3"		1.77	9.040		193	330	45.50	568.50	760	
3848	5" x 4"		1.48	10.811		610	400	54.50	1,064.50	1,325	
3850	6" x 3"	Q-16	2.19	10.959		485	420	37	942	1,200	
3852	6" x 4"		2.04	11.765		440	450	39.50	929.50	1,200	
3854	8" x 4"		1.46	16.438		810	625	55	1,490	1,900	
3856	10" x 8"		.69	34.783		2,250	1,325	117	3,692	4,600	
3858	12" x 10"		.55	43.636		2,775	1,675	147	4,597	5,700	
3950	Reducer, concentric, 3/4" x 1/2"	Q-15	11.85	1.350		25.50	49.50	6.80	81.80	111	
3952	1" x 3/4"		10.60	1.509		30	55.50	7.60	93.10	125	
3954	1-1/4" x 3/4"		10.19	1.570		85.50	58	7.90	151.40	190	
3956	1-1/4" x 1"		9.76	1.639		46	60.50	8.25	114.75	150	
3958	1-1/2" x 3/4"		9.88	1.619		53.50	59.50	8.15	121.15	157	
3960	1-1/2" x 1"		9.47	1.690		17.90	62	8.50	88.40	123	
3962	2" x 1"		8.65	1.850		21.50	68	9.30	98.80	136	
3964	2" x 1-1/2"		8.16	1.961		17.90	72	9.90	99.80	139	
3966	2-1/2" x 1"		5.71	2.802		78.50	103	14.10	195.60	257	
3968	2-1/2" x 2"		5.21	3.071		64.50	113	15.50	193	258	
3970	3" x 1"		4.88	3.279		85.50	121	16.50	223	293	
3972	3" x 1-1/2"		4.72	3.390		39.50	125	17.10	181.60	249	
3974	3" x 2"		4.51	3.548		32	130	17.90	179.90	251	
3976	4" x 2"		3.69	4.336		44	159	22	225	315	
3978	4" x 3"		2.77	5.776		38.50	212	29	279.50	395	
3980	5" x 3"		2.56	6.250		320	230	31.50	581.50	735	
3982	5" x 4"		2.27	7.048		257	259	35.50	551.50	710	
3984	6" x 3"	Q-16	3.57	6.723		122	256	22.50	400.50	545	
3986	6" x 4"		3.19	7.524		102	287	25.50	414.50	570	
3988	8" x 4"		2.44	9.836		385	375	33	793	1,025	
3990	8" x 6"		2.22	10.811		247	410	36.50	693.50	930	
3992	10" x 6"		1.91	12.565		515	480	42	1,037	1,325	
3994	10" x 8"		1.61	14.907		305	570	50	925	1,250	
3995	12" x 6"		1.63	14.724		770	560	49.50	1,379.50	1,750	
3996	12" x 8"		1.41	17.021		705	650	57	1,412	1,825	
3997	12" x 10"		1.27	18.898		425	720	63.50	1,208.50	1,600	

15107 | Metal Pipe & Fittings

		CREW	DAILY OUTPUT	LABOR-HOURS	UNIT	MAT.	LABOR	EQUIP.	TOTAL	TOTAL INCL O&P
6000	Threaded companion flange									
6010	Stainless steel, 150 lb., type 304									
6020	1/2" diam.	1 Plum	30	.267	Ea.	58	10.90		68.90	80
6030	3/4" diam.		28	.286		64.50	11.65		76.15	88.50
6040	1" diam.	↓	27	.296		72	12.10		84.10	97
6050	1-1/4" diam.	Q-1	44	.364		98	13.35		111.35	128
6060	1-1/2" diam.		40	.400		100	14.70		114.70	132
6070	2" diam.		36	.444		112	16.35		128.35	148
6080	2-1/2" diam.		28	.571		171	21		192	220
6090	3" diam.		20	.800		186	29.50		215.50	249
6110	4" diam.	↓	12	1.333		293	49		342	400
6130	6" diam.	Q-2	14	1.714		560	65.50		625.50	715
6140	8" diam.	"	12	2	↓	1,050	76.50		1,126.50	1,275
6150	For type 316 add					40%				
6260	Weld flanges, stainless steel, type 304									
6270	Slip on, 150 lb. (welded, front and back)									
6280	1/2" diam.	Q-15	12.70	1.260	Ea.	58	46.50	6.35	110.85	140
6290	3/4" diam.		11.11	1.440		64.50	53	7.25	124.75	159
6300	1" diam.		10.13	1.579		40	58	7.95	105.95	140
6310	1-1/4" diam.		9.41	1.700		98	62.50	8.55	169.05	211
6320	1-1/2" diam.		8.89	1.800		59	66	9.05	134.05	174
6330	2" diam.		7.55	2.119		77	78	10.70	165.70	213
6340	2-1/2" diam.		3.98	4.020		171	148	20.50	339.50	435
6350	3" diam.		3.21	4.984		186	183	25	394	510
6370	4" diam.	↓	2.44	6.557		162	241	33	436	575
6390	6" diam.	Q-16	1.89	12.698		255	485	42.50	782.50	1,050
6400	8" diam.	"	1.22	19.672	↓	1,050	750	66	1,866	2,350
6410	For type 316, add					40%				
6530	Weld neck 150 lb.									
6540	1/2" diam.	Q-15	25.40	.630	Ea.	37.50	23	3.17	63.67	80
6550	3/4" diam.		22.22	.720		42.50	26.50	3.63	72.63	91
6560	1" diam.		20.25	.790		50.50	29	3.98	83.48	103
6570	1-1/4" diam.		18.82	.850		61	31.50	4.28	96.78	119
6580	1-1/2" diam.		17.78	.900		71	33	4.54	108.54	133
6590	2" diam.		15.09	1.060		91	39	5.35	135.35	164
6600	2-1/2" diam.		7.96	2.010		152	74	10.15	236.15	289
6610	3" diam.		6.43	2.488		172	91.50	12.55	276.05	340
6630	4" diam.		4.88	3.279		223	121	16.50	360.50	445
6640	5" diam.	↓	4.26	3.756		296	138	18.95	452.95	555
6650	6" diam.	Q-16	5.66	4.240		345	162	14.25	521.25	640
6652	8" diam.		3.65	6.575		485	251	22	758	930
6654	10" diam.		2.88	8.333		665	320	28	1,013	1,250
6656	12" diam.	↓	2.28	10.526	↓	1,050	400	35.50	1,485.50	1,800
6670	For type 316 add					23%				
7000	Threaded joint, 150 lb., type 304									
7030	90° elbow									
7040	1/8"	1 Plum	13	.615	Ea.	16.60	25		41.60	56.50
7050	1/4"		13	.615		16.60	25		41.60	56.50
7070	3/8"		13	.615		18.95	25		43.95	59
7080	1/2"		12	.667		12.70	27		39.70	55
7090	3/4"	↓	11	.727		14.40	29.50		43.90	60.50
7100	1"		10	.800		19.15	32.50		51.65	70
7110	1-1/4"	Q-1	17	.941		39.50	34.50		74	95.50
7120	1-1/2"		16	1		32.50	37		69.50	91.50
7130	2"		14	1.143		47	42		89	115
7140	2-1/2"	↓	11	1.455	↓	174	53.50		227.50	272

Important: See the Reference Section for critical supporting data - Reference Nos., Crews, & City Cost Indexes

15107 | Metal Pipe & Fittings

		CREW	DAILY OUTPUT	LABOR-HOURS	UNIT	2005 BARE COSTS				TOTAL INCL O&P
						MAT.	LABOR	EQUIP.	TOTAL	
7150	3"	Q-1	8	2	Ea.	300	73.50		373.50	440
7160	4"	Q-2	11	2.182	↓	320	83		403	480
7180	45° elbow									
7190	1/8"	1 Plum	13	.615	Ea.	24.50	25		49.50	64.50
7200	1/4"		13	.615		24.50	25		49.50	64.50
7210	3/8"		13	.615		24	25		49	64.50
7220	1/2"		12	.667		18.20	27		45.20	61
7230	3/4"		11	.727		19.70	29.50		49.20	66
7240	1"	↓	10	.800		21.50	32.50		54	73
7250	1-1/4"	Q-1	17	.941		37.50	34.50		72	93
7260	1-1/2"		16	1		36	37		73	95.50
7270	2"		14	1.143		51.50	42		93.50	120
7280	2-1/2"		11	1.455		245	53.50		298.50	350
7290	3"	↓	8	2		360	73.50		433.50	505
7300	4"	Q-2	11	2.182	↓	485	83		568	660
7320	Tee, straight									
7330	1/8"	1 Plum	9	.889	Ea.	25.50	36.50		62	82.50
7340	1/4"		9	.889		25.50	36.50		62	82.50
7350	3/8"		9	.889		27	36.50		63.50	84
7360	1/2"		8	1		19.05	41		60.05	82.50
7370	3/4"		7	1.143		20.50	46.50		67	92.50
7380	1"	↓	6.50	1.231		24	50.50		74.50	102
7390	1-1/4"	Q-1	11	1.455		54	53.50		107.50	140
7400	1-1/2"		10	1.600		49.50	59		108.50	143
7410	2"		9	1.778		61.50	65.50		127	166
7420	2-1/2"		7	2.286		250	84		334	400
7430	3"	↓	5	3.200		380	118		498	595
7440	4"	Q-2	7	3.429	↓	705	131		836	970
7460	Coupling, straight									
7470	1/8"	1 Plum	19	.421	Ea.	6.85	17.20		24.05	33.50
7480	1/4"		19	.421		8.10	17.20		25.30	35
7490	3/8"		19	.421		9.70	17.20		26.90	36.50
7500	1/2"		19	.421		9.65	17.20		26.85	36.50
7510	3/4"		18	.444		17.15	18.15		35.30	46.50
7520	1"	↓	15	.533		21	22		43	56
7530	1-1/4"	Q-1	26	.615		32	22.50		54.50	69
7540	1-1/2"		24	.667		37.50	24.50		62	78
7550	2"		21	.762		59	28		87	107
7560	2-1/2"		18	.889		137	32.50		169.50	200
7570	3"	↓	14	1.143		186	42		228	268
7580	4"	Q-2	16	1.500		275	57		332	385
7600	Reducer, concentric, 1/2"	1 Plum	12	.667		10.35	27		37.35	52.50
7610	3/4"		11	.727		13.35	29.50		42.85	59
7612	1"	↓	10	.800		22	32.50		54.50	73.50
7614	1-1/4"	Q-1	17	.941		47	34.50		81.50	104
7616	1-1/2"		16	1		51.50	37		88.50	113
7618	2"		14	1.143		80.50	42		122.50	152
7620	2-1/2"		11	1.455		200	53.50		253.50	300
7622	3"	↓	8	2		222	73.50		295.50	355
7624	4"	Q-2	11	2.182	↓	370	83		453	530
7710	Union									
7720	1/8"	1 Plum	12	.667	Ea.	32	27		59	76
7730	1/4"		12	.667		32	27		59	76
7740	3/8"		12	.667		37	27		64	82
7750	1/2"		11	.727		34.50	29.50		64	82.50
7760	3/4"	↓	10	.800	↓	47.50	32.50		80	102

960

15107 | Metal Pipe & Fittings

		CREW	DAILY OUTPUT	LABOR-HOURS	UNIT	MAT.	LABOR	EQUIP.	TOTAL	TOTAL INCL O&P	
960	7770	1"	1 Plum	9	.889	Ea.	69	36.50		105.50	131	960
	7780	1-1/4"	Q-1	16	1		188	37		225	263	
	7790	1-1/2"		15	1.067		175	39		214	251	
	7800	2"		13	1.231		221	45.50		266.50	310	
	7810	2-1/2"		10	1.600		555	59		614	700	
	7820	3"		7	2.286		740	84		824	935	
	7830	4"	Q-2	10	2.400		1,025	91.50		1,116.50	1,275	
	7838	Caps										
	7840	1/2"	1 Plum	24	.333	Ea.	6.45	13.60		20.05	27.50	
	7841	3/4"		22	.364		9.40	14.85		24.25	33	
	7842	1"		20	.400		15	16.35		31.35	41	
	7843	1-1/2"	Q-1	32	.500		36	18.40		54.40	67	
	7844	2"	"	28	.571		45.50	21		66.50	81.50	
	7845	4"	Q-2	22	1.091		178	41.50		219.50	259	
	7850	For 150 lb., type 316, add					25%					

15108 | Plastic Pipe & Fittings

		CREW	DAILY OUTPUT	LABOR-HOURS	UNIT	MAT.	LABOR	EQUIP.	TOTAL	TOTAL INCL O&P	
520	0010	PIPE, PLASTIC R15100-070										520
	0020	Fiberglass reinforced, couplings 10' O.C., hangers 3 per 10'										
	0080	General service										
	0120	2" diameter	Q-1	59	.271	L.F.	9.60	9.95		19.55	25.50	
	0140	3" diameter		52	.308		12.10	11.30		23.40	30.50	
	0150	4" diameter		48	.333		13.95	12.25		26.20	34	
	0160	6" diameter		39	.410		21.50	15.10		36.60	46	
	0170	8" diameter	Q-2	49	.490		33.50	18.70		52.20	65	
	0180	10" diameter		41	.585		55.50	22.50		78	94.50	
	0190	12" diameter		36	.667		71	25.50		96.50	117	
	0200	High strength										
	0240	2" diameter	Q-1	58	.276	L.F.	9.20	10.15		19.35	25.50	
	0260	3" diameter		51	.314		13.10	11.55		24.65	32	
	0280	4" diameter		47	.340		15.85	12.50		28.35	36.50	
	0300	6" diameter		38	.421		24.50	15.50		40	50.50	
	0320	8" diameter	Q-2	48	.500		39	19.05		58.05	71	
	0340	10" diameter		40	.600		61.50	23		84.50	103	
	0360	12" diameter		36	.667		76	25.50		101.50	122	
	0550	To delete coupling & hangers, subtract										
	0560	2" diam. to 6" diam.					33%	56%				
	0570	8" diam. to 12" diam.					31%	52%				
	0600	PVC, high impact/pressure, cplgs. 10' O.C., hangers 3 per 10'										
	1020	Schedule 80										
	1040	1/4" diameter	1 Plum	58	.138	L.F.	1.40	5.65		7.05	10	
	1060	3/8" diameter		55	.145		1.59	5.95		7.54	10.70	
	1070	1/2" diameter		50	.160		1.28	6.55		7.83	11.20	
	1080	3/4" diameter		47	.170		1.46	6.95		8.41	12.05	
	1090	1" diameter		43	.186		1.76	7.60		9.36	13.35	
	1100	1-1/4" diameter		39	.205		2.17	8.40		10.57	15	
	1110	1-1/2" diameter		34	.235		2.43	9.60		12.03	17.15	
	1120	2" diameter	Q-1	55	.291		3.01	10.70		13.71	19.40	
	1130	2-1/2" diameter		52	.308		4.21	11.30		15.51	21.50	
	1140	3" diameter		50	.320		5.70	11.75		17.45	24	
	1150	4" diameter		46	.348		7.85	12.80		20.65	28	
	1160	5" diameter		42	.381		10.95	14		24.95	33	
	1170	6" diameter		38	.421		14.70	15.50		30.20	39.50	
	1180	8" diameter	Q-2	47	.511		23	19.45		42.45	54.50	
	1190	10" diameter		42	.571		39	22		61	76	

15108	Plastic Pipe & Fittings		CREW	DAILY OUTPUT	LABOR-HOURS	UNIT	2005 BARE COSTS				TOTAL INCL O&P
							MAT.	LABOR	EQUIP.	TOTAL	
520	1200	12" diameter	R15100 -070	Q-2	38	.632	L.F.	51	24		75	92
	1730	To delete coupling & hangers, subtract										
	1740	1/4" diam. to 1/2" diam.						62%	80%			
	1750	3/4" diam. to 1-1/4" diam.						58%	73%			
	1760	1-1/2" diam. to 6" diam.						40%	57%			
	1770	8" diam. to 12" diam.						34%	50%			
	1800	PVC, couplings 10' O.C., hangers 3 per 10'										
	1820	Schedule 40										
	1860	1/2" diameter		1 Plum	54	.148	L.F.	.98	6.05		7.03	10.20
	1870	3/4" diameter			51	.157		1.06	6.40		7.46	10.80
	1880	1" diameter			46	.174		1.17	7.10		8.27	12
	1890	1-1/4" diameter			42	.190		1.34	7.80		9.14	13.20
	1900	1-1/2" diameter			36	.222		1.44	9.10		10.54	15.25
	1910	2" diameter		Q-1	59	.271		1.63	9.95		11.58	16.80
	1920	2-1/2" diameter			56	.286		2.20	10.50		12.70	18.20
	1930	3" diameter			53	.302		2.87	11.10		13.97	19.85
	1940	4" diameter			48	.333		3.67	12.25		15.92	22.50
	1950	5" diameter			43	.372		4.86	13.70		18.56	26
	1960	6" diameter			39	.410		6.50	15.10		21.60	29.50
	1970	8" diameter		Q-2	48	.500		10.45	19.05		29.50	40
	1980	10" diameter			43	.558		26	21.50		47.50	61
	1990	12" diameter			42	.571		34.50	22		56.50	71
	2000	14" diameter			31	.774		55	29.50		84.50	105
	2010	16" diameter			23	1.043		69	40		109	136
	2340	To delete coupling & hangers, subtract										
	2360	1/2" diam. to 1-1/4" diam.						65%	74%			
	2370	1-1/2" diam. to 6" diam.						44%	57%			
	2380	8" diam. to 12" diam.						41%	53%			
	2390	14" diam. to 16" diam.						48%	45%			
	2420	Schedule 80										
	2440	1/4" diameter		1 Plum	58	.138	L.F.	.85	5.65		6.50	9.40
	2450	3/8" diameter			55	.145		.85	5.95		6.80	9.90
	2460	1/2" diameter			50	.160		1.02	6.55		7.57	10.90
	2470	3/4" diameter			47	.170		1.12	6.95		8.07	11.70
	2480	1" diameter			43	.186		1.26	7.60		8.86	12.80
	2490	1-1/4" diameter			39	.205		1.47	8.40		9.87	14.20
	2500	1-1/2" diameter			34	.235		1.60	9.60		11.20	16.20
	2510	2" diameter		Q-1	55	.291		1.87	10.70		12.57	18.15
	2520	2-1/2" diameter			52	.308		2.37	11.30		13.67	19.60
	2530	3" diameter			50	.320		3.35	11.75		15.10	21.50
	2540	4" diameter			46	.348		4.42	12.80		17.22	24
	2550	5" diameter			42	.381		7.15	14		21.15	29
	2560	6" diameter			38	.421		8.15	15.50		23.65	32.50
	2570	8" diameter		Q-2	47	.511		13	19.45		32.45	44
	2580	10" diameter			42	.571		31.50	22		53.50	68
	2590	12" diameter			38	.632		41	24		65	81.50
	2830	To delete coupling & hangers, subtract										
	2840	1/4" diam. to 1/2" diam.						66%	80%			
	2850	3/4" diam. to 1-1/4" diam.						61%	73%			
	2860	1-1/2" diam. to 6" diam.						41%	57%			
	2870	8" diam. to 12" diam.						31%	50%			
	2900	Schedule 120										
	2910	1/2" diameter		1 Plum	50	.160	L.F.	1.41	6.55		7.96	11.35
	2950	3/4" diameter			47	.170		1.62	6.95		8.57	12.25
	2960	1" diameter			43	.186		2	7.60		9.60	13.60
	2970	1-1/4" diameter			39	.205		2.51	8.40		10.91	15.35

15108	Plastic Pipe & Fittings	CREW	DAILY OUTPUT	LABOR-HOURS	UNIT	2005 BARE COSTS				TOTAL INCL O&P	
						MAT.	LABOR	EQUIP.	TOTAL		
2980	1-1/2" diameter	1 Plum	33	.242	L.F.	2.85	9.90		12.75	18.05	520
2990	2" diameter	Q-1	54	.296		3.64	10.90		14.54	20.50	
3000	2-1/2" diameter		52	.308		5.35	11.30		16.65	23	
3010	3" diameter		49	.327		6.95	12		18.95	25.50	
3020	4" diameter		45	.356		10.45	13.10		23.55	31	
3030	6" diameter		37	.432		19.60	15.90		35.50	45.50	
3240	To delete coupling & hangers, subtract										
3250	1/2" diam. to 1-1/4" diam.					52%	74%				
3260	1-1/2" diam. to 4" diam.					30%	57%				
3270	6" diam.					17%	50%				
3300	PVC, pressure, couplings 10' O.C., hangers 3 per 10'										
3310	SDR 26, 160 psi										
3350	1-1/4" diameter	1 Plum	42	.190	L.F.	1.17	7.80		8.97	13	
3360	1-1/2" diameter	"	36	.222		1.25	9.10		10.35	15.05	
3370	2" diameter	Q-1	59	.271		1.40	9.95		11.35	16.55	
3380	2-1/2" diameter		56	.286		1.83	10.50		12.33	17.80	
3390	3" diameter		53	.302		2.47	11.10		13.57	19.40	
3400	4" diameter		48	.333		3.27	12.25		15.52	22	
3420	6" diameter		39	.410		6.25	15.10		21.35	29.50	
3430	8" diameter	Q-2	48	.500		10.45	19.05		29.50	40	
3660	To delete coupling & hangers, subtract										
3670	1-1/4" diam.					63%	68%				
3680	1-1/2" diam. to 4" diam.					48%	57%				
3690	6" diam. to 8" diam.					60%	54%				
3720	SDR 21, 200 psi, 1/2" diameter	1 Plum	54	.148	L.F.	.94	6.05		6.99	10.15	
3740	3/4" diameter		51	.157		.98	6.40		7.38	10.75	
3750	1" diameter		46	.174		1.05	7.10		8.15	11.85	
3760	1-1/4" diameter		42	.190		1.20	7.80		9	13	
3770	1-1/2" diameter		36	.222		1.29	9.10		10.39	15.05	
3780	2" diameter	Q-1	59	.271		1.48	9.95		11.43	16.60	
3790	2-1/2" diameter		56	.286		2.29	10.50		12.79	18.30	
3800	3" diameter		53	.302		2.63	11.10		13.73	19.60	
3810	4" diameter		48	.333		3.53	12.25		15.78	22.50	
3830	6" diameter		39	.410		6.80	15.10		21.90	30	
3840	8" diameter	Q-2	48	.500		11.60	19.05		30.65	41.50	
4000	To delete coupling & hangers, subtract										
4010	1/2" diam. to 3/4" diam.					71%	77%				
4020	1" diam. to 1-1/4" diam.					63%	70%				
4030	1-1/2" diam. to 6" diam.					44%	57%				
4040	8" diam.					46%	54%				
4100	DWV type, schedule 40, couplings 10' O.C., hangers 3 per 10'										
4120	ABS										
4140	1-1/4" diameter	1 Plum	42	.190	L.F.	1.08	7.80		8.88	12.90	
4150	1-1/2" diameter	"	36	.222		1.09	9.10		10.19	14.85	
4160	2" diameter	Q-1	59	.271		1.19	9.95		11.14	16.30	
4170	3" diameter		53	.302		2.08	11.10		13.18	19	
4180	4" diameter		48	.333		2.78	12.25		15.03	21.50	
4190	6" diameter		39	.410		6.80	15.10		21.90	30	
4360	To delete coupling & hangers, subtract										
4370	1-1/4" diam.					64%	68%				
4380	1-1/2" diam. to 6" diam.					54%	57%				
4400	PVC										
4410	1-1/4" diameter	1 Plum	42	.190	L.F.	1.19	7.80		8.99	13	
4420	1-1/2" diameter	"	36	.222		1.20	9.10		10.30	14.95	
4460	2" diameter	Q-1	59	.271		1.35	9.95		11.30	16.50	
4470	3" diameter		53	.302		2.40	11.10		13.50	19.35	

R15100 -070

15108 | Plastic Pipe & Fittings

		CREW	DAILY OUTPUT	LABOR-HOURS	UNIT	MAT.	LABOR	EQUIP.	TOTAL	TOTAL INCL O&P
520							**2005 BARE COSTS**			**520**
4480	4" diameter R15100-070	Q-1	48	.333	L.F.	3.13	12.25		15.38	22
4490	6" diameter	↓	39	.410	↓	5.95	15.10		21.05	29
4500	8" diameter	Q-2	48	.500	↓	15.10	19.05		34.15	45
4750	To delete coupling & hangers, subtract									
4760	1-1/4" diam. to 1-1/2" diam.					71%	64%			
4770	2" diam. to 8" diam.					60%	57%			
4800	PVC, clear pipe, cplgs. 10' O.C., hangers 3 per 10', Sched. 40									
4840	1/4" diameter	1 Plum	59	.136	L.F.	1.34	5.55		6.89	9.85
4850	3/8" diameter		56	.143		1.54	5.85		7.39	10.45
4860	1/2" diameter		54	.148		1.92	6.05		7.97	11.20
4870	3/4" diameter		51	.157		2.33	6.40		8.73	12.20
4880	1" diameter		46	.174		3.12	7.10		10.22	14.15
4890	1-1/4" diameter		42	.190		3.97	7.80		11.77	16.05
4900	1-1/2" diameter	↓	36	.222		4.59	9.10		13.69	18.70
4910	2" diameter	Q-1	59	.271		5.95	9.95		15.90	21.50
4920	2-1/2" diameter		56	.286		9.10	10.50		19.60	26
4930	3" diameter		53	.302		11.75	11.10		22.85	29.50
4940	3-1/2" diameter		50	.320		15.80	11.75		27.55	35
4950	4" diameter	↓	48	.333	↓	17.30	12.25		29.55	37.50
5250	To delete coupling & hangers, subtract									
5260	1/4" diam. to 3/8" diam.					60%	81%			
5270	1/2" diam. to 3/4" diam.					41%	77%			
5280	1" diam. to 1-1/2" diam.					26%	67%			
5290	2" diam. to 4" diam.					16%	58%			
5360	CPVC, couplings 10' O.C., hangers 3 per 10'									
5380	Schedule 40									
5460	1/2" diameter	1 Plum	54	.148	L.F.	2.31	6.05		8.36	11.65
5470	3/4" diameter		51	.157		3.05	6.40		9.45	13
5480	1" diameter		46	.174		3.74	7.10		10.84	14.80
5490	1-1/4" diameter		42	.190		4.35	7.80		12.15	16.50
5500	1-1/2" diameter	↓	36	.222		4.86	9.10		13.96	19
5510	2" diameter	Q-1	59	.271		6	9.95		15.95	21.50
5520	2-1/2" diameter		56	.286		9.55	10.50		20.05	26.50
5530	3" diameter		53	.302		11.75	11.10		22.85	29.50
5540	4" diameter		48	.333		19.05	12.25		31.30	39.50
5550	6" diameter	↓	43	.372	↓	31	13.70		44.70	55
5730	To delete coupling & hangers, subtract									
5740	1/2" diam. to 3/4" diam.					37%	77%			
5750	1" diam. to 1-1/4" diam.					27%	70%			
5760	1-1/2" diam. to 3" diam.					21%	57%			
5770	4" diam. to 6" diam.					16%	57%			
5800	Schedule 80									
5860	1/2" diameter	1 Plum	50	.160	L.F.	2.73	6.55		9.28	12.80
5870	3/4" diameter		47	.170		3.26	6.95		10.21	14.05
5880	1" diameter		43	.186		4.02	7.60		11.62	15.80
5890	1-1/4" diameter		39	.205		4.79	8.40		13.19	17.85
5900	1-1/2" diameter	↓	34	.235		5.45	9.60		15.05	20.50
5910	2" diameter	Q-1	55	.291		6.90	10.70		17.60	23.50
5920	2-1/2" diameter		52	.308		10.75	11.30		22.05	29
5930	3" diameter		50	.320		13.50	11.75		25.25	32.50
5940	4" diameter		46	.348		22	12.80		34.80	43.50
5950	6" diameter	↓	38	.421		37.50	15.50		53	65
5960	8" diameter	Q-2	47	.511	↓	68	19.45		87.45	105
6060	To delete couplings & hangers, subtract									
6070	1/2" diam. to 3/4" diam.					44%	77%			
6080	1" diam. to 1-1/4" diam.					32%	71%			

15108	Plastic Pipe & Fittings		CREW	DAILY OUTPUT	LABOR-HOURS	UNIT	2005 BARE COSTS				TOTAL INCL O&P	
							MAT.	LABOR	EQUIP.	TOTAL		
520	6090	1-1/2" diam. to 4" diam.	R15100-070					25%	58%				520
	6100	6" diam. to 8" diam.						20%	53%			
	6240	CTS, 1/2" diameter		1 Plum	54	.148	L.F.	1.25	6.05		7.30	10.45
	6250	3/4" diameter			51	.157		1.70	6.40		8.10	11.50
	6260	1" diameter			46	.174		2.74	7.10		9.84	13.70
	6270	1 1/4"			42	.190		3.71	7.80		11.51	15.80
	6280	1 1/2" diameter		▼	36	.222		4.70	9.10		13.80	18.80
	6290	2" diameter		Q-1	59	.271	▼	7.45	9.95		17.40	23
	6370	To delete coupling & hangers, subtract										
	6380	1/2" diam.						51%	79%			
	6390	3/4" diam.						40%	76%			
	6392	1" thru 2" diam.						72%	68%			
	6500	Residential installation, plastic pipe										
	6510	Couplings 10' O.C., strap hangers 3 per 10'										
	6520	PVC, Schedule 40										
	6530	1/2" diameter		1 Plum	138	.058	L.F.	.29	2.37		2.66	3.88
	6540	3/4" diameter			128	.063		.35	2.55		2.90	4.22
	6550	1" diameter			119	.067		.45	2.75		3.20	4.63
	6560	1-1/4" diameter			111	.072		.57	2.94		3.51	5.05
	6570	1-1/2" diameter		▼	104	.077		.67	3.14		3.81	5.45
	6580	2" diameter		Q-1	197	.081		.85	2.99		3.84	5.45
	6590	2-1/2" diameter			162	.099		1.40	3.63		5.03	7
	6600	4" diameter		▼	123	.130	▼	6.85	4.78		11.63	14.75
	6700	PVC, DWV, Schedule 40										
	6720	1-1/4" diameter		1 Plum	100	.080	L.F.	.76	3.27		4.03	5.75
	6730	1-1/2" diameter		"	94	.085		.66	3.48		4.14	6
	6740	2" diameter		Q-1	178	.090		.84	3.31		4.15	5.90
	6760	4" diameter		"	110	.145	▼	6.75	5.35		12.10	15.50
	7280	Polyethylene, flexible, no couplings or hangers										
	7282	Note: For labor costs add 25% to the couplings and fittings labor total.										
	7300	SDR 15, 100 psi										
	7310	3/4" diameter					L.F.	.24			.24	.26
	7350	1" diameter						.32			.32	.35
	7360	1-1/4" diameter						.57			.57	.63
	7370	1-1/2" diameter						.69			.69	.76
	7380	2" diameter					▼	1.13			1.13	1.24
	7700	SIDR 9, 160 psi										
	7710	1/2" diameter					L.F.	.30			.30	.33
	7750	3/4" diameter						.31			.31	.34
	7760	1" diameter						.48			.48	.53
	7770	1-1/4" diameter						.84			.84	.92
	7780	1-1/2" diameter						1.17			1.17	1.29
	7790	2" diameter					▼	1.94			1.94	2.13
	8120	SDR 9, 200 psi										
	8150	3/4" diameter					L.F.	.28			.28	.31
	8160	1" diameter						.46			.46	.51
	8170	1-1/4" diameter					▼	.69			.69	.76
	8420	SDR 7, 250 psi										
	8440	3/4" diameter					L.F.	.40			.40	.44
	8450	1" diameter						.64			.64	.70
	8460	1-1/4" diameter					▼	1.13			1.13	1.24
	8800	PVC, type PSP, drain & sewer, belled end gasket jnt., no hngr.										
	8840	3" diameter					L.F.	1.25			1.25	1.38
	8850	4" diameter						1.55			1.55	1.71
	8860	6" diameter					▼	2.99			2.99	3.29
	9000	Perforated										

Important: See the Reference Section for critical supporting data - Reference Nos., Crews, & City Cost Indexes

15108 | Plastic Pipe & Fittings

		CREW	DAILY OUTPUT	LABOR-HOURS	UNIT	2005 BARE COSTS				TOTAL INCL O&P	
						MAT.	LABOR	EQUIP.	TOTAL		
520	9040	4″ diameter R15100-070				L.F.	1.55			1.55	1.71	520
560	0010	**PIPE, PLASTIC, FITTINGS**										560
	0030	Epoxy resin, fiberglass reinforced, general service										
	0090	Elbow, 90°, 2″	Q-1	33.10	.483	Ea.	45.50	17.80		63.30	76.50	
	0100	3″		20.80	.769		64.50	28.50		93	113	
	0110	4″		16.50	.970		88	35.50		123.50	151	
	0120	6″		10.10	1.584		134	58.50		192.50	236	
	0130	8″	Q-2	9.30	2.581		248	98.50		346.50	420	
	0140	10″		8.50	2.824		310	108		418	500	
	0150	12″		7.60	3.158		445	120		565	670	
	0160	45° Elbow, same as 90°										
	0170	Elbow, 90°, flanged										
	0172	2″	Q-1	23	.696	Ea.	100	25.50		125.50	149	
	0173	3″		16	1		116	37		153	183	
	0174	4″		13	1.231		151	45.50		196.50	235	
	0176	6″		8	2		274	73.50		347.50	410	
	0177	8″	Q-2	9	2.667		490	102		592	695	
	0178	10″		7	3.429		675	131		806	940	
	0179	12″		5	4.800		915	183		1,098	1,275	
	0186	Elbow, 45°, flanged										
	0188	2″	Q-1	23	.696	Ea.	101	25.50		126.50	150	
	0189	3″		16	1		116	37		153	183	
	0190	4″		13	1.231		151	45.50		196.50	234	
	0192	6″		8	2		276	73.50		349.50	415	
	0193	8″	Q-2	9	2.667		495	102		597	695	
	0194	10″		7	3.429		675	131		806	940	
	0195	12″		5	4.800		915	183		1,098	1,275	
	0290	Tee, 2″	Q-1	20	.800		55	29.50		84.50	105	
	0300	3″		13.90	1.151		64.50	42.50		107	134	
	0310	4″		11	1.455		76.50	53.50		130	165	
	0320	6″		6.70	2.388		205	88		293	360	
	0330	8″	Q-2	6.20	3.871		237	148		385	480	
	0340	10″		5.70	4.211		380	161		541	655	
	0350	12″		5.10	4.706		460	179		639	775	
	0352	Tee, flanged										
	0354	2″	Q-1	17	.941	Ea.	136	34.50		170.50	202	
	0355	3″		10	1.600		183	59		242	290	
	0356	4″		8	2		203	73.50		276.50	335	
	0358	6″		5	3.200		400	118		518	620	
	0359	8″	Q-2	6	4		685	153		838	985	
	0360	10″		5	4.800		995	183		1,178	1,375	
	0361	12″		4	6		1,275	229		1,504	1,750	
	0365	Wye, flanged										
	0367	2″	Q-1	17	.941	Ea.	350	34.50		384.50	435	
	0368	3″		10	1.600		390	59		449	520	
	0369	4″		8	2		485	73.50		558.50	645	
	0371	6″		5	3.200		580	118		698	815	
	0372	8″	Q-2	6	4		755	153		908	1,050	
	0373	10″		5	4.800		1,275	183		1,458	1,675	
	0374	12″		4	6		1,550	229		1,779	2,050	
	0380	Couplings										
	0410	2″	Q-1	33.10	.483	Ea.	9.85	17.80		27.65	37.50	
	0420	3″		20.80	.769		16.95	28.50		45.45	61	
	0430	4″		16.50	.970		22	35.50		57.50	78	
	0440	6″		10.10	1.584		42	58.50		100.50	134	

15108 | Plastic Pipe & Fittings

		CREW	DAILY OUTPUT	LABOR-HOURS	UNIT	2005 BARE COSTS				TOTAL INCL O&P
						MAT.	LABOR	EQUIP.	TOTAL	
560	0450	8"	Q-2	9.30	2.581	Ea.	64	98.50		162.50	219	560
	0460	10"		8.50	2.824		96	108		204	268
	0470	12"	↓	7.60	3.158	↓	129	120		249	325
	0473	High corrosion resistant couplings, add					30%				
	0474	Reducer, concentric, flanged									
	0475	2" x 1-1/2"	Q-1	30	.533	Ea.	145	19.60		164.60	190
	0476	3" x 2"		24	.667		165	24.50		189.50	219
	0477	4" x 3"		19	.842		174	31		205	238
	0479	6" x 4"	↓	15	1.067		227	39		266	310
	0480	8" x 6"	Q-2	16	1.500		320	57		377	435
	0481	10" x 8"		13	1.846		425	70.50		495.50	575
	0482	12" x 10"	↓	11	2.182	↓	610	83		693	795
	0486	Adapter, bell x male or female									
	0488	2"	Q-1	28	.571	Ea.	17.70	21		38.70	51
	0489	3"		20	.800		22.50	29.50		52	69
	0491	4"		17	.941		31	34.50		65.50	86
	0492	6"	↓	12	1.333		61.50	49		110.50	141
	0493	8"	Q-2	15	1.600		73	61		134	172
	0494	10"	"	11	2.182	↓	105	83		188	241
	0528	Flange									
	0532	2"	Q-1	46	.348	Ea.	19.15	12.80		31.95	40.50
	0533	3"		32	.500		23.50	18.40		41.90	53
	0534	4"		26	.615		32.50	22.50		55	69.50
	0536	6"	↓	16	1		55	37		92	116
	0537	8"	Q-2	18	1.333		90	51		141	176
	0538	10"		14	1.714		126	65.50		191.50	237
	0539	12"	↓	10	2.400	↓	177	91.50		268.50	330
	2100	PVC schedule 80, socket joint									
	2110	90° elbow, 1/2"	1 Plum	30.30	.264	Ea.	.59	10.80		11.39	16.85
	2130	3/4"		26	.308		.76	12.55		13.31	19.75
	2140	1"		22.70	.352		1.72	14.40		16.12	23.50
	2150	1-1/4"		20.20	.396		1.63	16.20		17.83	26.50
	2160	1-1/2"	↓	18.20	.440		1.74	17.95		19.69	29
	2170	2"	Q-1	33.10	.483		2.11	17.80		19.91	29
	2180	3"		20.80	.769		5.55	28.50		34.05	48.50
	2190	4"		16.50	.970		6.30	35.50		41.80	60.50
	2200	6"	↓	10.10	1.584		63.50	58.50		122	158
	2210	8"	Q-2	9.30	2.581		175	98.50		273.50	340
	2250	45° elbow, 1/2"	1 Plum	30.30	.264		1.12	10.80		11.92	17.45
	2270	3/4"		26	.308		1.70	12.55		14.25	21
	2280	1"		22.70	.352		2.55	14.40		16.95	24.50
	2290	1-1/4"		20.20	.396		3.25	16.20		19.45	28
	2300	1-1/2"	↓	18.20	.440		3.84	17.95		21.79	31
	2310	2"	Q-1	33.10	.483		4.97	17.80		22.77	32
	2320	3"		20.80	.769		12.70	28.50		41.20	56.50
	2330	4"		16.50	.970		23	35.50		58.50	78.50
	2340	6"	↓	10.10	1.584		29	58.50		87.50	119
	2350	8"	Q-2	9.30	2.581		166	98.50		264.50	330
	2400	Tee, 1/2"	1 Plum	20.20	.396		1.67	16.20		17.87	26.50
	2420	3/4"		17.30	.462		1.75	18.90		20.65	30.50
	2430	1"		15.20	.526		2.18	21.50		23.68	35
	2440	1-1/4"		13.50	.593		6	24		30	43
	2450	1-1/2"	↓	12.10	.661		6	27		33	47
	2460	2"	Q-1	20	.800		7.50	29.50		37	52.50
	2470	3"		13.90	1.151		8.15	42.50		50.65	72.50
	2480	4"	↓	11	1.455	↓	11.80	53.50		65.30	93.50

Important: See the Reference Section for critical supporting data - Reference Nos., Crews, & City Cost Indexes

15108	Plastic Pipe & Fittings	CREW	DAILY OUTPUT	LABOR-HOURS	UNIT	2005 BARE COSTS				TOTAL INCL O&P
						MAT.	LABOR	EQUIP.	TOTAL	
560 2490	6"	Q-1	6.70	2.388	Ea.	40.50	88		128.50	177 **560**
2500	8"	Q-2	6.20	3.871		237	148		385	480
2510	Flange, socket, 150 lb., 1/2"	1 Plum	55.60	.144		3.23	5.90		9.13	12.40
2514	3/4"		47.60	.168		3.46	6.85		10.31	14.10
2518	1"		41.70	.192		3.85	7.85		11.70	16.05
2522	1-1/2"		33.30	.240		4.05	9.80		13.85	19.20
2526	2"	Q-1	60.60	.264		5.40	9.70		15.10	20.50
2530	4"		30.30	.528		11.65	19.40		31.05	42
2534	6"		18.50	.865		18.30	32		50.30	68
2538	8"	Q-2	17.10	1.404		87	53.50		140.50	176
2550	Coupling, 1/2"	1 Plum	30.30	.264		1.07	10.80		11.87	17.40
2570	3/4"		26	.308		1.45	12.55		14	20.50
2580	1"		22.70	.352		1.49	14.40		15.89	23
2590	1-1/4"		20.20	.396		2.27	16.20		18.47	27
2600	1-1/2"		18.20	.440		2.44	17.95		20.39	29.50
2610	2"	Q-1	33.10	.483		2.62	17.80		20.42	29.50
2620	3"		20.80	.769		7.40	28.50		35.90	50.50
2630	4"		16.50	.970		9.25	35.50		44.75	63.50
2640	6"		10.10	1.584		19.95	58.50		78.45	110
2650	8"	Q-2	9.30	2.581		37.50	98.50		136	189
2660	10"		8.50	2.824		40.50	108		148.50	207
2670	12"		7.60	3.158		47	120		167	233
2700	PVC (white), schedule 40, socket joints									
2760	90° elbow, 1/2"	1 Plum	33.30	.240	Ea.	.29	9.80		10.09	15.05
2770	3/4"		28.60	.280		.32	11.45		11.77	17.50
2780	1"		25	.320		.57	13.05		13.62	20.50
2790	1-1/4"		22.20	.360		1	14.70		15.70	23
2800	1-1/2"		20	.400		1.07	16.35		17.42	25.50
2810	2"	Q-1	36.40	.440		1.68	16.15		17.83	26.50
2820	2-1/2"		26.70	.599		5.10	22		27.10	38.50
2830	3"		22.90	.699		6.10	25.50		31.60	45
2840	4"		18.20	.879		10.95	32.50		43.45	60.50
2850	5"		12.10	1.322		28.50	48.50		77	104
2860	6"		11.10	1.441		35	53		88	118
2870	8"	Q-2	10.30	2.330		89	89		178	232
2980	45° elbow, 1/2"	1 Plum	33.30	.240		.46	9.80		10.26	15.25
2990	3/4"		28.60	.280		.71	11.45		12.16	17.95
3000	1"		25	.320		.86	13.05		13.91	20.50
3010	1-1/4"		22.20	.360		1.20	14.70		15.90	23.50
3020	1-1/2"		20	.400		1.51	16.35		17.86	26
3030	2"	Q-1	36.40	.440		1.97	16.15		18.12	26.50
3040	2-1/2"		26.70	.599		5.10	22		27.10	38.50
3050	3"		22.90	.699		8.25	25.50		33.75	47.50
3060	4"		18.20	.879		14.25	32.50		46.75	64
3070	5"		12.10	1.322		28.50	48.50		77	104
3080	6"		11.10	1.441		35	53		88	118
3090	8"	Q-2	10.30	2.330		100	89		189	244
3180	Tee, 1/2"	1 Plum	22.20	.360		.35	14.70		15.05	22.50
3190	3/4"		19	.421		.40	17.20		17.60	26.50
3200	1"		16.70	.479		.75	19.55		20.30	30.50
3210	1-1/4"		14.80	.541		1.18	22		23.18	34.50
3220	1-1/2"		13.30	.601		1.43	24.50		25.93	38.50
3230	2"	Q-1	24.20	.661		2.07	24.50		26.57	39
3240	2-1/2"		17.80	.899		6.85	33		39.85	57
3250	3"		15.20	1.053		13.55	38.50		52.05	73
3260	4"		12.10	1.322		16.25	48.50		64.75	91

15108	Plastic Pipe & Fittings	CREW	DAILY OUTPUT	LABOR-HOURS	UNIT	2005 BARE COSTS				TOTAL INCL O&P
						MAT.	LABOR	EQUIP.	TOTAL	
3270	5"	Q-1	8.10	1.975	Ea.	39	72.50		111.50	152
3280	6"	↓	7.40	2.162		54.50	79.50		134	180
3290	8"	Q-2	6.80	3.529		65	135		200	274
3380	Coupling, 1/2"	1 Plum	33.30	.240		.19	9.80		9.99	14.95
3390	3/4"		28.60	.280		.26	11.45		11.71	17.45
3400	1"		25	.320		.44	13.05		13.49	20
3410	1-1/4"		22.20	.360		.61	14.70		15.31	22.50
3420	1-1/2"	↓	20	.400		.65	16.35		17	25
3430	2"	Q-1	36.40	.440		1.01	16.15		17.16	25.50
3440	2-1/2"		26.70	.599		2.22	22		24.22	35.50
3450	3"		22.90	.699		3.49	25.50		28.99	42.50
3460	4"		18.20	.879		5	32.50		37.50	54
3470	5"		12.10	1.322		12.25	48.50		60.75	86.50
3480	6"	↓	11.10	1.441		15.90	53		68.90	97
3490	8"	Q-2	10.30	2.330		39.50	89		128.50	178
3600	Cap, schedule 40, PVC socket 1/2"	1 Plum	60.60	.132		.26	5.40		5.66	8.40
3610	3/4"		51.90	.154		.30	6.30		6.60	9.80
3620	1"		45.50	.176		.47	7.20		7.67	11.30
3630	1-1/4"		40.40	.198		.65	8.10		8.75	12.85
3640	1-1/2"	↓	36.40	.220		.71	9		9.71	14.30
3650	2"	Q-1	66.10	.242		.86	8.90		9.76	14.35
3660	2-1/2"		48.50	.330		2.75	12.15		14.90	21.50
3670	3"		41.60	.385		3.01	14.15		17.16	25
3680	4"		33.10	.483		6.85	17.80		24.65	34
3690	6"	↓	20.20	.792		16.40	29		45.40	62
3700	8"	Q-2	18.60	1.290	↓	41	49		90	120
3710	Reducing insert, schedule 40, socket weld									
3712	3/4"	1 Plum	31.50	.254	Ea.	.29	10.35		10.64	15.90
3713	1"		27.50	.291		.53	11.90		12.43	18.45
3715	1-1/2"	↓	22	.364		.71	14.85		15.56	23.50
3716	2"	Q-1	40	.400		1.24	14.70		15.94	23.50
3717	4"		20	.800		6.60	29.50		36.10	51.50
3718	6"	↓	12.20	1.311		16.30	48		64.30	90.50
3719	8"	Q-2	11.30	2.124	↓	57	81		138	185
3730	Reducing insert, socket weld x female/male thread									
3732	1/2"	1 Plum	38.30	.209	Ea.	1.32	8.55		9.87	14.30
3733	3/4"		32.90	.243		.82	9.95		10.77	15.85
3734	1"		28.80	.278		1.14	11.35		12.49	18.30
3736	1-1/2"	↓	23	.348		2.06	14.20		16.26	24
3737	2"	Q-1	41.90	.382		2.20	14.05		16.25	23.50
3738	4"	"	20.90	.766	↓	6.15	28		34.15	49.50
3742	Male adapter, socket weld x male thread									
3744	1/2"	1 Plum	38.30	.209	Ea.	.26	8.55		8.81	13.15
3745	3/4"		32.90	.243		.29	9.95		10.24	15.25
3746	1"		28.80	.278		.50	11.35		11.85	17.60
3748	1-1/2"	↓	23	.348		.82	14.20		15.02	22.50
3749	2"	Q-1	41.90	.382		1.07	14.05		15.12	22
3750	4"	"	20.90	.766	↓	5.90	28		33.90	49
3754	Female adapter, socket weld x female thread									
3756	1/2"	1 Plum	38.30	.209	Ea.	.31	8.55		8.86	13.20
3757	3/4"		32.90	.243		.40	9.95		10.35	15.40
3758	1"		28.80	.278		.46	11.35		11.81	17.55
3760	1-1/2"	↓	23	.348		.82	14.20		15.02	22.50
3761	2"	Q-1	41.90	.382		1.10	14.05		15.15	22
3762	4"	"	20.90	.766	↓	6.15	28		34.15	49.50
3800	PVC, schedule 80, socket joints									

Important: See the Reference Section for critical supporting data - Reference Nos., Crews, & City Cost Indexes

		CREW	DAILY OUTPUT	LABOR-HOURS	UNIT	2005 BARE COSTS				TOTAL INCL O&P
						MAT.	LABOR	EQUIP.	TOTAL	
3810	Reducing insert									
3812	3/4"	1 Plum	28.60	.280	Ea.	.62	11.45		12.07	17.85
3813	1"		25	.320		2.40	13.05		15.45	22.50
3815	1-1/2"		20	.400		5.15	16.35		21.50	30
3816	2"	Q-1	36.40	.440		7.35	16.15		23.50	32.50
3817	4"		18.20	.879		28	32.50		60.50	79.50
3818	6"		11.10	1.441		39	53		92	123
3819	8"	Q-2	10.20	2.353		83.50	89.50		173	227
3830	Reducing insert, socket weld x female/male thread									
3832	1/2"	1 Plum	34.80	.230	Ea.	5.15	9.40		14.55	19.75
3833	3/4"		29.90	.268		3.92	10.95		14.87	21
3834	1"		26.10	.307		6.30	12.50		18.80	26
3836	1-1/2"		20.90	.383		6.35	15.65		22	30.50
3837	2"	Q-1	38	.421		12.45	15.50		27.95	37
3838	4"	"	19	.842		44	31		75	95
3844	Adapter, male socket x male thread									
3846	1/2"	1 Plum	34.80	.230	Ea.	2.68	9.40		12.08	17.05
3847	3/4"		29.90	.268		2.96	10.95		13.91	19.70
3848	1"		26.10	.307		5.15	12.50		17.65	24.50
3850	1-1/2"		20.90	.383		8.60	15.65		24.25	33
3851	2"	Q-1	38	.421		12.45	15.50		27.95	37
3852	4"	"	19	.842		28	31		59	77.50
3860	Adapter, female socket x female thread									
3862	1/2"	1 Plum	34.80	.230	Ea.	2.19	9.40		11.59	16.50
3863	3/4"		29.90	.268		3.62	10.95		14.57	20.50
3864	1"		26.10	.307		5.35	12.50		17.85	24.50
3866	1-1/2"		20.90	.383		10.60	15.65		26.25	35
3867	2"	Q-1	38	.421		18.55	15.50		34.05	44
3868	4"	"	19	.842		56.50	31		87.50	109
3872	Union, socket joints									
3874	1/2"	1 Plum	25.80	.310	Ea.	5.35	12.65		18	25
3875	3/4"		22.10	.362		6.80	14.80		21.60	29.50
3876	1"		19.30	.415		7.75	16.95		24.70	34
3878	1-1/2"		15.50	.516		17.45	21		38.45	50.50
3879	2"	Q-1	28.10	.569		23.50	21		44.50	57.50
3888	Cap									
3890	1/2"	1 Plum	54.50	.147	Ea.	2.56	6		8.56	11.80
3891	3/4"		46.70	.171		2.69	7		9.69	13.45
3892	1"		41	.195		4.78	7.95		12.73	17.25
3894	1-1/2"		32.80	.244		5.75	9.95		15.70	21.50
3895	2"	Q-1	59.50	.269		15.15	9.90		25.05	31.50
3896	4"		30	.533		46	19.60		65.60	80
3897	6"		18.20	.879		112	32.50		144.50	172
3898	8"	Q-2	16.70	1.437		143	55		198	241
4500	DWV, ABS, non pressure, socket joints									
4540	1/4 Bend, 1-1/4"	1 Plum	20.20	.396	Ea.	2.57	16.20		18.77	27.50
4560	1-1/2"	"	18.20	.440		1.97	17.95		19.92	29
4570	2"	Q-1	33.10	.483		3.04	17.80		20.84	30
4580	3"		20.80	.769		7.30	28.50		35.80	50.50
4590	4"		16.50	.970		10.95	35.50		46.45	65.50
4600	6"		10.10	1.584		58.50	58.50		117	152
4650	1/8 Bend, same as 1/4 Bend									
4800	Tee, sanitary									
4820	1-1/4"	1 Plum	13.50	.593	Ea.	2.91	24		26.91	39.50
4830	1-1/2"	"	12.10	.661		2.68	27		29.68	43.50
4840	2"	Q-1	20	.800		3.92	29.50		33.42	48.50

560

15108 | Plastic Pipe & Fittings

		CREW	DAILY OUTPUT	LABOR-HOURS	UNIT	2005 BARE COSTS				TOTAL INCL O&P
						MAT.	LABOR	EQUIP.	TOTAL	
4850	3"	Q-1	13.90	1.151	Ea.	8.80	42.50		51.30	73
4860	4"		11	1.455		16.10	53.50		69.60	98
4862	Tee, sanitary, reducing, 2" x 1-1/2"		22	.727		3.78	27		30.78	44
4864	3" x 2"		15.30	1.046		6.80	38.50		45.30	65.50
4868	4" x 3"		12.10	1.322		18	48.50		66.50	93
4870	Combination Y and 1/8 bend									
4872	1-1/2"	1 Plum	12.10	.661	Ea.	5.45	27		32.45	46.50
4874	2"	Q-1	20	.800		6.60	29.50		36.10	51.50
4876	3"		13.90	1.151		14.65	42.50		57.15	79.50
4878	4"		11	1.455		27.50	53.50		81	111
4880	3" x 1-1/2"		15.50	1.032		13.60	38		51.60	72
4882	4" x 3"		12.10	1.322		20.50	48.50		69	95.50
4900	Wye, 1-1/4"	1 Plum	13.50	.593		3.36	24		27.36	40
4902	1-1/2"	"	12.10	.661		3.43	27		30.43	44.50
4904	2"	Q-1	20	.800		4.82	29.50		34.32	49.50
4906	3"		13.90	1.151		11.95	42.50		54.45	76.50
4908	4"		11	1.455		21	53.50		74.50	104
4910	6"		6.70	2.388		72.50	88		160.50	212
4918	3" x 1-1/2"		15.50	1.032		8	38		46	66
4920	4" x 3"		12.10	1.322		15.30	48.50		63.80	90
4922	6" x 4"		6.90	2.319		60	85.50		145.50	194
4930	Double Wye, 1-1/2"	1 Plum	9.10	.879		7.95	36		43.95	63
4932	2"	Q-1	16.60	.964		9.35	35.50		44.85	64
4934	3"		10.40	1.538		24	56.50		80.50	112
4936	4"		8.25	1.939		49	71.50		120.50	161
4940	2" x 1-1/2"		16.80	.952		9.35	35		44.35	63
4942	3" x 2"		10.60	1.509		17.95	55.50		73.45	103
4944	4" x 3"		8.45	1.893		38.50	69.50		108	148
4946	6" x 4"		7.25	2.207		81	81		162	211
4950	Reducer bushing, 2" x 1-1/2"		36.40	.440		1.29	16.15		17.44	26
4952	3" x 1-1/2"		27.30	.586		4.33	21.50		25.83	37.50
4954	4" x 2"		18.20	.879		9.95	32.50		42.45	59.50
4956	6" x 4"		11.10	1.441		26.50	53		79.50	109
4960	Couplings, 1-1/2"	1 Plum	18.20	.440		.82	17.95		18.77	28
4962	2"	Q-1	33.10	.483		1.22	17.80		19.02	28
4963	3"		20.80	.769		3.13	28.50		31.63	46
4964	4"		16.50	.970		4.90	35.50		40.40	59
4966	6"		10.10	1.584		22	58.50		80.50	112
4970	2" x 1-1/2"		33.30	.480		2.12	17.65		19.77	29
4972	3" x 1-1/2"		21	.762		5.95	28		33.95	48.50
4974	4" x 3"		16.70	.958		10.55	35		45.55	64.50
4978	Closet flange, 4"	1 Plum	32	.250		5.20	10.20		15.40	21
4980	4" x 3"	"	34	.235		5.20	9.60		14.80	20
5000	DWV, PVC, schedule 40, socket joints									
5040	1/4 bend, 1-1/4"	1 Plum	20.20	.396	Ea.	3.25	16.20		19.45	28
5060	1-1/2"	"	18.20	.440		1.24	17.95		19.19	28.50
5070	2"	Q-1	33.10	.483		1.93	17.80		19.73	28.50
5080	3"		20.80	.769		5.55	28.50		34.05	48.50
5090	4"		16.50	.970		9.05	35.50		44.55	63.50
5100	6"		10.10	1.584		40.50	58.50		99	133
5105	8"	Q-2	9.30	2.581		82.50	98.50		181	239
5106	10"	"	8.50	2.824		75	108		183	245
5110	1/4 bend, long sweep, 1-1/2"	1 Plum	18.20	.440		3.05	17.95		21	30.50
5112	2"	Q-1	33.10	.483		2.93	17.80		20.73	29.50
5114	3"		20.80	.769		6.85	28.50		35.35	50
5116	4"		16.50	.970		12.80	35.50		48.30	67.50

560

Important: See the Reference Section for critical supporting data - Reference Nos., Crews, & City Cost Indexes

15108 | Plastic Pipe & Fittings

		CREW	DAILY OUTPUT	LABOR-HOURS	UNIT	2005 BARE COSTS				TOTAL INCL O&P	
						MAT.	LABOR	EQUIP.	TOTAL		
560	5150	1/8 bend, 1-1/4"	1 Plum	20.20	.396	Ea.	1.95	16.20		18.15	26.50	**560**
	5170	1-1/2"	"	18.20	.440		1.21	17.95		19.16	28.50	
	5180	2"	Q-1	33.10	.483		1.76	17.80		19.56	28.50	
	5190	3"		20.80	.769		4.81	28.50		33.31	48	
	5200	4"		16.50	.970		7.85	35.50		43.35	62	
	5210	6"		10.10	1.584		37.50	58.50		96	129	
	5215	8"	Q-2	9.30	2.581		66.50	98.50		165	221	
	5216	10"		8.50	2.824		109	108		217	282	
	5217	12"		7.60	3.158		150	120		270	345	
	5250	Tee, sanitary 1-1/4"	1 Plum	13.50	.593		3.36	24		27.36	40	
	5254	1-1/2"	"	12.10	.661		2.13	27		29.13	43	
	5255	2"	Q-1	20	.800		3.18	29.50		32.68	47.50	
	5256	3"		13.90	1.151		7.05	42.50		49.55	71.50	
	5257	4"		11	1.455		12.30	53.50		65.80	94	
	5259	6"		6.70	2.388		60.50	88		148.50	199	
	5261	8"	Q-2	6.20	3.871		181	148		329	420	
	5264	2" x 1-1/2"	Q-1	22	.727		2.94	27		29.94	43	
	5266	3" x 1-1/2"		15.50	1.032		4.47	38		42.47	62	
	5268	4" x 3"		12.10	1.322		19.20	48.50		67.70	94	
	5271	6" x 4"		6.90	2.319		65	85.50		150.50	200	
	5276	Tee, sanitary, reducing										
	5281	2" x 1-1/2" x 1-1/2"	Q-1	23	.696	Ea.	2.59	25.50		28.09	41.50	
	5282	2" x 1-1/2" x 2"		22	.727		3.56	27		30.56	44	
	5283	2" x 2" x 1-1/2"		22	.727		2.65	27		29.65	43	
	5284	3" x 3" x 1-1/2"		15.50	1.032		4.47	38		42.47	62	
	5285	3" x 3" x 2"		15.30	1.046		5.30	38.50		43.80	64	
	5286	4" x 4" x 1-1/2"		12.30	1.301		14.30	48		62.30	88	
	5287	4" x 4" x 2"		12.20	1.311		14.05	48		62.05	88	
	5288	4" x 4" x 3"		12.10	1.322		19.15	48.50		67.65	94	
	5291	6" x 6" x 4"		6.90	2.319		58.50	85.50		144	192	
	5294	Tee, double sanitary										
	5295	1-1/2"	1 Plum	9.10	.879	Ea.	4.32	36		40.32	59	
	5296	2"	Q-1	16.60	.964		6.30	35.50		41.80	60.50	
	5297	3"		10.40	1.538		17.70	56.50		74.20	104	
	5298	4"		8.25	1.939		28.50	71.50		100	139	
	5303	Wye, reducing										
	5304	2" x 1-1/2" x 1-1/2"	Q-1	23	.696	Ea.	5.05	25.50		30.55	44	
	5305	2" x 2" x 1-1/2"		22	.727		5.35	27		32.35	46	
	5306	3" x 3" x 2"		15.30	1.046		14.20	38.50		52.70	73.50	
	5307	4" x 4" x 2"		12.20	1.311		9.75	48		57.75	83	
	5309	4" x 4" x 3"		12.10	1.322		12.75	48.50		61.25	87	
	5314	Combination Y & 1/8 bend, 1-1/2"	1 Plum	12.10	.661		4.59	27		31.59	45.50	
	5315	2"	Q-1	20	.800		5.95	29.50		35.45	50.50	
	5317	3"		13.90	1.151		11.70	42.50		54.20	76.50	
	5318	4"		11	1.455		21	53.50		74.50	104	
	5324	Combination Y & 1/8 bend, reducing										
	5325	2" x 2" x 1-1/2"	Q-1	22	.727	Ea.	6.55	27		33.55	47.50	
	5327	3" x 3" x 1-1/2"		15.50	1.032		11.40	38		49.40	69.50	
	5328	3" x 3" x 2"		15.30	1.046		8.45	38.50		46.95	67.50	
	5329	4" x 4" x 2"		12.20	1.311		15.85	48		63.85	90	
	5331	Wye, 1-1/4"	1 Plum	13.50	.593		4.29	24		28.29	41	
	5332	1-1/2"	"	12.10	.661		2.86	27		29.86	43.50	
	5333	2"	Q-1	20	.800		3.65	29.50		33.15	48	
	5334	3"		13.90	1.151		9.90	42.50		52.40	74.50	
	5335	4"		11	1.455		16	53.50		69.50	98	
	5336	6"		6.70	2.388		62	88		150	200	

15108	Plastic Pipe & Fittings	CREW	DAILY OUTPUT	LABOR-HOURS	UNIT	2005 BARE COSTS				TOTAL INCL O&P
						MAT.	LABOR	EQUIP.	TOTAL	
5337	8"	Q-2	6.20	3.871	Ea.	73.50	148		221.50	305
5338	10"		5.70	4.211		143	161		304	400
5339	12"	▼	5.10	4.706		232	179		411	525
5341	2" x 1-1/2"	Q-1	22	.727		5.05	27		32.05	45.50
5342	3" x 1-1/2"		15.50	1.032		6.70	38		44.70	64.50
5343	4" x 3"		12.10	1.322		12.75	48.50		61.25	87
5344	6" x 4"	▼	6.90	2.319		44.50	85.50		130	177
5345	8" x 6"	Q-2	6.40	3.750		124	143		267	350
5347	Double wye, 1-1/2"	1 Plum	9.10	.879		6.10	36		42.10	60.50
5348	2"	Q-1	16.60	.964		7.85	35.50		43.35	62
5349	3"		10.40	1.538		20	56.50		76.50	107
5350	4"		8.25	1.939		41	71.50		112.50	152
5354	2" x 1-1/2"		16.80	.952		7.15	35		42.15	60.50
5355	3" x 2"		10.60	1.509		15.05	55.50		70.55	100
5356	4" x 3"		8.45	1.893		32.50	69.50		102	141
5357	6" x 4"	▼	7.25	2.207		67.50	81		148.50	197
5374	Coupling, 1-1/4"	1 Plum	20.20	.396		1.80	16.20		18	26.50
5376	1-1/2"	"	18.20	.440		.63	17.95		18.58	27.50
5378	2"	Q-1	33.10	.483		.78	17.80		18.58	27.50
5380	3"		20.80	.769		2.71	28.50		31.21	45.50
5390	4"		16.50	.970		4.45	35.50		39.95	58.50
5400	6"	▼	10.10	1.584		16.30	58.50		74.80	105
5402	8"	Q-2	9.30	2.581		36.50	98.50		135	189
5404	2" x 1-1/2"	Q-1	33.30	.480		1.60	17.65		19.25	28.50
5406	3" x 1-1/2"		21	.762		5.10	28		33.10	47.50
5408	4" x 3"		16.70	.958		9	35		44	63
5410	Reducer bushing, 2" x 1-1/4"		36.50	.438		.97	16.10		17.07	25
5411	2" x 1-1/2"		36.40	.440		1.08	16.15		17.23	25.50
5412	3" x 1-1/2"		27.30	.586		4.75	21.50		26.25	38
5413	3" x 2"		27.10	.590		2.34	21.50		23.84	35
5414	4" x 2"		18.20	.879		9.25	32.50		41.75	58.50
5415	4" x 3"		16.70	.958		5.10	35		40.10	58.50
5416	6" x 4"	▼	11.10	1.441		26.50	53		79.50	109
5418	8" x 6"	Q-2	10.20	2.353		53	89.50		142.50	194
5425	Closet flange 4"	Q-1	32	.500		5.40	18.40		23.80	33.50
5426	4" x 3"	"	34	.471	▼	8.25	17.30		25.55	35
5450	Solvent cement for PVC, industrial grade, per quart				Qt.	11.65			11.65	12.80
5500	CPVC, Schedule 80, threaded joints									
5540	90° Elbow, 1/4"	1 Plum	32	.250	Ea.	7.85	10.20		18.05	24
5560	1/2"		30.30	.264		4.55	10.80		15.35	21
5570	3/4"		26	.308		6.80	12.55		19.35	26.50
5580	1"		22.70	.352		9.55	14.40		23.95	32
5590	1-1/4"		20.20	.396		18.40	16.20		34.60	45
5600	1-1/2"	▼	18.20	.440		19.80	17.95		37.75	49
5610	2"	Q-1	33.10	.483		26.50	17.80		44.30	55.50
5620	2-1/2"		24.20	.661		78.50	24.50		103	123
5630	3"		20.80	.769		88.50	28.50		117	140
5640	4"		16.50	.970		138	35.50		173.50	206
5650	6"	▼	10.10	1.584	▼	163	58.50		221.50	267
5700	45° Elbow same as 90° Elbow									
5850	Tee, 1/4"	1 Plum	22	.364	Ea.	15.25	14.85		30.10	39.50
5870	1/2"		20.20	.396		15.25	16.20		31.45	41.50
5880	3/4"		17.30	.462		22	18.90		40.90	52.50
5890	1"		15.20	.526		23.50	21.50		45	58.50
5900	1-1/4"		13.50	.593		24	24		48	62.50
5910	1-1/2"	▼	12.10	.661	▼	25	27		52	68

Important: See the Reference Section for critical supporting data - Reference Nos., Crews, & City Cost Indexes

15108 | Plastic Pipe & Fittings

		CREW	DAILY OUTPUT	LABOR-HOURS	UNIT	2005 BARE COSTS MAT.	LABOR	EQUIP.	TOTAL	TOTAL INCL O&P	
560	5920	2"	Q-1	20	.800	Ea.	27.50	29.50		57	74.50	**560**
	5930	2-1/2"		16.20	.988		136	36.50		172.50	205	
	5940	3"		13.90	1.151		158	42.50		200.50	238	
	5950	4"		11	1.455		375	53.50		428.50	490	
	5960	6"		6.70	2.388		203	88		291	355	
	6000	Coupling, 1/4"	1 Plum	32	.250		10	10.20		20.20	26.50	
	6020	1/2"		30.30	.264		8.20	10.80		19	25.50	
	6030	3/4"		26	.308		13.30	12.55		25.85	33.50	
	6040	1"		22.70	.352		15.10	14.40		29.50	38	
	6050	1-1/4"		20.20	.396		16	16.20		32.20	42	
	6060	1-1/2"		18.20	.440		17.20	17.95		35.15	46	
	6070	2"	Q-1	33.10	.483		20.50	17.80		38.30	49	
	6080	2-1/2"		24.20	.661		36.50	24.50		61	76.50	
	6090	3"		20.80	.769		42.50	28.50		71	89	
	6100	4"		16.50	.970		86	35.50		121.50	149	
	6110	6"		10.10	1.584		124	58.50		182.50	224	
	6120	8"	Q-2	9.30	2.581		252	98.50		350.50	425	
	6200	CTS, 100 psi at 180°F, hot and cold water										
	6230	90° Elbow, 1/2"	1 Plum	20	.400	Ea.	.10	16.35		16.45	24.50	
	6250	3/4"		19	.421		.17	17.20		17.37	26	
	6251	1"		16	.500		.58	20.50		21.08	31	
	6252	1-1/4"		15	.533		.99	22		22.99	34	
	6253	1-1/2"		14	.571		1.59	23.50		25.09	37	
	6254	2"	Q-1	23	.696		3.43	25.50		28.93	42.50	
	6260	45° Elbow, 1/2"	1 Plum	20	.400		.19	16.35		16.54	24.50	
	6280	3/4"		19	.421		.23	17.20		17.43	26.50	
	6281	1"		16	.500		.53	20.50		21.03	31	
	6282	1-1/4"		15	.533		1.04	22		23.04	34	
	6283	1-1/2"		14	.571		1.65	23.50		25.15	37	
	6284	2"	Q-1	23	.696		3.48	25.50		28.98	42.50	
	6290	Tee, 1/2"	1 Plum	13	.615		.14	25		25.14	38	
	6310	3/4"		12	.667		.26	27		27.26	41.50	
	6311	1"		11	.727		1.20	29.50		30.70	46	
	6312	1-1/4"		10	.800		1.60	32.50		34.10	51	
	6313	1-1/2"		10	.800		2.60	32.50		35.10	52	
	6314	2"	Q-1	17	.941		4.22	34.50		38.72	56.50	
	6320	Coupling, 1/2"	1 Plum	22	.364		.10	14.85		14.95	22.50	
	6340	3/4"		21	.381		.14	15.55		15.69	23.50	
	6341	1"		18	.444		.50	18.15		18.65	28	
	6342	1-1/4"		17	.471		.62	19.20		19.82	29.50	
	6343	1-1/2"		16	.500		.88	20.50		21.38	31.50	
	6344	2"	Q-1	28	.571		1.75	21		22.75	33.50	
	6360	Solvent cement for CPVC, commercial grade, per quart				Qt.	17.85			17.85	19.65	
	7340	PVC flange, slip-on, Sch 80 std., 1/2"	1 Plum	22	.364	Ea.	8.20	14.85		23.05	31.50	
	7350	3/4"		21	.381		8.75	15.55		24.30	33	
	7360	1"		18	.444		9.75	18.15		27.90	38	
	7370	1-1/4"		17	.471		10.05	19.20		29.25	40	
	7380	1-1/2"		16	.500		10.25	20.50		30.75	42	
	7390	2"	Q-1	26	.615		13.65	22.50		36.15	49	
	7400	2-1/2"		24	.667		21	24.50		45.50	60	
	7410	3"		18	.889		23.50	32.50		56	74.50	
	7420	4"		15	1.067		29.50	39		68.50	91.50	
	7430	6"		10	1.600		48	59		107	142	
	7440	8"	Q-2	11	2.182		80.50	83		163.50	214	
	7550	Union, schedule 40, socket joints, 1/2"	1 Plum	19	.421		2.85	17.20		20.05	29	
	7560	3/4"		18	.444		3.25	18.15		21.40	31	

15108 | Plastic Pipe & Fittings

		CREW	DAILY OUTPUT	LABOR-HOURS	UNIT	2005 BARE COSTS MAT.	LABOR	EQUIP.	TOTAL	TOTAL INCL O&P	
560	7570	1"	1 Plum	15	.533	Ea.	3.34	22		25.34	36.50	**560**
	7580	1-1/4"		14	.571		10.75	23.50		34.25	47	
	7590	1-1/2"		13	.615		11.20	25		36.20	50.50	
	7600	2"	Q-1	20	.800		15.10	29.50		44.60	60.50	
	7990	Polybutyl/polyethyl pipe, for copper fittings see 15107-460-7000										
	8000	Compression type, PVC, 160 psi cold water										
	8010	Coupling, 3/4" CTS	1 Plum	21	.381	Ea.	2.36	15.55		17.91	26	
	8020	1" CTS		18	.444		2.92	18.15		21.07	30.50	
	8030	1-1/4" CTS		17	.471		4.08	19.20		23.28	33.50	
	8040	1-1/2" CTS		16	.500		5.60	20.50		26.10	36.50	
	8050	2" CTS		15	.533		7.85	22		29.85	41.50	
	8060	Female adapter, 3/4" FPT x 3/4" CTS		23	.348		3.76	14.20		17.96	25.50	
	8070	3/4" FPT x 1" CTS		21	.381		5.60	15.55		21.15	29.50	
	8080	1" FPT x 1" CTS		20	.400		5.65	16.35		22	30.50	
	8090	1-1/4" FPT x 1-1/4" CTS		18	.444		7.40	18.15		25.55	35.50	
	8100	1-1/2" FPT x 1-1/2" CTS		16	.500		8.65	20.50		29.15	40	
	8110	2" FPT x 2" CTS		13	.615		12.80	25		37.80	52	
	8130	Male adapter, 3/4" MPT x 3/4" CTS		23	.348		3.39	14.20		17.59	25	
	8140	3/4" MPT x 1" CTS		21	.381		4.17	15.55		19.72	28	
	8150	1" MPT x 1" CTS		20	.400		4.26	16.35		20.61	29	
	8160	1-1/4" MPT x 1-1/4" CTS		18	.444		6.55	18.15		24.70	34.50	
	8170	1-1/2" MPT x 1-1/2" CTS		16	.500		7.55	20.50		28.05	39	
	8180	2" MPT x 2" CTS		13	.615		11.75	25		36.75	51	
	8200	Spigot adapter, 3/4" IPS x 3/4" CTS		23	.348		3.39	14.20		17.59	25	
	8210	3/4" IPS x 1" CTS		21	.381		3.39	15.55		18.94	27	
	8220	1" IPS x 1" CTS		20	.400		3.66	16.35		20.01	28.50	
	8230	1-1/4" IPS x 1-1/4" CTS		18	.444		6.55	18.15		24.70	34.50	
	8240	1-1/2" IPS x 1-1/2" CTS		16	.500		7.55	20.50		28.05	39	
	8250	2" IPS x 2" CTS		13	.615		12.10	25		37.10	51.50	
	8270	Price includes insert stiffeners										
	8280	250 psi is same price as 160 psi										
	8300	Insert type, nylon, 160 & 250 psi, cold water										
	8310	Clamp ring stainless steel, 3/4" IPS	1 Plum	115	.070	Ea.	1.49	2.84		4.33	5.90	
	8320	1" IPS		107	.075		1.52	3.05		4.57	6.25	
	8330	1-1/4" IPS		101	.079		1.54	3.24		4.78	6.55	
	8340	1-1/2" IPS		95	.084		2.06	3.44		5.50	7.40	
	8350	2" IPS		85	.094		2.37	3.84		6.21	8.40	
	8370	Coupling, 3/4" IPS		22	.364		.71	14.85		15.56	23.50	
	8380	1" IPS		19	.421		.94	17.20		18.14	27	
	8390	1-1/4" IPS		18	.444		1.68	18.15		19.83	29.50	
	8400	1-1/2" IPS		17	.471		2.19	19.20		21.39	31.50	
	8410	2" IPS		16	.500		2.63	20.50		23.13	33.50	
	8430	Elbow, 90°, 3/4" IPS		22	.364		1.03	14.85		15.88	23.50	
	8440	1" IPS		19	.421		1.13	17.20		18.33	27	
	8450	1-1/4" IPS		18	.444		1.27	18.15		19.42	29	
	8460	1-1/2" IPS		17	.471		1.50	19.20		20.70	30.50	
	8470	2" IPS		16	.500		2.10	20.50		22.60	33	
	8490	Male adapter, 3/4" IPS x 3/4" MPT		25	.320		.71	13.05		13.76	20.50	
	8500	1" IPS x 1" MPT		21	.381		.94	15.55		16.49	24.50	
	8510	1-1/4" IPS x 1-1/4" MPT		20	.400		1.55	16.35		17.90	26	
	8520	1-1/2" IPS x 1-1/2" MPT		18	.444		2.05	18.15		20.20	30	
	8530	2" IPS x 2" MPT		15	.533		2.63	22		24.63	36	
	8550	Tee, 3/4" IPS		14	.571		1.34	23.50		24.84	36.50	
	8560	1" IPS		13	.615		2.21	25		27.21	40.50	
	8570	1-1/4" IPS		12	.667		3.96	27		30.96	45.50	
	8580	1-1/2" IPS		11	.727		4.97	29.50		34.47	50	

Important: See the Reference Section for critical supporting data - Reference Nos., Crews, & City Cost Indexes

15108	Plastic Pipe & Fittings	CREW	DAILY OUTPUT	LABOR-HOURS	UNIT	2005 BARE COSTS				TOTAL INCL O&P	
						MAT.	LABOR	EQUIP.	TOTAL		
560	8590	2" IPS	1 Plum	10	.800	Ea.	6.70	32.50		39.20	56.50	560
	8610	Insert type, PVC, 100 psi @ 180°F, hot & cold water										
	8620	Coupler, male, 3/8" CTS x 3/8" MPT	1 Plum	29	.276	Ea.	.40	11.25		11.65	17.40	
	8630	3/8" CTS x 1/2" MPT		28	.286		.47	11.65		12.12	18.05	
	8640	1/2" CTS x 1/2" MPT		27	.296		.43	12.10		12.53	18.65	
	8650	1/2" CTS x 3/4" MPT		26	.308		1.20	12.55		13.75	20	
	8660	3/4" CTS x 1/2" MPT		25	.320		1.20	13.05		14.25	21	
	8670	3/4" CTS x 3/4" MPT		25	.320		1.26	13.05		14.31	21	
	8700	Coupling, 3/8" CTS x 1/2" CTS		25	.320		2.15	13.05		15.20	22	
	8710	1/2" CTS		23	.348		2.50	14.20		16.70	24.50	
	8720	1/2" CTS x stub		23	.348		2.52	14.20		16.72	24.50	
	8730	3/4" CTS		22	.364		3.96	14.85		18.81	27	
	8750	Elbow 90°, 3/8" CTS		25	.320		2.55	13.05		15.60	22.50	
	8760	1/2" CTS		23	.348		3.01	14.20		17.21	25	
	8770	3/4" CTS		22	.364		4.61	14.85		19.46	27.50	
	8800	Rings, crimp, copper, 3/8" CTS		120	.067		.11	2.72		2.83	4.21	
	8810	1/2" CTS		117	.068		.12	2.79		2.91	4.33	
	8820	3/4" CTS		115	.070		.15	2.84		2.99	4.44	
	8850	Reducer tee, 3/8" x 3/8" x 1/2" CTS		17	.471		1.46	19.20		20.66	30.50	
	8860	1/2" x 3/8" x 1/2" CTS		15	.533		1.25	22		23.25	34.50	
	8870	3/4" x 1/2" x 1/2" CTS		14	.571		1.31	23.50		24.81	36.50	
	8890	3/4" x 3/4" x 1/2" CTS		14	.571		1.24	23.50		24.74	36.50	
	8900	3/4" x 1/2" x 3/8" CTS		14	.571		1.09	23.50		24.59	36	
	8930	Tee, 3/8" CTS		17	.471		1.05	19.20		20.25	30	
	8940	1/2" CTS		15	.533		1.19	22		23.19	34.50	
	8950	3/4" CTS		14	.571		1.29	23.50		24.79	36.50	
	8960	Copper rings included in fitting price										
	9000	Flare type, assembled, acetal, hot & cold water										
	9010	Coupling, 1/4" & 3/8" CTS	1 Plum	24	.333	Ea.	2.14	13.60		15.74	23	
	9020	1/2" CTS		22	.364		2.45	14.85		17.30	25	
	9030	3/4" CTS		21	.381		3.96	15.55		19.51	28	
	9040	1" CTS		18	.444		3.96	18.15		22.11	32	
	9050	Elbow 90°, 1/4" CTS		26	.308		2.38	12.55		14.93	21.50	
	9060	3/8" CTS		24	.333		2.55	13.60		16.15	23.50	
	9070	1/2" CTS		22	.364		3.01	14.85		17.86	26	
	9080	3/4" CTS		21	.381		4.61	15.55		20.16	28.50	
	9090	1" CTS		18	.444		5.80	18.15		23.95	34	
	9110	Tee ,1/4" & 3/8" CTS		15	.533		2.62	22		24.62	36	
	9120	1/2" CTS		14	.571		3.44	23.50		26.94	39	
	9130	3/4" CTS		13	.615		5.25	25		30.25	44	
	9140	1" CTS		12	.667		7	27		34	48.50	
	9550	For plastic hangers see 15060-300-8000										
	9560	For copper/brass fittings see 15107-460-7000										
590	0010	**PIPE, HIGH DENSITY POLYETHYLENE PLASTIC (HDPE)**										590
	0020	Not incl. hangers, trenching, backfill, hoisting or digging equipment.										
	0030	Standard length is 40', add a weld for each joint.										
	0040	Single wall										
	0050	Straight										
	0054	1" diameter DR 11				L.F.	.37			.37	.41	
	0058	1-1/2" diameter DR 11					.81			.81	.89	
	0062	2" diameter DR 11					1.05			1.05	1.16	
	0066	3" diameter DR 11					2.27			2.27	2.50	
	0070	3" diameter DR 17					1.52			1.52	1.67	
	0074	4" diameter DR 11					3.75			3.75	4.13	
	0078	4" diameter DR 17					2.52			2.52	2.77	

15108	Plastic Pipe & Fittings	CREW	DAILY OUTPUT	LABOR-HOURS	UNIT	2005 BARE COSTS				TOTAL INCL O&P
						MAT.	LABOR	EQUIP.	TOTAL	
0082	6" diameter DR 11				L.F.	8.10			8.10	8.95
0086	6" diameter DR 17					5.45			5.45	6
0090	8" diameter DR 11					13.75			13.75	15.15
0094	8" diameter DR 26					6.20			6.20	6.80
0098	10" diameter DR 11					21.50			21.50	23.50
0102	10" diameter DR 26					9.60			9.60	10.55
0106	12" diameter DR 11					30			30	33
0110	12" diameter DR 26					13.50			13.50	14.85
0114	16" diameter DR 11					44.50			44.50	49
0118	16" diameter DR 26					20			20	22
0122	18" diameter DR 11					56.50			56.50	62
0126	18" diameter DR 26					25.50			25.50	28
0130	20" diameter DR 11					70			70	77
0134	20" diameter DR 26					31.50			31.50	34.50
0138	22" diameter DR 11					84.50			84.50	93
0142	22" diameter DR 26					38			38	41.50
0146	24" diameter DR 11					101			101	111
0150	24" diameter DR 26					45			45	49.50
0154	28" diameter DR 17					92			92	101
0158	28" diameter DR 26					61.50			61.50	67.50
0162	30" diameter DR 21					86.50			86.50	95
0166	30" diameter DR 26					57			57	62.50
0170	36" diameter DR 26					57			57	62.50
0174	42" diameter DR 26					138			138	152
0178	48" diameter DR 26					175			175	192
0182	54" diameter DR 26				▼	228			228	251
0300	90° Elbow									
0304	1" diameter DR 11				Ea.	8.50			8.50	9.35
0308	1-1/2" diameter DR 11					14.20			14.20	15.60
0312	2" diameter DR 11					15.85			15.85	17.40
0316	3" diameter DR 11					30.50			30.50	33.50
0320	3" diameter DR 17					30.50			30.50	33.50
0324	4" diameter DR 11					35.50			35.50	39
0328	4" diameter DR 17					35.50			35.50	39
0332	6" diameter DR 11					104			104	115
0336	6" diameter DR 17					104			104	115
0340	8" diameter DR 11					345			345	380
0344	8" diameter DR 26					234			234	257
0348	10" diameter DR 11					850			850	935
0352	10" diameter DR 26					430			430	470
0356	12" diameter DR 11					850			850	935
0360	12" diameter DR 26					470			470	515
0364	16" diameter DR 11					950			950	1,050
0368	16" diameter DR 26					765			765	845
0372	18" diameter DR 11					995			995	1,100
0376	18" diameter DR 26					995			995	1,100
0380	20" diameter DR 11					1,475			1,475	1,625
0384	20" diameter DR 26					1,475			1,475	1,625
0388	22" diameter DR 11					1,850			1,850	2,025
0392	22" diameter DR 26					1,850			1,850	2,025
0396	24" diameter DR 11					2,125			2,125	2,350
0400	24" diameter DR 26					2,125			2,125	2,350
0404	28" diameter DR 17					3,225			3,225	3,550
0408	28" diameter DR 26					3,225			3,225	3,550
0412	30" diameter DR 17					2,975			2,975	3,275
0416	30" diameter DR 26				▼	2,975			2,975	3,275

Important: See the Reference Section for critical supporting data - Reference Nos., Crews, & City Cost Indexes

15108	Plastic Pipe & Fittings	CREW	DAILY OUTPUT	LABOR-HOURS	UNIT	MAT.	LABOR	EQUIP.	TOTAL	TOTAL INCL O&P
590							2005 BARE COSTS				590
0420	36" diameter DR 26				Ea.	4,550			4,550	5,000
0424	42" diameter DR 26					5,100			5,100	5,625
0428	48" diameter DR 26					5,400			5,400	5,925
0432	54" diameter DR 26					12,500			12,500	13,800
0500	45° Elbow									
0512	2" diameter DR 11				Ea.	31.50			31.50	35
0516	3" diameter DR 11					27.50			27.50	30.50
0520	3" diameter DR 17					45			45	49.50
0524	4" diameter DR 11					46			46	51
0528	4" diameter DR 17					55.50			55.50	61
0532	6" diameter DR 11					97.50			97.50	107
0536	6" diameter DR 17					92.50			92.50	102
0540	8" diameter DR 11					231			231	254
0544	8" diameter DR 26					139			139	152
0548	10" diameter DR 11					645			645	710
0552	10" diameter DR 26					187			187	206
0556	12" diameter DR 11					780			780	860
0560	12" diameter DR 26					296			296	325
0564	16" diameter DR 11					370			370	410
0568	16" diameter DR 26					370			370	410
0572	18" diameter DR 11					545			545	600
0576	18" diameter DR 26					545			545	600
0580	20" diameter DR 11					850			850	935
0584	20" diameter DR 26					850			850	935
0588	22" diameter DR 11					995			995	1,100
0592	22" diameter DR 26					995			995	1,100
0596	24" diameter DR 11					1,225			1,225	1,350
0600	24" diameter DR 26					1,225			1,225	1,350
0604	28" diameter DR 17					1,800			1,800	1,975
0608	28" diameter DR 26					1,800			1,800	1,975
0612	30" diameter DR 17					1,725			1,725	1,875
0616	30" diameter DR 26					1,725			1,725	1,875
0620	36" diameter DR 26					2,800			2,800	3,075
0624	42" diameter DR 26					3,425			3,425	3,775
0628	48" diameter DR 26					3,525			3,525	3,875
0632	54" diameter DR 26					5,750			5,750	6,325
0700	Tee									
0704	1" diameter DR 11				Ea.	8.50			8.50	9.35
0708	1-1/2" diameter DR 11					14.20			14.20	15.60
0712	2" diameter DR 11					19.80			19.80	22
0716	3" diameter DR 11					31.50			31.50	35
0720	3" diameter DR 17					31.50			31.50	35
0724	4" diameter DR 11					50			50	55
0728	4" diameter DR 17					50			50	55
0732	6" diameter DR 11					124			124	136
0736	6" diameter DR 17					124			124	136
0740	8" diameter DR 11					286			286	315
0744	8" diameter DR 17					260			260	286
0748	10" diameter DR 11					895			895	985
0752	10" diameter DR 17					320			320	350
0756	12" diameter DR 11					1,275			1,275	1,425
0760	12" diameter DR 17					365			365	400
0764	16" diameter DR 11					855			855	940
0768	16" diameter DR 17					855			855	940
0772	18" diameter DR 11					905			905	995
0776	18" diameter DR 17					905			905	995

15108	Plastic Pipe & Fittings	CREW	DAILY OUTPUT	LABOR-HOURS	UNIT	2005 BARE COSTS				TOTAL INCL O&P
						MAT.	LABOR	EQUIP.	TOTAL	
0780	20" diameter DR 11				Ea.	1,125			1,125	1,225
0784	20" diameter DR 17					1,125			1,125	1,225
0788	22" diameter DR 11					1,550			1,550	1,725
0792	22" diameter DR 17					1,550			1,550	1,725
0796	24" diameter DR 11					1,550			1,550	1,725
0800	24" diameter DR 17					1,550			1,550	1,725
0804	28" diameter DR 17					2,775			2,775	3,050
0812	30" diameter DR 17					3,075			3,075	3,400
0820	36" diameter DR 17					6,150			6,150	6,775
0824	42" diameter DR 26					6,775			6,775	7,450
0828	48" diameter DR 26				▼	7,925			7,925	8,725
1000	Flange adptr, w/back-up ring and 1/2 cost of plated bolt set									
1004	1" diameter DR 11				Ea.	78			78	85.50
1008	1-1/2" diameter DR 11					79.50			79.50	87.50
1012	2" diameter DR 11					44			44	48.50
1016	3" diameter DR 11					54.50			54.50	59.50
1020	3" diameter DR 17					54.50			54.50	59.50
1024	4" diameter DR 11					71			71	78
1028	4" diameter DR 17					71			71	78
1032	6" diameter DR 11					103			103	114
1036	6" diameter DR 17					103			103	114
1040	8" diameter DR 11					148			148	162
1044	8" diameter DR 26					148			148	162
1048	10" diameter DR 11					222			222	244
1052	10" diameter DR 26					222			222	244
1056	12" diameter DR 11					320			320	350
1060	12" diameter DR 26					320			320	350
1064	16" diameter DR 11					685			685	750
1068	16" diameter DR 26					685			685	750
1072	18" diameter DR 11					880			880	965
1076	18" diameter DR 26					880			880	965
1080	20" diameter DR 11					1,250			1,250	1,375
1084	20" diameter DR 26					1,250			1,250	1,375
1088	22" diameter DR 11					1,475			1,475	1,625
1092	22" diameter DR 26					1,475			1,475	1,625
1096	24" diameter DR 17					1,575			1,575	1,725
1100	24" diameter DR 32.5					1,575			1,575	1,725
1104	28" diameter DR 15.5					2,675			2,675	2,925
1108	28" diameter DR 32.5					2,600			2,600	2,875
1112	30" diameter DR 11					2,875			2,875	3,150
1116	30" diameter DR 21					2,500			2,500	2,750
1120	36" diameter DR 26					3,700			3,700	4,050
1124	42" diameter DR 26					4,175			4,175	4,575
1128	48" diameter DR 26					5,875			5,875	6,475
1132	54" diameter DR 26				▼	6,575			6,575	7,225
1200	Reducer									
1204	1-1/2" x 1" diameter DR 11				Ea.	9.10			9.10	10
1208	2" x 1-1/2" diameter DR 11					14.20			14.20	15.60
1212	3" x 2" diameter DR 11					18.50			18.50	20.50
1216	4" x 2" diameter DR 11					27.50			27.50	30.50
1220	4" x 3" diameter DR 11					27.50			27.50	30.50
1224	6" x 4" diameter DR 11					64.50			64.50	71
1228	8" x 6" diameter DR 11					114			114	125
1232	10" x 8" diameter DR 11					152			152	167
1236	12" x 8" diameter DR 11					284			284	310
1240	12" x 10" diameter DR 11				▼	203			203	224

Important: See the Reference Section for critical supporting data - Reference Nos., Crews, & City Cost Indexes

			DAILY	LABOR-		2005 BARE COSTS				TOTAL	
15108	**Plastic Pipe & Fittings**	CREW	OUTPUT	HOURS	UNIT	MAT.	LABOR	EQUIP.	TOTAL	INCL O&P	
590	1244	14″ x 12″ diameter DR 11				Ea.	228			228	251	590
	1248	16″ x 14″ diameter DR 11					275			275	300	
	1252	18″ x 16″ diameter DR 11					550			550	605	
	1256	20″ x 18″ diameter DR 11					745			745	820	
	1260	22″ x 20″ diameter DR 11					1,525			1,525	1,675	
	1264	24″ x 22″ diameter DR 11					975			975	1,075	
	1268	26″ x 24″ diameter DR 11					1,150			1,150	1,250	
	1272	28″ x 24″ diameter DR 11					1,300			1,300	1,450	
	1276	32″ x 28″ diameter DR 17					1,575			1,575	1,725	
	1280	36″ x 32″ diameter DR 17					1,875			1,875	2,050	
	4000	Welding labor per joint, not including welding machine										
	4010	Pipe joint size (cost based on thickest wall for each dia.)										
	4030	1″ pipe size	4 Skwk	273	.117	Ea.		4.09		4.09	6.35	
	4040	1-1/2″ pipe size		175	.183			6.35		6.35	9.90	
	4050	2″ pipe size		128	.250			8.70		8.70	13.55	
	4060	3″ pipe size		100	.320			11.15		11.15	17.35	
	4070	4″ pipe size		77	.416			14.50		14.50	22.50	
	4080	6″ pipe size	5 Skwk	63	.635			22		22	34.50	
	4090	8″ pipe size		48	.833			29		29	45	
	4100	10″ pipe size		40	1			35		35	54	
	4110	12″ pipe size	6 Skwk	41	1.171			41		41	63.50	
	4120	16″ pipe size		34	1.412			49		49	76.50	
	4130	18″ pipe size		32	1.500			52.50		52.50	81.50	
	4140	20″ pipe size	8 Skwk	37	1.730			60.50		60.50	94	
	4150	22″ pipe size		35	1.829			63.50		63.50	99	
	4160	24″ pipe size		34	1.882			65.50		65.50	102	
	4170	28″ pipe size		33	1.939			67.50		67.50	105	
	4180	30″ pipe size		32	2			69.50		69.50	108	
	4190	36″ pipe size		31	2.065			72		72	112	
	4200	42″ pipe size		30	2.133			74.50		74.50	116	
	4210	48″ pipe size	9 Skwk	33	2.182			76		76	118	
	4220	54″ pipe size	″	31	2.323			81		81	126	
	4300	Note: Cost for set up each time welder is moved.										
	4301	Add 50% of a weld cost										
	4310	Welder usually remains stationary with pipe moved through it.										
	4340	Weld machine, rental per day based on dia. capacity										
	4350	1″ thru 2″ diameter				Ea.			35	35	38.50	
	4360	3″ thru 4″ diameter							40	40	44	
	4370	6″ thru 8″ diameter							90	90	99	
	4380	10″ thru 12″ diameter							155	155	170.50	
	4390	16″ thru 18″ diameter							225	225	247.50	
	4400	20″ thru 24″ diameter							435	435	478.50	
	4410	28″ thru 32″ diameter							475	475	525	
	4420	36″ diameter							495	495	545	
	4430	42″ thru 54″ diameter							775	775	855	
	5000	Dual wall contained pipe										
	5040	Straight										
	5054	1″ DR 11 x 3″ DR 11				L.F.	7.60			7.60	8.35	
	5058	1″ DR 11 x 4″ DR 11					8.30			8.30	9.15	
	5062	1-1/2″ DR 11 x 4″ DR 17					8.80			8.80	9.65	
	5066	2″ DR 11 x 4″ DR 17					8.90			8.90	9.75	
	5070	2″ DR 11 x 6″ DR 17					13.85			13.85	15.25	
	5074	3″ DR 11 x 6″ DR 17					15.25			15.25	16.80	
	5078	3″ DR 11 x 6″ DR 26					12.25			12.25	13.45	
	5082	3″ DR 17 x 8″ DR 11					25			25	27.50	
	5086	3″ DR 17 x 8″ DR 17					19.90			19.90	22	

15108	Plastic Pipe & Fittings	CREW	DAILY OUTPUT	LABOR-HOURS	UNIT	2005 BARE COSTS				TOTAL INCL O&P
						MAT.	LABOR	EQUIP.	TOTAL	
590											590
5090	4" DR 11 x 8" DR 17				L.F.	22.50			22.50	25
5094	4" DR 17 x 8" DR 26					17.80			17.80	19.55
5098	6" DR 11 x 10" DR 17					36			36	39.50
5102	6" DR 17 x 10" DR 26					27.50			27.50	30
5106	6" DR 26 x 10" DR 26					25.50			25.50	28
5110	8" DR 17 x 12" DR 26					39.50			39.50	43.50
5114	8" DR 26 x 12" DR 32.5					33			33	36
5118	10" DR 17 x 14" DR 26					52			52	57
5122	10" DR 17 x 16" DR 26					58.50			58.50	64.50
5126	10" DR 26 x 16" DR 26					53			53	58.50
5130	12" DR 26 x 16" DR 26					60.50			60.50	66.50
5134	12" DR 17 x 18" DR 26					77			77	84.50
5138	12" DR 26 x 18" DR 26					69			69	76
5142	14" DR 26 x 20" DR 32.5					75.50			75.50	83
5146	16" DR 26 x 22" DR 32.5					114			114	126
5150	18" DR 26 x 24" DR 32.5					105			105	115
5154	20" DR 32.5 x 28" DR 32.5					136			136	150
5158	22" DR 32.5 x 30" DR 32.5					136			136	150
5162	24" DR 32.5 x 32" DR 32.5					173			173	191
5166	36" DR 32.5 x 42" DR 32.5				▼	293			293	320
5300	Force transfer coupling									
5354	1" DR 11 x 3" DR 11				Ea.	273			273	300
5358	1" DR 11 x 4" DR 17					287			287	315
5362	1-1/2" DR 11 x 4" DR 17					300			300	330
5366	2" DR 11 x 4" DR 17					315			315	345
5370	2" DR 11 x 6" DR 17					455			455	500
5374	3" DR 11 x 6" DR 17					455			455	500
5378	3" DR 11 x 6" DR 26					455			455	500
5382	3" DR 11 x 8" DR 11					480			480	530
5386	3" DR 11 x 8" DR 17					480			480	530
5390	4" DR 11 x 8" DR 17					480			480	530
5394	4" DR 17 x 8" DR 26					425			425	465
5398	6" DR 11 x 10" DR 17					610			610	675
5402	6" DR 17 x 10" DR 26					505			505	555
5406	6" DR 26 x 10" DR 26					505			505	555
5410	8" DR 17 x 12" DR 26					670			670	735
5414	8" DR 26 x 12" DR 32.5					540			540	595
5418	10" DR 17 x 14" DR 26					875			875	965
5422	10" DR 17 x 16" DR 26					875			875	965
5426	10" DR 26 x 16" DR 26					875			875	965
5430	12" DR 26 x 16" DR 26					1,100			1,100	1,200
5434	12" DR 17 x 18" DR 26					1,125			1,125	1,250
5438	12" DR 26 x 18" DR 26					1,125			1,125	1,250
5442	14" DR 26 x 20" DR 32.5					1,200			1,200	1,300
5446	16" DR 26 x 22" DR 32.5					1,225			1,225	1,350
5450	18" DR 26 x 24" DR 32.5					1,325			1,325	1,450
5454	20" DR 32.5 x 28" DR 32.5					1,625			1,625	1,775
5458	22" DR 32.5 x 30" DR 32.5					1,800			1,800	1,975
5462	24" DR 32.5 x 32" DR 32.5					1,925			1,925	2,125
5466	36" DR 32.5 x 42" DR 32.5				▼	3,200			3,200	3,525
5600	90° Elbow									
5654	1" DR 11 x 3" DR 11				Ea.	226			226	248
5658	1" DR 11 x 4" DR 17					216			216	238
5662	1-1/2" DR 11 x 4" DR 17					236			236	260
5666	2" DR 11 x 4" DR 17					250			250	275
5670	2" DR 11 x 6" DR 17				▼	320			320	350

		DAILY	LABOR-		2005 BARE COSTS				TOTAL	
15108 | Plastic Pipe & Fittings	CREW	OUTPUT	HOURS	UNIT	MAT.	LABOR	EQUIP.	TOTAL	INCL O&P	
590	5674	3" DR 11 x 6" DR 17				Ea.	365			365	400	590
5678	3" DR 17 x 6" DR 26					291			291	320
5682	3" DR 17 x 8" DR 11					590			590	645
5686	3" DR 17 x 8" DR 17					455			455	505
5690	4" DR 11 x 8" DR 17					540			540	590
5694	4" DR 17 x 8" DR 26					425			425	470
5698	6" DR 11 x 10" DR 17					715			715	790
5702	6" DR 17 x 10" DR 26					550			550	605
5706	6" DR 26 x 10" DR 26					515			515	565
5710	8" DR 17 x 12" DR 26					815			815	895
5714	8" DR 26 x 12" DR 32.5					755			755	830
5718	10" DR 17 x 14" DR 26					1,275			1,275	1,400
5722	10" DR 17 x 16" DR 26					1,225			1,225	1,350
5726	10" DR 26 x 16" DR 26					1,125			1,125	1,250
5730	12" DR 26 x 16" DR 26					1,250			1,250	1,400
5734	12" DR 17 x 18" DR 26					1,500			1,500	1,650
5738	12" DR 26 x 18" DR 26					1,375			1,375	1,525
5742	14" DR 26 x 20" DR 32.5					1,825			1,825	2,000
5746	16" DR 26 x 22" DR 32.5					1,825			1,825	2,025
5750	18" DR 26 x 24" DR 32.5					2,250			2,250	2,475
5754	20" DR 32.5 x 28" DR 32.5					2,600			2,600	2,850
5758	22" DR 32.5 x 30" DR 32.5					3,525			3,525	3,900
5762	24" DR 32.5 x 32" DR 32.5					3,425			3,425	3,775
5766	36" DR 32.5 x 42" DR 32.5				▼	5,575			5,575	6,150
5800	45° Elbow									
5804	1" DR 11 x 3" DR 11				Ea.	135			135	149
5808	1" DR 11 x 4" DR 17					134			134	148
5812	1-1/2" DR 11 x 4" DR 17					144			144	158
5816	2" DR 11 x 4" DR 17					157			157	173
5820	2" DR 11 x 6" DR 17					197			197	217
5824	3" DR 11 x 6" DR 17					219			219	241
5828	3" DR 17 x 6" DR 26					181			181	200
5832	3" DR 17 x 8" DR 11					335			335	370
5836	3" DR 17 x 8" DR 17					268			268	295
5840	4" DR 11 x 8" DR 17					315			315	345
5844	4" DR 17 x 8" DR 26					256			256	281
5848	6" DR 11 x 10" DR 17					415			415	455
5852	6" DR 17 x 10" DR 26					325			325	355
5856	6" DR 26 x 10" DR 26					305			305	335
5860	8" DR 17 x 12" DR 26					495			495	540
5864	8" DR 26 x 12" DR 32.5					460			460	505
5868	10" DR 17 x 14" DR 26					770			770	850
5872	10" DR 17 x 16" DR 26					755			755	830
5876	10" DR 26 x 16" DR 26					710			710	780
5880	12" DR 26 x 16" DR 26					800			800	880
5884	12" DR 17 x 18" DR 26					940			940	1,025
5888	12" DR 26 x 18" DR 26					875			875	965
5892	14" DR 26 x 20" DR 32.5					1,125			1,125	1,250
5896	16" DR 26 x 22" DR 32.5					1,175			1,175	1,300
5900	18" DR 26 x 24" DR 32.5					1,400			1,400	1,550
5904	20" DR 32.5 x 28" DR 32.5					1,725			1,725	1,875
5908	22" DR 32.5 x 30" DR 32.5					2,175			2,175	2,400
5912	24" DR 32.5 x 32" DR 32.5					2,175			2,175	2,375
5916	36" DR 32.5 x 42" DR 32.5				▼	3,375			3,375	3,700
6000	Access port with 4" riser									
6050	1" DR 11 x 4" DR 17				Ea.	295			295	325

15108	Plastic Pipe & Fittings	CREW	DAILY OUTPUT	LABOR-HOURS	UNIT	2005 BARE COSTS				TOTAL INCL O&P	
						MAT.	LABOR	EQUIP.	TOTAL		
590											**590**
6054	1-1/2" DR 11 x 4" DR 17				Ea.	300			300	330	
6058	2" DR 11 x 6" DR 17					365			365	405	
6062	3" DR 11 x 6" DR 17					370			370	405	
6066	3" DR 17 x 6" DR 26					360			360	400	
6070	3" DR 17 x 8" DR 11					415			415	455	
6074	3" DR 17 x 8" DR 17					395			395	435	
6078	4" DR 11 x 8" DR 17					410			410	450	
6082	4" DR 17 x 8" DR 26					390			390	430	
6086	6" DR 11 x 10" DR 17					510			510	565	
6090	6" DR 17 x 10" DR 26					470			470	520	
6094	6" DR 26 x 10" DR 26					460			460	510	
6098	8" DR 17 x 12" DR 26					495			495	545	
6102	8" DR 26 x 12" DR 32.5				▼	470			470	520	
6200	End termination with vent plug										
6204	1" DR 11 x 3" DR 11				Ea.	244			244	269	
6208	1" DR 11 x 4" DR 17					261			261	287	
6212	1-1/2" DR 11 x 4" DR 17					261			261	287	
6216	2" DR 11 x 4" DR 17					284			284	310	
6220	2" DR 11 x 6" DR 17					325			325	355	
6224	3" DR 11 x 6" DR 17					325			325	355	
6228	3" DR 17 x 6" DR 26					325			325	355	
6232	3" DR 17 x 8" DR 11					450			450	490	
6236	3" DR 17 x 8" DR 17					450			450	490	
6240	4" DR 11 x 8" DR 17					450			450	490	
6244	4" DR 17 x 8" DR 26					390			390	430	
6248	6" DR 11 x 10" DR 17					530			530	585	
6252	6" DR 17 x 10" DR 26					435			435	480	
6256	6" DR 26 x 10" DR 26					435			435	480	
6260	8" DR 17 x 12" DR 26					645			645	710	
6264	8" DR 26 x 12" DR 32.5					515			515	570	
6268	10" DR 17 x 14" DR 26					670			670	740	
6272	10" DR 17 x 16" DR 26					670			670	740	
6276	10" DR 26 x 16" DR 26					670			670	740	
6280	12" DR 26 x 16" DR 26					670			670	740	
6284	12" DR 17 x 18" DR 26					885			885	975	
6288	12" DR 26 x 18" DR 26					885			885	975	
6292	14" DR 26 x 20" DR 32.5					900			900	990	
6296	16" DR 26 x 22" DR 32.5					1,050			1,050	1,150	
6300	18" DR 26 x 24" DR 32.5					1,075			1,075	1,175	
6304	20" DR 32.5 x 28" DR 32.5					1,375			1,375	1,525	
6308	22" DR 32.5 x 30" DR 32.5					1,525			1,525	1,675	
6312	24" DR 32.5 x 32" DR 32.5					1,675			1,675	1,850	
6316	36" DR 32.5 x 42" DR 32.5				▼	2,450			2,450	2,700	
6600	Tee										
6604	1" DR 11 x 3" DR 11				Ea.	249			249	274	
6608	1" DR 11 x 4" DR 17					310			310	340	
6612	1-1/2" DR 11 x 4" DR 17					345			345	380	
6616	2" DR 11 x 4" DR 17					375			375	415	
6620	2" DR 11 x 6" DR 17					455			455	500	
6624	3" DR 11 x 6" DR 17					525			525	575	
6628	3" DR 17 x 6" DR 26					510			510	560	
6632	3" DR 17 x 8" DR 11					620			620	680	
6636	3" DR 17 x 8" DR 17					590			590	645	
6640	4" DR 11 x 8" DR 17					640			640	705	
6644	4" DR 17 x 8" DR 26					615			615	675	
6648	6" DR 11 x 10" DR 17				▼	800			800	880	

Important: See the Reference Section for critical supporting data - Reference Nos., Crews, & City Cost Indexes

			DAILY	LABOR-		\multicolumn{4}{c}{2005 BARE COSTS}			TOTAL
15108	**Plastic Pipe & Fittings**	CREW	OUTPUT	HOURS	UNIT	MAT.	LABOR	EQUIP.	TOTAL	INCL O&P
590 6652	6" DR 17 x 10" DR 26				Ea.	760			760	835	590
6656	6" DR 26 x 10" DR 26					750			750	825
6660	8" DR 17 x 12" DR 26					750			750	825
6664	8" DR 26 x 12" DR 32.5					980			980	1,075
6668	10" DR 17 x 14" DR 26					1,125			1,125	1,250
6672	10" DR 17 x 16" DR 26					1,200			1,200	1,300
6676	10" DR 26 x 16" DR 26					1,200			1,200	1,300
6680	12" DR 26 x 16" DR 26					1,275			1,275	1,400
6684	12" DR 17 x 18" DR 26					1,275			1,275	1,400
6688	12" DR 26 x 18" DR 26					1,425			1,425	1,575
6692	14" DR 26 x 20" DR 32.5					1,875			1,875	2,075
6696	16" DR 26 x 22" DR 32.5					2,300			2,300	2,525
6700	18" DR 26 x 24" DR 32.5					2,500			2,500	2,750
6704	20" DR 32.5 x 28" DR 32.5					3,150			3,150	3,475
6708	22" DR 32.5 x 30" DR 32.5					3,750			3,750	4,125
6712	24" DR 32.5 x 32" DR 32.5					4,500			4,500	4,950
6716	36" DR 32.5 x 42" DR 32.5				▼	9,150			9,150	10,100
6800	Wye									
6816	2" DR 11 x 4" DR 17				Ea.	475			475	520
6820	2" DR 11 x 6" DR 17					535			535	590
6824	3" DR 11 x 6" DR 17					545			545	595
6828	3" DR 17 x 6" DR 26					525			525	575
6832	3" DR 17 x 8" DR 11					660			660	725
6836	3" DR 17 x 8" DR 17					615			615	680
6840	4" DR 11 x 8" DR 17					670			670	740
6844	4" DR 17 x 8" DR 26					640			640	700
6848	6" DR 11 x 10" DR 17					830			830	915
6852	6" DR 17 x 10" DR 26					775			775	850
6856	6" DR 26 x 10" DR 26					770			770	845
6860	8" DR 17 x 12" DR 26					1,100			1,100	1,200
6864	8" DR 26 x 12" DR 32.5					1,075			1,075	1,175
6868	10" DR 17 x 14" DR 26					1,325			1,325	1,450
6872	10" DR 17 x 16" DR 26					1,400			1,400	1,550
6876	10" DR 26 x 16" DR 26					1,400			1,400	1,525
6880	12" DR 26 x 16" DR 26					1,550			1,550	1,725
6884	12" DR 17 x 18" DR 26					1,750			1,750	1,925
6888	12" DR 26 x 18" DR 26					1,775			1,775	1,950
6892	14" DR 26 x 20" DR 32.5					2,175			2,175	2,375
6896	16" DR 26 x 22" DR 32.5					3,125			3,125	3,425
6900	18" DR 26 x 24" DR 32.5					3,675			3,675	4,025
6904	20" DR 32.5 x 28" DR 32.5					4,600			4,600	5,050
6908	22" DR 32.5 x 30" DR 32.5					5,225			5,225	5,750
6912	24" DR 32.5 x 32" DR 32.5				▼	5,950			5,950	6,550
9000	Welding labor per joint, not including welding machine									
9010	Pipe joint size, outer pipe (cost based on the thickest walls)									
9020	Straight pipe									
9050	3" pipe size	4 Skwk	96	.333	Ea.		11.60		11.60	18.05
9060	4" pipe size	"	77	.416			14.50		14.50	22.50
9070	6" pipe size	5 Skwk	60	.667			23		23	36
9080	8" pipe size	"	40	1			35		35	54
9090	10" pipe size	6 Skwk	41	1.171			41		41	63.50
9100	12" pipe size		39	1.231			43		43	66.50
9110	14" pipe size		38	1.263			44		44	68.50
9120	16" pipe size		35	1.371			48		48	74.50
9130	18" pipe size	8 Skwk	45	1.422			49.50		49.50	77
9140	20" pipe size		42	1.524	▼		53		53	82.50

15108 | Plastic Pipe & Fittings

		CREW	DAILY OUTPUT	LABOR-HOURS	UNIT	2005 BARE COSTS				TOTAL INCL O&P
						MAT.	LABOR	EQUIP.	TOTAL	
590 9150	22" pipe size	8 Skwk	40	1.600	Ea.		56		56	86.50 **590**
9160	24" pipe size		38	1.684			58.50		58.50	91.50
9170	28" pipe size		37	1.730			60.50		60.50	94
9180	30" pipe size		36	1.778			62		62	96.50
9190	32" pipe size		35	1.829			63.50		63.50	99
9200	42" pipe size	↓	32	2	↓		69.50		69.50	108
9300	Note: Cost for set up each time welder is moved.									
9301	Add 100% of weld labor cost									
9310	For handling between fitting welds add 50% of weld labor cost									
9320	Welder usually remains stationary with pipe moved through it									
9360	Weld machine, rental per day based on dia. capacity									
9380	3" thru 4" diameter				Ea.			55	55	60.50
9390	6" thru 8" diameter							180	180	198
9400	10" thru 12" diameter							270	270	297
9410	14" thru 18" diameter							365	365	402
9420	20" thru 24" diameter							650	650	715
9430	28" thru 32" diameter							765	765	840
9440	42" thru 58" diameter				↓			790	790	870

15110 | Valves

		CREW	DAILY OUTPUT	LABOR-HOURS	UNIT	MAT.	LABOR	EQUIP.	TOTAL	TOTAL INCL O&P
100 0010	**VALVES, BRASS**									**100**
0030	For motorized valves, see Division 13838-200									
0500	Gas cocks, threaded									
0510	1/4"	1 Plum	26	.308	Ea.	5.95	12.55		18.50	25.50
0520	3/8"		24	.333		5.95	13.60		19.55	27
0530	1/2"		24	.333		5.75	13.60		19.35	27
0540	3/4"		22	.364		10.80	14.85		25.65	34.50
0550	1"		19	.421		11.35	17.20		28.55	38.50
0560	1-1/4"		15	.533		28	22		50	64
0570	1-1/2"		13	.615		29.50	25		54.50	70.50
0580	2"	↓	11	.727	↓	55.50	29.50		85	106
0670	For larger sizes use lubricated plug valve, Section 15110-600									
160 0010	**VALVES, BRONZE** R15100 -090									**160**
1020	Angle, 150 lb., rising stem, threaded									
1030	1/8"	1 Plum	24	.333	Ea.	49.50	13.60		63.10	75
1040	1/4"		24	.333		49.50	13.60		63.10	75
1050	3/8"		24	.333		49.50	13.60		63.10	75
1060	1/2"		22	.364		49.50	14.85		64.35	77
1070	3/4"		20	.400		67	16.35		83.35	98
1080	1"		19	.421		95.50	17.20		112.70	131
1090	1-1/4"		15	.533		124	22		146	169
1100	1-1/2"		13	.615		161	25		186	215
1110	2"	↓	11	.727	↓	260	29.50		289.50	330
1380	Ball, 150 psi, threaded									
1400	1/4"	1 Plum	24	.333	Ea.	7	13.60		20.60	28
1430	3/8"		24	.333		7	13.60		20.60	28
1450	1/2"		22	.364		7	14.85		21.85	30
1460	3/4"		20	.400		11.55	16.35		27.90	37
1470	1"		19	.421		14.55	17.20		31.75	42
1480	1-1/4"		15	.533		17.15	22		39.15	52
1490	1-1/2"		13	.615		26.50	25		51.50	67
1500	2"		11	.727		33	29.50		62.50	81
1510	2-1/2"		9	.889		156	36.50		192.50	227
1520	3"	↓	8	1	↓	255	41		296	345
1600	Butterfly, 175 psi, full port, solder or threaded ends									
1610	Stainless steel disc and stem									

Important: See the Reference Section for critical supporting data - Reference Nos., Crews, & City Cost Indexes

15110 | Valves

		CREW	DAILY OUTPUT	LABOR-HOURS	UNIT	2005 BARE COSTS				TOTAL INCL O&P
						MAT.	LABOR	EQUIP.	TOTAL	
1620	1/4"	1 Plum	24	.333	Ea.	6.25	13.60		19.85	27.50
1630	3/8"		24	.333		6.25	13.60		19.85	27.50
1640	1/2"		22	.364		6.65	14.85		21.50	30
1650	3/4"		20	.400		10.75	16.35		27.10	36.50
1660	1"		19	.421		13.15	17.20		30.35	40.50
1670	1-1/4"		15	.533		21	22		43	56.50
1680	1-1/2"		13	.615		27.50	25		52.50	68
1690	2"		11	.727		34	29.50		63.50	82
1750	Check, swing, class 150, regrinding disc, threaded									
1800	1/8"	1 Plum	24	.333	Ea.	25	13.60		38.60	48
1830	1/4"		24	.333		25	13.60		38.60	48
1840	3/8"		24	.333		26	13.60		39.60	49
1850	1/2"		24	.333		26	13.60		39.60	49
1860	3/4"		20	.400		35	16.35		51.35	63
1870	1"		19	.421		52	17.20		69.20	83
1880	1-1/4"		15	.533		73	22		95	113
1890	1-1/2"		13	.615		85.50	25		110.50	132
1900	2"		11	.727		125	29.50		154.50	183
1910	2-1/2"	Q-1	15	1.067		265	39		304	350
1920	3"	"	13	1.231		395	45.50		440.50	505
2000	For 200 lb, add					5%	10%			
2040	For 300 lb, add					15%	15%			
2060	Check swing, 300#, sweat, 3/8" size	1 Plum	24	.333	Ea.	16	13.60		29.60	38
2070	1/2"		24	.333		16	13.60		29.60	38
2080	3/4"		20	.400		20.50	16.35		36.85	47
2090	1"		19	.421		28.50	17.20		45.70	57.50
2100	1-1/4"		15	.533		37	22		59	73.50
2110	1-1/2"		13	.615		49	25		74	91.50
2120	2"		11	.727		73.50	29.50		103	126
2130	2-1/2"	Q-1	15	1.067		175	39		214	251
2140	3"	"	13	1.231		233	45.50		278.50	325
2850	Gate, N.R.S., soldered, 125 psi									
2900	3/8"	1 Plum	24	.333	Ea.	18.30	13.60		31.90	40.50
2920	1/2"		24	.333		16	13.60		29.60	38
2940	3/4"		20	.400		25	16.35		41.35	52
2950	1"		19	.421		34.50	17.20		51.70	64
2960	1-1/4"		15	.533		46.50	22		68.50	84
2970	1-1/2"		13	.615		52.50	25		77.50	95.50
2980	2"		11	.727		83.50	29.50		113	137
2990	2-1/2"	Q-1	15	1.067		195	39		234	274
3000	3"	"	13	1.231		256	45.50		301.50	350
3350	Threaded, class 150									
3410	1/4"	1 Plum	24	.333	Ea.	31	13.60		44.60	55
3420	3/8"		24	.333		31	13.60		44.60	55
3430	1/2"		24	.333		28.50	13.60		42.10	52
3440	3/4"		20	.400		32	16.35		48.35	59.50
3450	1"		19	.421		42	17.20		59.20	72
3460	1-1/4" size		15	.533		55.50	22		77.50	94
3470	1-1/2"		13	.615		71.50	25		96.50	117
3480	2"		11	.727		95	29.50		124.50	150
3490	2-1/2"	Q-1	15	1.067		222	39		261	305
3500	3" size	"	13	1.231		315	45.50		360.50	415
3600	Gate, flanged, 150 lb.									
3610	1"	1 Plum	7	1.143	Ea.	540	46.50		586.50	660
3620	1-1/2"		6	1.333		670	54.50		724.50	815
3630	2"		5	1.600		970	65.50		1,035.50	1,175

R15100
-090

160

259

15110 | Valves

		CREW	DAILY OUTPUT	LABOR-HOURS	UNIT	2005 BARE COSTS				TOTAL INCL O&P
						MAT.	LABOR	EQUIP.	TOTAL	
160 3634	2-1/2"	Q-1	5	3.200	Ea.	1,525	118		1,643	1,850	160
3640	3"	"	4.50	3.556		1,775	131		1,906	2,150
3850	Rising stem, soldered, 300 psi									
3900	3/8"	1 Plum	24	.333	Ea.	43	13.60		56.60	67.50
3920	1/2"		24	.333		38.50	13.60		52.10	62.50
3940	3/4"		20	.400		49	16.35		65.35	78
3950	1"		19	.421		59.50	17.20		76.70	91
3960	1-1/4"		15	.533		88	22		110	130
3970	1-1/2"		13	.615		107	25		132	156
3980	2"		11	.727		161	29.50		190.50	222
3990	2-1/2"	Q-1	15	1.067		390	39		429	485
4000	3"	"	13	1.231		455	45.50		500.50	570
4250	Threaded, class 150									
4310	1/4"	1 Plum	24	.333	Ea.	27	13.60		40.60	50.50
4320	3/8"		24	.333		27	13.60		40.60	50.50
4330	1/2"		24	.333		25.50	13.60		39.10	48.50
4340	3/4"		20	.400		30	16.35		46.35	57.50
4350	1"		19	.421		40.50	17.20		57.70	70.50
4360	1-1/4"		15	.533		53.50	22		75.50	91.50
4370	1-1/2"		13	.615		67.50	25		92.50	113
4380	2"		11	.727		91.50	29.50		121	146
4390	2-1/2"	Q-1	15	1.067		214	39		253	294
4400	3"	"	13	1.231		299	45.50		344.50	400
4500	For 300 psi, threaded, add					100%	15%			
4540	For chain operated type, add					15%				
4850	Globe, class 150, rising stem, threaded									
4920	1/4"	1 Plum	24	.333	Ea.	35.50	13.60		49.10	60
4940	3/8" size		24	.333		39	13.60		52.60	63.50
4950	1/2"		24	.333		39	13.60		52.60	63.50
4960	3/4"		20	.400		53.50	16.35		69.85	83
4970	1"		19	.421		83.50	17.20		100.70	118
4980	1-1/4"		15	.533		131	22		153	177
4990	1-1/2"		13	.615		160	25		185	214
5000	2"		11	.727		239	29.50		268.50	310
5010	2-1/2"	Q-1	15	1.067		480	39		519	590
5020	3"	"	13	1.231		605	45.50		650.50	735
5120	For 300 lb threaded, add					50%	15%			
5600	Relief, pressure & temperature, self-closing, ASME, threaded									
5640	3/4"	1 Plum	28	.286	Ea.	81.50	11.65		93.15	107
5650	1"		24	.333		118	13.60		131.60	151
5660	1-1/4"		20	.400		237	16.35		253.35	286
5670	1-1/2"		18	.444		455	18.15		473.15	530
5680	2"		16	.500		495	20.50		515.50	575
5950	Pressure, poppet type, threaded									
6000	1/2"	1 Plum	30	.267	Ea.	13.55	10.90		24.45	31.50
6040	3/4"	"	28	.286	"	45	11.65		56.65	67
6400	Pressure, water, ASME, threaded									
6440	3/4"	1 Plum	28	.286	Ea.	47	11.65		58.65	69.50
6450	1"		24	.333		93.50	13.60		107.10	124
6460	1-1/4"		20	.400		153	16.35		169.35	193
6470	1-1/2"		18	.444		211	18.15		229.15	260
6480	2"		16	.500		305	20.50		325.50	365
6490	2-1/2"		15	.533		1,550	22		1,572	1,725
6900	Reducing, water pressure									
6920	300 psi to 25-75 psi, threaded or sweat									
6940	1/2"	1 Plum	24	.333	Ea.	147	13.60		160.60	182

R15100 -090

Important: See the Reference Section for critical supporting data - Reference Nos., Crews, & City Cost Indexes

15110 | Valves

		CREW	DAILY OUTPUT	LABOR-HOURS	UNIT	2005 BARE COSTS MAT.	LABOR	EQUIP.	TOTAL	TOTAL INCL O&P
160 6950	3/4" R15100-090	1 Plum	20	.400	Ea.	147	16.35		163.35	186 **160**
6960	1"		19	.421		227	17.20		244.20	276
6970	1-1/4"		15	.533		405	22		427	480
6980	1-1/2"		13	.615		615	25		640	715
6990	2"	▼	11	.727		885	29.50		914.50	1,025
7100	For built-in by-pass or 10-35 psi, add				▼	7.30			7.30	8
7700	High capacity, 250 psi to 25-75 psi, threaded									
7740	1/2"	1 Plum	24	.333	Ea.	147	13.60		160.60	182
7780	3/4"		20	.400		147	16.35		163.35	186
7790	1"		19	.421		227	17.20		244.20	276
7800	1-1/4"		15	.533		405	22		427	480
7810	1-1/2"		13	.615		615	25		640	715
7820	2"		11	.727		890	29.50		919.50	1,025
7830	2-1/2"		9	.889		1,075	36.50		1,111.50	1,225
7840	3"	▼	8	1		1,175	41		1,216	1,350
7850	3" flanged (iron body)	Q-1	10	1.600		1,275	59		1,334	1,500
7860	4" flanged (iron body)	"	8	2	▼	2,125	73.50		2,198.50	2,450
7920	For higher pressure, add					25%				
8000	Silent check, bronze trim									
8010	Compact wafer type, for 125 or 150 lb. flanges									
8020	1-1/2"	1 Plum	11	.727	Ea.	198	29.50		227.50	263
8021	2"	"	9	.889		212	36.50		248.50	289
8022	2-1/2"	Q-1	9	1.778		224	65.50		289.50	345
8023	3"		8	2		250	73.50		323.50	385
8024	4"	▼	5	3.200		425	118		543	645
8025	5"	Q-2	6	4		590	153		743	875
8026	6"	"	5	4.800	▼	705	183		888	1,050
8050	For 250 or 300 lb flanges, thru 6" no change									
8060	Full flange wafer type, 150 lb.									
8063	1-1/2"	1 Plum	11	.727	Ea.	266	29.50		295.50	340
8064	2"	"	9	.889		355	36.50		391.50	450
8065	2-1/2"	Q-1	9	1.778		400	65.50		465.50	545
8066	3"		8	2		445	73.50		518.50	600
8067	4"	▼	5	3.200		660	118		778	900
8068	5"	Q-2	6	4		850	153		1,003	1,175
8069	6"	"	5	4.800		1,125	183		1,308	1,500
8080	For 300 lb., add				▼	40%	10%			
8100	Globe type, 150 lb.									
8110	2"	1 Plum	9	.889	Ea.	415	36.50		451.50	515
8111	2-1/2"	Q-1	9	1.778		515	65.50		580.50	665
8112	3"		8	2		615	73.50		688.50	785
8113	4"	▼	5	3.200		850	118		968	1,100
8114	5"	Q-2	6	4		1,050	153		1,203	1,375
8115	6"	"	5	4.800	▼	1,450	183		1,633	1,875
8130	For 300 lb., add					20%	10%			
8140	Screwed end type, 250 lb.									
8141	1/2"	1 Plum	24	.333	Ea.	37	13.60		50.60	61.50
8142	3/4"		20	.400		37	16.35		53.35	65.50
8143	1"		19	.421		43	17.20		60.20	73.50
8144	1-1/4"		15	.533		59	22		81	98
8145	1-1/2"		13	.615		66	25		91	111
8146	2"	▼	11	.727	▼	89	29.50		118.50	143
8350	Tempering, water, sweat connections									
8400	1/2"	1 Plum	24	.333	Ea.	52.50	13.60		66.10	78
8440	3/4"	"	20	.400	"	64	16.35		80.35	95
8650	Threaded connections									

15110 | Valves

		CREW	DAILY OUTPUT	LABOR-HOURS	UNIT	2005 BARE COSTS				TOTAL INCL O&P
						MAT.	LABOR	EQUIP.	TOTAL	
160	8700	1/2"	1 Plum	24	.333	Ea.	79	13.60		92.60	107	**160**
	8740	3/4"		20	.400		244	16.35		260.35	294
	8750	1"		19	.421		269	17.20		286.20	320
	8760	1-1/4"		15	.533		430	22		452	505
	8770	1-1/2"		13	.615		465	25		490	555
	8780	2"		11	.727		700	29.50		729.50	815
200	0010	**VALVES, IRON BODY**										**200**
	0020	For grooved joint, see Division 15107-690									
	0100	Angle, 125 lb.									
	0110	Flanged									
	0116	2"	1 Plum	5	1.600	Ea.	660	65.50		725.50	830
	0118	4"	Q-1	3	5.333		1,100	196		1,296	1,500
	0120	6"	Q-2	3	8		2,150	305		2,455	2,800
	0122	8"	"	2.50	9.600		3,825	365		4,190	4,775
	1020	Butterfly, wafer type, gear actuator, 200 lb.									
	1030	2"	1 Plum	14	.571	Ea.	113	23.50		136.50	159
	1040	2-1/2"	Q-1	9	1.778		116	65.50		181.50	227
	1050	3"		8	2		121	73.50		194.50	244
	1060	4"		5	3.200		151	118		269	345
	1070	5"	Q-2	5	4.800		182	183		365	475
	1080	6"		5	4.800		205	183		388	500
	1090	8"		4.50	5.333		274	203		477	605
	1100	10"		4	6		345	229		574	725
	1110	12"		3	8		480	305		785	985
	1200	Wafer type, lever actuator, 200 lb.									
	1220	2"	1 Plum	14	.571	Ea.	79.50	23.50		103	123
	1230	2-1/2"	Q-1	9	1.778		82.50	65.50		148	190
	1240	3"		8	2		87	73.50		160.50	207
	1250	4"		5	3.200		107	118		225	294
	1260	5"	Q-2	5	4.800		148	183		331	440
	1270	6"		5	4.800		178	183		361	470
	1280	8"		4.50	5.333		263	203		466	595
	1290	10"		4	6		475	229		704	865
	1300	12"		3	8		670	305		975	1,200
	1600	Gate, threaded, 125 lb.									
	1603	1-1/2"	1 Plum	13	.615	Ea.	236	25		261	298
	1604	2"		11	.727		625	29.50		654.50	730
	1605	2-1/2"		10	.800		675	32.50		707.50	790
	1606	3"		8	1		815	41		856	955
	1607	4"		5	1.600		1,125	65.50		1,190.50	1,350
	1650	Gate, 125 lb., N.R.S.									
	2150	Flanged									
	2200	2"	1 Plum	5	1.600	Ea.	320	65.50		385.50	450
	2240	2-1/2"	Q-1	5	3.200		330	118		448	535
	2260	3"		4.50	3.556		365	131		496	600
	2280	4"		3	5.333		525	196		721	875
	2290	5"	Q-2	3.40	7.059		895	269		1,164	1,400
	2300	6"		3	8		895	305		1,200	1,450
	2320	8"		2.50	9.600		1,525	365		1,890	2,250
	2340	10"		2.20	10.909		2,700	415		3,115	3,600
	2360	12"		1.70	14.118		3,725	540		4,265	4,900
	2420	For 250 lb flanged, add					200%	10%			
	3500	OS&Y, 125 lb., (225# noted), threaded									
	3504	3/4" (225 lb.)	1 Plum	18	.444	Ea.	385	18.15		403.15	455
	3505	1" (225 lb.)		16	.500		440	20.50		460.50	510
	3506	1-1/2" (225 lb.)		13	.615		640	25		665	740

R15100-090

Important: See the Reference Section for critical supporting data - Reference Nos., Crews, & City Cost Indexes

15110 | Valves

		CREW	DAILY OUTPUT	LABOR-HOURS	UNIT	2005 BARE COSTS				TOTAL INCL O&P
						MAT.	LABOR	EQUIP.	TOTAL	
200 3507	2" R15100-090	1 Plum	11	.727	Ea.	560	29.50		589.50	660	200
3508	2-1/2"		9	.889		575	36.50		611.50	685
3510	4"	▼	7	1.143	▼	925	46.50		971.50	1,100
3550	OS&Y, 125 lb., flanged									
3600	2"	1 Plum	5	1.600	Ea.	226	65.50		291.50	345
3640	2-1/2"	Q-1	5	3.200		234	118		352	435
3660	3"		4.50	3.556		263	131		394	485
3670	3-1/2"		3	5.333		410	196		606	745
3680	4"	▼	3	5.333		375	196		571	705
3690	5"	Q-2	3.40	7.059		615	269		884	1,075
3700	6"		3	8		615	305		920	1,150
3720	8"		2.50	9.600		1,100	365		1,465	1,750
3740	10"		2.20	10.909		2,000	415		2,415	2,825
3760	12"	▼	1.70	14.118	▼	2,675	540		3,215	3,750
3900	For 175 lb, flanged, add					200%	10%			
4350	Globe, OS&Y									
4540	Class 125, flanged									
4550	2"	1 Plum	5	1.600	Ea.	465	65.50		530.50	610
4560	2-1/2"	Q-1	5	3.200		490	118		608	715
4570	3"		4.50	3.556		570	131		701	820
4580	4"	▼	3	5.333		810	196		1,006	1,200
4590	5"	Q-2	3.40	7.059		1,475	269		1,744	2,025
4600	6"		3	8		1,475	305		1,780	2,075
4610	8"		2.50	9.600		2,900	365		3,265	3,725
4612	10"		2.20	10.909		4,525	415		4,940	5,600
4614	12"	▼	1.70	14.118		4,850	540		5,390	6,125
5040	Class 250, flanged									
5050	2"	1 Plum	4.50	1.778	Ea.	630	72.50		702.50	805
5060	2-1/2"	Q-1	4.50	3.556		820	131		951	1,100
5070	3"		4	4		850	147		997	1,150
5080	4"	▼	2.70	5.926		1,250	218		1,468	1,700
5090	5"	Q-2	3	8		2,225	305		2,530	2,900
5100	6"		2.70	8.889		2,225	340		2,565	2,950
5110	8"		2.20	10.909		3,775	415		4,190	4,775
5120	10"		2	12		6,175	460		6,635	7,500
5130	12"	▼	1.60	15		9,275	570		9,845	11,100
5240	Valve sprocket rim w/chain, for 2" valve	1 Stpi	30	.267		74.50	10.90		85.40	98.50
5250	2-1/2" valve		27	.296		74.50	12.15		86.65	100
5260	3-1/2" valve		25	.320		74.50	13.10		87.60	102
5270	6" valve		20	.400		74.50	16.40		90.90	107
5280	8" valve		18	.444		112	18.20		130.20	152
5290	12" valve		16	.500		112	20.50		132.50	155
5300	16" valve		12	.667		152	27.50		179.50	209
5310	20" valve		10	.800		152	33		185	217
5320	36" valve	▼	8	1	▼	283	41		324	370
5450	Swing check, 125 lb., threaded									
5470	1"	1 Plum	13	.615	Ea.	310	25		335	380
5500	2"	"	11	.727		400	29.50		429.50	485
5540	2-1/2"	Q-1	15	1.067		455	39		494	560
5550	3"		13	1.231		510	45.50		555.50	630
5560	4"	▼	10	1.600	▼	815	59		874	985
5950	Flanged									
5994	1"	1 Plum	7	1.143	Ea.	87	46.50		133.50	166
5998	1-1/2"		6	1.333		132	54.50		186.50	227
6000	2"	▼	5	1.600	▼	156	65.50		221.50	269
6040	2-1/2"	Q-1	5	3.200	▼	189	118		307	385

15110 | Valves

		CREW	DAILY OUTPUT	LABOR-HOURS	UNIT	2005 BARE COSTS MAT.	LABOR	EQUIP.	TOTAL	TOTAL INCL O&P	
200	6050	3" R15100-090	Q-1	4.50	3.556	Ea.	202	131		333	420	**200**
	6060	4"	↓	3	5.333		320	196		516	645	
	6070	6"	Q-2	3	8		545	305		850	1,050	
	6080	8"		2.50	9.600		1,025	365		1,390	1,675	
	6090	10"		2.20	10.909		1,750	415		2,165	2,550	
	6100	12"		1.70	14.118		2,725	540		3,265	3,800	
	6110	18"		1.30	18.462		18,500	705		19,205	21,500	
	6114	24"	↓	.75	32	↓	34,400	1,225		35,625	39,600	
	6160	For 250 lb flanged, add					200%	20%				
	6600	Silent check, bronze trim										
	6610	Compact wafer type, for 125 or 150 lb. flanges										
	6630	1-1/2"	1 Plum	11	.727	Ea.	91.50	29.50		121	146	
	6640	2"	"	9	.889		109	36.50		145.50	174	
	6650	2-1/2"	Q-1	9	1.778		119	65.50		184.50	230	
	6660	3"		8	2		128	73.50		201.50	252	
	6670	4"	↓	5	3.200		163	118		281	355	
	6680	5"	Q-2	6	4		214	153		367	465	
	6690	6"		6	4		286	153		439	545	
	6700	8"		4.50	5.333		510	203		713	865	
	6710	10"		4	6		885	229		1,114	1,325	
	6720	12"	↓	3	8	↓	1,675	305		1,980	2,300	
	6740	For 250 or 300 lb. flanges, thru 6" no change										
	6741	For 8" and 10", add				Ea.	11%	10%				
	6750	Twin disc										
	6752	2"	1 Plum	9	.889	Ea.	152	36.50		188.50	222	
	6754	4"	Q-1	5	3.200		248	118		366	450	
	6756	6"	Q-2	5	4.800		365	183		548	680	
	6758	8"		4.50	5.333		565	203		768	925	
	6760	10"		4	6		900	229		1,129	1,325	
	6762	12"		3	8		1,175	305		1,480	1,725	
	6764	18"		1.50	16		5,050	610		5,660	6,475	
	6766	24"	↓	.75	32	↓	7,100	1,225		8,325	9,650	
	6800	Full flange type, 150 lb.										
	6900	Globe type, 125 lb.										
	6911	2-1/2"	Q-1	9	1.778	Ea.	228	65.50		293.50	350	
	6912	3"		8	2		246	73.50		319.50	380	
	6913	4"	↓	5	3.200		330	118		448	540	
	6914	5"	Q-2	6	4		425	153		578	700	
	6915	6"		5	4.800		535	183		718	860	
	6916	8"		4.50	5.333		975	203		1,178	1,375	
	6917	10"		4	6		1,200	229		1,429	1,675	
	6918	12"	↓	3	8		2,050	305		2,355	2,700	
	6940	For 250 lb., add				↓	40%	10%				
	6980	Screwed end type, 125 lb.										
	6981	1"	1 Plum	19	.421	Ea.	39	17.20		56.20	69	
	6982	1-1/4"		15	.533		50.50	22		72.50	89	
	6983	1-1/2"		13	.615		62.50	25		87.50	107	
	6984	2"	↓	11	.727	↓	85	29.50		114.50	138	
300	0010	**VALVES, LINED, CORROSION RESISTANT/HIGH PURITY**										**300**
	2000	Butterfly 150 lb. ductile iron,										
	2010	Wafer type										
	2030	TFE lined 3" lever handle	Q-1	8	2	Ea.	1,025	73.50		1,098.50	1,225	
	2050	4" lever handle	"	5	3.200		1,425	118		1,543	1,750	
	2070	6" gear operated	Q-2	4.50	5.333		2,225	203		2,428	2,750	
	2080	8" gear operated		4	6		2,925	229		3,154	3,575	
	2090	10" gear operated	↓	3.50	6.857	↓	3,675	261		3,936	4,450	

Important: See the Reference Section for critical supporting data - Reference Nos., Crews, & City Cost Indexes

15110 | Valves

		CREW	DAILY OUTPUT	LABOR-HOURS	UNIT	2005 BARE COSTS MAT.	LABOR	EQUIP.	TOTAL	TOTAL INCL O&P
300 2100	12" gear operated	Q-2	2.50	9.600	Ea.	4,650	365		5,015	5,675
3500	Check lift, 125 lb., cast iron flanged									
3510	Horizontal PPL or SL lined									
3530	1"	1 Plum	14	.571	Ea.	435	23.50		458.50	515
3540	1-1/2"		11	.727		530	29.50		559.50	625
3550	2"		8	1		615	41		656	735
3560	2-1/2"	Q-1	5	3.200		805	118		923	1,050
3570	3"		4.50	3.556		1,000	131		1,131	1,300
3590	4"		3	5.333		1,325	196		1,521	1,750
3610	6"	Q-2	3	8		2,250	305		2,555	2,925
3620	8"	"	2.50	9.600		4,950	365		5,315	6,000
4250	Vertical PPL or SL lined									
4270	1"	1 Plum	14	.571	Ea.	540	23.50		563.50	625
4290	1-1/2"		11	.727		690	29.50		719.50	805
4300	2"		8	1		815	41		856	955
4310	2-1/2"	Q-1	5	3.200		1,025	118		1,143	1,300
4320	3"		4.50	3.556		1,300	131		1,431	1,625
4340	4"		3	5.333		1,600	196		1,796	2,050
4360	6"	Q-2	3	8		2,625	305		2,930	3,350
4370	8"	"	2.50	9.600		4,825	365		5,190	5,850
5000	Clamp type, ductile iron, 150 lb. flanged									
5010	TFE lined									
5030	1" size, lever handle	1 Plum	9	.889	Ea.	810	36.50		846.50	945
5050	1-1/2" size, lever handle		6	1.333		1,075	54.50		1,129.50	1,250
5060	2" size, lever handle		5	1.600		1,300	65.50		1,365.50	1,525
5080	3" size, lever handle	Q-1	4.50	3.556		1,625	131		1,756	2,000
5100	4" size, gear operated	"	3	5.333		2,050	196		2,246	2,550
5120	6" size, gear operated	Q-2	3	8		6,600	305		6,905	7,725
5130	8" size, gear operated	"	2.50	9.600		8,050	365		8,415	9,400
6000	Diaphragm type, cast iron, 125 lb. flanged									
6010	PTFE or VITON, lined									
6030	1" size, handwheel operated	1 Plum	9	.889	Ea.	174	36.50		210.50	246
6050	1-1/2" size, handwheel operated		6	1.333		222	54.50		276.50	325
6060	2" size, handwheel operated		5	1.600		280	65.50		345.50	410
6080	3" size, handwheel operated	Q-1	4.50	3.556		480	131		611	720
6100	4" size, handwheel operated	"	3	5.333		785	196		981	1,150
6120	6" size, handwheel operated	Q-2	3	8		1,475	305		1,780	2,075
6130	8" size, handwheel operated	"	2.50	9.600		3,150	365		3,515	4,025
500 0010	**VALVES, PLASTIC** R15100 -090									
1100	Angle, PVC, threaded									
1110	1/4"	1 Plum	26	.308	Ea.	49.50	12.55		62.05	73.50
1120	1/2"		26	.308		49.50	12.55		62.05	73.50
1130	3/4"		25	.320		58.50	13.05		71.55	84
1140	1"		23	.348		70	14.20		84.20	98.50
1150	Ball, PVC, socket or threaded, single union									
1200	1/4"	1 Plum	26	.308	Ea.	24	12.55		36.55	45.50
1220	3/8"		26	.308		24	12.55		36.55	45.50
1230	1/2"		26	.308		24	12.55		36.55	45.50
1240	3/4"		25	.320		28.50	13.05		41.55	51
1250	1"		23	.348		34.50	14.20		48.70	59.50
1260	1-1/4"		21	.381		57	15.55		72.55	86
1270	1-1/2"		20	.400		57	16.35		73.35	87
1280	2"		17	.471		82	19.20		101.20	119
1290	2-1/2"	Q-1	26	.615		204	22.50		226.50	258
1300	3"		24	.667		204	24.50		228.50	261
1310	4"		20	.800		278	29.50		307.50	350

		15110	Valves	CREW	DAILY OUTPUT	LABOR-HOURS	UNIT	2005 BARE COSTS				TOTAL INCL O&P
								MAT.	LABOR	EQUIP.	TOTAL	
500	1360		For PVC, flanged, add	R15100 -090			Ea.	100%	15%				**500**
	1450		Double union 1/2"	1 Plum	26	.308		13.80	12.55		26.35	34
	1460		3/4"		25	.320		16.20	13.05		29.25	37.50
	1470		1"		23	.348		17.75	14.20		31.95	41
	1480		1-1/4"		21	.381		27	15.55		42.55	53
	1490		1-1/2"		20	.400		30	16.35		46.35	57.50
	1500		2"		17	.471		42.50	19.20		61.70	75.50
	1650		CPVC, socket or threaded, single union									
	1700		1/2"	1 Plum	26	.308	Ea.	35	12.55		47.55	57.50
	1720		3/4"		25	.320		44	13.05		57.05	68
	1730		1"		23	.348		52.50	14.20		66.70	79.50
	1750		1-1/4"		21	.381		88	15.55		103.55	121
	1760		1-1/2"		20	.400		88	16.35		104.35	121
	1770		2"		17	.471		122	19.20		141.20	163
	1780		3"	Q-1	24	.667		365	24.50		389.50	435
	1840		For CPVC, flanged, add					65%	15%			
	1880		For true union, socket or threaded, add					50%	5%			
	2050		Polypropylene, threaded									
	2100		1/4"	1 Plum	26	.308	Ea.	36	12.55		48.55	58.50
	2120		3/8"		26	.308		36	12.55		48.55	58.50
	2130		1/2"		26	.308		36	12.55		48.55	58.50
	2140		3/4"		25	.320		45.50	13.05		58.55	69.50
	2150		1"		23	.348		53.50	14.20		67.70	80.50
	2160		1-1/4"		21	.381		77.50	15.55		93.05	109
	2170		1-1/2"		20	.400		89.50	16.35		105.85	123
	2180		2"		17	.471		122	19.20		141.20	164
	2190		3"	Q-1	24	.667		325	24.50		349.50	390
	2200		4"	"	20	.800		540	29.50		569.50	635
	2550		PVC, three way, socket or threaded									
	2600		1/2"	1 Plum	26	.308	Ea.	44.50	12.55		57.05	68
	2640		3/4"		25	.320		50.50	13.05		63.55	75
	2650		1"		23	.348		54.50	14.20		68.70	81.50
	2660		1-1/2"		20	.400		111	16.35		127.35	147
	2670		2"		17	.471		148	19.20		167.20	192
	2680		3"	Q-1	24	.667		320	24.50		344.50	390
	2740		For flanged, add					60%	15%			
	3150		Ball check, PVC, socket or threaded									
	3200		1/4"	1 Plum	26	.308	Ea.	30.50	12.55		43.05	52.50
	3220		3/8"		26	.308		30.50	12.55		43.05	52.50
	3240		1/2"		26	.308		30.50	12.55		43.05	52.50
	3250		3/4"		25	.320		34	13.05		47.05	57
	3260		1"		23	.348		43	14.20		57.20	68.50
	3270		1-1/4"		21	.381		71.50	15.55		87.05	103
	3280		1-1/2"		20	.400		71.50	16.35		87.85	104
	3290		2"		17	.471		170	19.20		189.20	216
	3310		3"	Q-1	24	.667		271	24.50		295.50	335
	3320		4"	"	20	.800		385	29.50		414.50	465
	3360		For PVC, flanged, add					50%	15%			
	3750		CPVC, socket or threaded									
	3800		1/2"	1 Plum	26	.308	Ea.	44.50	12.55		57.05	68
	3840		3/4"		25	.320		53.50	13.05		66.55	78
	3850		1"		23	.348		63	14.20		77.20	90.50
	3860		1-1/2"		20	.400		108	16.35		124.35	144
	3870		2"		17	.471		146	19.20		165.20	189
	3880		3"	Q-1	24	.667		395	24.50		419.50	470
	3920		4"	"	20	.800		530	29.50		559.50	630

Important: See the Reference Section for critical supporting data - Reference Nos., Crews, & City Cost Indexes

15110 | Valves

		CREW	DAILY OUTPUT	LABOR-HOURS	UNIT	2005 BARE COSTS				TOTAL INCL O&P
						MAT.	LABOR	EQUIP.	TOTAL	
500	3930	For CPVC, flanged, add	R15100 -090			Ea.	40%	15%				500
	4340	Polypropylene, threaded									
	4360	1/2"	1 Plum	26	.308	Ea.	29.50	12.55		42.05	51.50
	4400	3/4"		25	.320		34.50	13.05		47.55	57.50
	4440	1"		23	.348		44	14.20		58.20	70
	4450	1-1/2"		20	.400		85	16.35		101.35	118
	4460	2"		17	.471		106	19.20		125.20	146
	4500	For polypropylene flanged, add					200%	15%			
	4850	Foot valve, PVC, socket or threaded									
	4900	1/2"	1 Plum	34	.235	Ea.	49.50	9.60		59.10	69
	4930	3/4"		32	.250		56	10.20		66.20	77
	4940	1"		28	.286		73	11.65		84.65	97.50
	4950	1-1/4"		27	.296		140	12.10		152.10	172
	4960	1-1/2"		26	.308		140	12.55		152.55	173
	4970	2"		24	.333		162	13.60		175.60	199
	4980	3"		20	.400		385	16.35		401.35	450
	4990	4"		18	.444		680	18.15		698.15	775
	5000	For flanged, add					25%	10%			
	5050	CPVC, socket or threaded									
	5060	1/2"	1 Plum	34	.235	Ea.	48.50	9.60		58.10	67.50
	5070	3/4"		32	.250		55.50	10.20		65.70	77
	5080	1"		28	.286		69	11.65		80.65	93
	5090	1-1/4"		27	.296		110	12.10		122.10	139
	5100	1-1/2"		26	.308		110	12.55		122.55	140
	5110	2"		24	.333		141	13.60		154.60	176
	5120	3"		20	.400		288	16.35		304.35	340
	5130	4"		18	.444		525	18.15		543.15	605
	5140	For flanged, add					25%	10%			
	5280	Needle valve, PVC, threaded									
	5300	1/4"	1 Plum	26	.308	Ea.	35	12.55		47.55	57.50
	5340	3/8"		26	.308		41	12.55		53.55	64
	5360	1/2"		26	.308		41	12.55		53.55	64
	5380	For polypropylene, add					10%				
	5800	Y check, PVC, socket or threaded									
	5820	1/2"	1 Plum	26	.308	Ea.	55	12.55		67.55	79.50
	5840	3/4"		25	.320		59.50	13.05		72.55	85
	5850	1"		23	.348		65	14.20		79.20	93
	5860	1-1/4"		21	.381		102	15.55		117.55	136
	5870	1-1/2"		20	.400		111	16.35		127.35	147
	5880	2"		17	.471		138	19.20		157.20	180
	5890	2-1/2"		15	.533		201	22		223	254
	5900	3"	Q-1	24	.667		273	24.50		297.50	335
	5910	4"	"	20	.800		475	29.50		504.50	570
	5960	For PVC flanged, add					45%	15%			
	6350	Y sediment strainer, PVC, socket or threaded									
	6400	1/2"	1 Plum	26	.308	Ea.	36	12.55		48.55	58.50
	6440	3/4"		24	.333		39	13.60		52.60	63.50
	6450	1"		23	.348		47	14.20		61.20	73.50
	6460	1-1/4"		21	.381		77.50	15.55		93.05	109
	6470	1-1/2"		20	.400		77.50	16.35		93.85	110
	6480	2"		17	.471		95	19.20		114.20	134
	6490	2-1/2"		15	.533		233	22		255	289
	6500	3"	Q-1	24	.667		233	24.50		257.50	293
	6510	4"	"	20	.800		390	29.50		419.50	470
	6560	For PVC, flanged, add					55%	15%			

			DAILY	LABOR-		\multicolumn{4}{c}{2005 BARE COSTS}		TOTAL
---	---	CREW	OUTPUT	HOURS	UNIT	MAT.	LABOR	EQUIP.	TOTAL	INCL O&P
15110	**Valves**									
600 0010	**VALVES, SEMI-STEEL** R15100-090									600
1020	Lubricated plug valve, threaded, 200 psi									
1030	1/2"	1 Plum	18	.444	Ea.	61.50	18.15		79.65	95
1040	3/4"		16	.500		61.50	20.50		82	98
1050	1"		14	.571		79	23.50		102.50	122
1060	1-1/4"		12	.667		94.50	27		121.50	145
1070	1-1/2"		11	.727		102	29.50		131.50	157
1080	2"		8	1		120	41		161	194
1090	2-1/2"	Q-1	5	3.200		186	118		304	380
1100	3"	"	4.50	3.556		228	131		359	450
6990	Flanged, 200 psi									
7000	2"	1 Plum	8	1	Ea.	145	41		186	221
7010	2-1/2"	Q-1	5	3.200		218	118		336	415
7020	3"		4.50	3.556		263	131		394	485
7030	4"		3	5.333		335	196		531	660
7036	5"		2.50	6.400		500	235		735	905
7040	6"	Q-2	3	8		650	305		955	1,175
7050	8"		2.50	9.600		1,175	365		1,540	1,850
7060	10"		2.20	10.909		1,725	415		2,140	2,525
7070	12"		1.70	14.118		2,975	540		3,515	4,050
800 0010	**VALVES, STAINLESS STEEL** R15100-090									800
1700	Check, 200 lb., threaded									
1710	1/4"	1 Plum	24	.333	Ea.	88	13.60		101.60	117
1720	1/2"		22	.364		88	14.85		102.85	119
1730	3/4"		20	.400		96	16.35		112.35	130
1750	1"		19	.421		122	17.20		139.20	161
1760	1-1/2"		13	.615		235	25		260	296
1770	2"		11	.727		395	29.50		424.50	480
1800	150 lb., flanged									
1810	2-1/2"	Q-1	5	3.200	Ea.	945	118		1,063	1,225
1820	3"		4.50	3.556		945	131		1,076	1,250
1830	4"		3	5.333		1,400	196		1,596	1,850
1840	6"	Q-2	3	8		2,525	305		2,830	3,225
1850	8"	"	2.50	9.600		4,200	365		4,565	5,175
2100	Gate, OS&Y, 150 lb., flanged									
2120	1/2"	1 Plum	18	.444	Ea.	277	18.15		295.15	335
2140	3/4"		16	.500		296	20.50		316.50	355
2150	1"		14	.571		360	23.50		383.50	435
2155	1-1/4"		12	.667		505	27		532	600
2160	1-1/2"		11	.727		505	29.50		534.50	605
2170	2"		8	1		570	41		611	685
2180	2-1/2"	Q-1	5	3.200		940	118		1,058	1,200
2190	3"		4.50	3.556		940	131		1,071	1,225
2200	4"		3	5.333		1,375	196		1,571	1,825
2205	5"		2.80	5.714		2,400	210		2,610	2,975
2210	6"	Q-2	3	8		2,400	305		2,705	3,100
2220	8"		2.50	9.600		3,975	365		4,340	4,925
2230	10"		2.30	10.435		6,325	400		6,725	7,550
2240	12"		1.90	12.632		8,075	480		8,555	9,625
2260	For 300 lb., flanged, add					120%	15%			
2600	600 lb., flanged									
2620	1/2"	1 Plum	16	.500	Ea.	850	20.50		870.50	965
2640	3/4"		14	.571		925	23.50		948.50	1,050
2650	1"		12	.667		1,175	27		1,202	1,325
2660	1-1/2"		10	.800		1,650	32.50		1,682.50	1,850
2670	2"		7	1.143		2,350	46.50		2,396.50	2,675

Important: See the Reference Section for critical supporting data - Reference Nos., Crews, & City Cost Indexes

15110 | Valves

			CREW	DAILY OUTPUT	LABOR-HOURS	UNIT	MAT.	LABOR	EQUIP.	TOTAL	TOTAL INCL O&P	
800	2680	2-1/2"	Q-1	4	4	Ea.	3,650	147		3,797	4,250	80
	2690	3" R15100-090	"	3.60	4.444	↓	3,900	163		4,063	4,525	
	3100	Globe, OS&Y, 150 lb., flanged										
	3120	1/2"	1 Plum	18	.444	Ea.	370	18.15		388.15	435	
	3140	3/4"		16	.500		410	20.50		430.50	480	
	3150	1"		14	.571		505	23.50		528.50	590	
	3160	1-1/2"		11	.727		750	29.50		779.50	870	
	3170	2"	↓	8	1		910	41		951	1,050	
	3180	2-1/2"	Q-1	5	3.200		1,325	118		1,443	1,625	
	3190	3"		4.50	3.556		1,625	131		1,756	1,975	
	3200	4"	↓	3	5.333		2,475	196		2,671	3,025	
	3210	6"	Q-2	3	8	↓	3,100	305		3,405	3,850	

15120 | Piping Specialties

			CREW	DAILY OUTPUT	LABOR-HOURS	UNIT	MAT.	LABOR	EQUIP.	TOTAL	TOTAL INCL O&P	
120	0010	**AIR CONTROL**										12
	0030	Air separator, with strainer										
	0040	2" diameter	Q-5	6	2.667	Ea.	555	98.50		653.50	760	
	0080	2-1/2" diameter		5	3.200		630	118		748	870	
	0100	3" diameter	↓	4	4	↓	960	147		1,107	1,275	
	1000	Micro-bubble separator for total air removal, closed loop system										
	1010	Requires bladder type tank in system.										
	1020	Water (hot or chilled) or glycol system										
	1030	Threaded										
	1040	3/4" diameter	1 Stpi	20	.400	Ea.	69	16.40		85.40	100	
	1050	1" diameter		19	.421		77	17.25		94.25	111	
	1060	1-1/4" diameter		16	.500		107	20.50		127.50	148	
	1070	1-1/2" diameter		13	.615		139	25		164	191	
	1080	2" diameter	↓	11	.727		675	30		705	785	
	1090	2-1/2" diameter	Q-5	15	1.067		750	39.50		789.50	885	
	1100	3" diameter		13	1.231		1,075	45.50		1,120.50	1,250	
	1110	4" diameter	↓	10	1.600	↓	1,150	59		1,209	1,375	
140	0010	**AIR PURGING SCOOP** with tappings										14
	0020	for air vent and expansion tank connection										
	0100	1" pipe size, threaded	1 Stpi	19	.421	Ea.	13.90	17.25		31.15	41.50	
	0110	1-1/4" pipe size, threaded		15	.533		13.70	22		35.70	48	
	0120	1-1/2" pipe size, threaded		13	.615		30.50	25		55.50	71.50	
	0130	2" pipe size, threaded	↓	11	.727	↓	35	30		65	83.50	
160	0010	**AUTOMATIC AIR VENT**										16
	0020	Cast iron body, stainless steel internals, float type										
	0180	1/2" NPT inlet, 250 psi	1 Stpi	10	.800	Ea.	234	33		267	305	
	0220	3/4" NPT inlet, 250 psi	"	10	.800	"	234	33		267	305	
	0600	Forged steel body, stainless steel internals, float type										
	0640	1/2" NPT inlet, 750 psi	1 Stpi	12	.667	Ea.	725	27.50		752.50	835	
	0680	3/4" NPT inlet, 750 psi	"	12	.667	"	725	27.50		752.50	835	
	1100	Formed steel body, non corrosive										
	1110	1/8" NPT inlet 150 psi	1 Stpi	32	.250	Ea.	7.05	10.25		17.30	23	
	1120	1/4" NPT inlet 150 psi		32	.250		23.50	10.25		33.75	41.50	
	1130	3/4" NPT inlet 150 psi	↓	32	.250	↓	23.50	10.25		33.75	41.50	
	1300	Chrome plated brass, automatic/manual, for radiators										
	1310	1/8" NPT inlet, nickel plated brass	1 Stpi	32	.250	Ea.	4.10	10.25		14.35	19.90	
200	0010	**CIRCUIT SETTER** Balance valve										20
	0018	Threaded										
	0019	1/2" pipe size	1 Stpi	22	.364	Ea.	45	14.90		59.90	72	
	0020	3/4" pipe size	"	20	.400	"	48.50	16.40		64.90	77.50	

15120 | Piping Specialties

		CREW	DAILY OUTPUT	LABOR-HOURS	UNIT	2005 BARE COSTS				TOTAL INCL O&P
						MAT.	LABOR	EQUIP.	TOTAL	
220	0010	**COCKS, DRAINS & SPECIALTIES**									**220**
	1000	Boiler drain								
	1010	Pipe thread to hose								
	1020	Bronze								
	1030	1/2" size	1 Stpi	36	.222	Ea.	4.10	9.10		13.20	18.20
	1040	3/4" size	"	34	.235	"	4.44	9.65		14.09	19.40
	1100	Solder to hose								
	1110	Bronze								
	1120	1/2" size	1 Stpi	46	.174	Ea.	3.87	7.10		10.97	14.95
	1130	3/4" size	"	44	.182	"	3.68	7.45		11.13	15.25
	1600	With built-in vacuum breaker								
	1610	1/2" IP or solder	1 Stpi	36	.222	Ea.	16.45	9.10		25.55	32
	1630	With tamper proof vacuum breaker								
	1640	1/2" IP or solder	1 Stpi	36	.222	Ea.	19.50	9.10		28.60	35
	1650	3/4" IP or solder	"	34	.235	"	19.80	9.65		29.45	36.50
	3000	Cocks								
	3010	Air, lever or tee handle								
	3020	Bronze, single thread								
	3030	1/8" size	1 Stpi	52	.154	Ea.	5.25	6.30		11.55	15.25
	3040	1/4" size		46	.174		5.60	7.10		12.70	16.85
	3050	3/8" size		40	.200		5.65	8.20		13.85	18.50
	3060	1/2" size		36	.222		6.65	9.10		15.75	21
	3100	Bronze, double thread								
	3110	1/8" size	1 Stpi	26	.308	Ea.	6.60	12.60		19.20	26.50
	3120	1/4" size		22	.364		6.95	14.90		21.85	30
	3130	3/8" size		18	.444		7.20	18.20		25.40	35.50
	3140	1/2" size		15	.533		8.85	22		30.85	42.50
	4500	Gauge cock, brass								
	4510	1/4" FPT	1 Stpi	24	.333	Ea.	5.80	13.65		19.45	27
	4512	1/4" MPT	"	24	.333	"	7.30	13.65		20.95	28.50
	4600	Pigtail, steam syphon								
	4604	1/4"	1 Stpi	24	.333	Ea.	11.20	13.65		24.85	33
	4650	Snubber valve								
	4654	1/4"	1 Stpi	22	.364	Ea.	9.75	14.90		24.65	33.50
	4660	Nipple, black steel								
	4664	1/4" x 3"	1 Stpi	37	.216	Ea.	.58	8.85		9.43	13.95
250	0010	**DIELECTRIC UNIONS** Standard gaskets for water and air									**250**
	0020	250 psi maximum pressure								
	0280	Female IPT to sweat, straight								
	0300	1/2" pipe size	1 Plum	24	.333	Ea.	3.15	13.60		16.75	24
	0340	3/4" pipe size		20	.400		3.15	16.35		19.50	28
	0360	1" pipe size		19	.421		6.10	17.20		23.30	32.50
	0380	1-1/4" pipe size		15	.533		10	22		32	44
	0400	1-1/2" pipe size		13	.615		15	25		40	54.50
	0420	2" pipe size		11	.727		21	29.50		50.50	67.50
	0580	Female IPT to brass pipe thread, straight								
	0600	1/2" pipe size	1 Plum	24	.333	Ea.	5.95	13.60		19.55	27
	0640	3/4" pipe size		20	.400		7.60	16.35		23.95	33
	0660	1" pipe size		19	.421		11.40	17.20		28.60	38.50
	0680	1-1/4" pipe size		15	.533		17.05	22		39.05	52
	0700	1-1/2" pipe size		13	.615		23.50	25		48.50	64
	0720	2" pipe size		11	.727		30.50	29.50		60	78.50
	0780	Female IPT to female IPT, straight								
	0800	1/2" pipe size	1 Plum	24	.333	Ea.	4.75	13.60		18.35	26
	0840	3/4" pipe size		20	.400		5.30	16.35		21.65	30.50
	0860	1" pipe size		19	.421		7	17.20		24.20	33.50

Important: See the Reference Section for critical supporting data - Reference Nos., Crews, & City Cost Indexes

15120	Piping Specialties	CREW	DAILY OUTPUT	LABOR-HOURS	UNIT	2005 BARE COSTS				TOTAL INCL O&P
						MAT.	LABOR	EQUIP.	TOTAL	
250 0880	1-1/4" pipe size	1 Plum	15	.533	Ea.	9.75	22		31.75	44
0900	1-1/2" pipe size		13	.615		14.95	25		39.95	54.50
0920	2" pipe size	↓	11	.727	↓	21.50	29.50		51	68
2000	175 psi maximum pressure									
2180	Female IPT to sweat									
2200	1-1/2" pipe size	1 Plum	11	.727	Ea.	38.50	29.50		68	86.50
2240	2" pipe size	"	9	.889		44	36.50		80.50	103
2260	2-1/2" pipe size	Q-1	15	1.067		48	39		87	112
2280	3" pipe size		14	1.143		65.50	42		107.50	135
2300	4" pipe size	↓	11	1.455		174	53.50		227.50	273
2320	5" pipe size	Q-2	14	1.714		284	65.50		349.50	410
2340	6" pipe size		12	2		330	76.50		406.50	475
2360	8" pipe size	↓	10	2.400	↓	910	91.50		1,001.50	1,150
2480	Female IPT to brass pipe									
2500	1-1/2" pipe size	1 Plum	11	.727	Ea.	58	29.50		87.50	108
2540	2" pipe size	"	9	.889		69	36.50		105.50	131
2560	2-1/2" pipe size	Q-1	15	1.067		100	39		139	169
2580	3" pipe size		14	1.143		116	42		158	191
2600	4" pipe size	↓	11	1.455		199	53.50		252.50	300
2620	5" pipe size	Q-2	14	1.714		224	65.50		289.50	345
2640	6" pipe size		12	2		267	76.50		343.50	410
2680	8" pipe size	↓	10	2.400	↓	435	91.50		526.50	620
2880	Female IPT to female IPT									
2900	1-1/2" pipe size	1 Plum	11	.727	Ea.	36	29.50		65.50	84
2940	2" pipe size	"	9	.889		38.50	36.50		75	97
2960	2-1/2" pipe size	Q-1	15	1.067		48	39		87	112
2980	3" pipe size		14	1.143		60	42		102	130
3000	4" pipe size	↓	11	1.455		91	53.50		144.50	181
3020	5" pipe size	Q-2	14	1.714		105	65.50		170.50	215
3040	6" pipe size		12	2		141	76.50		217.50	270
3060	8" pipe size	↓	10	2.400	↓	192	91.50		283.50	350
3380	Copper to copper									
3400	1-1/2" pipe size	1 Plum	11	.727	Ea.	47.50	29.50		77	97
3440	2" pipe size	"	9	.889		54	36.50		90.50	114
3460	2-1/2" pipe size	Q-1	15	1.067		58	39		97	123
3480	3" pipe size		14	1.143		72.50	42		114.50	143
3500	4" pipe size	↓	11	1.455		230	53.50		283.50	335
3520	5" pipe size	Q-2	14	1.714		370	65.50		435.50	505
3540	6" pipe size		12	2		450	76.50		526.50	610
3560	8" pipe size	↓	10	2.400	↓	1,200	91.50		1,291.50	1,475
300 0010	**EXPANSION JOINTS**									
0100	Bellows type, neoprene cover, flanged spool									
0140	6" face to face, 1-1/4" diameter	1 Stpi	11	.727	Ea.	210	30		240	276
0160	1-1/2" diameter	"	10.60	.755		210	31		241	278
0180	2" diameter	Q-5	13.30	1.203		214	44.50		258.50	300
0190	2-1/2" diameter		12.40	1.290		222	47.50		269.50	315
0200	3" diameter		11.40	1.404		252	51.50		303.50	355
0210	4" diameter		8.40	1.905		269	70		339	400
0220	5" diameter		7.60	2.105		330	77.50		407.50	480
0230	6" diameter		6.80	2.353		340	86.50		426.50	505
0240	8" diameter		5.40	2.963		390	109		499	595
0250	10" diameter		5	3.200		520	118		638	745
0260	12" diameter		4.60	3.478		595	128		723	850
0480	10" face to face, 2" diameter		13	1.231		315	45.50		360.50	415
0500	2-1/2" diameter		12	1.333		330	49		379	440
0520	3" diameter	↓	11	1.455	↓	335	53.50		388.50	450

15120 | Piping Specialties

		CREW	DAILY OUTPUT	LABOR-HOURS	UNIT	2005 BARE COSTS				TOTAL INCL O&P	
						MAT.	LABOR	EQUIP.	TOTAL		
300	0540	4" diameter	Q-5	8	2	Ea.	375	73.50		448.50	525	**300**
	0560	5" diameter		7	2.286		455	84		539	625	
	0580	6" diameter		6	2.667		465	98.50		563.50	665	
	0600	8" diameter		5	3.200		530	118		648	755	
	0620	10" diameter		4.60	3.478		590	128		718	845	
	0640	12" diameter		4	4		675	147		822	965	
	0660	14" diameter		3.80	4.211		850	155		1,005	1,175	
	0680	16" diameter		2.90	5.517		975	203		1,178	1,375	
	0700	18" diameter		2.50	6.400		1,100	236		1,336	1,550	
	0720	20" diameter		2.10	7.619		1,150	281		1,431	1,700	
	0740	24" diameter		1.80	8.889		1,325	330		1,655	1,950	
	0760	26" diameter		1.40	11.429		1,475	420		1,895	2,225	
	0780	30" diameter		1.20	13.333		1,625	490		2,115	2,525	
	0800	36" diameter		1	16		2,200	590		2,790	3,275	
320	0010	**EXPANSION TANKS**										**320**
	1500	Plastic, corrosion resistant, see Div. 15220										
	1505	Fiberglass and steel single / double wall storage, see Div 13201										
	1510	Tank leak detection systems, see Div 13851-350										
	2000	Steel, liquid expansion, ASME, painted, 15 gallon capacity	Q-5	17	.941	Ea.	365	34.50		399.50	450	
	2020	24 gallon capacity		14	1.143		370	42		412	470	
	2040	30 gallon capacity		12	1.333		405	49		454	520	
	2060	40 gallon capacity		10	1.600		460	59		519	595	
	2360	Galvanized										
	2370	15 gallon capacity	Q-5	17	.941	Ea.	625	34.50		659.50	740	
	2380	24 gallon capacity		14	1.143		670	42		712	805	
	2390	30 gallon capacity		12	1.333		785	49		834	935	
	3000	Steel ASME expansion, rubber diaphragm, 19 gal. cap. accept.		12	1.333		1,425	49		1,474	1,650	
	3020	31 gallon capacity		8	2		1,600	73.50		1,673.50	1,850	
	3040	61 gallon capacity		6	2.667		2,250	98.50		2,348.50	2,625	
350	0010	**FLEXIBLE CONNECTORS**, Corrugated, 7/8" O.D., 1/2" I.D.										**350**
	0050	Gas, seamless brass, steel fittings										
	0200	12" long	1 Plum	36	.222	Ea.	11.95	9.10		21.05	27	
	0220	18" long		36	.222		14.85	9.10		23.95	30	
	0240	24" long		34	.235		17.55	9.60		27.15	34	
	0260	30" long		34	.235		18.95	9.60		28.55	35.50	
	0280	36" long		32	.250		21	10.20		31.20	38.50	
	0320	48" long		30	.267		26.50	10.90		37.40	46	
	0340	60" long		30	.267		31.50	10.90		42.40	51.50	
	0360	72" long		30	.267		36.50	10.90		47.40	56.50	
	2000	Water, copper tubing, dielectric separators										
	2100	12" long	1 Plum	36	.222	Ea.	9.40	9.10		18.50	24	
	2220	15" long		36	.222		10.45	9.10		19.55	25	
	2240	18" long		36	.222		11.40	9.10		20.50	26	
	2260	24" long		34	.235		13.95	9.60		23.55	30	
370	0010	**FLEXIBLE METAL HOSE** Connectors, standard lengths										**370**
	0100	Bronze braided, bronze ends										
	0120	3/8" diameter x 12"	1 Stpi	26	.308	Ea.	15.40	12.60		28	36	
	0140	1/2" diameter x 12"		24	.333		14.65	13.65		28.30	36.50	
	0160	3/4" diameter x 12"		20	.400		21.50	16.40		37.90	48	
	0180	1" diameter x 18"		19	.421		27.50	17.25		44.75	56.50	
	0200	1-1/2" diameter x 18"		13	.615		44	25		69	86.50	
	0220	2" diameter x 18"		11	.727		53	30		83	103	
	1000	Carbon steel ends										
	1020	1/4" diameter x 12"	1 Stpi	28	.286	Ea.	12.65	11.70		24.35	31.50	

Important: See the Reference Section for critical supporting data - Reference Nos., Crews, & City Cost Indexes

		CREW	DAILY OUTPUT	LABOR-HOURS	UNIT	2005 BARE COSTS				TOTAL INCL O&P
15120	**Piping Specialties**					MAT.	LABOR	EQUIP.	TOTAL	
370 1040	3/8" diameter x 12"	1 Stpi	26	.308	Ea.	13	12.60		25.60	33.50 370
1060	1/2" diameter x 12"		24	.333		12.65	13.65		26.30	34.50
1080	1/2" diameter x 24"		24	.333		16.20	13.65		29.85	38.50
1100	1/2" diameter x 36"		24	.333		19.80	13.65		33.45	42.50
1120	3/4" diameter x 12"		20	.400		16.80	16.40		33.20	43
1140	3/4" diameter x 24"		20	.400		20.50	16.40		36.90	47
1160	3/4" diameter x 36"		20	.400		24.50	16.40		40.90	51.50
1180	1" diameter x 18"		19	.421		22	17.25		39.25	50.50
1200	1" diameter x 30"		19	.421		26	17.25		43.25	54.50
1220	1" diameter x 36"		19	.421		28	17.25		45.25	57
1240	1-1/4" diameter x 18"		15	.533		22.50	22		44.50	58
1260	1-1/4" diameter x 36"		15	.533		30	22		52	66
1280	1-1/2" diameter x 18"		13	.615		24.50	25		49.50	64.50
1300	1-1/2" diameter x 36"		13	.615		32.50	25		57.50	73.50
1320	2" diameter x 24"		11	.727		32	30		62	80
1340	2" diameter x 36"		11	.727		39.50	30		69.50	88.50
1360	2-1/2" diameter x 24"		9	.889		45	36.50		81.50	104
1380	2-1/2" diameter x 36"		9	.889		53	36.50		89.50	113
1400	3" diameter x 24"		7	1.143		53.50	47		100.50	130
1420	3" diameter x 36"		7	1.143		62	47		109	139
2000	Carbon steel braid, carbon steel solid ends									
2100	1/2" diameter x 12"	1 Stpi	24	.333	Ea.	12.65	13.65		26.30	34.50
2120	3/4" diameter x 12"		20	.400		16.80	16.40		33.20	43
2140	1" diameter x 12"		19	.421		20	17.25		37.25	48
2160	1-1/4" diameter x 12"		15	.533		20	22		42	55
2180	1-1/2" diameter x 12"		13	.615		21.50	25		46.50	62
3000	Stainless steel braid, welded on carbon steel ends									
3100	1/2" diameter x 12"	1 Stpi	24	.333	Ea.	18.50	13.65		32.15	41
3120	3/4" diameter x 12"		20	.400		26.50	16.40		42.90	53.50
3140	3/4" diameter x 24"		20	.400		32	16.40		48.40	59.50
3160	3/4" diameter x 36"		20	.400		37.50	16.40		53.90	65.50
3180	1" diameter x 12"		19	.421		30	17.25		47.25	59
3200	1" diameter x 24"		19	.421		36	17.25		53.25	66
3220	1" diameter x 36"		19	.421		42.50	17.25		59.75	73
3240	1-1/4" diameter x 12"		15	.533		32.50	22		54.50	69
3260	1-1/4" diameter x 24"		15	.533		41.50	22		63.50	78.50
3280	1-1/4" diameter x 36"		15	.533		50	22		72	88
3300	1-1/2" diameter x 12"		13	.615		36	25		61	77.50
3320	1-1/2" diameter x 24"		13	.615		46	25		71	88.50
3340	1-1/2" diameter x 36"		13	.615		56	25		81	99.50
3400	Metal sst braid, over corrugated stainless steel, flanged ends									
3410	150 PSI									
3420	1/2" diameter x 12"	1 Stpi	24	.333	Ea.	30	13.65		43.65	53.50
3430	1" diameter x 12"		20	.400		40	16.40		56.40	68.50
3440	1-1/2" diameter x 12"		15	.533		50	22		72	88
3450	2-1/2" diameter x 9"		12	.667		54	27.50		81.50	101
3460	3" diameter x 9"		9	.889		65	36.50		101.50	126
3470	4" diameter x 9"		7	1.143		80	47		127	159
3480	4" diameter x 30"		5	1.600		200	65.50		265.50	320
3490	4" diameter x 36"		4.80	1.667		220	68.50		288.50	345
3500	6" diameter x 11"		5	1.600		132	65.50		197.50	244
3510	6" diameter x 36"		3.80	2.105		296	86		382	455
3520	8" diameter x 12"		4	2		281	82		363	435
3530	10" diameter x 13"		3	2.667		380	109		489	585
3540	12" diameter x 14"	Q-5	4	4		570	147		717	850
6000	Molded rubber with helical wire reinforcement									

15120 | Piping Specialties

		CREW	DAILY OUTPUT	LABOR-HOURS	UNIT	2005 BARE COSTS				TOTAL INCL O&P
						MAT.	LABOR	EQUIP.	TOTAL	
370	6010	150 PSI									**370**
	6020	1-1/2" diameter x 12"	1 Stpi	15	.533	Ea.	65	22		87	105
	6030	2" diameter x 12"		12	.667		162	27.50		189.50	219
	6040	3" diameter x 12"		8	1		176	41		217	256
	6050	4" diameter x 12"		6	1.333		225	54.50		279.50	330
	6060	6" diameter x 18"		4	2		330	82		412	490
	6070	8" diameter x 24"		3	2.667		455	109		564	665
	6080	10" diameter x 24"		2	4		545	164		709	845
	6090	12" diameter x 24"	Q-5	3	5.333		625	197		822	985
	7000	Molded teflon with stainless steel flanges									
	7010	150 PSI									
	7020	2-1/2" diameter x 3-3/16"	Q-1	7.80	2.051	Ea.	980	75.50		1,055.50	1,200
	7030	3" diameter x 3-5/8"		6.50	2.462		1,150	90.50		1,240.50	1,400
	7040	4" diameter x 3-5/8"		5	3.200		1,500	118		1,618	1,825
	7050	6" diameter x 4"		4.30	3.721		2,100	137		2,237	2,500
	7060	8" diameter x 6"		3.80	4.211		3,325	155		3,480	3,900
400	0010	**FLOAT VALVES**									**400**
	0020	With ball and bracket									
	0030	Single seat, threaded									
	0040	Brass body									
	0050	1/2"	1 Stpi	11	.727	Ea.	36.50	30		66.50	85.50
	0060	3/4"		9	.889		41.50	36.50		78	101
	0070	1"		7	1.143		61	47		108	138
	0080	1-1/2"		4.50	1.778		101	73		174	220
	0090	2"		3.60	2.222		102	91		193	249
	0300	For condensate receivers, CI, in-line mount									
	0320	1" inlet	1 Stpi	7	1.143	Ea.	104	47		151	185
	0360	For condensate receiver, CI, external float, flanged tank mount									
	0370	3/4" inlet	1 Stpi	5	1.600	Ea.	87	65.50		152.50	195
420	0010	**FLOW CHECK CONTROL**									**420**
	0100	Bronze body, soldered									
	0110	3/4" size	1 Stpi	20	.400	Ea.	36.50	16.40		52.90	64.50
	0120	1" size	"	19	.421	"	43.50	17.25		60.75	74
	0200	Cast iron body, threaded									
	0210	3/4" size	1 Stpi	20	.400	Ea.	29	16.40		45.40	56.50
	0220	1" size		19	.421		32.50	17.25		49.75	62
	0230	1-1/4" size		15	.533		39.50	22		61.50	76.50
	0240	1-1/2" size		13	.615		60.50	25		85.50	105
	0250	2" size		11	.727		87	30		117	141
520	0010	**HYDRONIC HEATING CONTROL VALVES**									**520**
	0050	Hot water, nonelectric, thermostatic									
	0100	Radiator supply, 1/2" diameter	1 Stpi	24	.333	Ea.	40	13.65		53.65	64.50
	0120	3/4" diameter	"	20	.400	"	40.50	16.40		56.90	69.50
	1000	Manual, radiator supply									
	1010	1/2" pipe size, angle union	1 Stpi	24	.333	Ea.	30	13.65		43.65	53.50
	1020	3/4" pipe size, angle union	"	20	.400	"	28	16.40		44.40	55.50
	1100	Radiator, balancing, straight, sweat connections									
	1110	1/2" pipe size	1 Stpi	24	.333	Ea.	8.50	13.65		22.15	30
	1120	3/4" pipe size	"	20	.400	"	11.85	16.40		28.25	37.50
	1200	Steam, radiator, supply									
	1210	1/2" pipe size, angle union	1 Stpi	24	.333	Ea.	22	13.65		35.65	44.50
	1220	3/4" pipe size, angle union	"	20	.400	"	23.50	16.40		39.90	50.50
	8000	System balancing and shut-off									
	8020	Butterfly, quarter turn, calibrated, threaded or solder									
	8040	Bronze, -30° F to +350° F, pressure to 175 psi									

Important: See the Reference Section for critical supporting data - Reference Nos., Crews, & City Cost Indexes

		CREW	DAILY OUTPUT	LABOR-HOURS	UNIT	2005 BARE COSTS				TOTAL INCL O&P	
						MAT.	LABOR	EQUIP.	TOTAL		
520	8060	1/2" size	1 Stpi	22	.364	Ea.	6	14.90		20.90	29	520
	8070	3/4" size	"	20	.400	"	9.70	16.40		26.10	35	
580	0010	**MIXING VALVE** Automatic water tempering										580
	0040	1/2" size	1 Stpi	19	.421	Ea.	395	17.25		412.25	460	
	0050	3/4" size		18	.444		395	18.20		413.20	465	
	0100	1" size		16	.500		630	20.50		650.50	725	
	0120	1-1/4" size		13	.615		820	25		845	940	
	0140	1-1/2" size		10	.800		980	33		1,013	1,125	
	0160	2" size		8	1		1,225	41		1,266	1,400	
	0170	2-1/2" size		6	1.333		1,225	54.50		1,279.50	1,425	
	0180	3" size		4	2		2,850	82		2,932	3,250	
	0190	4" size		3	2.667		2,850	109		2,959	3,300	
610	0010	**MONOFLOW TEE FITTING**										610
	1100	For one pipe hydronic, supply and return										
	1110	Copper, soldered										
	1120	3/4" x 1/2" size	1 Stpi	13	.615	Ea.	13.50	25		38.50	53	
670	0010	**PRESSURE REGULATOR**										670
	0100	Gas appliance regulators										
	0106	Main burner and pilot applications										
	0108	Rubber seat poppet type										
	0109	1/8" pipe size	1 Stpi	24	.333	Ea.	10.85	13.65		24.50	32.50	
	0110	1/4" pipe size		24	.333		12.60	13.65		26.25	34.50	
	0112	3/8" pipe size		24	.333		13.40	13.65		27.05	35.50	
	0113	1/2" pipe size		24	.333		14.80	13.65		28.45	37	
	0114	3/4" pipe size		20	.400		17.50	16.40		33.90	44	
	0122	Lever action type										
	0123	3/8" pipe size	1 Stpi	24	.333	Ea.	20.50	13.65		34.15	43	
	0124	1/2" pipe size		24	.333		20.50	13.65		34.15	43	
	0125	3/4" pipe size		20	.400		41.50	16.40		57.90	70.50	
	0126	1" pipe size		19	.421		41.50	17.25		58.75	72	
	0132	Double diaphragm type										
	0133	3/8" pipe size	1 Stpi	24	.333	Ea.	26	13.65		39.65	49	
	0134	1/2" pipe size		24	.333		38	13.65		51.65	62.50	
	0135	3/4" pipe size		20	.400		65	16.40		81.40	96	
	0136	1" pipe size		19	.421		79	17.25		96.25	113	
	0137	1-1/4" pipe size		15	.533		256	22		278	315	
	0138	1-1/2" pipe size		13	.615		475	25		500	565	
	0139	2" pipe size		11	.727		475	30		505	570	
	0140	2-1/2" pipe size	Q-5	15	1.067		930	39.50		969.50	1,075	
	0141	3" pipe size		13	1.231		930	45.50		975.50	1,100	
	0142	4" pipe size (flanged)		8	2		1,725	73.50		1,798.50	2,000	
	0160	Main burner only										
	0162	Straight-thru-flow design										
	0163	1/2" pipe size	1 Stpi	24	.333	Ea.	28.50	13.65		42.15	51.50	
	0164	3/4" pipe size		20	.400		39	16.40		55.40	67	
	0165	1" pipe size		19	.421		59	17.25		76.25	90.50	
	0166	1-1/4" pipe size		15	.533		59	22		81	97.50	
	0200	Oil, light, hot water, ordinary steam, threaded										
	0220	Bronze body, 1/4" size	1 Stpi	24	.333	Ea.	102	13.65		115.65	133	
	0230	3/8" size		24	.333		110	13.65		123.65	142	
	0240	1/2" size		24	.333		136	13.65		149.65	171	
	0250	3/4" size		20	.400		163	16.40		179.40	204	
	0260	1" size		19	.421		247	17.25		264.25	298	
	0270	1-1/4" size		15	.533		335	22		357	400	
	0320	Iron body, 1/4" size		24	.333		79.50	13.65		93.15	108	
	0330	3/8" size		24	.333		93.50	13.65		107.15	124	

		CREW	DAILY OUTPUT	LABOR-HOURS	UNIT	2005 BARE COSTS				TOTAL INCL 0&P	
						MAT.	LABOR	EQUIP.	TOTAL		
670	0340	1/2" size	1 Stpi	24	.333	Ea.	99.50	13.65		113.15	130	**670**
	0350	3/4" size		20	.400		121	16.40		137.40	158	
	0360	1" size		19	.421		157	17.25		174.25	199	
	0370	1-1/4" size		15	.533		218	22		240	273	
	9000	For water pressure regulators, see Div. 15110-160										
730	0010	**SLEEVES AND ESCUTCHEONS**										**730**
	0100	Pipe sleeve										
	0110	Steel, w/water stop, 12" long, with link seal										
	0120	2" diam. for 1/2" carrier pipe	1 Plum	8.40	.952	Ea.	26.50	39		65.50	87.50	
	0130	2-1/2" diam. for 3/4" carrier pipe		8	1		31.50	41		72.50	96	
	0140	2-1/2" diam. for 1" carrier pipe		8	1		29	41		70	93.50	
	0150	3" diam. for 1-1/4" carrier pipe		7.20	1.111		41	45.50		86.50	114	
	0160	3-1/2" diam. for 1-1/2" carrier pipe		6.80	1.176		42.50	48		90.50	119	
	0170	4" diam. for 2" carrier pipe		6	1.333		47	54.50		101.50	134	
	0180	4" diam. for 2-1/2" carrier pipe		6	1.333		46.50	54.50		101	133	
	0190	5" diam. for 3" carrier pipe		5.40	1.481		57	60.50		117.50	154	
	0200	6" diam. for 4" carrier pipe		4.80	1.667		68	68		136	177	
	0210	10" diam. for 6" carrier pipe	Q-1	8	2		124	73.50		197.50	247	
	0220	12" diam. for 8" carrier pipe		7.20	2.222		182	81.50		263.50	325	
	0230	14" diam. for 10" carrier pipe		6.40	2.500		192	92		284	350	
	0240	16" diam. for 12" carrier pipe		5.80	2.759		224	101		325	400	
	0250	18" diam. for 14" carrier pipe		5.20	3.077		246	113		359	440	
	0260	24" diam. for 18" carrier pipe		4	4		410	147		557	670	
	0270	24" diam. for 20" carrier pipe		4	4		330	147		477	585	
	0280	30" diam. for 24" carrier pipe		3.20	5		750	184		934	1,100	
	0500	Wall sleeve										
	0510	Ductile iron with rubber gasket seal										
	0520	3"	1 Plum	8.40	.952	Ea.	155	39		194	230	
	0530	4"		7.20	1.111		210	45.50		255.50	299	
	0540	6"		6	1.333		320	54.50		374.50	430	
	0550	8"		4	2		380	81.50		461.50	545	
	0560	10"		3	2.667		455	109		564	665	
	0570	12"		2.40	3.333		274	136		410	505	
	5000	Escutcheon										
	5100	Split ring, pipe										
	5110	Chrome plated										
	5120	1/2"	1 Plum	160	.050	Ea.	1.01	2.04		3.05	4.18	
	5130	3/4"		160	.050		1.05	2.04		3.09	4.23	
	5140	1"		135	.059		1.12	2.42		3.54	4.87	
	5150	1-1/2"		115	.070		1.69	2.84		4.53	6.15	
	5160	2"		100	.080		1.77	3.27		5.04	6.85	
	5170	4"		80	.100		3.93	4.09		8.02	10.45	
	5180	6"		68	.118		6.55	4.81		11.36	14.40	
790	0010	**STRAINERS, BASKET TYPE** Perforated stainless steel basket										**790**
	0100	Brass or monel available										
	2000	Simplex style										
	2300	Bronze body										
	2320	Screwed, 3/8" pipe size	1 Stpi	22	.364	Ea.	103	14.90		117.90	136	
	2340	1/2" pipe size		20	.400		105	16.40		121.40	141	
	2360	3/4" pipe size		17	.471		108	19.25		127.25	148	
	2380	1" pipe size		15	.533		141	22		163	188	
	2400	1-1/4" pipe size		13	.615		212	25		237	271	
	2420	1-1/2" pipe size		12	.667		214	27.50		241.50	277	
	2440	2" pipe size		10	.800		325	33		358	405	
	2460	2-1/2" pipe size	Q-5	15	1.067		465	39.50		504.50	570	

15120	Piping Specialties	CREW	DAILY OUTPUT	LABOR-HOURS	UNIT	2005 BARE COSTS				TOTAL INCL O&P
						MAT.	LABOR	EQUIP.	TOTAL	
790 2480	3" pipe size	Q-5	14	1.143	Ea.	670	42		712	805	790
2600	Flanged, 2" pipe size	1 Stpi	6	1.333		490	54.50		544.50	620
2620	2-1/2" pipe size	Q-5	4.50	3.556		770	131		901	1,050
2640	3" pipe size		3.50	4.571		865	168		1,033	1,200
2660	4" pipe size		3	5.333		1,475	197		1,672	1,925
2680	5" pipe size	Q-6	3.40	7.059		2,200	270		2,470	2,825
2700	6" pipe size		3	8		2,725	305		3,030	3,425
2710	8" pipe size		2.50	9.600		4,650	365		5,015	5,675
3600	Iron body									
3700	Screwed, 3/8" pipe size	1 Stpi	22	.364	Ea.	117	14.90		131.90	152
3720	1/2" pipe size		20	.400		117	16.40		133.40	154
3740	3/4" pipe size		17	.471		155	19.25		174.25	199
3760	1" pipe size		15	.533		155	22		177	203
3780	1-1/4" pipe size		13	.615		207	25		232	265
3800	1-1/2" pipe size		12	.667		227	27.50		254.50	290
3820	2" pipe size		10	.800		271	33		304	345
3840	2-1/2" pipe size	Q-5	15	1.067		390	39.50		429.50	485
3860	3" pipe size	"	14	1.143		440	42		482	550
4000	Flanged, 2" pipe size	1 Stpi	6	1.333		370	54.50		424.50	485
4020	2-1/2" pipe size	Q-5	4.50	3.556		500	131		631	745
4040	3" pipe size		3.50	4.571		530	168		698	840
4060	4" pipe size		3	5.333		800	197		997	1,175
4080	5" pipe size	Q-6	3.40	7.059		1,250	270		1,520	1,775
4100	6" pipe size		3	8		1,550	305		1,855	2,150
4120	8" pipe size		2.50	9.600		2,875	365		3,240	3,700
4140	10" pipe size		2.20	10.909		4,300	415		4,715	5,350
4160	12" pipe size		1.70	14.118		5,125	540		5,665	6,450
4180	14" pipe size		1.40	17.143		6,450	655		7,105	8,075
4200	16" pipe size		1	24		6,750	915		7,665	8,800
7000	Stainless steel body									
7200	Screwed, 1" pipe size	1 Stpi	15	.533	Ea.	415	22		437	490
7210	1-1/4" pipe size		13	.615		605	25		630	705
7220	1-1/2" pipe size		12	.667		605	27.50		632.50	705
7240	2" pipe size		10	.800		885	33		918	1,025
7260	2-1/2" pipe size	Q-5	15	1.067		1,475	39.50		1,514.50	1,675
7280	3" pipe size	"	14	1.143		1,950	42		1,992	2,225
7400	Flanged, 2" pipe size	1 Stpi	6	1.333		1,225	54.50		1,279.50	1,400
7420	2-1/2" pipe size	Q-5	4.50	3.556		2,200	131		2,331	2,625
7440	3" pipe size		3.50	4.571		2,275	168		2,443	2,750
7460	4" pipe size		3	5.333		3,550	197		3,747	4,200
7480	6" pipe size	Q-6	3	8		6,150	305		6,455	7,225
7500	8" pipe size	"	2.50	9.600		10,400	365		10,765	12,100
8100	Duplex style									
8200	Bronze body									
8240	Screwed, 3/4" pipe size	1 Stpi	16	.500	Ea.	1,100	20.50		1,120.50	1,225
8260	1" pipe size		14	.571		1,100	23.50		1,123.50	1,225
8280	1-1/4" pipe size		12	.667		2,150	27.50		2,177.50	2,425
8300	1-1/2" pipe size		11	.727		2,150	30		2,180	2,425
8320	2" pipe size		9	.889		3,400	36.50		3,436.50	3,800
8340	2-1/2" pipe size	Q-5	14	1.143		3,400	42		3,442	3,825
8420	Flanged, 2" pipe size	1 Stpi	6	1.333		3,850	54.50		3,904.50	4,325
8440	2-1/2" pipe size	Q-5	4.50	3.556		3,950	131		4,081	4,550
8460	3" pipe size		3.50	4.571		5,450	168		5,618	6,250
8480	4" pipe size		3	5.333		8,625	197		8,822	9,775
8500	5" pipe size	Q-6	3.40	7.059		15,000	270		15,270	16,900
8520	6" pipe size	"	3	8		18,900	305		19,205	21,300

15120 | Piping Specialties

		CREW	DAILY OUTPUT	LABOR-HOURS	UNIT	2005 BARE COSTS				TOTAL INCL O&P
						MAT.	LABOR	EQUIP.	TOTAL	
790	8700	Iron body									790
	8740	Screwed, 3/4" pipe size	1 Stpi	16	.500	Ea.	575	20.50		595.50	660
	8760	1" pipe size		14	.571		575	23.50		598.50	665
	8780	1-1/4" pipe size		12	.667		995	27.50		1,022.50	1,150
	8800	1-1/2" pipe size		11	.727		995	30		1,025	1,150
	8820	2" pipe size		9	.889		1,700	36.50		1,736.50	1,925
	8840	2-1/2" pipe size	Q-5	14	1.143		1,700	42		1,742	1,950
	9000	Flanged, 2" pipe size	1 Stpi	6	1.333		1,775	54.50		1,829.50	2,050
	9020	2-1/2" pipe size	Q-5	4.50	3.556		1,850	131		1,981	2,250
	9040	3" pipe size		3.50	4.571		2,375	168		2,543	2,850
	9060	4" pipe size		3	5.333		3,750	197		3,947	4,400
	9080	5" pipe size	Q-6	3.40	7.059		7,500	270		7,770	8,650
	9100	6" pipe size		3	8		7,500	305		7,805	8,700
	9120	8" pipe size		2.50	9.600		14,400	365		14,765	16,500
	9140	10" pipe size		2.20	10.909		20,100	415		20,515	22,700
	9160	12" pipe size		1.70	14.118		22,300	540		22,840	25,300
	9170	14" pipe size		1.40	17.143		27,000	655		27,655	30,700
	9180	16" pipe size		1	24		33,900	915		34,815	38,700
	9700	Stainless steel body									
	9740	Screwed, 1" pipe size	1 Stpi	14	.571	Ea.	2,900	23.50		2,923.50	3,225
	9760	1-1/2" pipe size		11	.727		4,400	30		4,430	4,900
	9780	2" pipe size		9	.889		7,100	36.50		7,136.50	7,875
	9860	Flanged, 2" pipe size		6	1.333		7,400	54.50		7,454.50	8,225
	9880	2-1/2" pipe size	Q-5	4.50	3.556		7,775	131		7,906	8,750
	9900	3" pipe size		3.50	4.571		12,200	168		12,368	13,700
	9920	4" pipe size		3	5.333		14,900	197		15,097	16,700
	9940	6" pipe size	Q-6	3	8		26,300	305		26,605	29,400
	9960	8" pipe size	"	2.50	9.600		48,700	365		49,065	54,000
820	0010	**STRAINERS, Y TYPE** Bronze body									820
	0050	Screwed, 150 lb., 1/4" pipe size	1 Stpi	24	.333	Ea.	12.05	13.65		25.70	34
	0070	3/8" pipe size		24	.333		15.90	13.65		29.55	38
	0100	1/2" pipe size		20	.400		15.90	16.40		32.30	42
	0120	3/4" pipe size		19	.421		17.40	17.25		34.65	45
	0140	1" pipe size		17	.471		18.70	19.25		37.95	49.50
	0150	1-1/4" pipe size		15	.533		42	22		64	79
	0160	1-1/2" pipe size		14	.571		40.50	23.50		64	79.50
	0180	2" pipe size		13	.615		53.50	25		78.50	97
	0182	3" pipe size		12	.667		340	27.50		367.50	415
	0200	300 lb., 2-1/2" pipe size	Q-5	17	.941		286	34.50		320.50	365
	0220	3" pipe size		16	1		565	37		602	680
	0240	4" pipe size		15	1.067		1,300	39.50		1,339.50	1,475
	0500	For 300 lb rating 1/4" thru 2", add					15%				
	1000	Flanged, 150 lb., 1-1/2" pipe size	1 Stpi	11	.727	Ea.	296	30		326	370
	1020	2" pipe size	"	8	1		345	41		386	440
	1030	2-1/2" pipe size	Q-5	5	3.200		530	118		648	760
	1040	3" pipe size		4.50	3.556		655	131		786	915
	1060	4" pipe size		3	5.333		990	197		1,187	1,400
	1080	5" pipe size	Q-6	3.40	7.059		2,025	270		2,295	2,625
	1100	6" pipe size		3	8		2,650	305		2,955	3,350
	1106	8" pipe size		2.60	9.231		2,750	355		3,105	3,550
	1500	For 300 lb rating, add					40%				
840	0010	**STRAINERS, Y TYPE** Iron body									840
	0050	Screwed, 250 lb., 1/4" pipe size	1 Stpi	20	.400	Ea.	6.65	16.40		23.05	32
	0070	3/8" pipe size		20	.400		6.65	16.40		23.05	32
	0100	1/2" pipe size		20	.400		6.65	16.40		23.05	32

Important: See the Reference Section for critical supporting data - Reference Nos., Crews, & City Cost Indexes

15120	Piping Specialties	CREW	DAILY OUTPUT	LABOR-HOURS	UNIT	2005 BARE COSTS				TOTAL INCL O&P
						MAT.	LABOR	EQUIP.	TOTAL	
840 0120	3/4" pipe size	1 Stpi	18	.444	Ea.	7.95	18.20		26.15	36.50 **84**
0140	1" pipe size		16	.500		10.75	20.50		31.25	43
0150	1-1/4" pipe size		15	.533		14.75	22		36.75	49
0160	1-1/2" pipe size		12	.667		17.75	27.50		45.25	60.50
0180	2" pipe size	▼	8	1		27.50	41		68.50	91.50
0200	2-1/2" pipe size	Q-5	12	1.333		140	49		189	228
0220	3" pipe size		11	1.455		151	53.50		204.50	247
0240	4" pipe size	▼	5	3.200	▼	255	118		373	455
0500	For galvanized body, add					50%				
1000	Flanged, 125 lb., 1-1/2" pipe size	1 Stpi	11	.727	Ea.	86	30		116	140
1020	2" pipe size	"	8	1		64	41		105	132
1030	2-1/2" pipe size	Q-5	5	3.200		72.50	118		190.50	257
1040	3" pipe size		4.50	3.556		84.50	131		215.50	290
1060	4" pipe size	▼	3	5.333		155	197		352	465
1080	5" pipe size	Q-6	3.40	7.059		242	270		512	670
1100	6" pipe size		3	8		295	305		600	785
1120	8" pipe size		2.50	9.600		495	365		860	1,100
1140	10" pipe size		2	12		950	460		1,410	1,750
1160	12" pipe size		1.70	14.118		1,425	540		1,965	2,375
1170	14" pipe size		1.30	18.462		2,625	705		3,330	3,950
1180	16" pipe size	▼	1	24	▼	3,725	915		4,640	5,475
1500	For 250 lb rating, add					20%				
2000	For galvanized body, add					50%				
2500	For steel body, add					40%				
870 0010	**SUCTION DIFFUSERS**									**87**
0100	Cast iron body with integral straightening vanes, strainer									
1000	Flanged									
1010	2" inlet, 1-1/2" pump side	1 Stpi	6	1.333	Ea.	224	54.50		278.50	330
1020	2" pump side	"	5	1.600		232	65.50		297.50	355
1030	3" inlet, 2" pump side	Q-5	6.50	2.462	▼	284	90.50		374.50	445
900 0010	**THERMOFLO INDICATOR** For balancing									**90**
1000	Sweat connections, 1-1/4" pipe size	1 Stpi	12	.667	Ea.	375	27.50		402.50	450
1020	1-1/2" pipe size	"	10	.800	"	380	33		413	470
920 0010	**VENTURI FLOW** Measuring device									**92**
0050	1/2" diameter	1 Stpi	24	.333	Ea.	175	13.65		188.65	214
0100	3/4" diameter		20	.400		163	16.40		179.40	204
0120	1" diameter		19	.421		169	17.25		186.25	212
0140	1-1/4" diameter		15	.533		203	22		225	256
0160	1-1/2" diameter		13	.615		206	25		231	265
0180	2" diameter	▼	11	.727		222	30		252	289
0200	2-1/2" diameter	Q-5	16	1		330	37		367	420
0220	3" diameter		14	1.143		380	42		422	485
0240	4" diameter	▼	11	1.455		595	53.50		648.50	735
0260	5" diameter	Q-6	4	6		720	229		949	1,125
0280	6" diameter		3.50	6.857		780	262		1,042	1,250
0300	8" diameter		3	8		1,075	305		1,380	1,625
0320	10" diameter		2	12		2,050	460		2,510	2,975
0330	12" diameter		1.80	13.333		3,400	510		3,910	4,500
0340	14" diameter		1.60	15		3,650	575		4,225	4,875
0350	16" diameter	▼	1.40	17.143		3,950	655		4,605	5,325
0500	For meter, add				▼	1,175			1,175	1,300
940 0010	**WATER SUPPLY METERS**									**94**
1000	Detector, serves dual systems such as fire and domestic or									
1020	process water, wide range cap., UL and FM approved									
1100	3" mainline x 2" by-pass, 400 GPM	Q-1	3.60	4.444	Ea.	5,150	163		5,313	5,900

15120 | Piping Specialties

		CREW	DAILY OUTPUT	LABOR-HOURS	UNIT	MAT.	LABOR	EQUIP.	TOTAL	TOTAL INCL O&P
						2005 BARE COSTS				
940 1140	4" mainline x 2" by-pass, 700 GPM	Q-1	2.50	6.400	Ea.	5,150	235		5,385	6,000
1180	6" mainline x 3" by-pass, 1600 GPM	Q-2	2.60	9.231		7,775	350		8,125	9,075
1220	8" mainline x 4" by-pass, 2800 GPM		2.10	11.429		11,500	435		11,935	13,300
1260	10" mainline x 6" by-pass, 4400 GPM		2	12		15,700	460		16,160	18,000
1300	10"x12" mainlines x 6" by-pass, 5400 GPM		1.70	14.118		20,700	540		21,240	23,600
2000	Domestic/commercial, bronze									
2020	Threaded									
2060	5/8" diameter, to 20 GPM	1 Plum	16	.500	Ea.	40	20.50		60.50	74.50
2080	3/4" diameter, to 30 GPM		14	.571		67.50	23.50		91	110
2100	1" diameter, to 50 GPM		12	.667		94	27		121	144
2300	Threaded/flanged									
2340	1-1/2" diameter, to 100 GPM	1 Plum	8	1	Ea.	330	41		371	425
2360	2" diameter, to 160 GPM	"	6	1.333	"	415	54.50		469.50	535
2600	Flanged, compound									
2640	3" diameter, 320 GPM	Q-1	3	5.333	Ea.	1,950	196		2,146	2,425
2660	4" diameter, to 500 GPM		1.50	10.667		3,025	390		3,415	3,925
2680	6" diameter, to 1,000 GPM		1	16		4,350	590		4,940	5,650
2700	8" diameter, to 1,800 GPM		.80	20		8,600	735		9,335	10,600
7000	Turbine									
7260	Flanged									
7300	2" diameter, to 160 GPM	1 Plum	7	1.143	Ea.	455	46.50		501.50	570
7320	3" diameter, to 450 GPM	Q-1	3.60	4.444		745	163		908	1,050
7340	4" diameter, to 650 GPM	"	2.50	6.400		1,350	235		1,585	1,825
7360	6" diameter, to 1800 GPM	Q-2	2.60	9.231		2,550	350		2,900	3,325
7380	8" diameter, to 2500 GPM		2.10	11.429		4,050	435		4,485	5,100
7400	10" diameter, to 5500 GPM		1.70	14.118		5,400	540		5,940	6,750

15140 | Domestic Water Piping

		CREW	DAILY OUTPUT	LABOR-HOURS	UNIT	MAT.	LABOR	EQUIP.	TOTAL	TOTAL INCL O&P
100 0010	**BACKFLOW PREVENTER** Includes valves									
0020	and four test cocks, corrosion resistant, automatic operation									
1000	Double check principle									
1010	Threaded, with ball valves									
1020	3/4" pipe size	1 Plum	16	.500	Ea.	127	20.50		147.50	170
1030	1" pipe size		14	.571		140	23.50		163.50	189
1040	1-1/2" pipe size		10	.800		271	32.50		303.50	350
1050	2" pipe size		7	1.143		315	46.50		361.50	415
1080	Threaded, with gate valves									
1100	3/4" pipe size	1 Plum	16	.500	Ea.	500	20.50		520.50	575
1120	1" pipe size		14	.571		500	23.50		523.50	590
1140	1-1/2" pipe size		10	.800		645	32.50		677.50	760
1160	2" pipe size		7	1.143		790	46.50		836.50	940
1200	Flanged, valves are gate									
1210	3" pipe size	Q-1	4.50	3.556	Ea.	1,375	131		1,506	1,700
1220	4" pipe size	"	3	5.333		1,675	196		1,871	2,150
1230	6" pipe size	Q-2	3	8		2,475	305		2,780	3,175
1240	8" pipe size		2	12		4,500	460		4,960	5,650
1250	10" pipe size		1	24		6,675	915		7,590	8,725
1300	Flanged, valves are OS&Y									
1370	1" pipe size	1 Plum	5	1.600	Ea.	545	65.50		610.50	700
1374	1-1/2" pipe size		5	1.600		695	65.50		760.50	865
1378	2" pipe size		4.80	1.667		840	68		908	1,025
1380	3" pipe size	Q-1	4.50	3.556		1,550	131		1,681	1,925
1400	4" pipe size	"	3	5.333		2,175	196		2,371	2,700
1420	6" pipe size	Q-2	3	8		3,425	305		3,730	4,225
1430	8" pipe size	"	2	12		6,350	460		6,810	7,700
4000	Reduced pressure principle									

15140	Domestic Water Piping	CREW	DAILY OUTPUT	LABOR-HOURS	UNIT	2005 BARE COSTS				TOTAL INCL O&P
						MAT.	LABOR	EQUIP.	TOTAL	
100 4100	Threaded, bronze, valves are ball									**100**
4120	3/4" pipe size	1 Plum	16	.500	Ea.	188	20.50		208.50	238
4140	1" pipe size		14	.571		203	23.50		226.50	258
4150	1-1/4" pipe size		12	.667		345	27		372	420
4160	1-1/2" pipe size		10	.800		380	32.50		412.50	470
4180	2" pipe size		7	1.143		425	46.50		471.50	540
5000	Flanged, valves are OS&Y									
5060	2-1/2" pipe size	Q-1	5	3.200	Ea.	2,175	118		2,293	2,575
5080	3" pipe size		4.50	3.556		2,300	131		2,431	2,725
5100	4" pipe size		3	5.333		2,900	196		3,096	3,475
5120	6" pipe size	Q-2	3	8		4,175	305		4,480	5,050
5200	Flanged, iron, valves are gate									
5210	2-1/2" pipe size	Q-1	5	3.200	Ea.	1,550	118		1,668	1,875
5220	3" pipe size		4.50	3.556		1,600	131		1,731	1,975
5230	4" pipe size		3	5.333		2,175	196		2,371	2,700
5240	6" pipe size	Q-2	3	8		3,075	305		3,380	3,825
5250	8" pipe size		2	12		5,500	460		5,960	6,750
5260	10" pipe size		1	24		7,750	915		8,665	9,900
5600	Flanged, iron, valves are OS&Y									
5660	2-1/2" pipe size	Q-1	5	3.200	Ea.	1,750	118		1,868	2,100
5680	3" pipe size		4.50	3.556		1,850	131		1,981	2,225
5700	4" pipe size		3	5.333		2,375	196		2,571	2,925
5720	6" pipe size	Q-2	3	8		3,350	305		3,655	4,125
5740	8" pipe size		2	12		5,875	460		6,335	7,175
5760	10" pipe size		1	24		7,875	915		8,790	10,000
600 0010	**VACUUM BREAKERS** Hot or cold water									**600**
1030	Anti-siphon, brass									
1040	1/4" size	1 Plum	24	.333	Ea.	21	13.60		34.60	43.50
1050	3/8" size		24	.333		21	13.60		34.60	43.50
1060	1/2" size		24	.333		16.35	13.60		29.95	38.50
1080	3/4" size		20	.400		28	16.35		44.35	55
1100	1" size		19	.421		43.50	17.20		60.70	74
1120	1-1/4" size		15	.533		76.50	22		98.50	117
1140	1-1/2" size		13	.615		89.50	25		114.50	137
1160	2" size		11	.727		140	29.50		169.50	199
1180	2-1/2" size		9	.889		400	36.50		436.50	495
1200	3" size		7	1.143		535	46.50		581.50	655
1300	For polished chrome, (1/4" thru 1"), add					50%				
1900	Vacuum relief, water service, bronze									
2000	1/2" size	1 Plum	30	.267	Ea.	20.50	10.90		31.40	39
2040	3/4" size	"	28	.286	"	25.50	11.65		37.15	46
700 0010	**VACUUM BREAKERS**									**700**
0011	See also backflow preventers 15140-100									
1000	Anti-siphon continuous pressure type									
1010	Max. 150 PSI - 210°F									
1020	Bronze body									
1030	1/2" size	1 Stpi	24	.333	Ea.	92.50	13.65		106.15	123
1040	3/4" size		20	.400		92.50	16.40		108.90	127
1050	1" size		19	.421		96	17.25		113.25	131
1060	1-1/4" size		15	.533		189	22		211	241
1070	1-1/2" size		13	.615		233	25		258	294
1080	2" size		11	.727		240	30		270	310
1200	Max. 125 PSI with atmospheric vent									
1210	Brass, in-line construction									
1220	1/4" size	1 Stpi	24	.333	Ea.	38	13.65		51.65	62.50

15140 | Domestic Water Piping

		CREW	DAILY OUTPUT	LABOR-HOURS	UNIT	2005 BARE COSTS				TOTAL INCL O&P
						MAT.	LABOR	EQUIP.	TOTAL	
700	1230	3/8" size	1 Stpi	24	.333	Ea.	38	13.65		51.65	62.50	700
	1260	For polished chrome finish, add				↓	13%				
	2000	Anti-siphon, non-continuous pressure type									
	2010	Hot or cold water 125 PSI - 210° F									
	2020	Bronze body									
	2030	1/4" size	1 Stpi	24	.333	Ea.	23	13.65		36.65	46
	2040	3/8" size		24	.333		23	13.65		36.65	46
	2050	1/2" size		24	.333		26	13.65		39.65	49
	2060	3/4" size		20	.400		31	16.40		47.40	58.50
	2070	1" size		19	.421		48	17.25		65.25	79
	2080	1-1/4" size		15	.533		84.50	22		106.50	126
	2090	1-1/2" size		13	.615		99	25		124	147
	2100	2" size		11	.727		154	30		184	215
	2110	2-1/2" size		8	1		445	41		486	550
	2120	3" size	↓	6	1.333	↓	590	54.50		644.50	730
	2150	For polished chrome finish, add					50%				
800	0010	**WATER HAMMER ARRESTORS / SHOCK ABSORBERS**										800
	0490	Copper									
	0500	3/4" male I.P.S. For 1 to 11 fixtures	1 Plum	12	.667	Ea.	14.50	27		41.50	57
	0600	1" male I.P.S., For 12 to 32 fixtures		8	1		37	41		78	103
	0700	1-1/4" male I.P.S. For 33 to 60 fixtures		8	1		43	41		84	109
	0800	1-1/2" male I.P.S. For 61 to 113 fixtures		8	1		58.50	41		99.50	126
	0900	2" male I.P.S. For 114 to 154 fixtures		8	1		94	41		135	165
	1000	2-1/2" male I.P.S. For 155 to 330 fixtures	↓	4	2	↓	266	81.50		347.50	415
	4000	Bellows type									
	4010	3/4" FNPT, to 11 fixture units	1 Plum	10	.800	Ea.	90.50	32.50		123	149
	4020	1" FNPT, to 32 fixture units		8.80	.909		182	37		219	256
	4030	1" FNPT, to 60 fixture units		8.80	.909		273	37		310	355
	4040	1" FNPT, to 113 fixture units		8.80	.909		685	37		722	805
	4050	1" FNPT, to 154 fixture units		6.60	1.212		770	49.50		819.50	925
	4060	1-1/2" FNPT, to 300 fixture units	↓	5	1.600	↓	950	65.50		1,015.50	1,150

15150 | Sanitary Waste and Vent Piping

		CREW	DAILY OUTPUT	LABOR-HOURS	UNIT	MAT.	LABOR	EQUIP.	TOTAL	TOTAL INCL O&P
100	0010	**BACKWATER VALVES**, C.I. body										100
	6980	Bronze gate and automatic flapper valves									
	7000	3" and 4" pipe size	Q-1	13	1.231	Ea.	800	45.50		845.50	950
	7100	5" and 6" pipe size	"	13	1.231	"	1,225	45.50		1,270.50	1,425
	7240	Bronze flapper valve, bolted cover									
	7260	2" pipe size	Q-1	16	1	Ea.	235	37		272	315
	7280	3" pipe size		14.50	1.103		365	40.50		405.50	465
	7300	4" pipe size	↓	13	1.231		450	45.50		495.50	565
	7320	5" pipe size	Q-2	18	1.333		545	51		596	675
	7340	6" pipe size	"	17	1.412		650	54		704	795
	7360	8" pipe size	Q-3	10	3.200		910	125		1,035	1,175
	7380	10" pipe size	"	9	3.556	↓	1,500	138		1,638	1,850
	7500	For threaded cover, same cost									
	7540	Revolving disk type, same cost as flapper type									
200	0010	**CLEANOUTS**										200
	0060	Floor type									
	0080	Round or square, scoriated nickel bronze top									
	0100	2" pipe size	1 Plum	10	.800	Ea.	77	32.50		109.50	134
	0120	3" pipe size		8	1		116	41		157	189
	0140	4" pipe size		6	1.333		116	54.50		170.50	209
	0160	5" pipe size	↓	4	2		147	81.50		228.50	285
	0180	6" pipe size	Q-1	6	2.667	↓	147	98		245	310

Important: See the Reference Section for critical supporting data - Reference Nos., Crews, & City Cost Indexes

15150 | Sanitary Waste and Vent Piping

		CREW	DAILY OUTPUT	LABOR-HOURS	UNIT	2005 BARE COSTS				TOTAL INCL O&P
						MAT.	LABOR	EQUIP.	TOTAL	
200	0200	8" pipe size	Q-1	4	4	Ea.	260	147		407	505
	0340	Recessed for tile, same price									
	0980	Round top, recessed for terrazzo									
	1000	2" pipe size	1 Plum	9	.889	Ea.	77	36.50		113.50	139
	1080	3" pipe size		6	1.333		116	54.50		170.50	209
	1100	4" pipe size		4	2		116	81.50		197.50	250
	1120	5" pipe size	Q-1	6	2.667		147	98		245	310
	1140	6" pipe size		5	3.200		147	118		265	340
	1160	8" pipe size		4	4		260	147		407	505
	2000	Round scoriated nickel bronze top, extra heavy duty									
	2060	2" pipe size	1 Plum	9	.889	Ea.	105	36.50		141.50	171
	2080	3" pipe size		6	1.333		144	54.50		198.50	240
	2100	4" pipe size		4	2		144	81.50		225.50	281
	2120	5" pipe size	Q-1	6	2.667		175	98		273	340
	2140	6" pipe size		5	3.200		175	118		293	370
	2160	8" pipe size		4	4		288	147		435	535
	4000	Wall type, square smooth cover, over wall frame									
	4060	2" pipe size	1 Plum	14	.571	Ea.	126	23.50		149.50	174
	4080	3" pipe size		12	.667		136	27		163	190
	4100	4" pipe size		10	.800		146	32.50		178.50	209
	4120	5" pipe size		9	.889		211	36.50		247.50	288
	4140	6" pipe size		8	1		235	41		276	320
	4160	8" pipe size	Q-1	11	1.455		315	53.50		368.50	425
	5000	Extension, C.I.;bronze countersunk plug, 8" long									
	5040	2" pipe size	1 Plum	16	.500	Ea.	66	20.50		86.50	103
	5060	3" pipe size		14	.571		76.50	23.50		100	119
	5080	4" pipe size		13	.615		78.50	25		103.50	124
	5100	5" pipe size		12	.667		116	27		143	169
	5120	6" pipe size		11	.727		148	29.50		177.50	208
250	0010	**CLEANOUT TEE**									
	0100	Cast iron, B&S, with countersunk plug									
	0200	2" pipe size	1 Plum	4	2	Ea.	105	81.50		186.50	239
	0220	3" pipe size		3.60	2.222		115	91		206	262
	0240	4" pipe size		3.30	2.424		143	99		242	305
	0260	5" pipe size	Q-1	5.50	2.909		310	107		417	500
	0280	6" pipe size	"	5	3.200		385	118		503	600
	0300	8" pipe size	Q-3	5	6.400		440	249		689	860
	0500	For round smooth access cover, same price									
	0600	For round scoriated access cover, same price									
	0700	For square smooth access cover, add					60%				
	2000	Cast iron, no hub									
	2010	Cleanout tee, with 2 couplings									
	2012	2"	Q-1	22	.727	Ea.	17	27		44	58.50
	2014	3"		19	.842		22.50	31		53.50	71.50
	2016	4"		16.50	.970		32	35.50		67.50	88.50
	2018	6"	Q-2	20	1.200		78.50	46		124.50	155
	2040	Cleanout plug, no hub, with 1 coupling									
	2042	2"	Q-1	44	.364	Ea.	11.55	13.35		24.90	32.50
	2046	3"		38	.421		16.10	15.50		31.60	41
	2048	4"		33	.485		23.50	17.85		41.35	53
	2050	6"	Q-2	40	.600		77	23		100	119
	2052	8"	"	33	.727		110	27.50		137.50	163
	4000	Plastic, tees and adapters. Add plugs									
	4010	ABS, DWW									
	4020	Cleanout tee, 1-1/2" pipe size	1 Plum	15	.533	Ea.	3.41	22		25.41	37

15150	Sanitary Waste and Vent Piping	CREW	DAILY OUTPUT	LABOR-HOURS	UNIT	2005 BARE COSTS				TOTAL INCL O&P	
						MAT.	LABOR	EQUIP.	TOTAL		
250											**250**
4030	2" pipe size	Q-1	27	.593	Ea.	4.50	22		26.50	38	
4040	3" pipe size		21	.762		9	28		37	52	
4050	4" pipe size		16	1		16.70	37		53.70	74	
4100	Cleanout plug, 1-1/2" pipe size	1 Plum	32	.250		.95	10.20		11.15	16.40	
4110	2" pipe size	Q-1	56	.286		1.33	10.50		11.83	17.25	
4120	3" pipe size		36	.444		3.77	16.35		20.12	28.50	
4130	4" pipe size		30	.533		2.48	19.60		22.08	32	
4180	Cleanout adapter fitting, 1-1/2" pipe size	1 Plum	32	.250		.95	10.20		11.15	16.40	
4190	2" pipe size	Q-1	56	.286		1.33	10.50		11.83	17.25	
4200	3" pipe size		36	.444		3.77	16.35		20.12	28.50	
4210	4" pipe size		30	.533		6.05	19.60		25.65	36	
5000	PVC, DWW										
5010	Cleanout tee, 1-1/2" pipe size	1 Plum	15	.533	Ea.	3.42	22		25.42	37	
5020	2" pipe size	Q-1	27	.593		4.60	22		26.60	38	
5030	3" pipe size		21	.762		9.20	28		37.20	52	
5040	4" pipe size		16	1		17.10	37		54.10	74.50	
5090	Cleanout plug, 1-1/2" pipe size	1 Plum	32	.250		.77	10.20		10.97	16.20	
5100	2" pipe size	Q-1	56	.286		.95	10.50		11.45	16.85	
5110	3" pipe size		36	.444		1.98	16.35		18.33	26.50	
5120	4" pipe size		30	.533		2.94	19.60		22.54	32.50	
5130	6" pipe size		24	.667		9	24.50		33.50	47	
5170	Cleanout adapter fitting, 1-1/2" pipe size	1 Plum	32	.250		1.11	10.20		11.31	16.55	
5180	2" pipe size	Q-1	56	.286		1.57	10.50		12.07	17.55	
5190	3" pipe size		36	.444		4.47	16.35		20.82	29.50	
5200	4" pipe size		30	.533		7.30	19.60		26.90	37.50	
5210	6" pipe size		24	.667		16.45	24.50		40.95	55	
5300	Cleanout tee, with plug										
5310	4"	Q-1	14	1.143	Ea.	18.10	42		60.10	83	
5320	6"	"	8	2		35.50	73.50		109	150	
5330	8"	Q-2	11	2.182		75	83		158	208	
5340	12"	"	10	2.400		98.50	91.50		190	246	
5400	Polypropylene, Schedule 40										
5410	Cleanout tee with plug										
5420	1-1/2"	1 Plum	10	.800	Ea.	34.50	32.50		67	87	
5430	2"	Q-1	17	.941		38	34.50		72.50	94	
5440	3"		11	1.455		80	53.50		133.50	169	
5450	4"		9	1.778		109	65.50		174.50	219	
300	0010 **FLOOR AND AREA DRAINS**										**300**
0400	Deck, auto park, C.I., 13" top										
0440	3", 4", 5", and 6" pipe size	Q-1	8	2	Ea.	565	73.50		638.50	730	
0480	For galvanized body, add				"	268			268	295	
0800	Promenade, heelproof grate, C.I., 14" top										
0840	2", 3", and 4" pipe size	Q-1	10	1.600	Ea.	217	59		276	330	
0860	5" and 6" pipe size		9	1.778		268	65.50		333.50	395	
0880	8" pipe size		8	2		320	73.50		393.50	460	
0940	For galvanized body, add					141			141	155	
0960	For polished bronze top, add					147			147	162	
1200	Promenade, heelproof grate, C.I., lateral, 14" top										
1240	2", 3" and 4" pipe size	Q-1	10	1.600	Ea.	280	59		339	400	
1260	5" and 6" pipe size		9	1.778		330	65.50		395.50	465	
1280	8" pipe size		8	2		380	73.50		453.50	530	
1340	For galvanized body, add					141			141	155	
1360	For polished bronze top, add					147			147	162	
1500	Promenade, slotted grate, C.I., 11" top										
1540	2", 3", 4", 5", and 6" pipe size	Q-1	12	1.333	Ea.	195	49		244	288	

			CREW	DAILY OUTPUT	LABOR-HOURS	UNIT	2005 BARE COSTS				TOTAL INCL O&P
							MAT.	LABOR	EQUIP.	TOTAL	
300	1600	For galvanized body, add				Ea.	79			79	87
	1640	For polished bronze top, add				↓	116			116	127
	2000	Floor, medium duty, C.I., deep flange, 7" dia top									
	2040	2" and 3" pipe size	Q-1	12	1.333	Ea.	79	49		128	161
	2080	For galvanized body, add					33.50			33.50	37
	2120	For polished bronze top, add				↓	37			37	41
	2160	Heavy duty, C.I., 12" dia anti-tilt grate									
	2180	2", 3", 4", 5" and 6" pipe size	Q-1	10	1.600	Ea.	229	59		288	340
	2220	For galvanized body, add					102			102	112
	2240	For polished bronze top, add				↓	116			116	127
	2300	X-Heavy duty, C.I., 15" antitilt grate									
	2320	4", 5", 6", and 8" pipe size	Q-1	8	2	Ea.	490	73.50		563.50	650
	2360	For galvanized body, add					167			167	183
	2380	For polished bronze top, add				↓	195			195	214
	2400	Heavy duty, with sediment bucket, C.I., 12" dia. loose grate									
	2420	2", 3", 4", 5", and 6" pipe size	Q-1	9	1.778	Ea.	272	65.50		337.50	400
	2440	For galvanized body, add					152			152	167
	2460	For polished bronze top, add				↓	112			112	123
	2500	Heavy duty, cleanout & trap w/bucket, C.I., 15" top									
	2540	2", 3", and 4" pipe size	Q-1	6	2.667	Ea.	2,525	98		2,623	2,925
	2560	For galvanized body, add					580			580	640
	2580	For polished bronze top, add				↓	204			204	225
	2600	Medium duty, with perforated SS basket, C.I., body,									
	2610	18" top for refuse container washing area									
	2620	2" thru 6" pipe size	Q-1	4	4	Ea.	1,525	147		1,672	1,900
	2630	Acid resistant									
	2638	PVC									
	2640	2", 3" and 4" pipe size	Q-1	16	1	Ea.	325	37		362	415
	2644	Cast iron, epoxy coated									
	2646	2", 3" and 4" pipe size	Q-1	14	1.143	Ea.	130	42		172	205
	2650	PVC or ABS thermoplastic									
	2660	3" and 4" pipe size	Q-1	16	1	Ea.	325	37		362	415
	2680	Extra heavy duty, oil intercepting, gas seal cone,									
	2690	with cleanout, loose grate, C.I., body 16" top									
	2700	3" and 4" diameter outlet, 4" slab depth	Q-1	4	4	Ea.	2,725	147		2,872	3,225
	2720	3" and 4" diameter outlet, 8" slab depth		3	5.333		2,950	196		3,146	3,550
	2740	3" & 4" dia. outlet, 10"-12" slab depth, 16"top		2	8		3,050	294		3,344	3,825
	2910	Prison cell, vandal-proof, 1-1/2", and 2" diam. pipe		12	1.333		174	49		223	266
	2920	3" pipe size	↓	10	1.600	↓	195	59		254	305
	2930	Trap drain, light duty, backwater valve C.I. top									
	2950	8" diameter top, 2" pipe size	Q-1	12	1.333	Ea.	152	49		201	241
	2960	10" diameter top, 3" pipe size		10	1.600		209	59		268	320
	2970	12" diameter top, 4" pipe size	↓	8	2	↓	293	73.50		366.50	435
350	0010	**FLOOR RECEPTORS** For connection to 2", 3" & 4" diameter pipe									
	0200	12-1/2" square top, 25 sq in open area	Q-1	10	1.600	Ea.	380	59		439	510
	0300	For grate with 4" diam. x 3-3/4" high funnel, add					68			68	74.50
	0400	For grate with 6" diameter x 6" high funnel, add					87.50			87.50	96.50
	0500	For full hinged grate with open center, add					63.50			63.50	70
	0600	For aluminum bucket, add					56.50			56.50	62.50
	0700	For acid-resisting bucket, add					108			108	119
	0800	For galvanized bucket, add					70			70	77
	0900	For stainless steel mesh bucket liner, add					84.50			84.50	93
	1000	For bronze antisplash dome strainer, add					40.50			40.50	44.50
	1100	For partial solid cover, add					20.50			20.50	22.50
	1200	For trap primer connection, add				↓	34.50			34.50	37.50

	15150	Sanitary Waste and Vent Piping	CREW	DAILY OUTPUT	LABOR-HOURS	UNIT	2005 BARE COSTS				TOTAL INCL O&P	
							MAT.	LABOR	EQUIP.	TOTAL		
350	2000	12-5/8" diameter top, 40 sq. in. open area	Q-1	10	1.600	Ea.	300	59		359	420	**350**
	2100	For options, add same prices as square top										
	3000	8" x 4" rectangular top, 7.5 sq. in. open area	Q-1	14	1.143	Ea.	293	42		335	390	
	3100	For trap primer connections, add					34.50			34.50	37.50	
	4000	24" x 16" rectangular top, 70 sq. in. open area	Q-1	4	4		1,700	147		1,847	2,100	
	4100	For trap primer connection, add					68			68	74.50	
400	0010	**INTERCEPTORS**										**400**
	0150	Grease, cast iron, 4 GPM, 8 lb. fat capacity	1 Plum	4	2	Ea.	615	81.50		696.50	805	
	0200	7 GPM, 14 lb. fat capacity		4	2		650	81.50		731.50	840	
	1000	10 GPM, 20 lb. fat capacity		4	2		760	81.50		841.50	965	
	1040	15 GPM, 30 lb. fat capacity		4	2		1,125	81.50		1,206.50	1,375	
	1060	20 GPM, 40 lb. fat capacity		3	2.667		1,375	109		1,484	1,700	
	1080	25 GPM, 50 lb. fat capacity	Q-1	3.50	4.571		1,550	168		1,718	1,950	
	1100	35 GPM, 70 lb. fat capacity		3	5.333		1,925	196		2,121	2,400	
	1110	50 GPM, 100 lb. fat capacity		2	8		1,925	294		2,219	2,550	
	1120	Fabricated steel, 50 GPM, 100 lb. fat capacity		2	8		2,550	294		2,844	3,250	
	1140	75 GPM, 150 lb. fat capacity		2	8		5,125	294		5,419	6,075	
	1160	100 GPM, 200 lb. fat capacity		2	8		5,750	294		6,044	6,775	
	1180	150 GPM, 300 lb. fat capacity		2	8		6,400	294		6,694	7,475	
	1200	200 GPM, 400 lb. fat capacity		1.50	10.667		8,825	390		9,215	10,300	
	1220	250 GPM, 500 lb. fat capacity		1.30	12.308		10,800	455		11,255	12,600	
	1240	300 GPM, 600 lb. fat capacity		1	16		12,700	590		13,290	14,900	
	1260	400 GPM, 800 lb. fat capacity	Q-2	1.20	20		15,600	765		16,365	18,300	
	1280	500 GPM, 1000 lb. fat capacity	"	1	24		18,400	915		19,315	21,700	
	1560	For chemical add-port, add					98			98	108	
	1580	For seepage pan, add					7%					
	3000	Hair, cast iron, 1-1/4" and 1-1/2" pipe connection	1 Plum	8	1	Ea.	175	41		216	255	
	3100	For chrome-plated cast iron, add					108			108	119	
	3200	For polished bronze, add					320			320	355	
	3400	Lint interceptor, fabricated steel										
	3410	Size based on 10 GPM per machine										
	3420	30 GPM, 2" pipe size	Q-1	3	5.333	Ea.	2,050	196		2,246	2,550	
	3430	70 GPM, 3" pipe size		2.50	6.400		2,400	235		2,635	2,975	
	3440	100 GPM, 4" pipe size		2	8		2,800	294		3,094	3,550	
	3450	200 GPM, 4" pipe size		1.50	10.667		3,400	390		3,790	4,350	
	3460	300 GPM, 6" pipe size		1	16		3,950	590		4,540	5,225	
	3470	400 GPM, 6" pipe size	Q-2	1.20	20		4,500	765		5,265	6,100	
	3480	500 GPM, 6" pipe size	"	1	24		5,075	915		5,990	6,950	
	4000	Oil, fabricated steel, 10 GPM, 2" pipe size	1 Plum	4	2		1,050	81.50		1,131.50	1,275	
	4100	15 GPM, 2" or 3" pipe size		4	2		1,425	81.50		1,506.50	1,700	
	4120	20 GPM, 2" or 3" pipe size		3	2.667		1,725	109		1,834	2,075	
	4140	25 GPM, 2" or 3" pipe size	Q-1	3.50	4.571		1,875	168		2,043	2,300	
	4160	35 GPM, 2", 3", or 4" pipe size		3	5.333		2,275	196		2,471	2,800	
	4180	50 GPM, 2", 3", or 4" pipe size		2	8		3,075	294		3,369	3,825	
	4200	75 GPM, 3" pipe size		2	8		5,300	294		5,594	6,300	
	4220	100 GPM, 3" pipe size		2	8		5,750	294		6,044	6,775	
	4240	150 GPM, 4" pipe size		2	8		7,125	294		7,419	8,300	
	4260	200 GPM, 4" pipe size		1.50	10.667		10,100	390		10,490	11,700	
	4280	250 GPM, 5" pipe size		1.30	12.308		11,700	455		12,155	13,500	
	4300	300 GPM, 5" pipe size		1	16		13,200	590		13,790	15,400	
	4320	400 GPM, 6" pipe size	Q-2	1.20	20		17,000	765		17,765	19,900	
	4340	500 GPM, 6" pipe size	"	1	24		21,400	915		22,315	25,000	
	5000	Sand interceptor, fabricated steel										
	5020	20 GPM, 4" pipe size	Q-1	3	5.333	Ea.	2,575	196		2,771	3,125	
	5030	50 GPM, 4" pipe size		2.50	6.400		4,350	235		4,585	5,150	
	5040	150 GPM, 4" pipe size		2	8		13,200	294		13,494	14,900	

Important: See the Reference Section for critical supporting data - Reference Nos., Crews, & City Cost Indexes

MEANS
CostWorks®
2005

Maximize your estimating & budgeting efforts quickly & efficiently . . .

Introducing Means CostWorks Online – your preeminent Cost Data resource is now available in an easy-to-use online format as well as on CD.

RS**Means**

Catalog No.	Means CostWorks 2005 CD	Price	Cost
65065	Assemblies Cost Data	$202.95	
65015	Building Construction Cost Data	$142.95	
65315	Building Construction Cost Data/Metric	$142.95	
65155	Open Shop Building Construction Cost Data	$142.95	
65115	Concrete & Masonry Cost Data	$142.95	
65035	Electrical Cost Data	$142.95	
65165	Heavy Construction Cost Data	$142.95	
65095	Interior Cost Data	$142.95	
65185	Light Commercial Cost Data	$177.95	
65025	Mechanical Cost Data	$142.95	
65215	Plumbing Cost Data	$142.95	
65285	Site Work & Landscape Cost Data	$142.95	
65055	Square Foot Costs	$202.95	
65605*	CostWorks Estimator	$299.95	
65205	Facilities Construction Cost Data	$329.95	
65045	Repair & Remodeling Cost Data	$142.95	
65365	Heavy Construction Cost Data/Metric	$142.95	
65175	Residential Cost Data	$164.95	
65305	Facilities Maintenance & Repair Cost Data	$329.95	

Packages*

Catalog No.	Title	Price	Cost
65515	Builder's Package	$380.95	
65525	Building Professional's Package	$445.95	
65535	Facility Manager's Package	$572.95	
65545	Design Professional's Package	$445.95	

*(Packages include Means Estimator FREE)

Subtotal		
MA residents please add 5% Sales Tax		
Shipping & Handling	**$6.00**	
Grand Total		

Name: _____

Company: _____

Street: _____

City: _____ State: _____ Zip: _____

Phone: (___) _____ Fax: (___) _____

E-Mail: _____

☐ Bill me   ☐ Please charge my credit card

☐ Visa   ☐ MasterCard   ☐ American Express   ☐ Discover

Card # _____ Expiration Date: _____

Cardholder Signature: _____

From the planning stages through the punch list ...

Maximize your estimating & budgeting efforts quickly & efficiently with *Means CostWorks 2005*

As building requirements grow more and more complex, the estimating and budgeting process has never been more demanding. Tighter deadlines and more stringent cost controls mean you must find ways to work more efficiently. Let *Means CostWorks 2005*, available in CD or Online formats, help you quickly and easily perform in-depth cost research ... and take your estimating and budgeting to the next level.

Purchase your titles separately or in value priced packages. Plus, with your purchase you receive access to regular cost updates, access to the Means Estimating HOTLINE, and the Wage Rate Utility (with selected titles).

Your 2005 Means CostWorks Subscription now includes, *Free:*

- Access to Means Estimating Hotline
- The *RSMeans Quarterly Update Service*
- Wage Rate Utility – allows you to reference Union and Open Shop rates (included with selected titles)
- Free Means Estimator with your purchase of a *Means CostWorks* package

The same Cost Data you know and trust... delivered electronically in your choice of CD or Online formats!

- Localize costs to your geographic area for better accuracy
- Easily search and organize detailed cost data
- Determine best-cost solutions for each project
- Change and calculate results automatically
- Receive free online cost factor updates with the *RSMeans Quarterly Update Service*

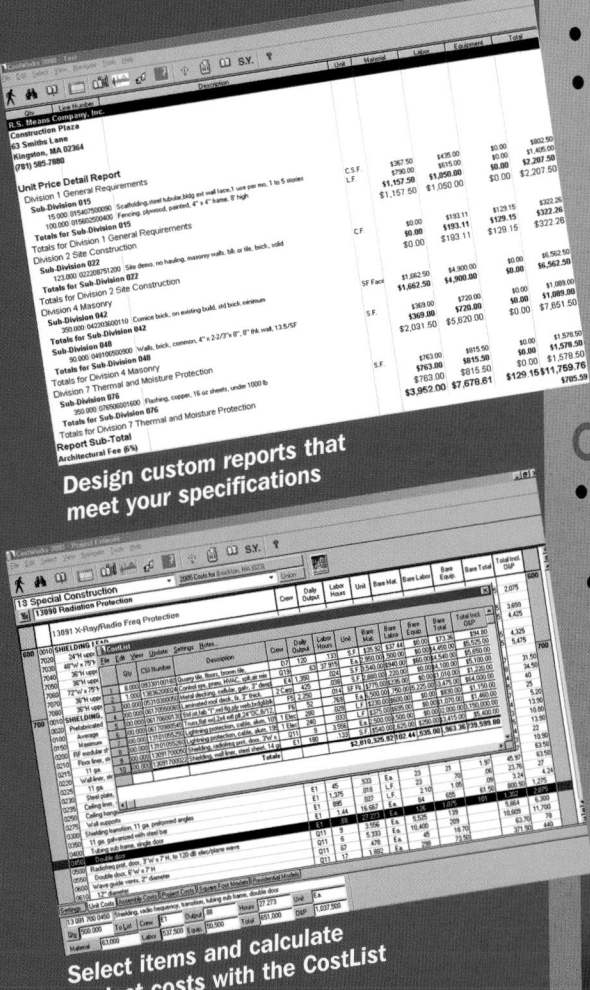

Design custom reports that meet your specifications

Select items and calculate project costs with the CostList

Quick and easy ... Get this powerful tool before you estimate your next project!

15150	Sanitary Waste and Vent Piping	CREW	DAILY OUTPUT	LABOR-HOURS	UNIT	2005 BARE COSTS				TOTAL INCL O&P
						MAT.	LABOR	EQUIP.	TOTAL	
400 5050	250 GPM, 6" pipe size	Q-1	1.30	12.308	Ea.	13,700	455		14,155	15,800
5060	500 GPM, 6" pipe size	Q-2	1	24		17,600	915		18,515	20,800
6000	Solids, precious metals recovery, C.I., 1-1/4" to 2" pipe	1 Plum	4	2		264	81.50		345.50	415
6100	Dental Lab., large, C.I., 1-1/2" to 2" pipe	"	3	2.667		925	109		1,034	1,200
500 0010	**SEPARATORS** Entrainment eliminator, steel body, 150 PSIG									
0100	1/4" size	1 Stpi	24	.333	Ea.	169	13.65		182.65	207
0120	1/2" size		24	.333		177	13.65		190.65	216
0140	3/4" size		20	.400		189	16.40		205.40	233
0160	1" size		19	.421		257	17.25		274.25	310
0180	1-1/4" size		15	.533		291	22		313	355
0200	1-1/2" size		13	.615		320	25		345	395
0220	2" size		11	.727		355	30		385	440
600 0010	**SHOWER DRAINS**									
2780	Shower, with strainer, uniform diam. trap, bronze top									
2800	2" and 3" pipe size	Q-1	8	2	Ea.	163	73.50		236.50	290
2820	4" pipe size	"	7	2.286		184	84		268	330
2840	For galvanized body, add					65			65	71.50
2860	With strainer, backwater valve, drum trap									
2880	1-1/2", 2",& 3" pipe size	Q-1	8	2	Ea.	150	73.50		223.50	276
2890	4" pipe size	"	7	2.286		209	84		293	355
2900	For galvanized body, add					65			65	71.50
700 0010	**SINK WASTE TREATMENT** System for commercial kitchens									
0100	includes clock timer, & fittings									
0200	System less chemical, wall mounted cabinet	1 Plum	16	.500	Ea.	430	20.50		450.50	505
2000	Chemical, 1 gallon, add					50			50	55
2100	6 gallons, add					485			485	535
2200	15 gallons, add					610			610	675
2300	30 gallons, add					1,150			1,150	1,250
2400	55 gallons, add					1,950			1,950	2,150
800 0010	**TRAPS**									
0030	Cast iron, service weight									
0050	Running P trap, without vent									
1100	2"	Q-1	16	1	Ea.	25	37		62	82.50
1140	3"		14	1.143		40.50	42		82.50	108
1150	4"		13	1.231		70	45.50		115.50	145
1160	6"	Q-2	17	1.412		325	54		379	435
1180	Running trap, single hub, with vent									
2080	3" pipe size, 3" vent	Q-1	14	1.143	Ea.	59	42		101	128
2120	4" pipe size, 4" vent	"	13	1.231		80	45.50		125.50	156
2140	5" pipe size, 4" vent	Q-2	11	2.182		128	83		211	266
2160	6" pipe size, 4" vent		10	2.400		345	91.50		436.50	520
2180	6" pipe size, 6" vent		8	3		375	114		489	580
2200	8" pipe size, 4" vent	Q-3	10	3.200		1,125	125		1,250	1,425
2220	8" pipe size, 6" vent	"	8	4		1,225	156		1,381	1,575
2300	For double hub, vent, add					10%	20%			
2800	S trap,									
2850	4" pipe size	Q-1	13	1.231	Ea.	79.50	45.50		125	156
3000	P trap, B&S, 2" pipe size		16	1		17.60	37		54.60	75
3040	3" pipe size		14	1.143		26.50	42		68.50	92
3060	4" pipe size		13	1.231		38	45.50		83.50	110
3080	5" pipe size	Q-2	18	1.333		80	51		131	165
3100	6" pipe size	"	17	1.412		117	54		171	210
3120	8" pipe size	Q-3	11	2.909		335	113		448	535
3130	10" pipe size	"	10	3.200		655	125		780	905
3150	P trap, no hub, 1-1/2" pipe size	Q-1	17	.941		10.35	34.50		44.85	63.50

15150 | Sanitary Waste and Vent Piping

800

		CREW	DAILY OUTPUT	LABOR-HOURS	UNIT	2005 BARE COSTS MAT.	LABOR	EQUIP.	TOTAL	TOTAL INCL O&P
3160	2" pipe size	Q-1	16	1	Ea.	9.90	37		46.90	66.50
3170	3" pipe size		14	1.143		22	42		64	87
3180	4" pipe size	↓	13	1.231	↓	38	45.50		83.50	110
3190	6" pipe size	Q-2	17	1.412	↓	92	54		146	182
3350	Deep seal trap, B&S									
3400	1-1/4" pipe size	Q-1	14	1.143	Ea.	29.50	42		71.50	95.50
3410	1-1/2" pipe size		14	1.143		29.50	42		71.50	95.50
3420	2" pipe size		14	1.143		20	42		62	85
3440	3" pipe size		12	1.333		32.50	49		81.50	109
3460	4" pipe size	↓	11	1.455		51.50	53.50		105	137
3500	For trap primer connection, add				↓	57.50			57.50	63
3540	For trap with floor cleanout, add					70%	5%			
3580	For trap with adjustable cleanout, add	Q-1	10	1.600	Ea.	120	59		179	221
4700	Copper, drainage, drum trap									
4800	3" x 5" solid, 1-1/2" pipe size	1 Plum	16	.500	Ea.	38	20.50		58.50	72.50
4840	3" x 6" swivel, 1-1/2" pipe size	"	16	.500	"	60.50	20.50		81	97
5100	P trap, standard pattern									
5200	1-1/4" pipe size	1 Plum	18	.444	Ea.	28	18.15		46.15	58.50
5240	1-1/2" pipe size		17	.471		27	19.20		46.20	59
5260	2" pipe size		15	.533		42	22		64	79
5280	3" pipe size	↓	11	.727	↓	101	29.50		130.50	156
5340	With cleanout and slip joint									
5360	1-1/4" pipe size	1 Plum	18	.444	Ea.	27	18.15		45.15	57
5400	1-1/2" pipe size		17	.471		41.50	19.20		60.70	74.50
5420	2" pipe size	↓	15	.533		67	22		89	107
5460	For swivel, add				↓	4.13			4.13	4.54
5750	Chromed brass, tubular, P trap, without cleanout, 20 Ga.									
5800	1-1/4" pipe size	1 Plum	18	.444	Ea.	13.75	18.15		31.90	42.50
5840	1-1/2" pipe size	"	17	.471	"	16.40	19.20		35.60	47
5900	With cleanout, 20 Ga.									
5940	1-1/4" pipe size	1 Plum	18	.444	Ea.	26.50	18.15		44.65	56.50
6000	1-1/2" pipe size	"	17	.471	"	33	19.20		52.20	65.50
6350	S trap, without cleanout, 20 Ga.									
6400	1-1/4" pipe size	1 Plum	18	.444	Ea.	29	18.15		47.15	59.50
6440	1-1/2" pipe size	"	17	.471	"	33	19.20		52.20	65.50
6550	With cleanout, 20 Ga.									
6600	1-1/4" pipe size	1 Plum	18	.444	Ea.	48	18.15		66.15	80.50
6640	1-1/2" pipe size	"	17	.471		52.50	19.20		71.70	86.50
6660	Corrosion resistant, glass, P trap, 1-1/2" pipe size	Q-1	17	.941		43.50	34.50		78	99.50
6670	2" pipe size		16	1		57	37		94	118
6680	3" pipe size		14	1.143		116	42		158	191
6690	4" pipe size	↓	13	1.231		174	45.50		219.50	259
6700	6" pipe size	Q-2	17	1.412	↓	665	54		719	815
6710	ABS DWV P trap, solvent weld joint									
6720	1-1/2" pipe size	1 Plum	18	.444	Ea.	4.31	18.15		22.46	32
6722	2" pipe size		17	.471		6.15	19.20		25.35	36
6724	3" pipe size		15	.533		22.50	22		44.50	57.50
6726	4" pipe size	↓	14	.571	↓	20	23.50		43.50	57
6732	PVC DWV P trap, solvent weld joint									
6733	1-1/2" pipe size	1 Plum	18	.444	Ea.	4.34	18.15		22.49	32.50
6734	2" pipe size		17	.471		6.45	19.20		25.65	36
6735	3" pipe size		15	.533		22.50	22		44.50	57.50
6736	4" pipe size		14	.571		51	23.50		74.50	91
6760	PP DWV, dilution trap, 1-1/2" pipe size		16	.500		190	20.50		210.50	240
6770	P trap, 1-1/2" pipe size		17	.471		51	19.20		70.20	85
6780	2" pipe size	↓	16	.500	↓	69.50	20.50		90	107

15150	Sanitary Waste and Vent Piping	CREW	DAILY OUTPUT	LABOR-HOURS	UNIT	2005 BARE COSTS				TOTAL INCL O&P
						MAT.	LABOR	EQUIP.	TOTAL	
800	6790	3" pipe size	1 Plum	14	.571	Ea.	161	23.50		184.50	212
	6800	4" pipe size		13	.615		205	25		230	263
	6810	Running trap, 1-1/2" pipe size		16	.500		28.50	20.50		49	62
	6820	2" pipe size		15	.533		40	22		62	77
	6830	S trap, 1-1/2" pipe size		16	.500		29	20.50		49.50	62.50
	6840	2" pipe size		15	.533		36.50	22		58.50	73
	6850	Universal trap, 1-1/2" pipe size		14	.571		82.50	23.50		106	126
	6860	PVC DWW hub x hub, basin trap, 1-1/4" pipe size		18	.444		6.10	18.15		24.25	34
	6870	Sink P trap, 1-1/2" pipe size		18	.444		6.10	18.15		24.25	34
	6880	Tubular S trap, 1-1/2" pipe size	↓	17	.471	↓	11.25	19.20		30.45	41.50
	6890	PVC sch. 40 DWV, drum trap									
	6900	1-1/2" pipe size	1 Plum	16	.500	Ea.	16.15	20.50		36.65	48.50
	6910	P trap, 1-1/2" pipe size		18	.444		4.34	18.15		22.49	32.50
	6920	2" pipe size		17	.471		6.45	19.20		25.65	36
	6930	3" pipe size		15	.533		22.50	22		44.50	57.50
	6940	4" pipe size		14	.571		51	23.50		74.50	91
	6950	P trap w/clean out, 1-1/2" pipe size		18	.444		7.40	18.15		25.55	35.50
	6960	2" pipe size		17	.471		12.55	19.20		31.75	43
	6970	P trap adjustable, 1-1/2" pipe size		17	.471		6.25	19.20		25.45	36
	6980	P trap adj. w/union & cleanout, 1-1/2" pipe size		16	.500		12.50	20.50		33	44.50
	7000	Trap primer, flow through type, 1/2" diameter		24	.333		39.50	13.60		53.10	64
	7100	With sediment strainer	↓	22	.364	↓	43.50	14.85		58.35	70
	7450	Trap primer distribution unit									
	7500	2 openings	1 Plum	18	.444	Ea.	22	18.15		40.15	51.50
	7540	3 opening		17	.471		22	19.20		41.20	53
	7560	4 opening	↓	16	.500	↓	22	20.50		42.50	54.50
	7850	Trap primer manifold									
	7900	2 outlet	1 Plum	18	.444	Ea.	50.50	18.15		68.65	83
	7940	4 outlet		16	.500		78.50	20.50		99	117
	7960	6 outlet		15	.533		98.50	22		120.50	141
	7980	8 outlet	↓	13	.615	↓	123	25		148	174
900	0010	**VENT FLASHING, CAPS**									
	0120	Vent caps									
	0140	Cast iron									
	0160	1-1/4" - 1-1/2" pipe	1 Plum	23	.348	Ea.	26.50	14.20		40.70	50.50
	0170	2" - 2-1/8" pipe		22	.364		30.50	14.85		45.35	56
	0180	2-1/2" - 3-5/8" pipe		21	.381		35	15.55		50.55	61.50
	0190	4" - 4-1/8" pipe		19	.421		41.50	17.20		58.70	72
	0200	5" - 6" pipe	↓	17	.471	↓	62	19.20		81.20	97
	0300	PVC									
	0320	1-1/4" - 1-1/2" pipe	1 Plum	24	.333	Ea.	19.80	13.60		33.40	42.50
	0330	2" - 2-1/8" pipe		23	.348		23	14.20		37.20	46.50
	0340	2-1/2" - 3-5/8" pipe		22	.364		26	14.85		40.85	51
	0350	4" - 4-1/8" pipe		20	.400		31.50	16.35		47.85	59
	0360	5" - 6" pipe	↓	18	.444	↓	46.50	18.15		64.65	78.50
	0900	Vent flashing									
	1000	Aluminum with lead ring									
	1020	1-1/4" pipe	1 Plum	20	.400	Ea.	6.65	16.35		23	32
	1030	1-1/2" pipe		20	.400		7.15	16.35		23.50	32.50
	1040	2" pipe		18	.444		7.15	18.15		25.30	35.50
	1050	3" pipe		17	.471		7.90	19.20		27.10	37.50
	1060	4" pipe	↓	16	.500	↓	9.55	20.50		30.05	41
	1350	Copper with neoprene ring									
	1400	1-1/4" pipe	1 Plum	20	.400	Ea.	14.05	16.35		30.40	40
	1430	1-1/2" pipe	↓	20	.400	↓	14.05	16.35		30.40	40

15150 | Sanitary Waste and Vent Piping

		CREW	DAILY OUTPUT	LABOR-HOURS	UNIT	2005 BARE COSTS				TOTAL INCL O&P
						MAT.	LABOR	EQUIP.	TOTAL	
900	1440	2" pipe	1 Plum	18	.444	Ea.	14.85	18.15		33	44	**900**
	1450	3" pipe		17	.471		17.40	19.20		36.60	48
	1460	4" pipe	▼	16	.500	▼	19.20	20.50		39.70	51.50
	2000	Galvanized with neoprene ring									
	2020	1-1/4" pipe	1 Plum	20	.400	Ea.	5.30	16.35		21.65	30.50
	2030	1-1/2" pipe		20	.400		5.30	16.35		21.65	30.50
	2040	2" pipe		18	.444		5.60	18.15		23.75	33.50
	2050	3" pipe		17	.471		6.10	19.20		25.30	35.50
	2060	4" pipe	▼	16	.500	▼	7.90	20.50		28.40	39
	2980	Neoprene, one piece									
	3000	1-1/4" pipe	1 Plum	24	.333	Ea.	5.15	13.60		18.75	26
	3030	1-1/2" pipe		24	.333		5.40	13.60		19	26.50
	3040	2" pipe		23	.348		5.40	14.20		19.60	27.50
	3050	3" pipe		21	.381		6.55	15.55		22.10	31
	3060	4" pipe	▼	20	.400	▼	8.90	16.35		25.25	34.50

15160 | Storm Drainage Piping

		CREW	DAILY OUTPUT	LABOR-HOURS	UNIT	MAT.	LABOR	EQUIP.	TOTAL	TOTAL INCL O&P
500	0010	**STORM AREA DRAINS**										**500**
	0140	Cornice, C.I., 45° or 90° outlet									
	0180	1-1/2" & 2" pipe size	Q-1	14	1.143	Ea.	95	42		137	168
	0200	3" and 4" pipe size	"	12	1.333		130	49		179	216
	0260	For galvanized body, add					26			26	28.50
	0280	For polished bronze dome, add				▼	22.50			22.50	24.50
	3860	Roof, flat metal deck, C.I. body, 12" C.I. dome									
	3880	2" pipe size	Q-1	15	1.067	Ea.	113	39		152	183
	3890	3" pipe size		14	1.143		155	42		197	233
	3900	4" pipe size		13	1.231		155	45.50		200.50	238
	3910	5" pipe size		12	1.333		268	49		317	370
	3920	6" pipe size	▼	10	1.600	▼	268	59		327	385
	4280	Integral expansion joint, C.I. body, 12" C.I. dome									
	4300	2" pipe size	Q-1	8	2	Ea.	232	73.50		305.50	365
	4320	3" pipe size		7	2.286		289	84		373	445
	4340	4" pipe size		6	2.667		299	98		397	475
	4360	5" pipe size		4	4		475	147		622	745
	4380	6" pipe size		3	5.333		510	196		706	855
	4400	8" pipe size	▼	3	5.333	▼	670	196		866	1,025
	4440	For galvanized body, add					90.50			90.50	99.50
	4480	Flexible neoprene bellows, no-hub connection									
	4500	3" and 4" pipe size	Q-1	5	3.200	Ea.	67	118		185	251
	4520	6" and 8" pipe size	"	4	4	"	149	147		296	385
	4620	Main, all aluminum, 12" low profile dome									
	4640	2", 3" and 4" pipe size	Q-1	14	1.143	Ea.	172	42		214	252
	4660	5" and 6" pipe size		13	1.231		226	45.50		271.50	315
	4680	8" pipe size		10	1.600		281	59		340	400
	4690	Main, CI body, 12" poly. dome, 2", 3", & 4" pipe		8	2		127	73.50		200.50	251
	4710	5" and 6" pipe size		6	2.667		182	98		280	345
	4720	8" pipe size		4	4		237	147		384	480
	4730	For underdeck clamp, add	▼	22	.727		88	27		115	137
	4740	For vandalproof dome, add					22.50			22.50	24.50
	4750	For galvanized body, add					152			152	167
	4760	Main, PVC body and dome, 2" pipe size	Q-1	14	1.143		89	42		131	161
	4780	3" pipe size		14	1.143		89	42		131	161
	4800	4" pipe size		14	1.143		89	42		131	161
	4820	For underdeck clamp, add	▼	24	.667	▼	19.30	24.50		43.80	58
	4900	Terrace planting area, with perforated overflow, C.I.									

15160 | Storm Drainage Piping

		CREW	DAILY OUTPUT	LABOR-HOURS	UNIT	2005 BARE COSTS				TOTAL INCL O&P
						MAT.	LABOR	EQUIP.	TOTAL	
500	4920	2", 3" and 4" pipe size	Q-1	8	2	Ea.	241	73.50		314.50	375
	4980	Scupper floor, oblique strainer, C.I.									
	5000	6" x 7" top, 2", 3" and 4" pipe size	Q-1	16	1	Ea.	111	37		148	178
	5100	8" x 12" top, 5" and 6" pipe size	"	14	1.143		216	42		258	300
	5160	For galvanized body, add					40%				
	5200	For polished bronze strainer, add					85%				
	5980	Trench, floor, hvy duty, modular, C.I., 12" x 12" top									
	6000	2", 3", 4", 5", & 6" pipe size	Q-1	8	2	Ea.	350	73.50		423.50	495
	6100	For polished bronze top, add					172			172	189
	6200	For 12" extension section, C.I. top, add					350			350	385
	6240	For 12" extension section, polished bronze top, add					165			165	181

15170 | Swimming Pool and Fountain Piping

		CREW	DAILY OUTPUT	LABOR-HOURS	UNIT	MAT.	LABOR	EQUIP.	TOTAL	TOTAL INCL O&P
500 0010	**SWIMMING POOL DRAINS**									
2990	Pool main discharge, polished bronze top, C.I. body									
3000	8" diam. grate, 2", 3" & 4" pipe size	Q-1	8	2	Ea.	340	73.50		413.50	485
3100	12" diam. grate, 4", 5" & 6" pipe size	"	8	2		575	73.50		648.50	740
3160	For galvanized body, add					137			137	151
3200	Gutter, bronze body, polished bronze top									
3220	9" grate, 1-1/2" & 2" pipe size	Q-1	8	2	Ea.	98	73.50		171.50	219
3240	10" grate, 2", 2-1/2" & 3" pipe size	"	7	2.286	"	102	84		186	238
3380	Overflow, 5" dome & 6" high pipe polished bronze,									
3400	cast iron body, 2", 3", & 4" pipe size	Q-1	10	1.600	Ea.	430	59		489	565
3580	Recirculating inlet, 6" sq face polished bronze,									
3600	C.I. body, 2", 2-1/2" & 3" pipe size	Q-1	10	1.600	Ea.	61.50	59		120.50	157
3620	For galvanized body, add				"	26.50			26.50	29.50

15180 | Heating and Cooling Piping

		CREW	DAILY OUTPUT	LABOR-HOURS	UNIT	MAT.	LABOR	EQUIP.	TOTAL	TOTAL INCL O&P
100 0010	**ANTI-FREEZE** Inhibited									
0900	Ethylene glycol concentrated									
1000	55 gallon drums, small quantities				Gal.	7.70			7.70	8.50
1200	Large quantities					7.25			7.25	8
2000	Propylene glycol, for solar heat, small quantities					8.75			8.75	9.65
2100	Large quantities					8.65			8.65	9.50
200 0010	**PUMPS, CIRCULATING** Heated or chilled water application									
0600	Bronze, sweat connections, 1/40 HP, in line									
0640	3/4" size	Q-1	16	1	Ea.	120	37		157	188
1000	Flange connection, 3/4" to 1-1/2" size									
1040	1/12 HP	Q-1	6	2.667	Ea.	325	98		423	505
1060	1/8 HP		6	2.667		560	98		658	760
1100	1/3 HP		6	2.667		630	98		728	835
1140	2" size, 1/6 HP		5	3.200		805	118		923	1,050
1180	2-1/2" size, 1/4 HP		5	3.200		1,050	118		1,168	1,325
1220	3" size, 1/4 HP		4	4		1,100	147		1,247	1,425
1260	1/3 HP		4	4		1,325	147		1,472	1,700
1300	1/2 HP		4	4		1,375	147		1,522	1,725
1340	3/4 HP		4	4		1,500	147		1,647	1,875
1380	1 HP		4	4		2,400	147		2,547	2,875
2000	Cast iron, flange connection									
2040	3/4" to 1-1/2" size, in line, 1/12 HP	Q-1	6	2.667	Ea.	213	98		311	380
2060	1/8 HP		6	2.667		355	98		453	535
2100	1/3 HP		6	2.667		395	98		493	580
2140	2" size, 1/6 HP		5	3.200		435	118		553	655
2180	2-1/2" size, 1/4 HP		5	3.200		570	118		688	800

15180	Heating and Cooling Piping	CREW	DAILY OUTPUT	LABOR-HOURS	UNIT	2005 BARE COSTS				TOTAL INCL O&P	
						MAT.	LABOR	EQUIP.	TOTAL		
200											**200**
2220	3" size, 1/4 HP	Q-1	4	4	Ea.	580	147		727	855	
2260	1/3 HP		4	4		780	147		927	1,075	
2300	1/2 HP		4	4		810	147		957	1,100	
2340	3/4 HP		4	4		935	147		1,082	1,250	
2380	1 HP		4	4		1,350	147		1,497	1,700	
2600	For non-ferrous impeller, add					3%					
3000	High head, bronze impeller										
3030	1-1/2" size 1/2 HP	Q-1	5	3.200	Ea.	680	118		798	920	
3040	1-1/2" size 3/4 HP		5	3.200		725	118		843	975	
3050	2" size 1 HP		4	4		870	147		1,017	1,175	
3090	2" size 1-1/2 HP		4	4		1,000	147		1,147	1,325	
4000	Close coupled, end suction, bronze impeller										
4040	1-1/2" size, 1-1/2 HP, to 40 GPM	Q-1	3	5.333	Ea.	1,125	196		1,321	1,550	
4090	2" size, 2 HP, to 50 GPM		3	5.333		1,150	196		1,346	1,550	
4100	2" size, 3 HP, to 90 GPM		2.30	6.957		1,250	256		1,506	1,750	
4190	2-1/2" size, 3 HP, to 150 GPM		2	8		1,350	294		1,644	1,925	
4300	3" size, 5 HP, to 225 GPM		1.80	8.889		1,475	325		1,800	2,100	
4410	3" size, 10 HP, to 350 GPM		1.60	10		2,125	370		2,495	2,875	
4420	4" size, 7-1/2 HP, to 350 GPM		1.60	10		2,050	370		2,420	2,800	
4520	4" size, 10 HP, to 600 GPM	Q-2	1.70	14.118		2,525	540		3,065	3,575	
4530	5" size, 15 HP, to 1000 GPM		1.70	14.118		3,025	540		3,565	4,125	
4610	5" size, 20 HP, to 1350 GPM		1.50	16		3,350	610		3,960	4,600	
4620	5" size, 25 HP, to 1550 GPM		1.50	16		3,975	610		4,585	5,300	
5000	Base mounted, bronze impeller, coupling guard										
5040	1-1/2" size, 1-1/2 HP, to 40 GPM	Q-1	2.30	6.957	Ea.	1,750	256		2,006	2,300	
5090	2" size, 2 HP, to 50 GPM		2.30	6.957		1,825	256		2,081	2,375	
5100	2" size, 3 HP, to 90 GPM		2	8		1,825	294		2,119	2,475	
5190	2-1/2" size, 3 HP, to 150 GPM		1.80	8.889		1,925	325		2,250	2,625	
5300	3" size, 5 HP, to 225 GPM		1.60	10		2,000	370		2,370	2,750	
5410	4" size, 5 HP, to 350 GPM		1.50	10.667		2,175	390		2,565	3,000	
5420	4" size, 7-1/2 HP, to 350 GPM		1.50	10.667		2,500	390		2,890	3,350	
5520	5" size, 10 HP, to 600 GPM	Q-2	1.60	15		3,175	570		3,745	4,350	
5530	5" size, 15 HP, to 1000 GPM		1.60	15		3,425	570		3,995	4,625	
5610	6" size, 20 HP, to 1350 GPM		1.40	17.143		3,800	655		4,455	5,150	
5620	6" size, 25 HP, to 1550 GPM		1.40	17.143		4,275	655		4,930	5,700	
5800	The above pump capacities are based on 1800 RPM,										
5810	at a 60 foot head. Increasing the RPM										
5820	or decreasing the head will increase the GPM.										

15190	Fuel Piping										
490											**490**
0010	**FUEL OIL SPECIALTIES**										
0020	Foot valve, single poppet, metal to metal construction										
0040	Bevel seat, 1/2" diameter	1 Stpi	20	.400	Ea.	36	16.40		52.40	64	
0060	3/4" diameter		18	.444		37.50	18.20		55.70	69	
1000	Oil filters, 3/8" IPT., 20 gal. per hour		20	.400		23	16.40		39.40	49.50	
1800	Pump and motor sets										
1810	Light fuel and diesel oils, 100 PSI										
1820	25 GPH 1/4 HP	Q-5	6	2.667	Ea.	795	98.50		893.50	1,025	
1850	160 GPH, 1/3 HP	"	4	4	"	885	147		1,032	1,200	
2000	Remote tank gauging system, self contained										
2100	Gage and one probe	1 Stpi	2.50	3.200	Ea.	2,600	131		2,731	3,050	
2120	Gage with two probes		2	4		3,500	164		3,664	4,100	
3000	Valve, ball check, globe type, 3/8" diameter		24	.333		9.05	13.65		22.70	30.50	
3500	Fusible, 3/8" diameter		24	.333		7.20	13.65		20.85	28.50	
3600	1/2" diameter		24	.333		18.65	13.65		32.30	41	
3610	3/4" diameter		20	.400		40	16.40		56.40	68	

Important: See the Reference Section for critical supporting data - Reference Nos., Crews, & City Cost Indexes

15100 | Building Services Piping

15190 | Fuel Piping

		CREW	DAILY OUTPUT	LABOR-HOURS	UNIT	MAT.	LABOR	EQUIP.	TOTAL	TOTAL INCL O&P	
490	3620	1" diameter	1 Stpi	19	.421	Ea.	118	17.25		135.25	155	**49**
	4000	Nonfusible, 3/8" diameter	↓	24	.333	↓	14.50	13.65		28.15	36.50	
	4500	Shutoff, gate type, lever handle, spring-fusible kit										
	4520	1/4" diameter	1 Stpi	14	.571	Ea.	20	23.50		43.50	57	
	4540	3/8" diameter		12	.667		21.50	27.50		49	64.50	
	4560	1/2" diameter		10	.800		27.50	33		60.50	79.50	
	4570	3/4" diameter	↓	8	1		48	41		89	115	
	5000	Vent alarm, whistling signal					26			26	28.50	
	5500	Vent protector/breather, 1-1/4" diameter	1 Stpi	32	.250	↓	7.20	10.25		17.45	23.50	
605	0010	**METERS**										**60**
	4000	Residential										
	4010	Gas meter, residential, 3/4" pipe size	1 Plum	14	.571	Ea.	117	23.50		140.50	163	
	4020	Gas meter, residential, 1" pipe size		12	.667		166	27		193	224	
	4030	Gas meter, residential, 1-1/4" pipe size	↓	10	.800	↓	176	32.50		208.50	243	

15200 | Process Piping

15210 | Process Air/Gas Piping

		CREW	DAILY OUTPUT	LABOR-HOURS	UNIT	MAT.	LABOR	EQUIP.	TOTAL	TOTAL INCL O&P	
100	0010	**AIR COMPRESSORS**										**10**
	5250	Air, reciprocating air cooled, splash lubricated, tank mounted										
	5300	Single stage, 1 phase, 140 psi										
	5303	1/2 HP, 30 gal tank	1 Stpi	3	2.667	Ea.	925	109		1,034	1,200	
	5305	3/4 HP, 30 gal tank		2.60	3.077		955	126		1,081	1,250	
	5307	1 HP, 30 gal tank	↓	2.20	3.636		1,175	149		1,324	1,525	
	5309	2 HP, 30 gal tank	Q-5	4	4		1,550	147		1,697	1,950	
	5310	3 HP, 30 gal tank		3.60	4.444		2,125	164		2,289	2,600	
	5314	3 HP, 60 gal tank		3.50	4.571		2,450	168		2,618	2,925	
	5320	5 HP, 60 gal tank		3.20	5		2,825	184		3,009	3,375	
	5330	5 HP, 80 gal tank		3	5.333		2,975	197		3,172	3,575	
	5340	7.5 HP, 80 gal tank	↓	2.60	6.154	↓	4,550	227		4,777	5,375	
	5600	2 stage pkg., 3 phase										
	5650	6 CFM at 125 psi 1-1/2 HP, 60 gal tank	Q-5	3	5.333	Ea.	2,050	197		2,247	2,575	
	5670	10.9 CFM at 125 psi, 3 HP, 80 gal tank		1.50	10.667		2,375	395		2,770	3,200	
	5680	38.7 CFM at 125 psi, 10 HP, 120 gal tank	↓	.60	26.667		4,775	985		5,760	6,725	
	5690	105 CFM at 125 psi, 25 HP, 250 gal tank	Q-6	.60	40	↓	9,200	1,525		10,725	12,400	
	5800	With single stage pump										
	5850	8.3 CFM at 125 psi, 2 HP, 80 gal tank	Q-6	3.50	6.857	Ea.	2,575	262		2,837	3,250	
	5860	38.7 CFM at 125 psi, 10 HP, 120 gal tank	"	.90	26.667	"	5,450	1,025		6,475	7,525	
	6000	Reciprocating, 2 stage, tank mtd, 3 Ph., Cap. rated @175 PSIG										
	6050	Pressure lubcatd, hvy duty, 9.7 CFM, 3 HP, 120 Gal. tank	Q-5	1.30	12.308	Ea.	4,050	455		4,505	5,125	
	6054	5 CFM, 1-1/2 HP, 80 gal tank		2.80	5.714		2,375	211		2,586	2,950	
	6056	6.4 CFM, 2 HP, 80 gal tank		2	8		2,500	295		2,795	3,200	
	6058	8.1 CFM, 3 HP, 80 gal tank		1.70	9.412		2,625	345		2,970	3,400	
	6059	14.8 CFM, 5 HP, 80 gal tank		1	16		2,700	590		3,290	3,850	
	6060	16.5 CFM, 5 HP, 120 gal. tank		1	16		4,100	590		4,690	5,375	
	6063	13 CFM, 6 HP, 80 gal tank		.90	17.778		4,325	655		4,980	5,725	
	6066	19.8 CFM, 7.5 HP, 80 gal tank		.80	20		4,550	735		5,285	6,100	
	6070	25.8 CFM, 7-1/2 HP, 120 gal. tank	↓	.80	20	↓	6,100	735		6,835	7,825	

15210	Process Air/Gas Piping	CREW	DAILY OUTPUT	LABOR-HOURS	UNIT	2005 BARE COSTS				TOTAL INCL O&P
						MAT.	LABOR	EQUIP.	TOTAL	
100 6078	34.8 CFM, 10 HP, 80 gal. tank	Q-5	.70	22.857	Ea.	5,775	840		6,615	7,625	**100**
6080	34.8 CFM, 10 HP, 120 gal. tank	↓	.60	26.667		6,075	985		7,060	8,150
6090	53.7 CFM, 15 HP, 120 gal. tank	Q-6	.80	30		6,300	1,150		7,450	8,650
6100	76.7 CFM, 20 HP, 120 gal. tank		.70	34.286		7,600	1,300		8,900	10,400
6104	76.7 CFM, 20 HP, 240 gal tank		.68	35.294		9,325	1,350		10,675	12,300
6110	90.1 CFM, 25 HP, 120 gal. tank		.63	38.095		9,400	1,450		10,850	12,500
6120	101 CFM, 30 HP, 120 gal. tank		.57	42.105		9,425	1,600		11,025	12,800
6130	101 CFM, 30 HP, 250 gal. tank	↓	.52	46.154		11,500	1,775		13,275	15,300
6200	Oil-less, 13.6 CFM, 5 HP, 120 gal. tank	Q-5	.88	18.182		12,600	670		13,270	14,900
6210	13.6 CFM, 5 HP, 250 gal. tank		.80	20		13,500	735		14,235	16,000
6220	18.2 CFM, 7.5 HP, 120 gal. tank		.73	21.918		12,800	810		13,610	15,300
6230	18.2 CFM, 7.5 HP, 250 gal. tank		.67	23.881		13,700	880		14,580	16,400
6250	30.5 CFM, 10 HP, 120 gal. tank		.57	28.070		15,000	1,025		16,025	18,000
6260	30.5 CFM, 10 HP, 250 gal. tank	↓	.53	30.189		15,800	1,100		16,900	19,100
6270	41.3 CFM, 15 HP, 120 gal. tank	Q-6	.70	34.286		16,200	1,300		17,500	19,800
6280	41.3 CFM, 15 HP, 250 gal. tank	"	.67	35.821	↓	17,100	1,375		18,475	20,900
200 0010	**COMPRESSOR ACCESSORIES**										**200**
4000	Couplers, air line, sleeve type									
4010	Female, connection size NPT									
4020	1/4"	1 Stpi	38	.211	Ea.	4.16	8.60		12.76	17.55
4030	3/8"		36	.222		5.20	9.10		14.30	19.40
4040	1/2"		35	.229		10.90	9.35		20.25	26
4050	3/4"	↓	34	.235	↓	12.40	9.65		22.05	28
4100	Male									
4110	1/4"	1 Stpi	38	.211	Ea.	4.11	8.60		12.71	17.45
4120	3/8"		36	.222		5.20	9.10		14.30	19.40
4130	1/2"		35	.229		10.25	9.35		19.60	25.50
4140	3/4"	↓	34	.235	↓	12.10	9.65		21.75	28
4150	Coupler, combined male and female halves									
4160	1/2"	1 Stpi	17	.471	Ea.	21	19.25		40.25	52.50
4170	3/4"	"	15	.533	"	24.50	22		46.50	60
500 0010	**MEDICAL GAS SYSTEM SPECIALTIES**										**500**
0100	Vacuum system									
0110	Vacuum outlet alarm panel	1 Plum	3.20	2.500	Ea.	810	102		912	1,050
1000	Nitrogen or oxygen system									
1010	Cylinder manifold									
1020	5 cylinder	1 Plum	.80	10	Ea.	4,350	410		4,760	5,400
1026	10 cylinder	"	.40	20	"	5,275	815		6,090	7,025
3000	Outlets and valves									
3010	Recessed, wall mounted									
3012	Single outlet	1 Plum	3.20	2.500	Ea.	51	102		153	210
3100	Ceiling outlet									
3200	Zone valve with box									
3210	2"	1 Plum	3.20	2.500	Ea.	360	102		462	555
4000	Alarm panel, medical gases and vacuum									
4010	Alarm panel	1 Plum	3.20	2.500	Ea.	810	102		912	1,050
900 0010	**VACUUM PUMPS**										**900**
0100	Medical, with receiver									
0110	Duplex									
0120	20 SCFM	Q-1	1.14	14.035	Ea.	14,200	515		14,715	16,400
0130	60 SCFM, 10 HP	Q-2	1.20	20	"	25,500	765		26,265	29,300
0200	Triplex									
0220	180 SCFM	Q-2	.86	27.907	Ea.	46,400	1,075		47,475	52,500
0300	Dental oral									

Important: See the Reference Section for critical supporting data - Reference Nos., Crews, & City Cost Indexes

15200 | Process Piping

			DAILY	LABOR-			2005 BARE COSTS			TOTAL
15210	**Process Air/Gas Piping**	CREW	OUTPUT	HOURS	UNIT	MAT.	LABOR	EQUIP.	TOTAL	INCL O&P
900 0310	Duplex									**90**
0330	165 SCFM with 77 Gal separator	Q-2	1.30	18.462	Ea.	42,000	705		42,705	47,300

15220	**Process Water/Waste Piping**									
100 0010	**LABORATORY EQUIPMENT** Corrosion resistant									**1**
7000	Tanks, covers included									
7800	Polyethylene liner, fiberglass casing									
7810	Continuous service to 220°F									
7830	2 gallon, 8" x 8" x 8"	Q-1	20	.800	Ea.	256	29.50		285.50	325
7850	7 gallon, 12" x 12" x 12"		20	.800		294	29.50		323.50	370
7870	15 gallon, 24" x 12" x 12"		17	.941		465	34.50		499.50	560
8010	30 gallon, 24" x 18" x 18"		12	1.333		535	49		584	665
8070	45 gallon, 24" x 18" x 24"		10	1.600		715	59		774	880
8080	90 gallon, 36" x 24" x 24"	▼	8	2	▼	1,350	73.50		1,423.50	1,600
8150	Polyethylene, heavy duty walls									
8160	Continuous service to 180°F									
8180	5 gallon, 11" I.D. x 15" deep	Q-1	20	.800	Ea.	121	29.50		150.50	177
8210	15 gallon, 14" I.D. x 27" deep		17	.941		183	34.50		217.50	253
8230	55 gallon, 22" I.D. x 36" deep		10	1.600		375	59		434	505
8250	100 gallon 28" I.D. x 42" deep		8	2		610	73.50		683.50	785
8270	200 gallon 36" I.D. x 48" deep		6	2.667		950	98		1,048	1,200
8290	360 gallon 48" I.D. x 48" deep	▼	5	3.200	▼	1,325	118		1,443	1,625

15230	**Industrial Process Piping**									
500 0010	**PUMPS, GENERAL UTILITY** With motor									**5**
0200	Multi-stage, horizontal split, for boiler feed applications									
0300	Two stage, 3" discharge x 4" suction, 75 HP	Q-7	.30	106	Ea.	15,700	4,175		19,875	23,500
0340	Four stage, 3" discharge x 4" suction, 150 HP	"	.18	177	"	25,800	6,950		32,750	38,700
2000	Single stage									
2060	End suction, 1"D. x 2"S., 3 HP	Q-1	.50	32	Ea.	3,825	1,175		5,000	5,975
2100	1-1/2"D. x 3"S., 10 HP	"	.40	40		4,125	1,475		5,600	6,725
2140	2"D. x 3"S., 15 HP	Q-2	.60	40		4,475	1,525		6,000	7,225
2180	3"D. x 4"S., 20 HP		.50	48		4,775	1,825		6,600	8,000
2220	4"D. x 6"S., 30 HP	▼	.40	60		5,175	2,300		7,475	9,150
3000	Double suction, 2"D. x 2-1/2"S., 10 HP	Q-1	.30	53.333		5,300	1,950		7,250	8,775
3060	3"D. x 4"S., 15 HP	Q-2	.46	52.174		5,475	2,000		7,475	9,000
3100	4"D. x 5"S., 30 HP		.40	60		7,425	2,300		9,725	11,600
3140	5"D. x 6"S., 50 HP	▼	.33	72.727		7,950	2,775		10,725	12,900
3180	6"D. x 8"S., 60 HP	Q-3	.30	106		8,900	4,150		13,050	16,100
3190	75 HP, to 2500 GPM		.28	114		11,100	4,450		15,550	18,900
3220	100 HP, to 3000 GPM		.26	123		14,100	4,800		18,900	22,700
3240	150 HP, to 4000 GPM	▼	.24	133	▼	18,900	5,200		24,100	28,600
4000	Centrifugal, end suction, mounted on base									
4010	Horizontal mounted, with drip proof motor, rated @ 100' head									
4020	Vertical split case, single stage									
4040	100 GPM, 5 HP, 1-1/2" discharge	Q-1	1.70	9.412	Ea.	2,025	345		2,370	2,750
4050	200 GPM, 10 HP, 2" discharge		1.30	12.308		2,450	455		2,905	3,375
4060	250 GPM, 10 HP, 3" discharge	▼	1.28	12.500		2,500	460		2,960	3,450
4070	300 GPM, 15 HP, 2" discharge	Q-2	1.56	15.385		2,800	585		3,385	3,950
4080	500 GPM, 20 HP, 4" discharge		1.44	16.667		3,275	635		3,910	4,550
4090	750 GPM, 30 HP, 4" discharge		1.20	20		4,375	765		5,140	5,975
4100	1050 GPM, 40 HP, 5" discharge		1	24		4,725	915		5,640	6,575
4110	1500 GPM, 60 HP, 6" discharge		.60	40		6,875	1,525		8,400	9,850
4120	2000 GPM, 75 HP, 6" discharge		.50	48		7,900	1,825		9,725	11,400
4130	3000 GPM, 100 HP, 8" discharge	▼	.40	60	▼	11,900	2,300		14,200	16,500
4200	Horizontal split case, single stage									

15230 | Industrial Process Piping

		CREW	DAILY OUTPUT	LABOR-HOURS	UNIT	MAT.	LABOR	EQUIP.	TOTAL	TOTAL INCL O&P
						2005 BARE COSTS				
500 4210	100 GPM, 7.5 HP, 1-1/2″ discharge	Q-1	1.70	9.412	Ea.	3,350	345		3,695	4,225
4220	250 GPM, 15 HP, 2-1/2″ discharge	″	1.30	12.308		4,450	455		4,905	5,550
4230	500 GPM, 20 HP, 4″ discharge	Q-2	1.60	15		5,375	570		5,945	6,750
4240	750 GPM, 25 HP, 5″ discharge		1.54	15.584		5,550	595		6,145	7,000
4250	1000 GPM, 40 HP, 5″ discharge		1.20	20		6,600	765		7,365	8,400
4260	1500 GPM, 50 HP, 6″ discharge	Q-3	1.42	22.535		8,675	880		9,555	10,900
4270	2000 GPM, 75 HP, 8″ discharge		1.14	28.070		12,400	1,100		13,500	15,300
4280	3000 GPM, 100 HP, 10″ discharge		.96	33.333		13,700	1,300		15,000	17,100
4290	3500 GPM, 150 HP, 10″ discharge		.86	37.209		17,100	1,450		18,550	21,000
4300	4000 GPM, 200 HP, 10″ discharge		.66	48.485		18,900	1,900		20,800	23,600
4330	Horizontal split case, two stage, 500′ head									
4340	100 GPM, 40 HP, 1-1/2″ discharge	Q-2	1.70	14.118	Ea.	4,400	540		4,940	5,625
4350	200 GPM, 50 HP, 1-1/2″ discharge	″	1.44	16.667		4,900	635		5,535	6,325
4360	300 GPM, 75 HP, 2″ discharge	Q-3	1.57	20.382		5,200	795		5,995	6,925
4370	400 GPM, 100 HP, 3″ discharge		1.14	28.070		6,550	1,100		7,650	8,850
4380	800 GPM, 200 HP, 4″ discharge		.86	37.209		15,300	1,450		16,750	19,100
5000	Centrifugal, in-line									
5006	Vertical mount, iron body, 125 lb. flgd, 3550 RPM TEFC mtr									
5010	Single stage									
5020	50 GPM, 3 HP, 1-1/2″ discharge	Q-1	2.30	6.957	Ea.	590	256		846	1,025
5030	75 GPM, 5 HP, 1-1/2″ discharge		1.60	10		995	370		1,365	1,650
5040	100 GPM, 7.5 HP, 1-1/2″ discharge		1.30	12.308		1,075	455		1,530	1,850
5050	125 GPM, 10 HP, 1-1/2″ discharge	Q-2	1.70	14.118		1,325	540		1,865	2,275
5060	150 GPM, 15 HP, 1-1/2″ discharge		1.60	15		1,725	570		2,295	2,725
5070	200 GPM, 30 HP, 1-1/2″ discharge		1.50	16		2,800	610		3,410	4,000
5080	250 GPM, 40 HP, 2″ discharge		1.40	17.143		3,700	655		4,355	5,050
5090	300 GPM, 50 HP, 3″ discharge		1.30	18.462		4,475	705		5,180	5,975
5100	400 GPM, 75 HP, 3″ discharge	Q-3	.60	53.333		7,300	2,075		9,375	11,200
5110	600 GPM, 100 HP, 3″ discharge	″	.50	64		9,800	2,500		12,300	14,600
8000	Vertical submerged, with non-submerged motor									
8060	1″D., 3 HP	Q-1	.50	32	Ea.	5,275	1,175		6,450	7,575
8100	1-1/2″D., 10 HP	″	.30	53.333		5,525	1,950		7,475	9,025
8140	2″D., 15 HP	Q-2	.40	60		5,675	2,300		7,975	9,700
8180	3″D., 25 HP		.30	80		5,800	3,050		8,850	11,000
8220	4″D., 30 HP		.20	120		6,500	4,575		11,075	14,000
600 0010	**PUMPS MISCELLANEOUS**									
0020	Water pump, portable, gasoline powered									
0100	6000 GPH, 2″ discharge	Q-1	11	1.455	Ea.	1,000	53.50		1,053.50	1,175
0110	8000 GPH, 2″ discharge		10.50	1.524		1,175	56		1,231	1,375
0120	10,000 GPH, 2″ discharge		10	1.600		1,800	59		1,859	2,075
0500	Pump, propylene body, housing and impeller									
0510	22 GPM, 1/3 HP, 40′ HD	1 Plum	5	1.600	Ea.	470	65.50		535.50	615
0520	33 GPM, 1/2 HP, 40′ HD	Q-1	5	3.200		480	118		598	700
0530	53 GPM, 3/4 HP, 40′ HD	″	4	4		700	147		847	990
0600	Rotary pump, CI									
0610	10 GPM, 1/2 HP, 1″ discharge	Q-1	5	3.200	Ea.	965	118		1,083	1,225
0614	10 GPM, 3/4 HP, 1″ discharge		4.50	3.556		965	131		1,096	1,250
0618	10 GPM 1 HP, 1″ discharge		4	4		1,000	147		1,147	1,325
0620	25 GPM, 1 HP, 1-1/4″ discharge		3.80	4.211		1,000	155		1,155	1,325
0624	25 GPM, 1.5 HP, 1-1/4″ discharge		3.60	4.444		1,100	163		1,263	1,450
0628	25 GPM, 2 HP, 1-1/4″ discharge		3.20	5		1,100	184		1,284	1,475
0700	Fuel oil pump, 2 stage, 3450 RPM									
0710	1/4 HP, 100 PSI	Q-1	8	2	Ea.	1,625	73.50		1,698.50	1,900
0800	Hydraulic oil pump									
0810	75 HP, 3 PH, 480 V	Q-3	1.20	26.667	Ea.	5,275	1,050		6,325	7,350

Important: See the Reference Section for critical supporting data - Reference Nos., Crews, & City Cost Indexes

15230	Industrial Process Piping	CREW	DAILY OUTPUT	LABOR-HOURS	UNIT	2005 BARE COSTS				TOTAL INCL O&P
						MAT.	LABOR	EQUIP.	TOTAL	
600	1000	Turbine pump, CI									60
	1010	50 GPM, 2 HP, 3" discharge	Q-1	.80	20	Ea.	1,100	735		1,835	2,325
	1020	100 GPM, 3 HP, 4" discharge	Q-2	.96	25		1,750	955		2,705	3,350
	1030	250 GPM, 15 HP, 6" discharge	"	.94	25.532		3,500	975		4,475	5,325
	1040	500 GPM, 25 HP, 6" discharge	Q-3	1.22	26.230		5,850	1,025		6,875	7,950
	1050	1000 GPM, 50 HP, 8" discharge		1.14	28.070		9,350	1,100		10,450	12,000
	1060	2000 GPM, 100 HP, 10" discharge		1	32		10,500	1,250		11,750	13,500
	1070	3000 GPM, 150 HP, 10" discharge		.80	40		11,000	1,550		12,550	14,500
	1080	4000 GPM, 200 HP, 12" discharge		.70	45.714		17,500	1,775		19,275	22,000
	1090	6000 GPM, 300 HP, 14" discharge		.60	53.333		17,900	2,075		19,975	22,800
	1100	10,000 GPM, 300 HP, 18" discharge		.58	55.172		19,500	2,150		21,650	24,600

15410	Plumbing Fixtures	CREW	DAILY OUTPUT	LABOR-HOURS	UNIT	2005 BARE COSTS				TOTAL INCL O&P
						MAT.	LABOR	EQUIP.	TOTAL	
040	0010	FIXTURES Includes trim fittings unless otherwise noted R15100 -410									04
	0080	For rough-in, supply, waste, and vent, see add for each type									
	0120	For electric water coolers, see division 15413									
	0160	For color, unless otherwise noted, add				Ea.	20%				
200	0010	CARRIERS/SUPPORTS For plumbing fixtures									20
	0500	Drinking fountain, wall mounted									
	0600	Plate type with studs, top back plate	1 Plum	7	1.143	Ea.	35	46.50		81.50	109
	0700	Top front and back plate		7	1.143		43.50	46.50		90	118
	0800	Top & bottom, front & back plates, w/bearing jacks		7	1.143		63	46.50		109.50	140
	3000	Lavatory, concealed arm									
	3050	Floor mounted, single									
	3100	High back fixture	1 Plum	6	1.333	Ea.	258	54.50		312.50	365
	3200	Flat slab fixture		6	1.333		300	54.50		354.50	410
	3220	Paraplegic		6	1.333		181	54.50		235.50	281
	3250	Floor mounted, back to back									
	3300	High back fixtures	1 Plum	5	1.600	Ea.	370	65.50		435.50	505
	3400	Flat slab fixtures		5	1.600		455	65.50		520.50	600
	3430	Paraplegic		5	1.600		241	65.50		306.50	365
	3500	Wall mounted, in stud or masonry									
	3600	High back fixture	1 Plum	6	1.333	Ea.	153	54.50		207.50	250
	3700	Flat slab fixture	"	6	1.333	"	196	54.50		250.50	298
	4000	Exposed arm type, floor mounted									
	4100	Single high back or flat slab fixture	1 Plum	6	1.333	Ea.	370	54.50		424.50	485
	4200	Back to back, high back or flat slab fixtures		5	1.600		580	65.50		645.50	735
	4300	Wall mounted, High back or flat slab lavatory		6	1.333		255	54.50		309.50	365
	4600	Sink, floor mounted									
	4650	Exposed arm system									
	4700	Single heavy fixture	1 Plum	5	1.600	Ea.	330	65.50		395.50	460
	4750	Single heavy sink with slab		5	1.600		266	65.50		331.50	390
	4800	Back to back, standard fixtures		5	1.600		241	65.50		306.50	365
	4850	Back to back, heavy fixtures		5	1.600		291	65.50		356.50	420
	4900	Back to back, heavy sink with slab		5	1.600		291	65.50		356.50	420
	4950	Exposed offset arm system									
	5000	Single heavy deep fixture	1 Plum	5	1.600	Ea.	410	65.50		475.50	550

15410	Plumbing Fixtures	CREW	DAILY OUTPUT	LABOR-HOURS	UNIT	2005 BARE COSTS				TOTAL INCL O&P
						MAT.	LABOR	EQUIP.	TOTAL	
5100	Plate type system									
5200	With bearing jacks, single fixture	1 Plum	5	1.600	Ea.	360	65.50		425.50	495
5300	With exposed arms, single heavy fixture		5	1.600		460	65.50		525.50	605
5400	Wall mounted, exposed arms, single heavy fixture		5	1.600		167	65.50		232.50	282
6000	Urinal, floor mounted, 2" or 3" coupling, blowout type		6	1.333		275	54.50		329.50	380
6100	With fixture or hanger bolts, blowout or washout		6	1.333		196	54.50		250.50	298
6200	With bearing plate		6	1.333		196	54.50		250.50	298
6300	Wall mounted, plate type system		6	1.333		128	54.50		182.50	223
6980	Water closet, siphon jet									
7000	Horizontal, adjustable, caulk									
7040	Single, 4" pipe size	1 Plum	5.33	1.501	Ea.	272	61.50		333.50	390
7050	4" pipe size, paraplegic		5.33	1.501		272	61.50		333.50	390
7060	5" pipe size		5.33	1.501		360	61.50		421.50	485
7100	Double, 4" pipe size		5	1.600		505	65.50		570.50	655
7110	4" pipe size, paraplegic		5	1.600		505	65.50		570.50	655
7120	5" pipe size		5	1.600		630	65.50		695.50	795
7160	Horizontal, adjustable, extended, caulk									
7180	Single, 4" pipe size	1 Plum	5.33	1.501	Ea.	340	61.50		401.50	465
7200	5" pipe size		5.33	1.501		485	61.50		546.50	625
7240	Double, 4" pipe size		5	1.600		565	65.50		630.50	720
7260	5" pipe size		5	1.600		805	65.50		870.50	990
7400	Vertical, adjustable, caulk or thread									
7440	Single, 4" pipe size	1 Plum	5.33	1.501	Ea.	272	61.50		333.50	390
7460	5" pipe size		5.33	1.501		360	61.50		421.50	485
7480	6" pipe size		5	1.600		420	65.50		485.50	565
7520	Double, 4" pipe size		5	1.600		505	65.50		570.50	655
7540	5" pipe size		5	1.600		630	65.50		695.50	795
7560	6" pipe size		4	2		695	81.50		776.50	890
7600	Vertical, adjustable, extended, caulk									
7620	Single, 4" pipe size	1 Plum	5.33	1.501	Ea.	272	61.50		333.50	390
7640	5" pipe size		5.33	1.501		360	61.50		421.50	485
7680	6" pipe size		5	1.600		450	65.50		515.50	600
7720	Double, 4" pipe size		5	1.600		505	65.50		570.50	655
7740	5" pipe size		5	1.600		630	65.50		695.50	795
7760	6" pipe size		4	2		675	81.50		756.50	870
7780	Water closet, blow out									
7800	Vertical offset, caulk or thread									
7820	Single, 4" pipe size	1 Plum	5.33	1.501	Ea.	370	61.50		431.50	500
7840	Double, 4" pipe size	"	5	1.600	"	635	65.50		700.50	800
7880	Vertical offset, extended, caulk									
7900	Single, 4" pipe size	1 Plum	5.33	1.501	Ea.	465	61.50		526.50	605
7920	Double, 4" pipe size	"	5	1.600	"	730	65.50		795.50	905
7960	Vertical, for floor mounted back-outlet									
7980	Single, 4" thread, 2" vent	1 Plum	5.33	1.501	Ea.	249	61.50		310.50	365
8000	Double, 4" thread, 2" vent	"	6	1.333	"	730	54.50		784.50	885
8040	Vertical, for floor mounted back-outlet, extended									
8060	Single, 4" caulk, 2" vent	1 Plum	6	1.333	Ea.	249	54.50		303.50	355
8080	Double, 4" caulk, 2" vent	"	6	1.333	"	730	54.50		784.50	885
8200	Water closet, residential									
8220	Vertical centerline, floor mount									
8240	Single, 3" caulk, 2" or 3" vent	1 Plum	6	1.333	Ea.	297	54.50		351.50	405
8260	4" caulk, 2" or 4" vent		6	1.333		385	54.50		439.50	500
8280	3" copper sweat, 3" vent		6	1.333		267	54.50		321.50	375
8300	4" copper sweat, 4" vent		6	1.333		325	54.50		379.50	435
8400	Vertical offset, floor mount									
8420	Single, 3" or 4" caulk, vent	1 Plum	4	2	Ea.	370	81.50		451.50	535

Important: See the Reference Section for critical supporting data - Reference Nos., Crews, & City Cost Indexes

			DAILY	LABOR-		2005 BARE COSTS				TOTAL	
15410	**Plumbing Fixtures**	CREW	OUTPUT	HOURS	UNIT	MAT.	LABOR	EQUIP.	TOTAL	INCL O&P	
200	8440	3" or 4" copper sweat, vent	1 Plum	5	1.600	Ea.	370	65.50		435.50	510	20
	8460	Double, 3" or 4" caulk, vent		4	2		635	81.50		716.50	825	
	8480	3" or 4" copper sweat, vent	↓	5	1.600	↓	635	65.50		700.50	800	
	9000	Water cooler (electric), floor mounted										
	9100	Plate type with bearing plate, single	1 Plum	6	1.333	Ea.	143	54.50		197.50	239	
	9120	Plate type with bearing plate, double		4	2		330	81.50		411.50	490	
	9140	Plate type with bearing plate, back to back	↓	4	2	↓	200	81.50		281.50	340	
300	0010	**FAUCETS/FITTINGS**										30
	0150	Bath, faucets, diverter spout combination, sweat	1 Plum	8	1	Ea.	69.50	41		110.50	138	
	0200	For integral stops, IPS unions, add					73			73	80.50	
	0300	Three valve combinations, spout, head, arm, flange, sweat	1 Plum	6	1.333	↓	69.50	54.50		124	159	
	0400	For integral stops, IPS unions, add				Pr.	86			86	94.50	
	0420	Bath, press-bal mix valve w/diverter, spout, shower hd, arm/flange	1 Plum	8	1	Ea.	114	41		155	187	
	0500	Drain, central lift, 1-1/2" IPS male		20	.400		38	16.35		54.35	66.50	
	0600	Trip lever, 1-1/2" IPS male		20	.400		38.50	16.35		54.85	67	
	0700	Pop up, 1-1/2" IPS male		18	.444		45.50	18.15		63.65	77.50	
	0800	Chain and stopper, 1-1/2" IPS male	↓	24	.333	↓	30.50	13.60		44.10	54	
	0810	Bidet										
	0812	Fitting, over the rim, swivel spray/pop-up drain	1 Plum	8	1	Ea.	138	41		179	214	
	0840	Flush valves, with vacuum breaker										
	0850	Water closet										
	0860	Exposed, rear spud	1 Plum	8	1	Ea.	112	41		153	185	
	0870	Top spud		8	1		103	41		144	175	
	0880	Concealed, rear spud		8	1		169	41		210	248	
	0890	Top spud		8	1		121	41		162	195	
	0900	Wall hung	↓	8	1	↓	130	41		171	205	
	0920	Urinal										
	0930	Exposed, stall	1 Plum	8	1	Ea.	103	41		144	175	
	0940	Wall, (washout)		8	1		103	41		144	175	
	0950	Pedestal, top spud		8	1		105	41		146	177	
	0960	Concealed, stall		8	1		118	41		159	192	
	0970	Wall (washout) ♿	↓	8	1	↓	118	41		159	192	
	0971	Automatic flush sensor and operator for										
	0972	urinals or water closets	1 Plum	5.33	1.501	Ea.	340	61.50		401.50	465	
	1000	Kitchen sink faucets, top mount, cast spout		10	.800		51.50	32.50		84	106	
	1100	For spray, add		24	.333		11.05	13.60		24.65	32.50	
	1110	For basket strainer w/tail piece, add		24	.333		12.15	13.60		25.75	34	
	1200	Wall type, swing tube spout	↓	10	.800		64	32.50		96.50	120	
	1240	For soap dish, add					4.60			4.60	5.05	
	1250	For basket strainer w/tail piece, add					12.15			12.15	13.35	
	2000	Laundry faucets, shelf type, IPS or copper unions	1 Plum	12	.667		42.50	27		69.50	87.50	
	2100	Lavatory faucet, centerset, without drain		10	.800		37.50	32.50		70	90	
	2200	With pop-up drain	↓	6.66	1.201		52	49		101	132	
	2250	For acrylic handles, add					4.56			4.56	5	
	2400	Concealed, 12" centers	1 Plum	10	.800		81.50	32.50		114	139	
	2450	With pop-up drain		6.66	1.201		96.50	49		145.50	180	
	2600	Shelfback, 4" to 6" centers, 17 Ga. tailpiece		10	.800		66	32.50		98.50	122	
	2650	With pop-up drain		6.66	1.201		81	49		130	163	
	2700	Shampoo faucet with supply tube		24	.333		40.50	13.60		54.10	65.50	
	2800	Self-closing, center set		10	.800		112	32.50		144.50	172	
	2810	Automatic sensor and operator, with faucet head ♿		6.15	1.301		300	53		353	410	
	2850	Medical, bedpan cleanser, with pedal valve,		12	.667		470	27		497	555	
	2860	With screwdriver stop valve		12	.667		244	27		271	310	
	2870	With self-closing spray valve ♿		12	.667		182	27		209	241	
	2900	Faucet, gooseneck spout, wrist handles, grid drain	↓	10	.800	↓	109	32.50		141.50	169	

	15410	Plumbing Fixtures	CREW	DAILY OUTPUT	LABOR-HOURS	UNIT	2005 BARE COSTS				TOTAL INCL O&P	
							MAT.	LABOR	EQUIP.	TOTAL		
300	2940	Mixing valve, knee action, screwdriver stops	1 Plum	4	2	Ea.	365	81.50		446.50	525	**300**
	3000	Service sink faucet, cast spout, pail hook, hose end		14	.571		72	23.50		95.50	114	
	4000	Shower by-pass valve with union		18	.444		50.50	18.15		68.65	83	
	4100	Shower arm with flange and head		22	.364		76	14.85		90.85	106	
	4140	Shower, hand held, pin mount, massage action, chrome		22	.364		47.50	14.85		62.35	75	
	4142	Polished brass		22	.364		91	14.85		105.85	123	
	4144	Shower, hand held, wall mtd, adj. spray, 2 wall mounts, chrome		20	.400		96	16.35		112.35	131	
	4146	Polished brass		20	.400		176	16.35		192.35	219	
	4148	Shower, hand held head, bar mounted 24", adj. spray, chrome		20	.400		119	16.35		135.35	156	
	4150	Polished brass		20	.400		176	16.35		192.35	219	
	4200	Shower thermostatic mixing valve, concealed	↓	8	1		242	41		283	330	
	4300	For inlet strainer, check, and stops, add					32			32	35.50	
	5000	Sillcock, compact, brass, IPS or copper to hose	1 Plum	24	.333	↓	4.74	13.60		18.34	25.50	
	6000	Stop and waste valves, bronze										
	6100	Angle, solder end 1/2"	1 Plum	24	.333	Ea.	4.18	13.60		17.78	25	
	6110	3/4"		20	.400		4.59	16.35		20.94	29.50	
	6300	Straightway, solder end 3/8"		24	.333		3.59	13.60		17.19	24.50	
	6310	1/2"		24	.333		2.32	13.60		15.92	23	
	6320	3/4"		20	.400		2.68	16.35		19.03	27.50	
	6330	1"		19	.421		4.04	17.20		21.24	30.50	
	6400	Straightway, threaded 3/8"		24	.333		3.30	13.60		16.90	24	
	6410	1/2"		24	.333		3.13	13.60		16.73	24	
	6420	3/4"		20	.400		3.73	16.35		20.08	28.50	
	6430	1"		19	.421		7.55	17.20		24.75	34.50	
	7800	Water closet, wax gasket		96	.083		.99	3.40		4.39	6.20	
	7820	Gasket toilet tank to bowl	↓	32	.250	↓	1.24	10.20		11.44	16.70	
	8000	Water supply stops, polished chrome plate										
	8200	Angle, 3/8"	1 Plum	24	.333	Ea.	4.73	13.60		18.33	25.50	
	8300	1/2"		22	.364		4.73	14.85		19.58	27.50	
	8400	Straight, 3/8"		26	.308		7.05	12.55		19.60	26.50	
	8500	1/2"		24	.333		7.05	13.60		20.65	28.50	
	8600	Water closet, angle, w/flex riser, 3/8"	↓	24	.333	↓	11.05	13.60		24.65	32.50	
	9100	Miscellaneous										
	9720	Teflon tape, 1/2" x 520" roll				Ea.	.52			.52	.57	
800	0010	**TOILET SEATS**										**800**
	0100	Molded composition, white										
	0150	Industrial, w/o cover, open front, regular bowl	1 Plum	24	.333	Ea.	14.85	13.60		28.45	37	
	0200	With self-sustaining hinge		24	.333		16.20	13.60		29.80	38.50	
	0220	With self-sustaining check hinge		24	.333		16.20	13.60		29.80	38.50	
	0240	Extra heavy, with check hinge	↓	24	.333	↓	15	13.60		28.60	37	
	0260	Elongated bowl, same price										
	0300	Junior size, w/o cover, open front	1 Plum	24	.333	Ea.	31	13.60		44.60	54.50	
	0320	Regular primary bowl, open front		24	.333		31	13.60		44.60	54.50	
	0340	Regular baby bowl, open front, check hinge		24	.333		29	13.60		42.60	52.50	
	0380	Open back & front, w/o cover, reg. or elongated bowl	↓	24	.333	↓	25.50	13.60		39.10	49	
	0400	Residential										
	0420	Regular bowl, w/cover, closed front	1 Plum	24	.333	Ea.	20.50	13.60		34.10	43.50	
	0440	Open front	"	24	.333	"	28.50	13.60		42.10	52	
	0460	Elongated bowl, add					25%					
	0500	Self-raising hinge, w/o cover, open front										
	0520	Regular bowl	1 Plum	24	.333	Ea.	93	13.60		106.60	124	
	0540	Elongated bowl	"	24	.333	"	80	13.60		93.60	109	
	0700	Molded wood, white, with cover										
	0720	Closed front, regular bowl, square back	1 Plum	24	.333	Ea.	9.85	13.60		23.45	31.50	
	0740	Extended back		24	.333		13.90	13.60		27.50	36	
	0780	Elongated bowl, square back	↓	24	.333	↓	12.70	13.60		26.30	34.50	

Important: See the Reference Section for critical supporting data - Reference Nos., Crews, & City Cost Indexes

15410		Plumbing Fixtures	CREW	DAILY OUTPUT	LABOR-HOURS	UNIT	MAT.	LABOR	EQUIP.	TOTAL	TOTAL INCL O&P	
800	0800	Open front	1 Plum	24	.333	Ea.	13.90	13.60		27.50	36	800
	0850	Decorator styles										
	0870	Cane top	1 Plum	24	.333	Ea.	23	13.60		36.60	46	
	0890	Vinyl top, patterned		24	.333		14.70	13.60		28.30	36.50	
	0900	Vinyl padded, plain colors, regular bowl		24	.333		14.70	13.60		28.30	36.50	
	0930	Elongated bowl		24	.333		19.80	13.60		33.40	42.50	
	1000	Solid plastic, white										
	1030	Industrial, w/o cover, open front, regular bowl	1 Plum	24	.333	Ea.	28.50	13.60		42.10	52	
	1080	Extra heavy, concealed check hinge		24	.333		15.10	13.60		28.70	37	
	1100	Self-sustaining hinge		24	.333		17.30	13.60		30.90	39.50	
	1150	Elongated bowl		24	.333		24	13.60		37.60	47	
	1170	Concealed check		24	.333		15.20	13.60		28.80	37	
	1190	Self-sustaining hinge, concealed check		24	.333		40.50	13.60		54.10	65	
	1220	Residential, with cover, closed front, regular bowl		24	.333		24	13.60		37.60	47	
	1240	Elongated bowl		24	.333		32	13.60		45.60	56	
	1260	Open front, regular bowl		24	.333		30.50	13.60		44.10	54.50	
	1280	Elongated bowl		24	.333		38.50	13.60		52.10	62.50	
900	0010	**WASHER DRYER ACCESSORIES**										900
	1020	Valves ball type single lever										
	1030	1/2" diam., IPS	1 Plum	21	.381	Ea.	27.50	15.55		43.05	53.50	
	1040	1/2" diam., solder	"	21	.381	"	27.50	15.55		43.05	53.50	
	1050	Recessed box, 16 ga., two hose valves and drain										
	1060	1/2" size, 1-1/2" drain	1 Plum	18	.444	Ea.	40	18.15		58.15	71.50	
	1070	1/2" size, 2" drain	"	17	.471	"	65	19.20		84.20	101	
	1080	With grounding electric receptacle										
	1090	1/2" size, 1-1/2" drain	1 Plum	18	.444	Ea.	75	18.15		93.15	110	
	1100	1/2" size, 2" drain	"	17	.471	"	54	19.20		73.20	88	
	1110	With grounding and dryer receptacle										
	1120	1/2" size, 1-1/2" drain	1 Plum	18	.444	Ea.	87	18.15		105.15	123	
	1130	1/2" size, 2" drain	"	17	.471	"	56.50	19.20		75.70	91.50	
	1140	Recessed box 16 ga., ball valves with single lever and drain										
	1150	1/2" size, 1-1/2" drain	1 Plum	19	.421	Ea.	102	17.20		119.20	139	
	1160	1/2" size, 2" drain	"	18	.444	"	103	18.15		121.15	141	
	1170	With grounding electric receptacle										
	1180	1/2" size, 1-1/2" drain	1 Plum	19	.421	Ea.	105	17.20		122.20	142	
	1190	1/2" size, 2" drain	"	18	.444	"	103	18.15		121.15	141	
	1200	With grounding and dryer receptacles										
	1210	1/2" size, 1-1/2" drain	1 Plum	19	.421	Ea.	118	17.20		135.20	155	
	1220	1/2" size, 2" drain	"	18	.444		123	18.15		141.15	164	
	1230	For duplex grounding receptacle add					22.50			22.50	25	
	1300	Recessed box, 20 ga., two hose valves and drain (economy type)										
	1310	1/2" size, 1-1/2" drain	1 Plum	19	.421	Ea.	50	17.20		67.20	81	
	1320	1/2" size, 2" drain		18	.444		50.50	18.15		68.65	83	
	1330	Box with drain only		24	.333		27	13.60		40.60	50	
	1340	1/2" size, 1-1/2" ABS/PVC drain		19	.421		20.50	17.20		37.70	48.50	
	1350	1/2" size, 2" ABS/PVC drain		18	.444		15.85	18.15		34	45	
	1352	Box with drain only		24	.333		46.50	13.60		60.10	71.50	
	1360	1/2" size, 2" drain ABS/PVC, 15 A receptacle		24	.333		54	13.60		67.60	79.50	
	1400	Wall mounted										
	1410	1/2" size, 1-1/2" plastic drain	1 Plum	19	.421	Ea.	18.60	17.20		35.80	46.50	
	1420	1/2" size, 2" plastic drain	"	18	.444	"	15.35	18.15		33.50	44.50	
	1500	Dryer vent kit										
	1510	8' flex duct, clamps and outside hood	1 Plum	20	.400	Ea.	6.70	16.35		23.05	32	
	1980	Rough-in, supply, waste, and vent for washer boxes	"	3.46	2.310	"	102	94.50		196.50	254	

15411	Commercial/Indust Fixtures	CREW	DAILY OUTPUT	LABOR-HOURS	UNIT	2005 BARE COSTS				TOTAL INCL O&P
						MAT.	LABOR	EQUIP.	TOTAL	
400	**0010**	**HOT WATER DISPENSERS**									**400**
0160	Commercial, 100 cup, 11.3 amp	1 Plum	14	.571	Ea.	291	23.50		314.50	355
3180	Household, 60 cup	"	14	.571	"	156	23.50		179.50	207
500	**0010**	**HYDRANTS**									**500**
0050	Wall type, moderate climate, bronze, encased									
0200	3/4" IPS connection	1 Plum	16	.500	Ea.	300	20.50		320.50	360
0300	1" IPS connection	"	14	.571		299	23.50		322.50	365
0500	Anti-siphon type					259			259	285
1000	Non-freeze, bronze, exposed									
1100	3/4" IPS connection, 4" to 9" thick wall	1 Plum	14	.571	Ea.	203	23.50		226.50	258
1120	10" to 14" thick wall		12	.667		220	27		247	283
1140	15" to 19" thick wall		12	.667		245	27		272	310
1160	20" to 24" thick wall		10	.800		265	32.50		297.50	340
1200	For 1" IPS connection, add					15%	10%			
1240	For 3/4" adapter type vacuum breaker, add				Ea.	25			25	27.50
1280	For anti-siphon type, add				"	54			54	59.50
2000	Non-freeze bronze, encased, anti-siphon type									
2100	3/4" IPS connection, 5" to 9" thick wall	1 Plum	14	.571	Ea.	271	23.50		294.50	335
2120	10" to 14" thick wall		12	.667		365	27		392	445
2140	15" to 19" thick wall		12	.667		395	27		422	470
2160	20" to 24" thick wall		10	.800		410	32.50		442.50	500
2200	For 1" IPS connection, add					10%	10%			
3000	Ground box type, bronze frame, 3/4" IPS connection									
3080	Non-freeze, all bronze, polished face, set flush									
3100	2 feet depth of bury	1 Plum	8	1	Ea.	375	41		416	470
3120	3 feet depth of bury		8	1		400	41		441	500
3140	4 feet depth of bury		8	1		430	41		471	530
3160	5 feet depth of bury		7	1.143		460	46.50		506.50	575
3180	6 feet depth of bury		7	1.143		490	46.50		536.50	605
3200	7 feet depth of bury		6	1.333		515	54.50		569.50	645
3220	8 feet depth of bury		5	1.600		545	65.50		610.50	700
3240	9 feet depth of bury		4	2		575	81.50		656.50	755
3260	10 feet depth of bury		4	2		600	81.50		681.50	785
3400	For 1" IPS connection, add					15%	10%			
3450	For 1-1/4" IPS connection, add					325%	14%			
3500	For 1-1/2" connection, add					370%	18%			
3550	For 2" connection, add					445%	24%			
3600	For tapped drain port in box, add					34.50			34.50	37.50
4000	Non-freeze, CI body, bronze frame & scoriated cover									
4010	with hose storage									
4100	2 feet depth of bury	1 Plum	7	1.143	Ea.	705	46.50		751.50	845
4120	3 feet depth of bury		7	1.143		735	46.50		781.50	880
4140	4 feet depth of bury		7	1.143		760	46.50		806.50	905
4160	5 feet depth of bury		6.50	1.231		775	50.50		825.50	925
4180	6 feet depth of bury		6	1.333		790	54.50		844.50	945
4200	7 feet depth of bury		5.50	1.455		815	59.50		874.50	985
4220	8 feet depth of bury		5	1.600		845	65.50		910.50	1,025
4240	9 feet depth of bury		4.50	1.778		870	72.50		942.50	1,075
4260	10 feet depth of bury		4	2		900	81.50		981.50	1,125
4280	For 1" IPS connection, add					155			155	171
4300	For tapped drain port in box, add					34.50			34.50	37.50
5000	Moderate climate, all bronze, polished face									
5020	and scoriated cover, set flush									
5100	3/4" IPS connection	1 Plum	16	.500	Ea.	265	20.50		285.50	325
5120	1" IPS connection	"	14	.571		265	23.50		288.50	325

Important: See the Reference Section for critical supporting data - Reference Nos., Crews, & City Cost Indexes

15411 | Commercial/Indust Fixtures

		CREW	DAILY OUTPUT	LABOR-HOURS	UNIT	2005 BARE COSTS				TOTAL INCL O&P	
						MAT.	LABOR	EQUIP.	TOTAL		
500	5200	For tapped drain port in box, add				Ea.	34.50			34.50	37.50	**500**
	6000	Ground post type, all non-freeze, all bronze, aluminum casing										
	6010	guard, exposed head, 3/4" IPS connection										
	6100	2 feet depth of bury	1 Plum	8	1	Ea.	375	41		416	470	
	6120	3 feet depth of bury		8	1		405	41		446	505	
	6140	4 feet depth of bury		8	1		435	41		476	540	
	6160	5 feet depth of bury		7	1.143		465	46.50		511.50	585	
	6180	6 feet depth of bury		7	1.143		500	46.50		546.50	620	
	6200	7 feet depth of bury		6	1.333		535	54.50		589.50	665	
	6220	8 feet depth of bury		5	1.600		565	65.50		630.50	725	
	6240	9 feet depth of bury		4	2		600	81.50		681.50	785	
	6260	10 feet depth of bury		4	2		635	81.50		716.50	825	
	6300	For 1" IPS connection, add					40%	10%				
	6350	For 1-1/4" IPS connection, add					140%	14%				
	6400	For 1-1/2" IPS connection, add					225%	18%				
	6450	For 2" IPS connection, add					315%	24%				
700	0010	**URINALS** D2010 210										**700**
	0100	For automatic flush see 15410-300-0972										
	3000	Wall hung, vitreous china, with hanger & self-closing valve										
	3100	Siphon jet type	Q-1	3	5.333	Ea.	298	196		494	625	
	3120	Blowout type		3	5.333		340	196		536	670	
	3300	Rough-in, supply, waste & vent		2.83	5.654		110	208		318	435	
	5000	Stall type, vitreous china, includes valve		2.50	6.400		510	235		745	915	
	5100	3" seam cover, add		12	1.333		146	49		195	235	
	5200	6" seam cover, add R15100 -410		12	1.333		161	49		210	251	
	6980	Rough-in, supply, waste and vent		1.99	8.040		152	296		448	615	
	8000	Waterless (no flush) urinal										
	8010	Wall hung										
	8020	Standard unit	Q-1	21.30	.751	Ea.	400	27.50		427.50	480	
	8030	ADA compliant unit	"	21.30	.751		385	27.50		412.50	460	
	8070	For solid color, add					60			60	66	
	8080	For 2" brass flange, (new const.), add	Q-1	96	.167		19.20	6.15		25.35	30	
	8090	Rough-in, supply, waste & vent	"	3.86	4.145		96	152		248	335	
	8100	Trap liquid										
	8110	1 quart				Ea.	12.80			12.80	14.10	
	8120	1 gallon				"	51			51	56.50	
840	0010	**WASH FOUNTAINS** D2010 610										**84**
	1900	Group, foot control										
	2000	Precast terrazzo, circular, 36" diam., 5 or 6 persons	Q-2	3	8	Ea.	2,775	305		3,080	3,500	
	2100	54" diameter for 8 or 10 persons		2.50	9.600		3,350	365		3,715	4,250	
	2400	Semi-circular, 36" diam. for 3 persons		3	8		2,475	305		2,780	3,175	
	2500	54" diam. for 4 or 5 persons		2.50	9.600		3,050	365		3,415	3,925	
	2700	Quarter circle (corner), 54" for 3 persons		3.50	6.857		3,050	261		3,311	3,775	
	3000	Stainless steel, circular, 36" diameter		3.50	6.857		3,175	261		3,436	3,900	
	3100	54" diameter		2.80	8.571		4,175	325		4,500	5,100	
	3400	Semi-circular, 36" diameter		3.50	6.857		2,750	261		3,011	3,425	
	3500	54" diameter		2.80	8.571		3,600	325		3,925	4,475	
	5000	Thermoplastic, pre-assembled, circular, 36" diameter		6	4		2,400	153		2,553	2,875	
	5100	54" diameter		4	6		2,825	229		3,054	3,475	
	5400	Semi-circular, 36" diameter		6	4		2,275	153		2,428	2,725	
	5600	54" diameter		4	6		2,700	229		2,929	3,300	
	5610	Group, infrared control, barrier free										
	5614	Precast terrazzo										
	5620	Semi-circular 36" diam. for 3 persons	Q-2	3	8	Ea.	3,650	305		3,955	4,450	
	5630	46" diam. for 4 persons		2.80	8.571		4,075	325		4,400	5,000	
	5640	Circular, 54" diam. for 8 persons, button control		2.50	9.600		5,975	365		6,340	7,125	

15411 | Commercial/Indust Fixtures

		CREW	DAILY OUTPUT	LABOR-HOURS	UNIT	2005 BARE COSTS MAT.	LABOR	EQUIP.	TOTAL	TOTAL INCL O&P	
840	5700	Rough-in, supply, waste and vent for above wash fountains	Q-1	1.82	8.791	Ea.	148	325		473	650	**840**
	6200	Duo for small washrooms, stainless steel		2	8		1,875	294		2,169	2,500	
	6400	Bowl with backsplash		2	8		1,575	294		1,869	2,200	
	6500	Rough-in, supply, waste & vent for duo fountains		2.02	7.921		83.50	291		374.50	530	
900	0010	**WATER CHILLERS REMOTE** 80° F inlet										**900**
	0100	Air cooled, 50° F outlet, 115V, 4.1 GPH	1 Plum	6	1.333	Ea.	555	54.50		609.50	690	
	0200	5.7 GPH		5.50	1.455		780	59.50		839.50	945	
	0300	8.0 GPH		5	1.600		575	65.50		640.50	735	
	0400	10.0 GPH		4.50	1.778		720	72.50		792.50	905	
	0500	13.4 GPH		4	2		900	81.50		981.50	1,125	
	0700	29 GPH	Q-1	5	3.200		1,350	118		1,468	1,675	
	1000	230V, 32 GPH	"	5	3.200		1,500	118		1,618	1,825	
	1200	For remote grill, add	1 Plum	16	.500		144	20.50		164.50	190	

15412 | Drinking Fountains

		CREW	DAILY OUTPUT	LABOR-HOURS	UNIT	2005 BARE COSTS MAT.	LABOR	EQUIP.	TOTAL	TOTAL INCL O&P	
200	0010	**DRINKING FOUNTAIN** For connection to cold water supply R15100-410										**200**
	0800	For remote water chiller, see division 15411										
	1000	Wall mounted, non-recessed R15100-430										
	1200	Aluminum,										
	1280	Dual bubbler type	1 Plum	3.20	2.500	Ea.	995	102		1,097	1,250	
	1400	Bronze, with no back		4	2		1,000	81.50		1,081.50	1,225	
	1600	Cast iron, enameled, low back, single bubbler		4	2		450	81.50		531.50	620	
	1640	Dual bubbler type		3.20	2.500		880	102		982	1,125	
	1680	Triple bubbler type		3.20	2.500		1,125	102		1,227	1,400	
	1800	Cast aluminum, enameled, for correctional institutions		4	2		625	81.50		706.50	810	
	2000	Fiberglass, 12" back, single bubbler unit		4	2		555	81.50		636.50	735	
	2040	Dual bubbler		3.20	2.500		810	102		912	1,050	
	2080	Triple bubbler		3.20	2.500		1,050	102		1,152	1,300	
	2200	Polymarble, no back, single bubbler		4	2		500	81.50		581.50	675	
	2240	Dual bubbler		3.20	2.500		855	102		957	1,100	
	2280	Triple bubbler		3.20	2.500		1,125	102		1,227	1,400	
	2400	Precast stone, no back		4	2		505	81.50		586.50	680	
	2700	Stainless steel, single bubbler, no back		4	2		825	81.50		906.50	1,025	
	2740	With back		4	2		980	81.50		1,061.50	1,200	
	2780	Dual handle & wheelchair projection type ♿		4	2		505	81.50		586.50	680	
	2820	Dual level for handicapped type ♿		3.20	2.500		1,025	102		1,127	1,275	
	2840	Vandal resistant type		4	2		430	81.50		511.50	600	
	3300	Vitreous china										
	3340	7" back	1 Plum	4	2	Ea.	420	81.50		501.50	585	
	3940	For vandal-resistant bottom plate, add					52.50			52.50	57.50	
	3960	For freeze-proof valve system, add	1 Plum	2	4		410	163		573	695	
	3980	For rough-in, supply and waste, add	"	2.21	3.620		65.50	148		213.50	294	
	4000	Wall mounted, semi-recessed										
	4200	Poly-marble, single bubbler	1 Plum	4	2	Ea.	680	81.50		761.50	870	
	4600	Stainless steel, satin finish, single bubbler		4	2		785	81.50		866.50	985	
	4900	Vitreous china, single bubbler		4	2		565	81.50		646.50	745	
	5980	For rough-in, supply and waste, add		1.83	4.372		65.50	179		244.50	340	
	6000	Wall mounted, fully recessed										
	6400	Poly-marble, single bubbler	1 Plum	4	2	Ea.	765	81.50		846.50	965	
	6440	For water glass filler, add					129			129	142	
	6800	Stainless steel, single bubbler	1 Plum	4	2		700	81.50		781.50	890	
	6900	Fountain and cuspidor combination		2	4		1,450	163		1,613	1,825	
	7560	For freeze-proof valve system, add		2	4		500	163		663	800	
	7580	For rough-in, supply and waste, add		1.83	4.372		65.50	179		244.50	340	
	7600	Floor mounted, pedestal type										

Important: See the Reference Section for critical supporting data - Reference Nos., Crews, & City Cost Indexes

15412 | Drinking Fountains

			CREW	DAILY OUTPUT	LABOR-HOURS	UNIT	2005 BARE COSTS				TOTAL INCL O&P	
							MAT.	LABOR	EQUIP.	TOTAL		
200	7700	Aluminum, architectural style, C.I. base	1 Plum	2	4	Ea.	460	163		623	750	200
	7780	Wheelchair handicap unit	R15100 -410	2	4		1,050	163		1,213	1,425	
	8000	Bronze, architectural style	R15100 -430	2	4		1,200	163		1,363	1,550	
	8040	Enameled steel cylindrical column style		2	4		1,000	163		1,163	1,350	
	8200	Precast stone/concrete, cylindrical column		1	8		770	325		1,095	1,325	
	8240	Wheelchair handicap unit		1	8		1,125	325		1,450	1,725	
	8400	Stainless steel, architectural style		2	4		1,200	163		1,363	1,550	
	8600	Enameled iron, heavy duty service, 2 bubblers		2	4		1,100	163		1,263	1,450	
	8660	4 bubblers		2	4		1,175	163		1,338	1,550	
	8880	For freeze-proof valve system, add		2	4		335	163		498	615	
	8900	For rough-in, supply and waste, add	↓	1.83	4.372	↓	65.50	179		244.50	340	
	9100	Deck mounted										
	9500	Stainless steel, circular receptor	1 Plum	4	2	Ea.	280	81.50		361.50	435	
	9540	14" x 9" receptor		4	2		252	81.50		333.50	400	
	9580	25" x 17" deep receptor, with water glass filler		3	2.667		208	109		317	395	
	9760	White enameled steel, 14" x 9" receptor		4	2		252	81.50		333.50	400	
	9860	White enameled cast iron, 24" x 16" receptor		3	2.667		242	109		351	430	
	9980	For rough-in, supply and waste, add	↓	1.83	4.372	↓	65.50	179		244.50	340	

15413 | Electric Water Coolers

			CREW	DAILY OUTPUT	LABOR-HOURS	UNIT	MAT.	LABOR	EQUIP.	TOTAL	TOTAL INCL O&P	
900	0010	**WATER COOLER**	D2010 820									900
	0030	See line 15413-900-9800 for rough-in, waste & vent										
	0040	for all water coolers	R15100 -430									
	0100	Wall mounted, non-recessed										
	0140	4 GPH	Q-1	4	4	Ea.	500	147		647	770	
	0160	8 GPH, barrier free, sensor operated		4	4		630	147		777	915	
	0180	8.2 GPH		4	4		560	147		707	835	
	0220	14.3 GPH	↓	4	4		645	147		792	930	
	0600	For hot and cold water, add					144			144	158	
	0640	For stainless steel cabinet, add					73.50			73.50	81	
	1000	Dual height, 8.2 GPH	Q-1	3.80	4.211		805	155		960	1,125	
	1040	14.3 GPH	"	3.80	4.211		855	155		1,010	1,175	
	1240	For stainless steel cabinet, add					112			112	123	
	2600	Wheelchair type, 8 GPH	Q-1	4	4		1,425	147		1,572	1,800	
	3000	Simulated recessed, 8 GPH		4	4		710	147		857	1,000	
	3040	11.5 GPH	↓	4	4		775	147		922	1,075	
	3200	For glass filler, add					73.50			73.50	81	
	3240	For stainless steel cabinet, add					51			51	56.50	
	3300	Semi-recessed, 8.1 GPH	Q-1	4	4		840	147		987	1,150	
	3320	12 GPH	"	4	4		885	147		1,032	1,200	
	3340	For glass filler, add					73.50			73.50	81	
	3360	For stainless steel cabinet, add					51			51	56.50	
	3400	Full recessed, stainless steel, 8 GPH	Q-1	3.50	4.571		1,325	168		1,493	1,700	
	3420	11.5 GPH	"	3.50	4.571		1,375	168		1,543	1,750	
	3460	For glass filler, add					134			134	148	
	3600	For mounting can only				↓	147			147	162	
	4600	Floor mounted, flush-to-wall										
	4640	4 GPH	1 Plum	3	2.667	Ea.	520	109		629	740	
	4680	8.2 GPH		3	2.667		575	109		684	795	
	4720	14.3 GPH		3	2.667		520	109		629	740	
	4960	14 GPH hot and cold water	↓	3	2.667		620	109		729	850	
	4980	For stainless steel cabinet, add					112			112	123	
	5000	Dual height, 8.2 GPH	1 Plum	2	4		870	163		1,033	1,200	
	5040	14.3 GPH	"	2	4		900	163		1,063	1,250	
	5120	For stainless steel cabinet, add					147			147	162	
	5600	Explosion Proof, 16 GPH	1 Plum	3	2.667		1,350	109		1,459	1,650	

15413 | Electric Water Coolers

		CREW	DAILY OUTPUT	LABOR-HOURS	UNIT	MAT.	LABOR	EQUIP.	TOTAL	TOTAL INCL O&P	
900	5640	For stainless steel cabinet, add				Ea.	96			96	106	**900**
	6000	Refrigerator Compartment Type, 4.5 GPH (R15100-430)	1 Plum	3	2.667		1,100	109		1,209	1,400	
	6600	Bottle Supply Type, 1.0 GPH		4	2		294	81.50		375.50	450	
	6640	Hot and cold, 1.0 GPH		4	2		400	81.50		481.50	565	
	9800	For supply, waste & vent, all coolers		2.21	3.620		65.50	148		213.50	294	

15414 | Emergency Fixtures

		CREW	DAILY OUTPUT	LABOR-HOURS	UNIT	MAT.	LABOR	EQUIP.	TOTAL	TOTAL INCL O&P	
200	0010	**INDUSTRIAL SAFETY FIXTURES** Rough-in not included										**200**
	1000	Eye wash fountain										
	1400	Plastic bowl, pedestal mounted	Q-1	4	4	Ea.	186	147		333	425	
	1600	Unmounted		4	4		132	147		279	365	
	1800	Wall mounted		4	4		182	147		329	420	
	2000	Stainless steel, pedestal mounted		4	4		278	147		425	525	
	2200	Unmounted		4	4		182	147		329	420	
	2400	Wall mounted		4	4		174	147		321	410	
	3000	Eye wash, portable, self-contained					375			375	410	
	4000	Eye and face wash, combination fountain										
	4200	Stainless steel, pedestal mounted	Q-1	4	4	Ea.	390	147		537	650	
	4400	Unmounted		4	4		300	147		447	550	
	4600	Wall mounted		4	4		305	147		452	555	
	5000	Shower, single head, drench, ball valve, pull, freestanding		4	4		255	147		402	500	
	5200	Horizontal or vertical supply		4	4		237	147		384	480	
	6000	Multi-nozzle, eye/face wash combination		4	4		480	147		627	750	
	6400	Multi-nozzle, 12 spray, shower only		4	4		1,175	147		1,322	1,500	
	6600	For freeze-proof, add		6	2.667		271	98		369	445	
	8000	Walk-thru decontamination with eye-face wash		2	8		1,975	294		2,269	2,625	
	8200	For freeze proof, add		4	4		445	147		592	710	

15417 | Institutional Fixtures

		CREW	DAILY OUTPUT	LABOR-HOURS	UNIT	MAT.	LABOR	EQUIP.	TOTAL	TOTAL INCL O&P	
700	0010	**PRISON/INSTITUTION FIXTURES,** Stainless steel										**700**
	1000	Lavatory, wall hung, push button filler valve										
	1100	Rectangular bowl	Q-1	8	2	Ea.	800	73.50		873.50	990	
	1200	Oval bowl		8	2		795	73.50		868.50	980	
	1240	Oval bowl, corner mount		8	2		915	73.50		988.50	1,100	
	1300	For lavatory rough-in, supply, waste and vent		1.50	10.667		107	390		497	710	
	1700	Service sink, with soap dish										
	1740	24" x 19" size	Q-1	3	5.333	Ea.	1,450	196		1,646	1,900	
	1790	For sink rough-in, supply, waste and vent	"	.89	17.978	"	244	660		904	1,275	
	1800	Shower cabinet, unitized										
	1840	36" x 36" x 88"	Q-1	2.20	7.273	Ea.	3,575	267		3,842	4,325	
	1900	Shower package for built-in										
	1940	Hot & cold valves, recessed soap dish	Q-1	6	2.667	Ea.	525	98		623	725	
	2000	Urinal, back supply and flush										
	2200	Wall hung	Q-1	4	4	Ea.	1,825	147		1,972	2,225	
	2240	Stall		2.50	6.400		1,325	235		1,560	1,800	
	2300	For urinal rough-in, supply, waste and vent		1.49	10.738		138	395		533	745	
	3000	Water closet, integral seat, back supply and flush										
	3300	Wall hung, wall outlet	Q-1	5.80	2.759	Ea.	820	101		921	1,050	
	3400	Floor mount, wall outlet		5.80	2.759		935	101		1,036	1,175	
	3440	Floor mount, floor outlet		5.80	2.759		1,025	101		1,126	1,275	
	3480	For recessed tissue holder, add					72			72	79	
	3500	For water closet rough-in, supply, waste and vent	Q-1	1.19	13.445		154	495		649	915	
	5000	Water closet and lavatory units, push button filler valves,										
	5010	soap & paper holders, seat										
	5300	Wall hung	Q-1	5	3.200	Ea.	1,850	118		1,968	2,200	

Important: See the Reference Section for critical supporting data - Reference Nos., Crews, & City Cost Indexes

15417 | Institutional Fixtures

		CREW	DAILY OUTPUT	LABOR-HOURS	UNIT	2005 BARE COSTS MAT.	LABOR	EQUIP.	TOTAL	TOTAL INCL O&P	
700	5400	Floor mount	Q-1	5	3.200	Ea.	1,850	118		1,968	2,200	700
	6300	For unit rough-in, supply, waste and vent	↓	1	16	↓	168	590		758	1,075	

15418 | Resi/Comm/Industrial Fixtures

			CREW	DAILY OUTPUT	LABOR-HOURS	UNIT	MAT.	LABOR	EQUIP.	TOTAL	TOTAL INCL O&P	
100	0010	**BATHS** D2010 510										100
	0100	Tubs, recessed porcelain enamel on cast iron, with trim										
	0140	42" x 37"	Q-1	5	3.200	Ea.	855	118		973	1,125	
	0180	48" x 42"		4	4		1,500	147		1,647	1,875	
	0220	72" x 36"		3	5.333		1,525	196		1,721	2,000	
	0300	Mat bottom, 4' long		5.50	2.909		1,025	107		1,132	1,275	
	0340	4'-6" long		5	3.200		765	118		883	1,025	
	0380	5' long		4.40	3.636		420	134		554	660	
	0420	5'-6" long		4	4		1,075	147		1,222	1,400	
	0480	Above floor drain, 5' long		4	4		655	147		802	945	
	0560	Corner 48" x 44"	↓	4.40	3.636	↓	1,500	134		1,634	1,850	
	0750	For color, add					25%					
	0760	For designer colors & trim, add					60%					
	2000	Enameled formed steel, 4'-6" long	Q-1	5.80	2.759	Ea.	330	101		431	520	
	2200	5' long		5.50	2.909		290	107		397	480	
	2300	Above floor drain, 5' long	↓	5.50	2.909	↓	270	107		377	460	
	2350	For color, add					10%					
	4000	Soaking, acrylic with pop-up drain, 60" x 32" x 21" deep	Q-1	5.50	2.909	Ea.	805	107		912	1,050	
	4100	60" x 48" x 18-1/2" deep		5	3.200		630	118		748	870	
	4200	72" x 45" x 21-1/2" deep	↓	4.80	3.333	↓	795	123		918	1,050	
	4310	For color, add					5%					
	4311	For designer colors & trim, add					20%					
	4600	Module tub & showerwall surround, molded fiberglass										
	4610	5' long x 34" wide x 76" high	Q-1	4	4	Ea.	495	147		642	765	
	4620	For color add					10%					
	4621	For designer colors and trim add				↓	25%					
	4750	Handicap with 1-1/2" OD grab bar, antiskid bottom										
	4760	60" x 32-3/4" x 72" high	Q-1	4	4	Ea.	670	147		817	955	
	4770	60" x 30" x 71" high with molded seat	"	3.50	4.571	"	830	168		998	1,175	
	5000	Hospital type, with trim (see 15410-300)										
	5050	Bathing pool, porcelain enamel on cast iron, grab bars										
	5060	pop-up drain, 72" x 36"	Q-1	3	5.333	Ea.	2,800	196		2,996	3,400	
	5100	Perineal (sitz), vitreous china		3	5.333		890	196		1,086	1,275	
	5120	Pedestal, vitreous china		8	2		135	73.50		208.50	260	
	5180	Pier tub, porcelain enamel on cast iron, 66-3/4" x 30"		3	5.333		2,225	196		2,421	2,750	
	5200	Base, porcelain enamel on cast iron		8	2		1,050	73.50		1,123.50	1,250	
	5300	Whirlpool, porcelain enamel on cast iron, 72" x 36"	↓	1	16		2,950	590		3,540	4,125	
	5310	For color add					5%					
	5311	For designer colors and trim add				↓	15%					
	6000	Whirlpool, bath with vented overflow, molded fiberglass										
	6100	66" x 48" x 24"	Q-1	1	16	Ea.	2,350	590		2,940	3,475	
	6400	72" x 36" x 24"		1	16		2,300	590		2,890	3,425	
	6500	60" x 30" x 21"		1	16		2,000	590		2,590	3,075	
	6600	72" x 42" x 22"		1	16		3,275	590		3,865	4,475	
	6700	83" x 65"	↓	.30	53.333		4,400	1,950		6,350	7,800	
	6710	For color add					10%					
	6711	For designer colors and trim add				↓	25%					
	7000	Redwood hot tub system										
	7050	4' diameter x 4' deep	Q-1	1	16	Ea.	1,575	590		2,165	2,600	
	7100	5' diameter x 4' deep		1	16		2,075	590		2,665	3,150	
	7150	6' diameter x 4' deep		.80	20		2,575	735		3,310	3,925	
	7200	8' diameter x 4' deep	↓	.80	20	↓	3,950	735		4,685	5,450	

15418 | Resi/Comm/Industrial Fixtures

			CREW	DAILY OUTPUT	LABOR-HOURS	UNIT	2005 BARE COSTS MAT.	LABOR	EQUIP.	TOTAL	TOTAL INCL O&P	
100	9600	Rough-in, supply, waste and vent, for all above tubs, add	Q-1	2.07	7.729	Ea.	142	284		426	580	100
200	0010	**BIDET**										200
	0180	Vitreous china, with trim on fixture	Q-1	5	3.200	Ea.	570	118		688	805	
	0200	With trim for wall mounting	"	5	3.200		570	118		688	805	
	9590	For color add					40%					
	9591	For designer colors and trim add					50%					
	9600	For rough-in, supply, waste and vent, add	Q-1	1.78	8.989		120	330		450	625	
400	0010	**LAUNDRY SINKS** With trim	D2010 420									400
	0020	Porcelain enamel on cast iron, black iron frame										
	0050	24" x 21", single compartment	Q-1	6	2.667	Ea.	345	98		443	525	
	0100	26" x 21", single compartment		6	2.667		269	98		367	445	
	0200	48" x 20", double compartment		5	3.200		262	118		380	465	
	2000	Molded stone, on wall hanger or legs										
	2020	22" x 23", single compartment	Q-1	6	2.667	Ea.	124	98		222	283	
	2100	45" x 21", double compartment	"	5	3.200	"	257	118		375	460	
	3000	Plastic, on wall hanger or legs										
	3020	18" x 23", single compartment	Q-1	6.50	2.462	Ea.	92	90.50		182.50	237	
	3100	20" x 24", single compartment		6.50	2.462		122	90.50		212.50	270	
	3200	36" x 23", double compartment		5.50	2.909		147	107		254	325	
	3300	40" x 24", double compartment		5.50	2.909		216	107		323	400	
	5000	Stainless steel, counter top, 22" x 17" single compartment		6	2.667		345	98		443	525	
	5100	22" x 22", single compartment		6	2.667		440	98		538	630	
	5200	33" x 22", double compartment		5	3.200		435	118		553	650	
	9600	Rough-in, supply, waste and vent, for all laundry sinks		2.14	7.477		104	275		379	530	
450	0010	**LAVATORIES** With trim, white unless noted otherwise	D2010 310									450
	0500	Vanity top, porcelain enamel on cast iron										
	0600	20" x 18"	R15100 -410 Q-1	6.40	2.500	Ea.	206	92		298	365	
	0640	33" x 19" oval		6.40	2.500		415	92		507	595	
	0680	20" x 17" oval		6.40	2.500		129	92		221	280	
	0720	19" round		6.40	2.500		221	92		313	380	
	0760	20" x 12" triangular bowl		6.40	2.500		215	92		307	375	
	0860	For color, add					25%					
	0861	For designer colors and trim, add					70%					
	1000	Cultured marble, 19" x 17", single bowl	Q-1	6.40	2.500	Ea.	153	92		245	305	
	1040	25" x 19", single bowl		6.40	2.500		153	92		245	305	
	1080	31" x 19", single bowl		6.40	2.500		168	92		260	325	
	1120	25" x 22", single bowl		6.40	2.500		153	92		245	305	
	1160	37" x 22", single bowl		6.40	2.500		178	92		270	335	
	1200	49" x 22", single bowl		6.40	2.500		205	92		297	365	
	1580	For color, same price										
	1900	Stainless steel, self-rimming, 25" x 22", single bowl, ledge	Q-1	6.40	2.500	Ea.	235	92		327	395	
	1960	17" x 22", single bowl		6.40	2.500		229	92		321	390	
	2040	18-3/4" round		6.40	2.500		455	92		547	640	
	2600	Steel, enameled, 20" x 17", single bowl		5.80	2.759		129	101		230	295	
	2660	19" round		5.80	2.759		127	101		228	293	
	2720	18" round		5.80	2.759		112	101		213	277	
	2860	For color, add					10%					
	2861	For designer colors and trim, add					20%					
	2900	Vitreous china, 20" x 16", single bowl	Q-1	5.40	2.963	Ea.	222	109		331	410	
	2960	20" x 17", single bowl		5.40	2.963		146	109		255	325	
	3020	19" round, single bowl		5.40	2.963		144	109		253	320	
	3080	19" x 16", single bowl		5.40	2.963		222	109		331	410	
	3140	17" x 14", single bowl		5.40	2.963		182	109		291	365	
	3200	22" x 13", single bowl		5.40	2.963		231	109		340	420	

Important: See the Reference Section for critical supporting data - Reference Nos., Crews, & City Cost Indexes

		15418	Resi/Comm/Industrial Fixtures	CREW	DAILY OUTPUT	LABOR-HOURS	UNIT	MAT.	LABOR	EQUIP.	TOTAL	TOTAL INCL O&P
450	3560		For color, add					50%				
	3561		For designer colors and trim, add					100%				
	3580		Rough-in, supply, waste and vent for all above lavatories	Q-1	2.30	6.957	Ea.	90.50	256		346.50	485
	4000		Wall hung									
	4040		Porcelain enamel on cast iron, 16″ x 14″, single bowl	Q-1	8	2	Ea.	320	73.50		393.50	460
	4060		18″ x 15″ single bowl		8	2		269	73.50		342.50	405
	4120		19″ x 17″, single bowl		8	2		269	73.50		342.50	405
	4180		20″ x 18″, single bowl		8	2		237	73.50		310.50	370
	4240		22″ x 19″, single bowl		8	2		405	73.50		478.50	555
	4580		For color, add					30%				
	4581		For designer colors and trim, add					75%				
	6000		Vitreous china, 18″ x 15″, single bowl with backsplash	Q-1	7	2.286	Ea.	198	84		282	345
	6060		19″ x 17″, single bowl		7	2.286		186	84		270	330
	6120		24″ x 20″, single bowl		7	2.286		410	84		494	575
	6180		19″ x 19″, corner style		8	2		420	73.50		493.50	575
	6210		27″ x 20″, wheelchair type		7	2.286		320	84		404	475
	6500		For color, add					30%				
	6501		For designer colors and trim, add					50%				
	6700		Hospital type, without trim (see 15410-300)									
	6710		20″ x 18″, contoured splash shield	Q-1	8	2	Ea.	114	73.50		187.50	236
	6720		26″ x 20″, patient,corner,side deck & back		7	2.286		375	84		459	535
	6730		28″ x 20″, surgeon, side decks		8	2		450	73.50		523.50	605
	6740		28″ x 22″, surgeon scrub-up, deep bowl		8	2		630	73.50		703.50	805
	6750		20″ x 27″, patient, wheelchair		7	2.286		305	84		389	460
	6760		30″ x 22″, all purpose		7	2.286		585	84		669	770
	6770		30″ x 22″, plaster work		7	2.286		585	84		669	770
	6820		20″ x 24″ clinic service, liquid/solid waste		6	2.667		735	98		833	955
	6960		Rough-in, supply, waste and vent for above lavatories		1.66	9.639		285	355		640	850
500	0010	**SHOWERS**										
	1500		Stall, with drain only. Add for valve and door/curtain									
	1510		Baked enamel, molded stone receptor, 30″ square	Q-1	2	8	Ea.	350	294		644	825
	1520		32″ square		2	8		325	294		619	800
	1530		36″ square		2	8		410	294		704	890
	1540		Terrazzo receptor, 32″ square		2	8		715	294		1,009	1,225
	1560		36″ square		1.80	8.889		875	325		1,200	1,450
	1580		36″ corner angle		1.80	8.889		800	325		1,125	1,375
	1600		For color, add					10%				
	1601		For designer colors and trim, add					15%				
	1604		For thermostatic valve add				Ea.	242			242	266
	3000		Fiberglass, one piece, with 3 walls, 32″ x 32″ square	Q-1	2.40	6.667		415	245		660	830
	3100		36″ x 36″ square	″	2.40	6.667		460	245		705	875
	3200		Handicap, 1-1/2″ O.D. grab bars, nonskid floor									
	3210		48″ x 34-1/2″ x 72″ corner seat	Q-1	2.40	6.667	Ea.	450	245		695	865
	3220		60″ x 34-1/2″ x 72″ corner seat		2	8		660	294		954	1,175
	3230		48″ x 34-1/2″ x 72″ fold up seat		2.20	7.273		625	267		892	1,100
	3250		64″ x 65-3/4″ x 81-1/2″ fold. seat, whlchr.		1.80	8.889		2,900	325		3,225	3,675
	3260		For thermostatic valve add					242			242	266
	4000		Polypropylene, stall only, w/ molded-stone floor, 30″ x 30″	Q-1	2	8		330	294		624	800
	4100		32″ x 32″	″	2	8		420	294		714	900
	4110		For thermostatic valve add					242			242	266
	4200		Rough-in, supply, waste and vent for above showers	Q-1	2.05	7.805		97	287		384	535
	4520		Showers, fiberglass receptor only	1 Plum	8	1		161	41		202	239
	5000		Built-in, head, arm, 4 GPM valve		4	2		91	81.50		172.50	223
	5200		Head, arm, by-pass, integral stops, handles		3.60	2.222		224	91		315	385
	5500		Head, water economizer		24	.333		65	13.60		78.60	92
	5800		Mixing valve, built-in		6	1.333		104	54.50		158.50	196

Notes: row 3560 reference box R15100-410. Row 0010 reference box D2010 710.

15418	Resi/Comm/Industrial Fixtures	CREW	DAILY OUTPUT	LABOR-HOURS	UNIT	2005 BARE COSTS				TOTAL INCL O&P
						MAT.	LABOR	EQUIP.	TOTAL	
500 5900	Exposed ♿	1 Plum	6	1.333	Ea.	505	54.50		559.50	635	**500**
5950	Module, handicap, SS panel, fixed & hand held head, control									
5960	valves, grab bar, curtain & rod, folding seat	1 Plum	4	2	Ea.	1,625	81.50		1,706.50	1,925
6000	Group, w/pressure balancing valve, rough-in and rigging not included									
6800	Column, 6 heads, no receptors, less partitions	Q-1	3	5.333	Ea.	2,025	196		2,221	2,525
6900	With stainless steel partitions		1	16		4,325	590		4,915	5,625
7600	5 heads, no receptors, less partitions ♿		3	5.333		1,550	196		1,746	2,000
7620	4 heads (1 handicap) no receptors, less partitions		3	5.333		2,175	196		2,371	2,675
7700	With stainless steel partitions		1	16		1,550	590		2,140	2,575
8000	Wall, 2 heads, no receptors, less partitions		4	4		895	147		1,042	1,200
8100	With stainless steel partitions		2	8		2,175	294		2,469	2,825
600 0010	**SINKS** With faucets and drain D2010 410										**600**
0050	Laboratory sinks, corrosion resistant									
1000	Polyethylene, single sink, bench mounted, with D2010 430									
1020	plug & waste fitting with 1-1/2" straight threads									
1030	2 drainboards, backnut & strainer									
1050	18-1/2" x 15-1/2" x 12-1/2" sink, 54" x 24" O.D.	Q-1	3	5.333	Ea.	815	196		1,011	1,200
1100	Single drainboard, backnut & strainer									
1130	18-1/2" x 15-1/2" x 12-1/2" sink, 47" x 24" O.D.	Q-1	3	5.333	Ea.	930	196		1,126	1,325
1150	18-1/2" x 15-1/2" x 12-1/2" sink, 70" x 24" O.D.	"	3	5.333	"	945	196		1,141	1,350
1290	Flanged 1-1/4" wide, rectangular with strainer									
1300	plug & waste fitting, 1-1/2" straight threads									
1320	12" x 12" x 8" sink, 14-1/2" x 14-1/2" O.D.	Q-1	4	4	Ea.	238	147		385	480
1340	16" x 16" x 8" sink, 18-1/2" x 18-1/2" O.D.		4	4		242	147		389	485
1360	21" x 18" x 10" sink, 23-1/2" x 20-1/2" O.D.		4	4		280	147		427	530
1490	For rough-in, supply, waste & vent, add		2.02	7.921		104	291		395	555
1600	Polypropylene									
1620	Cup sink, oval, integral strainers									
1640	6" x 3" I.D., 7" x 4" O.D.	Q-1	6	2.667	Ea.	114	98		212	272
1660	9" x 3" I.D., 10" x 4-1/2" O.D.	"	6	2.667	"	108	98		206	266
1720	Overflow standpipe									
1740	1-1/2" diam. x 11" long				Ea.	33.50			33.50	36.50
1980	For rough-in, supply, waste & vent, add	Q-1	1.70	9.412		102	345		447	630
2000	Kitchen, counter top style, P.E. on C.I., 24" x 21" single bowl		5.60	2.857		206	105		311	385
2100	31" x 22" single bowl		5.60	2.857		252	105		357	435
2200	32" x 21" double bowl		4.80	3.333		288	123		411	500
2300	42" x 21" double bowl		4.80	3.333		480	123		603	715
2310	For color, add					20%				
2311	For designer colors and trim, add					50%				
3000	Stainless steel, self rimming, 19" x 18" single bowl	Q-1	5.60	2.857	Ea.	335	105		440	525
3100	25" x 22" single bowl		5.60	2.857		370	105		475	565
3200	33" x 22" double bowl		4.80	3.333		530	123		653	770
3300	43" x 22" double bowl		4.80	3.333		615	123		738	860
3400	22" x 43" triple bowl		4.40	3.636		810	134		944	1,100
3500	Corner double bowl each 14" x 16"		4.80	3.333		520	123		643	755
4000	Steel, enameled, with ledge, 24" x 21" single bowl		5.60	2.857		110	105		215	279
4100	32" x 21" double bowl		4.80	3.333		147	123		270	345
4960	For color sinks except stainless steel, add					10%				
4961	For designer colors and trim add					20%				
4980	For rough-in, supply, waste and vent, counter top sinks	Q-1	2.14	7.477		104	275		379	530
5000	Kitchen, raised deck, P.E. on C.I.									
5100	32" x 21", dual level, double bowl	Q-1	2.60	6.154	Ea.	259	226		485	625
5200	42" x 21", double bowl & disposer well	"	2.20	7.273		615	267		882	1,075
5700	For color, add					20%				
5701	For designer colors and trim add					50%				

Important: See the Reference Section for critical supporting data - Reference Nos., Crews, & City Cost Indexes

15418 | Resi/Comm/Industrial Fixtures

		CREW	DAILY OUTPUT	LABOR-HOURS	UNIT	2005 BARE COSTS				TOTAL INCL O&P
						MAT.	LABOR	EQUIP.	TOTAL	
600 5790	For rough-in, supply, waste & vent, sinks	Q-1	1.85	8.649	Ea.	104	320		424	595	**600**
6650	Service, floor, corner, P.E. on C.I., 28" x 28"	"	4.40	3.636		530	134		664	785
6750	Vinyl coated rim guard, add					54			54	59.50
6760	Mop sink, molded stone, 24" x 36"	1 Plum	3.33	2.402		194	98		292	360
6770	Mop sink, molded stone, 24" x 36", w/rim 3 sides	"	3.33	2.402		218	98		316	390
6790	For rough-in, supply, waste & vent, floor service sinks	Q-1	1.64	9.756		263	360		623	830
7000	Service, wall, P.E. on C.I., roll rim, 22" x 18"	↓	4	4		445	147		592	710
7100	24" x 20"	↓	4	4		460	147		607	730
7600	For stainless steel rim guard, add					90.50			90.50	99.50
7800	For stainless steel rim guard, front only, add					35			35	38.50
8600	Vitreous china, 22" x 20"	Q-1	4	4		390	147		537	650
8960	For stainless steel rim guard, front or side, add					35			35	38.50
8980	For rough-in, supply, waste & vent, wall service sinks	Q-1	1.30	12.308	↓	410	455		865	1,125
900 0010	**WATER CLOSETS** D2010 110										**900**
0020	For seats, see 15410-800									
0030	For automatic flush, see 15410-300-0972									
0150	Tank type, vitreous china, incl. seat, supply pipe w/stop D2010 120									
0200	Wall hung, one piece R15100 -410	Q-1	5.30	3.019	Ea.	390	111		501	595
0400	Two piece, close coupled		5.30	3.019		480	111		591	690
0960	For rough-in, supply, waste, vent and carrier		2.73	5.861		335	216		551	690
1000	Floor mounted, one piece		5.30	3.019		495	111		606	710
1020	One piece, low profile		5.30	3.019		395	111		506	600
1050	One piece combination		5.30	3.019		680	111		791	910
1100	Two piece, close coupled		5.30	3.019		163	111		274	345
1150	With wall outlet		5.30	3.019		540	111		651	755
1200	With 18" high bowl	↓	5.30	3.019	↓	385	111		496	585
1960	For color, add					30%				
1961	For designer colors and trim, add					55%				
1980	For rough-in, supply, waste and vent	Q-1	3.05	5.246	Ea.	152	193		345	460
3000	Bowl only, with flush valve, seat									
3100	Wall hung	Q-1	5.80	2.759	Ea.	345	101		446	535
3150	Hospital type, slotted rim for bed pan									
3160	Elongated bowl	Q-1	5.80	2.759	Ea.	365	101		466	555
3200	For rough-in, supply, waste and vent, single WC		2.56	6.250		350	230		580	730
3300	Floor mounted		5.80	2.759		310	101		411	495
3350	With wall outlet	↓	5.80	2.759	↓	485	101		586	690
3360	Hospital type, slotted rim for bed pan									
3370	Elongated bowl	Q-1	5	3.200	Ea.	298	118		416	505
3380	Elongated bowl, 18" high		5	3.200		365	118		483	575
3400	For rough-in, supply, waste and vent, single WC	↓	2.84	5.634	↓	168	207		375	495
3500	Gang side by side carrier system, rough-in, supply, waste & vent									
3510	For single hook-up	Q-1	1.97	8.122	Ea.	420	299		719	915
3520	For each additional hook-up, add	"	2.14	7.477	"	385	275		660	840
3550	Gang back to back carrier system, rough-in, supply, waste & vent									
3560	For pair hook-up	Q-1	1.76	9.091	Pr.	670	335		1,005	1,250
3570	For each additional pair hook-up, add	"	1.81	8.840	"	635	325		960	1,200
4000	Water conserving systems									
4900	2 quart flush, residential	Q-1	5.40	2.963	Ea.	585	109		694	810
4980	For rough-in, supply, waste and vent		1.94	8.247		170	305		475	645
5100	2 quart flush		4.60	3.478		595	128		723	845
5200	For remote valve, add		24	.667		194	24.50		218.50	250
5300	For residential air compressor		6	2.667		705	98		803	920
5400	For light industrial air compressor		4	4		1,125	147		1,272	1,475
5500	For heavy duty industrial air compressor		1	16		1,675	590		2,265	2,725
5600	For dual compressor alternator	↓	20	.800	↓	565	29.50		594.50	665

15440 | Plumbing Pumps

		CREW	DAILY OUTPUT	LABOR-HOURS	UNIT	MAT.	LABOR	EQUIP.	TOTAL	TOTAL INCL O&P
240											**240**
0010	**PUMPS, PRESSURE BOOSTER SYSTEM**									
0200	Pump system, with diaphragm tank, control, press. switch									
0300	1 HP pump	Q-1	1.30	12.308	Ea.	4,000	455		4,455	5,075
0400	1-1/2 HP pump		1.25	12.800		4,050	470		4,520	5,150
0420	2 HP pump		1.20	13.333		4,150	490		4,640	5,300
0440	3 HP pump	↓	1.10	14.545		4,200	535		4,735	5,425
0460	5 HP pump	Q-2	1.50	16		4,650	610		5,260	6,050
0480	7-1/2 HP pump		1.42	16.901		5,175	645		5,820	6,675
0500	10 HP pump	↓	1.34	17.910	↓	5,425	685		6,110	7,000
1000	Pump/ energy storage system, diaphragm tank, 3 HP pump									
1100	motor, PRV, switch, gauge, control center, flow switch									
1200	125 lb. working pressure	Q-2	.70	34.286	Ea.	11,700	1,300		13,000	14,800
1300	250 lb. working pressure	"	.64	37.500	"	12,800	1,425		14,225	16,200
400											**400**
0010	**PUMPS, GRINDER SYSTEM** Complete, incl. check valve, tank, std.									
0020	controls incl. alarm/disconnect panel w/wire. Excavation not included									
0260	Simplex, 9 GPM at 60 PSIG, 70 gal. tank				Ea.	2,425			2,425	2,675
0300	For manway, 26" I.D., 18" high, add					310			310	340
0340	26" I.D., 36" high, add					425			425	470
0380	43" I.D., 4' high, add					475			475	525
0600	Simplex, 9 GPM at 60 PSIG, 150 gal. tank				↓	2,625			2,625	2,875
0660	For manway, add									
0700	26" I.D., 36" high, add				Ea.	565			565	625
0740	26" I.D., 4' high, add					645			645	710
2000	Duplex, 18 GPM at 60 PSIG, 150 gal. tank					1,725			1,725	1,900
2060	For manway 43" I.D., 4' high, add					1,425			1,425	1,550
2400	For core only				↓	1,300			1,300	1,425
800											**800**
0010	**PUMPS, SEWAGE EJECTOR** With operating and level controls									
0100	Simplex system incl. tank, cover, pump 15' head									
0500	37 gal PE tank, 12 GPM, 1/2 HP, 2" discharge	Q-1	3.20	5	Ea.	375	184		559	690
0510	3" discharge		3.10	5.161		405	190		595	730
0530	87 GPM, .7 HP, 2" discharge		3.20	5		570	184		754	905
0540	3" discharge		3.10	5.161		620	190		810	965
0600	45 gal. coated stl tank, 12 GPM, 1/2 HP, 2" discharge		3	5.333		670	196		866	1,025
0610	3" discharge		2.90	5.517		695	203		898	1,075
0630	87 GPM, .7 HP, 2" discharge		3	5.333		855	196		1,051	1,225
0640	3" discharge		2.90	5.517		905	203		1,108	1,300
0660	134 GPM, 1 HP, 2" discharge		2.80	5.714		925	210		1,135	1,350
0680	3" discharge		2.70	5.926		975	218		1,193	1,400
0700	70 gal. PE tank, 12 GPM, 1/2 HP, 2" discharge		2.60	6.154		735	226		961	1,150
0710	3" discharge		2.40	6.667		785	245		1,030	1,225
0730	87 GPM, 0.7 HP, 2" discharge		2.50	6.400		940	235		1,175	1,375
0740	3" discharge		2.30	6.957		1,000	256		1,256	1,475
0760	134 GPM, 1 HP, 2" discharge		2.20	7.273		1,025	267		1,292	1,525
0770	3" discharge		2	8		1,100	294		1,394	1,650
0800	75 gal. coated stl. tank, 12 GPM, 1/2 HP, 2 " discharge		2.40	6.667		760	245		1,005	1,200
0810	3" discharge		2.20	7.273		790	267		1,057	1,275
0830	87 GPM, .7 HP, 2" discharge		2.30	6.957		955	256		1,211	1,425
0840	3" discharge		2.10	7.619		1,000	280		1,280	1,525
0860	134 GPM, 1 HP, 2" discharge		2	8		1,025	294		1,319	1,575
0880	3" discharge	↓	1.80	8.889	↓	1,075	325		1,400	1,675
1040	Duplex system incl. tank, covers, pumps									
1060	110 gal. fiberglass tank, 24 GPM, 1/2 HP, 2" discharge	Q-1	1.60	10	Ea.	1,425	370		1,795	2,125
1080	3" discharge		1.40	11.429		1,500	420		1,920	2,275
1100	174 GPM, .7 HP, 2" discharge	↓	1.50	10.667	↓	1,875	390		2,265	2,650

15440 | Plumbing Pumps

		CREW	DAILY OUTPUT	LABOR-HOURS	UNIT	MAT.	2005 BARE COSTS LABOR	EQUIP.	TOTAL	TOTAL INCL O&P	
800	1120	3" discharge	Q-1	1.30	12.308	Ea.	1,950	455		2,405	2,825	80
	1140	268 GPM, 1 HP, 2" discharge		1.20	13.333		2,050	490		2,540	2,975	
	1160	3" discharge	▼	1	16		2,100	590		2,690	3,200	
	1260	135 gal. coated stl. tank, 24 GPM, 1/2 HP, 2" discharge	Q-2	1.70	14.118		1,500	540		2,040	2,450	
	2000	3" discharge		1.60	15		1,575	570		2,145	2,575	
	2640	174 GPM, .7 HP, 2" discharge		1.60	15		1,950	570		2,520	3,000	
	2660	3" discharge		1.50	16		2,050	610		2,660	3,175	
	2700	268 GPM, 1 HP, 2" discharge		1.30	18.462		2,125	705		2,830	3,375	
	3040	3" discharge		1.10	21.818		2,250	830		3,080	3,725	
	3060	275 gal. coated stl. tank, 24 GPM, 1/2 HP, 2" discharge		1.50	16		1,850	610		2,460	2,975	
	3080	3" discharge		1.40	17.143		1,875	655		2,530	3,050	
	3100	174 GPM, .7 HP, 2" discharge		1.40	17.143		2,400	655		3,055	3,625	
	3120	3" discharge		1.30	18.462		2,425	705		3,130	3,700	
	3140	268 GPM, 1 HP, 2" discharge		1.10	21.818		2,625	830		3,455	4,150	
	3160	3" discharge	▼	.90	26.667	▼	2,725	1,025		3,750	4,525	
	3260	Pump system accessories, add										
	3300	Alarm horn and lights, 115V mercury switch	Q-1	8	2	Ea.	82.50	73.50		156	202	
	3340	Switch, mag. contactor, alarm bell, light, 3 level control		5	3.200		590	118		708	825	
	3380	Alternator, mercury switch activated	▼	4	4	▼	750	147		897	1,050	
900	0010	**PUMPS, PEDESTAL SUMP** With float control										90
	0400	Molded PVC base, 21 GPM at 15' head, 1/3 HP	1 Plum	5	1.600	Ea.	87	65.50		152.50	194	
	0800	Iron base, 21 GPM at 15' head, 1/3 HP		5	1.600		110	65.50		175.50	219	
	1200	Solid brass, 21 GPM at 15' head, 1/3 HP	▼	5	1.600	▼	179	65.50		244.50	295	
940	0010	**PUMPS, SUBMERSIBLE** Sump										94
	7000	Sump pump, automatic										
	7100	Plastic, 1-1/4" discharge, 1/4 HP	1 Plum	6	1.333	Ea.	105	54.50		159.50	198	
	7140	1/3 HP		5	1.600		127	65.50		192.50	238	
	7160	1/2 HP		5	1.600		153	65.50		218.50	266	
	7180	1-1/2" discharge, 1/2 HP		4	2		171	81.50		252.50	310	
	7500	Cast iron, 1-1/4" discharge, 1/4 HP		6	1.333		122	54.50		176.50	216	
	7540	1/3 HP		6	1.333		144	54.50		198.50	240	
	7560	1/2 HP	▼	5	1.600	▼	170	65.50		235.50	285	

15450 | Potable Water Storage Tanks

		CREW	DAILY OUTPUT	LABOR-HOURS	UNIT	MAT.	LABOR	EQUIP.	TOTAL	TOTAL INCL O&P	
900	0010	**WATER HEATER STORAGE TANKS** 125 psi ASME										90
	2000	Galvanized steel, 15 gal., 14" diam., 26" LOA	1 Plum	12	.667	Ea.	535	27		562	630	
	2060	30 gal., 14" diam. x 49" LOA		11	.727		640	29.50		669.50	745	
	2080	80 gal., 20" diam. x 64" LOA		9	.889		990	36.50		1,026.50	1,150	
	2100	135 gal., 24" diam. x 75" LOA		6	1.333		2,150	54.50		2,204.50	2,450	
	2120	240 gal., 30" diam. x 86" LOA		4	2		2,525	81.50		2,606.50	2,900	
	2140	300 gal., 36" diam. x 76" LOA	▼	3	2.667		4,100	109		4,209	4,700	
	2160	400 gal., 36" diam. x 100" LOA	Q-1	4	4		5,050	147		5,197	5,775	
	2180	500 gal., 36" diam., x 126" LOA	"	3	5.333		6,250	196		6,446	7,175	
	3000	Glass lined, P.E., 80 gal., 20" diam. x 60" LOA	1 Plum	9	.889		1,725	36.50		1,761.50	1,925	
	3060	140 gal., 24" diam. x 80" LOA		6	1.333		2,450	54.50		2,504.50	2,775	
	3080	225 gal., 30" diam. x 78" LOA		4	2		2,800	81.50		2,881.50	3,200	
	3100	325 gal., 36" diam. x 81" LOA	▼	3	2.667		3,575	109		3,684	4,100	
	3120	460 gal., 42" diam. x 84" LOA	Q-1	4	4		3,750	147		3,897	4,350	
	3140	605 gal., 48" diam. x 87" LOA		3	5.333		7,175	196		7,371	8,200	
	3160	740 gal., 54" diam. x 91" LOA		3	5.333		8,075	196		8,271	9,200	
	3180	940 gal., 60" diam. x 93" LOA		2.50	6.400		9,325	235		9,560	10,700	
	3200	1330 gal., 66" diam. x 107" LOA		2	8		12,800	294		13,094	14,500	
	3220	1615 gal., 72" diam. x 110" LOA		1.50	10.667		14,500	390		14,890	16,600	
	3240	2285 gal., 84" diam. x 128" LOA	▼	1	16	▼	18,000	590		18,590	20,700	

15450	Potable Water Storage Tanks	CREW	DAILY OUTPUT	LABOR-HOURS	UNIT	2005 BARE COSTS				TOTAL INCL O&P	
						MAT.	LABOR	EQUIP.	TOTAL		
900	3260	3440 gal., 96" diam. x 157" LOA	Q-2	1.50	16	Ea.	26,600	610		27,210	30,200	900

15460 | Domestic Water Cond Equipment

			CREW	DAILY OUTPUT	LABOR-HOURS	UNIT	MAT.	LABOR	EQUIP.	TOTAL	INCL O&P	
900	0010	**WATER SOFTENER**										900
	5800	Softener systems, automatic, intermediate sizes										
	5820	available, may be used in multiples.										
	6000	Hardness capacity between regenerations and flow										
	6060	40,000 grains, 14 GPM	Q-1	4	4	Ea.	795	147		942	1,100	
	6070	50,000 grains, 17 GPM		3.60	4.444		920	163		1,083	1,250	
	6080	90,000 grains, 25 GPM		2.80	5.714		2,375	210		2,585	2,950	
	6100	150,000 grains, 37 GPM cont., 51 GPM peak		1.20	13.333		3,775	490		4,265	4,875	
	6200	300,000 grains, 81 GPM cont., 113 GPM peak		1	16		5,600	590		6,190	7,050	
	6300	750,000 grains, 160 GPM cont., 230 GPM peak		.80	20		7,300	735		8,035	9,150	
	6400	900,000 grains, 185 GPM cont., 270 GPM peak		.70	22.857		11,800	840		12,640	14,200	
	8000	Water treatment, salts, 50 lb. bag										
	8020	Salt, water softener, bag, pelletized				Lb.	.21			.21	.23	
	8030	Salt, water softener, bag, crystal rock salt				"	.18			.18	.20	

15470 | Domstc Water Filt Equip

			CREW	DAILY OUTPUT	LABOR-HOURS	UNIT	MAT.	LABOR	EQUIP.	TOTAL	INCL O&P	
400	0010	**WATER FILTERS** Purification and treatment										400
	1000	Cartridge style, dirt and rust type	1 Plum	12	.667	Ea.	29	27		56	73	
	1200	Replacement cartridge		32	.250		10.80	10.20		21	27.50	
	1600	Taste and odor type		12	.667		29.50	27		56.50	73.50	
	1700	Replacement cartridge		32	.250		25	10.20		35.20	43	
	3000	Central unit, dirt/rust/odor/taste/scale		4	2		665	81.50		746.50	855	
	3100	Replacement cartridge, standard		20	.400		131	16.35		147.35	169	
	3600	Replacement cartridge, heavy duty		20	.400		148	16.35		164.35	187	
	8000	Commercial, fully automatic or push button automatic										
	8200	Iron removal, 660 GPH, 1" pipe size	Q-1	1.50	10.667	Ea.	1,675	390		2,065	2,450	
	8240	1500 GPH, 1-1/4" pipe size		1	16		2,825	590		3,415	4,000	
	8280	2340 GPH, 1-1/2" pipe size		.80	20		3,100	735		3,835	4,525	
	8320	3420 GPH, 2" pipe size		.60	26.667		5,725	980		6,705	7,775	
	8360	4620 GPH, 2-1/2" pipe size		.50	32		9,100	1,175		10,275	11,800	
	8500	Neutralizer for acid water, 780 GPH, 1" pipe size		1.50	10.667		1,625	390		2,015	2,375	
	8540	1140 GPH, 1-1/4" pipe size		1	16		1,825	590		2,415	2,875	
	8580	1740 GPH, 1-1/2" pipe size		.80	20		2,650	735		3,385	4,025	
	8620	2520 GPH, 2" pipe size		.60	26.667		3,450	980		4,430	5,250	
	8660	3480 GPH, 2-1/2" pipe size		.50	32		5,725	1,175		6,900	8,075	
	8800	Sediment removal, 780 GPH, 1" pipe size		1.50	10.667		1,550	390		1,940	2,300	
	8840	1140 GPH, 1-1/4" pipe size		1	16		1,850	590		2,440	2,925	
	8880	1740 GPH, 1-1/2" pipe size		.80	20		2,450	735		3,185	3,800	
	8920	2520 GPH, 2" pipe size		.60	26.667		3,550	980		4,530	5,375	
	8960	3480 GPH, 2-1/2" pipe size		.50	32		5,575	1,175		6,750	7,900	
	9200	Taste and odor removal, 660 GPH, 1" pipe size		1.50	10.667		2,175	390		2,565	2,975	
	9240	1500 GPH, 1-1/4" pipe size		1	16		3,725	590		4,315	4,975	
	9280	2340 GPH, 1-1/2" pipe size		.80	20		3,575	735		4,310	5,025	
	9320	3420 GPH, 2" pipe size		.60	26.667		6,525	980		7,505	8,650	
	9360	4620 GPH, 2-1/2" pipe size		.50	32		10,000	1,175		11,175	12,800	

15480 | Domestic Water Heaters

			CREW	DAILY OUTPUT	LABOR-HOURS	UNIT	MAT.	LABOR	EQUIP.	TOTAL	INCL O&P	
200	0010	**WATER HEATERS**										200
	0020	For solar, see division 13600	D2020 210									
	1000	Residential, elec., glass lined tank, 5 yr, 10 gal., single elem.	D2020 220 / 1 Plum	2.30	3.478	Ea.	210	142		352	445	
	1040	20 gallon, single element		2.20	3.636		244	149		393	490	

Important: See the Reference Section for critical supporting data - Reference Nos., Crews, & City Cost Indexes

15480	Domestic Water Heaters		CREW	DAILY OUTPUT	LABOR-HOURS	UNIT	2005 BARE COSTS				TOTAL INCL O&P
							MAT.	LABOR	EQUIP.	TOTAL	
200 1060	30 gallon, double element	D2020 230	1 Plum	2.20	3.636	Ea.	280	149		429	535 **200**
1080	40 gallon, double element			2	4		300	163		463	575
1100	52 gallon, double element	D2020 240		2	4		355	163		518	635
1120	66 gallon, double element			1.80	4.444		475	182		657	800
1140	80 gallon, double element	D2020 350		1.60	5		535	204		739	895
1180	120 gallon, double element			1.40	5.714		770	233		1,003	1,200
2000	Gas fired, foam lined tank, 10 yr, vent not incl.,										
2040	30 gallon		1 Plum	2	4	Ea.	385	163		548	670
2060	40 gallon			1.90	4.211		405	172		577	705
2080	50 gallon			1.80	4.444		535	182		717	860
2100	75 gallon			1.50	5.333		745	218		963	1,150
2120	100 gallon			1.30	6.154		1,150	251		1,401	1,625
3000	Oil fired, glass lined tank, 5 yr, vent not included, 30 gal.	D2020 260		2	4		745	163		908	1,075
3040	50 gallon			1.80	4.444		1,375	182		1,557	1,775
3060	70 gallon			1.50	5.333		1,525	218		1,743	2,000
3080	85 gallon			1.40	5.714		4,300	233		4,533	5,075
4000	Commercial, 100° rise. NOTE: for each size tank, a range of										
4010	heaters between the ones shown are available										
4020	Electric										
4100	5 gal., 3 kW, 12 GPH, 208V		1 Plum	2	4	Ea.	1,625	163		1,788	2,050
4120	10 gal., 6 kW, 25 GPH, 208V			2	4		1,800	163		1,963	2,225
4140	50 gal., 9 kW, 37 GPH, 208V			1.80	4.444		2,475	182		2,657	3,000
4160	50 gal., 36 kW, 148 GPH, 208V			1.80	4.444		3,775	182		3,957	4,425
4180	80 gal., 12 kW, 49 GPH, 208V			1.50	5.333		3,050	218		3,268	3,675
4200	80 gal., 36 kW, 148 GPH, 208V			1.50	5.333		4,225	218		4,443	4,975
4220	100 gal., 36 kW, 148 GPH, 208V			1.20	6.667		4,400	272		4,672	5,225
4240	120 gal., 36 kW, 148 GPH, 208V			1.20	6.667		4,575	272		4,847	5,450
4260	150 gal., 15 kW , 61 GPH, 480V			1	8		10,900	325		11,225	12,500
4280	150 gal., 120 kW, 490 GPH, 480V			1	8		15,300	325		15,625	17,300
4300	200 gal., 15 kW, 61 GPH, 480V		Q-1	1.70	9.412		11,900	345		12,245	13,600
4320	200 gal., 120 kW , 490 GPH, 480V			1.70	9.412		16,300	345		16,645	18,400
4340	250 gal., 15 kW, 61 GPH, 480V			1.50	10.667		12,200	390		12,590	14,100
4360	250 gal., 150 kW, 615 GPH, 480V			1.50	10.667		17,900	390		18,290	20,200
4380	300 gal., 30 kW, 123 GPH, 480V			1.30	12.308		13,200	455		13,655	15,300
4400	300 gal., 180 kW, 738 GPH, 480V			1.30	12.308		19,200	455		19,655	21,800
4420	350 gal., 30 kW, 123 GPH, 480V			1.10	14.545		14,500	535		15,035	16,800
4440	350 gal., 180 kW, 738 GPH, 480V			1.10	14.545		19,700	535		20,235	22,500
4460	400 gal., 30 kW, 123 GPH, 480V			1	16		16,400	590		16,990	18,900
4480	400 gal., 210 kW, 860 GPH, 480V			1	16		22,800	590		23,390	26,000
4500	500 gal., 30 kW, 123 GPH, 480V			.80	20		18,900	735		19,635	21,900
4520	500 gal., 240 kW, 984 GPH, 480V			.80	20		27,200	735		27,935	31,000
4540	600 gal., 30 kW, 123 GPH, 480V		Q-2	1.20	20		21,800	765		22,565	25,200
4560	600 gal., 300 kW, 1230 GPH, 480V			1.20	20		32,800	765		33,565	37,200
4580	700 gal., 30 kW, 123 GPH, 480V			1	24		22,700	915		23,615	26,300
4600	700 gal., 300 kW, 1230 GPH, 480V			1	24		34,100	915		35,015	38,900
4620	800 gal., 60 kW, 245 GPH, 480V			.90	26.667		24,500	1,025		25,525	28,500
4640	800 gal., 300 kW, 1230 GPH, 480V			.90	26.667		34,900	1,025		35,925	39,800
4660	1000 gal., 60 kW, 245 GPH, 480V			.70	34.286		26,900	1,300		28,200	31,600
4680	1000 gal., 480 kW, 1970 GPH, 480V			.70	34.286		35,100	1,300		36,400	40,600
4700	1250 gal., 60 kW, 245 GPH, 480V			.60	40		30,300	1,525		31,825	35,700
4720	1250 gal., 480 kW, 1970 GPH, 480V			.60	40		46,100	1,525		47,625	53,000
4740	1500 gal., 60 kW, 245 GPH, 480V			.50	48		40,000	1,825		41,825	46,800
4760	1500 gal., 480 kW, 1970 GPH, 480V			.50	48		55,500	1,825		57,325	64,000
4770	2000 gal., 480 kW, 1970 GPH, 480V			.48	50		60,500	1,900		62,400	69,500
5400	Modulating step control, 2-5 steps		1 Elec	5.30	1.509		1,525	61.50		1,586.50	1,775
5440	6-10 steps			3.20	2.500		1,900	102		2,002	2,250

15480 | Domestic Water Heaters

		CREW	DAILY OUTPUT	LABOR-HOURS	UNIT	2005 BARE COSTS				TOTAL INCL O&P
						MAT.	LABOR	EQUIP.	TOTAL	
5460	11-15 steps	1 Elec	2.70	2.963	Ea.	2,350	121		2,471	2,750
5480	16-20 steps	↓	1.60	5	↓	2,775	204		2,979	3,350
6000	Gas fired, flush jacket, std. controls, vent not incl.									
6040	75 MBH input, 73 GPH	1 Plum	1.40	5.714	Ea.	1,475	233		1,708	1,975
6060	98 MBH input, 95 GPH		1.40	5.714		2,600	233		2,833	3,225
6080	120 MBH input, 110 GPH		1.20	6.667		2,825	272		3,097	3,500
6100	120 MBH input, 115 GPH		1.10	7.273		3,375	297		3,672	4,150
6120	140 MBH input, 130 GPH		1	8		3,775	325		4,100	4,650
6140	155 MBH input, 150 GPH		.80	10		4,100	410		4,510	5,150
6160	180 MBH input, 170 GPH		.70	11.429		4,275	465		4,740	5,400
6180	200 MBH input, 192 GPH		.60	13.333		4,300	545		4,845	5,575
6200	250 MBH input, 245 GPH	↓	.50	16		4,675	655		5,330	6,125
6220	260 MBH input, 250 GPH	Q-1	.80	20		5,900	735		6,635	7,575
6240	360 MBH input, 360 GPH		.80	20		6,550	735		7,285	8,300
6260	500 MBH input, 480 GPH		.70	22.857		8,200	840		9,040	10,300
6280	725 MBH input, 690 GPH	↓	.60	26.667		9,425	980		10,405	11,900
6900	For low water cutoff, add	1 Plum	8	1		220	41		261	305
6960	For bronze body hot water circulator, add	"	4	2	↓	1,150	81.50		1,231.50	1,400
8000	Oil fired, flush jacket, std. controls, vent not incl.									
8060	103 MBH gross output, 116 GPH	1 Plum	1.10	7.273	Ea.	980	297		1,277	1,525
8080	122 MBH gross output, 141 GPH		1	8		1,800	325		2,125	2,500
8100	137 MBH gross output, 161 GPH		.80	10		1,875	410		2,285	2,675
8120	168 MBH gross output, 192 GPH		.60	13.333		2,075	545		2,620	3,100
8140	195 MBH gross output, 224 GPH	↓	.50	16		2,250	655		2,905	3,450
8160	225 MBH gross output, 256 GPH	Q-1	.80	20		2,575	735		3,310	3,925
8180	262 MBH gross output, 315 GPH		.70	22.857		2,725	840		3,565	4,275
8200	315 MBH gross output, 409 GPH		.70	22.857		6,125	840		6,965	8,025
8220	420 MBH gross output, 504 GPH		.60	26.667		7,300	980		8,280	9,500
8240	525 MBH gross output, 630 GPH		.50	32		8,200	1,175		9,375	10,800
8260	630 MBH gross output, 756 GPH		.50	32		9,675	1,175		10,850	12,400
8280	735 MBH gross output, 880 GPH		.40	40		9,775	1,475		11,250	13,000
8300	840 MBH gross output, 1000 GPH	↓	.40	40		10,200	1,475		11,675	13,500
8320	1050 MBH gross output, 1260 GPH	Q-2	.60	40		11,200	1,525		12,725	14,600
8340	1365 MBH gross output, 1640 GPH		.50	48		13,200	1,825		15,025	17,300
8360	1680 MBH gross output, 2000 GPH		.40	60		14,300	2,300		16,600	19,200
8380	2310 MBH gross output, 2780 GPH		.30	80		18,800	3,050		21,850	25,300
8400	2835 MBH gross output, 3400 GPH		.30	80		21,500	3,050		24,550	28,300
8420	3150 MBH gross output, 3780 GPH	↓	.20	120	↓	24,400	4,575		28,975	33,700
8500	Oil fired, polymer lined									
8510	400 MBH, 125 gallon	Q-2	1	24	Ea.	15,700	915		16,615	18,700
8520	400 MBH, 600 gallon		.80	30		29,800	1,150		30,950	34,500
8530	800 MBH, 400 gallon		.67	35.821		27,100	1,375		28,475	31,900
8540	800 MBH, 600 gallon		.60	40		35,400	1,525		36,925	41,200
8550	1000 MBH, 600 gallon		.50	48		36,400	1,825		38,225	42,900
8560	1200 MBH, 900 gallon	↓	.40	60		42,000	2,300		44,300	49,700
8900	For low water cutoff, add	1 Plum	8	1		220	41		261	305
8960	For bronze body hot water circulator, add	"	4	2	↓	805	81.50		886.50	1,000
8965	Point of use, electric, glass lined									
8969	Energy saver									
8970	2.5 gal. single element	1 Plum	2.80	2.857	Ea.	171	117		288	365
8971	4 gal. single element		2.80	2.857		184	117		301	375
8974	6 gal. single element		2.50	3.200		198	131		329	415
8975	10 gal. single element		2.50	3.200		210	131		341	425
8976	15 gal. single element		2.40	3.333		223	136		359	450
8977	20 gal. single element		2.40	3.333		244	136		380	475
8978	30 gal. single element	↓	2.30	3.478	↓	271	142		413	510

Important: See the Reference Section for critical supporting data - Reference Nos., Crews, & City Cost Indexes

15480 | Domestic Water Heaters

		CREW	DAILY OUTPUT	LABOR-HOURS	UNIT	2005 BARE COSTS MAT.	LABOR	EQUIP.	TOTAL	TOTAL INCL O&P	
200	8979	40 gal. single element	1 Plum	2.20	3.636	Ea.	330	149		479	590	**200**
	8988	Commercial (ASHRAE energy std. 90)										
	8989	6 gallon	1 Plum	2.50	3.200	Ea.	375	131		506	605	
	8990	10 gallon		2.50	3.200		400	131		531	635	
	8991	15 gallon		2.40	3.333		425	136		561	670	
	8992	20 gallon		2.40	3.333		450	136		586	705	
	8993	30 gallon	↓	2.30	3.478	↓	950	142		1,092	1,275	
	8995	Under the sink, copper, w/bracket										
	8996	2.5 gallon	1 Plum	4	2	Ea.	209	81.50		290.50	350	
700	0010	**TEMPERATURE MAINTENANCE CABLE**										**700**
	0040	Components										
	0080	Heating cable										
	0100	208 V										
	0120	105 °F	Q-1	1,060.80	.015	L.F.	8.55	.55		9.10	10.25	
	0130	115 °F		1,060.80	.015		8.55	.55		9.10	10.25	
	0140	125 °F		1,060.80	.015		8.55	.55		9.10	10.25	
	0150	140 °F	↓	1,060.80	.015	↓	8.55	.55		9.10	10.25	
	0200	120 V										
	0220	125 °F	Q-1	1,060.80	.015	L.F.	5.50	.55		6.05	6.90	
	0300	Power kit w/1 end seal	1 Elec	48.80	.164	Ea.	48	6.70		54.70	63	
	0310	Splice kit		35.50	.225		57	9.20		66.20	76	
	0320	End seal		160	.050		7.65	2.04		9.69	11.45	
	0330	Tee kit w/1 end seal		26.80	.299		60.50	12.15		72.65	84.50	
	0340	Powered splice w/2 end seals		20	.400		57.50	16.30		73.80	87.50	
	0350	Powered tee kit w/3 end seals		18	.444		73.50	18.10		91.60	108	
	0360	Cross kit w/2 end seals	↓	18.60	.430	↓	76	17.55		93.55	110	
	0500	Recommended thickness of fiberglass insulation										
	0510	Pipe size										
	0520	1/2" - 1" use 1" insulation										
	0530	1-1/4" - 2" use 1-1/2" insulation										
	0540	2-1/2" - 6" use 2" insulation										
	0560	NOTE: For pipe sizes 1-1/4" and smaller use 1/4" larger diameter										
	0570	insulation to allow room for installation over cable.										
940	0010	**WATER HEATER PACKAGED SYSTEMS**										**940**
	1000	Car wash package, continuous duty, high recovery, gas fired										
	1040	100° rise, 180 MBH input, 174 GPH	1 Plum	3	2.667	Ea.	3,025	109		3,134	3,500	
	1060	280 MBH input, 270 GPH		2.50	3.200		3,225	131		3,356	3,750	
	1080	400 MBH input, 386 GPH		2.50	3.200		3,725	131		3,856	4,275	
	1100	480 MBH input, 464 GPH		2	4		3,925	163		4,088	4,550	
	1120	605 MBH input, 584 GPH		1.50	5.333		4,475	218		4,693	5,250	
	1140	700 MBH input, 676 GPH	↓	1	8		4,650	325		4,975	5,600	
	1160	1000 MBH input, 966 GPH	Q-1	1.60	10		5,175	370		5,545	6,250	
	1180	1200 MBH input, 1159 GPH		1.40	11.429		7,025	420		7,445	8,350	
	1200	1400 MBH input, 1353 GPH	↓	1.20	13.333	↓	7,675	490		8,165	9,150	
	3000	Combination dishwasher & general purpose, 2 temp. gas fired										
	3040	124 GPH @ 140° rise; 434 GPH @ 40° rise	1 Plum	2	4	Ea.	3,075	163		3,238	3,650	
	3060	193 GPH @ 140° rise; 677 GPH @ 40° rise		1.60	5		3,525	204		3,729	4,175	
	3080	276 GPH @ 140° rise; 969 GPH @ 40° rise		1.40	5.714		4,000	233		4,233	4,750	
	3100	330 GPH @ 140° rise; 1160 GPH @ 40°rise		1.20	6.667		4,250	272		4,522	5,075	
	3120	417 GPH @ 140° rise; 1460 GPH @ 40°rise	↓	1	8		4,800	325		5,125	5,775	
	3140	481 GPH @ 140° rise; 1685 GPH @ 40°rise	Q-1	1.40	11.429		5,550	420		5,970	6,725	
	3160	688 GPH @ 140° rise; 2410 GPH @ 40°rise		1.20	13.333		6,075	490		6,565	7,400	
	3180	828 GPH @ 140° rise; 2790 GPH @ 40°rise		1	16		7,950	590		8,540	9,600	
	3200	965 GPH @ 140° rise; 3370 GPH @ 40°rise	↓	.80	20		8,575	735		9,310	10,500	
	3960	For unit base, add				↓	237			237	261	

15480 | Domestic Water Heaters

		CREW	DAILY OUTPUT	LABOR-HOURS	UNIT	2005 BARE COSTS				TOTAL INCL O&P
						MAT.	LABOR	EQUIP.	TOTAL	
940	5000	Coin laundry units, gas fired, 100° rise									**940**
	5020	Single heater,								
	5040	280 MBH input, 270 GPH	1 Plum	1.60	5	Ea.	3,525	204		3,729	4,175
	5060	400 MBH input, 386 GPH		1.30	6.154		4,650	251		4,901	5,475
	5080	480 MBH input, 464 GPH	↓	1	8		4,900	325		5,225	5,875
	5100	605 MBH input, 584 GPH	Q-1	1.40	11.429		5,750	420		6,170	6,950
	5120	700 MBH input, 676 GPH		1.20	13.333		6,075	490		6,565	7,400
	5140	1000 MBH input, 966 GPH		1	16		6,600	590		7,190	8,125
	5160	1200 MBH input, 1159 GPH		.90	17.778		8,575	655		9,230	10,400
	5180	1400 MBH input, 1353 GPH	↓	.70	22.857	↓	9,500	840		10,340	11,800
	6000	Multiple heater									
	6040	560 MBH input, 540 GPH	1 Plum	1	8	Ea.	5,575	325		5,900	6,625
	6060	800 MBH input, 772 GPH	Q-1	1.30	12.308		7,225	455		7,680	8,625
	6080	960 MBH input, 928 GPH		1.20	13.333		8,150	490		8,640	9,700
	6100	1210 MBH input, 1168 GPH		1	16		9,175	590		9,765	11,000
	6120	1400 MBH input, 1352 GPH		1	16		9,600	590		10,190	11,500
	6140	1700 MBH input, 1642 GPH		.80	20		11,000	735		11,735	13,200
	6160	2000 MBH input, 1932 GPH	↓	.70	22.857		13,200	840		14,040	15,900
	6180	2400 MBH input, 2318 GPH	Q-2	.90	26.667		14,900	1,025		15,925	17,800
	6200	2600 MBH input, 2512 GPH		.80	30		16,600	1,150		17,750	19,900
	6220	2800 MBH input, 2706 GPH	↓	.70	34.286	↓	18,700	1,300		20,000	22,600
	6800	Standard system sizing is based on									
	6820	80% of the washers operating at one time.									

15490 | Pool and Fountain Equipment

		CREW	DAILY OUTPUT	LABOR-HOURS	UNIT	2005 BARE COSTS				TOTAL INCL O&P
100	0010	**FOUNTAINS/AERATORS**									**100**
	0100	Pump w/controls								
	0200	Single phase, 100' cord, 1/2 H.P. pump	2 Skwk	4.40	3.636	Ea.	2,675	127		2,802	3,150
	0300	3/4 H.P. pump		4.30	3.721		3,075	130		3,205	3,575
	0400	1 H.P. pump		4.20	3.810		3,100	133		3,233	3,600
	0500	1-1/2 H.P. pump		4.10	3.902		3,200	136		3,336	3,725
	0600	2 H.P. pump		4	4		3,250	139		3,389	3,800
	0700	Three phase, 200' cord, 5 H.P. pump		3.90	4.103		7,850	143		7,993	8,875
	0800	7-1/2 H.P. pump		3.80	4.211		8,875	147		9,022	9,975
	0900	10 H.P. pump		3.70	4.324		9,900	151		10,051	11,100
	1000	15 H.P. pump		3.60	4.444		11,700	155		11,855	13,100
	1100	Nozzles, minimum		8	2		197	69.50		266.50	325
	1200	Maximum		8	2		259	69.50		328.50	395
	1300	Lights w/mounting kits, 200 watt		18	.889		325	31		356	410
	1400	300 watt		18	.889		360	31		391	445
	1500	500 watt		18	.889		395	31		426	485
	1600	Color blender	↓	12	1.333	↓	305	46.50		351.50	410

15500 | Heat Generation Equipment

15510 | Heating Boilers and Accessories

		CREW	DAILY OUTPUT	LABOR-HOURS	UNIT	2005 BARE COSTS				TOTAL INCL O&P
						MAT.	LABOR	EQUIP.	TOTAL	
120	0010	**BURNERS**									**120**
	0990	Residential, conversion, gas fired, LP or natural									

Important: See the Reference Section for critical supporting data - Reference Nos., Crews, & City Cost Indexes

			CREW	DAILY OUTPUT	LABOR-HOURS	UNIT	2005 BARE COSTS				TOTAL INCL O&P	
15510	**Heating Boilers and Accessories**						MAT.	LABOR	EQUIP.	TOTAL		
120	1000	Gun type, atmospheric input 72 to 200 MBH	Q-1	2.50	6.400	Ea.	655	235		890	1,075	**120**
	1020	120 to 360 MBH	"	2	8	"	725	294		1,019	1,250	
	3000	Flame retention oil fired assembly, input										
	3020	.50 to 2.25 GPH	Q-1	2.40	6.667	Ea.	291	245		536	690	
	3040	2.0 to 5.0 GPH	"	2	8		525	294		819	1,025	
	4600	Gas safety, shut off valve, 3/4" threaded	1 Stpi	20	.400		165	16.40		181.40	207	
	4610	1" threaded		19	.421		166	17.25		183.25	209	
	4620	1-1/4" threaded		15	.533		181	22		203	232	
	4630	1-1/2" threaded		13	.615		184	25		209	240	
	4640	2" threaded		11	.727		219	30		249	286	
	4650	2-1/2" threaded	Q-1	15	1.067		251	39		290	335	
	4660	3" threaded		13	1.231		345	45.50		390.50	450	
	4670	4" flanged		3	5.333		2,375	196		2,571	2,900	
	4680	6" flanged	Q-2	3	8		5,275	305		5,580	6,250	
300	0010	**BOILERS, ELECTRIC, ASME** Standard controls and trim [D3020 102]										**300**
	1000	Steam, 6 KW, 20.5 MBH	Q-19	1.20	20	Ea.	2,725	765		3,490	4,150	
	1040	9 KW, 30.7 MBH		1.20	20		2,775	765		3,540	4,200	
	1060	18 KW, 61.4 MBH		1.20	20		2,850	765		3,615	4,300	
	1080	24 KW, 81.8 MBH		1.10	21.818		3,325	830		4,155	4,900	
	1120	36 KW, 123 MBH		1.10	21.818		3,650	830		4,480	5,275	
	2000	Hot water, 7.5 KW, 25.6 MBH		1.30	18.462		2,900	705		3,605	4,225	
	2020	15 KW, 51.2 MBH		1.30	18.462		2,925	705		3,630	4,275	
	2040	30 KW, 102 MBH		1.20	20		3,050	765		3,815	4,525	
	2060	45 KW, 164 MBH		1.20	20		3,450	765		4,215	4,925	
	2070	60 KW, 205 MBH		1.20	20		3,650	765		4,415	5,150	
	2080	75 KW, 256 MBH		1.10	21.818		3,900	830		4,730	5,550	
400	0010	**BOILERS, GAS FIRED** Natural or propane, standard controls										**400**
	1000	Cast iron, with insulated jacket										
	2000	Steam, gross output, 81 MBH	Q-7	1.40	22.857	Ea.	1,450	890		2,340	2,950	
	2020	102 MBH		1.30	24.615		1,650	960		2,610	3,250	
	2040	122 MBH		1	32		1,775	1,250		3,025	3,825	
	2060	163 MBH		.90	35.556		2,050	1,400		3,450	4,325	
	2080	203 MBH		.90	35.556		2,375	1,400		3,775	4,675	
	2100	240 MBH		.85	37.647		2,525	1,475		4,000	4,975	
	2120	280 MBH		.80	40		2,750	1,550		4,300	5,375	
	2140	320 MBH		.70	45.714		3,000	1,775		4,775	5,975	
	3000	Hot water, gross output, 80 MBH		1.46	21.918		1,550	855		2,405	2,975	
	3020	100 MBH		1.35	23.704		1,750	925		2,675	3,325	
	3040	122 MBH		1.10	29.091		1,875	1,125		3,000	3,750	
	3060	163 MBH		1	32		2,150	1,250		3,400	4,250	
	3080	203 MBH		1	32		2,200	1,250		3,450	4,300	
	3100	240 MBH		.95	33.684		2,375	1,325		3,700	4,600	
	3120	280 MBH		.90	35.556		2,625	1,400		4,025	4,975	
	3140	320 MBH		.80	40		2,800	1,550		4,350	5,425	
	6000	Hot water, including burner & one zone valve, gross output										
	6010	51.2 MBH	Q-6	2	12	Ea.	1,650	460		2,110	2,525	
	6020	72 MBH		2	12		2,025	460		2,485	2,925	
	6040	89 MBH		1.90	12.632		2,075	485		2,560	3,000	
	6060	105 MBH		1.80	13.333		2,325	510		2,835	3,325	
	6080	132 MBH		1.70	14.118		2,650	540		3,190	3,725	
	6100	155 MBH		1.50	16		3,075	610		3,685	4,300	
	6110	186 MBH		1.40	17.143		3,700	655		4,355	5,050	
	7000	For tankless water heater on smaller gas units, add					10%					
	7050	For additional zone valves up to 312 MBH add				Ea.	117			117	129	
	7990	Special feature gas fired boilers										
	8000	Pulse combustion, standard controls / trim										

	15510	Heating Boilers and Accessories	CREW	DAILY OUTPUT	LABOR-HOURS	UNIT	2005 BARE COSTS				TOTAL INCL O&P	
							MAT.	LABOR	EQUIP.	TOTAL		
400	8050	88,000 BTU	Q-5	1.40	11.429	Ea.	3,600	420		4,020	4,575	**400**
	8080	134,000 BTU	"	1.20	13.333	"	4,300	490		4,790	5,475	
500	0010	**BOILERS, OIL FIRED** Standard controls, flame retention burner										**500**
	1000	Cast iron, with insulated flush jacket										
	2000	Steam, gross output, 109 MBH	Q-7	1.20	26.667	Ea.	1,925	1,050		2,975	3,700	
	2020	144 MBH		1.10	29.091		1,975	1,125		3,100	3,875	
	2040	173 MBH		1	32		2,650	1,250		3,900	4,800	
	2060	207 MBH		.90	35.556		2,775	1,400		4,175	5,125	
	3000	Hot water, same price as steam										
	4000	For tankless coil in smaller sizes, add					15%					
	5000	Steel, insulated jacket, burner										
	6000	Steam, full water leg construction, gross output										
	6020	144 MBH	Q-6	1.33	18.045	Ea.	4,050	690		4,740	5,475	
	6040	198 MBH	"	1.20	20	"	4,400	765		5,165	6,000	
	6400	Larger sizes are same as steel, gas fired										
	7000	Hot water, gross output, 103 MBH	Q-6	1.60	15	Ea.	1,250	575		1,825	2,250	
	7020	122 MBH		1.45	16.506		2,475	630		3,105	3,675	
	7040	137 MBH		1.36	17.595		2,725	670		3,395	4,000	
	7060	168 MBH		1.30	18.405		3,125	705		3,830	4,475	
	7080	225 MBH		1.22	19.704		3,225	755		3,980	4,650	
	7340	For tankless coil in steam or hot water, add					7%					
880	0010	**SWIMMING POOL HEATERS** Not including wiring, external										**880**
	0020	piping, base or pad,										
	0060	Gas fired, input, 115 MBH	Q-6	3	8	Ea.	1,850	305		2,155	2,500	
	0100	135 MBH		2	12		2,100	460		2,560	3,000	
	0160	155 MBH		1.50	16		2,200	610		2,810	3,350	
	0180	175 MBH		1.30	18.462		2,450	705		3,155	3,750	
	0200	190 MBH		1	24		2,875	915		3,790	4,525	
	0220	235 MBH		.70	34.286		3,500	1,300		4,800	5,825	
	0240	285 MBH		.60	40		4,500	1,525		6,025	7,250	
	0260	365 MBH		.50	48		5,075	1,825		6,900	8,325	
	0280	500 MBH		.40	60		6,950	2,300		9,250	11,100	
	0300	600 MBH		.35	68.571		8,000	2,625		10,625	12,800	
	0320	800 MBH		.33	72.727		9,525	2,775		12,300	14,700	
	0360	1000 MBH		.22	109		11,100	4,175		15,275	18,600	
	0370	1,200 MBH		.21	114		13,900	4,375		18,275	21,800	
	0380	1,500 MBH		.19	126		17,100	4,825		21,925	26,100	
	0400	1,800 MBH		.14	171		19,400	6,550		25,950	31,200	
	0410	2,450 MBH		.13	184		25,800	7,050		32,850	39,000	
	0420	3,000 MBH		.11	218		31,500	8,350		39,850	47,100	
	0440	3,750 MBH		.09	266		39,100	10,200		49,300	58,500	
	1000	Oil fired, gross output, 97 MBH		1.70	14.118		2,125	540		2,665	3,125	
	1020	118 MBH		1.70	14.118		2,175	540		2,715	3,200	
	1040	134 MBH		1.60	15		2,300	575		2,875	3,400	
	1060	161 MBH		1.30	18.462		2,800	705		3,505	4,150	
	1080	187 MBH		1.20	20		2,975	765		3,740	4,400	
	1100	214 MBH		1.10	21.818		3,375	835		4,210	4,975	
	1120	262 MBH		.89	26.966		3,825	1,025		4,850	5,750	
	1140	340 MBH		.68	35.294		5,050	1,350		6,400	7,575	
	1160	420 MBH		.58	41.379		5,800	1,575		7,375	8,750	
	1180	525 MBH		.46	52.174		7,425	2,000		9,425	11,200	
	1200	630 MBH		.41	58.537		8,275	2,225		10,500	12,500	
	1220	735 MBH		.35	68.571		10,200	2,625		12,825	15,200	
	1240	840 MBH		.33	72.727		11,600	2,775		14,375	17,000	
	1260	1,050 MBH		.27	88.889		14,200	3,400		17,600	20,700	

Important: See the Reference Section for critical supporting data - Reference Nos., Crews, & City Cost Indexes

15510	Heating Boilers and Accessories	CREW	DAILY OUTPUT	LABOR-HOURS	UNIT	2005 BARE COSTS				TOTAL INCL O&P
						MAT.	LABOR	EQUIP.	TOTAL	
880	1280	1,365 MBH	Q-6	.21	114	Ea.	18,200	4,375		22,575	26,700
	1300	1,680 MBH		.18	133		21,400	5,100		26,500	31,300
	1320	2,310 MBH		.13	184		27,600	7,050		34,650	41,000
	1340	2,835 MBH		.11	218		34,400	8,350		42,750	50,500
	1360	3,150 MBH	▼	.09	266		38,600	10,200		48,800	58,000
	2000	Electric, 12 KW, 4,800 gallon pool	Q-19	3	8		1,725	305		2,030	2,350
	2020	15 KW, 7,200 gallon pool		2.80	8.571		1,725	325		2,050	2,400
	2040	24 KW, 9,600 gallon pool		2.40	10		2,325	380		2,705	3,150
	2060	30 KW, 12,000 gallon pool		2	12		2,425	460		2,885	3,325
	2080	35 KW, 14,400 gallon pool		1.60	15		2,425	570		2,995	3,525
	2100	55 KW, 24,000 gallon pool	▼	1.20	20	▼	3,325	765		4,090	4,800
	9000	To select pool heater: 12 BTUH x S.F. pool area									
	9010	X temperature differential =required output									
	9050	For electric, KW = gallons x 2.5 divided by 1000									
	9100	For family home type pool, double the									
	9110	Rated gallon capacity = 1/2°F rise per hour									

15550	Breechings, Chimneys & Stacks									
200	0010	**DRAFT CONTROL DEVICES**									
	5000	Vent damper, bi-metal, gas, 3" diameter	Q-9	24	.667	Ea.	23.50	24		47.50	62.50
	5010	4" diameter		24	.667		23.50	24		47.50	62.50
	5101	Electric, automatic, gas, 4" diameter		24	.667		145	24		169	196
	5250	Automatic, oil, 4" diameter	▼	24	.667	▼	162	24		186	215
440	0010	**VENT CHIMNEY** Prefab metal, U.L. listed									
	0020	Gas, double wall, galvanized steel									
	0080	3" diameter	Q-9	72	.222	V.L.F.	3.77	8.05		11.82	16.50
	0100	4" diameter		68	.235		4.71	8.50		13.21	18.30
	0120	5" diameter		64	.250		5.50	9.05		14.55	19.95
	0140	6" diameter		60	.267		6.40	9.65		16.05	22
	0160	7" diameter		56	.286		9.40	10.35		19.75	26.50
	0180	8" diameter		52	.308		10.45	11.15		21.60	28.50
	0200	10" diameter		48	.333		22	12.05		34.05	43
	0220	12" diameter		44	.364		29.50	13.15		42.65	52.50
	0240	14" diameter		42	.381		49.50	13.80		63.30	75.50
	0260	16" diameter		40	.400		67	14.45		81.45	96.50
	0280	18" diameter	▼	38	.421		86.50	15.25		101.75	119
	0300	20" diameter	Q-10	36	.667		102	25		127	151
	0320	22" diameter		34	.706		129	26.50		155.50	183
	0340	24" diameter	▼	32	.750	▼	159	28		187	219
	0600	For 4", 5" and 6" oval, add					50%				
	0650	Gas, double wall, galvanized steel, fittings									
	0660	Elbow 45°, 3" diameter	Q-9	36	.444	Ea.	9.35	16.10		25.45	35
	0670	4" diameter		34	.471		11.05	17.05		28.10	38
	0680	5" diameter		32	.500		12.95	18.10		31.05	42.50
	0690	6" diameter		30	.533		15.55	19.30		34.85	46.50
	0700	7" diameter		28	.571		22.50	20.50		43	56.50
	0710	8" diameter		26	.615		26	22.50		48.50	63
	0720	10" diameter		24	.667		64	24		88	108
	0730	12" diameter		22	.727		81	26.50		107.50	130
	0740	14" diameter		21	.762		113	27.50		140.50	167
	0750	16" diameter		20	.800		142	29		171	201
	0760	18" diameter	▼	19	.842		191	30.50		221.50	257
	0770	20" diameter	Q-10	18	1.333		216	50		266	315
	0780	22" diameter		17	1.412		325	53		378	435
	0790	24" diameter	▼	16	1.500	▼	415	56.50		471.50	540

15550	Breechings, Chimneys & Stacks	CREW	DAILY OUTPUT	LABOR-HOURS	UNIT	2005 BARE COSTS				TOTAL INCL O&P
						MAT.	LABOR	EQUIP.	TOTAL	
0916	Adjustable length									
0918	3" diameter, to 12"	Q-9	36	.444	Ea.	7.65	16.10		23.75	33
0920	4" diameter, to 12"		34	.471		9.05	17.05		26.10	36
0924	6" diameter, to 12"		30	.533		13.90	19.30		33.20	45
0928	8" diameter, to 12"		26	.615		17.15	22.50		39.65	53.50
0930	10" diameter, to 18"		24	.667		64.50	24		88.50	108
0932	12" diameter, to 18"		22	.727		78.50	26.50		105	127
0936	16" diameter, to 18"		20	.800		151	29		180	211
0938	18" diameter, to 18"	▼	19	.842		180	30.50		210.50	246
0944	24" diameter, to 18"	Q-10	16	1.500		325	56.50		381.50	440
0950	Elbow 90°, adjustable, 3" diameter	Q-9	36	.444		15.55	16.10		31.65	41.50
0960	4" diameter		34	.471		18.95	17.05		36	47
0970	5" diameter		32	.500		22	18.10		40.10	52.50
0980	6" diameter		30	.533		26.50	19.30		45.80	58.50
0990	7" diameter		28	.571		42	20.50		62.50	78
1010	8" diameter		26	.615		44.50	22.50		67	83.50
1020	Wall thimble, 4 to 7" adjustable, 3" diameter		36	.444		9.85	16.10		25.95	35.50
1022	4" diameter		34	.471		11	17.05		28.05	38
1024	5" diameter		32	.500		12.35	18.10		30.45	41.50
1026	6" diameter		30	.533		15.10	19.30		34.40	46
1028	7" diameter		28	.571		17.60	20.50		38.10	51.50
1030	8" diameter		26	.615		18.20	22.50		40.70	54.50
1040	Roof flashing, 3" diameter		36	.444		5.60	16.10		21.70	30.50
1050	4" diameter		34	.471		6.35	17.05		23.40	33
1060	5" diameter		32	.500		9.85	18.10		27.95	39
1070	6" diameter		30	.533		13.80	19.30		33.10	44.50
1080	7" diameter		28	.571		17.40	20.50		37.90	51
1090	8" diameter		26	.615		18.90	22.50		41.40	55.50
1100	10" diameter		24	.667		51.50	24		75.50	94
1110	12" diameter		22	.727		66	26.50		92.50	113
1120	14" diameter		20	.800		81	29		110	134
1130	16" diameter		18	.889		100	32		132	160
1140	18" diameter	▼	16	1		136	36		172	205
1150	20" diameter	Q-10	18	1.333		176	50		226	270
1160	22" diameter		14	1.714		223	64.50		287.50	345
1170	24" diameter	▼	12	2		268	75		343	410
1200	Tee, 3" diameter	Q-9	27	.593		23	21.50		44.50	58.50
1210	4" diameter		26	.615		24	22.50		46.50	61
1220	5" diameter		25	.640		26	23		49	64
1230	6" diameter		24	.667		29	24		53	68.50
1240	7" diameter		23	.696		31	25		56	72.50
1250	8" diameter		22	.727		42.50	26.50		69	87
1260	10" diameter		21	.762		115	27.50		142.50	169
1270	12" diameter		20	.800		138	29		167	197
1280	14" diameter		18	.889		225	32		257	298
1290	16" diameter		16	1		320	36		356	405
1300	18" diameter	▼	14	1.143		400	41.50		441.50	505
1310	20" diameter	Q-10	17	1.412		540	53		593	675
1320	22" diameter		13	1.846		670	69.50		739.50	845
1330	24" diameter	▼	12	2		790	75		865	980
1460	Tee cap, 3" diameter	Q-9	45	.356		1.63	12.85		14.48	21.50
1470	4" diameter		42	.381		1.85	13.80		15.65	23
1480	5" diameter		40	.400		2.47	14.45		16.92	25
1490	6" diameter		37	.432		2.86	15.65		18.51	27
1500	7" diameter		35	.457		4.65	16.55		21.20	30.50
1510	8" diameter	▼	34	.471		5.45	17.05		22.50	32

		CREW	DAILY OUTPUT	LABOR-HOURS	UNIT	2005 BARE COSTS				TOTAL INCL O&P	
						MAT.	LABOR	EQUIP.	TOTAL		
440	1520	10" diameter	Q-9	32	.500	Ea.	10.35	18.10		28.45	39.50	440
	1530	12" diameter		30	.533		13.85	19.30		33.15	45	
	1540	14" diameter		28	.571		28.50	20.50		49	63.50	
	1550	16" diameter		25	.640		33	23		56	72	
	1560	18" diameter		24	.667		37.50	24		61.50	78	
	1570	20" diameter	Q-10	27	.889		40	33.50		73.50	95.50	
	1580	22" diameter		22	1.091		46	41		87	114	
	1590	24" diameter		21	1.143		51.50	43		94.50	123	
	1750	Top, 3" diameter	Q-9	46	.348		8.65	12.60		21.25	29	
	1760	4" diameter		44	.364		9.05	13.15		22.20	30	
	1770	5" diameter		42	.381		12.35	13.80		26.15	34.50	
	1780	6" diameter		40	.400		15.85	14.45		30.30	40	
	1790	7" diameter		38	.421		23.50	15.25		38.75	49.50	
	1800	8" diameter		36	.444		30.50	16.10		46.60	58	
	1810	10" diameter		34	.471		71	17.05		88.05	105	
	1820	12" diameter		32	.500		107	18.10		125.10	146	
	1830	14" diameter		30	.533		123	19.30		142.30	165	
	1840	16" diameter		28	.571		221	20.50		241.50	275	
	1850	18" diameter		26	.615		390	22.50		412.50	465	
	1860	20" diameter	Q-10	28	.857		435	32		467	525	
	1870	22" diameter		22	1.091		590	41		631	715	
	1880	24" diameter		20	1.200		735	45		780	875	
	1882	26" diameter		19	1.263		1,100	47.50		1,147.50	1,300	
	1884	28" diameter		18	1.333		1,275	50		1,325	1,475	
	1886	30" diameter		17	1.412		1,425	53		1,478	1,625	
	1888	32" diameter		16	1.500		1,500	56.50		1,556.50	1,725	
	1890	34" diameter		15	1.600		1,600	60		1,660	1,850	
	1892	36" diameter		14	1.714		1,700	64.50		1,764.50	1,975	
	1900	Gas, double wall, galvanized steel, oval										
	1904	4" x 1'	Q-9	68	.235	V.L.F.	13.50	8.50		22	28	
	1906	5"		64	.250		31.50	9.05		40.55	48.50	
	1908	5"/6" x 1'		60	.267		29	9.65		38.65	47	
	1910	Oval fittings										
	1912	Adjustable length										
	1914	4" diameter to 12" lg	Q-9	34	.471	Ea.	12.80	17.05		29.85	40	
	1916	5" diameter to 12" lg		32	.500		35.50	18.10		53.60	67	
	1918	5"/6" diameter to 12"		30	.533		32.50	19.30		51.80	65	
	1920	Elbow 45°										
	1922	4"	Q-9	34	.471	Ea.	23.50	17.05		40.55	51.50	
	1924	5"		32	.500		44.50	18.10		62.60	76.50	
	1926	5"/6"		30	.533		42	19.30		61.30	75.50	
	1930	Elbow 45°, flat										
	1932	4"	Q-9	34	.471	Ea.	23.50	17.05		40.55	51.50	
	1934	5"		32	.500		44.50	18.10		62.60	76.50	
	1936	5"/6"		30	.533		42	19.30		61.30	75.50	
	1940	Top										
	1942	4"	Q-9	44	.364	Ea.	17.40	13.15		30.55	39	
	1944	5"		42	.381		42.50	13.80		56.30	67.50	
	1946	5"/6"		40	.400		26.50	14.45		40.95	51.50	
	1950	Adjustable flashing										
	1952	4"	Q-9	34	.471	Ea.	10.75	17.05		27.80	38	
	1954	5"		32	.500		31	18.10		49.10	62	
	1956	5"/6"		30	.533		21	19.30		40.30	52.50	
	1960	Tee										
	1962	4"	Q-9	26	.615	Ea.	34.50	22.50		57	72.50	
	1964	5"		25	.640		73	23		96	116	

| | | | DAILY | LABOR- | | 2005 BARE COSTS | | | | TOTAL |
15550	Breechings, Chimneys & Stacks	CREW	OUTPUT	HOURS	UNIT	MAT.	LABOR	EQUIP.	TOTAL	INCL O&P
440	**1966**	5"/6"	Q-9	24	.667	Ea.	67.50	24		91.50	111	**440**
1970	Tee with short snout									
1972	4"	Q-9	26	.615	Ea.	34.50	22.50		57	72.50
3200	All fuel, pressure tight, double wall, U.L. listed, 1400°F.									
3210	304 stainless steel liner, aluminized steel outer jacket									
3220	6" diameter	Q-9	60	.267	L.F.	36.50	9.65		46.15	55
3221	8" diameter		52	.308		45	11.15		56.15	66.50
3222	10" diameter		48	.333		50	12.05		62.05	73.50
3223	12" diameter		44	.364		57.50	13.15		70.65	83
3224	14" diameter		42	.381		65.50	13.80		79.30	93
3225	16" diameter		40	.400		69.50	14.45		83.95	99
3226	18" diameter		38	.421		78.50	15.25		93.75	110
3227	20" diameter	Q-10	36	.667		89.50	25		114.50	137
3228	24" diameter	"	32	.750		115	28		143	170
3260	For 316 stainless steel liner add					6%				
3280	All fuel, pressure tight, double wall fittings									
3284	304 stainless steel inner, aluminized steel jacket									
3288	Adjustable 18"/30" section									
3291	5" diameter	Q-9	32	.500	Ea.	94	18.10		112.10	132
3292	6" diameter		30	.533		119	19.30		138.30	161
3293	8" diameter		26	.615		137	22.50		159.50	185
3294	10" diameter		24	.667		154	24		178	206
3295	12" diameter		22	.727		175	26.50		201.50	233
3296	14" diameter		21	.762		202	27.50		229.50	265
3297	16" diameter		20	.800		210	29		239	276
3298	18" diameter		19	.842		259	30.50		289.50	330
3299	20" diameter	Q-10	18	1.333		298	50		348	405
3300	24" diameter	"	16	1.500		385	56.50		441.50	510
3326	For 316 stainless steel liner, add					15%				
3350	Elbow 90° fixed									
3353	5" diameter	Q-9	32	.500	Ea.	176	18.10		194.10	221
3354	6" diameter		30	.533		206	19.30		225.30	257
3355	8" diameter		26	.615		228	22.50		250.50	286
3356	10" diameter		24	.667		262	24		286	325
3357	12" diameter		22	.727		296	26.50		322.50	365
3358	14" diameter		21	.762		345	27.50		372.50	425
3359	16" diameter		20	.800		385	29		414	470
3360	18" diameter		19	.842		450	30.50		480.50	540
3361	20" diameter	Q-10	18	1.333		505	50		555	630
3362	24" diameter	"	16	1.500		625	56.50		681.50	770
3380	For 316 stainless steel liner, add					20%				
3400	Elbow 45°									
3404	6" diameter	Q-9	30	.533	Ea.	130	19.30		149.30	173
3405	8" diameter		26	.615		148	22.50		170.50	197
3406	10" diameter		24	.667		166	24		190	219
3407	12" diameter		22	.727		189	26.50		215.50	249
3408	14" diameter		21	.762		217	27.50		244.50	282
3409	16" diameter		20	.800		245	29		274	315
3410	18" diameter		19	.842		277	30.50		307.50	350
3411	20" diameter	Q-10	18	1.333		315	50		365	425
3412	24" diameter	"	16	1.500		405	56.50		461.50	530
3430	For 316 stainless steel liner, add					20%				
3450	Tee 90°									
3454	6" diameter	Q-9	24	.667	Ea.	136	24		160	187
3455	8" diameter		22	.727		157	26.50		183.50	213
3456	10" diameter		21	.762		212	27.50		239.50	277

Important: See the Reference Section for critical supporting data - Reference Nos., Crews, & City Cost Indexes

		CREW	DAILY OUTPUT	LABOR-HOURS	UNIT	2005 BARE COSTS				TOTAL INCL O&P
	15550 | Breechings, Chimneys & Stacks					MAT.	LABOR	EQUIP.	TOTAL	
440 3457	12" diameter	Q-9	20	.800	Ea.	214	29		243	280
3458	14" diameter		18	.889		253	32		285	330
3459	16" diameter		16	1		284	36		320	365
3460	18" diameter		14	1.143		335	41.50		376.50	435
3461	20" diameter	Q-10	17	1.412		375	53		428	495
3462	24" diameter	"	12	2		465	75		540	625
3480	For Tee Cap, add					35%	20%			
3500	For 316 stainless steel liner, add					22%				
3520	Plate support, galvanized									
3524	6" diameter	Q-9	26	.615	Ea.	74.50	22.50		97	117
3525	8" diameter		22	.727		84	26.50		110.50	133
3526	10" diameter		20	.800		92	29		121	146
3527	12" diameter		18	.889		95.50	32		127.50	155
3528	14" diameter		17	.941		116	34		150	180
3529	16" diameter		16	1		122	36		158	190
3530	18" diameter		15	1.067		128	38.50		166.50	201
3531	20" diameter	Q-10	16	1.500		136	56.50		192.50	237
3532	24" diameter	"	14	1.714		143	64.50		207.50	257
3570	Bellows, lined, 316 stainless steel only									
3574	6" diameter	Q-9	30	.533	Ea.	315	19.30		334.30	375
3575	8" diameter		26	.615		340	22.50		362.50	410
3576	10" diameter		24	.667		375	24		399	445
3577	12" diameter		22	.727		415	26.50		441.50	495
3578	14" diameter		21	.762		465	27.50		492.50	560
3579	16" diameter		20	.800		515	29		544	615
3580	18" diameter		19	.842		575	30.50		605.50	680
3581	20" diameter	Q-10	18	1.333		650	50		700	790
3600	Ventilated roof thimble, 304 stainless steel									
3620	6" diameter	Q-9	26	.615	Ea.	183	22.50		205.50	236
3624	8" diameter		22	.727		196	26.50		222.50	257
3625	10" diameter		20	.800		212	29		241	279
3626	12" diameter		18	.889		229	32		261	300
3627	14" diameter		17	.941		247	34		281	325
3628	16" diameter		16	1		262	36		298	345
3629	18" diameter		15	1.067		282	38.50		320.50	370
3630	20" diameter	Q-10	16	1.500		300	56.50		356.50	420
3631	24" diameter	"	14	1.714		340	64.50		404.50	475
3650	For 316 stainless steel, add					10%				
3670	Exit cone, 316 stainless steel only									
3674	6" diameter	Q-9	46	.348	Ea.	129	12.60		141.60	161
3675	8" diameter		42	.381		136	13.80		149.80	170
3676	10" diameter		40	.400		146	14.45		160.45	184
3677	12" diameter		38	.421		154	15.25		169.25	193
3678	14" diameter		37	.432		170	15.65		185.65	211
3679	16" diameter		36	.444		188	16.10		204.10	232
3680	18" diameter		35	.457		231	16.55		247.55	280
3681	20" diameter	Q-10	28	.857		275	32		307	350
3682	24" diameter	"	26	.923		355	34.50		389.50	445
3720	Roof support assy., incl. 30" pipe sect, 304 st. st.									
3724	6" diameter	Q-9	25	.640	Ea.	211	23		234	268
3725	8" diameter		21	.762		230	27.50		257.50	296
3726	10" diameter		19	.842		252	30.50		282.50	325
3727	12" diameter		17	.941		268	34		302	345
3728	14" diameter		16	1		300	36		336	385
3729	16" diameter		15	1.067		320	38.50		358.50	410
3730	18" diameter		14	1.143		345	41.50		386.50	445

		CREW	DAILY OUTPUT	LABOR-HOURS	UNIT	MAT.	LABOR	EQUIP.	TOTAL	TOTAL INCL O&P
15550	**Breechings, Chimneys & Stacks**							2005 BARE COSTS		

440

		CREW	DAILY OUTPUT	LABOR-HOURS	UNIT	MAT.	LABOR	EQUIP.	TOTAL	TOTAL INCL O&P
3731	20" diameter	Q-10	15	1.600	Ea.	365	60		425	495
3732	24" diameter	"	13	1.846		405	69.50		474.50	555
3750	For 316 stainless steel, add				↓	20%				
3770	Stack cap, 316 stainless steel only									
3773	5" diameter	Q-9	48	.333	Ea.	123	12.05		135.05	154
3774	6" diameter		46	.348		144	12.60		156.60	177
3775	8" diameter		42	.381		167	13.80		180.80	204
3776	10" diameter		40	.400		196	14.45		210.45	239
3777	12" diameter		38	.421		227	15.25		242.25	274
3778	14" diameter		37	.432		268	15.65		283.65	320
3779	16" diameter		36	.444		305	16.10		321.10	360
3780	18" diameter		35	.457		345	16.55		361.55	405
3781	20" diameter	Q-10	28	.857		395	32		427	485
3782	24" diameter	"	26	.923		475	34.50		509.50	580
3800	Round storm collar									
3806	5" diameter	Q-9	32	.500	Ea.	21.50	18.10		39.60	52
7800	All fuel, double wall, stainless steel, 6" diameter		60	.267	V.L.F.	32.50	9.65		42.15	50.50
7802	7" diameter		56	.286		42	10.35		52.35	62
7804	8" diameter		52	.308		49	11.15		60.15	71
7806	10" diameter		48	.333		71.50	12.05		83.55	97
7808	12" diameter		44	.364		96	13.15		109.15	125
7810	14" diameter		42	.381		126	13.80		139.80	159
8000	All fuel, double wall, stainless steel fittings									
8010	Roof support 6" diameter	Q-9	30	.533	Ea.	80.50	19.30		99.80	118
8020	7" diameter		28	.571		91	20.50		111.50	132
8030	8" diameter		26	.615		102	22.50		124.50	147
8040	10" diameter		24	.667		117	24		141	166
8050	12" diameter		22	.727		141	26.50		167.50	196
8060	14" diameter		21	.762		179	27.50		206.50	239
8100	Elbow 15°, 6" diameter		30	.533		129	19.30		148.30	172
8120	7" diameter		28	.571		145	20.50		165.50	192
8140	8" diameter		26	.615		165	22.50		187.50	217
8160	10" diameter		24	.667		216	24		240	275
8180	12" diameter		22	.727		262	26.50		288.50	330
8200	14" diameter		21	.762		315	27.50		342.50	390
8300	Insulated tee with insulated tee cap, 6" diameter		30	.533		122	19.30		141.30	164
8340	7" diameter		28	.571		160	20.50		180.50	208
8360	8" diameter		26	.615		180	22.50		202.50	233
8380	10" diameter		24	.667		253	24		277	315
8400	12" diameter		22	.727		355	26.50		381.50	430
8420	14" diameter		21	.762		465	27.50		492.50	560
8500	Joist shield, 6" diameter		30	.533		37	19.30		56.30	70.50
8510	7" diameter		28	.571		40.50	20.50		61	76.50
8520	8" diameter		26	.615		49.50	22.50		72	89
8530	10" diameter		24	.667		67	24		91	111
8540	12" diameter		22	.727		84	26.50		110.50	133
8550	14" diameter		21	.762		104	27.50		131.50	157
8600	Round top, 6" diameter		30	.533		41	19.30		60.30	75
8620	7" diameter		28	.571		55	20.50		75.50	92.50
8640	8" diameter		26	.615		115	22.50		137.50	161
8660	10" diameter		24	.667		136	24		160	186
8680	12" diameter		22	.727		194	26.50		220.50	254
8700	14" diameter		21	.762		259	27.50		286.50	330
8800	Adjustable roof flashing, 6" diameter		30	.533		49	19.30		68.30	83.50
8820	7" diameter		28	.571		56	20.50		76.50	93.50
8840	8" diameter		26	.615		61	22.50		83.50	102

440

Important: See the Reference Section for critical supporting data - Reference Nos., Crews, & City Cost Indexes

15550	Breechings, Chimneys & Stacks	CREW	DAILY OUTPUT	LABOR-HOURS	UNIT	2005 BARE COSTS				TOTAL INCL O&P
						MAT.	LABOR	EQUIP.	TOTAL	
440	8860	10" diameter	Q-9	24	.667	Ea.	78.50	24		102.50	123	440
8880	12" diameter		22	.727		101	26.50		127.50	152
8900	14" diameter		21	.762		126	27.50		153.50	182
9000	High temp. (2000° F), steel jacket, acid resist. refractory lining									
9010	11 ga. galvanized jacket, U.L. listed									
9020	Straight section, 48" long, 10" diameter	Q-10	13.30	1.805	Ea.	355	67.50		422.50	495
9030	12" diameter		11.20	2.143		385	80.50		465.50	550
9040	18" diameter		7.40	3.243		545	122		667	785
9050	24" diameter		4.60	5.217		745	196		941	1,125
9120	Tee section, 10" diameter		4.40	5.455		755	205		960	1,150
9130	12" diameter		3.70	6.486		795	243		1,038	1,250
9140	18" diameter		2.40	10		1,075	375		1,450	1,750
9150	24" diameter		1.50	16		1,475	600		2,075	2,550
9220	Cleanout pier section, 10" diameter		3.50	6.857		580	257		837	1,025
9230	12" diameter		2.50	9.600		610	360		970	1,225
9240	18" diameter		1.90	12.632		780	475		1,255	1,600
9250	24" diameter		1.30	18.462		990	695		1,685	2,175
9320	For drain, add					53%				
9330	Elbow, 30° and 45°, 10" diameter	Q-10	6.60	3.636		495	136		631	750
9340	12" diameter		5.60	4.286		530	161		691	825
9350	18" diameter		3.70	6.486		795	243		1,038	1,250
9360	24" diameter		2.30	10.435		1,200	390		1,590	1,900
9430	For 60° and 90° elbow, add					112%				
9440	End cap, 10" diameter	Q-10	26	.923		845	34.50		879.50	985
9450	12" diameter		22	1.091		965	41		1,006	1,125
9460	18" diameter		15	1.600		1,200	60		1,260	1,425
9470	24" diameter		9	2.667		1,475	100		1,575	1,750
9540	Increaser (1 diameter), 10" diameter		6.60	3.636		420	136		556	670
9550	12" diameter		5.60	4.286		450	161		611	740
9560	18" diameter		3.70	6.486		655	243		898	1,100
9570	24" diameter		2.30	10.435		940	390		1,330	1,625
9600	For expansion joints, add to straight section					7%				
9610	For 1/4" hot rolled steel jacket, add					157%				
9620	For 2950°F very high temperature, add					89%				
9630	26 ga. aluminized jacket, straight section, 48" long									
9640	10" diameter	Q-10	15.30	1.569	V.L.F.	210	59		269	320
9650	12" diameter		12.90	1.860		223	70		293	350
9660	18" diameter		8.50	2.824		325	106		431	520
9670	24" diameter		5.30	4.528		460	170		630	765
9700	Accessories (all models)									
9710	Guy band, 10" diameter	Q-10	32	.750	Ea.	65	28		93	115
9720	12" diameter		30	.800		69	30		99	122
9730	18" diameter		26	.923		89.50	34.50		124	152
9740	24" diameter		24	1		132	37.50		169.50	203
9810	Draw band, galv. stl., 11 gauge, 10" diameter		32	.750		57	28		85	106
9820	12" diameter		30	.800		61	30		91	113
9830	18" diameter		26	.923		76	34.50		110.50	138
9840	24" diameter		24	1		96.50	37.50		134	164
9910	Draw band, aluminized stl., 26 gauge, 10" diameter		32	.750		16.25	28		44.25	61.50
9920	12" diameter		30	.800		16.25	30		46.25	64
9930	18" diameter		26	.923		20.50	34.50		55	76
9940	24" diameter		24	1		24.50	37.50		62	84.50

15705 | Curbs/Pads/Stands Prefab

			CREW	DAILY OUTPUT	LABOR-HOURS	UNIT	MAT.	LABOR	EQUIP.	TOTAL	TOTAL INCL O&P	
600	0010	**CURBS/PADS PREFABRICATED**										600
	6000	Pad, fiberglass reinforced concrete with polystyrene foam core										
	6050	Condenser, 2" thick, 20" x 38"	1 Shee	8	1	Ea.	8.85	40		48.85	71.50	
	6220	30" x 36"	"	8	1		15.05	40		55.05	78.50	
	6340	36" x 54"	Q-9	6	2.667	▼	28.50	96.50		125	179	

15710 | Heat Exchangers

			CREW	DAILY OUTPUT	LABOR-HOURS	UNIT	MAT.	LABOR	EQUIP.	TOTAL	TOTAL INCL O&P	
900	0010	**HEAT EXCHANGERS**										900
	0016	Shell & tube type, 2 or 4 pass, 3/4" O.D. copper tubes,										
	0020	C.I. heads, C.I. tube sheet, steel shell										
	0100	Hot water 40°F to 180°F, by steam at 10 PSI										
	0120	8 GPM	Q-5	6	2.667	Ea.	960	98.50		1,058.50	1,200	
	0140	10 GPM		5	3.200		1,450	118		1,568	1,775	
	0160	40 GPM	▼	4	4	▼	2,250	147		2,397	2,700	
	0500	For bronze head and tube sheet, add					50%					
	1000	Hot water 40°F to 140°F, by water at 200°F										
	1020	7 GPM	Q-5	6	2.667	Ea.	1,175	98.50		1,273.50	1,450	
	1040	16 GPM		5	3.200		1,675	118		1,793	2,025	
	1060	34 GPM		4	4		2,525	147		2,672	3,000	
	1080	55 GPM		3	5.333		3,675	197		3,872	4,350	
	1100	74 GPM		1.50	10.667		4,600	395		4,995	5,650	
	1120	86 GPM	▼	1.40	11.429		6,150	420		6,570	7,375	
	1140	112 GPM	Q-6	2	12		7,650	460		8,110	9,100	
	1160	126 GPM		1.80	13.333		9,550	510		10,060	11,300	
	1180	152 GPM	▼	1	24	▼	12,200	915		13,115	14,800	
	3000	Plate type,										
	3100	400 GPM	Q-6	.80	30	Ea.	18,100	1,150		19,250	21,600	
	3120	800 GPM	"	.50	48		31,300	1,825		33,125	37,300	
	3140	1200 GPM	Q-7	.34	94.118		38,000	3,675		41,675	47,300	
	3160	1800 GPM	"	.24	133	▼	56,500	5,200		61,700	70,500	
	8000	Heat pipe type, glycol, 50% efficient										
	8010	100 MBH, 1700 CFM	1 Stpi	.80	10	Ea.	2,425	410		2,835	3,300	
	8020	160 MBH, 2700 CFM		.60	13.333		3,275	545		3,820	4,425	
	8030	620 MBH, 4000 CFM	▼	.40	20	▼	4,925	820		5,745	6,650	

15730 | Unitary Air Conditioning Equip

			CREW	DAILY OUTPUT	LABOR-HOURS	UNIT	MAT.	LABOR	EQUIP.	TOTAL	TOTAL INCL O&P	
500	0010	**PACKAGED TERMINAL AIR CONDITIONER** Cabinet, wall sleeve,										500
	0100	louver, electric heat, thermostat, manual changeover, 208 V										
	0200	6,000 BTUH cooling, 8800 BTU heat	Q-5	6	2.667	Ea.	955	98.50		1,053.50	1,200	
	0220	9,000 BTUH cooling, 13,900 BTU heat		5	3.200		1,000	118		1,118	1,275	
	0240	12,000 BTUH cooling, 13,900 BTU heat		4	4		1,100	147		1,247	1,450	
	0260	15,000 BTUH cooling, 13,900 BTU heat	▼	3	5.333		1,300	197		1,497	1,750	
800	0010	**WINDOW UNIT AIR CONDITIONERS**										800
	4000	Portable/window, 15 amp 125V grounded receptacle required										
	4060	5000 BTUH	1 Carp	8	1	Ea.	194	34.50		228.50	267	
	4340	6000 BTUH		8	1		204	34.50		238.50	278	
	4480	8000 BTUH		6	1.333		290	45.50		335.50	390	
	4500	10,000 BTUH	▼	6	1.333		390	45.50		435.50	495	
	4520	12,000 BTUH	L-2	8	2	▼	475	60		535	615	
	4600	Window/thru-the-wall, 15 amp 230V grounded receptacle required										
	4780	17,000 BTUH	L-2	6	2.667	Ea.	615	79.50		694.50	800	
	4940	25,000 BTUH		4	4		780	120		900	1,050	
	4960	29,000 BTUH	▼	4	4	▼	840	120		960	1,100	

Important: See the Reference Section for critical supporting data - Reference Nos., Crews, & City Cost Indexes

15740 | Heat Pumps

		DAILY OUTPUT	LABOR-HOURS	UNIT	2005 BARE COSTS				TOTAL INCL O&P
		CREW			MAT.	LABOR	EQUIP.	TOTAL	
100	0010	**AIR-SOURCE HEAT PUMPS** (Not including interconnecting tubing)								100
	1000	Air to air, split system, not including curbs, pads, or ductwork								
	1010	For curbs/pads see division 15705								
	1015	1.5 ton cooling, 7 MBH heat @ 0°F	Q-5	1.22	13.115	Ea.	1,375	485		1,860	2,225
	1020	2 ton cooling, 8.5 MBH heat @ 0°F		1.20	13.333		1,425	490		1,915	2,300
	1030	2.5 ton cooling, 10 MBH heat @ 0°F		1	16		1,575	590		2,165	2,625
	1040	3 ton cooling, 13 MBH heat @ 0°F		.80	20		1,725	735		2,460	3,000
	1050	3.5 ton cooling, 18 MBH heat @ 0°F		.75	21.333		1,975	785		2,760	3,350
	1054	4 ton cooling, 24 MBH heat @ 0°F		.60	26.667		2,150	985		3,135	3,850
	1060	5 ton cooling, 27 MBH heat @ 0°F	↓	.50	32	↓	2,450	1,175		3,625	4,475
	1300	Supplementary electric heat coil, included								
	1500	Single package, not including curbs, pads, or plenums								
	1502	1/2 ton cooling, supplementary heat not incl.	Q-5	8	2	Ea.	1,025	73.50		1,098.50	1,225
	1504	3/4 ton cooling, supplementary heat not incl.		6	2.667		1,100	98.50		1,198.50	1,350
	1506	1 ton cooling, supplementary heat not incl.		4	4		1,250	147		1,397	1,600
	1510	1.5 ton cooling, 5 MBH heat @ 0°F		1.55	10.323		1,925	380		2,305	2,700
	1520	2 ton cooling, 6.5 MBH heat @ 0°F		1.50	10.667		2,200	395		2,595	3,000
	1540	2.5 ton cooling, 8 MBH heat @ 0°F		1.40	11.429		2,375	420		2,795	3,250
	1560	3 ton cooling, 10 MBH heat @ 0°F		1.20	13.333		2,625	490		3,115	3,625
	1570	3.5 ton cooling, 11 MBH heat @ 0°F		1	16		2,850	590		3,440	4,025
	1580	4 ton cooling, 13 MBH heat @ 0°F		.96	16.667		3,075	615		3,690	4,325
	1620	5 ton cooling, 27 MBH heat @ 0°F		.65	24.615		3,325	905		4,230	5,025
	1640	7.5 ton cooling, 35 MBH heat @ 0°F	↓	.40	40	↓	5,075	1,475		6,550	7,825
800	0010	**WATER-SOURCE HEAT PUMPS** (Not including interconnecting tubing)									800
	2000	Water source to air, single package								
	2100	1 ton cooling, 13 MBH heat @ 75°F	Q-5	2	8	Ea.	990	295		1,285	1,550
	2120	1.5 ton cooling, 17 MBH heat @ 75°F		1.80	8.889		1,075	330		1,405	1,700
	2140	2 ton cooling, 19 MBH heat @ 75°F		1.70	9.412		1,125	345		1,470	1,750
	2160	2.5 ton cooling, 25 MBH heat @ 75°F		1.60	10		1,200	370		1,570	1,875
	2180	3 ton cooling, 27 MBH heat @ 75°F		1.40	11.429		1,275	420		1,695	2,050
	2190	3.5 ton cooling, 29 MBH heat @ 75°F		1.30	12.308		1,325	455		1,780	2,150
	2200	4 ton cooling, 31 MBH heat @ 75°F		1.20	13.333		1,475	490		1,965	2,375
	2220	5 ton cooling, 29 MBH heat @ 75°F	↓	.90	17.778		1,725	655		2,380	2,875
	3960	For supplementary heat coil, add				↓	10%				

15760 | Terminal Heating & Cooling Units

		CREW	DAILY OUTPUT	LABOR-HOURS	UNIT	MAT.	LABOR	EQUIP.	TOTAL	TOTAL INCL O&P
250	0010	**ELECTRIC HEATING**, not incl. conduit or feed wiring									250
	1100	Rule of thumb: Baseboard units, including control	1 Elec	4.40	1.818	kW	77	74		151	195
	1300	Baseboard heaters, 2' long, 375 watt		8	1	Ea.	29	41		70	92.50
	1400	3' long, 500 watt		8	1		34.50	41		75.50	98.50
	1600	4' long, 750 watt		6.70	1.194		41	48.50		89.50	118
	1800	5' long, 935 watt		5.70	1.404		48.50	57		105.50	139
	2000	6' long, 1125 watt		5	1.600		54	65		119	157
	2200	7' long, 1310 watt		4.40	1.818		59	74		133	175
	2400	8' long, 1500 watt		4	2		68	81.50		149.50	196
	2600	9' long, 1680 watt		3.60	2.222		76	90.50		166.50	219
	2800	10' long, 1875 watt	↓	3.30	2.424	↓	84.50	99		183.50	240
	2950	Wall heaters with fan, 120 to 277 volt								
	3160	Recessed, residential, 750 watt	1 Elec	6	1.333	Ea.	89.50	54.50		144	180
	3170	1000 watt		6	1.333		92	54.50		146.50	182
	3180	1250 watt		5	1.600		101	65		166	208
	3190	1500 watt		4	2		101	81.50		182.50	232
	3600	Thermostats, integral		16	.500	↓	18	20.50		38.50	50.50
	3800	Line voltage, 1 pole	↓	8	1	↓	22.50	41		63.50	85

		CREW	DAILY OUTPUT	LABOR-HOURS	UNIT	2005 BARE COSTS				TOTAL INCL O&P	
15760	**Terminal Heating & Cooling Units**					MAT.	LABOR	EQUIP.	TOTAL		
3810	2 pole	1 Elec	8	1	Ea.	29	41		70	92.50	250
3820	Low voltage, 1 pole	↓	8	1	↓	25	41		66	88	
4000	Heat trace system, 400 degree										
4020	115V, 2.5 watts per L.F.	1 Elec	530	.015	L.F.	6.70	.61		7.31	8.25	
4030	5 watts per L.F.		530	.015		6.70	.61		7.31	8.25	
4050	10 watts per L.F.		530	.015		6.70	.61		7.31	8.25	
4060	208V, 5 watts per L.F.		530	.015		6.70	.61		7.31	8.25	
4080	480V, 8 watts per L.F.	↓	530	.015	↓	6.70	.61		7.31	8.25	
4200	Heater raceway										
4260	Heat transfer cement										
4280	1 gallon				Ea.	53.50			53.50	58.50	
4300	5 gallon				"	218			218	240	
4320	Snap band, clamp										
4340	3/4" pipe size	1 Elec	470	.017	Ea.		.69		.69	1.03	
4360	1" pipe size		444	.018			.73		.73	1.09	
4380	1-1/4" pipe size		400	.020			.82		.82	1.21	
4400	1-1/2" pipe size		355	.023			.92		.92	1.37	
4420	2" pipe size		320	.025			1.02		1.02	1.52	
4440	3" pipe size		160	.050			2.04		2.04	3.03	
4460	4" pipe size		100	.080			3.26		3.26	4.85	
4480	Thermostat NEMA 3R, 22 amp, 0-150 Deg, 10' cap.		8	1		182	41		223	261	
4500	Thermostat NEMA 4X, 25 amp, 40 Deg, 5-1/2' cap.		7	1.143		182	46.50		228.50	270	
4520	Thermostat NEMA 4X, 22 amp, 25-325 Deg, 10' cap.		7	1.143		490	46.50		536.50	610	
4540	Thermostat NEMA 4X, 22 amp, 15-140 Deg,		6	1.333		410	54.50		464.50	535	
4580	Thermostat NEMA 4,7,9, 22 amp, 25-325 Deg, 10' cap.		3.60	2.222		605	90.50		695.50	800	
4600	Thermostat NEMA 4,7,9, 22 amp, 15-140 Deg,		3	2.667		600	109		709	820	
4720	Fiberglass application tape, 36 yard roll		11	.727		55	29.50		84.50	105	
5000	Radiant heating ceiling panels, 2' x 4', 500 watt		16	.500		203	20.50		223.50	254	
5050	750 watt		16	.500		223	20.50		243.50	277	
5200	For recessed plaster frame, add		32	.250		42	10.20		52.20	61.50	
5300	Infrared quartz heaters, 120 volts, 1000 watts		6.70	1.194		125	48.50		173.50	211	
5350	1500 watt		5	1.600		125	65		190	235	
5400	240 volts, 1500 watt		5	1.600		125	65		190	235	
5450	2000 watt		4	2		125	81.50		206.50	259	
5500	3000 watt		3	2.667		145	109		254	320	
5550	4000 watt		2.60	3.077		145	125		270	345	
5570	Modulating control	↓	.80	10	↓	72.50	410		482.50	685	
5600	Unit heaters, heavy duty, with fan & mounting bracket										
5650	Single phase, 208-240-277 volt, 3 kW	1 Elec	3.20	2.500	Ea.	335	102		437	515	
5750	5 kW		2.40	3.333		345	136		481	580	
5800	7 kW		1.90	4.211		530	172		702	840	
5850	10 kW		1.30	6.154		605	251		856	1,050	
5950	15 kW		.90	8.889		975	360		1,335	1,625	
6000	480 volt, 3 kW		3.30	2.424		345	99		444	525	
6020	4 kW		3	2.667		375	109		484	570	
6040	5 kW		2.60	3.077		385	125		510	605	
6060	7 kW		2	4		570	163		733	870	
6080	10 kW		1.40	5.714		625	233		858	1,025	
6100	13 kW		1.10	7.273		975	296		1,271	1,525	
6120	15 kW		1	8		975	325		1,300	1,550	
6140	20 kW		.90	8.889		1,275	360		1,635	1,950	
6300	3 phase, 208-240 volt, 5 kW		2.40	3.333		325	136		461	555	
6320	7 kW		1.90	4.211		500	172		672	805	
6340	10 kW		1.30	6.154		540	251		791	970	
6360	15 kW		.90	8.889		905	360		1,265	1,550	
6380	20 kW	↓	.70	11.429	↓	1,300	465		1,765	2,125	

Important: See the Reference Section for critical supporting data - Reference Nos., Crews, & City Cost Indexes

		DAILY	LABOR-		2005 BARE COSTS				TOTAL	
15760 | **Terminal Heating & Cooling Units**	CREW	OUTPUT	HOURS	UNIT	MAT.	LABOR	EQUIP.	TOTAL	INCL O&P	
250	6400	25 kW	1 Elec	.50	16	Ea.	1,525	650		2,175	2,650	**250**
	6500	480 volt, 5 kW		2.60	3.077		455	125		580	685	
	6520	7 kW		2	4		570	163		733	875	
	6540	10 kW		1.40	5.714		605	233		838	1,000	
	6560	13 kW		1.10	7.273		975	296		1,271	1,525	
	6580	15 kW		1	8		975	325		1,300	1,550	
	6600	20 kW	▼	.90	8.889	▼	1,075	360		1,435	1,750	
	6800	Vertical discharge heaters, with fan										
	6820	Single phase, 208-240-277 volt, 10 kW	1 Elec	1.30	6.154	Ea.	560	251		811	990	
	6840	15 kW		.90	8.889		940	360		1,300	1,575	
	6900	3 phase, 208-240 volt, 10 kW		1.30	6.154		540	251		791	970	
	6920	15 kW		.90	8.889		905	360		1,265	1,525	
	6940	20 kW		.70	11.429		1,300	465		1,765	2,125	
	7100	480 volt, 10 kW		1.40	5.714		605	233		838	1,000	
	7120	15 kW		1	8		975	325		1,300	1,550	
	7140	20 kW		.90	8.889		1,275	360		1,635	1,950	
	7160	25 kW	▼	.60	13.333	▼	1,525	545		2,070	2,475	
	7900	Cabinet convector heaters, 240 volt										
	7920	3' long, 2000 watt	1 Elec	5.30	1.509	Ea.	1,500	61.50		1,561.50	1,750	
	7940	3000 watt		5.30	1.509		1,575	61.50		1,636.50	1,825	
	7960	4000 watt		5.30	1.509		1,650	61.50		1,711.50	1,925	
	7980	6000 watt		4.60	1.739		1,725	71		1,796	2,000	
	8000	8000 watt		4.60	1.739		1,725	71		1,796	1,975	
	8020	4' long, 4000 watt		4.60	1.739		1,650	71		1,721	1,900	
	8040	6000 watt		4	2		1,725	81.50		1,806.50	2,025	
	8060	8000 watt		4	2		1,825	81.50		1,906.50	2,125	
	8080	10,000 watt	▼	4	2	▼	1,900	81.50		1,981.50	2,225	
	8100	Available also in 208 or 277 volt										
	8200	Cabinet unit heaters, 120 to 277 volt, 1 pole,										
	8220	wall mounted, 2 kW	1 Elec	4.60	1.739	Ea.	1,500	71		1,571	1,750	
	8230	3 kW		4.60	1.739		1,550	71		1,621	1,825	
	8240	4 kW		4.40	1.818		1,600	74		1,674	1,875	
	8250	5 kW		4.40	1.818		1,675	74		1,749	1,925	
	8260	6 kW		4.20	1.905		1,675	77.50		1,752.50	1,975	
	8270	8 kW		4	2		1,750	81.50		1,831.50	2,050	
	8280	10 kW		3.80	2.105		1,775	86		1,861	2,100	
	8290	12 kW		3.50	2.286		1,825	93		1,918	2,150	
	8300	13.5 kW		2.90	2.759		1,850	112		1,962	2,225	
	8310	16 kW		2.70	2.963		1,900	121		2,021	2,275	
	8320	20 kW		2.30	3.478		2,850	142		2,992	3,350	
	8330	24 kW		1.90	4.211		2,900	172		3,072	3,450	
	8350	Recessed, 2 kW		4.40	1.818		1,500	74		1,574	1,725	
	8370	3 kW		4.40	1.818		1,550	74		1,624	1,825	
	8380	4 kW		4.20	1.905		1,600	77.50		1,677.50	1,900	
	8390	5 kW		4.20	1.905		1,675	77.50		1,752.50	1,950	
	8400	6 kW		4	2		1,700	81.50		1,781.50	2,000	
	8410	8 kW		3.80	2.105		1,775	86		1,861	2,075	
	8420	10 kW		3.50	2.286		2,200	93		2,293	2,575	
	8430	12 kW		2.90	2.759		2,200	112		2,312	2,600	
	8440	13.5 kW		2.70	2.963		2,225	121		2,346	2,625	
	8450	16 kW		2.30	3.478		2,250	142		2,392	2,675	
	8460	20 kW		1.90	4.211		2,700	172		2,872	3,225	
	8470	24 kW		1.60	5		2,775	204		2,979	3,350	
	8490	Ceiling mounted, 2 kW		3.20	2.500		1,500	102		1,602	1,775	
	8510	3 kW		3.20	2.500		1,550	102		1,652	1,875	
	8520	4 kW	▼	3	2.667	▼	1,625	109		1,734	1,925	

			DAILY	LABOR-		2005 BARE COSTS				TOTAL	
15760	**Terminal Heating & Cooling Units**	CREW	OUTPUT	HOURS	UNIT	MAT.	LABOR	EQUIP.	TOTAL	INCL O&P	
250	8530	5 kW	1 Elec	3	2.667	Ea.	1,675	109		1,784	1,975	250
	8540	6 kW		2.80	2.857		1,675	116		1,791	2,025	
	8550	8 kW		2.40	3.333		1,750	136		1,886	2,125	
	8560	10 kW		2.20	3.636		1,775	148		1,923	2,200	
	8570	12 kW		2	4		1,775	163		1,938	2,225	
	8580	13.5 kW		1.50	5.333		1,850	217		2,067	2,375	
	8590	16 kW		1.30	6.154		1,900	251		2,151	2,475	
	8600	20 kW		.90	8.889		2,700	360		3,060	3,525	
	8610	24 kW	▼	.60	13.333	▼	2,750	545		3,295	3,825	
	8630	208 to 480 V, 3 pole										
	8650	Wall mounted, 2 kW	1 Elec	4.60	1.739	Ea.	1,500	71		1,571	1,725	
	8670	3 kW		4.60	1.739		1,550	71		1,621	1,825	
	8680	4 kW		4.40	1.818		1,600	74		1,674	1,875	
	8690	5 kW		4.40	1.818		1,675	74		1,749	1,925	
	8700	6 kW		4.20	1.905		1,675	77.50		1,752.50	1,975	
	8710	8 kW		4	2		1,750	81.50		1,831.50	2,050	
	8720	10 kW		3.80	2.105		1,775	86		1,861	2,100	
	8730	12 kW		3.50	2.286		1,775	93		1,868	2,125	
	8740	13.5 kW		2.90	2.759		1,850	112		1,962	2,225	
	8750	16 kW		2.70	2.963		2,050	121		2,171	2,425	
	8760	20 kW		2.30	3.478		2,700	142		2,842	3,175	
	8770	24 kW		1.90	4.211		2,750	172		2,922	3,275	
	8790	Recessed, 2 kW		4.40	1.818		1,500	74		1,574	1,725	
	8810	3 kW		4.40	1.818		1,550	74		1,624	1,825	
	8820	4 kW		4.20	1.905		1,600	77.50		1,677.50	1,900	
	8830	5 kW		4.20	1.905		1,675	77.50		1,752.50	1,950	
	8840	6 kW		4	2		1,675	81.50		1,756.50	1,975	
	8850	8 kW		3.80	2.105		1,750	86		1,836	2,050	
	8860	10 kW		3.50	2.286		1,775	93		1,868	2,100	
	8870	12 kW		2.90	2.759		1,775	112		1,887	2,125	
	8880	13.5 kW		2.70	2.963		1,900	121		2,021	2,275	
	8890	16 kW		2.30	3.478		1,900	142		2,042	2,300	
	8900	20 kW		1.90	4.211		2,700	172		2,872	3,225	
	8920	24 kW		1.60	5		2,750	204		2,954	3,325	
	8940	Ceiling mount, 2 kW		3.20	2.500		1,500	102		1,602	1,775	
	8950	3 kW		3.20	2.500		1,550	102		1,652	1,875	
	8960	4 kW		3	2.667		1,600	109		1,709	1,925	
	8970	5 kW		3	2.667		1,675	109		1,784	1,975	
	8980	6 kW		2.80	2.857		1,675	116		1,791	2,025	
	8990	8 kW		2.40	3.333		1,750	136		1,886	2,125	
	9000	10 kW		2.20	3.636		1,775	148		1,923	2,200	
	9020	13.5 kW		1.50	5.333		1,850	217		2,067	2,375	
	9030	16 kW		1.30	6.154		1,900	251		2,151	2,475	
	9040	20 kW		.90	8.889		2,700	360		3,060	3,525	
	9060	24 kW	▼	.60	13.333	▼	2,750	545		3,295	3,825	
600	0010	**HYDRONIC HEATING** Terminal units, not incl. main supply pipe										600
	1000	Radiation										
	1100	Panel, baseboard, C.I., including supports, no covers	Q-5	46	.348	L.F.	24.50	12.80		37.30	46.50	
	1310	Baseboard, pkgd, 1/2" copper tube, alum. fin, 7" high		60	.267		5.35	9.85		15.20	20.50	
	1320	3/4" copper tube, alum. fin, 7" high	▼	58	.276	▼	5.65	10.15		15.80	21.50	
	1381	Rough in baseboard panel & fin tube, supply & balance valves	1 Stpi	1.06	7.547	Ea.	107	310		417	585	
	1500	Note: fin tube may also require corners, caps, etc.										
	1990	Convector unit, floor recessed, flush, with trim										
	2000	for under large glass wall areas, no damper	Q-5	20	.800	L.F.	25.50	29.50		55	72.50	
	2100	For unit with damper				"	36			36	40	

Important: See the Reference Section for critical supporting data - Reference Nos., Crews, & City Cost Indexes

		15760	**Terminal Heating & Cooling Units**	CREW	DAILY OUTPUT	LABOR-HOURS	UNIT	2005 BARE COSTS				TOTAL INCL O&P
								MAT.	LABOR	EQUIP.	TOTAL	
600	3000		Radiators, cast iron									6
	3100		Free standing or wall hung, 6 tube, 25" high	Q-5	96	.167	Section	24	6.15		30.15	36
	3150		4 tube 25" high		96	.167		17.90	6.15		24.05	29
	3200		4 tube, 19" high	↓	96	.167	↓	16.65	6.15		22.80	27.50
	3250		Adj. brackets, 2 per wall radiator up to 30 sections	1 Stpi	32	.250	Ea.	17.10	10.25		27.35	34
	3500		Recessed, 20" high x 5" deep, without grille	Q-5	60	.267	Section	19.20	9.85		29.05	36
	3600		For inlet grille, add				"	2.06			2.06	2.27
	9500		To convert SFR to BTU rating: Hot water, 150 x SFR									
	9510		Forced hot water, 180 x SFR; steam, 240 x SFR									
700	0010		**INFRARED UNIT**									7
	0020		Gas fired, unvented, electric ignition, 100% shutoff.									
	0030		Piping and wiring not included									
	0060		Input, 15 MBH	Q-5	7	2.286	Ea.	360	84		444	525
	0100		30 MBH		6	2.667		375	98.50		473.50	560
	0120		45 MBH		5	3.200		430	118		548	650
	0140		50 MBH		4.50	3.556		450	131		581	690
	0160		60 MBH		4	4		465	147		612	730
	0180		75 MBH		3	5.333		540	197		737	890
	0200		90 MBH		2.50	6.400		560	236		796	970
	0220		105 MBH		2	8		685	295		980	1,200
	0240		120 MBH	↓	2	8	↓	895	295		1,190	1,425
	2000		Electric, single or three phase									
	2050		6 kW, 20,478 BTU	1 Elec	2.30	3.478	Ea.	370	142		512	620
	2100		13.5 KW, 40,956 BTU		2.20	3.636		585	148		733	865
	2150		24 KW, 81,912 BTU	↓	2	4	↓	1,250	163		1,413	1,625
	3000		Oil fired, pump, controls, fusible valve, oil supply tank									
	3050		91,000 BTU	Q-5	2.50	6.400	Ea.	5,700	236		5,936	6,625
	3080		105.000 BTU		2.25	7.111		5,975	262		6,237	6,950
	3110		119,000 BTU	↓	2	8	↓	6,550	295		6,845	7,650

		15770	**Floor-Heating & Snow-Melting Eq.**									
500	0010		**RADIANT FLOOR HEATING**									5
	0100		Tubing, PEX (cross-linked polyethylene)									
	0110		Oxygen barrier type for systems with ferrous materials									
	0120		1/2"	Q-5	800	.020	L.F.	.43	.74		1.17	1.58
	0130		3/4"		535	.030		.55	1.10		1.65	2.27
	0140		1"	↓	400	.040	↓	1.25	1.47		2.72	3.60
	0200		Non barrier type for ferrous free systems									
	0210		1/2"	Q-5	800	.020	L.F.	.23	.74		.97	1.36
	0220		3/4"		535	.030		.38	1.10		1.48	2.08
	0230		1"	↓	400	.040	↓	.66	1.47		2.13	2.95
	1000		Manifolds									
	1110		Brass									
	1120		Valved									
	1130		1", 2 circuit	Q-5	13	1.231	Ea.	38.50	45.50		84	110
	1140		1", 3 circuit		11	1.455		52	53.50		105.50	138
	1150		1", 4 circuit		9	1.778		70.50	65.50		136	176
	1160		1-1/4", 3 circuit		10	1.600		54	59		113	148
	1170		1-1/4", 4 circuit	↓	8	2	↓	72	73.50		145.50	190
	1300		Valveless									
	1310		1", 2 circuit	Q-5	14	1.143	Ea.	19.50	42		61.50	85
	1320		1", 3 circuit		12	1.333		26.50	49		75.50	103
	1330		1", 4 circuit		10	1.600		34	59		93	126
	1340		1-1/4", 3 circuit		11	1.455		32.50	53.50		86	116
	1350		1-1/4", 4 circuit	↓	9	1.778	↓	37	65.50		102.50	139

15770	Floor-Heating & Snow-Melting Eq.	CREW	DAILY OUTPUT	LABOR HOURS	UNIT	2005 BARE COSTS				TOTAL INCL O&P
						MAT.	LABOR	EQUIP.	TOTAL	
1400	Manifold connector kit									
1410	Mounting brackets, couplings, end cap, air vent, valve, plug									
1420	1"	Q-5	8	2	Ea.	66	73.50		139.50	184
1430	1-1/4"	"	7	2.286	"	79	84		163	214
1610	Copper, (cut to size)									
1620	1" x 1/2" x 72" Lg, 2" sweat drops, 24 circuit	Q-5	3.33	4.805	Ea.	69	177		246	340
1630	1-1/4" x 1/2" x 72" Lg, 2" sweat drops, 24 circuit		3.20	5		81	184		265	365
1640	1-1/2" x 1/2" x 72" Lg, 2" sweat drops, 24 circuit		3	5.333		99	197		296	405
1650	1-1/4" x 3/4" x 72" Lg, 2" sweat drops, 24 circuit		3.10	5.161		86.50	190		276.50	380
1660	1-1/2" x 3/4" x 72" Lg, 2" sweat drops, 24 circuit		2.90	5.517		104	203		307	420
1710	Copper, (modular)									
1720	1" x 1/2", 2" sweat drops, 2 circuit	Q-5	17.50	.914	Ea.	6.75	33.50		40.25	58
1730	1" x 1/2", 2" sweat drops, 3 circuit		14.50	1.103		9.75	40.50		50.25	72
1740	1" x 1/2", 2" sweat drops, 4 circuit		12.30	1.301		11.25	48		59.25	84.50
1750	1-1/4" x 1/2", 2" sweat drops, 2 circuit		17	.941		7.50	34.50		42	60.50
1760	1-1/4" x 1/2", 2" sweat drops, 3 circuit		14	1.143		11.25	42		53.25	76
1770	1-1/4" x 1/2", 2" sweat drops, 4 circuit		11.80	1.356		15	50		65	91.50
3000	Valves									
3110	Motorized zone valve operator, 24V	Q-5	30	.533	Ea.	40.50	19.65		60.15	74
3120	Motorized zone valve with operator complete, 24V									
3130	3/4"	Q-5	35	.457	Ea.	70	16.85		86.85	102
3140	1"		32	.500		79.50	18.45		97.95	115
3150	1-1/4"		29.60	.541		97.50	19.90		117.40	137
3200	Flow control/isolation valve, 1/2"		32	.500		21	18.45		39.45	50.50
3300	Thermostatic mixing valves									
3310	1/2"	Q-5	26.60	.601	Ea.	73	22		95	114
3320	3/4"		25	.640		88.50	23.50		112	133
3330	1"		21.40	.748		94.50	27.50		122	146
3400	3 Way mixing / diverting valve, manual, brass									
3410	1/2"	Q-5	21.30	.751	Ea.	96	27.50		123.50	148
3420	3/4"		20	.800		99	29.50		128.50	154
3430	1"		17.80	.899		115	33		148	176
3500	4 Way mixing / diverting valve, manual, brass									
3510	1/2"	Q-5	16	1	Ea.	101	37		138	167
3520	3/4"		15	1.067		112	39.50		151.50	182
3530	1"		13.30	1.203		125	44.50		169.50	205
3540	1-1/4"		11.40	1.404		137	51.50		188.50	229
3550	1-1/2"		10.60	1.509		185	55.50		240.50	287
3560	2"		10	1.600		199	59		258	310
3570	2-1/2"		8	2		550	73.50		623.50	710
3800	Motor control, 3 or 4 way, for valves up to 1-1/2"		30	.533		138	19.65		157.65	182
5000	Radiant floor heating, zone control panel									
5110	6 Zone control									
5120	For zone valves	Q-5	16	1	Ea.	131	37		168	200
5130	For circulators	"	16	1	"	185	37		222	260
7000	PEX tubing fittings, compression type									
7110	PEX tubing to brass valve conn., 1/2"	Q-5	54	.296	Ea.	2.25	10.90		13.15	18.90
7200	PEX x male NPT									
7210	1/2" x 1/2"	Q-5	54	.296	Ea.	4.37	10.90		15.27	21
7220	3/4" x 3/4"		46	.348		7.30	12.80		20.10	27.50
7230	1" x 1"		40	.400		17.65	14.75		32.40	41.50
7300	PEX coupling									
7310	1/2" x 1/2"	Q-5	50	.320	Ea.	5.25	11.80		17.05	23.50
7320	3/4" x 3/4"		42	.381		15.20	14.05		29.25	37.50
7330	1" x 1"		36	.444		20.50	16.40		36.90	47
7400	PEX x copper sweat									

Important: See the Reference Section for critical supporting data - Reference Nos., Crews, & City Cost Indexes

15700 | Heating/Ventilating/Air Conditioning Equipment

15770	Floor-Heating & Snow-Melting Eq.	CREW	DAILY OUTPUT	LABOR-HOURS	UNIT	2005 BARE COSTS				TOTAL INCL O&P	
						MAT.	LABOR	EQUIP.	TOTAL		
500 7410	1/2" x 1/2"	Q-5	38	.421	Ea.	3.88	15.50		19.38	28	500
7420	3/4" x 3/4"		34	.471		8.05	17.35		25.40	35	
7430	1" x 1"	↓	29	.552	↓	17.65	20.50		38.15	50	

15800 | Air Distribution

15830	Fans	CREW	DAILY OUTPUT	LABOR-HOURS	UNIT	2005 BARE COSTS				TOTAL INCL O&P	
						MAT.	LABOR	EQUIP.	TOTAL		
100 0010	**FANS**										100
2500	Ceiling fan, right angle, extra quiet, 0.10" S.P.										
2520	95 CFM	Q-20	20	1	Ea.	143	37		180	214	
2540	210 CFM	↓	19	1.053		169	39		208	246	
2560	385 CFM	↓	18	1.111		214	41		255	299	
2640	For wall or roof cap, add	1 Shee	16	.500		143	20		163	188	
2660	For straight thru fan, add					10%					
2680	For speed control switch, add	1 Elec	16	.500	↓	78	20.50		98.50	116	
6650	Residential, bath exhaust, grille, back draft damper										
6660	50 CFM	Q-20	24	.833	Ea.	30	31		61	80	
6670	110 CFM	↓	22	.909	↓	50	33.50		83.50	107	
6680	Light combination, squirrel cage, 100 watt, 70 CFM	↓	24	.833	↓	61.50	31		92.50	115	
6700	Light/heater combination, ceiling mounted										
6710	70 CFM, 1450 watt	Q-20	24	.833	Ea.	74.50	31		105.50	129	
6800	Heater combination, recessed, 70 CFM		24	.833		35.50	31		66.50	86	
6820	With 2 infrared bulbs	↓	23	.870	↓	53.50	32.50		86	109	
6940	Residential roof jacks and wall caps										
6944	Wall cap with back draft damper										
6946	3" & 4" dia. round duct	1 Shee	11	.727	Ea.	12.80	29		41.80	59	
6948	6" dia. round duct	"	11	.727	"	31	29		60	79	
6958	Roof jack with bird screen and back draft damper										
6960	3" & 4" dia. round duct	1 Shee	11	.727	Ea.	12.30	29		41.30	58.50	
6962	3-1/4" x 10" rectangular duct	"	10	.800	"	22.50	32		54.50	74.50	
6980	Transition										
6982	3-1/4" x 10" to 6" dia. round	1 Shee	20	.400	Ea.	13.85	16.10		29.95	40	

15860 | Air Cleaning Devices

		CREW	DAILY OUTPUT	LABOR-HOURS	UNIT	MAT.	LABOR	EQUIP.	TOTAL	TOTAL INCL O&P	
500 0010	**EXHAUST SYSTEMS** D3090 320										500
0500	Engine exhaust, garage, in-floor system										
0510	Single tube outlet assemblies										
0520	For transite pipe ducting, self-storing tube										
0530	3" tubing adapter plate	1 Shee	16	.500	Ea.	162	20		182	209	
0540	4" tubing adapter plate		16	.500		162	20		182	209	
0550	5" tubing adapter plate	↓	16	.500	↓	162	20		182	209	
0600	For vitrified tile ducting										
0610	3" tubing adapter plate, self-storing tube	1 Shee	16	.500	Ea.	162	20		182	209	
0620	4" tubing adapter plate, self-storing tube		16	.500		162	20		182	209	
0660	5" tubing adapter plate, self-storing tube	↓	16	.500	↓	162	20		182	209	
0800	Two tube outlet assemblies										
0810	For transite pipe ducting, self-storing tube										
0820	3" tubing, dual exhaust adapter plate	1 Shee	16	.500	Ea.	162	20		182	209	

	15860	Air Cleaning Devices	CREW	DAILY OUTPUT	LABOR-HOURS	UNIT	2005 BARE COSTS				TOTAL INCL O&P
							MAT.	LABOR	EQUIP.	TOTAL	
500	0850	For vitrified tile ducting										500
	0860	3" tubing, dual exhaust, self-storing tube	1 Shee	16	.500	Ea.	162	20		182	209
	0870	3" tubing, double outlet, non-storing tubes	"	16	.500	"	162	20		182	209
	0900	Accessories for metal tubing, (overhead systems also)									
	0910	Adapters, for metal tubing end									
	0920	3" tail pipe type				Ea.	36			36	39.50
	0930	4" tail pipe type					36			36	39.50
	0940	5" tail pipe type					36			36	39.50
	0990	5" diesel stack type					252			252	277
	1000	6" diesel stack type				↓	252			252	277
	1100	Bullnose (guide) required for in-floor assemblies									
	1110	3" tubing size				Ea.	18			18	19.80
	1120	4" tubing size					18			18	19.80
	1130	5" tubing size				↓	18			18	19.80
	1150	Plain rings, for tubing end									
	1160	3" tubing size				Ea.	8.55			8.55	9.40
	1170	4" tubing size				"	10.80			10.80	11.90
	1200	Tubing, galvanized, flexible, (for overhead systems also)									
	1210	3" ID				L.F.	4.82			4.82	5.30
	1220	4" ID					5.80			5.80	6.40
	1230	5" ID					6.75			6.75	7.45
	1240	6" ID				↓	7.75			7.75	8.50
	1250	Stainless steel, flexible, (for overhead sys, also)									
	1260	3" ID				L.F.	10			10	11
	1270	4" ID					13			13	14.30
	1280	5" ID					15.70			15.70	17.30
	1290	6" ID				↓	18.70			18.70	20.50
	1500	Engine exhaust, garage, ovrhd. compon., for neoprene tubing									
	1510	Alternate metal tubing & accessories see above									
	1550	Adapters, for neoprene tubing end									
	1560	3" tail pipe, adjustable, neoprene				Ea.	31.50			31.50	34.50
	1570	3" tail pipe, heavy wall neoprene					40.50			40.50	44.50
	1580	4" tail pipe, heavy wall neoprene					43			43	47.50
	1590	5" tail pipe, heavy wall neoprene				↓	58.50			58.50	64.50
	1650	Connectors, tubing									
	1660	3" interior, aluminum				Ea.	8.55			8.55	9.40
	1670	4" interior, aluminum					10.80			10.80	11.90
	1710	5" interior, neoprene					27			27	29.50
	1750	3" spiralock, neoprene					8.55			8.55	9.40
	1760	4" spiralock, neoprene					10.80			10.80	11.90
	1780	Y for 3" ID tubing, neoprene, dual exhaust					22.50			22.50	25
	1790	Y for 4" ID tubing, aluminum, dual exhaust				↓	40.50			40.50	44.50
	1850	Elbows, aluminum, splice into tubing for strap									
	1860	3" neoprene tubing size				Ea.	27			27	29.50
	1870	4" neoprene tubing size					31.50			31.50	34.50
	1900	Flange assemblies, connect tubing to overhead duct				↓	29.50			29.50	32.50
	2000	Hardware and accessories									
	2020	Cable, galvanized, 1/8" diameter				L.F.	.32			.32	.35
	2040	Cleat, tie down cable or rope				Ea.	3.60			3.60	3.96
	2060	Pulley					4.95			4.95	5.45
	2080	Pulley hook, universal				↓	3.60			3.60	3.96
	2100	Rope, nylon, 1/4" diameter				L.F.	.27			.27	.30
	2120	Winch, 1" diameter				Ea.	76.50			76.50	84
	2150	Lifting strap, mounts on neoprene									
	2160	3" tubing size				Ea.	18			18	19.80
	2170	4" tubing size				↓	18			18	19.80

Important: See the Reference Section for critical supporting data - Reference Nos., Crews, & City Cost Indexes

15860	Air Cleaning Devices	CREW	DAILY OUTPUT	LABOR-HOURS	UNIT	2005 BARE COSTS				TOTAL INCL O&P
						MAT.	LABOR	EQUIP.	TOTAL	
500	2180	5" tubing size				Ea.	18			18	19.80
	2190	6" tubing size				↓	18			18	19.80
	2200	Tubing, neoprene, 11' lengths									
	2210	3" ID				L.F.	5.40			5.40	5.95
	2220	4" ID					6.75			6.75	7.45
	2230	5" ID				↓	12.15			12.15	13.35
	2500	Engine exhaust, thru-door outlet									
	2510	3" tube size	1 Carp	16	.500	Ea.	31.50	17.15		48.65	61
	2530	4" tube size	"	16	.500	"	31.50	17.15		48.65	61
	3000	Tubing, exhaust, flex hose, with									
	3010	coupler, damper and tail pipe adapter									
	3020	Neoprene									
	3040	3" x 20'	1 Shee	6	1.333	Ea.	171	53.50		224.50	272
	3050	4" x 15'		5.40	1.481		171	59.50		230.50	281
	3060	4" x 20'		5	1.600		205	64.50		269.50	325
	3070	5" x 15'	↓	4.40	1.818	↓	289	73		362	430
	3100	Galvanized									
	3110	3" x 20'	1 Shee	6	1.333	Ea.	155	53.50		208.50	254
	3120	4" x 17'		5.60	1.429		162	57.50		219.50	267
	3130	4" x 20'		5	1.600		179	64.50		243.50	296
	3140	5" x 17'	↓	4.60	1.739	↓	199	70		269	325
	7500	Welding fume elimination accessories for garage exhaust systems									
	7600	Cut off (blast gate)									
	7610	3" tubing size, 3" x 6" opening	1 Shee	24	.333	Ea.	14.40	13.40		27.80	36.50
	7620	4" tubing size, 4" x 8" opening		24	.333		16.20	13.40		29.60	38.50
	7630	5" tubing size, 5" x 10" opening		24	.333		21.50	13.40		34.90	44.50
	7640	6" tubing size		24	.333		25	13.40		38.40	48
	7650	8" tubing size	↓	24	.333	↓	36	13.40		49.40	60
	7700	Hoods, magnetic, with handle & screen									
	7710	3" tubing size, 3" x 6" opening	1 Shee	24	.333	Ea.	67.50	13.40		80.90	95
	7720	4" tubing size, 4" x 8" opening		24	.333		67.50	13.40		80.90	95
	7730	5" tubing size, 5" x 10" opening	↓	24	.333	↓	67.50	13.40		80.90	95

15905	HVAC Instrumentation	CREW	DAILY OUTPUT	LABOR-HOURS	UNIT	2005 BARE COSTS				TOTAL INCL O&P
						MAT.	LABOR	EQUIP.	TOTAL	
960	0010	**WATER LEVEL CONTROLS**									
	3000	Low water cut-off for hot water boiler, 50 psi maximum									
	3100	1" top & bottom equalizing pipes, manual reset	1 Stpi	14	.571	Ea.	224	23.50		247.50	282
	3200	1" top & bottom equalizing pipes		14	.571		224	23.50		247.50	282
	3300	2-1/2" side connection for nipple-to-boiler	↓	14	.571	↓	202	23.50		225.50	257

	15955	HVAC Test/Adjust/Balance	CREW	DAILY OUTPUT	LABOR-HOURS	UNIT	2005 BARE COSTS				TOTAL INCL O&P	
							MAT.	LABOR	EQUIP.	TOTAL		
700	0010	**PIPING, TESTING**										700
	0100	Nondestructive testing										
	0110	Nondestructive hydraulic pressure test, isolate & 1 hr. hold										
	0120	1" - 4" pipe										
	0140	0 - 250 L.F.	1 Stpi	1.33	6.015	Ea.		246		246	370	
	0160	250 - 500 L.F.	"	.80	10			410		410	615	
	0180	500 - 1000 L.F.	Q-5	1.14	14.035			515		515	775	
	0200	1000 - 2000 L.F.	"	.80	20			735		735	1,100	
	0300	6" - 10" pipe										
	0320	0 - 250 L.F.	Q-5	1	16	Ea.		590		590	885	
	0340	250 - 500 L.F.		.73	21.918			810		810	1,225	
	0360	500 - 1000 L.F.		.53	30.189			1,100		1,100	1,675	
	0380	1000 - 2000 L.F.		.38	42.105			1,550		1,550	2,325	
	1000	Pneumatic pressure test, includes soaping joints										
	1120	1" - 4" pipe										
	1140	0 - 250 L.F.	Q-5	2.67	5.993	Ea.	5.10	221		226.10	335	
	1160	250 - 500 L.F.		1.33	12.030		10.25	445		455.25	675	
	1180	500 - 1000 L.F.		.80	20		15.35	735		750.35	1,125	
	1200	1000 - 2000 L.F.		.50	32		20.50	1,175		1,195.50	1,800	
	1300	6" - 10" pipe										
	1320	0 - 250 L.F.	Q-5	1.33	12.030	Ea.	5.10	445		450.10	670	
	1340	250 - 500 L.F.		.67	23.881		10.25	880		890.25	1,325	
	1360	500 - 1000 L.F.		.40	40		20.50	1,475		1,495.50	2,250	
	1380	1000 - 2000 L.F.		.25	64		25.50	2,350		2,375.50	3,575	
	2000	X-Ray of welds										
	2110	2" diam.	1 Stpi	8	1	Ea.	8.20	41		49.20	70.50	
	2120	3" diam.		8	1		8.20	41		49.20	70.50	
	2130	4" diam.		8	1		12.30	41		53.30	75	
	2140	6" diam.		8	1		12.30	41		53.30	75	
	2150	8" diam.		6.60	1.212		12.30	49.50		61.80	88	
	2160	10" diam.		6	1.333		16.40	54.50		70.90	100	
	3000	Liquid penetration of welds										
	3110	2" diam.	1 Stpi	14	.571	Ea.	2.30	23.50		25.80	37.50	
	3120	3" diam.		13.60	.588		2.30	24		26.30	38.50	
	3130	4" diam.		13.40	.597		2.30	24.50		26.80	39.50	
	3140	6" diam.		13.20	.606		2.30	25		27.30	40	
	3150	8" diam.		13	.615		3.45	25		28.45	42	
	3160	10" diam.		12.80	.625		3.45	25.50		28.95	42.50	

For information about Means Estimating Seminars, see yellow pages 12 and 13 in back of book

Division 16
Electrical

Estimating Tips

16060 Grounding & Bonding

- When taking off grounding system, identify separately the type and size of wire and list each unique type of ground connection.

16100 Wiring Methods

- Conduit should be taken off in three main categories:power distribution, branch power, and branch lighting, so the estimator can concentrate on systems and components, therefore making it easier to ensure all items have been accounted for.
- For cost modifications for elevated conduit installation, add the percentages to labor according to the height of installation and only the quantities exceeding the different height levels, not to the total conduit quantities.
- Remember that aluminum wiring of equal ampacity is larger in diameter than copper and may require larger conduit.
- If more than three wires at a time are being pulled, deduct percentages from the labor hours of that grouping of wires.

- The estimator should take the weights of materials into consideration when completing a takeoff. Topics to consider include:How will the materials be supported?What methods of support are available?How high will the support structure have to reach?Will the final support structure be able to withstand the total burden?Is the support material included or separate from the fixture, equipment and material specified?

16200 Electrical Power

- Do not overlook the costs for equipment used in the installation. If scaffolding or highlifts are available in the field, contractors may use them in lieu of the proposed ladders and rolling staging.

16400 Low-Voltage Distribution

- Supports and concrete pads may be shown on drawings for the larger equipment, or the support system may be just a piece of plywood for the back of a panelboard. In either case, it must be included in the costs.

16500 Lighting

- Fixtures should be taken off room by room, using the fixture schedule, specifications, and the ceiling plan. For large concentrations of lighting fixtures in the same area deduct the percentages from labor hours.

16700 Communications
16800 Sound & Video

- When estimating material costs for special systems, it is always prudent to obtain manufacturers' quotations for equipment prices and special installation requirements which will affect the total costs.

Reference Numbers

Reference numbers are shown in bold squares at the beginning of some major classifications. These numbers refer to related items in the Reference Section. The reference information may be an estimating procedure, an alternate pricing method or technical information.

Note: Not all subdivisions listed here necessarily appear in this publication.

Note: **i2 Trade Service,** *in part, has been used as a reference source for some of the material prices used in Division 16.*

16132	Conduit & Tubing	CREW	DAILY OUTPUT	LABOR-HOURS	UNIT	MAT.	LABOR	EQUIP.	TOTAL	TOTAL INCL O&P
							2005 BARE COSTS			
260	**0010 CUTTING AND DRILLING**									**260**
0100	Hole drilling to 10' high, concrete wall									
0110	8" thick, 1/2" pipe size	R-31	12	.667	Ea.	3.26	27	4.19	34.45	48.50
0120	3/4" pipe size		12	.667		3.26	27	4.19	34.45	48.50
0130	1" pipe size		9.50	.842		6.70	34.50	5.30	46.50	64
0140	1-1/4" pipe size		9.50	.842		6.70	34.50	5.30	46.50	64
0150	1-1/2" pipe size		9.50	.842		6.70	34.50	5.30	46.50	64
0160	2" pipe size		4.40	1.818		7.25	74	11.40	92.65	131
0170	2-1/2" pipe size		4.40	1.818		7.25	74	11.40	92.65	131
0180	3" pipe size		4.40	1.818		7.25	74	11.40	92.65	131
0190	3-1/2" pipe size		3.30	2.424		7.75	99	15.20	121.95	172
0200	4" pipe size		3.30	2.424		7.75	99	15.20	121.95	172
0500	12" thick, 1/2" pipe size		9.40	.851		4.98	34.50	5.35	44.83	63
0520	3/4" pipe size		9.40	.851		4.98	34.50	5.35	44.83	63
0540	1" pipe size		7.30	1.096		9.45	44.50	6.90	60.85	84.50
0560	1-1/4" pipe size		7.30	1.096		9.45	44.50	6.90	60.85	84.50
0570	1-1/2" pipe size		7.30	1.096		9.45	44.50	6.90	60.85	84.50
0580	2" pipe size		3.60	2.222		11.10	90.50	13.95	115.55	163
0590	2-1/2" pipe size		3.60	2.222		11.10	90.50	13.95	115.55	163
0600	3" pipe size		3.60	2.222		11.10	90.50	13.95	115.55	163
0610	3-1/2" pipe size		2.80	2.857		12.60	116	17.95	146.55	207
0630	4" pipe size		2.50	3.200		12.60	130	20	162.60	230
0650	16" thick, 1/2" pipe size		7.60	1.053		6.70	43	6.60	56.30	78.50
0670	3/4" pipe size		7	1.143		6.70	46.50	7.20	60.40	85
0690	1" pipe size		6	1.333		12.25	54.50	8.35	75.10	104
0710	1-1/4" pipe size		5.50	1.455		12.25	59.50	9.15	80.90	112
0730	1-1/2" pipe size		5.50	1.455		12.25	59.50	9.15	80.90	112
0750	2" pipe size		3	2.667		14.95	109	16.75	140.70	197
0770	2-1/2" pipe size		2.70	2.963		14.95	121	18.60	154.55	217
0790	3" pipe size		2.50	3.200		14.95	130	20	164.95	232
0810	3-1/2" pipe size		2.30	3.478		17.45	142	22	181.45	254
0830	4" pipe size		2	4		17.45	163	25	205.45	290
0850	20" thick, 1/2" pipe size		6.40	1.250		8.40	51	7.85	67.25	94
0870	3/4" pipe size		6	1.333		8.40	54.50	8.35	71.25	99.50
0890	1" pipe size		5	1.600		15	65	10.05	90.05	125
0910	1-1/4" pipe size		4.80	1.667		15	68	10.45	93.45	129
0930	1-1/2" pipe size		4.60	1.739		15	71	10.90	96.90	134
0950	2" pipe size		2.70	2.963		18.75	121	18.60	158.35	221
0970	2-1/2" pipe size		2.40	3.333		18.75	136	21	175.75	246
0990	3" pipe size		2.20	3.636		18.75	148	23	189.75	267
1010	3-1/2" pipe size		2	4		22.50	163	25	210.50	295
1030	4" pipe size		1.70	4.706		22.50	192	29.50	244	340
1050	24" thick, 1/2" pipe size		5.50	1.455		10.15	59.50	9.15	78.80	109
1070	3/4" pipe size		5.10	1.569		10.15	64	9.85	84	117
1090	1" pipe size		4.30	1.860		17.75	76	11.70	105.45	145
1110	1-1/4" pipe size		4	2		17.75	81.50	12.55	111.80	154
1130	1-1/2" pipe size		4	2		17.75	81.50	12.55	111.80	154
1150	2" pipe size		2.40	3.333		22.50	136	21	179.50	250
1170	2-1/2" pipe size		2.20	3.636		22.50	148	23	193.50	271
1190	3" pipe size		2	4		22.50	163	25	210.50	296
1210	3-1/2" pipe size		1.80	4.444		27	181	28	236	330
1230	4" pipe size		1.50	5.333		27	217	33.50	277.50	390
1500	Brick wall, 8" thick, 1/2" pipe size		18	.444		3.26	18.10	2.79	24.15	33.50
1520	3/4" pipe size		18	.444		3.26	18.10	2.79	24.15	33.50
1540	1" pipe size		13.30	.601		6.70	24.50	3.78	34.98	48
1560	1-1/4" pipe size		13.30	.601		6.70	24.50	3.78	34.98	48

Important: See the Reference Section for critical supporting data - Reference Nos., Crews, & City Cost Indexes

16132	Conduit & Tubing	CREW	DAILY OUTPUT	LABOR-HOURS	UNIT	2005 BARE COSTS				TOTAL INCL O&P
						MAT.	LABOR	EQUIP.	TOTAL	
260 1580	1-1/2" pipe size	R-31	13.30	.601	Ea.	6.70	24.50	3.78	34.98	48
1600	2" pipe size		5.70	1.404		7.25	57	8.80	73.05	103
1620	2-1/2" pipe size		5.70	1.404		7.25	57	8.80	73.05	103
1640	3" pipe size		5.70	1.404		7.25	57	8.80	73.05	103
1660	3-1/2" pipe size		4.40	1.818		7.75	74	11.40	93.15	131
1680	4" pipe size		4	2		7.75	81.50	12.55	101.80	143
1700	12" thick, 1/2" pipe size		14.50	.552		4.98	22.50	3.46	30.94	43
1720	3/4" pipe size		14.50	.552		4.98	22.50	3.46	30.94	43
1740	1" pipe size		11	.727		9.45	29.50	4.57	43.52	59.50
1760	1-1/4" pipe size		11	.727		9.45	29.50	4.57	43.52	59.50
1780	1-1/2" pipe size		11	.727		9.45	29.50	4.57	43.52	59.50
1800	2" pipe size		5	1.600		11.10	65	10.05	86.15	120
1820	2-1/2" pipe size		5	1.600		11.10	65	10.05	86.15	120
1840	3" pipe size		5	1.600		11.10	65	10.05	86.15	120
1860	3-1/2" pipe size		3.80	2.105		12.60	86	13.20	111.80	156
1880	4" pipe size		3.30	2.424		12.60	99	15.20	126.80	178
1900	16" thick, 1/2" pipe size		12.30	.650		6.70	26.50	4.08	37.28	51.50
1920	3/4" pipe size		12.30	.650		6.70	26.50	4.08	37.28	51.50
1940	1" pipe size		9.30	.860		12.25	35	5.40	52.65	71.50
1960	1-1/4" pipe size		9.30	.860		12.25	35	5.40	52.65	71.50
1980	1-1/2" pipe size		9.30	.860		12.25	35	5.40	52.65	71.50
2000	2" pipe size		4.40	1.818		14.95	74	11.40	100.35	139
2010	2-1/2" pipe size		4.40	1.818		14.95	74	11.40	100.35	139
2030	3" pipe size		4.40	1.818		14.95	74	11.40	100.35	139
2050	3-1/2" pipe size		3.30	2.424		17.45	99	15.20	131.65	183
2070	4" pipe size		3	2.667		17.45	109	16.75	143.20	200
2090	20" thick, 1/2" pipe size		10.70	.748		8.40	30.50	4.70	43.60	60
2110	3/4" pipe size		10.70	.748		8.40	30.50	4.70	43.60	60
2130	1" pipe size		8	1		15	41	6.30	62.30	84
2150	1-1/4" pipe size		8	1		15	41	6.30	62.30	84
2170	1-1/2" pipe size		8	1		15	41	6.30	62.30	84
2190	2" pipe size		4	2		18.75	81.50	12.55	112.80	155
2210	2-1/2" pipe size		4	2		18.75	81.50	12.55	112.80	155
2230	3" pipe size		4	2		18.75	81.50	12.55	112.80	155
2250	3-1/2" pipe size		3	2.667		22.50	109	16.75	148.25	205
2270	4" pipe size		2.70	2.963		22.50	121	18.60	162.10	225
2290	24" thick, 1/2" pipe size		9.40	.851		10.15	34.50	5.35	50	68.50
2310	3/4" pipe size		9.40	.851		10.15	34.50	5.35	50	68.50
2330	1" pipe size		7.10	1.127		17.75	46	7.10	70.85	96
2350	1-1/4" pipe size		7.10	1.127		17.75	46	7.10	70.85	96
2370	1-1/2" pipe size		7.10	1.127		17.75	46	7.10	70.85	96
2390	2" pipe size		3.60	2.222		22.50	90.50	13.95	126.95	175
2410	2-1/2" pipe size		3.60	2.222		22.50	90.50	13.95	126.95	175
2430	3" pipe size		3.60	2.222		22.50	90.50	13.95	126.95	175
2450	3-1/2" pipe size		2.80	2.857		27	116	17.95	160.95	223
2470	4" pipe size		2.50	3.200		27	130	20	177	246
3000	Knockouts to 8' high, metal boxes & enclosures									
3020	With hole saw, 1/2" pipe size	1 Elec	53	.151	Ea.		6.15		6.15	9.15
3040	3/4" pipe size		47	.170			6.95		6.95	10.30
3050	1" pipe size		40	.200			8.15		8.15	12.15
3060	1-1/4" pipe size		36	.222			9.05		9.05	13.50
3070	1-1/2" pipe size		32	.250			10.20		10.20	15.15
3080	2" pipe size		27	.296			12.05		12.05	17.95
3090	2-1/2" pipe size		20	.400			16.30		16.30	24.50
4010	3" pipe size		16	.500			20.50		20.50	30.50
4030	3-1/2" pipe size		13	.615			25		25	37.50

16132 | Conduit & Tubing

		DAILY OUTPUT	LABOR-HOURS	UNIT	2005 BARE COSTS				TOTAL INCL O&P
		CREW			MAT.	LABOR	EQUIP.	TOTAL	
260	4050	4" pipe size	1 Elec	11	.727	Ea.		29.50		29.50	44	**260**
	4070	With hand punch set, 1/2" pipe size		40	.200			8.15		8.15	12.15
	4090	3/4" pipe size		32	.250			10.20		10.20	15.15
	4110	1" pipe size		30	.267			10.85		10.85	16.15
	4130	1-1/4" pipe size		28	.286			11.65		11.65	17.35
	4150	1-1/2" pipe size		26	.308			12.55		12.55	18.65
	4170	2" pipe size		20	.400			16.30		16.30	24.50
	4190	2-1/2" pipe size		17	.471			19.20		19.20	28.50
	4200	3" pipe size		15	.533			21.50		21.50	32.50
	4220	3-1/2" pipe size		12	.667			27		27	40.50
	4240	4" pipe size		10	.800			32.50		32.50	48.50
	4260	With hydraulic punch, 1/2" pipe size		44	.182			7.40		7.40	11.05
	4280	3/4" pipe size		38	.211			8.60		8.60	12.75
	4300	1" pipe size		38	.211			8.60		8.60	12.75
	4320	1-1/4" pipe size		38	.211			8.60		8.60	12.75
	4340	1-1/2" pipe size		38	.211			8.60		8.60	12.75
	4360	2" pipe size		32	.250			10.20		10.20	15.15
	4380	2-1/2" pipe size		27	.296			12.05		12.05	17.95
	4400	3" pipe size		23	.348			14.15		14.15	21
	4420	3-1/2" pipe size		20	.400			16.30		16.30	24.50
	4440	4" pipe size	▼	18	.444	▼		18.10		18.10	27

16139 | Residential Wiring

		DAILY OUTPUT	LABOR-HOURS	UNIT	2005 BARE COSTS				TOTAL INCL O&P
		CREW			MAT.	LABOR	EQUIP.	TOTAL	
700	0010	**RESIDENTIAL WIRING**										**700**
	0020	20' avg. runs and #14/2 wiring incl. unless otherwise noted									
	1000	Service & panel, includes 24' SE-AL cable, service eye, meter,									
	1010	Socket, panel board, main bkr., ground rod, 15 or 20 amp									
	1020	1-pole circuit breakers, and misc. hardware									
	1100	100 amp, with 10 branch breakers	1 Elec	1.19	6.723	Ea.	445	274		719	895
	1110	With PVC conduit and wire		.92	8.696		485	355		840	1,050
	1120	With RGS conduit and wire		.73	10.959		675	445		1,120	1,400
	1150	150 amp, with 14 branch breakers		1.03	7.767		690	315		1,005	1,225
	1170	With PVC conduit and wire		.82	9.756		775	400		1,175	1,450
	1180	With RGS conduit and wire	▼	.67	11.940		1,125	485		1,610	1,975
	1200	200 amp, with 18 branch breakers	2 Elec	1.80	8.889		910	360		1,270	1,550
	1220	With PVC conduit and wire		1.46	10.959		990	445		1,435	1,775
	1230	With RGS conduit and wire	▼	1.24	12.903		1,450	525		1,975	2,375
	1800	Lightning surge suppressor for above services, add	1 Elec	32	.250	▼	41.50	10.20		51.70	60.50
	2000	Switch devices									
	2100	Single pole, 15 amp, Ivory, with a 1-gang box, cover plate,									
	2110	Type NM (Romex) cable	1 Elec	17.10	.468	Ea.	8.10	19.05		27.15	37.50
	2120	Type MC (BX) cable		14.30	.559		20.50	23		43.50	56.50
	2130	EMT & wire		5.71	1.401		29	57		86	117
	2150	3-way, #14/3, type NM cable		14.55	.550		12	22.50		34.50	46.50
	2170	Type MC cable		12.31	.650		29	26.50		55.50	71.50
	2180	EMT & wire		5	1.600		31	65		96	132
	2200	4-way, #14/3, type NM cable		14.55	.550		25.50	22.50		48	61.50
	2220	Type MC cable		12.31	.650		42.50	26.50		69	86.50
	2230	EMT & wire		5	1.600		45	65		110	146
	2250	S.P., 20 amp, #12/2, type NM cable		13.33	.600		14.05	24.50		38.55	52
	2270	Type MC cable		11.43	.700		25	28.50		53.50	70
	2280	EMT & wire		4.85	1.649		35	67		102	139
	2290	S.P. rotary dimmer, 600W, no wiring		17	.471		16.20	19.20		35.40	46.50
	2300	S.P. rotary dimmer, 600W, type NM cable		14.55	.550		19.65	22.50		42.15	55
	2320	Type MC cable	▼	12.31	.650	▼	32	26.50		58.50	75

16139 | Residential Wiring

		CREW	DAILY OUTPUT	LABOR-HOURS	UNIT	MAT.	LABOR	EQUIP.	TOTAL	TOTAL INCL O&P
700						2005 BARE COSTS					**700**
2330	EMT & wire	1 Elec	5	1.600	Ea.	41	65		106	143
2350	3-way rotary dimmer, type NM cable		13.33	.600		17.35	24.50		41.85	55.50
2370	Type MC cable		11.43	.700		30	28.50		58.50	75.50
2380	EMT & wire		4.85	1.649		39	67		106	143
2400	Interval timer wall switch, 20 amp, 1-30 min., #12/2									
2410	Type NM cable	1 Elec	14.55	.550	Ea.	31	22.50		53.50	67.50
2420	Type MC cable		12.31	.650		39	26.50		65.50	82
2430	EMT & wire		5	1.600		52	65		117	154
2500	Decorator style									
2510	S.P., 15 amp, type NM cable	1 Elec	17.10	.468	Ea.	11.65	19.05		30.70	41.50
2520	Type MC cable		14.30	.559		24	23		47	60.50
2530	EMT & wire		5.71	1.401		32.50	57		89.50	121
2550	3-way, #14/3, type NM cable		14.55	.550		15.55	22.50		38.05	50.50
2570	Type MC cable		12.31	.650		32.50	26.50		59	75.50
2580	EMT & wire		5	1.600		34.50	65		99.50	135
2600	4-way, #14/3, type NM cable		14.55	.550		29	22.50		51.50	65.50
2620	Type MC cable		12.31	.650		46	26.50		72.50	90.50
2630	EMT & wire		5	1.600		48.50	65		113.50	150
2650	S.P., 20 amp, #12/2, type NM cable		13.33	.600		17.60	24.50		42.10	56
2670	Type MC cable		11.43	.700		29	28.50		57.50	74
2680	EMT & wire		4.85	1.649		38.50	67		105.50	143
2700	S.P., slide dimmer, type NM cable		17.10	.468		27	19.05		46.05	58
2720	Type MC cable		14.30	.559		39.50	23		62.50	77.50
2730	EMT & wire		5.71	1.401		48.50	57		105.50	138
2750	S.P., touch dimmer, type NM cable		17.10	.468		23	19.05		42.05	54
2770	Type MC cable		14.30	.559		35.50	23		58.50	73
2780	EMT & wire		5.71	1.401		44.50	57		101.50	134
2800	3-way touch dimmer, type NM cable		13.33	.600		41	24.50		65.50	81.50
2820	Type MC cable		11.43	.700		53.50	28.50		82	102
2830	EMT & wire		4.85	1.649		62.50	67		129.50	169
3000	Combination devices									
3100	S.P. switch/15 amp recpt., Ivory, 1-gang box, plate									
3110	Type NM cable	1 Elec	11.43	.700	Ea.	16.35	28.50		44.85	60.50
3120	Type MC cable		10	.800		29	32.50		61.50	80.50
3130	EMT & wire		4.40	1.818		38	74		112	152
3150	S.P. switch/pilot light, type NM cable		11.43	.700		17	28.50		45.50	61
3170	Type MC cable		10	.800		29.50	32.50		62	81
3180	EMT & wire		4.43	1.806		38.50	73.50		112	153
3190	2-S.P. switches, 2-#14/2, no wiring		14	.571		6.75	23.50		30.25	42
3200	2-S.P. switches, 2-#14/2, type NM cables		10	.800		19.05	32.50		51.55	69.50
3220	Type MC cable		8.89	.900		40	36.50		76.50	98.50
3230	EMT & wire		4.10	1.951		39.50	79.50		119	162
3250	3-way switch/15 amp recpt., #14/3, type NM cable		10	.800		23.50	32.50		56	74
3270	Type MC cable		8.89	.900		40.50	36.50		77	99
3280	EMT & wire		4.10	1.951		42.50	79.50		122	165
3300	2-3 way switches, 2-#14/3, type NM cables		8.89	.900		31.50	36.50		68	89
3320	Type MC cable		8	1		61	41		102	128
3330	EMT & wire		4	2		47	81.50		128.50	173
3350	S.P. switch/20 amp recpt., #12/2, type NM cable		10	.800		27.50	32.50		60	78.50
3370	Type MC cable		8.89	.900		35.50	36.50		72	93.50
3380	EMT & wire		4.10	1.951		48.50	79.50		128	172
3400	Decorator style									
3410	S.P. switch/15 amp recpt., type NM cable	1 Elec	11.43	.700	Ea.	19.90	28.50		48.40	64.50
3420	Type MC cable		10	.800		32.50	32.50		65	84
3430	EMT & wire		4.40	1.818		41.50	74		115.50	156
3450	S.P. switch/pilot light, type NM cable		11.43	.700		20.50	28.50		49	65

16139 | Residential Wiring

	Description	CREW	DAILY OUTPUT	LABOR-HOURS	UNIT	MAT.	LABOR	EQUIP.	TOTAL	TOTAL INCL O&P
3470	Type MC cable	1 Elec	10	.800	Ea.	33	32.50		65.50	85
3480	EMT & wire		4.40	1.818		42	74		116	157
3500	2-S.P. switches, 2-#14/2, type NM cables		10	.800		22.50	32.50		55	73.50
3520	Type MC cable		8.89	.900		43.50	36.50		80	103
3530	EMT & wire		4.10	1.951		43	79.50		122.50	166
3550	3-way/15 amp recpt., #14/3, type NM cable		10	.800		27	32.50		59.50	78
3570	Type MC cable		8.89	.900		44	36.50		80.50	103
3580	EMT & wire		4.10	1.951		46	79.50		125.50	169
3650	2-3 way switches, 2-#14/3, type NM cables		8.89	.900		35	36.50		71.50	93
3670	Type MC cable		8	1		64.50	41		105.50	132
3680	EMT & wire		4	2		50.50	81.50		132	177
3700	S.P. switch/20 amp recpt., #12/2, type NM cable		10	.800		31	32.50		63.50	82.50
3720	Type MC cable		8.89	.900		39	36.50		75.50	97
3730	EMT & wire	▼	4.10	1.951	▼	52	79.50		131.50	175
4000	Receptacle devices									
4010	Duplex outlet, 15 amp recpt., Ivory, 1-gang box, plate									
4015	Type NM cable	1 Elec	14.55	.550	Ea.	6.45	22.50		28.95	40.50
4020	Type MC cable		12.31	.650		19	26.50		45.50	60.50
4030	EMT & wire		5.33	1.501		27	61		88	121
4050	With #12/2, type NM cable		12.31	.650		8.05	26.50		34.55	48.50
4070	Type MC cable		10.67	.750		19.20	30.50		49.70	66.50
4080	EMT & wire		4.71	1.699		29	69		98	135
4100	20 amp recpt., #12/2, type NM cable		12.31	.650		15.45	26.50		41.95	56.50
4120	Type MC cable		10.67	.750		26.50	30.50		57	75
4130	EMT & wire	▼	4.71	1.699	▼	36.50	69		105.50	143
4140	For GFI see line 4300 below									
4150	Decorator style, 15 amp recpt., type NM cable	1 Elec	14.55	.550	Ea.	10	22.50		32.50	44.50
4170	Type MC cable		12.31	.650		22.50	26.50		49	64.50
4180	EMT & wire		5.33	1.501		31	61		92	125
4200	With #12/2, type NM cable		12.31	.650		11.60	26.50		38.10	52.50
4220	Type MC cable		10.67	.750		23	30.50		53.50	70.50
4230	EMT & wire		4.71	1.699		32.50	69		101.50	139
4250	20 amp recpt. #12/2, type NM cable		12.31	.650		19	26.50		45.50	60.50
4270	Type MC cable		10.67	.750		30	30.50		60.50	78.50
4280	EMT & wire		4.71	1.699		40	69		109	147
4300	GFI, 15 amp recpt., type NM cable		12.31	.650		35	26.50		61.50	78
4320	Type MC cable		10.67	.750		47.50	30.50		78	98
4330	EMT & wire		4.71	1.699		56	69		125	165
4350	GFI with #12/2, type NM cable		10.67	.750		36.50	30.50		67	85.50
4370	Type MC cable		9.20	.870		47.50	35.50		83	105
4380	EMT & wire		4.21	1.900		57.50	77.50		135	179
4400	20 amp recpt., #12/2 type NM cable		10.67	.750		38	30.50		68.50	87.50
4420	Type MC cable		9.20	.870		49.50	35.50		85	107
4430	EMT & wire		4.21	1.900		59.50	77.50		137	180
4500	Weather-proof cover for above receptacles, add	▼	32	.250	▼	4.40	10.20		14.60	20
4550	Air conditioner outlet, 20 amp-240 volt recpt.									
4560	30' of #12/2, 2 pole circuit breaker									
4570	Type NM cable	1 Elec	10	.800	Ea.	44.50	32.50		77	97
4580	Type MC cable		9	.889		59	36		95	119
4590	EMT & wire		4	2		64.50	81.50		146	192
4600	Decorator style, type NM cable		10	.800		48	32.50		80.50	102
4620	Type MC cable		9	.889		63	36		99	123
4630	EMT & wire	▼	4	2	▼	68.50	81.50		150	196
4650	Dryer outlet, 30 amp-240 volt recpt., 20' of #10/3									
4660	2 pole circuit breaker									
4670	Type NM cable	1 Elec	6.41	1.248	Ea.	54	51		105	135

Important: See the Reference Section for critical supporting data - Reference Nos., Crews, & City Cost Indexes

16139	Residential Wiring	CREW	DAILY OUTPUT	LABOR-HOURS	UNIT	2005 BARE COSTS				TOTAL INCL O&P
						MAT.	LABOR	EQUIP.	TOTAL	
700	4680	Type MC cable	1 Elec	5.71	1.401	Ea.	65.50	57		122.50	157	700
	4690	EMT & wire	↓	3.48	2.299	↓	69	93.50		162.50	215
	4700	Range outlet, 50 amp-240 volt recpt., 30' of #8/3									
	4710	Type NM cable	1 Elec	4.21	1.900	Ea.	80.50	77.50		158	204
	4720	Type MC cable		4	2		128	81.50		209.50	262
	4730	EMT & wire		2.96	2.703		86	110		196	259
	4750	Central vacuum outlet, Type NM cable		6.40	1.250		48	51		99	129
	4770	Type MC cable		5.71	1.401		70	57		127	162
	4780	EMT & wire	↓	3.48	2.299	↓	75	93.50		168.50	222
	4800	30 amp-110 volt locking recpt., #10/2 circ. bkr.									
	4810	Type NM cable	1 Elec	6.20	1.290	Ea.	58.50	52.50		111	143
	4830	EMT & wire	"	3.20	2.500	"	82.50	102		184.50	243
	4900	Low voltage outlets									
	4910	Telephone recpt., 20' of 4/C phone wire	1 Elec	26	.308	Ea.	7.40	12.55		19.95	27
	4920	TV recpt., 20' of RG59U coax wire, F type connector	"	16	.500	"	12.65	20.50		33.15	44.50
	4950	Door bell chime, transformer, 2 buttons, 60' of bellwire									
	4970	Economy model	1 Elec	11.50	.696	Ea.	54.50	28.50		83	102
	4980	Custom model		11.50	.696		85.50	28.50		114	136
	4990	Luxury model, 3 buttons	↓	9.50	.842	↓	231	34.50		265.50	305
	6000	Lighting outlets									
	6050	Wire only (for fixture), type NM cable	1 Elec	32	.250	Ea.	4.42	10.20		14.62	20
	6070	Type MC cable		24	.333		14.10	13.60		27.70	35.50
	6080	EMT & wire		10	.800		21	32.50		53.50	72
	6100	Box (4"), and wire (for fixture), type NM cable		25	.320		9.40	13.05		22.45	30
	6120	Type MC cable		20	.400		19.10	16.30		35.40	45.50
	6130	EMT & wire	↓	11	.727	↓	26	29.50		55.50	73
	6200	Fixtures (use with lines 6050 or 6100 above)									
	6210	Canopy style, economy grade	1 Elec	40	.200	Ea.	26	8.15		34.15	40.50
	6220	Custom grade		40	.200		48	8.15		56.15	65
	6250	Dining room chandelier, economy grade		19	.421		78	17.15		95.15	112
	6260	Custom grade		19	.421		230	17.15		247.15	279
	6270	Luxury grade		15	.533		510	21.50		531.50	595
	6310	Kitchen fixture (fluorescent), economy grade		30	.267		52	10.85		62.85	73
	6320	Custom grade		25	.320		165	13.05		178.05	201
	6350	Outdoor, wall mounted, economy grade		30	.267		27.50	10.85		38.35	46.50
	6360	Custom grade		30	.267		104	10.85		114.85	130
	6370	Luxury grade		25	.320		235	13.05		248.05	278
	6410	Outdoor PAR floodlights, 1 lamp, 150 watt		20	.400		21.50	16.30		37.80	48
	6420	2 lamp, 150 watt each		20	.400		36	16.30		52.30	64
	6430	For infrared security sensor, add		32	.250		90	10.20		100.20	114
	6450	Outdoor, quartz-halogen, 300 watt flood		20	.400		39	16.30		55.30	67.50
	6600	Recessed downlight, round, pre-wired, 50 or 75 watt trim		30	.267		36	10.85		46.85	55.50
	6610	With shower light trim		30	.267		45	10.85		55.85	65.50
	6620	With wall washer trim		28	.286		54	11.65		65.65	77
	6630	With eye-ball trim	↓	28	.286		54	11.65		65.65	77
	6640	For direct contact with insulation, add					1.65			1.65	1.82
	6700	Porcelain lamp holder	1 Elec	40	.200		3.60	8.15		11.75	16.10
	6710	With pull switch		40	.200		3.86	8.15		12.01	16.40
	6750	Fluorescent strip, 1-20 watt tube, wrap around diffuser, 24"		24	.333		53	13.60		66.60	78.50
	6770	2-40 watt tubes, 48"		20	.400		79	16.30		95.30	112
	6780	With residential ballast		20	.400		89.50	16.30		105.80	123
	6800	Bathroom heat lamp, 1-250 watt		28	.286		35	11.65		46.65	56
	6810	2-250 watt lamps	↓	28	.286	↓	56	11.65		67.65	79
	6820	For timer switch, see line 2400									
	6900	Outdoor post lamp, incl. post, fixture, 35' of #14/2									
	6910	Type NMC cable	1 Elec	3.50	2.286	Ea.	184	93		277	340

16139 | Residential Wiring

		CREW	DAILY OUTPUT	LABOR-HOURS	UNIT	2005 BARE COSTS				TOTAL INCL O&P
						MAT.	LABOR	EQUIP.	TOTAL	
6920	Photo-eye, add	1 Elec	27	.296	Ea.	30	12.05		42.05	51
6950	Clock dial time switch, 24 hr., w/enclosure, type NM cable		11.43	.700		53.50	28.50		82	102
6970	Type MC cable		11	.727		66	29.50		95.50	117
6980	EMT & wire	↓	4.85	1.649	↓	74.50	67		141.50	182
7000	Alarm systems									
7050	Smoke detectors, box, #14/3, type NM cable	1 Elec	14.55	.550	Ea.	29.50	22.50		52	66
7070	Type MC cable		12.31	.650		43.50	26.50		70	87.50
7080	EMT & wire	↓	5	1.600		45.50	65		110.50	147
7090	For relay output to security system, add				↓	12.10			12.10	13.30
8000	Residential equipment									
8050	Disposal hook-up, incl. switch, outlet box, 3' of flex									
8060	20 amp-1 pole circ. bkr., and 25' of #12/2									
8070	Type NM cable	1 Elec	10	.800	Ea.	22.50	32.50		55	73
8080	Type MC cable		8	1		35.50	41		76.50	99.50
8090	EMT & wire	↓	5	1.600	↓	46.50	65		111.50	149
8100	Trash compactor or dishwasher hook-up, incl. outlet box,									
8110	3' of flex, 15 amp-1 pole circ. bkr., and 25' of #14/2									
8120	Type NM cable	1 Elec	10	.800	Ea.	15.90	32.50		48.40	66
8130	Type MC cable		8	1		31	41		72	94.50
8140	EMT & wire	↓	5	1.600		40.50	65		105.50	142
8150	Hot water sink dispensor hook-up, use line 8100									
8200	Vent/exhaust fan hook-up, type NM cable	1 Elec	32	.250	Ea.	4.42	10.20		14.62	20
8220	Type MC cable		24	.333		14.10	13.60		27.70	35.50
8230	EMT & wire	↓	10	.800	↓	21	32.50		53.50	72
8250	Bathroom vent fan, 50 CFM (use with above hook-up)									
8260	Economy model	1 Elec	15	.533	Ea.	22	21.50		43.50	56.50
8270	Low noise model		15	.533		31	21.50		52.50	66.50
8280	Custom model	↓	12	.667	↓	114	27		141	166
8300	Bathroom or kitchen vent fan, 110 CFM									
8310	Economy model	1 Elec	15	.533	Ea.	58	21.50		79.50	96.50
8320	Low noise model	"	15	.533	"	77	21.50		98.50	117
8350	Paddle fan, variable speed (w/o lights)									
8360	Economy model (AC motor)	1 Elec	10	.800	Ea.	103	32.50		135.50	162
8370	Custom model (AC motor)		10	.800		178	32.50		210.50	245
8380	Luxury model (DC motor)		8	1		350	41		391	445
8390	Remote speed switch for above, add	↓	12	.667	↓	25	27		52	68
8500	Whole house exhaust fan, ceiling mount, 36", variable speed									
8510	Remote switch, incl. shutters, 20 amp-1 pole circ. bkr.									
8520	30' of #12/2, type NM cable	1 Elec	4	2	Ea.	615	81.50		696.50	795
8530	Type MC cable		3.50	2.286		630	93		723	830
8540	EMT & wire	↓	3	2.667	↓	645	109		754	865
8600	Whirlpool tub hook-up, incl. timer switch, outlet box									
8610	3' of flex, 20 amp-1 pole GFI circ. bkr.									
8620	30' of #12/2, type NM cable	1 Elec	5	1.600	Ea.	85	65		150	191
8630	Type MC cable		4.20	1.905		93.50	77.50		171	219
8640	EMT & wire	↓	3.40	2.353	↓	103	96		199	257
8650	Hot water heater hook-up, incl. 1-2 pole circ. bkr., box;									
8660	3' of flex, 20' of #10/2, type NM cable	1 Elec	5	1.600	Ea.	25	65		90	125
8670	Type MC cable		4.20	1.905		40.50	77.50		118	161
8680	EMT & wire	↓	3.40	2.353		40.50	96		136.50	188
9000	Heating/air conditioning									
9050	Furnace/boiler hook-up, incl. firestat, local on-off switch									
9060	Emergency switch, and 40' of type NM cable	1 Elec	4	2	Ea.	45	81.50		126.50	171
9070	Type MC cable		3.50	2.286		66	93		159	212
9080	EMT & wire	↓	1.50	5.333	↓	82	217		299	415
9100	Air conditioner hook-up, incl. local 60 amp disc. switch									

16139 | Residential Wiring

		CREW	DAILY OUTPUT	LABOR-HOURS	UNIT	2005 BARE COSTS				TOTAL INCL O&P
						MAT.	LABOR	EQUIP.	TOTAL	
700 9110	3' sealtite, 40 amp, 2 pole circuit breaker									70
9130	40' of #8/2, type NM cable	1 Elec	3.50	2.286	Ea.	158	93		251	315
9140	Type MC cable		3	2.667		233	109		342	420
9150	EMT & wire	↓	1.30	6.154	↓	183	251		434	575
9200	Heat pump hook-up, 1-40 & 1-100 amp 2 pole circ. bkr.									
9210	Local disconnect switch, 3' sealtite									
9220	40' of #8/2 & 30' of #3/2									
9230	Type NM cable	1 Elec	1.30	6.154	Ea.	385	251		636	795
9240	Type MC cable		1.08	7.407		560	300		860	1,075
9250	EMT & wire	↓	.94	8.511	↓	425	345		770	985
9500	Thermostat hook-up, using low voltage wire									
9520	Heating only	1 Elec	24	.333	Ea.	5.25	13.60		18.85	26
9530	Heating/cooling	"	20	.400	"	6.35	16.30		22.65	31.50

16150 | Wiring Connections

		CREW	DAILY OUTPUT	LABOR-HOURS	UNIT	2005 BARE COSTS				TOTAL INCL O&P
						MAT.	LABOR	EQUIP.	TOTAL	
275 0010	**MOTOR CONNECTIONS**									27
0020	Flexible conduit and fittings, 115 volt, 1 phase, up to 1 HP motor	1 Elec	8	1	Ea.	4.96	41		45.96	66
0050	2 HP motor		6.50	1.231		5.05	50		55.05	80
0100	3 HP motor		5.50	1.455		8.40	59.50		67.90	97.50
0120	230 volt, 10 HP motor, 3 phase		4.20	1.905		9.05	77.50		86.55	126
0150	15 HP motor		3.30	2.424		17.35	99		116.35	166
0200	25 HP motor		2.70	2.963		19.50	121		140.50	202
0400	50 HP motor		2.20	3.636		54.50	148		202.50	281
0600	100 HP motor	↓	1.50	5.333	↓	139	217		356	480

16420 | Enclosed Controllers

		CREW	DAILY OUTPUT	LABOR-HOURS	UNIT	2005 BARE COSTS				TOTAL INCL O&P
						MAT.	LABOR	EQUIP.	TOTAL	
220 0010	**CONTROL STATIONS**									22
0050	NEMA 1, heavy duty, stop/start	1 Elec	8	1	Ea.	121	41		162	195
0100	Stop/start, pilot light		6.20	1.290		165	52.50		217.50	261
0200	Hand/off/automatic		6.20	1.290		90	52.50		142.50	177
0400	Stop/start/reverse		5.30	1.509		164	61.50		225.50	272
0500	NEMA 7, heavy duty, stop/start		6	1.333		199	54.50		253.50	300
0600	Stop/start, pilot light	↓	4	2	↓	244	81.50		325.50	390

16440 | Swbds, Panels & Control Centers

		CREW	DAILY OUTPUT	LABOR-HOURS	UNIT	2005 BARE COSTS				TOTAL INCL O&P
						MAT.	LABOR	EQUIP.	TOTAL	
660 0010	**MOTOR STARTERS & CONTROLS**									6
0050	Magnetic, FVNR, with enclosure and heaters, 480 volt									
0080	2 HP, size 00	1 Elec	3.50	2.286	Ea.	173	93		266	330
0100	5 HP, size 0		2.30	3.478		209	142		351	440
0200	10 HP, size 1	↓	1.60	5		235	204		439	565
0300	25 HP, size 2	2 Elec	2.20	7.273		440	296		736	925
0400	50 HP, size 3		1.80	8.889		720	360		1,080	1,325
0500	100 HP, size 4		1.20	13.333		1,600	545		2,145	2,550
0600	200 HP, size 5		.90	17.778		3,725	725		4,450	5,175
0610	400 HP, size 6	↓	.80	20	↓	10,900	815		11,715	13,100

For expanded coverage of these items see *Means Electrical Cost Data 2005*

	16440	Swbds, Panels & Control Centers	CREW	DAILY OUTPUT	LABOR-HOURS	UNIT	2005 BARE COSTS				TOTAL INCL O&P	
							MAT.	LABOR	EQUIP.	TOTAL		
660	0620	NEMA 7, 5 HP, size 0	1 Elec	1.60	5	Ea.	895	204		1,099	1,300	660
	0630	10 HP, size 1	"	1.10	7.273		930	296		1,226	1,475	
	0640	25 HP, size 2	2 Elec	1.80	8.889		1,500	360		1,860	2,200	
	0650	50 HP, size 3		1.20	13.333		2,225	545		2,770	3,250	
	0660	100 HP, size 4		.90	17.778		3,650	725		4,375	5,075	
	0670	200 HP, size 5	↓	.50	32		8,500	1,300		9,800	11,300	
	0700	Combination, with motor circuit protectors, 5 HP, size 0	1 Elec	1.80	4.444		680	181		861	1,025	
	0800	10 HP, size 1	"	1.30	6.154		705	251		956	1,150	
	0900	25 HP, size 2	2 Elec	2	8		990	325		1,315	1,550	
	1000	50 HP, size 3		1.32	12.121		1,425	495		1,920	2,300	
	1200	100 HP, size 4	↓	.80	20		3,100	815		3,915	4,650	
	1220	NEMA 7, 5 HP, size 0	1 Elec	1.30	6.154		1,575	251		1,826	2,125	
	1230	10 HP, size 1	"	1	8		1,625	325		1,950	2,250	
	1240	25 HP, size 2	2 Elec	1.32	12.121		2,150	495		2,645	3,100	
	1250	50 HP, size 3		.80	20		3,525	815		4,340	5,125	
	1260	100 HP, size 4		.60	26.667		5,500	1,075		6,575	7,675	
	1270	200 HP, size 5	↓	.40	40		11,900	1,625		13,525	15,500	
	1400	Combination, with fused switch, 5 HP, size 0	1 Elec	1.80	4.444		520	181		701	840	
	1600	10 HP, size 1	"	1.30	6.154		555	251		806	985	
	1800	25 HP, size 2	2 Elec	2	8		900	325		1,225	1,475	
	2000	50 HP, size 3		1.32	12.121		1,525	495		2,020	2,400	
	2200	100 HP, size 4	↓	.80	20	↓	2,650	815		3,465	4,150	
	3500	Magnetic FVNR with NEMA 12, enclosure & heaters, 480 volt										
	3600	5 HP, size 0	1 Elec	2.20	3.636	Ea.	197	148		345	440	
	3700	10 HP, size 1	"	1.50	5.333		297	217		514	650	
	3800	25 HP, size 2	2 Elec	2	8		555	325		880	1,100	
	3900	50 HP, size 3		1.60	10		855	410		1,265	1,550	
	4000	100 HP, size 4		1	16		2,050	650		2,700	3,225	
	4100	200 HP, size 5	↓	.80	20		4,900	815		5,715	6,625	
	4200	Combination, with motor circuit protectors, 5 HP, size 0	1 Elec	1.70	4.706		655	192		847	1,000	
	4300	10 HP, size 1	"	1.20	6.667		680	272		952	1,150	
	4400	25 HP, size 2	2 Elec	1.80	8.889		1,025	360		1,385	1,675	
	4500	50 HP, size 3		1.20	13.333		1,650	545		2,195	2,625	
	4600	100 HP, size 4	↓	.74	21.622		3,725	880		4,605	5,400	
	4700	Combination, with fused switch, 5 HP, size 0	1 Elec	1.70	4.706		630	192		822	980	
	4800	10 HP, size 1	"	1.20	6.667		660	272		932	1,125	
	4900	25 HP, size 2	2 Elec	1.80	8.889		1,000	360		1,360	1,650	
	5000	50 HP, size 3		1.20	13.333		1,600	545		2,145	2,575	
	5100	100 HP, size 4	↓	.74	21.622	↓	3,250	880		4,130	4,875	
	5200	Factory installed controls, adders to size 0 thru 5										
	5300	Start-stop push button	1 Elec	32	.250	Ea.	41.50	10.20		51.70	60.50	
	5400	Hand-off-auto-selector switch		32	.250		41.50	10.20		51.70	60.50	
	5500	Pilot light		32	.250		77.50	10.20		87.70	100	
	5600	Start-stop-pilot		32	.250		119	10.20		129.20	146	
	5700	Auxiliary contact, NO or NC		32	.250		57	10.20		67.20	77.50	
	5800	NO-NC	↓	32	.250	↓	114	10.20		124.20	140	

For information about Means Estimating Seminars, see yellow pages 12 and 13 in back of book

Important: See the Reference Section for critical supporting data - Reference Nos., Crews, & City Cost Indexes

Division 17
Square Foot & Cubic Foot Costs

Estimating Tips

- The cost figures in Division 17 were derived from approximately 11,200 projects contained in the Means database of completed construction projects, and include the contractor's overhead and profit, but do not generally include architectural fees or land costs. The figures have been adjusted to January of the current year. New projects are added to our files each year, and outdated projects are discarded. For this reason, certain costs may not show a uniform annual progression. In no case are all subdivisions of a project listed.

- These projects were located throughout the U.S. and reflect a tremendous variation in square foot (S.F.) and cubic foot (C.F.) costs. This is due to differences, not only in labor and material costs, but also in individual owners' requirements. For instance, a bank in a large city would have different features than one in a rural area. This is true of all the different types of buildings analyzed. Therefore, caution should be exercised when using Division 17 costs. For example, for court houses, costs in the database are local court house costs and will not apply to the larger, more elaborate federal court houses. As a general rule, the projects in the 1/4 column do not include any site work or equipment, while the projects in the 3/4 column may include both equipment and site work. The median figures do not generally include site work.

- None of the figures "go with" any others. All individual cost items were computed and tabulated separately. Thus the sum of the median figures for Plumbing, HVAC and Electrical will not normally total up to the total Mechanical and Electrical costs arrived at by separate analysis and tabulation of the projects.

- Each building was analyzed as to total and component costs and percentages. The figures were arranged in ascending order with the results tabulated as shown. The 1/4 column shows that 25% of the projects had lower costs, 75% higher. The 3/4 column shows that 75% of the projects had lower costs, 25% had higher. The median column shows that 50% of the projects had lower costs, 50% had higher.

- There are two times when square foot costs are useful. The first is in the conceptual stage when no details are available. Then square foot costs make a useful starting point. The second is after the bids are in and the costs can be worked back into their appropriate units for information purposes. As soon as details become available in the project design, the square foot approach should be discontinued and the project priced as to its particular components. When more precision is required or for estimating the replacement cost of specific buildings, the current edition of *Means Square Foot Costs* should be used.

- In using the figures in Division 17, it is recommended that the median column be used for preliminary figures if no additional information is available. The median figures, when multiplied by the total city construction cost index figures (see City Cost Indexes) and then multiplied by the project size modifier in Reference Number R17100-100, should present a fairly accurate base figure, which would then have to be adjusted in view of the estimator's experience, local economic conditions, code requirements and the owner's particular requirements. There is no need to factor the percentage figures, as these should remain constant from city to city. All tabulations mentioning air conditioning had at least partial air conditioning.

- The editors of this book would greatly appreciate receiving cost figures on one or more of your recent projects which would then be included in the averages for next year. All cost figures received will be kept confidential except that they will be averaged with other similar projects to arrive at S.F. and C.F. cost figures for next year's book. See the last page of the book for details and the discount available for submitting one or more of your projects.

17100 | S.F. & C.F. Costs

			UNIT	UNIT COSTS			% OF TOTAL		
				1/4	MEDIAN	3/4	1/4	MEDIAN	3/4
010	0010	**APARTMENTS** Low Rise (1 to 3 story) R17100-100	S.F.	53.50	67.50	90			
	0020	Total project cost	C.F.	4.80	6.35	7.85			
	2720	Plumbing	S.F.	4.17	5.35	6.80	6.65%	8.95%	10.05%
	2770	Heating, ventilating, air conditioning		2.66	3.28	4.82	4.20%	5.60%	7.60%
	2900	Electrical		3.10	4.06	5.55	5.20%	6.65%	8.40%
	3100	Total: Mechanical & Electrical	↓	10.75	13.70	17.05	15.90%	18.05%	23%
	9000	Per apartment unit, total cost	Apt.	49,900	75,500	112,500			
	9500	Total: Mechanical & Electrical	"	9,425	14,900	19,400			
020	0010	**APARTMENTS** Mid Rise (4 to 7 story) R17100-100	S.F.	71	85.50	105			
	0020	Total project costs	C.F.	5.55	7.65	10.45			
	2720	Plumbing	S.F.	4.17	6.65	7.40	5.70%	7.20%	8.95%
	2900	Electrical		4.90	6.70	7.80	6.65%	7.20%	8.95%
	3100	Total: Mechanical & Electrical	↓	14.15	18.80	23.50	18.85%	21%	25%
	9000	Per apartment unit, total cost	Apt.	80,500	95,000	157,000			
	9500	Total: Mechanical & Electrical	"	15,200	17,600	22,200			
030	0010	**APARTMENTS** High Rise (8 to 24 story) R17100-100	S.F.	80.50	97	119			
	0020	Total project costs	C.F.	6.85	9.55	11.65			
	2720	Plumbing	S.F.	5.95	7	9.80	6.80%	7.20%	10.45%
	2900	Electrical		5.55	7	9.45	6.45%	7.65%	8.80%
	3100	Total: Mechanical & Electrical	↓	16.35	21	25	17.95%	22.50%	24.50%
	9000	Per apartment unit, total cost	Apt.	83,500	92,000	128,000			
	9500	Total: Mechanical & Electrical	"	18,100	20,700	21,900			
040	0010	**AUDITORIUMS** R17100-100	S.F.	83	113	159			
	0020	Total project costs	C.F.	5.25	7.30	11			
	2720	Plumbing	S.F.	5.30	7.35	9.25	5.85%	7.20%	8.70%
	2900	Electrical		6.75	9.35	12.50	6.80%	8.95%	11.30%
	3100	Total: Mechanical & Electrical	↓	13	17.30	30	24.50%	30.50%	31.50%
050	0010	**AUTOMOTIVE SALES** R17100-100	S.F.	60.50	84	103			
	0020	Total project costs	C.F.	4.08	4.90	6.35			
	2720	Plumbing	S.F.	2.82	4.90	5.35	2.89%	6.05%	6.50%
	2770	Heating, ventilating, air conditioning		4.36	6.65	7.20	4.61%	10%	10.35%
	2900	Electrical		4.99	7.70	8.60	7.40%	9.95%	12.40%
	3100	Total: Mechanical & Electrical	↓	13.90	19.55	25	19.15%	20.50%	26%
060	0010	**BANKS** R17100-100	S.F.	121	151	189			
	0020	Total project costs	C.F.	8.70	11.80	15.55			
	2720	Plumbing	S.F.	3.81	5.45	7.95	2.82%	3.90%	4.93%
	2770	Heating, ventilating, air conditioning		7.25	9.65	12.90	4.86%	7.15%	8.50%
	2900	Electrical		11.50	15.35	19.95	8.25%	10.20%	12.20%
	3100	Total: Mechanical & Electrical	↓	27	36.50	44	17.75%	19.45%	23%
130	0010	**CHURCHES** R17100-100	S.F.	81.50	103	134			
	0020	Total project costs	C.F.	5.05	6.40	8.45			
	2720	Plumbing	S.F.	3.18	4.45	6.55	3.51%	4.96%	6.25%
	2770	Heating, ventilating, air conditioning		7.45	9.70	14.25	7.50%	10%	12%
	2900	Electrical		6.90	9.40	12.60	7.30%	8.75%	10.90%
	3100	Total: Mechanical & Electrical	↓	18.75	27	38	18.25%	21.50%	24%
150	0010	**CLUBS, COUNTRY** R17100-100	S.F.	87.50	106	133			
	0020	Total project costs	C.F.	7.05	8.60	11.90			
	2720	Plumbing	S.F.	5.65	7.85	17.90	5.60%	7.90%	10%
	2900	Electrical		6.90	9.85	13	7%	8.95%	11%
	3100	Total: Mechanical & Electrical	↓	21	36.50	46	19%	26.50%	29.50%
170	0010	**CLUBS, SOCIAL** Fraternal R17100-100	S.F.	70	100	131			
	0020	Total project costs	C.F.	4.38	6.65	7.90			
	2720	Plumbing	S.F.	4.40	5.50	6.65	5.60%	6.90%	8.55%
	2770	Heating, ventilating, air conditioning	↓	6.35	7.70	9.85	8.20%	9.25%	14.40%

17100		S.F. & C.F. Costs		UNIT	UNIT COSTS			% OF TOTAL		
					1/4	MEDIAN	3/4	1/4	MEDIAN	3/4
170	2900	Electrical		S.F.	5.55	8.65	10.45	6.50%	9.50%	10.55%	1
	3100	Total: Mechanical & Electrical		↓	15.55	27	29.50	21%	23%	37%	
180	0010	**CLUBS, Y.M.C.A.**	R17100 -100	S.F.	88	119	145				1
	0020	Total project costs		C.F.	4.05	6.80	10.10				
	2720	Plumbing		S.F.	5.55	11.05	12.40	5.65%	7.60%	10.85%	
	2900	Electrical			6.65	9.10	12.95	6.25%	8.65%	10.20%	
	3100	Total: Mechanical & Electrical		↓	19.70	25	33.50	18.40%	22.50%	29.50%	
190	0010	**COLLEGES** Classrooms & Administration	R17100 -100	S.F.	99	131	175				1
	0020	Total project costs		C.F.	7.15	10.10	15.90				
	2720	Plumbing		S.F.	4.81	9.65	17.30	5.10%	6.60%	8.95%	
	2900	Electrical			8.05	12.40	14.90	7.70%	9.85%	12%	
	3100	Total: Mechanical & Electrical		↓	19.80	35.50	49	24%	28%	31.50%	
210	0010	**COLLEGES** Science, Engineering, Laboratories	R17100 -100	S.F.	165	193	239				2
	0020	Total project costs		C.F.	9.45	13.80	15.70				
	2900	Electrical		S.F.	13.60	19.30	29.50	7.10%	9.40%	12.10%	
	3100	Total: Mechanical & Electrical		"	50.50	60	93	28.50%	31.50%	41%	
230	0010	**COLLEGES** Student Unions	R17100 -100	S.F.	105	148	173				2
	0020	Total project costs		C.F.	5.90	7.70	9.95				
	3100	Total: Mechanical & Electrical		S.F.	27.50	42.50	50.50	23.50%	26%	29%	
250	0010	**COMMUNITY CENTERS**	R17100 -100	"	82.50	107	144				2
	0020	Total project costs		C.F.	5.70	8.10	10.50				
	2720	Plumbing		S.F.	4.22	7.20	10.50	4.94%	7%	9.10%	
	2770	Heating, ventilating, air conditioning			6.95	10	13.75	6.95%	10.65%	13.05%	
	2900	Electrical			6.95	9.40	14.30	7.35%	9.10%	10.85%	
	3100	Total: Mechanical & Electrical		↓	26	31	44.50	22.50%	26.50%	32.50%	
280	0010	**COURT HOUSES**	R17100 -100	S.F.	124	144	166				2
	0020	Total project costs		C.F.	9.60	11.50	14.50				
	2720	Plumbing		S.F.	5.95	8.35	12	5.95%	7.45%	8.20%	
	2900	Electrical			11.70	14	17.50	8.55%	9.95%	11.50%	
	3100	Total: Mechanical & Electrical		↓	30	34	46.50	22.50%	29.50%	30.50%	
300	0010	**DEPARTMENT STORES**	R17100 -100	S.F.	46.50	62.50	79				3
	0020	Total project costs		C.F.	2.50	3.37	4.74				
	2720	Plumbing		S.F.	1.51	1.83	2.77	1.82%	4.21%	5.90%	
	2770	Heating, ventilating, air conditioning			4.22	6.50	9.80	8.20%	9.10%	14.80%	
	2900	Electrical			5.35	7.30	8.60	9.05%	12.15%	14.95%	
	3100	Total: Mechanical & Electrical		↓	9.35	12	21	13.20%	21.50%	50%	
310	0010	**DORMITORIES** Low Rise (1 to 3 story)	R17100 -100	S.F.	76	111	135				3
	0020	Total project costs		C.F.	5.25	8.05	12.05				
	2720	Plumbing		S.F.	5.30	7.10	8.95	8.05%	9%	9.65%	
	2770	Heating, ventilating, air conditioning			5.60	6.70	8.95	4.61%	8.05%	10%	
	2900	Electrical			5.65	8.55	10.75	6.55%	8.90%	9.55%	
	3100	Total: Mechanical & Electrical		↓	29.50	31	32.50	22%	26%	29%	
	9000	Per bed, total cost		Bed	37,300	41,400	89,000				
320	0010	**DORMITORIES** Mid Rise (4 to 8 story)	R17100 -100	S.F.	108	141	171				3
	0020	Total project costs		C.F.	11.90	13.10	15.65				
	2900	Electrical		S.F.	8	12.55	14.20	8.20%	9.45%	10.35%	
	3100	Total: Mechanical & Electrical		"	24	32	42	19.50%	30.50%	34.50%	
	9000	Per bed, total cost		Bed	15,400	35,100	72,500				
340	0010	**FACTORIES**	R17100 -100	S.F.	41	61	94				3
	0020	Total project costs		C.F.	2.62	3.91	6.50				
	2720	Plumbing		S.F.	2.29	4.10	6.80	3.73%	6.05%	8.10%	
	2770	Heating, ventilating, air conditioning		↓	4.29	6.15	8.30	5.25%	8.45%	11.35%	

17100 | S.F. & C.F. Costs

			UNIT	UNIT COSTS			% OF TOTAL			
				1/4	MEDIAN	3/4	1/4	MEDIAN	3/4	
340	2900	Electrical	S.F.	5.15	8.15	12.25	8.30%	10.55%	14.20%	**340**
	3100	Total: Mechanical & Electrical	↓	12.10	19.40	29.50	22%	28.50%	34.50%	
360	0010	**FIRE STATIONS** R17100-100	S.F.	78	107	144				**360**
	0020	Total project costs	C.F.	4.74	6.50	8.65				
	2720	Plumbing	S.F.	5.15	7.55	11.15	5.85%	7.35%	9.50%	
	2770	Heating, ventilating, air conditioning		4.42	7.05	11	4.86%	7.25%	9.25%	
	2900	Electrical		5.75	9.70	13.05	6.80%	8.75%	10.65%	
	3100	Total: Mechanical & Electrical	↓	28	33.50	39	18.75%	23%	27%	
370	0010	**FRATERNITY HOUSES** and Sorority Houses R17100-100	S.F.	81	104	143				**370**
	0020	Total project costs	C.F.	7.75	8.40	10.80				
	2720	Plumbing	S.F.	6.15	7	12.85	6.80%	8%	10.85%	
	2900	Electrical		5.35	11.55	14.15	6.60%	9.90%	10.65%	
	3100	Total: Mechanical & Electrical	↓	14.60	20.50	25	14.60%	15.10%	15.90%	
380	0010	**FUNERAL HOMES** R17100-100	S.F.	85.50	117	212				**380**
	0020	Total project costs	C.F.	8.75	9.70	18.65				
	2900	Electrical	S.F.	3.78	6.95	8.15	3.58%	4.44%	5.95%	
	3100	Total: Mechanical & Electrical	"	13.50	20	26.50	12.90%	17.80%	18.80%	
390	0010	**GARAGES, COMMERCIAL** (Service) R17100-100	S.F.	48	76	103				**390**
	0020	Total project costs	C.F.	3.26	4.71	6.85				
	2720	Plumbing	S.F.	3.36	5.15	9.45	5.45%	7.85%	10.65%	
	2730	Heating & ventilating		4.41	6	8.30	5.25%	6.85%	8.20%	
	2900	Electrical		4.67	7.15	10.20	7.15%	9.25%	10.85%	
	3100	Total: Mechanical & Electrical		10.45	19.40	29	13.60%	17.40%	27%	
400	0010	**GARAGES, MUNICIPAL** (Repair) R17100-100	S.F.	71	95	134				**400**
	0020	Total project costs	C.F.	4.44	5.65	9.20				
	2720	Plumbing	S.F.	3.19	6.15	11.55	3.59%	6.70%	7.95%	
	2730	Heating & ventilating		5.45	7.90	15.25	6.15%	7.45%	13.50%	
	2900	Electrical		5.25	8.25	11.90	6.65%	8.15%	11.15%	
	3100	Total: Mechanical & Electrical	↓	16.90	29.50	49	21.50%	25.50%	28.50%	
410	0010	**GARAGES, PARKING** R17100-100	S.F.	27.50	40	69.50				**410**
	0020	Total project costs	C.F.	2.60	3.53	5.15				
	2720	Plumbing	S.F.	.78	1.22	1.88	1.72%	2.70%	3.85%	
	2900	Electrical		1.52	1.84	2.87	4.33%	5.20%	6.25%	
	3100	Total: Mechanical & Electrical	↓	2.42	4.33	5.70	6.90%	8.80%	11.05%	
	9000	Per car, total cost	Car	11,700	14,700	18,700				
430	0010	**GYMNASIUMS** R17100-100	S.F.	77	102	125				**430**
	0020	Total project costs	C.F.	3.84	5.20	6.40				
	2720	Plumbing	S.F.	4.87	6	7.45	4.95%	6.75%	7.75%	
	2770	Heating, ventilating, air conditioning	"	5.25	8	16.05	5.80%	9.80%	11.10%	
	2900	Electrical	S.F.	5.30	7.80	9.90	6.60%	8.50%	10.35%	
	3100	Total: Mechanical & Electrical		21	29.50	33.50	19.60%	26%	29.50%	
460	0010	**HOSPITALS** R17100-100	↓	152	181	275				**460**
	0020	Total project costs	C.F.	11.20	13.95	19.95				
	2720	Plumbing	S.F.	12.75	17.80	23	7.60%	9.10%	10.85%	
	2770	Heating, ventilating, air conditioning		18.70	24	32.50	7.80%	12.95%	16.65%	
	2900	Electrical		16.15	21	32.50	9.85%	11.75%	14%	
	3100	Total: Mechanical & Electrical	↓	45.50	60	98.50	26.50%	33%	36.50%	
	9000	Per bed or person, total cost	Bed	76,500	171,500	223,500				
480	0010	**HOUSING** For the Elderly R17100-100	S.F.	72.50	92	113				**480**
	0020	Total project costs	C.F.	5.20	7.20	9.20				
	2720	Plumbing	S.F.	5.40	6.90	9	8.15%	9.55%	10.50%	
	2730	Heating, ventilating, air conditioning		2.77	3.92	5.85	3.30%	5.60%	7.25%	
	2900	Electrical		5.40	7.35	9.45	7.30%	8.50%	10.25%	
	3100	Total: Mechanical & Electrical	↓	18.65	22	29.50	18.10%	22.50%	29%	
	9000	Per rental unit, total cost	Unit	67,500	79,000	88,000				
	9500	Total: Mechanical & Electrical	"	14,900	17,300	20,200				

17100	S.F. & C.F. Costs		UNIT	UNIT COSTS			% OF TOTAL			
				1/4	MEDIAN	3/4	1/4	MEDIAN	3/4	
500	0010	**HOUSING** Public (Low Rise)	R17100 -100	S.F.	61	85	111				**500**
	0020	Total project costs		C.F.	5.45	6.80	8.45				
	2720	Plumbing		S.F.	4.41	5.85	7.35	7.15%	9.05%	11.60%	
	2730	Heating, ventilating, air conditioning			2.22	4.30	4.71	4.26%	6.05%	6.45%	
	2900	Electrical			3.70	5.50	7.65	5.10%	6.55%	8.25%	
	3100	Total: Mechanical & Electrical	↓		17.55	22.50	25.50	14.50%	17.55%	26.50%	
	9000	Per apartment, total cost		Apt.	67,000	76,500	96,000				
	9500	Total: Mechanical & Electrical		"	14,300	17,600	19,600				
510	0010	**ICE SKATING RINKS**	R17100 -100	S.F.	54.50	122	134				**510**
	0020	Total project costs		C.F.	3.84	3.93	4.53				
	2720	Plumbing		S.F.	1.95	3.66	3.74	3.12%	3.23%	5.65%	
	2900	Electrical			5.60	8.60	9.10	6.30%	10.15%	15.05%	
	3100	Total: Mechanical & Electrical	↓		9.35	13.20	16.50	18.95%	18.95%	18.95%	
520	0010	**JAILS**	R17100 -100	S.F.	159	205	265				**520**
	0020	Total project costs		C.F.	15	20	24.50				
	2720	Plumbing		S.F.	16.20	20.50	27	7%	8.90%	13.35%	
	2770	Heating, ventilating, air conditioning			14.35	19.15	37	7.50%	9.45%	17.75%	
	2900	Electrical			17.50	22	27.50	8.20%	11.45%	14.70%	
	3100	Total: Mechanical & Electrical			43.50	79	94	27.50%	30%	34%	
530	0010	**LIBRARIES**	R17100 -100	S.F.	99	126	165				**530**
	0020	Total project costs		C.F.	6.90	8.35	11				
	2720	Plumbing		S.F.	3.71	5.40	7.30	3.43%	4.60%	5.70%	
	2770	Heating, ventilating, air conditioning			8.20	13.90	18.10	7.80%	10.95%	12.80%	
	2900	Electrical			10.10	13	16.65	8.35%	10.40%	11.95%	
	3100	Total: Mechanical & Electrical	↓		30.50	38	47.50	19.65%	23%	26.50%	
540	0010	**LIVING, ASSISTED**	R17100 -100	S.F.	94	111	129				**540**
	0020	Total project costs		C.F.	7.85	9.15	10.40				
	2720	Plumbing		S.F.	7.80	10.45	10.80	6.05%	8.15%	10.60%	
	2770	Heating, ventilating, air conditioning			9.25	9.65	10.60	7.95%	9.35%	9.70%	
	2900	Electrical			9.15	10.25	11.65	9%	10.30%	10.70%	
	3100	Total: Mechanical & Electrical			26.50	30	34.50	26%	29%	31.50%	
550	0010	**MEDICAL CLINICS**	R17100 -100	S.F.	93	117	149				**550**
	0020	Total project costs		C.F.	6.95	9.40	12.10				
	2720	Plumbing		S.F.	6.30	8.85	12	6.15%	8.40%	10.10%	
	2770	Heating, ventilating, air conditioning			7.50	9.85	14.45	6.65%	8.85%	11.35%	
	2900	Electrical			8.15	11.50	15.05	8.10%	10%	12.20%	
	3100	Total: Mechanical & Electrical			26	36	48.50	22.50%	27.50%	33.50%	
570	0010	**MEDICAL OFFICES**	R17100 -100	S.F.	89	110	135				**570**
	0020	Total project costs		C.F.	6.65	9	12.15				
	2720	Plumbing		S.F.	4.92	7.60	10.20	5.60%	6.80%	8.50%	
	2770	Heating, ventilating, air conditioning			5.95	8.75	11.35	6.15%	8.05%	9.70%	
	2900	Electrical			7.10	10.35	14.50	7.65%	9.80%	11.70%	
	3100	Total: Mechanical & Electrical			19.20	27.50	41	19.35%	23.50%	30.50%	
590	0010	**MOTELS**	R17100 -100	S.F.	56	82.50	107				**590**
	0020	Total project costs		C.F.	5	6.70	10.95				
	2720	Plumbing		S.F.	5.70	7.25	8.65	9.45%	10.60%	12.55%	
	2770	Heating, ventilating, air conditioning			3.47	5.15	9.30	5.60%	5.60%	10%	
	2900	Electrical			5.35	6.80	8.65	7.45%	9.20%	10.80%	
	3100	Total: Mechanical & Electrical			18.05	22.50	38.50	18.50%	21%	25.50%	
	9000	Per rental unit, total cost		Unit	28,600	54,500	58,500				
	9500	Total: Mechanical & Electrical		"	5,575	8,425	9,800				
600	0010	**NURSING HOMES**	R17100 -100	S.F.	89	114	139				**600**
	0020	Total project costs		C.F.	6.90	8.65	11.80				
	2720	Plumbing		S.F.	7.85	11.45	13.80	8.75%	10.10%	12.70%	
	2770	Heating, ventilating, air conditioning			7.95	12.05	16	9.70%	11.45%	11.80%	
	2900	Electrical			8.75	11.10	14.85	9.50%	10.60%	12.50%	
	3100	Total: Mechanical & Electrical	↓		21	29	49	26%	29.50%	30.50%	

For expanded coverage of these items see *Means Square Foot Costs 2005*

17100 | S.F. & C.F. Costs

			UNIT	UNIT COSTS			% OF TOTAL		
				1/4	MEDIAN	3/4	1/4	MEDIAN	3/4
600	9000	Per bed or person, total cost	Bed	39,600	50,000	63,500			
610	0010	**OFFICES** Low Rise (1 to 4 story) R17100 -100	S.F.	73.50	95	124			
	0020	Total project costs	C.F.	5.30	7.30	9.60			
	2720	Plumbing	S.F.	2.65	4.10	5.95	3.66%	4.50%	6.10%
	2770	Heating, ventilating, air conditioning		5.85	8.30	11.95	7.20%	10.40%	11.70%
	2900	Electrical		6.05	8.60	12.15	7.45%	9.65%	11.40%
	3100	Total: Mechanical & Electrical	↓	15.80	22	33	18.25%	22.50%	27%
620	0010	**OFFICES** Mid Rise (5 to 10 story) R17100 -100	S.F.	78.50	95	129			
	0020	Total project costs	C.F.	5.55	7.10	10.05			
	2720	Plumbing	S.F.	2.37	3.83	5.30	2.83%	3.74%	4.50%
	2770	Heating, ventilating, air conditioning		5.95	8.55	13.60	7.65%	9.40%	11%
	2900	Electrical		5.85	7.45	9.90	6.35%	7.80%	10%
	3100	Total: Mechanical & Electrical	↓	14.90	18.80	37.50	19.15%	21%	26%
630	0010	**OFFICES** High Rise (11 to 20 story) R17100 -100	S.F.	96.50	122	150			
	0020	Total project costs	C.F.	6.75	8.45	12.10			
	2900	Electrical	S.F.	5.85	7.15	10.60	5.80%	7.85%	10.50%
	3100	Total: Mechanical & Electrical	"	18.85	25.50	42.50	16.90%	23.50%	34%
640	0010	**POLICE STATIONS** R17100 -100	S.F.	116	152	192			
	0020	Total project costs	C.F.	9.20	11.30	15.45			
	2720	Plumbing	S.F.	6.70	12.90	16.05	5.65%	6.90%	10.75%
	2770	Heating, ventilating, air conditioning		10.10	13.45	18.25	5.85%	10.55%	11.70%
	2900	Electrical		12.60	18.90	24	9.80%	11.85%	14.80%
	3100	Total: Mechanical & Electrical	↓	41.50	50	67.50	28.50%	32%	32.50%
650	0010	**POST OFFICES** R17100 -100	S.F.	87.50	113	140			
	0020	Total project costs	C.F.	5.50	6.95	8.40			
	2720	Plumbing	S.F.	4.11	5.10	6.45	4.24%	5.30%	5.60%
	2770	Heating, ventilating, air conditioning		6.45	7.95	8.85	6.65%	7.15%	9.35%
	2900	Electrical		7.55	10.60	12.60	7.25%	9%	11%
	3100	Total: Mechanical & Electrical	↓	22	28.50	32.50	16.25%	18.80%	22%
660	0010	**POWER PLANTS** R17100 -100	S.F.	635	835	1,550			
	0020	Total project costs	C.F.	17.45	38	81.50			
	2900	Electrical	S.F.	45	95	142	9.30%	12.75%	21.50%
	8100	Total: Mechanical & Electrical	"	112	365	815	32.50%	32.50%	52.50%
670	0010	**RELIGIOUS EDUCATION** R17100 -100	S.F.	73.50	94	118			
	0020	Total project costs	C.F.	4.09	5.85	7.55			
	2720	Plumbing	S.F.	3.08	4.36	6.15	4.40%	5.30%	7.10%
	2770	Heating, ventilating, air conditioning		7.80	8.80	12.45	10.05%	11.45%	12.35%
	2900	Electrical		5.85	7.70	10.80	7.60%	9.10%	10.35%
	3100	Total: Mechanical & Electrical	↓	20	31	33.50	22%	23%	26%
690	0010	**RESEARCH** Laboratories and facilities R17100 -100	S.F.	110	158	230			
	0020	Total project costs	C.F.	8.60	16.60	19.35			
	2720	Plumbing	S.F.	11.05	14.15	22.50	6.15%	8.30%	10.80%
	2770	Heating, ventilating, air conditioning		9.90	33.50	39.50	7.25%	16.50%	17.50%
	2900	Electrical		12.70	21.50	36	9.45%	11.15%	15.40%
	3100	Total: Mechanical & Electrical	↓	36	71	106	29.50%	37%	45.50%
700	0010	**RESTAURANTS** R17100 -100	S.F.	106	137	179			
	0020	Total project costs	C.F.	9	11.80	15.45			
	2720	Plumbing	S.F.	8.55	10.25	14.60	6.10%	8.15%	9%
	2770	Heating, ventilating, air conditioning		11.15	14.85	19.35	9.20%	12%	12.40%
	2900	Electrical		11.30	13.95	18.10	8.35%	10.55%	11.55%
	3100	Total: Mechanical & Electrical	↓	35	37	47.50	19.25%	24%	29.50%
	9000	Per seat unit, total cost	Seat	3,925	5,225	6,175			
	9500	Total: Mechanical & Electrical	"	980	1,300	1,550			
720	0010	**RETAIL STORES** R17100 -100	S.F.	50	67	88.50			
	0020	Total project costs	C.F.	3.38	4.82	6.70			
	2720	Plumbing	S.F.	1.81	3.02	5.15	3.26%	4.60%	6.80%
	2770	Heating, ventilating, air conditioning	↓	3.91	5.35	8.05	6.75%	8.75%	10.15%

See R17100-100 and City Cost Indexes in the Reference Section

17100		S.F. & C.F. Costs		UNIT	UNIT COSTS			% OF TOTAL		
					1/4	MEDIAN	3/4	1/4	MEDIAN	3/4
720	2900	Electrical		S.F.	4.48	6.25	8.90	7.30%	9.90%	11.65%
	3100	Total: Mechanical & Electrical		↓	11.95	15.35	20.50	17.10%	21.50%	23.50%
740	0010	**SCHOOLS** Elementary	R17100 -100	S.F.	80.50	99.50	120			
	0020	Total project costs		C.F.	5.35	6.80	8.80			
	2720	Plumbing		S.F.	4.67	6.60	8.85	5.70%	7.15%	9.35%
	2730	Heating, ventilating, air conditioning			7	11.20	15.60	8.15%	10.80%	15.20%
	2900	Electrical			7.55	9.90	12.50	8.45%	10.05%	11.80%
	3100	Total: Mechanical & Electrical		↓	27	33	39.50	24%	27.50%	30.50%
	9000	Per pupil, total cost		Ea.	7,850	14,000	41,400			
	9500	Total: Mechanical & Electrical		"	2,650	3,350	12,000			
760	0010	**SCHOOLS** Junior High & Middle	R17100 -100	S.F.	81.50	102	120			
	0020	Total project costs		C.F.	5.35	6.90	7.75			
	2720	Plumbing		S.F.	5.35	6.05	7.80	5.50%	6.90%	8.15%
	2770	Heating, ventilating, air conditioning			6.15	11.80	16.50	8.75%	12.75%	17.45%
	2900	Electrical			7.75	9.85	12.20	7.90%	9.30%	10.60%
	3100	Total: Mechanical & Electrical		↓	23.50	33	38.50	22.50%	25.50%	29.50%
	9000	Per pupil, total cost		Ea.	10,700	14,400	18,800			
780	0010	**SCHOOLS** Senior High	R17100 -100	S.F.	85.50	108	136			
	0020	Total project costs		C.F.	5.55	7.75	12.85			
	2720	Plumbing		S.F.	4.47	5.95	13.45	5.70%	7%	8.35%
	2770	Heating, ventilating, air conditioning			10	11.45	22	8.95%	11.60%	15%
	2900	Electrical			8.55	11.15	17.90	8.35%	10.10%	11.95%
	3100	Total: Mechanical & Electrical		↓	30	34	57	23%	26.50%	28.50%
	9000	Per pupil, total cost		Ea.	8,225	14,000	21,000			
800	0010	**SCHOOLS** Vocational	R17100 -100	S.F.	70.50	100	125			
	0020	Total project costs		C.F.	4.39	6.30	8.70			
	2720	Plumbing		S.F.	4.60	6.90	10.35	5.40%	6.90%	8.55%
	2770	Heating, ventilating, air conditioning			6.30	11.75	19.70	8.60%	11.90%	14.65%
	2900	Electrical			7.20	10.05	14.20	8.45%	10.95%	13.20%
	3100	Total: Mechanical & Electrical		↓	25.50	28.50	48.50	23.50%	29.50%	31%
	9000	Per pupil, total cost		Ea.	9,850	26,300	39,300			
830	0010	**SPORTS ARENAS**	R17100 -100	S.F.	60	82.50	127			
	0020	Total project costs		C.F.	3.35	6	7.75			
	2720	Plumbing		S.F.	3.58	5.45	11.50	4.35%	6.35%	9.40%
	2770	Heating, ventilating, air conditioning			7.70	9.15	12.65	8.80%	10.20%	13.55%
	2900	Electrical			6.50	8.75	11.30	8.60%	9.90%	12.25%
	3100	Total: Mechanical & Electrical		↓	16	28.50	37	21.50%	25%	27.50%
850	0010	**SUPERMARKETS**	R17100 -100	S.F.	56	66	80			
	0020	Total project costs		C.F.	3.18	3.84	5.80			
	2720	Plumbing		S.F.	3.19	4.10	4.67	5.40%	6%	7.45%
	2770	Heating, ventilating, air conditioning			4.69	6.25	7.60	8.60%	8.65%	9.60%
	2900	Electrical			7.15	8.20	9.70	10.40%	12.45%	13.60%
	3100	Total: Mechanical & Electrical		↓	18.35	19.95	28	20.50%	26.50%	31%
860	0010	**SWIMMING POOLS**	R17100 -100	S.F.	92.50	155	305			
	0020	Total project costs		C.F.	7.40	9.25	10.05			
	2720	Plumbing		S.F.	8.55	9.75	13.60	4.80%	9.70%	20.50%
	2900	Electrical			6.95	11.25	16.40	6.50%	7.25%	7.60%
	3100	Total: Mechanical & Electrical		↓	16.95	43	58.50	11.15%	14.10%	23.50%
870	0010	**TELEPHONE EXCHANGES**	R17100 -100	S.F.	123	180	228			
	0020	Total project costs		C.F.	7.65	12.25	16.85			
	2720	Plumbing		S.F.	5.20	8.20	11.70	4.52%	5.80%	6.90%
	2770	Heating, ventilating, air conditioning			12	24	30	11.80%	16.05%	18.40%
	2900	Electrical			12.50	19.80	35	10.90%	14%	17.85%
	3100	Total: Mechanical & Electrical		↓	37	70	99	29.50%	33.50%	44.50%

For expanded coverage of these items see *Means Square Foot Costs 2005*

17100 | S.F. & C.F. Costs

			UNIT	UNIT COSTS			% OF TOTAL			
				1/4	MEDIAN	3/4	1/4	MEDIAN	3/4	
910	0010	**THEATERS** R17100-100	S.F.	77	99	146				910
	0020	Total project costs	C.F.	3.56	5.25	7.75				
	2720	Plumbing	S.F.	2.57	2.79	11.40	2.92%	4.70%	6.80%	
	2770	Heating, ventilating, air conditioning		7.50	9.10	11.25	8%	12.25%	13.40%	
	2900	Electrical		6.75	9.10	18.55	8.05%	9.95%	12.25%	
	3100	Total: Mechanical & Electrical		17.35	26	53.50	23%	26.50%	27.50%	
940	0010	**TOWN HALLS** City Halls & Municipal Buildings R17100-100	S.F.	90.50	116	153				940
	0020	Total project costs	C.F.	7.85	9.70	13.25				
	2720	Plumbing	S.F.	3.60	6.75	12.40	4.31%	5.95%	7.95%	
	2770	Heating, ventilating, air conditioning		6.50	12.90	18.85	7.05%	9.05%	13.45%	
	2900	Electrical		8.20	11.95	15.95	8.05%	9.50%	12.05%	
	3100	Total: Mechanical & Electrical		27.50	32.50	42.50	22%	26.50%	31%	
970	0010	**WAREHOUSES** And Storage Buildings R17100-100	S.F.	34	49.50	68.50				970
	0020	Total project costs	C.F.	1.85	2.63	4.36				
	2720	Plumbing	S.F.	1.07	1.92	3.67	2.90%	4.80%	6.55%	
	2730	Heating, ventilating, air conditioning		1.29	3.45	4.63	2.41%	5%	8.90%	
	2900	Electrical		2.03	2.69	5.95	5.15%	7.20%	10.10%	
	3100	Total: Mechanical & Electrical		5.30	8.15	17.80	12.75%	18.90%	26%	
990	0010	**WAREHOUSE & OFFICES** Combination R17100-100	S.F.	39.50	52.50	73				990
	0020	Total project costs	C.F.	2.02	2.94	4.34				
	2720	Plumbing	S.F.	1.53	2.71	4.07	3.74%	4.76%	6.30%	
	2770	Heating, ventilating, air conditioning		2.41	3.77	5.30	5%	5.65%	10.05%	
	2900	Electrical	S.F.	2.65	3.93	6.20	5.85%	8%	10%	
	3100	Total: Mechanical & Electrical		7.30	10.45	17.25	14.40%	19.95%	24.50%	

For information about Means Estimating Seminars, see yellow pages 12 and 13 in back of book

See R17100-100 and City Cost Indexes in the Reference Section

Assemblies Section

Table of Contents

How to Use the Assemblies Cost Tables

The following is a detailed explanation of a sample Assemblies Cost Table. Most Assembly Tables are separated into three parts: 1) an illustration of the system to be estimated; 2) the components and related costs of a typical system; and 3) the costs for similar systems with dimensional and/or size variations. For costs of the components that comprise these systems or "assemblies" refer to the Unit Price Section. Next to each bold number below is the item being described with the appropriate component of the sample entry following in parenthesis. In most cases, if the work is to be subcontracted, the general contractor will need to add an additional markup (R.S. Means suggests using 10%) to the "Total" figures.

System/Line Numbers (D3010 510 1760)

Each Assemblies Cost Line has been assigned a unique identification number based on the UNIFORMAT II classification system.

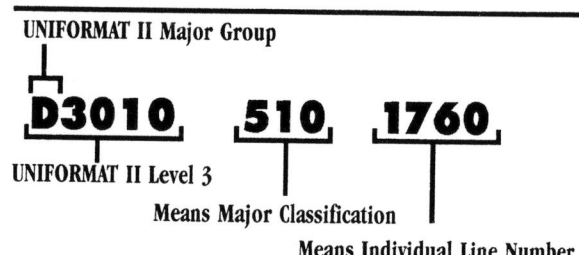

UNIFORMAT II Major Group

D3010 **510** **1760**

UNIFORMAT II Level 3

Means Major Classification

Means Individual Line Number

D30 HVAC

D3010 Energy Supply

Basis for Heat Loss Estimate, Apartment Type Structures:

1. Masonry walls and flat roof are insulated. U factor is assumed at .08.
2. Window glass area taken as BOCA minimum, 1/10th of floor area. Double insulating glass with 1/4" air space, U = .65.
3. Infiltration = 0.3 C.F. per hour per S.F. of net wall.
4. Concrete floor loss is 2 BTUH per S.F.
5. Temperature difference taken as 70° F.
6. Ventilating or makeup air has not been included and must be added if desired. Air shafts are not used.

System Components	QUANTITY	UNIT	COST EACH MAT.	COST EACH INST.	COST EACH TOTAL
SYSTEM D3010 510 1760					
HEATING SYSTEM, FIN TUBE RADIATION, FORCED HOT WATER					
1,000 S.F. AREA, 10,000 C.F. VOLUME					
Boiler, oil fired, CI, burner, ctrls/insul/breech/pipe/ftng/valves, 109 MBH	1.000	Ea.	3,718.75	2,756.25	6,475
Circulating pump, CI flange connection, 1/12 HP	1.000	Ea.	234	147	381
Expansion tank, painted steel, ASME 18 Gal capacity	1.000	Ea.	405	63.50	468.50
Storage tank, steel, above ground, 275 Gal capacity w/supports	1.000	Ea.			
Copper tubing type L, solder joint, hanger 10' OC, 3/4" diam	100.000	L.F.			
Radiation, 3/4" copper tube w/alum fin baseboard pkg, 7" high	30.000	L.F.	87.50		50
Pipe covering, calcium silicate w/cover, 1" wall, 3/4" diam	100.000	L.F.	240	496	736
TOTAL			5,367.25	4,743.75	10,111
COST PER S.F.			5.37	4.74	10.11

D3010 510	Apartment Building Heating - Fin Tube Radiation	COST PER S.F. MAT.	COST PER S.F. INST.	COST PER S.F. TOTAL
1740	Heating systems, fin tube radiation, forced hot water			
1760	1,000 S.F. area, 10,000 C.F. volume	5.38	4.75	10.13
1800	10,000 S.F. area, 100,000 C.F. volume	1.70	2.86	4.56

Illustration

At the top of most assembly pages is an illustration, a brief description, and the design criteria used to develop the cost.

System Components

The components of a typical system are listed separately to show what has been included in the development of the total system price. The table below contains prices for other similar systems with dimensional and/or size variations.

Quantity

This is the number of line item units required for one system unit. For example, we assume that it will take 30 linear feet of radiation for the 1000 S.F. area shown.

Unit of Measure for Each Item

The abbreviated designation indicates the unit of measure, as defined by industry standards, upon which the price of the component is based. For example, baseboard radiation is priced by the linear foot. For a complete listing of abbreviations, see the Reference Section.

Unit of Measure for Each System (Each)

Costs shown in the three right hand columns have been adjusted by the component quantity and unit of measure for the entire system. In this example, "Cost Each" is the unit of measure for this system or "assembly."

Materials (5,367.25)

This column contains the Materials Cost of each component. These cost figures are bare costs plus 10% for profit.

Installation (4,743.75)

Installation includes labor and equipment plus the installing contractor's overhead and profit. Equipment costs are the bare rental costs plus 10% for profit. The labor overhead and profit is defined on the inside back cover of this book.

Total (10,111.00)

The figure in this column is the sum of the material and installation costs.

Material Cost	+	Installation Cost	=	Total
$5,367.25	+	$4,743.75	=	$10,111.00

D SERVICES

D2010 Plumbing Fixtures

Systems are complete with trim seat and rough-in (supply, waste and vent) for connection to supply branches and waste mains.

One Piece Wall Hung   **Supply**   **Waste/Vent**   **Floor Mount**

System Components	QUANTITY	UNIT	COST EACH		
			MAT.	INST.	TOTAL
SYSTEM D2010 110 1840					
WATER CLOSET, VITREOUS CHINA, ELONGATED					
TANK TYPE, WALL HUNG, ONE PIECE					
Wtr closet tank type vit china wall hung 1 pc w/seat supply & stop	1.000	Ea.	430	167	597
Pipe steel galvanized, schedule 40, threaded, 2″ diam	4.000	L.F.	21.40	55.20	76.60
Pipe, CI soil, no hub, cplg 10′ OC, hanger 5′ OC, 4″ diam	2.000	L.F.	20	30.50	50.50
Pipe, coupling, standard coupling, CI soil, no hub, 4″ diam	2.000	Ea.	16.20	54	70.20
Copper tubing type L, solder joint, hanger 10′OC, 1/2″ diam	6.000	L.F.	9.90	36.30	46.20
Wrought copper 90° elbow for solder joints 1/2″ diam	2.000	Ea.	1.06	49	50.06
Wrought copper Tee for solder joints 1/2″ diam	1.000	Ea.	.89	38	38.89
Support/carrier, for water closet, siphon jet, horiz, single, 4″ waste	1.000	Ea.	299	92	391
TOTAL			798.45	522	1,320.45

D2010 110	Water Closet Systems		COST EACH		
			MAT.	INST.	TOTAL
1800	Water closet, vitreous china, elongated				
1840	Tank type, wall hung, one piece		800	520	1,320
1880	Close coupled two piece		895	520	1,415
1920	Floor mount, one piece	R15100 -410	735	555	1,290
1960	One piece low profile		625	555	1,180
2000	Two piece close coupled	R15100 -420	370	555	925
2040	Bowl only with flush valve				
2080	Wall hung		795	590	1,385
2120	Floor mount		545	565	1,110

D2010 Plumbing Fixtures

Systems are complete with trim, seat, flush valve and rough-in (supply, waste and vent) for connection to supply branches and waste mains.

Side by Side

Back to Back

Supply   Waste/Vent   Supply   Waste/Vent

System Components	QUANTITY	UNIT	COST EACH MAT.	COST EACH INST.	COST EACH TOTAL
SYSTEM D2010 120 1760					
WATER CLOSETS, BATTERY MOUNT, WALL HUNG, SIDE BY SIDE, FIRST CLOSET					
Water closet, bowl only w/flush valve, seat, wall hung	1.000	Ea.	380	153	533
Pipe, CI soil, no hub, cplg 10' OC, hanger 5' OC, 4" diam	3.000	L.F.	30	45.75	75.75
Coupling, standard, CI, soil, no hub, 4" diam	2.000	Ea.	16.20	54	70.20
Copper tubing type L, solder joints, hangers 10' OC, 1" diam	6.000	L.F.	20.46	43.20	63.66
Copper tubing, type DWV, solder joints, hangers 10'OC, 2" diam	6.000	L.F.	40.20	66.90	107.10
Wrought copper 90° elbow for solder joints 1" diam	1.000	Ea.	2.87	30.50	33.37
Wrought copper Tee for solder joints, 1" diam	1.000	Ea.	6.60	49	55.60
Support/carrier, siphon jet, horiz, adjustable single, 4" pipe	1.000	Ea.	299	92	391
Valve, gate, bronze, 125 lb, NRS, soldered 1" diam	1.000	Ea.	38	26	64
Wrought copper, DWV, 90° elbow, 2" diam	1.000	Ea.	10.35	49	59.35
TOTAL			843.68	609.35	1,453.03

D2010 120	Water Closets, Group		COST EACH MAT.	COST EACH INST.	COST EACH TOTAL
1760	Water closets, battery mount, wall hung, side by side, first closet	R15100 -410	845	610	1,455
1800	Each additional water closet, add		795	575	1,370
3000	Back to back, first pair of closets	R15100 -420	1,500	810	2,310
3100	Each additional pair of closets, back to back		1,450	795	2,245

D2010 Plumbing Fixtures

Systems are complete with trim, flush valve and rough-in (supply, waste and vent) for connection to supply branches and waste mains.

Stall Type

Supply   **Waste/Vent**

Wall Hung

System Components	QUANTITY	UNIT	COST EACH		
			MAT.	INST.	TOTAL
SYSTEM D2010 210 2000					
URINAL, VITREOUS CHINA, WALL HUNG					
Urinal, wall hung, vitreous china, incl. hanger	1.000	Ea.	330	295	625
Pipe, steel, galvanized, schedule 40, threaded, 1-1/2" diam	5.000	L.F.	20.40	55.25	75.65
Copper tubing type DWV, solder joint, hangers 10'OC, 2" diam	3.000	L.F.	20.10	33.45	53.55
Combination Y & 1/8 bend for CI soil pipe, no hub, 3" diam	1.000	Ea.	11.60		11.60
Pipe, CI, no hub, cplg 10' OC, hanger 5' OC, 3" diam	4.000	L.F.	31.60	55.20	86.80
Pipe coupling standard, CI soil, no hub, 3" diam	3.000	Ea.	13.70	47	60.70
Copper tubing type L, solder joint, hanger 10' OC 3/4" diam	5.000	L.F.	12.10	32.25	44.35
Wrought copper 90° elbow for solder joints 3/4" diam	1.000	Ea.	1.17	26	27.17
Wrought copper Tee for solder joints, 3/4" diam	1.000	Ea.	2.15	41	43.15
TOTAL			442.82	585.15	1,027.97

D2010 210	Urinal Systems		COST EACH		
			MAT.	INST.	TOTAL
2000	Urinal, vitreous china, wall hung	R15100 -410	445	585	1,030
2040	Stall type		730	700	1,430

D20 Plumbing

D2010 Plumbing Fixtures

Systems are complete with trim, flush valve and rough-in (supply, waste and vent) for connection to supply branches and waste mains.

Side by Side

Back to Back

Waste/Vent

Supply

Supply

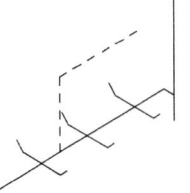

Waste/Vent

System Components	QUANTITY	UNIT	MAT.	INST.	TOTAL
SYSTEM D2010 220 1760					
URINALS, BATTERY MOUNT, WALL HUNG, SIDE BY SIDE, FIRST URINAL					
Urinal, wall hung, vitreous china, with hanger & trim	1.000	Ea.	330	295	625
No hub cast iron soil pipe, 3" diameter	4.000	L.F.	31.60	55.20	86.80
No hub cast iron sanitary tee, 3" diameter	1.000	Ea.	11.60		11.60
No hub coupling, 3" diameter	2.000	Ea.	13.70	47	60.70
Copper tubing, type L, 3/4" diameter	5.000	L.F.	12.10	32.25	44.35
Copper tubing, type DWV, 2" diameter	2.000	L.F.	13.40	22.30	35.70
Copper 90° elbow, 3/4" diameter	2.000	Ea.	2.34	52	54.34
Copper 90° elbow, type DWV, 2" diameter	2.000	Ea.	10.35	49	59.35
Galvanized steel pipe, 1-1/2" diameter	5.000	L.F.	20.40	55.25	75.65
Cast iron drainage elbow, 90°, 1-1/2" diameter	1.000	Ea.	2.15	41	43.15
TOTAL			447.64	649	1,096.64

D2010 220	Urinal Systems, Battery Mount		MAT.	INST.	TOTAL
1760	Urinals, battery mount, side by side, first urinal		450	650	1,100
1800	Each additional urinal, add		430	630	1,060
2000	Back to back, first pair of urinals	R15100 -410	890	1,050	1,940
2100	Each additional pair of urinals, back to back		765	840	1,605

D2010 Plumbing Fixtures

Systems are complete with trim and rough-in (supply, waste and vent) to connect to supply branches and waste mains.

Vanity Top

Supply   **Waste/Vent**

Wall Hung

System Components			COST EACH		
	QUANTITY	UNIT	MAT.	INST.	TOTAL
SYSTEM D2010 310 1560					
LAVATORY W/TRIM, VANITY TOP, P.E. ON C.I., 20″ X 18″					
Lavatory w/trim, PE on CI, white, vanity top, 20″ x 18″ oval	1.000	Ea.	227	138	365
Pipe, steel, galvanized, schedule 40, threaded, 1-1/4″ diam	4.000	L.F.	14.08	39.80	53.88
Copper tubing type DWV, solder joint, hanger 10′OC 1-1/4″ diam	4.000	L.F.	16.16	32.80	48.96
Wrought copper DWV, Tee, sanitary, 1-1/4″ diam	1.000	Ea.	9.45	54.50	63.95
P trap w/cleanout, 20 ga, 1-1/4″ diam	1.000	Ea.	31	27.50	58.50
Copper tubing type L, solder joint, hanger 10′ OC 1/2″ diam	10.000	L.F.	16.50	60.50	77
Wrought copper 90° elbow for solder joints 1/2″ diam	2.000	Ea.	1.06	49	50.06
Wrought copper Tee for solder joints, 1/2″ diam	2.000	Ea.	1.78	76	77.78
Stop, chrome, angle supply, 1/2″ diam	2.000	Ea.	10.40	45	55.40
TOTAL			327.43	523.10	850.53

D2010 310	Lavatory Systems		COST EACH		
			MAT.	INST.	TOTAL
1560	Lavatory w/trim, vanity top, PE on Cl, 20″ x 18″, Vanity top by others.		325	525	850
1600	19″ x 16″ oval		242	525	767
1640	18″ round		345	525	870
1680	Cultured marble, 19″ x 17″		269	525	794
1720	25″ x 19″		269	525	794
1760	Stainless, self-rimming, 25″ x 22″		360	525	885
1800	17″ x 22″		350	525	875
1840	Steel enameled, 20″ x 17″	R15100 -410	242	540	782
1880	19″ round		240	540	780
1920	Vitreous china, 20″ x 16″		345	550	895
1960	19″ x 16″		345	550	895
2000	22″ x 13″		355	550	905
2040	Wall hung, PE on Cl, 18″ x 15″		610	580	1,190
2080	19″ x 17″		610	580	1,190
2120	20″ x 18″		575	580	1,155
2160	Vitreous china, 18″ x 15″		530	595	1,125
2200	19″ x 17″		520	595	1,115
2240	24″ x 20″		765	595	1,360

Important: See the Reference Section for critical supporting data - Reference Numbers and City Cost Indexes

D20 Plumbing

D2010 Plumbing Fixtures

Systems are complete with trim, flush valve and rough-in (supply, waste and vent) for connection to supply branches and waste mains.

Side by Side

Back to Back

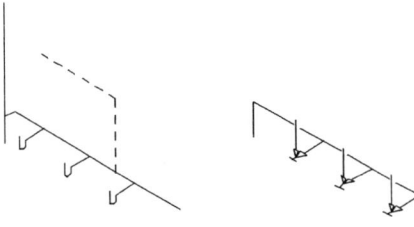

Waste/Vent

Supply
(Two supply systems required)

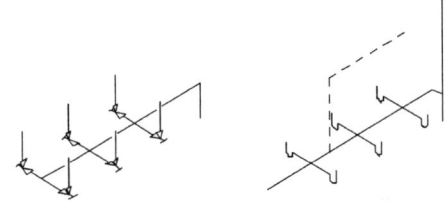

Supply
(Two supply systems required)

Waste/Vent

System Components	QUANTITY	UNIT	COST EACH		
			MAT.	INST.	TOTAL
SYSTEM D2010 320 1760					
LAVATORIES, BATTERY MOUNT, WALL HUNG, SIDE BY SIDE, FIRST LAVATORY					
Lavatory w/trim wall hung PE on CI 20" x 18"	1.000	Ea.	261	111	372
Stop, chrome, angle supply, 3/8" diameter	2.000	Ea.	10.40	41	51.40
Concealed arm support	1.000	Ea.	284	82	366
P trap w/cleanout, 20 ga. C.P., 1-1/4" diameter	1.000	Ea.	29	27.50	56.50
Copper tubing, type L, 1/2" diameter	10.000	L.F.	16.50	60.50	77
Copper tubing, type DWW, 1-1/4" diameter	4.000	L.F.	16.16	32.80	48.96
Copper 90° elbow, 1/2" diameter	2.000	Ea.	1.06	49	50.06
Copper tee, 1/2" diameter	2.000	Ea.	1.78	76	77.78
DWV copper sanitary tee, 1-1/4" diameter	2.000	Ea.	18.90	109	127.90
Galvanized steel pipe, 1-1/4" diameter	4.000	L.F.	14.08	39.80	53.88
Black cast iron 90° elbow, 1-1/4" diameter	1.000	Ea.	4.91	40	44.91
TOTAL			657.79	668.60	1,326.39

D2010 320	Lavatory Systems, Battery Mount	COST EACH		
		MAT.	INST.	TOTAL
1760	Lavatories, battery mount, side by side, first lavatory	660	670	1,330
1800	Each additional lavatory, add	620	465	1,085
2000	Back to back, first pair of lavatories	1,100	1,075	2,175
2100	Each additional pair of lavatories, back to back	1,075	880	1,955

D2010 Plumbing Fixtures

Systems are complete with trim and rough-in (supply, waste and vent) to connect to supply branches and waste mains.

Countertop Single Bowl

Supply

Waste/Vent

Countertop Double Bowl

System Components	QUANTITY	UNIT	COST EACH		
			MAT.	INST.	TOTAL
SYSTEM D2010 410 1720					
KITCHEN SINK W/TRIM, COUNTERTOP, P.E. ON C.I., 24″ X 21″, SINGLE BOWL					
Kitchen sink, counter top, PE on CI, 1 bowl, 24″ x 21″ OD	1.000	Ea.	227	158	385
Pipe, steel, galvanized, schedule 40, threaded, 1-1/4″ diam	4.000	L.F.	14.08	39.80	53.88
Copper tubing, type DWV, solder, hangers 10′ OC 1-1/2″ diam	6.000	L.F.	30	54.60	84.60
Wrought copper, DWV, Tee, sanitary, 1-1/2″ diam	1.000	Ea.	11.75	61.50	73.25
P trap, standard, copper, 1-1/2″ diam	1.000	Ea.	30	29	59
Copper tubing, type L, solder joints, hangers 10′ OC 1/2″ diam	10.000	L.F.	16.50	60.50	77
Wrought copper 90° elbow for solder joints 1/2″ diam	2.000	Ea.	1.06	49	50.06
Wrought copper Tee for solder joints, 1/2″ diam	2.000	Ea.	1.78	76	77.78
Stop, angle supply, chrome, 1/2″ CTS	2.000	Ea.	10.40	45	55.40
TOTAL			342.57	573.40	915.97

D2010 410	Kitchen Sink Systems		COST EACH		
			MAT.	INST.	TOTAL
1720	Kitchen sink w/trim, countertop, PE on CI, 24″x21″, single bowl		345	575	920
1760	30″ x 21″ single bowl		395	575	970
1800	32″ x 21″ double bowl		440	620	1,060
1840	42″ x 21″ double bowl	R15155 -210	660	625	1,285
1880	Stainless steel, 19″ x 18″ single bowl		480	575	1,055
1920	25″ x 22″ single bowl	R15155 -220	520	575	1,095
1960	33″ x 22″ double bowl		710	620	1,330
2000	43″ x 22″ double bowl	R15100 -120	805	625	1,430
2040	44″ x 22″ triple bowl		1,025	655	1,680
2080	44″ x 24″ corner double bowl		700	625	1,325
2120	Steel, enameled, 24″ x 21″ single bowl		237	575	812
2160	32″ x 21″ double bowl		287	620	907
2240	Raised deck, PE on CI, 32″ x 21″, dual level, double bowl		415	785	1,200
2280	42″ x 21″ dual level, triple bowl		810	850	1,660

Important: See the Reference Section for critical supporting data - Reference Numbers and City Cost Indexes

D2010 Plumbing Fixtures

Systems are complete with trim and rough-in (supply, waste and vent) to connect to supply branches and waste mains.

Single Compartment Sink   **Supply**   **Waste/Vent**   **Double Compartment Sink**

System Components	QUANTITY	UNIT	COST EACH		
			MAT.	INST.	TOTAL
SYSTEM D2010 420 1760					
LAUNDRY SINK W/TRIM, PE ON CI, BLACK IRON FRAME					
24" X 20" OD, SINGLE COMPARTMENT					
Laundry sink PE on CI w/trim & frame, 24" x 21" OD, 1 compartment	1.000	Ea.	380	147	527
Pipe, steel, galvanized, schedule 40, threaded, 1-1/4" diam	4.000	L.F.	14.08	39.80	53.88
Copper tubing, type DWV, solder joint, hanger 10' OC 1-1/2"diam	6.000	L.F.	30	54.60	84.60
Wrought copper, DWV, Tee, sanitary, 1-1/2" diam	1.000	Ea.	11.75	61.50	73.25
P trap, standard, copper, 1-1/2" diam	1.000	Ea.	30	29	59
Copper tubing type L, solder joints, hangers 10' OC, 1/2" diam	10.000	L.F.	16.50	60.50	77
Wrought copper 90° elbow for solder joints 1/2" diam	2.000	Ea.	1.06	49	50.06
Wrought copper Tee for solder joints, 1/2" diam	2.000	Ea.	1.78	76	77.78
Stop, angle supply, 1/2" diam	2.000	Ea.	10.40	45	55.40
TOTAL			495.57	562.40	1,057.97

D2010 420	Laundry Sink Systems		COST EACH		
			MAT.	INST.	TOTAL
1740	Laundry sink w/trim, PE on CI, black iron frame				
1760	24" x 20", single compartment		495	560	1,055
1800	24" x 23" single compartment	R15100 -410	410	560	970
1840	48" x 21" double compartment		415	610	1,025
1920	Molded stone, on wall, 22" x 21" single compartment		252	560	812
1960	45"x 21" double compartment		410	610	1,020
2040	Plastic, on wall or legs, 18" x 23" single compartment		217	550	767
2080	20" x 24" single compartment		250	550	800
2120	36" x 23" double compartment		288	595	883
2160	40" x 24" double compartment		365	595	960
2200	Stainless steel, countertop, 22" x 17" single compartment		495	560	1,055
2240	19" x 22" single compartment		600	560	1,160
2280	33" x 22" double compartment		600	610	1,210

D2010 Plumbing Fixtures

Corrosion resistant laboratory sink systems are complete with trim and rough-in (supply, waste and vent) to connect to supply branches and waste mains.

Laboratory Sink

Supply

Waste/Vent

Polypropylene Cup Sink

System Components

System Components	QUANTITY	UNIT	COST EACH MAT.	COST EACH INST.	COST EACH TOTAL
SYSTEM D2010 430 1600					
LABORATORY SINK W/TRIM, POLYETHYLENE, SINGLE BOWL					
DOUBLE DRAINBOARD, 54" X 24" OD					
Sink w/trim, polyethylene, 1 bowl, 2 drainboards 54" x 24" OD	1.000	Ea.	895	295	1,190
Pipe, polypropylene, schedule 40, acid resistant 1-1/2" diam	10.000	L.F.	12.40	130	142.40
Tee, sanitary, polypropylene, acid resistant, 1-1/2" diam	1.000	Ea.	19.15	49	68.15
P trap, polypropylene, acid resistant, 1-1/2" diam	1.000	Ea.	56	29	85
Copper tubing type L, solder joint, hanger 10' OC 1/2" diam	10.000	L.F.	16.50	60.50	77
Wrought copper 90° elbow for solder joints 1/2" diam	2.000	Ea.	1.06	49	50.06
Wrought copper Tee for solder joints, 1/2" diam	2.000	Ea.	1.78	76	77.78
Stop, angle supply, chrome, 1/2" diam	2.000	Ea.	10.40	45	55.40
TOTAL			1,012.29	733.50	1,745.79

D2010 430	Laboratory Sink Systems	COST EACH MAT.	COST EACH INST.	COST EACH TOTAL
1580	Laboratory sink w/trim, polyethylene, single bowl,			
1600	Double drainboard, 54" x 24" O.D.	1,000	735	1,735
1640	Single drainboard, 47" x 24"O.D.	1,150	735	1,885
1680	70" x 24" O.D.	1,175	735	1,910
1760	Flanged, 14-1/2" x 14-1/2" O.D.	380	660	1,040
1800	18-1/2" x 18-1/2" O.D.	385	660	1,045
1840	23-1/2" x 20-1/2" O.D.	425	660	1,085
1920	Polypropylene, cup sink, oval, 7" x 4" O.D.	237	580	817
1960	10" x 4-1/2" O.D.	231	580	811

Important: See the Reference Section for critical supporting data - Reference Numbers and City Cost Indexes

D2010 Plumbing Fixtures

Corrosion resistant laboratory sink systems are complete with trim and rough–in (supply, waste and vent) to connect to supply branches and waste mains.

Wall Hung

Supply

Waste/Vent

Corner, Floor

System Components	QUANTITY	UNIT	COST EACH		
			MAT.	INST.	TOTAL
SYSTEM D2010 440 4260					
SERVICE SINK, PE ON CI, CORNER FLOOR, 28"X28", W/RIM GUARD & TRIM					
Service sink, corner floor, PE on CI, 28" x 28", w/rim guard & trim	1.000	Ea.	585	201	786
Copper tubing type DWV, solder joint, hanger 10'OC 3" diam	6.000	L.F.	72	91.50	163.50
Copper tubing type DWV, solder joint, hanger 10'OC 2" diam	4.000	L.F.	26.80	44.60	71.40
Wrought copper DWV, Tee, sanitary, 3" diam	1.000	Ea.	55	126	181
P trap with cleanout & slip joint, copper 3" diam	1.000	Ea.	111	44.50	155.50
Copper tubing, type L, solder joints, hangers 10' OC, 1/2" diam	10.000	L.F.	16.50	60.50	77
Wrought copper 90° elbow for solder joints 1/2" diam	2.000	Ea.	1.06	49	50.06
Wrought copper Tee for solder joints, 1/2" diam	2.000	Ea.	1.78	76	77.78
Stop, angle supply, chrome, 1/2" diam	2.000	Ea.	10.40	45	55.40
TOTAL			879.54	738.10	1,617.64

D2010 440	Service Sink Systems		COST EACH		
			MAT.	INST.	TOTAL
4260	Service sink w/trim, PE on CI, corner floor, 28" x 28", w/rim guard	R15100 –410	880	740	1,620
4300	Wall hung w/rim guard, 22" x 18"		970	855	1,825
4340	24" x 20"		990	855	1,845
4380	Vitreous china, wall hung 22" x 20"		910	855	1,765

D2010 Plumbing Fixtures

Systems are complete with trim and rough-in (supply, waste and vent) to connect to supply branches and waste mains.

Recessed Bathtub   Supply   Waste/Vent   Corner Bathtub

System Components	QUANTITY	UNIT	COST EACH MAT.	COST EACH INST.	COST EACH TOTAL
SYSTEM D2010 510 1960					
BATHTUB RECESSED, PORCELAIN ENAMEL ON CAST IRON, 42" X 37"					
Bath tub, porcelain enamel on cast iron, w/fittings, 42" x 37"	1.000	Ea.	940	177	1,117
Pipe, steel, galvanized, schedule 40, threaded, 1-1/4" diam	4.000	L.F.	14.08	39.80	53.88
Pipe, CI, soil, no hub, w/coupling 10' OC, hanger 5' OC, 4" diam	3.000	L.F.	30	45.75	75.75
Combination Y and 1/8 bend for C.I. soil pipe, no hub, 4" pipe size	1.000	Ea.	26.50		26.50
Drum trap, 3" x 5", copper, 1-1/2" diam	1.000	Ea.	42	30.50	72.50
Copper tubing, type L, solder joints, hangers 10' OC 1/2" diam	10.000	L.F.	16.50	60.50	77
Wrought copper 90° elbow for solder joints, 1/2" diam	2.000	Ea.	1.06	49	50.06
Wrought copper Tee for solder joints, 1/2" diam	2.000	Ea.	1.78	76	77.78
Stop, angle supply, 1/2" diam	2.000	Ea.	10.40	45	55.40
Copper tubing type DWV, solder joint, hanger 10' OC 1-1/2" diam	3.000	L.F.	15	27.30	42.30
Pipe coupling, standard, C.I. soil no hub, 4" pipe size	2.000	Ea.	16.20	54	70.20
TOTAL			1,113.52	604.85	1,718.37

D2010 510	Bathtub Systems		COST EACH MAT.	COST EACH INST.	COST EACH TOTAL
1960	Bathtub, recessed, P.E. on CI., 42" x 37"		1,125	605	1,730
2000	48" x 42"		1,825	650	2,475
2040	72" x 36"		1,875	725	2,600
2080	Mat bottom, 5' long	R15100 -410	635	630	1,265
2120	5'-6" long		1,350	650	2,000
2160	Corner, 48" x 42"		1,825	630	2,455
2200	Formed steel, enameled, 4'-6" long		540	580	1,120
2240	5' long		495	590	1,085

Important: See the Reference Section for critical supporting data - Reference Numbers and City Cost Indexes

D2010 Plumbing Fixtures

Systems are complete with trim, flush valve and rough-in (supply, waste and vent) for connection to supply branches and waste mains.

Circular Fountain

Supply

Waste/Vent

Semi-Circular Fountain

System Components	QUANTITY	UNIT	COST EACH		
			MAT.	INST.	TOTAL
SYSTEM D2010 610 1760					
GROUP WASH FOUNTAIN, PRECAST TERRAZZO					
CIRCULAR, 36″ DIAMETER					
Wash fountain, group, precast terrazzo, foot control 36″ diam	1.000	Ea.	3,050	460	3,510
Copper tubing type DWV, solder joint, hanger 10'OC, 2″ diam	10.000	L.F.	67	111.50	178.50
P trap, standard, copper, 2″ diam	1.000	Ea.	46	33	79
Wrought copper, Tee, sanitary, 2″ diam	1.000	Ea.	13.65	70	83.65
Copper tubing type L, solder joint, hanger 10′ OC 1/2″ diam	20.000	L.F.	33	121	154
Wrought copper 90° elbow for solder joints 1/2″ diam	3.000	Ea.	1.59	73.50	75.09
Wrought copper Tee for solder joints, 1/2″ diam	2.000	Ea.	1.78	76	77.78
TOTAL			3,213.02	945	4,158.02

D2010 610	Group Wash Fountain Systems		COST EACH		
			MAT.	INST.	TOTAL
1740	Group wash fountain, precast terrazzo				
1760	Circular, 36″ diameter		3,225	945	4,170
1800	54″ diameter	R15100	3,875	1,025	4,900
1840	Semi-circular, 36″ diameter	-410	2,900	945	3,845
1880	54″ diameter		3,550	1,025	4,575
1960	Stainless steel, circular, 36″ diameter		3,675	880	4,555
2000	54″ diameter		4,775	975	5,750
2040	Semi-circular, 36″ diameter		3,200	880	4,080
2080	54″ diameter		4,150	975	5,125
2160	Thermoplastic, circular, 36″ diameter		2,825	715	3,540
2200	54″ diameter		3,300	830	4,130
2240	Semi-circular, 36″ diameter		2,675	715	3,390
2280	54″ diameter		3,125	830	3,955

D2010 Plumbing Fixtures

Systems are complete with trim and rough-in (supply, waste and vent) for connection to supply branches and waste mains.

Three Wall   **Supply**   **Waste/Vent**   **Corner Angle**

System Components

	QUANTITY	UNIT	COST EACH		
			MAT.	INST.	TOTAL
SYSTEM D2010 710 1560					
SHOWER, STALL, BAKED ENAMEL, MOLDED STONE RECEPTOR, 30" SQUARE					
Shower stall, enameled steel, molded stone receptor, 30" square	1.000	Ea.	385	440	825
Copper tubing type DWV, solder joints, hangers 10'OC, 2" diam	6.000	L.F.	30	54.60	84.60
Wrought copper DWV, Tee, sanitary, 2" diam	1.000	Ea.	11.75	61.50	73.25
Trap, standard, copper, 2" diam	1.000	Ea.	30	29	59
Copper tubing type L, solder joint, hanger 10' OC 1/2" diam	16.000	L.F.	26.40	96.80	123.20
Wrought copper 90° elbow for solder joints 1/2" diam	3.000	Ea.	1.59	73.50	75.09
Wrought copper Tee for solder joints, 1/2" diam	2.000	Ea.	1.78	76	77.78
Stop and waste, straightway, bronze, solder joint 1/2" diam	2.000	Ea.	5.10	41	46.10
TOTAL			491.62	872.40	1,364.02

D2010 710	Shower Systems	COST EACH		
		MAT.	INST.	TOTAL
1560	Shower, stall, baked enamel, molded stone receptor, 30" square	490	870	1,360
1600	32" square	465	870	1,335
1640	Terrazzo receptor, 32" square	890	870	1,760
1680	36" square	1,075	920	1,995
1720	36" corner angle	985	920	1,905
1800	Fiberglass one piece, three walls, 32" square	565	800	1,365
1840	36" square	610	800	1,410
1880	Polypropylene, molded stone receptor, 30" square	465	870	1,335
1920	32" square	565	870	1,435
1960	Built-in head, arm, bypass, stops and handles	110	227	337

R15100 -410

Important: See the Reference Section for critical supporting data - Reference Numbers and City Cost Indexes

D2010 Plumbing Fixtures

Systems are complete with trim and rough-in (supply, waste and vent) to connect to supply branches and waste mains.

Wall Mounted, No Back

Supply   **Waste/Vent**

Wall Mounted, Low Back

System Components	QUANTITY	UNIT	COST EACH MAT.	INST.	TOTAL
SYSTEM D2010 810 1800					
DRINKING FOUNTAIN, ONE BUBBLER, WALL MOUNTED					
NON RECESSED, BRONZE, NO BACK					
Drinking fountain, wall mount, bronze, 1 bubbler	1.000	Ea.	1,100	123	1,223
Copper tubing, type L, solder joint, hanger 10' OC 3/8" diam	5.000	L.F.	7	29.25	36.25
Stop, supply, straight, chrome, 3/8" diam	1.000	Ea.	3.95	20.50	24.45
Wrought copper 90° elbow for solder joints 3/8" diam	1.000	Ea.	1.64	22.50	24.14
Wrought copper Tee for solder joints, 3/8" diam	1.000	Ea.	2.78	35	37.78
Copper tubing, type DWV, solder joint, hanger 10' OC 1-1/4" diam	4.000	L.F.	16.16	32.80	48.96
P trap, standard, copper drainage, 1-1/4" diam	1.000	Ea.	31	27.50	58.50
Wrought copper, DWV, Tee, sanitary, 1-1/4" diam	1.000	Ea.	9.45	54.50	63.95
TOTAL			1,171.98	345.05	1,517.03

D2010 810	Drinking Fountain Systems		COST EACH MAT.	INST.	TOTAL
1740	Drinking fountain, one bubbler, wall mounted				
1760	Non recessed				
1800	Bronze, no back	R15100 -410	1,175	345	1,520
1840	Cast iron, enameled, low back		565	345	910
1880	Fiberglass, 12" back		680	345	1,025
1920	Stainless steel, no back		980	345	1,325
1960	Semi-recessed, poly marble		815	345	1,160
2040	Stainless steel		930	345	1,275
2080	Vitreous china		690	345	1,035
2120	Full recessed, poly marble		910	345	1,255
2200	Stainless steel		835	345	1,180
2240	Floor mounted, pedestal type, aluminum		575	470	1,045
2320	Bronze		1,375	470	1,845
2360	Stainless steel		1,375	470	1,845

D2010 Plumbing Fixtures

Systems are complete with trim and rough-in (supply, waste and vent) for connection to supply branches and waste mains.

Wall Hung   **Supply**   **Waste/Vent**   **Floor Mounted**

System Components	QUANTITY	UNIT	COST EACH		
			MAT.	INST.	TOTAL
SYSTEM D2010 820 1840					
WATER COOLER, ELECTRIC, SELF CONTAINED, WALL HUNG, 8.2 GPH					
Water cooler, wall mounted, 8.2 GPH	1.000	Ea.	615	221	836
Copper tubing type DWV, solder joint, hanger 10'OC 1-1/4" diam	4.000	L.F.	16.16	32.80	48.96
Wrought copper DWV, Tee, sanitary 1-1/4" diam	1.000	Ea.	9.45	54.50	63.95
P trap, copper drainage, 1-1/4" diam	1.000	Ea.	31	27.50	58.50
Copper tubing type L, solder joint, hanger 10' OC 3/8" diam	5.000	L.F.	7	29.25	36.25
Wrought copper 90° elbow for solder joints 3/8" diam	1.000	Ea.	1.64	22.50	24.14
Wrought copper Tee for solder joints, 3/8" diam	1.000	Ea.	2.78	35	37.78
Stop and waste, straightway, bronze, solder, 3/8" diam	1.000	Ea.	3.95	20.50	24.45
TOTAL			686.98	443.05	1,130.03

D2010 820	Water Cooler Systems		COST EACH		
			MAT.	INST.	TOTAL
1840	Water cooler, electric, wall hung, 8.2 GPH	R15100 -410	685	445	1,130
1880	Dual height, 14.3 GPH		1,000	455	1,455
1920	Wheelchair type, 7.5 G.P.H.		1,650	445	2,095
1960	Semi recessed, 8.1 G.P.H.		990	445	1,435
2000	Full recessed, 8 G.P.H.	R15100 -430	1,525	475	2,000
2040	Floor mounted, 14.3 G.P.H.		645	385	1,030
2080	Dual height, 14.3 G.P.H.		1,075	470	1,545
2120	Refrigerated compartment type, 1.5 G.P.H.		1,300	385	1,685

Important: See the Reference Section for critical supporting data - Reference Numbers and City Cost Indexes

D2010 Plumbing Fixtures

Two Fixture Bathroom Systems consisting of a lavatory, water closet, and rough-in service piping.

• Prices for plumbing and fixtures only.

*Common wall is with an adjacent bathroom

System Components	QUANTITY	UNIT	COST EACH		
			MAT.	INST.	TOTAL
SYSTEM D2010 920 1180					
BATHROOM, LAVATORY & WATER CLOSET, 2 WALL PLUMBING, STAND ALONE					
Water closet, 2 Pc close cpld vit china flr mntd w/seat, supply & stop	1.000	Ea.	179	167	346
Water closet, rough-in waste & vent	1.000	Set	168	290	458
Lavatory w/ftngs, wall hung, white, PE on CI, 20" x 18"	1.000	Ea.	261	111	372
Lavatory, rough-in waste & vent	1.000	Set	315	535	850
Copper tubing type L, solder joint, hanger 10' OC 1/2" diam	10.000	L.F.	16.50	60.50	77
Pipe, steel, galvanized, schedule 40, threaded, 2" diam	12.000	L.F.	64.20	165.60	229.80
Pipe, CI soil, no hub, coupling 10' OC, hanger 5' OC, 4" diam	7.000	L.F.	62.65	112.70	175.35
TOTAL			1,066.35	1,441.80	2,508.15

D2010 920	Two Fixture Bathroom, Two Wall Plumbing		COST EACH		
			MAT.	INST.	TOTAL
1180	Bathroom, lavatory & water closet, 2 wall plumbing, stand alone		1,075	1,450	2,525
1200	Share common plumbing wall*		995	1,250	2,245

D2010 922	Two Fixture Bathroom, One Wall Plumbing		COST EACH		
			MAT.	INST.	TOTAL
2220	Bathroom, lavatory & water closet, one wall plumbing, stand alone	R15100 -410	1,025	1,300	2,325
2240	Share common plumbing wall*		925	1,100	2,025
2260		R15100 -420			
2280					

D2010 Plumbing Fixtures

Three Fixture Bathroom Systems consisting of a lavatory, water closet, bathtub or shower and rough-in service piping.

• Prices for plumbing and fixtures only.

*Common wall is with an adjacent bathroom

System Components	QUANTITY	UNIT	COST EACH		
			MAT.	INST.	TOTAL
SYSTEM D2010 924 1170					
BATHROOM, LAVATORY, WATER CLOSET & BATHTUB					
ONE WALL PLUMBING, STAND ALONE					
Wtr closet, 2 pc close cpld vit china flr mntd w/seat supply & stop	1.000	Ea.	179	167	346
Water closet, rough-in waste & vent	1.000	Set	168	290	458
Lavatory w/ftngs, wall hung, white, PE on CI, 20" x 18"	1.000	Ea.	261	111	372
Lavatory, rough-in waste & vent	1.000	Set	315	535	850
Bathtub, white PE on CI, w/ftgs, mat bottom, recessed, 5' long	1.000	Ea.	460	201	661
Baths, rough-in waste and vent	1.000	Set	140.40	382.50	522.90
TOTAL			1,523.40	1,686.50	3,209.90

D2010 924	Three Fixture Bathroom, One Wall Plumbing		COST EACH		
			MAT.	INST.	TOTAL
1150	Bathroom, three fixture, one wall plumbing	R15100 -410			
1160	Lavatory, water closet & bathtub				
1170	Stand alone	R15100 -420	1,525	1,675	3,200
1180	Share common plumbing wall *		1,250	1,225	2,475

D2010 926	Three Fixture Bathroom, Two Wall Plumbing	COST EACH		
		MAT.	INST.	TOTAL
2130	Bathroom, three fixture, two wall plumbing			
2140	Lavatory, water closet & bathtub			
2160	Stand alone	1,525	1,700	3,225
2180	Long plumbing wall common *	1,325	1,350	2,675
3610	Lavatory, bathtub & water closet			
3620	Stand alone	1,625	1,950	3,575
3640	Long plumbing wall common *	1,550	1,750	3,300
4660	Water closet, corner bathtub & lavatory			
4680	Stand alone	2,725	1,725	4,450
4700	Long plumbing wall common *	2,575	1,300	3,875
6100	Water closet, stall shower & lavatory			
6120	Stand alone	1,500	2,200	3,700
6140	Long plumbing wall common *	1,425	2,050	3,475
7060	Lavatory, corner stall shower & water closet			
7080	Stand alone	1,900	2,025	3,925
7100	Short plumbing wall common *	1,725	1,450	3,175

D2010 Plumbing Fixtures

Four Fixture Bathroom Systems consisting of a lavatory, water closet, bathtub, shower and rough-in service piping.

- Prices for plumbing and fixtures only.

*Common wall is with an adjacent bathroom

System Components	QUANTITY	UNIT	COST EACH		
			MAT.	INST.	TOTAL
SYSTEM D2010 928 1160					
BATHROOM, BATHTUB, WATER CLOSET, STALL SHOWER & LAVATORY					
TWO WALL PLUMBING, STAND ALONE					
Wtr closet, 2 pc close cpld vit china flr mntd w/seat supply & stop	1.000	Ea.	179	167	346
Water closet, rough-in waste & vent	1.000	Set	168	290	458
Lavatory w/ftngs, wall hung, white PE on CI, 20" x 18"	1.000	Ea.	261	111	372
Lavatory, rough-in waste & vent	1.000	Set	31.50	53.50	85
Bathtub, white PE on CI, w/ftgs, mat bottom, recessed, 5' long	1.000	Ea.	460	201	661
Baths, rough-in waste and vent	1.000	Set	156	425	581
Shower stall, bkd enam, molded stone receptor, door & trim 32" sq.	1.000	Ea.	360	440	800
Shower stall, rough-in supply, waste & vent	1.000	Set	107	430	537
TOTAL			1,722.50	2,117.50	3,840

D2010 928	Four Fixture Bathroom, Two Wall Plumbing		COST EACH		
			MAT.	INST.	TOTAL
1140	Bathroom, four fixture, two wall plumbing	R15100 -410			
1150	Bathtub, water closet, stall shower & lavatory				
1160	Stand alone	R15100 -420	1,725	2,125	3,850
1180	Long plumbing wall common *		1,575	1,700	3,275
2260	Bathtub, lavatory, corner stall shower & water closet				
2280	Stand alone		2,250	2,175	4,425
2320	Long plumbing wall common *		2,075	1,750	3,825
3620	Bathtub, stall shower, lavatory & water closet				
3640	Stand alone		2,000	2,600	4,600
3660	Long plumbing wall (opp. door) common *		1,850	2,175	4,025

D2010 930	Four Fixture Bathroom, Three Wall Plumbing	COST EACH		
		MAT.	INST.	TOTAL
4680	Bathroom, four fixture, three wall plumbing			
4700	Bathtub, stall shower, lavatory & water closet			
4720	Stand alone	2,600	2,875	5,475
4760	Long plumbing wall (opposite door) common *	2,525	2,675	5,200

D2010 Plumbing Fixtures

Five Fixture Bathroom Systems consisting of two lavatories, a water closet, bathtub, shower and rough-in service piping.

- Prices for plumbing and fixtures only.

*Common wall is with an adjacent bathroom.

System Components				COST EACH		
		QUANTITY	UNIT	MAT.	INST.	TOTAL
SYSTEM D2010 932 1360						
BATHROOM, BATHTUB, WATER CLOSET, STALL SHOWER & TWO LAVATORIES						
TWO WALL PLUMBING, STAND ALONE						
Wtr closet, 2 pc close cpld vit china flr mntd incl seat, supply & stop		1.000	Ea.	179	167	346
Water closet, rough-in waste & vent		1.000	Set	168	290	458
Lavatory w/ftngs, wall hung, white PE on CI, 20" x 18"		2.000	Ea.	522	222	744
Lavatory, rough-in waste & vent		2.000	Set	630	1,070	1,700
Bathtub, white PE on CI, w/ftgs, mat bottom, recessed, 5' long		1.000	Ea.	460	201	661
Baths, rough-in waste and vent		1.000	Set	156	425	581
Shower stall, bkd enam molded stone receptor, door & ftng, 32" sq.		1.000	Ea.	360	440	800
Shower stall, rough-in supply, waste & vent,		1.000	Set	107	430	537
	TOTAL			2,582	3,245	5,827

D2010 932	Five Fixture Bathroom, Two Wall Plumbing		COST EACH		
			MAT.	INST.	TOTAL
1320	Bathroom, five fixture, two wall plumbing				
1340	Bathtub, water closet, stall shower & two lavatories	R15100 -410			
1360	Stand alone		2,575	3,250	5,825
1400	One short plumbing wall common *		2,425	2,825	5,250
1500	Bathtub, two lavatories, corner stall shower & water closet				
1520	Stand alone		3,100	3,300	6,400
1540	Long plumbing wall common*		2,700	2,625	5,325

D2010 934	Five Fixture Bathroom, Three Wall Plumbing	COST EACH		
		MAT.	INST.	TOTAL
2360	Bathroom, five fixture, three wall plumbing			
2380	Water closet, bathtub, two lavatories & stall shower			
2400	Stand alone	3,100	3,300	6,400
2440	One short plumbing wall common *	2,950	2,875	5,825

D2010 936	Five Fixture Bathroom, One Wall Plumbing	COST EACH		
		MAT.	INST.	TOTAL
4080	Bathroom, five fixture, one wall plumbing			
4100	Bathtub, two lavatories, corner stall shower & water closet			
4120	Stand alone	2,925	2,975	5,900
4160	Share common wall *	2,350	2,025	4,375

Important: See the Reference Section for critical supporting data - Reference Numbers and City Cost Indexes

D2010 Plumbing Fixtures

Example of Plumbing Cost Calculations: The bathroom system includes the individual fixtures such as bathtub, lavatory, shower and water closet. These fixtures are listed below as separate items merely as a checklist.

D2010 951	Plumbing Systems 20 Unit, 2 Story Apartment Building							
	FIXTURE	SYSTEM	LINE	QUANTITY	UNIT	COST EACH		
						MAT.	INST.	TOTAL
0440	Bathroom	D2010 953	3640	20	Ea.	31,000	35,100	66,100
0480	Bathtub							
0520	Booster pump[1]	not req'd.						
0560	Drinking fountain							
0600	Garbage disposal[1]	not incl.						
0660								
0680	Grease interceptor							
0720	Water heater	D2020 250	2140	1	Ea.	7,375	2,375	9,750
0760	Kitchen sink	D2010 410	1960	20	Ea.	14,200	12,400	26,600
0800	Laundry sink	D2010 420	1840	4	Ea.	1,650	2,450	4,100
0840	Lavatory							
0900								
0920	Roof drain, 1 floor	D2040 210	4200	2	Ea.	855	1,275	2,130
0960	Roof drain, add'l floor	D2040 210	4240	20	L.F.	179	320	499
1000	Service sink	D2010 420	4300	1	Ea.	970	855	1,825
1040	Sewage ejector[1]	not req'd.						
1080	Shower							
1100								
1160	Sump pump							
1200	Urinal							
1240	Water closet							
1320								
1360	**SUB TOTAL**					56,000	55,000	111,000
1480	Water controls	RD2010-031		10%[2]		5,625	5,475	11,100
1520	Pipe & fittings[3]	RD2010-031		30%[2]		16,900	16,400	33,300
1560	Other							
1600	Quality/complexity	RD2010-031		15%[2]		8,425	8,225	16,650
1680								
1720	**TOTAL**					87,500	85,000	172,500
1741								

[1]**Note:** Cost for items such as booster pumps, backflow preventers, sewage ejectors, water meters, etc., may be obtained from the unit price section in the front of this book.

Water controls, pipe and fittings, and the Quality/Complexity factors come from Table R15100-040.

[2]Percentage of subtotal.

[3]Long easily discernable runs of pipe would be more accurately priced from Section 15107. If this is done, reduce the miscellaneous percentage in proportion.

D2020 Domestic Water Distribution

Installation includes piping and fittings within 10′ of heater. Electric water heaters do not require venting.

1 Kilowatt hour will raise:			
Gallons of Water	Degrees F	Gallons of Water	Degrees F
4.1	100°	6.8	60°
4.5	90°	8.2	50°
5.1	80°	10.0	40°
5.9	70°		

System Components	QUANTITY	UNIT	COST EACH		
			MAT.	INST.	TOTAL
SYSTEM D2020 210 1780					
ELECTRIC WATER HEATER, RESIDENTIAL, 100° F RISE					
10 GALLON TANK, 7 GPH					
Water heater, residential electric, glass lined tank, 10 Gal	1.000	Ea.	231	214	445
Copper tubing, type L, solder joint, hanger 10′ OC 1/2″ diam	30.000	L.F.	49.50	181.50	231
Wrought copper 90° elbow for solder joints 1/2″ diam.	4.000	Ea.	2.12	98	100.12
Wrought copper Tee for solder joints, 1/2″ diam	2.000	Ea.	1.78	76	77.78
1/3 HP	2.000	Ea.	10.80	52	62.80
Valve, gate, bronze, 125 lb, NRS, soldered 1/2″ diam	2.000	Ea.	35.20	41	76.20
Relief valve, bronze, press & temp, self-close, 3/4″ IPS	1.000	Ea.	89.50	17.55	107.05
Wrought copper adapter, CTS to MPT 3/4″ IPS	1.000	Ea.	1.83	29	30.83
Copper tubing, type L, solder joints, 3/4″ diam	1.000	L.F.	2.42	6.45	8.87
Wrought copper 90° elbow for solder joints 3/4″ diam	1.000	Ea.	1.17	26	27.17
TOTAL			425.32	741.50	1,166.82

D2020 210	Electric Water Heaters - Residential Systems	COST EACH		
		MAT.	INST.	TOTAL
1760	Electric water heater, residential, 100° F rise			
1780	10 gallon tank, 7 GPH	425	740	1,165
1820	20 gallon tank, 7 GPH	515	790	1,305
1860	30 gallon tank, 7 GPH	610	830	1,440
1900	40 gallon tank, 8 GPH	710	915	1,625
1940	52 gallon tank, 10 GPH	770	920	1,690
1980	66 gallon tank, 13 GPH	1,150	1,050	2,200
2020	80 gallon tank, 16 GPH	1,225	1,100	2,325
2060	120 gallon tank, 23 GPH	1,825	1,275	3,100

R15100 -120

D2020 Domestic Water Distribution

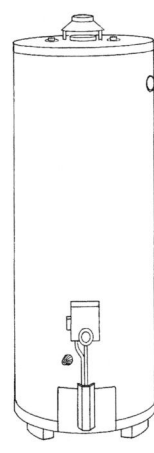

Installation includes piping and fittings within 10' of heater. Gas heaters require vent piping (not included with these units).

System Components	QUANTITY	UNIT	COST EACH		
			MAT.	INST.	TOTAL
SYSTEM D2020 220 2260					
GAS FIRED WATER HEATER, RESIDENTIAL, 100° F RISE					
30 GALLON TANK, 32 GPH					
Water heater, residential, gas, glass lined tank, 30 gallon	1.000	Ea.	425	246	671
Copper tubing, type L, solder joint, hanger 10' OC, 3/4" diam	33.000	L.F.	79.86	212.85	292.71
Wrought copper 90° elbow for solder joints, 3/4" diam	5.000	Ea.	5.85	130	135.85
Wrought copper Tee for solder joints, 3/4" diam	2.000	Ea.	4.30	82	86.30
Wrought copper union for soldered joints, 3/4" diam	2.000	Ea.	14.10	55	69.10
Valve bronze, 125 lb., NRS, soldered 3/4" diam	2.000	Ea.	55	49	104
Relief valve, press & temp, bronze, self-close, 3/4" diam	1.000	Ea.	89.50	17.55	107.05
Wrought copper, adapter, CTS to MPT 3/4" IPS	1.000	Ea.	1.83	29	30.83
Pipe steel black, schedule 40, threaded, 1/2" diam	10.000	L.F.	14.10	78	92.10
Pipe, 90° elbow, malleable iron black, 150 lb., threaded, 1/2" diam	2.000	Ea.	3.04	66	69.04
Pipe, union with brass seat, malleable iron black, 1/2" diam	1.000	Ea.	7.35	35	42.35
Valve, gas stop w/o check, brass, 1/2" IPS	1.000	Ea.	6.30	20.50	26.80
TOTAL			706.23	1,020.90	1,727.13

D2020 220	Gas Fired Water Heaters - Residential Systems		COST EACH		
			MAT.	INST.	TOTAL
2200	Gas fired water heater, residential, 100° F rise	R15100 -120			
2260	30 gallon tank, 32 GPH		705	1,025	1,730
2300	40 gallon tank, 32 GPH		810	1,150	1,960
2340	50 gallon tank, 63 GPH		955	1,150	2,105
2380	75 gallon tank, 63 GPH		1,300	1,275	2,575
2420	100 gallon tank, 63 GPH		1,725	1,350	3,075

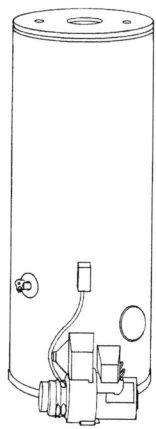

Installation includes piping and fittings within 10' of heater. Oil fired heaters require vent piping (not included in these prices).

System Components	QUANTITY	UNIT	COST EACH MAT.	COST EACH INST.	COST EACH TOTAL
SYSTEM D2020 230 2220					
OIL FIRED WATER HEATER, RESIDENTIAL, 100° F RISE					
30 GALLON TANK, 103 GPH					
Water heater, residential, oil glass lined tank, 30 Gal	1.000	Ea.	820	246	1,066
Copper tubing, type L, solder joint, hanger 10' O.C. 3/4" diam	33.000	L.F.	79.86	212.85	292.71
Wrought copper 90° elbow for solder joints 3/4" diam	5.000	Ea.	5.85	130	135.85
Wrought copper Tee for solder joints, 3/4" diam	2.000	Ea.	4.30	82	86.30
Wrought copper union for soldered joints, 3/4" diam	2.000	Ea.	14.10	55	69.10
Valve, gate, bronze, 125 lb, NRS, soldered 3/4" diam	2.000	Ea.	55	49	104
Relief valve, bronze, press & temp, self-close, 3/4" IPS	1.000	Ea.	89.50	17.55	107.05
Wrought copper adapter, CTS to MPT, 3/4" IPS	1.000	Ea.	1.83	29	30.83
Copper tubing, type L, solder joint, hanger 10' OC 3/8" diam	10.000	L.F.	14	58.50	72.50
Wrought copper 90° elbow for solder joints 3/8" diam	2.000	Ea.	3.28	45	48.28
Valve, globe, fusible, 3/8" diam	1.000	Ea.	7.90	20.50	28.40
TOTAL			1,095.62	945.40	2,041.02

D2020 230	Oil Fired Water Heaters - Residential Systems	COST EACH MAT.	COST EACH INST.	COST EACH TOTAL
2200	Oil fired water heater, residential, 100° F rise			
2220	30 gallon tank, 103 GPH	1,100	945	2,045
2260	50 gallon tank, 145 GPH	1,850	1,050	2,900
2300	70 gallon tank, 164 GPH R15100 -120	2,125	1,200	3,325
2340	85 gallon tank, 181 GPH	5,175	1,225	6,400

D2020 Domestic Water Distribution

Systems below include piping and fittings within 10′ of heater. Electric water heaters do not require venting.

System Components	QUANTITY	UNIT	COST EACH MAT.	COST EACH INST.	COST EACH TOTAL
SYSTEM D2020 240 1820					
ELECTRIC WATER HEATER, COMMERCIAL, 100° F RISE					
50 GALLON TANK, 9 KW, 37 GPH					
Water heater, commercial, electric, 50 Gal, 9 KW, 37 GPH	1.000	Ea.	2,725	273	2,998
Copper tubing, type L, solder joint, hanger 10′ OC, 3/4″ diam	34.000	L.F.	82.28	219.30	301.58
Wrought copper 90° elbow for solder joints 3/4″ diam	5.000	Ea.	5.85	130	135.85
Wrought copper Tee for solder joints, 3/4″ diam	2.000	Ea.	4.30	82	86.30
Wrought copper union for soldered joints, 3/4″ diam	2.000	Ea.	14.10	55	69.10
Valve, gate, bronze, 125 lb, NRS, soldered 3/4″ diam	2.000	Ea.	55	49	104
Relief valve, bronze, press & temp, self-close, 3/4″ IPS	1.000	Ea.	89.50	17.55	107.05
Wrought copper adapter, copper tubing to male, 3/4″ IPS	1.000	Ea.	1.83	29	30.83
TOTAL			2,977.86	854.85	3,832.71

D2020 240	Electric Water Heaters - Commercial Systems		COST EACH MAT.	COST EACH INST.	COST EACH TOTAL
1800	Electric water heater, commercial, 100° F rise				
1820	50 gallon tank, 9 KW 37 GPH		2,975	855	3,830
1860	80 gal, 12 KW 49 GPH	R15100 -110	3,775	1,050	4,825
1900	36 KW 147 GPH		5,150	1,150	6,300
1940	120 gal, 36 KW 147 GPH	R15100 -120	5,550	1,225	6,775
1980	150 gal, 120 KW 490 GPH		17,300	1,325	18,625
2020	200 gal, 120 KW 490 GPH		18,400	1,350	19,750
2060	250 gal, 150 KW 615 GPH		20,400	1,575	21,975
2100	300 gal, 180 KW 738 GPH		21,900	1,675	23,575
2140	350 gal, 30 KW 123 GPH		16,800	1,800	18,600
2180	180 KW 738 GPH		22,500	1,800	24,300
2220	500 gal, 30 KW 123 GPH		21,600	2,100	23,700
2260	240 KW 984 GPH		30,700	2,100	32,800
2300	700 gal, 30 KW 123 GPH		25,900	2,400	28,300
2340	300 KW 1230 GPH		38,500	2,400	40,900
2380	1000 gal, 60 KW 245 GPH		31,400	3,375	34,775
2420	480 KW 1970 GPH		40,400	3,375	43,775
2460	1500 gal, 60 KW 245 GPH		45,800	4,150	49,950
2500	480 KW 1970 GPH		63,000	4,150	67,150

D2020 Domestic Water Distribution

Units may be installed in multiples for increased capacity.

Included below is the heater with self-energizing gas controls, safety pilots, insulated jacket, hi-limit aquastat and pressure relief valve.

Installation includes piping and fittings within 10' of heater. Gas heaters require vent piping (not included in these prices).

System Components	QUANTITY	UNIT	COST EACH		
			MAT.	INST.	TOTAL
SYSTEM D2020 250 1780					
GAS FIRED WATER HEATER, COMMERCIAL, 100° F RISE					
75.5 MBH INPUT, 63 GPH					
Water heater, commercial, gas, 75.5 MBH, 63 GPH	1.000	Ea.	1,625	350	1,975
Copper tubing, type L, solder joint, hanger 10' OC, 1-1/4" diam	30.000	L.F.	140.40	253.50	393.90
Wrought copper 90° elbow for solder joints 1-1/4" diam	4.000	Ea.	17.32	132	149.32
Wrought copper Tee for solder joints, 1-1/4" diam	2.000	Ea.	19.10	109	128.10
Wrought copper union for soldered joints, 1-1/4" diam	2.000	Ea.	42	70	112
Valve, gate, bronze, 125 lb, NRS, soldered 1-1/4" diam	2.000	Ea.	102	66	168
Relief valve, bronze, press & temp, self-close, 3/4" IPS	1.000	Ea.	89.50	17.55	107.05
Copper tubing, type L, solder joints, 3/4" diam	8.000	L.F.	19.36	51.60	70.96
Wrought copper 90° elbow for solder joints 3/4" diam	1.000	Ea.	1.17	26	27.17
Wrought copper, adapter, CTS to MPT, 3/4" IPS	1.000	Ea.	1.83	29	30.83
Pipe steel black, schedule 40, threaded, 3/4" diam	10.000	L.F.	16.90	80.50	97.40
Pipe, 90° elbow, malleable iron black, 150 lb threaded, 3/4" diam	2.000	Ea.	3.68	70	73.68
Pipe, union with brass seat, malleable iron black, 3/4" diam	1.000	Ea.	7.70	38	45.70
Valve, gas stop w/o check, brass, 3/4" IPS	1.000	Ea.	11.90	22.50	34.40
TOTAL			2,097.86	1,315.65	3,413.51

D2020 250	Gas Fired Water Heaters - Commercial Systems		COST EACH		
			MAT.	INST.	TOTAL
1760	Gas fired water heater, commercial, 100° F rise				
1780	75.5 MBH input, 63 GPH		2,100	1,325	3,425
1860	100 MBH input, 91 GPH		3,575	1,375	4,950
1980	155 MBH input, 150 GPH	R15100 -110	5,000	1,600	6,600
2060	200 MBH input, 192 GPH		5,375	1,925	7,300
2140	300 MBH input, 278 GPH	R15100 -120	7,375	2,375	9,750
2180	390 MBH input, 374 GPH		8,250	2,400	10,650
2220	500 MBH input, 480 GPH		10,100	2,600	12,700
2260	600 MBH input, 576 GPH		11,500	2,800	14,300

D2020 Domestic Water Distribution

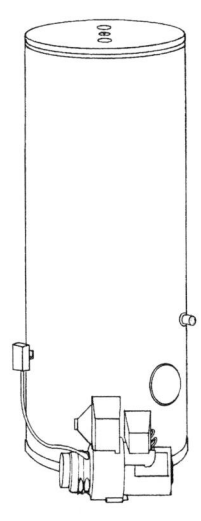

Units may be installed in multiples for increased capacity.

Included below is the heater, wired-in flame retention burners, cadmium cell primary controls, hi-limit controls, ASME pressure relief valves, draft controls, and insulated jacket.

Oil fired water heater systems include piping and fittings within 10′ of heater. Oil fired heaters require vent piping (not included in these systems).

System Components	QUANTITY	UNIT	COST EACH MAT.	INST.	TOTAL
SYSTEM D2020 260 1820					
OIL FIRED WATER HEATER, COMMERCIAL, 100° F RISE					
103 MBH OUTPUT, 116 GPH					
Water heater, commercial, oil, 103 MBH	1.000	Ea.	1,075	445	1,520
Copper tubing, type L, solder joint, hanger 10′ OC, 3/4″ diam	34.000	L.F.	82.28	219.30	301.58
Wrought copper 90° elbow for solder joints 3/4″ diam	5.000	Ea.	5.85	130	135.85
Wrought copper Tee for solder joints, 3/4″ diam	2.000	Ea.	4.30	82	86.30
Wrought copper union for soldered joints, 3/4″ diam	2.000	Ea.	14.10	55	69.10
Valve, bronze, 125 lb, NRS, soldered 3/4″ diam	2.000	Ea.	55	49	104
Relief valve, bronze, press & temp, self-close, 3/4″ IPS	1.000	Ea.	89.50	17.55	107.05
Wrought copper adapter, copper tubing to male, 3/4″ IPS	1.000	Ea.	1.83	29	30.83
Copper tubing, type L, solder joint, hanger 10′ OC, 3/8″ diam	10.000	L.F.	14	58.50	72.50
Wrought copper 90° elbow for solder joints 3/8″ diam	2.000	Ea.	3.28	45	48.28
Valve, globe, fusible, 3/8″ IPS	1.000	Ea.	7.90	20.50	28.40
TOTAL			1,353.04	1,150.85	2,503.89

D2020 260	Oil Fired Water Heaters - Commercial Systems		COST EACH MAT.	INST.	TOTAL
1800	Oil fired water heater, commercial, 100° F rise				
1820	103 MBH output, 116 GPH		1,350	1,150	2,500
1900	134 MBH output, 161 GPH	R15100 -110	2,500	1,475	3,975
1940	161 MBH output, 192 GPH		2,750	1,675	4,425
1980	187 MBH output, 224 GPH	R15100 -120	3,075	1,950	5,025
2060	262 MBH output, 315 GPH		3,600	2,250	5,850
2100	341 MBH output, 409 GPH		7,600	2,400	10,000
2140	420 MBH output, 504 GPH		8,875	2,600	11,475
2180	525 MBH output, 630 GPH		10,000	2,925	12,925
2220	630 MBH output, 756 GPH		11,600	2,925	14,525
2260	735 MBH output, 880 GPH		12,100	3,375	15,475
2300	840 MBH output, 1000 GPH		12,600	3,375	15,975
2340	1050 MBH output, 1260 GPH		13,600	3,475	17,075
2380	1365 MBH output, 1640 GPH		15,900	3,950	19,850
2420	1680 MBH output, 2000 GPH		17,700	4,975	22,675
2460	2310 MBH output, 2780 GPH		22,700	6,100	28,800
2500	2835 MBH output, 3400 GPH		25,700	6,100	31,800

D2020 Domestic Water Distribution

In this closed-loop indirect collection system, fluid with a low freezing temperature, such as propylene glycol, transports heat from the collectors to water storage. The transfer fluid is contained in a closed-loop consisting of collectors, supply and return piping, and a remote heat exchanger. The heat exchanger transfers heat energy from the fluid in the collector loop to potable water circulated in a storage loop. A typical two-or-three panel system contains 5 to 6 gallons of heat transfer fluid.

When the collectors become approximately 20° F warmer than the storage temperature, a controller activates the circulator on the collector and storage loops. The circulators will move the fluid and potable water through the heat exchanger until heat collection no longer occurs. At that point, the system shuts down. Since the heat transfer medium is a fluid with a very low freezing temperature, there is no need for it to be drained from the system between periods of collection.

D2020 Domestic Water Distribution

System Components	QUANTITY	UNIT	COST EACH		
			MAT.	INST.	TOTAL
SYSTEM D2020 265 2760					
SOLAR, CLOSED LOOP, ADD-ON HOT WATER SYS., EXTERNAL HEAT EXCHANGER					
3/4" TUBING, TWO 3'X7' BLACK CHROME COLLECTORS					
A,B,G,L,K,M Heat exchanger fluid-fluid pkg incl 2 circulators, expansion tank,					
Check valve, relief valve, controller, hi temp cutoff, & 2 sensors	1.000	Ea.	765	355	1,120
C Thermometer, 2" dial	3.000	Ea.	66	91.50	157.50
D, T Fill & drain valve, brass, 3/4" connection	1.000	Ea.	5.20	20.50	25.70
E Air vent, manual, 1/8" fitting	2.000	Ea.	4.92	30.70	35.62
F Air purger	1.000	Ea.	47.50	41	88.50
H Strainer, Y type, bronze body, 3/4" IPS	1.000	Ea.	19.15	26	45.15
I Valve, gate, bronze, NRS, soldered 3/4" diam	6.000	Ea.	165	147	312
J Neoprene vent flashing	2.000	Ea.	19.60	49	68.60
N-1, N Relief valve temp & press, 150 psi 210°F self-closing 3/4" IPS	1.000	Ea.	10	16.35	26.35
O Pipe covering, urethane, ultraviolet cover, 1" wall 3/4" diam	20.000	L.F.	38.20	91.80	130
P Pipe covering, fiberglass, all service jacket, 1" wall, 3/4" diam	50.000	L.F.	45.50	183.50	229
Q Collector panel solar energy blk chrome on copper, 1/8" temp glass 3'x7'	2.000	Ea.	1,200	186	1,386
Roof clamps for solar energy collector panels	2.000	Set	3.54	25.30	28.84
R Valve, swing check, bronze, regrinding disc, 3/4" diam	2.000	Ea.	77	49	126
S Pressure gauge, 60 psi, 2" dial	1.000	Ea.	24.50	15.35	39.85
U Valve, water tempering, bronze, sweat connections, 3/4" diam	1.000	Ea.	70.50	24.50	95
W-2, V Tank water storage w/heating element, drain, relief valve, existing	1.000	Ea.			
Copper tubing type L, solder joint, hanger 10' OC 3/4" diam	20.000	L.F.	48.40	129	177.40
Copper tubing, type M, solder joint, hanger 10' OC 3/4" diam	70.000	L.F.	133	441	574
Sensor wire, #22-2 conductor multistranded	.500	C.L.F.	5.68	24.25	29.93
Solar energy heat transfer fluid, propylene glycol anti-freeze	6.000	Gal.	58.20	105.30	163.50
Wrought copper fittings & solder, 3/4" diam	76.000	Ea.	88.92	1,976	2,064.92
TOTAL			2,895.81	4,028.05	6,923.86

D2020 265	Solar, Closed Loop, Add-On Hot Water Systems		COST EACH		
			MAT.	INST.	TOTAL
2550	Solar, closed loop, add-on hot water system, external heat exchanger				
2570	3/8" tubing, 3 ea. 4' x 4'-4" vacuum tube collectors		4,450	3,775	8,225
2580	1/2" tubing, 4 ea 4 x 4'-4" vacuum tube collectors		5,350	4,050	9,400
2600	2 ea 3'x7' black chrome collectors	R13600 -610	2,675	3,825	6,500
2620	3 ea 3'x7' black chrome collectors		3,275	3,925	7,200
2640	2 ea 3'x7' flat black collectors		2,425	3,850	6,275
2660	3 ea 3'x7' flat black collectors		2,900	3,950	6,850
2700	3/4" tubing, 3 ea 3'x7' black chrome collectors		3,500	4,125	7,625
2720	3 ea 3'x7' flat black absorber plate collectors		3,125	4,150	7,275
2740	2 ea 4'x9' flat black w/plastic glazing collectors		2,775	4,175	6,950
2760	2 ea 3'x7' black chrome collectors		2,900	4,025	6,925
2780	1" tubing, 4 ea 2'x9' plastic absorber & glazing collectors		4,175	4,650	8,825
2800	4 ea 3'x7' black chrome absorber collectors		4,400	4,675	9,075
2820	4 ea 3'x7' flat black absorber collectors		3,900	4,700	8,600

D2020 Domestic Water Distribution

In the drainback indirect-collection system, the heat transfer fluid is distilled water contained in a loop consisting of collectors, supply and return piping, and an unpressurized holding tank. A large heat exchanger containing incoming potable water is immersed in the holding tank. When a controller activates solar collection, the distilled water is pumped through the collectors and heated and pumped back down to the holding tank. When the temperature differential between the water in the collectors and water in storage is such that collection no longer occurs, the pump turns off and gravity causes the distilled water in the collector loop to drain back to the holding tank. All the loop piping is pitched so that the water can drain out of the collectors and piping and not freeze there. As hot water is needed in the home, incoming water first flows through the holding tank with the immersed heat exchanger and is warmed and then flows through a conventional heater for any supplemental heating that is necessary.

D2020 Domestic Water Distribution

System Components	QUANTITY	UNIT	COST EACH MAT.	COST EACH INST.	COST EACH TOTAL
SYSTEM D2020 270 2760					
SOLAR, DRAINBACK, ADD ON, HOT WATER, IMMERSED HEAT EXCHANGER					
3/4" TUBING, THREE EA 3'X7' BLACK CHROME COLLECTOR					
A, B Differential controller 2 sensors, thermostat, solar energy system	1.000	Ea.	92	41	133
C Thermometer 2" dial	3.000	Ea.	66	91.50	157.50
D, T Fill & drain valve, brass, 3/4" connection	1.000	Ea.	5.20	20.50	25.70
E-1 Automatic air vent 1/8" fitting	1.000	Ea.	11.25	15.35	26.60
H Strainer, Y type, bronze body, 3/4" IPS	1.000	Ea.	19.15	26	45.15
I Valve, gate, bronze, NRS, soldered 3/4" diam	2.000	Ea.	55	49	104
J Neoprene vent flashing	2.000	Ea.	19.60	49	68.60
L Circulator, solar heated liquid, 1/20 HP	1.000	Ea.	170	73.50	243.50
N Relief valve temp. & press. 150 psi 210°F self-closing 3/4" IPS	1.000	Ea.	10	16.35	26.35
O Pipe covering, urethane, ultraviolet cover, 1" wall, 3/4" diam	20.000	L.F.	38.20	91.80	130
P Pipe covering, fiberglass, all service jacket, 1" wall, 3/4" diam	50.000	L.F.	45.50	183.50	229
Q Collector panel solar energy blk chrome on copper, 1/8" temp glas 3'x7'	3.000	Ea.	1,800	279	2,079
Roof clamps for solar energy collector panels	3.000	Set	5.31	37.95	43.26
R Valve, swing check, bronze, regrinding disc, 3/4" diam	1.000	Ea.	38.50	24.50	63
U Valve, water tempering, bronze sweat connections, 3/4" diam	1.000	Ea.	70.50	24.50	95
V Tank, water storage w/heating element, drain, relief valve, existing	1.000	Ea.			
W Tank, water storage immersed heat exchr elec 2"x1/2# insul 120 gal	1.000	Ea.	1,025	350	1,375
X Valve, globe, bronze, rising stem, 3/4" diam, soldered	3.000	Ea.	175.50	73.50	249
Y Flow control valve	1.000	Ea.	64.50	22.50	87
Z Valve, ball, bronze, solder 3/4" diam, solar loop flow control	1.000	Ea.	12.70	24.50	37.20
Copper tubing, type L, solder joint, hanger 10' OC 3/4" diam	20.000	L.F.	48.40	129	177.40
Copper tubing, type M, solder joint, hanger 10' OC 3/4" diam	70.000	L.F.	133	441	574
Sensor wire, #22-2 conductor, multistranded	.500	C.L.F.	5.68	24.25	29.93
Wrought copper fittings & solder, 3/4" diam	76.000	Ea.	88.92	1,976	2,064.92
TOTAL			3,999.91	4,064.20	8,064.11

D2020 270	Solar, Drainback, Hot Water Systems		COST EACH MAT.	COST EACH INST.	COST EACH TOTAL
2550	Solar, drainback, hot water, immersed heat exchanger				
2560	3/8" tubing, 3 ea. 4' x 4'-4" vacuum tube collectors		4,850	3,675	8,525
2580	1/2" tubing, 4 ea 4' x 4'-4" vacuum tube collectors, 80 gal tank	R13600 -610	5,750	3,950	9,700
2600	120 gal tank		5,850	4,000	9,850
2640	2 ea. 3'x7' blk chrome collectors, 80 gal tank		3,075	3,725	6,800
2660	3 ea. 3'x7' blk chrome collectors, 120 gal tank		3,775	3,875	7,650
2700	2 ea. 3'x7' flat blk collectors, 120 gal tank		2,925	3,775	6,700
2720	3 ea. 3'x7' flat blk collectors, 120 gal tank		3,400	3,875	7,275
2760	3/4" tubing, 3 ea 3'x7' black chrome collectors, 120 gal tank		4,000	4,075	8,075
2780	3 ea. 3'x7' flat black absorber collectors, 120 gal tank		3,625	4,075	7,700
2800	2 ea. 4'x9' flat blk w/plastic glazing collectors 120 gal tank		3,275	4,100	7,375
2840	1" tubing, 4 ea. 2'x9' plastic absorber & glazing collectors, 120 gal tank		4,725	4,575	9,300
2860	4 ea. 3'x7' black chrome absorber collectors, 120 gal tank		4,950	4,600	9,550
2880	4 ea. 3'x7' flat black absorber collectors, 120 gal tank		4,450	4,625	9,075

In the draindown direct-collection system, incoming domestic water is heated in the collectors. When the controller activates solar collection, domestic water is first heated as it flows through the collectors and is then pumped to storage. When conditions are no longer suitable for heat collection, the pump shuts off and the water in the loop drains down and out of the system by means of solenoid valves and properly pitched piping.

D20 Plumbing

D2020 Domestic Water Distribution

System Components	QUANTITY	UNIT	COST EACH MAT.	COST EACH INST.	COST EACH TOTAL
SYSTEM D2020 275 2760					
SOLAR, DRAINDOWN, HOT WATER, DIRECT COLLECTION					
3/4″ TUBING, THREE 3′X7′ BLACK CHROME COLLECTORS					
A, B Differential controller, 2 sensors, thermostat, solar energy system	1.000	Ea.	92	41	133
A-1 Solenoid valve, solar heating loop, brass, 3/4″ diam, 24 volts	3.000	Ea.	699	163.50	862.50
B-1 Solar energy sensor, freeze prevention	1.000	Ea.	24	15.35	39.35
C Thermometer, 2″ dial	3.000	Ea.	66	91.50	157.50
E-1 Vacuum relief valve, 3/4″ diam	1.000	Ea.	29	15.35	44.35
F-1 Air vent, automatic, 1/8″ fitting	1.000	Ea.	11.25	15.35	26.60
H Strainer, Y type, bronze body, 3/4″ IPS	1.000	Ea.	19.15	26	45.15
I Valve, gate, bronze, NRS, soldered, 3/4″ diam	2.000	Ea.	55	49	104
J Vent flashing neoprene	2.000	Ea.	19.60	49	68.60
K Circulator, solar heated liquid, 1/25 HP	1.000	Ea.	162	63	225
N Relief valve temp & press 150 psi 210°F self-closing 3/4″ IPS	1.000	Ea.	10	16.35	26.35
O Pipe covering, urethane, ultraviolet cover, 1″ wall, 3/4″ diam	20.000	L.F.	38.20	91.80	130
P Pipe covering, fiberglass, all service jacket, 1″ wall, 3/4″ diam	50.000	L.F.	45.50	183.50	229
Roof clamps for solar energy collector panels	3.000	Set	5.31	37.95	43.26
Q Collector panel solar energy blk chrome on copper, 1/8″ temp glass 3′x7′	3.000	Ea.	1,800	279	2,079
R Valve, swing check, bronze, regrinding disc, 3/4″ diam, soldered	2.000	Ea.	77	49	126
T Drain valve, brass, 3/4″ connection	2.000	Ea.	10.40	41	51.40
U Valve, water tempering, bronze, sweat connections, 3/4″ diam	1.000	Ea.	70.50	24.50	95
W-2, W Tank, water storage elec elem 2″x1/2# insul 120 gal	1.000	Ea.	1,025	350	1,375
X Valve, globe, bronze, rising stem, 3/4″ diam, soldered	1.000	Ea.	58.50	24.50	83
Copper tubing, type L, solder joints, hangers 10′ OC 3/4″ diam	20.000	L.F.	48.40	129	177.40
Copper tubing, type M, solder joints, hangers 10′ OC 3/4″ diam	70.000	L.F.	133	441	574
Sensor wire, #22-2 conductor, multistranded	.500	C.L.F.	5.68	24.25	29.93
Wrought copper fittings & solder, 3/4″ diam	76.000	Ea.	88.92	1,976	2,064.92
TOTAL			4,593.41	4,196.90	8,790.31

D2020 275	Solar, Draindown, Hot Water Systems		COST EACH MAT.	COST EACH INST.	COST EACH TOTAL
2550	Solar, draindown, hot water				
2560	3/8″ tubing, 3 ea. 4′ x 4′-4″ vacuum tube collectors, 80 gal tank		5,400	3,975	9,375
2580	1/2″ tubing, 4 ea. 4′ x 4′-4″ vacuum tube collectors, 80 gal tank	R13600 -610	6,350	4,100	10,450
2600	120 gal tank		6,475	4,125	10,600
2640	2 ea 3′x7′ black chrome collectors, 80 gal tank		3,700	3,850	7,550
2660	3 ea 3′x7′ black chrome collectors, 120 gal tank		4,400	4,000	8,400
2700	2 ea 3′x7′ flat black collectors, 120 gal tank		3,550	3,925	7,475
2720	3 ea 3′x7′ flat black collectors, 120 gal tank		4,000	4,000	8,000
2760	3/4″ tubing, 3 ea 3′x7′ black chrome collectors, 120 gal tank		4,600	4,200	8,800
2780	3 ea 3′x7′ flat collectors, 120 gal tank		4,225	4,225	8,450
2800	2 ea. 4′x9′ flat black & plastic glazing collectors, 120 gal tank		3,875	4,225	8,100
2840	1″ tubing, 4 ea. 2′x9′ plastic absorber & glazing collectors, 120 gal tank		5,275	4,700	9,975
2860	4 ea 3′x7′ black chrome absorber collectors, 120 gal tank		5,500	4,725	10,225
2880	4 ea 3′x7′ flat black absorber collectors, 120 gal tank		5,000	4,750	9,750

In the recirculation system, (a direct-collection system), incoming domestic water is heated in the collectors. When the controller activates solar collection, domestic water is heated as it flows through the collectors and then it flows back to storage. When conditions are not suitable for heat collection, the pump shuts off and the flow of the water stops. In this type of system, water remains in the collector loop at all times. A "frost sensor" at the collector activates the circulation of warm water from storage through the collectors when protection from freezing is required.

Important: See the Reference Section for critical supporting data - Reference Numbers and City Cost Indexes

D2020 Domestic Water Distribution

System Components	QUANTITY	UNIT	COST EACH		
			MAT.	INST.	TOTAL
SYSTEM D2020 280 2820					
SOLAR, RECIRCULATION, HOT WATER					
3/4" TUBING, TWO 3'X7' BLACK CHROME COLLECTORS					
A, B Differential controller 2 sensors, thermostat, for solar energy system	1.000	Ea.	92	41	133
A-1 Solenoid valve, solar heating loop, brass, 3/4" IPS, 24 volts	2.000	Ea.	466	109	575
B-1 Solar energy sensor freeze prevention	1.000	Ea.	24	15.35	39.35
C Thermometer, 2" dial	3.000	Ea.	66	91.50	157.50
D Drain valve, brass, 3/4" connection	1.000	Ea.	5.20	20.50	25.70
F-1 Air vent, automatic, 1/8" fitting	1.000	Ea.	11.25	15.35	26.60
H Strainer, Y type, bronze body, 3/4" IPS	1.000	Ea.	19.15	26	45.15
I Valve, gate, bronze, 125 lb, soldered 3/4" diam	3.000	Ea.	82.50	73.50	156
J Vent flashing, neoprene	2.000	Ea.	19.60	49	68.60
L Circulator, solar heated liquid, 1/20 HP	1.000	Ea.	170	73.50	243.50
N Relief valve, temp & press 150 psi 210°F self-closing 3/4" IPS	2.000	Ea.	20	32.70	52.70
O Pipe covering, urethane, ultraviolet cover, 1" wall, 3/4" diam	20.000	L.F.	38.20	91.80	130
P Pipe covering, fiberglass, all service jacket, 1" wall 3/4" diam	50.000	L.F.	45.50	183.50	229
Q Collector panel solar energy blk chrome on copper, 1/8" temp glass 3'x7'	2.000	Ea.	1,200	186	1,386
Roof clamps for solar energy collector panels	2.000	Set	3.54	25.30	28.84
R Valve, swing check, bronze, 125 lb, regrinding disc, soldered 3/4" diam	2.000	Ea.	77	49	126
U Valve, water tempering, bronze, sweat connections, 3/4" diam	1.000	Ea.	70.50	24.50	95
V Tank, water storage, w/heating element, drain, relief valve, existing	1.000	Ea.			
X Valve, globe, bronze, 125 lb, 3/4" diam	2.000	Ea.	117	49	166
Copper tubing, type L, solder joints, hangers 10' OC 3/4" diam	20.000	L.F.	48.40	129	177.40
Copper tubing, type M, solder joints, hangers 10' OC 3/4" diam	70.000	L.F.	133	441	574
Wrought copper fittings & solder, 3/4" diam	76.000	Ea.	88.92	1,976	2,064.92
Sensor wire, #22-2 conductor, multistranded	.500	C.L.F.	5.68	24.25	29.93
TOTAL			2,803.44	3,726.75	6,530.19

D2020 280	Solar Recirculation, Domestic Hot Water Systems		COST EACH		
			MAT.	INST.	TOTAL
2550	Solar recirculation, hot water				
2560	3/8" tubing, 3 ea. 4' x 4'-4" vacuum tube collectors		4,350	3,475	7,825
2580	1/2" tubing, 4 ea. 4' x 4'-4" vacuum tube collectors	R13600 -610	5,250	3,775	9,025
2640	2 ea. 3'x7' black chrome collectors		2,575	3,525	6,100
2660	3 ea. 3'x7' black chrome collectors		3,200	3,625	6,825
2700	2 ea. 3'x7' flat black collectors		2,325	3,550	5,875
2720	3 ea. 3'x7' flat black collectors		2,825	3,650	6,475
2760	3/4" tubing, 3 ea. 3'x7' black chrome collectors		3,400	3,825	7,225
2780	3 ea. 3'x7' flat black absorber plate collectors		3,025	3,850	6,875
2800	2 ea. 4'x9' flat black w/plastic glazing collectors		2,675	3,875	6,550
2820	2 ea. 3'x7' black chrome collectors		2,800	3,725	6,525
2840	1" tubing, 4 ea. 2'x9' black plastic absorber & glazing collectors		4,125	4,350	8,475
2860	4 ea. 3'x7' black chrome absorber collectors		4,350	4,375	8,725
2880	4 ea. 3'x7' flat black absorber collectors		3,850	4,400	8,250

The thermosyphon domestic hot water system, a direct collection system, operates under city water pressure and does not require pumps for system operation. An insulated water storage tank is located above the collectors. As the sun heats the collectors, warm water in them rises by means of natural convection; the colder water in the storage tank flows into the collectors by means of gravity. As long as the sun is shining the water continues to flow through the collectors and to become warmer.

To prevent freezing, the system must be drained or the collectors covered with an insulated lid when the temperature drops below 32°F.

D2020 Domestic Water Distribution

System Components	QUANTITY	UNIT	COST EACH		
			MAT.	INST.	TOTAL
SYSTEM D2020 285 0960					
SOLAR, THERMOSYPHON, WATER HEATER					
3/4" TUBING, TWO 3'X7' BLACK CHROME COLLECTORS					
D-1 Framing lumber, fir, 2" x 6" x 8', tank cradle	.008	M.B.F.	4.72	5.48	10.20
F-1 Framing lumber, fir, 2" x 4" x 24', sleepers	.016	M.B.F.	8.96	16.40	25.36
I Valve, gate, bronze, 125 lb, soldered 1/2" diam	2.000	Ea.	35.20	41	76.20
J Vent flashing, neoprene	4.000	Ea.	39.20	98	137.20
O Pipe covering, urethane, ultraviolet cover, 1" wall, 1/2" diam	40.000	L.F.	58.40	180	238.40
P Pipe covering fiberglass all service jacket 1" wall 1/2" diam	160.000	L.F.	124.80	561.60	686.40
Q Collector panel solar, blk chrome on copper, 3/16" temp glass 3'-8" x 6'	2.000	Ea.	950	197	1,147
Y Flow control valve, globe, bronze, 125#, soldered, 1/2" diam	1.000	Ea.	43	20.50	63.50
U Valve, water tempering, bronze, sweat connections, 1/2" diam	1.000	Ea.	57.50	20.50	78
W Tank, water storage, solar energy system, 80 Gal, 2" x 1/2 lb insul	1.000	Ea.	920	305	1,225
Copper tubing type L, solder joints, hangers 10' OC 1/2" diam	150.000	L.F.	247.50	907.50	1,155
Copper tubing type M, solder joints, hangers 10' OC 1/2" diam	50.000	L.F.	65	292.50	357.50
Sensor wire, #22-2 gauge multistranded	.500	C.L.F.	5.68	24.25	29.93
Wrought copper fittings & solder, 1/2" diam	75.000	Ea.	39.75	1,837.50	1,877.25
TOTAL			2,599.71	4,507.23	7,106.94

D2020 285	Thermosyphon, Hot Water		MAT.	INST.	TOTAL
0960 0970	Solar, thermosyphon, hot water, two collector system	R13600 -610	2,600	4,500	7,100

This domestic hot water pre-heat system includes heat exchanger with a circulating pump, blower, air-to-water, coil and controls, mounted in the upper collector manifold. Heat from the hot air coming out of the collectors is transferred through the heat exchanger. For each degree of DHW preheating gained, one degree less heating is needed from the fuel fired water heater. The system is simple, inexpensive to operate and can provide a substantial portion of DHW requirements for modest additional cost.

D2020 Domestic Water Distribution

System Components	QUANTITY	UNIT	COST EACH		
			MAT.	INST.	TOTAL
SYSTEM D2020 290 2560					
SOLAR, HOT WATER, AIR TO WATER HEAT EXCHANGE					
THREE COLLECTORS, OPTICAL BLACK ON ALUMINUM, 10'X2',80GAL TANK					
A, B Differential controller, 2 sensors, thermostat, solar energy system	1.000	Ea.	92	41	133
C Thermometer, 2" dial	2.000	Ea.	44	61	105
C-1 Heat exchanger, air to fluid, up flow 70 MBH	1.000	Ea.	365	253	618
E Air vent, manual, for solar energy system 1/8" fitting	1.000	Ea.	2.46	15.35	17.81
F Air purger	1.000	Ea.	47.50	41	88.50
G Expansion tank	1.000	Ea.	67.50	15.35	82.85
H Strainer, Y type, bronze body, 1/2" IPS	1.000	Ea.	17.50	24.50	42
I Valve, gate, bronze, 125 lb, NRS, soldered 1/2" diam	5.000	Ea.	88	102.50	190.50
N Relief valve, temp & pressure solar 150 psi 210°F self-closing	1.000	Ea.	10	16.35	26.35
P Pipe covering, fiberglass, all service jacket, 1" wall, 1/2" diam	60.000	L.F.	46.80	210.60	257.40
Q Collector panel solar energy, air, black on alum plate, flush mount, 10'x2'	30.000	L.F.	1,710	279	1,989
B-1, R-1 Shutter damper wfor solar heater circulator	1.000	Ea.	43	74	117
B-1, R-1 Shutter motor for solar heater circulator	1.000	Ea.	102	55.50	157.50
R Backflow preventer for solar energy system, 1/2" diam	2.000	Ea.	128	61	189
T Drain valve, brass, 3/4" connection	1.000	Ea.	5.20	20.50	25.70
U Valve, water tempering, bronze, sweat connections, 1/2" diam	1.000	Ea.	57.50	20.50	78
W Tank, water storage, solar energy system, 80 Gal, 2" x 2 lb insul	1.000	Ea.	920	305	1,225
V Tank, water storage, w/heating element, drain, relief valve, existing	1.000	System			
X Valve, globe, bronze, 125 lb, rising stem, 1/2" diam	1.000	Ea.	43	20.50	63.50
Copper tubing type M, solder joints, hangers 10' OC 1/2" diam	50.000	L.F.	65	292.50	357.50
Copper tubing type L, solder joints, hangers 10' OC 1/2" diam	10.000	L.F.	16.50	60.50	77
Wrought copper fittings & solder, 1/2" diam	10.000	Ea.	5.30	245	250.30
Sensor wire, #22-2 conductor multistranded	.500	C.L.F.	5.68	24.25	29.93
Q-1, Q-2 Ductwork, fiberglass, aluminzed jacket, 1-1/2" thick, 8" diam	32.000	S.F.	158.08	158.40	316.48
Q-3 Manifold for flush mount solar energy collector panels, air	6.000	L.F.	177	33.30	210.30
TOTAL			4,217.02	2,430.60	6,647.62

D2020 290	Solar Hot Water, Air To Water Heat Exchange		COST EACH		
			MAT.	INST.	TOTAL
2550	Solar hot water, air to water heat exchange				
2560	Three collectors, optical black on aluminum, 10' x 2', 80 Gal tank		4,225	2,425	6,650
2580	Four collectors, optical black on aluminum, 10' x 2', 80 Gal tank	R13600 -610	4,900	2,575	7,475
2600	Four collectors, optical black on aluminum, 10' x 2', 120 Gal tank		5,000	2,625	7,625

In this closed-loop indirect collection system, fluid with a low freezing temperature, such as propylene glycol, transports heat from the collectors to water storage. The transfer fluid is contained in a closed-loop consisting of collectors, supply and return piping, and a heat exchanger immersed in the storage

tank. A typical two-or-three panel system contains 5 to 6 gallons of heat transfer fluid.

When the collectors become approximately 20° F warmer than the storage temperature, a controller activates the circulator. The circulator moves the fluid continuously through the collectors

until the temperature difference between the collectors and storage is such that heat collection no longer occurs; at that point, the circulator shuts off. Since the heat transfer fluid has a very low freezing temperature, there is no need for it to be drained from the collectors between periods of collection.

D2020 Domestic Water Distribution

System Components		QUANTITY	UNIT	COST EACH		
				MAT.	INST.	TOTAL
SYSTEM D2020 295 2760						
SOLAR, CLOSED LOOP, HOT WATER SYSTEM, IMMERSED HEAT EXCHANGER						
3/4″ TUBING, THREE 3′ X 7′ BLACK CHROME COLLECTORS						
A, B	Differential controller, 2 sensors, thermostat, solar energy system	1.000	Ea.	92	41	133
C	Thermometer 2″ dial	3.000	Ea.	66	91.50	157.50
D, T	Fill & drain valves, brass, 3/4″ connection	3.000	Ea.	15.60	61.50	77.10
E	Air vent, manual, 1/8″ fitting	1.000	Ea.	2.46	15.35	17.81
F	Air purger	1.000	Ea.	47.50	41	88.50
G	Expansion tank	1.000	Ea.	67.50	15.35	82.85
I	Valve, gate, bronze, NRS, soldered 3/4″ diam	3.000	Ea.	82.50	73.50	156
J	Neoprene vent flashing	2.000	Ea.	19.60	49	68.60
K	Circulator, solar heated liquid, 1/25 HP	1.000	Ea.	162	63	225
N-1, N	Relief valve, temp & press 150 psi 210°F self-closing 3/4″ IPS	2.000	Ea.	20	32.70	52.70
O	Pipe covering, urethane ultraviolet cover, 1″ wall, 3/4″ diam	20.000	L.F.	38.20	91.80	130
P	Pipe covering, fiberglass, all service jacket, 1″ wall, 3/4″ diam	50.000	L.F.	45.50	183.50	229
	Roof clamps for solar energy collector panel	3.000	Set	5.31	37.95	43.26
Q	Collector panel solar blk chrome on copper, 1/8″ temp glass, 3′x7′	3.000	Ea.	1,800	279	2,079
R-1	Valve, swing check, bronze, regrinding disc, 3/4″ diam, soldered	1.000	Ea.	38.50	24.50	63
S	Pressure gauge, 60 psi, 2-1/2″ dial	1.000	Ea.	24.50	15.35	39.85
U	Valve, water tempering, bronze, sweat connections, 3/4″ diam	1.000	Ea.	70.50	24.50	95
W-2, W	Tank, water storage immersed heat exchr elec elem 2″x2# insul 120 Gal	1.000	Ea.	1,025	350	1,375
X	Valve, globe, bronze, rising stem, 3/4″ diam, soldered	1.000	Ea.	58.50	24.50	83
	Copper tubing type L, solder joint, hanger 10′ OC 3/4″ diam	20.000	L.F.	48.40	129	177.40
	Copper tubing, type M, solder joint, hanger 10′ OC 3/4″ diam	70.000	L.F.	133	441	574
	Sensor wire, #22-2 conductor multistranded	.500	C.L.F.	5.68	24.25	29.93
	Solar energy heat transfer fluid, propylene glycol, anti-freeze	6.000	Gal.	58.20	105.30	163.50
	Wrought copper fittings & solder, 3/4″ diam	76.000	Ea.	88.92	1,976	2,064.92
	TOTAL			4,015.37	4,190.55	8,205.92

D2020 295	Solar, Closed Loop, Hot Water Systems		COST EACH		
			MAT.	INST.	TOTAL
2550	Solar, closed loop, hot water system, immersed heat exchanger				
2560	3/8″ tubing, 3 ea. 4′ x 4′-4″ vacuum tube collectors, 80 gal. tank		4,900	3,800	8,700
2580	1/2″ tubing, 4 ea. 4′ x 4′-4″ vacuum tube collectors, 80 gal. tank		5,775	4,075	9,850
2600	120 gal. tank		5,875	4,125	10,000
2640	2 ea. 3′x7′ black chrome collectors, 80 gal. tank	R13600 -610	3,150	3,900	7,050
2660	120 gal. tank		3,225	3,900	7,125
2700	2 ea. 3′x7′ flat black collectors, 120 gal. tank		2,975	3,900	6,875
2720	3 ea. 3′x7′ flat black collectors, 120 gal. tank		3,450	4,025	7,475
2760	3/4″ tubing, 3 ea. 3′x7′ black chrome collectors, 120 gal. tank		4,025	4,200	8,225
2780	3 ea. 3′x7′ flat black collectors, 120 gal. tank		3,650	4,200	7,850
2800	2 ea. 4′x9′ flat black w/plastic glazing collectors 120 gal. tank		3,300	4,225	7,525
2840	1″ tubing, 4 ea. 2′x9′ plastic absorber & glazing collectors 120 gal. tank		4,675	4,675	9,350
2860	4 ea. 3′x7′ black chrome collectors, 120 gal. tank		4,875	4,700	9,575
2880	4 ea. 3′x7′ flat black absorber collectors, 120 gal. tank		4,375	4,725	9,100

D2040   Rain Water Drainage

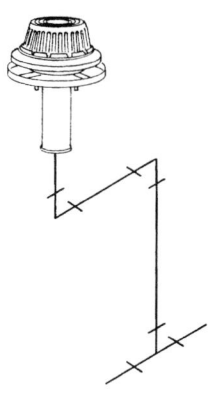

Design Assumptions: Vertical conductor size is based on a maximum rate of rainfall of 4" per hour. To convert roof area to other rates multiply "Max. S.F. Roof Area" shown by four and divide the result by desired local rate. The answer is the local roof area that may be handled by the indicated pipe diameter.

Basic cost is for roof drain, 10' of vertical leader and 10' of horizontal, plus connection to the main.

Pipe Dia.	Max. S.F. Roof Area	Gallons per Min.
2"	544	23
3"	1610	67
4"	3460	144
5"	6280	261
6"	10,200	424
8"	22,000	913

System Components			COST EACH		
	QUANTITY	UNIT	MAT.	INST.	TOTAL
SYSTEM D2040 210 1880					
ROOF DRAIN, DWV PVC PIPE, 2" DIAM., 10' HIGH					
Drain, roof, main, PVC, dome type 2" pipe size	1.000	Ea.	97.50	63	160.50
Clamp, roof drain, underdeck	1.000	Ea.	21	37	58
Pipe, Tee, PVC DWV, schedule 40, 2" pipe size	1.000	Ea.	3.50	44	47.50
Pipe, PVC, DWV, schedule 40, 2" diam.	20.000	L.F.	29.80	300	329.80
Pipe, elbow, PVC schedule 40, 2" diam.	2.000	Ea.	3.70	49	52.70
TOTAL			155.50	493	648.50

D2040 210	Roof Drain Systems	COST EACH		
		MAT.	INST.	TOTAL
1880	Roof drain, DWV PVC, 2" diam., piping, 10' high	156	495	651
1920	For each additional foot add	1.49	15	16.49
1960	3" diam., 10' high	192	575	767
2000	For each additional foot add	2.64	16.70	19.34
2040	4" diam., 10' high	225	645	870
2080	For each additional foot add	3.44	18.45	21.89
2120	5" diam., 10' high	525	745	1,270
2160	For each additional foot add	5.35	20.50	25.85
2200	6" diam., 10' high	570	825	1,395
2240	For each additional foot add	6.55	22.50	29.05
2280	8" diam., 10' high	1,350	1,400	2,750
2320	For each additional foot add	11.50	28.50	40
3940	C.I., soil, single hub, service wt., 2" diam. piping, 10' high	261	540	801
3980	For each additional foot add	5.10	14.05	19.15
4120	3" diam., 10' high	360	585	945
4160	For each additional foot add	6.95	14.75	21.70
4200	4" diam., 10' high	425	635	1,060
4240	For each additional foot add	8.95	16.10	25.05
4280	5" diam., 10' high	595	705	1,300
4320	For each additional foot add	9.05	18.10	27.15
4360	6" diam., 10' high	750	755	1,505
4400	For each additional foot add	15.15	18.85	34
4440	8" diam., 10' high	1,625	1,525	3,150
4480	For each additional foot add	23.50	31.50	55
6040	Steel galv. sch 40 threaded, 2" diam. piping, 10' high	330	520	850
6080	For each additional foot add	5.35	13.80	19.15
6120	3" diam., 10' high	680	750	1,430
6160	For each additional foot add	11.25	20.50	31.75

Important: See the Reference Section for critical supporting data - Reference Numbers and City Cost Indexes

D2040 Rain Water Drainage

D2040 210	Roof Drain Systems	COST EACH		
		MAT.	INST.	TOTAL
6200	4" diam., 10' high	1,050	965	2,015
6240	For each additional foot add	18.30	24.50	42.80
6280	5" diam., 10' high	1,925	1,225	3,150
6320	For each additional foot add	30.50	34	64.50
6360	6" diam, 10' high	2,450	1,500	3,950
6400	For each additional foot add	41.50	44.50	86
6440	8" diam., 10' high	5,275	2,050	7,325
6480	For each additional foot add	63	51	114

D3010 Energy Supply

Basis for Heat Loss Estimate, Apartment Type Structures:

1. Masonry walls and flat roof are insulated. U factor is assumed at .08.
2. Window glass area taken as BOCA minimum, 1/10th of floor area. Double insulating glass with 1/4″ air space, U = .65.
3. Infiltration = 0.3 C.F. per hour per S.F. of net wall.
4. Concrete floor loss is 2 BTUH per S.F.
5. Temperature difference taken as 70° F.
6. Ventilating or makeup air has not been included and must be added if desired. Air shafts are not used.

System Components	QUANTITY	UNIT	COST EACH		
			MAT.	INST.	TOTAL
SYSTEM D3010 510 1760					
HEATING SYSTEM, FIN TUBE RADIATION, FORCED HOT WATER					
1,000 S.F. AREA, 10,000 C.F. VOLUME					
Boiler, oil fired, CI, burner, ctrls/insul/breech/pipe/ftng/valves, 109 MBH	1.000	Ea.	3,718.75	2,756.25	6,475
Circulating pump, CI flange connection, 1/12 HP	1.000	Ea.	234	147	381
Expansion tank, painted steel, ASME 18 Gal capacity	1.000	Ea.	405	63.50	468.50
Storage tank, steel, above ground, 275 Gal capacity w/supports	1.000	Ea.	340	177	517
Copper tubing type L, solder joint, hanger 10′ OC, 3/4″ diam	100.000	L.F.	242	645	887
Radiation, 3/4″ copper tube w/alum fin baseboard pkg, 7″ high	30.000	L.F.	187.50	459	646.50
Pipe covering, calcium silicate w/cover, 1″ wall, 3/4″ diam	100.000	L.F.	240	496	736
TOTAL			5,367.25	4,743.75	10,111
COST PER S.F.			5.37	4.74	10.11

D3010 510	Apartment Building Heating - Fin Tube Radiation	COST PER S.F.		
		MAT.	INST.	TOTAL
1740	Heating systems, fin tube radiation, forced hot water			
1760	1,000 S.F. area, 10,000 C.F. volume	5.38	4.75	10.13
1800	10,000 S.F. area, 100,000 C.F. volume	1.70	2.86	4.56

Many styles of active solar energy systems exist. Those shown on the following page represent the majority of systems now being installed in different regions of the country.

The five active domestic hot water (DHW) systems which follow are typically specified as two or three panel systems with additional variations being the type of glazing and size of the storage tanks. Various combinations have been costed for the user's evaluation and comparison. The basic specifications from which the following systems were developed satisfy the construction detail requirements specified in the HUD Intermediate Minimum Property Standards (IMPS). If these standards are not complied with, the renewable energy system's costs could be significantly lower than shown.

To develop the system's specifications and costs it was necessary to make a number of assumptions about the systems. Certain systems are more appropriate to one climatic region than another or the systems may require modifications to be usable in particular locations. Specific instances in which the systems are not appropriate throughout the country as specified, or in which modification will be needed include the following:

- The freeze protection mechanisms provided in the DHW systems vary greatly. In harsh climates, where freeze is a major concern, a closed-loop indirect collection system may be more appropriate than a direct collection system.

- The thermosyphon water heater system described cannot be used when temperatures drop below 32° F.

- In warm climates it may be necessary to modify the systems installed to prevent overheating.

For each renewable resource (solar) system a schematic diagram and descriptive summary of the system is provided along with a list of all the components priced as part of the system. The costs were developed based on these specifications.

Considerations affecting costs which may increase or decrease beyond the estimates presented here include the following:

- Special structural qualities (allowance for earthquake, future expansion, high winds, and unusual spans or shapes);

- Isolated building site or rough terrain that would affect the transportation of personnel, material, or equipment;

- Unusual climatic conditions during the construction process;

- Substitution of other materials or system components for those used in the system specifications.

In this closed-loop indirect collection system, fluid with a low freezing temperature, propylene glycol, transports heat from the collectors to water storage. The transfer fluid is contained in a closed loop consisting of collectors, supply and return piping, and a heat exchanger immersed in the storage tank.

When the collectors become approximately 20° F warmer than the storage temperature, the controller activates the circulator. The circulator moves the fluid continuously until the temperature difference between fluid in the collectors and storage is such that the collection will no longer occur and then the circulator turns off. Since the heat transfer fluid has a very low freezing temperature, there is no need for it to be drained from the collectors between periods of collection.

D3010   Energy Supply

System Components		QUANTITY	UNIT	COST EACH		
				MAT.	INST.	TOTAL
	SYSTEM D3010 650 2750					
	SOLAR, CLOSED LOOP, SPACE/HOT WATER					
	1" TUBING, TEN 3'X7' BLK CHROME ON COPPER ABSORBER COLLECTORS					
A, B	Differential controller 2 sensors, thermostat, solar energy system	2.000	Ea.	184	82	266
C	Thermometer, 2" dial	10.000	Ea.	220	305	525
C-1	Heat exchanger, solar energy system, fluid to air, up flow, 80 MBH	1.000	Ea.	490	295	785
D, T	Fill & drain valves, brass, 3/4" connection	5.000	Ea.	26	102.50	128.50
D-1	Fan center	1.000	Ea.	119	108	227
E	Air vent, manual, 1/8" fitting	1.000	Ea.	2.46	15.35	17.81
E-2	Thermostat, 2 stage for sensing room temperature	1.000	Ea.	196	62	258
F	Air purger	2.000	Ea.	95	82	177
F-1	Controller, liquid temperature, solar energy system	1.000	Ea.	96	98	194
G	Expansion tank	2.000	Ea.	135	30.70	165.70
I	Valve, gate, bronze, 125 lb, soldered, 1" diam	5.000	Ea.	190	130	320
J	Vent flashing, neoprene	2.000	Ea.	19.60	49	68.60
K	Circulator, solar heated liquid, 1/25 HP	2.000	Ea.	324	126	450
L	Circulator, solar heated liquid, 1/20 HP	1.000	Ea.	170	73.50	243.50
N	Relief valve, temp & pressure 150 psi 210°F self-closing	4.000	Ea.	40	65.40	105.40
N-1	Relief valve, pressure poppet, bronze, 30 psi, 3/4" IPS	3.000	Ea.	148.50	52.65	201.15
O	Pipe covering, urethane, ultraviolet cover, 1" wall, 1" diam	50.000	L.F.	95.50	229.50	325
P	Pipe covering, fiberglass, all service jacket, 1" wall, 1" diam	60.000	L.F.	54.60	220.20	274.80
Q	Collector panel solar energy blk chrome on copper 1/8" temp glass 3'x7'	10.000	Ea.	6,000	930	6,930
	Roof clamps for solar energy collector panel	10.000	Set	17.70	126.50	144.20
R	Valve, swing check, bronze, 125 lb, regrinding disc, 3/4" & 1" diam	4.000	Ea.	228	104	332
S	Pressure gage, 0-60 psi, for solar energy system	2.000	Ea.	49	30.70	79.70
U	Valve, water tempering, bronze, sweat connections, 3/4" diam	1.000	Ea.	70.50	24.50	95
W-2, W-1, W	Tank, water storage immersed heat xchr elec elem 2"x1/2# ins 120 gal	4.000	Ea.	4,100	1,400	5,500
X	Valve, globe, bronze, 125 lb, rising stem, 1" diam	3.000	Ea.	274.50	78	352.50
Y	Valve, flow control	1.000	Ea.	64.50	22.50	87
	Copper tubing type M, solder joint, hanger 10' OC, 1" diam	110.000	L.F.	288.20	770	1,058.20
	Copper tubing, type L, solder joint, hanger 10' OC 3/4" diam	20.000	L.F.	48.40	129	177.40
	Wrought copper fittings & solder, 3/4" & 1" diam	121.000	Ea.	347.27	3,690.50	4,037.77
	Sensor, wire, #22-2 conductor, multistranded	.700	C.L.F.	7.95	33.95	41.90
	Ductwork, galvanized steel, for heat exchanger	8.000	Lb.	9.12	47.20	56.32
	Solar energy heat transfer fluid propylene glycol, anti-freeze	25.000	Gal.	242.50	438.75	681.25
	TOTAL			14,353.30	9,952.40	24,305.70

D3010 650	Solar, Closed Loop, Space/Hot Water Systems		COST EACH		
			MAT.	INST.	TOTAL
2540	Solar, closed loop, space/hot water				
2550	1/2" tubing, 12 ea. 4'x4'4" vacuum tube collectors		19,000	9,300	28,300
2600	3/4" tubing, 12 ea. 4'x4'4" vacuum tube collectors	R13600 -610	19,600	9,600	29,200
2650	10 ea. 3' x 7' black chrome absorber collectors		14,000	9,325	23,325
2700	10 ea. 3' x 7' flat black absorber collectors		12,700	9,375	22,075
2750	1" tubing, 10 ea. 3' x 7' black chrome absorber collectors		14,400	9,950	24,350
2800	10 ea. 3' x 7' flat black absorber collectors		13,100	10,000	23,100
2850	6 ea. 4' x 9' flat black w/plastic glazing collectors		11,600	9,950	21,550
2900	12 ea. 2' x 9' plastic absorber and glazing collectors		14,900	10,100	25,000

This draindown pool system uses a differential thermostat similar to those used in solar domestic hot water and space heating applications. To heat the pool, the pool water passes through the conventional pump-filter loop and then flows through the collectors. When collection is not possible, or when the pool temperature is reached, all water drains from the solar loop back to the pool through the existing piping. The modes are controlled by solenoid valves or other automatic valves in conjunction with a vacuum breaker relief valve, which facilitates draindown.

D3010 Energy Supply

System Components	QUANTITY	UNIT	COST EACH		
			MAT.	INST.	TOTAL
SYSTEM D3010 660 2640					
SOLAR SWIMMING POOL HEATER, ROOF MOUNTED COLLECTORS					
TEN 4' X 10' FULLY WETTED UNGLAZED PLASTIC ABSORBERS					
A Differential thermostat/controller, 110V, adj pool pump system	1.000	Ea.	305	246	551
A-1 Solenoid valve, PVC, normally 1 open 1 closed (included)	2.000	Ea.			
B Sensor, thermistor type (included)	2.000	Ea.			
E-1 Valve, vacuum relief	1.000	Ea.	29	15.35	44.35
Q Collector panel, solar energy, plastic, liquid full wetted, 4' x 10'	10.000	Ea.	2,150	1,770	3,920
R Valve, ball check, PVC, socket, 1-1/2" diam	1.000	Ea.	79	24.50	103.50
Z Valve, ball, PVC, socket, 1-1/2" diam	3.000	Ea.	187.50	73.50	261
Pipe, PVC, sch 40, 1-1/2" diam	80.000	L.F.	127.20	1,092	1,219.20
Pipe fittings, PVC sch 40, socket joint, 1-1/2" diam	10.000	Ea.	11.80	245	256.80
Sensor wire, #22-2 conductor, multistranded	.500	C.L.F.	5.68	24.25	29.93
Roof clamps for solar energy collector panels	10.000	Set	17.70	126.50	144.20
Roof strap, teflon for solar energy collector panels	26.000	L.F.	271.70	62.40	334.10
TOTAL			3,184.58	3,679.50	6,864.08

D3010 660	Solar Swimming Pool Heater Systems		COST EACH		
			MAT.	INST.	TOTAL
2530	Solar swimming pool heater systems, roof mounted collectors				
2540	10 ea. 3'x7' black chrome absorber, 1/8" temp. glass		7,025	2,850	9,875
2560	10 ea. 4'x8' black chrome absorber, 3/16" temp. glass	R13600	8,125	3,375	11,500
2580	10 ea. 3'8"x6' flat black absorber, 3/16" temp. glass	-610	5,775	2,900	8,675
2600	10 ea. 4'x9' flat black absorber, plastic glazing		6,425	3,525	9,950
2620	10 ea. 2'x9' rubber absorber, plastic glazing		12,000	3,800	15,800
2640	10 ea. 4'x10' fully wetted unglazed plastic absorber		3,175	3,675	6,850
2660	Ground mounted collectors				
2680	10 ea. 3'x7' black chrome absorber, 1/8" temp. glass		7,250	3,275	10,525
2700	10 ea. 4'x8' black chrome absorber, 3/16" temp glass		8,350	3,800	12,150
2720	10 ea. 3'8"x6' flat blk absorber, 3/16" temp. glass		6,000	3,325	9,325
2740	10 ea. 4'x9' flat blk absorber, plastic glazing		6,650	3,950	10,600
2760	10 ea. 2'x9' rubber absorber, plastic glazing		12,100	4,100	16,200
2780	10 ea. 4'x10' fully wetted unglazed plastic absorber		3,400	4,100	7,500

The complete Solar Air Heating System provides maximum savings of conventional fuel with both space heating and year-round domestic hot water heating. It allows for the home air conditioning to operate simultaneously and independently from the solar domestic water heating in summer. The system's modes of operation are:

Mode 1: The building is heated directly from the collectors with air circulated by the Solar Air Mover.

Mode 2: When heat is not needed in the building, the dampers change within the air mover to circulate the air from the collectors to the rock storage bin.

Mode 3: When heat is not available from the collector array and is available in rock storage, the air mover draws heated air from rock storage and directs it into the building. When heat is not available from the collectors or the rock storage bin, the auxiliary heating unit will provide heat for the building. The size of the collector array is typically 25% the size of the main floor area.

D3010 Energy Supply

System Components	QUANTITY	UNIT	COST EACH		
			MAT.	INST.	TOTAL
SYSTEM D3010 675 1210					
SOLAR, SPACE/HOT WATER, AIR TO WATER HEAT EXCHANGE					
A, B Differential controller 2 sensors thermos., solar energy sys liquid loop	1.000	Ea.	92	41	133
A-1, B Differential controller 2 sensors 6 station solar energy sys air loop	1.000	Ea.	239	164	403
B-1 Solar energy sensor, freeze prevention	1.000	Ea.	24	15.35	39.35
C Thermometer for solar energy system, 2" dial	2.000	Ea.	44	61	105
C-1 Heat exchanger, solar energy system, air to fluid, up flow, 70 MBH	1.000	Ea.	365	253	618
D Drain valve, brass, 3/4" connection	2.000	Ea.	10.40	41	51.40
E Air vent, manual, for solar energy system 1/8" fitting	1.000	Ea.	2.46	15.35	17.81
E-1 Thermostat, 2 stage for sensing room temperature	1.000	Ea.	196	62	258
F Air purger	1.000	Ea.	47.50	41	88.50
G Expansion tank, for solar energy system	1.000	Ea.	67.50	15.35	82.85
I Valve, gate, bronze, 125 lb, NRS, soldered 3/4" diam	2.000	Ea.	55	49	104
K Circulator, solar heated liquid, 1/25 HP	1.000	Ea.	162	63	225
N Relief valve temp & press 150 psi 210°F self-closing, 3/4" IPS	1.000	Ea.	10	16.35	26.35
N-1 Relief valve, pressure, poppet, bronze, 30 psi, 3/4" IPS	1.000	Ea.	49.50	17.55	67.05
P Pipe covering, fiberglass, all service jacket, 1" wall, 3/4" diam	60.000	L.F.	54.60	220.20	274.80
Q-3 Manifold for flush mount solar energy collector panels	20.000	L.F.	590	111	701
Q Collector panel solar energy, air, black on alum. plate, flush mount 10'x2'	100.000	L.F.	5,700	930	6,630
R Valve, swing check, bronze, 125 lb, regrinding disc, 3/4" diam	2.000	Ea.	77	49	126
S Pressure gage, 2" dial, for solar energy system	1.000	Ea.	24.50	15.35	39.85
U Valve, water tempering, bronze, sweat connections, 3/4" diam	1.000	Ea.	70.50	24.50	95
W-2, W Tank, water storage, solar, elec element 2"x1/2# insul, 80 Gal	1.000	Ea.	920	305	1,225
X Valve, globe, bronze, 125 lb, soldered, 3/4" diam	1.000	Ea.	58.50	24.50	83
Copper tubing type L, solder joints, hangers 10' OC 3/4" diam	10.000	L.F.	24.20	64.50	88.70
Copper tubing type M, solder joints, hangers 10' OC 3/4" diam	60.000	L.F.	114	378	492
Wrought copper fittings & solder, 3/4" diam	26.000	Ea.	30.42	676	706.42
Sensor wire, #22-2 conductor multistranded	1.200	C.L.F.	13.62	58.20	71.82
Q-1 Duct work, rigid fiberglass, rectangular	400.000	S.F.	292	1,584	1,876
Ductwork, transition pieces	8.000	Ea.	118.40	296	414.40
R-2 Shutter/damper for solar heater circulator	9.000	Ea.	387	666	1,053
K-1 Shutter motor for solar heater circulator blower	9.000	Ea.	918	499.50	1,417.50
K-1 Fan, solar energy heated air circulator, space & DHW system	1.000	Ea.	1,625	1,775	3,400
Z-1 Tank, solar energy air storage, 6'-3"H 7'x7' = 306 CF/2000 Gal	1.000	Ea.	10,400	1,150	11,550
Z-2 Crushed stone 1-1/2"	11.000	C.Y.	330	60.72	390.72
C-2 Thermometer, remote probe, 2" dial	1.000	Ea.	33.50	61.50	95
R-1 Solenoid valve	1.000	Ea.	233	54.50	287.50
V, R-4, R-3 Furnace, supply diffusers, return grilles, existing	1.000	Ea.	10,400	1,150	11,550
TOTAL			33,778.60	11,008.42	44,787.02

D3010 675	Air To Water Heat Exchange		COST EACH		
			MAT.	INST.	TOTAL
1210 1220	Solar, air to water heat exchange, for space / hot water heating	R13600 -610	33,800	11,000	44,800

D3020 Heat Generating Systems

Boiler   **Baseboard Radiation**

Small Electric Boiler
System Considerations:
1. Terminal units are fin tube baseboard radiation rated at 720 BTU/hr with 200° water temperature or 820 BTU/hr steam.
2. Primary use being for residential or smaller supplementary areas, the floor levels are based on 7-1/2' ceiling heights.
3. All distribution piping is copper for boilers through 205 MBH. All piping for larger systems is steel pipe.

System Components	QUANTITY	UNIT	COST EACH		
			MAT.	INST.	TOTAL
SYSTEM D3020 102 1120					
SMALL HEATING SYSTEM, HYDRONIC, ELECTRIC BOILER					
1,480 S.F., 61 MBH, STEAM, 1 FLOOR					
Boiler, electric steam, std cntrls, trim, ftngs and valves, 18 KW, 61.4 MBH	1.000	Ea.	3,465	1,265	4,730
Copper tubing type L, solder joint, hanger 10'OC, 1-1/4" diam	160.000	L.F.	748.80	1,352	2,100.80
Radiation, 3/4" copper tube w/alum fin baseboard pkg 7" high	60.000	L.F.	375	918	1,293
Rough in baseboard panel or fin tube with valves & traps	10.000	Set	1,180	4,650	5,830
Pipe covering, calcium silicate w/cover 1" wall 1-1/4" diam	160.000	L.F.	417.60	816	1,233.60
Low water cut-off, quick hookup, in gage glass tappings	1.000	Ea.	177	31	208
TOTAL			6,363.40	9,032	15,395.40
COST PER S.F.			4.30	6.10	10.40

D3020 102	Small Heating Systems, Hydronic, Electric Boilers	COST PER S.F.		
		MAT.	INST.	TOTAL
1100	Small heating systems, hydronic, electric boilers			
1120	Steam, 1 floor, 1480 S.F., 61 M.B.H.	4.31	6.10	10.41
1160	3,000 S.F., 123 M.B.H.	3.20	5.35	8.55
1200	5,000 S.F., 205 M.B.H.	2.73	4.92	7.65
1240	2 floors, 12,400 S.F., 512 M.B.H.	2.17	4.89	7.06
1280	3 floors, 24,800 S.F., 1023 M.B.H.	2.24	4.83	7.07
1360	Hot water, 1 floor, 1,000 S.F., 41 M.B.H.	7.25	3.41	10.66
1400	2,500 S.F., 103 M.B.H.	4.94	6.10	11.04
1440	2 floors, 4,850 S.F., 205 M.B.H.	4.61	7.35	11.96
1480	3 floors, 9,700 S.F., 410 M.B.H.	4.55	7.55	12.10

Important: See the Reference Section for critical supporting data - Reference Numbers and City Cost Indexes

D3090 Other HVAC Systems/Equip

Vitrified Clay Garage Exhaust System

Dual Exhaust System

System Components	QUANTITY	UNIT	COST EACH		
			MAT.	INST.	TOTAL
SYSTEM D3090 320 1040					
GARAGE, EXHAUST, SINGLE OUTLET, 3" EXHAUST, CARS & LIGHT TRUCKS					
A — Outlet top assy. for engine exhaust system with adapters and ftngs, 3" diam	1.000	Ea.	178	31	209
F — Bullnose (guide) for engine exhaust system, 3" diam	1.000	Ea.	19.80		19.80
G — Galvanized flexible tubing for engine exhaust system, 3" diam	8.000	L.F.	42.40		42.40
H — Adapter for metal tubing end of engine exhaust system, 3" tail pipe	1.000	Ea.	39.50		39.50
J — Pipe, sewer, vitrified clay, premium joint, 8" diam.	18.000	L.F.	79.74	136.26	216
Excavate utility trench w/chain trencher, 12 H.P., oper. walking	18.000	L.F.		23.22	23.22
Backfill utility trench by hand, incl. compaction, 8" wide 24" deep	18.000	L.F.		33.30	33.30
Stand for blower, concrete over polystyrene core, 6" high	1.000	Ea.	7.70	15.45	23.15
AC&V duct spiral reducer 10"x8"	1.000	Ea.	7.80	28	35.80
AC&V duct spiral reducer 12"x10"	1.000	Ea.	11.10	37	48.10
AC&V duct spiral preformed 45° elbow 8" diam	1.000	Ea.	3.64	32	35.64
AC&V utility fan, belt drive, 3 phase, 2000 CFM, 1 HP	2.000	Ea.	1,260	492	1,752
Safety switch, heavy duty fused, 240V, 3 pole, 30 amp	1.000	Ea.	138	152	290
TOTAL			1,787.68	980.23	2,767.91

D3090 320	Garage Exhaust Systems	COST PER BAY		
		MAT.	INST.	TOTAL
1040	Garage, single exhaust, 3" outlet, cars & light trucks, one bay	1,800	980	2,780
1060	Additional bays up to seven bays	325	138	463
1500	4" outlet, trucks, one bay	1,800	980	2,780
1520	Additional bays up to six bays	335	138	473
1600	5" outlet, diesel trucks, one bay	2,050	980	3,030
1650	Additional single bays up to six	590	160	750
1700	Two adjoining bays	2,050	980	3,030
2000	Dual exhaust, 3" outlets, pair of adjoining bays	2,100	1,075	3,175
2100	Additional pairs of adjoining bays	605	160	765

Dry Pipe System: A system employing automatic sprinklers attached to a piping system containing air under pressure, the release of which as from the opening of sprinklers permits the water pressure to open a valve known as a "dry pipe valve". The water then flows into the piping system and out the opened sprinklers.

All areas are assumed to be open.

System Components

System Components	QUANTITY	UNIT	COST EACH MAT.	COST EACH INST.	COST EACH TOTAL
SYSTEM D4010 310 0580					
DRY PIPE SPRINKLER, STEEL BLACK, SCH. 40 PIPE					
LIGHT HAZARD, ONE FLOOR, 2000 S.F.					
Valve, gate, iron body 125 lb, OS&Y, flanged, 4" pipe size	1.000	Ea.	307.50	221.25	528.75
Valve, swing check, bronze, 125 lb, regrinding disc, 2-1/2" pipe size	1.000	Ea.	218.25	44.25	262.50
Valve, angle, bronze, 150 lb, rising stem, threaded, 2" pipe size	1.000	Ea.	213.75	33.38	247.13
*Alarm valve, 2-1/2" pipe size	1.000	Ea.	701.25	221.25	922.50
Alarm, water motor, complete with gong,	1.000	Ea.	155.25	92.25	247.50
Fire alarm horn, electric	1.000	Ea.	30	54.38	84.38
Valve swing check w/balldrip CI with brass trim, 4" pipe size	1.000	Ea.	135	221.25	356.25
Pipe, steel, black, schedule 40, 4" diam	10.000	L.F.	99	198	297
Dry pipe valve, trim & gauges, 4" pipe size	1.000	Ea.	1,068.75	663.75	1,732.50
Pipe, steel, black, schedule 40, threaded, cplg & hngr 10'OC 2-1/2" diam	20.000	L.F.	115.50	265.50	381
Pipe, steel, black, schedule 40, threaded, cplg & hngr 10'OC 2" diam	12.500	L.F.	42.66	129.38	172.04
Pipe, steel, black, schedule 40, threaded, cplg & hngr 10'OC 1-1/4" diam	37.500	L.F.	84.94	279.84	364.78
Pipe, steel, black, schedule 40, threaded, cplg & hngr 10'OC 1" diam	112.000	L.F.	199.92	777	976.92
Pipe Tee, malleable iron black, 150 lb threaded, 4" pipe size	2.000	Ea.	196.50	331.50	528
Pipe Tee, malleable iron black, 150 lb threaded, 2-1/2" pipe size	2.000	Ea.	55.50	147.75	203.25
Pipe Tee, malleable iron black, 150 lb threaded, 2" pipe size	1.000	Ea.	12.83	60.38	73.21
Pipe Tee, malleable iron black, 150 lb threaded, 1-1/4" pipe size	5.000	Ea.	30.19	236.25	266.44
Pipe Tee, malleable iron black, 150 lb threaded, 1" pipe size	4.000	Ea.	14.91	184.50	199.41
Pipe 90° elbow malleable iron black, 150 lb threaded, 1" pipe size	6.000	Ea.	14.36	171	185.36
Sprinkler head dry 1/2" orifice 1" NPT, 3" to 4-3/4" length	12.000	Ea.	480	420	900
Air compressor, 200 Gal sprinkler system capacity, 1/3 HP	1.000	Ea.	592.50	285	877.50
*Standpipe connection, wall, flush, brs w/plug & chain 2-1/2"x2-1/2"	1.000	Ea.	73.13	132.75	205.88
Valve gate bronze, 300 psi, NRS, class 150, threaded, 1" pipe size	1.000	Ea.	34.50	19.50	54
TOTAL			4,876.19	5,190.11	10,066.30
COST PER S.F.			2.44	2.60	5.04

D4010 310	Dry Pipe Sprinkler Systems		COST PER S.F. MAT.	COST PER S.F. INST.	COST PER S.F. TOTAL
0520	Dry pipe sprinkler systems, steel, black, sch. 40 pipe				
0530	Light hazard, one floor, 500 S.F.		4.93	4.42	9.35
0560	1000 S.F.		2.74	2.55	5.29
0580	2000 S.F.	R10520 -110	2.44	2.60	5.04
0600	5000 S.F.		1.24	1.76	3
0620	10,000 S.F.	R10520 -120	.84	1.45	2.29
0640	50,000 S.F.		.63	1.28	1.91
0660	Each additional floor, 500 S.F.	R10520 -130	.94	2.13	3.07

Important: See the Reference Section for critical supporting data - Reference Numbers and City Cost Indexes

D4010 Sprinklers

D4010 310	Dry Pipe Sprinkler Systems	COST PER S.F.		
		MAT.	INST.	TOTAL
0680	1000 S.F.	.75	1.74	2.49
0700	2000 S.F.	.75	1.63	2.38
0720	5000 S.F.	.59	1.40	1.99
0740	10,000 S.F.	.53	1.29	1.82
0760	50,000 S.F.	.49	1.14	1.63
1000	Ordinary hazard, one floor, 500 S.F.	4.99	4.47	9.46
1020	1000 S.F.	2.77	2.58	5.35
1040	2000 S.F.	2.53	2.73	5.26
1060	5000 S.F.	1.41	1.87	3.28
1080	10,000 S.F.	1.10	1.91	3.01
1100	50,000 S.F.	1.03	1.87	2.90
1140	Each additional floor, 500 S.F.	1	2.18	3.18
1160	1000 S.F.	.84	1.96	2.80
1180	2000 S.F.	.88	1.81	2.69
1200	5000 S.F.	.80	1.53	2.33
1220	10,000 S.F.	.71	1.51	2.22
1240	50,000 S.F.	.73	1.39	2.12
1500	Extra hazard, one floor, 500 S.F.	6.75	5.55	12.30
1520	1000 S.F.	3.96	3.98	7.94
1540	2000 S.F.	2.74	3.47	6.21
1560	5000 S.F.	1.59	2.62	4.21
1580	10,000 S.F.	1.67	2.51	4.18
1600	50,000 S.F.	1.83	2.42	4.25
1660	Each additional floor, 500 S.F.	1.30	2.70	4
1680	1000 S.F.	1.28	2.57	3.85
1700	2000 S.F.	1.17	2.56	3.73
1720	5000 S.F.	.98	2.25	3.23
1740	10,000 S.F.	1.18	2.07	3.25
1760	50,000 S.F.	1.23	1.99	3.22
2020	Grooved steel, black, sch. 40 pipe, light hazard, one floor, 2000 S.F.	2.32	2.25	4.57
2060	10,000 S.F.	.74	1.24	1.98
2100	Each additional floor, 2000 S.F.	.67	1.33	2
2150	10,000 S.F.	.43	1.08	1.51
2200	Ordinary hazard, one floor, 2000 S.F.	2.39	2.36	4.75
2250	10,000 S.F.	.90	1.60	2.50
2300	Each additional floor, 2000 S.F.	.74	1.44	2.18
2350	10,000 S.F.	.59	1.44	2.03
2400	Extra hazard, one floor, 2000 S.F.	2.60	2.96	5.56
2450	10,000 S.F.	1.31	2.08	3.39
2500	Each additional floor, 2000 S.F.	1.04	2.10	3.14
2550	10,000 S.F.	.91	1.81	2.72
3050	Grooved steel black sch. 10 pipe, light hazard, one floor, 2000 S.F.	2.30	2.23	4.53
3100	10,000 S.F.	.77	1.23	2
3150	Each additional floor, 2000 S.F.	.65	1.31	1.96
3200	10,000 S.F.	.46	1.07	1.53
3250	Ordinary hazard, one floor, 2000 S.F.	2.37	2.36	4.73
3300	10,000 S.F.	.92	1.56	2.48
3350	Each additional floor, 2000 S.F.	.72	1.44	2.16
3400	10,000 S.F.	.61	1.40	2.01
3450	Extra hazard, one floor, 2000 S.F.	2.59	2.95	5.54
3500	10,000 S.F.	1.21	2.04	3.25
3550	Each additional floor, 2000 S.F.	1.03	2.09	3.12
3600	10,000 S.F.	.86	1.79	2.65
4050	Copper tubing, type M, light hazard, one floor, 2000 S.F.	2.48	2.19	4.67
4100	10,000 S.F.	.98	1.24	2.22
4150	Each additional floor, 2000 S.F.	.84	1.30	2.14
4200	10,000 S.F.	.67	1.09	1.76
4250	Ordinary hazard, one floor, 2000 S.F.	2.62	2.45	5.07

D4010 Sprinklers

D4010 310	Dry Pipe Sprinkler Systems	COST PER S.F.		
		MAT.	INST.	TOTAL
4300	10,000 S.F.	1.18	1.48	2.66
4350	Each additional floor, 2000 S.F.	1.12	1.51	2.63
4400	10,000 S.F.	.85	1.30	2.15
4450	Extra hazard, one floor, 2000 S.F.	2.94	2.98	5.92
4500	10,000 S.F.	2.28	2.22	4.50
4550	Each additional floor, 2000 S.F.	1.38	2.12	3.50
4600	10,000 S.F.	1.53	1.93	3.46
5050	Copper tubing, type M, T-drill system, light hazard, one floor			
5060	2000 S.F.	2.49	2.07	4.56
5100	10,000 S.F.	.94	1.04	1.98
5150	Each additional floor, 2000 S.F.	.85	1.18	2.03
5200	10,000 S.F.	.63	.89	1.52
5250	Ordinary hazard, one floor, 2000 S.F.	2.53	2.10	4.63
5300	10,000 S.F.	1.14	1.31	2.45
5350	Each additional floor, 2000 S.F.	.88	1.18	2.06
5400	10,000 S.F.	.78	1.09	1.87
5450	Extra hazard, one floor, 2000 S.F.	2.72	2.48	5.20
5500	10,000 S.F.	1.90	1.64	3.54
5550	Each additional floor, 2000 S.F.	1.16	1.62	2.78
5600	10,000 S.F.	1.15	1.35	2.50

D4010 Sprinklers

Pre-Action System: A system employing automatic sprinklers attached to a piping system containing air that may or may not be under pressure, with a supplemental heat responsive system of generally more sensitive characteristics than the automatic sprinklers themselves, installed in the same areas as the sprinklers. Actuation of the heat responsive system, as from a fire, opens a valve which permits water to flow into the sprinkler piping system and to be discharged from those sprinklers which were opened by heat from the fire.

All areas are assumed to be open.

System Components	QUANTITY	UNIT	COST EACH		
			MAT.	INST.	TOTAL
SYSTEM D4010 350 0580					
PREACTION SPRINKLER SYSTEM, STEEL BLACK SCH. 40 PIPE					
LIGHT HAZARD, 1 FLOOR, 2000 S.F.					
Valve, gate, iron body 125 lb, OS&Y, flanged, 4" pipe size	1.000	Ea.	307.50	221.25	528.75
*Valve, swing check w/ball drip CI with brass trim 4" pipe size	1.000	Ea.	135	221.25	356.25
Valve, swing check, bronze, 125 lb, regrinding disc, 2-1/2" pipe size	1.000	Ea.	218.25	44.25	262.50
Valve, angle, bronze, 150 lb, rising stem, threaded, 2" pipe size	1.000	Ea.	213.75	33.38	247.13
*Alarm valve, 2-1/2" pipe size	1.000	Ea.	701.25	221.25	922.50
Alarm, water motor, complete with gong	1.000	Ea.	155.25	92.25	247.50
Fire alarm horn, electric	1.000	Ea.	30	54.38	84.38
Thermostatic release for release line	2.000	Ea.	495	36.75	531.75
Pipe, steel, black, schedule 40, 4" diam	10.000	L.F.	99	198	297
Dry pipe valve, trim & gauges, 4" pipe size	1.000	Ea.	1,068.75	663.75	1,732.50
Pipe, steel, black, schedule 40, threaded, cplg & hngr 10'OC 2-1/2" diam	20.000	L.F.	115.50	265.50	381
Pipe steel black, schedule 40, threaded, cplg & hngr 10'OC 2" diam	12.500	L.F.	42.66	129.38	172.04
Pipe, steel, black, schedule 40, threaded, cplg & hngr 10'OC 1-1/4" diam	37.500	L.F.	84.94	279.84	364.78
Pipe, steel, black, schedule 40, threaded, cplg & hngr 10'OC 1" diam	112.000	L.F.	199.92	777	976.92
Pipe, Tee, malleable iron, black, 150 lb threaded, 4" diam	2.000	Ea.	196.50	331.50	528
Pipe, Tee, malleable iron, black, 150 lb threaded, 2-1/2" pipe size	2.000	Ea.	55.50	147.75	203.25
Pipe, Tee, malleable iron, black, 150 lb threaded, 2" pipe size	1.000	Ea.	12.83	60.38	73.21
Pipe, Tee, malleable iron, black, 150 lb threaded, 1-1/4" pipe size	5.000	Ea.	30.19	236.25	266.44
Pipe, Tee, malleable iron, black, 150 lb threaded, 1" pipe size	4.000	Ea.	14.91	184.50	199.41
Pipe, 90° elbow, malleable iron, blk, 150 lb threaded, 1" pipe size	6.000	Ea.	14.36	171	185.36
Sprinkler head, std spray, brass 135°-286°F 1/2" NPT, 3/8" orifice	12.000	Ea.	88.80	372	460.80
Air compressor auto complete 200 Gal sprinkler sys cap, 1/3 HP	1.000	Ea.	592.50	285	877.50
*Standpipe conn.,wall,flush,brass w/plug & chain 2-1/2" x 2-1/2"	1.000	Ea.	73.13	132.75	205.88
Valve, gate, bronze, 300 psi, NRS, class 150, threaded, 1" pipe size	1.000	Ea.	34.50	19.50	54
TOTAL			4,979.99	5,178.86	10,158.85
COST PER S.F.			2.49	2.59	5.08

*Not included in systems under 2000 S.F.

D4010 350	Preaction Sprinkler Systems		COST PER S.F.		
			MAT.	INST.	TOTAL
0520	Preaction sprinkler systems, steel, black, sch. 40 pipe				
0530	Light hazard, one floor, 500 S.F.		5.05	3.53	8.58
0560	1000 S.F.	R10520 -110	2.85	2.61	5.46
0580	2000 S.F.		2.49	2.59	5.08

D40 Fire Protection

D4010 Sprinklers

D4010 350	Preaction Sprinkler Systems	MAT.	INST.	TOTAL
0600	5000 S.F.	1.29	1.75	3.04
0620	10,000 S.F. R10520-120	.87	1.44	2.31
0640	50,000 S.F.	.66	1.27	1.93
0660	Each additional floor, 500 S.F. R10520-130	1.12	1.90	3.02
0680	1000 S.F.	.85	1.74	2.59
0700	2000 S.F.	.85	1.63	2.48
0720	5000 S.F.	.64	1.39	2.03
0740	10,000 S.F.	.56	1.28	1.84
0760	50,000 S.F.	.57	1.18	1.75
1000	Ordinary hazard, one floor, 500 S.F.	5.20	3.83	9.03
1020	1000 S.F.	2.82	2.58	5.40
1040	2000 S.F.	2.70	2.74	5.44
1060	5000 S.F.	1.41	1.86	3.27
1080	10,000 S.F.	1.07	1.90	2.97
1100	50,000 S.F.	.99	1.84	2.83
1140	Each additional floor, 500 S.F.	1.29	2.20	3.49
1160	1000 S.F.	.83	1.77	2.60
1180	2000 S.F.	.80	1.80	2.60
1200	5000 S.F.	.88	1.65	2.53
1220	10,000 S.F.	.76	1.74	2.50
1240	50,000 S.F.	.77	1.62	2.39
1500	Extra hazard, one floor, 500 S.F.	6.90	4.90	11.80
1520	1000 S.F.	3.88	3.63	7.51
1540	2000 S.F.	2.70	3.46	6.16
1560	5000 S.F.	1.65	2.83	4.48
1580	10,000 S.F.	1.64	2.78	4.42
1600	50,000 S.F.	1.77	2.69	4.46
1660	Each additional floor, 500 S.F.	1.50	2.71	4.21
1680	1000 S.F.	1.24	2.55	3.79
1700	2000 S.F.	1.13	2.55	3.68
1720	5000 S.F.	.93	2.26	3.19
1740	10,000 S.F.	1.07	2.09	3.16
1760	50,000 S.F.	1.08	1.96	3.04
2020	Grooved steel, black, sch. 40 pipe, light hazard, one floor, 2000 S.F.	2.42	2.25	4.67
2060	10,000 S.F.	.77	1.23	2
2100	Each additional floor of 2000 S.F.	.77	1.33	2.10
2150	10,000 S.F.	.46	1.07	1.53
2200	Ordinary hazard, one floor, 2000 S.F.	2.44	2.36	4.80
2250	10,000 S.F.	.87	1.59	2.46
2300	Each additional floor, 2000 S.F.	.79	1.44	2.23
2350	10,000 S.F.	.56	1.43	1.99
2400	Extra hazard, one floor, 2000 S.F.	2.56	2.95	5.51
2450	10,000 S.F.	1.21	2.05	3.26
2500	Each additional floor, 2000 S.F.	1	2.09	3.09
2550	10,000 S.F.	.79	1.78	2.57
3050	Grooved steel, black, sch. 10 pipe light hazard, one floor, 2000 S.F.	2.40	2.23	4.63
3100	10,000 S.F.	.80	1.22	2.02
3150	Each additional floor, 2000 S.F.	.75	1.31	2.06
3200	10,000 S.F.	.49	1.06	1.55
3250	Ordinary hazard, one floor, 2000 S.F.	2.39	2.21	4.60
3300	10,000 S.F.	.77	1.54	2.31
3350	Each additional floor, 2000 S.F.	.77	1.44	2.21
3400	10,000 S.F.	.58	1.39	1.97
3450	Extra hazard, one floor, 2000 S.F.	2.55	2.94	5.49
3500	10,000 S.F.	1.09	2.01	3.10
3550	Each additional floor, 2000 S.F.	.99	2.08	3.07
3600	10,000 S.F.	.74	1.76	2.50
4050	Copper tubing, type M, light hazard, one floor, 2000 S.F.	2.59	2.19	4.78

Important: See the Reference Section for critical supporting data - Reference Numbers and City Cost Indexes

D4010 Sprinklers

D4010 350	Preaction Sprinkler Systems	COST PER S.F.		
		MAT.	INST.	TOTAL
4100	10,000 S.F.	1.01	1.23	2.24
4150	Each additional floor, 2000 S.F.	.95	1.30	2.25
4200	10,000 S.F.	.58	1.07	1.65
4250	Ordinary hazard, one floor, 2000 S.F.	2.67	2.45	5.12
4300	10,000 S.F.	1.15	1.47	2.62
4350	Each additional floor, 2000 S.F.	.95	1.34	2.29
4400	10,000 S.F.	.74	1.19	1.93
4450	Extra hazard, one floor, 2000 S.F.	2.90	2.97	5.87
4500	10,000 S.F.	2.14	2.19	4.33
4550	Each additional floor, 2000 S.F.	1.34	2.11	3.45
4600	10,000 S.F.	1.41	1.90	3.31
5050	Copper tubing, type M, T-drill system, light hazard, one floor			
5060	2000 S.F.	2.59	2.07	4.66
5100	10,000 S.F.	.97	1.03	2
5150	Each additional floor, 2000 S.F.	.95	1.18	2.13
5200	10,000 S.F.	.66	.88	1.54
5250	Ordinary hazard, one floor, 2000 S.F.	2.58	2.10	4.68
5300	10,000 S.F.	1.11	1.30	2.41
5350	Each additional floor, 2000 S.F.	.93	1.19	2.12
5400	10,000 S.F.	.80	1.14	1.94
5450	Extra hazard, one floor, 2000 S.F.	2.68	2.47	5.15
5500	10,000 S.F.	1.76	1.61	3.37
5550	Each additional floor, 2000 S.F.	1.12	1.61	2.73
5600	10,000 S.F.	1.03	1.32	2.35

Deluge System: A system employing open sprinklers attached to a piping system connected to a water supply through a valve which is opened by the operation of a heat responsive system installed in the same areas as the sprinklers. When this valve opens, water flows into the piping system and discharges from all sprinklers attached thereto.

All areas are assumed to be open.

System Components

System Components	QUANTITY	UNIT	COST EACH MAT.	COST EACH INST.	COST EACH TOTAL
SYSTEM D4010 370 0580					
DELUGE SPRINKLER SYSTEM, STEEL BLACK SCH. 40 PIPE					
LIGHT HAZARD, 1 FLOOR, 2000 S.F.					
Valve, gate, iron body 125 lb, OS&Y, flanged, 4" pipe size	1.000	Ea.	307.50	221.25	528.75
Valve, swing check w/ball drip, CI w/brass ftngs, 4" pipe size	1.000	Ea.	135	221.25	356.25
Valve, swing check, bronze, 125 lb, regrinding disc, 2-1/2" pipe size	1.000	Ea.	218.25	44.25	262.50
Valve, angle, bronze, 150 lb, rising stem, threaded, 2" pipe size	1.000	Ea.	213.75	33.38	247.13
*Alarm valve, 2-1/2" pipe size	1.000	Ea.	701.25	221.25	922.50
Alarm, water motor, complete with gong	1.000	Ea.	155.25	92.25	247.50
Fire alarm horn, electric	1.000	Ea.	30	54.38	84.38
Thermostatic release for release line	2.000	Ea.	495	36.75	531.75
Pipe, steel, black, schedule 40, 4" diam	10.000	L.F.	99	198	297
Deluge valve trim, pressure relief, emergency release, gauge, 4" pipe size	1.000	Ea.	1,218.75	663.75	1,882.50
Deluge system pressured monitoring panel, 120V	1.000	Ea.	727.50	20.63	748.13
Pipe, steel, black, schedule 40, threaded, cplg & hngr 10'OC 2-1/2" diam	20.000	L.F.	115.50	265.50	381
Pipe, steel, black, schedule 40, threaded, cplg & hngr 10'OC 2" diam	12.500	L.F.	42.66	129.38	172.04
Pipe, steel, black, schedule 40, threaded, cplg & hngr 10'OC 1-1/4" diam	37.500	L.F.	84.94	279.84	364.78
Pipe, steel, black, schedule 40, threaded, cplg & hngr 10'OC 1" diam	112.000	L.F.	199.92	777	976.92
Pipe, Tee, malleable iron, black, 150 lb threaded, 4" pipe size	2.000	Ea.	196.50	331.50	528
Pipe, Tee, malleable iron, black, 150 lb threaded, 2-1/2" pipe size	2.000	Ea.	55.50	147.75	203.25
Pipe, Tee, malleable iron, black, 150 lb threaded, 2" pipe size	1.000	Ea.	12.83	60.38	73.21
Pipe, Tee, malleable iron, black, 150 lb threaded, 1-1/4" pipe size	5.000	Ea.	30.19	236.25	266.44
Pipe, Tee, malleable iron, black, 150 lb threaded, 1" pipe size	4.000	Ea.	14.91	184.50	199.41
Pipe, 90° elbow, malleable iron, black, 150 lb threaded 1" pipe size	6.000	Ea.	14.36	171	185.36
Sprinkler head, std spray, brass 135°-286°F 1/2" NPT, 3/8" orifice	9.720	Ea.	88.80	372	460.80
Air compressor, auto, complete, 200 Gal sprinkler sys cap, 1/3 HP	1.000	Ea.	592.50	285	877.50
*Standpipe connection, wall, flush w/plug & chain 2-1/2"x2-1/2"	1.000	Ea.	73.13	132.75	205.88
Valve, gate, bronze, 300 psi, NRS, class 150, threaded, 1" pipe size	1.000	Ea.	34.50	19.50	54
TOTAL			5,857.49	5,199.49	11,056.98
COST PER S.F.			2.93	2.60	5.53

*Not included in systems under 2000 S.F.

D4010 370	Deluge Sprinkler Systems	COST PER S.F. MAT.	COST PER S.F. INST.	COST PER S.F. TOTAL
0520	Deluge sprinkler systems, steel, black, sch. 40 pipe			
0530	Light hazard, one floor, 500 S.F.	6.10	3.57	9.67

D4010 Sprinklers

D4010 370	Deluge Sprinkler Systems		COST PER S.F.		
			MAT.	INST.	TOTAL
0560	1000 S.F.	R10520 -110	3.36	2.51	5.87
0580	2000 S.F.		2.93	2.60	5.53
0600	5000 S.F.	R10520 -120	1.47	1.75	3.22
0620	10,000 S.F.		.95	1.44	2.39
0640	50,000 S.F.	R10520 -130	.67	1.27	1.94
0660	Each additional floor, 500 S.F.		1.12	1.90	3.02
0680	1000 S.F.		.85	1.74	2.59
0700	2000 S.F.		.85	1.63	2.48
0720	5000 S.F.		.64	1.39	2.03
0740	10,000 S.F.		.56	1.28	1.84
0760	50,000 S.F.		.57	1.18	1.75
1000	Ordinary hazard, one floor, 500 S.F.		6.60	4.08	10.68
1020	1000 S.F.		3.36	2.60	5.96
1040	2000 S.F.		3.14	2.75	5.89
1060	5000 S.F.		1.59	1.86	3.45
1080	10,000 S.F.		1.15	1.90	3.05
1100	50,000 S.F.		1.03	1.88	2.91
1140	Each additional floor, 500 S.F.		1.29	2.20	3.49
1160	1000 S.F.		.83	1.77	2.60
1180	2000 S.F.		.80	1.80	2.60
1200	5000 S.F.		.80	1.52	2.32
1220	10,000 S.F.		.73	1.54	2.27
1240	50,000 S.F.		.73	1.49	2.22
1500	Extra hazard, one floor, 500 S.F.		7.95	4.94	12.89
1520	1000 S.F.		4.64	3.78	8.42
1540	2000 S.F.		3.14	3.47	6.61
1560	5000 S.F.		1.70	2.60	4.30
1580	10,000 S.F.		1.66	2.54	4.20
1600	50,000 S.F.		1.78	2.46	4.24
1660	Each additional floor, 500 S.F.		1.50	2.71	4.21
1680	1000 S.F.		1.24	2.55	3.79
1700	2000 S.F.		1.13	2.55	3.68
1720	5000 S.F.		.93	2.26	3.19
1740	10,000 S.F.		1.10	2.17	3.27
1760	50,000 S.F.		1.14	2.11	3.25
2000	Grooved steel, black, sch. 40 pipe, light hazard, one floor				
2020	2000 S.F.		2.87	2.26	5.13
2060	10,000 S.F.		.86	1.24	2.10
2100	Each additional floor, 2,000 S.F.		.77	1.33	2.10
2150	10,000 S.F.		.46	1.07	1.53
2200	Ordinary hazard, one floor, 2000 S.F.		2.44	2.36	4.80
2250	10,000 S.F.		.95	1.59	2.54
2300	Each additional floor, 2000 S.F.		.79	1.44	2.23
2350	10,000 S.F.		.56	1.43	1.99
2400	Extra hazard, one floor, 2000 S.F.		3	2.96	5.96
2450	10,000 S.F.		1.30	2.06	3.36
2500	Each additional floor, 2000 S.F.		1	2.09	3.09
2550	10,000 S.F.		.79	1.78	2.57
3000	Grooved steel, black, sch. 10 pipe, light hazard, one floor				
3050	2000 S.F.		2.53	2.14	4.67
3100	10,000 S.F.		.88	1.22	2.10
3150	Each additional floor, 2000 S.F.		.75	1.31	2.06
3200	10,000 S.F.		.49	1.06	1.55
3250	Ordinary hazard, one floor, 2000 S.F.		2.86	2.37	5.23
3300	10,000 S.F.		.85	1.54	2.39
3350	Each additional floor, 2000 S.F.		.77	1.44	2.21
3400	10,000 S.F.		.58	1.39	1.97
3450	Extra hazard, one floor, 2000 S.F.		2.99	2.95	5.94

D4010 Sprinklers

D4010 370	Deluge Sprinkler Systems	COST PER S.F.		
		MAT.	INST.	TOTAL
3500	10,000 S.F.	1.17	2.01	3.18
3550	Each additional floor, 2000 S.F.	.99	2.08	3.07
3600	10,000 S.F.	.74	1.76	2.50
4000	Copper tubing, type M, light hazard, one floor			
4050	2000 S.F.	3.02	2.20	5.22
4100	10,000 S.F.	1.09	1.23	2.32
4150	Each additional floor, 2000 S.F.	.94	1.30	2.24
4200	10,000 S.F.	.58	1.07	1.65
4250	Ordinary hazard, one floor, 2000 S.F.	3.11	2.46	5.57
4300	10,000 S.F.	1.23	1.47	2.70
4350	Each additional floor, 2000 S.F.	.95	1.34	2.29
4400	10,000 S.F.	.74	1.19	1.93
4450	Extra hazard, one floor, 2000 S.F.	3.34	2.98	6.32
4500	10,000 S.F.	2.25	2.20	4.45
4550	Each additional floor, 2000 S.F.	1.34	2.11	3.45
4600	10,000 S.F.	1.41	1.90	3.31
5000	Copper tubing, type M, T-drill system, light hazard, one floor			
5050	2000 S.F.	3.03	2.08	5.11
5100	10,000 S.F.	1.05	1.03	2.08
5150	Each additional floor, 2000 S.F.	.94	1.19	2.13
5200	10,000 S.F.	.66	.88	1.54
5250	Ordinary hazard, one floor, 2000 S.F.	3.02	2.11	5.13
5300	10,000 S.F.	1.19	1.30	2.49
5350	Each additional floor, 2000 S.F.	.93	1.18	2.11
5400	10,000 S.F.	.80	1.14	1.94
5450	Extra hazard, one floor, 2000 S.F.	3.12	2.48	5.60
5500	10,000 S.F.	1.84	1.61	3.45
5550	Each additional floor, 2000 S.F.	1.12	1.61	2.73
5600	10,000 S.F.	1.03	1.32	2.35

Important: See the Reference Section for critical supporting data - Reference Numbers and City Cost Indexes

D4010 Sprinklers

Firecycle is a fixed fire protection sprinkler system utilizing water as its extinguishing agent. It is a time delayed, recycling, preaction type which automatically shuts the water off when heat is reduced below the detector operating temperature and turns the water back on when that temperature is exceeded.

The system senses a fire condition through a closed circuit electrical detector system which controls water flow to the fire automatically. Batteries supply up to 90 hour emergency power supply for system operation. The piping system is dry (until water is required) and is monitored with pressurized air. Should any leak in the system piping occur, an alarm will sound, but water will not enter the system until heat is sensed by a Firecycle detector.

All areas are assumed to be open.

System Components	QUANTITY	UNIT	COST EACH MAT.	INST.	TOTAL
SYSTEM D4010 390 0580					
FIRECYCLE SPRINKLER SYSTEM, STEEL BLACK SCH. 40 PIPE					
LIGHT HAZARD, ONE FLOOR, 2000 S.F.					
Valve, gate, iron body 125 lb, OS&Y, flanged, 4″ pipe size	1.000	Ea.	307.50	221.25	528.75
Valve, angle, bronze, 150 lb, rising stem, threaded, 2″ pipe size	1.000	Ea.	213.75	33.38	247.13
Valve, swing check, bronze, 125 lb, regrinding disc, 2-1/2″ pipe size	1.000	Ea.	218.25	44.25	262.50
*Alarm valve, 2-1/2″ pipe size	1.000	Ea.	701.25	221.25	922.50
Alarm, water motor, complete with gong	1.000	Ea.	155.25	92.25	247.50
Pipe, steel, black, schedule 40, 4″ diam	10.000	L.F.	99	198	297
Fire alarm, horn, electric	1.000	Ea.	30	54.38	84.38
Pipe, steel, black, schedule 40, threaded, cplg & hngr 10′OC 2-1/2″ diam	20.000	L.F.	115.50	265.50	381
Pipe, steel, black, schedule 40, threaded, cplg & hngr 10′OC 2″ diam	12.500	L.F.	42.66	129.38	172.04
Pipe, steel, black, schedule 40, threaded, cplg & hngr 10′OC 1-1/4″ diam	37.500	L.F.	84.94	279.84	364.78
Pipe, steel, black, schedule 40, threaded, cplg & hngr 10′OC 1″ diam	112.000	L.F.	199.92	777	976.92
Pipe, Tee, malleable iron, black, 150 lb threaded, 4″ pipe size	2.000	Ea.	196.50	331.50	528
Pipe, Tee, malleable iron, black, 150 lb threaded, 2-1/2″ pipe size	2.000	Ea.	55.50	147.75	203.25
Pipe, Tee, malleable iron, black, 150 lb threaded, 2″ pipe size	1.000	Ea.	12.83	60.38	73.21
Pipe, Tee, malleable iron, black, 150 lb threaded, 1-1/4″ pipe size	5.000	Ea.	30.19	236.25	266.44
Pipe, Tee, malleable iron, black, 150 lb threaded, 1″ pipe size	4.000	Ea.	14.91	184.50	199.41
Pipe, 90° elbow, malleable iron, black, 150 lb threaded, 1″ pipe size	6.000	Ea.	14.36	171	185.36
Sprinkler head std spray, brass 135°-286°F 1/2″ NPT, 3/8″ orifice	12.000	Ea.	88.80	372	460.80
Firecycle controls, incls panel, battery, solenoid valves, press switches	1.000	Ea.	7,256.25	1,406.25	8,662.50
Detector, firecycle system	2.000	Ea.	532.50	46.50	579
Firecycle pkg, swing check & flow control valves w/trim 4″ pipe size	1.000	Ea.	1,893.75	663.75	2,557.50
Air compressor, auto, complete, 200 Gal sprinkler sys cap, 1/3 HP	1.000	Ea.	592.50	285	877.50
*Standpipe connection, wall, flush, brass w/plug & chain 2-1/2″x2-1/2″	1.000	Ea.	73.13	132.75	205.88
Valve, gate, bronze, 300 psi, NRS, class 150, threaded, 1″ diam	1.000	Ea.	34.50	19.50	54
TOTAL			12,963.74	6,373.61	19,337.35
COST PER S.F.			6.48	3.19	9.67

*Not included in systems under 2000 S.F.

D4010 Sprinklers

D4010 390	Firecycle Sprinkler Systems		COST PER S.F.		
			MAT.	INST.	TOTAL
0520	Firecycle sprinkler systems, steel black sch. 40 pipe				
0530	Light hazard, one floor, 500 S.F.		21	6.95	27.95
0560	1000 S.F.	R10520 -110	10.90	4.32	15.22
0580	2000 S.F.		6.50	3.19	9.69
0600	5000 S.F.	R10520 -120	2.89	2	4.89
0620	10,000 S.F.		1.71	1.57	3.28
0640	50,000 S.F.	R10520 -130	.85	1.30	2.15
0660	Each additional floor of 500 S.F.		1.16	1.91	3.07
0680	1000 S.F.		.87	1.74	2.61
0700	2000 S.F.		.73	1.62	2.35
0720	5000 S.F.		.65	1.40	2.05
0740	10,000 S.F.		.60	1.29	1.89
0760	50,000 S.F.		.59	1.18	1.77
1000	Ordinary hazard, one floor, 500 S.F.		21.50	7.25	28.75
1020	1000 S.F.		10.85	4.29	15.14
1040	2000 S.F.		6.60	3.32	9.92
1060	5000 S.F.		3.01	2.11	5.12
1080	10,000 S.F.		1.91	2.03	3.94
1100	50,000 S.F.		1.34	2.10	3.44
1140	Each additional floor, 500 S.F.		1.33	2.21	3.54
1160	1000 S.F.		.85	1.77	2.62
1180	2000 S.F.		.91	1.65	2.56
1200	5000 S.F.		.81	1.53	2.34
1220	10,000 S.F.		.72	1.51	2.23
1240	50,000 S.F.		.74	1.43	2.17
1500	Extra hazard, one floor, 500 S.F.		23	8.30	31.30
1520	1000 S.F.		11.80	5.35	17.15
1540	2000 S.F.		6.70	4.05	10.75
1560	5000 S.F.		3.12	2.85	5.97
1580	10,000 S.F.		2.45	2.89	5.34
1600	50,000 S.F.		2.18	3.18	5.36
1660	Each additional floor, 500 S.F.		1.54	2.72	4.26
1680	1000 S.F.		1.26	2.55	3.81
1700	2000 S.F.		1.15	2.55	3.70
1720	5000 S.F.		.94	2.27	3.21
1740	10,000 S.F.		1.11	2.10	3.21
1760	50,000 S.F.		1.14	2.05	3.19
2020	Grooved steel, black, sch. 40 pipe, light hazard, one floor				
2030	2000 S.F.		6.40	2.84	9.24
2060	10,000 S.F.		1.75	1.90	3.65
2100	Each additional floor, 2000 S.F.		.79	1.33	2.12
2150	10,000 S.F.		.50	1.08	1.58
2200	Ordinary hazard, one floor, 2000 S.F.		6.45	2.95	9.40
2250	10,000 S.F.		1.79	1.82	3.61
2300	Each additional floor, 2000 S.F.		.81	1.44	2.25
2350	10,000 S.F.		.60	1.44	2.04
2400	Extra hazard, one floor, 2000 S.F.		6.55	3.54	10.09
2450	10,000 S.F.		2.02	2.16	4.18
2500	Each additional floor, 2000 S.F.		1.02	2.09	3.11
2550	10,000 S.F.		.83	1.79	2.62
3050	Grooved steel, black, sch. 10 pipe light hazard, one floor,				
3060	2000 S.F.		6.40	2.82	9.22
3100	10,000 S.F.		1.64	1.35	2.99
3150	Each additional floor, 2000 S.F.		.77	1.31	2.08
3200	10,000 S.F.		.53	1.07	1.60
3250	Ordinary hazard, one floor, 2000 S.F.		6.40	2.95	9.35

D4010 Sprinklers

D4010 390	Firecycle Sprinkler Systems	COST PER S.F.		
		MAT.	INST.	TOTAL
3300	10,000 S.F.	1.73	1.68	3.41
3350	Each additional floor, 2000 S.F.	.79	1.44	2.23
3400	10,000 S.F.	.62	1.40	2.02
3450	Extra hazard, one floor, 2000 S.F.	6.55	3.53	10.08
3500	10,000 S.F.	1.92	2.12	4.04
3550	Each additional floor, 2000 S.F.	1.01	2.08	3.09
3600	10,000 S.F.	.78	1.77	2.55
4060	Copper tubing, type M, light hazard, one floor, 2000 S.F.	6.60	2.78	9.38
4100	10,000 S.F.	1.85	1.36	3.21
4150	Each additional floor, 2000 S.F.	.96	1.30	2.26
4200	10,000 S.F.	.74	1.09	1.83
4250	Ordinary hazard, one floor, 2000 S.F.	6.65	3.04	9.69
4300	10,000 S.F.	1.99	1.60	3.59
4350	Each additional floor, 2000 S.F.	.97	1.34	2.31
4400	10,000 S.F.	.77	1.18	1.95
4450	Extra hazard, one floor, 2000 S.F.	6.90	3.56	10.46
4500	10,000 S.F.	3.04	2.33	5.37
4550	Each additional floor, 2000 S.F.	1.36	2.11	3.47
4600	10,000 S.F.	1.45	1.91	3.36
5060	Copper tubing, type M, T-drill system, light hazard, one floor 2000 S.F.	6.60	2.66	9.26
5100	10,000 S.F.	1.81	1.16	2.97
5150	Each additional floor, 2000 S.F.	1.06	1.24	2.30
5200	10,000 S.F.	.70	.89	1.59
5250	Ordinary hazard, one floor, 2000 S.F.	6.60	2.69	9.29
5300	10,000 S.F.	1.95	1.43	3.38
5350	Each additional floor, 2000 S.F.	.95	1.18	2.13
5400	10,000 S.F.	.84	1.15	1.99
5450	Extra hazard, one floor, 2000 S.F.	6.70	3.06	9.76
5500	10,000 S.F.	2.59	1.72	4.31
5550	Each additional floor, 2000 S.F.	1.14	1.61	2.75
5600	10,000 S.F.	1.07	1.33	2.40

D4010 Sprinklers

Wet Pipe System. A system employing automatic sprinklers attached to a piping system containing water and connected to a water supply so that water discharges immediately from sprinklers opened by heat from a fire.

All areas are assumed to be open.

System Components	QUANTITY	UNIT	COST EACH MAT.	COST EACH INST.	COST EACH TOTAL
SYSTEM D4010 410 0580					
WET PIPE SPRINKLER, STEEL, BLACK, SCH. 40 PIPE					
LIGHT HAZARD, ONE FLOOR, 2000 S.F.					
Valve, gate, iron body, 125 lb, OS&Y, flanged, 4" diam	1.000	Ea.	307.50	221.25	528.75
Valve, swing check, bronze, 125 lb, regrinding disc, 2-1/2" pipe size	1.000	Ea.	218.25	44.25	262.50
Valve, angle, bronze, 150 lb, rising stem, threaded, 2" diam	1.000	Ea.	213.75	33.38	247.13
*Alarm valve, 2-1/2" pipe size	1.000	Ea.	701.25	221.25	922.50
Alarm, water motor, complete with gong	1.000	Ea.	155.25	92.25	247.50
Valve, swing check, w/balldrip CI with brass trim 4" pipe size	1.000	Ea.	135	221.25	356.25
Pipe, steel, black, schedule 40, 4" diam	10.000	L.F.	99	198	297
*Flow control valve, trim & gauges, 4" pipe size	1.000	Set	1,350	502.50	1,852.50
Fire alarm horn, electric	1.000	Ea.	30	54.38	84.38
Pipe, steel, black, schedule 40, threaded, cplg & hngr 10'OC, 2-1/2" diam	20.000	L.F.	115.50	265.50	381
Pipe, steel, black, schedule 40, threaded, cplg & hngr 10'OC, 2" diam	12.500	L.F.	42.66	129.38	172.04
Pipe, steel, black, schedule 40, threaded, cplg & hngr 10'OC, 1-1/4" diam	37.500	L.F.	84.94	279.84	364.78
Pipe steel, black, schedule 40, threaded cplg & hngr 10'OC, 1" diam	112.000	L.F.	199.92	777	976.92
Pipe Tee, malleable iron black, 150 lb threaded, 4" pipe size	2.000	Ea.	196.50	331.50	528
Pipe Tee, malleable iron black, 150 lb threaded, 2-1/2" pipe size	2.000	Ea.	55.50	147.75	203.25
Pipe Tee, malleable iron black, 150 lb threaded, 2" pipe size	1.000	Ea.	12.83	60.38	73.21
Pipe Tee, malleable iron black, 150 lb threaded, 1-1/4" pipe size	5.000	Ea.	30.19	236.25	266.44
Pipe Tee, malleable iron black, 150 lb threaded, 1" pipe size	4.000	Ea.	14.91	184.50	199.41
Pipe 90° elbow, malleable iron black, 150 lb threaded, 1" pipe size	6.000	Ea.	14.36	171	185.36
Sprinkler head, standard spray, brass 135°-286°F 1/2" NPT, 3/8" orifice	12.000	Ea.	88.80	372	460.80
Valve, gate, bronze, NRS, class 150, threaded, 1" pipe size	1.000	Ea.	34.50	19.50	54
*Standpipe connection, wall, single, flush w/plug & chain 2-1/2"x2-1/2"	1.000	Ea.	73.13	132.75	205.88
TOTAL			4,173.74	4,695.86	8,869.60
COST PER S.F.			2.09	2.35	4.44

*Not included in systems under 2000 S.F.

D4010 410	Wet Pipe Sprinkler Systems		COST PER S.F. MAT.	COST PER S.F. INST.	COST PER S.F. TOTAL
0520	Wet pipe sprinkler systems, steel, black, sch. 40 pipe				
0530	Light hazard, one floor, 500 S.F.		1.30	2.25	3.55
0560	1000 S.F.		1.94	2.30	4.24
0580	2000 S.F.	R10520 -110	2.09	2.35	4.44
0600	5000 S.F.		1.03	1.65	2.68
0620	10,000 S.F.	R10520 -120	.67	1.38	2.05

Important: See the Reference Section for critical supporting data - Reference Numbers and City Cost Indexes

D4010 Sprinklers

D4010 410	Wet Pipe Sprinkler Systems	COST PER S.F.		
		MAT.	INST.	TOTAL
0640	50,000 S.F. R10520 -130	.50	1.25	1.75
0660	Each additional floor, 500 S.F.	.64	1.91	2.55
0680	1000 S.F.	.61	1.77	2.38
0700	2000 S.F.	.60	1.61	2.21
0720	5000 S.F.	.44	1.38	1.82
0740	10,000 S.F.	.39	1.27	1.66
0760	50,000 S.F.	.33	.98	1.31
1000	Ordinary hazard, one floor, 500 S.F.	1.42	2.42	3.84
1020	1000 S.F.	1.91	2.27	4.18
1040	2000 S.F.	2.17	2.49	4.66
1060	5000 S.F.	1.15	1.76	2.91
1080	10,000 S.F.	.87	1.84	2.71
1100	50,000 S.F.	.81	1.80	2.61
1140	Each additional floor, 500 S.F.	.80	2.16	2.96
1160	1000 S.F.	.58	1.75	2.33
1180	2000 S.F.	.68	1.79	2.47
1200	5000 S.F.	.70	1.67	2.37
1220	10,000 S.F.	.59	1.73	2.32
1240	50,000 S.F.	.61	1.61	2.22
1500	Extra hazard, one floor, 500 S.F.	3.66	3.74	7.40
1520	1000 S.F.	2.44	3.21	5.65
1540	2000 S.F.	2.30	3.32	5.62
1560	5000 S.F.	1.49	2.89	4.38
1580	10,000 S.F.	1.44	2.75	4.19
1600	50,000 S.F.	1.58	2.66	4.24
1660	Each additional floor, 500 S.F.	1.01	2.67	3.68
1680	1000 S.F.	.99	2.53	3.52
1700	2000 S.F.	.88	2.53	3.41
1720	5000 S.F.	.73	2.25	2.98
1740	10,000 S.F.	.90	2.08	2.98
1760	50,000 S.F.	.92	1.99	2.91
2020	Grooved steel, black sch. 40 pipe, light hazard, one floor, 2000 S.F.	2.01	2.01	4.02
2060	10,000 S.F.	.70	1.22	1.92
2100	Each additional floor, 2000 S.F.	.52	1.31	1.83
2150	10,000 S.F.	.29	1.06	1.35
2200	Ordinary hazard, one floor, 2000 S.F.	2.03	2.12	4.15
2250	10,000 S.F.	.67	1.53	2.20
2300	Each additional floor, 2000 S.F.	.54	1.42	1.96
2350	10,000 S.F.	.39	1.42	1.81
2400	Extra hazard, one floor, 2000 S.F.	2.15	2.71	4.86
2450	10,000 S.F.	.99	1.99	2.98
2500	Each additional floor, 2000 S.F.	.75	2.07	2.82
2550	10,000 S.F.	.62	1.77	2.39
3050	Grooved steel black sch. 10 pipe, light hazard, one floor, 2000 S.F.	1.99	1.99	3.98
3100	10,000 S.F.	.60	1.16	1.76
3150	Each additional floor, 2000 S.F.	.50	1.29	1.79
3200	10,000 S.F.	.32	1.05	1.37
3250	Ordinary hazard, one floor, 2000 S.F.	2.01	2.12	4.13
3300	10,000 S.F.	.69	1.49	2.18
3350	Each additional floor, 2000 S.F.	.52	1.42	1.94
3400	10,000 S.F.	.41	1.38	1.79
3450	Extra hazard, one floor, 2000 S.F.	2.14	2.70	4.84
3500	10,000 S.F.	.89	1.95	2.84
3550	Each additional floor, 2000 S.F.	.74	2.06	2.80
3600	10,000 S.F.	.57	1.75	2.32
4050	Copper tubing, type M, light hazard, one floor, 2000 S.F.	2.17	1.95	4.12
4100	10,000 S.F.	.81	1.17	1.98
4150	Each additional floor, 2000 S.F.	.69	1.28	1.97

D4010 Sprinklers

D4010 410	Wet Pipe Sprinkler Systems	COST PER S.F.		
		MAT.	INST.	TOTAL
4200	10,000 S.F.	.53	1.07	1.60
4250	Ordinary hazard, one floor, 2000 S.F.	2.26	2.21	4.47
4300	10,000 S.F.	.95	1.41	2.36
4350	Each additional floor, 2000 S.F.	.80	1.44	2.24
4400	10,000 S.F.	.65	1.28	1.93
4450	Extra hazard, one floor, 2000 S.F.	2.49	2.73	5.22
4500	10,000 S.F.	1.94	2.13	4.07
4550	Each additional floor, 2000 S.F.	1.09	2.09	3.18
4600	10,000 S.F.	1.24	1.89	3.13
5050	Copper tubing, type M, T-drill system, light hazard, one floor			
5060	2000 S.F.	2.18	1.83	4.01
5100	10,000 S.F.	.77	.97	1.74
5150	Each additional floor, 2000 S.F.	.70	1.16	1.86
5200	10,000 S.F.	.49	.87	1.36
5250	Ordinary hazard, one floor, 2000 S.F.	2.17	1.86	4.03
5300	10,000 S.F.	.91	1.24	2.15
5350	Each additional floor, 2000 S.F.	.68	1.16	1.84
5400	10,000 S.F.	.63	1.13	1.76
5450	Extra hazard, one floor, 2000 S.F.	2.27	2.23	4.50
5500	10,000 S.F.	1.56	1.55	3.11
5550	Each additional floor, 2000 S.F.	.92	1.63	2.55
5600	10,000 S.F.	.86	1.31	2.17

D4020 Standpipes

Roof

Roof connections with hose gate valves (for combustible roof)

Hose connections on each floor (size based on class of service)

Check Valve

Siamese inlet connections (for fire department use)

System Components	QUANTITY	UNIT	COST PER FLOOR		
			MAT.	INST.	TOTAL
SYSTEM D4020 310 0560					
WET STANDPIPE RISER, CLASS I, STEEL, BLACK, SCH. 40 PIPE, 10' HEIGHT					
4" DIAMETER PIPE, ONE FLOOR					
Pipe, steel, black, schedule 40, threaded, 4" diam	20.000	L.F.	313	490	803
Pipe, Tee, malleable iron, black, 150 lb threaded, 4" pipe size	2.000	Ea.	262	442	704
Pipe, 90° elbow, malleable iron, black, 150 lb threaded 4" pipe size	1.000	Ea.	83.50	147	230.50
Pipe, nipple, steel, black, schedule 40, 2-1/2" pipe size x 3" long	2.000	Ea.	18.40	111	129.40
Fire valve, gate, 300 lb, brass w/handwheel, 2-1/2" pipe size	1.000	Ea.	112	70.50	182.50
Fire valve, pressure restricting, adj, rgh brs, 2-1/2" pipe size	1.000	Ea.	218	141	359
Valve, swing check, w/ball drip, CI w/brs ftngs, 4" pipe size	1.000	Ea.	180	295	475
Standpipe conn wall dble flush brs w/plugs & chains 2-1/2"x2-1/2"x4"	1.000	Ea.	325	177	502
Valve, swing check, bronze, 125 lb, regrinding disc, 2-1/2" pipe size	1.000	Ea.	291	59	350
Roof manifold, fire, w/valves & caps, horiz/vert brs 2-1/2"x2-1/2"x4"	1.000	Ea.	113	185	298
Fire, hydrolator, vent & drain, 2-1/2" pipe size	1.000	Ea.	46	41	87
Valve, gate, iron body 125 lb, OS&Y, threaded, 4" pipe size	1.000	Ea.	410	295	705
TOTAL			2,371.90	2,453.50	4,825.40

D4020 310	Wet Standpipe Risers, Class I		COST PER FLOOR		
			MAT.	INST.	TOTAL
0550	Wet standpipe risers, Class I, steel black sch. 40, 10' height				
0560	4" diameter pipe, one floor		2,375	2,450	4,825
0580	Additional floors		575	760	1,335
0600	6" diameter pipe, one floor	D4020 -410	4,225	4,325	8,550
0620	Additional floors	R10520 -310	1,100	1,200	2,300
0640	8" diameter pipe, one floor		6,400	5,225	11,625
0660	Additional floors	R10520 -320	1,550	1,475	3,025
0680					

D4020 310	Wet Standpipe Risers, Class II	COST PER FLOOR		
		MAT.	INST.	TOTAL
1030	Wet standpipe risers, Class II, steel black sch. 40, 10' height			
1040	2" diameter pipe, one floor	945	885	1,830
1060	Additional floors	310	340	650
1080	2-1/2" diameter pipe, one floor	1,325	1,300	2,625
1100	Additional floors	355	400	755
1120				

D40 Fire Protection

D4020 Standpipes

D4020 310	Wet Standpipe Risers, Class III	COST PER FLOOR		
		MAT.	INST.	TOTAL
1530	Wet standpipe risers, Class III, steel black sch. 40, 10' height			
1540	4" diameter pipe, one floor	2,450	2,450	4,900
1560	Additional floors	485	635	1,120
1580	6" diameter pipe, one floor	4,300	4,325	8,625
1600	Additional floors	1,125	1,200	2,325
1620	8" diameter pipe, one floor	6,475	5,225	11,700
1640	Additional floors	1,575	1,475	3,050

Important: See the Reference Section for critical supporting data - Reference Numbers and City Cost Indexes

D4020   Standpipes

Roof — Roof connections with hose gate valves (for combustible roof)

Hose connections on each floor (size based on class of service)

Check Valve —

Siamese inlet connections (for fire department use)

System Components	QUANTITY	UNIT	COST PER FLOOR		
			MAT.	INST.	TOTAL
SYSTEM D4020 330 0540					
DRY STANDPIPE RISER, CLASS I, PIPE, STEEL, BLACK, SCH 40, 10' HEIGHT					
4" DIAMETER PIPE, ONE FLOOR					
Pipe, steel, black, schedule 40, threaded, 4" diam	20.000	L.F.	313	490	803
Pipe, Tee, malleable iron, black, 150 lb threaded, 4" pipe size	2.000	Ea.	262	442	704
Pipe, 90° elbow, malleable iron, black, 150 lb threaded 4" pipe size	1.000	Ea.	83.50	147	230.50
Pipe, nipple, steel, black, schedule 40, 2-1/2" pipe size x 3" long	2.000	Ea.	18.40	111	129.40
Fire valve gate NRS 300 lb, brass w/handwheel, 2-1/2" pipe size	1.000	Ea.	112	70.50	182.50
Fire valve, pressure restricting, adj, rgh brs, 2-1/2" pipe size	1.000	Ea.	109	70.50	179.50
Standpipe conn wall dble flush brs w/plugs & chains 2-1/2"x2-1/2"x4"	1.000	Ea.	325	177	502
Valve swing check w/ball drip Cl w/brs ftngs, 4"pipe size	1.000	Ea.	180	295	475
Roof manifold, fire, w/valves & caps, horiz/vert brs 2-1/2"x2-1/2"x4"	1.000	Ea.	113	185	298
TOTAL			1,515.90	1,988	3,503.90

D4020 330	Dry Standpipe Risers, Class I		COST PER FLOOR		
			MAT.	INST.	TOTAL
0530	Dry standpipe riser, Class I, steel black sch. 40, 10' height				
0540	4" diameter pipe, one floor		1,525	2,000	3,525
0560	Additional floors		525	720	1,245
0580	6" diameter pipe, one floor	D4020 -410	3,150	3,450	6,600
0600	Additional floors	R10520 -310	1,050	1,175	2,225
0620	8" diameter pipe, one floor		4,800	4,200	9,000
0640	Additional floors	R10520 -320	1,500	1,450	2,950
0660					

D4020 330	Dry Standpipe Risers, Class II		COST PER FLOOR		
			MAT.	INST.	TOTAL
1030	Dry standpipe risers, Class II, steel black sch. 40, 10' height				
1040	2" diameter pipe, one floor		790	920	1,710
1060	Additional floor		264	300	564
1080	2-1/2" diameter pipe, one floor		1,025	1,075	2,100
1100	Additional floors		310	360	670
1120					

D4020 Standpipes

D4020 330	Dry Standpipe Risers, Class III	COST PER FLOOR		
		MAT.	INST.	TOTAL
1530	Dry standpipe risers, Class III, steel black sch. 40, 10' height			
1540	4" diameter pipe, one floor	1,525	1,950	3,475
1560	Additional floors	450	650	1,100
1580	6" diameter pipe, one floor	3,175	3,450	6,625
1600	Additional floors	1,075	1,175	2,250
1620	8" diameter pipe, one floor	4,850	4,200	9,050
1640	Additional floor	1,550	1,450	3,000

D4020 Standpipes

D4020 410	Fire Hose Equipment	COST EACH		
		MAT.	INST.	TOTAL
0100	Adapters, reducing, 1 piece, FxM, hexagon, cast brass, 2-1/2″ x 1-1/2″	34.50		34.50
0200	Pin lug, 1-1/2″ x 1″	28.50		28.50
0250	3″ x 2-1/2″	43		43
0300	For polished chrome, add 75% mat.			
0400	Cabinets, D.S. glass in door, recessed, steel box, not equipped			
0500	Single extinguisher, steel door & frame	68.50	111	179.50
0550	Stainless steel door & frame	142	111	253
0600	Valve, 2-1/2″ angle, steel door & frame	76.50	74	150.50
0650	Aluminum door & frame	101	74	175
0700	Stainless steel door & frame	133	74	207
0750	Hose rack assy, 2-1/2″ x 1-1/2″ valve & 100′ hose, steel door & frame	155	148	303
0800	Aluminum door & frame	229	148	377
0850	Stainless steel door & frame	299	148	447
0900	Hose rack assy,& extinguisher,2-1/2″x1-1/2″ valve & hose,steel door & frame	169	177	346
0950	Aluminum	292	177	469
1000	Stainless steel	320	177	497
1550	Compressor, air, dry pipe system, automatic, 200 gal., 1/3 H.P.	790	380	1,170
1600	520 gal., 1 H.P.	935	380	1,315
1650	Alarm, electric pressure switch (circuit closer)	163	18.95	181.95
2500	Couplings, hose, rocker lug, cast brass, 1-1/2″	32		32
2550	2-1/2″	42.50		42.50
3000	Escutcheon plate, for angle valves, polished brass, 1-1/2″	12		12
3050	2-1/2″	29		29
3500	Fire pump, electric, w/controller, fittings, relief valve			
3550	4″ pump, 30 H.P., 500 G.P.M.	15,600	2,750	18,350
3600	5″ pump, 40 H.P., 1000 G.P.M.	22,800	3,125	25,925
3650	5″ pump, 100 H.P., 1000 G.P.M.	25,500	3,475	28,975
3700	For jockey pump system, add	2,800	445	3,245
5000	Hose, per linear foot, synthetic jacket, lined,			
5100	300 lb. test, 1-1/2″ diameter	1.78	.34	2.12
5150	2-1/2″ diameter	2.97	.40	3.37
5200	500 lb. test, 1-1/2″ diameter	1.85	.34	2.19
5250	2-1/2″ diameter	3.20	.40	3.60
5500	Nozzle, plain stream, polished brass, 1-1/2″ x 10″	29.50		29.50
5550	2-1/2″ x 15″ x 13/16″ or 1-1/2″	53		53
5600	Heavy duty combination adjustable fog and straight stream w/handle 1-1/2″	279		279
5650	2-1/2″ direct connection	420		420
6000	Rack, for 1-1/2″ diameter hose 100 ft. long, steel	36.50	44.50	81
6050	Brass	56	44.50	100.50
6500	Reel, steel, for 50 ft. long 1-1/2″ diameter hose	83	63.50	146.50
6550	For 75 ft. long 2-1/2″ diameter hose	130	63.50	193.50
7050	Siamese, w/plugs & chains, polished brass, sidewalk, 4″ x 2-1/2″ x 2-1/2″	345	355	700
7100	6″ x 2-1/2″ x 2-1/2″	580	445	1,025
7200	Wall type, flush, 4″ x 2-1/2″ x 2-1/2″	325	177	502
7250	6″ x 2-1/2″ x 2-1/2″	445	193	638
7300	Projecting, 4″ x 2-1/2″ x 2-1/2″	300	177	477
7350	6″ x 2-1/2″ x 2-1/2″	495	193	688
7400	For chrome plate, add 15% mat.			
8000	Valves, angle, wheel handle, 300 Lb., rough brass, 1-1/2″	33	41	74
8050	2-1/2″	56.50	70.50	127
8100	Combination pressure restricting, 1-1/2″	52	41	93
8150	2-1/2″	110	70.50	180.50
8200	Pressure restricting, adjustable, satin brass, 1-1/2″	77.50	41	118.50
8250	2-1/2″	109	70.50	179.50
8300	Hydrolator, vent and drain, rough brass, 1-1/2″	46	41	87
8350	2-1/2″	46	41	87
8400	Cabinet assy, incls. adapter, rack, hose, and nozzle	485	268	753

D4090 Other Fire Protection Systems

General: Automatic fire protection (suppression) systems other than water sprinklers may be desired for special environments, high risk areas, isolated locations or unusual hazards. Some typical applications would include:

Paint dip tanks
Securities vaults
Electronic data processing
Tape and data storage
Transformer rooms
Spray booths
Petroleum storage
High rack storage

Piping and wiring costs are dependent on the individual application and must be added to the component costs shown below.

All areas are assumed to be open.

D4090 910	Unit Components	COST EACH		
		MAT.	INST.	TOTAL
0020	Detectors with brackets			
0040	Fixed temperature heat detector	31	60.50	91.50
0060	Rate of temperature rise detector	37	60.50	97.50
0080	Ion detector (smoke) detector	82.50	78.50	161
0200	Extinguisher agent			
0240	200 lb FM200, container	6,375	221	6,596
0280	75 lb carbon dioxide cylinder	1,050	147	1,197
0320	Dispersion nozzle			
0340	FM200 1-1/2" dispersion nozzle	55	35	90
0380	Carbon dioxide 3" x 5" dispersion nozzle	55	27.50	82.50
0420	Control station			
0440	Single zone control station with batteries	1,400	485	1,885
0470	Multizone (4) control station with batteries	2,700	970	3,670
0490				
0500	Electric mechanical release	138	244	382
0520				
0550	Manual pull station	49.50	82	131.50
0570				
0640	Battery standby power 10" x 10" x 17"	760	121	881
0700				
0740	Bell signalling device	54.50	60.50	115

D4090 920	FM200 Systems	COST PER C.F.		
		MAT.	INST.	TOTAL
0820	Average FM200 system, minimum			1.38
0840	Maximum			2.75

G BUILDING SITEWORK

G1030 Site Earthwork

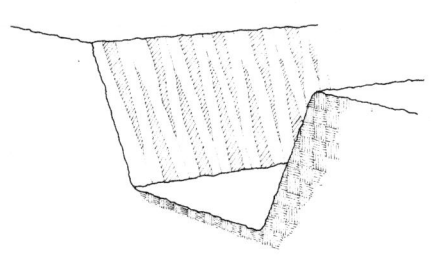

Trenching Systems are shown on a cost per linear foot basis. The systems include: excavation; backfill and removal of spoil; and compaction for various depths and trench bottom widths. The backfill has been reduced to accommodate a pipe of suitable diameter and bedding.

The slope for trench sides varies from 0:1 to 2:1.

The Expanded System Listing shows Trenching Systems that range from 2' to 12' in width. Depths range from 2' to 25'.

System Components				COST PER L.F.		
		QUANTITY	UNIT	EQUIP.	LABOR	TOTAL
SYSTEM G1030 805 1310						
TRENCHING, BACKHOE, 0 TO 1 SLOPE, 2' WIDE, 2' DP, 3/8 C.Y. BUCKET						
Excavation, trench, hyd. backhoe, track mtd., 3/8 C.Y. bucket		.174	C.Y.	.27	.87	1.14
Backfill and load spoil, from stockpile		.174	C.Y.	.09	.25	.34
Compaction by rammer tamper, 8" lifts, 4 passes		.014	C.Y.		.04	.04
Remove excess spoil, 6 C.Y. dump truck, 2 mile roundtrip		.160	C.Y.	.57	.54	1.11
	TOTAL			.93	1.70	2.63

G1030 805	Trenching	COST PER L.F.		
		EQUIP.	LABOR	TOTAL
1310	Trenching, backhoe, 0 to 1 slope, 2' wide, 2' deep, 3/8 C.Y. bucket	.93	1.70	2.63
1320	3' deep, 3/8 C.Y. bucket	1.15	2.49	3.64
1330	4' deep, 3/8 C.Y. bucket	1.36	3.28	4.64
1340	6' deep, 3/8 C.Y. bucket	1.69	4.18	5.87
1350	8' deep, 1/2 C.Y. bucket	1.98	5.30	7.28
1360	10' deep, 1 C.Y. bucket	3.01	6.30	9.31
1400	4' wide, 2' deep, 3/8 C.Y. bucket	1.98	3.42	5.40
1410	3' deep, 3/8 C.Y. bucket	2.80	5.10	7.90
1420	4' deep, 1/2 C.Y. bucket	2.79	5.10	7.89
1430	6' deep, 1/2 C.Y. bucket	3.56	7.75	11.31
1440	8' deep, 1/2 C.Y. bucket	5.60	10	15.60
1450	10' deep, 1 C.Y. bucket	6.25	12.40	18.65
1460	12' deep, 1 C.Y. bucket	7.95	15.90	23.85
1470	15' deep, 1-1/2 C.Y. bucket	6.75	13.70	20.45
1480	18' deep, 2-1/2 C.Y. bucket	9.30	19.50	28.80
1520	6' wide, 6' deep, 5/8 C.Y. bucket	7.25	10.75	18
1530	8' deep, 3/4 C.Y. bucket	8.85	12.90	21.75
1540	10' deep, 1 C.Y. bucket	9.50	15.50	25
1550	12' deep, 1-1/4 C.Y. bucket	8.35	12.30	20.65
1560	16' deep, 2 C.Y. bucket	10.95	14.05	25
1570	20' deep, 3-1/2 C.Y. bucket	14.90	25	39.90
1580	24' deep, 3-1/2 C.Y. bucket	24	41.50	65.50
1640	8' wide, 12' deep, 1-1/4 C.Y. bucket	12.40	16.80	29.20
1650	15' deep, 1-1/2 C.Y. bucket	14.60	20.50	35.10
1660	18' deep, 2-1/2 C.Y. bucket	17	21	38
1680	24' deep, 3-1/2 C.Y. bucket	33	56	89
1730	10' wide, 20' deep, 3-1/2 C.Y. bucket	27	42	69
1740	24' deep, 3-1/2 C.Y. bucket	42	70	112
1780	12' wide, 20' deep, 3-1/2 C.Y. bucket	32.50	50.50	83
1790	25' deep, bucket	56	93.50	149.50
1800	1/2 to 1 slope, 2' wide, 2' deep, 3/8 C.Y. bucket	1.27	2.18	3.45
1810	3' deep, 3/8 C.Y. bucket	1.75	3.60	5.35
1820	4' deep, 3/8 C.Y. bucket	2.22	5.25	7.47
1840	6' deep, 3/8 C.Y. bucket	3.23	8.10	11.33

G1030 Site Earthwork

G1030 805	Trenching	COST PER L.F.		
		EQUIP.	LABOR	TOTAL
1860	8' deep, 1/2 C.Y. bucket	4.49	12.35	16.84
1880	10' deep, 1 C.Y. bucket	8.35	17	25.35
2300	4' wide, 2' deep, 3/8 C.Y. bucket	2.11	3.63	5.74
2310	3' deep, 3/8 C.Y. bucket	3.48	6	9.48
2320	4' deep, 1/2 C.Y. bucket	3.66	6.25	9.91
2340	6' deep, 1/2 C.Y. bucket	5.10	10.85	15.95
2360	8' deep, 1/2 C.Y. bucket	8.65	14.70	23.35
2380	10' deep, 1 C.Y. bucket	11.35	21.50	32.85
2400	12' deep, 1 C.Y. bucket	11.95	23.50	35.45
2430	15' deep, 1-1/2 C.Y. bucket	15.20	29	44.20
2460	18' deep, 2-1/2 C.Y. bucket	21	35	56
2840	6' wide, 6' deep, 5/8 C.Y. bucket	8.60	11.85	20.45
2860	8' deep, 3/4 C.Y. bucket	12.25	17.75	30
2880	10' deep, 1 C.Y. bucket	14.30	23.50	37.80
2900	12' deep, 1-1/4 C.Y. bucket	13.45	19.55	33
2940	16' deep, 2 C.Y. bucket	21	31	52
2980	20' deep, 3-1/2 C.Y. bucket	32.50	53.50	86
3020	24' deep, 3-1/2 C.Y. bucket	58.50	100	158.50
3100	8' wide, 12' deep, 1-1/4 C.Y. bucket	18.05	24	42.05
3120	15' deep, 1-1/2 C.Y. bucket	23	32	55
3140	18' deep, 2-1/2 C.Y. bucket	29	35	64
3180	24' deep, 3-1/2 C.Y. bucket	67.50	112	179.50
3270	10' wide, 20' deep, 3-1/2 C.Y. bucket	42	67	109
3280	24' deep, 3-1/2 C.Y. bucket	77	126	203
3370	12' wide, 20' deep, 3-1/2 C.Y. bucket	47.50	74	121.50
3380	25' deep, 3-1/2 C.Y. bucket	90	147	237
3500	1 to 1 slope, 2' wide, 2' deep, 3/8 C.Y. bucket	1.48	2.55	4.03
3520	3' deep, 3/8 C.Y. bucket	2.27	4.46	6.73
3540	4' deep, 3/8 C.Y. bucket	2.92	6.70	9.62
3560	6' deep, 3/8 C.Y. bucket	4.40	10.70	15.10
3580	8' deep, 1/2 C.Y. bucket	6.35	16.85	23.20
3600	10' deep, 1 C.Y. bucket	13.90	27	40.90
3800	4' wide, 2' deep, 3/8 C.Y. bucket	2.52	4.35	6.87
3820	3' deep, 3/8 C.Y. bucket	4.44	7.65	12.09
3840	4' deep, 1/2 C.Y. bucket	4.45	7	11.45
3860	6' deep, 1/2 C.Y. bucket	6.35	13	19.35
3880	8' deep, 1/2 C.Y. bucket	11.65	19.40	31.05
3900	10' deep, 1 C.Y. bucket	15.15	28	43.15
3920	12' deep, 1 C.Y. bucket	22	40.50	62.50
3940	15' deep, 1-1/2 C.Y. bucket	21.50	38	59.50
3960	18' deep, 2-1/2 C.Y. bucket	30	45.50	75.50
4030	6' wide, 6' deep, 5/8 C.Y. bucket	10.45	14.40	24.85
4040	8' deep, 3/4 C.Y. bucket	15.30	23	38.30
4050	10' deep, 1 C.Y. bucket	18.45	32	50.45
4060	12' deep, 1-1/4 C.Y. bucket	17.95	29	46.95
4070	16' deep, 2 C.Y. bucket	28.50	41	69.50
4080	20' deep, 3-1/2 C.Y. bucket	46.50	85	131.50
4090	24' deep, 3-1/2 C.Y. bucket	87	165	252
4500	8' wide, 12' deep, 1-1/4 C.Y. bucket	23	32.50	55.50
4550	15' deep, 1-1/2 C.Y. bucket	30.50	46.50	77
4600	18' deep, 2-1/2 C.Y. bucket	39.50	54	93.50
4650	24' deep, 3-1/2 C.Y. bucket	96	176	272
4800	10' wide, 20' deep, 3-1/2 C.Y. bucket	58.50	98.50	157
4850	24' deep, 3-1/2 C.Y. bucket	104	187	291
4950	12' wide, 20' deep, 3-1/2 C.Y. bucket	63.50	105	168.50
4980	25' deep, 3-1/2 C.Y. bucket	126	227	353
5000	1-1/2 to 1 slope, 2' wide, 2' deep, 3/8 C.Y. bucket	1.71	2.95	4.66
5020	3' deep, 3/8 C.Y. bucket	2.77	5.25	8.02

G1030 Site Earthwork

G1030 805	Trenching	COST PER L.F.		
		EQUIP.	LABOR	TOTAL
5040	4' deep, 3/8 C.Y. bucket	3.58	7.85	11.43
5060	6' deep, 3/8 C.Y. bucket	5.45	12.50	17.95
5080	8' deep, 1/2 C.Y. bucket	7.90	19.80	27.70
5100	10' deep, 1 C.Y. bucket	15.55	28	43.55
5300	4' wide, 2' deep, 3/8 C.Y. bucket	2.39	4.10	6.49
5320	3' deep, 3/8 C.Y. bucket	4.32	7.45	11.77
5340	4' deep, 1/2 C.Y. bucket	5.30	8.20	13.50
5360	6' deep, 1/2 C.Y. bucket	7.55	14.75	22.30
5380	8' deep, 1/2 C.Y. bucket	14	22	36
5400	10' deep, 1 C.Y. bucket	18.40	31.50	49.90
5420	12' deep, 1 C.Y. bucket	27	47	74
5450	15' deep, 1-1/2 C.Y. bucket	26.50	42	68.50
5480	18' deep, 2-1/2 C.Y. bucket	37.50	49.50	87
5660	6' wide, 6' deep, 5/8 C.Y. bucket	12.10	16.10	28.20
5680	8' deep, 3/4 C.Y. bucket	17.90	25.50	43.40
5700	10' deep, 1 C.Y. bucket	22	35.50	57.50
5720	12' deep, 1-1/4 C.Y. bucket	21	30.50	51.50
5760	16' deep, 2 C.Y. bucket	34.50	43.50	78
5800	20' deep, 3-1/2 C.Y. bucket	57	96	153
5840	24' deep, 3-1/2 C.Y. bucket	109	187	296
6020	8' wide, 12' deep, 1-1/4 C.Y. bucket	27.50	36	63.50
6050	15' deep, 1-1/2 C.Y. bucket	36	51	87
6080	18' deep, 2-1/2 C.Y. bucket	48	58	106
6140	24' deep, 3-1/2 C.Y. bucket	118	198	316
6300	10' wide, 20' deep, 3-1/2 C.Y. bucket	70	109	179
6350	24' deep, 3-1/2 C.Y. bucket	125	207	332
6450	12' wide, 20' deep, 3-1/2 C.Y. bucket	76	116	192
6480	25' deep, 3-1/2 C.Y. bucket	152	250	402
6600	2 to 1 slope, 2' wide, 2' deep, 3/8 C.Y. bucket	2.70	2.68	5.38
6620	3' deep, 3/8 C.Y. bucket	3.19	5.85	9.04
6640	4' deep, 3/8 C.Y. bucket	4.10	8.75	12.85
6660	6' deep, 3/8 C.Y. bucket	5.85	13.25	19.10
6680	8' deep, 1/2 C.Y. bucket	9	22.50	31.50
6700	10' deep, 1 C.Y. bucket	17.65	31.50	49.15
6900	4' wide, 2' deep, 3/8 C.Y. bucket	2.38	4.08	6.46
6920	3' deep, 3/8 C.Y. bucket	4.44	7.65	12.09
6940	4' deep, 1/2 C.Y. bucket	5.95	8.75	14.70
6960	6' deep, 1/2 C.Y. bucket	8.45	16	24.45
6980	8' deep, 1/2 C.Y. bucket	15.55	24	39.55
7000	10' deep, 1 C.Y. bucket	20.50	35	55.50
7050	15' deep, 1-1/2 C.Y. bucket	29.50	47	76.50
7080	18' deep, 2-1/2 C.Y. bucket	42.50	56	98.50
7260	6' wide, 6' deep, 5/8 C.Y. bucket	13.35	17.20	30.55
7280	8' deep, 3/4 C.Y. bucket	19.70	27.50	47.20
7300	10' deep, 3/4 C.Y. bucket	26	39.50	65.50
7360	16' deep, 2 C.Y. bucket	38.50	48.50	87
7400	20' deep, 3-1/2 C.Y. bucket	64	107	171
7440	24' deep, 3-1/2 C.Y. bucket	122	210	332
7620	8' wide, 12' deep, 1-1/4 C.Y. bucket	30.50	39	69.50
7650	15' deep, 1-1/2 C.Y. bucket	40.50	56	96.50
7680	18' deep, 2-1/2 C.Y. bucket	53	64	117
7740	24' deep, 3-1/2 C.Y. bucket	132	220	352
7920	10' wide, 20' deep, 3-1/2 C.Y. bucket	77	120	197
7940	24' deep, 3-1/2 C.Y. bucket	139	229	368
8060	12' wide, 20' deep, 3-1/2 C.Y. bucket	84	126	210
8080	25' deep, 3-1/2 C.Y. bucket	168	274	442

Important: See the Reference Section for critical supporting data - Reference Numbers and City Cost Indexes

G1030 Site Earthwork

The Pipe Bedding System is shown for various pipe diameters. Compacted bank sand is used for pipe bedding and to fill 12″ over the pipe. No backfill is included. Various side slopes are shown to accommodate different soil conditions. Pipe sizes vary from 6″ to 84″ diameter.

System Components	QUANTITY	UNIT	COST PER L.F. MAT.	COST PER L.F. INST.	COST PER L.F. TOTAL
SYSTEM G1030 815 1440					
PIPE BEDDING, SIDE SLOPE 0 TO 1, 1′ WIDE, PIPE SIZE 6″ DIAMETER					
Borrow, bank sand, 2 mile haul, machine spread	.067	C.Y.	.30	.37	.67
Compaction, vibrating plate	.067	C.Y.		.12	.12
TOTAL			.30	.49	.79

G1030 815	Pipe Bedding	COST PER L.F. MAT.	COST PER L.F. INST.	COST PER L.F. TOTAL
1440	Pipe bedding, side slope 0 to 1, 1′ wide, pipe size 6″ diameter	.30	.49	.79
1460	2′ wide, pipe size 8″ diameter	.67	1.09	1.76
1480	Pipe size 10″ diameter	.69	1.11	1.80
1500	Pipe size 12″ diameter	.71	1.15	1.86
1520	3′ wide, pipe size 14″ diameter	1.17	1.89	3.06
1540	Pipe size 15″ diameter	1.18	1.90	3.08
1560	Pipe size 16″ diameter	1.19	1.93	3.12
1580	Pipe size 18″ diameter	1.22	1.97	3.19
1600	4′ wide, pipe size 20″ diameter	1.76	2.83	4.59
1620	Pipe size 21″ diameter	1.78	2.86	4.64
1640	Pipe size 24″ diameter	1.83	2.95	4.78
1660	Pipe size 30″ diameter	1.87	3.01	4.88
1680	6′ wide, pipe size 32″ diameter	3.26	5.25	8.51
1700	Pipe size 36″ diameter	3.35	5.40	8.75
1720	7′ wide, pipe size 48″ diameter	4.36	7.05	11.41
1740	8′ wide, pipe size 60″ diameter	5.45	8.80	14.25
1760	10′ wide, pipe size 72″ diameter	7.85	12.70	20.55
1780	12′ wide, pipe size 84″ diameter	10.70	17.25	27.95
2140	Side slope 1/2 to 1, 1′ wide, pipe size 6″ diameter	.64	1.02	1.66
2160	2′ wide, pipe size 8″ diameter	1.06	1.71	2.77
2180	Pipe size 10″ diameter	1.15	1.86	3.01
2200	Pipe size 12″ diameter	1.23	1.99	3.22
2220	3′ wide, pipe size 14″ diameter	1.77	2.84	4.61
2240	Pipe size 15″ diameter	1.81	2.92	4.73
2260	Pipe size 16″ diameter	1.87	3.01	4.88
2280	Pipe size 18″ diameter	1.97	3.18	5.15
2300	4′ wide, pipe size 20″ diameter	2.61	4.20	6.81
2320	Pipe size 21″ diameter	2.67	4.31	6.98
2340	Pipe size 24″ diameter	2.86	4.62	7.48
2360	Pipe size 30″ diameter	3.21	5.20	8.41
2380	6′ wide, pipe size 32″ diameter	4.73	7.65	12.38
2400	Pipe size 36″ diameter	5.05	8.15	13.20
2420	7′ wide, pipe size 48″ diameter	6.90	11.15	18.05
2440	8′ wide, pipe size 60″ diameter	9	14.50	23.50
2460	10′ wide, pipe size 72″ diameter	12.60	20.50	33.10
2480	12′ wide, pipe size 84″ diameter	16.75	27	43.75
2620	Side slope 1 to 1, 1′ wide, pipe size 6″ diameter	.98	1.58	2.56
2640	2′ wide, pipe size 8″ diameter	1.47	2.38	3.85

G1030 Site Earthwork

G1030 815	Pipe Bedding	COST PER L.F.		
		MAT.	INST.	TOTAL
2660	Pipe size 10" diameter	1.61	2.59	4.20
2680	Pipe size 12" diameter	1.77	2.84	4.61
2700	3' wide, pipe size 14" diameter	2.37	3.81	6.18
2720	Pipe size 15" diameter	2.45	3.95	6.40
2740	Pipe size 16" diameter	2.54	4.10	6.64
2760	Pipe size 18" diameter	2.73	4.40	7.13
2780	4' wide, pipe size 20" diameter	3.45	5.55	9
2800	Pipe size 21" diameter	3.55	5.75	9.30
2820	Pipe size 24" diameter	3.89	6.25	10.14
2840	Pipe size 30" diameter	4.54	7.30	11.84
2860	6' wide, pipe size 32" diameter	6.20	9.95	16.15
2880	Pipe size 36" diameter	6.75	10.90	17.65
2900	7' wide, pipe size 48" diameter	9.45	15.25	24.70
2920	8' wide, pipe size 60" diameter	12.55	20.50	33.05
2940	10' wide, pipe size 72" diameter	17.30	28	45.30
2960	12' wide, pipe size 84" diameter	23	37	60
3000	Side slope 1-1/2 to 1, 1' wide, pipe size 6" diameter	1.31	2.11	3.42
3020	2' wide, pipe size 8" diameter	1.86	3.01	4.87
3040	Pipe size 10" diameter	2.06	3.32	5.38
3060	Pipe size 12" diameter	2.29	3.69	5.98
3080	3' wide, pipe size 14" diameter	2.96	4.77	7.73
3100	Pipe size 15" diameter	3.08	4.98	8.06
3120	Pipe size 16" diameter	3.21	5.20	8.41
3140	Pipe size 18" diameter	3.49	5.65	9.14
3160	4' wide, pipe size 20" diameter	4.30	6.95	11.25
3180	Pipe size 21" diameter	4.44	7.15	11.59
3200	Pipe size 24" diameter	4.92	7.90	12.82
3220	Pipe size 30" diameter	5.90	9.50	15.40
3240	6' wide, pipe size 32" diameter	7.65	12.30	19.95
3260	Pipe size 36" diameter	8.45	13.65	22.10
3280	7' wide, pipe size 48" diameter	12	19.35	31.35
3300	8' wide, pipe size 60" diameter	16.10	26	42.10
3320	10' wide, pipe size 72" diameter	22	35.50	57.50
3340	12' wide, pipe size 84" diameter	29	46.50	75.50
3400	Side slope 2 to 1, 1' wide, pipe size 6" diameter	1.68	2.70	4.38
3420	2' wide, pipe size 8" diameter	2.26	3.64	5.90
3440	Pipe size 10" diameter	2.52	4.06	6.58
3460	Pipe size 12" diameter	2.81	4.53	7.34
3480	3' wide, pipe size 14" diameter	3.55	5.75	9.30
3500	Pipe size 15" diameter	3.70	5.95	9.65
3520	Pipe size 16" diameter	3.88	6.25	10.13
3540	Pipe size 18" diameter	4.24	6.85	11.09
3560	4' wide, pipe size 20" diameter	5.15	8.30	13.45
3580	Pipe size 21" diameter	5.35	8.60	13.95
3600	Pipe size 24" diameter	5.95	9.60	15.55
3620	Pipe size 30" diameter	7.25	11.70	18.95
3640	6' wide, pipe size 32" diameter	9.10	14.65	23.75
3660	Pipe size 36" diameter	10.15	16.40	26.55
3680	7' wide, pipe size 48" diameter	14.55	23.50	38.05
3700	8' wide, pipe size 60" diameter	19.65	31.50	51.15
3720	10' wide, pipe size 72" diameter	27	43	70
3740	12' wide, pipe size 84" diameter	35	56.50	91.50

Important: See the Reference Section for critical supporting data - Reference Numbers and City Cost Indexes

G2040 Site Development

Chain Link Fence

Concrete Sidewalk

Pool Equipment (Building not included in price)

12' x 30' Swimming Pool

Fence Gate

The Swimming Pool System is a complete package. Everything from excavation to deck hardware is included in system costs. Below are three basic types of pool systems: residential, motel, and municipal. Systems elements include: excavation, pool materials, installation, deck hardware, pumps and filters, sidewalk, and fencing.

The Expanded System Listing shows three basic types of pools with a variety of finishes and basic materials. These systems are either vinyl lined with metal sides; gunite shell with a cement plaster finish or tile finish; or concrete sided with vinyl lining. Pool sizes listed here vary from 12' x 30' to 60' x 82.5'. All costs are on a per unit basis.

System Components				COST EACH		
	QUANTITY	UNIT		MAT.	INST.	TOTAL
SYSTEM G2040 920 1000						
SWIMMING POOL, RESIDENTIAL, CONC. SIDES, VINYL LINED, 12'X 30'						
Swimming pool, residential, in ground including equipment	360.000	S.F.		4,194	3,776.40	7,970.40
4" thick reinforced concrete sidewalk, broom finish, no base	400.000	S.F.		596	764	1,360
Chain link fence, residential, 3' high	124.000	L.F.		488.56	281.48	770.04
Fence gate, chain link	1.000	Ea.		59	94.75	153.75
TOTAL				5,337.56	4,916.63	10,254.19

G2040 920	Swimming Pools	COST EACH		
		MAT.	INST.	TOTAL
1000	Swimming pool, residential class, concrete sides, vinyl lined, 12' x 30'	5,350	4,925	10,275
1100	16' x 32'	6,050	5,575	11,625
1200	20' x 40'	7,750	7,100	14,850
1500	Tile finish, 12' x 30'	11,500	21,100	32,600
1600	16' x 32'	12,100	22,100	34,200
1700	20' x 40'	17,700	31,200	48,900
2000	Metal sides, vinyl lined, 12' x 30'	4,650	3,175	7,825
2100	16' x 32'	5,250	3,600	8,850
2200	20' x 40'	6,725	4,550	11,275
3000	Gunite shell, cement plaster finish, 12' x 30'	8,100	8,975	17,075
3100	16' x 32'	10,200	11,700	21,900
3200	20' x 40'	13,900	11,700	25,600
4000	Motel class, concrete sides, vinyl lined, 20' x 40'	10,800	9,875	20,675
4100	28' x 60'	15,200	13,900	29,100
4500	Tile finish, 20' x 40'	20,300	36,100	56,400
4600	28' x 60'	33,000	58,500	91,500
5000	Metal sides, vinyl lined, 20' x 40'	11,200	7,175	18,375
5100	28' x 60'	15,100	9,650	24,750
6000	Gunite shell, cement plaster finish, 20' x 40'	25,500	28,700	54,200
6100	28' x 60'	38,100	42,900	81,000
7000	Municipal class, gunite shell, cement plaster finish, 42' x 75'	100,500	101,000	201,500
7100	60' x 82.5'	128,500	130,000	258,500
7500	Concrete walls, tile finish, 42' x 75'	120,500	136,500	257,000
7600	60' x 82.5'	158,000	181,000	339,000
7700	Tile finish and concrete gutter, 42' x 75'	133,000	136,500	269,500
7800	60' x 82.5'	173,500	181,000	354,500

G20 Site Improvements

G2040 Site Development

G2040 920	Swimming Pools	COST EACH		
		MAT.	INST.	TOTAL
7900	Tile finish and stainless gutter, 42' x 75'	158,000	136,500	294,500
8000	60'x 82.5'	231,000	209,500	440,500

Important: See the Reference Section for critical supporting data - Reference Numbers and City Cost Indexes

G2050 Landscaping

There are three basic types of Site Irrigation Systems: pop-up, riser mounted and quick coupling. Sprinkler heads are spray, impact or gear driven. Each system includes: the hardware for spraying the water; the pipe and fittings needed to deliver the water; and all other accessory equipment such as valves, couplings, nipples, and nozzles. Excavation heads and backfill costs are also included in the system.

The Expanded System Listing shows a wide variety of Site Irrigation Systems.

System Components	QUANTITY	UNIT	COST PER S.F. MAT.	COST PER S.F. INST.	COST PER S.F. TOTAL
SYSTEM G2050 710 1000					
SITE IRRIGATION, POP UP SPRAY, PLASTIC, 10' RADIUS, 1000 S.F., PVC PIPE					
Excavation, chain trencher	54.000	L.F.		30.78	30.78
Pipe, fittings & nipples, PVC Schedule 40, 1" dia	5.000	Ea.	32.21	1,056.83	1,089.04
Tees, PVC plastic, high pressure schedule, 40, 1" diameter	5.000	Ea.	20.30	295	315.30
Couplings PVC plastic, high pressure, 1" diameter	5.000	Ea.	2.40	98.25	100.65
Valves, bronze, globe, 125 lb. rising stem, threaded, 1" diameter	1.000	Ea.	91.50	26	117.50
Head & nozzle, pop-up spray, PVC plastic	5.000	Ea.	18.70	72.25	90.95
Backfill by hand with compaction	54.000	L.F.		25.38	25.38
Total cost per 1,000 S.F.			165.11	1,604.49	1,769.60
Total cost per S.F.			.17	1.60	1.77

G2050 710	Site Irrigation	COST PER S.F. MAT.	COST PER S.F. INST.	COST PER S.F. TOTAL
1000	Site irrigation, pop up spray, 10' radius, 1000 S.F., PVC pipe	.17	1.60	1.77
1100	Polyethylene pipe	.16	.57	.73
1200	14' radius, 8000 S.F., PVC pipe	.05	.29	.34
1300	Polyethylene pipe	.05	.26	.31
1400	18' square, 1000 S.F. PVC pipe	.14	.46	.60
1500	Polyethylene pipe	.14	.38	.52
1600	24' square, 8000 S.F., PVC pipe	.04	.18	.22
1700	Polyethylene pipe	.04	.17	.21
1800	4' x 30' strip, 200 S.F., PVC pipe	.67	1.90	2.57
1900	Polyethylene pipe	.65	1.71	2.36
2000	6' x 40' strip, 800 S.F., PVC pipe	.18	.63	.81
2100	Polyethylene pipe	.18	.56	.74
2200	Economy brass, 11' radius, 1000 S.F., PVC pipe	.18	.65	.83
2300	Polyethylene pipe	.18	.57	.75
2400	14' radius, 8000 S.F., PVC pipe	.06	.29	.35
2500	Polyethylene pipe	.06	.26	.32
2600	3' x 28' strip, 200 S.F., PVC pipe	.73	1.90	2.63
2700	Polyethylene pipe	.71	1.71	2.42
2800	7' x 36' strip, 1000 S.F., PVC pipe	.16	.52	.68
2900	Polyethylene pipe	.15	.46	.61
3000	Hd brass, 11' radius, 1000 S.F., PVC pipe	.21	.65	.86
3100	Polyethylene pipe	.21	.57	.78
3200	14' radius, 8000 S.F., PVC pipe	.07	.29	.36
3300	Polyethylene pipe	.07	.26	.33
3400	Riser mounted spray, plastic, 10'radius, 1000S.F., PVC pipe	.15	.65	.80
3500	Polyethylene pipe	.15	.57	.72

G2050 710	Site Irrigation	COST PER S.F.		
		MAT.	INST.	TOTAL
3600	12' radius, 5000 S.F., PVC pipe	.08	.38	.46
3700	Polyethylene pipe	.07	.34	.41
3800	19' square, 2000 S.F., PVC pipe	.07	.21	.28
3900	Polyethylene pipe	.07	.18	.25
4000	24' square, 8000 S.F., PVC pipe	.03	.18	.21
4100	Polyethylene pipe	.03	.17	.20
4200	5' x 32' strip, 300 S.F., PVC pipe	.43	1.27	1.70
4300	Polyethylene pipe	.42	1.14	1.56
4400	6' x 40' strip, 800 S.F., PVC pipe	.17	.63	.80
4500	Polyethylene pipe	.17	.56	.73
4600	Brass, 11' radius, 1000 S.F., PVC pipe	.16	.65	.81
4700	Polyethylene pipe	.16	.57	.73
4800	14' radius, 8000 S.F., PVC pipe	.05	.29	.34
4900	Polythylene pipe	.05	.26	.31
5000	Pop up gear drive stream type,plastic, 30'radius, 10,000 S.F.,PVC pipe	.08	.42	.50
5020	Polyethylene pipe	.07	.12	.19
5040	40,000 S.F., PVC pipe	.07	.41	.48
5060	Polyethylene pipe	.06	.11	.17
5080	Riser mounted gear drive stream type,plas.,30'rad.,10,000S.F.,PVC pipe	.08	.42	.50
5100	Polyethylene pipe	.07	.12	.19
5120	40,000 S.F., PVC pipe	.07	.41	.48
5140	Polyethylene pipe	.06	.11	.17
5200	Q.C. valve thread type w/impact head,brass,75'rad,20,000S.F.,PVC pipe	.04	.15	.19
5250	Polyethylene pipe	.03	.03	.06
5300	100,000 S.F., PVC pipe	.06	.26	.32
5350	Polyethylene pipe	.04	.06	.10
5400	Q.C. valve lug type w/impact head,brass, 75'rad, 20,000 S.F.,PVC pipe	.04	.15	.19
5450	Polyethylene pipe	.03	.03	.06
5500	100,000 S.F. PVC pipe	.06	.26	.32
5550	Polyethylene pipe	.04	.06	.10
6000	Site irrigation,pop up impact type,plastic,40'rad.,10000 S.F.,PVC pipe	.05	.27	.32
6100	Polyethylene pipe	.03	.07	.10
6200	45,000 S.F., PVC pipe	.05	.30	.35
6300	Polyethylene pipe	.03	.08	.11
6400	High medium volume brass, 40' radius, 10000 S.F., PVC pipe	.05	.27	.32
6500	Polyethylene pipe	.03	.07	.10
6600	45,000 S.F. PVC pipe	.05	.30	.35
6700	Polyethylene pipe	.03	.08	.11
6800	High volume brass, 60' radius, 25,000 S.F., PVC pipe	.05	.21	.26
6900	Polyethylene pipe	.04	.06	.10
7000	100,000 S.F., PVC pipe	.04	.21	.25
7100	Polyethylene pipe	.03	.05	.08
7200	Riser mounted part/full impact type,plas.,40'rad.,10,000 S.F.,PVC pipe	.04	.28	.32
7300	Polyethylene pipe	.03	.15	.18
7400	45,000 S.F., PVC pipe	.04	.31	.35
7500	Polyethylene	.02	.09	.11
7600	Low medium volume brass, 40' radius, 10,000 S.F., PVC pipe	.05	.28	.33
7700	Polyethylene pipe	.03	.08	.11
7800	45,000 S.F., PVC pipe	.05	.31	.36
7900	Polyethylene pipe	.03	.09	.12
8000	Medium volume brass, 50' radius, 30,000 S.F., PVC pipe	.04	.22	.26
8100	Polyethylene pipe	.03	.06	.09
8200	70,000 S.F., PVC pipe	.06	.36	.42
8300	Polyethylene pipe	.05	.08	.13
8400	Riser mounted full only impact type,plas.,40'rad.,10000 S.F., PVC pipe	.04	.28	.32
8500	Polyethylene pipe	.02	.08	.10
8600	45,000 S.F., PVC pipe	.04	.31	.35
8700	Polyethylene pipe	.02	.09	.11

Important: See the Reference Section for critical supporting data - Reference Numbers and City Cost Indexes

G2050 Landscaping

G2050 710	Site Irrigation	COST PER S.F.		
		MAT.	INST.	TOTAL
8800	Low medium volume brass, 40' radius, 10,000 S.F., PVC pipe	.05	.28	.33
8900	Polyethylene pipe	.03	.08	.11
9000	45,000 S.F., PVC pipe	.04	.31	.35
9100	Polyethylene pipe	.02	.09	.11
9200	High volume brass, 80' radius, 75,000 S.F., PVC pipe	.04	.14	.18
9300	Polyethylene pipe	.01	.02	.03
9400	150,000 S.F., PVC pipe	.03	.13	.16
9500	Polyethylene pipe	.01	.02	.03
9600	Very high volume brass, 100' radius, 100,000 S.F., PVC pipe	.03	.11	.14
9700	Polyethylene pipe	.01	.02	.03
9800	200,000 S.F., PVC pipe	.03	.11	.14
9900	Polyethylene pipe	.01	.02	.03

G2050 Landscaping

- Valve
- Backflow Preventer
- Pressure Regulator
- Filter
- Tubing Adapter
- Drip Tubing

Drip irrigation systems can be for either lawns or individual plants. Shown below are drip irrigation systems for lawns. These drip type systems are most cost effective where there are water restrictions in use or the cost of water is very high. The supply uses PVC piping with the laterals polyethylene drip lines.

System Components	QUANTITY	UNIT	COST PER S.F. MAT.	COST PER S.F. INST.	COST PER S.F. TOTAL
SYSTEM G2050 720 1000					
SITE IRRIGATION, DRIP SYSTEM 800 SF (20' X40')					
Excavte supply and header lines 4" x18" deep	50.000	L.F.		30	30
Excavate drip lines 4" x 4" deep	5.000	L.F.		509.60	509.60
Install supply manifold (stop valve, strainer, PRV, adapters)	1.000	Ea.	43.50	108	151.50
Install 3/4" PVC supply line and risers to 4 " drip level	75.000	L.F.	15	108.75	123.75
Install 3/4" PVC fittings for supply line	33.000	Ea.	13.20	358.05	371.25
Install PVC to drip line adapters	14.000	Ea.	30.52	151.90	182.42
Install 1/4 inch drip lines with hose clamps	560.000	L.F.	61.60	201.60	263.20
Backfill trenches	3.180	C.Y.		95.40	95.40
Cleanup area when completed	1.000	Ea.		83	83
Total cost per 800 S.F.			163.82	1,646.30	1,810.12
Total cost per S.F.			.20	2.06	2.26

G2050 720	Site Irrigation	COST PER S.F. MAT.	COST PER S.F. INST.	COST PER S.F. TOTAL
1000	Site irrigation, drip lawn watering system 20' x 40' area (800 SF)	.21	2.06	2.27
1100	Site irrigation, drip lawn watering system (20' x50') (1000 SF)	.18	1.84	2.02
1200	Site irrigation, drip lawn watering system (2 zone) 1600 SF (40' x40')	.23	1.97	2.20
1300	Site irrigation, drip lawn watering system (2 zone) 2000 SF (40' x 50')	.19	1.76	1.95
1400	Site irrigation, drip lawn watering system (3 zone) 2400 SF (60' x40')	.21	1.96	2.17
1500	Site irrigation, drip lawn watering system (3 zone) 3000 SF (60' x 50')	.17	1.76	1.93
1600	Site irrigation, drip lawn watering system (4 zone) 3200 SF (2 x 40' x 40')	.21	1.92	2.13
1700	Site irrigation, drip lawn watering system (4 zone) 4000 SF (2 x 40' x 50')	.18	1.73	1.91
1800	Site irri, drip lawn watering sys (6 zone) 4800 SF (2 x 60' x 40') w/contrl	.21	1.92	2.13
1900	Site irri, drip lawn watering sys (6 zone) 6000 SF (2 x 60' x 50') w/contrl	.18	1.74	1.92
2000	Site irri, drip lawn watering sys (8 zone) 6400 SF (4 x 40' x 40') w/contrl	.21	1.93	2.14
2100	Site irri, drip lawn watering sys (8 zone) 8000 SF (4 x 40' x 50') w/contrl	.18	1.75	1.93
2200	Site irri, drip lawn watering sys (4 zone) 3200 SF (2 x 40' x 40') w/contrl	.22	1.97	2.19
2300	Site irri, drip lawn watering sys (4 zone) 4000 SF (2 x 40' x 50') w/contrl	.19	1.77	1.96
2400	Site irri, drip lawn watering sys (3 zone) 2400 SF (60' x40') w/contrl	.23	2.03	2.26
2500	Site irri, drip lawn watering sys (3 zone) 3000 SF (60' x50') w/contrl	.18	1.81	1.99
2600	Site irri, drip lawn watering sys (6 zone) 4800 SF (2 x 60' x 40') manual	.20	1.90	2.10
2700	Site irri, drip lawn watering sys (6 zone) 6000 SF (2 x 60' x 50') manual	.17	1.71	1.88
2800	Site irrigation, drip lawn watering system 10' x 100' area (1000 SF)	.12	1.14	1.26
2900	Site irrigation, drip lawn watering system 2 x 10' x 100' area (2000 SF)	.13	1.09	1.22

G3010 Water Supply

The Water Service Systems are for copper service taps from 1″ to 2″ diameter into pressurized mains from 6″ to 8″ diameter. Costs are given for two offsets with depths varying from 2′ to 10′. Included in system components are excavation and backfill and required curb stops with boxes.

System Components	QUANTITY	UNIT	COST EACH MAT.	COST EACH INST.	COST EACH TOTAL
SYSTEM G3010 120 1000					
WATER SERVICE, 6″ MAIN, 1″ COPPER SERVICE, 10′ OFFSET, 2′ DEEP					
Trench excavation, 1/2 C.Y. backhoe, 1 laborer	2.220	C.Y.		11.46	11.46
Drill & tap pressurized main, 6″, 1″ to 2″ service	1.000	Ea.		295	295
Corporation, brass, 1″ diameter	1.000	Ea.	30.50	30.50	61
Saddle, cast iron, 1″ diameter	1.000	Ea.	37.50		37.50
Copper tubing, type K, 1″ diameter	11.000	L.F.	43.23	30.36	73.59
Curb stop, brass, 1″ diameter	1.000	Ea.	62.50	30.50	93
Curb box, cast iron, 1″ diameter	1.000	Ea.	56	41	97
Backfill by hand, no compaction, heavy soil	2.220	C.Y.		3.93	3.93
Compaction, vibrating plate	2.220	C.Y.		13.16	13.16
TOTAL			229.73	455.91	685.64

G3010 120	Water Service	COST EACH MAT.	COST EACH INST.	COST EACH TOTAL
1000	Water service, 6″ main, 1″ copper service, 10′ offset, 2′ deep	230	455	685
1040	4′ deep	230	500	730
1060	6′ deep	230	570	800
1080	8′ deep	230	655	885
1100	10′ deep	230	705	935
1200	20′ offset, 2′ deep	269	510	779
1240	4′ deep	269	610	879
1260	6′ deep	269	740	1,009
1280	8′ deep	269	915	1,184
1300	10′ deep	269	1,125	1,394
2000	1-1/2″ copper service, 10′ offset, 2′ deep	455	495	950
2040	4′ deep	455	545	1,000
2060	6′ deep	455	610	1,065
2080	8′ deep	455	700	1,155
2100	10′ deep	455	800	1,255
2200	20′ offset, 2′ deep	520	555	1,075
2240	4′ deep	520	655	1,175
2260	6′ deep	520	785	1,305
2280	8′ deep	520	960	1,480
2300	10′ deep	520	1,175	1,695
3000	6″ main, 2″ copper service, 10′ offset, 2′ deep	650	515	1,165
3040	4′ deep	650	565	1,215
3060	6′ deep	650	630	1,280
3080	8′ deep	650	715	1,365
3100	10′ deep	650	820	1,470
3200	20′ offset, 2′ deep	750	580	1,330
3240	4′ deep	750	680	1,430
3260	6′ deep	750	810	1,560

G3010 Water Supply

G3010 120	Water Service	COST EACH		
		MAT.	INST.	TOTAL
3280	8' deep	750	985	1,735
3300	10' deep	750	1,200	1,950
4000	8" main, 1" copper service, 10' offset, 2' deep	230	480	710
4040	4' deep	230	525	755
4060	6' deep	230	595	825
4080	8' deep	230	680	910
4100	10' deep	230	785	1,015
4200	20' offset, 2' deep	269	535	804
4240	4' deep	269	635	904
4260	6' deep	269	765	1,034
4280	8' deep	269	940	1,209
4300	10' deep	269	1,150	1,419
5000	1-1/2" copper service, 10' offset, 2' deep	455	520	975
5040	4' deep	455	570	1,025
5060	6' deep	455	635	1,090
5080	8' deep	455	725	1,180
5100	10' deep	455	825	1,280
5200	20' offset, 2' deep	520	580	1,100
5240	4' deep	520	680	1,200
5260	6' deep	520	810	1,330
5280	8' deep	520	985	1,505
5300	10' deep	520	1,200	1,720
6000	2" copper service, 10' offset, 2' deep	650	540	1,190
6040	4' deep	650	590	1,240
6060	6' deep	650	655	1,305
6080	8' deep	650	740	1,390
6100	10' deep	650	845	1,495
6200	20' offset, 2' deep	750	605	1,355
6240	4' deep	750	705	1,455
6260	6' deep	750	835	1,585
6280	8' deep	750	1,000	1,750
6300	10' deep	750	1,225	1,975

Important: See the Reference Section for critical supporting data - Reference Numbers and City Cost Indexes

G3010 Water Supply

The Fire Hydrant Systems include: four different hydrants with three different lengths of offsets, at several depths of trenching. Excavation and backfill is included with each system as well as thrust blocks as necessary and pipe bedding. Finally spreading of excess material and fine grading complete the components.

System Components	QUANTITY	UNIT	COST PER EACH		
			MAT.	INST.	TOTAL
SYSTEM G3010 410 1500					
HYDRANT, 4-1/2″ VALVE SIZE, TWO WAY, 10′ OFFSET, 2′ DEEP					
Excavation, trench, 1 C.Y. hydraulic backhoe	3.556	C.Y.		20.34	20.34
Sleeve, 12″ x 6″	1.000	Ea.	885	84.20	969.20
6″ diameter	8.000	L.F.	127.60	128.24	255.84
Gate valve, 6″	1.000	Ea.	510	75.70	585.70
Gate box	1.000	Ea.	76.50	56	132.50
Hydrant, incl. lower barrel and shoe	1.000	Ea.	1,200	151.45	1,351.45
Thrust blocks, at valve, shoe and sleeve	1.600	C.Y.	204.80	238.42	443.22
Borrow, crushed stone, 3/8″	4.089	C.Y.	116	22.08	138.08
Backfill and compact, dozer, air tamped	4.089	C.Y.		47.76	47.76
Spread excess excavated material	.204	C.Y.		6.12	6.12
Fine grade, hand	400.000	S.F.		192.50	192.50
TOTAL			3,119.90	1,022.81	4,142.71

G3010 410	Fire Hydrants	COST PER EACH		
		MAT.	INST.	TOTAL
1000	Hydrant, 4-1/2″ valve size, two way, 0′ offset, 2′ deep	3,000	625	3,625
1100	4′ deep	2,750	625	3,375
1200	6′ deep	3,125	695	3,820
1300	8′ deep	3,300	735	4,035
1400	10′ deep	3,400	785	4,185
1500	10′ offset, 2′ deep	3,125	1,025	4,150
1600	4′ deep	2,875	1,225	4,100
1700	6′ deep	3,250	1,600	4,850
1800	8′ deep	3,425	2,175	5,600
1900	10′ deep	3,525	3,325	6,850
2000	20′ offset, 2′ deep	3,275	1,450	4,725
2100	4′ deep	3,025	1,750	4,775
2200	6′ deep	3,400	2,350	5,750
2300	8′ deep	3,600	3,125	6,725
2400	10′ deep	3,700	4,250	7,950
2500	Hydrant, 4-1/2″ valve size, three way, 0′ offset, 2′ deep	3,075	640	3,715
2600	4′ deep	2,800	640	3,440
2700	6′ deep	3,200	705	3,905
2800	8′ deep	3,400	755	4,155
2900	10′ deep	3,525	805	4,330
3000	10′ offset, 2′ deep	3,200	1,025	4,225
3100	4′ deep	2,925	1,225	4,150
3200	6′ deep	3,350	1,600	4,950
3300	8′ deep	3,525	2,200	5,725

G3010 Water Supply

G3010 410	Fire Hydrants	COST PER EACH		
		MAT.	INST.	TOTAL
3400	10' deep	3,650	3,350	7,000
3500	20' offset, 2' deep	3,375	1,450	4,825
3600	4' deep	3,100	1,750	4,850
3700	6' deep	3,500	2,350	5,850
3800	8' deep	3,700	3,150	6,850
3900	10' deep	3,800	4,250	8,050
5000	5-1/4" valve size, two way, 0' offset, 2' deep	2,750	625	3,375
5100	4' deep	2,875	645	3,520
5200	6' deep	3,000	695	3,695
5300	8' deep	3,125	735	3,860
5400	10' deep	3,250	785	4,035
5500	10' offset, 2' deep	2,875	1,025	3,900
5600	4' deep	3,000	1,225	4,225
5700	6' deep	3,125	1,600	4,725
5800	8' deep	3,250	2,175	5,425
5900	10' deep	3,375	3,325	6,700
6000	20' offset, 2' deep	3,025	1,450	4,475
6100	4' deep	3,150	1,750	4,900
6200	6' deep	3,275	2,350	5,625
6300	8' deep	3,425	3,125	6,550
6400	10' deep	3,550	4,250	7,800
6500	Three way, 0' offset, 2' deep	2,800	640	3,440
6600	4' deep	2,950	655	3,605
6700	6' deep	3,075	705	3,780
6800	8' deep	3,225	755	3,980
6900	10' deep	3,350	805	4,155
7000	10' offset, 2' deep	2,925	1,025	3,950
7100	4' deep	3,075	1,250	4,325
7200	6' deep	3,200	1,600	4,800
7300	8' deep	3,350	2,200	5,550
7400	10' deep	3,475	3,350	6,825
7500	20' offset, 2' deep	3,100	1,450	4,550
7600	4' deep	3,225	1,575	4,800
7700	6' deep	3,375	2,350	5,725

G3020 Sanitary Sewer

The Septic System includes: a septic tank; leaching field; concrete distribution boxes; plus excavation and gravel backfill.

The Expanded System Listing shows systems with tanks ranging from 1000 gallons to 5000 gallons. Tanks are either concrete or fiberglass. Cost is on a complete unit basis.

System Components	QUANTITY	UNIT	COST EACH		
			MAT.	INST.	TOTAL
SYSTEM G3020 300 2010					
1000 GALLON TANK, LEACHING FIELD, 300 S.F., 50' FROM BLDG.					
Septic tank, precast, 1000 gallon	1.000	Ea.	615	190	805
Distribution box, precast, 5 outlets	1.000	Ea.	114	41.50	155.50
Sch. 40 sewer pipe, PVC, 4" diameter	40.000	L.F.	211	300	511
Sch. 40 fitting, wye, & tee, PVC, 4" diameter	4.000	Ea.	238	50	288
Excavation, hydraulic backhoe, backfill & load spoil	90.000	C.Y.		175.50	175.50
Backfill trench	90.000	C.Y.		179.10	179.10
Stone fill, 3/4" to 1-1/2"	15.000	C.Y.	307.50		307.50
Spread fill, 200 HP dozer	15.000	C.Y.		82.80	82.80
Delivery charge	15.000	C.Y.		307.50	307.50
Haul spoil, 12 C.Y. dump truck, 2 mile round trip	25.000	C.Y.		119.50	119.50
TOTAL			1,485.50	1,445.90	2,931.40

G3020 300	Septic Systems	COST EACH		
		MAT.	INST.	TOTAL
2010	1000 gal. septic tank, leaching field, 300 S.F., 50' from bldg.	1,475	1,450	2,925
2110	200' from bldg.	1,925	2,125	4,050
2210	6' deep precast pit, one at 6'-6" diameter, 50' from bldg.	1,800	1,675	3,475
2310	200' from bldg.	2,250	2,300	4,550
2410	4' deep precast galley, 12' long, 50' from bldg.	1,875	1,725	3,600
2510	200' from bldg.	2,300	2,375	4,675
3010	1250 gal. septic tank, leaching field, 400 S.F., 50' from bldg.	1,850	1,700	3,550
3110	200' from bldg.	2,275	2,325	4,600
3210	6' deep precast pit, one at 8'-0" diameter, 50' from bldg.	2,500	1,900	4,400
3310	200' from bldg.	2,950	2,550	5,500
3410	4' deep precast galley, 16' long, 50' from bldg.	2,400	2,000	4,400
3510	200' from bldg.	2,850	2,650	5,500
4010	1500 gal. septic tank, leaching field, 500 S.F., 50' from bldg.	1,700	1,850	3,550
4110	200' from bldg.	2,125	2,500	4,625
4210	6' deep precast pit, two at 6'-6" diameter, 50' from bldg.	3,175	2,775	5,950
4310	200' from bldg.	3,600	3,425	7,025
4410	4' deep precast galley, 20' long, 50' from bldg.	2,850	2,325	5,175
4510	200' from bldg.	3,275	2,950	6,225
5010	2000 gal. septic tank, leaching field, 600 S.F., 50' from bldg.	1,975	2,200	4,175
5110	200' from bldg.	2,400	2,850	5,250
5210	6' deep precast pit, two at 8'-0" diameter, 50' from bldg.	4,325	3,375	7,700
5310	200' from bldg.	4,750	4,050	8,800
5410	4' deep precast galley, 24 ft. long, 50' from bldg.	3,375	2,775	6,150
5510	200' from bldg.	3,800	3,425	7,225

G3030 Storm Sewer

Manhole   **Catch Basin**

The Manhole and Catch Basin System includes: excavation with a backhoe; a formed concrete footing; frame and cover; cast iron steps and compacted backfill.

The Expanded System Listing shows manholes that have a 4', 5' and 6' inside diameter riser. Depths range from 4' to 14'. Construction material shown is either concrete, concrete block, precast concrete, or brick.

System Components	QUANTITY	UNIT	COST PER EACH		
			MAT.	INST.	TOTAL
SYSTEM G3030 210 1920					
MANHOLE/CATCH BASIN, BRICK, 4' I.D. RISER, 4' DEEP					
Excavation, hydraulic backhoe, 3/8 C.Y. bucket	14.815	C.Y.		84.74	84.74
Trim sides and bottom of excavation	64.000	S.F.		45.44	45.44
Forms in place, manhole base, 4 uses	20.000	SFCA	11.80	78	89.80
Reinforcing in place footings, #4 to #7	.019	Ton	15.87	18.15	34.02
Concrete, 3000 psi	.925	C.Y.	82.33		82.33
Place and vibrate concrete, footing, direct chute	.925	C.Y.		35.58	35.58
Catch basin or MH, brick, 4' ID, 4' deep	1.000	Ea.	350	760	1,110
Catch basin or MH steps; heavy galvanized cast iron	1.000	Ea.	13.45	10.75	24.20
Catch basin or MH frame and cover	1.000	Ea.	212	166.50	378.50
Fill, granular	12.954	C.Y.	112.05		112.05
Backfill, spread with wheeled front end loader	12.954	C.Y.		22.93	22.93
Backfill compaction, 12" lifts, air tamp	12.954	C.Y.		83.55	83.55
TOTAL			797.50	1,305.64	2,103.14

G3030 210	Manholes & Catch Basins	COST PER EACH		
		MAT.	INST.	TOTAL
1920	Manhole/catch basin, brick, 4' I.D. riser, 4' deep	800	1,325	2,125
1940	6' deep	1,050	1,800	2,850
1960	8' deep	1,350	2,500	3,850
1980	10' deep	1,850	3,075	4,925
3000	12' deep	2,425	3,350	5,775
3020	14' deep	3,075	4,675	7,750
3200	Block, 4' I.D. riser, 4' deep	730	1,050	1,780
3220	6' deep	940	1,475	2,415
3240	8' deep	1,175	2,050	3,225
3260	10' deep	1,375	2,525	3,900
3280	12' deep	1,675	3,200	4,875
3300	14' deep	2,025	3,900	5,925
4620	Concrete, cast-in-place, 4' I.D. riser, 4' deep	940	1,800	2,740
4640	6' deep	1,275	2,425	3,700
4660	8' deep	1,750	3,525	5,275
4680	10' deep	2,100	4,375	6,475
4700	12' deep	2,550	5,400	7,950
4720	14' deep	3,025	6,475	9,500
5820	Concrete, precast, 4' I.D. riser, 4' deep	1,200	965	2,165
5840	6' deep	1,525	1,300	2,825

Important: See the Reference Section for critical supporting data - Reference Numbers and City Cost Indexes

G3030 Storm Sewer

G3030 210	Manholes & Catch Basins	COST PER EACH		
		MAT.	INST.	TOTAL
5860	8' deep	1,850	1,825	3,675
5880	10' deep	2,275	2,250	4,525
5900	12' deep	2,800	2,750	5,550
5920	14' deep	3,375	3,500	6,875
6000	5' I.D. riser, 4' deep	1,275	1,050	2,325
6020	6' deep	1,700	1,475	3,175
6040	8' deep	2,150	1,925	4,075
6060	10' deep	2,700	2,475	5,175
6080	12' deep	3,275	3,150	6,425
6100	14' deep	3,925	3,850	7,775
6200	6' I.D. riser, 4' deep	1,875	1,375	3,250
6220	6' deep	2,400	1,825	4,225
6240	8' deep	3,000	2,575	5,575
6260	10' deep	3,750	3,250	7,000
6280	12' deep	4,575	4,075	8,650
6300	14' deep	5,425	4,950	10,375

For information about Means Estimating Seminars, see yellow pages 12 and 13 in back of book

Reference Section

All the reference information is in one section making it easy to find what you need to know . . . and easy to use the book on a daily basis. This section is visually identified by a vertical gray bar on the edge of pages.

In the reference number information that follows, you'll see the background that relates to the "reference numbers" that appeared in the Unit Price Section. You'll find reference tables, explanations and estimating information that support how we arrived at the unit price data. Also included are alternate pricing methods, technical data and estimating procedures along with information on design and economy in construction.

Also in this Reference Section, we've included Change Orders, information on pricing changes to contract documents; Crew Listings, a full listing of all the crews, equipment and their costs; Historical Cost Indexes for cost comparisons over time; City Cost Indexes and Location Factors for adjusting costs to the region you are in; and an explanation of all abbreviations used in the book.

Table of Contents

Reference Numbers (cont.)

R01100-005 Tips for Accurate Estimating

1. Use pre-printed or columnar forms for orderly sequence of dimensions and locations and for recording telephone quotations.

2. Use only the front side of each paper or form except for certain pre-printed summary forms.

3. Be consistent in listing dimensions: For example, length x width x height. This helps in rechecking to ensure that, the total length of partitions is appropriate for the building area.

4. Use printed (rather than measured) dimensions where given.

5. Add up multiple printed dimensions for a single entry where possible.

6. Measure all other dimensions carefully.

7. Use each set of dimensions to calculate multiple related quantities.

8. Convert foot and inch measurements to decimal feet when listing. Memorize decimal equivalents to .01 parts of a foot (1/8" equals approximately .01').

9. Do not "round off" quantities until the final summary.

10. Mark drawings with different colors as items are taken off.

11. Keep similar items together, different items separate.

12. Identify location and drawing numbers to aid in future checking for completeness.

13. Measure or list everything on the drawings or mentioned in the specifications.

14. It may be necessary to list items not called for to make the job complete.

15. Be alert for: Notes on plans such as N.T.S. (not to scale); changes in scale throughout the drawings; reduced size drawings; discrepancies between the specifications and the drawings.

16. Develop a consistent pattern of performing an estimate. For example:
    a. Start the quantity takeoff at the lower floor and move to the next higher floor.
    b. Proceed from the main section of the building to the wings.
    c. Proceed from south to north or vice versa, clockwise or counterclockwise.
    d. Take off floor plan quantities first, elevations next, then detail drawings.

17. List all gross dimensions that can be either used again for different quantities, or used as a rough check of other quantities for verification (exterior perimeter, gross floor area, individual floor areas, etc.).

18. Utilize design symmetry or repetition (repetitive floors, repetitive wings, symmetrical design around a center line, similar room layouts, etc.). Note: Extreme caution is needed here so as not to omit or duplicate an area.

19. Do not convert units until the final total is obtained. For instance, when estimating concrete work, keep all units to the nearest cubic foot, then summarize and convert to cubic yards.

20. When figuring alternatives, it is best to total all items involved in the basic system, then total all items involved in the alternates. Therefore you work with positive numbers in all cases. When adds and deducts are used, it is often confusing whether to add or subtract a portion of an item; especially on a complicated or involved alternate.

R01100-040   Builder's Risk Insurance

Builder's Risk Insurance is insurance on a building during construction. Premiums are paid by the owner or the contractor. Blasting, collapse and underground insurance would raise total insurance costs above those listed. Floater policy for materials delivered to the job runs $.75 to $1.25 per $100 value. Contractor equipment insurance runs $.50 to $1.50 per $100 value. Insurance for miscellaneous tools to $1,500 value runs from $3.00 to $7.50 per $100 value.

Tabulated below are New England Builder's Risk insurance rates in dollars per $100 value for $1,000 deductible. For $25,000 deductible, rates can be reduced 13% to 34%. On contracts over $1,000,000, rates may be lower than those tabulated. Policies are written annually for the total completed value in place. For "all risk" insurance (excluding flood, earthquake and certain other perils) add $.025 to total rates below.

Coverage	Frame Construction (Class 1)		Average	Brick Construction (Class 4)		Average	Fire Resistive (Class 6)		Average
	Range			Range			Range		
Fire Insurance	$.350 to	$ .850	$.600	$.158 to	$.189	$.174	$.052 to	$.080	$.070
Extended Coverage	.115 to	.200	.158	.080 to	.105	.101	.081 to	.105	.100
Vandalism	.012 to	.016	.014	.008 to	.011	.011	.008 to	.011	.010
Total Annual Rate	$.477 to	$1.066	$.772	$.246 to	$.305	$.286	$.141 to	$.196	$.180

R01100-050   General Contractor's Overhead

There are two distinct types of overhead on a construction project: Project Overhead and Main Office Overhead. Project Overhead includes those costs at a construction site not directly associated with the installation of construction materials. Examples of Project Overhead costs include the following:

1. Superintendent
2. Construction office and storage trailers
3. Temporary sanitary facilities
4. Temporary utilities
5. Security fencing
6. Photographs
7. Clean up
8. Performance and payment bonds

The above Project Overhead items are also referred to as General Requirements and therefore are estimated in Division 1. Division 1 is the first division listed in the CSI MasterFormat but it is usually the last division estimated. The sum of the costs in Divisions 1 through 16 is referred to as the sum of the direct costs.

All construction projects also include indirect costs. The primary components of indirect costs are the contractor's Main Office Overhead and profit. The amount of the Main Office Overhead expense varies depending on the the following:

1. Owner's compensation
2. Project managers and estimator's wages
3. Clerical support wages
4. Office rent and utilities
5. Corporate legal and accounting costs
6. Advertising
7. Automobile expenses
8. Association dues
9. Travel and entertainment expenses

These costs are usually calculated as a percentage of annual sales volume. This percentage can range from 35% for a small contractor doing less than $500,000 to 5% for a large contractor with sales in excess of $100 million.

R01100-060 Workers' Compensation Insurance Rates by Trade

The table below tabulates the national averages for Workers' Compensation insurance rates by trade and type of building. The average "Insurance Rate" is multiplied by the "% of Building Cost" for each trade. This produces the "Workers' Compensation Cost" by % of total labor cost, to be added for each trade by building type to determine the weighted average Workers' Compensation rate for the building types analyzed.

Trade	Insurance Rate (% Labor Cost)			% of Building Cost			Workers' Compensation		
	Range		Average	Office Bldgs.	Schools & Apts.	Mfg.	Office Bldgs.	Schools & Apts.	Mfg.
Excavation, Grading, etc.	4.2 % to	19.5%	10.4%	4.8%	4.9%	4.5%	.50%	.51%	.47%
Piles & Foundations	7.7 to	76.7	22.5	7.1	5.2	8.7	1.60	1.17	1.96
Concrete	5.2 to	35.8	16.1	5.0	14.8	3.7	.81	2.38	.60
Masonry	5.1 to	28.3	15.1	6.9	7.5	1.9	1.04	1.13	.29
Structural Steel	7.8 to	103.8	38.6	10.7	3.9	17.6	4.13	1.51	6.79
Miscellaneous & Ornamental Metals	4.9 to	24.8	12.6	2.8	4.0	3.6	.35	.50	.45
Carpentry & Millwork	7.0 to	53.2	18.4	3.7	4.0	0.5	.68	.74	.09
Metal or Composition Siding	5.3 to	32.1	17.0	2.3	0.3	4.3	.39	.05	.73
Roofing	7.8 to	78.9	33.1	2.3	2.6	3.1	.76	.86	1.03
Doors & Hardware	4.5 to	24.9	11.3	0.9	1.4	0.4	.10	.16	.05
Sash & Glazing	4.6 to	33.7	13.9	3.5	4.0	1.0	.49	.56	.14
Lath & Plaster	4.3 to	40.2	14.9	3.3	6.9	0.8	.49	1.03	.12
Tile, Marble & Floors	2.4 to	31.8	9.7	2.6	3.0	0.5	.25	.29	.05
Acoustical Ceilings	2.4 to	29.7	10.9	2.4	0.2	0.3	.26	.02	.03
Painting	4.5 to	33.7	13.1	1.5	1.6	1.6	.20	.21	.21
Interior Partitions	7.0 to	53.2	18.4	3.9	4.3	4.4	.72	.79	.81
Miscellaneous Items	2.6 to	137.6	16.9	5.2	3.7	9.7	.88	.63	1.64
Elevators	2.8 to	42.8	7.6	2.1	1.1	2.2	.16	.08	.17
Sprinklers	3.2 to	22.1	8.8	0.5	—	2.0	.04	—	.18
Plumbing	2.7 to	12.4	8.0	4.9	7.2	5.2	.39	.58	.42
Heat., Vent., Air Conditioning	4.0 to	27.2	11.5	13.5	11.0	12.9	1.55	1.27	1.48
Electrical	2.6 to	12.7	6.5	10.1	8.4	11.1	.66	.55	.72
Total	2.4 % to	137.6%	—	100.0%	100.0%	100.0%	16.45%	15.02%	18.43%
				Overall Weighted Average	16.63%				

Workers' Compensation Insurance Rates by States

The table below lists the weighted average Workers' Compensation base rate for each state with a factor comparing this with the national average of 16.2%.

State	Weighted Average	Factor	State	Weighted Average	Factor	State	Weighted Average	Factor
Alabama	25.2%	156	Kentucky	17.4%	107	North Dakota	11.4%	70
Alaska	19.7	122	Louisiana	26.2	162	Ohio	12.8	79
Arizona	7.4	46	Maine	19.3	119	Oklahoma	13.7	85
Arkansas	13.7	85	Maryland	17.7	109	Oregon	12.6	78
California	17.4	107	Massachusetts	12.4	77	Pennsylvania	13.4	83
Colorado	12.3	76	Michigan	17.6	109	Rhode Island	19.6	121
Connecticut	23.1	143	Minnesota	25.1	155	South Carolina	12.9	80
Delaware	14.2	88	Mississippi	16.1	99	South Dakota	14.2	88
District of Columbia	17.3	107	Missouri	17.6	109	Tennessee	17.9	110
Florida	25.1	155	Montana	18.9	117	Texas	13.5	83
Georgia	21.3	131	Nebraska	18.8	116	Utah	11.4	70
Hawaii	15.7	97	Nevada	13.9	86	Vermont	18.6	115
Idaho	10.4	64	New Hampshire	21.3	131	Virginia	10.8	67
Illinois	17.2	106	New Jersey	10.5	65	Washington	10.4	64
Indiana	5.4	33	New Mexico	16.3	101	West Virginia	12.4	77
Iowa	11.8	73	New York	13.4	83	Wisconsin	15.2	94
Kansas	8.5	52	North Carolina	13.2	81	Wyoming	7.9	49
			Weighted Average for U.S. is	15.5% of payroll = 100%				

Rates in the following table are the base or manual costs per $100 of payroll for Workers' Compensation in each state. Rates are usually applied to straight time wages only and not to premium time wages and bonuses.

The weighted average skilled worker rate for 35 trades is 16.2%. For bidding purposes, apply the full value of Workers' Compensation directly to total labor costs, or if labor is 38%, materials 42% and overhead and profit 20% of total cost, carry 38/80 x 16.2% =7.7% of cost (before overhead and profit) into overhead. Rates vary not only from state to state but also with the experience rating of the contractor.

Rates are the most current available at the time of publication.

R01100-060 Workers' Compensation Insurance Rates by Trade and State (cont.)

State	Carpentry — 3 stories or less 5651	Carpentry — interior cab. work 5437	Carpentry — general 5403	Concrete Work — NOC 5213	Concrete Work — flat (flr., sdwk.) 5221	Electrical Wiring — inside 5190	Excavation — earth NOC 6217	Excavation — rock 6217	Glaziers 5462	Insulation Work 5479	Lathing 5443	Masonry 5022	Painting & Decorating 5474	Pile Driving 6003	Plastering 5480	Plumbing 5183	Roofing 5551	Sheet Metal Work (HVAC) 5538	Steel Erection — door & sash 5102	Steel Erection — inter., ornam. 5102	Steel Erection — structure 5040	Steel Erection — NOC 5057	Tile Work — (interior ceramic) 5348	Waterproofing 9014	Wrecking 5701
AL	26.93	15.28	34.89	12.39	9.90	8.95	13.69	13.69	33.66	17.12	12.78	27.57	33.72	34.63	40.23	11.91	60.85	27.15	23.11	23.11	51.16	39.09	13.16	6.65	51.16
AK	17.02	14.74	13.84	12.07	11.18	12.70	19.54	19.54	19.86	25.41	9.78	24.13	16.28	59.80	19.42	11.79	38.40	12.13	12.24	12.24	42.33	18.60	8.48	10.79	42.33
AZ	7.71	5.37	12.63	6.33	3.44	4.18	4.83	4.83	7.59	9.87	4.70	6.40	5.23	9.47	15.76	3.77	11.96	4.71	7.92	7.92	16.58	7.75	2.37	2.55	50.31
AR	13.04	11.01	16.99	13.77	6.64	6.11	9.42	9.42	15.49	33.95	8.65	11.34	10.50	12.82	12.93	5.69	21.95	10.28	7.70	7.70	37.63	28.19	7.10	3.75	37.63
CA	28.28	9.07	28.28	13.18	13.18	9.93	7.88	7.88	16.37	23.34	10.90	14.55	19.45	21.18	18.12	11.59	39.79	15.33	13.55	13.55	23.93	21.91	7.74	19.45	21.91
CO	17.36	9.17	12.24	14.36	7.69	5.32	10.44	10.44	9.28	13.68	5.57	13.41	10.77	17.70	10.81	8.51	24.47	12.11	7.39	7.39	31.30	15.53	7.99	5.78	31.30
CT	20.96	18.42	31.08	29.77	13.28	8.63	11.18	11.18	15.10	24.22	17.24	25.81	19.56	27.54	27.61	10.37	51.40	14.91	18.42	18.42	70.81	35.84	12.65	6.69	50.86
DE	15.39	15.39	12.92	13.03	7.13	6.79	10.03	10.03	11.96	12.92	13.83	12.33	16.85	21.38	13.83	7.94	27.84	10.69	13.02	13.02	29.92	13.02	10.51	12.33	29.92
DC	13.05	10.16	15.52	17.41	20.87	6.52	10.32	10.32	27.59	11.07	8.44	22.34	10.52	18.70	13.70	12.43	21.23	9.29	17.97	17.97	54.79	20.38	31.80	4.05	54.79
FL	30.55	21.73	31.10	30.26	15.05	10.74	13.79	13.79	23.24	22.05	13.08	23.21	21.38	58.93	36.28	10.71	46.17	18.10	14.78	14.78	58.02	39.86	11.26	9.25	58.02
GA	32.05	17.69	24.16	15.16	10.44	8.97	16.86	16.86	15.69	22.02	17.16	20.30	17.13	31.16	19.94	10.77	46.74	17.52	13.97	13.97	44.46	48.01	9.64	8.55	44.46
HI	17.38	11.62	28.20	13.74	12.27	7.10	7.73	7.73	20.55	21.19	10.94	17.87	11.33	20.15	15.61	6.06	33.24	8.02	11.11	11.11	31.71	21.70	9.78	11.54	31.71
ID	11.48	6.50	13.42	8.40	5.87	5.01	6.24	6.24	9.48	7.60	5.27	9.01	7.40	13.14	9.74	4.67	29.72	9.37	9.80	9.80	30.78	13.76	6.47	4.09	30.78
IL	16.38	13.48	17.63	29.67	11.05	8.51	10.49	10.49	15.50	11.58	9.74	16.53	11.34	28.71	11.92	9.79	28.32	14.36	14.87	14.87	52.98	26.69	14.11	4.36	52.98
IN	5.31	4.50	7.02	5.18	3.17	2.64	4.17	4.17	4.59	4.15	2.43	5.07	4.48	7.71	4.30	2.70	10.93	4.03	4.94	4.94	16.12	9.00	3.08	2.55	16.12
IA	11.67	5.80	10.47	12.55	6.82	4.18	5.98	5.98	13.66	9.22	5.45	8.79	7.73	10.70	9.26	5.92	18.71	7.17	9.92	9.92	51.91	36.28	5.29	4.20	30.67
KS	10.58	8.42	10.47	7.12	6.36	3.72	4.73	4.73	6.49	7.80	4.37	7.45	6.06	9.36	8.73	5.40	20.70	6.76	7.83	7.83	23.76	12.28	4.75	3.48	23.76
KY	15.79	12.91	19.21	20.50	6.16	7.64	17.08	17.08	21.66	24.14	11.20	10.81	11.40	18.49	14.90	6.85	48.37	13.64	11.61	11.61	40.21	30.64	12.56	4.72	40.21
LA	23.14	24.93	53.17	26.43	15.61	9.88	17.49	17.49	20.14	21.43	24.51	28.33	29.58	31.25	22.37	8.64	77.12	21.20	18.98	18.98	51.76	24.21	13.77	13.78	66.41
ME	13.43	10.42	43.59	24.29	11.12	5.03	12.19	12.19	13.35	15.39	14.03	17.26	16.07	31.44	17.95	7.32	32.07	8.82	15.08	15.08	37.84	61.81	11.66	6.06	37.84
MD	13.71	5.95	10.55	18.82	8.70	5.15	12.91	12.91	25.52	20.07	11.37	13.31	9.37	26.22	14.21	8.58	48.72	12.74	13.66	13.66	56.25	37.81	8.03	7.14	26.80
MA	9.93	6.07	16.09	17.95	8.10	3.69	6.49	6.49	7.55	13.48	5.76	13.04	7.54	14.31	5.34	4.55	38.25	7.47	12.35	12.35	35.13	27.39	9.24	3.32	28.82
MI	21.68	13.59	19.80	22.12	9.35	5.71	11.69	11.69	13.87	12.38	13.59	19.33	14.93	39.06	15.59	8.24	41.33	10.41	11.36	11.36	39.06	29.15	11.31	6.40	39.06
MN	21.84	21.96	43.39	16.70	14.64	6.94	16.68	16.68	17.42	11.95	21.02	20.28	17.69	26.83	21.02	10.70	78.89	12.72	14.64	14.64	103.82	38.11	14.59	6.96	132.92
MS	16.92	15.26	19.71	11.11	8.98	6.80	12.22	12.22	12.02	14.11	8.02	13.18	12.86	19.38	22.03	8.15	37.48	21.88	11.67	11.67	33.42	32.27	9.89	6.93	33.42
MO	21.84	12.24	16.70	15.66	11.37	7.09	10.05	10.05	11.75	18.49	12.47	16.14	13.64	21.93	16.25	9.37	33.06	14.20	12.51	12.51	59.20	40.85	9.59	7.02	59.20
MT	22.03	10.79	17.11	12.89	9.98	5.54	17.20	17.20	11.51	16.19	19.02	14.58	12.03	76.71	13.82	8.93	57.20	9.44	9.73	9.73	41.23	17.93	6.54	5.33	41.23
NE	23.15	10.97	20.95	28.60	9.65	8.32	12.20	12.20	15.50	24.47	11.30	18.50	12.32	23.35	14.75	11.17	36.72	13.05	13.15	13.15	52.27	38.35	9.97	6.15	50.40
NV	19.78	7.76	13.18	10.57	11.57	6.18	11.21	11.21	12.58	13.22	7.97	11.38	10.62	14.55	12.30	8.93	21.55	18.17	12.87	12.87	31.33	33.90	6.64	6.15	44.29
NH	21.98	14.43	24.90	35.84	11.10	6.02	17.16	17.16	14.70	33.70	8.60	24.98	15.72	17.80	18.65	10.22	60.89	14.43	14.01	14.01	54.76	25.50	17.84	6.36	54.76
NJ	11.76	8.69	11.76	9.82	7.77	4.22	7.74	7.74	6.91	11.95	10.39	12.07	9.66	13.10	10.39	6.14	29.59	6.69	9.94	9.94	19.67	10.61	4.42	4.65	23.42
NM	28.63	8.03	18.67	17.07	9.38	6.85	9.33	9.33	14.38	12.83	7.20	13.69	12.57	18.59	11.22	8.64	32.00	11.70	24.78	24.78	43.20	25.65	7.98	6.81	43.20
NY	17.66	6.98	15.91	17.16	13.57	6.32	10.02	10.02	12.02	9.09	5.11	19.20	13.75	16.81	8.43	8.53	33.44	17.47	9.63	9.63	18.80	19.14	9.43	6.09	11.29
NC	16.34	11.24	15.16	11.77	7.00	8.63	8.65	8.65	10.30	11.88	8.28	10.32	9.75	13.68	15.22	7.78	26.83	11.26	7.71	7.71	52.96	19.19	5.79	4.27	52.96
ND	9.60	9.60	9.60	5.66	5.66	4.00	5.39	5.39	9.60	9.60	9.04	8.10	6.27	20.12	9.04	5.35	20.74	5.35	20.12	20.12	20.12	20.12	9.60	20.24	14.48
OH	6.32	11.48	9.82	11.66	10.37	5.54	8.42	8.42	9.22	14.76	29.68	12.20	15.20	25.34	4.41	7.14	22.43	8.85	10.53	10.53	25.27	18.05	9.73	5.02	25.27
OK	13.44	9.11	12.70	12.65	6.63	5.82	10.25	10.25	16.10	19.50	8.46	10.98	9.09	20.29	12.21	7.24	22.32	9.49	13.78	13.78	39.78	24.01	6.95	5.88	39.78
OR	19.10	9.61	15.51	13.55	8.43	4.65	9.70	9.70	15.02	8.08	7.48	15.36	12.79	15.81	10.75	5.89	23.26	10.49	9.53	9.53	32.00	13.92	11.23	4.97	32.00
PA	13.85	13.85	12.80	14.90	9.67	6.74	8.44	8.44	9.60	12.80	11.67	12.18	14.33	17.72	11.67	7.51	27.83	8.19	15.17	15.17	26.43	15.17	8.22	12.18	26.43
RI	19.53	11.65	18.07	18.23	16.24	4.43	10.38	10.38	12.85	22.78	11.97	25.11	24.13	37.66	17.25	8.52	33.92	10.29	14.07	14.07	59.49	37.50	14.36	7.70	78.79
SC	19.10	13.03	19.24	12.88	6.19	7.02	8.32	8.32	12.22	10.69	7.27	9.55	11.19	16.64	18.89	6.88	28.98	11.42	9.49	9.49	21.80	23.66	6.23	4.65	21.80
SD	17.83	7.27	14.55	18.00	6.34	5.44	13.24	13.24	10.58	13.24	8.39	8.01	12.63	24.87	11.78	8.53	21.26	9.64	11.53	11.53	53.25	20.15	6.68	4.20	53.25
TN	31.79	14.64	24.26	19.40	9.95	8.72	14.06	14.06	12.09	14.84	13.05	17.02	14.59	17.43	17.47	11.33	38.38	14.51	13.38	13.38	45.26	25.19	9.73	6.57	45.26
TX	16.89	11.33	13.25	12.10	8.85	7.36	10.49	10.49	11.61	15.68	9.53	13.68	9.89	13.95	20.99	8.06	22.80	14.58	11.23	11.23	33.39	14.78	7.29	8.09	17.86
UT	9.45	6.23	10.52	7.68	7.99	7.05	6.38	6.38	9.55	10.95	12.32	13.75	15.29	17.24	10.60	6.41	26.27	6.30	8.99	8.99	23.51	23.51	6.06	5.83	26.53
VT	19.90	11.34	19.46	34.13	12.01	6.10	10.61	10.61	21.25	25.54	10.85	19.53	10.41	23.91	18.23	10.10	28.73	12.62	13.94	13.94	47.05	37.23	8.52	9.59	47.05
VA	11.97	7.81	9.72	10.47	5.32	4.43	6.53	6.53	7.60	7.96	13.79	8.17	9.95	13.55	7.16	5.52	21.62	8.30	11.34	11.34	39.01	16.86	6.32	3.09	39.01
WA	9.13	9.13	9.13	8.17	8.17	3.12	8.54	8.54	13.74	8.92	9.13	13.27	9.39	20.04	11.57	5.04	19.54	4.60	12.96	12.96	9.39	9.59	10.76	10.23	9.59
WV	12.30	12.30	12.30	27.37	27.37	5.86	8.90	8.90	6.99	13.15	13.15	11.94	12.35	8.90	12.35	5.48	14.63	6.99	10.89	10.89	10.89	10.89	13.15	13.15	48.62
WI	12.40	10.86	19.86	13.12	11.66	5.32	6.74	6.74	16.53	14.04	10.60	18.99	14.82	17.24	13.95	6.31	43.57	8.01	14.30	14.30	34.84	21.34	14.42	5.80	34.84
WY	7.77	7.77	7.77	7.77	7.77	7.77	7.77	7.77	7.77	7.77	7.77	7.77	7.77	7.77	7.77	7.77	7.77	7.77	7.77	7.77	7.77	7.77	7.77	7.77	7.77
AVG.	16.96	11.33	18.42	16.07	9.94	6.46	10.43	10.43	13.91	15.53	10.87	15.06	13.12	22.45	14.88	7.96	33.14	11.46	12.57	12.57	38.60	24.71	9.66	7.12	39.48

R01100-060 Workers' Compensation (cont.) (Canada in Canadian dollars)

Province		Alberta	British Columbia	Manitoba	Ontario	New Brunswick	Newfndld. & Labrador	Northwest Territories	Nova Scotia	Prince Edward Island	Quebec	Saskat-chewan	Yukon
Carpentry—3 stories or less	Rate	10.80	6.66	4.57	4.83	4.63	10.05	4.25	7.67	6.15	14.69	7.01	4.36
	Code	42143	721028	40102	723	422	4226	4-41	4226	401	80110	B1317	202
Carpentry—interior cab. work	Rate	2.59	5.94	4.57	4.83	4.21	6.35	4.25	5.74	3.41	14.69	3.65	4.36
	Code	42133	721021	40102	723	427	4270	4-41	4274	402	80110	B11-27	202
CARPENTRY—general	Rate	10.80	6.66	4.57	4.83	4.63	6.35	4.25	7.67	6.15	14.69	7.01	4.36
	Code	42143	721028	40102	723	422	4299	4-41	4226	401	80110	B1317	202
CONCRETE WORK—NOC	Rate	7.26	8.48	7.60	16.47	4.63	10.05	4.25	4.83	6.15	15.84	7.01	4.36
	Code	42104	721010	40110	748	422	4224	4-41	4224	401	80100	B13-14	203
CONCRETE WORK—flat (flr. sidewalk)	Rate	7.26	8.48	7.60	16.47	4.63	10.05	4.25	4.83	6.15	15.84	7.01	4.36
	Code	42104	721010	40110	748	422	4224	4-41	4224	401	80100	B13-14	203
ELECTRICAL Wiring—inside	Rate	2.86	2.67	2.52	3.03	2.43	3.19	3.00	2.23	3.41	7.29	3.65	4.36
	Code	42124	721019	40203	704	426	4261	4-46	4261	402	80170	B11-05	206
EXCAVATION—earth NOC	Rate	3.59	4.29	3.84	4.20	3.69	5.55	3.50	4.11	3.58	8.80	4.02	4.36
	Code	40604	721031	40706	711	421	4214	4-43	4214	404	80030	R11-06	207
EXCAVATION—rock	Rate	3.59	4.29	3.84	4.20	3.69	5.55	3.50	4.11	3.58	8.80	4.02	4.36
	Code	40604	721031	40706	711	421	4214	4-43	4214	404	80030	R11-06	207
GLAZIERS	Rate	4.84	3.00	4.57	8.12	6.12	5.97	4.25	7.67	3.41	14.59	7.01	4.36
	Code	42121	715020	40109	751	423	4233	4-41	4233	402	80150	B13-04	212
INSULATION WORK	Rate	3.67	6.94	4.57	8.12	6.12	5.97	4.25	7.67	6.15	14.69	6.20	4.36
	Code	42184	721029	40102	751	423	4234	4-41	4234	401	80110	B12-07	202
LATHING	Rate	8.30	11.20	4.57	4.83	4.21	6.35	4.25	5.74	3.41	14.69	7.01	4.36
	Code	42135	721033	40102	723	427	4279	4-41	4271	402	80110	B13-16	202
MASONRY	Rate	7.26	11.20	4.57	12.21	6.12	5.97	4.25	7.67	6.15	15.84	7.01	4.36
	Code	42102	721037	40102	741	423	4231	4-41	4231	401	80100	B13-18	202
PAINTING & DECORATING	Rate	5.64	6.94	3.45	6.83	4.21	6.35	4.25	5.74	3.41	14.69	6.20	4.36
	Code	42111	721041	40105	719	427	4275	4-41	4275	402	80110	B12-01	202
PILE DRIVING	Rate	7.26	13.07	3.84	5.84	4.63	10.05	3.50	4.83	6.15	8.80	7.01	4.36
	Code	42159	722004	40706	732	422	4221	4-43	4221	401	80030	B13-10	202
PLASTERING	Rate	8.30	11.20	5.13	6.83	4.21	6.35	4.25	5.74	3.41	14.69	6.20	4.36
	Code	42135	721042	40108	719	427	4271	4-41	4271	402	80110	B12-21	202
PLUMBING	Rate	2.86	5.48	2.88	3.83	3.74	3.89	3.00	2.23	3.41	7.75	3.65	4.36
	Code	42122	721043	40204	707	424	4241	4-46	4241	402	80160	B11-01	214
ROOFING	Rate	11.09	10.45	6.52	12.34	8.00	10.05	4.25	9.49	6.15	23.29	7.01	4.36
	Code	42118	721036	40403	728	430	4236	4-41	4236	401	80130	B13-20	202
SHEET METAL WORK (HVAC)	Rate	2.86	5.48	6.52	3.83	3.74	3.89	3.50	3.23	3.41	7.75	3.65	6.60
	Code	42117	721043	40402	707	424	4244	4-46	4244	402	80160	B11-07	208
STEEL ERECTION—door & sash	Rate	3.67	13.07	9.89	16.47	4.63	10.05	4.25	7.67	6.15	29.23	7.01	4.36
	Code	42106	722005	40502	748	422	4227	4-41	4227	401	80080	B13-22	202
STEEL ERECTION—inter., ornam.	Rate	3.67	13.07	9.89	16.47	4.63	10.05	4.25	7.67	6.15	29.23	7.01	4.36
	Code	42106	722005	40502	748	422	4227	4-41	4227	401	80080	B13-22	202
STEEL ERECTION—structure	Rate	3.67	13.07	9.89	16.47	4.63	10.05	4.25	7.67	6.15	29.23	7.01	4.36
	Code	42106	722005	40502	748	422	4227	4-41	4227	401	80080	B13-22	202
STEEL ERECTION—NOC	Rate	3.67	13.07	9.89	16.47	4.63	10.05	4.25	7.67	6.15	29.23	7.01	4.36
	Code	42106	722005	40502	748	422	4227	4-41	4227	401	80080	B13-22	202
TILE WORK—inter. (ceramic)	Rate	4.86	4.70	2.17	6.83	4.21	6.35	4.25	5.74	3.41	14.69	7.01	4.36
	Code	42113	721054	40103	719	427	4276	4-41	4276	402	80110	B13-01	202
WATERPROOFING	Rate	5.64	6.94	4.57	4.83	6.12	6.35	4.25	7.67	3.41	23.29	6.20	4.36
	Code	42139	721016	40102	723	423	4299	4-41	4239	402	80130	B12-17	202
WRECKING	Rate	3.59	6.53	6.46	16.47	3.69	5.55	3.50	4.11	6.15	14.69	7.01	4.36
	Code	40604	721005	40106	748	421	4211	4-43	4211	401	80110	B13-09	202

R01100-070   Contractor's Overhead & Profit

Below are the **average** installing contractor's percentage mark-ups applied to base labor rates to arrive at typical billing rates.

Column A: Labor rates are based on union wages averaged for 30 major U.S. cities. Base rates including fringe benefits are listed hourly and daily. These figures are the sum of the wage rate and employer-paid fringe benefits such as vacation pay, employer-paid health and welfare costs, pension costs, plus appropriate training and industry advancement funds costs.

Column B: Workers' Compensation rates are the national average of state rates established for each trade.

Column C: Column C lists average fixed overhead figures for all trades. Included are Federal and State Unemployment costs set at 6.2%; Social Security Taxes (FICA) set at 7.65%; Builder's Risk Insurance costs set at 0.44%; and Public Liability costs set at 2.02%. All the percentages except those for Social Security Taxes vary from state to state as well as from company to company.

Columns D and E: Percentages in Columns D and E are based on the presumption that the installing contractor has annual billing of $4,000,000 and up. Overhead percentages may increase with smaller annual billing. The overhead percentages for any given contractor may vary greatly and depend on a number of factors, such as the contractor's annual volume, engineering and logistical support costs, and staff requirements. The figures for overhead and profit will also vary depending on the type of job, the job location, and the prevailing economic conditions. All factors should be examined very carefully for each job.

Column F: Column F lists the total of Columns B, C, D, and E.

Column G: Column G is Column A (hourly base labor rate) multiplied by the percentage in Column F (O&P percentage).

Column H: Column H is the total of Column A (hourly base labor rate) plus Column G (Total O&P).

Column I: Column I is Column H multiplied by eight hours.

		A		B	C	D	E	F	G	H	I
		Base Rate Incl. Fringes		Workers' Comp. Ins.	Average Fixed Overhead	Over-head	Profit	Total Overhead & Profit		Rate with O & P	
Abbr.	Trade	Hourly	Daily					%	Amount	Hourly	Daily
Skwk	Skilled Workers Average (35 trades)	$34.85	$278.80	16.2%	16.3%	13.0%	10.0%	55.5%	$19.35	$54.20	$433.60
	Helpers Average (5 trades)	25.55	204.40	18.2		11.0		55.5	$14.20	39.75	318.00
	Foreman Average, Inside ($0.50 over trade)	35.35	282.80	16.2		13.0		55.5	19.60	54.95	439.60
	Foreman Average, Outside ($2.00 over trade)	36.85	294.80	16.2		13.0		55.5	20.45	57.30	458.40
Clab	Common Building Laborers	26.70	213.60	18.4		11.0		55.7	14.85	41.55	332.40
Asbe	Asbestos/Insulation Workers/Pipe Coverers	37.10	296.80	15.5		16.0		57.8	21.45	58.55	468.40
Boil	Boilermakers	42.90	343.20	13.7		16.0		56.0	24.00	66.90	535.20
Bric	Bricklayers	35.25	282.00	15.1		11.0		52.4	18.45	53.70	429.60
Brhe	Bricklayer Helpers	26.90	215.20	15.1		11.0		52.4	14.10	41.00	328.00
Carp	Carpenters	34.25	274.00	18.4		11.0		55.7	19.10	53.35	426.80
Cefi	Cement Finishers	32.85	262.80	9.9		11.0		47.2	15.50	48.35	386.80
Elec	Electricians	40.75	326.00	6.5		16.0		48.8	19.90	60.65	485.20
Elev	Elevator Constructors	45.05	360.40	7.6		16.0		49.9	22.50	67.55	540.40
Eqhv	Equipment Operators, Crane or Shovel	35.90	287.20	10.4		14.0		50.7	18.20	54.10	432.80
Eqmd	Equipment Operators, Medium Equipment	34.65	277.20	10.4		14.0		50.7	17.55	52.20	417.60
Eqlt	Equipment Operators, Light Equipment	33.05	264.40	10.4		14.0		50.7	16.75	49.80	398.40
Eqol	Equipment Operators, Oilers	30.10	240.80	10.4		14.0		50.7	15.25	45.35	362.80
Eqmm	Equipment Operators, Master Mechanics	36.20	289.60	10.4		14.0		50.7	18.35	54.55	436.40
Glaz	Glaziers	33.00	264.00	13.9		11.0		51.2	16.90	49.90	399.20
Lath	Lathers	31.45	251.60	10.9		11.0		48.2	15.15	46.60	372.80
Marb	Marble Setters	33.20	265.60	15.1		11.0		52.4	17.40	50.60	404.80
Mill	Millwrights	35.50	284.00	10.4		11.0		47.7	16.95	52.45	419.60
Mstz	Mosaic and Terrazzo Workers	32.65	261.20	9.7		11.0		47.0	15.35	48.00	384.00
Pord	Painters, Ordinary	30.60	244.80	13.1		11.0		50.4	15.40	46.00	368.00
Psst	Painters, Structural Steel	31.00	248.00	46.8		11.0		84.1	26.05	57.05	456.40
Pape	Paper Hangers	30.30	242.40	13.1		11.0		50.4	15.25	45.55	364.40
Pile	Pile Drivers	33.30	266.40	22.5		16.0		64.8	21.60	54.90	439.20
Plas	Plasterers	31.35	250.80	14.9		11.0		52.2	16.35	47.70	381.60
Plah	Plasterer Helpers	27.10	216.80	14.9		11.0		52.2	14.15	41.25	330.00
Plum	Plumbers	40.85	326.80	8.0		16.0		50.3	20.55	61.40	491.20
Rodm	Rodmen (Reinforcing)	37.95	303.60	24.7		14.0		65.0	24.65	62.60	500.80
Rofc	Roofers, Composition	29.45	235.60	33.1		11.0		70.4	20.75	50.20	401.60
Rots	Roofers, Tile and Slate	29.70	237.60	33.1		11.0		70.4	20.90	50.60	404.80
Rohe	Roofer Helpers (Composition)	21.60	172.80	33.1		11.0		70.4	15.20	36.80	294.40
Shee	Sheet Metal Workers	40.20	321.60	11.5		16.0		53.8	21.65	61.85	494.80
Spri	Sprinkler Installers	40.75	326.00	8.8		16.0		51.1	20.80	61.55	492.40
Stpi	Steamfitters or Pipefitters	40.95	327.60	8.0		16.0		50.3	20.60	61.55	492.40
Ston	Stone Masons	35.30	282.40	15.1		11.0		52.4	18.50	53.80	430.40
Sswk	Structural Steel Workers	38.15	305.20	38.6		14.0		78.9	30.10	68.25	546.00
Tilf	Tile Layers	32.70	261.60	9.7		11.0		47.0	15.35	48.05	384.40
Tilh	Tile Layer Helpers	25.35	202.80	9.7		11.0		47.0	11.90	37.25	298.00
Trlt	Truck Drivers, Light	26.80	214.40	15.2		11.0		52.5	14.05	40.85	326.80
Trhv	Truck Drivers, Heavy	27.55	220.40	15.2		11.0		52.5	14.45	42.00	336.00
Sswl	Welders, Structural Steel	38.15	305.20	38.6		14.0		78.9	30.10	68.25	546.00
Wrck	*Wrecking	26.70	213.60	39.5		11.0		76.8	20.50	47.20	377.60

*Not included in Averages.

R01100-080 Performance Bond

This table shows the cost of a Performance Bond for a construction job scheduled to be completed in 12 months. Add 1% of the premium cost per month for jobs requiring more than 12 months to complete. The rates are "standard" rates offered to contractors that the bonding company considers financially sound and capable of doing the work. Preferred rates are offered by some bonding companies based upon financial strength of the contractor. Actual rates vary from contractor to contractor and from bonding company to bonding company. Contractors should prequalify through a bonding agency before submitting a bid on a contract that requires a bond.

Contract Amount	Building Construction Class B Projects			Highways & Bridges					
				Class A New Construction			Class A-1 Highway Resurfacing		
First $ 100,000 bid	$25.00 per M			$15.00 per M			$9.40 per M		
Next 400,000 bid	$ 2,500	plus	$15.00 per M	$ 1,500	plus	$10.00 per M	$ 940	plus	$7.20 per M
Next 2,000,000 bid	8,500	plus	10.00 per M	5,500	plus	7.00 per M	3,820	plus	5.00 per M
Next 2,500,000 bid	28,500	plus	7.50 per M	19,500	plus	5.50 per M	15,820	plus	4.50 per M
Next 2,500,000 bid	47,250	plus	7.00 per M	33,250	plus	5.00 per M	28,320	plus	4.50 per M
Over 7,500,000 bid	64,750	plus	6.00 per M	45,750	plus	4.50 per M	39,570	plus	4.00 per M

R01100-090 Sales Tax by State

State sales tax on materials is tabulated below (5 states have no sales tax). Many states allow local jurisdictions, such as a county or city, to levy additional sales tax.

Some projects may be sales tax exempt, particularly those constructed with public funds.

State	Tax (%)	State	Tax (%)	State	Tax (%)	State	Tax (%)
Alabama	4	Illinois	6.25	Montana	0	Rhode Island	7
Alaska	0	Indiana	6	Nebraska	5.5	South Carolina	5
Arizona	5.6	Iowa	5	Nevada	6.5	South Dakota	4
Arkansas	5.125	Kansas	5.3	New Hampshire	0	Tennessee	7
California	7.25	Kentucky	6	New Jersey	6	Texas	6.25
Colorado	2.9	Louisiana	4	New Mexico	5	Utah	4.75
Connecticut	6	Maine	5	New York	4.25	Vermont	6
Delaware	0	Maryland	5	North Carolina	4.5	Virginia	5
District of Columbia	5.75	Massachusetts	5	North Dakota	5	Washington	6.5
Florida	6	Michigan	6	Ohio	6	West Virginia	6
Georgia	4	Minnesota	6.5	Oklahoma	4.5	Wisconsin	5
Hawaii	4	Mississippi	7	Oregon	0	Wyoming	4
Idaho	6	Missouri	4.225	Pennsylvania	6	Average	4.86 %

Sales Tax by Province (Canada)

GST - a value-added tax, which the government imposes on most goods and services provided in or imported into Canada. PST - a retail sales tax, which five of the provinces impose on the price of most goods and some services. QST - a value-added tax, similar to the federal GST, which Quebec imposes. HST - Three provinces have combined their retail sales tax with the federal GST into one harmonized tax.

Province	PST (%)	QST (%)	GST(%)	HST(%)
Alberta	0	0	7	0
British Columbia	7.5	0	7	0
Manitoba	7	0	7	0
New Brunswick	0	0	0	15
Newfoundland	0	0	0	15
Northwest Territories	0	0	7	0
Nova Scotia	0	0	0	15
Ontario	8	0	7	0
Prince Edward Island	10	0	7	0
Quebec	0	7.5	7	0
Saskatchewan	6	0	7	0
Yukon	0	0	7	0

R01100-100 Unemployment Taxes and Social Security Taxes

Mass. State Unemployment tax ranges from 1.12% to 10.96% plus an experience rating assessment the following year, on the first $14,000 of wages. Federal Unemployment tax is 6.2% of the first $7,000 of wages. This is reduced by a credit for payment to the state. The minimum Federal Unemployment tax is 0.8% after all credits.

Combined rates in Mass. thus vary from 1.92% to 11.76% of the first $14,000 of wages. Combined average U.S. rate is about 6.2% of the first $7,000. Contractors with permanent workers will pay less since the average annual wages for skilled workers is $34.85 x 2,000 hours or about $69,700 per year. The average combined rate for U.S. would thus be 6.2% x $7,000 ÷ $69,700 = 0.6% of total wages for permanent employees.

Rates vary not only from state to state but also with the experience rating of the contractor.

Social Security (FICA) for 2005 is estimated at time of publication to be 7.65% of wages up to $87,900.

R01100-110   Overtime

One way to improve the completion date of a project or eliminate negative float from a schedule is to compress activity duration times. This can be achieved by increasing the crew size or working overtime with the proposed crew.

To determine the costs of working overtime to compress activity duration times, consider the following examples. Below is an overtime efficiency and cost chart based on a five, six, or seven day week with an eight through twelve hour day. Payroll percentage increases for time and one half and double time are shown for the various working days.

Days per Week	Hours per Day	Production Efficiency					Payroll Cost Factors	
		1 Week	2 Weeks	3 Weeks	4 Weeks	Average 4 Weeks	@ 1-1/2 Times	@ 2 Times
	8	100%	100%	100%	100%	100 %	100 %	100 %
	9	100	100	95	90	96.25	105.6	111.1
5	10	100	95	90	85	91.25	110.0	120.0
	11	95	90	75	65	81.25	113.6	127.3
	12	90	85	70	60	76.25	116.7	133.3
	8	100	100	95	90	96.25	108.3	116.7
	9	100	95	90	85	92.50	113.0	125.9
6	10	95	90	85	80	87.50	116.7	133.3
	11	95	85	70	65	78.75	119.7	139.4
	12	90	80	65	60	73.75	122.2	144.4
	8	100	95	85	75	88.75	114.3	128.6
	9	95	90	80	70	83.75	118.3	136.5
7	10	90	85	75	65	78.75	121.4	142.9
	11	85	80	65	60	72.50	124.0	148.1
	12	85	75	60	55	68.75	126.2	152.4

R01100-710   Unit Gross Area Requirements

The figures in the table below indicate typical ranges in square feet as a function of the "occupant" unit. This table is best used in the preliminary design stages to help determine the probable size requirement for the total project. See R17100-100 for the typical total size ranges for various types of buildings.

Building Type	Unit	Gross Area in S.F.		
		1/4	Median	3/4
Apartments	Unit	660	860	1,100
Auditorium & Play Theaters	Seat	18	25	38
Bowling Alleys	Lane		940	
Churches & Synagogues	Seat	20	28	39
Dormitories	Bed	200	230	275
Fraternity & Sorority Houses	Bed	220	315	370
Garages, Parking	Car	325	355	385
Hospitals	Bed	685	850	1,075
Hotels	Rental Unit	475	600	710
Housing for the elderly	Unit	515	635	755
Housing, Public	Unit	700	875	1,030
Ice Skating Rinks	Total	27,000	30,000	36,000
Motels	Rental Unit	360	465	620
Nursing Homes	Bed	290	350	450
Restaurants	Seat	23	29	39
Schools, Elementary	Pupil	65	77	90
Junior High & Middle		85	110	129
Senior High		102	130	145
Vocational		110	135	195
Shooting Ranges	Point		450	
Theaters & Movies	Seat		15	

R01100-720 Floor Area Ratios

Table below lists commonly used gross to net area and net to gross area ratios expressed in % for various building types.

Building Type	Gross to Net Ratio	Net to Gross Ratio	Building Type	Gross to Net Ratio	Net to Gross Ratio
Apartment	156	64	School Buildings (campus type)		
Bank	140	72	Administrative	150	67
Church	142	70	Auditorium	142	70
Courthouse	162	61	Biology	161	62
Department Store	123	81	Chemistry	170	59
Garage	118	85	Classroom	152	66
Hospital	183	55	Dining Hall	138	72
Hotel	158	63	Dormitory	154	65
Laboratory	171	58	Engineering	164	61
Library	132	76	Fraternity	160	63
Office	135	75	Gymnasium	142	70
Restaurant	141	70	Science	167	60
Warehouse	108	93	Service	120	83
			Student Union	172	59

The gross area of a building is the total floor area based on outside dimensions.

The net area of a building is the usable floor area for the function intended and excludes such items as stairways, corridors and mechanical rooms. In the case of a commercial building, it might be considered as the "leasable area."

R01100-730 Occupancy Determinations

Description		S.F. Required per Person		
		BOCA	SBC	UBC
Assembly Areas	Fixed Seats	**	6	7
	Movable Seats		15	15
	Concentrated	7		
	Unconcentrated	15		
	Standing Space	3		
Educational	Unclassified			
	Classrooms	20	40	20
	Shop Areas	50	100	50
Institutional	Unclassified		125	
	In-Patient Areas	240		
	Sleeping Areas	120		
Mercantile	Basement	30	30	20
	Ground Floor	30	30	30
	Upper Floors	60	60	50
Office		100	100	100

BOCA=Building Officials & Code Administrators
SBC=Southern Building Code
UBC=Uniform Building Code

** The occupancy load for assembly area with fixed seats shall be determined by the number of fixed seats installed.

R01100-740 Weather Data and Design Conditions

City	Latitude (1) 0	Latitude (1) 1'	Winter Temperatures (1) Med. of Annual Extremes	Winter Temperatures (1) 99%	Winter Temperatures (1) 97½%	Winter Degree Days (2)	Summer (Design Dry Bulb) Temperatures and Relative Humidity 1%	Summer (Design Dry Bulb) Temperatures and Relative Humidity 2½%	Summer (Design Dry Bulb) Temperatures and Relative Humidity 5%
UNITED STATES									
Albuquerque, NM	35	0	5.1	12	16	4,400	96/61	94/61	92/61
Atlanta, GA	33	4	11.9	17	22	3,000	94/74	92/74	90/73
Baltimore, MD	39	2	7	14	17	4,600	94/75	91/75	89/74
Birmingham, AL	33	3	13	17	21	2,600	96/74	94/75	92/74
Bismarck, ND	46	5	-32	-23	-19	8,800	95/68	91/68	88/67
Boise, ID	43	3	1	3	10	5,800	96/65	94/64	91/64
Boston, MA	42	2	-1	6	9	5,600	91/73	88/71	85/70
Burlington, VT	44	3	-17	-12	-7	8,200	88/72	85/70	82/69
Charleston, WV	38	2	3	7	11	4,400	92/74	90/73	87/72
Charlotte, NC	35	1	13	18	22	3,200	95/74	93/74	91/74
Casper, WY	42	5	-21	-11	-5	7,400	92/58	90/57	87/57
Chicago, IL	41	5	-8	-3	2	6,600	94/75	91/74	88/73
Cincinnati, OH	39	1	0	1	6	4,400	92/73	90/72	88/72
Cleveland, OH	41	2	-3	1	5	6,400	91/73	88/72	86/71
Columbia, SC	34	0	16	20	24	2,400	97/76	95/75	93/75
Dallas, TX	32	5	14	18	22	2,400	102/75	100/75	97/75
Denver, CO	39	5	-10	-5	1	6,200	93/59	91/59	89/59
Des Moines, IA	41	3	-14	-10	-5	6,600	94/75	91/74	88/73
Detroit, MI	42	2	-3	3	6	6,200	91/73	88/72	86/71
Great Falls, MT	47	3	-25	-21	-15	7,800	91/60	88/60	85/59
Hartford, CT	41	5	-4	3	7	6,200	91/74	88/73	85/72
Houston, TX	29	5	24	28	33	1,400	97/77	95/77	93/77
Indianapolis, IN	39	4	-7	-2	2	5,600	92/74	90/74	87/73
Jackson, MS	32	2	16	21	25	2,200	97/76	95/76	93/76
Kansas City, MO	39	1	-4	2	6	4,800	99/75	96/74	93/74
Las Vegas, NV	36	1	18	25	28	2,800	108/66	106/65	104/65
Lexington, KY	38	0	-1	3	8	4,600	93/73	91/73	88/72
Little Rock, AR	34	4	11	15	20	3,200	99/76	96/77	94/77
Los Angeles, CA	34	0	36	41	43	2,000	93/70	89/70	86/69
Memphis, TN	35	0	10	13	18	3,200	98/77	95/76	93/76
Miami, FL	25	5	39	44	47	200	91/77	90/77	89/77
Milwaukee, WI	43	0	-11	-8	-4	7,600	90/74	87/73	84/71
Minneapolis, MN	44	5	-22	-16	-12	8,400	92/75	89/73	86/71
New Orleans, LA	30	0	28	29	33	1,400	93/78	92/77	90/77
New York, NY	40	5	6	11	15	5,000	92/74	89/73	87/72
Norfolk, VA	36	5	15	20	22	3,400	93/77	91/76	89/76
Oklahoma City, OK	35	2	4	9	13	3,200	100/74	97/74	95/73
Omaha, NE	41	2	-13	-8	-3	6,600	94/76	91/75	88/74
Philadelphia, PA	39	5	6	10	14	4,400	93/75	90/74	87/72
Phoenix, AZ	33	3	27	31	34	1,800	109/71	107/71	105/71
Pittsburgh, PA	40	3	-1	3	7	6,000	91/72	88/71	86/70
Portland, ME	43	4	-10	-6	-1	7,600	87/72	84/71	81/69
Portland, OR	45	4	18	17	23	4,600	89/68	85/67	81/65
Portsmouth, NH	43	1	-8	-2	2	7,200	89/73	85/71	83/70
Providence, RI	41	4	-1	5	9	6,000	89/73	86/72	83/70
Rochester, NY	43	1	-5	1	5	6,800	91/73	88/71	85/70
Salt Lake City, UT	40	5	0	3	8	6,000	97/62	95/62	92/61
San Francisco, CA	37	5	36	38	40	3,000	74/63	71/62	69/61
Seattle, WA	47	4	22	22	27	5,200	85/68	82/66	78/65
Sioux Falls, SD	43	4	-21	-15	-11	7,800	94/73	91/72	88/71
St. Louis, MO	38	4	-3	3	8	5,000	98/75	94/75	91/75
Tampa, FL	28	0	32	36	40	680	92/77	91/77	90/76
Trenton, NJ	40	1	4	11	14	5,000	91/75	88/74	85/73
Washington, DC	38	5	7	14	17	4,200	93/75	91/74	89/74
Wichita, KS	37	4	-3	3	7	4,600	101/72	98/73	96/73
Wilmington, DE	39	4	5	10	14	5,000	92/74	89/74	87/73
ALASKA									
Anchorage	61	1	-29	-23	-18	10,800	71/59	68/58	66/56
Fairbanks	64	5	-59	-51	-47	14,280	82/62	78/60	75/59
CANADA									
Edmonton, Alta.	53	3	-30	-29	-25	11,000	85/66	82/65	79/63
Halifax, N.S.	44	4	-4	1	5	8,000	79/66	76/65	74/64
Montreal, Que.	45	3	-20	-16	-10	9,000	88/73	85/72	83/71
Saskatoon, Sask.	52	1	-35	-35	-31	11,000	89/68	86/66	83/65
St. John's, N.F.	47	4	1	3	7	8,600	77/66	75/65	73/64
Saint John, N.B.	45	2	-15	-12	-8	8,200	80/67	77/65	75/64
Toronto, Ont.	43	4	-10	-5	-1	7,000	90/73	87/72	85/71
Vancouver, B.C.	49	1	13	15	19	6,000	79/67	77/66	74/65
Winnipeg, Man.	49	5	-31	-30	-27	10,800	89/73	86/71	84/70

(1) Handbook of Fundamentals, ASHRAE, Inc., NY 1989
(2) Local Climatological Annual Survey, USDC Env. Science Services
Administration, Asheville, NC

R01100-750 Metric Conversion Factors

Description: This table is primarily for converting customary U.S. units in the left hand column to SI metric units in the right hand column. In addition, conversion factors for some commonly encountered Canadian and non-SI metric units are included.

If You Know		Multiply By		To Find
Length				
Inches	x	25.4[a]	=	Millimeters
Feet	x	0.3048[a]	=	Meters
Yards	x	0.9144[a]	=	Meters
Miles (statute)	x	1.609	=	Kilometers
Area				
Square inches	x	645.2	=	Square millimeters
Square feet	x	0.0929	=	Square meters
Square yards	x	0.8361	=	Square meters
Volume (Capacity)				
Cubic inches	x	16,387	=	Cubic millimeters
Cubic feet	x	0.02832	=	Cubic meters
Cubic yards	x	0.7646	=	Cubic meters
Gallons (U.S. liquids)[b]	x	0.003785	=	Cubic meters[c]
Gallons (Canadian liquid)[b]	x	0.004546	=	Cubic meters[c]
Ounces (U.S. liquid)[b]	x	29.57	=	Milliliters[c, d]
Quarts (U.S. liquid)[b]	x	0.9464	=	Liters[c, d]
Gallons (U.S. liquid)[b]	x	3.785	=	Liters[c, d]
Force				
Kilograms force[d]	x	9.807	=	Newtons
Pounds force	x	4.448	=	Newtons
Pounds force	x	0.4536	=	Kilograms force[d]
Kips	x	4448	=	Newtons
Kips	x	453.6	=	Kilograms force[d]
Pressure, Stress, Strength (Force per unit area)				
Kilograms force per square centimeter[d]	x	0.09807	=	Megapascals
Pounds force per square inch (psi)	x	0.006895	=	Megapascals
Kips per square inch	x	6.895	=	Megapascals
Pounds force per square inch (psi)	x	0.07031	=	Kilograms force per square centimeter[d]
Pounds force per square foot	x	47.88	=	Pascals
Pounds force per square foot	x	4.882	=	Kilograms force per square meter[d]
Bending Moment Or Torque				
Inch-pounds force	x	0.01152	=	Meter-kilograms force[d]
Inch-pounds force	x	0.1130	=	Newton-meters
Foot-pounds force	x	0.1383	=	Meter-kilograms force[d]
Foot-pounds force	x	1.356	=	Newton-meters
Meter-kilograms force[d]	x	9.807	=	Newton-meters
Mass				
Ounces (avoirdupois)	x	28.35	=	Grams
Pounds (avoirdupois)	x	0.4536	=	Kilograms
Tons (metric)	x	1000	=	Kilograms
Tons, short (2000 pounds)	x	907.2	=	Kilograms
Tons, short (2000 pounds)	x	0.9072	=	Megagrams[e]
Mass per Unit Volume				
Pounds mass per cubic foot	x	16.02	=	Kilograms per cubic meter
Pounds mass per cubic yard	x	0.5933	=	Kilograms per cubic meter
Pounds mass per gallon (U.S. liquid)[b]	x	119.8	=	Kilograms per cubic meter
Pounds mass per gallon (Canadian liquid)[b]	x	99.78	=	Kilograms per cubic meter
Temperature				
Degrees Fahrenheit	(F-32)/1.8		=	Degrees Celsius
Degrees Fahrenheit	(F+459.67)/1.8		=	Degrees Kelvin
Degrees Celsius	C+273.15		=	Degrees Kelvin

[a]The factor given is exact
[b]One U.S. gallon = 0.8327 Canadian gallon
[c]1 liter = 1000 milliliters = 1000 cubic centimeters
1 cubic decimeter = 0.001 cubic meter

[d]Metric but not SI unit
[e]Called "tonne" in England and
"metric ton" in other metric countries

R01100-760 Weights and Measures

Measures of Length

1 Mile = 1760 Yards = 5280 Feet
1 Yard = 3 Feet = 36 inches
1 Foot = 12 Inches
1 Mil = 0.001 Inch
1 Fathom = 2 Yards = 6 Feet
1 Rod = 5.5 Yards = 16.5 Feet
1 Hand = 4 Inches
1 Span = 9 Inches
1 Micro-inch = One Millionth Inch or 0.000001 Inch
1 Micron = One Millionth Meter + 0.00003937 Inch

Surveyor's Measure

1 Mile = 8 Furlongs = 80 Chains
1 Furlong = 10 Chains = 220 Yards
1 Chain = 4 Rods = 22 Yards = 66 Feet = 100 Links
1 Link = 7.92 Inches

Square Measure

1 Square Mile = 640 Acres = 6400 Square Chains
1 Acre = 10 Square Chains = 4840 Square Yards = 43,560 Sq. Ft.
1 Square Chain = 16 Square Rods = 484 Square Yards = 4356 Sq. Ft.
1 Square Rod = 30.25 Square Yards = 272.25 Square Feet = 625 Square Lines
1 Square Yard = 9 Square Feet
1 Square Foot = 144 Square Inches
An Acre equals a Square 208.7 Feet per Side

Cubic Measure

1 Cubic Yard = 27 Cubic Feet
1 Cubic Foot = 1728 Cubic Inches
1 Cord of Wood = 4 x 4 x 8 Feet = 128 Cubic Feet
1 Perch of Masonry = 16½ x 1½ x 1 Foot = 24.75 Cubic Feet

Avoirdupois or Commercial Weight

1 Gross or Long Ton = 2240 Pounds
1 Net or Short ton = 2000 Pounds
1 Pound = 16 Ounces = 7000 Grains
1 Ounce = 16 Drachms = 437.5 Grains
1 Stone = 14 Pounds

Shipping Measure

For Measuring Internal Capacity of a Vessel:
1 Register Ton = 100 Cubic Feet

For Measurement of Cargo:
Approximately 40 Cubic Feet of Merchandise is considered a Shipping Ton, unless that bulk would weigh more than 2000 Pounds, in which case Freight Charge may be based upon weight.

40 Cubic Feet = 32.143 U.S. Bushels = 31.16 Imp. Bushels

Liquid Measure

1 Imperial Gallon = 1.2009 U.S. Gallon = 277.42 Cu. In.
1 Cubic Foot = 7.48 U.S. Gallons

R01107-030 Engineering Fees

Typical **Structural Engineering Fees** based on type of construction and total project size. These fees are included in Architectural Fees.

Type of Construction	Total Project Size (in thousands of dollars)			
	$500	$500-$1,000	$1,000-$5,000	Over $5000
Industrial buildings, factories & warehouses	Technical payroll times 2.0 to 2.5	1.60%	1.25%	1.00%
Hotels, apartments, offices, dormitories, hospitals, public buildings, food stores		2.00%	1.70%	1.20%
Museums, banks, churches and cathedrals		2.00%	1.75%	1.25%
Thin shells, prestressed concrete, earthquake resistive		2.00%	1.75%	1.50%
Parking ramps, auditoriums, stadiums, convention halls, hangars & boiler houses		2.50%	2.00%	1.75%
Special buildings, major alterations, underpinning & future expansion	↓	Add to above 0.5%	Add to above 0.5%	Add to above 0.5%

For complex reinforced concrete or unusually complicated structures, add 20% to 50%.

Typical **Mechanical and Electrical Engineering Fees** are based on the size of the subcontract. The fee structure for Mechanical Engineering is shown below; Electrical Engineering fees range from 2 to 5%. These fees are included in Architectural Fees.

Type of Construction	Subcontract Size							
	$25,000	$50,000	$100,000	$225,000	$350,000	$500,000	$750,000	$1,000,000
Simple structures	6.4%	5.7%	4.8%	4.5%	4.4%	4.3%	4.2%	4.1%
Intermediate structures	8.0	7.3	6.5	5.6	5.1	5.0	4.9	4.8
Complex structures	12.0	9.0	9.0	8.0	7.5	7.5	7.0	7.0

For renovations, add 15% to 25% to applicable fee.

R01250-010   Repair and Remodeling

Cost figures are based on new construction utilizing the most cost-effective combination of labor, equipment and material with the work scheduled in proper sequence to allow the various trades to accomplish their work in an efficient manner.

The costs for repair and remodeling work must be modified due to the following factors that may be present in any given repair and remodeling project.

1. Equipment usage curtailment due to the physical limitations of the project, with only hand-operated equipment being used.

2. Increased requirement for shoring and bracing to hold up the building while structural changes are being made and to allow for temporary storage of construction materials on above-grade floors.

3. Material handling becomes more costly due to having to move within the confines of an enclosed building. For multi-story construction, low capacity elevators and stairwells may be the only access to the upper floors.

4. Large amount of cutting and patching and attempting to match the existing construction is required. It is often more economical to remove entire walls rather than create many new door and window openings. This sort of trade-off has to be carefully analyzed.

5. Cost of protection of completed work is increased since the usual sequence of construction usually cannot be accomplished.

6. Economies of scale usually associated with new construction may not be present. If small quantities of components must be custom fabricated due to job requirements, unit costs will naturally increase. Also, if only small work areas are available at a given time, job scheduling between trades becomes difficult and subcontractor quotations may reflect the excessive start-up and shut-down phases of the job.

7. Work may have to be done on other than normal shifts and may have to be done around an existing production facility which has to stay in production during the course of the repair and remodeling.

8. Dust and noise protection of adjoining non-construction areas can involve substantial special protection and alter usual construction methods.

9. Job may be delayed due to unexpected conditions discovered during demolition or removal. These delays ultimately increase construction costs.

10. Piping and ductwork runs may not be as simple as for new construction. Wiring may have to be snaked through walls and floors.

11. Matching "existing construction" may be impossible because materials may no longer be manufactured. Substitutions may be expensive.

12. Weather protection of existing structure requires additional temporary structures to protect building at openings.

13. On small projects, because of local conditions, it may be necessary to pay a tradesman for a minimum of four hours for a task that is completed in one hour.

All of the above areas can contribute to increased costs for a repair and remodeling project. Each of the above factors should be considered in the planning, bidding and construction stage in order to minimize the increased costs associated with repair and remodeling jobs.

R01540-100   Steel Tubular Scaffolding

On new construction, tubular scaffolding is efficient up to 60′ high or five stories. Above this it is usually better to use a hung scaffolding if construction permits. Swing scaffolding operations may interfere with tenants. In this case, the tubular is more practical at all heights.

In repairing or cleaning the front of an existing building the cost of tubular scaffolding per S.F. of building front increases as the height increases above the first tier. The first tier cost is relatively high due to leveling and alignment.

The minimum efficient crew for erection is three workers. For heights over 50′, a crew of four is more efficient. Use two or more on top and two at the bottom for handing up or hoisting. Four workers can erect and dismantle about nine frames per hour up to five stories. From five to eight stories they will average six frames per hour. With 7′ horizontal spacing this will run about 400 S.F. and 265 S.F. of wall surface, respectively. Time for placing planks must be added to the above. On heights above 50′, five planks can be placed per labor-hour.

The table below shows the number of pieces required to erect tubular steel scaffolding for 1000 S.F. of building frontage. This area is made up of a scaffolding system that is 12 frames (11 bays) long by 2 frames high.

For jobs under twenty-five frames, add 50% to rental cost. Rental rates will be lower for jobs over three months duration. Large quantities for long periods can reduce rental rates by 20%.

Description of Component	CSI Line Item	Number of Pieces for 1000 S.F. of Building Front	Unit
5′ Wide Standard Frame, 6′-4″ High	01540-750-2200	24	Ea.
Leveling Jack & Plate	01540-750-2650	24	
Cross Brace	01540-750-2500	44	
Side Arm Bracket, 21″	01540-750-2700	12	
Guardrail Post	01540-750-2550	12	
Guardrail, 7′ section	01540-750-2600	22	
Stairway Section	01540-750-2900	2	
Stairway Starter Bar	01540-750-2910	1	
Stairway Inside Handrail	01540-750-2920	2	
Stairway Outside Handrail	01540-750-2930	2	
Walk-Thru Frame Guardrail	01540-750-2940	2	

Scaffolding is often used as falsework over 15′ high during construction of cast-in-place concrete beams and slabs. Two foot wide scaffolding is generally used for heavy beam construction. The span between frames depends upon the load to be carried with a maximum span of 5′.

Heavy duty shoring frames with a capacity of 10,000#/leg can be spaced up to 10′ O.C. depending upon form support design and loading.

Scaffolding used as horizontal shoring requires less than half the material required with conventional shoring.

On new construction, erection is done by carpenters.

Rolling towers supporting horizontal shores can reduce labor and speed the job. For maintenance work, catwalks with spans up to 70′ can be supported by the rolling towers.

R01590-100 Contractor Equipment

Rental Rates shown in Division 01590 pertain to late model high quality machines in excellent working condition, rented from equipment dealers. Rental rates from contractors may be substantially lower than the rental rates from equipment dealers depending upon economic conditions; for older, less productive machines, reduce rates by a maximum of 15%. Any overtime must be added to the base rates. For shift work, rates are lower. Usual rule of thumb is 150% of one shift rate for two shifts; 200% for three shifts.

For periods of less than one week, operated equipment is usually more economical to rent than renting bare equipment and hiring an operator.

Costs to move equipment to a job site (mobilization) or from a job site (demobilization) are not included in rental rates in Division 01590, nor in any Equipment costs on any Unit Price line items or crew listings. These costs can be found in section 02305-250. If a piece of equipment is already at a job site, it is not appropriate to utilize mob/demob costs in an estimate again.

Rental rates vary throughout the country with larger cities generally having lower rates. Lease plans for new equipment are available for periods in excess of six months with a percentage of payments applying toward purchase.

Monthly rental rates vary from 2% to 5% of the cost of the equipment depending on the anticipated life of the equipment and its wearing parts. Weekly rates are about 1/3 the monthly rates and daily rental rates about 1/3 the weekly rate.

The hourly operating costs for each piece of equipment include costs to the user such as fuel, oil, lubrication, normal expendables for the equipment, and a percentage of mechanic's wages chargeable to maintenance. The hourly operating costs listed do not include the operator's wages.

The daily cost for equipment used in the standard crews is figured by dividing the weekly rate by five, then adding eight times the hourly operating cost to give the total daily equipment cost, not including the operator. This figure is in the right hand column of Division 01590 under Crew Equipment Cost/Day.

Pile Driving rates shown for pile hammer and extractor do not include leads, crane, boiler or compressor. Vibratory pile driving requires an added field specialist during set-up and pile driving operation for the electric model. The hydraulic model requires a field specialist for set-up only. Up to 125 reuses of sheet piling are possible using vibratory drivers. For normal conditions, crane capacity for hammer type and size are as follows.

Crane Capacity	Hammer Type and Size		
	Air or Steam	Diesel	Vibratory
25 ton	to 8,750 ft.-lb.		70 H.P.
40 ton	15,000 ft.-lb.	to 32,000 ft.-lb.	170 H.P.
60 ton	25,000 ft.-lb.		300 H.P.
100 ton		112,000 ft.-lb.	

Cranes should be specified for the job by size, building and site characteristics, availability, performance characteristics, and duration of time required.

Backhoes & Shovels rent for about the same as equivalent size cranes but maintenance and operating expense is higher. Crane operators rate must be adjusted for high boom heights. Average adjustments: for 150' boom add 2% per hour; over 185', add 4% per hour; over 210', add 6% per hour; over 250', add 8% per hour and over 295', add 12% per hour.

Tower Cranes of the climbing or static type have jibs from 50' to 200' and capacities at maximum reach range from 4,000 to 14,000 pounds. Lifting capacities increase up to maximum load as the hook radius decreases.

Typical rental rates, based on purchase price are about 2% to 3% per month.

Erection and dismantling runs between 500 and 2000 labor hours. Climbing operation takes 10 labor hours per 20' climb. Crane dead time is about 5 hours per 40' climb. If crane is bolted to side of the building add cost of ties and extra mast sections. Climbing cranes have from 80' to 180' of mast while static cranes have 80' to 800' of mast.

Truck Cranes can be converted to tower cranes by using tower attachments. Mast heights over 400' have been used. See Division 01590-600 for rental rates of high boom cranes.

A single 100' high material **Hoist and Tower** can be erected and dismantled in about 400 labor hours; a double 100' high hoist and tower in about 600 labor hours. Erection times for additional heights are 3 and 4 labor hours

per vertical foot respectively up to 150', and 4 to 5 labor hours per vertical foot over 150' high. A 40' high portable Buck hoist takes about 160 labor hours to erect and dismantle. Additional heights take 2 labor hours per vertical foot to 80' and 3 labor hours per vertical foot for the next 100'. Most material hoists do not meet local code requirements for carrying personnel.

A 150' high **Personnel Hoist** requires about 500 to 800 labor hours to erect and dismantle. Budget erection time at 5 labor hours per vertical foot for all trades. Local code requirements or labor scarcity requiring overtime can add up to 50% to any of the above erection costs.

Earthmoving Equipment: The selection of earthmoving equipment depends upon the type and quantity of material, moisture content, haul distance, haul road, time available, and equipment available. Short haul cut and fill operations may require dozers only, while another operation may require excavators, a fleet of trucks, and spreading and compaction equipment. Stockpiled material and granular material are easily excavated with front end loaders. Scrapers are most economically used with hauls between 300' and 1-1/2 miles if adequate haul roads can be maintained. Shovels are often used for blasted rock and any material where a vertical face of 8' or more can be excavated. Special conditions may dictate the use of draglines, clamshells, or backhoes. Spreading and compaction equipment must be matched to the soil characteristics, the compaction required and the rate the fill is being supplied.

For information about Means Estimating Seminars, see yellow pages 12 and 13 in back of book

R02115-200 Underground Storage Tank Removal

Underground Storage Tank Removal can be divided into two categories: Non-Leaking and Leaking. Prior to removing an underground storage tank, tests should be made, with the proper authorities present, to determine whether a tank has been leaking or the surrounding soil has been contaminated.

To safely remove Liquid Underground Storage Tanks:
1. Excavate to the top of the tank.
2. Disconnect all piping.
3. Open all tank vents and access ports.
4. Remove all liquids and/or sludge.
5. Purge the tank with an inert gas.
6. Provide access to the inside of the tank and clean out the interior using proper personal protective equipment (PPE).
7. Excavate soil surrounding the tank using proper PPE for on-site personnel.
8. Pull and properly dispose of the tank.
9. Clean up the site of all contaminated material.
10. Install new tanks or close the excavation.

R02220-510 Demolition Defined

Whole Building Demolition (Division 02220-110) - Demolition of the whole building with no concern for any particular building element, component, or material type being demolished. This type of demolition is accomplished with large pieces of construction equipment that break up the structure, load it into trucks and haul it to a disposal site, but disposal or dump fees are not included (Divisions 02220-320 and 02220-330). Demolition of below-grade foundation elements, such as footings, foundation walls, grade beams, slabs on grade, etc. (Division 02220-130), is not included. Certain mechanical equipment containing flammable liquids or ozone-depleting refrigerants, electric lighting elements, communication equipment components, and other building elements may contain hazardous waste, and must be removed, either selectively or carefully, as hazardous waste before the building can be demolished (Division 13281).

Foundation Demolition (Division 02220-130) - Demolition of below-grade foundation footings, foundation walls, grade beams, and slabs on grade. This type of demolition is accomplished by hand or pneumatic hand tools, and does not include saw cutting (Division 02220-360), or handling, loading, hauling, or disposal of the debris (Divisions 02220-320, 02220-330 and 02220-350).

Gutting (Division 02220-340) - Removal of building interior finishes and electrical/mechanical systems down to the load-bearing and sub-floor elements of the rough building frame, with no concern for any particular building element, component, or material type being demolished. This type of demolition is accomplished by hand or pneumatic hand tools, and includes loading into trucks, but not hauling, disposal or dump fees (Divisions 02220-320 and 02220-330); scaffolding (01540-750); or shoring (03150-600). Certain mechanical equipment containing flammable liquids or ozone-depleting refrigerants, electric lighting elements, communication equipment components, and other building elements may contain hazardous waste, and must be removed, either selectively or carefully, as hazardous waste, before the building is gutted (Division 13281).

Selective Demolition - Demolition of a selected building element, component, or finish, with some concern for surrounding or adjacent elements, components, or finishes (see the first Subdivision (s) at the beginning of Divisions 3 through 16). This type of demolition is accomplished by hand or pneumatic hand tools, and does not include handling, loading, storing, hauling, or disposal of the debris (Divisions 02220-320, 02220-330

and 02220-350); scaffolding (Division 01540-750); or shoring (Division 03150-600). "Gutting" methods may be used in order to save time, but damage that was caused to surrounding or adjacent elements, components, or finishes may have to be repaired at a later time.

Careful Removal - Removal of a piece of service equipment, building element or component, or material type, with great concern for both the removed item and surrounding or adjacent elements, components or finishes. The purpose of careful removal may be to protect the removed item for later re-use, preserve a higher salvage value of the removed item, or replace an item while taking care to protect surrounding or adjacent elements, components, connections, or finishes from cosmetic and/or structural damage. An approximation of the time required to perform this type of removal is 1/3 to 1/2 the time it would take to install a new item of like kind (see Reference Numbers R15050-720 and R16055-300). This type of removal is accomplished by hand or pneumatic hand tools, and does not include loading, hauling, or storing the removed item (Divisions 01840-100 and 02220-350); scaffolding (01540-750); shoring (Division 03150-600); or lifting equipment (01590-600).

Cutout Demolition (Division 02220-310) - Demolition of a small quantity of floor, wall, roof, or other assembly, with concern for the appearance and structural integrity of the surrounding materials. This type of demolition is accomplished by hand or pneumatic hand tools, and does not include saw cutting (Division 02220-360); handling, loading, hauling, or disposal of debris (Divisions 02220-320, 02220-330 and 02220-350); scaffolding (Division 01540-750); or shoring (Division 03150-600).

Rubbish Handling (Division 02220-350) - Work activities that involve handling, loading or hauling of debris. Generally, the cost of rubbish handling must be added to the cost of all types of demolition, with the exception of whole building demolition.

Minor Site Demolition (Division 02220-240) - Demolition of site element outside the footprint of a building. This type of demolition is accomplished by hand or pneumatic hand tools, or with larger pieces of construction equipment, and may include loading a removed item onto a truck (check the Crew for equipment used). It does not include saw cutting (Division 02220-360), hauling or disposal of debris (Divisions 02220-320 and 02220-330), and, sometimes, handling or loading (Division 02220-350).

R02240-900 Wellpoints

A single stage wellpoint system is usually limited to dewatering an average 15′ depth below normal ground water level. Multi-stage systems are employed for greater depth with the pumping equipment installed only at the lowest header level. Ejectors, with unlimited lift capacity, can be economical when two or more stages of wellpoints can be replaced or when horizontal clearance is restricted, such as in deep trenches or tunneling projects, and where low water flows are expected. Wellpoints are usually spaced on 2-1/2′ to 10′ centers along a header pipe. Wellpoint spacing, header size, and pump size are all determined by the expected flow, as dictated by soil conditions.

In almost all soils encountered in wellpoint dewatering, the wellpoints may be jetted into place. Cemented soils and stiff clays may require sand wicks about 12″ in diameter around each wellpoint to increase efficiency and eliminate weeping into the excavation. These sand wicks require 1/2 to 3 C.Y. of washed filter sand and are installed by using a 12″ diameter steel casing and hole puncher jetted into the ground 2′ deeper than the wellpoint. Rock may require predrilled holes.

Labor required for the complete installation and removal of a single stage wellpoint system is in the range of 3/4 to 2 labor-hours per linear foot of header, depending upon jetting conditions, wellpoint spacing, etc.

Continuous pumping is necessary except in some free draining soil where temporary flooding is permissible (as in trenches which are backfilled after each day's work). Good practice requires provision of a stand-by pump during the continuous pumping operation.

Systems for continuous trenching below the water table should be installed three to four times the length of expected daily progress to insure uninterrupted digging, and header pipe size should not be changed during the job.

For pervious free draining soils, deep wells in place of wellpoints may be economical because of lower installation and maintenance costs. Daily production ranges between two to three wells per day, for 25′ to 40′ depths, to one well per day for depths over 50′.

Detailed analysis and estimating for any dewatering problem is available at no cost from wellpoint manufacturers. Major firms will quote "sufficient equipment" quotes or their affiliates offer lump sum proposals to cover complete dewatering responsibility.

	Description for 200′ System with 8″ Header	Quantities	1st Month		Thereafter	
			Unit	Total	Unit	Total
Equipment & Material	Wellpoints 25′ long, 2″ diameter @ 5′ O.C.	40 Each	$ 31.00	$ 1,240.00	$ 23.25	$ 930.00
	Header pipe, 8″ diameter	200 L.F.	5.90	1,180.00	2.95	590.00
	Discharge pipe, 8″ diameter	100 L.F.	3.96	396.00	2.57	257.00
	8″ valves	3 Each	151.00	453.00	98.15	294.45
	Combination Jetting & Wellpoint pump (standby)	1 Each	2,450.00	2,450.00	1,715.00	1,715.00
	Wellpoint pump, 8″ diameter	1 Each	2,400.00	2,400.00	1,680.00	1,680.00
	Transportation to and from site	1 Day	697.00	697.00	—	—
	Fuel 30 days x 60 gal./day	1800 Gallons	2.00	3,600.00	2.00	3,600.00
	Lubricants for 30 days x 16 lbs./day	480 Lbs.	1.50	720.00	1.50	720.00
	Sand for points	40 C.Y.	21.35	854.00	—	—
	Equipment & Materials Subtotal			$13,990.00		$ 9,786.45
Labor	Technician to supervise installation	1 Week	$ 36.20	$ 1,448.00	—	—
	Labor for installation and removal of system	300 Labor-hours	26.70	8,010.00	—	—
	4 Operators straight time 40 hrs./wk. for 4.33 wks.	693 Hrs.	33.05	22,903.65	$ 33.05	$22,903.65
	4 Operators overtime 2 hrs./wk. for 4.33 wks.	35 Hrs.	66.10	2,313.50	66.10	2,313.50
	Bare Labor Subtotal			$34,675.15		$25,217.15
	Total Bare Cost			$48,665.15		$35,003.60
	Monthly Bare Cost/L.F. Header			$ 243.33		$ 175.02

R02250-400   Wood Sheet Piling

Wood sheet piling may be used for depths to 20′ where there is no ground water. If moderate ground water is encountered Tongue & Groove sheeting will help to keep it out. When considerable ground water is present, steel sheeting must be used.

For estimating purposes on trench excavation, sizes are as follows:

Depth	Sheeting	Wales	Braces	B.F. per S.F.
To 8′	3 x 12's	6 x 8's, 2 line	6 x 8's, @ 10′	4.0 @ 8′
8′ x 12′	3 x 12's	10 x 10's, 2 line	10 x 10's, @ 9′	5.0 average
12′ to 20′	3 x 12's	12 x 12's, 3 line	12 x 12's, @ 8′	7.0 average

Sheeting to be toed in at least 2′ depending upon soil conditions. A five person crew with an air compressor and sheeting driver can drive and brace 440 SF/day at 8′ deep, 360 SF/day at 12′ deep, and 320 SF/day at 16′ deep. For normal soils, piling can be pulled in 1/3 the time to install. Pulling difficulty increases with the time in the ground. Production can be increased by high pressure jetting. Figures below assume 50% of lumber is salvaged and includes pulling costs. Some jurisdictions require an equipment operator in addition to Crew B-31.

Sheeting Pulled

Daily Cost Crew B-31	L.H./ Day	Hourly Cost	Daily Cost	8′ Depth, 440 S.F./Day		16′ Depth, 320 S.F./Day	
				To Drive (1 Day)	To Pull (1/3 Day)	To Drive (1 Day)	To Pull (1/3 Day)
1 Foreman	8	$28.70	$ 229.60	$ 229.60	$ 76.46	$ 229.60	$ 76.46
3 Laborers	24	26.70	640.80	640.80	213.39	640.80	213.39
1 Carpenter	8	34.25	274.00	274.00	91.24	274.00	91.24
1 Air Compressor			112.40	112.40	37.43	112.40	37.43
1 Sheeting Driver			7.20	7.20	2.40	7.20	2.40
2 -50 Ft. Air Hoses, 1-1/2″ Diam.			25.50	25.50	8.49	25.50	8.49
Lumber (50% salvage)				1.76 MBF 616.00		2.24 MBF 784.00	
Bare Total			$1,289.50	$1,905.50	$429.41	$2,073.50	$429.41
Bare Total/S.F.				$ 4.33	$ .98	$ 4.71	$ 1.34
Bare Total (Drive and Pull)/S.F.				$ 5.31		$ 6.05	

Sheeting Left in Place

Daily Cost	8′ Depth 440 S.F./Day	10′ Depth 400 S.F./Day	12′ Depth 360 S.F./Day	16′ Depth 320 S.F./Day	18′ Depth 305 S.F./Day	20′ Depth 280 S.F./Day
Crew B-31	$1,273.40	$1,273.40	$1,273.40	$1,273.40	$1,273.40	$1,273.40
Lumber	1.76 MBF 1,232.00	1.8 MBF 1,260.00	1.8 MBF 1,260.00	2.24 MBF 1,568.00	2.1 MBF 1,470.00	1.9 MBF 1,330.00
Bare Total in Place	$2,505.40	$2,533.40	$2,533.40	$2,841.40	$2,743.40	$2,603.40
Bare Total/S.F.	$ 5.69	$ 6.33	$ 7.04	$ 8.88	$ 8.99	$ 9.30
Bare Total/M.B.F.	$1,423.52	$1,407.44	$1,407.44	$1,268.48	$1,306.38	$1,370.21

R02250-450   Steel Sheet Piling

Limiting weights are 22 to 38#/S.F. of wall surface with 27#/S.F. average for usual types and sizes. (Weights of piles themselves are from 30.7#/L.F. to 57#/L.F. but they are 15" to 21" wide.) Lightweight sections 12" to 28" wide from 3 ga. to 12 ga. thick are also available for shallow excavations. Piles may be driven two at a time with an impact or vibratory hammer (use vibratory to pull) hung from a crane without leads. A reasonable estimate of the life of steel sheet piling is 10 uses with up to 125 uses possible if a vibratory hammer is used. Used piling costs from 50% to 80% of new piling depending on location and market conditions. Sheet piling and H piles can be rented for about 30% of the delivered mill price for the first month and 5% per month thereafter. Allow 1 labor-hour per pile for cleaning and trimming after driving. These costs increase with depth and hydrostatic head. Vibratory drivers are faster in wet granular soils and are excellent for pile extraction. Pulling difficulty increases with the time in the ground and may cost more than driving. It is often economical to abandon the sheet piling, especially if it can be used as the outer wall form. Allow about 1/3 additional length or more, for toeing into ground. Add bracing, waler and strut costs. Waler costs can equal the cost per ton of sheeting.

Cost of Sheet Piling & Production Rate by Ton & S.F.

Depth of Excavation	15' Depth		20' Depth		25' Depth	
Description of Pile	22 psf = 90.9 S.F./Ton		27 psf = 74 S.F./Ton		38 psf = 52.6 S.F./Ton	
Type of operation:	Drive &	Drive &	Drive &	Drive &	Drive &	Drive &
Left in Place or Removed	Left	Extract*	Left	Extract*	Left	Extract*
Labor & Equip. to Drive 1 ton	$ 412.06	$742.40	$ 343.97	$680.06	$ 234.44	$424.23
Piling (75% Salvage)	880.00	220.00	880.00	220.00	880.00	220.00
Bare Cost/Ton in Place	$1,292.06	$962.40	$1,223.97	$900.06	$1,114.44	$644.23
Production Rate, Tons/Day	10.81	6.00	12.95	6.55	19.00	10.50
Bare Cost, S.F. in Place (incl. 33% toe in)	$ 14.21	$ 10.59	$ 16.54	$ 12.16	$ 21.19	$ 12.25
Production Rate, S.F./Day (incl. 33% toe in)	983	545	959	485	1000	553

Crew B-40	Bare Costs		Inc. Subs O&P		Bare Installation Cost per Ton
	Hr.	Daily	Hr.	Daily	
1 Pile Driver Foreman (Out)	$35.30	$ 282.40	$58.15	$ 465.20	for 15' Deep Excavation plus 33% toe-in
4 Pile Drivers	33.30	1,065.60	54.90	1,756.80	$4,454 per day/10.81 tons = $412.03 per ton
2 Equip. Oper. (crane)	35.90	574.40	54.10	865.60	Bare Installation Cost per S.F.
1 Equip. Oper. Oiler	30.10	240.80	45.35	362.80	$\frac{2000\#}{22\#/S.F.}$ = 90.9 S.F. x 10.81 tons = 983 S.F. per day
1 Crane, 40 Ton		888.20		977.00	
Vibratory Hammer & Gen.		1,403.00		1,543.30	$4,454 / 983 S.F. per day = $4.53 per S.F.
64 L.H., Daily Totals		$4,454.40		$5,970.70	

*For driving & extracting, two mobilizations & demobilizations will be necessary

R02315-260 Trench Excavation and Pipe Bedding

						6′ Wide Bottom of Trench			
Example: Find the cost per LF for excavating, bedding, and backfilling a 36″ diameter pipe 10′ deep in damp sandy loam. **Method:** For trench bottom widths for various size pipe, Assembly G1030-815, a 36″ diameter pipe uses a 6′ wide trench bottom.						Slope 1/2 to 1	Slope 1 to 1	Slope 0	Slope 2 to 1
Excavation & Backfill System	For 1/2 to 1 slope see: 1 to 1 slope 1 2 to 1 slope	Trench Excavation ↓	System ↓	G1030-805 ↓	-2880 -4050 -5700 -7300	$37.80	$50.45	$57.50	$65.50
Pipe Bedding System	For 1/2 to 1 slope see: 1 to 1 slope 1 2 to 1 slope	Trench Pipe Bedding ↓	System ↓	G1030-815 ↓	-2400 -2880 -3260 -3660	$13.20	$17.65	$22.10	$26.55
Total Cost, including O&P, per LF (not incl. pipe)						$51.00	$68.10	$79.60	$92.05

R02450-660 Maximum Depth of Frost Penetration in Inches

THIS MAP IS REASONABLY ACCURATE FOR MOST PARTS OF THE UNITED STATES BUT IS NECESSARILY HIGHLY GENERALIZED, AND CONSEQUENTLY NOT TOO ACCURATE IN MOUNTAINOUS REGIONS, PARTICULARLY IN THE ROCKIES.

R02510-800   Piping Designations

There are several systems currently in use to describe pipe and fittings. The following paragraphs will help to identify and clarify classifications of piping systems used for water distribution.

Piping may be classified by schedule. Piping schedules include 5S, 10S, 10, 20, 30, Standard, 40, 60, Extra Strong, 80, 100, 120, 140, 160 and Double Extra Strong. These schedules are dependent upon the pipe wall thickness. The wall thickness of a particular schedule may vary with pipe size.

Ductile iron pipe for water distribution is classified by Pressure Classes such as Class 150, 200, 250, 300 and 350. These classes are actually the rated water working pressure of the pipe in pounds per square inch (psi). The pipe in these pressure classes is designed to withstand the rated water working pressure plus a surge allowance of 100 psi.

The American Water Works Association (AWWA) provides standards for various types of **plastic pipe**. C-900 is the specification for polyvinyl chloride (PVC) piping used for water distribution in sizes ranging from 4″ through 12″. C-901 is the specification for polyethylene (PE) pressure pipe, tubing and fittings used for water distribution in sizes ranging from 1/2″ through 3″. C-905 is the specification for PVC piping sizes 14″ and greater.

PVC pressure-rated pipe is identified using the standard dimensional ratio (SDR) method. This method is defined by the American Society for Testing and Materials (ASTM) Standard D 2241. This pipe is available in SDR numbers 64, 41, 32.5, 26, 21, 17, and 13.5. Pipe with an SDR of 64 will have the thinnest wall while pipe with an SDR of 13.5 will have the thickest wall. When the pressure rating (PR) of a pipe is given in psi, it is based on a line supplying water at 73 degrees F.

The National Sanitation Foundation (NSF) seal of approval is applied to products that can be used with potable water. These products have been tested to ANSI/NSF Standard 14.

Valves and strainers are classified by American National Standards Institute (ANSI) Classes. These Classes are 125, 150, 200, 250, 300, 400, 600, 900, 1500 and 2500. Within each class there is an operating pressure range dependent upon temperature. Design parameters should be compared to the appropriate material dependent, pressure-temperature rating chart for accurate valve selection.

R05090-520   Welded Structural Steel

Usual weight reductions with welded design run 10% to 20% compared with bolted or riveted connections. This amounts to about the same total cost compared with bolted structures since field welding is more expensive than bolts. For normal spans of 18′ to 24′ figure 6 to 7 connections per ton.

Trusses — For welded trusses add 4% to weight of main members for connections. Up to 15% less steel can be expected in a welded truss compared to one that is shop bolted. Cost of erection is the same whether shop bolted or welded.

General — Typical electrodes for structural steel welding are E6010, E6011, E60T and E70T. Typical buildings vary between 2# to 8# of weld rod per

ton of steel. Buildings utilizing continuous design require about three times as much welding as conventional welded structures. In estimating field erection by welding, it is best to use the average linear feet of weld per ton to arrive at the welding cost per ton. The type, size and position of the weld will have a direct bearing on the cost per linear foot. A typical field welder will deposit 1.8# to 2# of weld rod per hour manually. Using semiautomatic methods can increase production by as much as 50% to 75%.

R07800-030   Firestopping

Firestopping is the sealing of structural, mechanical, electrical and other penetrations through fire-rated assemblies. The basic components of firestop systems are safing insulation and firestop sealant on both sides of wall penetrations and the top side of floor penetrations.

Pipe penetrations are assumed to be through concrete, grout, or joint compound and can be sleeved or unsleeved. Costs for the penetrations and sleeves are not included. An annular space of 1″ is assumed. Escutcheons are not included.

Metallic pipe is assumed to be copper, aluminum, cast iron or similar metallic material. Insulated metallic pipe is assumed to be covered with a thermal insulating jacket of varying thickness and materials.

Non-metallic pipe is assumed to be PVC, CPVC, FR Polypropylene or similar plastic piping material. Intumescent firestop sealant or wrap strips are included. Collars on both sides of wall penetrations and a sheet metal plate on the underside of floor penetrations are included.

Ductwork is assumed to be sheet metal, stainless steel or similar metallic material. Duct penetrations are assumed to be through concrete, grout or joint compound. Costs for penetrations and sleeves are not included. An annular space of 1/2″ is assumed.

Multi-trade openings include costs for sheet metal forms, firestop mortar, wrap strips, collars and sealants as necessary.

Structural penetrations joints are assumed to be 1/2″ or less. CMU walls are assumed to be within 1-1/2″ of metal deck. Drywall walls are assumed to be tight to the underside of metal decking.

Metal panel, glass or curtain wall systems include a spandrel area of 5′ filled with mineral wool foil-faced insulation. Fasteners and stiffeners are included.

R10520-110 Sprinkler Systems (Automatic)

Sprinkler systems may be classified by type as follows:

1. **Wet Pipe System.** A system employing automatic sprinklers attached to a piping system containing water and connected to a water supply so that water discharges immediately from sprinklers opened by a fire.

2. **Dry Pipe System.** A system employing automatic sprinklers attached to a piping system containing air under pressure, the release of which as from the opening of sprinklers permits the water pressure to open a valve known as a "dry pipe valve". The water then flows into the piping system and out the opened sprinklers.

3. **Pre-Action System.** A system employing automatic sprinklers attached to a piping system containing air that may or may not be under pressure, with a supplemental heat responsive system of generally more sensitive characteristics than the automatic sprinklers themselves, installed in the same areas as the sprinklers; actuation of the heat responsive system, as from a fire, opens a valve which permits water to flow into the sprinkler piping system and to be discharged from any sprinklers which may be open.

4. **Deluge System.** A system employing open sprinklers attached to a piping system connected to a water supply through a valve which is opened by the operation of a heat responsive system installed in the same areas as the sprinklers. When this valve opens, water flows into the piping system and discharges from all sprinklers attached thereto.

5. **Combined Dry Pipe and Pre-Action Sprinkler System.** A system employing automatic sprinklers attached to a piping system containing air under pressure with a supplemental heat responsive system of generally more sensitive characteristics than the automatic sprinklers themselves, installed in the same areas as the sprinklers; operation of the heat responsive system, as from a fire, actuates tripping devices which open dry pipe valves simultaneously and without loss of air pressure in the system. Operation of the heat responsive system also opens approved air exhaust valves at the end of the feed main which facilitates the filling of the system with water which usually precedes the opening of sprinklers. The heat responsive system also serves as an automatic fire alarm system.

6. **Limited Water Supply System.** A system employing automatic sprinklers and conforming to these standards but supplied by a pressure tank of limited capacity.

7. **Chemical Systems.** Systems using halon, carbon dioxide, dry chemical or high expansion foam as selected for special requirements. Agent may extinguish flames by chemically inhibiting flame propagation, suffocate flames by excluding oxygen, interrupting chemical action of oxygen uniting with fuel or sealing and cooling the combustion center.

8. **Firecycle System.** Firecycle is a fixed fire protection sprinkler system utilizing water as its extinguishing agent. It is a time delayed, recycling, preaction type which automatically shuts the water off when heat is reduced below the detector operating temperature and turns the water back on when that temperature is exceeded. The system senses a fire condition through a closed circuit electrical detector system which controls water flow to the fire automatically. Batteries supply up to 90 hour emergency power supply for system operation. The piping system is dry (until water is required) and is monitored with pressurized air. Should any leak in the system piping occur, an alarm will sound, but water will not enter the system until heat is sensed by a firecycle detector.

Area coverage sprinkler systems may be laid out and fed from the supply in any one of several patterns as shown below. It is desirable, if possible, to utilize a central feed and achieve a shorter flow path from the riser to the furthest sprinkler. This permits use of the smallest sizes of pipe possible with resulting savings.

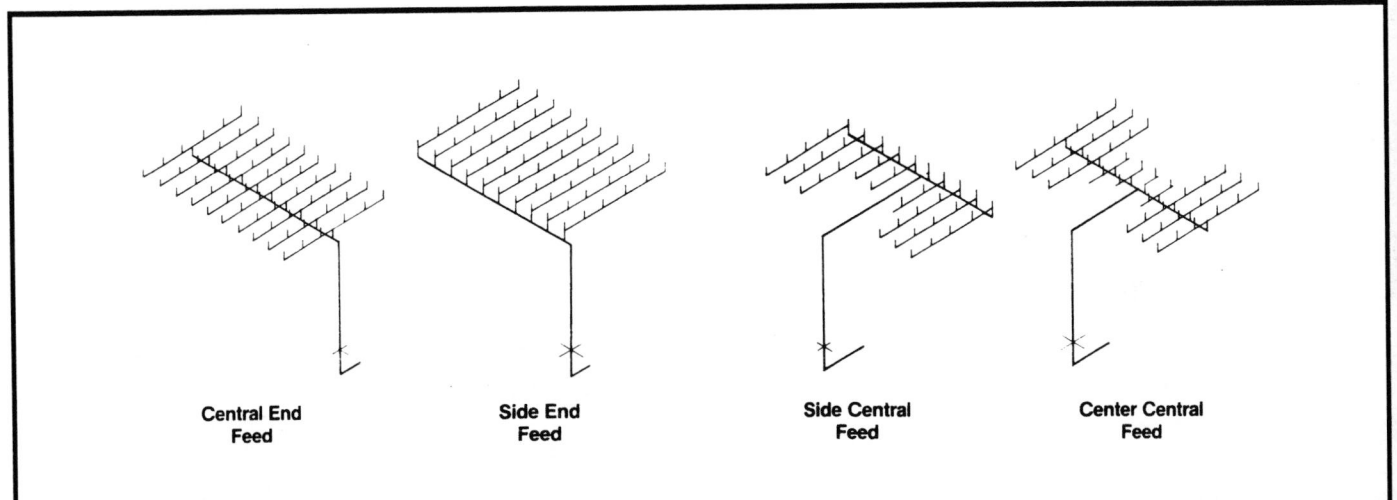

Central End Feed   Side End Feed   Side Central Feed   Center Central Feed

R10520-120 System Classification

System Classification

Rules for installation of sprinkler systems vary depending on the classification of occupancy falling into one of three categories as follows:

Light Hazard Occupancy

The protection area allotted per sprinkler should not exceed 225 S.F., with the maximum distance between lines and sprinklers on lines being 15'. The sprinklers do not need to be staggered. Branch lines should not exceed eight sprinklers on either side of a cross main. Each large area requiring more than 100 sprinklers and without a sub-dividing partition should be supplied by feed mains or risers sized for ordinary hazard occupancy.
Maximum system area = 52,000 S.F.

Included in this group are:

Churches	Nursing Homes
Clubs	Offices
Educational	Residential
Hospitals	Restaurants
Institutional	Theaters and Auditoriums
Libraries	(except stages and prosceniums)
(except large stack rooms)	Unused Attics
Museums	

Ordinary Hazard Occupancy

The protection area allotted per sprinkler shall not exceed 130 S.F. of noncombustible ceiling and 130 S.F. of combustible ceiling. The maximum allowable distance between sprinkler lines and sprinklers on line is 15'. Sprinklers shall be staggered if the distance between heads exceeds 12'. Branch lines should not exceed eight sprinklers on either side of a cross main.
Maximum system area = 52,000 S.F.

Included in this group are:

Group 1
- Automotive Parking and Showrooms
- Bakeries
- Beverage manufacturing
- Canneries
- Dairy Products Manufacturing/Processing
- Electronic Plans
- Glass and Glass Products Manufacturing
- Laundries
- Restaurant Service Areas

Group 2
- Cereal Mills
- Chemical Plants—Ordinary
- Confectionery Products
- Distilleries
- Dry Cleaners
- Feed Mills
- Horse Stables
- Leather Goods Manufacturing
- Libraries—Large Stack Room Areas
- Machine Shops
- Metal Working
- Mercantile
- Paper and Pulp Mills
- Paper Process Plants
- Piers and Wharves
- Post Offices
- Printing and Publishing
- Repair Garages
- Stages
- Textile Manufacturing
- Tire Manufacturing
- Tobacco Products Manufacturing
- Wood Machining
- Wood Product Assembly

Extra Hazard Occupancy

The protection area allotted per sprinkler shall not exceed 100 S.F. of noncombustible ceiling and 100 S.F. of combustible ceiling. The maximum allowable distance between lines and between sprinklers on lines is 12'. Sprinklers on alternate lines shall be staggered if the distance between sprinklers on lines exceeds 8'. Branch lines should not exceed six sprinklers on either side of a cross main.
Maximum system area:
  Design by pipe schedule = 25,000 S.F.
  Design by hydraulic calculation = 40,000 S.F.

Included in this group are:

Group 1
- Aircraft hangars
- Combustible Hydraulic Fluid Use Area
- Die Casting
- Metal Extruding
- Plywood/Particle Board Manufacturing
- Printing (inkls with flash points < 100 degrees F
- Rubber Reclaiming, Compounding, Drying, Milling, Vulcanizing
- Saw Mills
- Textile Picking, Opening, Blending, Garnetting, Carding, Combing of Cotton, Synthetics, Wood Shoddy, or Burlap
- Upholstering with Plastic Foams

Group 2
- Asphalt Saturating
- Flammable Liquids Spraying
- Flow Coating
- Manufactured/Modular Home Building Assemblies (where finished enclosure is present and has combustible interiors)
- Open Oil Quenching
- Plastics Processing
- Solvent Cleaning
- Varnish and Paint Dipping

R10520-130 Sprinkler Quantities for Various Sizes and Types of Pipe

Sprinkler Quantities: The table below lists the usual maximum number of sprinkler heads for each size of copper and steel pipe for both wet and dry systems. These quantities may be adjusted to meet individual structural needs or local code requirements. Maximum area on any one floor for one system is: light hazard and ordinary hazard 52,000 S.F., extra hazardous 25,000 S.F.

| Pipe Size | Light Hazard Occupancy | | Ordinary Hazard Occupancy | | Extra Hazard Occupancy | |
Diameter	Steel Pipe	Copper Pipe	Steel Pipe	Copper Pipe	Steel Pipe	Copper Pipe
1″	2 sprinklers	2 sprinklers	2 sprinklers	2 sprinklers	1 sprinklers	1 sprinklers
1-1/4″	3	3	3	3	2	2
1-1/2″	5	5	5	5	5	5
2″	10	12	10	12	8	8
2-1/2″	30	40	20	25	15	20
3″	60	65	40	45	27	30
3-1/2″	100	115	65	75	40	45
4″			100	115	55	65
5″			160	180	90	100
6″			275	300	150	170

Dry Pipe Systems: A dry pipe system should be installed where a wet pipe system is impractical, as in rooms or buildings which cannot be properly heated.

The use of an approved dry pipe system is more desirable than shutting off the water supply during cold weather.

Not more than 750 gallons of system capacity should be controlled by one dry pipe valve. Where two or more dry pipe valves are used, systems should preferably be divided horizontally.

R10520-140 Adjustment for Sprinkler/Standpipe Installations

Quality/Complexity Multiplier (For all installations)

Economy installation, add . . . . 0 to 5%
Good quality, medium complexity, add . . . . 5 to 15%
Above average quality and complexity, add . . . . 15 to 25%

R10520-310 Standpipe Systems

The basis for standpipe system design is National Fire Protection Association NFPA 14, however, the authority having jurisdiction should be consulted for special conditions, local requirements and approval.

Standpipe systems, properly designed and maintained, are an effective and valuable time saving aid for extinguishing fires, especially in the upper stories of tall buildings, the interior of large commercial or industrial malls, or other areas where construction features or access make the laying of temporary hose lines time consuming and/or hazardous. Standpipes are frequently installed with automatic sprinkler systems for maximum protection.

There are three general classes of service for standpipe systems:
Class I for use by fire departments and personnel with special training for heavy streams (2-1/2″ hose connections).
Class II for use by building occupants until the arrival of the fire department (1-1/2″ hose connector with hose).

Class III for use by either fire departments and trained personnel or by the building occupants (both 2-1/2″ and 1-1/2″ hose connections or one 2-1/2″ hose valve with an easily removable 2-1/2″ by 1-1/2″ adapter).

Standpipe systems are also classified by the way water is supplied to the system. The four basic types are:
Type 1: Wet standpipe system having supply valve open and water pressure maintained at all times.
Type 2: Standpipe system so arranged through the use of approved devices as to admit water to the system automatically by opening a hose valve.
Type 3: Standpipe system arranged to admit water to the system through manual operation of approved remote control devices located at each hose station.
Type 4: Dry standpipe having no permanent water supply.

R10520-320 NFPA 14 Basic Standpipe Design

Class	Design-Use	Pipe Size Minimums	Water Supply Minimums
Class I	2 1/2" hose connection on each floor All areas within 150' of an exit in every exit stairway Fire Department Trained Personnel	Height to 100', 4" dia. Heights above 100', 6" dia. (275' max. except with pressure regulators 400' max.)	For each standpipe riser 500 GPM flow For common supply pipe allow 500 GPM for first standpipe plus 250 GPM for each additional standpipe (2500 GPM max. total) 30 min. duration 65 PSI at 500 GPM
Class II	1 1/2" hose connection with hose on each floor All areas within 130' of hose connection measured along path of hose travel Occupant personnel	Height to 50', 2" dia. Height above 50', 2 1/2" dia.	For each standpipe riser 100 GPM flow For multiple riser common supply pipe 100 GPM 300 min. duration, 65 PSI at 100 GPM
Class III	Both of above. Class I valved connections will meet Class III with additional 2 1/2" by 1 1/2" adapter and 1 1/2" hose.	Same as Class I	Same as Class I

*Note: Where 2 or more standpipes are installed in the same building or section of building they shall be interconnected at the bottom.

Combined Systems

Combined systems are systems where the risers supply both automatic sprinklers and 2-1/2" hose connection outlets for fire department use. In such a system the sprinkler spacing pattern shall be in accordance with NFPA 13 while the risers and supply piping will be sized in accordance with NFPA 14. When the building is completely sprinklered the risers may be sized by hydraulic calculation. The minimum size riser for buildings not completely sprinklered is 6".

The minimum water supply of a completely sprinklered, light hazard, high-rise occupancy building will be 500 GPM while the supply required for other types of completely sprinklered high-rise buildings is 1000 GPM.

General System Requirements

1. Approved valves will be provided at the riser for controlling branch lines to hose outlets.
2. A hose valve will be provided at each outlet for attachment of hose.
3. Where pressure at any standpipe outlet exceeds 100 PSI a pressure reducer must be installed to limit the pressure to 100 PSI. Note that the pressure head due to gravity in 100' of riser is 43.4 PSI. This must be overcome by city pressure, fire pumps, or gravity tanks to provide adequate pressure at the top of the riser.
4. Each hose valve on a wet system having linen hose shall have an automatic drip connection to prevent valve leakage from entering the hose.
5. Each riser will have a valve to isolate it from the rest of the system.
6. One or more fire department connections as an auxiliary supply shall be provided for each Class I or Class III standpipe system. In buildings having two or more zones, a connection will be provided for each zone.
7. There will be no shutoff valve in the fire department connection, but a check valve will be located in the line before it joins the system.
8. All hose connections street side will be identified on a cast plate or fitting as to purpose.

For information about Means Estimating Seminars, see yellow pages 12 and 13 in back of book

R13128-520   Swimming Pools

Pool prices given per square foot of surface area include pool structure, filter and chlorination equipment, pumps, related piping, ladders/steps, maintenance kit, skimmer and vacuum system. Decks and electrical service to equipment are not included.

Residential in-ground pool construction can be divided into two categories: vinyl lined and gunite. Vinyl lined pool walls are constructed of different materials including wood, concrete, plastic or metal. The bottom is often graded with sand over which the vinyl liner is installed. Vermiculite or soil cement bottoms may be substituted for an added cost.

Gunite pool construction is used both in residential and municipal installations. These structures are steel reinforced for strength and finished with a white cement limestone plaster.

Municipal pools will have a higher cost because plumbing codes require more expensive materials, chlorination equipment and higher filtration rates.

Municipal pools greater than 1,800 S.F. require gutter systems to control waves. This gutter may be formed into the concrete wall. Often a vinyl/stainless steel gutter or gutter/wall system is specified, which will raise the pool cost.

Competition pools usually require tile bottoms and sides with contrasting lane striping, which will also raise the pool cost.

R13281-120   Asbestos Removal Process

Asbestos removal is accomplished by a specialty contractor who understands the federal and state regulations regarding the handling and disposal of the material. The process of asbestos removal is divided into many individual steps. An accurate estimate can be calculated only after all the steps have been priced.

The steps are generally as follows:

1. Obtain an asbestos abatement plan from an industrial hygienist.
2. Monitor the air quality in and around the removal area and along the path of travel between the removal area and transport area. This establishes the background contamination.
3. Construct a two part decontamination chamber at entrance to removal area.
4. Install a HEPA filter to create a negative pressure in the removal area.
5. Install wall, floor and ceiling protection as required by the plan, usually 2 layers of fireproof 6 mil polyethylene.
6. Industrial hygienist visually inspects work area to verify compliance with plan.
7. Provide temporary supports for conduit and piping affected by the removal process.
8. Proceed with asbestos removal and bagging process. Monitor air quality as described in Step #2. Discontinue operations when contaminate levels exceed applicable standards.
9. Document the legal disposal of materials in accordance with EPA standards.
10. Thoroughly clean removal area including all ledges, crevices and surfaces.
11. Post abatement inspection by industrial hygienist to verify plan compliance.
12. Provide a certificate from a licensed industrial hygienist attesting that contaminate levels are within acceptable standards before returning area to regular use.

R13600-610   Solar Heating (Space and Hot Water)

Collectors should face as close to due South as possible, however, variations of up to 20 degrees on either side of true South are acceptable. Local climate and collector type may influence the choice between east or west deviations. Obviously they should be located so they are not shaded from the sun's rays. Incline collectors at a slope of latitude minus 5 degrees for domestic hot water and latitude plus 15 degrees for space heating.

Flat plate collectors consist of a number of components as follows: Insulation to reduce heat loss through the bottom and sides of the collector. The enclosure which contains all the components in this assembly is usually weatherproof and prevents dust, wind and water from coming in contact with the absorber plate. The cover plate usually consists of one or more layers of a variety of glass or plastic and reduces the reradiation by creating an air space which traps the heat between the cover and the absorber plates.

The absorber plate must have a good thermal bond with the fluid passages. The absorber plate is usually metallic and treated with a surface coating which improves absorptivity. Black or dark paints or selective coatings are used for this purpose, and the design of this passage and plate combination helps determine a solar system's effectiveness.

Heat transfer fluid passage tubes are attached above and below or integral with an absorber plate for the purpose of transferring thermal energy from the absorber plate to a heat transfer medium. The heat exchanger is a device for transferring thermal energy from one fluid to another.

Piping and storage tanks should be well insulated to minimize heat losses.

Size domestic water heating storage tanks to hold 20 gallons of water per user, minimum, plus 10 gallons per dishwasher or washing machine. For domestic water heating an optimum collector size is approximately 3/4 square foot of area per gallon of water storage. For space heating of residences and small commercial applications the collector is commonly sized between 30% and 50% of the internal floor area. For space heating of large commercial applications, collector areas less than 30% of the internal floor area can still provide significant heat reductions.

A supplementary heat source is recommended for Northern states for December through February.

The solar energy transmission per square foot of collector surface varies greatly with the material used. Initial cost, heat transmittance and useful life are obviously interrelated.

R15050-710 Subcontractors

On the unit cost pages of the R.S. Means Cost Data books, the last column is entitled "Total Incl. O&P". This is normally the cost of the installing contractor. In Division 15, this is the cost of the mechanical contractor. If the particular work being estimated is to be performed by a sub to the mechanical contractor, the mechanical's profit and handling charge (usually 10%) is added to the total of the last column.

R15050-720 Demolition (Selective vs. Removal for Replacement)

Demolition can be divided into two basic categories.

One type of demolition involves the removal of material with no concern for its replacement. The labor-hours to estimate this work are found in Div. 15055 under "Selective Demolition". It is selective in that individual items or all the material installed as a system or trade grouping such as plumbing or heating systems are removed. This may be accomplished by the easiest way possible, such as sawing, torch cutting, or sledge hammer as well as simple unbolting.

The second type of demolition is the removal of some item for repair or replacement. This removal may involve careful draining, opening of unions,

disconnecting and tagging of electrical connections, capping of pipes/ducts to prevent entry of debris or leakage of the material contained as well as transport of the item away from its in-place location to a truck/dumpster. An approximation of the time required to accomplish this type of demolition is to use half of the time indicated as necessary to install a new unit. For example; installation of a new pump might be listed as requiring 6 labor-hours so if we had to estimate the removal of the old pump we would allow an additional 3 hours for a total of 9 hours. That is, the complete replacement of a defective pump with a new pump would be estimated to take 9 labor-hours.

R15100-040 Plumbing Approximations for Quick Estimating

Water Control

Water Meter; Backflow Preventer, Shock Absorbers; Vacuum Breakers; Mixer.	10 to 15% of Fixtures
Pipe And Fittings	30 to 60% of Fixtures

> **Note:** Lower percentage for compact buildings or larger buildings with plumbing in one area.
> Larger percentage for large buildings with plumbing spread out.
> In extreme cases pipe may be more than 100% of fixtures.
> Percentages **do not** include special purpose or process piping.

Plumbing Labor

1 & 2 Story Residential	Rough-in Labor = 80% of Materials
Apartment Buildings	Rough-in Labor = 90 to 100% of Materials

Labor for handling and placing fixtures is approximately 25 to 30% of fixtures

Quality/Complexity Multiplier (for all installations)

Economy installation, add	0 to 5%
Good quality, medium complexity, add	5 to 15%
Above average quality and complexity, add	15 to 25%

R15100-050 Pipe Material Considerations

1. Malleable fittings should be used for gas service.
2. Malleable fittings are used where there are stresses/strains due to expansion and vibration.
3. Cast fittings may be broken as an aid to disassembling of heating lines frozen by long use, temperature and minerals.
4. Cast iron pipe is extensively used for underground and submerged service.
5. Type M (light wall) copper tubing is available in hard temper only and is used for nonpressure and less severe applications than K and L.
6. Type L (medium wall) copper tubing, available hard or soft for interior service.
7. Type K (heavy wall) copper tubing, available in hard or soft temper for use where conditions are severe. For underground and interior service.
8. Hard drawn tubing requires fewer hangers or supports but should not be bent. Silver brazed fittings are recommended, however soft solder is normally used.
9. Type DMV (very light wall) copper tubing designed for drainage, waste and vent plus other non-critical pressure services.

Domestic/Imported Pipe and Fittings Cost

The prices shown in this publication for steel/cast iron pipe and steel, cast iron, malleable iron fittings are based on domestic production sold at the normal trade discounts. The above listed items of foreign manufacture may be available at prices of 1/3 to 1/2 those shown. Some imported items after minor machining or finishing operations are being sold as domestic to further complicate the system.

Caution: Most pipe prices in this book also include a coupling and pipe hangers which for the larger sizes can add significantly to the per foot cost and should be taken into account when comparing "book cost" with quoted supplier's cost.

R15100-070 Piping to 10' High

When taking off pipe, it is important to identify the different material types and joining procedures, as well as distances between supports and components required for proper support.

During the takeoff, measure through all fittings. Do not subtract the lengths of the fittings, valves, or strainers, etc. This added length plus the final rounding of the totals will compensate for nipples and waste.

When rounding off totals always increase the actual amount to correspond with manufacturer's shipping lengths.

A. Both red brass and yellow brass pipe are normally furnished in 12' lengths, plain end. Division 15107-220 includes in the linear foot costs two field threads and one coupling per 10' length. A carbon steel clevis type hanger assembly every 10' is also prorated into the linear foot costs, including both material and labor.

B. Cast iron soil pipe is furnished in either 5' or 10' lengths. For pricing purposes, Division 15107-320 in *Means Plumbing Cost Data* features 10' lengths with a joint and a carbon steel clevis hanger assembly every 5' prorated into the per foot costs of both material and labor.

   Three methods of joining are considered in Division 15107-320, lead and oakum poured joints, or push-on gasket type joints for the bell and spigot pipe, and a joint clamp for the no-hub soil pipe. The labor and material costs for each of these individual joining procedures are also prorated into the linear costs per foot.

C. Copper tubing in Division 15107-420 covers types K, L, M, and DWV which are furnished in 20' lengths. Means pricing data is based on a tubing cut each length and a coupling and two soft soldered joints every 10'. A carbon steel, clevis type hanger assembly every 10' is also prorated into the per foot costs. The prices for refrigeration tubing are for materials only. Labor for full lengths may be based on the type L labor but short cut measures in tight areas can increase the installation labor-hours from 20 to 40%.

D. Corrosion-resistant piping does not lend itself to one particular standard of hanging or support assembly due to its diversity of application and placement. The several varieties of corrosion-resistant piping do not include any material or labor costs for hanger assemblies (See Division 15060-300 for appropriate selection). The estimator should refer to Division 15107-560 in *Means Plumbing Cost Data* for the compatible corrosion-resistant fittings and piping.

E. Glass pipe, as shown in *Means Plumbing Cost Data* Division 15106-120, is furnished in standard lengths either 5' or 10' long, beaded on one end. Special orders for diverse lengths beaded on both ends are also available. For pricing purposes, R.S. Means features 10' lengths with a coupling and a carbon steel band hanger assembly every 10' prorated into the per foot linear costs.

   Glass pipe is also available with conical ends and standard lengths ranging from 6" through 3' in 6" increments, then up to 10' in 12" increments. Special lengths can be customized for particular installation requirements.

   For pricing purposes, Means has based the labor and material pricing on 10' lengths. Included in these costs per linear foot are the prorated costs for a flanged assembly every 10' consisting of two flanges, a gasket, two insertable seals, and the required number of bolts and nuts. A carbon steel band hanger assembly based on 10' center lines has also been prorated into the costs per foot for labor and materials.

F. Plastic pipe of several compositions and joining methods are considered in Division 15108-520. Fiberglass reinforced pipe (FRP) is priced, based on 10' lengths (20' lengths are also available), with coupling and epoxy joints every 10'. FRP is furnished in both "General Service" and "High Strength." A carbon steel clevis hanger assembly, 3 for every 10', is built into the prorated labor and material costs on a per foot basis.

   The PVC and CPVC pipe schedules 40, 80 and 120, plus SDR ratings are all based on 20' lengths with a coupling installed every 10', as well as a carbon steel clevis hanger assembly every 3'. The PVC and ABS type DWV piping is based on 10' lengths with solvent weld couplings every 10', and with carbon steel clevis hanger assemblies, 3 for every 10'. The rest of the plastic piping in this section is based on flexible 100' coils and does not include any coupling or supports.

This section ends with PVC drain and sewer piping based on 10' lengths with bell and spigot ends and 0-ring type, push-on joints. Material costs only are given, but the complete installation costs should be similar to those shown in line number 02620-270.

G. Stainless steel piping, Division 15107-900, includes both weld end and threaded piping, both in the type 304 and 316 specification and in the following schedules, 5, 10, 40, 80, and 160. Although this piping is usually furnished in 20' lengths, this cost grouping has a joint (either heli-arc butt-welded or threads and coupling) every 10'. A carbon steel clevis type hanger assembly is also included at 10' intervals and prorated into the linear foot costs.

H. Division 15107-620 covers carbon steel pipe, both black and galvanized. This section encompasses schedules 40 (standard) and 80 (extra heavy).

   Several common methods of joining steel pipe — such as thread and coupled, butt welded, and flanged (150 lb. weld neck flanges) are also included.

   For estimating purposes, it is assumed that the piping is purchased in 20' lengths and that a compatible joint is made up every 10'. These joints are prorated into the labor and material costs per linear foot. The following hanger and support assemblies every 10' are also included: carbon steel clevis for the T & C pipe, and single rod roll type for both the welded and flanged piping. All of these hangers are oversized to accommodate pipe insulation 3/4" thick through 5" pipe size and 1-1/2" thick from 6" through 12" pipe size.

I. Division 15107-690 prices grooved joint steel pipe, both black and galvanized, in schedules 10, 40, and 80, furnished in 20' lengths. This section describes two joining methods: cut groove and roll groove. The schedule 10 piping is roll-grooved, while the heavier schedules are cut-grooved. The labor and material costs are prorated into per linear foot prices, including a coupled joint every 10', as well as a carbon steel clevis hanger assembly.

Notes:

The pipe hanger assemblies mentioned in the preceding paragraphs include the described hanger; appropriately sized steel, box-type insert and nut; plus a threaded hanger rod. On average, the distance from the pipe center line to the insert face is 2'.

C clamps are used when the pipe is to be supported from steel shapes rather than anchored in the slab. C clamps are slightly less costly than inserts. However, to save time in estimating, it is advisable to use the given line number cost, rather than substituting a C clamp for the insert.

Add to piping labor for elevated installation:

10' to 15' high	10%	30' to 35' high	40%
15' to 20' high	20%	35' to 40' high	50%
20' to 25' high	25%	Over 40' high	55%
25' to 30' high	35%		

When using the percentage adds for elevated piping installations as shown above, bear in mind that the given heights are for the pipe supports, even though the insert, anchor, or clamp may be several feet higher than the pipe itself.

An allowance has been included in the piping installation time for testing and minor tightening of leaking joints, fittings, stuffing boxes, packing glands, etc. For extraordinary test requirements such as x-rays, prolonged pressure or demonstration tests, a percentage of the piping labor, based on the estimator's experience, must be added to the labor total. A testing service specializing in weld x-rays should be consulted for pricing if it is an estimate requirement. Equipment installation time includes start-up with associated adjustments.

R15100-080 Valve Materials

VALVE MATERIALS

Bronze:
Bronze is one of the oldest materials used to make valves. It is most commonly used in hot and cold water systems and other non-corrosive services. It is often used as a seating surface in larger iron body valves to ensure tight closure.

Carbon Steel:
Carbon steel is a high strength material. Therefore, valves made from this metal are used in higher pressure services, such as steam lines up to 600 psi at 850°F. Many steel valves are available with butt-weld ends for economy and are generally used in high pressure steam service as well as other higher pressure non-corrosive services.

Forged Steel:
Valves from tough carbon steel are used in service up to 2000 psi and temperatures up to 1000°F in Gate, Globe and Check valves.

Iron:
Valves are normally used in medium to large pipe lines to control non-corrosive fluid and gases, where pressures do not exceed 250 psi at 450° or 500 psi cold water, oil or gas.

Stainless Steel:
Developed steel alloys can be used in over 90% corrosive services.

Plastic PVC:
This is used in a great variety of valves generally in high corrosive service with lower temperatures and pressures.

VALVE SERVICE PRESSURES

Pressure ratings on valves provide an indication of the safe operating pressure for a valve at some elevated temperature. This temperature is dependent upon the materials used and the fabrication of the valve. When specific data is not available, a good "rule-of-thumb" to follow is the temperature of saturated steam on the primary rating indicated on the valve body. Example: The valve has the number 150S printed on the side indicating 150 psi and hence, a maximum operating temperature of 367°F (temperature of saturated steam and 150 psi).

DEFINITIONS

1. "WOG" – Water, oil, gas (cold working pressures).
2. "SWP" – Steam working pressure.
3. 100% area (full port) – means the area through the valve is equal to or greater than the area of standard pipe.
4. "Standard Opening" – means that the area through the valve is less than the area of standard pipe and therefore these valves should be used only where restriction of flow is unimportant.
5. "Round Port" – means the valve has a full round opening through the plug and body, of the same size and area as standard pipe.
6. "Rectangular Port" – valves have rectangular shaped ports through the plug body. The area of the port is either equal to 100% of the area of standard pipe, or restricted (standard opening). In either case it is clearly marked.
7. "ANSI" – American National Standards Institute.

R15100-090 Valve Selection Considerations

INTRODUCTION: In any piping application, valve performance is critical. Valves should be selected to give the best performance at the lowest cost.

The following is a list of performance characteristics generally expected of valves.

1. Stopping flow or starting it.
2. Throttling flow (Modulation).
3. Flow direction changing.
4. Checking backflow (Permitting flow in only one direction).
5. Relieving or regulating pressure.

In order to properly select the right valve, some facts must be determined.

A. What liquid or gas will flow through the valve?

B. Does the fluid contain suspended particles?

C. Does the fluid remain in liquid form at all times?

D. Which metals does fluid corrode?

E. What are the pressure and temperature limits? (As temperature and pressure rise, so will the price of the valve.)

F. Is there constant line pressure?

G. Is the valve merely an on-off valve?

H. Will checking of backflow be required?

I. Will the valve operate frequently or infrequently?

Valves are classified by design type into such classifications as Gate, Globe, Angle, Check, Ball, Butterfly and Plug. They are also classified by end connection, stem, pressure restrictions and material such as bronze, cast iron, etc. Each valve has a specific use. A quality valve used correctly will provide a lifetime of trouble-free service, but a high quality valve installed in the wrong service may require frequent attention.

STEM TYPES
(OS & Y)—Rising Stem-Outside Screw and Yoke

Offers a visual indication of whether the valve is open or closed. Recommended where high temperatures, corrosives, and solids in the line might cause damage to inside-valve stem threads. The stem threads are engaged by the yoke bushing so the stem rises through the hand wheel as it is turned.

(R.S.)—Rising Stem-Inside Screw

Adequate clearance for operation must be provided because both the hand wheel and the stem rise.

The valve wedge position is indicated by the position of the stem and hand wheel.

(N.R.S.)—Non-Rising Stem-Inside Screw

A minimum clearance is required for operating this type of valve. Excessive wear or damage to stem threads inside the valve may be caused by heat, corrosion, and solids. Because the hand wheel and stem do not rise, wedge position cannot be visually determined.

VALVE TYPES
Gate Valves

Provide full flow, minute pressure drop, minimum turbulence and minimum fluid trapped in the line.

They are normally used where operation is infrequent.

Globe Valves

Globe valves are designed for throttling and/or frequent operation with positive shut-off. Particular attention must be paid to the several types of seating materials available to avoid unnecessary wear. The seats must be compatible with the fluid in service and may be composition or metal. The configuration of the Globe valve opening causes turbulence which results in increased resistance. Most bronze Globe valves are rising stem-inside screw, but they are also available on O.S. & Y.

Angle Valves

The fundamental difference between the Angle valve and the Globe valve is the fluid flow through the Angle valve. It makes a 90° turn and offers less resistance to flow than the Globe valve while replacing an elbow. An angle valve thus reduces the number of joints and installation time.

Check Valves

Check valves are designed to prevent backflow by automatically seating when the direction of fluid is reversed.

Swing Check valves are generally installed with Gate-valves, as they provide comparable full flow.

Usually recommended for lines where flow velocities are low and should not be used on lines with pulsating flow. Recommended for horizontal installation, or in vertical lines only where flow is upward.

Lift Check Valves

These are commonly used with Globe and Angle valves since they have similar diaphragm seating arrangements and are recommended for preventing backflow of steam, air, gas and water, and on vapor lines with high flow velocities. For horizontal lines, horizontal lift checks should be used and vertical lift checks for vertical lines.

Ball Valves

Ball valves are light and easily installed, yet because of modern elastomeric seats, provide tight closure. Flow is controlled by rotating up to 90° a drilled ball which fits tightly against resilient seals. This ball seats with flow in either direction, and valve handle indicates the degree of opening. Recommended for frequent operation readily adaptable to automation, ideal for installation where space is limited.

Butterfly Valves

Butterfly valves provide bubble-tight closure with excellent throttling characteristics. They can be used for full-open, closed and for throttling applications.

The Butterfly valve consists of a disc within the valve body which is controlled by a shaft. In its closed position, the valve disc seals against a resilient seat. The disc position throughout the full 90° rotation is visually indicated by the position of the operator.

A Butterfly valve is only a fraction of the weight of a Gate valve and requires no gaskets between flanges in most cases. Recommended for frequent operation and adaptable to automation where space is limited.

Wafer and Lug type bodies when installed between two pipe flanges, can be easily removed from the line. The pressure of the bolted flanges holds the valve in place.

Locating lugs makes installation easier.

Plug Valves

Lubricated plug valves, because of the wide range of service to which they are adapted, may be classified as all purpose valves. They can be safely used at all pressure and vacuums, and at all temperatures up to the limits of available lubricants. They are the most satisfactory valves for the handling of gritty suspensions and many other destructive, erosive, corrosive and chemical solutions.

R15100-110 Hot Water Consumption Rates

Type of Building	Size Factor	Maximum Hourly Demand	Average Day Demand
Apartment Dwellings	No. of Apartments: Up to 20 21 to 50 51 to 75 76 to 100 101 to 200 201 up	12.0 Gal. per apt. 10.0 Gal. per apt. 8.5 Gal. per apt. 7.0 Gal. per apt. 6.0 Gal. per apt. 5.0 Gal. per apt.	42.0 Gal. per apt. 40.0 Gal. per apt. 38.0 Gal. per apt. 37.0 Gal. per apt. 36.0 Gal. per apt. 35.0 Gal. per apt.
Dormitories	Men Women	3.8 Gal. per man 5.0 Gal. per woman	13.1 Gal. per man 12.3 Gal. per woman
Hospitals	Per bed	23.0 Gal. per patient	90.0 Gal. per patient
Hotels	Single room with bath Double room with bath	17.0 Gal. per unit 27.0 Gal. per unit	50.0 Gal. per unit 80.0 Gal. per unit
Motels	No. of units: Up to 20 21 to 100 101 Up	6.0 Gal. per unit 5.0 Gal. per unit 4.0 Gal. per unit	20.0 Gal. per unit 14.0 Gal. per unit 10.0 Gal. per unit
Nursing Homes		4.5 Gal. per bed	18.4 Gal. per bed
Office buildings		0.4 Gal. per person	1.0 Gal. per person
Restaurants	Full meal type Drive-in snack type	1.5 Gal./max. meals/hr. 0.7 Gal./max. meals/hr.	2.4 Gal. per meal 0.7 Gal. per meal
Schools	Elementary Secondary & High	0.6 Gal. per student 1.0 Gal. per student	0.6 Gal. per student 1.8 Gal. per student

For evaluation purposes, recovery rate and storage capacity are inversely proportional. Water heaters should be sized so that the maximum hourly demand anticipated can be met in addition to allowance for the heat loss from the pipes and storage tank.

R15100-120 Fixture Demands in Gallons Per Fixture Per Hour

Table below is based on 140°F final temperature except for dishwashers in public places (*) where 180°F water is mandatory.

Fixture	Apartment House	Club	Gym	Hospital	Hotel	Indust. Plant	Office	Private Home	School
Bathtubs	20	20	30	20	20			20	
Dishwashers, automatic	15	50-150*		50-150*	50-200*	20-100*		15	20-100*
Kitchen sink	10	20		20	30	20	20	10	20
Laundry, stationary tubs	20	28		28	28			20	
Laundry, automatic wash	75	75		100	150			75	
Private lavatory	2	2	2	2	2	2	2	2	2
Public lavatory	4	6	8	6	8	12	6		15
Showers	30	150	225	75	75	225	30	30	225
Service sink	20	20		20	30	20	20	15	20
Demand factor	0.30	0.30	0.40	0.25	0.25	0.40	0.30	0.30	0.40
Storage capacity factor	1.25	0.90	1.00	0.60	0.80	1.00	2.00	0.70	1.00

To obtain the probable maximum demand multiply the total demands for the fixtures (gal./fixture/hour) by the demand factor. The heater should have a heating capacity in gallons per hour equal to this maximum. The storage tank should have a capacity in gallons equal to the probable maximum demand multiplied by the storage capacity factor.

R15100-410 Minimum Plumbing Fixture Requirements

Minimum Plumbing Fixture Requirements

Type of Building or Occupancy (2)	Water Closets (14) (Fixtures per Person)		Urinals (5,10) (Fixtures per Person)		Lavatories (Fixtures per Person)		Bathtubs or Showers (Fixtures per Person)	Drinking Fountains (Fixtures per Person) (3, 13)
	Male	Female	Male	Female	Male	Female		
Assembly Places- Theatres, Auditoriums, Convention Halls, etc.-for permanent employee use	1: 1 - 15 2: 16 - 35 3: 36 - 55 Over 55, add 1 fixture for each additional 40 persons	1: 1 - 15 2: 16 - 35 3: 36 - 55	0: 1 - 9 1: 10 - 50 Add one fixture for each additional 50 males		1 per 40	1 per 40		
Assembly Places- Theatres, Auditoriums, Convention Halls, etc. - for public use	1: 1 - 100 2: 101 - 200 3: 201 - 400 Over 400, add 1 fixture for each additional 500 males and 1 for each additional 125 females	3: 1 - 50 4: 51 - 100 8: 101 - 200 11: 201 - 400	1: 1 - 100 2: 101 - 200 3: 201 - 400 4: 401 - 600 Over 600, add 1 fixture for each additional 300 males		1: 1 - 200 2: 201 - 400 3: 401 - 750 Over 750, add 1 fixture for each additional 500 persons	1: 1 - 200 2: 201 - 400 3: 401 - 750		1: 1 - 150 2: 151 - 400 3: 401 - 750 Over 750, add one fixture for each additional 500 persons
Dormitories (9) School or Labor	1 per 10 Add 1 fixture for each additional 25 males (over 10) and 1 for each additional 20 females (over 8)	1 per 8	1 per 25 Over 150, add 1 fixture for each additional 50 males		1 per 12 Over 12 add 1 fixture for each additional 20 males and 1 for each 15 additional females	1 per 12	1 per 8 For females add 1 bathtub per 30. Over 150, add 1 per 20	1 per 150 (12)
Dormitories- for Staff Use	1: 1 - 15 2: 16 - 35 3: 36 - 55 Over 55, add 1 fixture for each additional 40 persons	1: 1 - 15 3: 16 - 35 4: 36 - 55	1 per 50		1 per 40	1 per 40	1 per 8	

Dwellings:
  Single Dwelling
  Multiple Dwelling
  or Apartment House
	1 per dwelling
	1 per dwelling or apartment unit
					1 per dwelling
					1 per dwelling or apartment unit
							1 per dwelling
							1 per dwelling or apartment unit	

Hospital Waiting rooms	1 per room				1 per room			1 per 150 (12)
Hospitals- for employee use	1: 1 - 15 2: 16 - 35 3: 36 - 55 Over 55, add 1 fixture for each additional 40 persons	1: 1 - 15 3: 16 - 35 4: 36 - 55	0: 1 - 9 1: 10 - 50 Add 1 fixture for each additional 50 males		1 per 40	1 per 40		

Hospitals:
  Individual Room
  Ward Room
	1 per room
	1 per 8 patients
					1 per room
					1 per 10 patients
							1 per room
							1 per 20 patients
								1 per 150 (12)

Industrial (6) Warehouses Workshops, Foundries and similar establishments- for employee use	1: 1 -10 2: 11 - 25 3: 26 - 50 4: 51 - 75 5: 76 - 100 Over 100, add 1 fixture for each additional 30 persons	1: 1 -10 2: 11 - 25 3: 26 - 50 4: 51 - 75 5: 76 - 100			Up to 100, per 10 persons Over 100, 1 per 15 persons (7, 8)		1 shower for each 15 persons exposed to excessive heat or to skin contamination with poisonous, infectious or irritating material	1 per 150 (12)
Institutional - Other than Hospitals or Penal Institutions (on each occupied floor)	1 per 25	1 per 20	0: 1 - 9 1: 10 - 50 Add 1 fixture for each additional 50 males		1 per 10	1 per 10	1 per 8	1 per 150 (12)
Institutional - Other than Hospitals or Penal Institutions (on each occupied floor)- for employee use	1: 1 - 15 2: 16 - 35 3: 36 - 55 Over 55, add 1 fixture for each additional 40 persons	1: 1 - 15 3: 16 - 35 4: 36 - 55	0: 1 - 9 1: 10 - 50 Add 1 fixture for each additional 50 males		1 per 40	1 per 40	1 per 8	1 per 150 (12)
Office or Public Buildings	1: 1 - 100 2: 101 - 200 3: 201 - 400 Over 400, add 1 fixture for each additional 500 males and 1 for each additional 150 females	3: 1 - 50 4: 51 - 100 8: 101 - 200 11: 201 - 400	1: 1 - 100 2: 101 - 200 3: 201 - 400 4: 401 - 600 Over 600, add 1 fixture for each additional 300 males		1: 1 - 200 2: 201 - 400 3: 401 - 750 Over 750, add 1 fixture for each additional 500 persons	1: 1 - 200 2: 201 - 400 3: 401 - 750		1 per 150 (12)
Office or Public Buildings - for employee use	1: 1 - 15 2: 16 - 35 3: 36 - 55 Over 55, add 1 fixture for each additional 40 persons	1: 1 - 15 3: 16 - 35 4: 36 - 55	0: 1 - 9 1: 10 - 50 Add 1 fixture for each additional 50 males		1 per 40	1 per 40		

R15100-410   Minimum Plumbing Fixture Requirements

Minimum Plumbing Fixture Requirements

Type of Building or Occupancy	Water Closets (14) (Fixtures per Person)		Urinals (5, 10) (Fixtures per Person)		Lavatories (Fixtures per Person)		Bathtubs or Showers (Fixtures per Person)	Drinking Fountains (Fixtures per Person) (3, 13)
	Male	Female	Male	Female	Male	Female		
Penal Institutions - for employee use	1: 1 - 15 2: 16 - 35 3: 36 - 55 Over 55, add 1 fixture for each additional 40 persons	1: 1 - 15 3: 16 - 35 4: 36 - 55	0: 1 - 9 1: 10 - 50 Add 1 fixture for each additional 50 males		1 per 40	1 per 40		1 per 150 (12)
Penal Institutions - for prison use Cell	1 per cell				1 per cell			1 per cellblock floor
Exercise room	1 per exercise room		1 per exercise room		1 per exercise room			1 per exercise room
Restaurants, Pubs and Lounges (11)	1: 1 - 50 2: 51 - 150 3: 151 - 300 Over 300, add 1 fixture for each additional 200 persons	1: 1 - 50 2: 51 - 150 4: 151 - 300	1: 1 - 150 Over 150, add 1 fixture for each additional 150 males		1: 1 - 150 2: 151 - 200 3: 201 - 400 Over 400, add 1 fixture for each additional 400 persons	1: 1 - 150 2: 151 - 200 3: 201 - 400		
Schools - for staff use All Schools	1: 1 - 15 2: 16 - 35 3: 36 - 55 Over 55, add 1 fixture for each additional 40 persons	1: 1 - 15 3: 16 - 35 4: 36 - 55	1 per 50		1 per 40	1 per 40		
Schools - for student use: Nursery	1: 1 - 20 2: 21 - 50 Over 50, add 1 fixture for each additional 50 persons	1: 1 - 20 2: 21 - 50			1: 1 - 25 2: 26 - 50 Over 50, add 1 fixture for each additional 50 persons	1: 1 - 25 2: 26 - 50		1 per 150 (12)
Elementary	1 per 30	1 per 25	1 per 75		1 per 35	1 per 35		1 per 150 (12)
Secondary	1 per 40	1 per 30	1 per 35		1 per 40	1 per 40		1 per 150 (12)
Others (Colleges, Universities, Adult Centers, etc.	1 per 40	1 per 30	1 per 35		1 per 40	1 per 40		1 per 150 (12)
Worship Places Educational and Activities Unit	1 per 150	1 per 75	1 per 150		1 per 2 water closets			1 per 150 (12)
Worship Places Principal Assembly Place	1 per 150	1 per 75	1 per 150		1 per 2 water closets			1 per 150 (12)

Notes:

1. The figures shown are based upon one (1) fixture being the minimum required for the number of persons indicated or any fraction thereof.
2. Building categories not shown on this table shall be considered separately by the Administrative Authority.
3. Drinking fountains shall not be installed in toilet rooms.
4. Laundry trays. One (1) laundry tray or one (1) automatic washer standpipe for each dwelling unit or one (1) laundry trays or one (1) automatic washer standpipes, or combination thereof, for each twelve (12) apartments. Kitchen sinks, one (1) for each dwelling or apartment unit.
5. For each urinal added in excess of the minimum required, one water closet may be deducted. The number of water closets shall not be reduced to less than two-thirds (2/3) of the minimum requirement.
6. As required by ANSI Z4.1-1968, Sanitation in Places of Employment.
7. Where there is exposure to skin contamination with poisonous, infectious, or irritating materials, provide one (1) lavatory for each five (5) persons.
8. Twenty-four (24) lineal inches of wash sink or eighteen (18) inches of a circular basin, when provided with water outlets for such space shall be considered equivalent to one (1) lavatory.
9. Laundry trays, one (1) for each fifty (50) persons. Service sinks, one (1) for each hundred (100) persons.
10. General. In applying this schedule of facilities, consideration shall be given to the accessibility of the fixtures. Conformity purely on a numerical basis may not result in an installation suited to the need of the individual establishment. For example, schools should be provided with toilet facilities on each floor having classrooms.
    a. Surrounding materials, wall and floor space to a point two (2) feet in front of urinal lip and four (4) feet above the floor, and at least two (2) feet to each side of the urinal shall be lined with non-absorbent materials.
    b. Trough urinals shall be prohibited.
11. A restaurant is defined as a business which sells food to be consumed on the premises.
    a. The number of occupants for a drive-in restaurant shall be considered as equal to the number of parking stalls.
    b. Employee toilet facilities shall not to be included in the above restaurant requirements. Hand washing facilities shall be available in the kitchen for employees.
12. Where food is consumed indoors, water stations may be substituted for drinking fountains. Offices, or public buildings for use by more than six (6) persons shall have one (1) drinking fountain for the first one hundred fifty (150) persons and one additional fountain for each three hundred (300) persons thereafter.
13. There shall be a minimum of one (1) drinking fountain per occupied floor in schools, theaters, auditoriums, dormitories, offices of public building.
14. The total number of water closets for females shall be at least equal to the total number of water closets and urinals required for males.

R15100-420   Plumbing Fixture Installation Time

Item	Rough-In	Set	Total Hours	Item	Rough-In	Set	Total Hours
Bathtub	5	5	10	Shower head only	2	1	3
Bathtub and shower, cast iron	6	6	12	Shower drain	3	1	4
Fire hose reel and cabinet	4	2	6	Shower stall, slate		15	15
Floor drain to 4 inch diameter	3	1	4	Slop sink	5	3	8
Grease trap, single, cast iron	5	3	8	Test 6 fixtures			14
Kitchen gas range		4	4	Urinal, wall	6	2	8
Kitchen sink, single	4	4	8	Urinal, pedestal or floor	6	4	10
Kitchen sink, double	6	6	12	Water closet and tank	4	3	7
Laundry tubs	4	2	6	Water closet and tank, wall hung	5	3	8
Lavatory wall hung	5	3	8	Water heater, 45 gals. gas, automatic	5	2	7
Lavatory pedestal	5	3	8	Water heaters, 65 gals. gas, automatic	5	2	7
Shower and stall	6	4	10	Water heaters, electric, plumbing only	4	2	6

Fixture prices in front of book are based on the cost per fixture set in place. The rough-in cost, which must be added for each fixture, includes carrier, if required, some supply, waste and vent pipe connecting fittings and stops. The lengths of rough-in pipe are nominal runs which would connect to the larger runs and stacks. The supply runs and DWV runs and stacks must be accounted for in separate entries. In the eastern half of the United States it is common for the plumber to carry these to a point 5′ outside the building.

R15100-430   Water Cooler Application

Type of Service	Requirement
Office, School or Hospital	12 persons per gallon per hour
Office, Lobby or Department Store	4 or 5 gallons per hour per fountain
Light manufacturing	7 persons per gallon per hour
Heavy manufacturing	5 persons per gallon per hour
Hot heavy manufacturing	4 persons per gallon per hour
Hotel	.08 gallons per hour per room
Theatre	1 gallon per hour per 100 seats

R15155-210 Drainage Requirements

Drainage lines must have a slope to maintain flow for proper operation. This slope should not be less than 1/4″ per foot for 3″ diameter or smaller pipe and not less than 1/8″ per foot for 4″ diameter or larger pipe. The capacity of building drainage systems is calculated on a basis of "drainage fixture units" (d.f.u.) as per the following chart.

Type of Fixture	d.f.u. Value	Type of Fixture	d.f.u. Value
Automatic clothes washer (2″ standpipe)	3	Service sink (trap standard)	3
Bathroom group (water closet, lavatory and		Service sink (P trap)	2
bathtub or shower) tank type closet	6	Urinal, pedestal, syphon jet blowout	6
Bathtub (with or without overhead shower)	2	Urinal, wall hung	4
Clinic sink	6	Urinal, stall washout	4
Combination sink & tray with food disposal	4	Wash sink (cir. or mult.) per faucet set	2
Dental unit or cuspidor	1	Water closet, tank operated	4
Dental lavatory	1	Water closet, valve operated	6
Drinking fountain	1/2	Fixtures not listed above	
Dishwasher, domestic	2	Trap size 1-1/4″ or smaller	1
Floor drains with 2″ waste	3	Trap size 1-1/2″	2
Kitchen sink, domestic with one 1-1/2″ trap	2	Trap size 2″	3
Kitchen sink, domestic with food disposal	2	Trap size 2-1/2″	4
Lavatory with 1-1/4″ waste	1	Trap size 3″	5
Laundry tray (1 or 2 compartment)	2	Trap size 4″	6
Shower stall, domestic	2		

For continuous or nearly continuous flow into the sytem from a pump, air conditioning equipment or other item, allow 2 fixture units for each gallon per minute of flow.

When the "drainage fixture units" (d.f.u.) for each horizontal branch or vertical stack is computed from the table above, the appropriate pipe size for each branch or stack is determined from the table below.

R15155-220 Allowable Fixture Units (d.f.u.) for Branches and Stacks

Pipe Diam.	Horiz. Branch (not incl. drains)	Stack Size for 3 Stories or 3 Levels	Stack size for Over 3 levels	Maximum for 1 Story building Stack
1-1/2″	3	4	8	2
2″	6	10	24	6
2-1/2″	12	20	42	9
3″	20*	48*	72*	20*
4″	160	240	500	90
5″	360	540	1100	200
6″	620	960	1900	350
8″	1400	2200	3600	600
10″	2500	3800	5600	1000
12″	3900	6000	8400	1500
15″	7000			

*Not more than two water closets or bathroom groups within each branch interval nor more than six water closets or bathroom groups on the stack.

Stacks sized for the total may be reduced as load decreases at each story t a minimum diameter of 1/2 the maximum diameter.

R17100-100 Square Foot Project Size Modifier

One factor that affects the S.F. cost of a particular building is the size. In general, for buildings built to the same specifications in the same locality, the larger building will have the lower S.F. cost. This is due mainly to the decreasing contribution of the exterior walls plus the economy of scale usually achievable in larger buildings. The Area Conversion Scale shown below will give a factor to convert costs for the typical size building to an adjusted cost for the particular project.

The Square Foot Base Size lists the median costs, most typical project size in our accumulated data and the range in size of the projects.

The Size Factor for your project is determined by dividing your project area in S.F. by the typical project size for the particular Building Type. With this factor, enter the Area Conversion Scale at the appropriate Size Factor and determine the appropriate cost multiplier for your building size.

Example: Determine the cost per S.F. for a 100,000 S.F. Mid-rise apartment building.

$$\frac{\text{Proposed building area} = 100,000 \text{ S.F.}}{\text{Typical size from below} = 50,000 \text{ S.F.}} = 2.00$$

Enter Area Conversion scale at 2.0, intersect curve, read horizontally the appropriate cost multiplier of .94. Size adjusted cost becomes .94 x $85.50 = $80.35 based on national average costs.

Note: For Size Factors less than .50, the Cost Multiplier is 1.1
For Size Factors greater than 3.5, the Cost Multiplier is .90

Building Type	Median Cost per S.F.	Typical Size Gross S.F.	Typical Range Gross S.F.		Building Type	Median Cost per S.F.	Typical Size Gross S.F.	Typical Range Gross S.F.	
Apartments, Low Rise	$ 67.50	21,000	9,700	37,200	Jails	$205.00	40,000	5,500	145,000
Apartments, Mid Rise	85.50	50,000	32,000	100,000	Libraries	126.00	12,000	7,000	31,000
Apartments, High Rise	97.00	145,000	95,000	600,000	Living, Assisted	111.00	32,300	23,500	50,300
Auditoriums	113.00	25,000	7,600	39,000	Medical Clinics	117.00	7,200	4,200	15,700
Auto Sales	84.00	20,000	10,800	28,600	Medical Offices	110.00	6,000	4,000	15,000
Banks	151.00	4,200	2,500	7,500	Motels	82.50	40,000	15,800	120,000
Churches	103.00	17,000	2,000	42,000	Nursing Homes	114.00	23,000	15,000	37,000
Clubs, Country	106.00	6,500	4,500	15,000	Offices, Low Rise	95.00	20,000	5,000	80,000
Clubs, Social	100.00	10,000	6,000	13,500	Offices, Mid Rise	95.00	120,000	20,000	300,000
Clubs, YMCA	119.00	28,300	12,800	39,400	Offices, High Rise	122.00	260,000	120,000	800,000
Colleges (Class)	131.00	50,000	15,000	150,000	Police Stations	152.00	10,500	4,000	19,000
Colleges (Science Lab)	193.00	45,600	16,600	80,000	Post Offices	113.00	12,400	6,800	30,000
College (Student Union)	148.00	33,400	16,000	85,000	Power Plants	835.00	7,500	1,000	20,000
Community Center	107.00	9,400	5,300	16,700	Religious Education	94.00	9,000	6,000	12,000
Court Houses	144.00	32,400	17,800	106,000	Research	158.00	19,000	6,300	45,000
Dept. Stores	62.50	90,000	44,000	122,000	Restaurants	137.00	4,400	2,800	6,000
Dormitories, Low Rise	111.00	25,000	10,000	95,000	Retail Stores	67.00	7,200	4,000	17,600
Dormitories, Mid Rise	141.00	85,000	20,000	200,000	Schools, Elementary	99.50	41,000	24,500	55,000
Factories	61.00	26,400	12,900	50,000	Schools, Jr. High	102.00	92,000	52,000	119,000
Fire Stations	107.00	5,800	4,000	8,700	Schools, Sr. High	108.00	101,000	50,500	175,000
Fraternity Houses	104.00	12,500	8,200	14,800	Schools, Vocational	100.00	37,000	20,500	82,000
Funeral Homes	117.00	10,000	4,000	20,000	Sports Arenas	82.50	15,000	5,000	40,000
Garages, Commercial	76.00	9,300	5,000	13,600	Supermarkets	66.00	44,000	12,000	60,000
Garages, Municipal	95.00	8,300	4,500	12,600	Swimming Pools	155.00	20,000	10,000	32,000
Garages, Parking	40.00	163,000	76,400	225,300	Telephone Exchange	180.00	4,500	1,200	10,600
Gymnasiums	102.00	19,200	11,600	41,000	Theaters	99.00	10,500	8,800	17,500
Hospitals	181.00	55,000	27,200	125,000	Town Halls	116.00	10,800	4,800	23,400
House (Elderly)	92.00	37,000	21,000	66,000	Warehouses	49.50	25,000	8,000	72,000
Housing (Public)	85.00	36,000	14,400	74,400	Warehouse & Office	52.50	25,000	8,000	72,000
Ice Rinks	122.00	29,000	27,200	33,600					

Square Foot Base Size

Change Orders

Change Order Considerations

A Change Order is a written document, usually prepared by the design professional, and signed by the owner, the architect/engineer and the contractor. A change order states the agreement of the parties to: an addition, deletion, or revision in the work; an adjustment in the contract sum, if any; or an adjustment in the contract time, if any. Change orders, or "extras" in the construction process occur after execution of the construction contract and impact architects/engineers, contractors and owners.

Change orders that are properly recognized and managed can ensure orderly, professional and profitable progress for all who are involved in the project. There are many causes for change orders and change order requests. In all cases, change orders or change order requests should be addressed promptly and in a precise and prescribed manner. The following paragraphs include information regarding change order pricing and procedures.

The Causes of Change Orders

Reasons for issuing change orders include:

- Unforeseen field conditions that require a change in the work
- Correction of design discrepancies, errors or omissions in the contract documents
- Owner-requested changes, either by design criteria, scope of work, or project objectives
- Completion date changes for reasons unrelated to the construction process
- Changes in building code interpretations, or other public authority requirements that require a change in the work
- Changes in availability of existing or new materials and products

Procedures

Properly written contract documents must include the correct change order procedures for all parties—owners, design professionals and contractors—to follow in order to avoid costly delays and litigation.

Being "in the right" is not always a sufficient or acceptable defense. The contract provisions requiring notification and documentation must be adhered to within a defined or reasonable time frame.

The appropriate method of handling change orders is by a written proposal and acceptance by all parties involved. Prior to starting work on a project, all parties should identify their authorized agents who may sign and accept change orders, as well as any limits placed on their authority.

Time may be a critical factor when the need for a change arises. For such cases, the contractor might be directed to proceed on a "time and materials" basis, rather than wait for all paperwork to be processed—a delay that could impede progress. In this situation, the contractor must still follow the prescribed change order procedures, including but not limited to, notification and documentation.

All forms used for change orders should be dated and signed by the proper authority. Lack of documentation can be very costly, especially if legal judgments are to be made and if certain field personnel are no longer available. For time and material change orders, the contractor should keep accurate daily records of all labor and material allocated to the change. Forms that can be used to document change order work are available in *Means Forms for Building Construction Professionals*.

Owners or awarding authorities who do considerable and continual building construction (such as the federal government) realize the inevitability of change orders for numerous reasons, both predictable and unpredictable. As a result, the federal government, the American Institute of Architects (AIA), the Engineers Joint Contract Documents Committee (EJCDC) and other contractor, legal and technical organizations have developed standards and procedures to be followed by all parties to achieve contract continuance and timely completion, while being financially fair to all concerned.

In addition to the change order standards put forth by industry associations, there are also many books available on the subject.

Pricing Change Orders

When pricing change orders, regardless of their cause, the most significant factor is *when* the change occurs. The need for a change may be perceived in the field or requested by the architect/engineer *before* any of the actual installation has begun, or may evolve or appear *during* construction when the item of work in question is partially installed. In the latter cases, the original sequence of construction is disrupted, along with all contiguous and supporting systems. Change orders cause the greatest impact when they occur *after* the installation has been completed and must be uncovered, or even replaced. Post-completion changes may be caused by necessary design changes, product failure, or changes in the owner's requirements that are not discovered until the building or the systems begin to function.

Specified procedures of notification and record keeping must be adhered to and enforced regardless of the stage of construction: *before, during,* or *after* installation. Some bidding documents anticipate change orders by requiring that unit prices including overhead and profit percentages—for additional as well as deductible changes—be listed. Generally these unit prices do not fully take into account the ripple effect, or impact on other trades, and should be used for general guidance only.

When pricing change orders, it is important to classify the time frame in which the change occurs. There are two basic time frames for change orders: *pre-installation change orders* which occur before the start of construction, and *post-installation change orders*, which involve reworking after the original installation. Change orders that occur between these stages may be priced according to the extent of work completed using a combination of techniques developed for pricing *pre-* and *post-installation* changes.

The following factors are the basis for a checklist to use when preparing a change order estimate.

Factors To Consider When Pricing Change Orders

As an estimator begins to prepare a change order, the following questions should be reviewed to determine their impact on the final price.

General

- Is the change order work *pre-installation* or *post-installation*?

    Change order work costs vary according to how much of the installation has been completed. Once workers have the project scoped in their mind, even though they have not started, it can be difficult to refocus.

Consequently they may spend more than the normal amount of time understanding the change. Also, modifications to work in place such as trimming or refitting usually take more time than was initially estimated. The greater the amount of work in place, the more reluctant workers are to change it. Psychologically they may resent the change and as a result the rework takes longer than normal. Post-installation change order estimates must include demolition of existing work as required to accomplish the change. If the work is performed at a later time, additional obstacles such as building finishes may be present which must be

protected. Regardless of whether the change occurs pre-installation or post-installation, attempt to isolate the identifiable factors and price them separately. For example, add shipping costs that may be required pre-installation or any demolition required post-installation. Then analyze the potential impact on productivity of psychological and/or learning curve factors and adjust the output rates accordingly. One approach is to break down the typical workday into segments and quantify the impact on each segment. The following chart may be useful as a guide:

Task	Activities (Productivity) Expressed as Percentages of a Workday		
	Means Mechanical Cost Data (for New Construction)	Pre-Installation Change Orders	Post-Installation Change Orders
1. Study plans	3%	6%	6%
2. Material procurement	3%	3%	3%
3. Receiving and storing	3%	3%	3%
4. Mobilization	5%	5%	5%
5. Site movement	5%	5%	8%
6. Layout and marking	8%	10%	12%
7. Actual installation	64%	59%	54%
8. Clean-up	3%	3%	3%
9. Breaks—non-productive	6%	6%	6%
Total	100%	100%	100%

Change Order Installation Efficiency

The labor-hours expressed (for new construction) are based on average installation time, using an efficiency level of approximately 60-65%. For change order situations, adjustments to this efficiency level should reflect the daily labor-hour allocation for that particular occurrence.

If any of the specific percentages expressed in the above chart do not apply to a particular project situation, then those percentage points should be reallocated to the appropriate task(s). Example: Using data for new construction, assume there is no new material being utilized. The percentages for Tasks 2 and 3 would therefore be reallocated to other tasks. If the time required for Tasks 2 and 3 can now be applied to installation, we can add the time allocated for *Material Procurement* and *Receiving and Storing* to the *Actual Installation* time for new construction, thereby increasing the Actual Installation percentage.

This chart shows that, due to reduced productivity, labor costs will be higher than those for new construction by 5% to 15% for pre-installation change orders and by 15% to 25% for post-installation change orders. Each job and change order is unique and must be examined individually. Many factors, covered elsewhere in this section, can each have a significant impact on productivity and change order costs. All such factors should be considered in every case.

- Will the change substantially delay the original completion date?

A significant change in the project may cause the original completion date to be extended. The extended schedule may subject the contractor to new wage rates dictated by relevant labor contracts. Project supervision and other project overhead must also be extended beyond the original completion date. The schedule extension may also put installation into a new weather season. For example, underground piping scheduled for October installation was delayed until January. As a result, frost penetrated the trench area, thereby changing the degree of difficulty of the task. Changes and delays may have a ripple effect throughout the project. This effect must be analyzed and negotiated with the owner.

- What is the net effect of a deduct change order?

In most cases, change orders resulting in a deduction or credit reflect only bare costs. The contractor may retain the overhead and profit based on the original bid.

Materials

- Will you have to pay more or less for the new material, required by the change order, than you paid for the original purchase?

The same material prices or discounts will usually apply to materials purchased for change orders as new construction. In some instances, however, the contractor may forfeit the advantages of competitive pricing for change orders. Consider the following example:

A contractor purchased over $20,000 worth of fan coil units for an installation, and obtained the maximum discount. Some time later it was determined the project required an additional matching unit. The contractor has to purchase this unit from the original supplier to ensure a match. The supplier at this time may not discount the unit because of the small quantity, and the fact that he is no longer in a competitive situation. The impact of quantity on purchase can add between 0% and 25% to material prices and/or subcontractor quotes.

- If materials have been ordered or delivered to the job site, will they be subject to a cancellation charge or restocking fee?

Check with the supplier to determine if ordered materials are subject to a cancellation charge. Delivered materials not used as result of a change order may be subject to a restocking fee if returned to the supplier. Common restocking charges run between 20% and 40%. Also, delivery charges to return the goods to the supplier must be added.

Labor

- How efficient is the existing crew at the actual installation?

Is the same crew that performed the initial work going to do the change order? Possibly the change consists of the installation of a unit identical to one already installed; therefore the change should take less time. Be sure to consider this potential productivity increase and modify the productivity rates accordingly.

- If the crew size is increased, what impact will that have on supervision requirements?

Under most bargaining agreements or management practices, there is a point at which a working foreman is replaced by a nonworking foreman. This replacement increases project overhead by adding a nonproductive worker. If additional workers are added to accelerate the project or to perform changes while maintaining the schedule, be sure to add additional supervision time if warranted. Calculate the hours involved and the additional cost directly if possible.

- What are the other impacts of increased crew size?

The larger the crew, the greater the potential for productivity to decrease. Some of the factors that cause this productivity loss are: overcrowding (producing restrictive conditions in the working space), and possibly a shortage of any special tools and equipment required. Such factors affect not only the crew working on the elements directly involved in the change order, but other crews whose movement may also be hampered.

As the crew increases, check its basic composition for changes by the addition or deletion of apprentices or nonworking foreman and quantify the potential effects of equipment shortages or other logistical factors.

- As new crews, unfamiliar with the project, are brought onto the site, how long will it take them to become oriented to the project requirements?

The orientation time for a new crew to become 100% effective varies with the site and type of project. Orientation is easiest at a new construction site, and most difficult at existing, very restrictive renovation sites. The type of work also affects orientation time. When all elements of the work are exposed, such as concrete or masonry work, orientation is decreased. When the work is concealed or less visible, such as existing electrical systems, orientation takes longer. Usually orientation can be accomplished in one day or less. Costs for added orientation should be itemized and added to the total estimated cost.

- How much actual production can be gained by working overtime?

Short term overtime can be used effectively to accomplish more work in a day. However as overtime is scheduled to run beyond several weeks, studies have shown marked decreases in output. The following chart shows the effect of long term overtime on worker efficiency. If the anticipated change requires extended overtime to keep the job on schedule, these factors can be used as a guide to predict the impact on time and cost. Add project overhead, particularly supervision, that may also be incurred.

Days per Week	Hours per Day	Production Efficiency					Payroll Cost Factors	
		1 Week	2 Weeks	3 Weeks	4 Weeks	Average 4 Weeks	@ 1-1/2 Times	@ 2 Times
5	8	100%	100%	100%	100%	100%	100%	100%
	9	100	100	95	90	96.25	105.6	111.1
	10	100	95	90	85	91.25	110.0	120.0
	11	95	90	75	65	81.25	113.6	127.3
	12	90	85	70	60	76.25	116.7	133.3
6	8	100	100	95	90	96.25	108.3	116.7
	9	100	95	90	85	92.50	113.0	125.9
	10	95	90	85	80	87.50	116.7	133.3
	11	95	85	70	65	78.75	119.7	139.4
	12	90	80	65	60	73.75	122.2	144.4
7	8	100	95	85	75	88.75	114.3	128.6
	9	95	90	80	70	83.75	118.3	136.5
	10	90	85	75	65	78.75	121.4	142.9
	11	85	80	65	60	72.50	124.0	148.1
	12	85	75	60	55	68.75	126.2	152.4

Effects of Overtime

Caution: Under many labor agreements, Sundays and holidays are paid at a higher premium than the normal overtime rate.

The use of long-term overtime is counterproductive on almost any construction job; that is, the longer the period of overtime, the lower the actual production rate. Numerous studies have been conducted, and while they have resulted in slightly different numbers, all reach the same conclusion. The figure above tabulates the effects of overtime work on efficiency.

As illustrated, there can be a difference between the *actual* payroll cost per hour and the *effective* cost per hour for overtime work. This is due to the reduced production efficiency with the increase in weekly hours beyond 40. This difference between actual and effective cost results from overtime work over a prolonged period. Short-term overtime work does not result in as great a reduction in efficiency, and in such cases, effective cost may not vary significantly from the actual payroll cost. As the total hours per week are increased on a regular basis, more time is lost because of fatigue, lowered morale, and an increased accident rate.

As an example, assume a project where workers are working 6 days a week, 10 hours per day. From the figure above (based on productivity studies), the average effective productive hours over a four-week period are:

$$0.875 \times 60 = 52.5$$

Depending upon the locale and day of week, overtime hours may be paid at time and a half or double time. For time and a half, the overall (average) *actual* payroll cost (including regular and overtime hours) is determined as follows:

$$\frac{40 \text{ reg. hrs.} + (20 \text{ overtime hrs. x } 1.5)}{60 \text{ hrs.}} = 1.167$$

Based on 60 hours, the payroll cost per hour will be 116.7% of the normal rate at 40 hours per week. However, because the effective production (efficiency) for 60 hours is reduced to the equivalent of 52.5 hours, the effective cost of overtime is calculated as follows:

For time and a half:

$$\frac{40 \text{ reg. hrs.} + (20 \text{ overtime hrs. x } 1.5)}{52.5 \text{ hrs.}} = 1.33$$

Installed cost will be 133% of the normal rate (for labor).

Thus, when figuring overtime, the actual cost per unit of work will be higher than the apparent overtime payroll dollar increase, due to the reduced productivity of the longer workweek. These efficiency calculations are true only for those cost factors determined by hours worked. Costs that are applied weekly or monthly, such as equipment rentals, will not be similarly affected.

Equipment

- What equipment is required to complete the change order?

Change orders may require extending the rental period of equipment already on the job site, or the addition of special equipment brought in to accomplish the change work. In either case, the additional rental charges and operator labor charges must be added.

Summary

The preceding considerations and others you deem appropriate should be analyzed and applied to a change order estimate. The impact of each should be quantified and listed on the estimate to form an audit trail.

Change orders that are properly identified, documented, and managed help to ensure the orderly, professional and profitable progress of the work. They also minimize potential claims or disputes at the end of the project.

Crew No.	Bare Costs		Incl. Subs O & P		Cost Per Labor-Hour	
Crew A-1	Hr.	Daily	Hr.	Daily	Bare Costs	Incl. O&P
1 Building Laborer	$26.70	$213.60	$41.55	$332.40	$26.70	$41.55
1 Concrete saw, gas manual		49.80		54.80	6.23	6.85
8 L.H., Daily Totals		$263.40		$387.20	$32.93	$48.40
Crew A-1A	Hr.	Daily	Hr.	Daily	Bare Costs	Incl. O&P
1 Skilled Worker	$34.85	$278.80	$54.20	$433.60	$34.85	$54.20
1 Shot Blaster, 20"		338.35		372.20	42.29	46.52
8 L.H., Daily Totals		$617.15		$805.80	$77.14	$100.72
Crew A-1B	Hr.	Daily	Hr.	Daily	Bare Costs	Incl. O&P
1 Building Laborer	$26.70	$213.60	$41.55	$332.40	$26.70	$41.55
1 Concr. saw, gas, self-prop.		114.40		125.85	14.30	15.73
8 L.H., Daily Totals		$328.00		$458.25	$41.00	$57.28
Crew A-1C	Hr.	Daily	Hr.	Daily	Bare Costs	Incl. O&P
1 Building Laborer	$26.70	$213.60	$41.55	$332.40	$26.70	$41.55
1 Brush saw		19.20		21.10	2.40	2.64
8 L.H., Daily Totals		$232.80		$353.50	$29.10	$44.19
Crew A-1D	Hr.	Daily	Hr.	Daily	Bare Costs	Incl. O&P
1 Building Laborer	$26.70	$213.60	$41.55	$332.40	$26.70	$41.55
1 Vibrating plate, gas, 18"		25.20		27.70	3.15	3.47
8 L.H., Daily Totals		$238.80		$360.10	$29.85	$45.02
Crew A-1E	Hr.	Daily	Hr.	Daily	Bare Costs	Incl. O&P
1 Building Laborer	$26.70	$213.60	$41.55	$332.40	$26.70	$41.55
1 Vibrating plate, gas, 21"		36.80		40.50	4.60	5.06
8 L.H., Daily Totals		$250.40		$372.90	$31.30	$46.61
Crew A-1F	Hr.	Daily	Hr.	Daily	Bare Costs	Incl. O&P
1 Building Laborer	$26.70	$213.60	$41.55	$332.40	$26.70	$41.55
1 Rammer/tamper, gas, 8"		34.80		38.30	4.35	4.79
8 L.H., Daily Totals		$248.40		$370.70	$31.05	$46.34
Crew A-1G	Hr.	Daily	Hr.	Daily	Bare Costs	Incl. O&P
1 Building Laborer	$26.70	$213.60	$41.55	$332.40	$26.70	$41.55
1 Rammer/tamper, gas, 8"		34.80		38.30	4.35	4.79
8 L.H., Daily Totals		$248.40		$370.70	$31.05	$46.34
Crew A-1H	Hr.	Daily	Hr.	Daily	Bare Costs	Incl. O&P
1 Building Laborer	$26.70	$213.60	$41.55	$332.40	$26.70	$41.55
1 Pressure washer		55.00		60.50	6.88	7.56
8 L.H., Daily Totals		$268.60		$392.90	$33.58	$49.11
Crew A-1J	Hr.	Daily	Hr.	Daily	Bare Costs	Incl. O&P
1 Building Laborer	$26.70	$213.60	$41.55	$332.40	$26.70	$41.55
1 Rototiller		84.50		92.95	10.56	11.61
8 L.H., Daily Totals		$298.10		$425.35	$37.26	$53.16
Crew A-1K	Hr.	Daily	Hr.	Daily	Bare Costs	Incl. O&P
1 Building Laborer	$26.70	$213.60	$41.55	$332.40	$26.70	$41.55
1 Lawn aerator		23.95		26.35	2.99	3.29
8 L.H., Daily Totals		$237.55		$358.75	$29.69	$44.84
Crew A-1L	Hr.	Daily	Hr.	Daily	Bare Costs	Incl. O&P
1 Building Laborer	$26.70	$213.60	$41.55	$332.40	$26.70	$41.55
1 Power blower/vacuum		23.95		26.35	2.99	3.29
8 L.H., Daily Totals		$237.55		$358.75	$29.69	$44.84
Crew A-1M	Hr.	Daily	Hr.	Daily	Bare Costs	Incl. O&P
1 Building Laborer	$26.70	$213.60	$41.55	$332.40	$26.70	$41.55
1 Snow blower		84.50		92.95	10.56	11.61
8 L.H., Daily Totals		$298.10		$425.35	$37.26	$53.16
Crew A-2	Hr.	Daily	Hr.	Daily	Bare Costs	Incl. O&P
2 Laborers	$26.70	$427.20	$41.55	$664.80	$26.73	$41.32
1 Truck Driver (light)	26.80	214.40	40.85	326.80		
1 Light Truck, 1.5 Ton		128.80		141.70	5.37	5.90
24 L.H., Daily Totals		$770.40		$1133.30	$32.10	$47.22
Crew A-2A	Hr.	Daily	Hr.	Daily	Bare Costs	Incl. O&P
2 Laborers	$26.70	$427.20	$41.55	$664.80	$26.73	$41.32
1 Truck Driver (light)	26.80	214.40	40.85	326.80		
1 Light Truck, 1.5 Ton		128.80		141.70		
1 Concrete Saw		114.40		125.85	10.13	11.15
24 L.H., Daily Totals		$884.80		$1259.15	$36.86	$52.47
Crew A-3	Hr.	Daily	Hr.	Daily	Bare Costs	Incl. O&P
1 Truck Driver (heavy)	$27.55	$220.40	$42.00	$336.00	$27.55	$42.00
1 Dump Truck, 12 Ton		326.60		359.25	40.83	44.91
8 L.H., Daily Totals		$547.00		$695.25	$68.38	$86.91
Crew A-3A	Hr.	Daily	Hr.	Daily	Bare Costs	Incl. O&P
1 Truck Driver (light)	$26.80	$214.40	$40.85	$326.80	$26.80	$40.85
1 Pickup Truck (4x4)		85.20		93.70	10.65	11.72
8 L.H., Daily Totals		$299.60		$420.50	$37.45	$52.57
Crew A-3B	Hr.	Daily	Hr.	Daily	Bare Costs	Incl. O&P
1 Equip. Oper. (medium)	$34.65	$277.20	$52.20	$417.60	$31.10	$47.10
1 Truck Driver (heavy)	27.55	220.40	42.00	336.00		
1 Dump Truck, 16 Ton		476.60		524.25		
1 F.E. Loader, 3 C.Y.		305.00		335.50	48.85	53.74
16 L.H., Daily Totals		$1279.20		$1613.35	$79.95	$100.84
Crew A-3C	Hr.	Daily	Hr.	Daily	Bare Costs	Incl. O&P
1 Equip. Oper. (light)	$33.05	$264.40	$49.80	$398.40	$33.05	$49.80
1 Wheeled Skid Steer Loader		197.80		217.60	24.73	27.20
8 L.H., Daily Totals		$462.20		$616.00	$57.78	$77.00
Crew A-3D	Hr.	Daily	Hr.	Daily	Bare Costs	Incl. O&P
1 Truck Driver, Light	$26.80	$214.40	$40.85	$326.80	$26.80	$40.85
1 Pickup Truck (4x4)		85.20		93.70		
1 Flatbed Trailer, 25 Ton		88.00		96.80	21.65	23.82
8 L.H., Daily Totals		$387.60		$517.30	$48.45	$64.67
Crew A-3E	Hr.	Daily	Hr.	Daily	Bare Costs	Incl. O&P
1 Equip. Oper. (crane)	$35.90	$287.20	$54.10	$432.80	$31.73	$48.05
1 Truck Driver (heavy)	27.55	220.40	42.00	336.00		
1 Pickup Truck (4x4)		85.20		93.70	5.33	5.86
16 L.H., Daily Totals		$592.80		$862.50	$37.06	$53.91
Crew A-3F	Hr.	Daily	Hr.	Daily	Bare Costs	Incl. O&P
1 Equip. Oper. (crane)	$35.90	$287.20	$54.10	$432.80	$31.73	$48.05
1 Truck Driver (heavy)	27.55	220.40	42.00	336.00		
1 Pickup Truck (4x4)		85.20		93.70		
1 Tractor, 6x2, 40 Ton Cap.		301.60		331.75		
1 Lowbed Trailer, 75 Ton		170.20		187.20	34.81	38.29
16 L.H., Daily Totals		$1064.60		$1381.45	$66.54	$86.34

Crew A-3G

Crew No.	Bare Costs Hr.	Daily	Incl. Subs O & P Hr.	Daily	Cost Per Labor-Hour Bare Costs	Incl. O&P
1 Equip. Oper. (crane)	$35.90	$287.20	$54.10	$432.80	$31.73	$48.05
1 Truck Driver (heavy)	27.55	220.40	42.00	336.00		
1 Pickup Truck (4x4)		85.20		93.70		
1 Tractor, 6x4, 45 Ton Cap.		353.60		388.95		
1 Lowbed Trailer, 75 Ton		170.20		187.20	38.06	41.87
16 L.H., Daily Totals		$1116.60		$1438.65	$69.79	$89.92

Crew A-4

Crew No.	Bare Costs Hr.	Daily	Incl. Subs O & P Hr.	Daily	Cost Per Labor-Hour Bare Costs	Incl. O&P
2 Carpenters	$34.25	$548.00	$53.35	$853.60	$33.03	$50.90
1 Painter, Ordinary	30.60	244.80	46.00	368.00		
24 L.H., Daily Totals		$792.80		$1221.60	$33.03	$50.90

Crew A-5

Crew No.	Bare Costs Hr.	Daily	Incl. Subs O & P Hr.	Daily	Cost Per Labor-Hour Bare Costs	Incl. O&P
2 Laborers	$26.70	$427.20	$41.55	$664.80	$26.71	$41.47
.25 Truck Driver (light)	26.80	53.60	40.85	81.70		
.25 Light Truck, 1.5 Ton		32.20		35.40	1.79	1.97
18 L.H., Daily Totals		$513.00		$781.90	$28.50	$43.44

Crew A-6

Crew No.	Bare Costs Hr.	Daily	Incl. Subs O & P Hr.	Daily	Cost Per Labor-Hour Bare Costs	Incl. O&P
1 Instrument Man	$34.85	$278.80	$54.20	$433.60	$33.93	$52.05
1 Rodman/Chainman	33.00	264.00	49.90	399.20		
1 Laser Transit/Level		60.75		66.85	3.79	4.17
16 L.H., Daily Totals		$603.55		$899.65	$37.72	$56.22

Crew A-7

Crew No.	Bare Costs Hr.	Daily	Incl. Subs O & P Hr.	Daily	Cost Per Labor-Hour Bare Costs	Incl. O&P
1 Chief Of Party	$42.90	$343.20	$66.90	$535.20	$36.92	$57.00
1 Instrument Man	34.85	278.80	54.20	433.60		
1 Rodman/Chainman	33.00	264.00	49.90	399.20		
1 Laser Transit/Level		60.75		66.85	2.53	2.78
24 L.H., Daily Totals		$946.75		$1434.85	$39.45	$59.78

Crew A-8

Crew No.	Bare Costs Hr.	Daily	Incl. Subs O & P Hr.	Daily	Cost Per Labor-Hour Bare Costs	Incl. O&P
1 Chief Of Party	$42.90	$343.20	$66.90	$535.20	$35.94	$55.23
1 Instrument Man	34.85	278.80	54.20	433.60		
2 Rodmen/Chainmen	33.00	528.00	49.90	798.40		
1 Laser Transit/Level		60.75		66.85	1.90	2.09
32 L.H., Daily Totals		$1210.75		$1834.05	$37.84	$57.32

Crew A-9

Crew No.	Bare Costs Hr.	Daily	Incl. Subs O & P Hr.	Daily	Cost Per Labor-Hour Bare Costs	Incl. O&P
1 Asbestos Foreman	$37.60	$300.80	$59.35	$474.80	$37.16	$58.65
7 Asbestos Workers	37.10	2077.60	58.55	3278.80		
64 L.H., Daily Totals		$2378.40		$3753.60	$37.16	$58.65

Crew A-10

Crew No.	Bare Costs Hr.	Daily	Incl. Subs O & P Hr.	Daily	Cost Per Labor-Hour Bare Costs	Incl. O&P
1 Asbestos Foreman	$37.60	$300.80	$59.35	$474.80	$37.16	$58.65
7 Asbestos Workers	37.10	2077.60	58.55	3278.80		
64 L.H., Daily Totals		$2378.40		$3753.60	$37.16	$58.65

Crew A-10A

Crew No.	Bare Costs Hr.	Daily	Incl. Subs O & P Hr.	Daily	Cost Per Labor-Hour Bare Costs	Incl. O&P
1 Asbestos Foreman	$37.60	$300.80	$59.35	$474.80	$37.27	$58.82
2 Asbestos Workers	37.10	593.60	58.55	936.80		
24 L.H., Daily Totals		$894.40		$1411.60	$37.27	$58.82

Crew A-10B

Crew No.	Bare Costs Hr.	Daily	Incl. Subs O & P Hr.	Daily	Cost Per Labor-Hour Bare Costs	Incl. O&P
1 Asbestos Foreman	$37.60	$300.80	$59.35	$474.80	$37.23	$58.75
3 Asbestos Workers	37.10	890.40	58.55	1405.20		
32 L.H., Daily Totals		$1191.20		$1880.00	$37.23	$58.75

Crew A-10C

Crew No.	Bare Costs Hr.	Daily	Incl. Subs O & P Hr.	Daily	Cost Per Labor-Hour Bare Costs	Incl. O&P
3 Asbestos Workers	$37.10	$890.40	$58.55	$1405.20	$37.10	$58.55
1 Flatbed Truck		128.80		141.70	5.37	5.90
24 L.H., Daily Totals		$1019.20		$1546.90	$42.47	$64.45

Crew A-10D

Crew No.	Bare Costs Hr.	Daily	Incl. Subs O & P Hr.	Daily	Cost Per Labor-Hour Bare Costs	Incl. O&P
2 Asbestos Workers	$37.10	$593.60	$58.55	$936.80	$35.05	$54.14
1 Equip. Oper. (crane)	35.90	287.20	54.10	432.80		
1 Equip. Oper. Oiler	30.10	240.80	45.35	362.80		
1 Hydraulic Crane, 33 Ton		641.60		705.75	20.05	22.06
32 L.H., Daily Totals		$1763.20		$2438.15	$55.10	$76.20

Crew A-11

Crew No.	Bare Costs Hr.	Daily	Incl. Subs O & P Hr.	Daily	Cost Per Labor-Hour Bare Costs	Incl. O&P
1 Asbestos Foreman	$37.60	$300.80	$59.35	$474.80	$37.16	$58.65
7 Asbestos Workers	37.10	2077.60	58.55	3278.80		
2 Chipping Hammers		32.00		35.20	.50	.55
64 L.H., Daily Totals		$2410.40		$3788.80	$37.66	$59.20

Crew A-12

Crew No.	Bare Costs Hr.	Daily	Incl. Subs O & P Hr.	Daily	Cost Per Labor-Hour Bare Costs	Incl. O&P
1 Asbestos Foreman	$37.60	$300.80	$59.35	$474.80	$37.16	$58.65
7 Asbestos Workers	37.10	2077.60	58.55	3278.80		
1 Large Prod. Vac. Loader		480.00		528.00	7.50	8.25
64 L.H., Daily Totals		$2858.40		$4281.60	$44.66	$66.90

Crew A-13

Crew No.	Bare Costs Hr.	Daily	Incl. Subs O & P Hr.	Daily	Cost Per Labor-Hour Bare Costs	Incl. O&P
1 Equip. Oper. (light)	$33.05	$264.40	$49.80	$398.40	$33.05	$49.80
1 Large Prod. Vac. Loader		480.00		528.00	60.00	66.00
8 L.H., Daily Totals		$744.40		$926.40	$93.05	$115.80

Crew B-1

Crew No.	Bare Costs Hr.	Daily	Incl. Subs O & P Hr.	Daily	Cost Per Labor-Hour Bare Costs	Incl. O&P
1 Labor Foreman (outside)	$28.70	$229.60	$44.70	$357.60	$27.37	$42.60
2 Laborers	26.70	427.20	41.55	664.80		
24 L.H., Daily Totals		$656.80		$1022.40	$27.37	$42.60

Crew B-1A

Crew No.	Bare Costs Hr.	Daily	Incl. Subs O & P Hr.	Daily	Cost Per Labor-Hour Bare Costs	Incl. O&P
1 Laborer Foreman	$28.70	$229.60	$44.70	$357.60	$27.37	$42.60
2 Laborers	26.70	427.20	41.55	664.80		
2 Cutting Torches		36.00		39.60		
2 Gases		129.60		142.55	6.90	7.59
24 L.H., Daily Totals		$822.40		$1204.55	$34.27	$50.19

Crew B-1B

Crew No.	Bare Costs Hr.	Daily	Incl. Subs O & P Hr.	Daily	Cost Per Labor-Hour Bare Costs	Incl. O&P
1 Laborer Foreman	$28.70	$229.60	$44.70	$357.60	$29.50	$45.48
2 Laborers	26.70	427.20	41.55	664.80		
1 Equip. Oper. (crane)	35.90	287.20	54.10	432.80		
2 Cutting Torches		36.00		39.60		
2 Gases		129.60		142.55		
1 Hyd. Crane, 12 Ton		602.60		662.85	24.01	26.41
32 L.H., Daily Totals		$1712.20		$2300.20	$53.51	$71.89

Crew B-2

Crew No.	Bare Costs Hr.	Daily	Incl. Subs O & P Hr.	Daily	Cost Per Labor-Hour Bare Costs	Incl. O&P
1 Labor Foreman (outside)	$28.70	$229.60	$44.70	$357.60	$27.10	$42.18
4 Laborers	26.70	854.40	41.55	1329.60		
40 L.H., Daily Totals		$1084.00		$1687.20	$27.10	$42.18

Crew No.	Bare Costs		Incl. Subs O & P		Cost Per Labor-Hour	
	Hr.	Daily	Hr.	Daily	Bare Costs	Incl. O&P
Crew B-3	Hr.	Daily	Hr.	Daily	Bare Costs	Incl. O&P
1 Labor Foreman (outside)	$28.70	$229.60	$44.70	$357.60	$28.64	$44.00
2 Laborers	26.70	427.20	41.55	664.80		
1 Equip. Oper. (med.)	34.65	277.20	52.20	417.60		
2 Truck Drivers (heavy)	27.55	440.80	42.00	672.00		
1 F.E. Loader, T.M., 2.5 C.Y.		788.00		866.80		
2 Dump Trucks, 16 Ton		953.20		1048.50	36.28	39.90
48 L.H., Daily Totals		$3116.00		$4027.30	$64.92	$83.90
Crew B-3A	Hr.	Daily	Hr.	Daily	Bare Costs	Incl. O&P
4 Laborers	$26.70	$854.40	$41.55	$1329.60	$28.29	$43.68
1 Equip. Oper. (med.)	34.65	277.20	52.20	417.60		
1 Hyd. Excavator, 1.5 C.Y.		720.40		792.45	18.01	19.81
40 L.H., Daily Totals		$1852.00		$2539.65	$46.30	$63.49
Crew B-3B	Hr.	Daily	Hr.	Daily	Bare Costs	Incl. O&P
2 Laborers	$26.70	$427.20	$41.55	$664.80	$28.90	$44.33
1 Equip. Oper. (med.)	34.65	277.20	52.20	417.60		
1 Truck Driver (heavy)	27.55	220.40	42.00	336.00		
1 Backhoe Loader, 80 H.P.		256.40		282.05		
1 Dump Truck, 16 Ton		476.60		524.25	22.91	25.20
32 L.H., Daily Totals		$1657.80		$2224.70	$51.81	$69.53
Crew B-3C	Hr.	Daily	Hr.	Daily	Bare Costs	Incl. O&P
3 Laborers	$26.70	$640.80	$41.55	$997.20	$28.69	$44.21
1 Equip. Oper. (med.)	34.65	277.20	52.20	417.60		
1 F.E. Crawler Ldr, 4 C.Y.		1094.00		1203.40	34.19	37.61
32 L.H., Daily Totals		$2012.00		$2618.20	$62.88	$81.82
Crew B-4	Hr.	Daily	Hr.	Daily	Bare Costs	Incl. O&P
1 Labor Foreman (outside)	$28.70	$229.60	$44.70	$357.60	$27.18	$42.15
4 Laborers	26.70	854.40	41.55	1329.60		
1 Truck Driver (heavy)	27.55	220.40	42.00	336.00		
1 Tractor, 4 x 2, 195 H.P.		208.20		229.00		
1 Platform Trailer		119.40		131.35	6.83	7.51
48 L.H., Daily Totals		$1632.00		$2383.55	$34.01	$49.66
Crew B-5	Hr.	Daily	Hr.	Daily	Bare Costs	Incl. O&P
1 Labor Foreman (outside)	$28.70	$229.60	$44.70	$357.60	$29.26	$45.04
4 Laborers	26.70	854.40	41.55	1329.60		
2 Equip. Oper. (med.)	34.65	554.40	52.20	835.20		
1 Air Compr., 250 C.F.M.		112.40		123.65		
2 Air Tools & Accessories		19.60		21.55		
2-50 Ft. Air Hoses, 1.5" Dia.		9.40		10.35		
1 F.E. Loader, T.M., 2.5 C.Y.		788.00		866.80	16.60	18.26
56 L.H., Daily Totals		$2567.80		$3544.75	$45.86	$63.30
Crew B-5A	Hr.	Daily	Hr.	Daily	Bare Costs	Incl. O&P
1 Foreman	$28.70	$229.60	$44.70	$357.60	$28.86	$44.35
6 Laborers	26.70	1281.60	41.55	1994.40		
2 Equip. Oper. (med.)	34.65	554.40	52.20	835.20		
1 Equip. Oper. (light)	33.05	264.40	49.80	398.40		
2 Truck Drivers (heavy)	27.55	440.80	42.00	672.00		
1 Air Compr. 365 C.F.M.		146.80		161.50		
2 Pavement Breakers		19.60		21.55		
8 Air Hoses w/Coup.,1"		29.20		32.10		
2 Dump Trucks, 12 Ton		653.20		718.50	8.84	9.73
96 L.H., Daily Totals		$3619.60		$5191.25	$37.70	$54.08

Crew No.	Bare Costs		Incl. Subs O & P		Cost Per Labor-Hour	
	Hr.	Daily	Hr.	Daily	Bare Costs	Incl. O&P
Crew B-5B	Hr.	Daily	Hr.	Daily	Bare Costs	Incl. O&P
1 Powderman	$34.85	$278.80	$54.20	$433.60	$31.13	$47.43
2 Equip. Oper. (med.)	34.65	554.40	52.20	835.20		
3 Truck Drivers (heavy)	27.55	661.20	42.00	1008.00		
1 F.E. Ldr. 2-1/2 CY		305.00		335.50		
3 Dump Trucks, 16 Ton		1429.80		1572.80		
1 Air Compr. 365 C.F.M.		146.80		161.50	39.20	43.12
48 L.H., Daily Totals		$3376.00		$4346.60	$70.33	$90.55
Crew B-5C	Hr.	Daily	Hr.	Daily	Bare Costs	Incl. O&P
3 Laborers	$26.70	$640.80	$41.55	$997.20	$29.48	$45.04
1 Equip. Oper. (medium)	34.65	277.20	52.20	417.60		
2 Truck Drivers (heavy)	27.55	440.80	42.00	672.00		
1 Equip. Oper. (crane)	35.90	287.20	54.10	432.80		
1 Equip. Oper. Oiler	30.10	240.80	45.35	362.80		
2 Dump Trucks, 16 Ton		953.20		1048.50		
1 F.E. Crawler Ldr, 4 C.Y.		1094.00		1203.40		
1 Hyd. Crane, 25 Ton		575.00		632.50	40.97	45.07
64 L.H., Daily Totals		$4509.00		$5766.80	$70.45	$90.11
Crew B-6	Hr.	Daily	Hr.	Daily	Bare Costs	Incl. O&P
2 Laborers	$26.70	$427.20	$41.55	$664.80	$28.82	$44.30
1 Equip. Oper. (light)	33.05	264.40	49.80	398.40		
1 Backhoe Loader, 48 H.P.		215.80		237.40	8.99	9.89
24 L.H., Daily Totals		$907.40		$1300.60	$37.81	$54.19
Crew B-6A	Hr.	Daily	Hr.	Daily	Bare Costs	Incl. O&P
.5 Labor Foreman (outside)	$28.70	$114.80	$44.70	$178.80	$30.28	$46.44
1 Laborer	26.70	213.60	41.55	332.40		
1 Equip. Oper. (med.)	34.65	277.20	52.20	417.60		
1 Vacuum Trk.,5000 Gal.		331.50		364.65	16.58	18.23
20 L.H., Daily Totals		$937.10		$1293.45	$46.86	$64.67
Crew B-6B	Hr.	Daily	Hr.	Daily	Bare Costs	Incl. O&P
2 Labor Foreman (out)	$28.70	$459.20	$44.70	$715.20	$27.37	$42.60
4 Laborers	26.70	854.40	41.55	1329.60		
1 Winch Truck		317.00		348.70		
1 Flatbed Truck		128.80		141.70		
1 Butt Fusion Machine		432.80		476.10	18.30	20.13
48 L.H., Daily Totals		$2192.20		$3011.30	$45.67	$62.73
Crew B-7	Hr.	Daily	Hr.	Daily	Bare Costs	Incl. O&P
1 Labor Foreman (outside)	$28.70	$229.60	$44.70	$357.60	$28.36	$43.85
4 Laborers	26.70	854.40	41.55	1329.60		
1 Equip. Oper. (med.)	34.65	277.20	52.20	417.60		
1 Chipping Machine		164.80		181.30		
1 F.E. Loader, T.M., 2.5 C.Y.		788.00		866.80		
2 Chain Saws, 36"		66.80		73.50	21.24	23.37
48 L.H., Daily Totals		$2380.80		$3226.40	$49.60	$67.22
Crew B-7A	Hr.	Daily	Hr.	Daily	Bare Costs	Incl. O&P
2 Laborers	$26.70	$427.20	$41.55	$664.80	$28.82	$44.30
1 Equip. Oper. (light)	33.05	264.40	49.80	398.40		
1 Rake w/Tractor		191.50		210.65		
2 Chain Saws, 18"		38.40		42.25	9.58	10.54
24 L.H., Daily Totals		$921.50		$1316.10	$38.40	$54.84

Crew B-8

Crew No.	Bare Costs Hr.	Daily	Incl. Subs O & P Hr.	Daily	Bare Costs	Incl. O&P
1 Labor Foreman (outside)	$28.70	$229.60	$44.70	$357.60	$29.58	$45.19
2 Laborers	26.70	427.20	41.55	664.80		
2 Equip. Oper. (med.)	34.65	554.40	52.20	835.20		
1 Equip. Oper. Oiler	30.10	240.80	45.35	362.80		
2 Truck Drivers (heavy)	27.55	440.80	42.00	672.00		
1 Hyd. Crane, 25 Ton		616.80		678.50		
1 F.E. Loader, T.M., 2.5 C.Y.		788.00		866.80		
2 Dump Trucks, 16 Ton		953.20		1048.50	36.84	40.53
64 L.H., Daily Totals		$4250.80		$5486.20	$66.42	$85.72

Crew B-9

Crew No.	Bare Costs Hr.	Daily	Incl. Subs O & P Hr.	Daily	Bare Costs	Incl. O&P
1 Labor Foreman (outside)	$28.70	$229.60	$44.70	$357.60	$27.10	$42.18
4 Laborers	26.70	854.40	41.55	1329.60		
1 Air Compr., 250 C.F.M.		112.40		123.65		
2 Air Tools & Accessories		19.60		21.55		
2-50 Ft. Air Hoses, 1.5" Dia.		9.40		10.35	3.54	3.89
40 L.H., Daily Totals		$1225.40		$1842.75	$30.64	$46.07

Crew B-9A

Crew No.	Bare Costs Hr.	Daily	Incl. Subs O & P Hr.	Daily	Bare Costs	Incl. O&P
2 Laborers	$26.70	$427.20	$41.55	$664.80	$26.98	$41.70
1 Truck Driver (heavy)	27.55	220.40	42.00	336.00		
1 Water Tanker		116.80		128.50		
1 Tractor		208.20		229.00		
2-50 Ft. Disch. Hoses		5.00		5.50	13.75	15.13
24 L.H., Daily Totals		$977.60		$1363.80	$40.73	$56.83

Crew B-9B

Crew No.	Bare Costs Hr.	Daily	Incl. Subs O & P Hr.	Daily	Bare Costs	Incl. O&P
2 Laborers	$26.70	$427.20	$41.55	$664.80	$26.98	$41.70
1 Truck Driver (heavy)	27.55	220.40	42.00	336.00		
2-50 Ft. Disch. Hoses		5.00		5.50		
1 Water Tanker		116.80		128.50		
1 Tractor		208.20		229.00		
1 Pressure Washer		46.20		50.80	15.68	17.24
24 L.H., Daily Totals		$1023.80		$1414.60	$42.66	$58.94

Crew B-9C

Crew No.	Bare Costs Hr.	Daily	Incl. Subs O & P Hr.	Daily	Bare Costs	Incl. O&P
1 Labor Foreman (outside)	$28.70	$229.60	$44.70	$357.60	$27.10	$42.18
4 Laborers	26.70	854.40	41.55	1329.60		
1 Air Compr., 250 C.F.M.		112.40		123.65		
2-50 Ft. Air Hoses, 1.5" Dia.		9.40		10.35		
2 Breaker, Pavement, 60 lb.		19.60		21.55	3.54	3.89
40 L.H., Daily Totals		$1225.40		$1842.75	$30.64	$46.07

Crew B-9D

Crew No.	Bare Costs Hr.	Daily	Incl. Subs O & P Hr.	Daily	Bare Costs	Incl. O&P
1 Labor Foreman (Outside)	$28.70	$229.60	$44.70	$357.60	$27.10	$42.18
4 Common Laborers	26.70	854.40	41.55	1329.60		
1 Air Compressor, 250 CFM		112.40		123.65		
2 Air hoses, 1.5" x 50'		9.40		10.35		
2 Air tamper		55.30		60.85	4.43	4.87
40 L.H., Daily Totals		$1261.10		$1882.05	$31.53	$47.05

Crew B-10

Crew No.	Bare Costs Hr.	Daily	Incl. Subs O & P Hr.	Daily	Bare Costs	Incl. O&P
1 Equip. Oper. (med.)	$34.65	$277.20	$52.20	$417.60	$32.00	$48.65
.5 Laborer	26.70	106.80	41.55	166.20		
12 L.H., Daily Totals		$384.00		$583.80	$32.00	$48.65

Crew B-10A

Crew No.	Bare Costs Hr.	Daily	Incl. Subs O & P Hr.	Daily	Bare Costs	Incl. O&P
1 Equip. Oper. (med.)	$34.65	$277.20	$52.20	$417.60	$32.00	$48.65
.5 Laborer	26.70	106.80	41.55	166.20		
1 Walk behind compactor, 7.5 HP		122.80		135.10	10.23	11.26
12 L.H., Daily Totals		$506.80		$718.90	$42.23	$59.91

Crew B-10B

Crew No.	Bare Costs Hr.	Daily	Incl. Subs O & P Hr.	Daily	Bare Costs	Incl. O&P
1 Equip. Oper. (med.)	$34.65	$277.20	$52.20	$417.60	$32.00	$48.65
.5 Laborer	26.70	106.80	41.55	166.20		
1 Dozer, 200 H.P.		919.60		1011.55	76.63	84.30
12 L.H., Daily Totals		$1303.60		$1595.35	$108.63	$132.95

Crew B-10C

Crew No.	Bare Costs Hr.	Daily	Incl. Subs O & P Hr.	Daily	Bare Costs	Incl. O&P
1 Equip. Oper. (med.)	$34.65	$277.20	$52.20	$417.60	$32.00	$48.65
.5 Laborer	26.70	106.80	41.55	166.20		
1 Dozer, 200 H.P.		919.60		1011.55		
1 Vibratory Roller, Towed		591.60		650.75	125.93	138.53
12 L.H., Daily Totals		$1895.20		$2246.10	$157.93	$187.18

Crew B-10D

Crew No.	Bare Costs Hr.	Daily	Incl. Subs O & P Hr.	Daily	Bare Costs	Incl. O&P
1 Equip. Oper. (med.)	$34.65	$277.20	$52.20	$417.60	$32.00	$48.65
.5 Laborer	26.70	106.80	41.55	166.20		
1 Dozer, 200 H.P.		919.60		1011.55		
1 Sheepsft. Roller, Towed		610.20		671.20	127.48	140.23
12 L.H., Daily Totals		$1913.80		$2266.55	$159.48	$188.88

Crew B-10E

Crew No.	Bare Costs Hr.	Daily	Incl. Subs O & P Hr.	Daily	Bare Costs	Incl. O&P
1 Equip. Oper. (med.)	$34.65	$277.20	$52.20	$417.60	$32.00	$48.65
.5 Laborer	26.70	106.80	41.55	166.20		
1 Tandem Roller, 5 Ton		107.40		118.15	8.95	9.85
12 L.H., Daily Totals		$491.40		$701.95	$40.95	$58.50

Crew B-10F

Crew No.	Bare Costs Hr.	Daily	Incl. Subs O & P Hr.	Daily	Bare Costs	Incl. O&P
1 Equip. Oper. (med.)	$34.65	$277.20	$52.20	$417.60	$32.00	$48.65
.5 Laborer	26.70	106.80	41.55	166.20		
1 Tandem Roller, 10 Ton		179.20		197.10	14.93	16.43
12 L.H., Daily Totals		$563.20		$780.90	$46.93	$65.08

Crew B-10G

Crew No.	Bare Costs Hr.	Daily	Incl. Subs O & P Hr.	Daily	Bare Costs	Incl. O&P
1 Equip. Oper. (med.)	$34.65	$277.20	$52.20	$417.60	$32.00	$48.65
.5 Laborer	26.70	106.80	41.55	166.20		
1 Sheepsft. Roll., 130 H.P.		794.80		874.30	66.23	72.86
12 L.H., Daily Totals		$1178.80		$1458.10	$98.23	$121.51

Crew B-10H

Crew No.	Bare Costs Hr.	Daily	Incl. Subs O & P Hr.	Daily	Bare Costs	Incl. O&P
1 Equip. Oper. (med.)	$34.65	$277.20	$52.20	$417.60	$32.00	$48.65
.5 Laborer	26.70	106.80	41.55	166.20		
1 Diaphr. Water Pump, 2"		50.60		55.65		
1-20 Ft. Suction Hose, 2"		2.55		2.80		
2-50 Ft. Disch. Hoses, 2"		3.80		4.20	4.75	5.22
12 L.H., Daily Totals		$440.95		$646.45	$36.75	$53.87

Crew B-10I

Crew No.	Bare Costs Hr.	Daily	Incl. Subs O & P Hr.	Daily	Bare Costs	Incl. O&P
1 Equip. Oper. (med.)	$34.65	$277.20	$52.20	$417.60	$32.00	$48.65
.5 Laborer	26.70	106.80	41.55	166.20		
1 Diaphr. Water Pump, 4"		74.40		81.85		
1-20 Ft. Suction Hose, 4"		5.05		5.55		
2-50 Ft. Disch. Hoses, 4"		7.10		7.80	7.21	7.93
12 L.H., Daily Totals		$470.55		$679.00	$39.21	$56.58

Crew B-10J

Crew No.	Bare Costs Hr.	Daily	Incl. Subs O & P Hr.	Daily	Bare Costs	Incl. O&P
1 Equip. Oper. (med.)	$34.65	$277.20	$52.20	$417.60	$32.00	$48.65
.5 Laborer	26.70	106.80	41.55	166.20		
1 Centr. Water Pump, 3"		53.60		58.95		
1-20 Ft. Suction Hose, 3"		4.25		4.70		
2-50 Ft. Disch. Hoses, 3"		5.00		5.50	5.24	5.76
12 L.H., Daily Totals		$446.85		$652.95	$37.24	$54.41

Crew B-10K

	Bare Costs Hr.	Daily	Incl. Subs O & P Hr.	Daily	Cost Per Labor-Hour Bare Costs	Incl. O&P
1 Equip. Oper. (med.)	$34.65	$277.20	$52.20	$417.60	$32.00	$48.65
.5 Laborer	26.70	106.80	41.55	166.20		
1 Centr. Water Pump, 6"		217.00		238.70		
1-20 Ft. Suction Hose, 6"		12.80		14.10		
2-50 Ft. Disch. Hoses, 6"		17.80		19.60	20.63	22.70
12 L.H., Daily Totals		$631.60		$856.20	$52.63	$71.35

Crew B-10L

	Bare Costs Hr.	Daily	Incl. Subs O & P Hr.	Daily	Cost Per Labor-Hour Bare Costs	Incl. O&P
1 Equip. Oper. (med.)	$34.65	$277.20	$52.20	$417.60	$32.00	$48.65
.5 Laborer	26.70	106.80	41.55	166.20		
1 Dozer, 80 H.P.		314.80		346.30	26.23	28.86
12 L.H., Daily Totals		$698.80		$930.10	$58.23	$77.51

Crew B-10M

	Bare Costs Hr.	Daily	Incl. Subs O & P Hr.	Daily	Cost Per Labor-Hour Bare Costs	Incl. O&P
1 Equip. Oper. (med.)	$34.65	$277.20	$52.20	$417.60	$32.00	$48.65
.5 Laborer	26.70	106.80	41.55	166.20		
1 Dozer, 300 H.P.		1195.00		1314.50	99.58	109.54
12 L.H., Daily Totals		$1579.00		$1898.30	$131.58	$158.19

Crew B-10N

	Bare Costs Hr.	Daily	Incl. Subs O & P Hr.	Daily	Cost Per Labor-Hour Bare Costs	Incl. O&P
1 Equip. Oper. (med.)	$34.65	$277.20	$52.20	$417.60	$32.00	$48.65
.5 Laborer	26.70	106.80	41.55	166.20		
1 F.E. Loader, T.M., 1.5 C.Y		312.60		343.85	26.05	28.66
12 L.H., Daily Totals		$696.60		$927.65	$58.05	$77.31

Crew B-10O

	Bare Costs Hr.	Daily	Incl. Subs O & P Hr.	Daily	Cost Per Labor-Hour Bare Costs	Incl. O&P
1 Equip. Oper. (med.)	$34.65	$277.20	$52.20	$417.60	$32.00	$48.65
.5 Laborer	26.70	106.80	41.55	166.20		
1 F.E. Loader, T.M., 2.25 C.Y.		551.80		607.00	45.98	50.58
12 L.H., Daily Totals		$935.80		$1190.80	$77.98	$99.23

Crew B-10P

	Bare Costs Hr.	Daily	Incl. Subs O & P Hr.	Daily	Cost Per Labor-Hour Bare Costs	Incl. O&P
1 Equipment Oper. (med.)	$34.65	$277.20	$52.20	$417.60	$32.00	$48.65
.5 Laborer	26.70	106.80	41.55	166.20		
1 F.E. Loader, T.M., 2.5 C.Y.		788.00		866.80	65.67	72.23
12 L.H., Daily Totals		$1172.00		$1450.60	$97.67	$120.88

Crew B-10Q

	Bare Costs Hr.	Daily	Incl. Subs O & P Hr.	Daily	Cost Per Labor-Hour Bare Costs	Incl. O&P
1 Equip. Oper. (med.)	$34.65	$277.20	$52.20	$417.60	$32.00	$48.65
.5 Laborer	26.70	106.80	41.55	166.20		
1 F.E. Loader, T.M., 5 C.Y.		1094.00		1203.40	91.17	100.28
12 L.H., Daily Totals		$1478.00		$1787.20	$123.17	$148.93

Crew B-10R

	Bare Costs Hr.	Daily	Incl. Subs O & P Hr.	Daily	Cost Per Labor-Hour Bare Costs	Incl. O&P
1 Equip. Oper. (med.)	$34.65	$277.20	$52.20	$417.60	$32.00	$48.65
.5 Laborer	26.70	106.80	41.55	166.20		
1 F.E. Loader, W.M., 1 C.Y.		193.80		213.20	16.15	17.77
12 L.H., Daily Totals		$577.80		$797.00	$48.15	$66.42

Crew B-10S

	Bare Costs Hr.	Daily	Incl. Subs O & P Hr.	Daily	Cost Per Labor-Hour Bare Costs	Incl. O&P
1 Equip. Oper. (med.)	$34.65	$277.20	$52.20	$417.60	$32.00	$48.65
.5 Laborer	26.70	106.80	41.55	166.20		
1 F.E. Loader, W.M., 1.5 C.Y.		241.00		265.10	20.08	22.09
12 L.H., Daily Totals		$625.00		$848.90	$52.08	$70.74

Crew B-10T

	Bare Costs Hr.	Daily	Incl. Subs O & P Hr.	Daily	Cost Per Labor-Hour Bare Costs	Incl. O&P
1 Equip. Oper. (med.)	$34.65	$277.20	$52.20	$417.60	$32.00	$48.65
.5 Laborer	26.70	106.80	41.55	166.20		
1 F.E. Loader, W.M.,2.5 C.Y.		305.00		335.50	25.42	27.96
12 L.H., Daily Totals		$689.00		$919.30	$57.42	$76.61

Crew B-10U

	Bare Costs Hr.	Daily	Incl. Subs O & P Hr.	Daily	Cost Per Labor-Hour Bare Costs	Incl. O&P
1 Equip. Oper. (med.)	$34.65	$277.20	$52.20	$417.60	$32.00	$48.65
.5 Laborer	26.70	106.80	41.55	166.20		
1 F.E. Loader, W.M., 5.5 C.Y.		700.40		770.45	58.37	64.20
12 L.H., Daily Totals		$1084.40		$1354.25	$90.37	$112.85

Crew B-10V

	Bare Costs Hr.	Daily	Incl. Subs O & P Hr.	Daily	Cost Per Labor-Hour Bare Costs	Incl. O&P
1 Equip. Oper. (med.)	$34.65	$277.20	$52.20	$417.60	$32.00	$48.65
.5 Laborer	26.70	106.80	41.55	166.20		
1 Dozer, 700 H.P.		3139.00		3452.90	261.58	287.74
12 L.H., Daily Totals		$3523.00		$4036.70	$293.58	$336.39

Crew B-10W

	Bare Costs Hr.	Daily	Incl. Subs O & P Hr.	Daily	Cost Per Labor-Hour Bare Costs	Incl. O&P
1 Equip. Oper. (med.)	$34.65	$277.20	$52.20	$417.60	$32.00	$48.65
.5 Laborer	26.70	106.80	41.55	166.20		
1 Dozer, 105 H.P.		453.80		499.20	37.82	41.60
12 L.H., Daily Totals		$837.80		$1083.00	$69.82	$90.25

Crew B-10X

	Bare Costs Hr.	Daily	Incl. Subs O & P Hr.	Daily	Cost Per Labor-Hour Bare Costs	Incl. O&P
1 Equip. Oper. (med.)	$34.65	$277.20	$52.20	$417.60	$32.00	$48.65
.5 Laborer	26.70	106.80	41.55	166.20		
1 Dozer, 410 H.P.		1524.00		1676.40	127.00	139.70
12 L.H., Daily Totals		$1908.00		$2260.20	$159.00	$188.35

Crew B-10Y

	Bare Costs Hr.	Daily	Incl. Subs O & P Hr.	Daily	Cost Per Labor-Hour Bare Costs	Incl. O&P
1 Equip. Oper. (med.)	$34.65	$277.20	$52.20	$417.60	$32.00	$48.65
.5 Laborer	26.70	106.80	41.55	166.20		
1 Vibratory Drum Roller		344.80		379.30	28.73	31.61
12 L.H., Daily Totals		$728.80		$963.10	$60.73	$80.26

Crew B-11A

	Bare Costs Hr.	Daily	Incl. Subs O & P Hr.	Daily	Cost Per Labor-Hour Bare Costs	Incl. O&P
1 Equipment Oper. (med.)	$34.65	$277.20	$52.20	$417.60	$30.68	$46.88
1 Laborer	26.70	213.60	41.55	332.40		
1 Dozer, 200 H.P.		919.60		1011.55	57.48	63.22
16 L.H., Daily Totals		$1410.40		$1761.55	$88.16	$110.10

Crew B-11B

	Bare Costs Hr.	Daily	Incl. Subs O & P Hr.	Daily	Cost Per Labor-Hour Bare Costs	Incl. O&P
1 Equipment Oper. (light)	$33.05	$264.40	$49.80	$398.40	$29.88	$45.68
1 Laborer	26.70	213.60	41.55	332.40		
1 Air Powered Tamper		27.65		30.40		
1 Air Compr. 365 C.F.M.		146.80		161.50		
2-50 Ft. Air Hoses, 1.5" Dia.		9.40		10.35	11.49	12.64
16 L.H., Daily Totals		$661.85		$933.05	$41.37	$58.32

Crew B-11C

	Bare Costs Hr.	Daily	Incl. Subs O & P Hr.	Daily	Cost Per Labor-Hour Bare Costs	Incl. O&P
1 Equipment Oper. (med.)	$34.65	$277.20	$52.20	$417.60	$30.68	$46.88
1 Laborer	26.70	213.60	41.55	332.40		
1 Backhoe Loader, 48 H.P.		215.80		237.40	13.49	14.84
16 L.H., Daily Totals		$706.60		$987.40	$44.17	$61.72

Crew B-11J

	Bare Costs Hr.	Daily	Incl. Subs O & P Hr.	Daily	Cost Per Labor-Hour Bare Costs	Incl. O&P
1 Equipment Oper. (med.)	$34.65	$277.20	$52.20	$417.60	$30.68	$46.88
1 Laborer	26.70	213.60	41.55	332.40		
1 Grader, 30,000 Lbs.		456.60		502.25		
1 Ripper, beam & 1 shank		71.40		78.55	33.00	36.30
16 L.H., Daily Totals		$1018.80		$1330.80	$63.68	$83.18

Crew B-11K

Crew No.	Bare Costs Hr.	Daily	Incl. Subs O & P Hr.	Daily	Bare Costs	Incl. O&P
1 Equipment Oper. (med.)	$34.65	$277.20	$52.20	$417.60	$30.68	$46.88
1 Laborer	26.70	213.60	41.55	332.40		
1 Trencher, 8' D., 16" W.		1420.00		1562.00	88.75	97.63
16 L.H., Daily Totals		$1910.80		$2312.00	$119.43	$144.51

Crew B-11L

Crew No.	Hr.	Daily	Hr.	Daily	Bare Costs	Incl. O&P
1 Equipment Oper. (med.)	$34.65	$277.20	$52.20	$417.60	$30.68	$46.88
1 Laborer	26.70	213.60	41.55	332.40		
1 Grader, 30,000 Lbs.		456.60		502.25	28.54	31.39
16 L.H., Daily Totals		$947.40		$1252.25	$59.22	$78.27

Crew B-11M

Crew No.	Hr.	Daily	Hr.	Daily	Bare Costs	Incl. O&P
1 Equipment Oper. (med.)	$34.65	$277.20	$52.20	$417.60	$30.68	$46.88
1 Laborer	26.70	213.60	41.55	332.40		
1 Backhoe Loader, 80 H.P.		256.40		282.05	16.03	17.63
16 L.H., Daily Totals		$747.20		$1032.05	$46.71	$64.51

Crew B-11N

Crew No.	Hr.	Daily	Hr.	Daily	Bare Costs	Incl. O&P
1 Labor Foreman	$28.70	$229.60	$44.70	$357.60	$29.26	$44.57
2 Equipment Operators (med.)	34.65	554.40	52.20	835.20		
6 Truck Drivers (hvy.)	27.55	1322.40	42.00	2016.00		
1 F.E. Loader, 5.5 C.Y.		700.40		770.45		
1 Dozer, 400 H.P.		1524.00		1676.40		
6 Off Hwy. Tks. 50 Ton		7086.00		7794.60	129.31	142.24
72 L.H., Daily Totals		$11416.80		$13450.25	$158.57	$186.81

Crew B-11Q

Crew No.	Hr.	Daily	Hr.	Daily	Bare Costs	Incl. O&P
1 Equipment Operator (med.)	$34.65	$277.20	$52.20	$417.60	$32.00	$48.65
.5 Laborer	26.70	106.80	41.55	166.20		
1 Dozer, 140 H.P.		572.00		629.20	47.67	52.43
12 L.H., Daily Totals		$956.00		$1213.00	$79.67	$101.08

Crew B-11R

Crew No.	Hr.	Daily	Hr.	Daily	Bare Costs	Incl. O&P
1 Equipment Operator (med.)	$34.65	$277.20	$52.20	$417.60	$32.00	$48.65
.5 Laborer	26.70	106.80	41.55	166.20		
1 Dozer, 215 H.P.		919.60		1011.55	76.63	84.30
12 L.H., Daily Totals		$1303.60		$1595.35	$108.63	$132.95

Crew B-11S

Crew No.	Hr.	Daily	Hr.	Daily	Bare Costs	Incl. O&P
1 Equipment Operator (med.)	$34.65	$277.20	$52.20	$417.60	$32.00	$48.65
.5 Laborer	26.70	106.80	41.55	166.20		
1 Dozer, 300 H.P.		1195.00		1314.50		
1 Ripper, Beam & 1 Shank		71.40		78.55	105.53	116.09
12 L.H., Daily Totals		$1650.40		$1976.85	$137.53	$164.74

Crew B-11T

Crew No.	Hr.	Daily	Hr.	Daily	Bare Costs	Incl. O&P
1 Equipment Operator (med.)	$34.65	$277.20	$52.20	$417.60	$32.00	$48.65
.5 Laborer	26.70	106.80	41.55	166.20		
1 Dozer, 410 H.P.		1524.00		1676.40		
1 Ripper, Beam & 2 Shanks		80.20		88.20	133.68	147.05
12 L.H., Daily Totals		$1988.20		$2348.40	$165.68	$195.70

Crew B-11U

Crew No.	Hr.	Daily	Hr.	Daily	Bare Costs	Incl. O&P
1 Equipment Operator (med.)	$34.65	$277.20	$52.20	$417.60	$32.00	$48.65
.5 Laborer	26.70	106.80	41.55	166.20		
1 Dozer, 520 H.P.		2035.00		2238.50	169.58	186.54
12 L.H., Daily Totals		$2419.00		$2822.30	$201.58	$235.19

Crew B-11V

Crew No.	Hr.	Daily	Hr.	Daily	Bare Costs	Incl. O&P
3 Laborers	$26.70	$640.80	$41.55	$997.20	$26.70	$41.55
1 Walk behind compactor, 7.5 HP		122.80		135.10	5.12	5.63
24 L.H., Daily Totals		$763.60		$1132.30	$31.82	$47.18

Crew B-11W

Crew No.	Hr.	Daily	Hr.	Daily	Bare Costs	Incl. O&P
1 Equipment Operator (med.)	$34.65	$277.20	$52.20	$417.60	$28.07	$42.81
1 Common Laborer	26.70	213.60	41.55	332.40		
10 Truck Drivers, Heavy	27.55	2204.00	42.00	3360.00		
1 Dozer, 200 H.P.		919.60		1011.55		
1 Vib. roller, smth, towed, 23 Ton		591.60		650.75		
10 Dump Truck, 10 Ton		3266.00		3592.60	49.76	54.74
96 L.H., Daily Totals		$7472.00		$9364.90	$77.83	$97.55

Crew B-11Y

Crew No.	Hr.	Daily	Hr.	Daily	Bare Costs	Incl. O&P
1 Labor Foreman (Outside)	$28.70	$229.60	$44.70	$357.60	$29.57	$45.45
5 Common Laborers	26.70	1068.00	41.55	1662.00		
3 Equipment Operator (med.)	34.65	831.60	52.20	1252.80		
1 Dozer, 80 H.P.		314.80		346.30		
2 Walk behind compactor, 7.5 HP		245.60		270.15		
4 Vibratory plate, gas, 21"		147.20		161.90	9.83	10.81
72 L.H., Daily Totals		$2836.80		$4050.75	$39.40	$56.26

Crew B-12A

Crew No.	Hr.	Daily	Hr.	Daily	Bare Costs	Incl. O&P
1 Equip. Oper. (crane)	$35.90	$287.20	$54.10	$432.80	$31.30	$47.83
1 Laborer	26.70	213.60	41.55	332.40		
1 Hyd. Excavator, 1 C.Y.		557.80		613.60	34.86	38.35
16 L.H., Daily Totals		$1058.60		$1378.80	$66.16	$86.18

Crew B-12B

Crew No.	Hr.	Daily	Hr.	Daily	Bare Costs	Incl. O&P
1 Equip. Oper. (crane)	$35.90	$287.20	$54.10	$432.80	$31.30	$47.83
1 Laborer	26.70	213.60	41.55	332.40		
1 Hyd. Excavator, 1.5 C.Y.		720.40		792.45	45.03	49.53
16 L.H., Daily Totals		$1221.20		$1557.65	$76.33	$97.36

Crew B-12C

Crew No.	Hr.	Daily	Hr.	Daily	Bare Costs	Incl. O&P
1 Equip. Oper. (crane)	$35.90	$287.20	$54.10	$432.80	$31.30	$47.83
1 Laborer	26.70	213.60	41.55	332.40		
1 Hyd. Excavator, 2 C.Y.		909.80		1000.80	56.86	62.55
16 L.H., Daily Totals		$1410.60		$1766.00	$88.16	$110.38

Crew B-12D

Crew No.	Hr.	Daily	Hr.	Daily	Bare Costs	Incl. O&P
1 Equip. Oper. (crane)	$35.90	$287.20	$54.10	$432.80	$31.30	$47.83
1 Laborer	26.70	213.60	41.55	332.40		
1 Hyd. Excavator, 3.5 C.Y.		2020.00		2222.00	126.25	138.88
16 L.H., Daily Totals		$2520.80		$2987.20	$157.55	$186.71

Crew B-12E

Crew No.	Hr.	Daily	Hr.	Daily	Bare Costs	Incl. O&P
1 Equip. Oper. (crane)	$35.90	$287.20	$54.10	$432.80	$31.30	$47.83
1 Laborer	26.70	213.60	41.55	332.40		
1 Hyd. Excavator, .5 C.Y.		332.40		365.65	20.78	22.85
16 L.H., Daily Totals		$833.20		$1130.85	$52.08	$70.68

Crew B-12F

Crew No.	Hr.	Daily	Hr.	Daily	Bare Costs	Incl. O&P
1 Equip. Oper. (crane)	$35.90	$287.20	$54.10	$432.80	$31.30	$47.83
1 Laborer	26.70	213.60	41.55	332.40		
1 Hyd. Excavator, .75 C.Y.		474.00		521.40	29.63	32.59
16 L.H., Daily Totals		$974.80		$1286.60	$60.93	$80.42

Crew No.	Bare Costs		Incl. Subs O & P		Cost Per Labor-Hour	

Crew B-12G

	Hr.	Daily	Hr.	Daily	Bare Costs	Incl. O&P
1 Equip. Oper. (crane)	$35.90	$287.20	$54.10	$432.80	$31.30	$47.83
1 Laborer	26.70	213.60	41.55	332.40		
1 Power Shovel, .5 C.Y.		508.10		558.90		
1 Clamshell Bucket, .5 C.Y.		33.80		37.20	33.87	37.25
16 L.H., Daily Totals		$1042.70		$1361.30	$65.17	$85.08

Crew B-12H

	Hr.	Daily	Hr.	Daily	Bare Costs	Incl. O&P
1 Equip. Oper. (crane)	$35.90	$287.20	$54.10	$432.80	$31.30	$47.83
1 Laborer	26.70	213.60	41.55	332.40		
1 Power Shovel, 1 C.Y.		833.20		916.50		
1 Clamshell Bucket, 1 C.Y.		43.40		47.75	54.79	60.27
16 L.H., Daily Totals		$1377.40		$1729.45	$86.09	$108.10

Crew B-12I

	Hr.	Daily	Hr.	Daily	Bare Costs	Incl. O&P
1 Equip. Oper. (crane)	$35.90	$287.20	$54.10	$432.80	$31.30	$47.83
1 Laborer	26.70	213.60	41.55	332.40		
1 Power Shovel, .75 C.Y.		630.45		693.50		
1 Dragline Bucket, .75 C.Y.		18.80		20.70	40.58	44.63
16 L.H., Daily Totals		$1150.05		$1479.40	$71.88	$92.46

Crew B-12J

	Hr.	Daily	Hr.	Daily	Bare Costs	Incl. O&P
1 Equip. Oper. (crane)	$35.90	$287.20	$54.10	$432.80	$31.30	$47.83
1 Laborer	26.70	213.60	41.55	332.40		
1 Gradall, 3 Ton, .5 C.Y.		837.80		921.60	52.36	57.60
16 L.H., Daily Totals		$1338.60		$1686.80	$83.66	$105.43

Crew B-12K

	Hr.	Daily	Hr.	Daily	Bare Costs	Incl. O&P
1 Equip. Oper. (crane)	$35.90	$287.20	$54.10	$432.80	$31.30	$47.83
1 Laborer	26.70	213.60	41.55	332.40		
1 Gradall, 3 Ton, 1 C.Y.		971.00		1068.10	60.69	66.76
16 L.H., Daily Totals		$1471.80		$1833.30	$91.99	$114.59

Crew B-12L

	Hr.	Daily	Hr.	Daily	Bare Costs	Incl. O&P
1 Equip. Oper. (crane)	$35.90	$287.20	$54.10	$432.80	$31.30	$47.83
1 Laborer	26.70	213.60	41.55	332.40		
1 Power Shovel, .5 C.Y.		508.10		558.90		
1 F.E. Attachment, .5 C.Y.		47.40		52.15	34.72	38.19
16 L.H., Daily Totals		$1056.30		$1376.25	$66.02	$86.02

Crew B-12M

	Hr.	Daily	Hr.	Daily	Bare Costs	Incl. O&P
1 Equip. Oper. (crane)	$35.90	$287.20	$54.10	$432.80	$31.30	$47.83
1 Laborer	26.70	213.60	41.55	332.40		
1 Power Shovel, .75 C.Y.		630.45		693.50		
1 F.E. Attachment, .75 C.Y.		51.80		57.00	42.64	46.90
16 L.H., Daily Totals		$1183.05		$1515.70	$73.94	$94.73

Crew B-12N

	Hr.	Daily	Hr.	Daily	Bare Costs	Incl. O&P
1 Equip. Oper. (crane)	$35.90	$287.20	$54.10	$432.80	$31.30	$47.83
1 Laborer	26.70	213.60	41.55	332.40		
1 Power Shovel, 1 C.Y.		833.20		916.50		
1 F.E. Attachment, 1 C.Y.		58.60		64.45	55.74	61.31
16 L.H., Daily Totals		$1392.60		$1746.15	$87.04	$109.14

Crew B-12O

	Hr.	Daily	Hr.	Daily	Bare Costs	Incl. O&P
1 Equip. Oper. (crane)	$35.90	$287.20	$54.10	$432.80	$31.30	$47.83
1 Laborer	26.70	213.60	41.55	332.40		
1 Power Shovel, 1.5 C.Y.		888.20		977.00		
1 F.E. Attachment, 1.5 C.Y.		67.80		74.60	59.75	65.73
16 L.H., Daily Totals		$1456.80		$1816.80	$91.05	$113.56

Crew B-12P

	Hr.	Daily	Hr.	Daily	Bare Costs	Incl. O&P
1 Equip. Oper. (crane)	$35.90	$287.20	$54.10	$432.80	$31.30	$47.83
1 Laborer	26.70	213.60	41.55	332.40		
1 Crawler Crane, 40 Ton		888.20		977.00		
1 Dragline Bucket, 1.5 C.Y.		30.60		33.65	57.43	63.17
16 L.H., Daily Totals		$1419.60		$1775.85	$88.73	$111.00

Crew B-12Q

	Hr.	Daily	Hr.	Daily	Bare Costs	Incl. O&P
1 Equip. Oper. (crane)	$35.90	$287.20	$54.10	$432.80	$31.30	$47.83
1 Laborer	26.70	213.60	41.55	332.40		
1 Hyd. Excavator, 5/8 C.Y.		432.40		475.65	27.03	29.73
16 L.H., Daily Totals		$933.20		$1240.85	$58.33	$77.56

Crew B-12R

	Hr.	Daily	Hr.	Daily	Bare Costs	Incl. O&P
1 Equip. Oper. (crane)	$35.90	$287.20	$54.10	$432.80	$31.30	$47.83
1 Laborer	26.70	213.60	41.55	332.40		
1 Hyd. Excavator, 1.5 C.Y.		720.40		792.45	45.03	49.53
16 L.H., Daily Totals		$1221.20		$1557.65	$76.33	$97.36

Crew B-12S

	Hr.	Daily	Hr.	Daily	Bare Costs	Incl. O&P
1 Equip. Oper. (crane)	$35.90	$287.20	$54.10	$432.80	$31.30	$47.83
1 Laborer	26.70	213.60	41.55	332.40		
1 Hyd. Excavator, 2.5 C.Y.		1215.00		1336.50	75.94	83.53
16 L.H., Daily Totals		$1715.80		$2101.70	$107.24	$131.36

Crew B-12T

	Hr.	Daily	Hr.	Daily	Bare Costs	Incl. O&P
1 Equip. Oper. (crane)	$35.90	$287.20	$54.10	$432.80	$31.30	$47.83
1 Laborer	26.70	213.60	41.55	332.40		
1 Crawler Crane, 75 Ton		1170.00		1287.00		
1 F.E. Attachment, 3 C.Y.		90.20		99.20	78.76	86.64
16 L.H., Daily Totals		$1761.00		$2151.40	$110.06	$134.47

Crew B-12V

	Hr.	Daily	Hr.	Daily	Bare Costs	Incl. O&P
1 Equip. Oper. (crane)	$35.90	$287.20	$54.10	$432.80	$31.30	$47.83
1 Laborer	26.70	213.60	41.55	332.40		
1 Crawler Crane, 75 Ton		1170.00		1287.00		
1 Dragline Bucket, 3 C.Y.		47.40		52.15	76.09	83.70
16 L.H., Daily Totals		$1718.20		$2104.35	$107.39	$131.53

Crew B-13

	Hr.	Daily	Hr.	Daily	Bare Costs	Incl. O&P
1 Labor Foreman (outside)	$28.70	$229.60	$44.70	$357.60	$28.79	$44.34
4 Laborers	26.70	854.40	41.55	1329.60		
1 Equip. Oper. (crane)	35.90	287.20	54.10	432.80		
1 Equip. Oper. Oiler	30.10	240.80	45.35	362.80		
1 Hyd. Crane, 25 Ton		616.80		678.50	11.01	12.12
56 L.H., Daily Totals		$2228.80		$3161.30	$39.80	$56.46

Crew B-13A

	Hr.	Daily	Hr.	Daily	Bare Costs	Incl. O&P
1 Foreman	$28.70	$229.60	$44.70	$357.60	$29.50	$45.17
2 Laborers	26.70	427.20	41.55	664.80		
2 Equipment Operators	34.65	554.40	52.20	835.20		
2 Truck Drivers (heavy)	27.55	440.80	42.00	672.00		
1 Crane, 75 Ton		1170.00		1287.00		
1 F.E. Lder, 3.75 C.Y.		1094.00		1203.40		
2 Dump Trucks, 12 Ton		653.20		718.50	52.09	57.30
56 L.H., Daily Totals		$4569.20		$5738.50	$81.59	$102.47

Crew No.	Bare Costs		Incl. Subs O & P		Cost Per Labor-Hour	

Crew B-13B	Hr.	Daily	Hr.	Daily	Bare Costs	Incl. O&P
1 Labor Foreman (outside)	$28.70	$229.60	$44.70	$357.60	$28.79	$44.34
4 Laborers	26.70	854.40	41.55	1329.60		
1 Equip. Oper. (crane)	35.90	287.20	54.10	432.80		
1 Equip. Oper. Oiler	30.10	240.80	45.35	362.80		
1 Hyd. Crane, 55 Ton		895.40		984.95	15.99	17.59
56 L.H., Daily Totals		$2507.40		$3467.75	$44.78	$61.93

Crew B-13C	Hr.	Daily	Hr.	Daily	Bare Costs	Incl. O&P
1 Labor Foreman (outside)	$28.70	$229.60	$44.70	$357.60	$28.79	$44.34
4 Laborers	26.70	854.40	41.55	1329.60		
1 Equip. Oper. (crane)	35.90	287.20	54.10	432.80		
1 Equip. Oper. Oiler	30.10	240.80	45.35	362.80		
1 Crawler Crane, 100 Ton		1519.00		1670.90	27.13	29.84
56 L.H., Daily Totals		$3131.00		$4153.70	$55.92	$74.18

Crew B-14	Hr.	Daily	Hr.	Daily	Bare Costs	Incl. O&P
1 Labor Foreman (outside)	$28.70	$229.60	$44.70	$357.60	$28.09	$43.45
4 Laborers	26.70	854.40	41.55	1329.60		
1 Equip. Oper. (light)	33.05	264.40	49.80	398.40		
1 Backhoe Loader, 48 H.P.		215.80		237.40	4.50	4.95
48 L.H., Daily Totals		$1564.20		$2323.00	$32.59	$48.40

Crew B-15	Hr.	Daily	Hr.	Daily	Bare Costs	Incl. O&P
1 Equipment Oper. (med.)	$34.65	$277.20	$52.20	$417.60	$29.46	$44.85
.5 Laborer	26.70	106.80	41.55	166.20		
2 Truck Drivers (heavy)	27.55	440.80	42.00	672.00		
2 Dump Trucks, 16 Ton		953.20		1048.50		
1 Dozer, 200 H.P.		919.60		1011.55	66.89	73.57
28 L.H., Daily Totals		$2697.60		$3315.85	$96.35	$118.42

Crew B-16	Hr.	Daily	Hr.	Daily	Bare Costs	Incl. O&P
1 Labor Foreman (outside)	$28.70	$229.60	$44.70	$357.60	$27.41	$42.45
2 Laborers	26.70	427.20	41.55	664.80		
1 Truck Driver (heavy)	27.55	220.40	42.00	336.00		
1 Dump Truck, 16 Ton		476.60		524.25	14.89	16.38
32 L.H., Daily Totals		$1353.80		$1882.65	$42.30	$58.83

Crew B-17	Hr.	Daily	Hr.	Daily	Bare Costs	Incl. O&P
2 Laborers	$26.70	$427.20	$41.55	$664.80	$28.50	$43.73
1 Equip. Oper. (light)	33.05	264.40	49.80	398.40		
1 Truck Driver (heavy)	27.55	220.40	42.00	336.00		
1 Backhoe Loader, 48 H.P.		215.80		237.40		
1 Dump Truck, 12 Ton		326.60		359.25	16.95	18.65
32 L.H., Daily Totals		$1454.40		$1995.85	$45.45	$62.38

Crew B-17A	Hr.	Daily	Hr.	Daily	Bare Costs	Incl. O&P
2 Laborer Foremen	$28.70	$459.20	$44.70	$715.20	$28.93	$45.02
6 Laborers	26.70	1281.60	41.55	1994.40		
1 Skilled Worker Foreman	36.85	294.80	57.30	458.40		
1 Skilled Worker	34.85	278.80	54.20	433.60		
80 L.H., Daily Totals		$2314.40		$3601.60	$28.93	$45.02

Crew B-18	Hr.	Daily	Hr.	Daily	Bare Costs	Incl. O&P
1 Labor Foreman (outside)	$28.70	$229.60	$44.70	$357.60	$27.37	$42.60
2 Laborers	26.70	427.20	41.55	664.80		
1 Vibrating Compactor		36.80		40.50	1.53	1.69
24 L.H., Daily Totals		$693.60		$1062.90	$28.90	$44.29

Crew B-19	Hr.	Daily	Hr.	Daily	Bare Costs	Incl. O&P
1 Pile Driver Foreman	$35.30	$282.40	$58.15	$465.20	$33.80	$53.91
4 Pile Drivers	33.30	1065.60	54.90	1756.80		
2 Equip. Oper. (crane)	35.90	574.40	54.10	865.60		
1 Equip. Oper. Oiler	30.10	240.80	45.35	362.80		
1 Crane, 40 Ton & Access.		888.20		977.00		
60 L.F. Pile Leads		75.00		82.50		
1 Hammer, Diesel, 22k Ft-Lb		564.00		620.40	23.91	26.30
64 L.H., Daily Totals		$3690.40		$5130.30	$57.71	$80.21

Crew B-19A	Hr.	Daily	Hr.	Daily	Bare Costs	Incl. O&P
1 Pile Driver Foreman	$35.30	$282.40	$58.15	$465.20	$33.80	$53.91
4 Pile Drivers	33.30	1065.60	54.90	1756.80		
2 Equip. Oper. (crane)	35.90	574.40	54.10	865.60		
1 Equip. Oper. Oiler	30.10	240.80	45.35	362.80		
1 Crawler Crane, 75 Ton		1170.00		1287.00		
60 L.F. Leads, 25K Ft. Lbs.		120.00		132.00		
1 Hammer, Diesel, 41k Ft-Lb		662.60		728.85	30.51	33.56
64 L.H., Daily Totals		$4115.80		$5598.25	$64.31	$87.47

Crew B-20	Hr.	Daily	Hr.	Daily	Bare Costs	Incl. O&P
1 Labor Foreman (out)	$28.70	$229.60	$44.70	$357.60	$30.08	$46.82
1 Skilled Worker	34.85	278.80	54.20	433.60		
1 Laborer	26.70	213.60	41.55	332.40		
24 L.H., Daily Totals		$722.00		$1123.60	$30.08	$46.82

Crew B-20A	Hr.	Daily	Hr.	Daily	Bare Costs	Incl. O&P
1 Labor Foreman	$28.70	$229.60	$44.70	$357.60	$32.24	$49.20
1 Laborer	26.70	213.60	41.55	332.40		
1 Plumber	40.85	326.80	61.40	491.20		
1 Plumber Apprentice	32.70	261.60	49.15	393.20		
32 L.H., Daily Totals		$1031.60		$1574.40	$32.24	$49.20

Crew B-21	Hr.	Daily	Hr.	Daily	Bare Costs	Incl. O&P
1 Labor Foreman (out)	$28.70	$229.60	$44.70	$357.60	$30.91	$47.86
1 Skilled Worker	34.85	278.80	54.20	433.60		
1 Laborer	26.70	213.60	41.55	332.40		
.5 Equip. Oper. (crane)	35.90	143.60	54.10	216.40		
.5 S.P. Crane, 5 Ton		158.50		174.35	5.66	6.23
28 L.H., Daily Totals		$1024.10		$1514.35	$36.57	$54.09

Crew B-21A	Hr.	Daily	Hr.	Daily	Bare Costs	Incl. O&P
1 Labor Foreman	$28.70	$229.60	$44.70	$357.60	$32.97	$50.18
1 Laborer	26.70	213.60	41.55	332.40		
1 Plumber	40.85	326.80	61.40	491.20		
1 Plumber Apprentice	32.70	261.60	49.15	393.20		
1 Equip. Oper. (crane)	35.90	287.20	54.10	432.80		
1 S.P. Crane, 12 Ton		505.80		556.40	12.65	13.91
40 L.H., Daily Totals		$1824.60		$2563.60	$45.62	$64.09

Crew B-21B	Hr.	Daily	Hr.	Daily	Bare Costs	Incl. O&P
1 Laborer Foreman	$28.70	$229.60	$44.70	$357.60	$28.94	$44.69
3 Laborers	26.70	640.80	41.55	997.20		
1 Equip. Oper. (crane)	35.90	287.20	54.10	432.80		
1 Hyd. Crane, 12 Ton		602.60		662.85	15.07	16.57
40 L.H., Daily Totals		$1760.20		$2450.45	$44.01	$61.26

Crew No.	Bare Costs		Incl. Subs O & P		Cost Per Labor-Hour	

Crew B-21C	Hr.	Daily	Hr.	Daily	Bare Costs	Incl. O&P
1 Laborer Foreman	$28.70	$229.60	$44.70	$357.60	$28.79	$44.34
4 Laborers	26.70	854.40	41.55	1329.60		
1 Equip. Oper. (crane)	35.90	287.20	54.10	432.80		
1 Equip. Oper. Oiler	30.10	240.80	45.35	362.80		
2 Cutting Torches		36.00		39.60		
2 Gases		129.60		142.55		
1 Crane, 90 Ton		1325.00		1457.50	26.62	29.28
56 L.H., Daily Totals		$3102.60		$4122.45	$55.41	$73.62

Crew B-22	Hr.	Daily	Hr.	Daily	Bare Costs	Incl. O&P
1 Labor Foreman (out)	$28.70	$229.60	$44.70	$357.60	$31.25	$48.27
1 Skilled Worker	34.85	278.80	54.20	433.60		
1 Laborer	26.70	213.60	41.55	332.40		
.75 Equip. Oper. (crane)	35.90	215.40	54.10	324.60		
.75 S.P. Crane, 5 Ton		237.75		261.50	7.93	8.72
30 L.H., Daily Totals		$1175.15		$1709.70	$39.18	$56.99

Crew B-22A	Hr.	Daily	Hr.	Daily	Bare Costs	Incl. O&P
1 Labor Foreman (out)	$28.70	$229.60	$44.70	$357.60	$30.29	$46.86
1 Skilled Worker	34.85	278.80	54.20	433.60		
2 Laborers	26.70	427.20	41.55	664.80		
.75 Equipment Oper. (crane)	35.90	215.40	54.10	324.60		
.75 Crane, 5 Ton		237.75		261.50		
1 Generator, 5 KW		32.20		35.40		
1 Butt Fusion Machine		432.80		476.10	18.49	20.34
38 L.H., Daily Totals		$1853.75		$2553.60	$48.78	$67.20

Crew B-22B	Hr.	Daily	Hr.	Daily	Bare Costs	Incl. O&P
1 Skilled Worker	$34.85	$278.80	$54.20	$433.60	$30.78	$47.88
1 Laborer	26.70	213.60	41.55	332.40		
1 Electro Fusion Machine		171.80		189.00	10.74	11.81
16 L.H., Daily Totals		$664.20		$955.00	$41.52	$59.69

Crew B-23	Hr.	Daily	Hr.	Daily	Bare Costs	Incl. O&P
1 Labor Foreman (outside)	$28.70	$229.60	$44.70	$357.60	$27.10	$42.18
4 Laborers	26.70	854.40	41.55	1329.60		
1 Drill Rig, Wells		3179.00		3496.90		
1 Light Truck, 3 Ton		176.20		193.80	83.88	92.27
40 L.H., Daily Totals		$4439.20		$5377.90	$110.98	$134.45

Crew B-23A	Hr.	Daily	Hr.	Daily	Bare Costs	Incl. O&P
1 Labor Foreman (outside)	$28.70	$229.60	$44.70	$357.60	$30.02	$46.15
1 Laborer	26.70	213.60	41.55	332.40		
1 Equip. Operator (medium)	34.65	277.20	52.20	417.60		
1 Drill Rig, Wells		3179.00		3496.90		
1 Pickup Truck, 3/4 Ton		78.00		85.80	135.71	149.28
24 L.H., Daily Totals		$3977.40		$4690.30	$165.73	$195.43

Crew B-23B	Hr.	Daily	Hr.	Daily	Bare Costs	Incl. O&P
1 Labor Foreman (outside)	$28.70	$229.60	$44.70	$357.60	$30.02	$46.15
1 Laborer	26.70	213.60	41.55	332.40		
1 Equip. Operator (medium)	34.65	277.20	52.20	417.60		
1 Drill Rig, Wells		3179.00		3496.90		
1 Pickup Truck, 3/4 Ton		78.00		85.80		
1 Pump, Cntfgl, 6"		217.00		238.70	144.75	159.23
24 L.H., Daily Totals		$4194.40		$4929.00	$174.77	$205.38

Crew B-24	Hr.	Daily	Hr.	Daily	Bare Costs	Incl. O&P
1 Cement Finisher	$32.85	$262.80	$48.35	$386.80	$31.27	$47.63
1 Laborer	26.70	213.60	41.55	332.40		
1 Carpenter	34.25	274.00	53.35	426.80		
24 L.H., Daily Totals		$750.40		$1146.00	$31.27	$47.63

Crew B-25	Hr.	Daily	Hr.	Daily	Bare Costs	Incl. O&P
1 Labor Foreman	$28.70	$229.60	$44.70	$357.60	$29.05	$44.74
7 Laborers	26.70	1495.20	41.55	2326.80		
3 Equip. Oper. (med.)	34.65	831.60	52.20	1252.80		
1 Asphalt Paver, 130 H.P.		1590.00		1749.00		
1 Tandem Roller, 10 Ton		179.20		197.10		
1 Roller, Pneumatic Wheel		241.80		266.00	22.85	25.14
88 L.H., Daily Totals		$4567.40		$6149.30	$51.90	$69.88

Crew B-25B	Hr.	Daily	Hr.	Daily	Bare Costs	Incl. O&P
1 Labor Foreman	$28.70	$229.60	$44.70	$357.60	$29.52	$45.36
7 Laborers	26.70	1495.20	41.55	2326.80		
4 Equip. Oper. (medium)	34.65	1108.80	52.20	1670.40		
1 Asphalt Paver, 130 H.P.		1590.00		1749.00		
2 Rollers, Steel Wheel		358.40		394.25		
1 Roller, Pneumatic Wheel		241.80		266.00	22.81	25.10
96 L.H., Daily Totals		$5023.80		$6764.05	$52.33	$70.46

Crew B-25C	Hr.	Daily	Hr.	Daily	Bare Costs	Incl. O&P
1 Labor Foreman	$28.70	$229.60	$44.70	$357.60	$29.68	$45.63
3 Laborers	26.70	640.80	41.55	997.20		
2 Equip. Oper. (medium)	34.65	554.40	52.20	835.20		
1 Asphalt Paver, 130 H.P.		1590.00		1749.00		
1 Rollers, Steel Wheel		179.20		197.10	36.86	40.54
48 L.H., Daily Totals		$3194.00		$4136.10	$66.54	$86.17

Crew B-26	Hr.	Daily	Hr.	Daily	Bare Costs	Incl. O&P
1 Labor Foreman (outside)	$28.70	$229.60	$44.70	$357.60	$29.91	$46.27
6 Laborers	26.70	1281.60	41.55	1994.40		
2 Equip. Oper. (med.)	34.65	554.40	52.20	835.20		
1 Rodman (reinf.)	37.95	303.60	62.60	500.80		
1 Cement Finisher	32.85	262.80	48.35	386.80		
1 Grader, 30,000 Lbs.		456.60		502.25		
1 Paving Mach. & Equip.		1910.00		2101.00	26.89	29.58
88 L.H., Daily Totals		$4998.60		$6678.05	$56.80	$75.85

Crew B-27	Hr.	Daily	Hr.	Daily	Bare Costs	Incl. O&P
1 Labor Foreman (outside)	$28.70	$229.60	$44.70	$357.60	$27.20	$42.34
3 Laborers	26.70	640.80	41.55	997.20		
1 Berm Machine		216.20		237.80	6.76	7.43
32 L.H., Daily Totals		$1086.60		$1592.60	$33.96	$49.77

Crew B-28	Hr.	Daily	Hr.	Daily	Bare Costs	Incl. O&P
2 Carpenters	$34.25	$548.00	$53.35	$853.60	$31.73	$49.42
1 Laborer	26.70	213.60	41.55	332.40		
24 L.H., Daily Totals		$761.60		$1186.00	$31.73	$49.42

Crew B-29	Hr.	Daily	Hr.	Daily	Bare Costs	Incl. O&P
1 Labor Foreman (outside)	$28.70	$229.60	$44.70	$357.60	$28.79	$44.34
4 Laborers	26.70	854.40	41.55	1329.60		
1 Equip. Oper. (crane)	35.90	287.20	54.10	432.80		
1 Equip. Oper. Oiler	30.10	240.80	45.35	362.80		
1 Gradall, 3 Ton, 1/2 C.Y.		837.80		921.60	14.96	16.46
56 L.H., Daily Totals		$2449.80		$3404.40	$43.75	$60.80

Crew No.	Bare Costs		Incl. Subs O & P		Cost Per Labor-Hour	
	Hr.	Daily	Hr.	Daily	Bare Costs	Incl. O&P
Crew B-30	Hr.	Daily	Hr.	Daily	Bare Costs	Incl. O&P
1 Equip. Oper. (med.)	$34.65	$277.20	$52.20	$417.60	$29.92	$45.40
2 Truck Drivers (heavy)	27.55	440.80	42.00	672.00		
1 Hyd. Excavator, 1.5 C.Y.		720.40		792.45		
2 Dump Trucks, 16 Ton		953.20		1048.50	69.73	76.71
24 L.H., Daily Totals		$2391.60		$2930.55	$99.65	$122.11
Crew B-31	Hr.	Daily	Hr.	Daily	Bare Costs	Incl. O&P
1 Labor Foreman (outside)	$28.70	$229.60	$44.70	$357.60	$28.61	$44.54
3 Laborers	26.70	640.80	41.55	997.20		
1 Carpenter	34.25	274.00	53.35	426.80		
1 Air Compr., 250 C.F.M.		112.40		123.65		
1 Sheeting Driver		7.20		7.90		
2-50 Ft. Air Hoses, 1.5" Dia.		9.40		10.35	3.23	3.55
40 L.H., Daily Totals		$1273.40		$1923.50	$31.84	$48.09
Crew B-32	Hr.	Daily	Hr.	Daily	Bare Costs	Incl. O&P
1 Laborer	$26.70	$213.60	$41.55	$332.40	$32.66	$49.54
3 Equip. Oper. (med.)	34.65	831.60	52.20	1252.80		
1 Grader, 30,000 Lbs.		456.60		502.25		
1 Tandem Roller, 10 Ton		179.20		197.10		
1 Dozer, 200 H.P.		919.60		1011.55	48.61	53.47
32 L.H., Daily Totals		$2600.60		$3296.10	$81.27	$103.01
Crew B-32A	Hr.	Daily	Hr.	Daily	Bare Costs	Incl. O&P
1 Laborer	$26.70	$213.60	$41.55	$332.40	$32.00	$48.65
2 Equip. Oper. (medium)	34.65	554.40	52.20	835.20		
1 Grader, 30,000 Lbs.		456.60		502.25		
1 Roller, Vibratory, 29,000 Lbs.		433.20		476.50	37.08	40.78
24 L.H., Daily Totals		$1657.80		$2146.35	$69.08	$89.43
Crew B-32B	Hr.	Daily	Hr.	Daily	Bare Costs	Incl. O&P
1 Laborer	$26.70	$213.60	$41.55	$332.40	$32.00	$48.65
2 Equip. Oper. (medium)	34.65	554.40	52.20	835.20		
1 Dozer, 200 H.P.		919.60		1011.55		
1 Roller, Vibratory, 29,000 Lbs.		433.20		476.50	56.37	62.00
24 L.H., Daily Totals		$2120.80		$2655.65	$88.37	$110.65
Crew B-32C	Hr.	Daily	Hr.	Daily	Bare Costs	Incl. O&P
1 Labor Foreman	$28.70	$229.60	$44.70	$357.60	$31.01	$47.40
2 Laborers	26.70	427.20	41.55	664.80		
3 Equip. Oper. (medium)	34.65	831.60	52.20	1252.80		
1 Grader, 30,000 Lbs.		456.60		502.25		
1 Roller, Steel Wheel		179.20		197.10		
1 Dozer, 200 H.P.		919.60		1011.55	32.40	35.64
48 L.H., Daily Totals		$3043.80		$3986.10	$63.41	$83.04
Crew B-33A	Hr.	Daily	Hr.	Daily	Bare Costs	Incl. O&P
1 Equip. Oper. (med.)	$34.65	$277.20	$52.20	$417.60	$32.38	$49.16
.5 Laborer	26.70	106.80	41.55	166.20		
.25 Equip. Oper. (med.)	34.65	69.30	52.20	104.40		
1 Scraper, Towed, 7 C.Y.		150.75		165.80		
1.25 Dozer, 300 H.P.		1493.75		1643.15	117.47	129.21
14 L.H., Daily Totals		$2097.80		$2497.15	$149.85	$178.37
Crew B-33B	Hr.	Daily	Hr.	Daily	Bare Costs	Incl. O&P
1 Equip. Oper. (med.)	$34.65	$277.20	$52.20	$417.60	$32.38	$49.16
.5 Laborer	26.70	106.80	41.55	166.20		
.25 Equip. Oper. (med.)	34.65	69.30	52.20	104.40		
1 Scraper, Towed, 10 C.Y.		169.20		186.10		
1.25 Dozer, 300 H.P.		1493.75		1643.15	118.78	130.66
14 L.H., Daily Totals		$2116.25		$2517.45	$151.16	$179.82

Crew No.	Bare Costs		Incl. Subs O & P		Cost Per Labor-Hour	
	Hr.	Daily	Hr.	Daily	Bare Costs	Incl. O&P
Crew B-33C	Hr.	Daily	Hr.	Daily	Bare Costs	Incl. O&P
1 Equip. Oper. (med.)	$34.65	$277.20	$52.20	$417.60	$32.38	$49.16
.5 Laborer	26.70	106.80	41.55	166.20		
.25 Equip. Oper. (med.)	34.65	69.30	52.20	104.40		
1 Scraper, Towed, 12 C.Y.		169.20		186.10		
1.25 Dozer, 300 H.P.		1493.75		1643.15	118.78	130.66
14 L.H., Daily Totals		$2116.25		$2517.45	$151.16	$179.82
Crew B-33D	Hr.	Daily	Hr.	Daily	Bare Costs	Incl. O&P
1 Equip. Oper. (med.)	$34.65	$277.20	$52.20	$417.60	$32.38	$49.16
.5 Laborer	26.70	106.80	41.55	166.20		
.25 Equip. Oper. (med.)	34.65	69.30	52.20	104.40		
1 S.P. Scraper, 14 C.Y.		1552.00		1707.20		
.25 Dozer, 300 H.P.		298.75		328.65	132.20	145.42
14 L.H., Daily Totals		$2304.05		$2724.05	$164.58	$194.58
Crew B-33E	Hr.	Daily	Hr.	Daily	Bare Costs	Incl. O&P
1 Equip. Oper. (med.)	$34.65	$277.20	$52.20	$417.60	$32.38	$49.16
.5 Laborer	26.70	106.80	41.55	166.20		
.25 Equip. Oper. (med.)	34.65	69.30	52.20	104.40		
1 S.P. Scraper, 24 C.Y.		2317.00		2548.70		
.25 Dozer, 300 H.P.		298.75		328.65	186.84	205.52
14 L.H., Daily Totals		$3069.05		$3565.55	$219.22	$254.68
Crew B-33F	Hr.	Daily	Hr.	Daily	Bare Costs	Incl. O&P
1 Equip. Oper. (med.)	$34.65	$277.20	$52.20	$417.60	$32.38	$49.16
.5 Laborer	26.70	106.80	41.55	166.20		
.25 Equip. Oper. (med.)	34.65	69.30	52.20	104.40		
1 Elev. Scraper, 11 C.Y.		819.40		901.35		
.25 Dozer, 300 H.P.		298.75		328.65	79.87	87.85
14 L.H., Daily Totals		$1571.45		$1918.20	$112.25	$137.01
Crew B-33G	Hr.	Daily	Hr.	Daily	Bare Costs	Incl. O&P
1 Equip. Oper. (med.)	$34.65	$277.20	$52.20	$417.60	$32.38	$49.16
.5 Laborer	26.70	106.80	41.55	166.20		
.25 Equip. Oper. (med.)	34.65	69.30	52.20	104.40		
1 Elev. Scraper, 20 C.Y.		1648.00		1812.80		
.25 Dozer, 300 H.P.		298.75		328.65	139.05	152.96
14 L.H., Daily Totals		$2400.05		$2829.65	$171.43	$202.12
Crew B-33H	Hr.	Daily	Hr.	Daily	Bare Costs	Incl. O&P
.25 Laborer	$26.70	$53.40	$41.55	$83.10	$33.28	$50.36
1 Equipment Operator (med.)	34.65	277.20	52.20	417.60		
.2 Equipment Operator (med.)	34.65	55.44	52.20	83.52		
1 Scraper, 32-44 C.Y.		2728.00		3000.80		
.2 Dozer, 400 H.P.		304.80		335.30	261.45	287.59
11. L.H., Daily Totals		$3418.84		$3920.32	$294.73	$337.95
Crew B-33J	Hr.	Daily	Hr.	Daily	Bare Costs	Incl. O&P
1 Equipment Operator (med.)	$34.65	$277.20	$52.20	$417.60	$34.65	$52.20
1 Scraper 17 C.Y.		1552.00		1707.20	194.00	213.40
8 L.H., Daily Totals		$1829.20		$2124.80	$228.65	$265.60
Crew B-34A	Hr.	Daily	Hr.	Daily	Bare Costs	Incl. O&P
1 Truck Driver (heavy)	$27.55	$220.40	$42.00	$336.00	$27.55	$42.00
1 Dump Truck, 12 Ton		326.60		359.25	40.83	44.91
8 L.H., Daily Totals		$547.00		$695.25	$68.38	$86.91

Crew No.	Bare Costs		Incl. Subs O & P		Cost Per Labor-Hour	

Crew B-34B	Hr.	Daily	Hr.	Daily	Bare Costs	Incl. O&P
1 Truck Driver (heavy)	$27.55	$220.40	$42.00	$336.00	$27.55	$42.00
1 Dump Truck, 16 Ton		476.60		524.25	59.58	65.53
8 L.H., Daily Totals		$697.00		$860.25	$87.13	$107.53

Crew B-34C	Hr.	Daily	Hr.	Daily	Bare Costs	Incl. O&P
1 Truck Driver (heavy)	$27.55	$220.40	$42.00	$336.00	$27.55	$42.00
1 Truck Tractor, 40 Ton		301.60		331.75		
1 Dump Trailer, 16.5 C.Y.		103.00		113.30	50.58	55.63
8 L.H., Daily Totals		$625.00		$781.05	$78.13	$97.63

Crew B-34D	Hr.	Daily	Hr.	Daily	Bare Costs	Incl. O&P
1 Truck Driver (heavy)	$27.55	$220.40	$42.00	$336.00	$27.55	$42.00
1 Truck Tractor, 40 Ton		301.60		331.75		
1 Dump Trailer, 20 C.Y.		115.80		127.40	52.18	57.39
8 L.H., Daily Totals		$637.80		$795.15	$79.73	$99.39

Crew B-34E	Hr.	Daily	Hr.	Daily	Bare Costs	Incl. O&P
1 Truck Driver (heavy)	$27.55	$220.40	$42.00	$336.00	$27.55	$42.00
1 Truck, Off Hwy., 25 Ton		912.20		1003.40	114.03	125.43
8 L.H., Daily Totals		$1132.60		$1339.40	$141.58	$167.43

Crew B-34F	Hr.	Daily	Hr.	Daily	Bare Costs	Incl. O&P
1 Truck Driver (heavy)	$27.55	$220.40	$42.00	$336.00	$27.55	$42.00
1 Truck, Off Hwy., 22 C.Y.		933.40		1026.75	116.68	128.34
8 L.H., Daily Totals		$1153.80		$1362.75	$144.23	$170.34

Crew B-34G	Hr.	Daily	Hr.	Daily	Bare Costs	Incl. O&P
1 Truck Driver (heavy)	$27.55	$220.40	$42.00	$336.00	$27.55	$42.00
1 Truck, Off Hwy., 34 C.Y.		1181.00		1299.10	147.63	162.39
8 L.H., Daily Totals		$1401.40		$1635.10	$175.18	$204.39

Crew B-34H	Hr.	Daily	Hr.	Daily	Bare Costs	Incl. O&P
1 Truck Driver (heavy)	$27.55	$220.40	$42.00	$336.00	$27.55	$42.00
1 Truck, Off Hwy., 42 C.Y.		1276.00		1403.60	159.50	175.45
8 L.H., Daily Totals		$1496.40		$1739.60	$187.05	$217.45

Crew B-34J	Hr.	Daily	Hr.	Daily	Bare Costs	Incl. O&P
1 Truck Driver (heavy)	$27.55	$220.40	$42.00	$336.00	$27.55	$42.00
1 Truck, Off Hwy., 60 C.Y.		1649.00		1813.90	206.13	226.74
8 L.H., Daily Totals		$1869.40		$2149.90	$233.68	$268.74

Crew B-34K	Hr.	Daily	Hr.	Daily	Bare Costs	Incl. O&P
1 Truck Driver (heavy)	$27.55	$220.40	$42.00	$336.00	$27.55	$42.00
1 Truck Tractor, 240 H.P.		353.60		388.95		
1 Low Bed Trailer		170.20		187.20	65.48	72.02
8 L.H., Daily Totals		$744.20		$912.15	$93.03	$114.02

Crew B-34N	Hr.	Daily	Hr.	Daily	Bare Costs	Incl. O&P
1 Truck Driver (heavy)	$27.55	$220.40	$42.00	$336.00	$27.55	$42.00
1 Dump Truck, 12 Ton		326.60		359.25		
1 Flatbed Trailer, 40 Ton		119.40		131.35	55.75	61.33
8 L.H., Daily Totals		$666.40		$826.60	$83.30	$103.33

Crew B-34P	Hr.	Daily	Hr.	Daily	Bare Costs	Incl. O&P
1 Pipe Fitter	$40.95	$327.60	$61.55	$492.40	$34.13	$51.53
1 Truck Driver (light)	26.80	214.40	40.85	326.80		
1 Equip. Oper. (medium)	34.65	277.20	52.20	417.60		
1 Flatbed Truck, 3 Ton		176.20		193.80		
1 Backhoe Loader, 48 H.P.		215.80		237.40	16.33	17.97
24 L.H., Daily Totals		$1211.20		$1668.00	$50.46	$69.50

Crew B-34Q	Hr.	Daily	Hr.	Daily	Bare Costs	Incl. O&P
1 Pipe Fitter	$40.95	$327.60	$61.55	$492.40	$34.55	$52.17
1 Truck Driver (light)	26.80	214.40	40.85	326.80		
1 Eqip. Oper. (crane)	35.90	287.20	54.10	432.80		
1 Flatbed Trailer, 25 Ton		88.00		96.80		
1 Dump Truck, 12 Ton		326.60		359.25		
1 Hyd. Crane, 25 Ton		616.80		678.50	42.98	47.27
24 L.H., Daily Totals		$1860.60		$2386.55	$77.53	$99.44

Crew B-34R	Hr.	Daily	Hr.	Daily	Bare Costs	Incl. O&P
1 Pipe Fitter	$40.95	$327.60	$61.55	$492.40	$34.55	$52.17
1 Truck Driver (light)	26.80	214.40	40.85	326.80		
1 Eqip. Oper. (crane)	35.90	287.20	54.10	432.80		
1 Flatbed Trailer, 25 Ton		88.00		96.80		
1 Dump Truck, 12 Ton		326.60		359.25		
1 Hyd. Crane, 25 Ton		616.80		678.50		
1 Hyd. Excavator, 1 C.Y.		557.80		613.60	66.22	72.84
24 L.H., Daily Totals		$2418.40		$3000.15	$100.77	$125.01

Crew B-34S	Hr.	Daily	Hr.	Daily	Bare Costs	Incl. O&P
2 Pipe Fitters	$40.95	$655.20	$61.55	$984.80	$36.34	$54.80
1 Truck Driver (heavy)	27.55	220.40	42.00	336.00		
1 Eqip. Oper. (crane)	35.90	287.20	54.10	432.80		
1 Flatbed Trailer, 40 Ton		119.40		131.35		
1 Truck Tractor, 40 Ton		301.60		331.75		
1 Truck Crane, 80 Ton		995.00		1094.50		
1 Hyd. Excavator, 2 C.Y.		909.80		1000.80	72.68	79.95
32 L.H., Daily Totals		$3488.60		$4312.00	$109.02	$134.75

Crew B-34T	Hr.	Daily	Hr.	Daily	Bare Costs	Incl. O&P
2 Pipe Fitters	$40.95	$655.20	$61.55	$984.80	$36.34	$54.80
1 Truck Driver (heavy)	27.55	220.40	42.00	336.00		
1 Eqip. Oper. (crane)	35.90	287.20	54.10	432.80		
1 Flatbed Trailer, 40 Ton		119.40		131.35		
1 Truck Tractor, 40 Ton		301.60		331.75		
1 Truck Crane, 80 Ton		995.00		1094.50	44.25	48.68
32 L.H., Daily Totals		$2578.80		$3311.20	$80.59	$103.48

Crew B-35	Hr.	Daily	Hr.	Daily	Bare Costs	Incl. O&P
1 Laborer Foreman (out)	$28.70	$229.60	$44.70	$357.60	$32.85	$50.22
1 Skilled Worker	34.85	278.80	54.20	433.60		
1 Welder (plumber)	40.85	326.80	61.40	491.20		
1 Laborer	26.70	213.60	41.55	332.40		
1 Equip. Oper. (crane)	35.90	287.20	54.10	432.80		
1 Equip. Oper. Oiler	30.10	240.80	45.35	362.80		
1 Electric Welding Mach.		80.65		88.70		
1 Hyd. Excavator, .75 C.Y.		474.00		521.40	11.56	12.71
48 L.H., Daily Totals		$2131.45		$3020.50	$44.41	$62.93

Crew B-35A

Crew No.	Bare Costs Hr.	Bare Costs Daily	Incl. Subs O&P Hr.	Incl. Subs O&P Daily	Bare Costs	Incl. O&P
1 Laborer Foreman (out)	$28.70	$229.60	$44.70	$357.60	$31.97	$48.98
2 Laborers	26.70	427.20	41.55	664.80		
1 Skilled Worker	34.85	278.80	54.20	433.60		
1 Welder (plumber)	40.85	326.80	61.40	491.20		
1 Equip. Oper. (crane)	35.90	287.20	54.10	432.80		
1 Equip. Oper. Oiler	30.10	240.80	45.35	362.80		
1 Welder, 300 amp		81.20		89.30		
1 Crane, 75 Ton		1170.00		1287.00	22.34	24.58
56 L.H., Daily Totals		$3041.60		$4119.10	$54.31	$73.56

Crew B-36

Crew No.	Bare Costs Hr.	Bare Costs Daily	Incl. Subs O&P Hr.	Incl. Subs O&P Daily	Bare Costs	Incl. O&P
1 Labor Foreman (outside)	$28.70	$229.60	$44.70	$357.60	$30.28	$46.44
2 Laborers	26.70	427.20	41.55	664.80		
2 Equip. Oper. (med.)	34.65	554.40	52.20	835.20		
1 Dozer, 200 H.P.		919.60		1011.55		
1 Aggregate Spreader		35.00		38.50		
1 Tandem Roller, 10 Ton		179.20		197.10	28.35	31.18
40 L.H., Daily Totals		$2345.00		$3104.75	$58.63	$77.62

Crew B-36A

Crew No.	Bare Costs Hr.	Bare Costs Daily	Incl. Subs O&P Hr.	Incl. Subs O&P Daily	Bare Costs	Incl. O&P
1 Labor Foreman (outside)	$28.70	$229.60	$44.70	$357.60	$31.53	$48.09
2 Laborers	26.70	427.20	41.55	664.80		
4 Equip. Oper. (med.)	34.65	1108.80	52.20	1670.40		
1 Dozer, 200 H.P.		919.60		1011.55		
1 Aggregate Spreader		35.00		38.50		
1 Roller, Steel Wheel		179.20		197.10		
1 Roller, Pneumatic Wheel		241.80		266.00	24.56	27.02
56 L.H., Daily Totals		$3141.20		$4205.95	$56.09	$75.11

Crew B-36B

Crew No.	Bare Costs Hr.	Bare Costs Daily	Incl. Subs O&P Hr.	Incl. Subs O&P Daily	Bare Costs	Incl. O&P
1 Labor Foreman (outside)	$28.70	$229.60	$44.70	$357.60	$31.03	$47.33
2 Laborers	26.70	427.20	41.55	664.80		
4 Equip. Oper. (medium)	34.65	1108.80	52.20	1670.40		
1 Truck Driver, Heavy	27.55	220.40	42.00	336.00		
1 Grader, 30,000 Lbs.		456.60		502.25		
1 F.E. Loader, crl. 1.5 C.Y.		368.80		405.70		
1 Dozer, 300 H.P.		1195.00		1314.50		
1 Roller, Vibratory		433.20		476.50		
1 Truck, Tractor, 240 H.P.		353.60		388.95		
1 Water Tanker, 5000 Gal.		116.80		128.50	45.69	50.26
64 L.H., Daily Totals		$4910.00		$6245.20	$76.72	$97.59

Crew B-36C

Crew No.	Bare Costs Hr.	Bare Costs Daily	Incl. Subs O&P Hr.	Incl. Subs O&P Daily	Bare Costs	Incl. O&P
1 Labor Foreman (outside)	$28.70	$229.60	$44.70	$357.60	$32.04	$48.66
3 Equip. Oper. (medium)	34.65	831.60	52.20	1252.80		
1 Truck Driver, Heavy	27.55	220.40	42.00	336.00		
1 Grader, 30,000 Lbs.		456.60		502.25		
1 Dozer, 300 H.P.		1195.00		1314.50		
1 Roller, Vibratory		433.20		476.50		
1 Truck, Tractor, 240 H.P.		353.60		388.95		
1 Water Tanker, 5000 Gal.		116.80		128.50	63.88	70.27
40 L.H., Daily Totals		$3836.80		$4757.10	$95.92	$118.93

Crew B-37

Crew No.	Bare Costs Hr.	Bare Costs Daily	Incl. Subs O&P Hr.	Incl. Subs O&P Daily	Bare Costs	Incl. O&P
1 Labor Foreman (outside)	$28.70	$229.60	$44.70	$357.60	$28.09	$43.45
4 Laborers	26.70	854.40	41.55	1329.60		
1 Equip. Oper. (light)	33.05	264.40	49.80	398.40		
1 Tandem Roller, 5 Ton		107.40		118.15	2.24	2.46
48 L.H., Daily Totals		$1455.80		$2203.75	$30.33	$45.91

Crew B-38

Crew No.	Bare Costs Hr.	Bare Costs Daily	Incl. Subs O&P Hr.	Incl. Subs O&P Daily	Bare Costs	Incl. O&P
1 Labor Foreman (outside)	$28.70	$229.60	$44.70	$357.60	$29.96	$45.96
2 Laborers	26.70	427.20	41.55	664.80		
1 Equip. Oper. (light)	33.05	264.40	49.80	398.40		
1 Equip. Oper. (medium)	34.65	277.20	52.20	417.60		
1 Backhoe Loader, 48 H.P.		215.80		237.40		
1 Hyd.Hammer, (1200 lb)		112.00		123.20		
1 F.E. Loader (170 H.P.)		445.60		490.15		
1 Pavt. Rem. Bucket		51.80		57.00	20.63	22.69
40 L.H., Daily Totals		$2023.60		$2746.15	$50.59	$68.65

Crew B-39

Crew No.	Bare Costs Hr.	Bare Costs Daily	Incl. Subs O&P Hr.	Incl. Subs O&P Daily	Bare Costs	Incl. O&P
1 Labor Foreman (outside)	$28.70	$229.60	$44.70	$357.60	$28.09	$43.45
4 Laborers	26.70	854.40	41.55	1329.60		
1 Equip. Oper. (light)	33.05	264.40	49.80	398.40		
1 Air Compr., 250 C.F.M.		112.40		123.65		
2 Air Tools & Accessories		19.60		21.55		
2-50 Ft. Air Hoses, 1.5" Dia.		9.40		10.35	2.95	3.24
48 L.H., Daily Totals		$1489.80		$2241.15	$31.04	$46.69

Crew B-40

Crew No.	Bare Costs Hr.	Bare Costs Daily	Incl. Subs O&P Hr.	Incl. Subs O&P Daily	Bare Costs	Incl. O&P
1 Pile Driver Foreman (out)	$35.30	$282.40	$58.15	$465.20	$33.80	$53.91
4 Pile Drivers	33.30	1065.60	54.90	1756.80		
2 Equip. Oper. (crane)	35.90	574.40	54.10	865.60		
1 Equip. Oper. Oiler	30.10	240.80	45.35	362.80		
1 Crane, 40 Ton		888.20		977.00		
1 Vibratory Hammer & Gen.		1403.00		1543.30	35.80	39.38
64 L.H., Daily Totals		$4454.40		$5970.70	$69.60	$93.29

Crew B-40B

Crew No.	Bare Costs Hr.	Bare Costs Daily	Incl. Subs O&P Hr.	Incl. Subs O&P Daily	Bare Costs	Incl. O&P
1 Laborer Foreman	$28.70	$229.60	$44.70	$357.60	$29.13	$44.80
3 Laborers	26.70	640.80	41.55	997.20		
1 Equip. Oper. (crane)	35.90	287.20	54.10	432.80		
1 Equip. Oper. Oiler	30.10	240.80	45.35	362.80		
1 Crane, 40 Ton		951.90		1047.10	19.83	21.81
48 L.H., Daily Totals		$2350.30		$3197.50	$48.96	$66.61

Crew B-41

Crew No.	Bare Costs Hr.	Bare Costs Daily	Incl. Subs O&P Hr.	Incl. Subs O&P Daily	Bare Costs	Incl. O&P
1 Labor Foreman (outside)	$28.70	$229.60	$44.70	$357.60	$27.64	$42.87
4 Laborers	26.70	854.40	41.55	1329.60		
.25 Equip. Oper. (crane)	35.90	71.80	54.10	108.20		
.25 Equip. Oper. Oiler	30.10	60.20	45.35	90.70		
.25 Crawler Crane, 40 Ton		222.05		244.25	5.05	5.55
44 L.H., Daily Totals		$1438.05		$2130.35	$32.69	$48.42

Crew B-42

Crew No.	Bare Costs Hr.	Bare Costs Daily	Incl. Subs O&P Hr.	Incl. Subs O&P Daily	Bare Costs	Incl. O&P
1 Labor Foreman (outside)	$28.70	$229.60	$44.70	$357.60	$29.96	$47.33
4 Laborers	26.70	854.40	41.55	1329.60		
1 Equip. Oper. (crane)	35.90	287.20	54.10	432.80		
1 Equip. Oper. Oiler	30.10	240.80	45.35	362.80		
1 Welder	38.15	305.20	68.25	546.00		
1 Hyd. Crane, 25 Ton		616.80		678.50		
1 Gas Welding Machine		81.20		89.30		
1 Horz. Boring Csg. Mch.		372.40		409.65	16.73	18.40
64 L.H., Daily Totals		$2987.60		$4206.25	$46.69	$65.73

Crew No.	Bare Costs		Incl. Subs O & P		Cost Per Labor-Hour	
Crew B-43	Hr.	Daily	Hr.	Daily	Bare Costs	Incl. O&P
1 Labor Foreman (outside)	$28.70	$229.60	$44.70	$357.60	$29.13	$44.80
3 Laborers	26.70	640.80	41.55	997.20		
1 Equip. Oper. (crane)	35.90	287.20	54.10	432.80		
1 Equip. Oper. Oiler	30.10	240.80	45.35	362.80		
1 Drill Rig & Augers		3179.00		3496.90	66.23	72.85
48 L.H., Daily Totals		$4577.40		$5647.30	$95.36	$117.65

Crew B-44	Hr.	Daily	Hr.	Daily	Bare Costs	Incl. O&P
1 Pile Driver Foreman	$35.30	$282.40	$58.15	$465.20	$33.38	$53.44
4 Pile Drivers	33.30	1065.60	54.90	1756.80		
2 Equip. Oper. (crane)	35.90	574.40	54.10	865.60		
1 Laborer	26.70	213.60	41.55	332.40		
1 Crane, 40 Ton, & Access.		888.20		977.00		
45 L.F. Leads, 15K Ft. Lbs.		56.25		61.90	14.79	16.27
64 L.H., Daily Totals		$3080.45		$4458.90	$48.17	$69.71

Crew B-45	Hr.	Daily	Hr.	Daily	Bare Costs	Incl. O&P
1 Equip. Oper. (med.)	$34.65	$277.20	$52.20	$417.60	$31.10	$47.10
1 Truck Driver (heavy)	27.55	220.40	42.00	336.00		
1 Dist. Tank Truck, 3K Gal.		233.80		257.20		
1 Tractor, 4 x 2, 250 H.P.		291.60		320.75	32.84	36.12
16 L.H., Daily Totals		$1023.00		$1331.55	$63.94	$83.22

Crew B-46	Hr.	Daily	Hr.	Daily	Bare Costs	Incl. O&P
1 Pile Driver Foreman	$35.30	$282.40	$58.15	$465.20	$30.33	$48.77
2 Pile Drivers	33.30	532.80	54.90	878.40		
3 Laborers	26.70	640.80	41.55	997.20		
1 Chain Saw, 36" Long		33.40		36.75	.70	.77
48 L.H., Daily Totals		$1489.40		$2377.55	$31.03	$49.54

Crew B-47	Hr.	Daily	Hr.	Daily	Bare Costs	Incl. O&P
1 Blast Foreman	$28.70	$229.60	$44.70	$357.60	$29.48	$45.35
1 Driller	26.70	213.60	41.55	332.40		
1 Equip. Oper. (light)	33.05	264.40	49.80	398.40		
1 Crawler Type Drill, 4"		664.40		730.85		
1 Air Compr., 600 C.F.M.		297.80		327.60		
2-50 Ft. Air Hoses, 3" Dia.		35.40		38.95	41.57	45.72
24 L.H., Daily Totals		$1705.20		$2185.80	$71.05	$91.07

Crew B-47A	Hr.	Daily	Hr.	Daily	Bare Costs	Incl. O&P
1 Drilling Foreman	$28.70	$229.60	$44.70	$357.60	$31.57	$48.05
1 Equip. Oper. (heavy)	35.90	287.20	54.10	432.80		
1 Oiler	30.10	240.80	45.35	362.80		
1 Quarry Drill		822.60		904.85	34.28	37.70
24 L.H., Daily Totals		$1580.20		$2058.05	$65.85	$85.75

Crew B-47C	Hr.	Daily	Hr.	Daily	Bare Costs	Incl. O&P
1 Laborer	$26.70	$213.60	$41.55	$332.40	$29.88	$45.68
1 Equip. Oper. (light)	33.05	264.40	49.80	398.40		
1 Air Compressor, 750 CFM		317.40		349.15		
2-50' Air Hose, 3"		35.40		38.95		
1 Air Track Drill, 4"		664.40		730.85	63.58	69.93
16 L.H., Daily Totals		$1495.20		$1849.75	$93.46	$115.61

Crew B-47E	Hr.	Daily	Hr.	Daily	Bare Costs	Incl. O&P
1 Laborer Foreman	$28.70	$229.60	$44.70	$357.60	$27.20	$42.34
3 Laborers	26.70	640.80	41.55	997.20		
1 Truck, Flatbed, 3 Ton		176.20		193.80	5.51	6.06
32 L.H., Daily Totals		$1046.60		$1548.60	$32.71	$48.40

Crew B-47G	Hr.	Daily	Hr.	Daily	Bare Costs	Incl. O&P
1 Laborer Foreman	$28.70	$229.60	$44.70	$357.60	$28.79	$44.40
2 Laborers	26.70	427.20	41.55	664.80		
1 Equip. Oper. (light)	33.05	264.40	49.80	398.40		
1 Air Track Drill, 4"		664.40		730.85		
1 Air Compr., 600 C.F.M.		297.80		327.60		
2-50 Ft. Air Hoses, 3" Dia.		35.40		38.95		
1 Grout Pump		101.40		111.55	34.34	37.78
32 L.H., Daily Totals		$2020.20		$2629.75	$63.13	$82.18

Crew B-48	Hr.	Daily	Hr.	Daily	Bare Costs	Incl. O&P
1 Labor Foreman (outside)	$28.70	$229.60	$44.70	$357.60	$29.69	$45.51
3 Laborers	26.70	640.80	41.55	997.20		
1 Equip. Oper. (crane)	35.90	287.20	54.10	432.80		
1 Equip. Oper. Oiler	30.10	240.80	45.35	362.80		
1 Equip. Oper. (light)	33.05	264.40	49.80	398.40		
1 Centr. Water Pump, 6"		217.00		238.70		
1-20 Ft. Suction Hose, 6"		12.80		14.10		
1-50 Ft. Disch. Hose, 6"		8.90		9.80		
1 Drill Rig & Augers		3179.00		3496.90	61.03	67.13
56 L.H., Daily Totals		$5080.50		$6308.30	$90.72	$112.64

Crew B-49	Hr.	Daily	Hr.	Daily	Bare Costs	Incl. O&P
1 Labor Foreman (outside)	$28.70	$229.60	$44.70	$357.60	$30.95	$47.99
3 Laborers	26.70	640.80	41.55	997.20		
2 Equip. Oper. (crane)	35.90	574.40	54.10	865.60		
2 Equip. Oper. Oilers	30.10	481.60	45.35	725.60		
1 Equip. Oper. (light)	33.05	264.40	49.80	398.40		
2 Pile Drivers	33.30	532.80	54.90	878.40		
1 Hyd. Crane, 25 Ton		616.80		678.50		
1 Centr. Water Pump, 6"		217.00		238.70		
1-20 Ft. Suction Hose, 6"		12.80		14.10		
1-50 Ft. Disch. Hose, 6"		8.90		9.80		
1 Drill Rig & Augers		3179.00		3496.90	45.85	50.43
88 L.H., Daily Totals		$6758.10		$8660.80	$76.80	$98.42

Crew B-50	Hr.	Daily	Hr.	Daily	Bare Costs	Incl. O&P
2 Pile Driver Foremen	$35.30	$564.80	$58.15	$930.40	$32.31	$51.71
6 Pile Drivers	33.30	1598.40	54.90	2635.20		
2 Equip. Oper. (crane)	35.90	574.40	54.10	865.60		
1 Equip. Oper. Oiler	30.10	240.80	45.35	362.80		
3 Laborers	26.70	640.80	41.55	997.20		
1 Crane, 40 Ton		888.20		977.00		
60 L.F. Leads, 15K Ft. Lbs.		75.00		82.50		
1 Hammer, 15K Ft. Lbs.		359.80		395.80		
1 Air Compr., 600 C.F.M.		297.80		327.60		
2-50 Ft. Air Hoses, 3" Dia.		35.40		38.95		
1 Chain Saw, 36" Long		33.40		36.75	15.11	16.62
112 L.H., Daily Totals		$5308.80		$7649.80	$47.42	$68.33

Crew B-51	Hr.	Daily	Hr.	Daily	Bare Costs	Incl. O&P
1 Labor Foreman (outside)	$28.70	$229.60	$44.70	$357.60	$27.05	$41.96
4 Laborers	26.70	854.40	41.55	1329.60		
1 Truck Driver (light)	26.80	214.40	40.85	326.80		
1 Light Truck, 1.5 Ton		128.80		141.70	2.68	2.95
48 L.H., Daily Totals		$1427.20		$2155.70	$29.73	$44.91

Crew No.	Bare Costs		Incl. Subs O & P		Cost Per Labor-Hour	

Crew B-52	Hr.	Daily	Hr.	Daily	Bare Costs	Incl. O&P
1 Carpenter Foreman	$36.25	$290.00	$56.45	$451.60	$31.39	$48.55
1 Carpenter	34.25	274.00	53.35	426.80		
3 Laborers	26.70	640.80	41.55	997.20		
1 Cement Finisher	32.85	262.80	48.35	386.80		
.5 Rodman (reinf.)	37.95	151.80	62.60	250.40		
.5 Equip. Oper. (med.)	34.65	138.60	52.20	208.80		
.5 F.E. Ldr., T.M., 2.5 C.Y.		394.00		433.40	7.04	7.74
56 L.H., Daily Totals		$2152.00		$3155.00	$38.43	$56.29

Crew B-53	Hr.	Daily	Hr.	Daily	Bare Costs	Incl. O&P
1 Equip. Oper. (light)	$33.05	$264.40	$49.80	$398.40	$33.05	$49.80
1 Trencher, Chain, 12 H.P.		48.00		52.80	6.00	6.60
8 L.H., Daily Totals		$312.40		$451.20	$39.05	$56.40

Crew B-54	Hr.	Daily	Hr.	Daily	Bare Costs	Incl. O&P
1 Equip. Oper. (light)	$33.05	$264.40	$49.80	$398.40	$33.05	$49.80
1 Trencher, Chain, 40 H.P.		213.40		234.75	26.68	29.34
8 L.H., Daily Totals		$477.80		$633.15	$59.73	$79.14

Crew B-54A	Hr.	Daily	Hr.	Daily	Bare Costs	Incl. O&P
.17 Labor Foreman (outside)	$28.70	$39.03	$44.70	$60.79	$33.79	$51.11
1 Equipment Operator (med.)	34.65	277.20	52.20	417.60		
1 Wheel Trencher, 67 H.P.		816.20		897.80	87.20	95.92
9.36 L.H., Daily Totals		$1132.43		$1376.19	$120.99	$147.03

Crew B-54B	Hr.	Daily	Hr.	Daily	Bare Costs	Incl. O&P
.25 Labor Foreman (outside)	$28.70	$57.40	$44.70	$89.40	$33.46	$50.70
1 Equipment Operator (med.)	34.65	277.20	52.20	417.60		
1 Wheel Trencher, 150 H.P.		1460.00		1606.00	146.00	160.60
10 L.H., Daily Totals		$1794.60		$2113.00	$179.46	$211.30

Crew B-55	Hr.	Daily	Hr.	Daily	Bare Costs	Incl. O&P
2 Laborers	$26.70	$427.20	$41.55	$664.80	$26.73	$41.32
1 Truck Driver (light)	26.80	214.40	40.85	326.80		
1 Auger, 4" to 36" Dia		604.60		665.05		
1 Flatbed 3 Ton Truck		176.20		193.80	32.53	35.79
24 L.H., Daily Totals		$1422.40		$1850.45	$59.26	$77.11

Crew B-56	Hr.	Daily	Hr.	Daily	Bare Costs	Incl. O&P
1 Laborer	$26.70	$213.60	$41.55	$332.40	$29.88	$45.68
1 Equip. Oper. (light)	33.05	264.40	49.80	398.40		
1 Crawler Type Drill, 4"		664.40		730.85		
1 Air Compr., 600 C.F.M.		297.80		327.60		
1-50 Ft. Air Hose, 3" Dia.		17.70		19.45	61.24	67.37
16 L.H., Daily Totals		$1457.90		$1808.70	$91.12	$113.05

Crew B-57	Hr.	Daily	Hr.	Daily	Bare Costs	Incl. O&P
1 Labor Foreman (outside)	$28.70	$229.60	$44.70	$357.60	$30.19	$46.18
2 Laborers	26.70	427.20	41.55	664.80		
1 Equip. Oper. (crane)	35.90	287.20	54.10	432.80		
1 Equip. Oper. (light)	33.05	264.40	49.80	398.40		
1 Equip. Oper. Oiler	30.10	240.80	45.35	362.80		
1 Power Shovel, 1 C.Y.		833.20		916.50		
1 Clamshell Bucket, 1 C.Y.		43.40		47.75		
1 Centr. Water Pump, 6"		217.00		238.70		
1-20 Ft. Suction Hose, 6"		12.80		14.10		
20-50 Ft. Disch. Hoses, 6"		178.00		195.80	26.76	29.43
48 L.H., Daily Totals		$2733.60		$3629.25	$56.95	$75.61

Crew B-58	Hr.	Daily	Hr.	Daily	Bare Costs	Incl. O&P
2 Laborers	$26.70	$427.20	$41.55	$664.80	$28.82	$44.30
1 Equip. Oper. (light)	33.05	264.40	49.80	398.40		
1 Backhoe Loader, 48 H.P.		215.80		237.40		
1 Small Helicopter, w/pilot		2153.00		2368.30	98.70	108.57
24 L.H., Daily Totals		$3060.40		$3668.90	$127.52	$152.87

Crew B-59	Hr.	Daily	Hr.	Daily	Bare Costs	Incl. O&P
1 Truck Driver (heavy)	$27.55	$220.40	$42.00	$336.00	$27.55	$42.00
1 Truck, 30 Ton		208.20		229.00		
1 Water tank, 6000 Gal.		116.80		128.50	40.63	44.69
8 L.H., Daily Totals		$545.40		$693.50	$68.18	$86.69

Crew B-59A	Hr.	Daily	Hr.	Daily	Bare Costs	Incl. O&P
2 Laborers	$26.70	$427.20	$41.55	$664.80	$26.98	$41.70
1 Truck Driver (heavy)	27.55	220.40	42.00	336.00		
1 Water tank, 5K w/pum		116.80		128.50		
1 Truck, 30 Ton		208.20		229.00	13.54	14.90
24 L.H., Daily Totals		$972.60		$1358.30	$40.52	$56.60

Crew B-60	Hr.	Daily	Hr.	Daily	Bare Costs	Incl. O&P
1 Labor Foreman (outside)	$28.70	$229.60	$44.70	$357.60	$30.60	$46.69
2 Laborers	26.70	427.20	41.55	664.80		
1 Equip. Oper. (crane)	35.90	287.20	54.10	432.80		
2 Equip. Oper. (light)	33.05	528.80	49.80	796.80		
1 Equip. Oper. Oiler	30.10	240.80	45.35	362.80		
1 Crawler Crane, 40 Ton		888.20		977.00		
45 L.F. Leads, 15K Ft. Lbs.		56.25		61.90		
1 Backhoe Loader, 48 H.P.		215.80		237.40	20.76	22.83
56 L.H., Daily Totals		$2873.85		$3891.10	$51.36	$69.52

Crew B-61	Hr.	Daily	Hr.	Daily	Bare Costs	Incl. O&P
1 Labor Foreman (outside)	$28.70	$229.60	$44.70	$357.60	$28.37	$43.83
3 Laborers	26.70	640.80	41.55	997.20		
1 Equip. Oper. (light)	33.05	264.40	49.80	398.40		
1 Cement Mixer, 2 C.Y.		162.00		178.20		
1 Air Compr., 160 C.F.M.		87.40		96.15	6.24	6.86
40 L.H., Daily Totals		$1384.20		$2027.55	$34.61	$50.69

Crew B-62	Hr.	Daily	Hr.	Daily	Bare Costs	Incl. O&P
2 Laborers	$26.70	$427.20	$41.55	$664.80	$28.82	$44.30
1 Equip. Oper. (light)	33.05	264.40	49.80	398.40		
1 Loader, Skid Steer		153.40		168.75	6.39	7.03
24 L.H., Daily Totals		$845.00		$1231.95	$35.21	$51.33

Crew B-63	Hr.	Daily	Hr.	Daily	Bare Costs	Incl. O&P
4 Laborers	$26.70	$854.40	$41.55	$1329.60	$27.97	$43.20
1 Equip. Oper. (light)	33.05	264.40	49.80	398.40		
1 Loader, Skid Steer		153.40		168.75	3.84	4.22
40 L.H., Daily Totals		$1272.20		$1896.75	$31.81	$47.42

Crew B-64	Hr.	Daily	Hr.	Daily	Bare Costs	Incl. O&P
1 Laborer	$26.70	$213.60	$41.55	$332.40	$26.75	$41.20
1 Truck Driver (light)	26.80	214.40	40.85	326.80		
1 Power Mulcher (small)		103.20		113.50		
1 Light Truck, 1.5 Ton		128.80		141.70	14.50	15.95
16 L.H., Daily Totals		$660.00		$914.40	$41.25	$57.15

Crew No.	Bare Costs		Incl. Subs O & P		Cost Per Labor-Hour	
Crew B-65	Hr.	Daily	Hr.	Daily	Bare Costs	Incl. O&P
1 Laborer	$26.70	$213.60	$41.55	$332.40	$26.75	$41.20
1 Truck Driver (light)	26.80	214.40	40.85	326.80		
1 Power Mulcher (large)		184.00		202.40		
1 Light Truck, 1.5 Ton		128.80		141.70	19.55	21.51
16 L.H., Daily Totals		$740.80		$1003.30	$46.30	$62.71
Crew B-66	Hr.	Daily	Hr.	Daily	Bare Costs	Incl. O&P
1 Equip. Oper. (light)	$33.05	$264.40	$49.80	$398.40	$33.05	$49.80
1 Backhoe Ldr. w/Attchmt.		173.20		190.50	21.65	23.82
8 L.H., Daily Totals		$437.60		$588.90	$54.70	$73.62
Crew B-67	Hr.	Daily	Hr.	Daily	Bare Costs	Incl. O&P
1 Millwright	$35.50	$284.00	$52.45	$419.60	$34.28	$51.13
1 Equip. Oper. (light)	33.05	264.40	49.80	398.40		
1 Forklift		233.00		256.30	14.56	16.02
16 L.H., Daily Totals		$781.40		$1074.30	$48.84	$67.15
Crew B-68	Hr.	Daily	Hr.	Daily	Bare Costs	Incl. O&P
2 Millwrights	$35.50	$568.00	$52.45	$839.20	$34.68	$51.57
1 Equip. Oper. (light)	33.05	264.40	49.80	398.40		
1 Forklift		233.00		256.30	9.71	10.68
24 L.H., Daily Totals		$1065.40		$1493.90	$44.39	$62.25
Crew B-69	Hr.	Daily	Hr.	Daily	Bare Costs	Incl. O&P
1 Labor Foreman (outside)	$28.70	$229.60	$44.70	$357.60	$29.13	$44.80
3 Laborers	26.70	640.80	41.55	997.20		
1 Equip Oper. (crane)	35.90	287.20	54.10	432.80		
1 Equip Oper. Oiler	30.10	240.80	45.35	362.80		
1 Truck Crane, 80 Ton		995.00		1094.50	20.73	22.80
48 L.H., Daily Totals		$2393.40		$3244.90	$49.86	$67.60
Crew B-69A	Hr.	Daily	Hr.	Daily	Bare Costs	Incl. O&P
1 Labor Foreman	$28.70	$229.60	$44.70	$357.60	$29.38	$44.93
3 Laborers	26.70	640.80	41.55	997.20		
1 Equip. Oper. (medium)	34.65	277.20	52.20	417.60		
1 Concrete Finisher	32.85	262.80	48.35	386.80		
1 Curb Paver		582.80		641.10	12.14	13.36
48 L.H., Daily Totals		$1993.20		$2800.30	$41.52	$58.29
Crew B-69B	Hr.	Daily	Hr.	Daily	Bare Costs	Incl. O&P
1 Labor Foreman	$28.70	$229.60	$44.70	$357.60	$29.38	$44.93
3 Laborers	26.70	640.80	41.55	997.20		
1 Equip. Oper. (medium)	34.65	277.20	52.20	417.60		
1 Cement Finisher	32.85	262.80	48.35	386.80		
1 Curb/Gutter Paver		725.00		797.50	15.10	16.61
48 L.H., Daily Totals		$2135.40		$2956.70	$44.48	$61.54
Crew B-70	Hr.	Daily	Hr.	Daily	Bare Costs	Incl. O&P
1 Labor Foreman (outside)	$28.70	$229.60	$44.70	$357.60	$30.39	$46.56
3 Laborers	26.70	640.80	41.55	997.20		
3 Equip. Oper. (med.)	34.65	831.60	52.20	1252.80		
1 Motor Grader, 30,000 Lb.		456.60		502.25		
1 Grader Attach., Ripper		71.40		78.55		
1 Road Sweeper, S.P.		445.00		489.50		
1 F.E. Loader, 1-3/4 C.Y.		241.00		265.10	21.68	23.85
56 L.H., Daily Totals		$2916.00		$3943.00	$52.07	$70.41

Crew No.	Bare Costs		Incl. Subs O & P		Cost Per Labor-Hour	
Crew B-71	Hr.	Daily	Hr.	Daily	Bare Costs	Incl. O&P
1 Labor Foreman (outside)	$28.70	$229.60	$44.70	$357.60	$30.39	$46.56
3 Laborers	26.70	640.80	41.55	997.20		
3 Equip. Oper. (med.)	34.65	831.60	52.20	1252.80		
1 Pvmt. Profiler, 750 H.P.		4266.00		4692.60		
1 Road Sweeper, S.P.		445.00		489.50		
1 F.E. Loader, 1-3/4 C.Y.		241.00		265.10	88.43	97.27
56 L.H., Daily Totals		$6654.00		$8054.80	$118.82	$143.83
Crew B-72	Hr.	Daily	Hr.	Daily	Bare Costs	Incl. O&P
1 Labor Foreman (outside)	$28.70	$229.60	$44.70	$357.60	$30.93	$47.27
3 Laborers	26.70	640.80	41.55	997.20		
4 Equip. Oper. (med.)	34.65	1108.80	52.20	1670.40		
1 Pvmt. Profiler, 750 H.P.		4266.00		4692.60		
1 Hammermill, 250 H.P.		1344.00		1478.40		
1 Windrow Loader		791.20		870.30		
1 Mix Paver 165 H.P.		1656.00		1821.60		
1 Roller, Pneu. Tire, 12 T.		241.80		266.00	129.67	142.64
64 L.H., Daily Totals		$10278.20		$12154.10	$160.60	$189.91
Crew B-73	Hr.	Daily	Hr.	Daily	Bare Costs	Incl. O&P
1 Labor Foreman (outside)	$28.70	$229.60	$44.70	$357.60	$31.92	$48.60
2 Laborers	26.70	427.20	41.55	664.80		
5 Equip. Oper. (med.)	34.65	1386.00	52.20	2088.00		
1 Road Mixer, 310 H.P.		1546.00		1700.60		
1 Roller, Tandem, 12 Ton		179.20		197.10		
1 Hammermill, 250 H.P.		1344.00		1478.40		
1 Motor Grader, 30,000 Lb.		456.60		502.25		
.5 F.E. Loader, 1-3/4 C.Y.		120.50		132.55		
.5 Truck, 30 Ton		104.10		114.50		
.5 Water Tank 5000 Gal.		58.40		64.25	59.51	65.46
64 L.H., Daily Totals		$5851.60		$7300.05	$91.43	$114.06
Crew B-74	Hr.	Daily	Hr.	Daily	Bare Costs	Incl. O&P
1 Labor Foreman (outside)	$28.70	$229.60	$44.70	$357.60	$31.14	$47.38
1 Laborer	26.70	213.60	41.55	332.40		
4 Equip. Oper. (med.)	34.65	1108.80	52.20	1670.40		
2 Truck Drivers (heavy)	27.55	440.80	42.00	672.00		
1 Motor Grader, 30,000 Lb.		456.60		502.25		
1 Grader Attach., Ripper		71.40		78.55		
2 Stabilizers, 310 H.P.		2090.00		2299.00		
1 Flatbed Truck, 3 Ton		176.20		193.80		
1 Chem. Spreader, Towed		54.60		60.05		
1 Vibr. Roller, 29,000 Lb.		433.20		476.50		
1 Water Tank 5000 Gal.		116.80		128.50		
1 Truck, 30 Ton		208.20		229.00	56.36	62.00
64 L.H., Daily Totals		$5599.80		$7000.05	$87.50	$109.38
Crew B-75	Hr.	Daily	Hr.	Daily	Bare Costs	Incl. O&P
1 Labor Foreman (outside)	$28.70	$229.60	$44.70	$357.60	$31.65	$48.15
1 Laborer	26.70	213.60	41.55	332.40		
4 Equip. Oper. (med.)	34.65	1108.80	52.20	1670.40		
1 Truck Driver (heavy)	27.55	220.40	42.00	336.00		
1 Motor Grader, 30,000 Lb.		456.60		502.25		
1 Grader Attach., Ripper		71.40		78.55		
2 Stabilizers, 310 H.P.		2090.00		2299.00		
1 Dist. Truck, 3000 Gal.		233.80		257.20		
1 Vibr. Roller, 29,000 Lb.		433.20		476.50	58.66	64.53
56 L.H., Daily Totals		$5057.40		$6309.90	$90.31	$112.68

Crew B-76

Crew No.	Bare Costs Hr.	Daily	Incl. Subs O & P Hr.	Daily	Cost Per Labor-Hour Bare Costs	Incl. O&P
1 Dock Builder Foreman	$35.30	$282.40	$58.15	$465.20	$33.74	$54.02
5 Dock Builders	33.30	1332.00	54.90	2196.00		
2 Equip. Oper. (crane)	35.90	574.40	54.10	865.60		
1 Equip. Oper. Oiler	30.10	240.80	45.35	362.80		
1 Crawler Crane, 50 Ton		1206.00		1326.60		
1 Barge, 400 Ton		273.20		300.50		
1 Hammer, 15K Ft. Lbs.		359.80		395.80		
60 L.F. Leads, 15K Ft. Lbs.		75.00		82.50		
1 Air Compr., 600 C.F.M.		297.80		327.60		
2-50 Ft. Air Hoses, 3" Dia.		35.40		38.95	31.25	34.38
72 L.H., Daily Totals		$4676.80		$6361.55	$64.99	$88.40

Crew B-76A

Crew No.	Bare Costs Hr.	Daily	Incl. Subs O & P Hr.	Daily	Cost Per Labor-Hour Bare Costs	Incl. O&P
1 Laborer Foreman	$28.70	$229.60	$44.70	$357.60	$28.53	$43.99
5 Laborers	26.70	1068.00	41.55	1662.00		
1 Equip. Oper. (crane)	35.90	287.20	54.10	432.80		
1 Equip. Oper. Oiler	30.10	240.80	45.35	362.80		
1 Crawler Crane, 50 Ton		1206.00		1326.60		
1 Barge, 400 Ton		273.20		300.50	23.11	25.42
64 L.H., Daily Totals		$3304.80		$4442.30	$51.64	$69.41

Crew B-77

Crew No.	Bare Costs Hr.	Daily	Incl. Subs O & P Hr.	Daily	Cost Per Labor-Hour Bare Costs	Incl. O&P
1 Labor Foreman	$28.70	$229.60	$44.70	$357.60	$27.12	$42.04
3 Laborers	26.70	640.80	41.55	997.20		
1 Truck Driver (light)	26.80	214.40	40.85	326.80		
1 Crack Cleaner, 25 H.P.		44.40		48.85		
1 Crack Filler, Trailer Mtd.		157.00		172.70		
1 Flatbed Truck, 3 Ton		176.20		193.80	9.44	10.38
40 L.H., Daily Totals		$1462.40		$2096.95	$36.56	$52.42

Crew B-78

Crew No.	Bare Costs Hr.	Daily	Incl. Subs O & P Hr.	Daily	Cost Per Labor-Hour Bare Costs	Incl. O&P
1 Labor Foreman	$28.70	$229.60	$44.70	$357.60	$27.05	$41.96
4 Laborers	26.70	854.40	41.55	1329.60		
1 Truck Driver (light)	26.80	214.40	40.85	326.80		
1 Paint Striper, S.P.		134.20		147.60		
1 Flatbed Truck, 3 Ton		176.20		193.80		
1 Pickup Truck, 3/4 Ton		78.00		85.80	8.09	8.90
48 L.H., Daily Totals		$1686.80		$2441.20	$35.14	$50.86

Crew B-79

Crew No.	Bare Costs Hr.	Daily	Incl. Subs O & P Hr.	Daily	Cost Per Labor-Hour Bare Costs	Incl. O&P
1 Labor Foreman	$28.70	$229.60	$44.70	$357.60	$27.12	$42.04
3 Laborers	26.70	640.80	41.55	997.20		
1 Truck Driver (light)	26.80	214.40	40.85	326.80		
1 Thermo. Striper, T.M.		549.20		604.10		
1 Flatbed Truck, 3 Ton		176.20		193.80		
2 Pickup Truck, 3/4 Ton		156.00		171.60	22.04	24.24
40 L.H., Daily Totals		$1966.20		$2651.10	$49.16	$66.28

Crew B-80

Crew No.	Bare Costs Hr.	Daily	Incl. Subs O & P Hr.	Daily	Cost Per Labor-Hour Bare Costs	Incl. O&P
1 Labor Foreman	$28.70	$229.60	$44.70	$357.60	$28.81	$44.23
1 Laborer	26.70	213.60	41.55	332.40		
1 Truck Driver (light)	26.80	214.40	40.85	326.80		
1 Equip. Oper. (light)	33.05	264.40	49.80	398.40		
1 Flatbed Truck, 3 Ton		176.20		193.80		
1 Fence Post Auger, T.M.		357.40		393.15	16.68	18.34
32 L.H., Daily Totals		$1455.60		$2002.15	$45.49	$62.57

Crew B-80A

Crew No.	Bare Costs Hr.	Daily	Incl. Subs O & P Hr.	Daily	Cost Per Labor-Hour Bare Costs	Incl. O&P
3 Laborers	$26.70	$640.80	$41.55	$997.20	$26.70	$41.55
1 Flatbed Truck, 3 Ton		176.20		193.80	7.34	8.08
24 L.H., Daily Totals		$817.00		$1191.00	$34.04	$49.63

Crew B-80B

Crew No.	Bare Costs Hr.	Daily	Incl. Subs O & P Hr.	Daily	Cost Per Labor-Hour Bare Costs	Incl. O&P
3 Laborers	$26.70	$640.80	$41.55	$997.20	$28.29	$43.61
1 Equip. Oper. (light)	33.05	264.40	49.80	398.40		
1 Crane, Flatbed Mnt.		209.60		230.55	6.55	7.21
32 L.H., Daily Totals		$1114.80		$1626.15	$34.84	$50.82

Crew B-80C

Crew No.	Bare Costs Hr.	Daily	Incl. Subs O & P Hr.	Daily	Cost Per Labor-Hour Bare Costs	Incl. O&P
2 Laborers	$26.70	$427.20	$41.55	$664.80	$26.73	$41.32
1 Truck Driver (light)	26.80	214.40	40.85	326.80		
1 Light Truck, 1.5 Ton		128.80		141.70		
1 Manual fence post auger, gas		5.00		5.50	5.58	6.13
24 L.H., Daily Totals		$775.40		$1138.80	$32.31	$47.45

Crew B-81

Crew No.	Bare Costs Hr.	Daily	Incl. Subs O & P Hr.	Daily	Cost Per Labor-Hour Bare Costs	Incl. O&P
1 Laborer	$26.70	$213.60	$41.55	$332.40	$29.63	$45.25
1 Equip. Oper. (med.)	34.65	277.20	52.20	417.60		
1 Truck Driver (heavy)	27.55	220.40	42.00	336.00		
1 Hydromulcher, T.M.		196.00		215.60		
1 Tractor Truck, 4x2		208.20		229.00	16.84	18.53
24 L.H., Daily Totals		$1115.40		$1530.60	$46.47	$63.78

Crew B-82

Crew No.	Bare Costs Hr.	Daily	Incl. Subs O & P Hr.	Daily	Cost Per Labor-Hour Bare Costs	Incl. O&P
1 Laborer	$26.70	$213.60	$41.55	$332.40	$29.88	$45.68
1 Equip. Oper. (light)	33.05	264.40	49.80	398.40		
1 Horiz. Borer, 6 H.P.		62.60		68.85	3.91	4.30
16 L.H., Daily Totals		$540.60		$799.65	$33.79	$49.98

Crew B-83

Crew No.	Bare Costs Hr.	Daily	Incl. Subs O & P Hr.	Daily	Cost Per Labor-Hour Bare Costs	Incl. O&P
1 Tugboat Captain	$34.65	$277.20	$52.20	$417.60	$30.68	$46.88
1 Tugboat Hand	26.70	213.60	41.55	332.40		
1 Tugboat, 250 H.P.		440.60		484.65	27.54	30.29
16 L.H., Daily Totals		$931.40		$1234.65	$58.22	$77.17

Crew B-84

Crew No.	Bare Costs Hr.	Daily	Incl. Subs O & P Hr.	Daily	Cost Per Labor-Hour Bare Costs	Incl. O&P
1 Equip. Oper. (med.)	$34.65	$277.20	$52.20	$417.60	$34.65	$52.20
1 Rotary Mower/Tractor		228.80		251.70	28.60	31.46
8 L.H., Daily Totals		$506.00		$669.30	$63.25	$83.66

Crew B-85

Crew No.	Bare Costs Hr.	Daily	Incl. Subs O & P Hr.	Daily	Cost Per Labor-Hour Bare Costs	Incl. O&P
3 Laborers	$26.70	$640.80	$41.55	$997.20	$28.46	$43.77
1 Equip. Oper. (med.)	34.65	277.20	52.20	417.60		
1 Truck Driver (heavy)	27.55	220.40	42.00	336.00		
1 Aerial Lift Truck, 80'		496.80		546.50		
1 Brush Chipper, 130 H.P.		164.80		181.30		
1 Pruning Saw, Rotary		8.75		9.65	16.76	18.43
40 L.H., Daily Totals		$1808.75		$2488.25	$45.22	$62.20

Crew B-86

Crew No.	Bare Costs Hr.	Daily	Incl. Subs O & P Hr.	Daily	Cost Per Labor-Hour Bare Costs	Incl. O&P
1 Equip. Oper. (med.)	$34.65	$277.20	$52.20	$417.60	$34.65	$52.20
1 Stump Chipper, S.P.		70.70		77.75	8.84	9.72
8 L.H., Daily Totals		$347.90		$495.35	$43.49	$61.92

Crew B-86A

Crew No.	Bare Costs Hr.	Daily	Incl. Subs O & P Hr.	Daily	Cost Per Labor-Hour Bare Costs	Incl. O&P
1 Equip. Oper. (medium)	$34.65	$277.20	$52.20	$417.60	$34.65	$52.20
1 Grader, 30,000 Lbs.		456.60		502.25	57.08	62.78
8 L.H., Daily Totals		$733.80		$919.85	$91.73	$114.98

Crew B-86B

Crew No.	Bare Costs Hr.	Daily	Incl. Subs O & P Hr.	Daily	Cost Per Labor-Hour Bare Costs	Incl. O&P
1 Equip. Oper. (medium)	$34.65	$277.20	$52.20	$417.60	$34.65	$52.20
1 Dozer, 200 H.P.		919.60		1011.55	114.95	126.45
8 L.H., Daily Totals		$1196.80		$1429.15	$149.60	$178.65

Crew B-87

Crew B-87	Hr.	Daily	Hr.	Daily	Bare Costs	Incl. O&P
1 Laborer	$26.70	$213.60	$41.55	$332.40	$33.06	$50.07
4 Equip. Oper. (med.)	34.65	1108.80	52.20	1670.40		
2 Feller Bunchers, 50 H.P.		864.80		951.30		
1 Log Chipper, 22" Tree		1062.00		1168.20		
1 Dozer, 105 H.P.		453.80		499.20		
1 Chainsaw, Gas, 36" Long		33.40		36.75	60.35	66.39
40 L.H., Daily Totals		$3736.40		$4658.25	$93.41	$116.46

Crew B-88

Crew B-88	Hr.	Daily	Hr.	Daily	Bare Costs	Incl. O&P
1 Laborer	$26.70	$213.60	$41.55	$332.40	$33.51	$50.68
6 Equip. Oper. (med.)	34.65	1663.20	52.20	2505.60		
2 Feller Bunchers, 50 H.P.		864.80		951.30		
1 Log Chipper, 22" Tree		1062.00		1168.20		
2 Log Skidders, 50 H.P.		1490.00		1639.90		
1 Dozer, 105 H.P.		453.80		499.20		
1 Chainsaw, Gas, 36" Long		33.40		36.75	69.73	76.70
56 L.H., Daily Totals		$5781.60		$7133.35	$103.24	$127.38

Crew B-89

Crew B-89	Hr.	Daily	Hr.	Daily	Bare Costs	Incl. O&P
1 Equip. Oper. (light)	$33.05	$264.40	$49.80	$398.40	$29.93	$45.33
1 Truck Driver (light)	26.80	214.40	40.85	326.80		
1 Truck, Stake Body, 3 Ton		176.20		193.80		
1 Concrete Saw		114.40		125.85		
1 Water Tank, 65 Gal.		13.60		14.95	19.01	20.91
16 L.H., Daily Totals		$783.00		$1059.80	$48.94	$66.24

Crew B-89A

Crew B-89A	Hr.	Daily	Hr.	Daily	Bare Costs	Incl. O&P
1 Skilled Worker	$34.85	$278.80	$54.20	$433.60	$30.78	$47.88
1 Laborer	26.70	213.60	41.55	332.40		
1 Core Drill (large)		103.05		113.35	6.44	7.08
16 L.H., Daily Totals		$595.45		$879.35	$37.22	$54.96

Crew B-89B

Crew B-89B	Hr.	Daily	Hr.	Daily	Bare Costs	Incl. O&P
1 Equip. Oper. (light)	$33.05	$264.40	$49.80	$398.40	$29.93	$45.33
1 Truck Driver, Light	26.80	214.40	40.85	326.80		
1 Wall Saw, Hydraulic, 10 H.P.		73.30		80.65		
1 Generator, Diesel, 100 KW		176.60		194.25		
1 Water Tank, 65 Gal.		13.60		14.95		
1 Flatbed Truck, 3 Ton		176.20		193.80	27.48	30.23
16 L.H., Daily Totals		$918.50		$1208.85	$57.41	$75.56

Crew B-90

Crew B-90	Hr.	Daily	Hr.	Daily	Bare Costs	Incl. O&P
1 Labor Foreman (outside)	$28.70	$229.60	$44.70	$357.60	$28.75	$44.12
3 Laborers	26.70	640.80	41.55	997.20		
2 Equip. Oper. (light)	33.05	528.80	49.80	796.80		
2 Truck Drivers (heavy)	27.55	440.80	42.00	672.00		
1 Road Mixer, 310 H.P.		1546.00		1700.60		
1 Dist. Truck, 2000 Gal.		204.80		225.30	27.36	30.09
64 L.H., Daily Totals		$3590.80		$4749.50	$56.11	$74.21

Crew B-90A

Crew B-90A	Hr.	Daily	Hr.	Daily	Bare Costs	Incl. O&P
1 Labor Foreman	$28.70	$229.60	$44.70	$357.60	$31.53	$48.09
2 Laborers	26.70	427.20	41.55	664.80		
4 Equip. Oper. (medium)	34.65	1108.80	52.20	1670.40		
2 Graders, 30,000 Lbs.		913.20		1004.50		
1 Roller, Steel Wheel		179.20		197.10		
1 Roller, Pneumatic Wheel		241.80		266.00	23.83	26.21
56 L.H., Daily Totals		$3099.80		$4160.40	$55.36	$74.30

Crew B-90B

Crew B-90B	Hr.	Daily	Hr.	Daily	Bare Costs	Incl. O&P
1 Labor Foreman	$28.70	$229.60	$44.70	$357.60	$31.01	$47.40
2 Laborers	26.70	427.20	41.55	664.80		
3 Equip. Oper. (medium)	34.65	831.60	52.20	1252.80		
1 Roller, Steel Wheel		179.20		197.10		
1 Roller, Pneumatic Wheel		241.80		266.00		
1 Road Mixer, 310 H.P.		1546.00		1700.60	40.98	45.08
48 L.H., Daily Totals		$3455.40		$4438.90	$71.99	$92.48

Crew B-91

Crew B-91	Hr.	Daily	Hr.	Daily	Bare Costs	Incl. O&P
1 Labor Foreman (outside)	$28.70	$229.60	$44.70	$357.60	$31.03	$47.33
2 Laborers	26.70	427.20	41.55	664.80		
4 Equip. Oper. (med.)	34.65	1108.80	52.20	1670.40		
1 Truck Driver (heavy)	27.55	220.40	42.00	336.00		
1 Dist. Truck, 3000 Gal.		233.80		257.20		
1 Aggreg. Spreader, S.P.		771.20		848.30		
1 Roller, Pneu. Tire, 12 Ton		241.80		266.00		
1 Roller, Steel, 10 Ton		179.20		197.10	22.28	24.51
64 L.H., Daily Totals		$3412.00		$4597.40	$53.31	$71.84

Crew B-92

Crew B-92	Hr.	Daily	Hr.	Daily	Bare Costs	Incl. O&P
1 Labor Foreman (outside)	$28.70	$229.60	$44.70	$357.60	$27.20	$42.34
3 Laborers	26.70	640.80	41.55	997.20		
1 Crack Cleaner, 25 H.P.		44.40		48.85		
1 Air Compressor		55.40		60.95		
1 Tar Kettle, T.M.		40.95		45.05		
1 Flatbed Truck, 3 Ton		176.20		193.80	9.90	10.90
32 L.H., Daily Totals		$1187.35		$1703.45	$37.10	$53.24

Crew B-93

Crew B-93	Hr.	Daily	Hr.	Daily	Bare Costs	Incl. O&P
1 Equip. Oper. (med.)	$34.65	$277.20	$52.20	$417.60	$34.65	$52.20
1 Feller Buncher, 50 H.P.		432.40		475.65	54.05	59.46
8 L.H., Daily Totals		$709.60		$893.25	$88.70	$111.66

Crew B-94A

Crew B-94A	Hr.	Daily	Hr.	Daily	Bare Costs	Incl. O&P
1 Laborer	$26.70	$213.60	$41.55	$332.40	$26.70	$41.55
1 Diaph. Water Pump, 2"		50.60		55.65		
1-20 Ft. Suction Hose, 2"		2.55		2.80		
2-50 Ft. Disch. Hoses, 2"		3.80		4.20	7.12	7.83
8 L.H., Daily Totals		$270.55		$395.05	$33.82	$49.38

Crew B-94B

Crew B-94B	Hr.	Daily	Hr.	Daily	Bare Costs	Incl. O&P
1 Laborer	$26.70	$213.60	$41.55	$332.40	$26.70	$41.55
1 Diaph. Water Pump, 4"		74.40		81.85		
1-20 Ft. Suction Hose, 4"		5.05		5.55		
2-50 Ft. Disch. Hoses, 4"		7.10		7.80	10.82	11.90
8 L.H., Daily Totals		$300.15		$427.60	$37.52	$53.45

Crew B-94C

Crew B-94C	Hr.	Daily	Hr.	Daily	Bare Costs	Incl. O&P
1 Laborer	$26.70	$213.60	$41.55	$332.40	$26.70	$41.55
1 Centr. Water Pump, 3"		53.60		58.95		
1-20 Ft. Suction Hose, 3"		4.25		4.70		
2-50 Ft. Disch. Hoses, 3"		5.00		5.50	7.86	8.64
8 L.H., Daily Totals		$276.45		$401.55	$34.56	$50.19

Crew B-94D

Crew B-94D	Hr.	Daily	Hr.	Daily	Bare Costs	Incl. O&P
1 Laborer	$26.70	$213.60	$41.55	$332.40	$26.70	$41.55
1 Centr. Water Pump, 6"		217.00		238.70		
1-20 Ft. Suction Hose, 6"		12.80		14.10		
2-50 Ft. Disch. Hoses, 6"		17.80		19.60	30.95	34.05
8 L.H., Daily Totals		$461.20		$604.80	$57.65	$75.60

Crew B-95A

Crew No.	Bare Costs Hr.	Daily	Incl. Subs O & P Hr.	Daily	Cost Per Labor-Hour Bare Costs	Incl. O&P
1 Equip. Oper. (crane)	$35.90	$287.20	$54.10	$432.80	$31.30	$47.83
1 Laborer	26.70	213.60	41.55	332.40		
1 Hyd. Excavator, 5/8 C.Y.		432.40		475.65	27.03	29.73
16 L.H., Daily Totals		$933.20		$1240.85	$58.33	$77.56

Crew B-95B

Crew No.	Bare Costs Hr.	Daily	Incl. Subs O & P Hr.	Daily	Cost Per Labor-Hour Bare Costs	Incl. O&P
1 Equip. Oper. (crane)	$35.90	$287.20	$54.10	$432.80	$31.30	$47.83
1 Laborer	26.70	213.60	41.55	332.40		
1 Hyd. Excavator, 1.5 C.Y.		720.40		792.45	45.03	49.53
16 L.H., Daily Totals		$1221.20		$1557.65	$76.33	$97.36

Crew B-95C

Crew No.	Bare Costs Hr.	Daily	Incl. Subs O & P Hr.	Daily	Cost Per Labor-Hour Bare Costs	Incl. O&P
1 Equip. Oper. (crane)	$35.90	$287.20	$54.10	$432.80	$31.30	$47.83
1 Laborer	26.70	213.60	41.55	332.40		
1 Hyd. Excavator, 2.5 C.Y.		1215.00		1336.50	75.94	83.53
16 L.H., Daily Totals		$1715.80		$2101.70	$107.24	$131.36

Crew C-1

Crew No.	Bare Costs Hr.	Daily	Incl. Subs O & P Hr.	Daily	Cost Per Labor-Hour Bare Costs	Incl. O&P
3 Carpenters	$34.25	$822.00	$53.35	$1280.40	$32.36	$50.40
1 Laborer	26.70	213.60	41.55	332.40		
32 L.H., Daily Totals		$1035.60		$1612.80	$32.36	$50.40

Crew C-2

Crew No.	Bare Costs Hr.	Daily	Incl. Subs O & P Hr.	Daily	Cost Per Labor-Hour Bare Costs	Incl. O&P
1 Carpenter Foreman (out)	$36.25	$290.00	$56.45	$451.60	$33.33	$51.90
4 Carpenters	34.25	1096.00	53.35	1707.20		
1 Laborer	26.70	213.60	41.55	332.40		
48 L.H., Daily Totals		$1599.60		$2491.20	$33.33	$51.90

Crew C-2A

Crew No.	Bare Costs Hr.	Daily	Incl. Subs O & P Hr.	Daily	Cost Per Labor-Hour Bare Costs	Incl. O&P
1 Carpenter Foreman (out)	$36.25	$290.00	$56.45	$451.60	$33.09	$51.01
3 Carpenters	34.25	822.00	53.35	1280.40		
1 Cement Finisher	32.85	262.80	48.35	386.80		
1 Laborer	26.70	213.60	41.55	332.40		
48 L.H., Daily Totals		$1588.40		$2451.20	$33.09	$51.01

Crew C-3

Crew No.	Bare Costs Hr.	Daily	Incl. Subs O & P Hr.	Daily	Cost Per Labor-Hour Bare Costs	Incl. O&P
1 Rodman Foreman	$39.95	$319.60	$65.90	$527.20	$34.78	$56.15
4 Rodmen (reinf.)	37.95	1214.40	62.60	2003.20		
1 Equip. Oper. (light)	33.05	264.40	49.80	398.40		
2 Laborers	26.70	427.20	41.55	664.80		
3 Stressing Equipment		40.80		44.90		
.5 Grouting Equipment		76.40		84.05	1.83	2.01
64 L.H., Daily Totals		$2342.80		$3722.55	$36.61	$58.16

Crew C-4

Crew No.	Bare Costs Hr.	Daily	Incl. Subs O & P Hr.	Daily	Cost Per Labor-Hour Bare Costs	Incl. O&P
1 Rodman Foreman	$39.95	$319.60	$65.90	$527.20	$38.45	$63.43
3 Rodmen (reinf.)	37.95	910.80	62.60	1502.40		
3 Stressing Equipment		40.80		44.90	1.28	1.40
32 L.H., Daily Totals		$1271.20		$2074.50	$39.73	$64.83

Crew C-5

Crew No.	Bare Costs Hr.	Daily	Incl. Subs O & P Hr.	Daily	Cost Per Labor-Hour Bare Costs	Incl. O&P
1 Rodman Foreman	$39.95	$319.60	$65.90	$527.20	$36.82	$59.39
4 Rodmen (reinf.)	37.95	1214.40	62.60	2003.20		
1 Equip. Oper. (crane)	35.90	287.20	54.10	432.80		
1 Equip. Oper. Oiler	30.10	240.80	45.35	362.80		
1 Hyd. Crane, 25 Ton		616.80		678.50	11.01	12.12
56 L.H., Daily Totals		$2678.80		$4004.50	$47.83	$71.51

Crew C-6

Crew No.	Bare Costs Hr.	Daily	Incl. Subs O & P Hr.	Daily	Cost Per Labor-Hour Bare Costs	Incl. O&P
1 Labor Foreman (outside)	$28.70	$229.60	$44.70	$357.60	$28.06	$43.15
4 Laborers	26.70	854.40	41.55	1329.60		
1 Cement Finisher	32.85	262.80	48.35	386.80		
2 Gas Engine Vibrators		48.00		52.80	1.00	1.10
48 L.H., Daily Totals		$1394.80		$2126.80	$29.06	$44.25

Crew C-7

Crew No.	Bare Costs Hr.	Daily	Incl. Subs O & P Hr.	Daily	Cost Per Labor-Hour Bare Costs	Incl. O&P
1 Labor Foreman (outside)	$28.70	$229.60	$44.70	$357.60	$28.87	$44.22
5 Laborers	26.70	1068.00	41.55	1662.00		
1 Cement Finisher	32.85	262.80	48.35	386.80		
1 Equip. Oper. (med.)	34.65	277.20	52.20	417.60		
1 Equip. Oper. (oiler)	30.10	240.80	45.35	362.80		
2 Gas Engine Vibrators		48.00		52.80		
1 Concrete Bucket, 1 C.Y.		16.00		17.60		
1 Hyd. Crane, 55 Ton		895.40		984.95	13.33	14.66
72 L.H., Daily Totals		$3037.80		$4242.15	$42.20	$58.88

Crew C-7A

Crew No.	Bare Costs Hr.	Daily	Incl. Subs O & P Hr.	Daily	Cost Per Labor-Hour Bare Costs	Incl. O&P
1 Labor Foreman (outside)	$28.70	$229.60	$44.70	$357.60	$27.16	$42.06
5 Laborers	26.70	1068.00	41.55	1662.00		
2 Truck Drivers (Heavy)	27.55	440.80	42.00	672.00		
2 Conc. Transit Mixers		1480.80		1628.90	23.14	25.45
64 L.H., Daily Totals		$3219.20		$4320.50	$50.30	$67.51

Crew C-7B

Crew No.	Bare Costs Hr.	Daily	Incl. Subs O & P Hr.	Daily	Cost Per Labor-Hour Bare Costs	Incl. O&P
1 Labor Foreman (outside)	$28.70	$229.60	$44.70	$357.60	$28.53	$43.99
5 Laborers	26.70	1068.00	41.55	1662.00		
1 Equipment Operator (heavy)	35.90	287.20	54.10	432.80		
1 Equipment Oiler	30.10	240.80	45.35	362.80		
1 Conc. Bucket, 2 C.Y.		24.60		27.05		
1 Truck Crane, 165 Ton		1912.00		2103.20	30.26	33.29
64 L.H., Daily Totals		$3762.20		$4945.45	$58.79	$77.28

Crew C-7C

Crew No.	Bare Costs Hr.	Daily	Incl. Subs O & P Hr.	Daily	Cost Per Labor-Hour Bare Costs	Incl. O&P
1 Labor Foreman (outside)	$28.70	$229.60	$44.70	$357.60	$28.94	$44.61
5 Laborers	26.70	1068.00	41.55	1662.00		
2 Equipment Operator (medium)	34.65	554.40	52.20	835.20		
2 Wheel Loader, 4 C.Y.		891.20		980.30	13.93	15.32
64 L.H., Daily Totals		$2743.20		$3835.10	$42.87	$59.93

Crew C-7D

Crew No.	Bare Costs Hr.	Daily	Incl. Subs O & P Hr.	Daily	Cost Per Labor-Hour Bare Costs	Incl. O&P
1 Labor Foreman (outside)	$28.70	$229.60	$44.70	$357.60	$28.12	$43.52
5 Laborers	26.70	1068.00	41.55	1662.00		
1 Equipment Operator (med.)	34.65	277.20	52.20	417.60		
1 Concrete Conveyer		152.80		168.10	2.73	3.00
56 L.H., Daily Totals		$1727.60		$2605.30	$30.85	$46.52

Crew C-8

Crew No.	Bare Costs Hr.	Daily	Incl. Subs O & P Hr.	Daily	Cost Per Labor-Hour Bare Costs	Incl. O&P
1 Labor Foreman (outside)	$28.70	$229.60	$44.70	$357.60	$29.88	$45.36
3 Laborers	26.70	640.80	41.55	997.20		
2 Cement Finishers	32.85	525.60	48.35	773.60		
1 Equip. Oper. (med.)	34.65	277.20	52.20	417.60		
1 Concrete Pump (small)		704.20		774.60	12.58	13.83
56 L.H., Daily Totals		$2377.40		$3320.60	$42.46	$59.19

Crew C-8A

Crew No.	Bare Costs Hr.	Daily	Incl. Subs O & P Hr.	Daily	Cost Per Labor-Hour Bare Costs	Incl. O&P
1 Labor Foreman (outside)	$28.70	$229.60	$44.70	$357.60	$29.08	$44.23
3 Laborers	26.70	640.80	41.55	997.20		
2 Cement Finishers	32.85	525.60	48.35	773.60		
48 L.H., Daily Totals		$1396.00		$2128.40	$29.08	$44.23

Crew No.	Bare Costs		Incl. Subs O & P		Cost Per Labor-Hour	
Crew C-8B	Hr.	Daily	Hr.	Daily	Bare Costs	Incl. O&P
1 Labor Foreman (outside)	$28.70	$229.60	$44.70	$357.60	$28.69	$44.31
3 Laborers	26.70	640.80	41.55	997.20		
1 Equipment Operator	34.65	277.20	52.20	417.60		
1 Vibrating Screed		43.05		47.35		
1 Vibratory Roller		433.20		476.50		
1 Dozer, 200 H.P.		919.60		1011.55	34.90	38.38
40 L.H., Daily Totals		$2543.45		$3307.80	$63.59	$82.69
Crew C-8C	Hr.	Daily	Hr.	Daily	Bare Costs	Incl. O&P
1 Labor Foreman (outside)	$28.70	$229.60	$44.70	$357.60	$29.38	$44.93
3 Laborers	26.70	640.80	41.55	997.20		
1 Cement Finisher	32.85	262.80	48.35	386.80		
1 Equipment Operator (med.)	34.65	277.20	52.20	417.60		
1 Shotcrete Rig, 12 CY/hr		223.80		246.20	4.66	5.13
48 L.H., Daily Totals		$1634.20		$2405.40	$34.04	$50.06
Crew C-8D	Hr.	Daily	Hr.	Daily	Bare Costs	Incl. O&P
1 Labor Foreman (outside)	$28.70	$229.60	$44.70	$357.60	$30.33	$46.01
1 Laborer	26.70	213.60	41.55	332.40		
1 Cement Finisher	32.85	262.80	48.35	386.80		
1 Equipment Operator (light)	33.05	264.40	49.80	398.40		
1 Compressor, 250 CFM		112.40		123.65		
2 Hoses, 1", 50'		7.30		8.05	3.74	4.11
32 L.H., Daily Totals		$1090.10		$1606.90	$34.07	$50.12
Crew C-8E	Hr.	Daily	Hr.	Daily	Bare Costs	Incl. O&P
1 Labor Foreman (outside)	$28.70	$229.60	$44.70	$357.60	$30.33	$46.01
1 Laborer	26.70	213.60	41.55	332.40		
1 Cement Finisher	32.85	262.80	48.35	386.80		
1 Equipment Operator (light)	33.05	264.40	49.80	398.40		
1 Compressor, 250 CFM		112.40		123.65		
2 Hoses, 1", 50'		7.30		8.05		
1 Concrete Pump (small)		704.20		774.60	25.75	28.32
32 L.H., Daily Totals		$1794.30		$2381.50	$56.08	$74.33
Crew C-10	Hr.	Daily	Hr.	Daily	Bare Costs	Incl. O&P
1 Laborer	$26.70	$213.60	$41.55	$332.40	$30.80	$45.85
2 Cement Finishers	32.85	525.60	48.35	773.60		
24 L.H., Daily Totals		$739.20		$1106.00	$30.80	$45.85
Crew C-10B	Hr.	Daily	Hr.	Daily	Bare Costs	Incl. O&P
3 Laborers	$26.70	$640.80	$41.55	$997.20	$29.16	$44.13
2 Cement Finishers	32.85	525.60	48.35	773.60		
1 Concrete mixer, 10 CF		123.20		135.50		
2 Concrete finisher, 48" dia		49.20		54.10	4.31	4.74
40 L.H., Daily Totals		$1338.80		$1960.40	$33.47	$48.87
Crew C-11	Hr.	Daily	Hr.	Daily	Bare Costs	Incl. O&P
1 Struc. Steel Foreman	$40.15	$321.20	$71.85	$574.80	$37.23	$64.53
6 Struc. Steel Workers	38.15	1831.20	68.25	3276.00		
1 Equip. Oper. (crane)	35.90	287.20	54.10	432.80		
1 Equip. Oper. Oiler	30.10	240.80	45.35	362.80		
1 Truck Crane, 150 Ton		1445.00		1589.50	20.07	22.08
72 L.H., Daily Totals		$4125.40		$6235.90	$57.30	$86.61

Crew No.	Bare Costs		Incl. Subs O & P		Cost Per Labor-Hour	
Crew C-12	Hr.	Daily	Hr.	Daily	Bare Costs	Incl. O&P
1 Carpenter Foreman (out)	$36.25	$290.00	$56.45	$451.60	$33.60	$52.03
3 Carpenters	34.25	822.00	53.35	1280.40		
1 Laborer	26.70	213.60	41.55	332.40		
1 Equip. Oper. (crane)	35.90	287.20	54.10	432.80		
1 Hyd. Crane, 12 Ton		602.60		662.85	12.55	13.81
48 L.H., Daily Totals		$2215.40		$3160.05	$46.15	$65.84
Crew C-13	Hr.	Daily	Hr.	Daily	Bare Costs	Incl. O&P
1 Struc. Steel Worker	$38.15	$305.20	$68.25	$546.00	$36.85	$63.28
1 Welder	38.15	305.20	68.25	546.00		
1 Carpenter	34.25	274.00	53.35	426.80		
1 Gas Welding Machine		81.20		89.30	3.38	3.72
24 L.H., Daily Totals		$965.60		$1608.10	$40.23	$67.00
Crew C-14	Hr.	Daily	Hr.	Daily	Bare Costs	Incl. O&P
1 Carpenter Foreman (out)	$36.25	$290.00	$56.45	$451.60	$33.21	$51.96
5 Carpenters	34.25	1370.00	53.35	2134.00		
4 Laborers	26.70	854.40	41.55	1329.60		
4 Rodmen (reinf.)	37.95	1214.40	62.60	2003.20		
2 Cement Finishers	32.85	525.60	48.35	773.60		
1 Equip. Oper. (crane)	35.90	287.20	54.10	432.80		
1 Equip. Oper. Oiler	30.10	240.80	45.35	362.80		
1 Crane, 80 Ton, & Tools		995.00		1094.50	6.91	7.60
144 L.H., Daily Totals		$5777.40		$8582.10	$40.12	$59.56
Crew C-14A	Hr.	Daily	Hr.	Daily	Bare Costs	Incl. O&P
1 Carpenter Foreman (out)	$36.25	$290.00	$56.45	$451.60	$34.28	$53.75
16 Carpenters	34.25	4384.00	53.35	6828.80		
4 Rodmen (reinf.)	37.95	1214.40	62.60	2003.20		
2 Laborers	26.70	427.20	41.55	664.80		
1 Cement Finisher	32.85	262.80	48.35	386.80		
1 Equip. Oper. (med.)	34.65	277.20	52.20	417.60		
1 Gas Engine Vibrator		24.00		26.40		
1 Concrete Pump (small)		704.20		774.60	3.64	4.01
200 L.H., Daily Totals		$7583.80		$11553.80	$37.92	$57.76
Crew C-14B	Hr.	Daily	Hr.	Daily	Bare Costs	Incl. O&P
1 Carpenter Foreman (out)	$36.25	$290.00	$56.45	$451.60	$34.22	$53.53
16 Carpenters	34.25	4384.00	53.35	6828.80		
4 Rodmen (reinf.)	37.95	1214.40	62.60	2003.20		
2 Laborers	26.70	427.20	41.55	664.80		
2 Cement Finishers	32.85	525.60	48.35	773.60		
1 Equip. Oper. (med.)	34.65	277.20	52.20	417.60		
1 Gas Engine Vibrator		24.00		26.40		
1 Concrete Pump (small)		704.20		774.60	3.50	3.85
208 L.H., Daily Totals		$7846.60		$11940.60	$37.72	$57.38
Crew C-14C	Hr.	Daily	Hr.	Daily	Bare Costs	Incl. O&P
1 Carpenter Foreman (out)	$36.25	$290.00	$56.45	$451.60	$32.66	$51.14
6 Carpenters	34.25	1644.00	53.35	2560.80		
2 Rodmen (reinf.)	37.95	607.20	62.60	1001.60		
4 Laborers	26.70	854.40	41.55	1329.60		
1 Cement Finisher	32.85	262.80	48.35	386.80		
1 Gas Engine Vibrator		24.00		26.40	.21	.24
112 L.H., Daily Totals		$3682.40		$5756.80	$32.87	$51.38

Crew C-14D

Crew No.	Bare Costs Hr.	Bare Costs Daily	Incl. Subs O & P Hr.	Incl. Subs O & P Daily	Cost Per L.H. Bare Costs	Cost Per L.H. Incl. O&P
1 Carpenter Foreman (out)	$36.25	$290.00	$56.45	$451.60	$33.98	$53.01
18 Carpenters	34.25	4932.00	53.35	7682.40		
2 Rodmen (reinf.)	37.95	607.20	62.60	1001.60		
2 Laborers	26.70	427.20	41.55	664.80		
1 Cement Finisher	32.85	262.80	48.35	386.80		
1 Equip. Oper. (med.)	34.65	277.20	52.20	417.60		
1 Gas Engine Vibrator		24.00		26.40		
1 Concrete Pump (small)		704.20		774.60	3.64	4.01
200 L.H., Daily Totals		$7524.60		$11405.80	$37.62	$57.02

Crew C-14E

Crew No.	Bare Costs Hr.	Bare Costs Daily	Incl. Subs O & P Hr.	Incl. Subs O & P Daily	Cost Per L.H. Bare Costs	Cost Per L.H. Incl. O&P
1 Carpenter Foreman (out)	$36.25	$290.00	$56.45	$451.60	$33.59	$53.29
2 Carpenters	34.25	548.00	53.35	853.60		
4 Rodmen (reinf.)	37.95	1214.40	62.60	2003.20		
3 Laborers	26.70	640.80	41.55	997.20		
1 Cement Finisher	32.85	262.80	48.35	386.80		
1 Gas Engine Vibrator		24.00		26.40	.27	.30
88 L.H., Daily Totals		$2980.00		$4718.80	$33.86	$53.59

Crew C-14F

Crew No.	Bare Costs Hr.	Bare Costs Daily	Incl. Subs O & P Hr.	Incl. Subs O & P Daily	Cost Per L.H. Bare Costs	Cost Per L.H. Incl. O&P
1 Laborer Foreman (out)	$28.70	$229.60	$44.70	$357.60	$31.02	$46.20
2 Laborers	26.70	427.20	41.55	664.80		
6 Cement Finishers	32.85	1576.80	48.35	2320.80		
1 Gas Engine Vibrator		24.00		26.40	.33	.37
72 L.H., Daily Totals		$2257.60		$3369.60	$31.35	$46.57

Crew C-14G

Crew No.	Bare Costs Hr.	Bare Costs Daily	Incl. Subs O & P Hr.	Incl. Subs O & P Daily	Cost Per L.H. Bare Costs	Cost Per L.H. Incl. O&P
1 Laborer Foreman (out)	$28.70	$229.60	$44.70	$357.60	$30.50	$45.69
2 Laborers	26.70	427.20	41.55	664.80		
4 Cement Finishers	32.85	1051.20	48.35	1547.20		
1 Gas Engine Vibrator		24.00		26.40	.43	.47
56 L.H., Daily Totals		$1732.00		$2596.00	$30.93	$46.16

Crew C-14H

Crew No.	Bare Costs Hr.	Bare Costs Daily	Incl. Subs O & P Hr.	Incl. Subs O & P Daily	Cost Per L.H. Bare Costs	Cost Per L.H. Incl. O&P
1 Carpenter Foreman (out)	$36.25	$290.00	$56.45	$451.60	$33.71	$52.55
2 Carpenters	34.25	548.00	53.35	853.60		
1 Rodman (reinf.)	37.95	303.60	62.60	500.80		
1 Laborer	26.70	213.60	41.55	332.40		
1 Cement Finisher	32.85	262.80	48.35	386.80		
1 Gas Engine Vibrator		24.00		26.40	.50	.55
48 L.H., Daily Totals		$1642.00		$2551.60	$34.21	$53.10

Crew C-15

Crew No.	Bare Costs Hr.	Bare Costs Daily	Incl. Subs O & P Hr.	Incl. Subs O & P Daily	Cost Per L.H. Bare Costs	Cost Per L.H. Incl. O&P
1 Carpenter Foreman (out)	$36.25	$290.00	$56.45	$451.60	$32.06	$49.60
2 Carpenters	34.25	548.00	53.35	853.60		
3 Laborers	26.70	640.80	41.55	997.20		
2 Cement Finishers	32.85	525.60	48.35	773.60		
1 Rodman (reinf.)	37.95	303.60	62.60	500.80		
72 L.H., Daily Totals		$2308.00		$3576.80	$32.06	$49.60

Crew C-16

Crew No.	Bare Costs Hr.	Bare Costs Daily	Incl. Subs O & P Hr.	Incl. Subs O & P Daily	Cost Per L.H. Bare Costs	Cost Per L.H. Incl. O&P
1 Labor Foreman (outside)	$28.70	$229.60	$44.70	$357.60	$31.67	$49.19
3 Laborers	26.70	640.80	41.55	997.20		
2 Cement Finishers	32.85	525.60	48.35	773.60		
1 Equip. Oper. (med.)	34.65	277.20	52.20	417.60		
2 Rodmen (reinf.)	37.95	607.20	62.60	1001.60		
1 Concrete Pump (small)		704.20		774.60	9.78	10.76
72 L.H., Daily Totals		$2984.60		$4322.20	$41.45	$59.95

Crew C-17

Crew No.	Bare Costs Hr.	Bare Costs Daily	Incl. Subs O & P Hr.	Incl. Subs O & P Daily	Cost Per L.H. Bare Costs	Cost Per L.H. Incl. O&P
2 Skilled Worker Foremen	$36.85	$589.60	$57.30	$916.80	$35.25	$54.82
8 Skilled Workers	34.85	2230.40	54.20	3468.80		
80 L.H., Daily Totals		$2820.00		$4385.60	$35.25	$54.82

Crew C-17A

Crew No.	Bare Costs Hr.	Bare Costs Daily	Incl. Subs O & P Hr.	Incl. Subs O & P Daily	Cost Per L.H. Bare Costs	Cost Per L.H. Incl. O&P
2 Skilled Worker Foremen	$36.85	$589.60	$57.30	$916.80	$35.26	$54.81
8 Skilled Workers	34.85	2230.40	54.20	3468.80		
.125 Equip. Oper. (crane)	35.90	35.90	54.10	54.10		
.125 Crane, 80 Ton, & Tools		124.38		136.80	1.54	1.69
81 L.H., Daily Totals		$2980.28		$4576.50	$36.80	$56.50

Crew C-17B

Crew No.	Bare Costs Hr.	Bare Costs Daily	Incl. Subs O & P Hr.	Incl. Subs O & P Daily	Cost Per L.H. Bare Costs	Cost Per L.H. Incl. O&P
2 Skilled Worker Foremen	$36.85	$589.60	$57.30	$916.80	$35.27	$54.80
8 Skilled Workers	34.85	2230.40	54.20	3468.80		
.25 Equip. Oper. (crane)	35.90	71.80	54.10	108.20		
.25 Crane, 80 Ton, & Tools		248.75		273.65		
.25 Walk Behind Power Tools		6.15		6.75	3.11	3.42
82 L.H., Daily Totals		$3146.70		$4774.20	$38.38	$58.22

Crew C-17C

Crew No.	Bare Costs Hr.	Bare Costs Daily	Incl. Subs O & P Hr.	Incl. Subs O & P Daily	Cost Per L.H. Bare Costs	Cost Per L.H. Incl. O&P
2 Skilled Worker Foremen	$36.85	$589.60	$57.30	$916.80	$35.27	$54.79
8 Skilled Workers	34.85	2230.40	54.20	3468.80		
.375 Equip. Oper. (crane)	35.90	107.70	54.10	162.30		
.375 Crane, 80 Ton & Tools		373.13		410.45	4.50	4.95
83 L.H., Daily Totals		$3300.83		$4958.35	$39.77	$59.74

Crew C-17D

Crew No.	Bare Costs Hr.	Bare Costs Daily	Incl. Subs O & P Hr.	Incl. Subs O & P Daily	Cost Per L.H. Bare Costs	Cost Per L.H. Incl. O&P
2 Skilled Worker Foremen	$36.85	$589.60	$57.30	$916.80	$35.28	$54.79
8 Skilled Workers	34.85	2230.40	54.20	3468.80		
.5 Equip. Oper. (crane)	35.90	143.60	54.10	216.40		
.5 Crane, 80 Ton & Tools		497.50		547.25	5.92	6.51
84 L.H., Daily Totals		$3461.10		$5149.25	$41.20	$61.30

Crew C-17E

Crew No.	Bare Costs Hr.	Bare Costs Daily	Incl. Subs O & P Hr.	Incl. Subs O & P Daily	Cost Per L.H. Bare Costs	Cost Per L.H. Incl. O&P
2 Skilled Worker Foremen	$36.85	$589.60	$57.30	$916.80	$35.25	$54.82
8 Skilled Workers	34.85	2230.40	54.20	3468.80		
1 Hyd. Jack with Rods		79.00		86.90	.99	1.09
80 L.H., Daily Totals		$2899.00		$4472.50	$36.24	$55.91

Crew C-18

Crew No.	Bare Costs Hr.	Bare Costs Daily	Incl. Subs O & P Hr.	Incl. Subs O & P Daily	Cost Per L.H. Bare Costs	Cost Per L.H. Incl. O&P
.125 Labor Foreman (out)	$28.70	$28.70	$44.70	$44.70	$26.92	$41.90
1 Laborer	26.70	213.60	41.55	332.40		
1 Concrete Cart, 10 C.F.		49.80		54.80	5.53	6.09
9 L.H., Daily Totals		$292.10		$431.90	$32.45	$47.99

Crew C-19

Crew No.	Bare Costs Hr.	Bare Costs Daily	Incl. Subs O & P Hr.	Incl. Subs O & P Daily	Cost Per L.H. Bare Costs	Cost Per L.H. Incl. O&P
.125 Labor Foreman (out)	$28.70	$28.70	$44.70	$44.70	$26.92	$41.90
1 Laborer	26.70	213.60	41.55	332.40		
1 Concrete Cart, 18 C.F.		76.80		84.50	8.53	9.39
9 L.H., Daily Totals		$319.10		$461.60	$35.45	$51.29

Crew C-20

Crew No.	Bare Costs Hr.	Bare Costs Daily	Incl. Subs O & P Hr.	Incl. Subs O & P Daily	Cost Per L.H. Bare Costs	Cost Per L.H. Incl. O&P
1 Labor Foreman (outside)	$28.70	$229.60	$44.70	$357.60	$28.71	$44.08
5 Laborers	26.70	1068.00	41.55	1662.00		
1 Cement Finisher	32.85	262.80	48.35	386.80		
1 Equip. Oper. (med.)	34.65	277.20	52.20	417.60		
2 Gas Engine Vibrators		48.00		52.80		
1 Concrete Pump (small)		704.20		774.60	11.75	12.93
64 L.H., Daily Totals		$2589.80		$3651.40	$40.46	$57.01

Crew No.	Bare Costs		Incl. Subs O & P		Cost Per Labor-Hour	

Crew C-21	Hr.	Daily	Hr.	Daily	Bare Costs	Incl. O&P
1 Labor Foreman (outside)	$28.70	$229.60	$44.70	$357.60	$28.71	$44.08
5 Laborers	26.70	1068.00	41.55	1662.00		
1 Cement Finisher	32.85	262.80	48.35	386.80		
1 Equip. Oper. (med.)	34.65	277.20	52.20	417.60		
2 Gas Engine Vibrators		48.00		52.80		
1 Concrete Conveyer		152.80		168.10	3.14	3.45
64 L.H., Daily Totals		$2038.40		$3044.90	$31.85	$47.53

Crew C-22	Hr.	Daily	Hr.	Daily	Bare Costs	Incl. O&P
1 Rodman Foreman	$39.95	$319.60	$65.90	$527.20	$38.10	$62.62
4 Rodmen (reinf.)	37.95	1214.40	62.60	2003.20		
.125 Equip. Oper. (crane)	35.90	35.90	54.10	54.10		
.125 Equip. Oper. Oiler	30.10	30.10	45.35	45.35		
.125 Hyd. Crane, 25 Ton		77.10		84.80	1.84	2.02
42 L.H., Daily Totals		$1677.10		$2714.65	$39.94	$64.64

Crew C-23	Hr.	Daily	Hr.	Daily	Bare Costs	Incl. O&P
2 Skilled Worker Foremen	$36.85	$589.60	$57.30	$916.80	$34.88	$53.93
6 Skilled Workers	34.85	1672.80	54.20	2601.60		
1 Equip. Oper. (crane)	35.90	287.20	54.10	432.80		
1 Equip. Oper. Oiler	30.10	240.80	45.35	362.80		
1 Crane, 90 Ton		1325.00		1457.50	16.56	18.22
80 L.H., Daily Totals		$4115.40		$5771.50	$51.44	$72.15

Crew C-23A	Hr.	Daily	Hr.	Daily	Bare Costs	Incl. O&P
1 Labor Foreman (outside)	$28.70	$229.60	$44.70	$357.60	$29.62	$45.45
2 Laborers	26.70	427.20	41.55	664.80		
1 Equip. Oper. (crane)	35.90	287.20	54.10	432.80		
1 Equip. Oper. Oiler	30.10	240.80	45.35	362.80		
1 Crane, 100 ton capacity		1519.00		1670.90		
3 Conc. bucket, 8 C.Y.		465.60		512.15	49.62	54.58
40 L.H., Daily Totals		$3169.40		$4001.05	$79.24	$100.03

Crew C-24	Hr.	Daily	Hr.	Daily	Bare Costs	Incl. O&P
2 Skilled Worker Foremen	$36.85	$589.60	$57.30	$916.80	$34.88	$53.93
6 Skilled Workers	34.85	1672.80	54.20	2601.60		
1 Equip. Oper. (crane)	35.90	287.20	54.10	432.80		
1 Equip. Oper. Oiler	30.10	240.80	45.35	362.80		
1 Truck Crane, 150 Ton		1445.00		1589.50	18.06	19.87
80 L.H., Daily Totals		$4235.40		$5903.50	$52.94	$73.80

Crew C-25	Hr.	Daily	Hr.	Daily	Bare Costs	Incl. O&P
2 Rodmen (reinf.)	$37.95	$607.20	$62.60	$1001.60	$29.78	$49.70
2 Rodmen Helpers	21.60	345.60	36.80	588.80		
32 L.H., Daily Totals		$952.80		$1590.40	$29.78	$49.70

Crew C-27	Hr.	Daily	Hr.	Daily	Bare Costs	Incl. O&P
2 Cement Finishers	$32.85	$525.60	$48.35	$773.60	$32.85	$48.00
1 Concrete Saw		114.40		125.85	7.15	7.87
16 L.H., Daily Totals		$640.00		$899.45	$40.00	$55.87

Crew C-28	Hr.	Daily	Hr.	Daily	Bare Costs	Incl. O&P
1 Cement Finisher	$32.85	$262.80	$48.35	$386.80	$32.85	$48.00
1 Portable Air Compressor		17.30		19.05	2.16	2.37
8 L.H., Daily Totals		$280.10		$405.85	$35.01	$50.37

Crew D-1	Hr.	Daily	Hr.	Daily	Bare Costs	Incl. O&P
1 Bricklayer	$35.25	$282.00	$53.70	$429.60	$31.08	$47.35
1 Bricklayer Helper	26.90	215.20	41.00	328.00		
16 L.H., Daily Totals		$497.20		$757.60	$31.08	$47.35

Crew D-2	Hr.	Daily	Hr.	Daily	Bare Costs	Incl. O&P
3 Bricklayers	$35.25	$846.00	$53.70	$1288.80	$32.12	$49.05
2 Bricklayer Helpers	26.90	430.40	41.00	656.00		
.5 Carpenter	34.25	137.00	53.35	213.40		
44 L.H., Daily Totals		$1413.40		$2158.20	$32.12	$49.05

Crew D-3	Hr.	Daily	Hr.	Daily	Bare Costs	Incl. O&P
3 Bricklayers	$35.25	$846.00	$53.70	$1288.80	$32.02	$48.85
2 Bricklayer Helpers	26.90	430.40	41.00	656.00		
.25 Carpenter	34.25	68.50	53.35	106.70		
42 L.H., Daily Totals		$1344.90		$2051.50	$32.02	$48.85

Crew D-4	Hr.	Daily	Hr.	Daily	Bare Costs	Incl. O&P
1 Bricklayer	$35.25	$282.00	$53.70	$429.60	$30.53	$46.38
2 Bricklayer Helpers	26.90	430.40	41.00	656.00		
1 Equip. Oper. (light)	33.05	264.40	49.80	398.40		
1 Grout Pump, 50 C.F./hr		107.75		118.55		
1 Hoses & Hopper		15.20		16.70		
1 Accessories		11.90		13.10	4.22	4.64
32 L.H., Daily Totals		$1111.65		$1632.35	$34.75	$51.02

Crew D-5	Hr.	Daily	Hr.	Daily	Bare Costs	Incl. O&P
1 Bricklayer	$35.25	$282.00	$53.70	$429.60	$35.25	$53.70
8 L.H., Daily Totals		$282.00		$429.60	$35.25	$53.70

Crew D-6	Hr.	Daily	Hr.	Daily	Bare Costs	Incl. O&P
3 Bricklayers	$35.25	$846.00	$53.70	$1288.80	$31.20	$47.59
3 Bricklayer Helpers	26.90	645.60	41.00	984.00		
.25 Carpenter	34.25	68.50	53.35	106.70		
50 L.H., Daily Totals		$1560.10		$2379.50	$31.20	$47.59

Crew D-7	Hr.	Daily	Hr.	Daily	Bare Costs	Incl. O&P
1 Tile Layer	$32.70	$261.60	$48.05	$384.40	$29.03	$42.65
1 Tile Layer Helper	25.35	202.80	37.25	298.00		
16 L.H., Daily Totals		$464.40		$682.40	$29.03	$42.65

Crew D-8	Hr.	Daily	Hr.	Daily	Bare Costs	Incl. O&P
3 Bricklayers	$35.25	$846.00	$53.70	$1288.80	$31.91	$48.62
2 Bricklayer Helpers	26.90	430.40	41.00	656.00		
40 L.H., Daily Totals		$1276.40		$1944.80	$31.91	$48.62

Crew D-9	Hr.	Daily	Hr.	Daily	Bare Costs	Incl. O&P
3 Bricklayers	$35.25	$846.00	$53.70	$1288.80	$31.08	$47.35
3 Bricklayer Helpers	26.90	645.60	41.00	984.00		
48 L.H., Daily Totals		$1491.60		$2272.80	$31.08	$47.35

Crew D-10	Hr.	Daily	Hr.	Daily	Bare Costs	Incl. O&P
1 Bricklayer Foreman	$37.25	$298.00	$56.75	$454.00	$32.44	$49.31
1 Bricklayer	35.25	282.00	53.70	429.60		
2 Bricklayer Helpers	26.90	430.40	41.00	656.00		
1 Equip. Oper. (crane)	35.90	287.20	54.10	432.80		
1 Truck Crane, 12.5 Ton		505.80		556.40	12.65	13.91
40 L.H., Daily Totals		$1803.40		$2528.80	$45.09	$63.22

Crew D-11	Hr.	Daily	Hr.	Daily	Bare Costs	Incl. O&P
1 Bricklayer Foreman	$37.25	$298.00	$56.75	$454.00	$33.13	$50.48
1 Bricklayer	35.25	282.00	53.70	429.60		
1 Bricklayer Helper	26.90	215.20	41.00	328.00		
24 L.H., Daily Totals		$795.20		$1211.60	$33.13	$50.48

Crew No.	Bare Costs Hr.	Daily	Incl. Subs O & P Hr.	Daily	Cost Per Labor-Hour Bare Costs	Incl. O&P
Crew D-12	Hr.	Daily	Hr.	Daily	Bare Costs	Incl. O&P
1 Bricklayer Foreman	$37.25	$298.00	$56.75	$454.00	$31.58	$48.11
1 Bricklayer	35.25	282.00	53.70	429.60		
2 Bricklayer Helpers	26.90	430.40	41.00	656.00		
32 L.H., Daily Totals		$1010.40		$1539.60	$31.58	$48.11
Crew D-13	Hr.	Daily	Hr.	Daily	Bare Costs	Incl. O&P
1 Bricklayer Foreman	$37.25	$298.00	$56.75	$454.00	$32.74	$49.98
1 Bricklayer	35.25	282.00	53.70	429.60		
2 Bricklayer Helpers	26.90	430.40	41.00	656.00		
1 Carpenter	34.25	274.00	53.35	426.80		
1 Equip. Oper. (crane)	35.90	287.20	54.10	432.80		
1 Truck Crane, 12.5 Ton		505.80		556.40	10.54	11.59
48 L.H., Daily Totals		$2077.40		$2955.60	$43.28	$61.57
Crew E-1	Hr.	Daily	Hr.	Daily	Bare Costs	Incl. O&P
1 Welder Foreman	$40.15	$321.20	$71.85	$574.80	$37.12	$63.30
1 Welder	38.15	305.20	68.25	546.00		
1 Equip. Oper. (light)	33.05	264.40	49.80	398.40		
1 Gas Welding Machine		81.20		89.30	3.38	3.72
24 L.H., Daily Totals		$972.00		$1608.50	$40.50	$67.02
Crew E-2	Hr.	Daily	Hr.	Daily	Bare Costs	Incl. O&P
1 Struc. Steel Foreman	$40.15	$321.20	$71.85	$574.80	$36.96	$63.47
4 Struc. Steel Workers	38.15	1220.80	68.25	2184.00		
1 Equip. Oper. (crane)	35.90	287.20	54.10	432.80		
1 Equip. Oper. Oiler	30.10	240.80	45.35	362.80		
1 Crane, 90 Ton		1325.00		1457.50	23.66	26.03
56 L.H., Daily Totals		$3395.00		$5011.90	$60.62	$89.50
Crew E-3	Hr.	Daily	Hr.	Daily	Bare Costs	Incl. O&P
1 Struc. Steel Foreman	$40.15	$321.20	$71.85	$574.80	$38.82	$69.45
1 Struc. Steel Worker	38.15	305.20	68.25	546.00		
1 Welder	38.15	305.20	68.25	546.00		
1 Gas Welding Machine		81.20		89.30	3.38	3.72
24 L.H., Daily Totals		$1012.80		$1756.10	$42.20	$73.17
Crew E-4	Hr.	Daily	Hr.	Daily	Bare Costs	Incl. O&P
1 Struc. Steel Foreman	$40.15	$321.20	$71.85	$574.80	$38.65	$69.15
3 Struc. Steel Workers	38.15	915.60	68.25	1638.00		
1 Gas Welding Machine		81.20		89.30	2.54	2.79
32 L.H., Daily Totals		$1318.00		$2302.10	$41.19	$71.94
Crew E-5	Hr.	Daily	Hr.	Daily	Bare Costs	Incl. O&P
2 Struc. Steel Foremen	$40.15	$642.40	$71.85	$1149.60	$37.52	$65.27
5 Struc. Steel Workers	38.15	1526.00	68.25	2730.00		
1 Equip. Oper. (crane)	35.90	287.20	54.10	432.80		
1 Welder	38.15	305.20	68.25	546.00		
1 Equip. Oper. Oiler	30.10	240.80	45.35	362.80		
1 Crane, 90 Ton		1325.00		1457.50		
1 Gas Welding Machine		81.20		89.30	17.58	19.34
80 L.H., Daily Totals		$4407.80		$6768.00	$55.10	$84.61

Crew No.	Bare Costs Hr.	Daily	Incl. Subs O & P Hr.	Daily	Cost Per Labor-Hour Bare Costs	Incl. O&P
Crew E-6	Hr.	Daily	Hr.	Daily	Bare Costs	Incl. O&P
3 Struc. Steel Foremen	$40.15	$963.60	$71.85	$1724.40	$37.56	$65.46
9 Struc. Steel Workers	38.15	2746.80	68.25	4914.00		
1 Equip. Oper. (crane)	35.90	287.20	54.10	432.80		
1 Welder	38.15	305.20	68.25	546.00		
1 Equip. Oper. Oiler	30.10	240.80	45.35	362.80		
1 Equip. Oper. (light)	33.05	264.40	49.80	398.40		
1 Crane, 90 Ton		1325.00		1457.50		
1 Gas Welding Machine		81.20		89.30		
1 Air Compr., 160 C.F.M.		87.40		96.15		
2 Impact Wrenches		24.00		26.40	11.86	13.04
128 L.H., Daily Totals		$6325.60		$10047.75	$49.42	$78.50
Crew E-7	Hr.	Daily	Hr.	Daily	Bare Costs	Incl. O&P
1 Struc. Steel Foreman	$40.15	$321.20	$71.85	$574.80	$37.52	$65.27
4 Struc. Steel Workers	38.15	1220.80	68.25	2184.00		
1 Equip. Oper. (crane)	35.90	287.20	54.10	432.80		
1 Equip. Oper. Oiler	30.10	240.80	45.35	362.80		
1 Welder Foreman	40.15	321.20	71.85	574.80		
2 Welders	38.15	610.40	68.25	1092.00		
1 Crane, 90 Ton		1325.00		1457.50		
2 Gas Welding Machines		162.40		178.65	18.59	20.45
80 L.H., Daily Totals		$4489.00		$6857.35	$56.11	$85.72
Crew E-8	Hr.	Daily	Hr.	Daily	Bare Costs	Incl. O&P
1 Struc. Steel Foreman	$40.15	$321.20	$71.85	$574.80	$37.27	$64.53
4 Struc. Steel Workers	38.15	1220.80	68.25	2184.00		
1 Welder Foreman	40.15	321.20	71.85	574.80		
4 Welders	38.15	1220.80	68.25	2184.00		
1 Equip. Oper. (crane)	35.90	287.20	54.10	432.80		
1 Equip. Oper. Oiler	30.10	240.80	45.35	362.80		
1 Equip. Oper. (light)	33.05	264.40	49.80	398.40		
1 Crane, 90 Ton		1325.00		1457.50		
4 Gas Welding Machines		324.80		357.30	15.86	17.45
104 L.H., Daily Totals		$5526.20		$8526.40	$53.13	$81.98
Crew E-9	Hr.	Daily	Hr.	Daily	Bare Costs	Incl. O&P
2 Struc. Steel Foremen	$40.15	$642.40	$71.85	$1149.60	$37.56	$65.46
5 Struc. Steel Workers	38.15	1526.00	68.25	2730.00		
1 Welder Foreman	40.15	321.20	71.85	574.80		
5 Welders	38.15	1526.00	68.25	2730.00		
1 Equip. Oper. (crane)	35.90	287.20	54.10	432.80		
1 Equip. Oper. Oiler	30.10	240.80	45.35	362.80		
1 Equip. Oper. (light)	33.05	264.40	49.80	398.40		
1 Crane, 90 Ton		1325.00		1457.50		
5 Gas Welding Machines		406.00		446.60	13.52	14.88
128 L.H., Daily Totals		$6539.00		$10282.50	$51.08	$80.34
Crew E-10	Hr.	Daily	Hr.	Daily	Bare Costs	Incl. O&P
1 Welder Foreman	$40.15	$321.20	$71.85	$574.80	$39.15	$70.05
1 Welder	38.15	305.20	68.25	546.00		
1 Gas Welding Machines		81.20		89.30		
1 Truck, 3 Ton		176.20		193.80	16.09	17.70
16 L.H., Daily Totals		$883.80		$1403.90	$55.24	$87.75
Crew E-11	Hr.	Daily	Hr.	Daily	Bare Costs	Incl. O&P
2 Painters, Struc. Steel	$31.00	$496.00	$57.05	$912.80	$30.44	$51.36
1 Building Laborer	26.70	213.60	41.55	332.40		
1 Equip. Oper. (light)	33.05	264.40	49.80	398.40		
1 Air Compressor 250 C.F.M.		112.40		123.65		
1 Sand Blaster		15.20		16.70		
1 Sand Blasting Accessories		11.90		13.10	4.36	4.80
32 L.H., Daily Totals		$1113.50		$1797.05	$34.80	$56.16

Crew E-12

Crew E-12	Hr.	Daily	Hr.	Daily	Bare Costs	Incl. O&P
1 Welder Foreman	$40.15	$321.20	$71.85	$574.80	$36.60	$60.83
1 Equip. Oper. (light)	33.05	264.40	49.80	398.40		
1 Gas Welding Machine		81.20		89.30	5.08	5.58
16 L.H., Daily Totals		$666.80		$1062.50	$41.68	$66.41

Crew E-13

Crew E-13	Hr.	Daily	Hr.	Daily	Bare Costs	Incl. O&P
1 Welder Foreman	$40.15	$321.20	$71.85	$574.80	$37.78	$64.50
.5 Equip. Oper. (light)	33.05	132.20	49.80	199.20		
1 Gas Welding Machine		81.20		89.30	6.77	7.44
12 L.H., Daily Totals		$534.60		$863.30	$44.55	$71.94

Crew E-14

Crew E-14	Hr.	Daily	Hr.	Daily	Bare Costs	Incl. O&P
1 Welder Foreman	$40.15	$321.20	$71.85	$574.80	$40.15	$71.85
1 Gas Welding Machine		81.20		89.30	10.15	11.17
8 L.H., Daily Totals		$402.40		$664.10	$50.30	$83.02

Crew E-16

Crew E-16	Hr.	Daily	Hr.	Daily	Bare Costs	Incl. O&P
1 Welder Foreman	$40.15	$321.20	$71.85	$574.80	$39.15	$70.05
1 Welder	38.15	305.20	68.25	546.00		
1 Gas Welding Machine		81.20		89.30	5.08	5.58
16 L.H., Daily Totals		$707.60		$1210.10	$44.23	$75.63

Crew E-17

Crew E-17	Hr.	Daily	Hr.	Daily	Bare Costs	Incl. O&P
1 Structural Steel Foreman	$40.15	$321.20	$71.85	$574.80	$39.15	$70.05
1 Structural Steel Worker	38.15	305.20	68.25	546.00		
1 Power Tool		5.20		5.70	.33	.36
16 L.H., Daily Totals		$631.60		$1126.50	$39.48	$70.41

Crew E-18

Crew E-18	Hr.	Daily	Hr.	Daily	Bare Costs	Incl. O&P
1 Structural Steel Foreman	$40.15	$321.20	$71.85	$574.80	$37.85	$65.76
3 Structural Steel Workers	38.15	915.60	68.25	1638.00		
1 Equipment Operator (med.)	34.65	277.20	52.20	417.60		
1 Crane, 20 Ton		797.10		876.80	19.93	21.92
40 L.H., Daily Totals		$2311.10		$3507.20	$57.78	$87.68

Crew E-19

Crew E-19	Hr.	Daily	Hr.	Daily	Bare Costs	Incl. O&P
1 Structural Steel Worker	$38.15	$305.20	$68.25	$546.00	$37.12	$63.30
1 Structural Steel Foreman	40.15	321.20	71.85	574.80		
1 Equip. Oper. (light)	33.05	264.40	49.80	398.40		
1 Power Tool		5.20		5.70		
1 Crane, 20 Ton		797.10		876.80	33.43	36.77
24 L.H., Daily Totals		$1693.10		$2401.70	$70.55	$100.07

Crew E-20

Crew E-20	Hr.	Daily	Hr.	Daily	Bare Costs	Incl. O&P
1 Structural Steel Foreman	$40.15	$321.20	$71.85	$574.80	$37.11	$64.07
5 Structural Steel Workers	38.15	1526.00	68.25	2730.00		
1 Equip. Oper. (crane)	35.90	287.20	54.10	432.80		
1 Oiler	30.10	240.80	45.35	362.80		
1 Power Tool		5.20		5.70		
1 Crane, 40 Ton		951.90		1047.10	14.95	16.45
64 L.H., Daily Totals		$3332.30		$5153.20	$52.06	$80.52

Crew E-22

Crew E-22	Hr.	Daily	Hr.	Daily	Bare Costs	Incl. O&P
1 Skilled Worker Foreman	$36.85	$294.80	$57.30	$458.40	$35.52	$55.23
2 Skilled Workers	34.85	557.60	54.20	867.20		
24 L.H., Daily Totals		$852.40		$1325.60	$35.52	$55.23

Crew E-24

Crew E-24	Hr.	Daily	Hr.	Daily	Bare Costs	Incl. O&P
3 Structural Steel Workers	$38.15	$915.60	$68.25	$1638.00	$37.28	$64.24
1 Equipment Operator (medium)	34.65	277.20	52.20	417.60		
1-25 Ton Crane		616.80		678.50	19.28	21.20
32 L.H., Daily Totals		$1809.60		$2734.10	$56.56	$85.44

Crew E-25

Crew E-25	Hr.	Daily	Hr.	Daily	Bare Costs	Incl. O&P
1 Welder Foreman	$40.15	$321.20	$71.85	$574.80	$40.15	$71.85
1 Cutting Torch		18.00		19.80		
1 Gases		64.80		71.30	10.35	11.39
8 L.H., Daily Totals		$404.00		$665.90	$50.50	$83.24

Crew F-3

Crew F-3	Hr.	Daily	Hr.	Daily	Bare Costs	Incl. O&P
4 Carpenters	$34.25	$1096.00	$53.35	$1707.20	$34.58	$53.50
1 Equip. Oper. (crane)	35.90	287.20	54.10	432.80		
1 Hyd. Crane, 12 Ton		602.60		662.85	15.07	16.57
40 L.H., Daily Totals		$1985.80		$2802.85	$49.65	$70.07

Crew F-4

Crew F-4	Hr.	Daily	Hr.	Daily	Bare Costs	Incl. O&P
4 Carpenters	$34.25	$1096.00	$53.35	$1707.20	$33.83	$52.14
1 Equip. Oper. (crane)	35.90	287.20	54.10	432.80		
1 Equip. Oper. Oiler	30.10	240.80	45.35	362.80		
1 Hyd. Crane, 55 Ton		895.40		984.95	18.65	20.52
48 L.H., Daily Totals		$2519.40		$3487.75	$52.48	$72.66

Crew F-5

Crew F-5	Hr.	Daily	Hr.	Daily	Bare Costs	Incl. O&P
1 Carpenter Foreman	$36.25	$290.00	$56.45	$451.60	$34.75	$54.13
3 Carpenters	34.25	822.00	53.35	1280.40		
32 L.H., Daily Totals		$1112.00		$1732.00	$34.75	$54.13

Crew F-6

Crew F-6	Hr.	Daily	Hr.	Daily	Bare Costs	Incl. O&P
2 Carpenters	$34.25	$548.00	$53.35	$853.60	$31.56	$48.78
2 Building Laborers	26.70	427.20	41.55	664.80		
1 Equip. Oper. (crane)	35.90	287.20	54.10	432.80		
1 Hyd. Crane, 12 Ton		602.60		662.85	15.07	16.57
40 L.H., Daily Totals		$1865.00		$2614.05	$46.63	$65.35

Crew F-7

Crew F-7	Hr.	Daily	Hr.	Daily	Bare Costs	Incl. O&P
2 Carpenters	$34.25	$548.00	$53.35	$853.60	$30.48	$47.45
2 Building Laborers	26.70	427.20	41.55	664.80		
32 L.H., Daily Totals		$975.20		$1518.40	$30.48	$47.45

Crew G-1

Crew G-1	Hr.	Daily	Hr.	Daily	Bare Costs	Incl. O&P
1 Roofer Foreman	$31.45	$251.60	$53.60	$428.80	$27.49	$46.86
4 Roofers, Composition	29.45	942.40	50.20	1606.40		
2 Roofer Helpers	21.60	345.60	36.80	588.80		
1 Application Equipment		146.60		161.25		
1 Tar Kettle/Pot		52.50		57.75		
1 Crew Truck		98.80		108.70	5.32	5.85
56 L.H., Daily Totals		$1837.50		$2951.70	$32.81	$52.71

Crew G-2

Crew G-2	Hr.	Daily	Hr.	Daily	Bare Costs	Incl. O&P
1 Plasterer	$31.35	$250.80	$47.70	$381.60	$28.38	$43.50
1 Plasterer Helper	27.10	216.80	41.25	330.00		
1 Building Laborer	26.70	213.60	41.55	332.40		
1 Grouting Equipment		107.75		118.55	4.49	4.94
24 L.H., Daily Totals		$788.95		$1162.55	$32.87	$48.44

Crew No.	Bare Costs Hr.	Daily	Incl. Subs O & P Hr.	Daily	Cost Per Labor-Hour Bare Costs	Incl. O&P
Crew G-3	Hr.	Daily	Hr.	Daily	Bare Costs	Incl. O&P
2 Sheet Metal Workers	$40.20	$643.20	$61.85	$989.60	$33.45	$51.70
2 Building Laborers	26.70	427.20	41.55	664.80		
32 L.H., Daily Totals		$1070.40		$1654.40	$33.45	$51.70
Crew G-4	Hr.	Daily	Hr.	Daily	Bare Costs	Incl. O&P
1 Labor Foreman (outside)	$28.70	$229.60	$44.70	$357.60	$27.37	$42.60
2 Building Laborers	26.70	427.20	41.55	664.80		
1 Light Truck, 1.5 Ton		128.80		141.70		
1 Air Compr., 160 C.F.M.		87.40		96.15	9.01	9.91
24 L.H., Daily Totals		$873.00		$1260.25	$36.38	$52.51
Crew G-5	Hr.	Daily	Hr.	Daily	Bare Costs	Incl. O&P
1 Roofer Foreman	$31.45	$251.60	$53.60	$428.80	$26.71	$45.52
2 Roofers, Composition	29.45	471.20	50.20	803.20		
2 Roofer Helpers	21.60	345.60	36.80	588.80		
1 Application Equipment		146.60		161.25	3.67	4.03
40 L.H., Daily Totals		$1215.00		$1982.05	$30.38	$49.55
Crew G-6A	Hr.	Daily	Hr.	Daily	Bare Costs	Incl. O&P
2 Roofers Composition	$29.45	$471.20	$50.20	$803.20	$29.45	$50.20
1 Small Compressor		12.15		13.35		
2 Pneumatic Nailers		39.00		42.90	3.19	3.51
16 L.H., Daily Totals		$522.35		$859.45	$32.64	$53.71
Crew G-7	Hr.	Daily	Hr.	Daily	Bare Costs	Incl. O&P
1 Carpenter	$34.25	$274.00	$53.35	$426.80	$34.25	$53.35
1 Small Compressor		12.15		13.35		
1 Pneumatic Nailer		19.50		21.45	3.94	4.34
8 L.H., Daily Totals		$305.65		$461.60	$38.19	$57.69
Crew H-1	Hr.	Daily	Hr.	Daily	Bare Costs	Incl. O&P
2 Glaziers	$33.00	$528.00	$49.90	$798.40	$35.58	$59.08
2 Struc. Steel Workers	38.15	610.40	68.25	1092.00		
32 L.H., Daily Totals		$1138.40		$1890.40	$35.58	$59.08
Crew H-2	Hr.	Daily	Hr.	Daily	Bare Costs	Incl. O&P
2 Glaziers	$33.00	$528.00	$49.90	$798.40	$30.90	$47.12
1 Building Laborer	26.70	213.60	41.55	332.40		
24 L.H., Daily Totals		$741.60		$1130.80	$30.90	$47.12
Crew H-3	Hr.	Daily	Hr.	Daily	Bare Costs	Incl. O&P
1 Glazier	$33.00	$264.00	$49.90	$399.20	$29.28	$44.83
1 Helper	25.55	204.40	39.75	318.00		
16 L.H., Daily Totals		$468.40		$717.20	$29.28	$44.83
Crew J-1	Hr.	Daily	Hr.	Daily	Bare Costs	Incl. O&P
3 Plasterers	$31.35	$752.40	$47.70	$1144.80	$29.65	$45.12
2 Plasterer Helpers	27.10	433.60	41.25	660.00		
1 Mixing Machine, 6 C.F.		103.00		113.30	2.58	2.83
40 L.H., Daily Totals		$1289.00		$1918.10	$32.23	$47.95
Crew J-2	Hr.	Daily	Hr.	Daily	Bare Costs	Incl. O&P
3 Plasterers	$31.35	$752.40	$47.70	$1144.80	$29.95	$45.37
2 Plasterer Helpers	27.10	433.60	41.25	660.00		
1 Lather	31.45	251.60	46.60	372.80		
1 Mixing Machine, 6 C.F.		103.00		113.30	2.15	2.36
48 L.H., Daily Totals		$1540.60		$2290.90	$32.10	$47.73

Crew No.	Bare Costs Hr.	Daily	Incl. Subs O & P Hr.	Daily	Cost Per Labor-Hour Bare Costs	Incl. O&P
Crew J-3	Hr.	Daily	Hr.	Daily	Bare Costs	Incl. O&P
1 Terrazzo Worker	$32.65	$261.20	$48.00	$384.00	$29.38	$43.18
1 Terrazzo Helper	26.10	208.80	38.35	306.80		
1 Terrazzo Grinder, Electric		70.00		77.00		
1 Terrazzo Mixer		139.60		153.55	13.10	14.41
16 L.H., Daily Totals		$679.60		$921.35	$42.48	$57.59
Crew J-4	Hr.	Daily	Hr.	Daily	Bare Costs	Incl. O&P
1 Tile Layer	$32.70	$261.60	$48.05	$384.40	$29.03	$42.65
1 Tile Layer Helper	25.35	202.80	37.25	298.00		
16 L.H., Daily Totals		$464.40		$682.40	$29.03	$42.65
Crew K-1	Hr.	Daily	Hr.	Daily	Bare Costs	Incl. O&P
1 Carpenter	$34.25	$274.00	$53.35	$426.80	$30.53	$47.10
1 Truck Driver (light)	26.80	214.40	40.85	326.80		
1 Truck w/Power Equip.		176.20		193.80	11.01	12.11
16 L.H., Daily Totals		$664.60		$947.40	$41.54	$59.21
Crew K-2	Hr.	Daily	Hr.	Daily	Bare Costs	Incl. O&P
1 Struc. Steel Foreman	$40.15	$321.20	$71.85	$574.80	$35.03	$60.32
1 Struc. Steel Worker	38.15	305.20	68.25	546.00		
1 Truck Driver (light)	26.80	214.40	40.85	326.80		
1 Truck w/Power Equip.		176.20		193.80	7.34	8.08
24 L.H., Daily Totals		$1017.00		$1641.40	$42.37	$68.40
Crew L-1	Hr.	Daily	Hr.	Daily	Bare Costs	Incl. O&P
1 Electrician	$40.75	$326.00	$60.65	$485.20	$40.80	$61.03
1 Plumber	40.85	326.80	61.40	491.20		
16 L.H., Daily Totals		$652.80		$976.40	$40.80	$61.03
Crew L-2	Hr.	Daily	Hr.	Daily	Bare Costs	Incl. O&P
1 Carpenter	$34.25	$274.00	$53.35	$426.80	$29.90	$46.55
1 Carpenter Helper	25.55	204.40	39.75	318.00		
16 L.H., Daily Totals		$478.40		$744.80	$29.90	$46.55
Crew L-3	Hr.	Daily	Hr.	Daily	Bare Costs	Incl. O&P
1 Carpenter	$34.25	$274.00	$53.35	$426.80	$37.36	$57.30
.5 Electrician	40.75	163.00	60.65	242.60		
.5 Sheet Metal Worker	40.20	160.80	61.85	247.40		
16 L.H., Daily Totals		$597.80		$916.80	$37.36	$57.30
Crew L-3A	Hr.	Daily	Hr.	Daily	Bare Costs	Incl. O&P
1 Carpenter Foreman (outside)	$36.25	$290.00	$56.45	$451.60	$37.57	$58.25
.5 Sheet Metal Worker	40.20	160.80	61.85	247.40		
12 L.H., Daily Totals		$450.80		$699.00	$37.57	$58.25
Crew L-4	Hr.	Daily	Hr.	Daily	Bare Costs	Incl. O&P
2 Skilled Workers	$34.85	$557.60	$54.20	$867.20	$31.75	$49.38
1 Helper	25.55	204.40	39.75	318.00		
24 L.H., Daily Totals		$762.00		$1185.20	$31.75	$49.38
Crew L-5	Hr.	Daily	Hr.	Daily	Bare Costs	Incl. O&P
1 Struc. Steel Foreman	$40.15	$321.20	$71.85	$574.80	$38.11	$66.74
5 Struc. Steel Workers	38.15	1526.00	68.25	2730.00		
1 Equip. Oper. (crane)	35.90	287.20	54.10	432.80		
1 Hyd. Crane, 25 Ton		616.80		678.50	11.01	12.12
56 L.H., Daily Totals		$2751.20		$4416.10	$49.12	$78.86

Crew No.	Bare Costs		Incl. Subs O & P		Cost Per Labor-Hour	

Crew L-5A	Hr.	Daily	Hr.	Daily	Bare Costs	Incl. O&P
1 Structural Steel Foreman	$40.15	$321.20	$71.85	$574.80	$38.09	$65.61
2 Structural Steel Workers	38.15	610.40	68.25	1092.00		
1 Equip. Oper. (crane)	35.90	287.20	54.10	432.80		
1 Crane, SP, 25 Ton		575.00		632.50	17.97	19.77
32 L.H., Daily Totals		$1793.80		$2732.10	$56.06	$85.38

Crew L-6	Hr.	Daily	Hr.	Daily	Bare Costs	Incl. O&P
1 Plumber	$40.85	$326.80	$61.40	$491.20	$40.82	$61.15
.5 Electrician	40.75	163.00	60.65	242.60		
12 L.H., Daily Totals		$489.80		$733.80	$40.82	$61.15

Crew L-7	Hr.	Daily	Hr.	Daily	Bare Costs	Incl. O&P
2 Carpenters	$34.25	$548.00	$53.35	$853.60	$33.02	$51.02
1 Building Laborer	26.70	213.60	41.55	332.40		
.5 Electrician	40.75	163.00	60.65	242.60		
28 L.H., Daily Totals		$924.60		$1428.60	$33.02	$51.02

Crew L-8	Hr.	Daily	Hr.	Daily	Bare Costs	Incl. O&P
2 Carpenters	$34.25	$548.00	$53.35	$853.60	$35.57	$54.96
.5 Plumber	40.85	163.40	61.40	245.60		
20 L.H., Daily Totals		$711.40		$1099.20	$35.57	$54.96

Crew L-9	Hr.	Daily	Hr.	Daily	Bare Costs	Incl. O&P
1 Labor Foreman (inside)	$27.20	$217.60	$42.35	$338.80	$30.92	$49.78
2 Building Laborers	26.70	427.20	41.55	664.80		
1 Struc. Steel Worker	38.15	305.20	68.25	546.00		
.5 Electrician	40.75	163.00	60.65	242.60		
36 L.H., Daily Totals		$1113.00		$1792.20	$30.92	$49.78

Crew L-10	Hr.	Daily	Hr.	Daily	Bare Costs	Incl. O&P
1 Structural Steel Foreman	$40.15	$321.20	$71.85	$574.80	$38.07	$64.73
1 Structural Steel Worker	38.15	305.20	68.25	546.00		
1 Equip. Oper. (crane)	35.90	287.20	54.10	432.80		
1 Hyd. Crane, 12 Ton		602.60		662.85	25.11	27.62
24 L.H., Daily Totals		$1516.20		$2216.45	$63.18	$92.35

Crew L-11	Hr.	Daily	Hr.	Daily	Bare Costs	Incl. O&P
2 Wreckers	$26.70	$427.20	$47.20	$755.20	$30.59	$49.58
1 Equip. Oper. (crane)	35.90	287.20	54.10	432.80		
1 Equip. Oper. (light)	33.05	264.40	49.80	398.40		
1 Hyd. Excavator, 2.5 C.Y.		1215.00		1336.50		
1 Skid steer loader		197.80		217.60	44.15	48.57
32 L.H., Daily Totals		$2391.60		$3140.50	$74.74	$98.15

Crew M-1	Hr.	Daily	Hr.	Daily	Bare Costs	Incl. O&P
3 Elevator Constructors	$45.05	$1081.20	$67.55	$1621.20	$42.80	$64.18
1 Elevator Apprentice	36.05	288.40	54.05	432.40		
5 Hand Tools		63.00		69.30	1.97	2.17
32 L.H., Daily Totals		$1432.60		$2122.90	$44.77	$66.35

Crew M-3	Hr.	Daily	Hr.	Daily	Bare Costs	Incl. O&P
1 Electrician Foreman (out)	$42.75	$342.00	$63.60	$508.80	$37.46	$56.42
1 Common Laborer	26.70	213.60	41.55	332.40		
.25 Equipment Operator, Medium	34.65	69.30	52.20	104.40		
1 Elevator Constructor	45.05	360.40	67.55	540.40		
1 Elevator Apprentice	36.05	288.40	54.05	432.40		
.25 Crane, SP, 4 x 4, 20 ton		145.35		159.90	4.28	4.70
34 L.H., Daily Totals		$1419.05		$2078.30	$41.74	$61.12

Crew M-4	Hr.	Daily	Hr.	Daily	Bare Costs	Incl. O&P
1 Electrician Foreman (out)	$42.75	$342.00	$63.60	$508.80	$37.12	$55.91
1 Common Laborer	26.70	213.60	41.55	332.40		
.25 Equipment Operator, Crane	35.90	71.80	54.10	108.20		
.25 Equipment Operator, Oiler	30.10	60.20	45.35	90.70		
1 Elevator Constructor	45.05	360.40	67.55	540.40		
1 Elevator Apprentice	36.05	288.40	54.05	432.40		
.25 Crane, Hyd, SP, 4WD, 40 Ton		217.40		239.15	6.04	6.64
36 L.H., Daily Totals		$1553.80		$2252.05	$43.16	$62.55

Crew Q-1	Hr.	Daily	Hr.	Daily	Bare Costs	Incl. O&P
1 Plumber	$40.85	$326.80	$61.40	$491.20	$36.78	$55.28
1 Plumber Apprentice	32.70	261.60	49.15	393.20		
16 L.H., Daily Totals		$588.40		$884.40	$36.78	$55.28

Crew Q-1C	Hr.	Daily	Hr.	Daily	Bare Costs	Incl. O&P
1 Plumber	$40.85	$326.80	$61.40	$491.20	$36.07	$54.25
1 Plumber Apprentice	32.70	261.60	49.15	393.20		
1 Equip. Oper. (medium)	34.65	277.20	52.20	417.60		
1 Trencher, Chain		1420.00		1562.00	59.17	65.08
24 L.H., Daily Totals		$2285.60		$2864.00	$95.24	$119.33

Crew Q-2	Hr.	Daily	Hr.	Daily	Bare Costs	Incl. O&P
2 Plumbers	$40.85	$653.60	$61.40	$982.40	$38.13	$57.32
1 Plumber Apprentice	32.70	261.60	49.15	393.20		
24 L.H., Daily Totals		$915.20		$1375.60	$38.13	$57.32

Crew Q-3	Hr.	Daily	Hr.	Daily	Bare Costs	Incl. O&P
1 Plumber Foreman (inside)	$41.35	$330.80	$62.15	$497.20	$38.94	$58.53
2 Plumbers	40.85	653.60	61.40	982.40		
1 Plumber Apprentice	32.70	261.60	49.15	393.20		
32 L.H., Daily Totals		$1246.00		$1872.80	$38.94	$58.53

Crew Q-4	Hr.	Daily	Hr.	Daily	Bare Costs	Incl. O&P
1 Plumber Foreman (inside)	$41.35	$330.80	$62.15	$497.20	$38.94	$58.53
1 Plumber	40.85	326.80	61.40	491.20		
1 Welder (plumber)	40.85	326.80	61.40	491.20		
1 Plumber Apprentice	32.70	261.60	49.15	393.20		
1 Electric Welding Mach.		80.65		88.70	2.52	2.77
32 L.H., Daily Totals		$1326.65		$1961.50	$41.46	$61.30

Crew Q-5	Hr.	Daily	Hr.	Daily	Bare Costs	Incl. O&P
1 Steamfitter	$40.95	$327.60	$61.55	$492.40	$36.85	$55.38
1 Steamfitter Apprentice	32.75	262.00	49.20	393.60		
16 L.H., Daily Totals		$589.60		$886.00	$36.85	$55.38

Crew Q-6	Hr.	Daily	Hr.	Daily	Bare Costs	Incl. O&P
2 Steamfitters	$40.95	$655.20	$61.55	$984.80	$38.22	$57.43
1 Steamfitter Apprentice	32.75	262.00	49.20	393.60		
24 L.H., Daily Totals		$917.20		$1378.40	$38.22	$57.43

Crew Q-7	Hr.	Daily	Hr.	Daily	Bare Costs	Incl. O&P
1 Steamfitter Foreman (inside)	$41.45	$331.60	$62.30	$498.40	$39.03	$58.65
2 Steamfitters	40.95	655.20	61.55	984.80		
1 Steamfitter Apprentice	32.75	262.00	49.20	393.60		
32 L.H., Daily Totals		$1248.80		$1876.80	$39.03	$58.65

Crew No.	Bare Costs Hr.	Daily	Incl. Subs O & P Hr.	Daily	Cost Per Labor-Hour Bare Costs	Incl. O&P
Crew Q-8	**Hr.**	**Daily**	**Hr.**	**Daily**	**Bare Costs**	**Incl. O&P**
1 Steamfitter Foreman (inside)	$41.45	$331.60	$62.30	$498.40	$39.03	$58.65
1 Steamfitter	40.95	327.60	61.55	492.40		
1 Welder (steamfitter)	40.95	327.60	61.55	492.40		
1 Steamfitter Apprentice	32.75	262.00	49.20	393.60		
1 Electric Welding Mach.		80.65		88.70	2.52	2.77
32 L.H., Daily Totals		$1329.45		$1965.50	$41.55	$61.42
Crew Q-9	**Hr.**	**Daily**	**Hr.**	**Daily**	**Bare Costs**	**Incl. O&P**
1 Sheet Metal Worker	$40.20	$321.60	$61.85	$494.80	$36.18	$55.65
1 Sheet Metal Apprentice	32.15	257.20	49.45	395.60		
16 L.H., Daily Totals		$578.80		$890.40	$36.18	$55.65
Crew Q-10	**Hr.**	**Daily**	**Hr.**	**Daily**	**Bare Costs**	**Incl. O&P**
2 Sheet Metal Workers	$40.20	$643.20	$61.85	$989.60	$37.52	$57.72
1 Sheet Metal Apprentice	32.15	257.20	49.45	395.60		
24 L.H., Daily Totals		$900.40		$1385.20	$37.52	$57.72
Crew Q-11	**Hr.**	**Daily**	**Hr.**	**Daily**	**Bare Costs**	**Incl. O&P**
1 Sheet Metal Foreman (inside)	$40.70	$325.60	$62.60	$500.80	$38.31	$58.94
2 Sheet Metal Workers	40.20	643.20	61.85	989.60		
1 Sheet Metal Apprentice	32.15	257.20	49.45	395.60		
32 L.H., Daily Totals		$1226.00		$1886.00	$38.31	$58.94
Crew Q-12	**Hr.**	**Daily**	**Hr.**	**Daily**	**Bare Costs**	**Incl. O&P**
1 Sprinkler Installer	$40.75	$326.00	$61.55	$492.40	$36.68	$55.40
1 Sprinkler Apprentice	32.60	260.80	49.25	394.00		
16 L.H., Daily Totals		$586.80		$886.40	$36.68	$55.40
Crew Q-13	**Hr.**	**Daily**	**Hr.**	**Daily**	**Bare Costs**	**Incl. O&P**
1 Sprinkler Foreman (inside)	$41.25	$330.00	$62.35	$498.80	$38.84	$58.68
2 Sprinkler Installers	40.75	652.00	61.55	984.80		
1 Sprinkler Apprentice	32.60	260.80	49.25	394.00		
32 L.H., Daily Totals		$1242.80		$1877.60	$38.84	$58.68
Crew Q-14	**Hr.**	**Daily**	**Hr.**	**Daily**	**Bare Costs**	**Incl. O&P**
1 Asbestos Worker	$37.10	$296.80	$58.55	$468.40	$33.40	$52.70
1 Asbestos Apprentice	29.70	237.60	46.85	374.80		
16 L.H., Daily Totals		$534.40		$843.20	$33.40	$52.70
Crew Q-15	**Hr.**	**Daily**	**Hr.**	**Daily**	**Bare Costs**	**Incl. O&P**
1 Plumber	$40.85	$326.80	$61.40	$491.20	$36.78	$55.28
1 Plumber Apprentice	32.70	261.60	49.15	393.20		
1 Electric Welding Mach.		80.65		88.70	5.04	5.54
16 L.H., Daily Totals		$669.05		$973.10	$41.82	$60.82
Crew Q-16	**Hr.**	**Daily**	**Hr.**	**Daily**	**Bare Costs**	**Incl. O&P**
2 Plumbers	$40.85	$653.60	$61.40	$982.40	$38.13	$57.32
1 Plumber Apprentice	32.70	261.60	49.15	393.20		
1 Electric Welding Mach.		80.65		88.70	3.36	3.70
24 L.H., Daily Totals		$995.85		$1464.30	$41.49	$61.02
Crew Q-17	**Hr.**	**Daily**	**Hr.**	**Daily**	**Bare Costs**	**Incl. O&P**
1 Steamfitter	$40.95	$327.60	$61.55	$492.40	$36.85	$55.38
1 Steamfitter Apprentice	32.75	262.00	49.20	393.60		
1 Electric Welding Mach.		80.65		88.70	5.04	5.54
16 L.H., Daily Totals		$670.25		$974.70	$41.89	$60.92

Crew No.	Bare Costs Hr.	Daily	Incl. Subs O & P Hr.	Daily	Cost Per Labor-Hour Bare Costs	Incl. O&P
Crew Q-17A	**Hr.**	**Daily**	**Hr.**	**Daily**	**Bare Costs**	**Incl. O&P**
1 Steamfitter	$40.95	$327.60	$61.55	$492.40	$36.53	$54.95
1 Steamfitter Apprentice	32.75	262.00	49.20	393.60		
1 Equip. Oper. (crane)	35.90	287.20	54.10	432.80		
1 Truck Crane, 12 Ton		602.60		662.85		
1 Electric Welding Mach.		80.65		88.70	28.47	31.32
24 L.H., Daily Totals		$1560.05		$2070.35	$65.00	$86.27
Crew Q-18	**Hr.**	**Daily**	**Hr.**	**Daily**	**Bare Costs**	**Incl. O&P**
2 Steamfitters	$40.95	$655.20	$61.55	$984.80	$38.22	$57.43
1 Steamfitter Apprentice	32.75	262.00	49.20	393.60		
1 Electric Welding Mach.		80.65		88.70	3.36	3.70
24 L.H., Daily Totals		$997.85		$1467.10	$41.58	$61.13
Crew Q-19	**Hr.**	**Daily**	**Hr.**	**Daily**	**Bare Costs**	**Incl. O&P**
1 Steamfitter	$40.95	$327.60	$61.55	$492.40	$38.15	$57.13
1 Steamfitter Apprentice	32.75	262.00	49.20	393.60		
1 Electrician	40.75	326.00	60.65	485.20		
24 L.H., Daily Totals		$915.60		$1371.20	$38.15	$57.13
Crew Q-20	**Hr.**	**Daily**	**Hr.**	**Daily**	**Bare Costs**	**Incl. O&P**
1 Sheet Metal Worker	$40.20	$321.60	$61.85	$494.80	$37.09	$56.65
1 Sheet Metal Apprentice	32.15	257.20	49.45	395.60		
.5 Electrician	40.75	163.00	60.65	242.60		
20 L.H., Daily Totals		$741.80		$1133.00	$37.09	$56.65
Crew Q-21	**Hr.**	**Daily**	**Hr.**	**Daily**	**Bare Costs**	**Incl. O&P**
2 Steamfitters	$40.95	$655.20	$61.55	$984.80	$38.85	$58.24
1 Steamfitter Apprentice	32.75	262.00	49.20	393.60		
1 Electrician	40.75	326.00	60.65	485.20		
32 L.H., Daily Totals		$1243.20		$1863.60	$38.85	$58.24
Crew Q-22	**Hr.**	**Daily**	**Hr.**	**Daily**	**Bare Costs**	**Incl. O&P**
1 Plumber	$40.85	$326.80	$61.40	$491.20	$36.78	$55.28
1 Plumber Apprentice	32.70	261.60	49.15	393.20		
1 Truck Crane, 12 Ton		602.60		662.85	37.66	41.43
16 L.H., Daily Totals		$1191.00		$1547.25	$74.44	$96.71
Crew Q-22A	**Hr.**	**Daily**	**Hr.**	**Daily**	**Bare Costs**	**Incl. O&P**
1 Plumber	$40.85	$326.80	$61.40	$491.20	$34.04	$51.55
1 Plumber Apprentice	32.70	261.60	49.15	393.20		
1 Laborer	26.70	213.60	41.55	332.40		
1 Equip. Oper. (crane)	35.90	287.20	54.10	432.80		
1 Truck Crane, 12 Ton		602.60		662.85	18.83	20.71
32 L.H., Daily Totals		$1691.80		$2312.45	$52.87	$72.26
Crew Q-23	**Hr.**	**Daily**	**Hr.**	**Daily**	**Bare Costs**	**Incl. O&P**
1 Plumber Foreman	$42.85	$342.80	$64.40	$515.20	$39.45	$59.33
1 Plumber	40.85	326.80	61.40	491.20		
1 Equip. Oper. (medium)	34.65	277.20	52.20	417.60		
1 Power Tools		5.20		5.70		
1 Crane, 20 Ton		797.10		876.80	33.43	36.77
24 L.H., Daily Totals		$1749.10		$2306.50	$72.88	$96.10
Crew R-1	**Hr.**	**Daily**	**Hr.**	**Daily**	**Bare Costs**	**Incl. O&P**
1 Electrician Foreman	$41.25	$330.00	$61.40	$491.20	$35.77	$53.81
3 Electricians	40.75	978.00	60.65	1455.60		
2 Helpers	25.55	408.80	39.75	636.00		
48 L.H., Daily Totals		$1716.80		$2582.80	$35.77	$53.81

Crew No.	Bare Costs		Incl. Subs O & P		Cost Per Labor-Hour	

Crew R-1A	Hr.	Daily	Hr.	Daily	Bare Costs	Incl. O&P
1 Electrician	$40.75	$326.00	$60.65	$485.20	$33.15	$50.20
1 Helper	25.55	204.40	39.75	318.00		
16 L.H., Daily Totals		$530.40		$803.20	$33.15	$50.20

Crew R-2	Hr.	Daily	Hr.	Daily	Bare Costs	Incl. O&P
1 Electrician Foreman	$41.25	$330.00	$61.40	$491.20	$35.79	$53.85
3 Electricians	40.75	978.00	60.65	1455.60		
2 Helpers	25.55	408.80	39.75	636.00		
1 Equip. Oper. (crane)	35.90	287.20	54.10	432.80		
1 S.P. Crane, 5 Ton		317.00		348.70	5.66	6.23
56 L.H., Daily Totals		$2321.00		$3364.30	$41.45	$60.08

Crew R-3	Hr.	Daily	Hr.	Daily	Bare Costs	Incl. O&P
1 Electrician Foreman	$41.25	$330.00	$61.40	$491.20	$39.98	$59.64
1 Electrician	40.75	326.00	60.65	485.20		
.5 Equip. Oper. (crane)	35.90	143.60	54.10	216.40		
.5 S.P. Crane, 5 Ton		158.50		174.35	7.93	8.72
20 L.H., Daily Totals		$958.10		$1367.15	$47.91	$68.36

Crew R-4	Hr.	Daily	Hr.	Daily	Bare Costs	Incl. O&P
1 Struc. Steel Foreman	$40.15	$321.20	$71.85	$574.80	$39.07	$67.45
3 Struc. Steel Workers	38.15	915.60	68.25	1638.00		
1 Electrician	40.75	326.00	60.65	485.20		
1 Gas Welding Machine		81.20		89.30	2.03	2.23
40 L.H., Daily Totals		$1644.00		$2787.30	$41.10	$69.68

Crew R-5	Hr.	Daily	Hr.	Daily	Bare Costs	Incl. O&P
1 Electrician Foreman	$41.25	$330.00	$61.40	$491.20	$35.27	$53.12
4 Electrician Linemen	40.75	1304.00	60.65	1940.80		
2 Electrician Operators	40.75	652.00	60.65	970.40		
4 Electrician Groundmen	25.55	817.60	39.75	1272.00		
1 Crew Truck		98.80		108.70		
1 Tool Van		122.15		134.35		
1 Pickup Truck, 3/4 Ton		78.00		85.80		
.2 Crane, 55 Ton		179.08		197.00		
.2 Crane, 12 Ton		120.52		132.55		
.2 Auger, Truck Mtd.		635.80		699.40		
1 Tractor w/Winch		260.00		286.00	16.98	18.68
88 L.H., Daily Totals		$4597.95		$6318.20	$52.25	$71.80

Crew R-6	Hr.	Daily	Hr.	Daily	Bare Costs	Incl. O&P
1 Electrician Foreman	$41.25	$330.00	$61.40	$491.20	$35.27	$53.12
4 Electrician Linemen	40.75	1304.00	60.65	1940.80		
2 Electrician Operators	40.75	652.00	60.65	970.40		
4 Electrician Groundmen	25.55	817.60	39.75	1272.00		
1 Crew Truck		98.80		108.70		
1 Tool Van		122.15		134.35		
1 Pickup Truck, 3/4 Ton		78.00		85.80		
.2 Crane, 55 Ton		179.08		197.00		
.2 Crane, 12 Ton		120.52		132.55		
.2 Auger, Truck Mtd.		635.80		699.40		
1 Tractor w/Winch		260.00		286.00		
3 Cable Trailers		490.65		539.70		
.5 Tensioning Rig		160.00		176.00		
.5 Cable Pulling Rig		934.00		1027.40	34.99	38.49
88 L.H., Daily Totals		$6182.60		$8061.30	$70.26	$91.61

Crew R-7	Hr.	Daily	Hr.	Daily	Bare Costs	Incl. O&P
1 Electrician Foreman	$41.25	$330.00	$61.40	$491.20	$28.17	$43.36
5 Electrician Groundmen	25.55	1022.00	39.75	1590.00		
1 Crew Truck		98.80		108.70	2.06	2.26
48 L.H., Daily Totals		$1450.80		$2189.90	$30.23	$45.62

Crew R-8	Hr.	Daily	Hr.	Daily	Bare Costs	Incl. O&P
1 Electrician Foreman	$41.25	$330.00	$61.40	$491.20	$35.77	$53.81
3 Electrician Linemen	40.75	978.00	60.65	1455.60		
2 Electrician Groundmen	25.55	408.80	39.75	636.00		
1 Pickup Truck, 3/4 Ton		78.00		85.80		
1 Crew Truck		98.80		108.70	3.68	4.05
48 L.H., Daily Totals		$1893.60		$2777.30	$39.45	$57.86

Crew R-9	Hr.	Daily	Hr.	Daily	Bare Costs	Incl. O&P
1 Electrician Foreman	$41.25	$330.00	$61.40	$491.20	$33.21	$50.29
1 Electrician Lineman	40.75	326.00	60.65	485.20		
2 Electrician Operators	40.75	652.00	60.65	970.40		
4 Electrician Groundmen	25.55	817.60	39.75	1272.00		
1 Pickup Truck, 3/4 Ton		78.00		85.80		
1 Crew Truck		98.80		108.70	2.76	3.04
64 L.H., Daily Totals		$2302.40		$3413.30	$35.97	$53.33

Crew R-10	Hr.	Daily	Hr.	Daily	Bare Costs	Incl. O&P
1 Electrician Foreman	$41.25	$330.00	$61.40	$491.20	$38.30	$57.29
4 Electrician Linemen	40.75	1304.00	60.65	1940.80		
1 Electrician Groundman	25.55	204.40	39.75	318.00		
1 Crew Truck		98.80		108.70		
3 Tram Cars		345.75		380.35	9.26	10.19
48 L.H., Daily Totals		$2282.95		$3239.05	$47.56	$67.48

Crew R-11	Hr.	Daily	Hr.	Daily	Bare Costs	Incl. O&P
1 Electrician Foreman	$41.25	$330.00	$61.40	$491.20	$38.12	$57.09
4 Electricians	40.75	1304.00	60.65	1940.80		
1 Equip. Oper. (crane)	35.90	287.20	54.10	432.80		
1 Common Laborer	26.70	213.60	41.55	332.40		
1 Crew Truck		98.80		108.70		
1 Crane, 12 Ton		602.60		662.85	12.53	13.78
56 L.H., Daily Totals		$2836.20		$3968.75	$50.65	$70.87

Crew R-12	Hr.	Daily	Hr.	Daily	Bare Costs	Incl. O&P
1 Carpenter Foreman	$34.75	$278.00	$54.10	$432.80	$31.94	$50.38
4 Carpenters	34.25	1096.00	53.35	1707.20		
4 Common Laborers	26.70	854.40	41.55	1329.60		
1 Equip. Oper. (med.)	34.65	277.20	52.20	417.60		
1 Steel Worker	38.15	305.20	68.25	546.00		
1 Dozer, 200 H.P.		919.60		1011.55		
1 Pickup Truck, 3/4 Ton		78.00		85.80	11.34	12.47
88 L.H., Daily Totals		$3808.40		$5530.55	$43.28	$62.85

Crew R-13	Hr.	Daily	Hr.	Daily	Bare Costs	Incl. O&P
1 Electrician Foreman	$41.25	$330.00	$61.40	$491.20	$38.59	$57.57
3 Electricians	40.75	978.00	60.65	1455.60		
.25 Equip. Oper. (crane)	35.90	71.80	54.10	108.20		
1 Equipment Oiler	30.10	240.80	45.35	362.80		
.25-1 Hyd. Crane, 33 Ton		160.40		176.45	3.82	4.20
42 L.H., Daily Totals		$1781.00		$2594.25	$42.41	$61.77

Crew No.	Bare Costs		Incl. Sub O & P		Cost Per Labor-Hour	
Crew R-15	Hr.	Daily	Hr.	Daily	Bare Costs	Incl. O&P
1 Electrician Foreman	$41.25	$330.00	$61.40	$491.20	$39.55	$58.97
4 Electricians	40.75	1304.00	60.65	1940.80		
1 Equipment Operator	33.05	264.40	49.80	398.40		
1 Aerial Lift Truck		252.60		277.85	5.26	5.79
48 L.H., Daily Totals		$2151.00		$3108.25	$44.81	$64.76
Crew R-18	Hr.	Daily	Hr.	Daily	Bare Costs	Incl. O&P
.25 Electrician Foreman	$41.25	$82.50	$61.40	$122.80	$31.43	$47.85
1 Electrician	40.75	326.00	60.65	485.20		
2 Helpers	25.55	408.80	39.75	636.00		
26 L.H., Daily Totals		$817.30		$1244.00	$31.43	$47.85
Crew R-19	Hr.	Daily	Hr.	Daily	Bare Costs	Incl. O&P
.5 Electrician Foreman	$41.25	$165.00	$61.40	$245.60	$40.85	$60.80
2 Electricians	40.75	652.00	60.65	970.40		
20 L.H., Daily Totals		$817.00		$1216.00	$40.85	$60.80
Crew R-21	Hr.	Daily	Hr.	Daily	Bare Costs	Incl. O&P
1 Electrician Foreman	$41.25	$330.00	$61.40	$491.20	$40.72	$60.63
3 Electricians	40.75	978.00	60.65	1455.60		
.1 Equip. Oper. (med.)	34.65	27.72	52.20	41.76		
.1 Hyd. Crane 25 Ton		57.50		63.25	1.75	1.93
32. L.H., Daily Totals		$1393.22		$2051.81	$42.47	$62.56
Crew R-22	Hr.	Daily	Hr.	Daily	Bare Costs	Incl. O&P
.66 Electrician Foreman	$41.25	$217.80	$61.40	$324.19	$34.30	$51.79
2 Helpers	25.55	408.80	39.75	636.00		
2 Electricians	40.75	652.00	60.65	970.40		
37.28 L.H., Daily Totals		$1278.60		$1930.59	$34.30	$51.79
Crew R-30	Hr.	Daily	Hr.	Daily	Bare Costs	Incl. O&P
.25 Electrician	$42.75	$85.50	$63.60	$127.20	$32.26	$49.12
1 Electrician	40.75	326.00	60.65	485.20		
2 Laborers, (Semi-Skilled)	26.70	427.20	41.55	664.80		
26 L.H., Daily Totals		$838.70		$1277.20	$32.26	$49.12
Crew R-31	Hr.	Daily	Hr.	Daily	Bare Costs	Incl. O&P
1 Electrician	$40.75	$326.00	$60.65	$485.20	$40.75	$60.65
1 Core Drill, Elec, 2.5 HP		50.30		55.35	6.28	6.91
8 L.H., Daily Totals		$376.30		$540.55	$47.03	$67.56
Crew W-41E	Hr.	Daily	Hr.	Daily	Bare Costs	Incl. O&P
1 Laborers, (Semi-Skilled)	$26.70	$213.60	$41.55	$332.40	$35.59	$54.06
1 Plumber	40.85	326.80	61.40	491.20		
.5 Plumber	42.85	171.40	64.40	257.60		
20 L.H., Daily Totals		$711.80		$1081.20	$35.59	$54.06

Historical Cost Indexes

The table below lists both the Means Historical Cost Index based on Jan. 1, 1993 = 100 as well as the computed value of an index based on Jan. 1, 2005 costs. Since the Jan. 1, 2005 figure is estimated, space is left to write in the actual index figures as they become available through either the quarterly "Means Construction Cost Indexes" or as printed in the "Engineering News-Record." To compute the actual index based on Jan. 1, 2005 = 100, divide the Historical Cost Index for a particular year by the actual Jan. 1, 2005 Construction Cost Index. Space has been left to advance the index figures as the year progresses.

Year	Historical Cost Index Jan. 1, 1993 = 100		Current Index Based on Jan. 1, 2005 = 100		Year	Historical Cost Index Jan. 1, 1993 = 100	Current Index Based on Jan. 1, 2005 = 100		Year	Historical Cost Index Jan. 1, 1993 = 100	Current Index Based on Jan. 1, 2005 = 100	
	Est.	Actual	Est.	Actual		Actual	Est.	Actual		Actual	Est.	Actual
Oct 2005					July 1990	94.3	63.5		July 1972	34.8	23.4	
July 2005					1989	92.1	62.0		1971	32.1	21.6	
April 2005					1988	89.9	60.5		1970	28.7	19.3	
Jan 2005	148.5		100.0	100.0	1987	87.7	59.0		1969	26.9	18.1	
July 2004		143.7	96.8		1986	84.2	56.7		1968	24.9	16.8	
2003		132.0	88.9		1985	82.6	55.6		1967	23.5	15.8	
2002		128.7	86.7		1984	82.0	55.2		1966	22.7	15.3	
2001		125.1	84.2		1983	80.2	54.0		1965	21.7	14.6	
2000		120.9	81.4		1982	76.1	51.3		1964	21.2	14.3	
1999		117.6	79.2		1981	70.0	47.1		1963	20.7	13.9	
1998		115.1	77.5		1980	62.9	42.4		1962	20.2	13.6	
1997		112.8	76.0		1979	57.8	38.9		1961	19.8	13.3	
1996		110.2	74.2		1978	53.5	36.0		1960	19.7	13.3	
1995		107.6	72.5		1977	49.5	33.3		1959	19.3	13.0	
1994		104.4	70.3		1976	46.9	31.6		1958	18.8	12.7	
1993		101.7	68.5		1975	44.8	30.2		1957	18.4	12.4	
1992		99.4	67.0		1974	41.4	27.9		1956	17.6	11.9	
1991		96.8	65.2		1973	37.7	25.4		1955	16.6	11.2	

Adjustments to Costs

The Historical Cost Index can be used to convert National Average building costs at a particular time to the approximate building costs for some other time.

Example:

Estimate and compare construction costs for different years in the same city.

To estimate the National Average construction cost of a building in 1970, knowing that it cost $900,000 in 2005:

INDEX in 1970 = 28.7

INDEX in 2005 = 148.5

Note: The City Cost Indexes for Canada can be used to convert U.S. National averages to local costs in Canadian dollars.

Time Adjustment using the Historical Cost Indexes:

$$\frac{\text{Index for Year A}}{\text{Index for Year B}} \times \text{Cost in Year B} = \text{Cost in Year A}$$

$$\frac{\text{INDEX 1970}}{\text{INDEX 2005}} \times \text{Cost 2005} = \text{Cost 1970}$$

$$\frac{28.7}{148.5} \times \$900,000 = .193 \times \$900,000 = \$173,700$$

The construction cost of the building in 1970 is $173,700.

How to Use the City Cost Indexes

What you should know before you begin

Means City Cost Indexes (CCI) are an extremely useful tool to use when you want to compare costs from city to city and region to region.

This publication contains average construction cost indexes for 316 major U.S. and Canadian cities and Location Factors covering over 930 three-digit zip code locations.

Keep in mind that a City Cost Index number is a *percentage ratio* of a specific city's cost to the national average cost of the same item at a stated time period.

In other words, these index figures represent relative construction *factors* (or, if you prefer, multipliers) for Material and Installation costs, as well as the weighted average for Total In Place costs for each CSI MasterFormat division. Installation costs include both labor and equipment rental costs. When estimating equipment rental rates only, for a specific location, use 01590 EQUIPMENT RENTAL index.

The 30 City Average Index is the average of 30 major U.S. cities and serves as a National Average.

Index figures for both material and installation are based on the 30 major city average of 100 and represent the cost relationship as of July 1, 2003. The index for each division is computed from representative material and labor quantities for that division. The weighted average for each city is a weighted total of the components listed above it, but does not include relative productivity between trades or cities.

As changes occur in local material prices, labor rates and equipment rental rates, the impact of these changes should be accurately measured by the change in the City Cost Index for each particular city (as compared to the 30 City Average).

Therefore, if you know (or have estimated) building costs in one city today, you can easily convert those costs to expected building costs in another city.

In addition, by using the Historical Cost Index, you can easily convert National Average building costs at a particular time to the approximate building costs for some other time. The City Cost Indexes can then be applied to calculate the costs for a particular city.

Quick Calculations

Location Adjustment Using the City Cost Indexes:

$$\frac{\text{Index for City A}}{\text{Index for City B}} \times \text{Cost in City B} = \text{Cost in City A}$$

Time Adjustment for the National Average Using the Historical Cost Index:

$$\frac{\text{Index for Year A}}{\text{Index for Year B}} \times \text{Cost in Year B} = \text{Cost in Year A}$$

Adjustment from the National Average:

$$\frac{\text{Index for City A}}{100} \times \text{National Average Cost} = \text{Cost in City A}$$

Since each of the other RSMeans publications contains many different items, any *one* item multiplied by the particular city index may give incorrect results. However, the larger the number of items compiled, the closer the results should be to actual costs for that particular city.

The City Cost Indexes for Canadian cities are calculated using Canadian material and equipment prices and labor rates, in Canadian dollars. Therefore, indexes for Canadian cities can be used to convert U.S. National Average prices to local costs in Canadian dollars.

How to use this section

1. Compare costs from city to city.

In using the Means Indexes, remember that an index number is not a fixed number but a *ratio:* It's a percentage ratio of a building component's cost at any stated time to the National Average cost of that same component at the same time period. Put in the form of an equation:

$$\frac{\text{Specific City Cost}}{\text{National Average Cost}} \times 100 = \text{City Index Number}$$

Therefore, when making cost comparisons between cities, do not subtract one city's index number from the index number of another city and read the result as a percentage difference. Instead, divide one city's index number by that of the other city. The resulting number may then be used as a multiplier to calculate cost differences from city to city.

The formula used to find cost differences between cities for the purpose of comparison is as follows:

$$\frac{\text{City A Index}}{\text{City B Index}} \times \text{City B Cost (Known)} = \text{City A Cost (Unknown)}$$

In addition, you can use *Means CCI* to calculate and compare costs division by division between cities using the same basic formula. (Just be sure that you're comparing similar divisions.)

2. Compare a specific city's construction costs with the National Average.

When you're studying construction location feasibility, it's advisable to compare a prospective project's cost index with an index of the National Average cost.

For example, divide the weighted average index of construction costs of a specific city by that of the 30 City Average, which = 100.

$$\frac{\text{City Index}}{100} = \text{\% of National Average}$$

As a result, you get a ratio that indicates the relative cost of construction in that city in comparison with the National Average.

3. Convert U.S. National Average to actual costs in Canadian City.

$$\frac{\text{Index for Canadian City}}{100} \times \text{National Average Cost} = \text{Cost in Canadian City in \$ CAN}$$

4. Adjust construction cost data based on a National Average.

When you use a source of construction cost data which is based on a National Average (such as *Means cost data publications*), it is necessary to adjust those costs to a specific location.

$$\frac{\text{City Index}}{100} \quad \text{x} \quad \begin{array}{c}\text{``Book'' Cost Based on} \\ \text{National Average Costs}\end{array} = \begin{array}{c}\text{City Cost} \\ \text{(Unknown)}\end{array}$$

5. When applying the City Cost Indexes to demolition projects, use the appropriate division index. For example, for removal of existing doors and windows, use the Division 8 index.

What you might like to know about how we developed the Indexes

To create a reliable index, RSMeans researched the building type most often constructed in the United States and Canada. Because it was concluded that no one type of building completely represented the building construction industry, nine different types of buildings were combined to create a composite model.

The exact material, labor and equipment quantities are based on detailed analysis of these nine building types, then each quantity is weighted in proportion to expected usage. These various material items, labor hours, and equipment rental rates are thus combined to form a composite building representing as closely as possible the actual usage of materials, labor and equipment used in the North American Building Construction Industry.

The following structures were chosen to make up that composite model:

1. Factory, 1 story
2. Office, 2–4 story
3. Store, Retail
4. Town Hall, 2–3 story
5. High School, 2–3 story
6. Hospital, 4–8 story
7. Garage, Parking
8. Apartment, 1–3 story
9. Hotel/Motel, 2–3 story

For the purposes of ensuring the timeliness of the data, the components of the index for the composite model have been streamlined. They currently consist of:

- specific quantities of 66 commonly used construction materials;
- specific labor-hours for 21 building construction trades; and
- specific days of equipment rental for 6 types of construction equipment (normally used to install the 66 material items by the 21 trades.)

A sophisticated computer program handles the updating of all costs for each city on a quarterly basis. Material and equipment price quotations are gathered quarterly from over 316 cities in the United States and Canada. These prices and the latest negotiated labor wage rates for 21 different building trades are used to compile the quarterly update of the City Cost Index.

The 30 major U.S. cities used to calculate the National Average are:

Atlanta, GA	Memphis, TN
Baltimore, MD	Milwaukee, WI
Boston, MA	Minneapolis, MN
Buffalo, NY	Nashville, TN
Chicago, IL	New Orleans, LA
Cincinnati, OH	New York, NY
Cleveland, OH	Philadelphia, PA
Columbus, OH	Phoenix, AZ
Dallas, TX	Pittsburgh, PA
Denver, CO	St. Louis, MO
Detroit, MI	San Antonio, TX
Houston, TX	San Diego, CA
Indianapolis, IN	San Francisco, CA
Kansas City, MO	Seattle, WA
Los Angeles, CA	Washington, DC

F.Y.I.: The CSI MasterFormat Divisions

1. General Requirements
2. Site Construction
3. Concrete
4. Masonry
5. Metals
6. Wood & Plastics
7. Thermal & Moisture Protection
8. Doors & Windows
9. Finishes
10. Specialties
11. Equipment
12. Furnishings
13. Special Construction
14. Conveying Systems
15. Mechanical
16. Electrical

The information presented in the CCI is organized according to the Construction Specifications Institute (CSI) MasterFormat.

What the CCI does not indicate

The weighted average for each city is a total of the components listed above weighted to reflect typical usage, but it does *not* include the productivity variations between trades or cities.

In addition, the CCI does not take into consideration factors such as the following:

- managerial efficiency
- competitive conditions
- automation
- restrictive union practices
- unique local requirements
- regional variations due to specific building codes

DIVISION	UNITED STATES 30 CITY AVERAGE			BIRMINGHAM			ALABAMA HUNTSVILLE			MOBILE			MONTGOMERY			TUSCALOOSA		
	MAT.	INST.	TOTAL	MAT.	INST.	TOTAL	MAT.	INST.	TOTAL	MAT.	INST.	TOTAL	MAT.	INST.	TOTAL	MAT.	INST.	TOTAL
01590 EQUIPMENT RENTAL	.0	100.0	100.0	.0	101.3	101.3	.0	101.2	101.2	.0	97.7	97.7	.0	97.7	97.7	.0	101.2	101.2
02 SITE CONSTRUCTION	100.0	100.0	100.0	84.6	93.1	90.9	82.7	92.9	90.2	93.7	86.1	88.0	94.1	86.6	88.6	83.2	91.9	89.6
03100 CONCRETE FORMS & ACCESSORIES	100.0	100.0	100.0	95.0	75.6	78.2	97.1	68.6	72.4	97.1	53.0	58.9	95.9	48.1	54.5	97.0	39.7	47.3
03200 CONCRETE REINFORCEMENT	100.0	100.0	100.0	92.1	88.2	90.1	92.1	77.4	84.7	95.0	53.9	74.2	95.0	86.7	90.8	92.1	87.0	89.5
03300 CAST-IN-PLACE CONCRETE	100.0	100.0	100.0	92.8	68.1	82.5	87.8	66.2	78.8	92.7	54.7	76.8	94.3	49.4	75.6	91.4	47.1	72.9
03 CONCRETE	100.0	100.0	100.0	90.9	76.4	83.7	88.7	70.7	79.7	91.5	55.4	73.5	92.2	57.5	74.9	90.4	53.9	71.8
04 MASONRY	100.0	100.0	100.0	85.9	76.0	79.7	85.8	65.2	72.8	86.3	52.3	65.0	86.9	36.3	55.2	86.0	38.6	56.3
05 METALS	100.0	100.0	100.0	94.6	95.5	94.9	96.0	91.0	94.4	94.8	79.6	89.8	94.6	92.1	93.8	95.2	92.7	94.4
06 WOOD & PLASTICS	100.0	100.0	100.0	96.7	76.0	85.9	96.7	68.4	81.9	96.7	52.7	73.7	95.3	48.3	70.7	96.7	38.3	66.2
07 THERMAL & MOISTURE PROTECTION	100.0	100.0	100.0	95.7	80.9	88.4	95.4	75.1	85.4	95.4	69.5	82.7	95.1	63.0	79.3	95.3	61.8	78.8
08 DOORS & WINDOWS	100.0	100.0	100.0	98.5	78.5	93.3	98.5	65.5	89.9	98.5	53.2	86.8	98.5	58.2	88.0	98.5	58.0	88.0
09200 PLASTER & GYPSUM BOARD	100.0	100.0	100.0	103.1	75.9	85.9	100.5	68.1	80.0	100.5	52.0	69.7	100.5	47.4	66.9	100.5	37.2	60.4
095,098 CEILINGS & ACOUSTICAL TREATMENT	100.0	100.0	100.0	102.7	75.9	86.6	102.7	68.1	81.9	102.7	52.0	72.3	102.7	47.4	69.6	102.7	37.2	63.4
09600 FLOORING	100.0	100.0	100.0	104.4	50.5	90.1	104.4	54.2	91.1	114.7	54.7	98.8	112.9	28.3	90.5	104.4	41.5	87.7
097,099 WALL FINISHES, PAINTS & COATINGS	100.0	100.0	100.0	95.3	70.8	80.4	95.3	64.2	76.4	100.1	56.9	73.9	95.3	56.0	71.4	95.3	48.2	66.7
09 FINISHES	100.0	100.0	100.0	100.8	70.1	84.8	100.3	65.0	81.9	104.8	53.1	77.9	104.0	43.9	72.8	100.2	39.2	68.5
10-14 TOTAL DIV. 10000-14000	100.0	100.0	100.0	100.0	86.6	97.1	100.0	84.5	96.7	100.0	69.0	93.4	100.0	77.1	95.1	100.0	74.5	94.6
15 MECHANICAL	100.0	100.0	100.0	100.3	69.6	86.9	100.3	69.8	86.9	99.8	70.8	87.2	99.8	41.7	74.4	99.8	36.7	72.2
16 ELECTRICAL	100.0	100.0	100.0	96.3	66.6	81.8	96.4	70.5	83.8	96.4	51.3	74.4	96.3	69.1	83.0	96.4	66.6	81.9
01-16 WEIGHTED AVERAGE	100.0	100.0	100.0	96.4	76.5	87.4	96.3	73.3	85.8	97.1	63.0	81.6	97.0	58.4	79.5	96.3	56.4	78.2

DIVISION	ALASKA ANCHORAGE			FAIRBANKS			JUNEAU			ARIZONA FLAGSTAFF			MESA/TEMPE			PHOENIX		
	MAT.	INST.	TOTAL	MAT.	INST.	TOTAL	MAT.	INST.	TOTAL	MAT.	INST.	TOTAL	MAT.	INST.	TOTAL	MAT.	INST.	TOTAL
01590 EQUIPMENT RENTAL	.0	118.4	118.4	.0	118.4	118.4	.0	118.4	118.4	.0	94.9	94.9	.0	98.2	98.2	.0	98.8	98.8
02 SITE CONSTRUCTION	143.5	133.8	136.3	127.1	133.8	132.1	139.1	133.8	135.1	83.5	100.9	96.4	88.3	103.9	99.9	88.8	104.7	100.6
03100 CONCRETE FORMS & ACCESSORIES	132.4	113.9	116.4	134.1	119.7	121.6	133.8	113.9	116.6	102.8	61.5	67.1	97.3	63.1	67.7	98.4	69.8	73.6
03200 CONCRETE REINFORCEMENT	146.2	106.6	126.2	122.9	106.7	114.7	108.7	106.6	107.7	105.5	73.5	89.3	104.8	72.9	88.7	103.2	74.2	88.5
03300 CAST-IN-PLACE CONCRETE	195.7	114.6	161.8	163.1	115.1	143.0	196.5	114.6	162.3	91.8	78.3	86.2	98.3	71.4	87.1	98.4	78.9	90.3
03 CONCRETE	153.5	112.2	132.9	129.8	115.0	122.4	147.9	112.2	130.1	116.7	69.5	93.1	99.2	67.8	83.5	98.8	73.6	86.2
04 MASONRY	217.0	121.0	156.7	210.3	121.0	154.3	219.4	121.0	157.7	102.9	56.2	73.6	108.2	52.8	73.4	96.0	68.1	78.5
05 METALS	130.1	102.2	121.0	130.1	102.5	121.1	130.1	102.2	121.0	93.8	67.9	85.4	94.5	69.2	86.2	96.0	70.9	87.8
06 WOOD & PLASTICS	117.8	111.5	114.5	118.0	118.9	118.5	117.8	111.5	114.5	108.6	60.1	83.3	99.6	68.7	83.5	100.6	70.5	84.9
07 THERMAL & MOISTURE PROTECTION	198.5	115.8	157.8	194.6	118.6	157.2	195.2	115.8	156.2	107.3	66.9	87.4	105.4	63.8	85.0	105.3	70.8	88.4
08 DOORS & WINDOWS	126.2	108.8	121.7	123.3	112.9	120.6	123.3	108.8	119.6	102.2	64.2	92.3	98.9	65.2	90.2	100.0	69.9	92.2
09200 PLASTER & GYPSUM BOARD	132.6	111.6	119.3	132.6	119.3	124.1	132.6	111.6	119.3	90.8	59.0	70.6	91.6	67.7	76.4	93.9	69.6	78.5
095,098 CEILINGS & ACOUSTICAL TREATMENT	132.1	111.6	119.8	132.1	119.3	124.4	132.1	111.6	119.8	101.3	59.0	75.9	95.6	67.7	78.8	102.7	69.6	82.8
09600 FLOORING	164.9	127.0	154.9	164.9	127.0	154.9	164.9	127.0	154.9	95.2	51.8	83.8	97.3	64.6	88.7	97.6	67.5	89.7
097,099 WALL FINISHES, PAINTS & COATINGS	168.1	111.4	133.7	168.1	125.0	141.9	168.1	111.4	133.7	94.6	48.8	66.8	105.0	53.1	73.5	105.0	60.4	77.9
09 FINISHES	154.8	116.0	134.6	152.6	121.9	136.6	153.4	116.0	134.0	96.7	57.6	76.4	97.2	62.0	78.9	99.5	68.0	83.1
10-14 TOTAL DIV. 10000-14000	100.0	112.8	102.7	100.0	113.7	102.9	100.0	112.8	102.7	100.0	72.6	94.2	100.0	69.0	93.4	100.0	74.4	94.6
15 MECHANICAL	100.9	108.3	104.2	100.9	115.9	107.4	100.9	108.3	104.2	100.2	77.4	90.2	100.2	67.9	86.1	100.2	77.5	90.2
16 ELECTRICAL	149.1	112.3	131.2	150.9	112.3	132.1	150.9	112.3	132.1	96.9	48.0	73.1	90.6	60.3	75.8	99.6	65.7	83.1
01-16 WEIGHTED AVERAGE	134.5	113.8	125.1	130.7	116.9	124.4	133.6	113.8	124.6	100.5	67.6	85.5	97.8	67.7	84.1	98.8	74.3	87.7

DIVISION	ARIZONA PRESCOTT			TUCSON			ARKANSAS FORT SMITH			JONESBORO			LITTLE ROCK			PINE BLUFF		
	MAT.	INST.	TOTAL	MAT.	INST.	TOTAL	MAT.	INST.	TOTAL	MAT.	INST.	TOTAL	MAT.	INST.	TOTAL	MAT.	INST.	TOTAL
01590 EQUIPMENT RENTAL	.0	94.9	94.9	.0	98.2	98.2	.0	86.0	86.0	.0	107.8	107.8	.0	86.0	86.0	.0	86.0	86.0
02 SITE CONSTRUCTION	71.2	100.3	92.7	85.1	104.4	99.4	77.6	83.8	82.2	100.5	99.2	99.6	77.4	83.8	82.1	79.6	83.8	82.7
03100 CONCRETE FORMS & ACCESSORIES	98.2	56.0	61.6	97.9	69.2	73.1	97.7	42.0	49.4	84.5	47.8	52.7	91.8	59.2	63.5	77.2	58.9	61.4
03200 CONCRETE REINFORCEMENT	105.5	72.5	88.8	86.6	73.5	80.0	96.9	74.6	85.6	92.7	49.6	70.9	97.1	72.2	84.5	97.1	72.1	84.5
03300 CAST-IN-PLACE CONCRETE	91.7	64.2	80.2	101.1	78.7	91.8	89.8	66.8	80.2	85.6	57.1	73.7	89.8	67.0	80.3	82.4	66.9	75.9
03 CONCRETE	102.6	62.0	82.3	97.4	73.1	85.3	87.4	57.4	72.4	84.0	53.0	68.6	87.0	64.7	75.9	85.0	64.5	74.8
04 MASONRY	103.3	58.8	75.4	97.8	56.2	71.7	97.5	53.3	69.8	92.8	46.9	64.0	95.9	53.3	69.2	118.2	53.3	77.5
05 METALS	93.8	66.7	85.0	95.3	68.7	86.6	95.7	71.3	87.7	90.0	75.9	85.4	91.6	70.8	84.8	94.3	70.6	86.6
06 WOOD & PLASTICS	103.8	54.6	78.1	99.8	70.5	84.5	100.4	40.6	69.2	86.4	48.8	66.8	97.4	63.3	79.6	78.9	63.3	70.7
07 THERMAL & MOISTURE PROTECTION	105.6	61.7	84.0	106.6	64.2	85.8	99.4	48.4	74.3	109.1	51.6	80.8	97.9	50.7	74.7	98.0	50.7	74.8
08 DOORS & WINDOWS	102.2	58.3	90.8	96.0	69.9	89.2	97.2	45.8	83.9	98.8	47.9	85.6	97.2	58.4	87.1	92.5	58.4	83.7
09200 PLASTER & GYPSUM BOARD	87.9	53.3	66.0	92.0	69.6	77.8	85.4	39.6	56.4	92.3	47.7	64.0	85.4	62.9	71.1	77.7	62.9	68.3
095,098 CEILINGS & ACOUSTICAL TREATMENT	99.6	53.3	71.8	95.6	69.6	80.0	91.2	39.6	60.3	89.0	47.7	64.2	91.2	62.9	74.2	86.8	62.9	72.4
09600 FLOORING	93.6	51.6	82.5	96.8	51.8	84.9	115.8	70.0	103.7	77.3	45.6	68.9	117.1	70.0	104.6	104.8	70.0	95.6
097,099 WALL FINISHES, PAINTS & COATINGS	94.6	48.8	66.8	102.6	48.8	70.0	96.4	58.3	73.3	84.9	52.4	65.2	96.4	59.9	74.2	96.4	59.9	74.2
09 FINISHES	94.3	53.8	73.2	96.9	63.8	79.6	96.4	48.4	71.4	87.7	47.9	67.0	96.7	61.9	78.6	91.1	61.9	75.9
10-14 TOTAL DIV. 10000-14000	100.0	71.1	93.9	100.0	74.4	94.6	100.0	63.4	92.2	100.0	54.9	90.4	100.0	66.4	92.8	100.0	66.4	92.8
15 MECHANICAL	100.2	73.6	88.6	100.1	69.2	86.6	100.1	43.2	75.2	100.2	43.1	75.2	100.1	58.4	81.8	100.1	45.6	76.3
16 ELECTRICAL	96.7	46.2	72.1	92.5	60.9	77.1	95.6	64.4	80.4	102.7	47.8	75.9	95.3	70.0	83.0	95.0	70.0	82.8
01-16 WEIGHTED AVERAGE	98.2	64.7	83.0	97.0	69.8	84.6	96.0	56.8	78.2	95.5	55.0	77.1	95.2	64.3	81.2	95.3	61.7	80.0

DIVISION		ARKANSAS			CALIFORNIA														
		TEXARKANA			ANAHEIM			BAKERSFIELD			FRESNO			LOS ANGELES			OAKLAND		
		MAT.	INST.	TOTAL	MAT.	INST.	TOTAL	MAT.	INST.	TOTAL	MAT.	INST.	TOTAL	MAT.	INST.	TOTAL	MAT.	INST.	TOTAL
01590	EQUIPMENT RENTAL	.0	86.8	86.8	.0	102.1	102.1	.0	99.5	99.5	.0	99.5	99.5	.0	98.5	98.5	.0	103.3	103.3
02	SITE CONSTRUCTION	95.3	84.6	87.4	103.2	109.0	107.5	108.1	106.0	106.6	109.6	106.0	106.9	98.1	107.8	105.3	155.8	104.6	117.9
03100	CONCRETE FORMS & ACCESSORIES	85.0	39.6	45.7	105.0	121.5	119.3	97.5	121.1	117.9	101.2	127.2	123.7	105.5	121.4	119.3	107.4	133.5	130.0
03200	CONCRETE REINFORCEMENT	96.6	48.1	72.1	95.6	112.5	104.1	108.9	112.5	110.7	92.2	112.8	102.6	113.0	112.7	112.9	103.1	113.4	108.3
03300	CAST-IN-PLACE CONCRETE	89.8	44.6	70.9	107.2	117.2	111.4	102.3	116.1	108.1	112.0	113.5	112.6	87.3	115.4	99.1	139.5	118.9	130.9
03	CONCRETE	84.3	43.9	64.1	107.8	117.4	112.6	107.1	116.8	112.0	109.3	118.7	114.0	103.1	116.7	109.9	126.4	123.6	125.0
04	MASONRY	97.5	32.8	56.9	89.0	111.4	103.1	107.5	111.2	109.8	111.1	111.2	111.2	97.7	115.3	108.8	148.4	122.5	132.1
05	METALS	86.9	59.4	77.9	110.9	99.8	107.2	105.4	99.0	103.3	111.1	100.2	107.6	112.8	98.4	108.1	105.1	104.9	105.0
06	WOOD & PLASTICS	88.2	42.3	64.3	94.9	121.2	108.7	86.0	121.3	104.4	99.1	129.8	115.1	91.8	120.9	107.0	101.9	134.8	119.1
07	THERMAL & MOISTURE PROTECTION	98.9	42.2	71.0	121.2	115.0	118.1	105.5	108.9	107.2	101.3	110.5	105.8	111.6	117.0	114.3	113.0	124.0	118.4
08	DOORS & WINDOWS	97.6	41.1	83.0	102.8	116.2	106.2	101.6	113.3	104.6	101.9	118.6	106.2	96.7	116.0	101.7	106.4	125.7	111.4
09200	PLASTER & GYPSUM BOARD	83.3	41.4	56.7	98.1	121.7	113.1	98.1	121.7	113.1	96.1	130.4	117.9	101.6	121.7	114.4	94.7	135.1	120.3
095,098	CEILINGS & ACOUSTICAL TREATMENT	93.9	41.4	62.4	119.6	121.7	120.9	119.6	121.7	120.9	119.6	130.4	126.1	119.8	121.7	121.0	114.8	135.1	127.0
09600	FLOORING	107.6	43.5	90.6	123.1	103.7	118.0	117.3	93.4	111.0	133.8	138.6	135.1	114.0	103.7	111.3	116.1	122.8	117.9
097,099	WALL FINISHES, PAINTS & COATINGS	96.4	36.8	60.2	111.4	117.0	111.0	110.9	92.5	99.7	134.6	100.1	113.7	99.6	110.7	106.4	115.6	136.3	128.1
09	FINISHES	95.0	40.1	66.4	114.3	117.4	115.9	114.5	113.6	114.0	120.9	127.5	124.3	110.8	117.1	114.1	115.9	132.6	124.6
10-14	TOTAL DIV. 10000 - 14000	100.0	39.3	87.1	100.0	114.8	103.2	100.0	119.2	104.1	100.0	128.8	106.1	100.0	114.1	103.0	100.0	132.3	106.9
15	MECHANICAL	100.1	38.1	73.0	100.2	108.4	103.8	100.2	99.6	99.9	100.2	111.6	105.2	100.1	108.4	103.7	100.2	127.0	111.9
16	ELECTRICAL	96.8	39.8	69.0	88.6	107.9	98.0	89.5	93.6	91.5	87.5	97.1	92.2	96.5	112.9	104.5	102.6	134.4	118.1
01-16	WEIGHTED AVERAGE	94.7	45.3	72.3	102.9	111.1	106.6	102.6	106.2	104.2	104.3	111.9	107.8	102.7	111.9	106.8	110.6	123.9	116.6

DIVISION		CALIFORNIA																	
		OXNARD			REDDING			RIVERSIDE			SACRAMENTO			SAN DIEGO			SAN FRANCISCO		
		MAT.	INST.	TOTAL	MAT.	INST.	TOTAL	MAT.	INST.	TOTAL	MAT.	INST.	TOTAL	MAT.	INST.	TOTAL	MAT.	INST.	TOTAL
01590	EQUIPMENT RENTAL	.0	98.1	98.1	.0	99.2	99.2	.0	100.6	100.6	.0	102.9	102.9	.0	97.1	97.1	.0	108.6	108.6
02	SITE CONSTRUCTION	109.7	103.8	105.3	114.4	105.2	107.6	100.8	106.7	105.1	119.5	110.9	113.2	102.2	100.6	101.0	158.2	111.2	123.4
03100	CONCRETE FORMS & ACCESSORIES	103.4	121.5	119.1	103.1	126.9	123.7	105.4	121.4	119.3	105.9	127.6	124.7	106.2	111.0	110.3	107.8	134.8	131.2
03200	CONCRETE REINFORCEMENT	108.9	112.2	110.6	105.6	112.6	109.1	107.9	112.5	110.2	95.3	112.9	104.2	105.6	112.4	109.0	117.5	113.9	115.7
03300	CAST-IN-PLACE CONCRETE	108.6	116.4	111.9	123.0	113.3	119.0	106.1	117.2	110.7	120.2	114.3	117.8	110.8	104.9	108.3	139.4	120.7	131.6
03	CONCRETE	110.5	117.0	113.8	120.2	118.4	119.3	109.3	117.4	113.3	115.8	119.1	117.5	113.0	108.4	110.7	128.7	124.9	126.8
04	MASONRY	112.3	105.9	108.2	114.4	108.2	110.5	85.6	111.0	101.5	119.0	112.2	114.7	98.1	106.4	103.3	148.8	129.0	136.4
05	METALS	105.2	99.0	103.1	110.3	99.4	106.7	111.1	99.6	107.4	100.4	99.8	100.2	108.8	98.8	105.6	111.2	107.1	109.9
06	WOOD & PLASTICS	94.0	121.3	108.3	94.2	129.8	112.8	94.9	121.2	108.7	96.7	130.0	114.1	100.3	109.2	104.9	101.9	135.0	119.2
07	THERMAL & MOISTURE PROTECTION	110.5	110.0	110.3	111.3	108.5	109.9	118.9	112.0	115.5	123.8	112.8	118.4	113.8	104.1	109.0	113.0	130.8	121.7
08	DOORS & WINDOWS	100.4	116.2	104.5	103.2	119.8	107.5	102.8	116.2	106.2	117.7	120.8	118.5	103.9	107.7	104.9	110.8	128.8	115.5
09200	PLASTER & GYPSUM BOARD	98.1	121.7	113.1	96.8	130.4	118.1	97.0	121.7	112.7	92.1	130.4	116.4	97.6	109.2	104.9	97.6	135.1	121.4
095,098	CEILINGS & ACOUSTICAL TREATMENT	119.6	121.7	120.9	126.7	130.4	128.9	115.1	121.7	119.1	114.8	130.4	124.2	103.7	109.2	107.0	123.5	135.1	130.5
09600	FLOORING	117.3	103.7	113.7	116.6	106.8	114.0	121.8	103.7	117.0	119.0	115.6	118.1	110.7	103.5	108.8	116.1	128.7	119.5
097,099	WALL FINISHES, PAINTS & COATINGS	110.3	102.7	105.7	110.3	112.3	111.5	107.8	110.7	109.6	112.9	116.2	114.9	109.4	105.0	106.7	115.6	151.7	137.5
09	FINISHES	114.3	116.5	115.5	116.1	123.1	119.8	112.1	117.4	114.8	114.5	125.5	120.2	107.8	109.1	108.5	118.6	136.0	127.7
10-14	TOTAL DIV. 10000 - 14000	100.0	115.1	103.2	100.0	128.9	106.1	100.0	114.8	103.1	100.0	129.6	106.3	100.0	112.5	102.7	100.0	133.1	107.0
15	MECHANICAL	100.2	108.4	103.8	100.2	105.7	102.6	100.1	108.4	103.7	100.2	111.2	105.0	100.3	106.8	103.1	100.2	151.6	122.7
16	ELECTRICAL	94.6	106.1	100.2	97.2	102.7	99.9	88.7	102.2	95.3	97.9	102.8	100.3	96.0	98.7	97.4	102.6	152.1	126.8
01-16	WEIGHTED AVERAGE	103.9	109.5	106.4	106.8	110.4	108.4	102.6	109.9	105.9	106.9	113.1	109.7	103.9	105.0	104.4	112.4	133.8	122.2

DIVISION		CALIFORNIA												COLORADO					
		SAN JOSE			SANTA BARBARA			STOCKTON			VALLEJO			COLORADO SPRINGS			DENVER		
		MAT.	INST.	TOTAL	MAT.	INST.	TOTAL	MAT.	INST.	TOTAL	MAT.	INST.	TOTAL	MAT.	INST.	TOTAL	MAT.	INST.	TOTAL
01590	EQUIPMENT RENTAL	.0	100.0	100.0	.0	99.5	99.5	.0	99.2	99.2	.0	103.5	103.5	.0	95.2	95.2	.0	100.4	100.4
02	SITE CONSTRUCTION	148.5	99.7	112.4	109.6	106.1	107.0	107.6	105.3	105.9	116.2	111.1	112.4	95.0	97.1	96.5	93.6	106.4	103.0
03100	CONCRETE FORMS & ACCESSORIES	105.7	133.8	130.0	104.1	121.4	119.1	103.8	126.9	123.8	107.0	131.7	128.4	92.6	85.9	86.8	101.9	87.1	89.1
03200	CONCRETE REINFORCEMENT	96.2	113.6	105.0	108.9	112.5	110.7	109.5	112.6	111.1	102.7	113.5	108.2	113.0	84.6	98.6	113.0	84.7	98.7
03300	CAST-IN-PLACE CONCRETE	128.5	118.6	124.4	108.2	116.3	111.6	108.2	113.4	110.3	125.1	115.4	121.1	98.0	88.1	93.9	91.7	88.5	90.3
03	CONCRETE	118.1	123.8	120.9	110.4	117.0	113.7	110.4	118.5	114.5	119.4	121.4	120.4	108.6	86.5	97.5	102.6	87.2	94.5
04	MASONRY	147.2	125.5	133.6	107.9	111.1	109.9	111.9	108.2	109.5	84.4	121.6	107.7	108.7	82.6	92.3	108.4	85.8	94.2
05	METALS	104.5	106.9	105.3	105.7	99.3	103.6	107.1	99.8	104.7	103.8	101.4	103.0	97.9	88.1	94.7	100.2	88.4	96.1
06	WOOD & PLASTICS	101.0	134.5	118.5	94.0	121.3	108.3	95.9	129.8	113.6	95.1	134.5	115.6	93.9	88.3	91.0	103.3	88.8	95.1
07	THERMAL & MOISTURE PROTECTION	107.2	129.5	118.2	106.8	108.9	107.8	110.8	107.5	109.2	126.1	121.6	123.9	103.8	85.8	95.0	103.2	84.3	93.?
08	DOORS & WINDOWS	92.8	126.7	101.6	101.6	116.2	105.4	98.5	130.4	106.0	119.2	128.6	121.6	98.5	90.6	96.5	99.9	91.5	97.?
09200	PLASTER & GYPSUM BOARD	97.9	135.1	121.5	98.1	121.7	113.1	98.5	130.4	118.7	94.9	135.1	120.4	87.6	88.1	87.9	97.6	88.8	92.0
095,098	CEILINGS & ACOUSTICAL TREATMENT	106.2	135.1	123.5	119.6	121.7	120.9	119.6	130.4	126.1	123.7	135.1	130.5	99.2	88.1	92.5	98.3	88.8	92.?
09600	FLOORING	114.3	128.7	118.1	117.3	100.4	112.9	117.3	110.9	115.6	123.7	128.7	125.0	107.8	89.1	102.9	108.8	101.0	106.?
097,099	WALL FINISHES, PAINTS & COATINGS	112.0	139.4	128.6	110.3	102.7	105.7	110.3	101.4	104.9	112.1	136.0	126.6	106.7	53.6	74.5	106.7	77.7	89.?
09	FINISHES	111.8	134.0	123.3	114.6	116.0	115.3	114.4	122.6	118.7	116.0	132.8	124.8	98.2	83.0	90.3	99.6	88.8	94.?
10-14	TOTAL DIV. 10000 - 14000	100.0	131.7	106.7	100.0	115.1	103.2	100.0	128.9	106.1	100.0	129.9	106.4	100.0	91.3	98.1	100.0	91.5	98.?
15	MECHANICAL	100.2	141.3	118.2	100.2	108.4	103.8	100.2	101.3	100.7	100.2	118.4	108.2	100.2	81.3	91.9	100.2	86.4	94.?
16	ELECTRICAL	102.5	142.9	122.2	85.8	107.2	96.3	97.0	109.6	103.2	92.2	123.1	107.2	95.3	90.8	93.1	97.1	95.4	96.?
01-16	WEIGHTED AVERAGE	107.3	128.4	116.9	102.7	110.3	106.1	104.5	110.5	107.2	105.6	120.4	112.3	100.1	86.6	94.0	100.3	90.3	95.?

		COLORADO												CONNECTICUT					
	DIVISION	FORT COLLINS			GRAND JUNCTION			GREELEY			PUEBLO			BRIDGEPORT			BRISTOL		
		MAT.	INST.	TOTAL	MAT.	INST.	TOTAL	MAT.	INST.	TOTAL	MAT.	INST.	TOTAL	MAT.	INST.	TOTAL	MAT.	INST.	TOTAL
01590	EQUIPMENT RENTAL	.0	96.8	96.8	.0	100.0	100.0	.0	96.8	96.8	.0	96.9	96.9	.0	101.6	101.6	.0	101.6	101.6
02	SITE CONSTRUCTION	104.3	99.5	100.8	124.7	102.1	108.0	91.7	98.2	96.5	116.8	96.5	101.7	102.3	104.4	103.8	101.4	104.3	103.6
03100	CONCRETE FORMS & ACCESSORIES	101.3	80.4	83.2	106.7	81.0	84.4	98.7	49.5	56.1	103.3	86.1	88.4	101.0	110.7	109.4	101.0	110.3	109.1
03200	CONCRETE REINFORCEMENT	113.9	83.0	98.3	113.7	83.0	98.2	113.8	82.4	97.9	109.5	84.6	96.9	108.2	122.7	115.5	108.2	122.6	115.5
03300	CAST-IN-PLACE CONCRETE	106.4	83.5	96.9	113.2	84.6	101.2	89.0	60.8	77.2	101.5	89.3	96.4	105.2	117.7	110.4	98.6	117.6	106.5
03	CONCRETE	115.0	82.1	98.6	115.3	82.6	99.0	100.0	60.5	80.3	105.2	87.1	96.1	108.8	115.3	112.0	105.6	115.1	110.3
04	MASONRY	122.8	60.9	84.0	148.3	63.4	95.0	116.4	40.4	68.7	107.1	82.6	91.7	108.2	117.3	113.9	99.9	117.3	110.8
05	METALS	96.1	83.4	92.0	96.9	82.6	92.3	96.1	81.7	91.4	98.4	89.2	95.4	101.2	120.2	107.4	101.2	119.9	107.3
06	WOOD & PLASTICS	103.0	83.6	92.9	106.4	83.8	94.6	100.2	48.3	73.1	103.1	88.7	95.6	99.4	108.0	103.8	99.4	108.0	103.8
07	THERMAL & MOISTURE PROTECTION	103.8	72.5	88.4	104.9	68.6	87.0	102.9	60.7	82.2	104.2	85.5	95.0	101.6	117.5	109.4	101.8	113.6	107.6
08	DOORS & WINDOWS	95.9	87.9	93.8	100.8	88.0	97.5	95.9	68.7	88.8	95.2	90.8	94.1	107.1	121.0	110.7	107.1	113.0	108.6
09200	PLASTER & GYPSUM BOARD	96.0	83.4	88.0	106.7	83.4	91.9	94.7	47.1	64.5	85.2	88.1	87.0	104.4	107.2	106.2	104.4	107.2	106.2
095,098	CEILINGS & ACOUSTICAL TREATMENT	92.1	83.4	86.9	96.0	83.4	88.4	92.1	47.1	65.1	107.6	88.1	95.9	101.1	107.2	104.8	101.1	107.2	104.8
09600	FLOORING	108.6	71.5	98.8	116.1	71.5	104.4	107.2	71.5	97.8	112.5	101.0	109.5	99.8	115.1	103.9	99.8	115.1	103.9
097,099	WALL FINISHES, PAINTS & COATINGS	106.7	53.2	74.2	117.0	77.7	93.2	106.7	33.1	62.1	117.1	49.6	76.2	92.8	109.6	103.0	92.8	109.6	103.0
09	FINISHES	98.3	75.9	86.7	106.3	79.5	92.4	97.1	50.2	72.7	104.6	85.3	94.6	102.7	110.7	106.9	102.7	110.7	106.9
10 - 14	TOTAL DIV. 10000 - 14000	100.0	89.1	97.7	100.0	90.5	98.0	100.0	80.3	95.8	100.0	92.3	98.4	100.0	102.1	100.5	100.0	102.1	102.1
15	MECHANICAL	100.2	82.5	92.4	100.1	72.5	88.0	100.2	76.7	89.9	100.1	71.3	87.5	100.1	104.0	101.8	100.1	103.9	101.8
16	ELECTRICAL	92.3	93.0	92.6	88.5	65.4	77.3	92.3	93.0	92.6	89.3	82.6	86.0	100.3	105.8	103.0	100.3	108.4	104.2
01 - 16	WEIGHTED AVERAGE	101.0	82.7	92.7	103.8	77.6	91.9	98.5	71.1	86.0	100.0	83.9	92.6	102.7	110.8	106.4	101.9	110.6	105.9

		CONNECTICUT																	
	DIVISION	HARTFORD			NEW BRITAIN			NEW HAVEN			NORWALK			STAMFORD			WATERBURY		
		MAT.	INST.	TOTAL	MAT.	INST.	TOTAL	MAT.	INST.	TOTAL	MAT.	INST.	TOTAL	MAT.	INST.	TOTAL	MAT.	INST.	TOTAL
01590	EQUIPMENT RENTAL	.0	101.6	101.6	.0	101.6	101.6	.0	102.1	102.1	.0	101.6	101.6	.0	101.6	101.6	.0	101.6	101.6
02	SITE CONSTRUCTION	101.8	104.3	103.7	101.6	104.3	103.6	101.5	105.2	104.2	102.0	104.4	103.8	102.7	104.4	103.9	101.8	104.4	103.7
03100	CONCRETE FORMS & ACCESSORIES	99.9	110.3	108.9	101.2	110.4	109.1	100.8	110.6	109.3	101.0	110.7	109.4	101.0	110.9	109.6	101.0	110.6	109.3
03200	CONCRETE REINFORCEMENT	108.2	122.6	115.5	108.2	122.6	115.5	108.2	122.7	115.5	108.2	122.8	115.6	108.2	122.8	115.6	108.2	122.7	115.5
03300	CAST-IN-PLACE CONCRETE	99.7	117.6	107.2	100.1	117.6	107.4	101.9	117.7	108.5	103.5	119.3	110.1	105.2	119.4	111.1	105.2	117.7	110.4
03	CONCRETE	106.1	115.1	110.6	106.4	115.1	110.7	121.6	115.2	118.4	107.9	115.9	111.9	108.8	116.0	112.4	108.8	115.2	112.0
04	MASONRY	100.0	117.3	110.9	100.3	117.3	110.9	100.3	117.3	110.9	100.3	119.0	112.1	100.4	119.0	112.1	100.4	117.3	111.0
05	METALS	106.3	119.9	110.7	97.5	119.9	104.8	97.7	120.2	105.1	101.2	120.4	107.5	101.2	120.7	107.5	101.2	120.1	107.4
06	WOOD & PLASTICS	99.4	108.0	103.8	99.4	108.0	103.8	99.4	108.0	103.8	99.4	108.0	103.8	99.4	108.0	103.8	99.4	108.0	103.8
07	THERMAL & MOISTURE PROTECTION	100.3	113.6	106.9	101.8	115.2	108.4	101.9	115.2	108.5	101.8	118.2	109.9	101.8	118.2	109.9	101.8	115.2	108.4
08	DOORS & WINDOWS	107.1	113.0	108.6	107.1	113.0	108.6	107.1	121.0	110.7	107.1	121.0	110.7	107.1	121.0	110.7	107.1	121.0	110.7
09200	PLASTER & GYPSUM BOARD	104.4	107.2	106.2	104.4	107.2	106.2	104.4	107.2	106.2	104.4	107.2	106.2	104.4	107.2	106.2	104.4	107.2	106.2
095,098	CEILINGS & ACOUSTICAL TREATMENT	101.1	107.2	104.8	101.1	107.2	104.8	101.1	107.2	104.8	101.1	107.2	104.8	101.1	107.2	104.8	101.1	107.2	104.8
09600	FLOORING	99.8	115.1	103.9	99.8	115.1	103.9	99.8	115.1	103.9	99.8	115.1	103.9	99.8	115.1	103.9	99.8	115.1	103.9
097,099	WALL FINISHES, PAINTS & COATINGS	92.8	109.6	103.0	92.8	109.6	103.0	92.8	109.6	103.0	92.8	109.6	103.0	92.8	109.6	103.0	92.8	109.6	103.0
09	FINISHES	102.7	110.7	106.9	102.7	110.7	106.9	102.8	110.7	106.9	102.7	110.7	106.9	102.8	110.7	106.9	102.6	110.7	106.8
10 - 14	TOTAL DIV. 10000 - 14000	100.0	110.1	102.1	100.0	110.1	102.1	100.0	110.1	102.1	100.0	110.1	102.1	100.0	110.2	102.2	100.0	110.1	102.1
15	MECHANICAL	100.1	103.9	101.8	100.1	103.9	101.8	100.1	104.0	101.8	100.1	104.0	101.8	100.1	104.0	101.8	100.1	104.0	101.8
16	ELECTRICAL	99.9	103.1	101.5	100.3	108.4	104.3	100.2	108.4	104.2	100.3	108.4	104.2	100.3	137.7	118.5	99.8	105.8	102.7
01 - 16	WEIGHTED AVERAGE	102.7	109.9	106.0	101.4	110.7	105.6	103.2	111.2	106.8	102.2	111.5	106.4	102.3	115.6	108.4	102.2	110.7	106.1

		D.C.			DELAWARE			FLORIDA											
	DIVISION	WASHINGTON			WILMINGTON			DAYTONA BEACH			FORT LAUDERDALE			JACKSONVILLE			MELBOURNE		
		MAT.	INST.	TOTAL	MAT.	INST.	TOTAL	MAT.	INST.	TOTAL	MAT.	INST.	TOTAL	MAT.	INST.	TOTAL	MAT.	INST.	TOTAL
01590	EQUIPMENT RENTAL	.0	102.3	102.3	.0	118.4	118.4	.0	97.7	97.7	.0	89.3	89.3	.0	97.7	97.7	.0	97.7	97.7
02	SITE CONSTRUCTION	100.3	89.5	92.3	87.5	111.4	105.1	115.6	86.7	94.2	101.3	73.8	81.0	115.7	87.2	94.6	123.2	87.2	96.6
03100	CONCRETE FORMS & ACCESSORIES	98.0	80.7	83.0	96.7	101.4	100.7	98.2	69.8	73.6	95.6	68.4	72.0	98.0	53.3	59.3	92.5	73.7	76.2
03200	CONCRETE REINFORCEMENT	105.6	89.5	97.5	102.2	99.0	100.6	95.0	85.5	90.2	95.0	72.9	83.8	95.0	51.2	72.9	96.1	85.5	90.8
03300	CAST-IN-PLACE CONCRETE	114.7	88.1	103.6	78.8	94.4	85.3	89.9	70.5	81.8	94.3	69.4	83.9	90.8	58.7	77.4	104.5	75.8	92.5
03	CONCRETE	111.2	86.2	98.7	97.8	99.5	98.6	90.4	74.2	82.3	92.3	70.9	81.6	90.8	56.5	73.7	99.6	77.8	88.7
04	MASONRY	92.2	82.0	85.8	108.5	88.8	96.1	86.7	64.7	72.9	86.9	67.6	74.8	86.2	51.4	64.4	83.9	73.6	77.4
05	METALS	107.4	107.8	107.5	101.8	115.1	106.1	96.4	95.9	96.2	96.3	89.9	94.2	96.0	80.2	90.8	105.3	96.4	102.4
06	WOOD & PLASTICS	95.0	80.9	87.7	95.3	103.2	99.4	97.8	72.5	84.6	93.3	65.8	79.0	97.8	52.1	74.0	91.6	72.5	81.7
07	THERMAL & MOISTURE PROTECTION	98.3	81.7	90.2	100.6	104.3	102.4	95.4	69.7	82.7	95.4	74.7	85.2	95.7	58.0	77.2	95.6	76.3	86.1
08	DOORS & WINDOWS	102.6	90.7	99.5	94.4	106.2	97.4	100.9	69.9	92.8	98.5	63.3	89.4	100.9	49.7	87.6	100.0	75.4	93.7
09200	PLASTER & GYPSUM BOARD	101.1	80.5	88.0	112.3	103.1	106.5	100.5	72.3	82.6	100.1	65.5	78.1	100.5	51.3	69.3	96.2	72.3	81.1
095,098	CEILINGS & ACOUSTICAL TREATMENT	99.3	80.5	88.0	103.6	103.1	103.3	102.7	72.3	84.5	102.7	65.5	80.4	102.7	51.3	71.9	97.3	72.3	82.3
09600	FLOORING	113.7	102.8	110.8	83.7	98.3	87.6	118.6	73.5	106.7	118.6	61.0	103.4	118.6	50.1	100.5	115.3	73.5	104.2
097,099	WALL FINISHES, PAINTS & COATINGS	113.0	87.8	97.7	90.7	97.0	94.5	111.2	72.1	87.5	107.4	49.9	72.5	111.2	48.1	72.9	111.2	72.6	99.9
09	FINISHES	101.2	85.3	92.9	102.2	100.2	101.2	108.4	70.8	88.9	106.4	64.5	84.6	108.4	51.5	78.8	106.0	75.4	90.0
10 - 14	TOTAL DIV. 10000 - 14000	100.0	97.5	99.5	100.0	100.0	100.0	100.0	82.5	96.3	100.0	89.4	97.8	100.0	77.5	95.2	100.0	86.0	97.0
15	MECHANICAL	100.0	90.9	96.0	100.3	116.6	107.4	99.8	72.6	87.9	99.8	69.7	86.7	99.8	51.4	78.6	99.8	77.7	90.2
16	ELECTRICAL	97.7	97.6	97.6	97.2	104.8	100.9	96.3	67.0	82.0	96.3	75.2	86.0	96.1	67.2	82.0	96.2	73.9	85.3
01 - 16	WEIGHTED AVERAGE	102.0	90.8	96.9	99.5	105.7	102.3	98.4	74.5	87.5	97.7	72.3	86.2	98.3	61.0	81.4	100.5	79.0	90.7

FLORIDA

DIVISION		MIAMI			ORLANDO			PANAMA CITY			PENSACOLA			ST. PETERSBURG			TALLAHASSEE		
		MAT.	INST.	TOTAL	MAT.	INST.	TOTAL	MAT.	INST.	TOTAL	MAT.	INST.	TOTAL	MAT.	INST.	TOTAL	MAT.	INST.	TOTA
01590	EQUIPMENT RENTAL	.0	89.3	89.3	.0	97.7	97.7	.0	97.7	97.7	.0	97.7	97.7	.0	97.7	97.7	.0	97.7	97.
02	SITE CONSTRUCTION	100.6	73.6	80.6	116.2	86.4	94.1	130.2	83.9	95.9	127.8	86.3	97.1	116.5	85.7	93.7	117.0	85.8	93.
03100	CONCRETE FORMS & ACCESSORIES	95.8	68.4	72.0	98.0	71.3	74.9	97.0	27.7	37.0	87.4	51.1	55.9	95.5	47.2	53.7	97.9	39.0	46.
03200	CONCRETE REINFORCEMENT	95.0	72.9	83.8	95.0	81.8	88.3	99.2	50.1	74.4	101.7	50.4	75.8	98.5	57.6	77.8	95.0	50.6	72.
03300	CAST-IN-PLACE CONCRETE	91.9	69.6	82.5	97.7	70.4	86.3	95.4	34.6	70.0	95.4	54.9	78.5	101.7	55.3	82.3	94.1	48.4	75.
03	CONCRETE	91.1	70.9	81.0	92.0	74.2	83.1	99.0	36.1	67.6	98.0	54.0	76.0	96.4	53.7	75.1	92.3	46.5	69.
04	MASONRY	84.8	68.0	74.3	90.9	64.7	74.5	91.0	27.6	51.2	88.1	49.9	64.1	134.0	49.3	80.8	89.1	38.8	57.
05	METALS	101.7	89.3	97.6	102.5	94.1	99.8	96.2	65.6	86.2	96.2	79.6	90.8	99.6	80.6	93.4	89.4	78.5	85.
06	WOOD & PLASTICS	93.3	65.8	79.0	97.8	75.2	86.0	96.6	27.8	60.7	86.3	51.4	68.1	95.0	47.2	70.0	96.2	37.4	65.
07	THERMAL & MOISTURE PROTECTION	99.3	71.4	85.6	95.7	69.9	83.1	96.0	31.4	64.2	95.7	52.8	74.6	95.3	49.6	72.8	95.7	46.5	71.
08	DOORS & WINDOWS	98.5	63.9	89.5	100.9	69.6	92.8	98.4	26.7	79.9	98.4	50.2	85.9	99.5	46.5	85.8	99.5	41.0	84.
09200	PLASTER & GYPSUM BOARD	100.1	65.5	78.1	103.1	75.0	85.3	98.4	26.4	52.7	94.0	50.6	66.5	97.6	46.3	65.1	100.5	36.2	59.
095,098	CEILINGS & ACOUSTICAL TREATMENT	102.7	65.5	80.4	102.7	75.0	86.1	96.5	26.4	54.4	96.5	50.6	69.0	96.5	46.3	66.4	102.7	36.2	62.
09600	FLOORING	126.0	61.7	109.0	118.6	73.5	106.7	118.0	19.2	91.9	112.2	53.2	96.6	116.9	53.0	100.0	118.6	38.9	97.
097,099	WALL FINISHES, PAINTS & COATINGS	107.4	49.9	72.5	111.2	55.2	77.2	111.2	24.6	58.6	111.2	55.9	77.6	111.2	46.5	71.9	111.2	39.6	67.
09	FINISHES	108.6	64.6	85.7	108.8	70.4	88.9	107.5	25.4	64.8	105.0	51.8	77.3	106.0	48.0	75.8	108.6	38.2	72.
10 - 14	TOTAL DIV. 10000 - 14000	100.0	89.4	97.8	100.0	82.7	96.3	100.0	46.0	88.5	100.0	51.5	89.7	100.0	53.9	90.2	100.0	63.9	92.
15	MECHANICAL	99.8	74.2	88.6	99.8	63.8	84.1	99.8	25.2	67.2	99.8	50.0	78.1	99.8	49.9	78.0	99.8	39.9	73.
16	ELECTRICAL	96.7	73.3	85.3	96.7	44.6	71.3	95.3	33.4	65.1	98.4	58.8	79.1	96.6	48.1	73.0	96.7	41.3	69.
01 - 16	WEIGHTED AVERAGE	98.7	72.9	87.0	99.8	69.3	86.0	99.5	37.9	71.5	99.2	58.1	80.5	101.6	55.8	80.8	97.6	49.2	75.

		FLORIDA			GEORGIA														
DIVISION		TAMPA			ALBANY			ATLANTA			AUGUSTA			COLUMBUS			MACON		
		MAT.	INST.	TOTAL	MAT.	INST.	TOTAL	MAT.	INST.	TOTAL	MAT.	INST.	TOTAL	MAT.	INST.	TOTAL	MAT.	INST.	TOTA
01590	EQUIPMENT RENTAL	.0	97.7	97.7	.0	90.1	90.1	.0	92.8	92.8	.0	92.1	92.1	.0	90.1	90.1	.0	102.6	102
02	SITE CONSTRUCTION	116.7	86.4	94.3	101.6	76.2	82.8	101.2	95.6	97.0	97.8	92.6	94.0	101.5	76.6	83.1	102.3	93.8	96
03100	CONCRETE FORMS & ACCESSORIES	99.4	74.9	78.2	97.1	45.9	52.8	96.0	78.8	81.1	93.1	51.9	57.4	97.0	58.7	63.8	96.4	55.5	61.
03200	CONCRETE REINFORCEMENT	95.0	90.6	92.8	94.6	91.7	93.1	98.7	93.5	96.1	99.8	77.3	88.4	95.0	92.7	93.9	96.3	91.9	94.
03300	CAST-IN-PLACE CONCRETE	99.5	61.6	83.7	96.0	46.2	75.2	99.6	74.8	89.2	94.1	48.9	75.2	95.7	55.0	78.7	94.4	45.8	74.
03	CONCRETE	95.0	74.3	84.7	93.0	56.1	74.6	98.5	80.0	89.3	94.2	56.1	75.2	92.9	65.1	79.0	92.5	60.4	76.
04	MASONRY	87.7	75.7	80.2	88.9	39.9	58.1	90.9	70.6	78.1	91.2	40.2	59.2	88.9	57.2	69.0	102.8	38.4	62.
05	METALS	99.7	95.6	98.4	95.6	88.8	93.4	93.3	80.4	89.1	92.2	69.6	84.8	95.2	92.3	94.3	90.9	90.8	90
06	WOOD & PLASTICS	99.3	77.4	87.9	96.7	42.5	68.4	96.4	80.9	88.3	93.4	53.3	72.5	96.7	58.6	76.8	104.4	57.5	79.
07	THERMAL & MOISTURE PROTECTION	95.7	62.3	79.3	95.5	54.1	75.1	92.3	75.5	84.0	91.7	49.8	71.1	95.1	61.9	78.8	93.9	56.8	75.
08	DOORS & WINDOWS	100.9	71.2	93.2	98.5	50.1	85.9	100.0	77.5	94.2	94.1	54.7	83.9	98.5	60.9	88.7	96.9	59.3	87.
09200	PLASTER & GYPSUM BOARD	100.5	77.4	85.8	100.2	41.5	63.0	113.1	80.8	92.7	111.9	52.5	74.2	100.2	58.0	73.5	106.2	56.9	75.
095,098	CEILINGS & ACOUSTICAL TREATMENT	102.7	77.4	87.5	101.8	41.5	65.6	107.9	80.8	91.7	108.9	52.5	75.1	101.8	58.0	75.6	97.7	56.9	73
09600	FLOORING	118.6	53.0	101.3	118.6	32.5	95.8	84.0	78.0	82.4	82.8	41.4	71.9	118.6	54.8	101.7	92.5	38.2	78.
097,099	WALL FINISHES, PAINTS & COATINGS	111.2	46.5	71.9	107.4	41.4	67.3	90.3	84.5	86.8	90.3	39.3	59.4	107.4	49.1	72.0	109.1	48.4	72
09	FINISHES	108.5	68.3	87.6	106.2	41.5	72.6	96.2	79.3	87.4	95.7	48.2	71.0	106.1	56.2	80.2	94.1	51.1	71
10 - 14	TOTAL DIV. 10000 - 14000	100.0	85.3	96.9	100.0	77.5	95.2	100.0	84.6	96.7	100.0	67.6	93.1	100.0	79.6	95.7	100.0	77.6	95
15	MECHANICAL	99.8	73.9	88.5	99.8	49.7	77.9	100.0	80.4	91.5	100.0	66.3	85.3	99.8	50.1	78.1	99.8	41.9	74
16	ELECTRICAL	95.7	48.1	72.5	91.9	60.4	76.5	96.6	82.1	89.5	96.7	49.3	73.6	92.3	50.9	72.1	90.0	59.6	75
01 - 16	WEIGHTED AVERAGE	99.4	72.8	87.3	97.3	56.8	78.9	97.4	80.7	89.8	96.0	59.2	79.3	97.2	61.7	81.1	95.7	59.2	79

		GEORGIA						HAWAII			IDAHO								
DIVISION		SAVANNAH			VALDOSTA			HONOLULU			BOISE			LEWISTON			POCATELLO		
		MAT.	INST.	TOTAL	MAT.	INST.	TOTAL	MAT.	INST.	TOTAL	MAT.	INST.	TOTAL	MAT.	INST.	TOTAL	MAT.	INST.	TOTA
01590	EQUIPMENT RENTAL	.0	91.2	91.2	.0	90.1	90.1	.0	99.3	99.3	.0	101.7	101.7	.0	94.6	94.6	.0	101.7	101
02	SITE CONSTRUCTION	102.1	77.1	83.6	111.8	76.4	85.6	139.1	106.5	115.0	79.9	102.7	96.8	84.1	96.9	93.6	81.1	102.6	97
03100	CONCRETE FORMS & ACCESSORIES	97.4	52.8	58.8	84.2	45.7	50.8	106.8	144.6	139.5	97.7	82.3	84.4	111.5	67.5	73.4	97.8	81.8	84.
03200	CONCRETE REINFORCEMENT	96.0	77.5	86.6	97.0	44.0	70.2	108.9	126.2	117.6	108.2	78.0	92.9	115.6	97.7	106.5	108.5	77.5	92
03300	CAST-IN-PLACE CONCRETE	92.7	51.6	75.5	94.0	52.2	76.5	199.7	129.5	170.3	95.9	91.0	93.9	104.0	86.7	96.8	95.2	90.8	93
03	CONCRETE	91.7	58.4	75.1	97.0	49.3	73.2	154.0	134.6	144.3	104.2	84.4	94.3	111.9	80.0	96.0	100.9	84.0	92
04	MASONRY	91.9	53.9	68.0	94.8	49.0	66.1	132.4	128.9	130.8	137.5	74.5	98.0	138.3	86.6	105.8	132.9	66.5	91
05	METALS	91.2	84.3	89.0	94.9	72.5	87.6	145.3	111.2	134.1	99.7	77.6	92.5	92.6	88.0	91.1	107.8	76.4	97
06	WOOD & PLASTICS	110.2	50.2	78.9	82.3	41.5	61.0	97.7	148.7	124.3	95.7	81.9	88.5	103.9	60.5	81.2	95.7	81.9	88
07	THERMAL & MOISTURE PROTECTION	95.5	53.0	74.6	95.2	57.4	76.7	113.0	130.0	121.4	96.9	80.0	88.6	167.0	80.7	124.6	97.1	70.5	84
08	DOORS & WINDOWS	99.6	50.9	87.0	93.8	38.7	79.5	105.8	140.1	114.7	94.7	77.8	90.3	114.8	67.5	102.6	94.7	71.0	88
09200	PLASTER & GYPSUM BOARD	100.5	49.4	68.1	91.2	40.4	59.0	116.7	149.8	137.7	83.2	81.2	81.9	134.0	59.3	86.7	83.2	81.2	81
095,098	CEILINGS & ACOUSTICAL TREATMENT	102.7	49.4	70.7	96.5	40.4	62.8	119.6	149.8	137.7	107.6	81.2	91.7	128.4	59.3	87.0	107.6	81.2	91
09600	FLOORING	118.6	48.8	100.1	110.4	39.0	91.6	168.4	134.9	159.5	99.8	56.3	88.3	137.7	93.5	126.1	100.1	56.3	88
097,099	WALL FINISHES, PAINTS & COATINGS	108.1	49.1	72.3	107.4	35.8	64.0	104.9	148.5	131.4	103.7	52.2	72.5	129.6	74.3	96.0	103.7	54.3	73
09	FINISHES	106.7	51.4	77.9	101.8	42.7	71.1	134.0	144.6	139.5	98.5	74.5	86.0	157.0	71.7	112.6	98.6	74.7	86
10 - 14	TOTAL DIV. 10000 - 14000	100.0	70.6	93.8	100.0	69.4	93.5	100.0	121.1	104.5	100.0	86.4	97.1	100.0	88.3	97.5	100.0	86.4	97
15	MECHANICAL	99.8	47.8	77.1	99.8	42.1	74.6	100.2	117.8	107.9	100.1	77.1	90.1	101.3	85.9	94.6	100.1	77.1	90
16	ELECTRICAL	93.8	58.9	76.8	90.6	30.2	61.1	108.0	124.1	115.9	93.5	78.3	86.1	79.1	85.1	82.0	88.7	73.9	81
01 - 16	WEIGHTED AVERAGE	97.1	58.8	79.7	97.0	49.1	75.2	121.0	125.7	123.1	100.2	80.5	91.2	108.1	83.3	96.8	100.3	78.3	90

Section 1

DIVISION		IDAHO TWIN FALLS			ILLINOIS CHICAGO			DECATUR			EAST ST. LOUIS			JOLIET			PEORIA		
		MAT.	INST.	TOTAL	MAT.	INST.	TOTAL	MAT.	INST.	TOTAL	MAT.	INST.	TOTAL	MAT.	INST.	TOTAL	MAT.	INST.	TOTAL
01590	EQUIPMENT RENTAL	.0	101.7	101.7	.0	91.5	91.5	.0	102.4	102.4	.0	109.0	109.0	.0	89.3	89.3	.0	101.6	101.6
02	SITE CONSTRUCTION	87.8	100.6	97.2	96.9	90.9	92.5	87.2	96.0	93.7	103.5	96.3	98.2	96.8	89.9	91.7	95.9	95.0	95.3
03100	CONCRETE FORMS & ACCESSORIES	98.5	35.2	43.7	101.6	140.3	135.2	97.1	103.2	102.4	91.0	102.4	100.9	102.5	131.8	127.8	95.4	108.0	106.3
03200	CONCRETE REINFORCEMENT	110.4	49.1	79.4	100.7	139.3	120.2	97.0	99.8	98.4	98.9	102.5	100.7	100.7	128.0	114.5	97.0	99.8	98.5
03300	CAST-IN-PLACE CONCRETE	97.7	45.2	75.7	113.2	136.6	123.0	98.9	103.5	100.8	92.6	108.3	99.2	113.1	121.2	116.4	96.6	106.9	100.9
03	CONCRETE	108.4	42.3	75.4	111.4	137.9	124.6	97.7	103.1	100.4	88.6	105.5	97.1	111.4	126.6	118.9	96.6	106.4	101.5
04	MASONRY	136.3	39.0	75.2	93.4	134.3	119.1	68.6	91.8	83.2	73.9	107.9	95.2	94.9	121.5	111.6	112.3	108.7	110.1
05	METALS	107.8	62.2	92.9	96.2	122.1	104.7	97.9	106.4	100.7	95.6	118.6	103.1	94.1	114.3	100.7	97.9	107.4	101.0
06	WOOD & PLASTICS	96.4	35.1	64.4	103.3	139.5	122.2	98.8	102.1	100.5	94.9	99.9	97.5	104.9	132.2	119.2	98.8	104.3	101.7
07	THERMAL & MOISTURE PROTECTION	98.2	46.7	72.9	100.0	131.3	115.4	97.2	94.0	95.6	92.3	101.5	96.8	99.8	123.0	111.2	97.1	105.3	101.1
08	DOORS & WINDOWS	98.0	36.0	82.0	103.4	139.3	112.7	96.3	101.4	97.6	86.1	107.7	91.7	101.4	132.5	109.4	96.3	102.6	98.0
09200	PLASTER & GYPSUM BOARD	82.9	33.1	51.4	102.8	140.5	126.7	104.4	102.0	102.9	100.5	99.8	100.1	100.0	133.1	121.0	104.4	104.2	104.3
095,098	CEILINGS & ACOUSTICAL TREATMENT	100.4	33.1	60.1	106.3	140.5	126.8	97.1	102.0	100.1	90.0	99.8	95.9	106.3	133.1	122.3	97.1	104.2	101.4
09600	FLOORING	101.5	56.3	89.5	83.2	127.6	94.9	96.4	84.6	93.3	105.1	104.5	104.9	82.7	116.2	91.6	96.4	99.7	97.3
097,099	WALL FINISHES, PAINTS & COATINGS	103.7	32.0	60.2	80.5	129.5	110.2	91.2	97.6	95.1	99.3	90.4	93.9	78.3	115.5	100.9	91.2	96.3	94.3
09	FINISHES	97.7	38.4	66.9	93.1	137.4	116.1	95.8	99.4	97.7	94.5	100.2	97.5	92.5	127.9	110.9	95.9	105.4	100.8
10 - 14	TOTAL DIV. 10000 - 14000	100.0	48.6	89.1	100.0	126.9	105.7	100.0	93.5	98.6	100.0	94.8	98.9	100.0	123.2	104.9	100.0	95.3	99.0
15	MECHANICAL	100.1	34.6	72.2	100.0	123.2	110.1	100.0	101.4	100.6	100.0	95.8	98.2	100.1	117.7	107.8	100.0	102.1	101.0
16	ELECTRICAL	81.0	31.5	56.9	98.0	119.9	108.7	98.8	91.3	95.1	96.1	91.7	94.0	96.3	106.3	101.2	98.1	88.7	93.5
01 - 16	WEIGHTED AVERAGE	100.8	45.8	75.8	99.7	125.8	111.6	96.5	98.5	97.4	94.2	101.1	97.3	99.0	117.2	107.3	98.7	101.7	100.1

Section 2

DIVISION		ILLINOIS ROCKFORD			SPRINGFIELD			INDIANA ANDERSON			BLOOMINGTON			EVANSVILLE			FORT WAYNE		
		MAT.	INST.	TOTAL	MAT.	INST.	TOTAL	MAT.	INST.	TOTAL	MAT.	INST.	TOTAL	MAT.	INST.	TOTAL	MAT.	INST.	TOTAL
01590	EQUIPMENT RENTAL	.0	101.6	101.6	.0	102.4	102.4	.0	96.7	96.7	.0	85.9	85.9	.0	121.3	121.3	.0	96.7	96.7
02	SITE CONSTRUCTION	95.0	95.6	95.4	93.3	96.0	95.3	87.6	96.6	94.2	77.1	95.6	90.8	82.1	130.1	117.6	88.6	96.5	94.4
03100	CONCRETE FORMS & ACCESSORIES	99.7	113.7	111.9	99.0	103.8	103.2	96.1	75.6	78.3	99.9	76.1	79.3	92.6	75.8	78.0	94.7	75.0	77.6
03200	CONCRETE REINFORCEMENT	97.0	130.2	113.8	97.0	99.8	98.4	93.8	81.2	87.4	83.9	81.0	82.5	92.0	81.2	86.6	93.8	78.0	85.8
03300	CAST-IN-PLACE CONCRETE	98.9	98.8	98.9	91.9	104.2	97.0	102.4	80.8	93.4	101.6	77.3	91.5	97.0	90.4	94.2	108.8	76.2	95.2
03	CONCRETE	97.9	112.0	104.9	94.5	103.6	99.0	96.7	79.1	87.9	103.1	77.2	90.2	103.4	82.1	92.8	99.6	76.6	88.2
04	MASONRY	83.1	110.8	100.5	69.8	92.8	84.3	91.3	77.5	82.6	90.6	75.9	81.4	86.8	81.2	83.3	91.6	77.7	82.9
05	METALS	97.9	123.4	106.2	100.5	106.6	102.5	90.9	89.1	90.3	93.4	75.1	87.4	86.2	86.3	86.2	90.9	87.1	89.7
06	WOOD & PLASTICS	98.8	112.0	105.7	101.5	102.1	101.8	113.2	74.1	92.8	118.9	74.7	95.8	97.7	72.2	84.4	112.9	73.6	92.4
07	THERMAL & MOISTURE PROTECTION	97.1	111.2	104.1	96.3	98.5	97.6	101.0	76.1	88.7	91.5	80.6	86.1	94.8	84.1	89.5	100.7	82.4	91.7
08	DOORS & WINDOWS	96.3	114.8	101.1	96.3	101.4	97.6	99.3	78.1	93.8	103.2	78.4	96.8	95.6	75.8	90.5	99.3	74.7	92.9
09200	PLASTER & GYPSUM BOARD	104.4	112.1	109.3	104.4	102.0	102.9	96.7	74.0	82.3	98.2	74.7	83.3	91.8	70.6	78.4	95.9	73.5	81.7
095,098	CEILINGS & ACOUSTICAL TREATMENT	97.1	112.1	106.1	97.1	102.0	100.1	89.4	74.0	80.2	81.7	74.7	77.5	86.4	70.6	77.0	89.4	73.5	79.9
09600	FLOORING	96.4	93.3	95.6	96.6	84.7	93.4	87.3	88.6	87.6	103.9	88.6	99.9	97.0	83.6	93.5	87.3	79.9	85.3
097,099	WALL FINISHES, PAINTS & COATINGS	91.2	108.8	101.9	91.2	90.2	90.6	92.1	67.6	77.3	89.8	86.2	87.7	96.5	88.1	91.4	92.1	74.7	81.6
09	FINISHES	95.9	109.3	102.9	95.9	98.8	97.4	88.9	76.8	82.6	94.7	79.2	86.7	92.9	77.9	85.1	88.7	75.6	81.8
10 - 14	TOTAL DIV. 10000 - 14000	100.0	103.9	100.8	100.0	93.9	98.7	100.0	81.7	96.1	100.0	81.2	96.0	100.0	84.4	96.7	100.0	83.6	96.5
15	MECHANICAL	100.0	105.7	102.5	100.0	103.7	101.7	99.9	81.0	91.7	99.6	82.9	92.3	99.8	85.0	93.3	99.9	78.6	90.6
16	ELECTRICAL	98.1	110.0	103.9	98.7	89.1	94.0	85.2	92.8	88.9	102.4	80.5	91.7	95.5	83.7	89.7	86.3	80.8	83.6
01 - 16	WEIGHTED AVERAGE	97.4	109.4	102.8	96.8	98.9	97.8	94.6	83.3	89.5	98.2	80.7	90.3	95.2	86.8	91.4	95.1	80.6	88.5

Section 3

DIVISION		INDIANA GARY			INDIANAPOLIS			MUNCIE			SOUTH BEND			TERRE HAUTE			IOWA CEDAR RAPIDS		
		MAT.	INST.	TOTAL	MAT.	INST.	TOTAL	MAT.	INST.	TOTAL	MAT.	INST.	TOTAL	MAT.	INST.	TOTAL	MAT.	INST.	TOTAL
01590	EQUIPMENT RENTAL	.0	96.7	96.7	.0	91.5	91.5	.0	96.2	96.2	.0	106.8	106.8	.0	121.3	121.3	.0	95.5	95.5
02	SITE CONSTRUCTION	88.2	98.3	95.7	87.5	100.4	97.0	77.2	95.8	90.9	87.7	96.1	93.9	83.4	130.0	117.9	87.6	94.6	92.8
03100	CONCRETE FORMS & ACCESSORIES	96.2	100.6	100.0	96.6	86.8	88.1	90.5	75.3	77.3	100.3	77.6	80.6	94.1	77.7	79.9	103.6	78.2	81.6
03200	CONCRETE REINFORCEMENT	93.8	99.6	96.7	93.2	85.4	89.2	92.9	81.1	86.9	93.8	76.4	85.0	92.0	80.7	86.3	93.3	82.6	87.9
03300	CAST-IN-PLACE CONCRETE	107.0	101.5	104.7	98.5	87.6	93.9	106.8	80.1	95.7	99.6	82.4	92.4	94.0	90.4	92.5	107.8	80.1	96.2
03	CONCRETE	98.9	100.9	99.9	98.8	86.4	92.6	101.8	78.7	90.3	91.4	80.5	86.0	106.3	82.9	94.6	99.7	80.4	90.1
04	MASONRY	92.7	93.1	93.0	98.2	86.6	90.9	92.8	77.5	83.2	89.8	77.4	82.0	93.7	78.2	84.0	106.2	79.6	89.5
05	METALS	90.9	101.1	94.2	92.8	80.1	88.7	95.1	88.8	93.0	90.9	100.9	94.2	86.9	86.4	86.8	87.2	92.2	88.8
06	WOOD & PLASTICS	111.2	100.9	105.8	110.1	86.4	97.7	111.0	73.9	91.6	113.1	77.0	94.3	99.9	76.2	87.5	109.6	76.9	92.5
07	THERMAL & MOISTURE PROTECTION	100.0	98.0	99.0	99.9	85.9	93.0	93.1	76.4	84.9	99.7	82.3	91.1	94.9	81.5	88.3	100.0	80.3	90.3
08	DOORS & WINDOWS	99.3	102.5	100.1	106.9	86.3	101.6	98.2	78.0	93.0	92.7	77.0	88.6	96.2	79.0	91.7	99.3	80.2	94.4
09200	PLASTER & GYPSUM BOARD	91.4	101.5	97.8	94.8	86.5	89.5	91.5	74.0	80.4	96.3	77.0	84.0	91.8	74.7	81.0	104.4	76.6	86.8
095,098	CEILINGS & ACOUSTICAL TREATMENT	89.4	101.5	96.7	93.9	86.5	89.4	82.5	74.0	77.4	89.4	77.0	82.0	86.4	74.7	79.4	112.4	76.6	90.9
09600	FLOORING	87.3	99.4	90.5	87.0	93.4	88.7	96.2	88.6	94.2	87.3	77.4	84.7	97.0	87.3	94.5	123.0	58.2	105.9
097,099	WALL FINISHES, PAINTS & COATINGS	92.1	101.8	98.0	92.1	89.3	90.4	89.8	67.6	76.4	92.1	81.0	85.4	96.5	87.7	91.2	107.7	77.4	89.3
09	FINISHES	88.1	100.5	94.6	90.0	88.4	89.1	91.4	76.7	83.7	88.8	77.8	83.1	92.9	80.2	86.3	111.8	73.5	91.8
10 - 14	TOTAL DIV. 10000 - 14000	100.0	98.5	99.7	100.0	85.1	96.8	100.0	81.2	96.0	100.0	77.6	95.2	100.0	85.4	96.9	100.0	81.8	96.1
15	MECHANICAL	99.9	94.7	97.6	99.9	88.6	95.0	99.6	81.0	91.4	99.9	76.9	89.8	99.8	78.6	90.5	100.4	83.6	93.0
16	ELECTRICAL	93.4	102.6	97.9	105.1	92.8	99.1	90.1	76.4	83.4	99.8	83.4	91.8	94.1	80.6	87.5	97.0	82.1	89.7
01 - 16	WEIGHTED AVERAGE	95.8	98.7	97.1	98.8	88.7	94.2	96.1	80.8	89.2	95.0	82.6	89.4	96.0	85.3	91.1	98.8	82.7	91.5

IOWA

DIVISION		COUNCIL BLUFFS			DAVENPORT			DES MOINES			DUBUQUE			SIOUX CITY			WATERLOO		
		MAT.	INST.	TOTAL	MAT.	INST.	TOTAL	MAT.	INST.	TOTAL	MAT.	INST.	TOTAL	MAT.	INST.	TOTAL	MAT.	INST.	TOTAL
01590	EQUIPMENT RENTAL	.0	95.2	95.2	.0	99.5	99.5	.0	101.5	101.5	.0	94.1	94.1	.0	99.5	99.5	.0	99.5	99.5
02	SITE CONSTRUCTION	92.8	91.2	91.6	85.8	98.4	95.1	79.3	99.9	94.6	85.7	91.5	90.0	94.7	95.4	95.2	86.3	94.5	92.4
03100	CONCRETE FORMS & ACCESSORIES	81.5	58.4	61.5	103.3	90.8	92.5	105.3	78.1	81.8	83.3	67.3	69.5	103.6	62.2	67.7	104.2	49.3	56.6
03200	CONCRETE REINFORCEMENT	95.2	78.5	86.8	93.3	95.7	94.5	93.3	79.3	86.2	92.0	82.3	87.1	93.3	67.3	80.2	93.3	82.1	87.6
03300	CAST-IN-PLACE CONCRETE	112.1	74.3	96.3	103.8	100.9	102.6	108.1	78.8	95.8	105.5	87.9	98.2	106.9	57.2	86.1	107.8	48.4	83.0
03	CONCRETE	101.5	68.8	85.2	97.8	95.7	96.7	98.8	79.4	89.1	96.1	78.2	87.1	99.3	62.5	80.9	99.7	56.6	78.2
04	MASONRY	105.8	73.9	85.8	103.8	87.6	93.7	100.0	80.3	87.6	107.2	72.0	85.1	99.1	59.5	74.3	100.7	59.5	74.8
05	METALS	92.4	89.5	91.5	87.2	101.6	91.9	87.3	92.3	89.0	85.8	91.4	87.6	87.2	83.7	86.0	87.2	90.2	88.2
06	WOOD & PLASTICS	85.2	54.5	69.2	109.6	89.1	98.9	111.1	77.0	93.3	87.2	64.7	75.5	109.6	60.7	84.1	110.3	49.3	78.5
07	THERMAL & MOISTURE PROTECTION	99.2	66.8	83.2	99.4	88.3	93.9	100.3	78.5	89.6	99.5	69.5	84.7	99.4	57.4	78.8	99.1	53.6	76.7
08	DOORS & WINDOWS	98.3	61.6	88.8	99.3	91.9	97.4	99.3	81.1	94.6	98.4	75.4	92.4	99.3	62.3	89.7	94.9	61.5	86.3
09200	PLASTER & GYPSUM BOARD	93.7	53.5	68.3	104.4	88.9	94.6	101.3	76.4	85.5	94.1	64.1	75.1	104.4	59.6	76.0	104.4	48.0	68.6
095,098	CEILINGS & ACOUSTICAL TREATMENT	107.0	53.5	74.9	112.4	88.9	98.3	110.6	76.4	90.1	107.0	64.1	81.3	112.4	59.6	80.7	112.4	48.0	73.7
09600	FLOORING	98.1	47.5	84.7	109.4	82.3	102.2	109.2	43.4	91.8	112.0	40.5	93.1	109.9	61.3	97.0	111.1	61.6	98.0
097,099	WALL FINISHES, PAINTS & COATINGS	99.2	72.1	82.8	103.5	94.9	98.3	103.5	78.5	88.3	106.5	77.4	88.9	104.7	63.0	79.4	104.7	36.8	63.5
09	FINISHES	100.8	56.4	77.7	107.4	89.4	98.0	105.6	71.0	87.6	105.5	62.2	83.0	108.6	61.7	84.2	108.0	49.5	77.6
10 - 14	TOTAL DIV. 10000 - 14000	100.0	74.3	94.5	100.0	86.3	97.1	100.0	82.5	96.3	100.0	78.7	95.5	100.0	77.7	95.2	100.0	72.1	94.1
15	MECHANICAL	100.4	76.9	90.1	100.4	94.9	98.0	100.4	80.7	91.7	100.4	78.9	91.0	100.4	82.0	92.3	100.4	46.9	77.0
16	ELECTRICAL	101.7	81.3	91.7	95.5	89.1	92.4	97.5	83.2	90.5	100.4	70.3	85.7	97.0	77.6	87.6	97.0	63.8	80.8
01 - 16	WEIGHTED AVERAGE	99.1	74.7	88.0	97.8	93.1	95.7	97.7	82.4	90.7	97.7	76.5	88.1	98.3	73.3	87.0	97.7	62.0	81.5

KANSAS / KENTUCKY

DIVISION		DODGE CITY			KANSAS CITY			SALINA			TOPEKA			WICHITA			BOWLING GREEN		
		MAT.	INST.	TOTAL	MAT.	INST.	TOTAL	MAT.	INST.	TOTAL	MAT.	INST.	TOTAL	MAT.	INST.	TOTAL	MAT.	INST.	TOTAL
01590	EQUIPMENT RENTAL	.0	103.5	103.5	.0	100.1	100.1	.0	103.5	103.5	.0	101.6	101.6	.0	103.5	103.5	.0	96.0	96.0
02	SITE CONSTRUCTION	109.9	92.7	97.1	91.0	90.9	90.9	99.9	93.0	94.8	92.9	90.4	91.1	94.1	93.4	93.6	69.5	100.3	92.3
03100	CONCRETE FORMS & ACCESSORIES	93.4	43.4	50.1	99.0	91.9	92.8	89.2	46.0	51.8	98.5	45.2	52.3	95.3	58.1	63.1	83.8	82.9	83.0
03200	CONCRETE REINFORCEMENT	105.1	58.5	81.5	99.8	82.4	91.0	104.5	80.0	92.1	97.0	94.5	95.8	97.0	81.1	89.0	83.3	74.5	78.9
03300	CAST-IN-PLACE CONCRETE	114.8	57.4	90.8	90.0	91.6	90.7	99.6	49.1	78.5	91.0	53.6	75.4	86.8	57.6	74.6	88.6	82.0	85.8
03	CONCRETE	114.8	52.6	83.8	96.2	90.5	93.3	102.0	55.1	78.6	94.1	59.0	76.6	91.9	63.5	77.7	94.9	81.1	88.0
04	MASONRY	104.2	42.2	65.3	103.5	87.5	93.5	118.5	40.6	69.7	98.1	64.0	76.7	91.8	65.8	75.5	93.9	79.6	84.9
05	METALS	95.3	78.2	89.7	100.1	94.5	98.3	95.1	88.2	92.8	100.5	96.9	99.3	100.5	90.3	97.2	91.2	80.0	87.5
06	WOOD & PLASTICS	91.7	43.4	66.5	98.5	92.3	95.2	87.9	46.7	66.4	95.7	41.3	67.3	93.7	56.8	74.4	93.3	82.3	87.5
07	THERMAL & MOISTURE PROTECTION	98.2	49.2	74.1	96.1	90.1	93.2	97.5	54.7	76.5	97.4	71.3	84.6	97.1	65.8	81.7	84.7	80.2	82.5
08	DOORS & WINDOWS	96.2	47.0	83.5	95.1	89.4	93.6	96.1	52.5	84.8	96.3	60.2	87.0	96.3	63.5	87.8	95.3	76.0	90.3
09200	PLASTER & GYPSUM BOARD	100.8	41.7	63.3	97.0	91.9	93.8	100.0	45.1	65.2	102.3	39.5	62.5	102.3	55.4	72.6	87.4	82.2	84.1
095,098	CEILINGS & ACOUSTICAL TREATMENT	88.2	41.7	60.3	89.1	91.9	90.8	88.2	45.1	62.3	89.1	39.5	59.4	89.1	55.4	68.9	86.4	82.2	83.9
09600	FLOORING	95.7	51.2	83.9	86.7	93.1	88.4	93.7	37.4	78.8	97.2	47.5	84.1	96.4	79.3	91.9	92.9	63.0	85.0
097,099	WALL FINISHES, PAINTS & COATINGS	91.2	49.9	66.2	99.3	89.5	93.4	91.2	39.3	59.7	91.2	69.5	78.0	91.2	63.4	74.3	96.5	75.6	83.8
09	FINISHES	93.9	45.1	68.5	92.2	91.2	91.7	92.6	42.9	66.7	93.8	46.0	68.9	93.6	62.2	77.3	91.1	78.5	84.5
10 - 14	TOTAL DIV. 10000 - 14000	100.0	50.2	89.4	100.0	84.4	96.7	100.0	66.1	92.8	100.0	70.0	93.6	100.0	70.7	93.8	100.0	64.2	92.4
15	MECHANICAL	100.0	47.9	77.2	99.9	88.2	94.8	100.0	36.3	72.1	100.0	70.6	87.1	100.0	67.9	86.0	99.8	88.6	94.9
16	ELECTRICAL	99.7	50.3	75.6	104.4	102.1	103.3	99.4	72.6	86.3	103.3	75.1	89.5	101.5	72.6	87.4	94.5	84.2	89.5
01 - 16	WEIGHTED AVERAGE	100.3	54.8	79.6	98.7	91.6	95.5	99.0	57.2	80.1	98.5	69.3	85.2	97.7	71.1	85.6	94.4	83.6	89.5

KENTUCKY / LOUISIANA

DIVISION		LEXINGTON			LOUISVILLE			OWENSBORO			ALEXANDRIA			BATON ROUGE			LAKE CHARLES		
		MAT.	INST.	TOTAL	MAT.	INST.	TOTAL	MAT.	INST.	TOTAL	MAT.	INST.	TOTAL	MAT.	INST.	TOTAL	MAT.	INST.	TOTAL
01590	EQUIPMENT RENTAL	.0	103.5	103.5	.0	96.0	96.0	.0	121.3	121.3	.0	86.8	86.8	.0	86.2	86.2	.0	86.2	86.2
02	SITE CONSTRUCTION	77.3	102.4	95.8	67.0	100.2	91.5	82.1	129.6	117.2	102.8	85.7	90.1	109.3	85.3	91.5	110.9	85.0	91.8
03100	CONCRETE FORMS & ACCESSORIES	97.6	70.3	74.0	92.5	82.6	83.9	88.7	64.7	67.9	80.0	44.3	49.1	96.9	48.4	54.9	97.4	49.2	55.7
03200	CONCRETE REINFORCEMENT	92.0	96.6	94.3	92.0	97.8	94.9	83.4	93.9	88.7	98.4	63.8	80.9	93.6	51.2	72.2	93.6	51.2	72.2
03300	CAST-IN-PLACE CONCRETE	95.7	80.1	89.2	92.7	80.5	87.6	91.4	76.7	85.3	93.9	44.2	73.2	85.9	52.3	71.9	91.0	53.8	75.5
03	CONCRETE	97.3	79.0	88.2	95.5	84.8	90.1	104.3	74.8	89.5	89.0	49.1	69.1	93.5	51.1	72.3	96.0	52.0	74.0
04	MASONRY	90.8	48.1	64.0	92.4	81.8	85.8	90.8	52.0	66.4	115.3	54.0	76.9	96.6	55.6	70.9	94.4	56.6	70.7
05	METALS	92.6	89.3	91.5	100.0	90.3	96.8	83.1	87.4	84.5	85.7	74.4	82.0	95.7	65.5	85.8	90.1	65.7	82.1
06	WOOD & PLASTICS	104.3	73.3	88.1	102.8	82.3	92.1	93.5	63.6	77.9	82.5	43.9	62.3	103.8	48.3	74.8	102.0	49.0	74.3
07	THERMAL & MOISTURE PROTECTION	95.3	78.7	87.2	85.0	82.0	83.5	94.8	67.5	81.4	99.4	54.2	77.2	97.4	55.2	76.7	100.7	55.3	78.4
08	DOORS & WINDOWS	96.2	80.1	92.0	96.2	87.1	93.6	93.6	75.3	88.8	99.3	50.0	86.6	104.3	48.6	89.9	104.3	50.6	90.4
09200	PLASTER & GYPSUM BOARD	94.8	71.8	80.2	95.9	82.2	87.2	86.5	61.8	70.9	79.1	43.0	56.2	99.1	47.5	66.4	99.1	48.2	66.8
095,098	CEILINGS & ACOUSTICAL TREATMENT	91.8	71.8	79.8	91.8	82.2	86.0	75.7	61.8	67.4	90.3	43.0	61.9	96.7	47.5	67.2	96.7	48.2	67.6
09600	FLOORING	97.9	43.8	83.6	96.4	74.8	90.7	95.3	63.0	86.8	105.0	68.9	95.5	109.4	67.6	98.4	109.7	52.8	94.6
097,099	WALL FINISHES, PAINTS & COATINGS	96.5	59.7	74.2	96.5	84.5	89.2	96.5	99.6	98.4	96.4	42.1	63.5	98.7	48.3	68.1	98.7	44.5	65.8
09	FINISHES	95.0	63.7	78.7	94.4	81.1	87.5	89.0	67.4	77.8	93.2	48.1	69.7	102.1	51.8	75.9	102.2	49.0	74.5
10 - 14	TOTAL DIV. 10000 - 14000	100.0	81.9	96.2	100.0	85.3	96.9	100.0	84.3	96.7	100.0	71.6	94.0	100.0	75.3	94.7	100.0	75.5	94.8
15	MECHANICAL	99.8	48.8	77.5	99.8	87.3	94.3	99.8	53.4	79.5	100.1	38.5	73.1	100.0	43.7	75.4	100.0	59.3	82.2
16	ELECTRICAL	94.9	50.8	73.4	94.9	84.2	89.7	94.0	82.3	88.3	92.4	56.4	74.8	92.8	50.4	72.1	92.8	63.1	78.4
01 - 16	WEIGHTED AVERAGE	95.9	66.9	82.7	96.3	86.4	91.8	94.2	74.3	85.2	95.6	54.7	77.0	98.4	55.1	78.7	97.8	60.0	80.7

Block 1

DIVISION		LOUISIANA MONROE MAT.	INST.	TOTAL	NEW ORLEANS MAT.	INST.	TOTAL	SHREVEPORT MAT.	INST.	TOTAL	MAINE AUGUSTA MAT.	INST.	TOTAL	BANGOR MAT.	INST.	TOTAL	LEWISTON MAT.	INST.	TOTAL
01590	EQUIPMENT RENTAL	.0	86.8	86.8	.0	87.4	87.4	.0	86.8	86.8	.0	101.6	101.6	.0	101.6	101.6	.0	101.6	101.6
02	SITE CONSTRUCTION	102.8	85.2	89.8	113.4	87.8	94.5	101.5	85.1	89.3	83.4	102.3	97.4	83.3	100.7	96.2	82.0	100.7	95.8
03100	CONCRETE FORMS & ACCESSORIES	79.3	44.9	49.5	97.2	69.5	73.2	98.6	47.6	54.4	99.6	64.9	69.6	93.9	80.3	82.1	80.3	80.3	82.9
03200	CONCRETE REINFORCEMENT	97.4	64.0	80.5	93.6	65.0	79.2	96.9	63.7	80.1	88.1	112.7	100.5	88.1	113.0	100.7	108.2	113.0	110.7
03300	CAST-IN-PLACE CONCRETE	93.9	48.1	74.8	90.1	60.0	77.5	92.6	51.8	75.5	81.5	63.2	73.9	81.5	65.1	74.7	92.8	65.1	81.2
03	CONCRETE	88.8	50.6	69.7	95.5	65.8	80.7	88.8	53.0	70.9	98.6	73.3	86.0	97.7	80.7	89.2	102.8	80.7	91.7
04	MASONRY	110.2	58.1	77.5	96.3	64.3	76.2	102.4	51.1	70.2	96.1	52.3	68.6	113.7	60.2	80.1	97.9	60.2	74.2
05	METALS	85.7	73.0	81.5	103.1	75.1	94.0	82.1	72.2	78.8	87.9	84.9	86.9	87.5	83.1	86.1	90.7	83.1	88.2
06	WOOD & PLASTICS	81.7	44.3	62.2	98.4	72.7	85.0	102.2	48.2	74.0	98.5	63.4	80.2	92.3	82.6	87.2	98.5	82.6	90.2
07	THERMAL & MOISTURE PROTECTION	99.4	57.0	78.5	101.6	65.2	83.7	98.3	55.6	77.3	102.1	52.5	77.7	102.0	59.5	81.1	101.8	59.5	81.0
08	DOORS & WINDOWS	99.3	55.0	87.9	104.9	69.4	95.7	97.2	52.5	85.6	103.9	62.1	93.1	103.8	74.3	96.2	107.1	74.3	98.6
09200	PLASTER & GYPSUM BOARD	78.7	43.4	56.3	97.9	72.5	81.8	85.4	47.4	61.3	102.8	61.4	76.6	98.9	81.2	87.6	105.2	81.2	89.9
095,098	CEILINGS & ACOUSTICAL TREATMENT	90.3	43.4	62.2	95.7	72.5	81.8	91.2	47.4	65.0	92.2	61.4	73.7	89.5	81.2	84.5	101.1	81.2	89.1
09600	FLOORING	104.6	43.6	88.5	110.1	51.8	94.7	115.6	64.6	102.1	99.8	43.9	85.0	97.3	54.7	86.1	99.8	54.7	87.9
097,099	WALL FINISHES, PAINTS & COATINGS	96.4	50.3	68.4	100.5	63.9	78.3	96.4	42.1	63.5	92.8	36.4	58.6	92.8	33.5	56.8	92.8	33.5	56.8
09	FINISHES	93.0	44.3	67.7	102.3	65.8	83.3	97.4	49.7	72.6	99.1	57.1	77.3	97.2	70.9	83.5	101.4	70.9	85.5
10-14	TOTAL DIV. 10000 - 14000	100.0	65.4	92.6	100.0	80.3	95.8	100.0	65.7	92.7	100.0	63.9	92.3	100.0	83.0	96.4	100.0	83.0	96.4
15	MECHANICAL	100.1	52.0	79.1	100.0	65.9	85.1	100.1	54.6	80.2	100.1	72.3	87.6	100.1	71.5	87.6	100.1	71.5	87.6
16	ELECTRICAL	94.1	55.9	75.4	94.0	70.5	82.6	92.9	67.0	80.3	100.2	80.9	90.8	98.5	80.9	89.9	100.3	80.9	90.8
01-16	WEIGHTED AVERAGE	95.5	57.5	78.2	100.2	69.8	86.4	94.7	59.7	78.8	97.7	72.3	86.2	98.0	76.8	88.3	99.2	76.8	89.0

Block 2

DIVISION		MAINE PORTLAND MAT.	INST.	TOTAL	MARYLAND BALTIMORE MAT.	INST.	TOTAL	HAGERSTOWN MAT.	INST.	TOTAL	MASSACHUSETTS BOSTON MAT.	INST.	TOTAL	BROCKTON MAT.	INST.	TOTAL	FALL RIVER MAT.	INST.	TOTAL
01590	EQUIPMENT RENTAL	.0	101.6	101.6	.0	103.4	103.4	.0	99.2	99.2	.0	108.4	108.4	.0	103.8	103.8	.0	105.0	105.0
02	SITE CONSTRUCTION	80.8	100.7	95.5	97.6	93.3	94.4	87.4	88.9	88.5	88.4	108.9	103.6	86.1	104.8	100.0	85.2	105.0	99.8
03100	CONCRETE FORMS & ACCESSORIES	98.8	80.3	82.7	101.7	74.8	78.4	90.3	76.1	78.0	105.1	135.6	131.5	104.7	120.9	118.8	104.7	121.2	119.0
03200	CONCRETE REINFORCEMENT	108.2	113.0	110.7	95.9	87.8	91.8	84.8	76.3	80.5	107.1	143.7	125.6	108.2	142.8	125.7	108.2	125.6	117.0
03300	CAST-IN-PLACE CONCRETE	84.5	65.1	76.4	98.3	78.7	90.1	83.6	61.7	74.4	105.2	143.5	121.2	100.2	129.2	112.3	96.9	141.6	115.5
03	CONCRETE	98.8	80.7	89.7	99.5	79.7	89.6	85.3	72.2	78.8	111.7	138.9	125.3	109.0	127.2	118.1	107.4	128.4	117.9
04	MASONRY	95.4	60.2	73.3	92.8	74.6	81.4	97.0	75.9	83.8	116.7	144.8	134.3	112.6	127.3	121.9	112.4	140.0	129.7
05	METALS	92.0	83.1	89.1	100.3	99.3	99.9	96.5	91.4	94.0	103.3	125.9	110.7	100.0	121.6	107.1	100.0	115.7	105.2
06	WOOD & PLASTICS	98.5	82.6	90.2	97.9	75.7	86.3	85.9	75.6	80.5	102.3	136.3	120.0	101.3	120.4	111.3	101.3	120.8	111.4
07	THERMAL & MOISTURE PROTECTION	101.6	59.5	80.9	94.2	81.7	88.1	93.4	71.6	82.7	102.4	141.9	121.8	102.1	130.1	115.9	102.1	127.9	114.8
08	DOORS & WINDOWS	107.1	74.3	98.6	93.8	83.3	91.1	91.1	77.1	87.5	103.1	135.4	111.4	101.0	123.2	106.8	101.0	119.9	105.9
09200	PLASTER & GYPSUM BOARD	105.2	81.2	89.9	108.7	75.3	87.5	103.5	75.3	85.6	107.8	136.5	126.0	102.5	120.1	113.6	102.5	120.1	113.6
095,098	CEILINGS & ACOUSTICAL TREATMENT	101.1	81.2	89.1	104.1	75.3	86.8	105.1	75.3	87.2	100.1	136.5	121.9	102.1	120.1	112.9	102.1	120.1	112.9
09600	FLOORING	99.8	54.7	87.9	92.8	82.2	90.0	87.9	78.8	85.5	100.4	153.6	114.5	100.9	153.6	114.8	100.7	153.6	114.7
097,099	WALL FINISHES, PAINTS & COATINGS	92.8	33.5	56.8	96.0	71.1	80.9	96.0	41.1	62.7	95.9	150.7	129.1	95.3	119.2	109.8	95.3	119.2	109.8
09	FINISHES	101.4	70.9	85.6	97.3	75.2	85.8	94.7	72.9	83.4	101.7	141.0	122.1	101.5	127.0	114.8	101.5	127.2	114.9
10-14	TOTAL DIV. 10000 - 14000	100.0	83.0	96.4	100.0	87.2	97.3	100.0	89.7	97.8	100.0	122.5	104.8	100.0	118.7	104.0	100.0	119.6	104.2
15	MECHANICAL	100.1	71.5	87.6	100.0	85.7	93.7	100.0	85.7	93.7	100.1	125.8	111.4	100.1	105.5	102.5	100.1	105.4	102.4
16	ELECTRICAL	100.4	80.9	90.9	107.1	90.7	99.1	103.6	82.1	93.1	99.3	123.3	111.0	98.3	96.3	97.3	97.8	102.2	99.9
01-16	WEIGHTED AVERAGE	98.8	76.8	88.8	99.4	84.8	92.7	96.0	80.8	89.1	102.9	130.5	115.4	101.4	115.5	107.8	101.1	117.0	108.4

Block 3 — MASSACHUSETTS

DIVISION		HYANNIS MAT.	INST.	TOTAL	LAWRENCE MAT.	INST.	TOTAL	LOWELL MAT.	INST.	TOTAL	NEW BEDFORD MAT.	INST.	TOTAL	PITTSFIELD MAT.	INST.	TOTAL	SPRINGFIELD MAT.	INST.	TOTAL
01590	EQUIPMENT RENTAL	.0	103.8	103.8	.0	103.8	103.8	.0	101.6	101.6	.0	105.0	105.0	.0	101.6	101.6	.0	101.6	101.6
02	SITE CONSTRUCTION	82.7	104.8	99.1	86.7	104.8	100.1	85.8	104.8	99.9	84.8	105.0	99.7	86.9	103.6	99.2	86.3	103.7	99.2
03100	CONCRETE FORMS & ACCESSORIES	95.4	120.9	117.5	104.5	121.4	119.1	101.2	121.6	118.9	104.7	121.2	119.0	101.2	101.2	101.2	101.4	109.8	108.7
03200	CONCRETE REINFORCEMENT	86.8	125.6	106.4	107.4	130.6	119.1	108.2	130.7	119.6	108.2	125.6	117.0	90.2	105.9	98.1	108.2	124.8	116.6
03300	CAST-IN-PLACE CONCRETE	91.4	140.8	112.0	101.0	139.5	117.1	91.8	139.6	111.8	94.0	141.6	113.9	100.2	116.0	106.8	95.5	117.3	104.6
03	CONCRETE	98.6	127.9	113.3	109.2	128.7	118.9	100.0	128.6	114.3	106.1	128.4	117.2	101.0	106.7	103.9	101.8	114.5	108.2
04	MASONRY	111.5	140.1	129.4	111.9	138.4	128.5	97.8	138.5	123.3	112.2	140.0	129.6	98.4	112.1	107.0	98.1	114.1	108.2
05	METALS	96.3	115.2	102.5	97.3	117.6	103.9	97.3	114.9	103.0	100.0	115.7	105.2	97.1	99.4	97.8	99.9	107.5	102.4
06	WOOD & PLASTICS	91.3	120.4	106.5	101.3	120.4	111.3	100.4	120.4	110.9	101.3	120.8	111.4	100.4	100.4	100.4	100.4	110.6	105.7
07	THERMAL & MOISTURE PROTECTION	101.6	128.8	115.0	102.1	134.4	118.0	101.8	134.4	117.8	102.1	127.9	114.8	101.9	109.1	105.4	101.8	111.0	106.3
08	DOORS & WINDOWS	97.2	119.7	103.0	101.0	119.8	105.9	107.1	119.8	110.4	101.0	119.9	105.9	107.1	103.9	106.2	107.1	114.5	109.0
09200	PLASTER & GYPSUM BOARD	94.2	120.1	110.6	105.2	120.1	114.6	105.2	120.1	114.6	102.5	120.1	113.6	105.2	99.4	101.5	105.2	110.0	108.2
095,098	CEILINGS & ACOUSTICAL TREATMENT	90.5	120.1	108.2	101.1	120.1	112.5	101.1	120.1	112.5	102.1	120.1	112.9	101.1	99.4	100.1	101.1	110.0	106.4
09600	FLOORING	96.9	153.6	111.9	99.8	153.6	114.0	99.8	153.6	114.0	100.7	153.6	114.7	100.0	126.7	107.1	99.7	126.7	106.9
097,099	WALL FINISHES, PAINTS & COATINGS	95.3	119.2	109.8	92.9	119.2	108.8	92.8	119.2	108.8	95.3	119.2	109.8	92.8	99.0	96.6	94.6	99.0	97.3
09	FINISHES	96.1	127.0	112.1	101.2	127.0	114.6	101.2	127.0	114.6	101.4	127.2	114.8	101.2	105.6	103.5	101.3	112.1	106.9
10-14	TOTAL DIV. 10000 - 14000	100.0	118.7	104.0	100.0	118.8	104.0	100.0	118.8	104.0	100.0	119.6	104.2	100.0	102.7	100.6	100.0	104.8	101.0
15	MECHANICAL	100.1	105.3	102.4	100.1	109.6	104.3	100.1	122.8	110.0	100.1	105.4	102.4	100.1	95.3	98.0	100.1	98.4	99.4
16	ELECTRICAL	96.5	102.2	99.3	99.8	123.3	111.3	100.3	123.3	111.5	99.3	102.2	100.7	100.3	88.7	94.7	100.3	88.0	94.3
01-16	WEIGHTED AVERAGE	98.3	116.9	106.7	101.1	121.0	110.2	100.0	123.5	110.7	101.1	117.0	108.4	100.2	101.1	100.6	100.7	105.1	102.7

		MASSACHUSETTS			MICHIGAN														
	DIVISION	WORCESTER			ANN ARBOR			DEARBORN			DETROIT			FLINT			GRAND RAPIDS		
		MAT.	INST.	TOTAL	MAT.	INST.	TOTAL	MAT.	INST.	TOTAL	MAT.	INST.	TOTAL	MAT.	INST.	TOTAL	MAT.	INST.	TOTAL
01590	EQUIPMENT RENTAL	.0	101.6	101.6	.0	112.2	112.2	.0	112.2	112.2	.0	97.7	97.7	.0	112.2	112.2	.0	105.1	105.1
02	SITE CONSTRUCTION	85.8	104.8	99.8	78.1	95.2	90.7	77.9	95.5	90.9	91.5	97.2	95.8	69.0	94.2	87.6	82.2	87.8	86.3
03100	CONCRETE FORMS & ACCESSORIES	101.8	121.0	118.5	97.6	110.4	108.7	97.4	123.0	119.6	99.0	123.1	119.9	101.1	93.5	94.5	97.2	79.1	81.5
03200	CONCRETE REINFORCEMENT	108.2	142.5	125.6	97.1	124.3	110.9	97.1	125.1	111.3	96.5	125.1	110.9	97.1	124.0	110.7	92.0	86.7	89.3
03300	CAST-IN-PLACE CONCRETE	91.8	139.0	111.5	94.3	111.2	101.4	92.1	122.0	104.6	101.2	122.0	109.9	94.9	93.6	94.4	95.9	99.9	97.6
03	CONCRETE	100.1	130.4	115.2	93.0	113.9	103.4	91.9	123.3	107.6	96.2	121.9	109.1	93.5	100.2	96.9	97.3	87.1	92.2
04	MASONRY	97.8	138.0	123.0	98.5	106.2	103.3	98.3	122.1	113.2	97.3	122.1	112.9	98.5	92.6	94.8	94.9	56.0	70.5
05	METALS	100.0	119.1	106.2	100.6	126.0	108.9	100.7	127.9	109.6	105.2	105.7	105.4	100.7	123.3	108.1	92.9	80.6	88.9
06	WOOD & PLASTICS	100.9	120.4	111.1	101.4	109.6	105.7	101.4	123.0	112.7	102.3	123.0	113.1	104.9	94.1	99.3	98.2	79.7	88.5
07	THERMAL & MOISTURE PROTECTION	101.8	129.0	115.2	97.7	109.2	103.4	96.6	124.2	110.2	95.2	124.2	109.5	95.9	92.5	94.2	91.7	62.4	77.3
08	DOORS & WINDOWS	107.1	126.5	112.1	96.5	111.1	100.3	96.5	118.4	102.2	98.0	120.3	103.8	96.5	100.2	97.4	94.2	71.0	88.2
09200	PLASTER & GYPSUM BOARD	105.2	120.1	114.6	102.4	108.6	106.3	102.4	122.3	115.1	102.4	122.3	115.1	104.0	92.6	96.8	93.2	74.2	81.2
095,098	CEILINGS & ACOUSTICAL TREATMENT	101.1	120.1	112.5	95.6	108.6	103.4	95.6	122.3	111.6	96.6	122.3	112.0	95.6	92.6	93.8	91.8	74.2	81.2
09600	FLOORING	99.8	153.6	114.0	87.7	109.8	93.5	87.2	123.8	96.9	87.3	123.8	97.0	87.4	74.9	84.1	96.6	44.2	82.8
097,099	WALL FINISHES, PAINTS & COATINGS	92.8	119.2	108.8	86.5	103.3	96.7	86.5	114.0	103.1	88.3	114.0	103.9	86.5	88.1	87.5	96.5	42.7	63.8
09	FINISHES	101.2	127.0	114.6	93.9	109.5	102.0	93.8	122.5	108.7	94.8	122.5	109.2	93.6	89.0	91.2	94.3	68.5	80.9
10 - 14	TOTAL DIV. 10000 - 14000	100.0	108.3	101.8	100.0	111.3	102.4	100.0	114.9	103.2	100.0	114.9	103.2	100.0	96.0	99.2	100.0	101.4	100.3
15	MECHANICAL	100.1	108.8	103.9	99.9	103.3	101.4	99.9	116.9	107.3	99.9	119.7	108.6	99.9	93.6	97.1	99.8	57.0	81.1
16	ELECTRICAL	100.3	104.5	102.3	94.1	87.9	91.1	94.1	116.4	105.0	95.9	116.3	105.8	94.1	88.9	91.5	94.6	61.3	78.4
01 - 16	WEIGHTED AVERAGE	100.5	118.4	108.6	96.9	105.7	100.9	96.7	118.3	106.5	98.7	116.9	107.0	96.7	96.2	96.5	95.8	70.2	84.2

		MICHIGAN												MINNESOTA					
	DIVISION	KALAMAZOO			LANSING			MUSKEGON			SAGINAW			DULUTH			MINNEAPOLIS		
		MAT.	INST.	TOTAL	MAT.	INST.	TOTAL	MAT.	INST.	TOTAL	MAT.	INST.	TOTAL	MAT.	INST.	TOTAL	MAT.	INST.	TOTAL
01590	EQUIPMENT RENTAL	.0	105.1	105.1	.0	112.2	112.2	.0	105.1	105.1	.0	112.2	112.2	.0	101.8	101.8	.0	106.1	106.1
02	SITE CONSTRUCTION	82.7	88.1	86.7	84.1	94.1	91.5	80.5	88.0	86.1	71.2	93.9	88.0	82.7	102.5	97.4	82.6	108.8	102.0
03100	CONCRETE FORMS & ACCESSORIES	96.7	88.5	89.6	101.1	91.5	92.8	97.4	85.8	87.3	97.6	92.2	92.9	103.5	123.7	121.0	104.2	137.5	133.0
03200	CONCRETE REINFORCEMENT	92.0	88.3	90.1	97.1	123.8	110.6	92.5	88.1	90.3	97.1	123.4	110.4	95.3	109.0	102.2	95.4	130.0	112.9
03300	CAST-IN-PLACE CONCRETE	97.6	100.4	98.8	94.3	94.7	94.5	95.3	98.3	96.5	93.1	93.5	93.3	109.6	104.2	107.3	107.9	126.9	115.8
03	CONCRETE	100.5	91.8	96.2	93.2	99.7	96.4	95.4	89.8	92.6	92.4	99.5	95.9	100.2	114.7	107.0	101.0	132.8	116.9
04	MASONRY	97.9	84.3	89.4	91.9	94.6	93.6	95.1	80.1	85.7	100.0	86.1	91.2	107.9	113.8	111.6	107.9	136.5	125.9
05	METALS	91.5	84.5	89.2	99.1	122.9	106.8	89.2	84.1	87.5	100.7	121.9	107.7	93.9	125.5	104.2	96.7	139.9	110.8
06	WOOD & PLASTICS	99.9	88.2	93.8	104.3	89.6	96.6	96.7	85.2	90.7	97.4	92.4	94.8	114.7	127.0	121.2	115.2	136.5	126.3
07	THERMAL & MOISTURE PROTECTION	91.3	84.4	87.9	96.7	92.5	94.6	90.4	78.7	84.7	96.5	88.9	92.8	101.5	116.6	108.9	101.3	136.3	118.5
08	DOORS & WINDOWS	90.9	83.1	88.9	96.5	97.7	96.8	90.1	82.0	88.0	96.0	98.2	96.6	96.8	124.6	104.0	99.9	143.7	111.2
09200	PLASTER & GYPSUM BOARD	93.2	83.0	86.8	105.1	87.9	94.2	84.1	79.9	81.4	102.4	90.9	95.1	102.3	128.3	118.8	102.5	137.8	124.9
095,098	CEILINGS & ACOUSTICAL TREATMENT	91.8	83.0	86.5	95.6	87.9	91.0	93.6	79.9	85.3	95.6	90.9	92.8	93.4	128.3	114.3	94.2	137.8	120.4
09600	FLOORING	96.6	70.1	89.6	95.8	83.0	92.4	95.8	81.5	92.0	87.7	59.5	80.2	105.8	121.7	110.0	103.2	126.4	109.3
097,099	WALL FINISHES, PAINTS & COATINGS	96.5	86.6	90.5	100.4	94.4	96.8	95.7	61.9	75.2	86.5	81.1	83.2	93.3	103.0	99.2	99.6	127.1	116.3
09	FINISHES	94.3	84.7	89.3	98.2	89.7	93.8	92.7	82.1	87.2	93.7	84.4	88.9	100.6	122.4	111.9	100.4	134.8	118.3
10 - 14	TOTAL DIV. 10000 - 14000	100.0	103.7	100.8	100.0	97.1	99.4	100.0	103.0	100.6	100.0	96.5	99.2	100.0	99.3	99.8	100.0	109.4	102.0
15	MECHANICAL	99.8	86.3	93.9	99.9	93.2	97.0	99.7	89.7	95.4	99.9	91.0	96.0	100.2	94.8	97.8	100.1	122.2	109.8
16	ELECTRICAL	94.2	71.8	83.2	92.3	88.2	90.3	94.7	67.3	81.4	93.5	83.2	88.5	105.8	95.1	100.6	107.7	112.4	110.0
01 - 16	WEIGHTED AVERAGE	95.7	84.9	90.8	96.7	96.1	96.4	94.3	83.6	89.5	96.5	93.2	95.0	99.7	108.9	103.8	100.7	126.8	112.6

		MINNESOTA									MISSISSIPPI								
	DIVISION	ROCHESTER			SAINT PAUL			ST. CLOUD			BILOXI			GREENVILLE			JACKSON		
		MAT.	INST.	TOTAL	MAT.	INST.	TOTAL	MAT.	INST.	TOTAL	MAT.	INST.	TOTAL	MAT.	INST.	TOTAL	MAT.	INST.	TOTAL
01590	EQUIPMENT RENTAL	.0	101.8	101.8	.0	101.8	101.8	.0	101.5	101.5	.0	98.3	98.3	.0	98.3	98.3	.0	98.3	98.3
02	SITE CONSTRUCTION	81.8	101.7	96.5	84.7	103.3	98.5	78.3	105.2	98.2	103.3	86.2	90.6	107.4	86.0	91.5	99.9	86.0	89.6
03100	CONCRETE FORMS & ACCESSORIES	104.0	104.5	104.4	94.5	130.9	126.0	84.5	126.8	121.1	95.5	42.5	49.6	80.3	35.1	41.1	95.3	40.0	47.4
03200	CONCRETE REINFORCEMENT	95.3	129.2	112.4	91.9	129.9	111.1	94.3	129.2	111.9	95.0	62.4	78.6	103.0	44.9	73.6	95.0	48.7	71.6
03300	CAST-IN-PLACE CONCRETE	105.8	98.2	102.6	108.9	126.0	116.0	97.8	124.0	108.7	104.5	46.3	80.2	104.9	40.5	78.0	102.5	43.0	77.7
03	CONCRETE	98.4	108.1	103.2	102.5	129.5	116.0	90.0	126.7	108.4	97.0	49.5	73.3	100.2	40.9	70.6	96.0	44.7	70.4
04	MASONRY	107.2	109.5	108.6	117.7	136.5	129.5	109.2	122.2	117.4	89.3	38.1	57.2	131.8	39.5	73.9	91.3	39.5	58.8
05	METALS	93.7	137.1	107.9	93.2	139.2	108.3	90.0	136.2	105.1	92.1	83.2	89.2	90.4	74.9	85.4	92.1	76.8	87.1
06	WOOD & PLASTICS	115.2	103.7	109.2	104.4	128.1	116.8	91.7	124.5	108.8	96.6	42.4	68.3	79.0	34.4	55.7	96.7	40.8	67.5
07	THERMAL & MOISTURE PROTECTION	101.3	103.7	102.5	101.2	133.9	117.3	98.6	117.3	107.8	95.2	47.3	71.6	95.3	42.4	69.3	95.0	43.6	69.8
08	DOORS & WINDOWS	96.8	125.9	104.3	94.1	139.1	105.8	91.6	137.2	103.4	98.5	48.4	85.5	97.8	39.2	82.7	98.9	43.8	84.6
09200	PLASTER & GYPSUM BOARD	102.0	104.3	103.5	96.3	129.4	117.3	87.0	125.8	111.6	103.7	41.3	64.2	91.2	33.1	54.4	103.7	39.8	63.2
095,098	CEILINGS & ACOUSTICAL TREATMENT	92.5	104.3	99.6	90.7	129.4	113.9	72.1	125.8	104.3	102.7	41.3	65.9	96.5	33.1	58.5	102.7	39.8	65.0
09600	FLOORING	105.6	82.1	99.4	98.6	126.4	106.0	97.4	126.4	105.1	118.6	41.4	98.2	109.1	35.4	89.6	118.6	40.9	98.1
097,099	WALL FINISHES, PAINTS & COATINGS	95.7	103.5	100.4	99.6	120.5	112.3	103.4	127.1	117.8	107.4	36.5	64.4	107.4	35.9	64.0	107.4	35.9	64.0
09	FINISHES	100.3	100.6	100.5	97.5	129.1	113.9	90.9	127.0	109.7	106.9	41.1	72.7	101.1	34.9	66.7	106.9	39.9	72.0
10 - 14	TOTAL DIV. 10000 - 14000	100.0	97.4	99.4	100.0	107.9	101.7	100.0	102.8	100.6	100.0	55.2	90.5	100.0	53.4	90.1	100.0	54.3	90.3
15	MECHANICAL	100.1	99.2	99.7	100.1	113.2	105.8	99.7	120.5	108.8	99.8	55.9	80.6	99.8	35.6	71.7	99.8	35.3	71.6
16	ELECTRICAL	105.7	89.4	97.8	106.5	106.1	106.3	102.5	106.1	104.3	96.7	61.1	79.3	96.1	38.1	67.8	97.5	38.1	68.5
01 - 16	WEIGHTED AVERAGE	99.4	105.1	102.0	99.8	122.0	109.8	95.6	120.5	106.9	97.9	56.6	79.1	99.4	45.9	75.1	97.9	47.5	75.0

DIVISION		MISSISSIPPI			MISSOURI														
		MERIDIAN			CAPE GIRARDEAU			COLUMBIA			JOPLIN			KANSAS CITY			SPRINGFIELD		
		MAT.	INST.	TOTAL	MAT.	INST.	TOTAL	MAT.	INST.	TOTAL	MAT.	INST.	TOTAL	MAT.	INST.	TOTAL	MAT.	INST.	TOTAL
01590	EQUIPMENT RENTAL	.0	98.3	98.3	.0	108.3	108.3	.0	109.0	109.0	.0	106.9	106.9	.0	104.8	104.8	.0	102.7	102.7
02	SITE CONSTRUCTION	98.7	86.0	89.3	93.6	92.5	92.8	101.5	94.5	96.3	100.7	98.6	99.1	94.3	98.0	97.0	95.3	93.2	93.7
03100	CONCRETE FORMS & ACCESSORIES	78.8	34.1	40.1	84.4	74.9	76.2	85.0	72.4	74.0	101.2	65.4	70.2	100.2	105.2	104.5	100.2	65.6	70.2
03200	CONCRETE REINFORCEMENT	101.7	48.1	74.6	93.9	84.7	89.2	100.7	112.5	106.7	111.3	73.6	92.2	105.8	113.4	109.6	97.0	111.7	104.4
03300	CAST-IN-PLACE CONCRETE	99.0	43.0	75.6	84.6	84.5	84.6	88.5	68.9	80.3	101.2	73.1	89.5	94.1	107.5	99.7	101.5	63.6	85.6
03	CONCRETE	93.0	41.9	67.5	85.5	81.8	83.7	85.9	80.5	83.2	99.9	70.8	85.4	95.9	107.9	101.9	99.2	74.7	87.0
04	MASONRY	89.0	29.4	51.6	114.0	75.8	90.0	131.1	83.4	101.2	98.6	60.4	74.6	103.0	105.7	104.7	87.4	65.4	73.6
05	METALS	90.4	75.8	85.6	92.1	109.3	97.7	92.7	121.1	102.0	99.8	89.8	96.5	107.5	115.0	109.9	96.6	105.6	99.5
06	WOOD & PLASTICS	77.7	33.0	54.4	80.2	72.3	76.1	88.7	68.3	78.1	100.7	63.6	81.3	100.1	104.4	102.4	99.1	66.6	82.1
07	THERMAL & MOISTURE PROTECTION	94.7	40.2	67.9	95.5	79.8	87.7	92.4	81.4	87.0	98.6	63.8	81.5	97.8	107.1	102.4	96.5	70.1	83.5
08	DOORS & WINDOWS	97.8	37.9	82.3	93.9	75.0	89.1	90.9	91.7	91.1	91.9	71.0	86.5	98.0	107.0	100.3	96.3	78.2	91.6
09200	PLASTER & GYPSUM BOARD	91.2	31.7	53.5	96.1	71.4	80.4	98.1	67.3	78.6	103.9	62.3	77.5	100.3	104.3	102.8	102.5	65.5	79.0
095,098	CEILINGS & ACOUSTICAL TREATMENT	96.5	31.7	57.6	85.5	71.4	77.0	90.0	67.3	76.4	87.6	62.3	72.4	93.9	104.3	100.1	90.0	65.5	75.3
09600	FLOORING	108.0	28.7	87.0	92.4	78.2	88.7	101.9	90.5	98.9	111.7	52.6	96.1	91.7	90.5	91.4	104.9	52.6	91.1
097,099	WALL FINISHES, PAINTS & COATINGS	107.4	35.4	63.7	100.9	78.5	87.3	99.3	74.8	84.5	92.3	46.4	64.4	97.2	118.8	110.3	93.6	66.4	77.1
09	FINISHES	100.2	32.6	65.0	99.0	74.9	82.1	93.1	73.3	82.8	99.0	60.6	79.0	95.3	103.6	99.6	96.6	62.6	78.9
10 - 14	TOTAL DIV. 10000 - 14000	100.0	53.4	90.1	100.0	70.2	93.7	100.0	91.0	98.1	100.0	79.8	95.7	100.0	93.8	98.7	100.0	80.5	95.9
15	MECHANICAL	99.8	35.3	71.6	100.1	93.8	97.4	100.0	92.2	96.6	100.1	57.9	81.6	100.0	107.1	103.1	100.0	67.7	85.9
16	ELECTRICAL	95.7	60.2	78.4	102.4	108.8	105.5	99.4	82.4	91.1	97.6	63.6	81.1	109.0	108.1	108.6	102.6	68.7	86.1
01 - 16	WEIGHTED AVERAGE	96.0	47.8	74.1	96.1	89.3	93.0	96.9	88.1	92.9	98.7	69.0	85.2	101.1	106.3	103.4	98.2	74.5	87.4

DIVISION		MISSOURI						MONTANA											
		ST. JOSEPH			ST. LOUIS			BILLINGS			BUTTE			GREAT FALLS			HELENA		
		MAT.	INST.	TOTAL	MAT.	INST.	TOTAL	MAT.	INST.	TOTAL	MAT.	INST.	TOTAL	MAT.	INST.	TOTAL	MAT.	INST.	TOTAL
01590	EQUIPMENT RENTAL	.0	103.3	103.3	.0	109.5	109.5	.0	98.5	98.5	.0	98.2	98.2	.0	98.2	98.2	.0	98.2	98.2
02	SITE CONSTRUCTION	95.8	93.5	94.1	93.3	96.5	95.7	85.2	96.1	93.3	92.2	94.7	94.0	95.7	95.5	95.5	97.5	95.3	95.9
03100	CONCRETE FORMS & ACCESSORIES	100.1	86.0	87.9	99.6	105.7	104.9	98.1	65.3	69.7	85.4	62.9	65.9	99.4	66.5	70.9	99.5	65.9	70.4
03200	CONCRETE REINFORCEMENT	104.5	97.3	100.9	85.8	115.6	100.9	93.2	75.9	84.5	101.2	76.0	88.4	93.3	75.9	84.5	96.6	65.9	81.0
03300	CAST-IN-PLACE CONCRETE	94.0	100.6	96.8	84.6	108.2	94.5	119.0	68.6	97.9	122.5	68.7	100.0	129.4	58.6	99.9	131.9	67.1	104.8
03	CONCRETE	95.6	93.9	94.8	85.3	109.4	97.3	104.6	69.4	87.0	104.6	68.4	86.5	109.7	66.5	88.1	111.4	67.2	89.3
04	MASONRY	102.4	81.0	89.0	95.2	111.2	105.2	120.1	66.4	86.4	120.1	67.4	87.0	123.7	71.5	90.9	119.2	65.7	85.6
05	METALS	103.3	104.0	103.6	96.7	126.0	106.3	105.7	86.8	99.5	98.5	86.5	94.6	102.2	86.6	97.1	101.3	81.2	94.8
06	WOOD & PLASTICS	100.7	85.9	93.0	96.2	103.6	100.0	101.8	65.1	82.6	88.9	63.6	75.7	104.5	65.3	84.0	104.6	65.3	84.1
07	THERMAL & MOISTURE PROTECTION	98.4	90.3	94.4	95.4	106.2	100.7	99.9	68.5	84.5	99.2	67.4	83.6	100.1	68.5	84.5	100.0	68.2	84.4
08	DOORS & WINDOWS	96.8	92.9	95.8	92.9	112.3	97.9	99.1	64.1	90.0	95.8	63.2	87.4	99.3	64.2	90.2	98.8	61.6	89.2
09200	PLASTER & GYPSUM BOARD	105.3	85.2	92.6	105.2	103.5	104.1	104.4	64.5	79.1	94.8	63.0	74.6	104.4	64.7	79.2	103.7	64.7	79.0
095,098	CEILINGS & ACOUSTICAL TREATMENT	93.0	85.2	88.3	90.9	103.5	98.4	112.4	64.5	83.6	109.7	63.0	81.7	112.4	64.7	83.8	109.7	64.7	82.7
09600	FLOORING	94.0	84.2	91.4	99.7	98.5	99.4	107.8	51.2	92.9	99.1	40.8	83.7	107.8	55.3	94.0	107.8	53.6	93.5
097,099	WALL FINISHES, PAINTS & COATINGS	92.9	80.2	85.2	100.9	108.7	105.6	99.2	61.9	76.6	99.2	42.1	64.6	99.2	47.4	67.8	99.2	52.6	71.0
09	FINISHES	96.3	84.5	90.2	94.9	104.1	99.7	106.6	61.9	83.4	101.8	56.0	78.0	106.7	61.9	83.4	106.0	62.0	83.1
10 - 14	TOTAL DIV. 10000 - 14000	100.0	87.3	97.3	100.0	101.1	100.2	100.0	72.0	94.0	100.0	70.8	93.8	100.0	73.3	94.3	100.0	59.2	91.3
15	MECHANICAL	100.1	90.2	95.8	100.1	104.4	102.0	100.4	77.8	90.5	100.4	70.5	87.3	100.4	79.6	91.3	100.4	74.7	89.1
16	ELECTRICAL	105.9	83.4	94.9	105.3	114.1	109.6	95.0	74.7	85.1	101.6	69.6	86.0	95.0	75.1	85.3	95.0	74.1	84.8
01 - 16	WEIGHTED AVERAGE	100.1	89.6	95.3	96.7	108.6	102.1	102.0	74.4	89.5	101.0	71.1	87.4	102.6	74.9	90.0	102.4	72.3	88.7

DIVISION		MONTANA			NEBRASKA												NEVADA		
		MISSOULA			GRAND ISLAND			LINCOLN			NORTH PLATTE			OMAHA			CARSON CITY		
		MAT.	INST.	TOTAL	MAT.	INST.	TOTAL	MAT.	INST.	TOTAL	MAT.	INST.	TOTAL	MAT.	INST.	TOTAL	MAT.	INST.	TOTAL
01590	EQUIPMENT RENTAL	.0	98.2	98.2	.0	101.6	101.6	.0	101.6	101.6	.0	101.6	101.6	.0	90.8	90.8	.0	101.7	101.7
02	SITE CONSTRUCTION	75.6	94.6	89.6	98.1	91.0	92.9	89.5	91.1	90.6	98.8	89.8	92.2	78.9	89.7	86.9	63.0	104.1	93.4
03100	CONCRETE FORMS & ACCESSORIES	89.1	59.0	63.1	96.6	50.7	56.9	101.3	46.8	54.1	96.5	47.3	53.8	96.1	72.4	75.5	94.3	103.0	101.8
03200	CONCRETE REINFORCEMENT	102.9	78.4	90.5	105.6	74.5	89.9	97.0	75.4	86.1	107.1	74.2	90.5	101.6	76.0	88.7	113.7	118.6	116.2
03300	CAST-IN-PLACE CONCRETE	90.2	65.1	79.7	115.5	58.7	91.8	104.3	61.2	86.3	115.5	52.5	89.2	108.3	75.9	94.8	111.6	85.8	100.8
03	CONCRETE	83.6	65.8	74.7	108.5	59.3	83.9	100.6	58.5	79.6	108.7	55.6	82.2	102.0	74.4	88.2	109.3	99.8	104.5
04	MASONRY	145.0	60.5	91.9	105.5	49.9	70.6	95.8	66.7	77.5	91.1	40.5	59.4	100.9	75.4	84.9	135.0	80.6	100.9
05	METALS	95.3	86.6	92.5	92.8	83.3	89.7	97.9	84.8	93.6	93.0	83.2	89.8	97.3	76.1	90.3	93.4	104.5	97.0
06	WOOD & PLASTICS	93.4	59.8	75.9	95.5	45.5	69.4	100.1	39.8	68.6	95.4	45.5	69.3	94.7	73.3	83.5	91.2	106.4	99.1
07	THERMAL & MOISTURE PROTECTION	98.3	70.9	84.8	96.9	57.5	77.5	97.3	60.3	79.1	96.8	47.4	72.5	92.5	70.4	81.6	102.6	89.4	96.1
08	DOORS & WINDOWS	95.9	62.5	87.2	90.0	50.1	79.7	95.5	49.7	83.6	89.3	53.2	80.0	98.8	67.0	90.6	94.4	112.7	99.1
09200	PLASTER & GYPSUM BOARD	96.9	59.1	72.9	100.0	43.8	64.4	104.4	37.9	62.3	100.5	43.8	64.5	106.9	73.0	85.4	81.3	106.4	97.2
095,098	CEILINGS & ACOUSTICAL TREATMENT	109.7	59.1	79.3	88.2	43.8	61.6	97.1	37.9	61.6	90.0	43.8	62.3	107.9	73.0	87.0	100.4	106.4	104.0
09600	FLOORING	101.3	63.7	91.3	94.6	36.3	79.2	96.4	43.7	82.5	94.5	37.8	79.5	122.7	48.4	103.1	101.2	66.7	92.1
097,099	WALL FINISHES, PAINTS & COATINGS	99.2	47.4	67.8	91.2	42.8	61.9	91.2	43.2	62.1	91.2	51.8	67.3	148.5	73.6	103.0	103.7	81.6	90.3
09	FINISHES	102.0	58.2	79.2	93.1	45.7	68.4	96.3	43.9	69.0	93.6	45.3	68.4	114.6	67.9	90.3	96.1	94.0	95.0
10 - 14	TOTAL DIV. 10000 - 14000	100.0	55.2	90.5	100.0	73.9	94.4	100.0	73.2	94.3	100.0	55.2	90.5	100.0	75.9	94.9	100.0	114.9	103.2
15	MECHANICAL	100.4	70.2	87.2	100.0	77.5	90.2	100.0	77.6	90.2	100.0	73.4	88.4	99.8	78.4	90.4	100.1	92.9	96.9
16	ELECTRICAL	99.7	72.3	86.3	91.2	66.6	79.2	96.6	66.6	81.9	93.0	44.0	69.1	90.7	82.1	86.5	88.6	88.0	88.3
01 - 16	WEIGHTED AVERAGE	98.7	70.3	85.8	97.2	66.2	83.1	98.0	67.7	84.2	96.8	59.9	80.0	99.1	76.7	89.0	98.4	95.6	97.2

DIVISION		NEVADA						NEW HAMPSHIRE									NEW JERSEY		
		LAS VEGAS			RENO			MANCHESTER			NASHUA			PORTSMOUTH			CAMDEN		
		MAT.	INST.	TOTAL	MAT.	INST.	TOTAL	MAT.	INST.	TOTAL	MAT.	INST.	TOTAL	MAT.	INST.	TOTAL	MAT.	INST.	TOTA
01590	EQUIPMENT RENTAL	.0	101.7	101.7	.0	101.7	101.7	.0	101.6	101.6	.0	101.6	101.6	.0	101.6	101.6	.0	99.4	99.
02	SITE CONSTRUCTION	62.5	106.3	94.9	62.7	104.1	93.3	86.7	100.2	96.7	88.0	100.2	97.0	82.2	99.2	94.8	91.3	105.0	101.
03100	CONCRETE FORMS & ACCESSORIES	93.2	107.9	105.9	94.3	102.9	101.8	100.9	63.9	68.9	101.4	63.9	68.9	88.8	59.4	63.3	101.4	123.2	120.
03200	CONCRETE REINFORCEMENT	108.1	118.7	113.4	108.1	118.6	113.4	108.2	91.9	100.0	108.2	91.9	100.0	86.6	91.8	89.2	108.2	104.9	106.
03300	CAST-IN-PLACE CONCRETE	105.3	108.6	106.7	113.7	85.8	102.1	101.0	97.4	99.5	92.8	97.4	94.7	88.0	90.9	89.2	80.7	128.7	100.
03	CONCRETE	105.3	109.8	107.6	109.4	99.7	104.6	106.8	81.0	93.9	102.9	81.0	91.9	93.8	76.6	85.2	97.1	120.6	108.
04	MASONRY	125.8	89.8	103.2	134.3	80.6	100.6	97.9	92.8	94.7	98.1	92.8	94.8	93.5	82.2	86.4	90.3	125.0	112.
05	METALS	94.0	105.9	97.9	93.8	104.5	97.3	100.0	87.1	95.8	99.9	87.1	95.7	96.3	85.0	92.6	99.8	97.9	99.
06	WOOD & PLASTICS	90.9	106.4	99.0	91.2	106.4	99.1	100.4	55.9	77.2	100.4	55.9	77.2	86.2	55.9	70.4	100.4	122.9	112.
07	THERMAL & MOISTURE PROTECTION	105.1	98.7	102.0	102.6	89.4	96.1	101.6	89.0	95.4	102.1	89.0	95.6	101.5	92.1	96.9	101.4	122.1	111
08	DOORS & WINDOWS	94.7	112.7	99.4	94.7	111.1	99.0	107.1	65.4	96.3	107.1	65.4	96.3	108.0	59.4	95.4	107.1	113.8	108.
09200	PLASTER & GYPSUM BOARD	81.9	106.4	97.5	83.2	106.4	97.9	105.2	53.8	72.6	105.2	53.8	72.6	94.0	53.8	68.5	105.2	122.6	116.
095,098	CEILINGS & ACOUSTICAL TREATMENT	105.8	106.4	106.2	107.6	106.4	106.9	101.1	53.8	72.7	101.1	53.8	72.7	89.5	53.8	68.1	101.1	122.6	114
09600	FLOORING	101.2	79.5	95.4	101.2	66.7	92.1	100.0	103.7	101.0	99.8	103.7	100.8	94.4	103.7	96.8	99.8	132.6	108.
097,099	WALL FINISHES, PAINTS & COATINGS	103.7	105.1	104.6	103.7	81.6	90.3	92.8	90.6	91.4	92.8	90.6	91.4	92.8	40.3	61.0	92.8	119.7	109.
09	FINISHES	97.5	101.9	99.8	98.2	94.0	96.0	101.3	72.5	86.4	101.6	72.5	86.5	95.1	64.2	79.0	101.8	125.0	113.
10 - 14	TOTAL DIV. 10000 - 14000	100.0	110.2	102.2	100.0	114.9	103.2	100.0	80.0	95.7	100.0	80.0	95.7	100.0	75.8	94.9	100.0	117.8	103
15	MECHANICAL	100.1	111.1	104.9	100.1	92.9	96.9	100.1	86.1	94.0	100.1	86.1	94.0	100.1	80.4	91.5	100.1	109.6	104.
16	ELECTRICAL	90.5	105.5	97.8	88.6	88.0	88.3	100.4	74.5	87.8	100.2	74.5	87.7	98.5	74.5	86.8	100.6	117.6	108.
01 - 16	WEIGHTED AVERAGE	98.1	105.5	101.5	98.7	95.5	97.3	101.2	82.8	92.9	100.9	82.8	92.7	98.0	78.4	89.1	100.0	115.2	106.

DIVISION		NEW JERSEY															NEW MEXICO		
		ELIZABETH			JERSEY CITY			NEWARK			PATERSON			TRENTON			ALBUQUERQUE		
		MAT.	INST.	TOTAL	MAT.	INST.	TOTAL	MAT.	INST.	TOTAL	MAT.	INST.	TOTAL	MAT.	INST.	TOTAL	MAT.	INST.	TOTA
01590	EQUIPMENT RENTAL	.0	101.6	101.6	.0	99.4	99.4	.0	101.6	101.6	.0	101.6	101.6	.0	98.9	98.9	.0	116.3	116
02	SITE CONSTRUCTION	105.3	105.3	105.3	91.3	104.9	101.3	109.8	105.3	106.5	103.7	104.8	104.5	93.0	105.4	102.1	80.2	111.5	103
03100	CONCRETE FORMS & ACCESSORIES	111.5	122.7	121.2	101.4	123.7	120.7	99.8	122.7	119.7	100.4	122.6	119.6	100.3	123.4	120.3	96.1	69.0	72
03200	CONCRETE REINFORCEMENT	83.4	118.6	101.2	108.2	118.6	113.5	108.2	118.6	113.5	108.2	118.6	113.5	108.2	114.7	111.5	108.8	67.2	87
03300	CAST-IN-PLACE CONCRETE	88.9	128.5	105.4	80.7	130.0	101.3	94.2	128.6	108.5	102.8	128.5	113.6	95.5	119.9	105.7	104.5	76.4	92
03	CONCRETE	99.4	123.1	111.2	97.1	123.8	110.5	103.4	123.1	113.2	107.6	123.0	115.3	104.1	119.4	111.8	105.2	72.2	88
04	MASONRY	112.4	123.0	119.0	90.3	125.5	112.4	100.1	123.0	114.5	95.8	123.0	112.9	89.9	116.0	106.3	122.3	66.3	87
05	METALS	96.0	108.6	100.1	99.9	105.6	101.7	99.8	108.6	102.7	94.7	108.5	99.2	94.6	101.2	96.8	100.2	88.0	96
06	WOOD & PLASTICS	115.3	122.9	119.3	100.4	122.9	112.1	102.2	122.9	113.0	102.2	122.9	113.0	100.4	122.8	112.1	95.7	70.1	82
07	THERMAL & MOISTURE PROTECTION	102.4	126.0	114.0	101.4	127.1	114.1	101.7	126.0	113.7	102.2	119.7	110.8	100.0	116.0	107.9	101.6	72.6	87
08	DOORS & WINDOWS	108.0	118.2	110.7	107.1	118.2	109.9	113.1	118.2	114.4	113.1	118.2	114.4	107.1	116.4	109.5	94.7	72.5	89
09200	PLASTER & GYPSUM BOARD	108.9	122.6	117.6	105.2	122.6	116.2	105.2	122.6	116.2	105.2	122.6	116.2	105.2	122.6	116.2	85.4	68.8	74
095,098	CEILINGS & ACOUSTICAL TREATMENT	89.5	122.6	109.3	101.1	122.6	114.0	101.1	122.6	114.0	101.1	122.6	114.0	101.1	122.6	114.0	105.8	68.8	83
09600	FLOORING	104.5	132.6	111.9	99.8	132.6	108.5	100.0	132.6	108.6	99.8	132.6	108.5	100.0	117.0	104.5	101.2	65.0	91
097,099	WALL FINISHES, PAINTS & COATINGS	92.7	119.7	109.1	92.8	119.7	109.1	92.7	119.7	109.1	92.7	119.7	109.1	92.8	119.7	109.1	103.7	60.7	77
09	FINISHES	101.6	124.1	113.3	101.8	124.8	113.8	102.4	124.1	113.7	102.0	124.1	113.5	101.9	121.8	112.3	98.5	67.2	82
10 - 14	TOTAL DIV. 10000 - 14000	100.0	124.0	105.1	100.0	125.1	105.3	100.0	124.1	105.1	100.0	124.0	105.1	100.0	117.7	103.8	100.0	76.4	95
15	MECHANICAL	100.1	121.4	109.4	100.1	126.8	111.8	100.1	122.5	109.9	100.1	125.4	111.2	100.1	126.1	111.5	100.1	72.9	88
16	ELECTRICAL	98.5	128.9	113.3	102.2	132.5	117.0	102.1	132.5	116.9	102.2	128.9	115.2	100.8	129.8	114.9	81.8	76.3	79
01 - 16	WEIGHTED AVERAGE	101.1	120.7	110.0	100.2	122.5	110.3	102.5	121.4	111.1	101.8	121.2	110.6	99.9	119.1	108.6	98.3	76.8	88

DIVISION		NEW MEXICO												NEW YORK					
		FARMINGTON			LAS CRUCES			ROSWELL			SANTA FE			ALBANY			BINGHAMTON		
		MAT.	INST.	TOTAL	MAT.	INST.	TOTAL	MAT.	INST.	TOTAL	MAT.	INST.	TOTAL	MAT.	INST.	TOTAL	MAT.	INST.	TOT/
01590	EQUIPMENT RENTAL	.0	116.3	116.3	.0	86.7	86.7	.0	116.3	116.3	.0	116.3	116.3	.0	115.7	115.7	.0	115.8	115
02	SITE CONSTRUCTION	85.7	111.5	104.8	91.2	86.7	87.9	90.3	111.4	105.9	79.9	111.5	103.3	71.8	106.5	97.5	93.8	89.8	90
03100	CONCRETE FORMS & ACCESSORIES	96.1	69.0	72.6	93.5	67.2	70.7	96.1	68.9	72.5	96.1	69.0	72.6	98.7	92.1	93.0	103.9	76.5	80
03200	CONCRETE REINFORCEMENT	118.7	67.2	92.7	112.8	58.8	85.5	117.9	59.3	88.3	116.6	67.2	91.6	103.1	89.8	96.4	102.0	87.8	94
03300	CAST-IN-PLACE CONCRETE	105.1	76.4	93.1	91.4	66.7	81.1	96.8	76.4	88.3	98.8	76.4	89.4	88.7	100.5	93.6	105.9	86.7	97
03	CONCRETE	109.0	72.2	90.6	87.3	66.2	76.7	109.3	70.7	90.0	103.8	72.2	88.0	101.3	95.5	98.4	101.1	84.2	92
04	MASONRY	127.3	66.3	89.1	110.9	60.7	79.4	127.4	66.3	89.1	119.4	66.3	86.1	89.3	97.0	94.1	106.0	79.0	89
05	METALS	97.9	88.0	94.7	97.8	75.2	90.4	97.9	84.1	93.4	97.9	88.0	94.6	96.9	104.9	99.5	93.4	114.1	100
06	WOOD & PLASTICS	95.7	70.1	82.4	87.4	68.7	77.6	95.7	70.1	82.4	95.7	70.1	82.4	98.3	91.3	94.6	107.3	75.0	90
07	THERMAL & MOISTURE PROTECTION	102.5	72.6	87.8	88.7	65.5	77.3	102.8	72.6	87.9	101.9	72.6	87.5	90.8	92.8	91.8	101.5	80.3	91
08	DOORS & WINDOWS	97.6	72.5	91.1	87.1	69.2	82.5	93.6	70.0	87.5	93.7	72.5	88.2	95.7	85.0	92.9	90.7	73.4	86
09200	PLASTER & GYPSUM BOARD	80.4	68.8	73.0	82.7	68.8	73.9	80.4	68.8	73.0	80.4	68.8	73.0	111.0	90.8	98.2	116.3	73.7	89
095,098	CEILINGS & ACOUSTICAL TREATMENT	96.9	68.8	80.0	96.8	68.8	80.0	96.9	68.8	80.0	96.9	68.8	80.0	99.8	90.8	94.4	99.8	73.7	84
09600	FLOORING	101.2	65.0	91.6	133.4	65.2	115.4	101.2	65.0	91.6	101.2	65.0	91.6	86.8	94.1	88.7	97.6	80.2	93
097,099	WALL FINISHES, PAINTS & COATINGS	103.7	60.7	77.6	97.1	60.7	75.0	103.7	60.7	77.6	103.7	60.7	77.6	84.9	78.0	80.7	90.1	79.1	83
09	FINISHES	95.9	67.2	81.0	111.0	66.2	87.7	96.3	67.2	81.2	95.7	67.2	80.9	97.5	91.0	94.1	98.8	77.0	87
10 - 14	TOTAL DIV. 10000 - 14000	100.0	76.4	95.0	100.0	72.3	94.1	100.0	76.4	95.0	100.0	76.4	95.0	100.0	87.3	97.3	100.0	89.3	97
15	MECHANICAL	100.1	72.9	88.2	100.3	72.4	88.1	100.1	72.8	88.2	100.1	72.9	88.2	100.4	88.9	95.3	100.6	82.4	92
16	ELECTRICAL	80.2	76.3	78.3	82.5	60.4	71.7	80.9	76.3	78.7	81.8	76.3	79.1	103.1	89.4	96.4	101.1	80.2	90
01 - 16	WEIGHTED AVERAGE	98.6	76.8	88.7	95.8	69.1	83.7	98.5	76.1	88.4	97.3	76.8	88.0	97.9	93.9	96.1	98.6	84.5	92

NEW YORK

DIVISION		BUFFALO MAT.	INST.	TOTAL	HICKSVILLE MAT.	INST.	TOTAL	NEW YORK MAT.	INST.	TOTAL	RIVERHEAD MAT.	INST.	TOTAL	ROCHESTER MAT.	INST.	TOTAL	SCHENECTADY MAT.	INST.	TOTAL
01590	EQUIPMENT RENTAL	.0	93.1	93.1	.0	117.2	117.2	.0	117.6	117.6	.0	117.2	117.2	.0	116.5	116.5	.0	115.7	115.7
02	SITE CONSTRUCTION	96.1	93.4	94.1	112.1	130.3	125.6	134.2	128.5	129.9	112.3	130.4	125.7	73.9	107.2	98.6	71.5	106.3	97.3
03100	CONCRETE FORMS & ACCESSORIES	100.3	113.7	111.9	90.5	149.8	141.9	113.8	180.5	171.6	94.6	149.9	142.5	100.2	94.2	95.0	103.5	91.0	92.7
03200	CONCRETE REINFORCEMENT	104.0	105.9	105.0	103.3	184.3	144.2	111.2	187.4	149.7	105.3	184.3	145.2	105.0	86.5	95.7	101.8	89.8	95.7
03300	CAST-IN-PLACE CONCRETE	118.7	119.3	118.9	99.1	151.6	121.0	121.6	160.8	137.9	97.2	151.7	120.0	112.9	100.5	107.8	97.3	99.0	98.0
03	CONCRETE	108.2	113.3	110.7	104.6	155.5	130.0	120.2	172.8	146.5	103.7	155.5	129.6	113.6	96.0	104.8	105.5	94.5	100.0
04	MASONRY	104.9	119.8	114.2	109.5	152.4	136.4	105.3	161.1	140.3	116.0	152.4	138.9	103.0	98.0	99.8	90.8	94.1	92.9
05	METALS	100.4	94.8	98.5	103.5	137.5	114.6	108.1	140.9	118.8	103.6	137.9	114.8	96.8	107.3	100.2	97.0	104.9	99.6
06	WOOD & PLASTICS	104.3	112.9	108.8	88.5	151.2	121.2	111.4	187.9	151.4	92.4	151.2	123.1	101.1	93.8	97.3	103.9	91.3	97.3
07	THERMAL & MOISTURE PROTECTION	98.9	110.7	104.7	104.0	145.1	124.2	105.5	160.3	132.4	104.0	145.1	124.2	95.6	96.9	96.2	91.0	91.6	91.3
08	DOORS & WINDOWS	92.7	103.3	95.5	88.3	153.6	105.2	97.7	176.6	118.1	88.4	153.6	105.2	96.0	86.2	93.5	95.7	85.0	92.9
09200	PLASTER & GYPSUM BOARD	102.4	113.0	109.1	102.1	152.7	134.2	115.8	189.9	162.8	103.0	152.7	134.5	104.6	93.6	97.6	111.0	90.8	98.2
095,098	CEILINGS & ACOUSTICAL TREATMENT	94.6	113.0	105.6	79.3	152.7	123.3	109.4	189.9	157.7	79.3	152.7	123.3	99.1	93.6	95.8	99.8	90.8	94.4
09600	FLOORING	97.7	121.0	103.9	97.4	90.1	95.5	97.4	171.3	117.0	98.7	90.1	96.5	86.8	85.8	86.6	86.8	94.1	88.7
097,099	WALL FINISHES, PAINTS & COATINGS	94.8	114.8	106.9	112.6	146.7	133.3	97.2	150.2	129.3	112.6	146.7	133.3	92.5	98.4	96.1	84.9	78.0	80.7
09	FINISHES	95.7	115.9	106.2	104.0	137.5	121.4	108.9	176.8	144.2	104.4	137.5	121.6	96.8	93.1	94.9	97.3	90.3	93.6
10-14	TOTAL DIV. 10000 - 14000	100.0	108.0	101.7	100.0	120.4	104.3	100.0	151.7	111.0	100.0	120.4	104.3	100.0	91.5	98.2	100.0	86.2	97.1
15	MECHANICAL	99.8	97.4	98.8	99.8	144.9	119.6	100.3	163.2	127.8	99.8	144.9	119.6	99.8	88.7	95.0	100.4	87.4	94.7
16	ELECTRICAL	99.7	97.6	98.7	101.6	152.1	126.3	107.8	178.5	142.3	102.2	152.1	126.6	103.4	89.1	96.4	100.5	89.4	95.1
01-16	WEIGHTED AVERAGE	99.9	104.7	102.1	101.2	145.0	121.1	106.6	163.5	132.4	101.6	145.1	121.4	100.1	94.9	97.7	98.2	93.0	95.8

NEW YORK / NORTH CAROLINA

DIVISION		SYRACUSE MAT.	INST.	TOTAL	UTICA MAT.	INST.	TOTAL	WATERTOWN MAT.	INST.	TOTAL	WHITE PLAINS MAT.	INST.	TOTAL	YONKERS MAT.	INST.	TOTAL	ASHEVILLE MAT.	INST.	TOTAL
01590	EQUIPMENT RENTAL	.0	115.7	115.7	.0	115.7	115.7	.0	115.7	115.7	.0	117.3	117.3	.0	117.3	117.3	.0	93.3	93.3
02	SITE CONSTRUCTION	92.6	106.6	103.0	69.6	105.6	96.2	77.5	107.6	99.8	123.8	125.2	124.9	133.5	125.1	127.3	101.6	72.3	79.9
03100	CONCRETE FORMS & ACCESSORIES	102.7	87.4	89.5	104.0	83.3	86.1	85.8	90.6	90.0	113.4	136.1	133.0	113.6	136.3	133.3	94.3	43.0	49.8
03200	CONCRETE REINFORCEMENT	103.1	88.4	95.7	103.1	86.9	94.9	103.7	88.4	96.0	104.0	183.5	144.2	108.1	183.5	146.2	95.1	39.5	67.0
03300	CAST-IN-PLACE CONCRETE	98.3	93.7	96.4	89.9	91.2	90.4	104.7	85.7	96.8	107.1	134.2	118.4	119.9	134.3	125.9	96.3	47.3	75.8
03	CONCRETE	104.4	90.8	97.6	102.3	87.8	95.1	115.2	89.5	102.4	108.1	143.1	125.6	119.0	143.3	131.1	99.5	45.8	72.7
04	MASONRY	96.9	92.4	94.1	89.4	89.1	89.2	90.6	94.0	92.7	100.0	134.1	121.4	104.8	134.1	123.2	79.1	37.1	52.7
05	METALS	96.8	104.2	99.3	94.9	103.2	97.6	94.9	103.9	97.9	94.8	134.3	107.7	104.0	134.6	114.0	93.0	74.5	87.0
06	WOOD & PLASTICS	103.9	86.8	95.0	103.9	82.8	92.9	84.1	91.8	88.2	112.3	135.7	124.5	112.2	135.7	124.5	93.1	44.6	67.8
07	THERMAL & MOISTURE PROTECTION	99.1	93.2	96.2	91.0	92.0	91.5	91.2	93.6	92.4	106.2	137.5	121.6	106.6	137.5	121.8	96.8	42.2	70.0
08	DOORS & WINDOWS	93.6	80.7	90.3	95.7	77.8	91.0	95.7	82.0	92.1	92.2	145.1	105.9	95.8	147.6	109.2	92.2	41.3	79.0
09200	PLASTER & GYPSUM BOARD	111.0	86.2	95.3	111.0	82.0	92.6	102.6	91.4	95.5	108.4	136.3	126.1	115.3	136.3	128.6	100.7	42.9	64.0
095,098	CEILINGS & ACOUSTICAL TREATMENT	99.8	86.2	91.6	99.8	82.0	89.1	99.8	91.4	94.7	79.9	136.3	113.7	107.6	136.3	124.8	91.9	42.9	62.5
09600	FLOORING	88.2	89.4	88.6	86.8	89.5	87.5	79.2	89.5	81.9	93.6	158.9	110.9	93.2	158.9	110.6	101.3	41.4	85.5
097,099	WALL FINISHES, PAINTS & COATINGS	90.3	86.8	88.2	84.9	84.6	84.8	84.9	73.4	78.0	94.8	146.7	126.3	94.8	146.7	126.3	108.5	39.1	66.4
09	FINISHES	98.6	87.6	92.9	97.3	84.3	90.5	94.3	88.3	91.2	98.7	141.6	121.0	107.0	141.6	125.0	96.4	42.3	68.3
10-14	TOTAL DIV. 10000 - 14000	100.0	97.7	99.5	100.0	84.1	96.6	100.0	83.7	96.5	100.0	117.6	103.7	100.0	141.6	108.9	100.0	62.5	92.0
15	MECHANICAL	100.4	84.3	93.3	100.4	80.8	91.8	100.4	77.0	90.1	100.6	127.0	112.1	100.6	127.0	112.1	100.0	43.1	75.1
16	ELECTRICAL	100.5	82.5	91.7	97.1	81.7	89.6	100.5	79.0	90.0	97.3	143.1	119.6	104.3	143.1	123.2	97.9	40.4	69.9
01-16	WEIGHTED AVERAGE	99.0	90.5	95.2	97.0	87.8	92.8	98.7	88.4	94.0	99.9	135.4	116.0	105.1	136.2	119.2	96.3	48.3	74.5

NORTH CAROLINA

DIVISION		CHARLOTTE MAT.	INST.	TOTAL	DURHAM MAT.	INST.	TOTAL	FAYETTEVILLE MAT.	INST.	TOTAL	GREENSBORO MAT.	INST.	TOTAL	RALEIGH MAT.	INST.	TOTAL	WILMINGTON MAT.	INST.	TOTAL
01590	EQUIPMENT RENTAL	.0	93.3	93.3	.0	99.7	99.7	.0	99.7	99.7	.0	99.7	99.7	.0	99.7	99.7	.0	93.3	93.3
02	SITE CONSTRUCTION	101.6	72.3	80.0	101.6	82.2	87.2	99.7	82.2	86.7	101.5	82.2	87.2	102.6	82.2	87.5	102.5	72.3	80.2
03100	CONCRETE FORMS & ACCESSORIES	101.7	43.0	50.9	96.2	43.3	50.4	92.0	43.3	49.8	96.3	43.3	50.4	99.0	43.3	50.7	95.8	43.3	50.3
03200	CONCRETE REINFORCEMENT	95.5	39.5	67.2	95.5	54.7	74.9	94.5	54.7	74.4	95.5	54.7	74.9	95.5	54.7	74.9	95.8	54.7	75.0
03300	CAST-IN-PLACE CONCRETE	98.7	47.4	77.3	96.5	47.5	76.0	93.2	47.5	74.1	95.7	47.5	75.6	102.2	47.5	79.3	95.9	47.5	75.6
03	CONCRETE	100.5	45.8	73.2	99.1	48.8	74.0	96.7	48.8	72.8	98.7	48.8	73.8	102.0	48.8	75.4	99.4	48.8	74.2
04	MASONRY	84.0	37.1	54.6	80.8	37.1	53.4	84.4	37.1	54.7	80.6	37.1	53.3	85.5	37.1	55.1	69.7	37.1	49.2
05	METALS	95.5	74.5	88.7	98.1	81.1	92.6	98.1	81.1	92.5	98.7	81.1	92.9	95.9	81.1	91.1	92.5	81.1	88.8
06	WOOD & PLASTICS	102.0	44.6	72.0	95.4	44.6	68.9	90.4	44.6	66.5	95.4	44.6	68.9	98.6	44.6	70.4	94.9	44.6	68.6
07	THERMAL & MOISTURE PROTECTION	96.9	43.2	70.5	97.5	42.8	70.6	97.2	42.8	70.4	97.5	42.8	70.6	97.3	42.8	70.6	96.9	42.8	70.3
08	DOORS & WINDOWS	96.3	41.3	82.1	96.3	47.1	83.6	92.3	47.1	80.6	96.3	47.1	83.6	93.0	47.1	81.1	92.3	47.1	80.6
09200	PLASTER & GYPSUM BOARD	107.2	42.9	66.4	107.2	42.9	66.4	101.0	42.9	64.1	107.2	42.9	66.4	107.2	42.9	66.4	101.8	42.9	64.4
095,098	CEILINGS & ACOUSTICAL TREATMENT	98.1	42.9	65.0	98.1	42.9	65.0	92.8	42.9	62.8	98.1	42.9	65.0	98.1	42.9	65.0	92.8	42.9	62.8
09600	FLOORING	105.1	41.4	88.3	105.3	41.4	88.4	101.4	41.4	85.6	105.3	41.4	88.4	105.3	41.4	88.4	102.3	41.4	86.2
097,099	WALL FINISHES, PAINTS & COATINGS	108.5	39.1	66.4	108.5	39.1	66.4	108.5	39.1	66.4	108.5	39.1	66.4	108.5	39.1	66.4	108.5	39.1	66.4
09	FINISHES	100.0	42.3	70.0	100.1	42.3	70.0	96.7	42.3	68.4	100.1	42.3	70.0	100.1	42.3	70.1	97.0	42.3	68.6
10-14	TOTAL DIV. 10000 - 14000	100.0	62.5	92.0	100.0	65.2	92.6	100.0	65.2	92.6	100.0	62.5	92.0	100.0	65.2	92.6	100.0	65.2	92.6
15	MECHANICAL	100.0	43.1	75.1	100.0	43.3	75.2	100.0	43.3	75.2	100.0	43.3	75.2	100.0	43.3	75.2	100.0	43.3	75.2
16	ELECTRICAL	99.7	36.3	68.8	97.4	38.6	68.8	94.8	38.6	67.4	98.0	38.6	69.1	98.0	38.6	69.1	98.5	38.6	69.3
01-16	WEIGHTED AVERAGE	98.2	47.8	75.3	97.9	50.3	76.3	96.7	50.3	75.6	98.0	50.3	76.3	97.9	50.3	76.3	96.0	49.4	74.8

City Cost Indexes

NORTH CAROLINA / NORTH DAKOTA / OHIO

DIVISION		WINSTON-SALEM MAT.	INST.	TOTAL	BISMARCK MAT.	INST.	TOTAL	FARGO MAT.	INST.	TOTAL	GRAND FORKS MAT.	INST.	TOTAL	MINOT MAT.	INST.	TOTAL	AKRON MAT.	INST.	TOTA
01590	EQUIPMENT RENTAL	.0	99.7	99.7	.0	98.2	98.2	.0	98.2	98.2	.0	98.2	98.2	.0	98.2	98.2	.0	97.4	97.
02	SITE CONSTRUCTION	101.9	82.2	87.3	88.4	96.0	94.0	87.8	96.0	93.9	96.3	94.1	94.7	93.9	96.0	95.5	100.8	106.8	105.
03100	CONCRETE FORMS & ACCESSORIES	97.6	43.1	50.4	96.4	51.0	57.0	97.4	51.4	57.5	93.1	44.8	51.3	88.1	57.8	61.9	98.1	99.0	98.
03200	CONCRETE REINFORCEMENT	95.5	39.5	67.2	101.2	75.5	88.2	93.3	74.6	83.8	99.6	81.6	90.5	103.1	75.6	89.2	94.1	98.0	96.
03300	CAST-IN-PLACE CONCRETE	98.7	47.4	77.3	104.9	56.3	84.6	109.9	58.6	88.5	107.4	54.4	85.2	107.4	54.9	85.5	99.1	106.2	102.
03	CONCRETE	100.2	45.8	73.1	99.1	58.5	78.9	109.5	59.3	84.4	105.1	56.1	80.6	103.7	61.1	82.4	99.4	100.2	99.
04	MASONRY	80.8	37.1	53.4	104.0	66.9	80.7	106.3	50.0	71.0	105.6	66.2	80.9	106.7	68.2	82.5	90.0	102.4	97.
05	METALS	95.9	74.6	88.9	92.0	81.2	88.5	94.4	79.8	89.6	92.0	81.0	88.4	92.3	81.3	88.7	88.2	82.6	86.
06	WOOD & PLASTICS	95.4	44.6	68.9	88.9	46.2	66.6	88.9	46.8	66.9	85.3	42.4	62.9	80.3	55.4	67.3	92.2	97.8	95.
07	THERMAL & MOISTURE PROTECTION	97.5	42.8	70.6	100.5	56.9	79.1	101.2	54.2	78.1	101.2	57.1	79.5	100.9	58.1	79.9	104.0	100.4	102.
08	DOORS & WINDOWS	96.3	43.1	82.5	99.5	50.5	86.8	99.4	50.9	86.9	99.5	47.1	85.9	99.6	55.6	88.2	104.2	98.9	102.
09200	PLASTER & GYPSUM BOARD	107.2	42.9	66.4	115.0	45.0	70.6	115.0	45.7	71.1	113.0	41.2	67.5	111.4	54.5	75.4	95.2	97.3	96.
095,098	CEILINGS & ACOUSTICAL TREATMENT	98.1	42.9	65.0	134.3	45.0	80.7	134.3	45.7	81.2	134.3	41.2	78.5	134.3	54.5	86.5	91.9	97.3	95.
09600	FLOORING	105.3	41.4	88.4	108.4	78.4	100.4	108.2	46.9	91.9	106.6	46.9	90.8	103.9	81.0	97.8	103.2	91.7	100.
097,099	WALL FINISHES, PAINTS & COATINGS	108.5	39.1	66.4	99.2	38.9	62.6	99.2	73.6	83.7	99.2	29.7	57.1	99.2	32.6	58.8	107.1	122.7	116.
09	FINISHES	100.1	42.3	70.0	113.9	53.8	82.6	113.7	51.7	81.4	113.6	42.6	76.7	112.3	59.0	84.6	99.5	100.2	99.
10 - 14	TOTAL DIV. 10000 - 14000	100.0	62.5	92.0	100.0	74.2	94.5	100.0	74.2	94.5	100.0	48.3	89.0	100.0	75.4	94.8	100.0	92.6	98.
15	MECHANICAL	100.0	43.1	75.1	100.6	64.2	84.7	100.6	69.7	87.0	100.6	46.9	77.1	100.6	61.8	83.6	99.8	101.4	100.
16	ELECTRICAL	98.0	38.6	69.1	94.4	70.1	82.5	94.2	69.8	82.3	96.8	71.4	84.4	99.7	71.4	85.9	97.9	90.5	94.
01 - 16	WEIGHTED AVERAGE	97.8	49.1	75.7	99.2	66.9	84.5	100.8	65.9	85.0	100.4	60.7	82.4	100.5	68.2	85.8	97.7	98.0	97.

OHIO

DIVISION		CANTON MAT.	INST.	TOTAL	CINCINNATI MAT.	INST.	TOTAL	CLEVELAND MAT.	INST.	TOTAL	COLUMBUS MAT.	INST.	TOTAL	DAYTON MAT.	INST.	TOTAL	LORAIN MAT.	INST.	TOTA
01590	EQUIPMENT RENTAL	.0	97.4	97.4	.0	102.5	102.5	.0	97.7	97.7	.0	96.4	96.4	.0	96.8	96.8	.0	97.4	97.
02	SITE CONSTRUCTION	100.9	106.9	105.4	74.9	108.7	99.9	100.8	106.8	105.3	86.5	101.8	97.8	73.7	108.2	99.2	100.2	106.0	104.
03100	CONCRETE FORMS & ACCESSORIES	98.1	86.1	87.7	95.9	90.9	91.6	98.3	105.6	104.6	98.6	86.7	88.3	95.8	80.2	82.3	98.2	106.2	105.
03200	CONCRETE REINFORCEMENT	94.1	84.1	89.0	98.7	86.7	92.6	94.6	98.3	96.5	76.1	86.5	81.4	98.7	82.1	90.3	94.1	98.1	96.
03300	CAST-IN-PLACE CONCRETE	100.1	103.4	101.5	84.8	91.5	87.6	97.2	114.1	104.3	89.6	93.7	91.3	77.3	89.2	82.3	94.3	108.9	100.
03	CONCRETE	99.8	91.0	95.4	92.9	90.4	91.6	98.6	106.1	102.3	92.2	88.7	90.5	89.3	83.4	86.4	97.1	104.4	100.
04	MASONRY	90.7	93.3	92.4	84.9	95.9	91.8	94.2	110.6	104.5	95.4	93.8	94.4	84.3	85.3	84.9	87.0	100.5	95.
05	METALS	88.2	76.0	84.2	86.8	89.6	87.7	89.6	85.8	88.3	96.5	82.0	91.8	86.1	77.3	83.3	88.7	83.5	87.
06	WOOD & PLASTICS	92.6	84.1	88.2	96.4	89.5	92.8	91.5	102.4	97.2	104.3	84.8	94.1	97.5	78.2	87.4	92.2	108.3	100.
07	THERMAL & MOISTURE PROTECTION	104.7	95.9	100.3	88.9	97.8	93.3	102.9	114.0	108.3	99.4	95.6	97.5	94.5	87.6	91.1	104.6	108.1	106.
08	DOORS & WINDOWS	98.6	77.9	93.3	95.9	87.6	93.8	95.0	101.3	96.7	99.5	84.1	95.5	96.2	78.1	91.5	98.6	104.6	100.
09200	PLASTER & GYPSUM BOARD	96.0	83.2	87.9	101.2	89.5	93.8	94.3	102.0	99.2	99.3	84.2	89.8	101.2	77.9	86.5	95.2	108.1	103
095,098	CEILINGS & ACOUSTICAL TREATMENT	91.9	83.2	86.7	95.5	89.5	91.9	90.1	102.0	97.2	96.2	84.2	89.0	96.5	77.9	85.3	91.9	108.1	101.
09600	FLOORING	103.4	83.5	98.1	107.2	101.8	105.8	103.0	112.2	105.4	97.6	91.0	95.9	110.0	84.0	103.1	103.4	109.5	105.
097,099	WALL FINISHES, PAINTS & COATINGS	107.1	85.8	94.2	104.8	96.1	99.5	107.1	122.4	116.4	99.3	99.6	99.5	104.8	92.2	97.1	107.1	122.4	116.
09	FINISHES	99.7	84.6	91.8	98.1	93.4	95.6	98.9	108.3	103.8	98.9	88.4	93.5	99.1	81.6	90.0	99.5	109.2	104.
10 - 14	TOTAL DIV. 10000 - 14000	100.0	80.1	95.8	100.0	94.7	98.9	100.0	108.1	101.7	100.0	94.5	98.8	100.0	91.6	98.2	100.0	105.5	101.
15	MECHANICAL	99.8	85.4	93.5	99.8	91.6	96.2	99.8	108.1	103.5	99.9	92.6	96.7	100.7	84.4	93.6	99.8	89.4	95.
16	ELECTRICAL	97.1	90.1	93.7	99.5	83.0	91.4	98.0	109.7	103.7	99.7	87.2	93.6	97.4	80.7	89.3	97.2	90.8	94.
01 - 16	WEIGHTED AVERAGE	97.2	88.5	93.2	94.7	92.3	93.6	97.1	106.0	101.1	97.8	90.5	94.5	94.4	84.9	90.1	96.8	98.1	97

OHIO / OKLAHOMA

DIVISION		SPRINGFIELD MAT.	INST.	TOTAL	TOLEDO MAT.	INST.	TOTAL	YOUNGSTOWN MAT.	INST.	TOTAL	ENID MAT.	INST.	TOTAL	LAWTON MAT.	INST.	TOTAL	MUSKOGEE MAT.	INST.	TOTA
01590	EQUIPMENT RENTAL	.0	96.8	96.8	.0	99.4	99.4	.0	97.4	97.4	.0	77.1	77.1	.0	78.2	78.2	.0	86.8	86.
02	SITE CONSTRUCTION	74.2	107.3	98.7	85.8	102.9	98.4	100.7	107.2	105.5	108.8	89.1	94.3	104.2	90.8	94.3	94.8	85.2	87.
03100	CONCRETE FORMS & ACCESSORIES	95.8	85.1	86.6	98.6	98.5	98.5	98.1	92.4	93.2	93.7	40.4	47.5	97.5	55.4	61.0	99.0	36.6	44.
03200	CONCRETE REINFORCEMENT	98.7	82.1	90.3	76.1	95.6	85.9	94.1	98.0	96.0	96.7	82.7	89.6	96.9	82.7	89.7	96.7	39.3	67.
03300	CAST-IN-PLACE CONCRETE	81.0	89.8	84.7	89.6	108.4	97.5	98.2	105.1	101.1	92.6	51.6	75.5	89.6	51.7	73.7	83.3	39.7	65
03	CONCRETE	91.1	85.8	88.4	92.2	100.9	96.5	98.9	97.0	97.9	90.6	52.4	71.5	87.3	59.1	73.2	83.1	39.3	61.
04	MASONRY	84.6	86.4	85.7	105.3	104.7	104.9	90.0	98.1	95.1	103.5	60.2	76.3	97.5	60.2	74.1	113.4	43.2	69.
05	METALS	86.1	77.2	83.2	96.3	89.5	94.1	88.2	83.0	86.5	89.2	67.3	82.0	92.6	67.4	84.4	89.1	59.1	79.
06	WOOD & PLASTICS	98.7	84.4	91.2	104.3	97.2	100.6	92.2	90.2	91.2	97.4	37.5	66.1	100.4	57.6	78.1	102.4	37.1	68.
07	THERMAL & MOISTURE PROTECTION	94.4	88.7	91.6	101.0	109.3	105.1	104.8	100.9	102.9	99.8	60.3	80.4	99.5	62.4	81.2	99.3	45.5	72.
08	DOORS & WINDOWS	94.2	81.4	90.9	97.5	93.6	96.5	96.8	96.3	96.8	95.6	51.5	84.2	97.2	62.5	88.2	95.6	36.6	80.
09200	PLASTER & GYPSUM BOARD	101.2	84.2	90.5	99.3	97.0	97.8	95.2	89.5	91.6	82.7	36.6	53.5	85.4	57.3	67.6	86.6	36.1	54.
095,098	CEILINGS & ACOUSTICAL TREATMENT	96.5	84.2	89.1	96.2	97.0	96.7	91.9	89.5	90.4	81.4	36.6	54.6	91.2	57.3	70.9	91.2	36.1	58.
09600	FLOORING	110.0	84.0	103.1	96.8	100.8	97.8	103.4	97.9	102.0	112.9	53.7	97.2	115.8	53.7	99.4	117.1	43.3	97.
097,099	WALL FINISHES, PAINTS & COATINGS	104.8	92.2	97.1	99.4	110.4	106.1	107.1	103.6	105.0	96.4	57.1	72.6	96.4	57.1	72.6	96.4	36.3	59.
09	FINISHES	99.1	85.5	92.1	98.6	100.1	99.4	99.6	94.0	96.7	94.6	42.8	67.6	97.8	54.6	75.3	97.7	37.7	66.
10 - 14	TOTAL DIV. 10000 - 14000	100.0	92.9	98.5	100.0	93.4	98.6	100.0	91.9	98.3	100.0	63.9	92.3	100.0	66.5	92.9	100.0	63.3	92.
15	MECHANICAL	100.7	85.0	93.8	99.9	99.9	99.9	99.8	90.7	95.8	100.1	63.7	84.2	100.1	63.7	84.2	100.1	30.3	69.
16	ELECTRICAL	97.4	84.0	90.9	99.5	99.4	99.4	97.2	95.4	96.3	95.3	68.5	82.3	96.8	68.5	83.0	95.0	35.1	65.
01 - 16	WEIGHTED AVERAGE	94.4	86.7	90.9	98.0	99.6	98.7	97.1	94.8	96.0	96.1	61.6	80.4	96.5	65.0	82.2	95.7	43.7	72.

City Cost Indexes

OKLAHOMA / OREGON

DIVISION		OKLAHOMA CITY MAT.	INST.	TOTAL	TULSA MAT.	INST.	TOTAL	EUGENE MAT.	INST.	TOTAL	MEDFORD MAT.	INST.	TOTAL	PORTLAND MAT.	INST.	TOTAL	SALEM MAT.	INST.	TOTAL
01590	EQUIPMENT RENTAL	.0	78.5	78.5	.0	86.8	86.8	.0	99.5	99.5	.0	99.5	99.5	.0	99.5	99.5	.0	99.5	99.5
02	SITE CONSTRUCTION	105.5	91.4	95.0	101.2	86.2	90.1	108.8	104.8	105.8	118.4	104.8	108.3	109.9	104.8	106.1	108.4	104.8	105.7
03100	CONCRETE FORMS & ACCESSORIES	98.6	47.7	54.5	98.7	48.2	54.9	103.5	105.8	105.5	98.3	105.6	104.6	104.9	106.3	106.1	104.8	106.0	105.9
03200	CONCRETE REINFORCEMENT	96.9	82.7	89.7	96.9	82.6	89.7	106.0	99.5	102.7	103.4	99.4	101.4	106.9	99.9	103.3	107.0	99.9	103.4
03300	CAST-IN-PLACE CONCRETE	96.9	55.0	79.4	90.8	50.4	73.9	106.0	107.9	106.8	109.5	107.8	108.7	113.8	108.0	111.4	109.9	107.9	109.1
03	CONCRETE	90.9	56.8	73.8	87.9	56.4	72.2	108.8	105.0	106.9	115.6	104.8	110.2	112.8	105.3	109.0	110.9	105.1	108.0
04	MASONRY	99.3	62.4	76.1	96.8	62.4	75.2	112.0	101.4	105.3	108.9	101.4	104.2	113.2	107.4	109.8	116.7	107.8	111.1
05	METALS	94.5	67.3	85.6	92.0	81.9	88.7	94.4	97.0	95.3	94.0	96.8	94.9	95.6	98.0	96.4	94.8	97.7	95.8
06	WOOD & PLASTICS	102.2	46.2	73.0	101.1	47.6	73.1	94.1	105.7	100.1	87.8	105.7	97.1	95.3	105.7	100.7	95.3	105.7	100.7
07	THERMAL & MOISTURE PROTECTION	98.5	62.1	80.6	98.8	59.9	79.7	110.3	92.6	101.6	111.1	89.6	100.5	110.1	102.6	106.4	110.1	98.3	104.3
08	DOORS & WINDOWS	97.2	56.2	86.6	97.2	56.5	86.7	98.8	106.3	100.7	101.5	106.3	102.7	96.3	106.3	98.9	98.1	106.3	100.2
09200	PLASTER & GYPSUM BOARD	85.4	45.5	60.1	85.4	46.8	60.9	94.3	105.7	101.5	92.5	105.7	100.9	92.3	105.7	100.8	92.3	105.7	100.8
095,098	CEILINGS & ACOUSTICAL TREATMENT	91.2	45.5	63.8	91.2	46.8	64.6	105.3	105.7	105.5	115.1	105.7	109.5	105.3	105.7	105.5	105.3	105.7	105.5
09600	FLOORING	115.8	53.7	99.4	115.6	55.9	99.8	115.8	103.1	112.4	113.3	103.1	110.6	115.8	103.1	112.4	115.8	103.1	112.4
097,099	WALL FINISHES, PAINTS & COATINGS	96.4	57.1	72.6	96.4	53.6	70.4	115.8	74.6	90.8	113.3	62.8	83.6	115.8	74.6	90.8	115.8	74.6	90.8
09	FINISHES	97.9	48.4	72.1	97.3	49.3	72.3	110.2	102.1	106.0	112.3	100.7	106.3	110.0	102.1	105.9	109.8	102.1	105.8
10 - 14	TOTAL DIV. 10000 - 14000	100.0	65.9	92.7	100.0	66.8	92.9	100.0	108.2	101.7	100.0	108.1	101.7	100.0	108.2	101.7	100.0	108.2	101.7
15	MECHANICAL	100.1	64.9	84.7	100.1	62.6	83.7	100.2	103.9	101.8	100.2	103.9	101.8	100.2	109.5	104.2	100.2	104.0	101.8
16	ELECTRICAL	96.3	68.5	82.7	96.8	45.8	71.9	97.3	97.7	97.5	100.6	87.7	94.3	97.8	105.8	101.7	97.4	97.7	97.6
01 - 16	WEIGHTED AVERAGE	97.3	64.0	82.1	96.3	61.2	80.4	101.7	102.1	101.9	103.4	100.3	102.0	102.2	105.4	103.7	102.2	103.0	102.5

PENNSYLVANIA

DIVISION		ALLENTOWN MAT.	INST.	TOTAL	ALTOONA MAT.	INST.	TOTAL	ERIE MAT.	INST.	TOTAL	HARRISBURG MAT.	INST.	TOTAL	PHILADELPHIA MAT.	INST.	TOTAL	PITTSBURGH MAT.	INST.	TOTAL
01590	EQUIPMENT RENTAL	.0	115.7	115.7	.0	115.7	115.7	.0	115.7	115.7	.0	114.9	114.9	.0	94.7	94.7	.0	115.5	115.5
02	SITE CONSTRUCTION	91.4	106.4	102.5	95.3	106.6	103.7	92.0	107.1	103.2	82.5	104.9	99.1	101.1	94.8	96.4	98.5	109.1	106.3
03100	CONCRETE FORMS & ACCESSORIES	101.9	105.2	104.8	83.9	86.3	86.0	101.2	96.9	97.4	94.4	85.2	86.4	99.9	134.0	129.5	101.5	98.8	99.1
03200	CONCRETE REINFORCEMENT	103.1	102.0	102.5	100.0	97.1	98.5	102.0	94.3	98.1	103.1	96.9	100.0	104.0	135.7	120.0	99.3	106.2	102.8
03300	CAST-IN-PLACE CONCRETE	89.0	94.9	91.5	99.3	86.9	94.1	97.6	84.9	92.3	98.3	89.2	94.5	97.1	128.4	110.2	99.8	97.3	98.7
03	CONCRETE	98.8	102.1	100.5	94.0	90.0	92.0	93.1	93.5	93.3	100.9	90.3	95.6	107.0	131.9	119.4	95.9	100.9	98.4
04	MASONRY	94.0	90.7	91.9	96.6	80.7	86.6	86.4	92.3	90.1	94.6	84.3	88.1	97.0	129.3	117.2	89.6	97.4	94.5
05	METALS	97.3	120.2	104.8	91.0	117.0	99.5	91.2	116.7	99.5	99.3	117.4	105.2	104.5	127.7	112.1	95.6	123.0	104.6
06	WOOD & PLASTICS	103.2	108.5	105.9	80.4	88.2	84.4	99.6	98.1	98.8	96.8	84.4	90.4	97.9	133.8	116.6	100.1	98.0	99.0
07	THERMAL & MOISTURE PROTECTION	99.1	109.8	104.3	97.6	92.0	94.9	97.8	96.5	97.2	102.8	103.3	103.0	101.2	132.8	116.6	97.5	101.1	99.2
08	DOORS & WINDOWS	93.6	105.3	96.6	88.1	95.2	90.0	88.3	94.3	89.8	93.6	93.1	93.5	96.7	139.5	107.8	93.8	107.6	97.4
09200	PLASTER & GYPSUM BOARD	108.4	108.4	108.4	100.7	87.6	92.4	108.4	97.8	101.7	108.4	83.7	92.8	105.1	134.6	123.8	99.3	97.7	98.3
095,098	CEILINGS & ACOUSTICAL TREATMENT	90.0	108.4	101.1	94.5	87.6	90.3	90.0	97.8	94.7	90.0	83.7	86.2	101.9	134.6	121.5	89.8	97.7	94.5
09600	FLOORING	88.2	83.4	87.0	82.3	59.5	76.3	90.2	88.6	89.8	88.5	78.4	85.8	85.7	134.9	98.7	96.3	106.2	98.9
097,099	WALL FINISHES, PAINTS & COATINGS	90.3	93.3	92.1	85.4	111.7	101.4	96.9	88.6	91.8	90.3	87.0	88.3	92.5	141.7	122.3	96.3	117.6	109.2
09	FINISHES	95.8	100.0	98.0	94.2	83.3	88.5	97.2	94.5	95.8	94.8	83.4	88.9	101.2	135.1	118.9	96.0	101.6	98.9
10 - 14	TOTAL DIV. 10000 - 14000	100.0	96.8	99.3	100.0	99.7	99.9	100.0	104.0	100.9	100.0	98.0	99.6	100.0	126.5	105.6	100.0	104.8	101.0
15	MECHANICAL	100.4	99.8	100.1	99.8	84.7	93.2	99.8	92.9	96.8	100.4	90.3	95.9	100.1	132.3	114.2	100.1	100.6	100.3
16	ELECTRICAL	100.1	92.4	96.3	89.3	102.0	95.5	90.7	86.4	88.6	99.1	83.7	91.6	97.2	134.2	115.2	95.4	102.0	98.6
01 - 16	WEIGHTED AVERAGE	98.0	101.2	99.4	94.3	93.3	93.8	94.3	96.2	95.1	98.1	92.3	95.5	100.8	128.9	113.6	96.7	103.8	99.9

PENNSYLVANIA / PUERTO RICO / RHODE ISLAND / SOUTH CAROLINA

DIVISION		READING MAT.	INST.	TOTAL	SCRANTON MAT.	INST.	TOTAL	YORK MAT.	INST.	TOTAL	SAN JUAN MAT.	INST.	TOTAL	PROVIDENCE MAT.	INST.	TOTAL	CHARLESTON MAT.	INST.	TOTAL
01590	EQUIPMENT RENTAL	.0	118.2	118.2	.0	115.7	115.7	.0	114.9	114.9	.0	89.1	89.1	.0	103.4	103.4	.0	99.3	99.3
02	SITE CONSTRUCTION	98.7	110.7	107.6	91.9	106.3	102.6	81.6	104.8	98.8	115.6	92.1	98.2	80.9	104.0	98.0	95.2	81.2	84.9
03100	CONCRETE FORMS & ACCESSORIES	99.4	86.2	88.0	102.0	84.5	86.9	82.4	79.7	80.1	94.1	20.1	30.0	103.5	115.4	113.8	96.3	38.4	46.1
03200	CONCRETE REINFORCEMENT	102.2	103.1	102.7	103.1	101.9	102.5	102.0	96.8	99.4	194.4	12.5	102.5	108.2	117.3	112.8	95.5	65.6	80.4
03300	CAST-IN-PLACE CONCRETE	71.0	91.3	79.5	93.0	88.7	91.2	87.6	83.9	86.1	105.2	31.4	74.4	86.1	113.2	97.4	85.1	49.3	70.2
03	CONCRETE	94.3	92.7	93.5	100.7	90.8	95.8	98.9	86.0	92.5	113.9	23.3	68.7	102.3	114.6	108.4	93.7	49.1	71.4
04	MASONRY	98.2	87.2	91.3	94.3	92.8	93.4	93.9	81.7	86.2	219.3	17.3	92.6	107.7	126.5	119.5	88.6	36.1	55.7
05	METALS	100.9	120.7	107.4	99.4	119.5	106.0	96.8	117.0	103.4	115.2	31.3	87.8	100.0	110.3	103.3	93.3	81.2	89.4
06	WOOD & PLASTICS	98.1	84.4	90.9	103.2	81.9	92.1	87.4	78.0	82.5	91.1	20.4	54.2	101.2	113.9	107.8	95.4	37.7	65.3
07	THERMAL & MOISTURE PROTECTION	101.2	105.4	103.3	98.9	99.1	99.0	97.2	98.7	97.9	167.7	23.8	97.0	100.6	114.4	107.4	97.1	43.8	70.9
08	DOORS & WINDOWS	94.4	92.5	93.9	93.6	93.2	93.5	90.5	89.6	90.3	153.8	17.1	118.5	101.0	115.1	104.7	96.3	42.6	82.4
09200	PLASTER & GYPSUM BOARD	108.7	83.8	92.9	111.0	81.2	92.1	104.4	77.1	87.1	252.1	18.0	103.7	102.0	113.4	109.2	107.2	35.7	61.9
095,098	CEILINGS & ACOUSTICAL TREATMENT	90.2	83.8	86.4	99.8	81.2	88.6	91.8	77.1	83.0	343.0	18.0	148.1	100.3	113.4	108.2	98.1	35.7	60.7
09600	FLOORING	83.9	87.6	84.9	88.2	82.1	86.6	84.2	83.8	84.1	200.8	17.1	152.3	100.7	133.9	109.5	105.3	41.0	88.3
097,099	WALL FINISHES, PAINTS & COATINGS	90.7	98.8	95.6	90.3	95.3	93.3	90.3	76.4	81.8	201.8	15.8	89.0	95.3	113.4	106.3	108.5	39.5	66.7
09	FINISHES	98.3	86.8	92.3	98.5	84.4	91.2	93.5	79.0	86.0	251.5	19.3	130.7	100.9	119.0	110.3	100.3	38.4	68.1
10 - 14	TOTAL DIV. 10000 - 14000	100.0	92.1	98.3	100.0	93.0	98.5	100.0	96.6	99.3	100.0	21.0	83.2	100.0	102.8	100.6	100.0	63.8	92.3
15	MECHANICAL	100.3	107.3	103.4	100.4	87.4	94.7	100.4	89.5	95.6	102.4	15.9	64.5	100.1	112.5	105.5	100.0	44.6	75.8
16	ELECTRICAL	98.9	90.9	95.0	100.1	84.5	92.5	94.0	71.7	83.1	133.6	15.7	76.1	99.0	101.4	100.1	97.5	33.2	66.1
01 - 16	WEIGHTED AVERAGE	98.6	98.5	98.6	98.8	92.9	96.1	96.2	88.6	92.7	136.6	26.3	86.5	100.2	112.4	105.8	96.8	48.9	75.1

SOUTH CAROLINA / SOUTH DAKOTA

DIVISION		COLUMBIA MAT.	INST.	TOTAL	FLORENCE MAT.	INST.	TOTAL	GREENVILLE MAT.	INST.	TOTAL	SPARTANBURG MAT.	INST.	TOTAL	ABERDEEN MAT.	INST.	TOTAL	PIERRE MAT.	INST.	TOTAL
01590	EQUIPMENT RENTAL	.0	99.3	99.3	.0	99.3	99.3	.0	99.3	99.3	.0	99.3	99.3	.0	98.2	98.2	.0	98.2	98.2
02	SITE CONSTRUCTION	94.5	81.2	84.7	105.8	81.2	87.6	100.8	80.8	86.0	100.7	80.8	86.0	84.2	93.6	91.1	82.5	93.6	90.7
03100	CONCRETE FORMS & ACCESSORIES	100.9	40.7	48.7	82.0	40.6	46.1	96.1	40.3	47.8	99.8	40.3	48.3	97.0	40.1	47.7	95.3	41.6	48.8
03200	CONCRETE REINFORCEMENT	95.5	65.6	80.4	95.2	65.6	80.2	95.1	40.0	67.2	95.1	40.0	67.2	99.8	49.3	74.3	99.4	61.2	80.1
03300	CAST-IN-PLACE CONCRETE	80.3	49.0	67.2	72.5	49.2	62.8	72.5	49.1	62.8	72.5	49.1	62.8	102.3	51.2	81.0	99.4	46.9	77.5
03	CONCRETE	91.7	50.0	70.9	94.3	50.0	72.2	93.0	45.2	69.1	93.2	45.2	69.2	98.4	47.1	72.8	96.1	48.5	72.4
04	MASONRY	87.3	37.1	55.8	73.3	36.1	50.0	71.2	36.1	49.2	73.3	36.1	50.0	107.7	58.0	76.5	104.2	56.2	74.0
05	METALS	93.3	81.1	89.3	92.3	81.0	88.6	92.3	71.6	85.5	92.3	71.6	85.5	98.3	70.8	89.3	98.3	75.9	91.0
06	WOOD & PLASTICS	101.0	40.9	69.6	79.5	40.9	59.4	95.2	40.9	66.9	99.5	40.9	68.9	101.8	38.2	68.6	99.8	40.3	68.8
07	THERMAL & MOISTURE PROTECTION	97.0	43.7	70.8	97.4	44.2	71.2	97.4	44.2	71.2	97.4	44.2	71.3	99.0	52.2	76.0	99.2	49.9	75.0
08	DOORS & WINDOWS	96.3	44.4	82.8	92.3	44.4	79.9	92.2	38.4	78.3	92.2	38.4	78.3	95.0	41.1	81.1	98.3	45.7	84.7
09200	PLASTER & GYPSUM BOARD	107.2	39.1	64.1	95.8	39.1	59.8	102.4	39.1	62.3	104.4	39.1	63.0	106.3	36.8	62.3	104.4	39.0	63.0
095,098	CEILINGS & ACOUSTICAL TREATMENT	98.1	39.1	62.7	92.8	39.1	60.6	91.9	39.1	60.2	91.9	39.1	60.2	112.0	36.8	66.9	109.4	39.0	67.2
09600	FLOORING	105.1	41.0	88.2	95.4	41.0	81.0	102.7	41.8	86.6	104.4	41.8	87.9	108.7	58.7	95.5	107.8	41.8	90.4
097,099	WALL FINISHES, PAINTS & COATINGS	108.5	39.5	66.7	108.5	39.5	66.7	108.5	39.5	66.7	108.5	39.5	66.7	99.2	35.6	60.6	99.2	40.7	63.7
09	FINISHES	100.3	40.4	69.1	95.3	40.4	66.7	97.9	40.5	68.1	98.7	40.5	68.4	106.8	42.6	73.4	105.5	41.0	72.0
10 - 14	TOTAL DIV. 10000 - 14000	100.0	64.2	92.4	100.0	64.2	92.4	100.0	64.2	92.4	100.0	64.2	92.4	100.0	57.5	91.0	100.0	68.4	93.3
15	MECHANICAL	100.0	37.6	72.7	100.0	37.6	72.7	100.0	37.5	72.7	100.0	37.5	72.7	100.2	40.1	73.9	100.2	38.4	73.2
16	ELECTRICAL	98.0	35.0	67.3	95.7	20.0	58.8	97.6	31.8	65.5	97.6	31.8	65.5	97.3	60.8	79.5	93.7	51.9	73.3
01 - 16	WEIGHTED AVERAGE	96.6	48.3	74.7	95.0	46.1	72.8	95.2	46.0	72.9	95.5	46.0	73.0	99.4	54.6	79.0	98.7	53.7	78.2

SOUTH DAKOTA / TENNESSEE

DIVISION		RAPID CITY MAT.	INST.	TOTAL	SIOUX FALLS MAT.	INST.	TOTAL	CHATTANOOGA MAT.	INST.	TOTAL	JACKSON MAT.	INST.	TOTAL	JOHNSON CITY MAT.	INST.	TOTAL	KNOXVILLE MAT.	INST.	TOTAL
01590	EQUIPMENT RENTAL	.0	98.2	98.2	.0	99.4	99.4	.0	104.5	104.5	.0	104.9	104.9	.0	97.4	97.4	.0	97.4	97.4
02	SITE CONSTRUCTION	82.7	93.3	90.6	83.7	95.5	92.4	103.2	96.5	98.2	100.6	95.7	97.0	113.1	85.2	92.5	89.9	85.3	86.5
03100	CONCRETE FORMS & ACCESSORIES	104.4	37.0	46.0	95.4	42.6	49.6	98.7	47.1	54.0	88.7	38.4	45.1	83.5	47.9	52.7	97.8	47.9	54.6
03200	CONCRETE REINFORCEMENT	93.3	61.3	77.1	93.3	61.3	77.1	87.9	44.6	66.0	89.6	44.3	66.7	88.5	44.8	66.4	87.9	44.8	66.1
03300	CAST-IN-PLACE CONCRETE	98.7	44.3	76.0	102.8	49.8	80.7	100.8	49.4	79.3	100.5	42.7	76.4	81.0	54.1	69.8	94.6	49.6	75.8
03	CONCRETE	95.4	45.6	70.6	95.6	50.0	72.8	92.6	49.3	71.0	93.9	43.1	68.5	97.6	51.3	74.5	89.9	49.9	69.9
04	MASONRY	104.1	52.2	71.6	101.5	56.2	73.0	96.9	44.4	64.0	111.4	34.5	63.2	108.8	44.2	68.3	75.1	44.2	55.7
05	METALS	100.1	76.2	92.3	100.2	76.3	92.4	95.4	77.7	89.6	92.8	75.9	87.3	92.7	77.3	87.7	96.0	77.6	90.0
06	WOOD & PLASTICS	106.1	36.1	69.5	99.8	41.0	69.1	102.3	48.1	74.0	87.2	38.2	61.6	76.1	49.2	62.0	91.5	49.2	69.4
07	THERMAL & MOISTURE PROTECTION	99.6	48.1	74.3	99.1	49.5	74.7	100.3	51.7	76.4	99.3	40.6	70.5	96.1	49.3	73.1	93.7	49.5	72.0
08	DOORS & WINDOWS	99.3	43.4	84.8	99.3	46.0	85.5	101.4	48.2	87.6	102.1	40.6	86.2	97.2	49.3	84.8	93.7	49.3	82.2
09200	PLASTER & GYPSUM BOARD	106.3	34.6	60.9	106.3	39.7	64.1	84.2	47.2	60.8	89.3	37.1	56.2	90.1	48.4	63.6	98.0	48.4	66.6
095,098	CEILINGS & ACOUSTICAL TREATMENT	116.5	34.6	67.4	116.5	39.7	70.4	98.1	47.2	67.6	95.1	37.1	60.3	93.4	48.4	66.4	96.1	48.4	67.5
09600	FLOORING	107.8	73.6	98.8	107.8	72.1	98.4	104.0	51.6	90.1	92.8	25.0	74.8	98.0	53.1	86.1	103.6	53.1	90.2
097,099	WALL FINISHES, PAINTS & COATINGS	99.2	40.7	63.7	99.2	40.7	63.7	109.4	47.4	71.8	96.5	32.1	57.5	107.2	55.7	75.9	107.2	55.7	75.9
09	FINISHES	107.6	43.7	74.4	107.6	47.5	76.4	98.6	47.9	72.3	95.6	35.2	64.2	99.2	49.6	73.4	94.6	49.6	71.2
10 - 14	TOTAL DIV. 10000 - 14000	100.0	66.0	92.8	100.0	68.6	93.3	100.0	55.4	90.5	100.0	52.3	89.8	100.0	63.7	92.3	100.0	63.7	92.3
15	MECHANICAL	100.2	35.8	72.1	100.2	37.4	72.7	99.8	46.5	76.5	100.0	57.6	81.4	99.8	61.3	83.0	99.8	61.3	83.0
16	ELECTRICAL	94.2	51.9	73.6	93.6	73.7	83.9	103.3	54.1	79.3	100.7	52.3	77.1	91.6	52.4	72.4	96.8	61.2	79.4
01 - 16	WEIGHTED AVERAGE	99.3	52.5	78.0	99.1	57.8	80.3	98.8	55.8	79.3	98.5	53.3	78.0	97.6	58.2	79.7	94.9	59.2	78.7

TENNESSEE / TEXAS

DIVISION		MEMPHIS MAT.	INST.	TOTAL	NASHVILLE MAT.	INST.	TOTAL	ABILENE MAT.	INST.	TOTAL	AMARILLO MAT.	INST.	TOTAL	AUSTIN MAT.	INST.	TOTAL	BEAUMONT MAT.	INST.	TOTAL
01590	EQUIPMENT RENTAL	.0	102.2	102.2	.0	105.5	105.5	.0	86.8	86.8	.0	86.8	86.8	.0	85.9	85.9	.0	88.7	88.7
02	SITE CONSTRUCTION	95.7	91.8	92.8	95.0	99.2	98.1	101.5	85.4	89.6	101.9	86.3	90.3	89.8	85.3	86.5	96.5	87.4	89.7
03100	CONCRETE FORMS & ACCESSORIES	99.3	67.1	71.4	97.8	65.2	69.5	95.4	43.9	50.8	97.6	53.2	59.1	94.1	56.6	61.6	103.1	54.5	61.0
03200	CONCRETE REINFORCEMENT	102.2	70.5	86.2	87.1	67.0	76.9	94.1	53.6	73.6	94.1	56.1	74.9	98.7	54.0	76.1	92.5	45.7	68.8
03300	CAST-IN-PLACE CONCRETE	88.9	71.2	81.5	94.4	69.7	84.1	94.2	43.6	73.0	99.0	49.9	78.5	90.0	50.8	73.6	86.2	56.3	73.7
03	CONCRETE	89.6	70.7	80.2	90.0	68.6	79.3	88.9	46.7	67.8	91.3	53.4	72.4	81.8	54.7	68.3	85.6	54.3	70.0
04	MASONRY	84.4	78.0	80.4	85.7	65.9	73.3	100.4	54.4	71.5	102.4	52.9	71.3	94.7	56.4	70.7	104.9	59.1	76.1
05	METALS	97.0	96.1	96.7	96.7	91.9	95.1	95.7	67.7	86.5	95.7	68.9	86.9	94.0	66.7	85.1	97.1	65.7	86.8
06	WOOD & PLASTICS	97.4	67.7	81.9	98.4	66.2	81.6	97.1	43.4	69.0	98.4	55.7	76.1	94.6	59.8	76.4	110.9	55.1	81.7
07	THERMAL & MOISTURE PROTECTION	96.6	74.9	85.9	97.0	66.7	82.1	99.4	51.0	75.6	102.0	50.1	76.5	96.1	55.4	76.1	103.3	59.6	81.8
08	DOORS & WINDOWS	100.9	69.8	92.9	98.5	67.5	90.5	92.9	46.4	80.9	92.9	51.4	82.2	94.5	59.2	85.4	97.6	51.0	85.7
09200	PLASTER & GYPSUM BOARD	95.3	67.3	77.5	92.3	65.5	75.4	85.4	42.5	58.2	85.4	55.1	66.2	90.7	59.3	70.8	90.5	54.5	67.7
095,098	CEILINGS & ACOUSTICAL TREATMENT	106.3	67.3	82.9	96.5	65.7	78.0	91.2	42.5	62.0	91.2	55.1	69.6	88.0	59.3	70.8	93.0	54.5	69.9
09600	FLOORING	98.3	47.1	84.7	106.8	77.5	99.1	115.8	69.1	103.4	115.6	66.7	102.7	101.4	51.2	88.1	110.1	72.6	100.1
097,099	WALL FINISHES, PAINTS & COATINGS	99.0	60.6	75.7	112.7	70.1	86.9	95.2	56.6	71.8	95.2	39.2	61.3	92.2	44.0	63.0	91.4	52.9	68.1
09	FINISHES	97.3	62.1	79.0	104.0	68.0	85.3	97.3	49.6	72.5	97.3	54.3	74.9	92.1	54.2	72.4	91.8	57.5	74.1
10 - 14	TOTAL DIV. 10000 - 14000	100.0	79.1	95.6	100.0	70.8	93.8	100.0	73.9	94.5	100.0	66.0	92.8	100.0	66.2	92.8	100.0	78.2	95.0
15	MECHANICAL	99.8	76.6	89.7	99.8	73.6	88.3	100.1	46.8	76.8	100.1	52.9	79.5	99.9	56.9	81.1	100.0	64.4	84.4
16	ELECTRICAL	95.8	78.5	87.4	102.9	68.2	86.0	96.9	50.0	74.0	97.4	61.5	79.9	99.0	69.4	84.6	95.0	65.3	80.6
01 - 16	WEIGHTED AVERAGE	96.6	77.2	87.8	98.0	74.0	87.1	96.8	54.5	77.6	97.3	59.1	79.9	94.9	61.8	79.8	96.6	63.5	81.7

TEXAS

DIVISION		CORPUS CHRISTI			DALLAS			EL PASO			FORT WORTH			HOUSTON			LAREDO		
		MAT.	INST.	TOTAL	MAT.	INST.	TOTAL	MAT.	INST.	TOTAL	MAT.	INST.	TOTAL	MAT.	INST.	TOTAL	MAT.	INST.	TOTAL
01590	EQUIPMENT RENTAL	.0	94.7	94.7	.0	98.1	98.1	.0	86.8	86.8	.0	86.8	86.8	.0	98.9	98.9	.0	85.9	85.9
02	SITE CONSTRUCTION	124.9	80.6	92.1	128.3	86.3	97.2	102.0	85.0	89.4	101.4	85.7	89.8	123.8	85.8	95.7	90.3	85.1	86.5
03100	CONCRETE FORMS & ACCESSORIES	97.7	39.9	47.6	93.6	59.3	63.9	94.8	48.5	54.7	96.3	58.6	63.6	91.4	66.7	70.0	91.7	40.2	47.1
03200	CONCRETE REINFORCEMENT	97.8	52.0	74.7	96.5	58.4	77.2	94.1	50.7	72.2	94.1	58.3	76.0	94.2	64.8	79.3	98.7	52.3	75.3
03300	CAST-IN-PLACE CONCRETE	106.7	47.0	81.8	97.1	57.4	80.5	96.5	38.2	72.1	92.6	52.4	75.8	88.8	69.5	80.8	82.1	63.3	74.3
03	CONCRETE	91.6	46.6	69.1	88.2	60.0	74.2	89.9	46.2	68.1	88.2	57.1	72.7	84.9	68.9	76.9	81.6	51.4	66.6
04	MASONRY	86.0	49.4	63.1	102.7	61.3	76.8	99.4	49.4	68.0	95.7	61.3	74.1	100.6	65.3	78.4	95.0	55.4	70.1
05	METALS	93.5	79.9	89.1	93.0	84.3	90.2	95.5	63.7	85.1	92.4	70.0	85.1	101.5	90.9	98.1	94.9	65.6	85.3
06	WOOD & PLASTICS	107.2	39.3	71.8	96.8	60.2	77.7	97.1	52.1	73.6	103.8	60.0	80.9	96.2	66.4	80.6	90.5	39.2	63.7
07	THERMAL & MOISTURE PROTECTION	99.1	47.8	73.9	94.3	60.9	77.9	98.9	52.1	75.9	99.8	55.5	78.0	94.7	67.6	81.3	95.1	52.6	74.2
08	DOORS & WINDOWS	101.6	41.4	86.0	105.6	56.7	93.0	92.9	47.5	81.2	87.5	56.6	79.5	107.1	66.0	96.5	95.1	42.0	81.4
09200	PLASTER & GYPSUM BOARD	90.7	38.2	57.4	91.7	59.5	71.3	85.4	51.4	63.9	85.4	59.5	69.0	88.6	66.0	74.3	87.4	38.2	56.2
095,098	CEILINGS & ACOUSTICAL TREATMENT	88.0	38.2	58.1	94.6	59.5	73.6	91.2	51.4	67.4	91.2	59.5	72.2	94.6	66.0	77.4	88.0	38.2	58.1
09600	FLOORING	114.7	50.6	97.7	105.1	62.0	93.7	115.8	66.4	102.7	150.1	49.6	123.5	98.2	68.2	90.3	97.9	50.6	85.4
097,099	WALL FINISHES, PAINTS & COATINGS	105.7	43.1	67.7	104.7	56.1	75.2	95.2	37.7	60.3	96.4	55.9	71.9	99.8	64.0	78.1	92.2	57.5	71.2
09	FINISHES	98.0	41.4	68.6	99.9	59.3	78.8	97.3	51.2	73.3	107.8	56.6	81.2	96.8	66.5	81.0	90.8	43.0	65.9
10 - 14	TOTAL DIV. 10000 - 14000	100.0	74.7	94.6	100.0	78.8	95.5	100.0	64.3	92.4	100.0	78.3	95.4	100.0	83.2	96.4	100.0	64.4	92.4
15	MECHANICAL	99.9	47.8	77.1	99.8	67.0	85.4	100.1	35.7	71.9	100.1	61.0	83.0	100.0	73.2	88.3	99.8	43.1	75.0
16	ELECTRICAL	97.2	53.6	75.9	98.7	64.7	82.1	96.2	56.9	77.0	96.1	62.4	79.6	97.7	68.6	83.5	98.9	63.9	81.8
01 - 16	WEIGHTED AVERAGE	97.7	54.1	77.8	98.7	67.1	84.3	96.7	52.4	76.6	96.4	63.3	81.4	99.1	72.8	87.2	94.9	55.0	76.8

TEXAS / UTAH

DIVISION		LUBBOCK			ODESSA			SAN ANTONIO			WACO			WICHITA FALLS			LOGAN		
		MAT.	INST.	TOTAL	MAT.	INST.	TOTAL	MAT.	INST.	TOTAL	MAT.	INST.	TOTAL	MAT.	INST.	TOTAL	MAT.	INST.	TOTAL
01590	EQUIPMENT RENTAL	.0	96.9	96.9	.0	86.8	86.8	.0	88.8	88.8	.0	86.8	86.8	.0	86.8	86.8	.0	100.7	100.7
02	SITE CONSTRUCTION	131.4	83.5	96.0	101.7	85.8	89.9	90.0	89.8	89.9	100.4	85.7	89.5	101.1	85.4	89.5	92.0	100.8	98.5
03100	CONCRETE FORMS & ACCESSORIES	94.9	43.2	50.1	95.3	41.1	48.3	91.8	57.9	62.4	96.0	41.9	49.2	96.0	44.2	51.1	101.0	61.5	66.8
03200	CONCRETE REINFORCEMENT	95.4	53.5	74.2	94.1	53.4	73.6	105.4	54.3	79.6	94.1	53.8	73.8	94.1	53.8	73.8	108.5	80.7	94.5
03300	CAST-IN-PLACE CONCRETE	94.3	50.0	75.8	94.2	45.8	73.9	80.6	70.6	76.4	85.0	54.0	72.1	90.8	49.5	73.5	88.3	72.8	81.8
03	CONCRETE	88.0	49.5	68.8	88.9	46.1	67.5	82.0	62.3	72.2	84.5	49.4	67.0	87.3	48.8	68.1	108.4	69.3	88.9
04	MASONRY	99.8	51.2	69.3	100.4	51.5	69.7	94.9	65.3	76.3	97.2	59.9	73.8	97.6	58.0	72.7	121.3	62.3	84.3
05	METALS	99.2	82.5	93.8	95.1	66.9	85.8	95.6	69.6	87.1	95.6	67.5	86.4	95.5	68.3	86.6	99.5	77.4	92.3
06	WOOD & PLASTICS	97.4	43.5	69.2	97.1	41.2	67.9	90.5	56.5	72.8	102.8	38.3	69.1	102.8	43.3	71.8	88.1	59.7	73.3
07	THERMAL & MOISTURE PROTECTION	90.6	52.3	71.8	99.4	47.6	73.9	95.1	66.0	80.8	100.1	51.4	76.1	100.1	55.0	77.9	102.2	68.7	85.8
08	DOORS & WINDOWS	104.2	44.6	88.8	92.9	42.9	80.0	97.0	56.7	86.6	87.5	40.2	75.3	87.5	46.3	76.9	89.3	61.8	82.2
09200	PLASTER & GYPSUM BOARD	86.1	42.4	58.4	85.4	40.2	56.8	87.4	55.9	67.5	85.4	37.3	54.9	85.4	42.4	58.1	80.6	58.4	66.5
095,098	CEILINGS & ACOUSTICAL TREATMENT	93.9	42.4	63.0	91.2	40.2	60.6	88.0	55.9	68.8	91.2	37.3	58.8	91.2	42.4	61.9	97.8	58.4	74.1
09600	FLOORING	107.2	42.1	90.0	115.8	41.5	96.2	97.9	57.8	87.3	150.3	38.6	120.7	150.7	75.3	130.8	101.2	62.3	90.9
097,099	WALL FINISHES, PAINTS & COATINGS	106.8	35.2	63.4	95.2	35.2	58.8	92.2	57.5	71.2	96.4	36.3	59.9	98.2	57.5	73.5	103.7	49.6	70.9
09	FINISHES	99.3	41.3	69.1	97.3	39.9	67.4	90.8	57.1	73.3	107.8	39.4	72.2	108.0	50.9	78.3	97.0	59.6	77.5
10 - 14	TOTAL DIV. 10000 - 14000	100.0	73.6	94.4	100.0	63.5	92.2	100.0	70.7	93.8	100.0	75.5	94.8	100.0	64.4	92.4	100.0	75.8	94.9
15	MECHANICAL	99.7	50.1	78.0	100.1	39.1	73.4	99.8	69.5	86.6	100.1	56.6	81.1	100.1	51.0	78.6	100.1	71.3	87.5
16	ELECTRICAL	95.5	47.0	71.9	97.0	43.8	71.0	99.0	63.9	81.9	97.0	63.2	80.5	100.1	57.4	79.3	88.6	76.4	82.7
01 - 16	WEIGHTED AVERAGE	98.8	54.9	78.9	96.7	49.9	75.4	95.2	66.8	82.3	96.6	57.8	79.0	97.4	57.2	79.1	98.9	72.1	86.7

UTAH / VERMONT / VIRGINIA

DIVISION		OGDEN			PROVO			SALT LAKE CITY			BURLINGTON			RUTLAND			ALEXANDRIA		
		MAT.	INST.	TOTAL	MAT.	INST.	TOTAL	MAT.	INST.	TOTAL	MAT.	INST.	TOTAL	MAT.	INST.	TOTAL	MAT.	INST.	TOTAL
01590	EQUIPMENT RENTAL	.0	100.7	100.7	.0	99.4	99.4	.0	100.7	100.7	.0	101.6	101.6	.0	101.6	101.6	.0	101.2	101.2
02	SITE CONSTRUCTION	80.1	100.8	95.4	88.4	98.8	96.1	79.8	100.7	95.3	79.6	99.1	94.0	79.6	99.1	94.0	112.3	86.1	92.9
03100	CONCRETE FORMS & ACCESSORIES	101.0	61.5	66.8	102.5	61.5	67.0	99.5	61.5	66.6	91.4	51.9	57.2	101.8	52.0	58.6	91.7	75.0	77.2
03200	CONCRETE REINFORCEMENT	113.4	80.7	96.9	117.4	80.7	98.9	110.7	80.7	95.6	108.2	56.0	81.8	108.2	56.0	81.8	84.8	81.8	83.3
03300	CAST-IN-PLACE CONCRETE	89.6	72.8	82.6	88.4	72.8	81.9	97.2	72.8	87.0	97.2	64.9	83.7	92.6	64.9	81.1	99.6	82.7	92.6
03	CONCRETE	99.2	69.3	84.3	108.4	69.4	88.9	116.9	69.4	93.2	104.3	57.7	81.0	102.8	57.8	80.3	101.8	80.0	90.9
04	MASONRY	114.6	62.3	81.8	127.3	62.3	86.5	132.9	62.3	88.6	98.7	62.0	75.7	86.5	62.0	71.1	85.8	74.6	78.8
05	METALS	99.9	77.4	92.6	97.8	77.5	91.1	103.1	77.5	94.7	101.7	67.6	90.5	99.9	67.7	89.4	102.4	96.0	100.3
06	WOOD & PLASTICS	88.1	59.7	73.3	89.7	59.7	74.0	88.3	59.7	73.4	88.6	50.1	68.5	100.9	50.1	74.4	94.5	75.4	84.5
07	THERMAL & MOISTURE PROTECTION	100.7	68.7	85.0	104.7	68.7	87.0	103.8	68.7	86.6	101.4	56.9	79.5	101.2	58.1	80.0	96.0	78.9	87.6
08	DOORS & WINDOWS	89.3	61.8	82.2	93.7	61.8	85.4	89.3	61.8	82.2	107.1	46.9	91.5	107.1	46.9	91.5	96.3	77.5	91.4
09200	PLASTER & GYPSUM BOARD	80.6	58.4	66.5	81.0	58.4	66.7	80.6	58.4	66.5	103.5	47.8	68.2	103.5	47.8	68.2	107.2	74.6	86.5
095,098	CEILINGS & ACOUSTICAL TREATMENT	97.8	58.4	74.1	97.8	58.4	74.1	97.8	58.4	74.1	94.9	47.8	66.6	94.9	47.8	66.6	98.1	74.6	84.0
09600	FLOORING	101.2	62.3	90.9	102.0	62.3	91.5	100.8	62.3	90.6	99.8	72.0	92.4	99.8	72.0	92.4	105.3	91.2	101.6
097,099	WALL FINISHES, PAINTS & COATINGS	103.7	49.6	70.9	103.7	63.3	79.2	103.7	63.3	79.2	92.8	41.1	61.4	92.8	41.1	61.4	119.6	86.4	99.4
09	FINISHES	95.9	59.6	77.0	97.6	61.1	78.6	96.2	61.1	77.9	98.8	53.8	75.4	98.7	53.8	75.4	100.7	79.1	89.5
10 - 14	TOTAL DIV. 10000 - 14000	100.0	75.8	94.9	100.0	75.8	94.9	100.0	75.8	94.9	100.0	84.8	96.8	100.0	84.8	96.8	100.0	88.0	97.4
15	MECHANICAL	100.1	71.3	87.5	100.1	71.3	87.5	100.2	71.3	87.6	100.1	68.5	86.3	100.1	68.5	86.3	100.0	85.8	93.8
16	ELECTRICAL	88.7	76.4	82.7	88.8	76.4	82.8	92.5	76.4	84.6	100.9	67.4	84.6	100.2	67.4	84.2	98.5	94.4	96.5
01 - 16	WEIGHTED AVERAGE	97.2	72.1	85.8	99.5	72.1	87.1	101.2	72.3	88.1	100.8	66.0	85.0	99.8	66.1	84.5	99.6	84.6	92.8

VIRGINIA

DIVISION		ARLINGTON			NEWPORT NEWS			NORFOLK			PORTSMOUTH			RICHMOND			ROANOKE		
		MAT.	INST.	TOTAL	MAT.	INST.	TOTAL	MAT.	INST.	TOTAL	MAT.	INST.	TOTAL	MAT.	INST.	TOTAL	MAT.	INST.	TOTAL
01590	EQUIPMENT RENTAL	.0	99.7	99.7	.0	104.9	104.9	.0	105.6	105.6	.0	104.8	104.8	.0	104.8	104.8	.0	99.7	99.7
02	SITE CONSTRUCTION	122.9	82.7	93.2	105.5	84.5	90.0	104.8	85.6	90.6	104.1	84.1	89.3	106.1	84.9	90.4	103.2	81.0	86.8
03100	CONCRETE FORMS & ACCESSORIES	90.5	71.9	74.4	96.4	58.0	63.2	100.7	58.3	64.0	84.5	58.2	61.7	97.3	57.3	62.7	96.2	36.9	44.8
03200	CONCRETE REINFORCEMENT	95.7	65.1	80.3	95.5	72.0	83.6	95.5	72.0	83.6	95.2	72.0	83.5	95.5	71.2	83.2	95.5	64.6	79.9
03300	CAST-IN-PLACE CONCRETE	96.9	74.3	87.5	97.6	56.2	80.3	100.5	56.3	82.0	96.6	56.0	79.7	103.8	56.0	83.8	109.6	47.7	83.8
03	CONCRETE	106.7	72.6	89.7	99.6	61.5	80.6	101.3	61.7	81.5	98.3	61.5	79.9	102.6	61.0	81.8	105.3	47.7	76.6
04	MASONRY	98.5	68.6	79.8	89.6	55.3	68.1	96.2	55.3	70.6	94.6	55.3	69.9	88.5	50.8	64.9	90.0	40.0	58.6
05	METALS	101.1	87.7	96.7	102.4	89.6	98.2	101.5	89.8	97.7	101.4	89.1	97.4	104.4	89.6	99.5	102.2	82.0	95.6
06	WOOD & PLASTICS	91.0	75.4	82.9	95.4	61.5	77.7	100.7	61.5	80.2	82.2	61.5	71.4	96.5	59.5	77.2	95.4	36.4	64.6
07	THERMAL & MOISTURE PROTECTION	98.1	72.7	85.6	97.1	52.3	75.1	96.8	52.4	75.0	97.0	52.4	75.1	96.6	54.3	75.8	97.0	45.9	71.9
08	DOORS & WINDOWS	94.2	72.6	88.6	96.3	58.8	86.6	96.3	61.5	87.3	96.3	61.5	87.3	96.3	56.6	86.0	96.3	43.6	82.6
09200	PLASTER & GYPSUM BOARD	103.4	74.6	85.1	107.2	59.4	76.9	107.2	59.4	76.9	98.8	59.4	73.8	107.2	57.4	75.6	107.2	34.4	61.1
095,098	CEILINGS & ACOUSTICAL TREATMENT	92.8	74.6	81.9	98.1	59.4	74.9	98.1	59.4	74.9	98.1	59.4	74.9	98.1	57.4	73.7	98.1	34.4	59.9
09600	FLOORING	103.5	70.5	94.8	105.3	41.1	88.4	105.1	41.1	88.2	96.5	41.1	81.9	105.1	65.4	94.6	105.3	36.9	87.3
097,099	WALL FINISHES, PAINTS & COATINGS	119.6	86.4	99.4	108.5	43.5	69.1	108.5	64.1	81.6	108.5	64.1	81.6	108.5	56.9	77.2	108.5	34.0	63.3
09	FINISHES	99.3	73.5	85.9	100.1	53.5	75.9	100.1	55.8	77.1	96.3	55.8	75.2	100.0	58.9	78.6	100.0	36.0	66.7
10-14	TOTAL DIV. 10000 - 14000	100.0	72.8	94.2	100.0	70.0	93.6	100.0	70.0	93.6	100.0	70.0	93.6	100.0	74.9	94.7	100.0	62.9	92.1
15	MECHANICAL	100.0	82.2	92.2	100.0	58.5	81.9	100.0	62.3	83.5	100.0	62.3	83.5	100.0	62.3	83.5	100.0	40.7	74.1
16	ELECTRICAL	96.0	92.1	94.1	97.9	60.6	79.7	98.0	62.5	80.7	96.3	62.5	79.8	98.5	69.6	84.4	97.9	35.0	67.2
01-16	WEIGHTED AVERAGE	100.2	79.4	90.7	99.2	63.6	83.0	99.6	65.2	84.0	98.4	65.0	83.2	99.9	65.9	84.5	99.8	48.4	76.5

WASHINGTON

DIVISION		EVERETT			RICHLAND			SEATTLE			SPOKANE			TACOMA			VANCOUVER		
		MAT.	INST.	TOTAL	MAT.	INST.	TOTAL	MAT.	INST.	TOTAL	MAT.	INST.	TOTAL	MAT.	INST.	TOTAL	MAT.	INST.	TOTAL
01590	EQUIPMENT RENTAL	.0	104.0	104.0	.0	90.0	90.0	.0	103.8	103.8	.0	90.0	90.0	.0	104.0	104.0	.0	97.2	97.2
02	SITE CONSTRUCTION	94.3	115.2	109.8	104.4	89.5	93.3	98.4	113.3	109.4	105.2	89.5	93.6	97.1	115.5	110.7	109.6	100.9	103.1
03100	CONCRETE FORMS & ACCESSORIES	106.4	101.3	102.0	112.9	81.2	85.4	97.1	103.0	102.2	119.8	81.4	86.5	97.1	102.5	101.8	97.9	97.0	97.1
03200	CONCRETE REINFORCEMENT	107.9	93.6	100.7	103.1	88.0	95.5	106.7	93.9	100.2	103.8	88.1	95.8	106.7	93.7	100.1	107.5	93.5	100.4
03300	CAST-IN-PLACE CONCRETE	97.8	103.4	100.2	117.4	85.4	104.1	102.7	109.9	105.7	121.6	85.5	106.5	100.6	109.7	104.4	112.3	102.6	108.3
03	CONCRETE	98.5	100.0	99.2	108.0	83.8	95.9	101.3	103.2	102.2	110.7	83.9	97.3	100.3	102.7	101.5	110.5	98.0	104.2
04	MASONRY	130.9	99.1	110.9	113.3	84.1	95.0	125.6	103.9	112.0	114.9	84.8	96.0	125.4	101.3	110.3	125.9	89.3	108.6
05	METALS	105.4	87.3	99.4	94.3	82.5	90.4	107.0	90.6	101.6	96.9	82.8	92.3	107.0	88.2	100.8	104.2	89.6	99.5
06	WOOD & PLASTICS	97.2	101.4	99.4	95.3	80.5	87.6	87.7	101.4	94.8	104.6	80.5	92.0	86.7	101.4	94.3	81.0	96.7	89.2
07	THERMAL & MOISTURE PROTECTION	107.2	98.5	102.9	172.6	80.1	127.1	107.2	102.1	104.7	169.8	81.7	126.5	107.0	100.0	103.6	109.6	91.1	100.5
08	DOORS & WINDOWS	98.6	98.2	98.5	112.0	75.9	102.6	100.8	99.1	100.3	111.7	77.3	102.8	99.2	99.1	99.2	96.2	94.5	95.8
09200	PLASTER & GYPSUM BOARD	104.4	101.3	102.5	119.1	79.8	94.2	97.8	101.3	100.0	123.9	79.8	95.9	100.0	101.3	100.8	98.3	96.9	97.4
095,098	CEILINGS & ACOUSTICAL TREATMENT	114.7	101.3	106.7	107.1	79.8	90.7	118.0	101.3	108.0	107.1	79.8	90.7	118.2	101.3	108.1	113.0	96.9	103.4
09600	FLOORING	120.1	101.6	115.2	110.8	45.2	93.5	112.0	105.8	110.4	114.4	76.1	104.3	112.6	101.6	109.7	118.5	83.1	109.2
097,099	WALL FINISHES, PAINTS & COATINGS	113.4	81.4	94.0	113.5	74.1	89.6	113.4	95.6	102.6	113.5	76.8	91.3	113.4	95.6	102.6	121.5	70.5	90.6
09	FINISHES	114.8	99.3	106.8	122.8	73.1	96.9	112.5	102.4	107.2	124.6	79.7	101.2	113.0	101.5	107.0	110.9	91.4	100.8
10-14	TOTAL DIV. 10000 - 14000	100.0	95.7	99.1	100.0	88.9	97.6	100.0	105.6	101.2	100.0	88.9	97.6	100.0	105.5	101.2	100.0	85.6	96.9
15	MECHANICAL	99.9	95.7	98.1	100.5	94.5	97.9	99.9	114.1	106.1	100.5	84.6	93.5	100.0	95.1	97.9	100.1	100.2	100.2
16	ELECTRICAL	104.4	95.3	100.0	94.9	86.7	90.9	104.2	109.1	106.6	93.2	83.3	88.4	104.2	97.1	100.8	115.3	103.5	109.5
01-16	WEIGHTED AVERAGE	104.0	98.3	101.4	105.5	85.1	96.2	104.3	105.8	105.0	106.2	83.7	96.0	104.1	99.7	102.1	106.0	97.2	102.0

DIVISION		WASHINGTON			WEST VIRGINIA												WISCONSIN		
		YAKIMA			CHARLESTON			HUNTINGTON			PARKERSBURG			WHEELING			EAU CLAIRE		
		MAT.	INST.	TOTAL	MAT.	INST.	TOTAL	MAT.	INST.	TOTAL	MAT.	INST.	TOTAL	MAT.	INST.	TOTAL	MAT.	INST.	TOTAL
01590	EQUIPMENT RENTAL	.0	104.0	104.0	.0	99.7	99.7	.0	99.7	99.7	.0	99.7	99.7	.0	99.7	99.7	.0	100.5	100.5
02	SITE CONSTRUCTION	100.5	114.0	110.5	101.0	85.9	89.8	102.4	86.6	90.7	107.1	85.8	91.3	107.6	85.6	91.3	84.7	102.4	97.8
03100	CONCRETE FORMS & ACCESSORIES	97.5	96.2	96.4	103.9	93.8	95.1	97.3	95.2	95.5	87.0	88.2	88.1	88.8	88.2	88.3	99.8	94.8	95.5
03200	CONCRETE REINFORCEMENT	107.2	92.5	99.8	95.5	88.0	91.7	95.5	88.8	92.1	94.1	90.2	92.1	93.5	92.7	93.1	91.8	100.2	96.1
03300	CAST-IN-PLACE CONCRETE	107.5	83.5	97.5	95.5	108.2	100.8	103.0	108.8	105.4	96.2	95.3	95.8	96.2	99.9	97.7	100.1	94.8	97.9
03	CONCRETE	105.2	90.6	97.9	99.1	98.2	98.7	102.2	99.2	100.7	102.3	91.8	97.1	102.3	93.8	98.1	96.4	96.2	96.3
04	MASONRY	117.4	73.2	89.7	87.2	93.7	91.3	89.5	94.6	92.7	76.5	85.5	82.1	99.6	90.5	93.9	92.3	96.2	94.7
05	METALS	105.3	82.9	98.0	102.4	100.2	101.7	102.4	100.5	101.8	101.0	100.3	100.8	101.2	101.9	101.4	91.3	102.5	94.9
06	WOOD & PLASTICS	87.2	101.4	94.6	103.6	93.4	98.3	95.4	94.7	95.0	84.6	86.7	85.7	86.4	86.2	86.3	112.8	93.9	102.9
07	THERMAL & MOISTURE PROTECTION	107.2	78.2	93.0	97.0	90.7	93.9	97.4	91.2	94.4	97.4	86.0	91.8	97.6	89.0	93.4	98.8	88.2	93.5
08	DOORS & WINDOWS	98.8	85.4	95.3	97.5	86.0	94.5	96.3	86.9	93.8	96.9	80.6	92.7	97.8	85.4	94.6	100.7	93.5	98.9
09200	PLASTER & GYPSUM BOARD	99.8	101.3	100.8	106.8	93.0	98.1	105.6	94.4	98.5	99.3	86.2	91.0	99.7	85.6	90.8	106.3	94.1	98.5
095,098	CEILINGS & ACOUSTICAL TREATMENT	112.9	101.3	106.0	105.1	104.2	104.9	105.1	109.0	106.1	99.8	95.1	98.5	100.8	105.0	101.9	90.9	105.2	94.7
09600	FLOORING	114.0	67.9	101.8	105.1	104.2	104.9	105.1	109.0	106.1	99.8	95.1	98.5	100.8	105.0	101.9	90.9	105.2	94.7
097,099	WALL FINISHES, PAINTS & COATINGS	113.4	76.8	91.2	108.5	96.6	101.3	108.5	93.2	99.2	108.5	100.2	103.4	108.5	89.0	96.7	89.5	84.8	86.7
09	FINISHES	112.2	89.5	100.4	99.6	96.3	97.9	98.1	97.9	98.0	95.9	90.7	93.2	96.3	91.1	93.6	97.3	95.8	96.5
10-14	TOTAL DIV. 10000 - 14000	100.0	100.7	100.2	100.0	101.4	100.3	100.0	102.0	100.4	100.0	100.3	100.1	100.0	103.6	100.8	100.0	92.8	98.5
15	MECHANICAL	100.0	93.6	97.2	100.0	85.4	93.6	100.0	82.2	92.3	100.0	90.0	95.6	100.0	94.9	97.8	100.2	86.7	94.3
16	ELECTRICAL	106.9	86.7	97.0	97.9	89.4	93.8	97.9	81.9	90.1	98.3	93.5	96.0	95.7	98.7	97.2	102.3	85.2	93.9
01-16	WEIGHTED AVERAGE	104.2	90.0	97.8	99.1	92.0	95.9	99.3	90.9	95.5	98.3	90.6	94.8	99.3	93.7	96.8	97.7	93.3	95.

WISCONSIN

DIVISION		GREEN BAY			KENOSHA			LA CROSSE			MADISON			MILWAUKEE			RACINE		
		MAT.	INST.	TOTAL	MAT.	INST.	TOTAL	MAT.	INST.	TOTAL	MAT.	INST.	TOTAL	MAT.	INST.	TOTAL	MAT.	INST.	TOTAL
01590	EQUIPMENT RENTAL	.0	98.2	98.2	.0	97.7	97.7	.0	100.5	100.5	.0	100.0	100.0	.0	86.1	86.1	.0	100.0	100.0
02	SITE CONSTRUCTION	87.2	98.3	95.4	92.8	101.6	99.3	78.9	102.4	96.3	87.5	104.6	100.1	88.8	93.3	92.1	87.9	105.5	100.9
03100	CONCRETE FORMS & ACCESSORIES	110.9	94.0	96.2	108.5	101.2	102.2	83.9	94.3	92.9	99.2	94.9	95.5	101.5	108.6	107.7	98.7	101.3	100.9
03200	CONCRETE REINFORCEMENT	90.1	89.5	89.8	89.3	97.9	93.7	91.5	88.2	89.9	89.5	88.5	89.0	89.5	98.3	93.9	89.5	97.9	93.7
03300	CAST-IN-PLACE CONCRETE	103.5	96.7	100.7	115.2	98.3	108.1	90.0	94.2	91.7	103.8	97.3	101.1	103.8	104.4	104.1	103.8	98.0	101.4
03	CONCRETE	97.6	94.4	96.0	103.3	99.7	101.5	87.9	93.5	90.7	97.3	94.7	96.0	97.7	104.5	101.1	97.2	99.6	98.4
04	MASONRY	123.1	95.2	105.6	98.3	106.4	103.4	91.5	96.2	94.4	102.4	99.1	100.4	101.9	115.1	110.2	101.9	106.4	104.7
05	METALS	93.8	97.5	95.0	93.2	100.8	95.7	91.2	96.8	93.0	94.3	95.2	94.6	96.0	92.0	94.7	94.3	100.8	96.4
06	WOOD & PLASTICS	118.8	93.9	105.8	114.7	99.9	107.0	95.1	93.9	94.5	109.5	93.8	101.3	112.1	108.0	110.0	109.7	99.9	104.6
07	THERMAL & MOISTURE PROTECTION	100.7	86.4	93.7	96.8	101.3	99.0	98.0	87.3	92.7	96.5	95.2	95.9	95.8	105.6	100.6	97.2	100.8	99.0
08	DOORS & WINDOWS	99.1	87.0	96.0	96.9	100.5	97.9	100.7	83.7	96.3	101.8	93.7	99.7	104.0	105.0	104.3	101.8	100.5	101.5
09200	PLASTER & GYPSUM BOARD	97.9	94.1	95.5	94.4	100.4	98.2	99.0	94.1	95.9	103.3	94.1	97.4	104.2	108.5	106.9	103.3	100.4	101.4
095,098	CEILINGS & ACOUSTICAL TREATMENT	91.2	94.1	92.9	84.3	100.4	93.9	95.4	94.1	94.6	90.2	94.1	92.5	93.8	108.5	102.6	90.2	100.4	96.3
09600	FLOORING	108.8	105.2	107.9	110.0	110.6	110.1	83.5	105.2	89.3	93.6	106.0	96.9	96.3	110.2	100.0	93.6	110.6	98.1
097,099	WALL FINISHES, PAINTS & COATINGS	99.2	75.8	85.0	103.6	91.9	96.5	89.5	69.8	77.6	93.3	91.7	92.4	95.7	102.6	99.9	93.3	88.0	90.1
09	FINISHES	100.9	94.5	97.6	98.5	102.2	100.4	93.2	94.1	93.7	95.6	97.0	96.3	97.6	108.8	103.4	95.6	101.7	98.8
10-14	TOTAL DIV. 10000 - 14000	100.0	92.4	98.4	100.0	91.7	98.2	100.0	92.8	98.5	100.0	92.6	98.2	100.0	94.4	98.8	100.0	91.8	98.2
15	MECHANICAL	100.6	83.2	93.0	100.1	94.2	97.5	100.2	86.2	94.1	99.8	89.7	95.4	99.8	95.7	98.0	99.8	94.2	97.4
16	ELECTRICAL	96.9	84.8	91.0	95.8	85.9	91.0	102.5	85.2	94.1	96.6	85.2	91.1	96.9	101.8	99.3	95.7	93.9	94.8
01-16	WEIGHTED AVERAGE	99.5	90.9	95.6	98.2	97.8	98.0	96.0	91.7	94.0	97.9	93.9	96.1	98.7	101.6	100.0	97.8	99.2	98.4

WYOMING / CANADA

DIVISION		CASPER			CHEYENNE			ROCK SPRINGS			CALGARY, ALBERTA			EDMONTON, ALBERTA			HALIFAX, NOVA SCOTIA		
		MAT.	INST.	TOTAL	MAT.	INST.	TOTAL	MAT.	INST.	TOTAL	MAT.	INST.	TOTAL	MAT.	INST.	TOTAL	MAT.	INST.	TOTAL
01590	EQUIPMENT RENTAL	.0	101.7	101.7	.0	101.7	101.7	.0	101.7	101.7	.0	107.0	107.0	.0	107.0	107.0	.0	98.6	98.6
02	SITE CONSTRUCTION	80.9	100.3	95.2	80.7	100.2	95.2	83.8	99.8	95.6	113.8	105.1	107.3	117.5	105.1	108.3	97.2	95.8	96.1
03100	CONCRETE FORMS & ACCESSORIES	97.8	43.6	50.8	98.5	43.4	50.8	99.3	30.6	39.8	122.1	76.4	82.5	119.8	76.4	82.2	91.0	66.3	69.6
03200	CONCRETE REINFORCEMENT	115.6	52.0	83.5	108.5	52.1	80.0	117.9	51.8	84.5	161.3	66.7	113.5	161.3	66.7	113.5	148.1	49.7	98.3
03300	CAST-IN-PLACE CONCRETE	99.4	58.0	82.1	99.4	57.9	82.1	99.9	43.6	76.4	191.6	87.1	147.9	206.9	87.1	156.9	178.4	68.3	132.4
03	CONCRETE	104.0	50.9	77.5	102.9	50.8	76.9	104.7	40.2	72.5	160.8	78.9	119.9	168.0	78.9	123.5	151.7	64.6	108.2
04	MASONRY	109.8	39.3	65.5	107.0	41.4	65.9	174.9	32.1	85.3	180.3	74.1	113.7	181.6	74.1	114.1	172.3	69.6	107.9
05	METALS	97.3	63.0	86.1	98.1	63.0	86.6	95.1	61.6	84.1	135.7	85.6	119.3	136.5	85.6	119.9	120.8	72.5	105.0
06	WOOD & PLASTICS	95.5	41.8	67.5	95.5	41.8	67.5	98.4	29.6	62.5	116.3	76.0	95.3	112.9	76.0	93.6	85.0	65.6	74.8
07	THERMAL & MOISTURE PROTECTION	102.4	54.3	78.8	102.4	54.9	79.1	103.0	41.3	72.7	125.9	79.9	103.3	126.2	79.9	103.4	105.4	67.7	86.9
08	DOORS & WINDOWS	93.5	44.4	80.8	94.3	44.4	81.4	100.6	37.7	84.4	91.8	73.2	87.0	91.8	73.2	87.0	82.5	61.1	77.0
09200	PLASTER & GYPSUM BOARD	80.4	40.0	54.8	80.4	40.0	54.8	86.4	27.5	49.1	151.0	74.8	102.7	153.8	74.8	103.7	166.0	64.7	101.7
095,098	CEILINGS & ACOUSTICAL TREATMENT	96.9	40.0	62.8	96.9	40.0	62.8	96.9	27.5	55.3	107.3	74.8	87.8	125.1	74.8	94.9	108.1	64.7	82.1
09600	FLOORING	101.2	43.1	85.8	101.2	69.2	92.7	103.5	42.2	87.3	132.6	86.2	120.4	132.6	86.2	120.4	110.0	62.6	97.4
097,099	WALL FINISHES, PAINTS & COATINGS	103.7	56.9	75.3	103.6	56.9	75.3	103.7	34.6	61.8	109.1	76.5	89.3	109.2	76.5	89.4	109.2	60.4	79.6
09	FINISHES	95.9	43.9	68.3	95.8	49.2	71.6	97.4	32.3	63.6	121.0	78.0	98.7	125.9	78.0	101.0	115.7	65.1	89.4
10-14	TOTAL DIV. 10000 - 14000	100.0	75.8	94.8	100.0	75.7	94.8	100.0	50.6	89.5	140.0	85.6	128.4	140.0	85.6	128.4	140.0	62.3	123.5
15	MECHANICAL	100.1	68.4	86.2	100.1	44.8	75.9	100.1	31.6	70.1	101.5	77.6	91.1	101.7	77.6	91.2	101.3	69.0	87.1
16	ELECTRICAL	88.2	62.8	75.8	88.2	52.6	70.8	87.6	45.3	67.0	117.3	86.3	102.2	118.2	86.3	102.6	127.3	70.0	99.3
01-16	WEIGHTED AVERAGE	97.6	60.1	80.5	97.5	54.7	78.0	101.4	44.8	75.7	124.6	82.0	105.2	126.2	82.0	106.2	119.3	70.3	97.0

CANADA

DIVISION		HAMILTON, ONTARIO			KITCHENER, ONTARIO			LAVAL, QUEBEC			LONDON, ONTARIO			MONTREAL, QUEBEC			OSHAWA, ONTARIO		
		MAT.	INST.	TOTAL	MAT.	INST.	TOTAL	MAT.	INST.	TOTAL	MAT.	INST.	TOTAL	MAT.	INST.	TOTAL	MAT.	INST.	TOTAL
01590	EQUIPMENT RENTAL	.0	103.1	103.1	.0	103.0	103.0	.0	99.2	99.2	.0	103.2	103.2	.0	101.0	101.0	.0	103.0	103.0
02	SITE CONSTRUCTION	110.0	103.5	105.2	99.2	102.9	101.9	92.2	96.8	95.6	110.0	103.2	105.0	91.9	97.1	95.7	111.5	103.1	105.3
03100	CONCRETE FORMS & ACCESSORIES	121.5	86.8	91.4	115.8	79.5	84.4	130.1	81.3	87.8	127.2	82.4	88.4	130.4	81.6	88.1	122.7	83.2	88.5
03200	CONCRETE REINFORCEMENT	174.6	89.1	131.4	110.3	89.0	99.5	156.8	83.7	119.9	123.3	87.5	105.2	159.4	83.8	121.2	174.6	89.7	131.7
03300	CAST-IN-PLACE CONCRETE	157.2	94.9	131.2	146.7	76.5	117.3	140.1	89.5	119.0	157.2	92.5	130.2	137.5	91.1	118.1	169.7	81.8	133.0
03	CONCRETE	146.6	90.1	118.4	124.4	80.5	102.5	136.2	84.8	110.5	138.6	87.0	112.9	135.4	85.5	110.5	152.6	84.1	118.4
04	MASONRY	167.2	91.2	119.5	162.8	87.1	115.3	160.7	81.5	111.0	167.3	88.4	117.8	163.9	81.5	112.2	165.8	90.5	118.5
05	METALS	139.4	89.2	123.0	116.4	88.7	107.3	105.6	85.4	99.0	124.5	88.8	112.8	125.1	85.9	112.3	107.2	89.0	101.2
06	WOOD & PLASTICS	117.7	85.6	100.9	111.2	77.9	93.8	131.7	81.2	105.3	117.7	80.6	98.3	131.7	81.5	105.5	118.6	81.4	99.2
07	THERMAL & MOISTURE PROTECTION	110.7	88.3	99.7	108.8	83.9	96.5	105.4	85.7	95.7	114.1	85.6	100.5	106.2	86.4	96.5	110.1	84.4	97.5
08	DOORS & WINDOWS	91.8	85.7	90.2	83.2	79.9	82.4	91.8	72.3	86.8	92.7	82.0	90.0	91.8	72.4	86.8	90.9	83.1	88.9
09200	PLASTER & GYPSUM BOARD	190.9	85.2	123.9	157.4	77.3	106.6	156.3	80.7	108.4	191.4	80.2	120.9	159.4	80.7	109.5	159.9	81.0	109.9
095,098	CEILINGS & ACOUSTICAL TREATMENT	112.6	85.2	96.2	107.3	77.3	89.3	95.7	80.7	86.7	114.4	80.2	93.9	107.3	80.7	91.3	100.1	81.0	88.6
09600	FLOORING	132.6	91.9	121.9	128.3	91.9	118.7	132.6	92.6	122.1	134.7	91.9	123.4	132.6	92.6	122.1	132.6	94.3	122.5
097,099	WALL FINISHES, PAINTS & COATINGS	109.2	99.2	103.2	109.2	90.9	98.1	109.2	86.3	95.3	109.2	99.2	103.2	109.2	86.3	95.3	109.2	104.3	106.2
09	FINISHES	127.7	89.2	107.7	119.7	83.0	100.6	118.0	84.3	100.4	128.9	85.9	106.5	121.3	84.5	102.1	120.2	87.6	103.2
10-14	TOTAL DIV. 10000 - 14000	140.0	91.5	129.7	140.0	89.3	129.2	140.0	79.2	127.1	140.0	90.2	129.4	140.0	80.0	127.2	140.0	90.7	129.5
15	MECHANICAL	101.7	85.5	94.7	101.3	81.9	92.8	101.3	74.6	89.6	101.8	82.1	93.2	101.8	74.7	89.9	101.3	83.5	93.5
16	ELECTRICAL	130.4	93.9	112.6	125.1	90.7	108.3	121.9	73.3	98.2	130.7	90.9	111.3	120.9	73.3	97.7	126.2	91.4	109.2
01-16	WEIGHTED AVERAGE	124.7	90.6	109.2	115.6	86.3	102.3	115.4	81.1	99.8	121.8	88.0	106.5	118.8	81.4	101.8	118.9	88.4	105.1

CANADA

DIVISION		OTTAWA, ONTARIO			QUEBEC, QUEBEC			REGINA, SASKATCHEWAN			SASKATOON, SASKATCHEWAN			ST CATHARINES, ONTARIO			ST JOHNS, NEWFOUNDLAND		
		MAT.	INST.	TOTAL	MAT.	INST.	TOTAL	MAT.	INST.	TOTAL	MAT.	INST.	TOTAL	MAT.	INST.	TOTAL	MAT.	INST.	TOTAL
01590	EQUIPMENT RENTAL	.0	103.0	103.0	.0	101.4	101.4	.0	98.6	98.6	.0	98.6	98.6	.0	100.4	100.4	.0	98.6	98.
02	SITE CONSTRUCTION	110.3	103.1	105.0	93.7	97.2	96.2	110.0	94.7	98.7	105.2	94.8	97.5	99.6	98.8	99.0	112.5	94.5	99.
03100	CONCRETE FORMS & ACCESSORIES	120.2	83.2	88.2	130.2	81.8	88.3	102.4	56.0	62.2	102.4	55.9	62.1	113.3	81.2	85.5	100.6	56.9	62.
03200	CONCRETE REINFORCEMENT	174.6	87.5	130.6	148.1	83.8	115.6	113.8	65.8	89.6	118.9	65.8	92.1	111.2	89.0	100.0	148.1	51.5	99.
03300	CAST-IN-PLACE CONCRETE	159.6	92.4	131.5	151.5	91.6	126.5	162.8	65.7	122.2	148.0	65.6	113.6	140.1	80.9	115.4	179.6	64.7	131.
03	CONCRETE	147.6	87.4	117.6	140.2	85.7	113.0	131.6	61.9	96.8	125.3	61.8	93.6	121.2	82.8	102.1	164.6	59.5	112.
04	MASONRY	167.4	88.0	117.6	164.9	81.5	112.6	166.9	60.7	100.3	167.1	60.7	100.3	162.3	84.5	113.5	160.7	60.0	97.
05	METALS	116.5	88.8	107.4	120.7	86.2	109.4	105.7	72.8	94.9	105.7	72.7	94.9	106.6	88.8	100.8	108.6	70.3	96.
06	WOOD & PLASTICS	117.2	82.2	99.0	132.3	81.6	105.8	95.6	54.5	74.2	94.2	54.5	73.5	108.5	82.0	94.7	95.0	56.0	74.
07	THERMAL & MOISTURE PROTECTION	110.9	85.1	98.2	105.7	86.6	96.3	106.5	61.8	84.5	104.9	60.8	83.2	108.8	85.1	97.1	109.9	59.4	85.
08	DOORS & WINDOWS	91.8	82.8	89.5	91.8	79.6	88.7	86.7	53.7	78.2	85.7	53.7	77.5	91.8	82.5	87.7	98.1	53.8	86.
09200	PLASTER & GYPSUM BOARD	231.0	81.8	136.4	198.1	80.7	123.7	166.3	53.3	94.7	147.7	53.3	87.9	143.3	81.6	104.2	172.3	54.8	97.
095,098	CEILINGS & ACOUSTICAL TREATMENT	106.4	81.8	91.6	97.4	80.7	87.4	123.3	53.3	81.3	123.3	53.3	81.3	100.1	81.6	89.0	106.4	54.8	75.
09600	FLOORING	132.6	90.6	121.5	132.6	92.6	122.1	120.0	58.9	103.9	120.0	58.9	103.9	126.5	91.9	117.4	114.9	53.3	98.
097,099	WALL FINISHES, PAINTS & COATINGS	109.2	92.9	99.3	109.2	86.3	95.3	109.2	62.9	81.1	109.2	53.6	75.5	109.2	92.9	99.3	109.2	58.2	78.
09	FINISHES	132.0	85.7	107.9	124.5	84.5	103.7	123.6	57.0	88.9	120.7	55.9	87.0	115.3	84.7	99.4	118.7	56.3	86.
10-14	TOTAL DIV. 10000 - 14000	140.0	88.2	129.0	140.0	80.2	127.3	140.0	60.0	123.0	140.0	60.1	123.0	140.0	68.0	124.7	140.0	60.8	123.
15	MECHANICAL	101.8	82.4	93.3	101.7	74.7	89.9	101.5	62.0	84.2	101.4	62.0	84.2	101.3	80.3	92.1	101.7	58.7	82.
16	ELECTRICAL	121.9	91.9	107.3	125.7	73.3	100.2	126.3	63.4	95.6	126.5	63.4	95.7	128.2	91.9	110.5	120.7	58.9	90.
01-16	WEIGHTED AVERAGE	120.7	88.1	105.9	119.6	81.8	102.5	116.0	64.9	92.8	114.7	64.7	92.0	113.6	85.6	100.9	120.0	62.8	94.

CANADA

DIVISION		THUNDER BAY, ONTARIO			TORONTO, ONTARIO			VANCOUVER, B C			WINDSOR, ONTARIO			WINNIPEG, MANITOBA					
		MAT.	INST.	TOTAL	MAT.	INST.	TOTAL	MAT.	INST.	TOTAL	MAT.	INST.	TOTAL	MAT.	INST.	TOTAL	MAT.	INST.	TOTAL
01590	EQUIPMENT RENTAL	.0	100.4	100.4	.0	103.2	103.2	.0	111.3	111.3	.0	100.4	100.4	.0	105.3	105.3	.0	.0	.0
02	SITE CONSTRUCTION	105.3	98.9	100.6	111.2	103.9	105.8	115.5	107.5	109.6	95.6	99.1	98.2	109.6	100.3	102.7	.0	.0	.0
03100	CONCRETE FORMS & ACCESSORIES	122.7	82.3	87.7	123.1	93.3	97.3	114.6	84.9	88.9	122.7	82.9	88.2	122.3	67.3	74.7	.0	.0	.0
03200	CONCRETE REINFORCEMENT	99.4	88.4	93.8	170.9	90.0	130.0	161.3	81.7	121.1	109.0	87.6	98.2	161.3	56.5	108.4	.0	.0	.0
03300	CAST-IN-PLACE CONCRETE	154.2	92.6	128.5	163.2	102.1	137.7	153.2	94.6	128.7	143.5	93.7	122.7	157.5	72.5	122.0	.0	.0	.0
03	CONCRETE	130.0	87.1	108.6	149.0	95.7	122.4	146.0	88.0	117.0	123.1	87.7	105.5	144.4	67.8	106.1	.0	.0	.0
04	MASONRY	163.0	85.4	114.3	186.3	99.1	131.6	175.3	88.8	121.0	162.4	91.7	118.1	177.7	63.4	106.0	.0	.0	.0
05	METALS	106.4	87.9	100.4	130.5	91.1	117.6	141.7	90.9	125.1	106.5	89.1	100.8	139.3	75.8	118.5	.0	.0	.0
06	WOOD & PLASTICS	118.6	82.4	99.7	118.6	91.7	104.6	114.3	82.7	97.8	118.6	81.0	99.0	116.3	68.0	91.1	.0	.0	.0
07	THERMAL & MOISTURE PROTECTION	109.1	84.3	96.9	111.0	95.0	103.1	114.4	89.5	102.2	108.8	88.1	98.6	106.1	70.2	88.5	.0	.0	.0
08	DOORS & WINDOWS	81.7	82.1	81.8	90.9	90.8	90.9	94.3	81.2	90.9	91.8	81.8	81.5	91.8	61.3	83.9	.0	.0	.0
09200	PLASTER & GYPSUM BOARD	169.3	82.0	113.9	173.3	91.5	121.5	147.7	81.7	105.9	164.0	80.5	111.1	153.3	66.6	98.4	.0	.0	.0
095,098	CEILINGS & ACOUSTICAL TREATMENT	95.7	82.0	87.5	120.6	91.5	103.2	109.0	81.7	92.7	95.7	80.5	86.6	107.3	66.6	82.9	.0	.0	.0
09600	FLOORING	132.6	52.9	111.6	132.6	97.5	123.3	132.6	89.2	121.2	132.6	92.6	122.1	132.6	65.3	114.8	.0	.0	.0
097,099	WALL FINISHES, PAINTS & COATINGS	109.2	94.9	100.5	112.2	104.3	107.4	109.1	98.7	102.8	109.2	94.5	100.3	109.2	55.3	76.5	.0	.0	.0
09	FINISHES	120.1	79.1	98.8	127.5	95.4	110.8	121.6	86.7	103.4	119.1	85.9	101.8	121.5	65.9	92.6	.0	.0	.0
10-14	TOTAL DIV. 10000 - 14000	140.0	69.0	124.9	140.0	93.7	130.2	140.0	89.3	129.2	140.0	70.3	125.2	140.0	63.9	123.8	.0	.0	.0
15	MECHANICAL	101.3	81.0	92.4	101.7	91.4	97.2	101.7	84.6	94.2	101.3	83.6	93.5	101.7	68.5	87.1	.0	.0	.0
16	ELECTRICAL	125.1	90.1	108.0	129.8	94.6	112.6	132.9	82.8	108.5	133.5	91.9	113.2	130.9	70.3	101.3	.0	.0	.0
01-16	WEIGHTED AVERAGE	114.8	85.4	101.5	124.4	95.0	111.0	125.6	88.2	108.6	114.7	88.0	102.6	124.3	71.0	100.1	.0	.0	.0

Costs shown in *Means cost data publications* are based on National averages for materials and installation. To adjust these costs to a specific location, simply multiply the base cost by the factor and divide by 100 for that city. The data is arranged alphabetically by state and postal zip code numbers. For a city not listed, use the factor for a nearby city with similar economic characteristics.

STATE/ZIP	CITY	MAT.	INST.	TOTAL
ALABAMA				
350-352	Birmingham	96.4	76.5	87.4
354	Tuscaloosa	96.3	56.4	78.2
355	Jasper	96.7	53.5	77.1
356	Decatur	96.3	57.6	78.8
357-358	Huntsville	96.3	73.3	85.8
359	Gadsden	96.3	60.0	79.8
360-361	Montgomery	97.0	58.4	79.5
362	Anniston	96.0	47.7	74.1
363	Dothan	96.5	50.5	75.6
364	Evergreen	95.8	51.8	75.8
365-366	Mobile	97.1	63.0	81.6
367	Selma	96.1	53.0	76.5
368	Phenix City	96.8	56.4	78.5
369	Butler	96.2	50.5	75.5
ALASKA				
995-996	Anchorage	134.5	113.8	125.1
997	Fairbanks	130.7	116.9	124.4
998	Juneau	133.6	113.8	124.6
999	Ketchikan	143.2	113.4	129.6
ARIZONA				
850,853	Phoenix	98.8	74.3	87.7
852	Mesa/Tempe	97.8	67.7	84.1
855	Globe	98.2	62.9	82.2
856-857	Tucson	97.0	69.8	84.6
859	Show Low	98.3	66.9	84.1
860	Flagstaff	100.5	67.6	85.5
863	Prescott	98.2	64.7	83.0
864	Kingman	96.8	65.6	82.6
865	Chambers	96.8	62.5	81.2
ARKANSAS				
716	Pine Bluff	95.3	61.7	80.0
717	Camden	93.4	40.1	69.2
718	Texarkana	94.7	45.3	72.3
719	Hot Springs	92.6	39.6	68.5
720-722	Little Rock	95.2	64.3	81.2
723	West Memphis	95.1	55.0	76.9
724	Jonesboro	95.5	55.0	77.1
725	Batesville	93.5	49.9	73.7
726	Harrison	94.8	49.9	74.4
727	Fayetteville	92.1	48.1	72.1
728	Russellville	93.4	48.0	72.8
729	Fort Smith	96.0	56.8	78.2
CALIFORNIA				
900-902	Los Angeles	102.7	111.9	106.8
903-905	Inglewood	98.5	108.2	102.9
906-908	Long Beach	100.0	108.3	103.8
910-912	Pasadena	100.5	108.3	104.0
913-916	Van Nuys	103.8	108.3	105.8
917-918	Alhambra	102.8	108.2	105.3
919-921	San Diego	103.9	105.0	104.4
922	Palm Springs	100.5	105.8	102.9
923-924	San Bernardino	98.3	106.7	102.2
925	Riverside	102.6	109.9	105.9
926-927	Santa Ana	100.4	106.7	103.2
928	Anaheim	102.9	111.1	106.6
930	Oxnard	103.9	109.5	106.4
931	Santa Barbara	102.7	110.3	106.1
932-933	Bakersfield	102.6	106.2	104.2
934	San Luis Obispo	103.8	106.3	105.0
935	Mojave	100.8	103.7	102.1
936-938	Fresno	104.3	111.9	107.8
939	Salinas	104.5	116.1	109.8
940-941	San Francisco	112.6	133.8	122.2
942,956-958	Sacramento	106.9	113.1	109.7
943	Palo Alto	105.5	125.9	114.8
944	San Mateo	108.4	125.1	116.0
945	Vallejo	105.6	120.4	112.3
946	Oakland	110.6	123.9	116.6
947	Berkeley	110.1	122.5	115.7
948	Richmond	109.5	122.8	115.6
949	San Rafael	110.5	122.3	115.9
950	Santa Cruz	109.6	116.1	112.6

STATE/ZIP	CITY	MAT.	INST.	TOTAL
CALIFORNIA (CONT'D)				
951	San Jose	107.3	128.4	116.9
952	Stockton	104.5	110.5	107.2
953	Modesto	104.4	111.6	107.7
954	Santa Rosa	104.7	121.5	112.4
955	Eureka	106.3	103.7	105.1
959	Marysville	105.3	113.4	109.0
960	Redding	106.8	110.4	108.4
961	Susanville	106.1	111.1	108.4
COLORADO				
800-802	Denver	100.3	90.3	95.8
803	Boulder	97.6	86.7	92.7
804	Golden	99.9	86.2	93.7
805	Fort Collins	101.0	82.7	92.7
806	Greeley	98.5	71.1	86.0
807	Fort Morgan	98.4	85.9	92.7
808-809	Colorado Springs	100.1	86.6	94.0
810	Pueblo	100.0	83.9	92.6
811	Alamosa	101.6	81.3	92.4
812	Salida	101.6	81.1	92.3
813	Durango	102.0	79.2	91.6
814	Montrose	100.6	79.0	90.8
815	Grand Junction	103.8	77.6	91.9
816	Glenwood Springs	101.4	84.6	93.8
CONNECTICUT				
060	New Britain	101.4	110.7	105.6
061	Hartford	102.7	109.9	106.0
062	Willimantic	102.2	109.6	105.6
063	New London	98.6	110.7	104.1
064	Meriden	100.6	111.1	105.4
065	New Haven	103.2	111.2	106.8
066	Bridgeport	102.7	110.8	106.4
067	Waterbury	102.2	110.7	106.1
068	Norwalk	102.2	111.5	106.4
069	Stamford	102.3	115.6	108.4
D.C.				
200-205	Washington	102.0	90.8	96.9
DELAWARE				
197	Newark	100.5	105.7	102.8
198	Wilmington	99.5	105.7	102.3
199	Dover	100.6	105.7	102.9
FLORIDA				
320,322	Jacksonville	98.3	61.0	81.4
321	Daytona Beach	98.4	74.5	87.5
323	Tallahassee	97.6	49.2	75.6
324	Panama City	99.5	37.9	71.5
325	Pensacola	99.2	58.1	80.5
326,344	Gainesville	99.7	58.7	81.1
327-328,347	Orlando	99.8	69.3	86.0
329	Melbourne	100.5	79.0	90.7
330-332,340	Miami	98.7	72.9	87.0
333	Fort Lauderdale	97.7	72.3	86.2
334,349	West Palm Beach	96.6	68.5	83.8
335-336,346	Tampa	99.4	72.8	87.3
337	St. Petersburg	101.6	55.8	80.8
338	Lakeland	98.5	72.3	86.6
339,341	Fort Myers	97.9	64.3	82.6
342	Sarasota	99.6	66.7	84.6
GEORGIA				
300-303,399	Atlanta	97.4	80.7	89.8
304	Statesboro	97.1	45.0	73.4
305	Gainesville	95.8	57.4	78.4
306	Athens	95.2	62.1	80.1
307	Dalton	97.5	50.9	76.3
308-309	Augusta	96.0	59.2	79.3
310-312	Macon	95.7	59.2	79.1
313-314	Savannah	97.1	58.8	79.7
315	Waycross	97.2	51.6	76.5
316	Valdosta	97.0	49.1	75.2
317	Albany	97.3	56.8	78.9
318-319	Columbus	97.2	61.7	81.1

STATE/ZIP	CITY	MAT.	INST.	TOTAL
HAWAII				
967	Hilo	115.3	125.7	120.0
968	Honolulu	121.0	125.7	123.1
STATES & POSS.				
969	Guam	193.9	54.3	130.5
IDAHO				
832	Pocatello	100.3	78.3	90.3
833	Twin Falls	100.8	45.8	75.8
834	Idaho Falls	98.2	55.3	78.7
835	Lewiston	108.1	83.3	96.8
836-837	Boise	100.2	80.5	91.2
838	Coeur d'Alene	107.3	56.8	84.4
ILLINOIS				
600-603	North Suburban	99.1	116.3	106.9
604	Joliet	99.0	117.2	107.3
605	South Suburban	99.1	116.3	106.9
606-608	Chicago	99.7	125.8	111.6
609	Kankakee	95.4	105.3	99.9
610-611	Rockford	97.4	109.4	102.8
612	Rock Island	95.1	97.7	96.3
613	La Salle	96.5	102.7	99.3
614	Galesburg	96.2	100.8	98.3
615-616	Peoria	98.7	101.7	100.1
617	Bloomington	95.5	101.2	98.1
618-619	Champaign	99.0	100.0	99.5
620-622	East St. Louis	94.2	101.1	97.3
623	Quincy	95.6	95.4	95.5
624	Effingham	94.9	97.3	96.0
625	Decatur	96.5	98.5	97.4
626-627	Springfield	96.8	98.9	97.8
628	Centralia	93.1	98.0	95.3
629	Carbondale	92.8	97.8	95.1
INDIANA				
460	Anderson	94.6	83.3	89.5
461-462	Indianapolis	98.8	88.7	94.2
463-464	Gary	95.8	98.7	97.1
465-466	South Bend	95.0	82.6	89.4
467-468	Fort Wayne	95.1	80.6	88.5
469	Kokomo	92.9	82.2	88.0
470	Lawrenceburg	92.2	80.1	86.7
471	New Albany	93.7	74.8	85.1
472	Columbus	96.3	80.2	89.0
473	Muncie	96.1	80.8	89.2
474	Bloomington	98.2	80.7	90.3
475	Washington	94.1	85.2	90.1
476-477	Evansville	95.2	86.8	91.4
478	Terre Haute	96.0	85.3	91.1
479	Lafayette	95.9	79.5	88.4
IOWA				
500-503,509	Des Moines	97.7	82.4	90.7
504	Mason City	96.0	63.3	81.2
505	Fort Dodge	96.2	59.4	79.5
506-507	Waterloo	97.7	62.0	81.5
508	Creston	96.6	66.3	82.8
510-511	Sioux City	98.3	73.3	87.0
512	Sibley	97.0	52.9	77.0
513	Spencer	98.8	51.0	77.1
514	Carroll	95.8	57.0	78.2
515	Council Bluffs	99.1	74.7	88.0
516	Shenandoah	96.1	52.2	76.2
520	Dubuque	97.7	76.5	88.1
521	Decorah	96.8	54.5	77.6
522-524	Cedar Rapids	98.8	82.7	91.5
525	Ottumwa	96.9	70.7	85.0
526	Burlington	96.0	73.7	85.9
527-528	Davenport	97.8	93.1	95.7
KANSAS				
660-662	Kansas City	98.7	91.6	95.5
664-666	Topeka	98.5	69.3	85.2
667	Fort Scott	97.5	66.8	83.6
668	Emporia	97.4	59.4	80.1
669	Belleville	99.2	54.8	79.0
670-672	Wichita	97.7	71.1	85.6
673	Independence	99.0	53.0	78.1
674	Salina	99.0	57.2	80.1
675	Hutchinson	94.5	51.1	74.7
676	Hays	98.6	54.8	78.7
677	Colby	99.2	54.8	79.0

STATE/ZIP	CITY	MAT.	INST.	TOTAL
KANSAS (CONT'D)				
678	Dodge City	100.3	54.8	79.6
679	Liberal	98.4	45.6	74.4
KENTUCKY				
400-402	Louisville	96.3	86.4	91.8
403-405	Lexington	95.9	66.9	82.7
406	Frankfort	96.1	68.5	83.6
407-409	Corbin	93.3	44.6	71.2
410	Covington	94.0	93.6	93.8
411-412	Ashland	92.6	98.6	95.3
413-414	Campton	94.2	45.1	71.9
415-416	Pikeville	95.1	62.1	80.1
417-418	Hazard	93.5	44.8	71.4
420	Paducah	92.2	88.8	90.7
421-422	Bowling Green	94.4	83.6	89.5
423	Owensboro	94.2	74.3	85.2
424	Henderson	91.9	89.2	90.7
425-426	Somerset	91.5	46.5	71.0
427	Elizabethtown	91.1	82.8	87.3
LOUISIANA				
700-701	New Orleans	100.2	69.8	86.4
703	Thibodaux	98.1	64.4	82.8
704	Hammond	95.4	63.9	81.1
705	Lafayette	97.7	55.0	78.3
706	Lake Charles	97.8	60.0	80.7
707-708	Baton Rouge	98.4	55.1	78.7
710-711	Shreveport	94.7	59.7	78.8
712	Monroe	95.5	57.5	78.2
713-714	Alexandria	95.6	54.7	77.0
MAINE				
039	Kittery	95.6	72.4	85.1
040-041	Portland	98.8	76.8	88.8
042	Lewiston	99.2	76.8	89.0
043	Augusta	97.7	72.3	86.2
044	Bangor	98.0	76.8	88.3
045	Bath	96.7	72.4	85.7
046	Machias	96.3	71.8	85.2
047	Houlton	96.5	75.5	87.0
048	Rockland	95.5	70.6	84.2
049	Waterville	96.8	67.3	83.4
MARYLAND				
206	Waldorf	98.0	73.1	86.7
207-208	College Park	98.0	80.6	90.1
209	Silver Spring	97.3	78.2	88.6
210-212	Baltimore	99.4	84.8	92.7
214	Annapolis	98.8	79.9	90.2
215	Cumberland	95.5	80.8	88.8
216	Easton	96.9	45.4	73.5
217	Hagerstown	96.0	80.8	89.1
218	Salisbury	97.4	54.8	78.0
219	Elkton	94.6	67.2	82.2
MASSACHUSETTS				
010-011	Springfield	100.7	105.1	102.7
012	Pittsfield	100.2	101.1	100.6
013	Greenfield	98.3	103.3	100.5
014	Fitchburg	97.0	116.9	106.0
015-016	Worcester	100.5	118.4	108.6
017	Framingham	96.4	122.8	108.4
018	Lowell	100.0	123.5	110.7
019	Lawrence	101.1	121.0	110.2
020-022, 024	Boston	102.9	130.5	115.4
023	Brockton	101.4	115.5	107.8
025	Buzzards Bay	95.8	116.9	105.4
026	Hyannis	98.3	116.9	106.7
027	New Bedford	101.1	117.0	108.4
MICHIGAN				
480,483	Royal Oak	94.9	106.2	100.0
481	Ann Arbor	96.9	105.7	100.9
482	Detroit	98.7	116.9	107.0
484-485	Flint	96.7	96.2	96.5
486	Saginaw	96.5	93.2	95.0
487	Bay City	96.4	93.3	95.0
488-489	Lansing	96.7	96.1	96.4
490	Battle Creek	95.4	88.1	92.1
491	Kalamazoo	95.7	84.9	90.8
492	Jackson	93.9	94.7	94.3
493,495	Grand Rapids	95.8	70.2	84.2
494	Muskegon	94.3	83.6	89.5

Location Factors

STATE/ZIP	CITY	MAT.	INST.	TOTAL
MICHIGAN (CONT'D)				
496	Traverse City	93.4	73.9	84.6
497	Gaylord	94.6	77.9	87.0
498-499	Iron Mountain	96.5	87.1	92.2
MINNESOTA				
550-551	Saint Paul	99.8	122.0	109.8
553-555	Minneapolis	100.7	126.8	112.6
556-558	Duluth	99.7	108.9	103.8
559	Rochester	99.4	105.1	102.0
560	Mankato	96.3	102.1	99.0
561	Windom	95.1	81.4	88.9
562	Willmar	94.5	87.7	91.4
563	St. Cloud	95.6	120.5	106.9
564	Brainerd	96.1	101.0	98.3
565	Detroit Lakes	98.0	99.1	98.5
566	Bemidji	97.3	97.7	97.5
567	Thief River Falls	96.4	94.0	95.3
MISSISSIPPI				
386	Clarksdale	96.0	31.7	66.8
387	Greenville	99.4	45.9	75.1
388	Tupelo	97.4	38.4	70.6
389	Greenwood	97.3	34.3	68.7
390-392	Jackson	97.9	47.5	75.0
393	Meridian	96.0	47.8	74.1
394	Laurel	97.3	34.4	68.7
395	Biloxi	97.9	56.6	79.1
396	McComb	95.8	53.3	76.5
397	Columbus	97.2	37.8	70.2
MISSOURI				
630-631	St. Louis	96.7	108.6	102.1
633	Bowling Green	95.6	89.4	92.8
634	Hannibal	94.4	85.1	90.2
635	Kirksville	96.7	78.3	88.3
636	Flat River	96.6	93.7	95.3
637	Cape Girardeau	96.1	89.3	93.0
638	Sikeston	94.3	84.3	89.8
639	Poplar Bluff	93.9	85.0	89.8
640-641	Kansas City	101.1	106.3	103.4
644-645	St. Joseph	100.1	89.6	95.3
646	Chillicothe	97.0	70.7	85.1
647	Harrisonville	96.7	97.7	97.2
648	Joplin	98.7	69.0	85.2
650-651	Jefferson City	95.9	87.3	92.0
652	Columbia	96.9	88.1	92.9
653	Sedalia	96.1	84.3	90.7
654-655	Rolla	95.0	77.8	87.2
656-658	Springfield	98.2	74.5	87.4
MONTANA				
590-591	Billings	102.0	74.4	89.5
592	Wolf Point	100.8	70.5	87.0
593	Miles City	98.8	71.0	86.2
594	Great Falls	102.6	74.9	90.0
595	Havre	99.6	70.7	86.5
596	Helena	102.4	72.3	88.7
597	Butte	101.0	71.1	87.4
598	Missoula	98.7	70.3	85.8
599	Kalispell	97.8	69.8	85.1
NEBRASKA				
680-681	Omaha	99.1	76.7	89.0
683-685	Lincoln	98.0	67.7	84.2
686	Columbus	96.0	51.2	75.6
687	Norfolk	97.7	62.5	81.7
688	Grand Island	97.2	66.2	83.1
689	Hastings	97.0	58.5	79.5
690	Mccook	96.8	48.8	75.0
691	North Platte	96.8	59.9	80.0
692	Valentine	99.2	39.7	72.2
693	Alliance	98.9	37.0	70.8
NEVADA				
889-891	Las Vegas	98.1	105.5	101.5
893	Ely	98.7	81.2	90.7
894-895	Reno	98.7	95.5	97.3
897	Carson City	98.4	95.6	97.2
898	Elko	97.5	87.9	93.1
NEW HAMPSHIRE				
030	Nashua	100.9	82.8	92.7
031	Manchester	101.2	82.8	92.9

STATE/ZIP	CITY	MAT.	INST.	TOTAL
NEW HAMPSHIRE (CONT'D)				
032-033	Concord	98.6	82.8	91.4
034	Keene	97.6	52.5	77.1
035	Littleton	97.7	62.0	81.5
036	Charleston	97.0	49.3	75.4
037	Claremont	96.3	49.3	75.0
038	Portsmouth	98.0	78.4	89.1
NEW JERSEY				
070-071	Newark	102.5	121.4	111.1
072	Elizabeth	101.1	120.7	110.0
073	Jersey City	100.2	122.5	110.3
074-075	Paterson	101.8	121.2	110.6
076	Hackensack	99.7	121.7	109.7
077	Long Branch	99.4	120.0	108.7
078	Dover	100.0	121.5	109.8
079	Summit	100.1	120.7	109.4
080,083	Vineland	97.8	117.7	106.8
081	Camden	100.0	115.2	106.9
082,084	Atlantic City	98.6	115.7	106.4
085-086	Trenton	99.9	119.1	108.6
087	Point Pleasant	99.8	116.8	107.5
088-089	New Brunswick	100.3	119.4	108.9
NEW MEXICO				
870-872	Albuquerque	98.3	76.8	88.6
873	Gallup	98.3	76.8	88.6
874	Farmington	98.6	76.8	88.7
875	Santa Fe	97.3	76.8	88.0
877	Las Vegas	96.8	76.8	87.7
878	Socorro	96.4	76.8	87.5
879	Truth/Consequences	97.1	71.6	85.5
880	Las Cruces	95.8	69.1	83.7
881	Clovis	96.9	76.1	87.4
882	Roswell	98.5	76.1	88.4
883	Carrizozo	99.0	76.8	88.9
884	Tucumcari	97.8	76.1	87.9
NEW YORK				
100-102	New York	106.6	163.5	132.4
103	Staten Island	102.1	155.2	126.2
104	Bronx	100.2	155.2	125.2
105	Mount Vernon	100.2	135.4	116.2
106	White Plains	99.9	135.4	116.0
107	Yonkers	105.1	136.2	119.2
108	New Rochelle	100.7	135.4	116.5
109	Suffern	100.5	123.7	111.0
110	Queens	101.4	155.0	125.8
111	Long Island City	103.1	155.0	126.7
112	Brooklyn	103.4	156.5	127.5
113	Flushing	103.6	155.0	127.0
114	Jamaica	101.7	155.2	126.0
115,117,118	Hicksville	101.2	145.0	121.1
116	Far Rockaway	103.7	155.0	127.0
119	Riverhead	101.6	145.1	121.4
120-122	Albany	97.9	93.9	96.1
123	Schenectady	98.2	93.0	95.8
124	Kingston	101.1	114.3	107.1
125-126	Poughkeepsie	100.3	119.0	108.8
127	Monticello	99.6	113.8	106.1
128	Glens Falls	92.8	89.9	91.5
129	Plattsburgh	97.0	82.8	90.6
130-132	Syracuse	99.0	90.5	95.2
133-135	Utica	97.0	87.8	92.8
136	Watertown	98.7	88.4	94.0
137-139	Binghamton	98.6	84.5	92.2
140-142	Buffalo	99.9	104.7	102.1
143	Niagara Falls	97.6	102.8	100.0
144-146	Rochester	100.1	94.9	97.7
147	Jamestown	96.6	85.4	91.5
148-149	Elmira	96.5	84.8	91.2
NORTH CAROLINA				
270,272-274	Greensboro	98.0	50.3	76.3
271	Winston-Salem	97.8	49.1	75.7
275-276	Raleigh	97.9	50.3	76.3
277	Durham	97.9	50.3	76.3
278	Rocky Mount	95.4	36.2	68.5
279	Elizabeth City	96.1	38.6	70.0
280	Gastonia	97.2	47.5	74.6
281-282	Charlotte	98.2	47.8	75.3
283	Fayetteville	96.7	50.3	75.6
284	Wilmington	96.0	49.4	74.8
285	Kinston	94.2	35.8	67.6

STATE/ZIP	CITY	MAT.	INST.	TOTAL
NORTH CAROLINA (CONT'D)				
286	Hickory	94.5	34.9	67.5
287-288	Asheville	96.3	48.3	74.5
289	Murphy	95.5	34.0	67.6
NORTH DAKOTA				
580-581	Fargo	100.8	65.9	85.0
582	Grand Forks	100.4	60.7	82.4
583	Devils Lake	100.0	61.0	82.3
584	Jamestown	100.0	54.2	79.2
585	Bismarck	99.2	66.9	84.5
586	Dickinson	100.8	62.3	83.3
587	Minot	100.5	68.2	85.8
588	Williston	99.3	61.7	82.2
OHIO				
430-432	Columbus	97.8	90.5	94.5
433	Marion	94.4	88.3	91.6
434-436	Toledo	98.0	99.6	98.7
437-438	Zanesville	94.8	83.9	89.8
439	Steubenville	95.7	94.3	95.1
440	Lorain	96.8	98.1	97.4
441	Cleveland	97.1	106.0	101.1
442-443	Akron	97.7	98.0	97.8
444-445	Youngstown	97.1	94.8	96.0
446-447	Canton	97.2	88.5	93.2
448-449	Mansfield	94.5	93.4	94.0
450	Hamilton	94.3	90.4	92.5
451-452	Cincinnati	94.7	92.3	93.6
453-454	Dayton	94.4	84.9	90.1
455	Springfield	94.4	86.7	90.9
456	Chillicothe	93.5	92.4	93.0
457	Athens	96.5	82.3	90.1
458	Lima	97.1	87.9	92.9
OKLAHOMA				
730-731	Oklahoma City	97.3	64.0	82.1
734	Ardmore	94.2	63.3	80.2
735	Lawton	96.5	65.0	82.2
736	Clinton	95.7	61.6	80.2
737	Enid	96.1	61.6	80.4
738	Woodward	94.5	61.7	79.6
739	Guymon	95.6	33.9	67.6
740-741	Tulsa	96.3	61.2	80.4
743	Miami	93.3	67.1	81.4
744	Muskogee	95.7	43.7	72.1
745	Mcalester	92.9	56.1	76.2
746	Ponca City	93.4	61.5	78.9
747	Durant	93.4	61.7	79.0
748	Shawnee	94.9	60.1	79.1
749	Poteau	92.5	64.3	79.7
OREGON				
970-972	Portland	102.2	105.4	103.7
973	Salem	102.2	103.0	102.5
974	Eugene	101.7	102.1	101.9
975	Medford	103.4	100.3	102.0
976	Klamath Falls	103.7	100.1	102.1
977	Bend	102.5	102.2	102.4
978	Pendleton	96.6	100.7	98.5
979	Vale	94.4	93.7	94.1
PENNSYLVANIA				
150-152	Pittsburgh	96.7	103.8	99.9
153	Washington	93.8	102.1	97.6
154	Uniontown	94.0	99.6	96.5
155	Bedford	95.0	91.8	93.5
156	Greensburg	95.1	99.6	97.1
157	Indiana	93.9	97.3	95.4
158	Dubois	95.2	95.7	95.5
159	Johnstown	94.9	96.2	95.5
160	Butler	92.3	101.6	96.5
161	New Castle	92.3	100.2	95.9
162	Kittanning	92.8	103.2	97.5
163	Oil City	92.3	95.8	93.9
164-165	Erie	94.3	96.2	95.1
166	Altoona	94.3	93.3	93.8
167	Bradford	95.9	93.3	94.7
168	State College	95.5	93.7	94.7
169	Wellsboro	96.5	87.4	92.4
170-171	Harrisburg	98.1	92.3	95.5
172	Chambersburg	96.1	87.2	92.0
173-174	York	96.2	88.6	92.7
175-176	Lancaster	94.9	87.5	91.5

STATE/ZIP	CITY	MAT.	INST.	TOTAL
PENNSYLVANIA (CONT'D)				
177	Williamsport	93.4	81.8	88.2
178	Sunbury	95.5	87.9	92.0
179	Pottsville	94.6	89.3	92.2
180	Lehigh Valley	96.0	102.7	99.0
181	Allentown	98.0	101.2	99.4
182	Hazleton	95.3	92.4	94.0
183	Stroudsburg	95.3	95.8	95.5
184-185	Scranton	98.8	92.9	96.1
186-187	Wilkes-Barre	95.3	91.4	93.5
188	Montrose	94.9	92.0	93.6
189	Doylestown	94.8	116.5	104.6
190-191	Philadelphia	100.8	128.9	113.6
193	Westchester	97.6	115.6	105.7
194	Norristown	96.5	116.5	105.6
195-196	Reading	98.6	98.5	98.6
PUERTO RICO				
009	San Juan	136.6	26.3	86.5
RHODE ISLAND				
028	Newport	99.9	112.4	105.6
029	Providence	100.2	112.4	105.8
SOUTH CAROLINA				
290-292	Columbia	96.6	48.3	74.7
293	Spartanburg	95.5	46.0	73.0
294	Charleston	96.8	48.9	75.1
295	Florence	95.0	46.1	72.8
296	Greenville	95.2	46.0	72.9
297	Rock Hill	95.0	33.8	67.2
298	Aiken	95.9	70.4	84.3
299	Beaufort	96.7	37.4	69.8
SOUTH DAKOTA				
570-571	Sioux Falls	99.1	57.8	80.3
572	Watertown	97.9	52.0	77.0
573	Mitchell	96.9	51.6	76.3
574	Aberdeen	99.4	54.6	79.0
575	Pierre	98.7	53.7	78.2
576	Mobridge	97.5	51.9	76.8
577	Rapid City	99.3	52.5	78.0
TENNESSEE				
370-372	Nashville	98.0	74.0	87.1
373-374	Chattanooga	98.8	55.8	79.3
375,380-381	Memphis	96.6	77.2	87.8
376	Johnson City	97.6	58.2	79.7
377-379	Knoxville	94.9	59.2	78.7
382	Mckenzie	96.5	49.7	75.3
383	Jackson	98.5	53.3	78.0
384	Columbia	95.0	56.9	77.7
385	Cookeville	96.3	51.4	75.9
TEXAS				
750	Mckinney	98.2	57.1	79.5
751	Waxahackie	98.1	57.8	79.8
752-753	Dallas	98.7	67.1	84.3
754	Greenville	98.3	41.9	72.7
755	Texarkana	97.2	53.0	77.2
756	Longview	97.5	43.9	73.1
757	Tyler	98.1	56.4	79.2
758	Palestine	94.1	44.3	71.5
759	Lufkin	95.1	48.1	73.7
760-761	Fort Worth	96.4	63.3	81.4
762	Denton	96.9	53.2	77.0
763	Wichita Falls	97.4	57.2	79.1
764	Eastland	95.9	43.3	72.0
765	Temple	94.8	51.4	75.1
766-767	Waco	96.6	57.8	79.0
768	Brownwood	96.9	40.7	71.4
769	San Angelo	96.6	49.0	75.0
770-772	Houston	99.1	72.8	87.2
773	Huntsville	97.6	41.6	72.1
774	Wharton	98.8	46.5	75.1
775	Galveston	97.0	70.7	85.0
776-777	Beaumont	96.6	63.5	81.6
778	Bryan	94.5	65.2	81.2
779	Victoria	99.0	49.1	76.4
780	Laredo	94.9	55.0	76.8
781-782	San Antonio	95.2	66.8	82.3
783-784	Corpus Christi	97.7	54.1	77.8
785	Mc Allen	97.8	48.2	75.2
786-787	Austin	94.9	61.8	79.8

Location Factors

STATE/ZIP	CITY	MAT.	INST.	TOTAL		STATE/ZIP	CITY	MAT.	INST.	TOTAL
TEXAS (CONT'D)						**(CONT'D)**				
788	Del Rio	97.1	34.2	68.6		538	Lancaster	95.6	91.4	93.7
789	Giddings	94.5	43.4	71.3		539	Portage	94.2	93.3	93.8
790-791	Amarillo	97.3	59.1	79.9		540	New Richmond	95.5	94.6	95.1
792	Childress	96.5	53.5	76.9		541-543	Green Bay	99.5	90.9	95.6
793-794	Lubbock	98.8	54.9	78.9		544	Wausau	94.9	90.9	93.1
795-796	Abilene	96.8	54.5	77.6		545	Rhinelander	98.0	90.7	94.7
797	Midland	99.1	51.4	77.4		546	La Crosse	96.0	91.7	94.0
798-799,885	El Paso	96.7	52.4	76.6		547	Eau Claire	97.7	93.3	95.7
						548	Superior	95.4	94.8	95.1
UTAH						549	Oshkosh	95.5	90.5	93.3
840-841	Salt Lake City	101.2	72.3	88.1						
842,844	Ogden	97.2	72.1	85.8		**WYOMING**				
843	Logan	98.9	72.1	86.7		820	Cheyenne	97.5	54.7	78.0
845	Price	99.6	51.1	77.6		821	Yellowstone Nat'l Park	96.7	49.1	75.1
846-847	Provo	99.5	72.1	87.1		822	Wheatland	98.1	48.0	75.4
						823	Rawlins	99.4	44.1	74.3
VERMONT						824	Worland	97.3	44.8	73.5
050	White River Jct.	99.0	47.9	75.8		825	Riverton	98.4	47.3	75.2
051	Bellows Falls	97.5	50.3	76.0		826	Casper	97.6	60.1	80.5
052	Bennington	97.8	50.6	76.4		827	Newcastle	97.2	44.1	73.1
053	Brattleboro	98.2	50.8	76.7		828	Sheridan	98.1	52.8	77.5
054	Burlington	100.8	66.0	85.0		829-831	Rock Springs	101.4	44.8	75.7
056	Montpelier	97.7	66.1	83.3						
057	Rutland	99.8	66.1	84.5		**CANADIAN FACTORS** (reflect Canadian currency)				
058	St. Johnsbury	99.2	50.6	77.1						
059	Guildhall	97.8	50.1	76.1		**ALBERTA**				
							Calgary	124.6	82.0	105.2
VIRGINIA							Edmonton	126.2	82.0	106.2
220-221	Fairfax	99.2	81.2	91.0			Fort McMurray	116.8	81.6	100.8
222	Arlington	100.2	79.4	90.7			Lethbridge	117.7	81.0	101.1
223	Alexandria	99.6	84.6	92.8			Lloydminster	116.9	81.6	100.8
224-225	Fredericksburg	97.8	68.0	84.3			Medicine Hat	117.0	81.0	100.6
226	Winchester	98.5	55.4	78.9			Red Deer	117.4	81.0	100.9
227	Culpeper	98.3	56.6	79.4						
228	Harrisonburg	98.6	48.1	75.7		**BRITISH COLUMBIA**				
229	Charlottesville	98.9	60.5	81.5			Kamloops	117.8	84.3	102.6
230-232	Richmond	99.9	65.9	84.5			Prince George	119.1	84.3	103.3
233-235	Norfolk	99.6	65.2	84.0			Vancouver	125.6	88.2	108.6
236	Newport News	99.2	63.6	83.0			Victoria	119.1	84.8	103.6
237	Portsmouth	98.4	65.0	83.2						
238	Petersburg	98.6	65.9	83.8		**MANITOBA**				
239	Farmville	98.1	42.3	72.8			Brandon	117.1	70.7	96.0
240-241	Roanoke	99.8	48.4	76.5			Portage la Prairie	117.1	70.7	96.0
242	Bristol	97.8	48.7	75.5			Winnipeg	124.3	71.0	100.1
243	Pulaski	97.5	43.7	73.1						
244	Staunton	98.2	46.5	74.8		**NEW BRUNSWICK**				
245	Lynchburg	98.4	51.0	76.8			Bathurst	115.2	62.8	91.4
246	Grundy	97.8	45.8	74.2			Dalhousie	115.2	62.8	91.4
							Fredericton	117.1	66.9	94.3
WASHINGTON							Moncton	115.5	62.8	91.6
980-981,987	Seattle	104.3	105.8	105.0			Newcastle	115.2	62.8	91.4
982	Everett	104.0	98.3	101.4			Saint John	118.3	66.9	94.9
983-984	Tacoma	104.1	99.7	102.1						
985	Olympia	102.8	99.7	101.4		**NEWFOUNDLAND**				
986	Vancouver	106.0	97.2	102.0			Corner Brook	120.1	62.8	94.1
988	Wenatchee	104.3	84.0	95.1			St. John's	120.0	62.8	94.0
989	Yakima	104.2	90.0	97.8						
990-992	Spokane	106.2	83.7	96.0		**NORTHWEST TERRITORIES**				
993	Richland	105.5	85.1	96.2			Yellowknife	114.3	79.8	98.6
994	Clarkston	104.7	82.3	94.6						
						NOVA SCOTIA				
WEST VIRGINIA							Dartmouth	117.7	69.2	95.7
247-248	Bluefield	96.6	80.2	89.1			Halifax	119.3	70.3	97.0
249	Lewisburg	98.1	83.9	91.6			New Glasgow	115.8	69.2	94.6
250-253	Charleston	99.1	92.0	95.9			Sydney	113.3	69.2	93.3
254	Martinsburg	98.0	80.0	89.8			Yarmouth	115.6	69.2	94.5
255-257	Huntington	99.3	90.9	95.5						
258-259	Beckley	96.4	89.5	93.3		**ONTARIO**				
260	Wheeling	99.3	93.7	96.8			Barrie	120.0	86.7	104.9
261	Parkersburg	98.3	90.6	94.8			Brantford	119.2	90.3	106.1
262	Buckhannon	97.9	93.2	95.8			Cornwall	119.1	86.9	104.5
263-264	Clarksburg	98.3	92.2	95.5			Hamilton	124.7	90.6	109.2
265	Morgantown	98.4	93.2	96.0			Kingston	120.1	87.3	105.2
266	Gassaway	97.7	92.3	95.3			Kitchener	115.6	86.3	102.3
267	Romney	97.7	86.4	92.6			London	121.8	88.0	106.5
268	Petersburg	97.7	89.1	93.8			North Bay	119.2	85.0	103.6
							Oshawa	118.9	88.4	105.1
WISCONSIN							Ottawa	120.7	88.1	105.9
530,532	Milwaukee	98.7	101.6	100.0			Owen Sound	120.3	85.0	104.2
531	Kenosha	98.2	97.8	98.0			Peterborough	119.2	86.7	104.4
534	Racine	97.8	99.2	98.4			Sarnia	119.4	90.7	106.3
535	Beloit	97.7	96.4	97.1			St. Catharines	113.6	85.6	100.9
537	Madison	97.9	93.9	96.1			Sudbury	113.6	85.2	100.7

Location Factors

STATE/ZIP	CITY	MAT.	INST.	TOTAL
ONTARIO (CONT'D)				
	Thunder Bay	114.8	85.4	101.5
	Toronto	124.4	95.0	111.0
	Windsor	114.7	88.0	102.6
PRINCE EDWARD ISLAND				
	Charlottetown	117.8	58.7	91.0
	Summerside	117.5	58.7	90.8
QUEBEC				
	Cap-de-la-Madeleine	116.0	81.4	100.3
	Charlesbourg	116.0	81.4	100.3
	Chicoutimi	115.0	81.1	99.6
	Gatineau	115.4	81.1	99.9
	Laval	115.4	81.1	99.8
	Montreal	118.8	81.4	101.8
	Quebec	119.6	81.8	102.5
	Sherbrooke	115.8	81.1	100.1
	Trois Rivieres	116.3	81.4	100.4
SASKATCHEWAN				
	Moose Jaw	114.2	64.8	91.8
	Prince Albert	113.6	64.7	91.4
	Regina	116.0	64.9	92.8
	Saskatoon	114.7	64.7	92.0
YUKON				
	Whitehorse	114.0	63.8	91.2

Abbr.	Meaning
A	Area Square Feet; Ampere
ABS	Acrylonitrile Butadiene Stryrene; Asbestos Bonded Steel
A.C.	Alternating Current; Air-Conditioning; Asbestos Cement; Plywood Grade A & C
A.C.I.	American Concrete Institute
AD	Plywood, Grade A & D
Addit.	Additional
Adj.	Adjustable
af	Audio-frequency
A.G.A.	American Gas Association
Agg.	Aggregate
A.H.	Ampere Hours
A hr.	Ampere-hour
A.H.U.	Air Handling Unit
A.I.A.	American Institute of Architects
AIC	Ampere Interrupting Capacity
Allow.	Allowance
alt.	Altitude
Alum.	Aluminum
a.m.	Ante Meridiem
Amp.	Ampere
Anod.	Anodized
Approx.	Approximate
Apt.	Apartment
Asb.	Asbestos
A.S.B.C.	American Standard Building Code
Asbe.	Asbestos Worker
A.S.H.R.A.E.	American Society of Heating, Refrig. & AC Engineers
A.S.M.E.	American Society of Mechanical Engineers
A.S.T.M.	American Society for Testing and Materials
Attchmt.	Attachment
Avg.	Average
A.W.G.	American Wire Gauge
AWWA	American Water Works Assoc.
Bbl.	Barrel
B&B	Grade B and Better; Balled & Burlapped
B.&S.	Bell and Spigot
B.&W.	Black and White
b.c.c.	Body-centered Cubic
B.C.Y.	Bank Cubic Yards
BE	Bevel End
B.F.	Board Feet
Bg. cem.	Bag of Cement
BHP	Boiler Horsepower; Brake Horsepower
B.I.	Black Iron
Bit.; Bitum.	Bituminous
Bk.	Backed
Bkrs.	Breakers
Bldg.	Building
Blk.	Block
Bm.	Beam
Boil.	Boilermaker
B.P.M.	Blows per Minute
BR	Bedroom
Brg.	Bearing
Brhe.	Bricklayer Helper
Bric.	Bricklayer
Brk.	Brick
Brng.	Bearing
Brs.	Brass
Brz.	Bronze
Bsn.	Basin
Btr.	Better
BTU	British Thermal Unit
BTUH	BTU per Hour
B.U.R.	Built-up Roofing
BX	Interlocked Armored Cable
c	Conductivity, Copper Sweat
C	Hundred; Centigrade
C/C	Center to Center, Cedar on Cedar
Cab.	Cabinet
Cair.	Air Tool Laborer
Calc	Calculated
Cap.	Capacity
Carp.	Carpenter
C.B.	Circuit Breaker
C.C.A.	Chromate Copper Arsenate
C.C.F.	Hundred Cubic Feet
cd	Candela
cd/sf	Candela per Square Foot
CD	Grade of Plywood Face & Back
CDX	Plywood, Grade C & D, exterior glue
Cefi.	Cement Finisher
Cem.	Cement
CF	Hundred Feet
C.F.	Cubic Feet
CFM	Cubic Feet per Minute
c.g.	Center of Gravity
CHW	Chilled Water; Commercial Hot Water
C.I.	Cast Iron
C.I.P.	Cast in Place
Circ.	Circuit
C.L.	Carload Lot
Clab.	Common Laborer
Clam	Common maintenance laborer
C.L.F.	Hundred Linear Feet
CLF	Current Limiting Fuse
CLP	Cross Linked Polyethylene
cm	Centimeter
CMP	Corr. Metal Pipe
C.M.U.	Concrete Masonry Unit
CN	Change Notice
Col.	Column
CO_2	Carbon Dioxide
Comb.	Combination
Compr.	Compressor
Conc.	Concrete
Cont.	Continuous; Continued
Corr.	Corrugated
Cos	Cosine
Cot	Cotangent
Cov.	Cover
C/P	Cedar on Paneling
CPA	Control Point Adjustment
Cplg.	Coupling
C.P.M.	Critical Path Method
CPVC	Chlorinated Polyvinyl Chloride
C.Pr.	Hundred Pair
CRC	Cold Rolled Channel
Creos.	Creosote
Crpt.	Carpet & Linoleum Layer
CRT	Cathode-ray Tube
CS	Carbon Steel, Constant Shear Bar Joist
Csc	Cosecant
C.S.F.	Hundred Square Feet
CSI	Construction Specifications Institute
C.T.	Current Transformer
CTS	Copper Tube Size
Cu	Copper, Cubic
Cu. Ft.	Cubic Foot
cw	Continuous Wave
C.W.	Cool White; Cold Water
Cwt.	100 Pounds
C.W.X.	Cool White Deluxe
C.Y.	Cubic Yard (27 cubic feet)
C.Y./Hr.	Cubic Yard per Hour
Cyl.	Cylinder
d	Penny (nail size)
D	Deep; Depth; Discharge
Dis.;Disch.	Discharge
Db.	Decibel
Dbl.	Double
DC	Direct Current
DDC	Direct Digital Control
Demob.	Demobilization
d.f.u.	Drainage Fixture Units
D.H.	Double Hung
DHW	Domestic Hot Water
Diag.	Diagonal
Diam.	Diameter
Distrib.	Distribution
Dk.	Deck
D.L.	Dead Load; Diesel
DLH	Deep Long Span Bar Joist
Do.	Ditto
Dp.	Depth
D.P.S.T.	Double Pole, Single Throw
Dr.	Driver
Drink.	Drinking
D.S.	Double Strength
D.S.A.	Double Strength A Grade
D.S.B.	Double Strength B Grade
Dty.	Duty
DWV	Drain Waste Vent
DX	Deluxe White, Direct Expansion
dyn	Dyne
e	Eccentricity
E	Equipment Only; East
Ea.	Each
E.B.	Encased Burial
Econ.	Economy
E.C.Y	Embankment Cubic Yards
EDP	Electronic Data Processing
EIFS	Exterior Insulation Finish System
E.D.R.	Equiv. Direct Radiation
Eq.	Equation
Elec.	Electrician; Electrical
Elev.	Elevator; Elevating
EMT	Electrical Metallic Conduit; Thin Wall Conduit
Eng.	Engine, Engineered
EPDM	Ethylene Propylene Diene Monomer
EPS	Expanded Polystyrene
Eqhv.	Equip. Oper., Heavy
Eqlt.	Equip. Oper., Light
Eqmd.	Equip. Oper., Medium
Eqmm.	Equip. Oper., Master Mechanic
Eqol.	Equip. Oper., Oilers
Equip.	Equipment
ERW	Electric Resistance Welded
E.S.	Energy Saver
Est.	Estimated
esu	Electrostatic Units
E.W.	Each Way
EWT	Entering Water Temperature
Excav.	Excavation
Exp.	Expansion, Exposure
Ext.	Exterior
Extru.	Extrusion
f.	Fiber stress
F	Fahrenheit; Female; Fill
Fab.	Fabricated
FBGS	Fiberglass
F.C.	Footcandles
f.c.c.	Face-centered Cubic
f'c.	Compressive Stress in Concrete; Extreme Compressive Stress
F.E.	Front End
FEP	Fluorinated Ethylene Propylene (Teflon)
F.G.	Flat Grain
F.H.A.	Federal Housing Administration
Fig.	Figure
Fin.	Finished
Fixt.	Fixture
Fl. Oz.	Fluid Ounces
Flr.	Floor
F.M.	Frequency Modulation; Factory Mutual
Fmg.	Framing
Fndtn.	Foundation

Abbr.	Meaning	Abbr.	Meaning	Abbr.	Meaning
Fori.	Foreman, Inside	I.W.	Indirect Waste	M.C.F.	Thousand Cubic Feet
Foro.	Foreman, Outside	J	Joule	M.C.F.M.	Thousand Cubic Feet per Minute
Fount.	Fountain	J.I.C.	Joint Industrial Council	M.C.M.	Thousand Circular Mils
FPM	Feet per Minute	K	Thousand; Thousand Pounds;	M.C.P.	Motor Circuit Protector
FPT	Female Pipe Thread		Heavy Wall Copper Tubing, Kelvin	MD	Medium Duty
Fr.	Frame	K.A.H.	Thousand Amp. Hours	M.D.O.	Medium Density Overlaid
F.R.	Fire Rating	KCMIL	Thousand Circular Mils	Med.	Medium
FRK	Foil Reinforced Kraft	KD	Knock Down	MF	Thousand Feet
FRP	Fiberglass Reinforced Plastic	K.D.A.T.	Kiln Dried After Treatment	M.F.B.M.	Thousand Feet Board Measure
FS	Forged Steel	kg	Kilogram	Mfg.	Manufacturing
FSC	Cast Body; Cast Switch Box	kG	Kilogauss	Mfrs.	Manufacturers
Ft.	Foot; Feet	kgf	Kilogram Force	mg	Milligram
Ftng.	Fitting	kHz	Kilohertz	MGD	Million Gallons per Day
Ftg.	Footing	Kip.	1000 Pounds	MGPH	Thousand Gallons per Hour
Ft. Lb.	Foot Pound	KJ	Kiljoule	MH, M.H.	Manhole; Metal Halide; Man-Hour
Furn.	Furniture	K.L.	Effective Length Factor	MHz	Megahertz
FVNR	Full Voltage Non-Reversing	K.L.F.	Kips per Linear Foot	Mi.	Mile
FXM	Female by Male	Km	Kilometer	MI	Malleable Iron; Mineral Insulated
Fy.	Minimum Yield Stress of Steel	K.S.F.	Kips per Square Foot	mm	Millimeter
g	Gram	K.S.I.	Kips per Square Inch	Mill.	Millwright
G	Gauss	kV	Kilovolt	Min., min.	Minimum, minute
Ga.	Gauge	kVA	Kilovolt Ampere	Misc.	Miscellaneous
Gal.	Gallon	K.V.A.R.	Kilovar (Reactance)	ml	Milliliter, Mainline
Gal./Min.	Gallon per Minute	KW	Kilowatt	M.L.F.	Thousand Linear Feet
Galv.	Galvanized	KWh	Kilowatt-hour	Mo.	Month
Gen.	General	L	Labor Only; Length; Long;	Mobil.	Mobilization
G.F.I.	Ground Fault Interrupter		Medium Wall Copper Tubing	Mog.	Mogul Base
Glaz.	Glazier	Lab.	Labor	MPH	Miles per Hour
GPD	Gallons per Day	lat	Latitude	MPT	Male Pipe Thread
GPH	Gallons per Hour	Lath.	Lather	MRT	Mile Round Trip
GPM	Gallons per Minute	Lav.	Lavatory	ms	Millisecond
GR	Grade	lb.; #	Pound	M.S.F.	Thousand Square Feet
Gran.	Granular	L.B.	Load Bearing; L Conduit Body	Mstz.	Mosaic & Terrazzo Worker
Grnd.	Ground	L. & E.	Labor & Equipment	M.S.Y.	Thousand Square Yards
H	High; High Strength Bar Joist;	lb./hr.	Pounds per Hour	Mtd.	Mounted
	Henry	lb./L.F.	Pounds per Linear Foot	Mthe.	Mosaic & Terrazzo Helper
H.C.	High Capacity	lbf/sq.in.	Pound-force per Square Inch	Mtng.	Mounting
H.D.	Heavy Duty; High Density	L.C.L.	Less than Carload Lot	Mult.	Multi; Multiply
H.D.O.	High Density Overlaid	L.C.Y.	Loose Cubic Yard	M.V.A.	Million Volt Amperes
Hdr.	Header	Ld.	Load	M.V.A.R.	Million Volt Amperes Reactance
Hdwe.	Hardware	LE	Lead Equivalent	MV	Megavolt
Help.	Helper Average	LED	Light Emitting Diode	MW	Megawatt
HEPA	High Efficiency Particulate Air	L.F.	Linear Foot	MXM	Male by Male
	Filter	Lg.	Long; Length; Large	MYD	Thousand Yards
Hg	Mercury	L & H	Light and Heat	N	Natural; North
HIC	High Interrupting Capacity	LH	Long Span Bar Joist	nA	Nanoampere
HM	Hollow Metal	L.H.	Labor Hours	NA	Not Available; Not Applicable
H.O.	High Output	L.L.	Live Load	N.B.C.	National Building Code
Horiz.	Horizontal	L.L.D.	Lamp Lumen Depreciation	NC	Normally Closed
H.P.	Horsepower; High Pressure	lm	Lumen	N.E.M.A.	National Electrical Manufacturers
H.P.F.	High Power Factor	lm/sf	Lumen per Square Foot		Assoc.
Hr.	Hour	lm/W	Lumen per Watt	NEHB	Bolted Circuit Breaker to 600V.
Hrs./Day	Hours per Day	L.O.A.	Length Over All	N.L.B.	Non-Load-Bearing
HSC	High Short Circuit	log	Logarithm	NM	Non-Metallic Cable
Ht.	Height	L-O-L	Lateralolet	nm	Nanometer
Htg.	Heating	L.P.	Liquefied Petroleum; Low Pressure	No.	Number
Htrs.	Heaters	L.P.F.	Low Power Factor	NO	Normally Open
HVAC	Heating, Ventilation & Air-	LR	Long Radius	N.O.C.	Not Otherwise Classified
	Conditioning	L.S.	Lump Sum	Nose.	Nosing
Hvy.	Heavy	Lt.	Light	N.P.T.	National Pipe Thread
HW	Hot Water	Lt. Ga.	Light Gauge	NQOD	Combination Plug-on/Bolt on
Hyd.;Hydr.	Hydraulic	L.T.L.	Less than Truckload Lot		Circuit Breaker to 240V.
Hz.	Hertz (cycles)	Lt. Wt.	Lightweight	N.R.C.	Noise Reduction Coefficient
I.	Moment of Inertia	L.V.	Low Voltage	N.R.S.	Non Rising Stem
I.C.	Interrupting Capacity	M	Thousand; Material; Male;	ns	Nanosecond
ID	Inside Diameter		Light Wall Copper Tubing	nW	Nanowatt
I.D.	Inside Dimension; Identification	M²CA	Meters Squared Contact Area	OB	Opposing Blade
I.F.	Inside Frosted	m/hr; M.H.	Man-hour	OC	On Center
I.M.C.	Intermediate Metal Conduit	mA	Milliampere	OD	Outside Diameter
In.	Inch	Mach.	Machine	O.D.	Outside Dimension
Incan.	Incandescent	Mag. Str.	Magnetic Starter	ODS	Overhead Distribution System
Incl.	Included; Including	Maint.	Maintenance	O.G.	Ogee
Int.	Interior	Marb.	Marble Setter	O.H.	Overhead
Inst.	Installation	Mat; Mat'l.	Material	O&P	Overhead and Profit
Insul.	Insulation/Insulated	Max.	Maximum	Oper.	Operator
I.P.	Iron Pipe	MBF	Thousand Board Feet	Opng.	Opening
I.P.S.	Iron Pipe Size	MBH	Thousand BTU's per hr.	Orna.	Ornamental
I.P.T.	Iron Pipe Threaded	MC	Metal Clad Cable	OSB	Oriented Strand Board

O.S.&Y.	Outside Screw and Yoke
Ovhd.	Overhead
OWG	Oil, Water or Gas
Oz.	Ounce
P.	Pole; Applied Load; Projection
p.	Page
Pape.	Paperhanger
P.A.P.R.	Powered Air Purifying Respirator
PAR	Parabolic Reflector
Pc., Pcs.	Piece, Pieces
P.C.	Portland Cement; Power Connector
P.C.F.	Pounds per Cubic Foot
P.C.M.	Phase Contract Microscopy
P.E.	Professional Engineer; Porcelain Enamel; Polyethylene; Plain End
Perf.	Perforated
Ph.	Phase
P.I.	Pressure Injected
Pile.	Pile Driver
Pkg.	Package
Pl.	Plate
Plah.	Plasterer Helper
Plas.	Plasterer
Pluh.	Plumbers Helper
Plum.	Plumber
Ply.	Plywood
p.m.	Post Meridiem
Pntd.	Painted
Pord.	Painter, Ordinary
pp	Pages
PP; PPL	Polypropylene
P.P.M.	Parts per Million
Pr.	Pair
P.E.S.B.	Pre-engineered Steel Building
Prefab.	Prefabricated
Prefin.	Prefinished
Prop.	Propelled
PSF; psf	Pounds per Square Foot
PSI; psi	Pounds per Square Inch
PSIG	Pounds per Square Inch Gauge
PSP	Plastic Sewer Pipe
Pspr.	Painter, Spray
Psst.	Painter, Structural Steel
P.T.	Potential Transformer
P. & T.	Pressure & Temperature
Ptd.	Painted
Ptns.	Partitions
Pu	Ultimate Load
PVC	Polyvinyl Chloride
Pvmt.	Pavement
Pwr.	Power
Q	Quantity Heat Flow
Quan.; Qty.	Quantity
Q.C.	Quick Coupling
r	Radius of Gyration
R	Resistance
R.C.P.	Reinforced Concrete Pipe
Rect.	Rectangle
Reg.	Regular
Reinf.	Reinforced
Req'd.	Required
Res.	Resistant
Resi.	Residential
Rgh.	Rough
RGS	Rigid Galvanized Steel
R.H.W.	Rubber, Heat & Water Resistant; Residential Hot Water
rms	Root Mean Square
Rnd.	Round
Rodm.	Rodman
Rofc.	Roofer, Composition
Rofp.	Roofer, Precast
Rohe.	Roofer Helpers (Composition)
Rots.	Roofer, Tile & Slate
R.O.W.	Right of Way
RPM	Revolutions per Minute
R.S.	Rapid Start
Rsr	Riser
RT	Round Trip
S.	Suction; Single Entrance; South
SCFM	Standard Cubic Feet per Minute
Scaf.	Scaffold
Sch.; Sched.	Schedule
S.C.R.	Modular Brick
S.D.	Sound Deadening
S.D.R.	Standard Dimension Ratio
S.E.	Surfaced Edge
Sel.	Select
S.E.R.; S.E.U.	Service Entrance Cable
S.F.	Square Foot
S.F.C.A.	Square Foot Contact Area
S.F. Flr.	Square Foot of Floor
S.F.G.	Square Foot of Ground
S.F. Hor.	Square Foot Horizontal
S.F.R.	Square Feet of Radiation
S.F. Shlf.	Square Foot of Shelf
S4S	Surface 4 Sides
Shee.	Sheet Metal Worker
Sin.	Sine
Skwk.	Skilled Worker
SL	Saran Lined
S.L.	Slimline
Sldr.	Solder
SLH	Super Long Span Bar Joist
S.N.	Solid Neutral
S-O-L	Socketolet
sp	Standpipe
S.P.	Static Pressure; Single Pole; Self-Propelled
Spri.	Sprinkler Installer
spwg	Static Pressure Water Gauge
S.P.D.T.	Single Pole, Double Throw
SPF	Spruce Pine Fir
S.P.S.T.	Single Pole, Single Throw
SPT	Standard Pipe Thread
Sq.	Square; 100 Square Feet
Sq. Hd.	Square Head
Sq. In.	Square Inch
S.S.	Single Strength; Stainless Steel
S.S.B.	Single Strength B Grade
sst	Stainless Steel
Sswk.	Structural Steel Worker
Sswl.	Structural Steel Welder
St.; Stl.	Steel
S.T.C.	Sound Transmission Coefficient
Std.	Standard
STK	Select Tight Knot
STP	Standard Temperature & Pressure
Stpi.	Steamfitter, Pipefitter
Str.	Strength; Starter; Straight
Strd.	Stranded
Struct.	Structural
Sty.	Story
Subj.	Subject
Subs.	Subcontractors
Surf.	Surface
Sw.	Switch
Swbd.	Switchboard
S.Y.	Square Yard
Syn.	Synthetic
S.Y.P.	Southern Yellow Pine
Sys.	System
t.	Thickness
T	Temperature; Ton
Tan	Tangent
T.C.	Terra Cotta
T & C	Threaded and Coupled
T.D.	Temperature Difference
T.E.M.	Transmission Electron Microscopy
TFE	Tetrafluoroethylene (Teflon)
T. & G.	Tongue & Groove; Tar & Gravel
Th.; Thk.	Thick
Thn.	Thin
Thrded	Threaded
Tilf.	Tile Layer, Floor
Tilh.	Tile Layer, Helper
THHN	Nylon Jacketed Wire
THW.	Insulated Strand Wire
THWN;	Nylon Jacketed Wire
T.L.	Truckload
T.M.	Track Mounted
Tot.	Total
T-O-L	Threadolet
T.S.	Trigger Start
Tr.	Trade
Transf.	Transformer
Trhv.	Truck Driver, Heavy
Trlr	Trailer
Trlt.	Truck Driver, Light
TV	Television
T.W.	Thermoplastic Water Resistant Wire
UCI	Uniform Construction Index
UF	Underground Feeder
UGND	Underground Feeder
U.H.F.	Ultra High Frequency
U.L.	Underwriters Laboratory
Unfin.	Unfinished
URD	Underground Residential Distribution
US	United States
USP	United States Primed
UTP	Unshielded Twisted Pair
V	Volt
V.A.	Volt Amperes
V.C.T.	Vinyl Composition Tile
VAV	Variable Air Volume
VC	Veneer Core
Vent.	Ventilation
Vert.	Vertical
V.F.	Vinyl Faced
V.G.	Vertical Grain
V.H.F.	Very High Frequency
VHO	Very High Output
Vib.	Vibrating
V.L.F.	Vertical Linear Foot
Vol.	Volume
VRP	Vinyl Reinforced Polyester
W	Wire; Watt; Wide; West
w/	With
W.C.	Water Column; Water Closet
W.F.	Wide Flange
W.G.	Water Gauge
Wldg.	Welding
W. Mile	Wire Mile
W-O-L	Weldolet
W.R.	Water Resistant
Wrck.	Wrecker
W.S.P.	Water, Steam, Petroleum
WT., Wt.	Weight
WWF	Welded Wire Fabric
XFER	Transfer
XFMR	Transformer
XHD	Extra Heavy Duty
XHHW; XLPE	Cross-Linked Polyethylene Wire Insulation
XLP	Cross-linked Polyethylene
Y	Wye
yd	Yard
yr	Year
Δ	Delta
%	Percent
~	Approximately
$\emptyset$	Phase
@	At
#	Pound; Number
<	Less Than
>	Greater Than

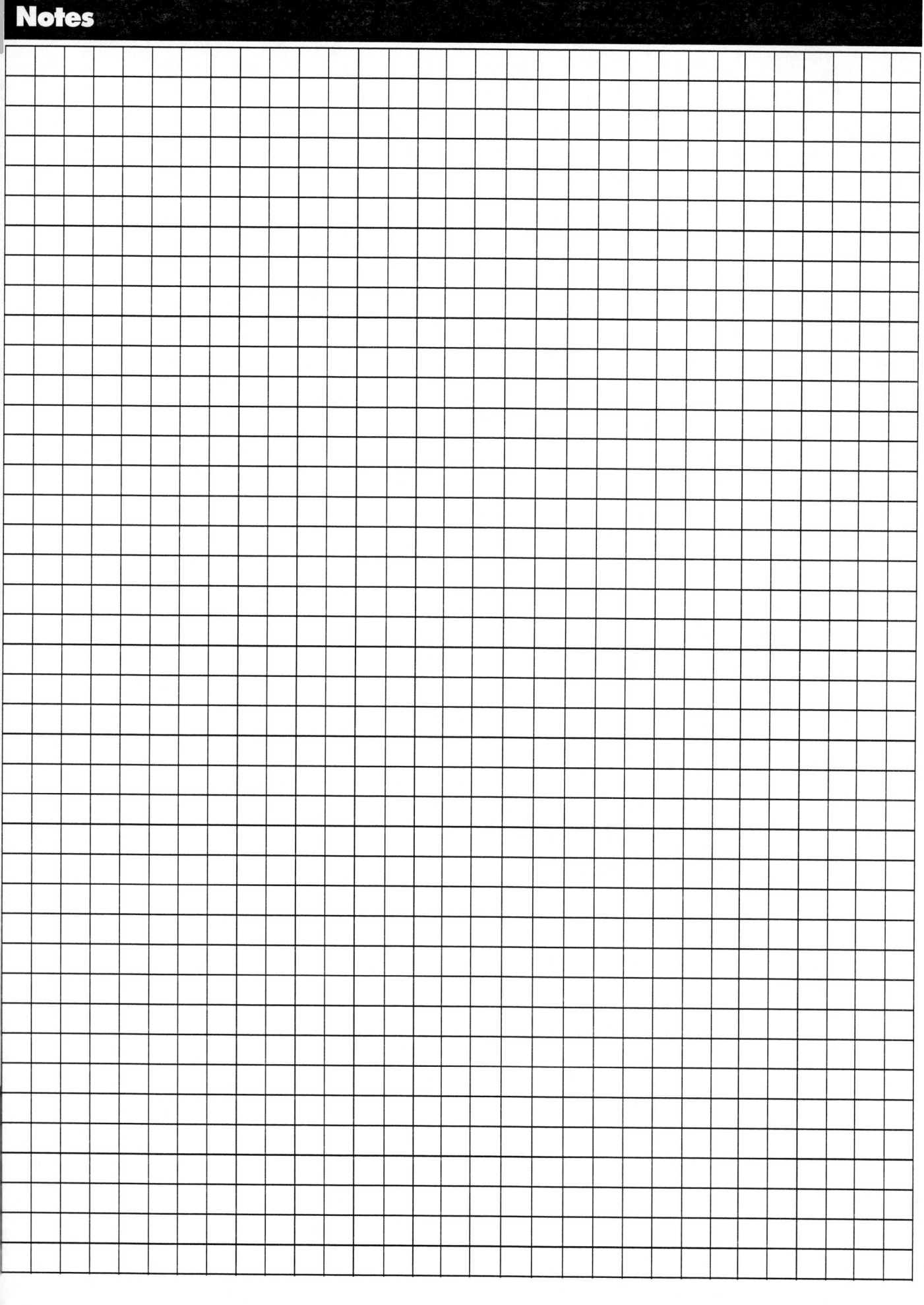

Reed Construction Data, Inc.

Reed Construction Data, Inc., a leading worldwide provider of total construction information solutions, is comprised of three main product groups designed specifically to help construction professionals advance their businesses with timely, accurate and actionable project, product, and cost data. Reed Construction Data is a division of Reed Business Information, a member of the Reed Elsevier plc group of companies.

The *Project, Product, and Cost & Estimating* divisions offer a variety of innovative products and services designed for the full spectrum of design, construction, and manufacturing professionals. Through it's *International* companies, Reed Construction Data's reputation for quality construction market data is growing worldwide.

Cost Information
RSMeans, the undisputed market leader and authority on construction costs, publishes current cost and estimating information in annual cost books and on the CostWorks CD-ROM. RSMeans furnishes the construction industry with a rich library of complementary reference books and a series of professional seminars that are designed to sharpen professional skills and maximize the effective use of cost estimating and management tools. RSMeans also provides construction cost consulting for Owners, Manufacturers, Designers, and Contractors.

Project Data
Reed Construction Data provides complete, accurate and relevant project information through all stages of construction. Customers are supplied industry data through leads, project reports, contact lists, plans and specifications surveys, market penetration analyses and sales evaluation reports. Any of these products can pinpoint a county, look at a state, or cover the country. Data is delivered via paper, e-mail, CD-ROM or the Internet.

Building Product Information
The First Source suite of products is the only integrated building product information system offered to the commercial construction industry for comparing and specifying building products. These print and online resources include *First Source*, CSI's SPEC-DATA™, CSI's MANU-SPEC™, First Source CAD, and Manufacturer Catalogs. Written by industry professionals and organized using CSI's MasterFormat™, construction professionals use this information to make better design decisions.

FirstSourceONL.com combines Reed Construction Data's project, product and cost data with news and information from Reed Business Information's *Building Design & Construction* and *Consulting-Specifying Engineer,* this industry-focused site offers easy and unlimited access to vital information for all construction professionals.

International
BIMSA/Mexico provides construction project news, product information, cost-data, seminars and consulting services to construction professionals in Mexico. Its subsidiary, PRISMA, provides job costing software.

Byggfakta Scandinavia AB, founded in 1936, is the parent company for the leaders of customized construction market data for Denmark, Estonia, Finland, Norway and Sweden. Each company fully covers the local construction market and provides information across several platforms including subscription, ad-hoc basis, electronically and on paper.

Reed Construction Data Canada serves the Canadian construction market with reliable and comprehensive project and product information services that cover all facets of construction. Core services include: *BuildSource, BuildSpec, BuildSelect,* product selection and specification tools available in print and on the Internet; Building Reports, a national construction project lead service; CanaData, statistical and forecasting information; *Daily Commercial News,* a construction newspaper reporting on news and projects in Ontario; and *Journal of Commerce,* reporting news in British Columbia and Alberta.

Cordell Building Information Services, with its complete range of project and cost and estimating services, is Australia's specialist in the construction information industry. Cordell provides in-depth and historical information on all aspects of construction projects and estimation, including several customized reports, construction and sales leads, and detailed cost information among others.

For more information, please visit our Web site at www.reedconstructiondata.com.

Reed Construction Data, Inc., Corporate Office
30 Technology Parkway South
Norcross, GA 30092-2912
(800) 322-6996
(800) 895-8661 (fax)
info@reedbusiness.com
www.reedconstructiondata.com

 Reed Construction Data

Means Project Cost Report

By filling out and returning the Project Description, you can receive a discount of $20.00 off any one of the Means products advertised in the following pages. The cost information required includes all items marked (✔) except those where no costs occurred. The sum of all major items should equal the Total Project Cost.

$20.00 Discount per product for each report you submit.

DISCOUNT PRODUCTS AVAILABLE—FOR U.S. CUSTOMERS ONLY—STRICTLY CONFIDENTIAL

Project Description (No remodeling projects, please.)

✔ Type Building _____

✔ Location _____

 Capacity _____

✔ Frame _____

✔ Exterior _____

✔ Basement: full ☐ partial ☐ none ☐ crawl ☐

✔ Height in Stories _____

✔ Total Floor Area _____

 Ground Floor Area _____

✔ Volume in C.F. _____

 % Air Conditioned _____ Tons _____

 Comments _____

 Owner _____

 Architect _____

 General Contractor _____

✔ Bid Date _____

 Typical Bay Size _____

✔ Labor Force: _____ % Union _____ % Non-Union

✔ Project Description (Circle one number in each line)
 1. Economy 2. Average 3. Custom 4. Luxury
 1. Square 2. Rectangular 3. Irregular 4. Very Irregular

	✔ **Total Project Cost**			$	
A	✔ **General Conditions**			$	
B	✔ **Site Work**			$	
BS	Site Clearing & Improvement				
BE	Excavation	(		C.Y.)	
BF	Caissons & Piling	(		L.F.)	
BU	Site Utilities				
BP	Roads & Walks Exterior Paving	(		S.Y.)	
C	✔ **Concrete**			$	
C	Cast in Place	(		C.Y.)	
CP	Precast	(		S.F.)	
D	✔ **Masonry**			$	
DB	Brick	(		M)	
DC	Block	(		M)	
DT	Tile	(		S.F.)	
DS	Stone	(		S.F.)	
E	✔ **Metals**			$	
ES	Structural Steel	(		Tons)	
EM	Misc. & Ornamental Metals				
F	✔ **Wood & Plastics**			$	
FR	Rough Carpentry	(		MBF)	
FF	Finish Carpentry				
FM	Architectural Millwork				
G	✔ **Thermal & Moisture Protection**			$	
GW	Waterproofing-Dampproofing	(		S.F.)	
GN	Insulation	(		S.F.)	
GR	Roofing & Flashing	(		S.F.)	
GM	Metal Siding/Curtain Wall	(		S.F.)	
H	✔ **Doors and Windows**			$	
HD	Doors	(		Ea.)	
HW	Windows	(		S.F.)	
HH	Finish Hardware				
HG	Glass & Glazing	(		S.F.)	
HS	Storefronts	(		S.F.)	

J	✔ **Finishes**				$
JL	Lath & Plaster	(		S.Y.)	
JD	Drywall	(		S.F.)	
JM	Tile & Marble	(		S.F.)	
JT	Terrazzo	(		S.F.)	
JA	Acoustical Treatment	(		S.F.)	
JC	Carpet	(		S.Y.)	
JF	Hard Surface Flooring	(		S.F.)	
JP	Painting & Wall Covering	(		S.F.)	
K	✔ **Specialties**				$
KB	Bathroom Partitions & Access.	(		S.F.)	
KF	Other Partitions	(		S.F.)	
KL	Lockers	(		Ea.)	
L	✔ **Equipment**				$
LK	Kitchen				
LS	School				
LO	Other				
M	✔ **Furnishings**				$
MW	Window Treatment				
MS	Seating	(		Ea.)	
N	✔ **Special Construction**				$
NA	Acoustical	(		S.F.)	
NB	Prefab. Bldgs.	(		S.F.)	
NO	Other				
P	✔ **Conveying Systems**				$
PE	Elevators	(		Ea.)	
PS	Escalators	(		Ea.)	
PM	Material Handling				
Q	✔ **Mechanical**				$
QP	Plumbing (No. of fixtures)				
QS	Fire Protection (Sprinklers)				
QF	Fire Protection (Hose Standpipes)				
QB	Heating, Ventilating & A.C.				
QH	Heating & Ventilating (BTU Output)				
QA	Air Conditioning	(		Tons)	
R	✔ **Electrical**				$
RL	Lighting	(		S.F.)	
RP	Power Service				
RD	Power Distribution				
RA	Alarms				
RG	Special Systems				
S	✔ **Mech./Elec. Combined**				$

Product Name _____

Product Number _____

Your Name _____

Title _____

Company _____

☐ Company

☐ Home Street Address _____

City, State, Zip _____

☐ Please send _____ forms.

Please specify the Means product you wish to receive. Complete the address information as requested and return this form with your check (product cost less $20.00) to address below.

RSMeans Company, Inc.,
Square Foot Costs Department
P.O. Box 800
Kingston, MA 02364-9988

For more information
visit Means Web S...
at www.rsmeans.c...

Reed Construction Data/RSMeans . . . a tradition of excellence in Construction Cost Information and Services since 1942.

Table of Contents

Book Selection Guide

The following table provides definitive information on the content of each cost data publication. The number of lines of data provided in each unit price or assemblies division, as well as the number of reference tables and crews is listed for each book. The presence of other elements such as an historical cost index, city cost indexes, square foot models or cross-referenced index is also indicated. You can use the table to help select the Means' book that has the quantity and type of information you most need in your work

Unit Cost Divisions	Building Construction Costs	Mechanical	Electrical	Repair & Remodel.	Square Foot	Site Work Landsc.	Assemblies	Interior	Concrete Masonry	Open Shop	Heavy Construc.	Light Commercial	Facil. Construc.	Plumbing	Western Construction Costs	Re-den
1	1104	822	907	1013		1045		847	1014	1102	1051	752	1528	874	1102	7
2	2505	1468	460	1450		7794		547	1387	2467	4978	692	4305	1584	2490	7
3	1401	113	100	726		1259		201	1782	1396	1404	229	1312	82	1399	2
4	835	18	0	647		657		574	1056	811	594	407	1058	0	821	3
5	1820	239	193	957		768		927	691	1788	1023	805	1802	316	1802	1
6	1489	82	78	1475		452		1406	318	1481	608	1626	1596	47	1826	17
7	1264	158	74	1241		468		489	408	1264	349	956	1318	168	1264	7
8	1839	28	0	1907		313		1673	645	1821	51	1227	2034	0	1841	11
9	1595	47	0	1424		233		1688	360	1547	159	1337	1809	47	1586	12
10	861	47	25	498		193		700	170	863	0	394	906	233	861	2
11	1018	322	169	502		137		813	44	925	107	218	1173	291	925	1
12	307	0	0	47		210		1416	27	298	0	62	1435	0	298	
13	1153	1005	375	494		388		884	75	1136	279	484	1783	928	1119	1
14	345	36	0	258		36		292	0	344	30	12	343	35	343	
15	1987	13138	636	1779		1569		1174	59	1996	1766	1217	10865	9695	2017	8
16	1277	470	10063	1002		739		1108	55	1265	760	1081	9834	415	1226	5
17	427	354	427	0		0		0	0	427	0	0	427	356	427	
Totals	21227	18347	13507	15420		16261		14739	8091	20961	13159	11499	43528	15071	21347	96

Assembly Divisions	Building Construction Costs	Mechanical	Electrical	Repair & Remodel.	Square Foot	Site Work Landsc.	Assemblies	Interior	Concrete Masonry	Open Shop	Heavy Construc.	Light Commercial	Facil. Construc.	Plumbing	Western Construction Costs	Asm Div
A		19	0	192	150	540	612	0	550		542	149	24	0		1
B		0	0	809	2480	0	5590	333	1914		0	2024	144	0		2
C		0	0	635	862	0	1220	1568	145		0	767	238	0		3
D		1031	780	693	1823	0	2439	753	0		0	1310	1027	896		4
E		0	0	85	255	0	292	5	0		0	255	5	0		5
F		0	0	0	123	0	126	0	0		0	123	3	0		6
G		465	172	332	111	1856	584	0	482		432	110	113	560		7
																8
																9
																10
																11
																12
Totals		1515	952	2746	5804	2396	10863	2659	3091		974	4738	1554	1456		

Reference Section	Building Construction Costs	Mechanical	Electrical	Repair & Remodel.	Square Foot	Site Work Landsc.	Assemblies	Interior	Concrete Masonry	Open Shop	Heavy Construc.	Light Commercial	Facil. Construc.	Plumbing	Western Construction Costs	Re-den
Tables	130	45	85	69	4	80	219	46	71	130	61	57	104	51	131	
Models					102							43				
Crews	410	410	410	391		410		410	410	393	410	393	391	410	410	3
City Cost Indexes	yes	yes	yes	yes	yes	yes	yes	yes	yes	yes	yes	yes	yes	yes	yes	
Historical Cost Indexes	yes	yes	yes	yes	yes	yes	yes	yes	yes	yes	yes	yes	yes	yes	yes	

Annual Cost Guides

For more information
visit Means Web Site
at www.rsmeans.com

Means Building Construction Cost Data 2005

Available in Both Softbound and Looseleaf Editions

The "Bible" of the industry comes in the standard softcover edition or the looseleaf edition.

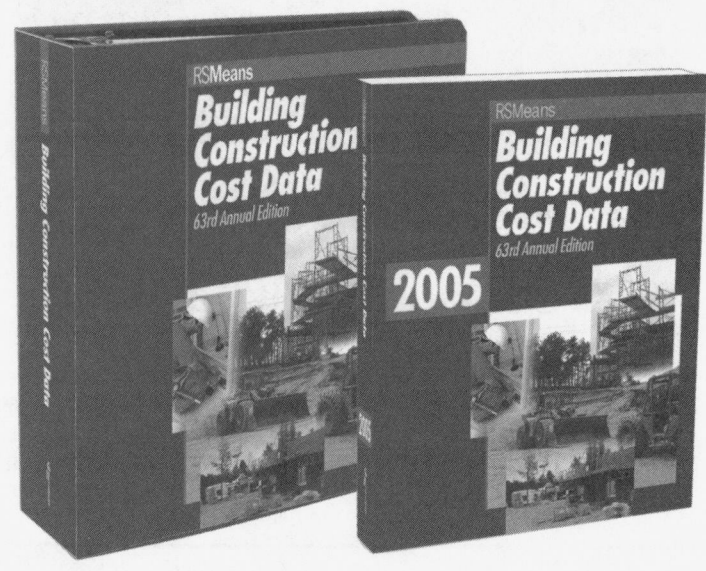

Many customers enjoy the convenience and flexibility of the looseleaf binder, which increases the usefulness of *Means Building Construction Cost Data 2005* by making it easy to add and remove pages. You can insert your own cost information pages, so everything is in one place. Copying pages for faxing is easier also. Whichever edition you prefer, softbound or the convenient looseleaf edition, you'll be eligible to receive *The Change Notice* FREE. Current subscribers receive *The Change Notice* via e-mail.

$115.95 per copy, Softbound
Catalog No. 60015

$144.95 per copy, Looseleaf
Catalog No. 61015

Means Building Construction Cost Data 2005

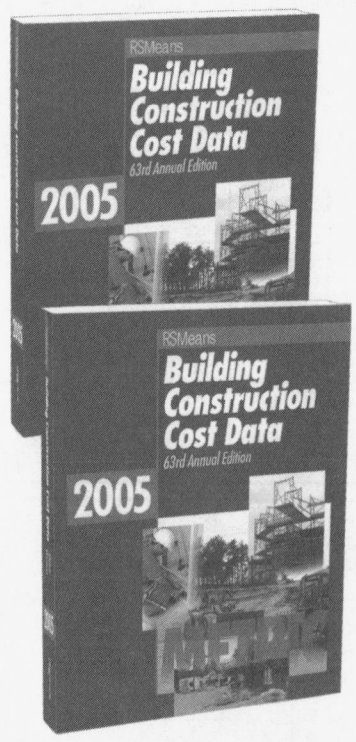

Offers you unchallenged unit price reliability in an easy-to-use arrangement. Whether used for complete, finished estimates or for periodic checks, it supplies more cost facts better and faster than any comparable source. Over 23,000 unit prices for 2005. The City Cost Indexes cover over 930 areas, for indexing to any project location in North America. Order and get *The Change Notice* FREE. You'll have year-long access to the Means Estimating **HOTLINE** FREE with your subscription. Expert assistance when using Means data is just a phone call away.

$115.95 per copy
Over 700 pages, illustrated, available Oct. 2004
Catalog No. 60015

Means Building Construction Cost Data 2005

Metric Version

The Federal Government has stated that all federal construction projects must now use metric documentation. The *Metric Version* of *Means Building Construction Cost Data 2005* is presented in metric measurements covering all construction areas. Don't miss out on these billion dollar opportunities. Make the switch to metric today.

$115.95 per copy
Over 700 pages, illus., available Dec. 2004
Catalog No. 63015

Annual Cost Guides

Means Mechanical Cost Data 2005

• **HVAC** • **Controls**

Total unit and systems price guidance for mechanical construction...materials, parts, fittings, and complete labor cost information. Includes prices for piping, heating, air conditioning, ventilation, and all related construction.

Plus new 2005 unit costs for:
• Over 2500 installed HVAC/controls assemblies
• "On Site" Location Factors for over 930 cities and towns in the U.S. and Canada
• Crews, labor and equipment

$115.95 per copy
Over 600 pages, illustrated, available Oct. 2004
Catalog No. 60025

Means Plumbing Cost Data 2005

Comprehensive unit prices and assemblies for plumbing, irrigation systems, commercial and residential fire protection, point-of-use water heaters, and the latest approved materials. This publication and its companion, *Means Mechan Cost Data*, provide full-range cost estimating coverage for all the mechanical trades.

$115.95 per copy
Over 550 pages, illustrated, available Oct. 2004
Catalog No. 60215

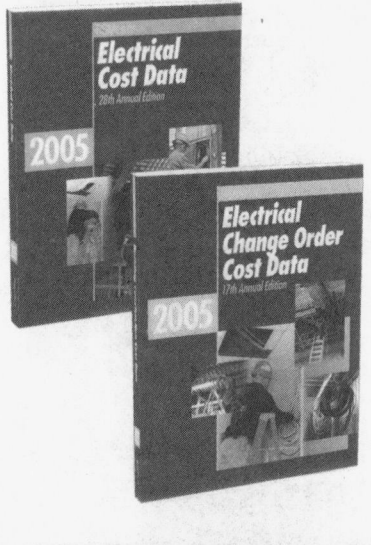

Means Electrical Cost Data 2005

Pricing information for every part of electrical cost planning: More than 15,000 unit and systems costs with design tables; clear specifications and drawings; engineering guides and illustrated estimating procedures; complete labor-hour and materials costs for better scheduling and procurement; the latest electrical products and construction methods.
• A Variety of Special Electrical Systems including Cathodic Protection
• Costs for maintenance, demolition, HVAC/ mechanical, specialties, equipment, and more

$115.95 per copy
Over 450 pages, illustrated, available Oct. 2004
Catalog No. 60035

Means Electrical Change Order Cost Data 2005

You are provided with electrical unit prices exclusively for pricing change orders based on recent, direct experience of contractors and suppliers. Analyze and check your own chang order estimates against the experience others had doing the same work. It also covers productivity analysis and change order cost justifications. With useful information for calculating the effects of change orders and dealing with their administration.

$115.95 per copy
Over 450 pages, available Oct. 2004
Catalog No. 60235

Means Facilities Maintenance & Repair Cost Data 2005

Published in a looseleaf format, *Means Facilities Maintenance & Repair Cost Data* gives you a complete system to manage and plan your facility repair and maintenance costs and budget efficiently. Guidelines for auditing a facility and developing an annual maintenance plan. Budgeting is included, along with reference tables on cost and management and information on frequency and productivity of maintenance operations.

The only nationally recognized source of maintenance and repair costs. Developed in cooperation with the Army Corps of Engineers.

$252.95 per copy
Over 600 pages, illustrated, available Dec. 2004
Catalog No. 60305

Means Square Foot Costs 2005

It's Accurate and Easy To Use!

• **Updated 2005 price information,** based on nationwide figures from suppliers, estimator labor experts and contractors.
• "How-to-Use" Sections, with **clear example** commercial, residential, industrial, and institutional structures.
• Realistic graphics, offering true-to-life illustrations of building projects.
• Extensive information on using square foot data, including **sample estimates** and **altern pricing methods.**

$126.95 per copy
Over 450 pages, illustrated, available Nov. 2004
Catalog No. 60055

Annual Cost Guides

For more information
visit Means Web Site
at www.rsmeans.com

Means Repair & Remodeling Cost Data 2005

Commercial/Residential

You can use this valuable tool to estimate commercial and residential renovation and remodeling.

Includes: New costs for hundreds of unique methods, materials and conditions that only come up in repair and remodeling. PLUS:

- Unit costs for over 16,000 construction components
- Installed costs for over 90 assemblies
- Costs for 300+ construction crews
- Over 930 "On Site" localization factors for the U.S. and Canada.

$99.95 per copy
Over 650 pages, illustrated, available Nov. 2004
Catalog No. 60045

Means Facilities Construction Cost Data 2005

For the maintenance and construction of commercial, industrial, municipal, and institutional properties. Costs are shown for new and remodeling construction and are broken down into materials, labor, equipment, overhead, and profit. Special emphasis is given to sections on mechanical, electrical, furnishings, site work, building maintenance, finish work, and demolition. More than 45,000 unit costs, plus assemblies and reference sections are included.

$279.95 per copy
Over 1200 pages, illustrated, available Nov. 2004
Catalog No. 60205

Means Residential Cost Data 2005

Contains square foot costs for 30 basic home models with the look of today, plus hundreds of custom additions and modifications you can quote right off the page. With costs for the 100 residential systems you're most likely to use in the year ahead. Complete with blank estimating forms, sample estimates and step-by-step instructions.

$99.95 per copy
Over 600 pages, illustrated, available Oct. 2004
Catalog No. 60175

Means Light Commercial Cost Data 2005

Specifically addresses the light commercial market, which is an increasingly specialized niche in the industry. Aids you, the owner/designer/contractor, in preparing all types of estimates, from budgets to detailed bids. Includes new advances in methods and materials. Assemblies section allows you to evaluate alternatives in the early stages of design/planning.

Over 13,000 unit costs for 2005 ensure you have the prices you need...when you need them.

$99.95 per copy
Over 650 pages, illustrated, available Nov. 2004
Catalog No. 60185

Means Assemblies Cost Data 2005

Means Assemblies Cost Data 2005 takes the guesswork out of preliminary or conceptual estimates. Now you don't have to try to calculate an assembled cost by working up individual components costs. We've done all the work for you.

Presents detailed illustrations, descriptions, specifications and costs for every conceivable building assembly—240 types in all—arranged in the easy-to-use UNIFORMAT II system. Each illustrated "assembled" cost includes a complete grouping of materials and associated installation costs including the installing contractor's overhead and profit.

$189.95 per copy
Over 600 pages, illustrated, available Oct. 2004
Catalog No. 60065

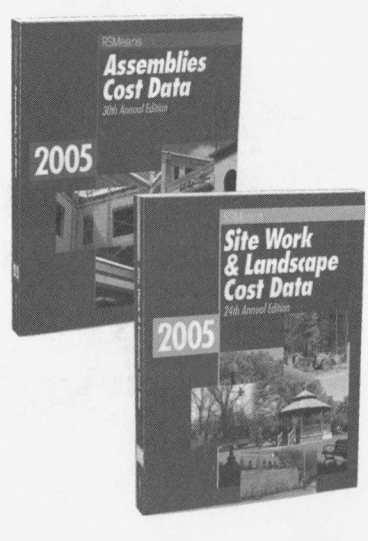

Means Site Work & Landscape Cost Data 2005

Means Site Work & Landscape Cost Data 2005 is organized to assist you in all your estimating needs. Hundreds of fact-filled pages help you make accurate cost estimates efficiently.

Updated for 2005!

- Demolition features—including ceilings, doors, electrical, flooring, HVAC, millwork, plumbing, roofing, walls and windows
- State-of-the-art segmental retaining walls
- Flywheel trenching costs and details
- Updated Wells section
- Landscape materials, flowers, shrubs and trees

$115.95 per copy
Over 600 pages, illustrated, available Nov. 2004
Catalog No. 60285

Building Security: Strategies & Costs
By David Owen

Unauthorized systems access, terrorism, hurricanes and tornadoes, sabotage, vandalism, fire, explosions, and other threats… All are considerations as building owners, facility managers, and design and construction professionals seek ways to meet security needs.

This comprehensive resource will help you evaluate your facility's security needs — and design and budget for the materials and devices needed to fulfill them. The text and cost data will help you to:

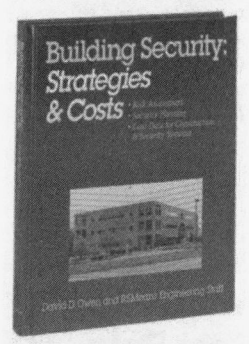

- Identify threats, probability of occurrence, and the potential losses— to determine and address your real vulnerabilities.
- Perform a detailed risk assessment of an existing facility, and prioritize and budget for security enhancement.
- Evaluate and price security systems and construction solutions, so you can make cost-effective choices.

Includes over 130 pages of Means Cost Data for installation of security systems and materials, plus a review of more than 50 security devices and construction solutions—how they work, and how they compare.

9.95 per copy
er 400 pages, Hardcover
talog No. 37339

Green Building: Project Planning & Cost Estimating
By RSMeans and Contributing Authors

Written by a team of leading experts in sustainable design, this new book is a complete guide to planning and estimating green building projects, a growing trend in building design and construction – commercial, industrial, institutional and residential. It explains:

- All the different criteria for "green-ness"
- What criteria your building needs to meet to get a LEED, Energy Star, or other recognized rating for green buildings
- How the project team works differently on a green versus a traditional building project
- How to select and specify green products
- How to evaluate the cost and value of green products versus conventional ones — not only for their first (installation) cost, but their cost over time (in maintenance and operation).

Features an extensive Green Building Cost Data section, which details the available products, how they are specified, and how much they cost.

9.95 per copy
pages, illustrated, Hardcover
talog No. 67338

Life Cycle Costing for Facilities
By Alphonse Dell'Isola and Dr. Steven Kirk

Guidance for achieving higher quality design and construction projects at lower costs!

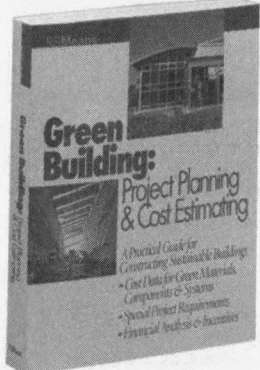

acility designers and owners are frustrated with cost-cutting efforts that yield the cheapest product, but acrifice quality. Life Cycle Costing, properly done, enables them to achieve both—high quality, incorporating innovative design, and costs that meet their budgets.

he authors show how LCC can work for a broad variety of projects – from several types of buildings, to roads nd bridges, to HVAC and electrical upgrades, to materials and equipment procurement. Case studies include:

Health care and nursing facilities

College campus and high schools

Office buildings, courthouses, and banks

Exterior walls, elevators, lighting, HVAC, and more

he book's extensive cost section provides maintenance and replacement costs for facility elements.

95 per copy
pages, Hardcover
alog No. 67341

Reference Books

Value Engineering: Practical Applications

. . . For Design, Construction, Maintenance & Operations

By Alphonse Dell'Isola, PE

A tool for immediate application—for engineers, architects, facility managers, owners, and contractors. Includes: making the case for VE—the management briefing, integrating VE into planning and budgeting, conducting life cycle costing, integrating VE into the design process, using VE methodology in design review and consultant selection, case studies, VE workbook, and a life cycle costing program on disk.

$79.95 per copy
Over 450 pages, illustrated, Hardcover
Catalog No. 67319

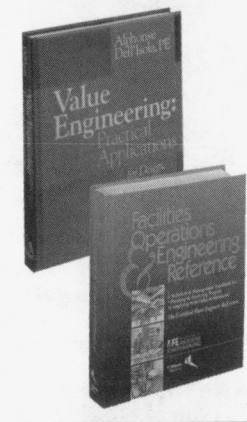

Facilities Operations & Engineering Reference

By the Association for Facilities Engineering and RSMeans

An all-in-one technical referance for planning and managi facility projects and solving day-to-day operations proble Selected as the official Certified Plant Engineer referenc handbook covers financial analysis, maintenance, HVAC a energy efficiency, and more.

$109.95 per copy
Over 700 pages, illustrated, Hardcover
Catalog No. 67318

The Building Professional's Guide to Contract Documents

3rd Edition

By Waller S. Poage, AIA, CSI, CVS

This comprehensive treatment of Contract Documents is an important reference for owners, design professionals, contractors, and students.

- Structure your Documents for Maximum Efficiency
- Effectively communicate construction requirements to all concerned
- Understand the Roles and Responsibilities of Construction Professionals
- Improve Methods of Project Delivery

$64.95 per copy, 400 pages
Diagrams and construction forms, Hardcover
Catalog No. 67261A

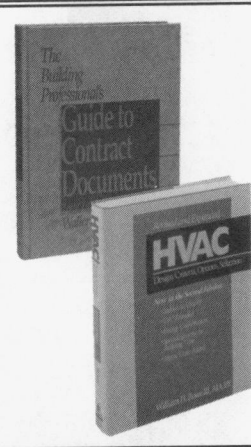

HVAC: Design Criteria, Options, Selection

Expanded 2nd Edition

By William H. Rowe III, AIA, PE

Includes Indoor Air Quality, CFC Removal, Energy Effici Systems, and Special Systems by Building Type. Helps yo solve a wide range of HVAC system design and selection problems effectively and economically. Gives you clear explanations of the latest ASHRAE standards.

$84.95 per copy
Over 600 pages, illustrated, Hardcover
Catalog No. 67306

Cost Planning & Estimating for Facilities Maintenance

In this unique book, a team of facilities management authorities shares their expertise at:

- Evaluating and budgeting maintenance operations
- Maintaining & repairing key building components
- Applying *Means Facilities Maintenance & Repair Cost Data* to your estimating

Covers special maintenance requirements of the 10 major building types.

$89.95 per copy
Over 475 pages, Hardcover
Catalog No. 67314

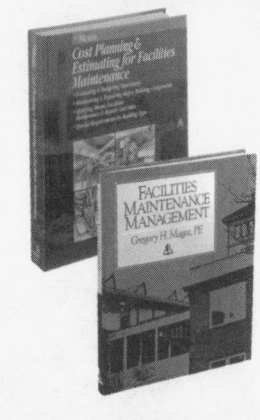

Facilities Maintenance Management

By Gregory H. Magee, PE

Now you can get successful management methods and techniques for all aspects of facilities maintenance. This comprehensive reference explains and demonstrates suc management techniques for all aspects of maintenance, and improvements for buildings, machinery, equipment, grounds. Plus, guidance for outsourcing and managing in staffs.

$86.95 per copy
Over 280 pages with illustrations, Hardcover
Catalog No. 67249

Builder's Essentials: Advanced Framing Methods

By Scot Simpson

A highly illustrated, "framer-friendly" approach to advanced framing elements. Provides expert, but easy-to-interpret, instruction for laying out and framing complex walls, roofs, and stairs, and special requirements for earthquake and hurricane protection. Also helps bring framers up to date on the latest building code changes, and provides tips on the lead framer's role and responsibilities, how to prepare for a job, and how to get the crew started.

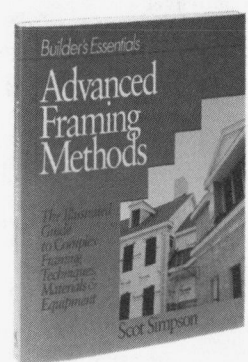

$24.95 per copy
250 pages, illustrated, Softcover
Catalog No. 67330

Reference Books

For more information
visit Means Web Site
at www.rsmeans.com

terior Home Improvement Costs,

w 9th Edition

mates for the most popular remodeling and repair projects—
small, do-it-yourself jobs—to major renovations and new
struction. Includes: Kitchens & Baths; New Living Space from
Attic, Basement or Garage; New Floors, Paint & Wallpaper;
ing Out or Building New Walls; Closets, Stairs & Fireplaces;
Energy-Saving Improvements, Home Theatres, and More!

95 per copy
pages, illustrated, Softcover
alog No. 67308E

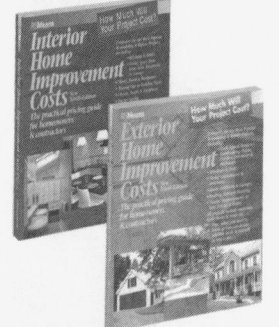

Exterior Home Improvement Costs

New 9th Edition

Estimates for the most popular remodeling and repair
projects—from small, do-it-yourself jobs, to major renovations
and new construction. Includes: Curb Appeal Projects—
Landscaping, Patios, Porches, Driveways and Walkways; New
Windows and Doors; Decks, Greenhouses, and Sunrooms; Room
Additions and Garages; Roofing, Siding, and Painting; "Green"
Improvements to Save Energy & Water

$24.95 per copy
Over 275 pages, illustrated, softcover
Catalog No. 67309E

ilder's Essentials: Plan ading & Material Takeoff

Wayne J. DelPico

Residential and Light Commercial Construction

luable tool for understanding plans and specs, and
rately calculating material quantities.
-by-step instructions and takeoff procedures based on a
set of working drawings.

95 per copy
420 pages, Softcover
log No. 67307

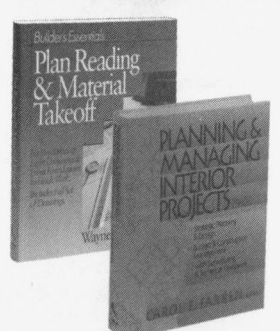

Planning & Managing Interior Projects

2nd Edition

By Carol E. Farren, CFM

Addresses changes in technology and business, guiding you
through commercial design and construction from initial client
meetings to post-project administration. Includes: evaluating
space requirements, alternative work models,
telecommunications and data management, and environmental
issues.

$69.95 per copy
Over 400 pages, illustrated, Hardcover
Catalog No. 67245A

ilder's Essentials: Best Business actices for Builders & Remodelers:

Easy-to-Use Checklist System

Thomas N. Frisby

mprehensive guide covering all aspects of running a
truction business, with more than 40 user-friendly checklists.
book provides expert guidance on: increasing your revenue
keeping more of your profit; planning for long-term growth;
ing good employees and managing subcontractors.

95 per copy
220 pages, Softcover
log No. 67329

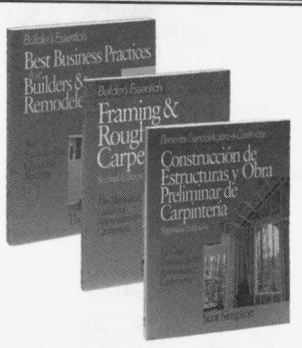

Builder's Essentials: Framing & Rough Carpentry 2nd Edition

By Scot Simpson

A complete training manual for apprentice and experienced
carpenters. Develop and improve your skills with "framer-
friendly," easy-to-follow instructions, and step-by-step
illustrations. Learn proven techniques for framing walls, floors,
roofs, stairs, doors, and windows. Updated guidance on
standards, building codes, safety requirements, and more.
Also available in Spanish!

$24.95 per copy
Over 150 pages, Softcover
Catalog No. 67298A   Spanish Catalog No. 67298AS

w to Estimate with Means Data ostWorks

RSMeans and Saleh Mubarak

w 2nd Edition!

estimating techniques using Means cost data.
des an instructional version of Means CostWorks
ROM with Sample Building Plans.
tep-by-step guide takes you through all the major
ruction items. Over 300 sample estimating problems
cluded.

5 per copy
190 pages, Softcover
log No. 67324A

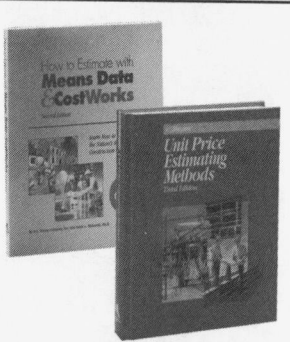

Unit Price Estimating Methods

New 3rd Edition

This new edition includes up-to-date cost data and estimating
examples, updated to reflect changes to the CSI numbering
system and new features of Means cost data. It describes the
most productive, universally accepted ways to estimate, and
uses checklists and forms to illustrate shortcuts and
timesavers. A model estimate demonstrates procedures. A
new chapter explores computer estimating alternatives.

$59.95 per copy
Over 350 pages, illustrated, Hardcover
Catalog No. 67303A

ans Landscape Estimating Methods

Edition

ylvia H. Fee

evised edition offers expert guidance for preparing accurate
ates for new landscape construction and grounds maintenance.
es a complete project estimate featuring the latest equipment
ethods, and **two chapters on Life Cycle Costing, and Landscape
enance Estimating.**

5 per copy
300 pages, illustrated, Hardcover
og No. 67295B

Means Environmental Remediation Estimating Methods, 2nd Edition

By Richard R. Rast

Guidelines for estimating 50 standard remediation technologies.
Use it to prepare preliminary budgets, develop estimates,
compare costs and solutions, estimate liability, review quotes,
negotiate settlements.

A valuable support tool for *Means Environmental Remediation
Unit Price* and *Assemblies* books.

$99.95 per copy
Over 750 pages, illustrated, Hardcover
Catalog No. 64777A

For more information
visit Means Web Site
at www.rsmeans.com

Reference Books

Means Illustrated Construction Dictionary, Condensed Edition, 2nd Edition
By RSMeans

Recognized in the industry as the best resource of its kind, this has been further enhanced with updates to existing terms and the addition of hundreds of new terms and illustrations . . . in keeping with the most recent developments in the industry.
The best portable dictionary for office or field use—an essential tool for contractors, architects, insurance and real estate personnel, homeowners, and anyone who needs quick, clear definitions for construction terms.

$59.95 per copy
Over 500 pages
Catalog No. 67282A

Means Repair and Remodeling Estimating
New 4th Edition
By Edward B. Wetherill & RSMeans

Focuses on the unique problems of estimating renovations of existing structures. It helps you determine the true costs of remodeling through careful evaluation of architectural details a site visit.
New section on disaster restoration costs.

$69.95 per copy
Over 450 pages, illustrated, Hardcover
Catalog No. 67265B

Facilities Planning & Relocation
New, lower price and user-friendly format.
By David D. Owen

A complete system for planning space needs and managing relocations. Includes step-by-step manual, over 50 forms, and extensive reference section on materials and furnishings.

$89.95 per copy
Over 450 pages, Softcover
Catalog No. 67301

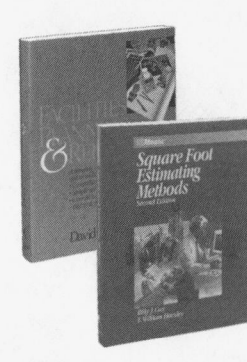

Means Square Foot & Assemblies Estimating Methods
3rd Edition!

Develop realistic Square Foot and Assemblies Costs for budge and construction funding. The new edition features updated guidance on square foot and assemblies estimating using UNIFORMAT II. An essential reference for anyone who perfor conceptual estimates.

$69.95 per copy
Over 300 pages, illustrated, Hardcover
Catalog No. 67145B

Means Electrical Estimating Methods 3rd Edition

Expanded new edition includes sample estimates and cost information in keeping with the latest version of the CSI MasterFormat and UNIFORMAT II. Complete coverage of Fiber Optic and Uninterruptible Power Supply electrical systems, broken down by components and explained in detail. Includes a new chapter on computerized estimating methods. A practical companion to *Means Electrical Cost Data*.

$64.95 per copy
Over 325 pages, Hardcover
Catalog No. 67230A

Means Mechanical Estimatin Methods 3rd Edition

This guide assists you in making a review of plans, specs, and packages with suggestions for takeoff procedures, listings, substitutions and pre-bid scheduling. Includes suggestions for budgeting labor and equipment usage. Compares materials ar construction methods to allow you to select the best option.

$64.95 per copy
Over 350 pages, illustrated, Hardcover
Catalog No. 67294A

Means Spanish/English Construction Dictionary
By RSMeans, The International Conference of Building Officials (ICBO), and Rolf Jensen & Associates (RJA)

Designed to facilitate communication among Spanish- and English-speaking construction personnel—improving performance and job-site safety. Features the most common words and phrases used in the construction industry, with easy-to-follow pronunciations. Includes extensive building systems and tools illustrations.

$22.95 per copy
250 pages, illustrated, Softcover
Catalog No. 67327

Project Scheduling & Management for Constructio
New 3rd Edition
By David R. Pierce, Jr.

A comprehensive yet easy-to-follow guide to construction p scheduling and control—from vital project management prin through the latest scheduling, tracking, and controlling tech niques. The author is a leading authority on scheduling with of field and teaching experience at leading academic institu Spend a few hours with this book and come away with a soli understanding of this essential management topic.

$64.95 per copy
Over 300 pages, illustrated, Hardcover
Catalog No. 67247B

Reference Books

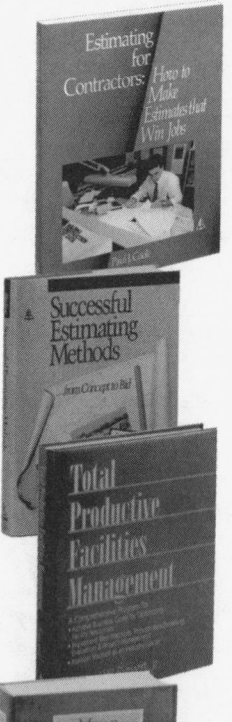

Concrete Repair and Maintenance Illustrated
By Peter H. Emmons
$9.95 per copy
Catalog No. 67146

Superintending for Contractors:
How to Bring Jobs in On-Time, On-Budget
By Paul J. Cook
$5.95 per copy
Catalog No. 67233

HVAC Systems Evaluation
By Harold R. Colen, PE
$4.95 per copy
Catalog No. 67281

Basics for Builders: How to Survive and Prosper in Construction
By Thomas N. Frisby
$9.95 per copy
Catalog No. 67273

Successful Interior Projects Through Effective Contract Documents
By Joel Downey & Patricia K. Gilbert
Now $24.98 per copy, limited quantity
Catalog No. 67313

Building Spec Homes Profitably
By Kenneth V. Johnson
$9.95 per copy
Catalog No. 67312

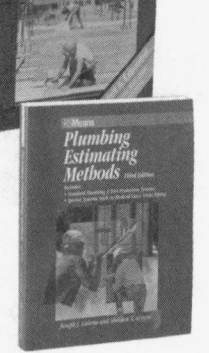

Estimating for Contractors
How to Make Estimates that Win Jobs
By Paul J. Cook
$35.95 per copy
Catalog No. 67160

Successful Estimating Methods:
From Concept to Bid
By John D. Bledsoe, PhD, PE
$32.48 per copy
Catalog No. 67287

Total Productive Facilities Management
By Richard W. Sievert, Jr.
$29.98 per copy
Over 270 pages, illustrated, Hardcover
Catalog No. 67321

Means Productivity Standards for Construction
Expanded Edition (Formerly Man-Hour Standards)
$49.98 per copy
Over 800 pages, Hardcover
Catalog No. 67236A

Means Plumbing Estimating Methods, 3rd Edition
By Joseph Galeno and Sheldon Greene
Now $29.98 per copy
Catalog No. 67283B

Reference Books

Preventive Maintenance Guidelines for School Facilities

By John C. Maciha

A complete PM program for K-12 schools that ensures sustained security, safety, property integrity, user satisfaction, and reasonable ongoing expenditures.

Includes schedules for weekly, monthly, semiannual, and annual maintenance with hard copy and electronic forms.

$149.95 per copy
Over 225 pages, Hardcover
Catalog No. 67326

Preventive Maintenance for Higher Education Facilities

By Applied Management Engineering, Inc.

An easy-to-use system to help facilities professionals establish value of PM, and to develop and budget for an appropriate PM program for their college or university. Features interactive campus building models typical of those found in different-size higher education facilities, and PM checklists linked to each of equipment or system in hard copy and electronic format.

$149.95 per copy
150 pages, Hardcover
Catalog No. 67337

Historic Preservation: Project Planning & Estimating

By Swanke Hayden Connell Architects

Expert guidance on managing historic restoration, rehabilitation, and preservation building projects and determining and controlling their costs. Includes:

- How to determine whether a structure qualifies as historic
- Where to obtain funding and other assistance
- How to evaluate and repair more than 75 historic building materials

$99.95 per copy
Over 675 pages, Hardcover
Catalog No. 67323

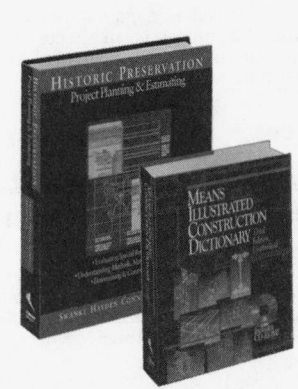

Means Illustrated Construction Dictionary, 3rd Edition

Long regarded as the Industry's finest, the Means *Illustrated Construction Dictionary* is now even better.
With the addition of over 1,000 new terms and hundreds of illustrations, it is the clear choice for the most comprehensive and current information. **The companion CD-ROM that comes with this new edition adds many extra features: larger graphics, expanded definitions, and links to both CSI MasterFormat numbers and product information.**

$99.95 per copy
Over 790 pages, Illustrated, Hardcover
Catalog No. 67292A

Designing & Building with the IBC, 2nd Edition

By Rolf Jensen & Associates, Inc.

This updated comprehensive guide helps building professionals make the transition to the 2003 *International Building Code®*. Includes a side-by-side code comparison of the IBC 2003 to the IBC 2000 and the three primary model codes, a quick-find index, and professional code commentary. With illustrations, abbreviations key, and an extensive Resource section.

$99.95 per copy
Over 875 pages
Catalog No. 67328A

Residential & Light Commercial Construction Standards, 2nd Edition

By RSMeans and Contributing Authors

New, updated second edition of this unique collection of industry standards that define quality construction. For contractors, subcontractors, owners, developers, architects, engineers, attorneys, and insurance personnel, this book provides authoritative requirements and recommendations compiled from the nation's leading professional associations, industry publications, and building code organizations.

$59.95 per copy
600 pages, illustrated, Softcover
Catalog No. 67322A

Means Estimating Handbook, 2nd Edition

By RSMeans

Updated new Second Edition answers virtually any estimating technical question - all organized by CSI MasterFormat. This comprehensive reference covers the full spectrum of technical data required to estimate construction costs. The book includes information on sizing, productivity, equipment requirements, code-mandated specifications, design standards and engineering factors.

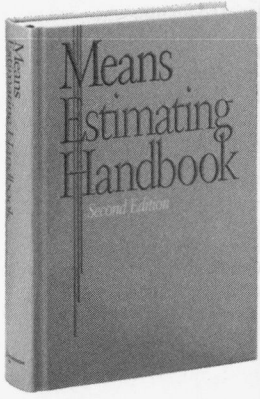

$99.95 per copy
Over 900 pages, Hardcover
Catalog No. 67276A

For more information
visit Means Web Site
www.rsmeans.com

Seminars

Means CostWorks Training

This one-day seminar course has been designed with the intention of assisting both new and existing users to become more familiar with CostWorks program. The class is broken into two unique sections: (1) A one-half day presentation on the function of each option; and each student will be shown how to use the software to develop a cost estimate. (2) Hands-on estimating exercises that will ensure that each student thoroughly understands how to use CostWorks. You must bring your own laptop computer to this course.

CostWorks Benefits/Features:
- Estimate in your own spreadsheet format
- Power of Means National Database
- Database automatically regionalized
- Save time with keyword searches
- Save time by establishing common estimate items in "Bookmark" files
- Customize your spreadsheet template
- Hot key to Product Manufacturers' listings and specs
- Merge capability for networking environments
- View crews and assembly components
- AutoSave capability
- Enhanced sorting capability

Unit Price Estimating

This interactive two-day seminar teaches attendees how to interpret project information and process it into final, detailed estimates with the greatest accuracy level.

The single most important credential an estimator can take to the job is the ability to visualize construction in the mind's eye, and thereby estimate accurately.

Some Of What You'll Learn:
- Interpreting the design in terms of cost
- The most detailed, time tested methodology for accurate "pricing"
- Key cost drivers—material, labor, equipment, staging and subcontracts
- Understanding direct and indirect costs for accurate job cost accounting and change order management

Who Should Attend: Corporate and government estimators and purchasers, architects, engineers…and others needing to produce accurate project estimates.

Square Foot and Assemblies Cost Estimating

This two-day course teaches attendees how to quickly deliver accurate square foot estimates using limited budget and design information.

Some Of What You'll Learn:
- How square foot costing gets the estimate done faster
- Taking advantage of a "systems" or "assemblies" format
- The Means "building assemblies/square foot cost approach"
- How to create a very reliable preliminary and systems estimate using bare-bones design information

Who Should Attend: Facilities managers, facilities engineers, estimators, planners, developers, construction finance professionals…and others needing to make quick, accurate construction cost estimates at commercial, government, educational and medical facilities.

Repair and Remodeling Estimating

This two-day seminar emphasizes all the underlying considerations unique to repair/remodeling estimating and presents the correct methods for generating accurate, reliable R&R project costs using the unit price and assemblies methods.

Some Of What You'll Learn:
- Estimating considerations—like labor-hours, building code compliance, working within existing structures, purchasing materials in smaller quantities, unforeseen deficiencies
- Identifies problems and provides solutions to estimating building alterations
- Rules for factoring in minimum labor costs, accurate productivity estimates and allowances for project contingencies
- R&R estimating examples are calculated using unit prices and assemblies data

Who Should Attend: Facilities managers, plant engineers, architects, contractors, estimators, builders…and others who are concerned with the proper preparation and/or evaluation of repair and remodeling estimates.

Mechanical and Electrical Estimating

This two-day course teaches attendees how to prepare more accurate and complete mechanical/electrical estimates, avoiding the pitfalls of omission and double-counting, while understanding the composition and rationale within the Means Mechanical/Electrical database.

Some Of What You'll Learn:
- The unique way mechanical and electrical systems are interrelated
- M&E estimates, conceptual, planning, budgeting and bidding stages
- Order of magnitude, square foot, assemblies and unit price estimating
- Comparative cost analysis of equipment and design alternatives

Who Should Attend: Architects, engineers, facilities managers, mechanical and electrical contractors…and others needing a highly reliable method for developing, understanding and evaluating mechanical and electrical contracts.

Plan Reading and Material Takeoff

This two-day program teaches attendees to read and understand construction documents and to use them in the preparation of material takeoffs.

Some of What You'll Learn:
- Skills necessary to read and understand typical contract documents—blueprints and specifications
- Details and symbols used by architects and engineers
- Construction specifications' importance in conjunction with blueprints
- Accurate takeoff of construction materials and industry-accepted takeoff methods

Who Should Attend: Facilities managers, construction supervisors, office managers…and other responsible for the execution and administration of a construction project including government, medical, commercial, educational or retail facilities.

Facilities Maintenance and Repair Estimating

This two-day course teaches attendees how to plan, budget, and estimate the cost of ongoing and preventive maintenance and repair for existing buildings and grounds.

Some Of What You'll Learn:
- The most financially favorable maintenance, repair and replacement scheduling and estimating
- Auditing and value engineering facilities
- Preventive planning and facilities upgrading
- Determining both in-house and contract-out service costs; annual, asset-protecting M&R plan

Who Should Attend: Facility managers, maintenance supervisors, buildings and grounds superintendents, plant managers, planners, estimators…and others involved in facilities planning and budgeting.

Scheduling and Project Management

This two-day course teaches attendees the most current and proven scheduling and management techniques needed to bring projects in on time and on budget.

Some Of What You'll Learn:
- Crucial phases of planning and scheduling
- How to establish project priorities, develop realistic schedules and management techniques
- Critical Path and Precedence Methods
- Special emphasis on cost control

Who Should Attend: Construction project managers, supervisors, engineers, estimators, contractors…and others who want to improve their project planning, scheduling and management skills.

Advanced Project Management

This two-day seminar will teach you how to effectively manage and control the entire design-build process and allow you to take home tangible skills that will be immediately applicable on existing projects.

Some Of What You'll Learn:
- Value engineering, bonding, fast-tracking and bid package creation
- How estimates and schedules can be integrated to provide advanced project management tools
- Cost engineering, quality control, productivity measurement and improvement
- Front loading a project and predicting its cash flow

Who Should Attend: Owners, project managers, architectural and engineering managers, construction managers, contractors…and anyone else who is responsible for the timely design and completion of construction projects.

For more informat
visit Means Web
at www.rsmeans.

Seminars

2005
Means
Seminar
Schedule

Note: Call for exact dates and details.

Location	Dates
Las Vegas, NV	March
Washington, DC	April
Phoenix, AZ	April
Denver, CO	May
San Francisco, CA	June
Philadelphia, PA	June
Washington, DC	September
Dallas, TX	September
Las Vegas, NV	October
Orlando, FL	November
Atlantic City, NJ	November
San Diego, CA	December

Registration Information

Register Early... Save up to $100! Register 30 days before the start date of a seminar and save $100 off your total fee. *Note: This discount can be applied only once per order. It cannot be applied to team discount registrations or any other special offer.*

How to Register Register by phone today! Means toll-free number for making reservations is: **1-800-334-3509.**

Individual Seminar Registration Fee $895. Individual CostWorks Training Registration Fee $349. To register by mail, complete the registration form and return with your full fee to: Seminar Division, Reed Construction Data, RSMeans Seminars, 63 Smiths Lane, Kingston, MA 02364.

Federal Government Pricing All Federal Government employees save 25% off regular seminar price. Other promotional discounts cannot be combined with Federal Government discount.

Team Discount Program Two to four seminar registrations: Call for pricing.

Multiple Course Discounts When signing up for two or more courses, call for pricing.

Refund Policy Cancellations will be accepted up to ten days prior to the seminar start. There are no refunds for cancellations received later than ten working days prior to the first day of the seminar. A $150 processing fee will be applied for all cancellations. Written notice of cancellation is required . Substitutions can be made at anytime before the session starts. **No-shows are subject to the full seminar fee.**

AACE Approved Courses The RSMeans Construction Estimating and Management Seminars described and offered to you here have each been approved for 14 hours (1.4 recertification credits) of credit by the AACE International Certification Board toward meeting the continuing education requirements for re-certification as a Certified Cost Engineer/Certified Cost Consultant.

AIA Continuing Education We are registered with the AIA Continuing Education System (AIA/CES) and are committed to developing quality learning activities in accordance with the CES criteria. RSMeans seminars meet the AIA CES criteria for Quality Level 2. AIA members will receive (14) learning units (I for each two-day RSMeans Course.

NASBA CPE Sponsor Credits We are part of the National Registry of CPE Sponsors. Attendees may be eligable for (16) CPE credits.

Daily Course Schedule The first day of each seminar session begins at 8:3 A.M. and ends at 4:30 P.M. The second day is 8:00 A.M.–4:00 P.M. Participants urged to bring a hand-held calculator since many actual problems will be worke in each session.

Continental Breakfast Your registration includes the cost of a continental breakfast, a morning coffee break, and an afternoon break. These informal segm will allow you to discuss topics of mutual interest with other members of the seminar. (You are free to make your own lunch and dinner arrangements.)

Hotel/Transportation Arrangements RSMeans has arranged to hold a bl of rooms at each hotel hosting a seminar. To take advantage of special group ra when making your reservation, be sure to mention that you are attending the M Seminar. You are, of course, free to stay at the lodging place of your choice. (H **reservations and transportation arrangements should be made directly by semi attendees.)**

Important Class sizes are limited, so please register as soon as possible.

Note: Pricing subject to change.

Registration Form   Call 1-800-334-3509 to register or FAX 1-800-632-6732. Visit our Web site www.rsmeans.com

Please register the following people for the Means Construction Seminars as shown here. Full payment or deposit is enclosed, and we understand that we must make our own hotel reservations if overnight stays are necessary.

☐ Full payment of $ _____ enclosed.

☐ Bill me

Name of Registrant(s)
(To appear on certificate of completion)

_____

_____

_____

P.O. #: _____
GOVERNMENT AGENCIES MUST SUPPLY PURCHASE ORDER NUMBER

Firm Name _____

Address _____

City/State/Zip _____

Telephone No. fax No. _____

E-mail Address _____

Charge our registration(s) to: ☐ MasterCard ☐ VISA ☐ American Express ☐ Discover

Account No. _____ Exp. Date _____

Cardholder's Signature _____

Seminar Name   City   Dates

_____

Please mail check to: Seminar Division, Reed Construction Data, RSMeans Seminars, 63 Smiths Lane, P.O.Box 800, Kingston, MA 02364

MeansData™

CONSTRUCTION COSTS FOR SOFTWARE APPLICATIONS

Your construction estimating software is only as good as your cost data.

A proven construction cost database is a mandatory part of any estimating package. The following list of softwa providers can offer you MeansData™ as an added feature for their estimating systems. See the table below for w types of products and services they offer (match their numbers). Visit online at **www.rsmeans.com/demosource/** for more information and free demos. Or call their numbers listed below.

1. **3D International**
713-871-7000
venegas@3di.com

2. **4Clicks-Solutions, LLC**
719-574-7721
mbrown@4clicks-solutions.com

3. **Aepco, Inc.**
301-670-4642
blueworks@aepco.com

4. **Applied Flow Technology**
800-589-4943
info@aft.com

5. **ArenaSoft Estimating**
888-370-8806
info@arenasoft.com

6. **ARES Corporation**
925-299-6700
sales@arescorporation.com

7. **BSD - Building Systems Design, Inc.**
888-273-7638
bsd@bsdsoftlink.com

8. **CMS - Computerized Micro Solutions**
800-255-7407
cms@proest.com

9. **Corecon Technologies, Inc.**
714-895-7222
sales@corecon.com

10. **CorVet Systems**
301-622-9069
sales@corvetsys.com

11. **Discover Software**
727-559-0161
sales@discoversoftware.com

12. **Estimating Systems, Inc.**
800-967-8572
esipulsar@adelphia.net

13. **MAESTRO Estimator Schwaab Technology Solutions, Inc.**
281-578-3039
Stefan@schwaabtech.com

14. **Magellan K-12**
936-447-1744
sam.wilson@magellan-K12.com

15. **MC² - Management Computer**
800-225-5622
vkeys@mc2-ice.com

16. **Maximus Asset Solutions**
800-659-9001
assetsolutions@maximus.com

17. **Prime Time**
610-964-8200
dick@primetime.com

18. **Prism Computer Corp.**
800-774-7622
famis@prismcc.com

19. **Quest Solutions, Inc.**
800-452-2342
info@questsolutions.com

20. **RIB Software (Americas), Inc.**
800-945-7093
san@rib-software.com

21. **Shaw Beneco Enterprises, Inc.**
877-719-4748
inquire@beneco.com

22. **Timberline Software Corp.**
800-628-6583
product.info@timberline.com

23. **TMA SYSTEMS, Inc.**
800-862-1130
sales@tmasys.com

24. **US Cost, Inc.**
800-372-4003
sales@uscost.com

25. **Vanderweil Facility Advisors**
617-451-5100
info@VFA.com

26. **Vertigraph, Inc.**
800-989-4243
info-request@vertigraph.com

27. **WinEstimator, Inc.**
800-950-2374
sales@winest.com

TYPE	1	2	3	4	5	6	7	8	9	10	11	12	13	14	15	16	17	18	19	20	21	22	23	24	25	26	27
BID					●		●	●		●	●				●		●		●	●	●					●	●
Estimating		●			●	●	●	●	●	●	●	●	●		●				●	●	●	●		●		●	●
DOC/JOC/SABER		●			●		●				●		●				●				●	●					●
IDIQ		●									●		●				●				●						●
Asset Mgmt.											●			●		●									●		●
Facility Mgmt.	●		●								●		●	●		●	●	●					●		●		
Project Mgmt.	●	●					●				●		●	●			●			●	●	●					
TAKE-OFF					●			●	●	●	●				●		●		●	●				●		●	●
EARTHWORK									●		●				●		●	●								●	
Pipe Flow				●							●						●										
HVAC/Plumbing					●			●			●						●										
Roofing					●						●						●										
Design	●				●						●						●				●						●
Other Offers/Links:																											
Accounting/HR		●			●						●									●	●	●					
Scheduling					●						●	●									●	●		●		●	●
CAD											●			●								●					●
PDA																						●					●
Lt. Versions		●						●			●											●		●			
Consulting	●	●			●			●			●				●	●						●			●		●
Training		●			●		●	●	●		●				●	●	●	●	●		●	●					●

Reseller applications now being accepted. Call Carol Polio Ext. 5107.

FOR MORE INFORMATION
CALL 1-800-448-8182, EXT. 5107 OR FAX 1-800-632-6732

For more information
visit Means Web Site
at www.rsmeans.com

New Titles

Builder's Essentials:
Estimating Building Costs for the Residential & Light Commercial Contractor

By Wayne J. DelPico

Step-by-step estimating methods for residential and light commercial contractors. Includes a detailed look at every construction specialty—explaining all the components, takeoff units, and labor needed for well-organized, complete estimates covers:

Correctly interpreting plans and specifications.

Developing accurate and complete labor and material costs.

Understanding direct and indirect overhead costs... and accounting for time-sensitive costs.

Using historical cost data to generate new project budgets.

Plus hard-to-find, professional guidance on what to consider so you can allocate the right amount for profit and contingencies.

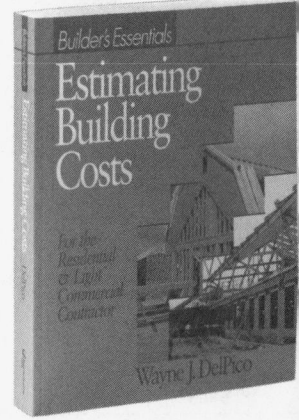

$29.95 per copy
Over 400 pages, illustrated, softcover
Catalog No. 67343

Building & Renovating Schools

This all-inclusive guide covers every step of the school construction process—from initial planning, needs assessment, and design, right through moving into the new facility. A must-have resource for anyone concerned with new school construction or renovation, including architects and engineers, contractors and project managers, facility managers, school administrators and school board members, building committees, community leaders, and anyone else who wants to ensure that the project meets the schools' needs in a cost-effective, timely manner. With square foot cost models for elementary, middle, and high school facilities and real-life case studies of recently completed school projects.

The contributors to this book – architects, construction project managers, contractors, and estimators who specialize in school construction – provide start-to-finish, expert guidance on the process.

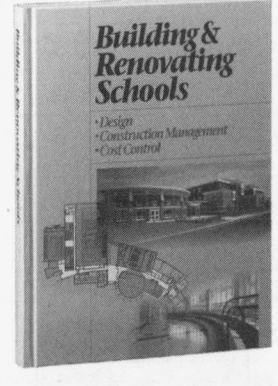

$99.95 per copy
Over 425 pages
Catalog No. 67342

Means ADA Compliance Pricing Guide, New Second Edition
By Adaptive Environments and RSMeans

Completely updated and revised to the new 2004 Americans with Disabilities Act Accessible Guidelines, this book features more than 70 of the most commonly needed modifications for ADA compliance—their design requirements, suggestions, and final cost. Projects range from installing ramps and walkways, widening doorways and entryways, and installing and refitting elevators, to relocating light switches and signage, and remodeling bathrooms and kitchens. Also provided are:

Detailed cost estimates for budgeting modification projects, including estimates for each of 260 alternates.

An assembly estimate for every project, with detailed cost breakdown including materials, labor hours, and contractor's overhead.

3,000 Additional ADA compliance-related unit cost line items.

Costs that are easily adjusted to over 900 cities and towns.

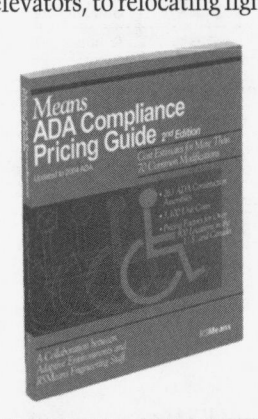

Over 350 pages
$79.99 per copy
Catalog No. 67310A

Qty.	Book No.	COST ESTIMATING BOOKS	Unit Price	Total
	60065	Assemblies Cost Data 2005	$189.95	
	60015	Building Construction Cost Data 2005	115.95	
	61015	Building Const. Cost Data–Looseleaf Ed. 2005	144.95	
	63015	Building Const. Cost Data–Metric Version 2005	115.95	
	60225	Building Const. Cost Data–Western Ed. 2005	115.95	
	60115	Concrete & Masonry Cost Data 2005	105.95	
	60145	Construction Cost Indexes 2005	251.95	
	60145A	Construction Cost Index–January 2005	63.00	
	60145B	Construction Cost Index–April 2005	63.00	
	60145C	Construction Cost Index–July 2005	63.00	
	60145D	Construction Cost Index–October 2005	63.00	
	60345	Contr. Pricing Guide: Resid. R & R Costs 2005	39.95	
	60335	Contr. Pricing Guide: Resid. Detailed 2005	39.95	
	60325	Contr. Pricing Guide: Resid. Sq. Ft. 2005	39.95	
	64025	ECHOS Assemblies Cost Book 2005	189.95	
	64015	ECHOS Unit Cost Book 2005	126.95	
	54005	ECHOS (Combo set of both books)	251.95	
	60235	Electrical Change Order Cost Data 2005	115.95	
	60035	Electrical Cost Data 2005	115.95	
	60205	Facilities Construction Cost Data 2005	279.95	
	60305	Facilities Maintenance & Repair Cost Data 2005	252.95	
	60165	Heavy Construction Cost Data 2005	115.95	
	63165	Heavy Const. Cost Data–Metric Version 2005	115.95	
	60095	Interior Cost Data 2005	115.95	
	60125	Labor Rates for the Const. Industry 2005	253.95	
	60185	Light Commercial Cost Data 2005	99.95	
	60025	Mechanical Cost Data 2005	115.95	
	60155	Open Shop Building Const. Cost Data 2005	115.95	
	60215	Plumbing Cost Data 2005	115.95	
	60045	Repair and Remodeling Cost Data 2005	99.95	
	60175	Residential Cost Data 2005	99.95	
	60285	Site Work & Landscape Cost Data 2005	115.95	
	60055	Square Foot Costs 2005	126.95	
		REFERENCE BOOKS		
	67147A	ADA in Practice	59.98	
	67310A	ADA Compliance Pricing Guide, 2nd Ed.	79.95	
	67273	Basics for Builders: How to Survive and Prosper	34.95	
	67330	Bldrs Essentials: Adv. Framing Methods	24.95	
	67329	Bldrs Essentials: Best Bus. Practices for Bldrs	29.95	
	67298A	Bldrs Essentials: Framing/Carpentry 2nd Ed.	24.95	
	67298AS	Bldrs Essentials: Framing/Carpentry Spanish	24.95	
	67307	Bldrs Essentials: Plan Reading & Takeoff	35.95	
	67261A	Bldg. Prof. Guide to Contract Documents 3rd Ed.	64.95	
	67342	Building & Renovating Schools	99.95	
	67339	Building Security: Strategies & Costs	89.95	
	67312	Building Spec Homes Profitably	29.95	
	67146	Concrete Repair & Maintenance Illustrated	69.95	
	67314	Cost Planning & Est. for Facil. Maint.	89.95	
	67317A	Cyberplaces: The Internet Guide 2nd Ed.	59.95	
	67328A	Designing & Building with the IBC, 2nd Ed.	99.95	
	67230B	Electrical Estimating Methods 3rd Ed.	64.95	

Qty.	Book No.	REFERENCE BOOKS (Cont.)	Unit Price	Total
	64777A	Environmental Remediation Est. Methods 2nd Ed.	$ 99.95	
	67160	Estimating for Contractors	35.95	
	67276A	Estimating Handbook 2nd Ed.	99.95	
	67249	Facilities Maintenance Management	86.95	
	67246	Facilities Maintenance Standards	79.95	
	67318	Facilities Operations & Engineering Reference	109.95	
	67301	Facilities Planning & Relocation	89.95	
	67231	Forms for Building Const. Professional	47.48	
	67260	Fundamentals of the Construction Process	34.98	
	67323	Historic Preservation: Proj. Planning & Est.	99.95	
	67308E	Home Improvement Costs–Int. Projects 9th Ed.	29.95	
	67309E	Home Improvement Costs–Ext. Projects 9th Ed.	29.95	
	67324A	How to Est. w/Means Data & CostWorks 2nd Ed.	59.95	
	67304	How to Estimate with Metric Units	9.98	
	67306	HVAC: Design Criteria, Options, Select. 2nd Ed.	84.95	
	67281	HVAC Systems Evaluation	84.95	
	67282A	Illustrated Construction Dictionary, Condensed	59.95	
	67292A	Illustrated Construction Dictionary, w/CD-ROM	99.95	
	67295B	Landscape Estimating 4th Ed.	62.95	
	67341	Life Cycle Costing	99.95	
	67302	Managing Construction Purchasing	19.98	
	67294A	Mechanical Estimating 3rd Ed.	64.95	
	67245A	Planning and Managing Interior Projects 2nd Ed.	69.95	
	67283A	Plumbing Estimating Methods 2nd Ed.	59.95	
	67326	Preventive Maint. Guidelines for School Facil.	149.95	
	67337	Preventive Maint. for Higher Education Facilities	149.95	
	67236A	Productivity Standards for Constr.–3rd Ed.	49.98	
	67247B	Project Scheduling & Management for Constr. 3rd Ed.	64.95	
	67265B	Repair & Remodeling Estimating 4th Ed.	69.95	
	67322A	Resi. & Light Commercial Const. Stds. 2nd Ed.	59.95	
	67327	Spanish/English Construction Dictionary	22.95	
	67145A	Sq. Ft. & Assem. Estimating Methods 3rd Ed.	69.95	
	67287	Successful Estimating Methods	32.48	
	67313	Successful Interior Projects	24.98	
	67233	Superintending for Contractors	35.95	
	67321	Total Productive Facilities Management	29.98	
	67284	Understanding Building Automation Systems	29.98	
	67303A	Unit Price Estimating Methods 3rd Ed.	59.95	
	67319	Value Engineering: Practical Applications	79.95	

MA residents add 5% state sales tax		
Shipping & Handling**		
Total (U.S. Funds)*		

Prices are subject to change and are for U.S. delivery only. *Canadian customers may call for current prices. **Shipping & handling charges: Add 7% of total order for check and credit card payments. Add 9% of total order for invoiced orders.

Send Order To: ADDV-1001

Name (Please Print) _____

Company _____

☐ **Company**
☐ **Home** Address _____

_____

City/State/Zip _____

Phone # _____ P.O. # _____

(Must accompany all orders being billed)